T0269257

This is for Aurea.

Contents

Foreword

Commutative Algebra, flushed with its great successes in dealing with the difficult homological questions that originated in the 60's, has flourished recently around many clusters of related activities. To name a few, in no particular order of significance: linkage theory, Cohen–Macaulay approximation, theory of Hilbert functions, the structure of free resolutions, tight closure, and a growing web of relationships to computer algebra and combinatorics. All this, while strengthening its connections to algebraic geometry and other parts of algebra and retaining the vitality of its established domains.

This is a panorama that begs for systematic accounts of these developments to guide a possible reader through a bountiful but scattered literature. Several efforts have begun to address these needs and there are definite stirrings of an accelerating pace.

Another of these nuclei of activity is the theory and practice of the so-called Rees or blowup algebras. Through its many variants it interacts in some measure with all the areas listed above. Furthermore, it is generating questions at a much faster clip than they are being dealt with. It became then a likely target to be written about.

We decided to assemble some recent results and methods on these algebras in a manner which would be comprehensible/useful to a reader conversant with basic commutative algebra. It was the original intent to frame these notes into a second course in commutative algebra, requiring the core part of [192], along with a good foundation in homological algebra. The task made easier because so much of the technical background, particularly on local cohomology, is already dealt with in the excellent [103]. The reader would also be provided with several entry points along the text to make access to different parts of the notes more direct. For example, chapters 9 and 10 are nearly independent of the rest of the notes. Along the way, part of this intent decayed: The sheer task of bringing together and integrating properly the material became too demanding, and was replaced by the more modest aim of presenting some significant results and methods, while providing gateways to other developments, some even richer in content.

One early ground rule, the emphasis on bringing in the computer as a basic tool in most aspects of the discussion, did not change. In fact, it became clear that its role would have to be strengthened and even took a life of its own. A result has been the lengthy Chapter 10, with a potpourri of methods and recipes to deal with basic constructions in commutative algebra.

These notes owe much to numerous conversations with many colleagues in the course of the past few years. We are particularly grateful to Jürgen Herzog, Craig Huneke, Aron Simis, Bernd Ulrich, and Rafael Villarreal; their published (and unpublished) work, lectures and comments influenced considerably the choice of topics. Several others, colleagues and friends, unsuspectingly will see their own words, in forgotten letters or messages, stare back at them! We are also thankful to Alberto Corso, Susan Morey, Claudia Polini, Maria Vaz Pinto and two anonymous reviewers for pointing out many inaccuracies in an early version of the notes and for offering helpful suggestions.

For financial support during its writing, we thank the National Science Foundation.

Introduction

There is a class of rings, collectively designated **blowup algebras**, that appear in many constructions in Commutative Algebra and Algebraic Geometry. They represent fibrations of a variety with fibers which are often affine spaces; a polynomial ring $R[T_1, \ldots, T_n]$ is the notorious example of such algebras. The algebras arise from module and categorical theoretic constructions over a base ring R, which has led to the terminology of algebras of linear type. Its uses include counterexamples to Hilbert's 14th Problem, the determination of the minimal number of equations needed to define algebraic varieties, the computation of some invariants of Lie groups, and several others. The main impetus for their systematic study has been the long list of beautiful Cohen–Macaulay algebras produced by the various processes. Finally, they provide a testing ground for several computational methods in Commutative Algebra.

Schematically these rings are:

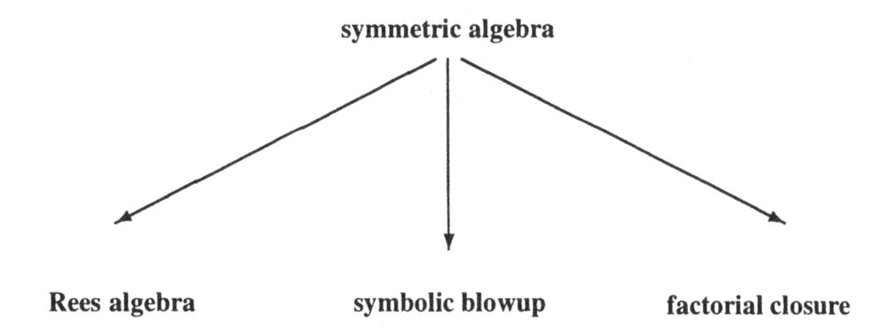

They will be introduced individually, beginning with the most ubiquitous. A multiplicative $\mathbb{N}$–filtration $\mathcal{F}$, of a commutative ring R, is a sequence of subgroups $\{R_n,\ n \in \mathbb{N}\}$ of R with the property

$$R_m \cdot R_n \subset R_{m+n}.$$

The Rees algebra of $\mathcal{F}$ is the subring of the ring of polynomials $R[t]$

$$R(\mathcal{F}) := \sum_{n \in \mathbb{N}} R_n t^n.$$

If the filtration is decreasing, $R_{n+1} \subset R_n$, there are two other algebras attached to it: The extended Rees algebra $R(\mathcal{F})[t^{-1}]$, and the associated graded ring

$$\mathrm{gr}_{\mathcal{F}}(R) := \bigoplus_{n=0}^{\infty} R_n/R_{n+1}.$$

A major example is the I–adic filtration of an ideal I: $R_n = I^n$, $n \geq 0$. Its **Rees algebra**, which will be denoted by $\mathcal{R}(I)$ or $R[It]$, has its significance centered on the fact that it provides an algebraic realization for the classical notion of blowing–up a variety along a subvariety, and plays an important role in the birational study of algebraic varieties, particularly in the study of desingularization.

Although one of our purposes here is to study the algebraic properties of these rings, an equal effort will be placed on their ancestors, **symmetric algebras**, that has several other interesting descendants. Given a commutative ring R and an R–module E, the symmetric algebra of E is an R–algebra $S(E)$ which together with a R–module homomorphism

$$\pi : E \longrightarrow S(E)$$

solves the following universal problem. For a commutative R-algebra B and any R–module homomorphism $\varphi : E \longrightarrow B$, there exists a unique R–algebra homomorphism $\Phi : S(E) \longrightarrow B$ such that the diagram

$$\begin{array}{ccc} E & \xrightarrow{\varphi} & B \\ \pi \downarrow & \nearrow \Phi & \\ S(E) & & \end{array}$$

is commutative. Thus, if E is a free module, $S(E)$ is a polynomial ring $R[T_1, \ldots, T_n]$, one variable for each element in a given basis of E. More generally, when E is given by the presentation

$$R^m \xrightarrow{\varphi} R^n \longrightarrow E \rightarrow 0, \; \varphi = (a_{ij}),$$

its symmetric algebra is the quotient of the polynomial ring $R[T_1, \cdots, T_n]$ by the ideal $J(E)$ generated by the 1-forms

$$f_j = a_{1j}T_1 + \cdots + a_{nj}T_n, \; j = 1, \ldots, m.$$

Conversely, any quotient ring of a polynomial ring $R[T_1, \ldots, T_n]/J$, with J generated by 1–forms in the T_i's, is the symmetric algebra of a module. Like the classical blowup, the morphism

$$\mathrm{Spec}(S(E)) \longrightarrow \mathrm{Spec}(R)$$

is a fibration of Spec(R) by a family of hyperplanes. The case of a vector bundle, when E is a projective module, already warrants interest.

The other algebras are derived from $S(E)$ by effecting modifications on its components, some rather mild but others brutal. To show how this comes about, consider first the case of ideals. For an ideal $I \subset R$, there is a canonical surjection

$$\alpha : S(I) \to \mathcal{R}(I).$$

If, further, R is an integral domain, the kernel of α is just the R–torsion submodule of $S(I)$. This suggests the definition of the Rees algebra $\mathcal{R}(E)$ of an R–module as $S(E)/T$, with T the (prime) ideal of the R–torsion elements of $S(E)$.

Another filtration is that associated to the symbolic powers of the ideal I. If I is a prime ideal, its nth symbolic power is the I–primary component of I^n. (There is a more general definition if I is not prime.) Its Rees algebra

$$\mathcal{R}_s(I) := \sum_{n \geq 0} I^{(n)} t^n,$$

the **symbolic Rees algebra** of I, which also represents a blowup, inherits more readily the divisorial properties of R, but has its usefulness limited because it is not always Noetherian. The presence of Noetherianess in $\mathcal{R}_s(I)$ is loosely linked to the number of equations necessary to define set–theoretically the subvariety $V(I)$. In turn, the lack of Noetherianess of certain cases has been used to construct counterexamples to Hilbert's 14th Problem.

Wedged between the Rees algebra of an ideal I and its symbolic blowup lies the integral closure of $\mathcal{R}(I)$. It is also defined by a filtration but we shall be unable to deal with it other than algorithmically. On the other hand, there will be methods to ascertain whether a symmetric algebra is normal and to compute its divisor class group.

The last algebra attached to a module E is the graded bi–dual of $S(E)$

$$B(E) := \sum_{i \geq 0} (S_i(E))^{**},$$

where $S_i(E)^{**}$ is the R–module bi–dual of the ith graded component of $S(E)$. If R is a Noetherian factorial domain and E is finitely generated, then $B(E)$ is a factorial domain. It is called the **factorial closure** of $S(E)$.

A common thread of the three algebras derived from $S(E)$ is that each is obtained by the same process of taking the ring of global sections of Spec$(S(E))$ on an appropriate affine open set.

To study the ideal theory of any class of algebras is, to a great extent, to make comparisons between two series of numbers: those arising from expressions of gross size—such as Krull dimension—and numbers measuring embeddings into regular (*i.e.* canonical) objects—depth is an example—and which have a more delicate geometric flavor. The running theme proper here is the ideal theory—Krull dimension, integrality, normality and factorization—of symmetric algebras, and its modifications. The aim is to highlight some of the known results, oftentimes technical points of their proofs, and to list significant open problems. At the same time, and this is emphasized, it seeks to provide gateways to other related developments in commutative algebra, but constraints of time and space leave this task barely started.

We shall now give an overall description of the contents. Brief introductions in each chapter have a more technical explanation of its results.

The first chapter focuses on the most primitive of the measures of an algebra: The analysis of the Krull dimension of a symmetric algebra $S(E)$. It is fairly complete in that there is an abstract general formula, and, based on it, a constructive method for computing the dimension in terms of a presentation of the module. It depends on the sizing of determinantal ideals, an exposition of which is also given. Two notions that will be visible throughout are introduced. First, we consider the family of conditions $\mathcal{F}_k$ on the Fitting ideals of a module. It will be used as a vehicle to read Krull dimension, and later of finer invariants such as divisor class groups. The other device is the technique of Jacobian duals, a process that attaches modules over different rings to certain ideals generated by quadrics. An important unresolved issue is when these special quadrics generate prime ideals.

The next two chapters deal with several generalizations of regular sequences, and there will be quite a diversity of them! A dominating theme is that of deciding when an ideal generated by linear forms in a partial set of variables is prime. Some interesting classes of examples are discussed in Chapter 2.

Chapter 3 constructs the **approximation complexes**, a family of chain complexes derived from Koszul complexes. They were introduced to make up for the lack of methods to find projective resolutions of the high symmetric powers of modules. Although they are not complexes of free modules, they permit one to ascertain the Cohen–Macaulayness of many blowup algebras. They work well when considerable information is know about the homology of the ordinary Koszul complexes is available. These complexes are the underpinnings for all the remaining exposition.

The next chapter is a primer on linkage theory, and highlights the connections between Koszul homology and residual intersections. It describes invariants of ideals in the linkage class of complete intersections, sketches out some differences between the odd and even linkage classes and in greater detail considers the Koszul homology of ideals associated to graphs. Hopefully it can be used as a hook to the beautiful work of Huneke and Ulrich on the structure of linkage and its multiple uses.

Chapter 5 has a fundamental character as it deals with normality of Rees algebras and basic constructions to generate new algebras from old ones, and effects a bridge between the special theory of ideals of linear type with more general ideals. It exploits the Noether like normalization expressed by a reduction of an ideal. Several recent results on the Hilbert function of primary ideals are explained away. New classes of Cohen–Macaulay rings are constructed.

There is a simple form of Serre's criterion, and an accompanying method to calculate the divisor class of normal algebras. It also explores some interesting relationships between subrings generated by monomials and their Rees algebras.There are a number of elementary techniques to produce Rees algebras with predictable properties from simpler algebras, along with computational tests of normality. Finally it introduces a novel technique (we almost said, technology) that seeks to apply the methods used to study algebras of linear type to much more general Rees algebras.

The next focus is the canonical module of a symmetric algebra. It occurs prominently in an unsettled conjecture purporting to describe all such algebras which are factorial: that they are complete intersections. Some cases are affirmed, and they can be viewed as assertions of homological rigidity. Other related results include the proof of the *Zariski–Lipman conjecture* for symmetric algebras over polynomial rings. It ends with a detailed examination of the structure of the canonical module.

Chapters 7 aims at finding ideal transforms of symmetric algebras. This includes the factorial closure of these algebras and the computation of symbolic blowups. A counterexample by Roberts to Hilbert's 14th Problem can be framed in this context. On the other hand, modules with linear presentation provide a vigorous set of fresh problems.

The next chapter seeks ways to determine the equations of the Rees algebras of ideals that are not necessarily of linear type. It is heavily mediated by homological algebra. There are several *ad hoc* approaches including computer–assisted methods. (There is even a little proof by computer!)

Chapter 9 is about commuting varieties of algebras, particularly Lie algebras. It is focused on two things: proving that certain ideals generated by quadrics are prime and bringing some invariant theory into the picture. It is very promising in applications, or at least looks so to this non–expert.

The last chapter discusses computer algebra methods in Commutative Algebra. It introduces the Buchberger's algorithm and traces out its role in the practice of computation in polynomial ideal theory. Because almost any application of these algorithms seems to lie on the brink of combinatorial explosion, greater emphasis is put on what might be called second generation methods, whereby theoretical information has to be fed into the computation, in order to beat the beast of complexity. (As it will be seen, they are still crude artifacts.) Among these methods are: Noether normalization, various forms of the *Nullstellensatz*, elements of primary decomposition, primality testing, computation of the integral closure of affine domains and of ideal transforms. There is another application domain for some of these techniques, the development of algebraic *solvers*: Programs to compute the solutions of systems of polynomial equations. Because of their need for *practicality* they tend to be integrated with numerical routines.

There are a great number of related topics that were not treated. The bountiful bibliography is intended to make up for that fault. Actually, at the time of this writing, several developments are leading to a very deep understanding of which Rees algebras are Cohen–Macaulay. Regretfully these will go unreported.

Chapter 1

Krull Dimension

This chapter has for aim the development of techniques to determine the most basic measure of a symmetric algebra—its Krull dimension. It requires an exposition of the classical Fitting ideals of modules, with estimates of codimensions of determinantal ideals.

The Krull dimension of a symmetric algebra $S(E)$ of a module E turns out to be connected to the invariant $b(E)$ of the module E introduced by Forster [71], a quarter of a century ago, that bounds the number of generators of E. This was shown by Huneke and Rossi [150]; furthermore, it was accomplished in a manner that makes the search for dimension formulas for $S(E)$ much easier. Based on slightly different ideas, [261] gives another proof of that result in terms of the heights of the Fitting ideals of E. It makes for an often effective way of determining the Krull dimension of $S(E)$.

The Fitting ideals are introduced from various directions, and their divisorial properties are noted. It is to be expected that these ideals play such an important role since they code the symmetric algebra of the module. In later chapters, their primary components will be used in expressing the divisor class group of normal algebras.

One topic which is central to this study, that of ascertaining when a symmetric algebra is an integral domain will have a very brief development here but will recur under various circumstances in several other chapters. One has been unable to capture the general properties that lead to symmetric algebras whose prime spectrum is irreducible.

At the end of the chapter we introduce a formalism—*the Jacobian dual*—that permits viewing some ideals generated by quadrics as ideals of definition of two distinct modules over different rings. It has proven itself to be an useful technique.

1.1 Fitting Ideals

The Fitting ideals of a module E generalize the classical invariants of finitely generated modules over Dedekind domains. In general, they do not fully determine the module. Nevertheless they convey many of its properties and are literally read off the presentation of the module.

Determinantal ideals

Given a presentation of an R–module E

$$R^m \xrightarrow{\varphi} R^n \longrightarrow E \longrightarrow 0,$$

its Fitting invariants are the various ideals generated by the minors of a matrix representation of φ.

We recall some notation in the following twin definitions:

Definition 1.1.1 Given a $n \times m$ matrix φ with entries in the commutative ring R, we set $I_t(\varphi)$ for the ideal generated by the $t \times t$ minors of the matrix φ. The ideal $I_1(\varphi)$ will be called the content of φ.

Definition 1.1.2 For an integer $0 \leq r < n$ the ideal $f_r(E)$ generated by the minors of order $n-r$ of the matrix φ is the rth Fitting ideal of E. If $r \geq n$ one puts $f_r(E) = R$.

The shift in the order of the minors makes the definition independent of the presentation of E; this was already shown by Fitting [69]. It was an early form of what became known as the Schanuel's lemma of Homological Algebra. When working from a fixed presentation of the module dealing with the ideals $I_t(\varphi)$ will be more convenient than using the notation of the Fitting ideals.

Another way of defining $f_r(E)$ is through the following broader notion.

Definition 1.1.3 Let $\varphi\colon F \to G$ be a homomorphism of R–modules. The *order ideal* of φ is the ideal

$$o(\varphi) = \sum f(\varphi(F)),\ f \in \operatorname{Hom}_R(G, R).$$

One could then put the notion above in the form

$$f_r(E) = o(\wedge^{n-r}\varphi)$$

where $\wedge^{n-r}\varphi$ denotes the $(n-r)$th exterior power of the presentation matrix of E. The advantage of the previous definition is its accessibility. In fact, when working out from a fixed presentation of E, we shall often state conditions directly in terms of the $I_t(\varphi)$.

Remark 1.1.4 If $\psi\colon R \to S$ is a ring homomorphism and E is a finitely generated module, one has from the definition that $f_r(E \otimes_R S) = \psi(f_r(E)) \cdot S$.

Elementary properties

We shall later be concerned with estimates of the heights of the Fitting ideals. Due to their lengthy expression, it is convenient to have other ways of describing them at least up to radicals. We briefly discuss two of them.

First observe that if $a_1(E)$ denotes the annihilator of E then

$$a_1(E)^n \subset f_0(E) \subset a_1(E).$$

More generally, the *invariant factors* of E are

$$a_r(E) = a_1(\wedge^r E),\ r \geq 1.$$

In addition, one also has the *Kaplansky invariants* of E: If $\mathcal{E}_r$ denotes the set of submodules of E that can be generated by r elements, then

$$k_r(E) = \sum a_1(E/F),\ F \in \mathcal{E}_r.$$

The relationship between them is summed up in:

Proposition 1.1.5 *Let E be a finitely generated R–module. Then*

$$\sqrt{f_r(E)} = \sqrt{k_r(E)} = \sqrt{a_{r+1}(E)}.$$

Proof. For a prime ideal P of R, let E_P be the localization of E at P. Denote by $\nu(E_P)$ the minimal number of generators of E_P. It follows easily that

$$\nu(\wedge^r E_P) = \binom{\nu(E_P)}{r}.$$

We then have:

(i) $a_r(E) \subset P$ if and only if $\nu(E_P) \geq r$.

(ii) If $k_r(E) \not\subset P$, there is $x \in k_r(E) \setminus P$ such that $x \cdot (E/F) = 0$, for some submodule $F \in \mathcal{E}_r$. Localizing at P one has $E_P = F_P$ and $\nu(E_P) \leq r$. Conversely, if $\nu(E_P) \leq r$ there is a submodule F generated by r elements and $y \notin P$, $y \cdot E \subset F$. Thus

$$k_r(E) \subset P \text{ if and only if } \nu(E_P) > r \text{ and } \sqrt{k_r(E)} = \sqrt{a_{r+1}(E)}.$$

(iii) Finally, assume $f_r(E) \not\subset P$. Choosing a minimal presentation of E_P this means $\nu(E_P) \leq r$. Since the converse is clear, altogether we have that $f_r(E) \subset P$ if and only if $\nu(E_P) \geq r+1$. □

Divisors and factoriality

A classical application of these invariants is in developing a theory of divisors, that is, an association of torsion modules with a class of ideals with good multiplicative properties.

To simplify the discussion, let us assume that R is an integral domain (or that, at appropriate places, ideals contain regular elements). First, one singles out the following set of prime ideals (*cf.* [71]). Let $\mathcal{P}$ be the collection of prime ideals that are minimal over an ideal of the type $(a):_R b = \{r \in R \mid rb \in (a)\}$. In other words, $P \in \mathcal{P}$ if and only if R_P has depth one.

Some of the properties of $\mathcal{P}$ are the following. A fractionary ideal I of R is a submodule of the field of fractions K of R for which there exists $0 \neq x \in R$, such that $xI \subset R$. The set $\{x \in K \mid xI \subset R\}$ will be denoted by I^{-1}. The ideal I is called *reflexive* or *divisorial* if $(I^{-1})^{-1} = I$, that is, I is reflexive as an R–module.

Proposition 1.1.6 *Let I be a fractionary ideal of R.*

(a) *If I is not contained in any $P \in \mathcal{P}$, $I^{-1} = R$.*

(b) *I is a reflexive ideal if and only if $I = \bigcap_{P \in \mathcal{P}} I_P$.*

Proof. It is an easy exercise that is left to the reader. □

There is a composition on the set $Div(R)$ of all divisorial ideals of R. For two divisorial ideals I and J define

$$I \circ J = ((I \cdot J)^{-1})^{-1}.$$

The submonoid formed by the invertible ideals will be denoted by $Inv(R)$. They compose, with other elements of $Div(R)$, through ordinary multiplication of ideals.

We come to one of our aims, the definition of a function from torsion modules of finite projective dimension into $Inv(R)$.

Definition 1.1.7 Let

$$F_1 \longrightarrow F_0 \longrightarrow E \longrightarrow 0$$

be a projective presentation of the torsion module E; put $I = f_0(E)$. The ideal $\mathbf{d}(E) = (I^{-1})^{-1}$ is the *determinantal divisor* of E.

On modules of finite projective dimension the function $\mathbf{d}(\cdot)$ has the following remarkable property ([190]):

Theorem 1.1.8 *Let E be a torsion module of finite projective dimension. Then $\mathbf{d}(E)$ is an invertible ideal.*

Lemma 1.1.9 *If* proj dim $(E) = 1$, *then $\mathbf{d}(E)$ is an invertible ideal of R.*

Proof. We prove more generally the following statement. Let E be a finitely generated module over a commutative ring R, annihilated by a regular element. Then $\text{proj dim}_R(E) \leq 1$ if and only if $f_0(E)$ is an invertible ideal of R. For that, let

$$0 \longrightarrow G \stackrel{\varphi}{\longrightarrow} F \longrightarrow E \longrightarrow 0$$

be exact with F a finitely generated free module. If $\text{proj dim}_R(E) \leq 1$, G is a finitely generated projective module that locally has the same rank as F. This means that locally $f_0(E)$ is generated by the determinant of φ.

For the converse one may assume, after localization at a prime ideal, that $f_0(E)$ is given by an specific minor of φ. It is then easy to see that G is generated by the corresponding column vectors that enter in the minor (see [190] for additional details). □

Lemma 1.1.10 $\mathbf{d}(\cdot)$ *is an additive function on the category of torsion modules which have finite projective dimension when localized at the primes in* $\mathcal{P}$. *That is, if*

$$0 \rightarrow E \longrightarrow F \longrightarrow G \rightarrow 0$$

is an exact sequence of such modules then

$$\mathbf{d}(E) \circ \mathbf{d}(G) = \mathbf{d}(F).$$

Proof. Construct a projective presentation of the exact sequence

$$\begin{array}{ccccccccc}
0 & \rightarrow & E_1 & \longrightarrow & F_1 & \longrightarrow & G_1 & \rightarrow & 0 \\
 & & \alpha\downarrow & & \beta\downarrow & & \gamma\downarrow & & \\
0 & \rightarrow & E_0 & \longrightarrow & F_0 & \longrightarrow & G_0 & \rightarrow & 0 \\
 & & \downarrow & & \downarrow & & \downarrow & & \\
0 & \rightarrow & E & \longrightarrow & F & \longrightarrow & G & \rightarrow & 0 \\
 & & \downarrow & & \downarrow & & \downarrow & & \\
 & & 0 & & 0 & & 0 & &
\end{array}$$

The matrix β is given in terms of the others by

$$\begin{bmatrix} \alpha & \star \\ 0 & \gamma \end{bmatrix}.$$

To show the assertion it is enough to verify it at the prime ideals of $\mathcal{P}$, but then α and γ are square matrices and $\det(\beta) = \det(\alpha) \cdot \det(\gamma)$. □

Proof of Theorem. We may assume that $\text{proj dim}_R(E) > 1$. Let x be a regular element in the annihilator of E. If

$$F \longrightarrow E \rightarrow 0$$

is a presentation of E with F free, consider the exact sequence

$$0 \rightarrow G \longrightarrow F/xF \longrightarrow E \rightarrow 0.$$

Then proj $\dim_R(G) =$ proj $\dim_R(E) - 1$. The previous lemma says that

$$\mathbf{d}(E) \circ \mathbf{d}(G) = (x^n)$$

where n is the rank of F. By an induction hypothesis however, $\mathbf{d}(G)$ is an invertible ideal of R, thus showing that $\mathbf{d}(E)$ is invertible as well. □

Corollary 1.1.11 *Let I be an ideal containing a regular element and having finite projective dimension. Then there exists a factorization $I = \mathbf{d}(R/I)J$, where J is an ideal satisfying $J^{-1} = R$.*

Proof. It suffices to apply the theorem to the module $E = R/I$. □

Corollary 1.1.12 (Factoriality of regular local rings) *A regular local ring R is factorial.*

Proof. Use the previous corollary on the prime ideals in $\mathcal{P}$. □

Krull–Serre criterion

Let us recall the conditions (S_k) of Serre. Let A be a Noetherian ring and let E be a finitely generated A–module.

Definition 1.1.13 Let k be a non–negative integer. E satisfies (S_k) if for every prime ideal $\wp$ of A the length of the longest E–regular sequence contained in $\wp$, which we denote by $\wp$–depth E, satisfies

$$\wp\text{-depth } E \geq \inf\{k, \text{height } \wp\}.$$

Thus the ring A satisfies (S_1) if it has no embedded primes. Further, A has (S_2) if height$(\wp) = 1$ for all $\wp \in \text{Ass}(A/aA)$ for any regular element $a \in A$.

Let A be an integral domain. Then $A = \bigcap A_P$, where P runs over the prime ideals associated to principal ideals [165, Theorem 53]. This representation has two immediate consequences: First, it follows that A will be integrally closed if each such localization A_P is normal. When A is Noetherian this observation along with its converse is Krull–Serre normality criterion.

Theorem 1.1.14 *Let A be a Noetherian integral domain. A is integrally closed if and only if the following conditions hold:*

(a) *For each prime $\wp$ of codimension one, $A_\wp$ is a discrete valuation domain.*

(b) *Every ideal I of codimension two contains a regular sequence on A with two elements.*

Condition (b) is normally referred to as the condition R_1 of Serre.

There is a related result for modules we are going to use later ([241]):

Proposition 1.1.15 *Let R be an integral domain that satisfies (S_2) and which is Gorenstein in codimension 1. Let M be a finitely generated R–module. Then*

$$M^{**} = \bigcap M_{\wp}, \text{ height } \wp = 1.$$

The intersection is taken in the vector space $V = M \otimes R_{(0)}$. In particular, if M is a finite algebra over R then M^{**} is also a finite R–algebra.

Proof. The proof is the same as the case of a normal domain in [241]. □

Sizes of determinantal ideals

Let R be a Noetherian ring and let φ be a $m \times n$ matrix with entries in R. Estimating the sizes of the determinantal ideals $I_t(\varphi)$ will become important for the rest of this chapter. This is made more difficult by the large number of generators they have.

The classical bound for the sizes $EN(m, n; t)$ of these ideals is the theorem of Eagon and Northcott ([53]):

Theorem 1.1.16 *The ideals $I_t(\varphi)$ satisfy*

$$\text{height } I_t(\varphi) \leq (m - t + 1)(n - t + 1),$$

with equality reached when φ is a generic matrix in $m \cdot n$ indeterminates.

We are going to present some extensions of this result, discovered by Bruns [26]. Along the way, we shall prove his generalization of Krull's principal ideal theorem. See also the results of Faltings [67] and Huneke [147].

The following terminology shall be used. Given a Noetherian ring R and a finitely generated R–module E

$$\text{rank } (E) = \sup\{\nu(E_P) \mid P = \text{minimal prime of} R\}.$$

On the other hand, the order ideal of an element $e \in E$

$$E^*(e) = \{f(e) \mid f \in \text{Hom}_R(E, R)\}.$$

This is the order ideal of the natural mapping induced by evaluation at "e" on the dual module.

If E is a free module then $E^*(e)$ is the ideal generated by the coordinates of e in any given basis. The following result was conjectured in [59] and proved in [26] (the case E free is the classical theorem of Krull). We give the proof in [26].

Theorem 1.1.17 *Let R be a Noetherian ring and let E be a finitely generated R–module. Let $e \in E$ and assume there is a prime ideal P with $e \in PE_P$. Then*

$$\text{height } E^*(e) \leq \text{rank } (E).$$

Proof. Note the following properties of these numbers, with respect to localization. If Q is a prime ideal, then height $E^*(e) \leq$ height $E^*_Q(e)$, while rank $E \geq$ rank E_P. We may also assume that $(R, \mathfrak{m})$ is a local ring and pass to the m–adic completion of R. Furthermore, we can mod out a minimal prime Q of R for which

$$\text{height } E^*(e) = \text{height } (E^*(e) + Q)/Q.$$

This ensures that the rank will not increase, and height $(E/QE)^*(e)$ will not decrease; one may assume that R is a local complete domain. We shall use the fact that such rings are universally catenarian [192].

The elements are ready for the main argument. Through a change of rings let us reduce to the case where E is free. We have that

$$e = \sum_{i=1}^{n} a_i e_i, \ a_i \in \mathfrak{m}.$$

Let S be the localization of $R[T_1, \ldots, T_n]$ at $(\mathfrak{m}, T_1, \ldots, T_n)$. The ideal generated by the indeterminates $Q = (a_1 + T_1, \ldots, a_m + T_m)$ is prime and $Q \cap R = 0$. Thus

$$(E \otimes S)_Q = E_{(0)} \otimes S_Q$$

is a free S_Q–module ($E_{(0)}$ is a vector space over the field of fractions of R). The element

$$f = (a_1 + T_1)e_1 + \cdots + (a_m + T_m)e_m$$

is contained in $Q \cdot (E \otimes S)$. From the earlier interpretation of Krull's theorem, we must have height $(E \otimes S)^*(f) \leq$ rank $(E \otimes S)$.

Because S is catenarian, there must exist a prime J containing $(E \otimes S)^*(f)$ and the indeterminates $T_1, \ldots, T_m$, with height $(J) \leq$ rank $(E) + m$. But J contains $(E \otimes S)^*(e)$ as well and therefore one of its minimal primes, say $\mathfrak{q}$; these however are all extended from R. This means that we have a chain of primes in S of length m:

$$\mathfrak{q} \subset (\mathfrak{q}, T_1) \subset \cdots \subset (\mathfrak{q}, T_1, \ldots, T_m)$$

which shows that height $(\mathfrak{q}) \leq$ rank (E). □

Given an $m \times n$ matrix φ, the the classical bound for the height of the ideal $I_t(\varphi)$ is the Eagon–Northcott formula:

$$\text{height } I_t(\varphi) \leq EN(m, n; t) = (m - t + 1)(n - t + 1).$$

It can be obtained by an iteration of the next result. (Observe $EN(m, n; t + 1) - EN(m, n; t) = m + n - 2t + 1$.)

Theorem 1.1.18 *Let R be a Noetherian ring and let φ be a $m \times n$ matrix over R. If $I_t(\varphi) \neq 0$ and $I_{t+1}(\varphi) = 0$, then*

$$\text{height } I_t(\varphi) \leq m + n - 2t + 1.$$

Proof. We may assume that R is a local, complete, domain whose maximal ideal $\mathfrak{m}$ is minimal over $I_t(\varphi)$. On the other hand, if an entry of φ is a unit, an elementary transformation changes the matrix into

$$\begin{bmatrix} 1 & 0 \\ 0 & \psi \end{bmatrix},$$

thus permitting a reduction to the case $t-1$. Furthermore, we may also assume that $\mathfrak{m}$ is not minimal over $I_t(\varphi')$ where φ' is the submatrix obtained by deleting the last column of φ, as otherwise we could induct on m.

We claim that height $(\varphi') \leq m-t$. Let $P \neq \mathfrak{m}$ be a minimal prime of $I_t(\varphi')$. Pick a basis $e_1, \ldots, e_m$ of R^m and view $\varphi\colon R^n \longrightarrow R^m$ and put $E = \text{coker}(\varphi)$ and correspondingly $E' = \text{coker}(\varphi')$. One has an exact sequence

$$0 \to Re'_m \longrightarrow E \longrightarrow E' \to 0$$

Since $I_t(\varphi) \not\subset P$, E_P is a free module of rank $m-t$. But $e'_m \neq 0$ implies that if E'_P could be generated by fewer than $n-t$ elements it would be free, a condition that would violate $I_t(\varphi') \subset P$. Altogether we have $e'_m \in PE_P$, so that by Theorem 1.1.17 height $E^*(e'_m) \leq \text{rank}\,(E) = m-t$.

Finally, since $I_{t+1}(\varphi) = 0$, the determinantal relations of the columns of φ' can be viewed as elements of $(R^m)^*$ which vanish on the image of φ, and therefore $I_t(\varphi') \subset E^*(e'_m)$, to establish the claim.

To complete the proof, since R is catenarian, it suffices to show that in the ring $R/I_t(\varphi')$, height $I_t(\varphi)/I_t(\varphi') \leq n-t+1$. We formulate this into an independent assertion.

Lemma 1.1.19 *Let $(R, \mathfrak{m})$ be a local ring, and φ a $m \times n$ matrix over R, whose last row consists of entries in $\mathfrak{m}$. Let φ' be the matrix formed by the first $m-1$ rows of φ. If $I_t(\varphi') = 0$, then* height $I_t(\varphi) \leq n-t+1$.

Proof. The following trick reduces the proof to the argument above. Replace $\varphi = (a_{ij})$ by the matrix

$$\psi = \begin{bmatrix} & \varphi & & \\ 0 & \cdots & 0 & -1 \end{bmatrix}$$

with φ the transpose of φ. The relation between ψ and φ is similar to that between the latter and φ', with

$$e'_{n+1} = \sum_{i=1}^{m} a_{mi} e'_i.$$

Exercise 1.1.20 Let E be a finitely generated module over a Noetherian ring R and let $f_r(E)$ be the first non-vanishing Fitting ideal of E. If $f_r(E)$ is invertible prove that the torsion part of E is a direct summand and E has projective dimension at most 1 (see [183]).

Exercise 1.1.21 Let A be a Noetherian integral domain and let B be a finite extension of A with the same field of fractions. Prove that $\text{proj dim}_A B < \infty$ if and only if $A = B$ (see [213]).

1.2 The Forster–Swan Number

Let R be a commutative, Noetherian ring of finite Krull dimension, and let E be a finitely generated R–module. For a prime ideal P of R, denote by $\nu(E_P)$ the minimal number of generators of the localization of E at P. It is the same as the torsion free rank of the module E/PE over the ring R/P. The Forster–Swan number of E is:

$$b(E) = \sup_{\wp \in \mathrm{Spec}(R)} \{\dim(R/P) + \nu(E_\wp)\}.$$

The original result of Forster was that E can be globally generated by $b(E)$ elements. Later it was refined, at the hands of Swan and others, by appropriately restricting the set of primes. It has to this day played a role in algebraic K–theory. The theorem of Huneke–Rossi ([150, Theorem 2.6]) explains the nature of this number.

Theorem 1.2.1 $\dim S(E) = b(E)$.

To give the proof of this we need a few preliminary observations about dimension formulas for graded rings. M. Kühl also noticed these formulas.

Lemma 1.2.2 *Let B be a Noetherian integral domain that is finitely generated over a subring A. Suppose there exists a prime ideal Q of B such that $B = A+Q$, $A \cap Q = 0$. Then*

$$\dim(B) = \dim(A) + \mathrm{height}(Q) = \dim(A) + \mathrm{tr.deg.}_A(B).$$

Proof. We may assume that $\dim A$ is finite; $\dim(B) \geq \dim(A) + \mathrm{height}(Q)$ by our assumption. On the other hand, by the standard dimension formula of [192], for any prime ideal P of B, $\mathbf{p} = P \cap A$, we have

$$\mathrm{height}(P) \leq \mathrm{height}(\mathbf{p}) + \mathrm{tr.deg.}_A(B) - \mathrm{tr.deg.}_{k(\mathbf{p})} k(P).$$

The inequality

$$\dim(B) \leq \dim(A) + \mathrm{height}(Q)$$

follows from this formula and reduction to the affine algebra obtained by localizing B at the zero ideal of A. □

There are two cases of interest here. If B is a Noetherian graded ring and A denotes its degree 0 component then:

$$\dim(B/P) = \dim(A/\mathbf{p}) + \mathrm{tr.deg.}_{k(\mathbf{p})} k(P)$$

and

$$\dim(B_\mathbf{p}) = \dim(A_\mathbf{p}) + \mathrm{tr.deg.}_A(B).$$

We shall need to identify the prime ideals of $S(E)$ that correspond to the extended primes in case E is a free module. It is based on the observation that if R is an integral domain, then the R–torsion submodule T of a symmetric algebra $S(E)$ is a prime ideal of $S(E)$. This is clear from the embedding $S(E)/T \hookrightarrow S(E) \otimes K$ (K = field of quotients of R) and the latter being a polynomial ring over K.

Let $\mathbf{p}$ be a prime ideal of R; denote by $T(\mathbf{p})$ the $R/\mathbf{p}$–torsion submodule of

$$S(E) \otimes R/\mathbf{p} = S_{R/\mathbf{p}}(E/\mathbf{p}E).$$

The torsion submodule of $S(E)$ is just $T(0)$.

Proof of theorem. By the formula above we have

$$\begin{aligned}\dim S(E)/T(\mathbf{p}) &= \dim(R/\mathbf{p}) + \text{tr.deg.}_{R/\mathbf{p}} S(E)/T(\mathbf{p}) \\ &= \dim(R/\mathbf{p}) + \nu(E_{\mathbf{p}})\end{aligned}$$

and it follows that $\dim S(E) \geq b(E)$.

Conversely, let P be a prime of $S(E)$ and put $\mathbf{p} = P \cap R$. It is clear that $T(\mathbf{p}) \subset P$,

$$\dim S(E)/P \leq \dim S(E)/T(\mathbf{p}),$$

and $\dim S(E) \leq b(E)$ as desired. □

Corollary 1.2.3 *Let R be a local domain of dimension d, and let E be a finitely generated module that is free on the punctured spectrum of R. Then* $\dim S(E) = \sup\{\nu(E), d + \text{rank}(E)\}$.

Valuative dimension

We shall now give another derivation of the dimension formula for $S(E)$, based on the notion of valuative dimension of a ring.

If A is an integral domain of field of quotients K, its valuative dimension is the supremum of the Krull dimension of all valuation rings of K that contain A ([80]). If A is a Noetherian domain this is the same as its Krull dimension.

Proof of 1.2.1: The Krull dimension of the Noetherian ring $S(E)/T(P)$ is the supremum of the Krull dimension of all of its valuation rings. If A is a valuation domain and V is its restriction to a valuation of R/P, then the subring of A generated by V and the degree one component M of $S(E)/T(P)$ is a polynomial ring over V on as many variables as the rank of M over R/P, since VM is a free V–module. □

We use the same technique to obtain the Krull dimension of the ordinary Rees algebra (see [283]).

Theorem 1.2.4 *Let R be a Noetherian ring and I be an ideal of R. For each minimal prime $\wp$ of R set $c(\wp) = 1$ if $I \not\subset \wp$, and $c(\wp) = 0$ otherwise. Then*

$$\dim R[It] = \sup\{\dim R/\wp + c(\wp)\}.$$

In particular, $\dim R[It] \leq \dim R + 1$, *and equality will hold when I contains regular elements.*

Proof. We may assume that $\dim R < \infty$. Let N be the nilradical of R. Since $N[t]$ is the nilradical of $R[t]$, it follows that $N[t] \cap R[It]$ is the nilradical of $R[It]$, so that we may assume that R is reduced and the minimal primes of $R[It]$ have the form $P_i = \wp_i R[t] \cap R[It]$. In turn, this implies that to determine $\dim R[It]$ we may assume that R is a domain. In case $I = 0$, then $R[It] = R$. If $I \neq 0$, let V be any valuation ring of R. Since I is finitely generated, $V \cdot R[It] = V[at]$, which has Krull dimension $\dim V + 1$. □

Graphs and ideals

Let us give an application, due to R. Villarreal [305], of the dimension formula. Given a graph G on a vertex set $V = \{x_1, \ldots, x_n\}$, he attaches the ideal $I(G)$ of the polynomial ring $k[x_1, \ldots, x_n]$, generated by the monomials $x_i x_j$ defined by its edges. This is not the same as the Stanley–Reisner ideal of the simplicial complex defined by G. Instead, it is related to its complementary simplex G': if the latter has no triangles, then the two ideals coincide.

Theorem 1.2.5 *Let G be a graph with n vertices and q edges, and let $I = I(G)$. If G is connected, then the Krull dimension of $S(I)$ is* $\sup\{n + 1, q\}$.

Proof. By Theorem 1.2.1, it is clear that the Krull dimension of $S(I)$ is at least the given bound. To prove the converse, we recall the notion of a minimal vertex cover of a graph. It is simply a subset A of G such that every edge of G is incident with a vertex of A and admits no proper subset with this property. There is a one–to–one correspondence between the minimal covers of G and the minimal primes of $I(G)$.

Let P be a prime ideal of height $n - i$, containing I; set

$$B = \{v \in V \mid v \notin P\}$$

Let $Q = (A)$ be a minimal prime of I contained in P, where A is a minimal vertex cover of G. Define now

$$C = \{x \in A \mid x \text{ is adjacent to some vertex in } B\}$$

and

$$Y = \{\{x, y\} \in X(G) \mid \{x, y\} \bigcap X(G)\} = \emptyset,$$

where $X(G)$ is the edge set of G. Notice that $\nu(I_P) \leq |C| + |Y|$. Since $|B| \geq i$, we obtain

$$\nu(I_P) + \dim(R/P) \leq |C| + |Y| + |B|.$$

The formula is now a consequence of the following:

Proposition 1.2.6 *Let G be a connected (n, q)–graph with vertex set $V = V(G)$ and edge set $X = X(G)$. Let A be a given minimal vertex cover of G, and let B be a subset of $V \setminus A$. Then $|B| + |C| + |Y| \leq \sup\{n + 1, q\}$.*

Proof. Let Y' be the set of edges covered by C, that is, $Y' = X(G) \setminus Y$. Consider the subgraph G' of G with edge set equal to Y' and vertex set

$$V(G') = \{z \in V(G) \mid z \text{ lies in some edge in } Y'\}.$$

If $Y = \emptyset$, then $|B| + |C| + |Y| \leq n$. Assume $Y \neq \emptyset$, and denote the connected components of Y' by $G_1, \ldots, G_m$. Set $B_i = B \cap V(G_i)$ and $C_i = C \bigcap V(G_i)$; notice $B = \cup B_i$ and $C = \cup C_i$.

We claim that $|B_i| + |C_i| \leq n_i - 1$, for all i, where $n_i = |V(G_i)|$. For that, fix an edge $\{x, y\} \in Y$. If $x \in V(G_i)$, $x \notin B_i \cup C_i$; using that B_i and C_i are disjoint we get the asserted inequality. On the other hand, if $x \notin V(G_i)$, we choose a vertex $z \in V(G_i)$ at a minimum distance from x. This yields a path $\{x = x_0, x_1, \cdots, x_r = z\}$. As $x_{r-1} \notin V(G_i)$, $z \notin B_i \cup C_i$, which gives $|B_i| + |C_i| \leq n_i - 1$. Altogether we have

$$|B| + |C| + |Y| \leq \sum_{i=1}^{m}(n_i - 1) + |Y| = \sum_{i=1}^{m}(n_i - 1) + (q - |Y'|).$$

But Y' is the disjoint union of $X(G_i)$ for $i = 1, \ldots, m$, and thus $|Y'| = \sum_{i=1}^{m} q_i$, $q_i = |X(G_i)|$. This permits writing the last inequality as

$$|B| + |C| + |Y| \leq \sum_{i=1}^{m}(n_i - 1) + q - \sum_{i=1}^{m} q_i = \sum_{i=1}^{m}(n_i - q_i - 1) + q.$$

Since G_i is a connected (n_i, q_i)–graph we have $n_i - 1 \leq q_i$. Therefore $|B| + |C| + |Y| \leq q$, establishing the claim. □

1.3 Dimension Formulas

Let R be a Noetherian domain and let

$$\varphi : R^m \longrightarrow R^n$$

be a presentation of the R-module E. We intend to express $\dim S(E)$ in terms of the sizes of the determinantal ideals of the matrix φ.

The condition $\mathcal{F}_k$

There is a set of conditions on the sizes of the Fitting ideals of E that keep recurring.

Definition 1.3.1 Let φ be a matrix with entries in R, defining the R–module E. For an integer k, φ (or E) satisfies the condition $\mathcal{F}_k$ if:

$$\text{height } I_t(\varphi) \geq \text{rank}(\varphi) - t + 1 + k, \quad 1 \leq t \leq \text{rank}(\varphi).$$

Remark 1.3.2 If E is an ideal, $\mathcal{F}_1$ is the condition G_∞ of [6]. There are constraints as to which $\mathcal{F}_k$ condition a module may support; for instance, for an ideal the condition $\mathcal{F}_2$ would contradict Krull's principal ideal theorem.

One can rephrase $\mathcal{F}_k$ in terms of how the image L of φ embeds into R^n. It means that for each prime ideal P where the localization E_P is not a free module then

$$\nu(E_P) \leq \text{rank}(E) + \text{height}(P) - k.$$

This says that for each prime ideal P of R, L_P decomposes into a summand of R^n_P and a submodule K of rank at most height$(P) - k$ (cf. [116], [261]).

Example 1.3.3 W. Bruns indicated the following method to obtain examples of torsion–free modules with $\mathcal{F}_k$, for various values of k.

Let R be a local domain of dimension d, and let G be the module of global sections of a vector bundle on the punctured spectrum of R, of rank $d-k$. By the remark above, from a presentation

$$0 \rightarrow L \longrightarrow R^n \xrightarrow{\varphi} G \rightarrow 0$$

of G, the module E = coker (φ^*) satisfies $\mathcal{F}_k$.

For example, let G be the module of first–order syzygies of the ideal of $R = k[a, b, c, d, e]$ (char $k \neq 2, 5$) generated by

$$5abcde - a^5 - b^5 - c^5 - d^5 - e^5$$
$$ab^3c + bc^3d + a^3be + cd^3e + ade^3$$
$$a^2bc^2 + b^2cd^2 + a^2d^2e + ab^2e^2 + c^2de^2$$

According to Bruns, G is the module of global sections of the Horrocks–Mumford bundle. The corresponding module E has $\mathcal{F}_3$.

Example 1.3.4 Given a regular local ring $(R, \mathfrak{m})$ of dimension $n \geq 4$, Vetter [302] constructs an indecomposable vector bundle on the punctured spectrum of R, of rank $n - 2$. Its module of global sections E has a presentation:

$$0 \rightarrow R \xrightarrow{\psi} R^n \xrightarrow{\varphi} R^{2n-3} \longrightarrow E \rightarrow 0.$$

φ is a matrix of linear forms in a regular system of parameters $x_1, \ldots, x_n$ of R, and

$$\psi = \begin{bmatrix} x_1 \\ \vdots \\ x_n \end{bmatrix}.$$

The module E satisfies $\mathcal{F}_1$. The cases n odd or even exhibit significant differences, as will be seen later.

Dimension formulas

We now return to the derivation of a dimension formula for $S(E)$. A lower bound for $b(E)$ is $b_0(E) = \dim R + \text{rank}(E)$, the value corresponding to the generic prime ideal in the definition of $b(E)$. We show that the correction from $b_0(E)$ can be explained by how deeply the condition $\mathcal{F}_0$ is violated. Set $m_0 = \text{rank}(\varphi)$, so that $\text{rank}(E) = n - m_0$. Without loss of generality we assume that R is a local ring, $\dim R = d$. Consider the descending chain of affine closed sets:

$$V(I_{m_0}(\varphi)) \supseteq \cdots \supseteq V(I_1(\varphi)) \supseteq V(1).$$

Let $P \in \text{Spec}(R)$; if $I_{m_0}(\varphi) \not\subseteq P$, $\text{rank}(E/PE) = n - m_0$, and therefore

$$\dim(R/P) + \text{rank}(E/PE) \leq b_0(E).$$

On the other hand, if $P \in V(I_t(\varphi)) \backslash V(I_{t-1}(\varphi))$, we have $\text{rank}(E/PE) = n - t + 1$; if $\mathcal{F}_0$ holds at t, the height of P is at least $m_0 - t + 1$ and again $\dim(R/P) + \text{rank}(E/PE) \leq b_0(E)$.

Define the following integer valued function on $[1, \text{rank}(\varphi)]$:

$$d(t) = \begin{cases} m_0 - t + 1 - \text{height } I_t(\varphi) & \text{if } \mathcal{F}_0 \text{ is violated at t} \\ 0 & \text{otherwise.} \end{cases}$$

Finally, if we put $d(E) = \sup_t \{d(t)\}$, we have the following dimension formula ([261, Theorem 1.1.2]):

Theorem 1.3.5 *Let R be an equi–dimensional catenarian domain and let E be a finitely generated R-module. Then*

$$b(E) = b_0(E) + d(E).$$

Proof. Assume that $\mathcal{F}_0$ fails at t, and let P be a prime as above. We have

$$\text{height}(P) \geq m_0 - t + 1 - d(t),$$

and thus

$$\dim(R/P) + \text{rank}(E/PE) \leq (d - (m_0 - t + 1 - d(t))) + (n - t + 1) = b_0(E) + d(t).$$

Conversely, pick t to be an integer where the *l a r g e s t* deficit $d(t)$ occurs. Let P be a prime ideal minimal over $I_t(\varphi)$ of height exactly $m_0 - t + 1 - d(t)$. From the choice of t it follows that $P \notin V(I_{t-1}(\varphi))$, as otherwise the deficit at $t - 1$ would be even higher. Since R is catenarian the last displayed expression gives the desired equality. □

Corollary 1.3.6 *Let R be an equi–dimensional catenarian domain, and let E be a finitely generated R–module. Then* $\dim S(E) = \dim R + \text{rank } E$ *if and only if E satisfies $\mathcal{F}_0$.*

Remark 1.3.7 If R is not an integral domain, let $p_1, \ldots, p_n$ be its minimal primes. It is clear from Theorem 1.2.1 that

$$\dim S_R(E) = \sup_{i=1}^n \{\dim S_{R/p_i}(E/p_i E)\}.$$

Assume that for each p_i, R/P_i is equi–dimensional. If we put

$$b_i(E) = b_0(E/p_i E), \quad d_i(E) = d(E/p_i E),$$

then

$$\dim S(E) = \sup_{i=1}^n \{b_i(E) + d_i(E)\}.$$

Example 1.3.8 Here is a simple illustration: If $E = \text{coker}(\varphi)$

$$\varphi = \begin{bmatrix} 0 & x & x \\ y & z & 0 \\ x & 0 & x \end{bmatrix},$$

$m_0 = 3, d(1) = d(3) = 0,$ but $d(2) = 1$, so $\dim S(E) = 3 + 1$.

One can define the condition $\mathcal{F}_k$ for negative integers as well (*cf.* [254]). The number $d(E)$, if positive, determines the most strict of the conditions satisfied by E: If $k = -d(E)$, $\mathcal{F}_k$ holds but $\mathcal{F}_{k+1}$ does not.

We leave to the reader the proof of another reformulation of the condition $\mathcal{F}_k$.

Remark 1.3.9 We can express $\mathcal{F}_k$ in terms of Krull dimension: If E has $\mathcal{F}_k$ then for any ideal I of R, of height at least k, height $I \cdot S(E) \geq k$.

Remark 1.3.10 We shall now observe that the formula fails to hold for non-catenarian domains. Pick a local domain of dimension three with two saturated chains of primes as in the graph (see [208, Example 2]):

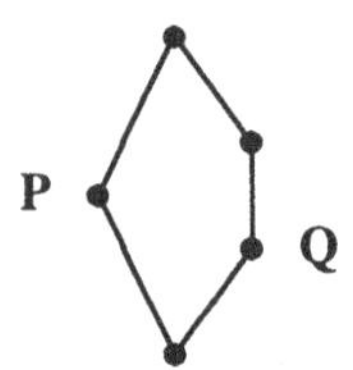

Let E be the module $R/P \oplus R/P \oplus R/Q$. A simple calculation shows that $b(E) = b_0(E) = 3$. If we present it as

$$\varphi \colon R^m \to R^3,$$

$I_2(\varphi) = P(P + Q)$ has height 1, so that $d(1) = 3 - 2 + 1 - 1 = 1$.

Exercise 1.3.11 Let E be a module defined by a square skew-symmetric matrix φ. Write out the requirements for E to satisfy the condition $\mathcal{F}_1$.

Exercise 1.3.12 Let R be a Noetherian local ring and let I be an ideal containing regular elements. Prove that

$$\dim S(I) = \dim R + 1$$

if and only if I satisfies $\mathcal{F}_0$.

Exercise 1.3.13 Let R be a Noetherian local ring and let I be an ideal which is generically a complete intersection. Prove that $\dim S_{R/I}(I/I^2) = \dim R$ if and only if I satisfies $\mathcal{F}_1$.

Exercise 1.3.14 Let E be a submodule of a free module R^n. One way to define the Rees algebra of E is the following. Pick a basis e_i, $1 \leq i \leq n$ of R^n. To each element $e = \sum_i r_i e_i$ of E associate the form $\varphi(e) = \sum_i r_i T_i$ of the polynomial ring $R[T_1, \ldots, T_n]$. Define $\mathcal{R}(E)$ to be the R–subalgebra of $R[T_1, \ldots, T_n]$ generated by all the $\varphi(e)$'s.

- Find a formula for Krull dimension of $\mathcal{R}(E)$.
- For each Fitting ideal of the module R^n/E estimate the Krull dimension of $\mathcal{R}(E) \times R/I$.

Problem 1.3.15 Let R be a regular local ring and let E and F be finitely generated R–modules. Find the required conditions for the equality

$$\dim S(E \oplus F) = \dim R + \text{rank } E + \text{rank } F$$

to hold.

1.4 Irreducibility and Reducedness

The conditions under which the symmetric algebra $S(E)$ is an integral domain have been a source of interest. The situation is well understood for modules of projective dimension one (*cf.* [11], [139], [260]), for several classes of ideals (see [116] for a survey and Chapters 2 and 3) and has been an incentive to the development of several generalizations of the notion of regular sequences and the theory of the approximation complexes.

Throughout this section R will be a Cohen-Macaulay integral domain, although the full strength of this condition will not always be used. The search for a general condition on E that leads $S(E)$ to be an integral domain is complicated by the many diverse ways it occurs. Nevertheless, the following result (*cf.* [20, Proposition 2.2]) identifies a basic ingredient.

Proposition 1.4.1 *Let R be a catenarian domain whose maximal ideals have the same height and let E be a finitely generated R–module. Then* Spec($S(E)$) *is irreducible if and only if E satisfies $\mathcal{F}_1$ and all the minimal primes of $S(E)$ have the same dimension.*

Proof. One of the minimal prime ideals of a symmetric algebra $S(E)$ is the R–torsion submodule T of $S(E)$. Since the Krull dimension of $S(E)/T$ is $\dim R+$ rank E by the proof of Theorem 1.2.1, it follows that if $S(E)$ is equi–dimensional then by Theorem 1.3.5 the condition $\mathcal{F}_0$ is automatically satisfied. Thus both conditions in the assertion imply $\mathcal{F}_0$.

Suppose Spec$(S(E))$ is irreducible; then, for each nonzero element x of R, as $x \notin T$, $\dim S(E) \otimes R/(x) = \dim S(E) - 1$. But $S_R(E) \otimes R/(x) \cong S_{R/(x)}(E/xE)$, so that the algebra $S(E/xE)$ will satisfy the condition of Theorem 1.3.5 if $R/(x)$ is reduced. It is not difficult to see that we can pick a square-free element x contained in all the associated primes of the $I_t(\varphi)$'s. The condition $\mathcal{F}_1$ will follow.

Conversely, suppose $\mathcal{F}_1$ holds and M is a minimal prime of $S(E)$ other than T. If $P = M \bigcap R$, then M is just $T(P)$. If however $P \neq 0$, reducing E modulo a prime element x of P would, in the presence of $\mathcal{F}_1$, yield an $R/(x)$–module E/xE whose $R/(x)$–rank is still rank E, so that the dimension of $S(E/xE)$, by Theorem 1.3.5, is one less than that of $S(E)$. The equi–dimensionality hypothesis on $S(E)$ rules this out. □

Corollary 1.4.2 *Let R be an integral domain. If E satisfies $\mathcal{F}_1$ and $S(E)$ is Cohen–Macaulay, then $S(E)$ is an integral domain.*

Proof. At each localization of R, $T(0)$ is the only minimal prime of $S(E)$ by the result above. Since the assumption implies that it is the only associated prime as well, the assertion ensues. □

Although we tend to emphasize the case of an integral domain, it will be clear that the arguments in this section also apply to the algebras of the following kind.

Definition 1.4.3 Let R be an integral domain and let E be a finitely generated R-module. E is said to be of *analytic type* if Spec$(S(E))$ is irreducible.

The advantage of this condition is that the condition $\mathcal{F}_1$ still holds for E, but in many cases it is much easier to prove irreducibility than integrality. Note that both imply $\mathcal{F}_1$ for E, and the former is (in the presence of this condition) characterized by equi–dimensionality (*cf.* [20]). It would be of interest to express it in ideal–theoretic terms.

Exercise 1.4.4 Let R be a Cohen–Macaulay integral domain, and let E be a module of rank e. Suppose I is a perfect ideal of R such that $f_e(E)$, the eth Fitting ideal of E, is not contained in any associated prime of I. Prove that if $S(E)$ is Cohen–Macaulay, then $S_{R/I}(E/IE)$ is Cohen–Macaulay if and only if for each prime ideal $\wp \supset I + f_e(E)$,

$$\nu(E_\wp) \leq \text{height}(\wp/I) + e.$$

1.5 Modules with Linear Presentation

Symmetric algebras of modules defined by matrices with linear entries have many unique properties. Because many Cohen–Macaulay varieties defined by quadrics will exhibit this feature, we introduce a duality formalism for them. Several of the most significant modules discussed later will belong to this class.

The most typical situation is that of a variety defined by an ideal $I = (f_1, \ldots, f_m)$ of a polynomial ring $k[x_1, \ldots, x_d]$, generated by quadrics, all monomials of which are square–free. We can attach to it a graph (the Newton graph) with the indeterminates as vertices and edges that correspond to the monomials that occur with nonzero coefficients among the f_j. If this graph is bipartite, then I defines two interacting fibrations over affine spaces, which arise in the following manner.

Let R be a ring and let E be a module with a presentation

$$R^m \xrightarrow{\varphi} R^n \longrightarrow E \to 0.$$

The equations of the symmetric algebra of E, $S = S(E)$, are

$$\mathbf{f} = [f_1, \ldots, f_m] = [T_1, \ldots, T_n] \cdot \varphi.$$

There is a naive duality for S that results from re–writing the equations in **f** (see [291]). Let

$$\mathbf{x} = [x_1, \ldots, x_d]$$

be an ideal containing the entries of φ. Then we can write

$$\mathbf{f} = \mathbf{x} \cdot B(\varphi),$$

where $B = B(\varphi)$ is a $d \times n$ matrix of linear forms in the T_i's variables. In general there will be several choices for B. When the entries of φ are 1–forms in the x_i's, with coefficients in some subring k, then B is well–defined. In fact, while φ is the Jacobian matrix of **f** with respect to **x**, $B(\varphi)$ is similarly defined with regard to **T**. By abuse of terminology, we shall say that $B(\varphi)$ defines the *Jacobian dual* of E, even in the case when the entries of φ are not linear forms (see [119], [257], [286], [293]).

To simplify matters, assume that $R = k[\mathbf{x}] = k[x_1, \ldots, x_d]$ is a polynomial ring, and E is an R–module with a presentation

$$R^m \xrightarrow{\varphi} R^n \longrightarrow E \to 0$$

where the entries of φ are homogeneous linear forms. Now write $Q = k[\mathbf{T}] = k[T_1, \ldots, T_n]$, and let B be a matrix of linear forms in Q such that

$$\mathbf{T} \cdot \varphi = \mathbf{x} \cdot B,$$

and consider the cokernel F of B

$$k[\mathbf{T}]^m \xrightarrow{B} k[\mathbf{T}]^d \longrightarrow F \longrightarrow 0.$$

The following encapsulates several technical aspects of the construction of the Jacobian dual.

Proposition 1.5.1 *Let E be a module with a linear presentation and let F be its Jacobian dual. Then*

(a) $S_{k[\mathbf{x}]}(E) \simeq S_{k[\mathbf{T}]}(F)$.

(b) *If $S(E)$ is unmixed then* $d + \text{rank } E = n + \text{rank } F$.

(c) *Let $\wp = (\mathbf{x})S(E)$. Then $S(E)_\wp$ is a regular local ring. Moreover, if $S(E)$ is unmixed then* $\dim S(E)_\wp = \text{rank } F$.

Proof. We only have to prove (c). The first assertion is clear since $S(E)_\wp$ is defined by linear equations. Next, note that

$$\text{height } (\mathbf{x})S(E) = d + \text{rank } E - n = \text{rank } F.$$

But B is a matrix of rank $d - \text{rank } F$, so that one of its minors of that order does not belong to $\wp$. This means that the system of linear equations in $S(E)_\wp$

$$\mathbf{x} \cdot B = 0,$$

has rank F independent solutions and therefore $\wp S(E)_\wp$ can be generated by rank F elements. □

This duality permits toggling between representations. Thus a question over a d–dimensional ring turns into the same question on a d–generated module over the other ring.

Another way to express this construction is as follows. Let

$$\psi : U \times V \rightarrow W$$

be a bilinear mapping of vector spaces over a field k. Picking bases, ψ is defined by a set of bihomogeneous quadratic equations which can be seen as defining symmetric algebras of modules over $k[U]$ or over $k[V]$.

Let us illustrate with a simple example. Consider the ideals associated with graphs in the manner of Theorem 1.2.5.

Example 1.5.2 Let $\mathcal{G}$ be a connected, bipartite graph. Partition its vertices into two subsets

$$V(\mathcal{G}) = X \cup Y,\ X = \{x_1, \ldots, x_m\},\ Y = \{y_1, \ldots, y_n\},$$

so that all the neighbours of elements in X lie in Y and vice–versa. It follows that $I(\mathcal{G})$ is the defining ideal of the symmetric algebras of two different modules

$$k[X]^p \longrightarrow k[X]^n \longrightarrow E_X \rightarrow 0,$$

and

$$k[Y]^p \longrightarrow k[Y]^m \longrightarrow E_Y \rightarrow 0.$$

It is easy to see that E_X (respectively, E_Y) has projective dimension equal to the supremum of the degrees of the elements in Y (respectively, in X).

Under certain circumstances, more can be said about the matrix B.

Proposition 1.5.3 *Let E be a module over $k[x_1, \ldots, x_d]$ with linear presentation φ, of size $n \times d$ whose rows are syzygies of the variables $x_1, \ldots, x_d$. Then B is skew symmetric.*

Proof. It is enough to observe that

$$\mathbf{x} \cdot B \cdot \mathbf{x}^t = \mathbf{T} \cdot \varphi \cdot \mathbf{x}^t = 0,$$

and the matrix B defines the trivial quadratic form over $k(\mathbf{T})$. □

Sometimes the assertion on B does not require that the entries of φ be linear. To explain this further we first need the following lemma ([61]).

Lemma 1.5.4 *Let R be a commutative ring, let $\mathbf{x} = (x_1, \ldots, x_d)$ be elements in R, and let $\mathbf{a} = (a_1, \ldots, a_d)$ be a first boundary in the Koszul complex on $\mathbf{x}$. Then there exists a skew symmetric $d \times d$ matrix A with entries in R such that*

$$\mathbf{a} = \mathbf{x} \cdot A.$$

Proof. Since a linear combination of alternating matrices is again alternating, it suffices to consider the case where $\mathbf{a}$ is a Koszul relation. We may even assume

$$(a_1, \ldots, a_d) = (x_2, -x_1, 0, \ldots, 0).$$

Now take

$$A = \begin{pmatrix} 0 & -1 & \\ 1 & 0 & \\ & & \\ & & O \end{pmatrix}$$

Proposition 1.5.5 *Let R be a Noetherian local ring, let $\mathbf{x} = (x_1, \ldots, x_d)$ be an R–regular sequence, let E be an R–module with a presentation*

$$R^d \xrightarrow{\varphi} R^n \longrightarrow E \to 0$$

such that $\varphi \cdot \mathbf{x}^t = 0$, and write $S = R[\mathbf{T}] = R[T_1, \ldots, T_n]$. Then there exists a skew symmetric $d \times d$ matrix B over S with linear entries in the variables $\mathbf{T}$ such that

$$\mathbf{T} \cdot \varphi = \mathbf{x} \cdot B.$$

Proof. Let φ_i be the ith row of φ. Since $\varphi_i \cdot \mathbf{x}^t = 0$ and $\mathbf{x}$ is a regular sequence, it follows that φ_i is a first boundary in the Koszul complex of $\mathbf{x}$. Hence by Lemma 1.5.4, there exists a skew symmetric $d \times d$ matrix A_i with entries in R such that $\varphi_i = \mathbf{x} \cdot A_i$. Now take $B = \sum_{i=1}^n A_i T_i$. □

Exercise 1.5.6 Let k be a field and let U be a finite dimensional k–algebra. The multiplication of U, $U \times U \to U$, gives rise to a module E over the ring of polynomials $A = S_k(U)$ obtained as follows. Let $e_1, \ldots, e_n$ be a basis of U as a vector space over k, and let c_{ijk} be the associated structure constants. Let $\sum_{i=1}^n x_i e_i$ and $\sum_{i=1}^n y_i e_i$ be two generic elements and set

$$(\sum_{i=1}^n x_i e_i) \cdot (\sum_{i=1}^n y_i e_i) = \sum_{i=1}^n f_i e_i.$$

The f_i's are quadratic polynomials that can be made explicit through the c_{ijk}'s. Let E be the module over A whose presentation matrix is the Jacobian matrix of the f_i's relative to the y_j's.

If k is perfect and U is commutative, show that E satisfies $\mathcal{F}_0$ if and only if U is a semi–simple algebra.

Exercise 1.5.7 (Bernd Ulrich) Let Y be a $n \times g$ matrix with entries in $R = k[\mathbf{x}]$, $\mathbf{x} = [x_1, \ldots, x_n]$. Suppose that the ideal $I_1(\mathbf{x} \cdot Y)$ has codimension g, and that

$$(I_1(\mathbf{x} \cdot Y), I_g(Y))$$

is $(\mathbf{x})$–primary. Show that $I_g(Y)$ has maximal codimension, i.e. $n - g + 1$. (*Hint*: View this ideal as the specialization of the equations defining the symmetric algebra of the $R/I_g(Y)$–module presented by the matrix Y; then use Theorem 1.2.1 to compute its Krull dimension.)

Chapter 2

Syzygetic Sequences

In this chapter we begin the task of studying blowup algebras through the comparison of the Rees algebra and the symmetric algebra of an ideal. Let I be an ideal of a commutative ring R. There is a canonical mapping

$$\alpha\colon S(I) \to \mathcal{R}(I)$$

from the symmetric algebra of I onto its Rees algebra. If I is generated by $\mathbf{x} = x_1, \ldots, x_n$, the homomorphism α can be more concretely described in the following manner. Let

$$0 \to Z_1 \longrightarrow R^n \stackrel{\varphi}{\longrightarrow} I \to 0.$$

φ induces a surjection from the ring of polynomials $R[T_1, \ldots, T_n]$ onto $S(I)$. Its kernel Q is the ideal determined by Z_1, that is, Q is generated by the linear forms $\sum r_i T_i$ such that $\sum r_i a_i = 0$. The Rees algebra $\mathcal{R}(I)$ is also a homomorphic image of $R[T_1, \ldots, T_n]$, with T_i mapping to $a_i t$. The kernel of this homomorphism is the ideal Q_∞ generated by all forms $f(T_1, \ldots, T_n)$ such that $f(a_1, \ldots, a_n) = 0$. We have the diagram

$$\begin{array}{ccccc}
Q_\infty & & \hookleftarrow & & Q \\
& \searrow & & \swarrow & \\
& & R[T_1, \ldots, T_n] & & \\
& \swarrow & & \searrow & \\
S(I) & & \stackrel{\alpha}{\longrightarrow} & & \mathcal{R}(I)
\end{array}$$

with $\mathcal{A} = \ker \alpha = Q_\infty / Q$.

It is apparent that $\mathcal{A}$ gives a measure of the difficulty in studying the algebra $\mathcal{R}(I)$. The ideal I is said to be of *linear type* if α is an isomorphism. The earliest significant example of an ideal of linear type is found in [198]: Every ideal generated by a regular sequence has this property (see [117] for further details). This result has led to a number of far-reaching generalizations.

The terminology itself was introduced by L. Robbiano and G. Valla. More recently, [104] used *geometrically* of linear type when $\ker(\alpha)$ is nilpotent. There is even an intermediate notion: I is *projectively* of linear type if the morphism

$$\mathrm{Proj}(\alpha)\colon \mathrm{Proj}(\mathcal{R}(I)) \to \mathrm{Proj}(\mathcal{S}(I))$$

is an isomorphism (see [171, Example 1.4]). The definition of linear type is further refined as follows. We will say that I is of rth *relation type*, if $\mathcal{A}$ is generated by its components of degree at most r.

This chapter considers several generalizations of regular sequences, and of their role in studying ideals of linear type. There is a much more detailed discussion in [116].

2.1 Syzygetic Ideals

In this section we begin the study of the comparison between the symmetric algebra and Rees algebra of an ideal I. It must start with the examination of the kernel of the canonical surjection

$$0 \to \delta(I) \longrightarrow S_2(I) \longrightarrow I^2 \to 0.$$

We shall be particularly interested in the class of ideals that satisfy the following property.

Definition 2.1.1 The ideal I is said to be *syzygetic* if $\delta(I) = 0$.

Let us approach the description of $\delta(I)$ from the point of view of the syzygies of a set of generators of the ideal. Let R be a commutative ring and let I be a finitely generated ideal. Let $\mathbf{a} = \langle a_1, \ldots, a_n \rangle$ be a set of generators of I and consider the corresponding module of syzygies

$$0 \longrightarrow Z \longrightarrow R^n \longrightarrow I \to 0,$$

where the ith base element e_i of R^n is mapped to a_i. Tensoring with R/I, induces the exact sequence

$$0 \to (Z \cap IR^n)/IZ \longrightarrow Z/IZ \longrightarrow R^n/IR^n \longrightarrow I/I^2 \to 0.$$

Note that $Z/(Z \cap IR^n)$ is the module of syzygies of the conormal module I/I^2, and we want to view it also as a component in the exact sequence

$$0 \to (Z \cap IR^n)/B \longrightarrow H_1 \longrightarrow Z/(Z \cap IR^n) \to 0.$$

Here H_1 (resp. B) is the first homology module (resp. the 1–boundaries) of the Koszul complex on $\mathbf{a}$. We set $\widetilde{H}_1 = Z/(Z \cap IR^n)$, and call it the *syzygy part* of H_1 (*cf.* [255]).

In this last exact sequence, H_1 and $\widetilde{H}_1$ may depend on $\mathbf{a}$; the module $\delta^*(I) = (Z \cap IR^n)/B$ is solely dependent on I according to ([259]):

Proposition 2.1.2 *There exists a natural homomorphism*

$$\alpha : \wedge^2 I \to \mathrm{Tor}_1^R(I, R/I)$$

and $\delta^*(I) = \mathrm{coker}(\alpha)$.

Proof. The mapping α is the degree two component of an algebra homomorphism from $\wedge I$, the exterior algebra of I, into $\mathrm{Tor}_*^R(R/I, R/I)$ ([91]). As the case here is more elementary, we give its full details.

Since I annihilates H_1, we have the exact sequence

$$0 \to B/IZ \longrightarrow (Z \cap IR^n)/IZ \longrightarrow \delta^*(I) \to 0,$$

where the mid module is just $\mathrm{Tor}_1^R(I, R/I)$. But $\mathrm{Tor}_1^R(I, R/I)$ is also the kernel of the multiplication mapping $I \otimes I \to I^2$, a fact we now exploit. The antisymmetrization mapping, $\wedge^2 I \to I \otimes I$, that sends $u \wedge v$ to $u \otimes v - v \otimes u$, defines a mapping from $\wedge^2 I$ to $\mathrm{Tor}_1^R(I, R/I) = (Z \cap IR^n)/IZ$. Thus $\alpha(a_i \wedge a_j) = a_j e_i - a_i e_j, \mathrm{mod} IZ$, which belongs to B/IZ, completing the proof. □

Note that since the image of α is annihilated by I, α factors through $\wedge^2 I \otimes R/I = \wedge^2 I/I^2$. If 2 is a unit of R, then α gives a splitting of $\mathrm{Tor}_1^R(I, R/I)$.

Proposition 2.1.3 $\delta(I) = \delta^*(I)$.

Proof. It follows from the commutative diagram

$$\begin{array}{ccccccc} & & \wedge^2 I & \longrightarrow & I \otimes I & \longrightarrow & S_2(I) & \to & 0 \\ & & \downarrow & & \| & & \downarrow & & \\ 0 & \to & \mathrm{Tor}_1^R(I, R/I) & \longrightarrow & I \otimes I & \longrightarrow & I^2 & \to & 0 \end{array}$$

of natural maps. □

There is an even more fundamental interpretation of $\delta(I)$ as a special homology module in the André–Quillen homology theory; see [110] for these and several other facts.

Under appropriate conditions, $\delta(I) = 0$ turns out to be very stringent.

Corollary 2.1.4 *Let $(R, \wp)$ be a local ring. Then $\delta(\wp) = 0$ if and only if R is a regular local ring.*

Proof. If R is a regular local ring, $\wp$ can be generated by a regular sequence $\mathbf{a}$, so that $H_1 = 0$, and $\delta(\wp) = 0$ along with it. Conversely, choosing a minimal generating set for $\wp$, $Z \subset \wp R^n$, so that $\delta(\wp) = (Z \cap \wp R^n)/B = 0$, implies $H_1 = 0$; thus $\wp$ is generated by a regular sequence. □

Another simple case is that of a principal ideal. If $I = (a)$, then $\delta(I) = 0$ means that

$$\mathrm{Tor}_1^R(I, R/I) = (0 \colon a)/((a) \cap (0 \colon a)) = 0,$$

in other words, $0\colon a = 0\colon a^2$. Another way to show $\delta(I) = 0$, which will be used later, requires a more detailed knowledge of the depth properties of both $\wedge^2 I$ and $\mathrm{Tor}_1^R(I, R/I)$.

It follows from the remark above that, if R is an Artinian local ring, then it admits no singly generated syzygetic ideal. One expects this phenomenon to be more widespread.

Proposition 2.1.5 *Let R be a Gorenstein local ring of dimension 0. Then R has no syzygetic ideal generated by two elements.*

Proof. For an ideal I generated by two elements, $I = (a, b)$, there is a canonical exact sequence

$$0 \to 0\colon I \longrightarrow R \xrightarrow{\varphi} I \oplus I \longrightarrow S_2(I) \to 0,$$

where $\varphi(1) = (a, -b)$. Denote by $\ell(\cdot)$ the length function for R–modules. If I is syzygetic, $S_2(I) = I^2$. On the other hand, $0\colon I = \mathrm{Hom}_R(R/I, R)$ has by duality the same length as R/I. This yields $\ell(I) = \ell(I^2)$, and therefore $I = 0$ by Nakayama's lemma. □

Exercise 2.1.6 Let R be a Cohen–Macaulay local ring of dimension 2 and let I be a perfect ideal. Show that I is syzygetic if and only if it is a complete intersection.

Exercise 2.1.7 Let A be an algebra of finite type over a field k and let

$$0 \to \mathbb{D} \longrightarrow A^e = A \otimes_k A \longrightarrow A \to 0$$

be the multiplication mapping. Describe $\delta(\mathbb{D})$ and study when it vanishes.

2.2 Residual Conditions

Denote by α the canonical homomorphism from the symmetric algebra of I, $S(I)$, onto its Rees algebra, $\mathcal{R}(I)$. We examine here the relationship between α and the induced homomorphism obtained by reducing it modulo I:

$$\beta : S(I) \otimes R/I = S_{R/I}(I/I^2) \longrightarrow \mathcal{R}(I) \otimes R/I = \mathrm{gr}_I(R).$$

Downgrading homomorphism

The mapping α is an isomorphism in degrees 0 an 1, and in order to examine more closely this comparison, we define an auxiliary R–module homomorphism of $S(I)$. We put

$$S(I) = \bigoplus_{t \geq 0} S_t(I),$$

the decomposition of the algebra into its graded components. The *downgrading* homomorphism, is defined by the formula

$$\forall\, a_1, \ldots, a_{t+1} \in I,\ \lambda(a_1 \cdot a_2 \cdots a_{t+1}) = a_1(a_2 \cdot a_3 \cdots a_{t+1}) \in S_t(I)$$

and extended by linearity. It is clear that λ is well-defined, and that it does not make a difference which element is pulled out of the monomial product. These maps together give rise to the diagram

$$\begin{array}{ccccccccc} 0 & \to & \mathcal{A}_{t+1} & \longrightarrow & S_{t+1}(I) & \longrightarrow & I^{t+1} & \to & 0 \\ & & \downarrow & & \lambda\downarrow & & \downarrow & & \\ 0 & \to & \mathcal{A}_t & \longrightarrow & S_t(I) & \longrightarrow & I^t & \to & 0 \end{array}$$

The first result is the surprising ([281]):

Theorem 2.2.1 *Let R be a Noetherian ring and let I be an ideal. Then α is an isomorphism if and only if β is an isomorphism.*

Proof. It is clear that β is an isomorphism along with α. For the converse, from the snake lemma applied to the diagram above, we obtain the exact sequence:

$$0 \to \mathcal{A}_t/\lambda(\mathcal{A}_{t+1}) \longrightarrow S_t(I/I^2) \longrightarrow \mathrm{gr}_I(R)_t \to 0.$$

By hypothesis we have $\mathcal{A}_t = \lambda(\mathcal{A}_{t+1})$ for $t \geq 2$. On the other hand, since $\mathcal{A}$ is a finitely generated ideal of $S(I)$, there exists an integer $s \geq 2$ such that

$$\mathcal{A}_{t+1} = S_1(I) \cdot \mathcal{A}_t, \quad t \geq s.$$

Applying λ to this equation, we get

$$\mathcal{A}_t = \lambda(\mathcal{A}_{t+1}) = \lambda(S_1(I) \cdot \mathcal{A}_t) = I\mathcal{A}_t.$$

By a localization argument and Nakayama's lemma, we get $\mathcal{A}_t = 0$ for $t \geq s$. The vanishing of the other components follows by a descending induction. □

Analytic independence

We look briefly into the geometric version of the property of being of linear type.

Theorem 2.2.2 $\mathcal{A} = \ker(\alpha)$ *is nilpotent if and only if* $\mathcal{B} = \ker(\beta)$ *is nilpotent.*

Lemma 2.2.3 *There exists an integer ℓ such that $I^\ell \cdot \mathcal{A} = 0$.*

Proof. For each prime ideal $I \not\subset P$, the localization α_P is an isomorphism. Thus for each homogeneous component $\mathcal{A}_t$ of $\mathcal{A}$ the closed set $V(I)$ determined by I contains the support of $\mathcal{A}_t$. As $\mathcal{A}$ is finitely generated, there exists a power of I that annihilates the ideal $\mathcal{A}$. □

Proof of Theorem. We only have to show that if $\ker(\beta)$ is nilpotent, then $\ker(\alpha)$ is nilpotent as well. Making use of the exact sequence above, as $\ker(\beta)$ is finitely generated, there exists an integer s such that

$$(\mathcal{A}_t/\lambda(\mathcal{A}_{t+1}))^s = 0, \quad \forall\, t \geq 0.$$

This means that $\mathcal{A}_t{}^s \subset \lambda(\mathcal{A}_{st+1}) \subset I.\mathcal{A}_{ts}$. Now we use the integer ℓ provided by the lemma to get $\mathcal{A}_t{}^{s\ell} = (\mathcal{A}_t{}^s)^\ell \subset I^\ell \cdot \mathcal{A}_{ts}{}^\ell = 0$. □

The meaning of this condition $\ker(\alpha) =$ nilpotent is the following. Let R be a local ring, of maximal ideal $\mathfrak{m}$. The elements $a_1, \ldots, a_n \in R$ are said to be *analytically independent* if any homogeneous polynomial $f(X_1, \ldots, X_n)$ for which $f(a_1, \ldots, a_n) = 0$ has all of its coefficients in $\mathfrak{m}$. If $I = (a_1, \ldots, a_n)$, this means that the ring $\mathcal{R}(I) \otimes (R/\mathfrak{m})$ is a polynomial ring in n indeterminates over $R/\mathfrak{m}$.

The following elementary observation gives an interpretation of this definition:

Proposition 2.2.4 *The ideal $\mathcal{A} = \ker(\alpha)$ is nilpotent if and only if I is locally generated by analytically independent elements.*

Proof. In the proof of Theorem 1.2.1 the so–called extended primes of a symmetric algebra of a module were identified: If $\wp$ is a prime of R, $T(\wp)$ is the kernel of the canonical mapping

$$S(I) \longrightarrow S(I) \otimes k(\wp),\ k(\wp) = R_\wp/\wp_\wp.$$

Every minimal prime of $S(I)$ is one of the $T(\wp)$'s. It follows that $\mathcal{A}$ is nilpotent if and only if, for every prime $\wp$ of R

$$\mathcal{A}_\wp \subset T(\wp)_\wp = \wp S(I)_\wp,$$

in other words $\mathcal{A} \subset T(\wp)$. □

Corollary 2.2.5 *Let R be a Noetherian ring and I be an ideal. If I is of linear type then it satisfies the condition $\mathcal{F}_1$.*

2.3 Proper Sequences and d-Sequences

The following notions play fundamental roles in the study of symmetric algebras of ideals and modules. In the next chapter we shall construct complexes whose acyclicity is related to these sequences.

Definition 2.3.1 Let $\mathbf{x} = \{x_1, \ldots, x_n\}$ be a sequence of elements in a ring R generating the ideal I. $\mathbf{x}$ is called a

(a) *d–sequence* if

1. $\mathbf{x}$ is a minimal generating system of the ideal I.
2. $(x_1, \ldots, x_i) \colon x_{i+1}x_k = (x_1, \ldots, x_i) \colon x_k$ for $i = 0, \ldots, n-1$ and $k \geq i+1$.

(b) *relative regular sequence* if $((x_1, \ldots, x_i) \colon x_{i+1}) \cap I = (x_1, \ldots, x_i)$ for $i = 0, \ldots, n-1$.

(c) *proper sequence* if $x_{i+1} \cdot H_j(x_1, \ldots, x_i; R) = 0$ for $i = 0, \ldots, n-1, j > 0$, where $H_j(x_1, \ldots, x_i; R)$ denotes the Koszul homology associated to the initial subsequence $\{x_1, \ldots, x_i\}$.

These conditions, which are visible in (a) and (b), provide a mechanism to examine algebraic relations among the elements of the sequence. In brief, the relationships between these sequences are:

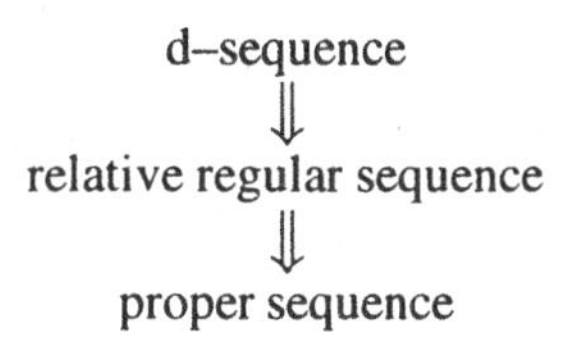

The similarity between *d–sequences* and *regular sequences* is further enhanced when the former are interpreted in terms of the vanishing of certain complexes (see Theorem 3.3.27) derived from Koszul complexes. In (c) it suffices to consider $j = 1$ ([171]). There are several other notions of sequences sharing in the properties of regular sequences ([117]).

The relationship between (a) and (c) is, broadly speaking, as follows: Each *d*–sequence is a proper sequence, and the linear forms in the symmetric algebra of the ideal generated by a proper sequence $\mathbf{x}$ generate a *d*–sequence relative to the ring $S(\mathbf{x})$ ([116, Theorem 12.10]).

d–sequences were introduced by Huneke as a mechanism to study the powers of ideals, particularly the *d*epths of *d*eterminantal ideals [137]. (The "*d*" arose from one of these *d*–words.) It plays in our study the same role as regular sequences in the classical theory of depth. The other notions are specialized aspects of *d*–sequences. The idea of a relative regular sequence is due to Fiorentini [68] and has served to focus attention on the coefficients of the syzygies of the elements in the sequence. The last notion, proper sequences, were defined in [114] to provide a tool to study the acyclicity of the approximation complexes.

To show as early as possible its significance, we prove the following ([141], [281]):

Theorem 2.3.2 *Every ideal generated by a d–sequence is of linear type.*

We make use of the following ascent method ([281]; see also [116, Proposition 3.3]):

Proposition 2.3.3 *Let I be an ideal of the ring R, and let f be an element of I such that for each integer t, $I^t \cap (0\colon f) = I^t(0\colon f)$. Then if $I/(f)$ is of linear type as an ideal of $R/(f)$, then I is of linear type as well.*

Proof. The inclusion $(f) \subset I$ of R–modules induces the exact sequence of maps of their symmetric algebras into the corresponding Rees rings

$$\begin{array}{ccccccccc} 0 & \to & (f)\cdot S(I) & \longrightarrow & S(I)_+ & \longrightarrow & S(I/(f))_+ & \to & 0 \\ & & \downarrow & & \alpha\downarrow & & \downarrow & & \\ 0 & \to & f\mathcal{R}(I)_+ & \longrightarrow & \mathcal{R}(I)_+ & \longrightarrow & \mathcal{R}(I/(f))_+ & \to & 0 \end{array}$$

and where all the vertical maps are induced by α. By assumption the mapping on the right is an isomorphism, so the bottom row is exact as well.

Assume that $0 \neq w$ is a homogeneous element of least degree in $\mathcal{A} = \ker \alpha$. From the exactness of the top row, $w = f \cdot w'$, $w' \in S_{t-1}(I)$. Since $\alpha(w) = 0 = f\alpha(w')$, $\alpha(w') \in I^{t-1} \cap (0\colon f) = I^{t-1}(0\colon f)$ and thus $w' \in (0\colon f)S_{t-1}(I)$ since $\ker(\alpha)_{t-1} = 0$. It follows that $w = f \cdot w' = 0$. □

The proof of the theorem will follow from

Proposition 2.3.4 *If* $\mathbf{x} = x_1, \ldots, x_n$ *is a d-sequence, then*

(a) *The images of* $x_2, \ldots, x_n$ *in* $R/(x_1)$ *form a d-sequence.*

(b) $(0\colon x_1) \cap (x_1, \ldots, x_n) = 0$.

Proof. (a) is clear from the definition of a d–sequence. To prove (b), we induct on n; we may assume that $n \geq 2$. Let $\sum r_i x_i \in 0\colon x_1$. Since $0\colon x_1 \subset 0\colon x_1 x_n = 0\colon x_n$, $r_n x_n^2 \in (x_1, \ldots, x_{n-1})$, we have $r_n x_n \in (x_1, \ldots, x_{n-1})$, as required. □

One aspect of d–sequences, that we shall make use later, is that they satisfy a sharp form of the Artin–Rees Lemma.

Proposition 2.3.5 *Let* $J \subset I$ *be ideals of the ring* R. *Suppose that* I *and* I/J *are ideals of linear type with respect to* R *and* R/J *respectively. Then* $J \cap I^m = J \cdot I^{m-1}$ *for* $m > 0$.

Proof. Consider the diagram of symmetric and Rees algebras associated to the ideals J and I:

$$\begin{array}{ccccccc} 0 \longrightarrow & J\cdot S(I) & \longrightarrow & S(I)_+ & \longrightarrow & S(I/J)_+ & \longrightarrow 0 \\ & \downarrow & & \downarrow & & \downarrow & \\ 0 \longrightarrow & Jt\cdot\mathcal{R}(I)_+ & \longrightarrow & \mathcal{R}(I)_+ & \longrightarrow & \mathcal{R}(I/J)_+ & \longrightarrow 0. \end{array}$$

Since the top sequence is exact and the vertical maps are isomorphisms, the bottom sequence is also exact. This means that for each m, $(I/J)^m = (I^m + J)/J \simeq I^m/JI^{m-1}$. □

Exercise 2.3.6 Let $(R, \mathfrak{m})$ be a Noetherian local ring and I an ideal generated by a d–sequence $x_1, \ldots, x_n$. Prove the following assertions:

1. If grade $I = g$, then $x_1, \ldots, x_g$ form a regular sequence.

2. If $R/\mathfrak{m}$ is infinite then I can be generated by a sequence that is a d–sequence in any order.

2.4 Determinantal Ideals of Linear Type

There are many other known classes of ideals of linear type. The determinantal ideals associated to a generic matrix that are of linear type have been fully described in Huneke [140]. More recently, B. Kotsev has proved that the ideal generated by the submaximal minors of a generic symmetric matrix is of linear type [169].

The circular complex

A complex of R–modules is called *circular* if it is periodic, of period 2:

$$\cdots \xrightarrow{\varphi} A \xrightarrow{\psi} B \xrightarrow{\varphi} A \xrightarrow{\psi} \cdots;$$

it is denoted by $\{A, B, \varphi, \psi\}$.

The quintessential circular complex was studied by Strickland [267]: A and B are free modules of rank n and φ and ψ are generic mappings. More precisely, let k be a ring and let $X = [x_{ij}]$ and $Y = [y_{ij}]$ be generic $n \times n$ matrices over k. She studied the variety of pairs of matrices satisfying the equations

$$X \cdot Y = Y \cdot X = 0,$$

and a deformation of it.

Consider the ring of polynomials $k[X, Y, t] = k[x_{ij}, y_{ij}, t]$ in $2n^2 + 1$ indeterminates and the ideal J generated by $I_1(X \cdot Y), I_1(Y \cdot X)$ and the polynomials

$$\sum_{k=1}^{n} x_{ik} y_{ki} - t, \quad i = 1, \ldots, n$$

$$\sum_{k=1}^{n} y_{ik} x_{ki} - t, \quad i = 1, \ldots, n.$$

For simplicity we state her result for the case of a field.

Theorem 2.4.1 *$k[X, Y, t]/J$ is a Cohen–Macaulay ring and t generates a radical ideal of height 1.*

It shall have several consequences, a significant for us being that

$$I = I_1(X \cdot Y) + I_1(Y \cdot X)$$

defines a symmetric algebra. It was necessary however to prove the more detailed theorem above. One of its corollaries is (see [267]):

Corollary 2.4.2 *Let $R = k[X, Y]/I$ and denote by φ and ψ the matrices X and Y taken modulo I. The complex $\{A = B = R^n, \varphi, \psi\}$ is acyclic.*

We shall make some comments about circular complexes attached to a square matrix. Let R be a ring and let φ be a $n \times n$ matrix with entries in R. Suppose $f = \det(\varphi)$ is a regular element. This means that the complex

$$0 \to R^n \xrightarrow{\varphi} R^n \longrightarrow E = \text{coker}(\varphi) \to 0$$

is acyclic. Tensoring with $S = R/(f)$ we obtain the exact sequence

$$0 \to E \longrightarrow S^n \xrightarrow{\overline{\varphi}} S^n \longrightarrow E \to 0,$$

since E is annihilated by f.

Let us assume that height $I_{n-1}(\varphi) \geq 2$. Let ψ be the adjoint matrix to $\overline{\varphi}$. It gives rise to the circular complex $\{S^n, \overline{\varphi}, \psi\}$; note that image $\psi \subset E$. Since ψ has rank 1 we may identify its image with the projection on one of the coordinates. In other words, there is an ideal J, given as the $(n-1)$–sized minors of the matrix obtained by deleting one column of φ, and an injection $\alpha\colon J/(f) \hookrightarrow E$.

Let us consider two special cases, when φ is either a generic matrix or a generic symmetric matrix. In both cases, $S = k[x_{ij}]/(f)$ is a Krull domain and E is a divisorial ideal. It is clear that the mapping α above is an isomorphism in codimension one. Since $J/(f)$ and E are divisorial, α must be an isomorphism.

We have the following consequence, a result originally observed and systematically exploited by Ferrand, and Valla [282]; see also [54] and [85].

Theorem 2.4.3 *Let φ be a symmetric $n \times n$ matrix and let J be the ideal generated by the maximal minors of a $n \times (n-1)$ submatrix. Then J is set–theoretically generated by 2 elements.*

Proof. It suffices to consider the generic symmetric case. Dualizing the exact sequence above, we get the sequence

$$0 \to E^* \longrightarrow (S^n)^* \xrightarrow{\overline{\varphi^*}} (S^n)^* \longrightarrow E^* \to 0,$$

which is exact since S is a Gorenstein ring. Because $\overline{\varphi} = \overline{\varphi^*}$, it follows that $E = E^*$. This implies that the divisor class of E^2 is trivial and therefore $E^2 = u \cdot L$, where L is an ideal of codimension at least 2. Since E is divisorial, we must have $\sqrt{E} = \sqrt{(u)}$, which completes the proof. (By specialization the result remains valid in all cases.) □

Submaximal minors

The generic determinantal ideals are also noteworthy because they cannot be generated by *d–sequences*. It is not known what happens in the symmetric case—that is, whether they can be generated by a *d–sequence*.

In this section we give a proof of the main result of [140]:

Theorem 2.4.4 *Let R be any Noetherian domain and let $X = (x_{ij})$ be an $n \times n$ generic matrix over R. Set $T = R[X] = R[x_{ij}]$ and $I = I_{n-1}(X)$. Then I is an ideal of linear type.*

Proof. A resolution of I over T is given by the *Scandinavian complex* of [96]; its details are as follows. Let Δ_{ij} be the $n-1 \times n-1$ minor of X determined by deleting the jth row and ith column, and multiply it by $(-1)^{i+j}$. The Δ_{ij}'s are the generators of I. Then all the relations on the Δ_{ij} follow from the cofactor expansion

$$X \cdot (\Delta_{ij}) = (\Delta_{ij}) \cdot X = \det(X) \cdot Id.$$

The equations arise from setting the off–diagonal entries equal to zero, and setting any two diagonal entries equal. Therefore the symmetric algebra of I is isomorphic to $R[X, Y]/J$, where J is generated by the sets of equations

$$\begin{cases} \sum_{j=1}^{n} x_{ij}y_{jk}, \ i \neq k \\ \sum_{j=1}^{n} y_{ij}x_{jk}, \ i \neq k \\ \sum_{j=1}^{n} x_{1j}y_{j1} - \sum_{j=1}^{n} x_{ij}y_{ji}, \ 2 \leq i \leq n \\ \sum_{j=1}^{n} y_{1j}x_{j1} - \sum_{j=1}^{n} y_{ij}x_{ji}, \ 2 \leq i \leq n \\ \sum_{j=1}^{n} x_{1j}y_{j1} - \sum_{j=1}^{n} x_{j1}y_{1j}. \end{cases}$$

Let $a \in S(I)$ be the image of $\sum_{j=1}^{n} x_{1j}y_{j1}$. From the equations above it is clear that in $S(I)$ we have the equations

$$\overline{X} \cdot \overline{Y} = \overline{X} \cdot \overline{Y} = a \cdot Id,$$

where '$^{-}$' denotes the map from $R[X, Y]$ to $S(I)$. Therefore,

$$S(I)/(a) = R[X, Y]/(I_1(X \cdot Y) + I_1(Y \cdot X))$$

which, by Strickland's theorem, is a reduced ring.

To complete the proof we need the following general fact:

Lemma 2.4.5 *Let $A = \sum_{n \geq 0} A_n$ be a non–negatively graded ring. Suppose A_{red} is a domain and there is a form x of positive degree such that $A/(x)$ is reduced and x is not nilpotent. Then A is a domain.*

Proof. Let $N = \text{rad}(A)$. As $A/(x)$ is reduced, $N \subset (x)$. Let y be a form in N. Then $y = rx$ for some r. However, as $x \notin N$ and N is prime, $r \in N$. Thus $N = xN$. It follows that $N = 0$. □

It remains to prove that $S(I)_{red}$ is a domain. From Proposition 2.2.4 it suffices to show that every localization $I_{\wp}$ is generated by analytically independent elements. This is achieved by inducting on n, as in the argument of [48]. Suppose it has been shown that $I_{t-1}(X')_Q$ is generated by analytically independent elements if $t < n$ and X' is a generic $t \times t$ matrix and Q is a prime ideal of $R'[X']$, where R' is any

Noetherian domain. Let $P \in \operatorname{Spec}(R[X])$. If $(x_{ij}) \not\subset P$, then $R[X]_P$ is a localization of $R[X]_{x_{11}}$ (we may assume $x_{11} \notin P$), $I_{n-1}(X)_{x_{11}}$ is generated by the $n-2$ sized minors of a generic $n-1 \times n-1$ matrix over the ring $R[x_{11}, x_{11}^{-1}]$. The induction shows that $(I_{n-1}(X)_{x_{11}})_P$ is generated by analytically independent elements. Thus we may assume that $(x_{ij}) \subset P$. We may also invert all nonzero elements of R. We must show that the elements Δ_{ij} are analytically independent in the ring $K[x_{ij}]_{(x_{ij})}$, K the field of quotients of R. In this case the ring $\bigoplus_{n\geq 0} I^n/\mathrm{m}I^n$ is isomorphic to $K[\Delta_{ij}] \subset K[x_{ij}]$. From Cramer's rule it follows however that $K[x_{ij}]$ lies in the field of quotients of $K[\Delta_{ij}, \det(X)]$. Since $\det(X)^{n-1} = \det(\Delta_{ij})$ it follows that $K[x_{ij}]$ is algebraic over $K[\Delta_{ij}]$. □

Corollary 2.4.6 *Let X be a $r \times s$ generic matrix ($r \leq s$), and let $I = I_t(X) \subset \mathbb{Z}[X]$. Then I is of linear type exactly in the following cases:*

(i) $t = 1$,

(ii) $t = r = s$,

(iii) $t = r$, $s = r + 1$,

(iv) $t = r - 1$, $r = s$.

Proof. The cases (i) and (ii) are obvious, (iv) is proved above, and (iii) will be proved in the next chapter.

On the other hand, any Plücker relation on the $t \times t$ minors of X corresponds to an element in the kernel of the map from $S_2(I)$ onto I^2. But there are such relations if $s \geq t + 2$ (if $t > 1$). □

Remark 2.4.7 The other class of linear type ideals of submaximal minors is that derived from symmetric matrices [169]. In one respect they may differ sharply. While both are of linear type, in the generic case the ideal I cannot be generated by a d–sequence. Indeed, according to [289, Theorem 3.7], if a Gorenstein ideal I of codimension four is generated by a d–sequence, then its conormal module, I/I^2, must be a reflexive R/I–module; but I^2 and the symbolic square $I^{(2)}$ already differ.

Exercise 2.4.8 Let R be a ring and let A be the symmetric algebra of the R–module M. Let $I = A_+$ be the augmentation ideal of the algebra A. Prove that I is an ideal of linear type.

Chapter 3

Approximation Complexes

Approximation complexes are, broadly speaking, differential, graded complexes lying over blowup algebras. Their main use is as surrogates for missing projective resolutions. At their most typical, approximation complexes are projective resolutions of the symmetric algebras of modules. Given a projective presentation of an R–module E

$$F \xrightarrow{\varphi} G \longrightarrow E \rightarrow 0,$$

the induced sequence

$$F \otimes S(G) \longrightarrow S(G) \longrightarrow S(E) \rightarrow 0$$

gives the beginning of a resolution of $S(E)$, the symmetric algebra of E, as a module over the polynomial ring $S(G)$. It is not known how to proceed many steps further, even for modules with uncomplicated free resolutions. There are nevertheless many significant cases where it can be fully resolved. These include modules whose symmetric algebras are complete intersection and the resolutions of low symmetric powers of many modules. In addition, there is a growing catalog of modules of projective dimension two for which a satisfactory situation has been achieved.

Another approach to the problem consists in replacing projective resolutions of $S(E)$ by finite differential graded complexes

$$0 \rightarrow B_n \longrightarrow B_{n-1} \longrightarrow \cdots \longrightarrow B_0 \longrightarrow S(E) \rightarrow 0, \tag{3.1}$$

where $B_0 = S(G)$, and the B_i are $S(G)$–modules that while no longer free modules still have good grade properties. It permits ascertaining many such algebras are Cohen–Macaulay.

The most important case for us will be when E is an ideal of R. The modules B_i's in these complexes are then intimately related to the cycles and homology of the ordinary Koszul complex built on a generating set of I. It will be the depth of these modules, in many cases of interest, that will lead to the acyclicity of the complexes (3.1) and estimates for the depth of $S(I)$.

In this chapter we construct several such complexes, introduced in [259] and systematically developed in [114], [115] and [116]. This last reference is an in-depth survey of the structural properties of the complexes. Here we only develop those properties that directly affect the applications.

3.1 Generalized Koszul Complexes

The basic Koszul complex $\mathbb{K}(\varphi)$ associated to a homomorphism

$$\varphi : E \longrightarrow R$$

is built on the exterior algebra of the module E having for differential the interior multiplication induced by φ:

$$\partial_\varphi : \wedge^r E \longrightarrow \wedge^{r-1} E$$

$$\partial_\varphi(e_1 \wedge \cdots \wedge e_r) = \sum_{i=1}^{r} (-1)^{i-1} \varphi(e_i)(e_1 \wedge \cdots \widehat{e_i} \cdots \wedge e_r).$$

More general complexes, associated to a homomorphism of R-modules $\varphi: F \longrightarrow G$, can be similarly constructed. There are two ways to go about it. First, one that emphasizes the R-module structure of the complexes. For each positive integer s there is a complex $K_s(\varphi)$:

$$0 \to \wedge^s F \longrightarrow \wedge^{s-1} F \otimes G \longrightarrow \cdots \longrightarrow F \otimes S_{s-1}(G) \longrightarrow S_s(G) \to 0,$$

with differential

$$\partial : \wedge^r F \otimes S_{s-r}(G) \longrightarrow \wedge^{r-1} F \otimes S_{s-r+1}(G)$$

$$\partial(e_1 \wedge \cdots \wedge e_r \otimes w) = \sum_{i=1}^{r} (-1)^{i-1} e_1 \wedge \cdots \widehat{e_i} \cdots \wedge e_r \otimes \varphi(e_i) \cdot w.$$

The other approach makes use of the induced homomorphism of $S(G)$-modules

$$\Phi : F \otimes S(G) \longrightarrow S(G)$$

obtained by the composition

$$F \otimes S(G) \longrightarrow G \otimes S(G) \longrightarrow S(G),$$

where the last mapping is the ordinary multiplication in the symmetric algebra $S(G)$. The basic construction of the Koszul complex is now applied to Φ.

The relationship between these complexes is clear. If $S(G)$ is graded in the usual manner, Φ is a homomorphism of graded $S(G)$-modules and $\mathbb{K}(\varphi)$ has an associated graded structure. Its component of degree s is $K_s(\varphi)$:

$$\mathbb{K}(\varphi) = \bigoplus_{s \geq 0} K_s(\varphi).$$

For example, if

$$\varphi: R^m \longrightarrow R^n, \ \varphi = (a_{ij}),$$

then $\mathbb{K}(\varphi)$ is the ordinary Koszul complex on the linear forms

$$f_j = \sum_{i=1}^{n} a_{ij} T_i, \ j = 1, \ldots, m,$$

of the ring of polynomials $S(R^n) = R[T_1, \ldots, T_n]$.

A major drawback of these complexes is that, except for the case when G is a free module and $F = R$, there are not many convenient criteria of acyclicity. For this reason, one often limits its usage to the case when both F and G are free modules. Because the 0–dimensional homology of these complexes are the symmetric powers of coker(φ), they may provide useful information about symmetric algebras of arbitrary modules.

The following summarizes several basic facts about the homology of these complexes (see [116]). For simplicity, the notation $K(\varphi)$ will be used for $K_s(\varphi)$ as well.

Proposition 3.1.1 *Let $\varphi: G \longrightarrow F$ be a homomorphism of R–modules.*

(a) *If $\Phi = \varphi + 0: G \oplus R \longrightarrow F$, then $H_*(\mathbb{K}(\Phi)) = H_*(\mathbb{K}(\varphi))[z]$, where z is a variable of degree 1 (i.e. $z^2 = 0$).*

(b) *If $\Phi = \varphi \oplus 1: G \oplus R \longrightarrow F \oplus R$, then $H_*(\mathbb{K}(\Phi)) = H_*(\mathbb{K}(\varphi))$.*

Lemma 3.1.2 *Let $\Phi: G = G' \oplus (R \simeq Re) \longrightarrow F$ and denote by φ its restriction to G'. There exists an exact sequence of complexes*

$$0 \to \mathbb{K}(\varphi) \longrightarrow \mathbb{K}(\Phi) \longrightarrow \mathbb{K}(\varphi)[-1] \to 0.$$

The connecting homomorphism is induced by multiplication by $\Phi(e)$.

Proof. It suffices to observe that

$$\mathbb{K}(\Phi)_{r+1} = \wedge^{r+1} G \otimes S(F) = \wedge^{r+1} G' \otimes S(F) \oplus (\wedge^r G' \otimes Re) \otimes S(F),$$

and its differential ∂ is related to the differential d of $\mathbb{K}(\varphi)$ in the following manner

$$\partial_r = \begin{bmatrix} d_r & \Phi_r(e) \\ 0 & -d_{r-1} \otimes E \end{bmatrix},$$

where E is the identity on Re and

$$\Phi_r(e) : (\wedge^r G' \otimes Re) \otimes S(F) \longrightarrow \wedge^r G' \otimes S(F)$$

sends $\omega \otimes e$ to $\omega \otimes \Phi(e)$. (In other words, $\mathbb{K}(\Phi)$ is the mapping cone induced by multiplication by $\Phi(e)$ [192, p. 135].) □

Proof of Proposition. (a) This follows directly from the lemma.

(b) We denote $S(F)$ by S and $S(F \oplus R) = S[T]$. $\mathbb{K}(\Phi)$ is then the Koszul complex associated to the mapping

$$(G \otimes S) \otimes S[T] \oplus S[T] \longrightarrow S[T],$$

with the generator of $S[T]$ mapped to T. The mapping on the first summand, as in the discussion above, is induced by φ. But the complex on $G \otimes S[T]$ is just $\mathbb{K}(\varphi) \otimes S[T]$. We then use flatness and the lemma. □

For ease of reference we state the acyclicity lemma of Peskine–Szpiro ([34]) in a convenient form for us:

Lemma 3.1.3 (Lemme d'acyclicité) *Let R be a local ring, and let $\mathcal{C}$ be a complex*

$$0 \to C_s \longrightarrow \cdots \longrightarrow C_1 \longrightarrow C_0 \to 0$$

of finitely generated R–modules.

(a) *If* depth $C_j \geq j$, *and* depth $H_j(\mathcal{C}) = 0$ *for* $j \geq 1$, *then $\mathcal{C}$ is acyclic.*

(b) *Furthermore, if for some positive integer t,* depth $C_j \geq t + j$ *for* $j \geq 0$, *then* depth $H_0(\mathcal{C}) \geq t$.

It is a consequence of the following basic result ([66, Lemma 1.1], [165, p.149]) on chasing depths in exact sequences:

Lemma 3.1.4 *Let R be a nonnegatively graded Noetherian ring such that R_0 is local, and denote by N its irrelevant maximal ideal. Let*

$$0 \to A \longrightarrow B \longrightarrow C \to 0$$

be an exact sequence of finitely generated R–modules where the maps are all homogeneous. Then, with depths taken relative to N, either

(a) depth$(A) \geq$ depth$(B) =$ depth(C), *or*

(b) depth$(B) \geq$ depth$(A) =$ depth$(C) + 1$, *or*

(c) depth$(C) >$ depth$(A) =$ depth(B).

Complete intersections

We look into the vanishing of the Koszul complex of a mapping

$$G = R^m \xrightarrow{\varphi} R^n = F,$$

in terms of the determinantal ideals of the matrix $\varphi = (a_{ij})$.

We begin by recalling the notion of the rank of a module. Let E be a module defined by a matrix φ, $E = \text{coker}(\varphi)$.

Definition 3.1.5 E is said to have generic rank r_0 if any of the following equivalent conditions hold:

1. $E \otimes K$ is K-free of rank r_0, where K = total ring of fractions of R.
2. If $m_0 = n - r_0$, then $I_{m_0}(\varphi)$ is faithful and $I_t(\varphi) = 0$ for $t > m_0$.

Let us describe the symmetric algebras which are complete intersections (*cf.* [11], [139] and [261]).

Theorem 3.1.6 *Let E be a module with a presentation*

$$R^m \xrightarrow{\varphi} R^n \longrightarrow E \to 0.$$

Then $S(E)$ is a complete intersection on the forms

$$f_j = \sum_{i=1}^{n} a_{ij} T_i,\ j = 1, \ldots, m$$

if and only if

$$\text{grade } I_t(\varphi) \geq \text{rank}(\varphi) - t + 1,\ 1 \leq t \leq \text{rank}(\varphi).$$

Note that these are the conditions embodied in the Fitting condition $\mathcal{F}_0$, with grade instead of height.

Proof. We make two observations before beginning the proof. If the polynomials f_j, $j = 1, \ldots, m$ form a regular sequence, their relations are of degree 1 and higher, so they are linearly independent over R. On the other hand, the required grade implies the independence of the forms. One may thus assume that E has projective dimension at most one.

Suppose that the f_j form a regular sequence. Consider the complexes $K_s(\varphi)$, for $s \leq m$:

$$0 \to \wedge^s R^m \xrightarrow{\varphi_s} \wedge^{s-1} R^m \otimes R \to \cdots \to R^m \otimes S_{s-1}(R^n) \to S_s(R^n) \to E \to 0.$$

By assumption each of these complexes is acyclic, so that in particular the map φ_s must satisfy the condition of the *lemme d'acyclicité* ([34]; see also Lemma 3.1.3): that is, the grade of the ideal of maximal minors of φ_s must be at least s. If we note that φ_s is the presentation matrix for of the module $\text{coker}(\wedge^s \varphi^t)$, the assertion will follow from the following computation.

Lemma 3.1.7 *Let E be a module defined by a presentation matrix*

$$R^m \xrightarrow{\varphi} R^n \longrightarrow E \to 0.$$

Then $\operatorname{Supp} R/I_r(\varphi) = \operatorname{Supp} R/\operatorname{ann}(\wedge^{m-r+1} E)\ \forall r.$

Proof. It suffices to note the equivalencies: For any prime ideal $\wp$,

$$\begin{aligned} I_r(\varphi) \subset \wp &\Leftrightarrow \text{rank}\ \varphi\ (\text{mod}\,\wp) < r \\ &\Leftrightarrow \nu(E_\wp) \geq m - r + 1 \\ &\Leftrightarrow (\wedge^{m-r+1} E)_\wp \neq 0 \\ &\Leftrightarrow \operatorname{ann}(\wedge^{m-r+1} E) \subset \wp \end{aligned}$$

The converse is a consequence of Proposition 3.1.1 and the Lemma 3.1.3, once the condition $\mathcal{F}_0$ is used. □

Lebelt–Weyman complexes

Let R be a ring and consider a complex

$$\mathbb{F}\ : 0 \to F_n \xrightarrow{f_n} \cdots \longrightarrow F_2 \xrightarrow{f_2} F_1 \xrightarrow{f_1} F_0 \to 0$$

of finitely generated free R–modules. Lebelt [179] and Weyman [307] have defined associated complexes of free modules over all the symmetric and exterior powers of $E = \operatorname{coker}(f_1)$.

Their elegant construction is founded on the notion of algebra of divided powers, a concept that can be used in both situations (see [95, Section 1.7]). Throughout the discussion, R will be a commutative ring and $\mathcal{D}$ a graded, skew–commutative algebra; the latter means that if x and y are homogeneous elements of $\mathcal{D}$ then

$$\begin{aligned} x \cdot y &= (-1)^{\deg(x)\cdot\deg(y)} y \cdot x \\ x \cdot x &= 0 \text{ if } x \text{ is homogeneous of odd degree.} \end{aligned}$$

Definition 3.1.8 Let R be a commutative ring. A graded algebra $\mathcal{D}$ admits a *system of divided powers* if to every element $x \in \mathcal{D}$ of even positive degree there is an associated sequence of elements $x^{(k)} \in \mathcal{D}$, $k \in \mathbb{N}$, satisfying:

1. $x^{(0)} = 1$, $x^{(1)} = x$, $\deg x^{(k)} = k \deg x$.
2. $x^{(h)} x^{(k)} = (h, k) x^{(h+k)}$, where $(h, k) = \binom{h+k}{h}$.
3. $(x + y)^{(k)} = \sum_{i+j=k} x^{(i)} y^{(j)}$.
4. $(x^{(h)})^{(k)} = [h, k] x^{(hk)}$ for $k \geq 0,\ h \geq 1$, where $[h, k] = \frac{(hk)!}{k!(h!)^k}$.
5. For $k \geq 2$

$$(xy)^{(k)} = \begin{cases} 0 & \text{if the degrees are odd} \\ x^{(k)} y^{(k)} & \text{if the degrees are even and one is nonzero.} \end{cases}$$

Such algebras will be simply called *divided powers algebras*. Their homomorphisms must preserve the divided powers structure: $f(x^{(k)}) = (f(x))^{(k)}$. This formalizes properties of a polynomial ring $R[x_1, \ldots, x_n]$, containing the rational numbers and graded so that each $\deg x_i$ is even. Indeed, for each form x one can define

$$x^{(k)} = \frac{1}{k!}x^k,$$

and the conditions are all realized.

We want to define divided powers algebras attached to free R–modules. At its simplest, let $E = R \cdot e$ be a free, graded R–module of rank 1. If $\deg e$ is even, the module $S(E)$ has a natural system of divided powers; we call it the divided polynomial algebra. If $\deg e$ is odd, it is clear that the exterior algebra of E, $\wedge(E)$, also supports a system of divided powers. They are the building blocks for free divided power algebras.

More generally, if $E = Re_1 \oplus \cdots \oplus Re_m \simeq R^n$ is a graded module, with e_i homogeneous, then

$$\mathcal{D}(E) = \mathcal{D}(Re_1) \otimes \cdots \otimes \mathcal{D}(Re_m),$$

where $\mathcal{D}(Re_i)$ is one of the basic algebras above, each chosen according to the degree of the generator e_i.

We return now to the complex $\mathbb{F}$, which we view as a differential graded module

$$\mathbb{F} = \bigoplus_{i=0}^{n} F_i,$$

where F_i is its component of degree i, with differential

$$\partial = \bigoplus_{i=1}^{n} f_i.$$

Definition 3.1.9 The *Lebelt–Weyman complexes* associated to $\mathbb{F}$ are:

$$\begin{aligned} \mathcal{S}(\mathbb{F}) &= \mathcal{D}(\mathbb{F}), \\ \mathcal{L}(\mathbb{F}) &= \mathcal{D}(\mathbb{F}[-1]). \end{aligned}$$

They will be turned into differential graded algebras once ∂ is fully extended. For each generator it is going to be required to satisfy

$$\partial(x^{(k)}) = x^{(k-1)}\partial(x).$$

$\mathcal{D}(\mathbb{F})$ is naturally bigraded: For each sequence $(a_0, a_1, \ldots, a_n)$ of natural numbers, set

$$\begin{aligned} \mathcal{S}_r(\mathbb{F}) &= \bigoplus_s \mathcal{D}(\mathbb{F})_{r,s} \\ &= \bigoplus_s \Big(\bigoplus_{(a_0, \ldots a_n)} \mathcal{D}_{a_0}(F_0) \otimes \wedge^{a_1} F_1 \otimes \cdots, \sum a_i = r, \sum i \cdot a_i = s \Big). \end{aligned}$$

∂ will have bi–degree $(-1,0)$, and

$$H_0(\mathcal{S}_r(\mathbb{F})) = S_r(E).$$

Similarly,

$$\begin{aligned}\mathcal{L}_r(\mathbb{F}) &= \bigoplus_s \mathcal{D}(\mathbb{F}[-1])_{r,s} \\ &= \bigoplus_s \Big(\bigoplus_{(a_0,\ldots a_n)} \wedge^{a_0} F_0 \otimes \mathcal{D}_{a_1} F_1 \otimes \cdots, \sum a_i = r, \sum i\cdot a_i = s\Big),\end{aligned}$$

and now

$$H_0(\mathcal{L}_r(\mathbb{F})) = \wedge^r E.$$

We will not go into the construction of the extension of ∂, other than to point out the similarity with the Koszul maps we encountered already (see [179] and [307] for fuller details). In fact, they are broad based extensions of Koszul complexes.

Remark 3.1.10 The lengths of these complexes are:

$$\begin{aligned}\text{length}(\mathcal{S}_r(\mathbb{F})) &= nr & &\text{for } n \text{ even.}\\ &= \min\{(n-1)r + \text{rank } F_n, nr\} & &\text{for } n \text{ odd.}\\ \text{length}(\mathcal{L}_r(\mathbb{F})) &= nr & &\text{for } n \text{ odd.}\\ &= \min\{(n-1)r + \text{rank } F_n, nr\} & &\text{for } n \text{ even.}\end{aligned}$$

It follows that if $n \geq 2$ and all the maps f_i are minimal in some sense (e.g. $(R, \mathfrak{m})$ is a local ring and the entries of each f_i lie in $\mathfrak{m}$) then the complexes cannot be acyclic for all values of r. Nevertheless, they may still provide information that can be fed into the complexes to be defined in the next section.

Symmetric squares

We shall attempt to pinpoint some general difficulties in the use of these complexes. Let

$$\mathbb{F}: \qquad 0 \to R^p \xrightarrow{\psi} R^m \xrightarrow{\varphi} R^n \longrightarrow E \to 0$$

be a free resolution of the module E.

For $S(E)$ to be a domain, the symmetric powers $S_r(E)$ must be torsion–free modules. For $r = 2$ one has the complex $\mathcal{S}_2(\mathbb{F})$:

$$0 \to D_2(R^p) \to R^p \otimes R^m \to R^p \otimes R^n \oplus \wedge^2 R^m \to R^m \otimes R^n \to S_2(R^n) \to 0.$$

To be acyclic, in the presence of $\mathcal{F}_1$, requires height$(I_p(\psi)) \geq 4$, cf. [307]. If in addition E is a reflexive module, it will be satisfied. Consequently we obtain:

Proposition 3.1.11 *Let R be a four-dimensional domain and let E be a reflexive module of projective dimension two. Then $S(E)$ is not an integral domain.*

In Chapter 9 we shall present another complex over the symmetric square of modules that mimic properties of Gorenstein ideals of codimension three.

3.2 Construction of the Approximation Complexes

There are various complexes that have been dubbed *approximation complexes*. Perhaps the broadest template for them is the following double Koszul complex. Let F and G be R–modules and suppose there are two mappings:

$$\begin{array}{ccc} F & \stackrel{\psi}{\longrightarrow} & G \\ \varphi \downarrow & & \\ R & & \end{array} \tag{3.2}$$

The algebra $\bigwedge F \otimes S(G)$ is a double complex with differentials

$$d_\varphi = \partial\colon \wedge^r F \otimes S_t(G) \longrightarrow \wedge^{r-1} F \otimes S_t(G),$$

which is the Koszul complex associated to the mapping φ and coefficients in $S(G)$, while

$$d_\psi = \partial'\colon \wedge^r F \otimes S_t(G) \longrightarrow \wedge^{r-1} F \otimes S_{t+1}(G)$$

defines the Koszul mapping of ψ. It is a straightforward computation to verify that

$$\partial \cdot \partial' + \partial' \cdot \partial = 0.$$

The resulting double complex will be denoted by $\mathcal{L} = \mathcal{L}(\varphi, \psi)$. The case of interest here is $\varphi\colon F = R^n \to R$, the mapping associated to a sequence $\mathbf{x} = \{x_1, \ldots, x_n\}$ of elements of R.

Definition 3.2.1 The complex $\mathcal{L} = \mathcal{L}(\mathbf{x}) = \mathcal{L}(\varphi, \text{identity})$ will be called the double Koszul complex of $\mathbf{x}$.

The complex $\mathcal{L}(\partial)$ is the Koszul complex associated to the sequence $\mathbf{x}$ in the polynomial ring $S = S(R^n) = R[T_1, \ldots, T_n]$. $\mathcal{L}(\partial')$, on the other hand, is also an ordinary Koszul complex but constructed over the sequence $\mathbf{T} = T_1, \ldots, T_n$. Thus we have a grading $\mathcal{L}(\partial')$ by subcomplexes of R–modules, $\mathcal{L} = \sum \mathcal{L}_t$

$$\mathcal{L}_t = \sum_{r+s=t} \wedge^r R^n \otimes S_s(R^n),$$

and the $\mathcal{L}_t$ are exact for $t > 0$. Attaching coefficients from an R–module E extends the construction to the complex $\mathcal{L}(\mathbf{x}, E) = E \otimes \bigwedge R^n \otimes S(R^n)$.

The approximation complexes

We can now define several new complexes. Denote by $\mathcal{Z} = \mathcal{Z}(\mathbf{x}; E)$ and $\mathcal{B} = \mathcal{B}(\mathbf{x}; E)$ the cycles and boundaries of the complex $\mathcal{L}(\partial)$. Since ∂ and ∂' skew commute, ∂' restricts to a differential on $\mathcal{Z}$ and $\mathcal{B}$.

More precisely, if we denote by $Z = Z(\mathbf{x}; E)$ the module of cycles of $\mathbb{K}(\mathbf{x}; E)$, by $B = B(\mathbf{x}; E)$ its boundaries, and by $H = H(\mathbf{x}; E)$ its homology, we have the basic approximation complexes:

- $\mathcal{Z}$–complex : $\mathcal{Z} = \mathcal{Z}(\mathbf{x}; E) = \{Z \otimes S, \partial'\}$
- $\mathcal{B}$–complex : $\mathcal{B} = \mathcal{B}(\mathbf{x}; E) = \{B \otimes S, \partial'\}$
- $\mathcal{M}$–complex : $\mathcal{M} = \mathcal{M}(\mathbf{x}; E) = \{H \otimes S, \partial'\}$

The $\mathcal{M}$–complex, as well as the $\mathcal{Z}$–complex, are graded complexes over the polynomial ring $S = R[T_1, \ldots, T_n]$. The tth homogeneous component

$$\mathcal{M}_t = \bigoplus_{r+s=t} H_r \otimes S_s(R^n)$$

of $\mathcal{M}$ is a complex of finitely generated R–modules

$$0 \to H_n \otimes S_{t-n} \longrightarrow \cdots \longrightarrow H_1 \otimes S_{t-1} \longrightarrow H_0 \otimes S_t \to 0.$$

For certain uses however we shall view them as modules over $R[T_1, \ldots, T_n]$, which for the $\mathcal{Z}$–complex is

$$0 \to Z_n \otimes S[-n] \longrightarrow \cdots \longrightarrow Z_1 \otimes S[-1] \longrightarrow Z_0 \otimes S \to 0.$$

The significance of these complexes lies in the fact that $H_0(\mathcal{Z}(\mathbf{x}; R)) = S(I)$ and $H_0(\mathcal{M}(\mathbf{x}; R)) = S(I/I^2)$, and if $\mathcal{M}(\mathbf{x}; R)$ is acyclic then I is of linear type. Thus in some loose sense, the homology of these complexes gives a measure of how close the symmetric algebra and Rees algebra of I are to one another. We refer to $\mathcal{Z}(\mathbf{x}; R)$ and $\mathcal{M}(\mathbf{x}; R)$ as the Z–complex and the M–complex of the ideal generated by $\mathbf{x}$.

Elementary properties

Even before we discuss other approximation complexes, we want to show their usefulness by examining the close relationship between the $\mathcal{Z}$ and $\mathcal{M}$ complexes.

Proposition 3.2.2 *For each integer $r > 0$ there exists an exact sequence of R–modules:*

$$\cdots \to H_r(\mathcal{Z}_{t+1}) \to H_r(\mathcal{Z}_t) \to H_r(\mathcal{M}_t) \to H_{r-1}(\mathcal{Z}_{t+1}) \to \cdots. \quad (3.3)$$

Proof. Consider the exact sequences of complexes

$$0 \to \mathcal{Z}_t \longrightarrow \mathcal{L}_t \longrightarrow \mathcal{B}_{t-1}[-1] \to 0$$

and

$$0 \to \mathcal{B}_t \longrightarrow \mathcal{Z}_t \longrightarrow \mathcal{M}_t \to 0.$$

As already observed, $\mathcal{L}(\partial')$ is the Koszul complex associated to the indeterminates of $B = R[T_1, \ldots, T_n]$. Note the homological shift in the complex $\mathcal{B}_{t-1}$ to make homogeneous the maps of the first sequence, whose differential is ∂. Thus

$$H_r(\mathcal{L}_t) = \begin{cases} R & r = t = 0 \\ 0 & \text{otherwise,} \end{cases}$$

which taken into the two long homology sequences gives the assertion. □

Corollary 3.2.3 *The following are equivalent for the complexes $\mathcal{M}$ and $\mathcal{Z}$:*
(a) *$\mathcal{M}$ is acyclic.*
(b) *$\mathcal{Z}$ is acyclic and λ is injective.*

Proof. This follows by examining how the tail of the homology exact sequence (3.3) is connected to the comparison between $S(I)$ and $\mathcal{R}(I)$:

$$H_1(\mathcal{M}_t) \xrightarrow{\sigma} H_0(\mathcal{Z}_{t+1}) = S_{t+1}(I) \xrightarrow{\lambda} H_0(\mathcal{Z}_t) = S_t(I) \rightarrow H_0(\mathcal{M}_t) = S_t(I/I^2) \rightarrow 0$$

where σ is the connecting homomorphism, while λ is the downgrading mapping defined earlier. □

Corollary 3.2.4 *Let I be a finitely generated ideal of the ring R. If the $\mathcal{M}$–complex of I is acyclic then I is of linear type.*

Proof. Consider the commutative exact diagram

$$\begin{array}{ccccccccc}
H_1(\mathcal{M}_1) & \xrightarrow{\sigma} & H_0(\mathcal{Z}_{t+1}) & \xrightarrow{\lambda} & H_0(\mathcal{Z}_t) & \longrightarrow & H_0(\mathcal{M}_t) & \rightarrow & 0 \\
 & & \downarrow & & \downarrow \alpha_t & & \downarrow \beta_t & & \\
0 & \longrightarrow & I^{t+1} & \longrightarrow & I^t & \longrightarrow & I^t/I^{t+1} & \rightarrow & 0
\end{array}$$

Since α and β are isomorphisms in degrees 0 and 1, the assertion follows by induction and the exactness of both complexes. □

Remark 3.2.5 In two sharp ways the approximation complexes differ from the ordinary Koszul complexes. First, their homology modules are independent of the generating sets. Then, they are not rigid in the sense that if the homology vanishes in one dimension it need not vanish thereafter.

We are going to sketch the proof of the first of these assertions because it will play a role in our discussion. It will suffice to deal with the complex $\mathcal{M}$.

Proposition 3.2.6 *Suppose $\varphi : G = G' \oplus Re \rightarrow R$ is such that $\varphi(e) = 0$. Pick a basis for G' and add e to obtain a basis for G. Let $\mathcal{M}$ and $\mathcal{M}'$ be the approximation complexes of the sequences defined by φ on the basis elements of G and of G'. Then*

$$H_\bullet(\mathcal{M}_t) = H_\bullet(\mathcal{M}'_t).$$

Proof. We use the notation K and K' for the ordinary Koszul complexes defined by the map φ and its restriction to G'. According to the definition of $\mathcal{M}_t$, its component $\mathcal{M}_{r,s}$ is given by

$$H_r(K) \otimes S_s(G) = (H_r(K') \oplus H_{r-1}(K') \otimes Re) \otimes (\sum_{j=0}^{s} S_j(G')e^{s-j})$$

which we are going to rearrange as the sum of 3 subcomplexes:

$$H_r(K') \otimes S_s(G') \oplus (H_r(K') \otimes (\sum_{j=0}^{s-1} S_j(G')e^{s-j})) \oplus (H_{r-1}(K') \otimes (\sum_{j=0}^{s} S_j(G')e^{s-j})).$$

The first component is $\mathcal{M}'_{r,s}$; we denote the other components by $\mathcal{A}_{r,s}$ and $\mathcal{B}_{r,s}$. Observe the isomorphism $\mathcal{A}_{r,s} = \mathcal{B}_{r+1,s-1}$ induced by multiplication by e. Furthermore

$$\mathcal{C}_{r,s} = \mathcal{A}_{r,s} \oplus \mathcal{B}_{r,s}$$

is the mapping cone induced by multiplication by e and is therefore acyclic, which proves our assertion. □

Corollary 3.2.7 *Let I be an ideal generated by the sequence $\mathbf{x}$. The homology of the complex $\mathcal{M}(\mathbf{x})$ depends on I alone.*

Proof. If $\mathbf{y}$ is another set of generators of I we form the M–complex associated to the sequence $\{\mathbf{x}, \mathbf{y}\}$ and repeatedly use the proposition. □

The $\mathcal{Z}$–complex of a module

There are several other variants ([116]), of which we point out the following. We denote by $Z^*(\mathbf{x}; E)$ the cycles of the Koszul complex $\mathbb{K}(\mathbf{x}; E)$ contained in $I \cdot \mathbb{K}(\mathbf{x}; E)$, and denote $H^*(\mathbf{x}; E) = Z(\mathbf{x}; E)/Z^*(\mathbf{x}; E)$. This is the so–called syzygy part of the Koszul homology I which we have already encountered in Chapter 2. It defines

- $\mathcal{M}^*$–complex : $\mathcal{M}^* = \mathcal{M}^*(\mathbf{x}; E) = \{H^* \otimes S, \partial'\}$

In the complexes constructed above the role of modules is fundamentally as coefficients. There is a natural extension where the role is intrinsic.

Let R be a Noetherian ring and let E be a finitely generated module. The following notation shall be used: $S = S(E)$ and $S_+ = S(E)_+ = \sum_{t \geq 1} S_t(E)$. We also single out a generating set of E, that is a surjection $\varphi \colon F = R^n \longrightarrow E$; the ring $S(R^n) = R[T_1, \ldots, T_n]$ will be denoted by B.

Definition 3.2.8 $\mathcal{Z}(E) = \mathcal{M}^*(S_+; S)$ is the $\mathcal{Z}$–complex of the module E.

In other words, $\mathcal{Z}(E)$ is a complex of finitely generated B–modules whose ith component is

$$\mathcal{Z}(E)_i = H_i(S_+; S)_i \otimes B[-i],$$

where $H_i(S_+; S)_i$ denotes the ith graded part of the Koszul homology of S with respect to a set of linear generators of S_+, that is, of E itself. Note that while the total Koszul homology may turn out to be cumbersome, its ith graded part has a natural description

$$H_i(S_+; S)_i = \ker(\wedge^i F \xrightarrow{\partial} \wedge^{i-1} F \otimes E),$$

$$\partial(a_1 \wedge \cdots \wedge a_i) = \sum (-1)^j (a_1 \wedge \cdots \widehat{a_j} \cdots \wedge a_i) \otimes \varphi(a_j).$$

Note that $H_1(S_+; S)_1 = \ker \varphi =$ first syzygy module of E, which explains the notation $\mathcal{Z}(E)$. If we write $H_i(S_+; S)_i = Z_i(E) = Z_i$, the complex $\mathcal{Z}(E)$ has the form

$$\mathcal{Z}(E) : 0 \to Z_n \otimes B[-n] \to \cdots \to Z_1 \otimes B[-1] \to B \to H_0(\mathcal{Z}(E)) = S(E) \to 0.$$

The differential of $\mathcal{Z}(E)$ is that induced by ∂'. An interesting point is its length; if E has a rank, say rank $E = e$, then $Z_i = 0$ for $i > n - e$. Furthermore, its homology is independent of the chosen mapping φ.

An advantage of considering $\mathcal{M}^*(S_+; S)$ rather than the full $\mathcal{M}(S_+; S)$ lies in its simplicity: the higher symmetric powers of E do not get directly involved. Moreover, one has

Proposition 3.2.9 *$\mathcal{M}(S_+; S(E))$ is acyclic $\Leftrightarrow$ $\mathcal{Z}(E)$ is acyclic, in which case, the complexes coincide.*

Proof. See [116, Corollary 11.9] and [117, Lemma 3.2] for details. □

Modified Koszul complexes

We introduce the last of the approximation complexes. Let

$$0 \to L \xrightarrow{\psi} R^n = F \longrightarrow E \to 0$$

be a presentation of the module E. Let $\mathbb{K}(E) = \mathbb{K}(\psi) = \bigwedge L \otimes S(F)$ be the Koszul complex associated to the mapping ψ. Since $L = Z_1(E)$, from the skew–commutative structure of $\mathcal{Z}(E)$, and the fact that the various differentials are also derivations, there is an induced homomorphism $\mathbb{K}(E) \to \mathcal{Z}(E)$ arising out of the mappings $\wedge^i L \to Z_i(E)$.

Let us assume that E is a torsion–free R–module, in which case each $Z_i(E)$ is a second syzygy module and thus reflexive. Replacing $\wedge^i L$ by its double dual $(\wedge^i L)^{**}$, we get a chain mapping $\mathbb{K}(E)^{**} \to \mathcal{Z}(E)$.

These modifications are actual identifications in the following cases.

Proposition 3.2.10 *Let E be a finitely generated R–module.*

(a) *If E is a torsion–free at the prime ideals $\wp$,* depth $R_\wp \leq 1$, *then* $\mathbb{K}(E)^{**} = \mathcal{Z}(E)$.

(b) *Let R be a Cohen–Macaulay ring, and assume E has a resolution*

$$0 \to R^m \longrightarrow R^n \longrightarrow E \to 0.$$

If E satisfies $\mathcal{F}_0$ then $\mathbb{K}(E) = \mathcal{Z}(E)$. In this case $\mathcal{Z}(E)$ is a free B–resolution of $S(E)$.

To point out certain duality features of $\mathcal{Z}(E)$ we recast Theorem 1.1.8:

Theorem 3.2.11 *Let M be a finitely generated module of finite projective dimension. Assume $M_\wp$ is free for each prime ideal with depth $R_\wp \leq 1$. If $r =$ rank M, then* $\det(M) := (\wedge^r M)^{**}$ *is an invertible ideal.*

To apply this result to exhibit duality in the complex $\mathcal{Z}(E)$, suppose that E is a torsion–free module with proj dim $E < \infty$. Then the module L is reflexive and $(\wedge^\ell L)^{**}$ is an invertible ideal ($\ell = \text{rank } L$). Thus, for any integer $r \leq \ell$, there is a pairing

$$\alpha\colon \wedge^r L \otimes \wedge^{\ell-r} L \longrightarrow \wedge^\ell L \longrightarrow (\wedge^\ell L)^{**} = \det(L)$$

that yields a mapping $\wedge^r L \rightarrow \operatorname{Hom}_R(\wedge^{\ell-r} L, \det(L))$, which is an isomorphism in depth $R_\wp \leq 1$.

Corollary 3.2.12 *Let E be a torsion–free module of finite projective dimension. If $\det E = R$ (e.g. R is factorial, or E admits a finite free resolution), then $Z_r(E) = Z_{\ell-r}(E)^*$.*

Low codimension

A significant point of the construction above is its length, equal to the torsion–free rank of L. If, for instance, E satisfies $\mathcal{F}_k$ then $\operatorname{rank}(L) \leq \dim R - k$. In general, if E satisfies $\mathcal{F}_0$ the ideal of definition of $S(E)$ is height unmixed.

Algebras of low codimension, that is when $\operatorname{rank}(L) \leq 4$, are easier to analyze. In the simplest case, when E is torsion-free and $\operatorname{rank}(L) = 2$, we have the approximation complex:

$$0 \rightarrow (\wedge^2 L)^{**} \otimes S_{t-2}(R^n) \rightarrow L \otimes S_{t-1}(R^n) \rightarrow S_t(R^n) \rightarrow S_t(E) \rightarrow 0.$$

If E has finite projective dimension, not necessarily two–or, if R is a factorial domain–$(\wedge^2 L)^{**} \cong R$. It follows that if E satisfies $\mathcal{F}_1$ then $S(E)$ is a domain. In fact, if R is a Cohen-Macaulay ring, then $S(E)$ is a codimension two Cohen-Macaulay domain if and only if L has projective dimension at most one.

This short complex typifies some of the differences between approximation complexes and projective resolutions of symmetric powers.

If $\operatorname{rank}(L) = 3$, under similar conditions, the approximation complex is still acyclic. To get depth information we need to find out the depth of $(\wedge^2 L^{**})$. By duality this module is isomorphic to L^*. Let us examine a simple case.

Proposition 3.2.13 *Assume that the module E is torsion-free and has a presentation*

$$0 \rightarrow C \xrightarrow{\psi} R^n \xrightarrow{\varphi} R^n \longrightarrow E \rightarrow 0,$$

where φ is either symmetric or skew symmetric. Then $L \simeq L^*$.

Proof. Dualizing the presentation we find that the image of φ^*, which we can identify to L, embeds into the kernel of ψ^* which is nothing but L^*. Since this embedding of reflexive modules is an isomorphism in codimension one, they must be isomorphic. □

Problem 3.2.14 Find necessary and sufficient conditions for the integrality of $S(E)$ in codimensions three and four.

3.3 Acyclicity Criteria

There are two basic approaches to the acyclicity of the approximation complexes. One emphasizes the similarity between the complexes and the ordinary Koszul complexes, and expresses acyclicity in sequential terms, that is, in the language of the syzygetic sequences introduced in the previous chapter. The other approach is more utilitarian as it seeks conditions on the coefficient modules of the complexes that ensure both acyclicity and high homological depth for the blowup algebras.

The exposition here will be tilted towards the second method for two reasons. First, [116] carries already a detailed account of the role of d–sequences in the theory of the approximation complexes. Furthermore, the case of ideals having been so thoroughly looked at there and elsewhere, for reasons of space only those aspects which show sharp differences when one passes from ideals to modules of higher rank are stressed.

d–sequences and acyclicity

The acyclicity of the complexes $\mathcal{Z}(I;E)$ and $\mathcal{M}(I;E)$ bears a striking resemblance to that of an ordinary Koszul complex $\mathbb{K}(I)$, with the role of regular sequences being played by the proper sequences and d–sequences.

A number of characterizations are found in [116] and [171], among which we single out ([116, Theorem 12.9]):

Theorem 3.3.1 *Let $(R, \mathfrak{m})$ be a local ring with infinite residue field. Let I be an ideal of R, and let E be a finitely generated R–module. Consider the following statements:*

(a) $\mathcal{M}(I;E)$ *is acyclic.*

(b) I *is generated by a d–sequence on* E.

(c) $\mathcal{Z}(I;E)$ *is acyclic.*

(d) I *is generated by a proper sequence on* E.

Then (a) $\Leftrightarrow$ (b) *and* (c) $\Leftrightarrow$ (d).

There is a formulation of this result that is particularly appropriate for the $\mathcal{Z}$–complex of a module E (see [116] and [114]). It will afford a framework to construct modules whose symmetric algebras are domains.

Theorem 3.3.2 *Let $(R, \mathfrak{m})$ be a local ring with infinite residue field, and let E be a finitely generated R–module. Denote by S_+ the irrelevant ideal of $S(E)$. The following are equivalent:*

(a) $\mathcal{Z}(E)$ *is acyclic.*

(b) S_+ *is generated by a d–sequence of* 1*–forms of* $S(E)$.

Let us now construct modules with the properties of the theorem (cf. [114]). If R is an integral domain, a Bourbaki sequence is an exact sequence

$$0 \to F \longrightarrow E \longrightarrow I \to 0,$$

where F is a free module and I is an ideal. Let us assume that I is generated by a d–sequence, so that the symmetric algebra of I is an integral domain. But $S(I) = S(E)/(F)$, (F) being the ideal of $S(E)$ generated by the 1–forms in F. (F) is a prime ideal of height equal to the rank of F; by [50] any basis of F will generate a regular sequence of $S(E)$, and $S(E)$ will be an integral domain. The generating set of E obtained from a basis of F and elements mapping to a d–sequence in I will provide a d–sequence in $S(E)$. As a consequence, $\mathcal{Z}(E)$ will be acyclic.

We also remark that if $S(I)$ is integrally closed and R is a Japanese ring, then $S(E)$ will be integrally closed as well; it follows from Hironaka's lemma [208, Theorem 36.10].

The aim is to ensure that there is always an underlying Bourbaki sequence construction whenever $\mathcal{Z}(E)$ is acyclic. Assume that $(R, \mathfrak{m})$ is a local ring with infinite residue field and let E be a finitely generated R–module.

Theorem 3.3.3 *Suppose E is a torsion–free module and either R is a normal domain or* proj dim $E < \infty$. *The following conditions are equivalent:*

(a) *$\mathcal{Z}(E)$ is acyclic.*

(b) *E admits a Bourbaki sequence $0 \to F \longrightarrow E \longrightarrow I \to 0$ such that:*

(b_1) *F is generated by elements which form a regular sequence on $S(E)$, and*

(b_2) *I is generated by a proper sequence.*

As preliminary to the proof, we shall need the following lemmas. E will denote an R–module and S_+ will stand for the irrelevant ideal of $S(E)$.

Lemma 3.3.4 *Let $x_1, \ldots, x_n$ be 1–forms of S_+. The following conditions are equivalent:*

(a) *$\{x_1, \ldots, x_n\}$ is a d–sequence.*

(b) *$H_1((x_1, \ldots, x_i); S(E))_2 = 0$ for $i = 1, \ldots, n$.*

(c) *$H_1((x_1, \ldots, x_i); S(E))_k = 0$ for $i = 1, \ldots, n$, and $k \geq 2$.*

Proof. (a) $\Leftrightarrow$ (c) is [116, Lemma 12.7].

(b) $\Leftrightarrow$ (c): We proceed by induction on k. Assume $k > 2$ and suppose

$$H_1((x_1, \ldots, x_i); S(E))_{k-1} = 0 \text{ for } i = 1, \ldots, n.$$

From the long exact sequence of Koszul complexes

$$H_1((x_1, \ldots, x_i); S(E))_{k-1} \stackrel{x_{i+1}}{\to} H_1((x_1, \ldots, x_i); S(E))_k \to H_1((x_1, \ldots, x_{i+1}); S(E))_k,$$

we inductively get that for each i the mapping

$$H_1((x_1, \ldots, x_i); S(E))_k \longrightarrow H_1((x_1, \ldots, x_n); S(E))_k$$

is injective. But the last module always vanishes for $k > 1$ [171]. □

Let E be a module of rank $E = e$. It is easy to see that grade $S_+ \leq e$.

Lemma 3.3.5 *If the complex $\mathcal{Z}(E)$ is acyclic, then* grade $S_+ = e$. *In particular, if $x_1, \ldots, x_n$ is a d–sequence of 1–forms generating S_+, then $x_1, \ldots, x_e$ is a regular sequence on $S(E)$.*

Proof. Map the polynomial ring $B = R[\mathbf{T}] = R[T_1, \ldots, T_n]$ onto a generating set of 1–forms of $S(E)$. The complex $\mathcal{Z}(E)$ has length $n - e$, and for each component

$$\mathrm{Tor}_i^B(B/(\mathbf{T}), Z_r(E) \otimes B[-r]) = 0 \text{ for } i > 0.$$

We may then read the grade of S_+ as

$$\mathbf{T}\text{–depth } S(E) = n - \sup\{i \mid \mathrm{Tor}_i^B(B/(\mathbf{T}), S(E)) \neq 0\} = n - (n - e) = e.$$

The last assertion follows from the definition of d–sequence. □

Lemma 3.3.6 *Let $x_1, \ldots, x_n$ be a sequence of 1–forms in S_+. Suppose the first r elements form a regular sequence. The following conditions are equivalent:*

(a) *$x_1, \ldots, x_n$ is a d–sequence.*

(b) *${x_{r+1}}^*, \ldots, {x_n}^*$ is a d–sequence on $S^* = S(E)/(x_1, \ldots, x_r)$.*

Proof. The assertion follows from Lemma 3.3.4 and the isomorphism of graded modules

$$H_1(({x_{r+1}}^*, \ldots, {x_i}^*); S^*) \simeq H_1((x_1, \ldots, x_i); S(E)),$$

valid for all $i > r$. □

Proof of Theorem 3.3.3. (b) ⇒ (a): The module E in the statement of the theorem has a well–defined rank, say rank $E = e$. Pick generators of F, $x_1, \ldots, x_{e-1}$ of F which form a regular sequence on $S(E)$, and pick elements $x_e, \ldots, x_n$ in E whose images in I generate the ideal and form a proper sequence of R; $x_1, \ldots, x_n$ is a system of generators of S_+. By [116, Theorem 12.10], the elements $x_e, \ldots, x_n$ form a d–sequence on $S(I)$; hence, by Lemma 3.3.6, $x_1, \ldots, x_n$ is a d–sequence on $S(E)$. The assertion now follows from Theorem 3.3.2.

(a) ⇒ (b): We argue by induction on e. If $e = 1$, E is isomorphic to an ideal and Theorem 3.3.2 applies. Suppose then $e > 1$. Since $\mathcal{Z}(E)$ is acyclic, we have by Lemma 3.3.5 that grade $S_+ = e$. Hence we can find $x \in E$, which is a nonzero divisor on $S(E)$. We claim we can find an element x with the additional property that E/Rx is torsion–free. We will then be able to apply the induction hypothesis to E/Rx and the theorem will be proved.

Denote by x^* the image of an element x in the vector space $E/\mathfrak{m}E$. To find x which is regular on $S(E)$ amounts to finding x such that $x^* \notin Y$, where $Y \subset E/\mathfrak{m}E$ is a finite union of proper subspaces which are determined by $\mathrm{Ass}(S(E))$.

To complete the proof of the theorem, all we need is the following elementary observation. Recall that an element x of an R–module is said to be $\wp$–basic, for some prime $\wp$, if the image of x in the localization $E_\wp$ is a minimal generator.

Lemma 3.3.7 *Suppose E satisfies the conditions of the theorem and* rank $E = e > 1$. *Let Y be a finite union of proper subspaces of $E/\mathfrak{m}E$. There exists $x \in E$ such that*

(i) *E/Rx is torsion–free.*

(ii) $x^* \notin Y$.

Proof. (i) is satisfied if x is $\wp$–basic for all primes $\wp$ of R with depth $R_\wp \leq 1$. (ii) will be satisfied if $\lambda_i(x)$ is $\mathfrak{m}$–basic for $i = 1, \ldots, k$, where $\lambda_i \colon E \to L_i$ are epimorphisms and $Y = \cup_{i=1}^k L_i$. Such elements always exist by [29, Theorem 2.4]. □

Strongly Cohen–Macaulay ideals

To make use of the approximation complexes, their coefficient modules must have relatively high depths. This makes possible the use of standard acyclicity results and enhances their role as a substitute for free resolutions. We shall define two classes of ideals that meet these requirements.

Definition 3.3.8 Let I be an ideal of the ring R of dimension d. Let $\mathbb{K}$ be the Koszul complex on the set $\mathbf{a} = \{a_1, \ldots, a_n\}$ of generators of I, and let k be a positive integer. I has *sliding depth* condition $\mathcal{SD}_k$ if the homology modules of $\mathbb{K}$ satisfy

$$\text{depth } H_i(\mathbb{K}) \geq d - n + i + k, \text{ for all } i.$$

The unspecified sliding depth is the case $k = 0$. We refer to it simply as $\mathcal{SD}$.

There is a more stringent condition on the depth of the Koszul homology modules ([142], [143]):

Definition 3.3.9 I is *strongly Cohen–Macaulay*, SCM for short, if the Koszul homology modules of I with respect to one (and then to any) generating set are Cohen–Macaulay.

Example 3.3.10 Let I be the ideal generated by the minors of order 2 of the generic symmetric matrix

$$\begin{bmatrix} x_1 & x_2 & x_3 \\ x_2 & x_4 & x_5 \\ x_3 & x_5 & x_6 \end{bmatrix}.$$

I satisfies sliding depth but it is not strongly Cohen–Macaulay.

We make some observations about the elementary properties of these notions, and the extent to which they may differ.

(i) First we remark that $\mathcal{SD}_k$ is independent of the generating set $\mathbf{a}$. Indeed, it is enough to compare $\mathcal{SD}_k$ for two generating sets $\mathbf{a}$ and $\mathbf{a}' = \{\mathbf{a}, 0\}$. Suppose $\mathcal{SD}_k$ holds for $\mathbf{a}$; then for each i, as in Proposition 3.1.1, we have

$$H_i(\mathbf{a}') = H_i(\mathbf{a}) \oplus H_{i-1}(\mathbf{a}),$$

and therefore depth $H_i(\mathbf{a}') \geq d - n + (i-1) + k$, as required. Conversely, if $\mathcal{SD}_k$ holds for $\mathbf{a}'$ but not for $\mathbf{a}$, let i be the highest integer for which the condition fails, that is, suppose depth $H_i(\mathbf{a}) \leq d - n + i - 1 + k$. But writing the direct sum above for $i+1$ would then lead to a contradiction.

(ii) If R is a Cohen–Macaulay ring, the localizations of I will have $\mathcal{SD}_k$ if it holds true for the maximal ideals. More generally, let $(R, \mathfrak{m})$ be a Cohen–Macaulay local ring of dimension d and let G be a finitely generated module. Assume depth $G \geq d-r$. Then for any prime ideal P, depth $G_P \geq$ height $P - r$. This is clear if R is a regular local ring, since $d -$ depth G is then the codimension of G, which does not increase under localization. But the underlying idea also works in the general case: Assume first that R admits a canonical module ω. In this case

$$d - \text{depth}\, G = \sup\{j \mid \text{Ext}^j_R(G, \omega) \neq 0\}.$$

Since the localization ω_P is a canonical module for R_P, the assertion again follows. For general Cohen–Macaulay rings a simple argument with the $\mathfrak{m}$–adic completion yields the same conclusion.

(iii) In the case of an R–module E, we define $\mathcal{SD}_k$ on $S(E)$ as

$$\text{depth}\, H_i(S_+; S(E))_i \geq d - n + i + k,\ i \geq 0,$$

and in all cases, unqualified sliding depth refers to the case $k =$ rank E.

It will be convenient to rephrase the condition $\mathcal{SD}$ for an ideal I in terms of the depths of the modules of cycles and boundaries of the associated Koszul complex. Assume that R is a Cohen–Macaulay local ring of dimension d and that I is generated by the sequence $\mathbf{x} = \{x_1, \ldots, x_n\}$; put $g =$ height I. Denote by Z_i and B_i the modules of cycles and boundaries of the associated Koszul complex $\mathbb{K}$.

Proposition 3.3.11 *We have*

$$\text{depth}\, Z_i \geq \begin{cases} \min\{d, d-n+i+1\}, & \text{for } \mathcal{SD} \text{ if } g \geq 1 \\ \min\{d, d-g+2\}, & \text{for } \mathcal{SCM}. \end{cases}$$

Proof. It is an immediate translation from the defining exact sequences

$$0 \to Z_{i+1} \longrightarrow K_{i+1} \longrightarrow B_i \to 0$$

$$0 \to B_i \longrightarrow Z_i \longrightarrow H_i \to 0,$$

and chasing depths. □

Corollary 3.3.12 *Let R be a Cohen–Macaulay ring and let I be an almost complete intersection. I has sliding depth if and only if* depth $R/I \geq \dim R/I - 1$.

Examining the dimension $i = n - g$ will be important to us. From now on we assume that R is a Gorenstein ring.

Proposition 3.3.13 *Let R be a Gorenstein local ring of dimension d and let I be a Cohen–Macaulay ideal of height g generated by n elements. Then* depth $Z_{n-g} \geq \min\{d, d-g+2\}$.

Proof. If $g = 0$, $Z_n = 0 \colon I = \mathrm{Hom}(R/I, R)$ is the canonical module of R/I, and therefore it is Cohen–Macaulay since R/I is a Cohen–Macaulay module. If $g = 1$, the exact sequence

$$0 \to B_{n-1} \longrightarrow Z_{n-1} \longrightarrow H_{n-1} \to 0$$

yields (E^* denotes the dual module $\mathrm{Hom}(E, R)$):

$$0 \to Z^*_{n-1} \longrightarrow B^*_{n-1} \longrightarrow \mathrm{Ext}^1_R(H_{n-1}, R) \longrightarrow \mathrm{Ext}^1_R(Z_{n-1}, R) \to 0.$$

Since $B^*_{n-1} = R$ and $\mathrm{Ext}^1_R(H_{n-1}, R) = R/I$ by duality, we get the exact sequence

$$0 \to R/Z^*_{n-1} \xrightarrow{\varphi} R/I \longrightarrow \mathrm{Ext}^1_R(Z_{n-1}, R) \to 0.$$

But as Z_{n-1} is a second syzygy module, the last module has support at primes of height greater than two. In the identification $B^*_{n-1} = R$, Z^*_{n-1} maps exactly onto I: To see this it suffices to localize at any prime P (necessarily of height 1) associated to either Z_{n-1} or I. Thus φ is essentially the multiplication of R/I into itself via a regular element of the Cohen–Macaulay ring R/I. By the remark above on the support of $\mathrm{Ext}^1_R(Z_{n-1}, R)$, φ is an isomorphism.

If $g > 1$, consider the sequence

$$0 \to B_{n-g} \longrightarrow Z_{n-g} \longrightarrow H_{n-g} \to 0.$$

Here B_{n-g} has depth $d - g + 1$ while H_{n-g} has depth $d - g$, being the canonical module of R/I. The exact sequence says that depth $Z_{n-g} \geq d - g$. We test the vanishing of the modules $\mathrm{Ext}^i_R(Z_{n-g}, R)$ for $i = g, g-1$. From above we obtain the homology sequence

$$\begin{array}{l} \mathrm{Ext}^{g-1}_R(H_{n-g}, R) \longrightarrow \mathrm{Ext}^{g-1}_R(Z_{n-g}, R) \longrightarrow \mathrm{Ext}^{g-1}_R(B_{n-g}, R) \longrightarrow \\ \mathrm{Ext}^g_R(H_{n-g}, R) \longrightarrow \mathrm{Ext}^g_R(Z_{n-g}, R) \longrightarrow \mathrm{Ext}^g_R(B_{n-g}, R). \end{array}$$

$\mathrm{Ext}^{g-1}_R(B_{n-g}, R) = R/I$ from the exactness of the tail of the Koszul complex. On the other hand $\mathrm{Ext}^g_R(B_{n-g}, R) = \mathrm{Ext}^{g-1}_R(H_{n-g}, R) = 0$, while $\mathrm{Ext}^g_R(H_{n-g}, R) = R/I$ since R is a Gorenstein ring. Thus we have the exact sequence

$$0 \to \mathrm{Ext}^{g-1}_R(Z_{n-g}, R) \longrightarrow R/I \xrightarrow{\varphi} R/I \longrightarrow \mathrm{Ext}^g_R(Z_{n-g}, R) \to 0.$$

Localizing at primes of height g and $g+1$, we get that φ is an isomorphism since Z_{n-g} is a second syzygy module. □

The following was one of the earliest results on the Cohen–Macaulayness of Koszul homology ([12]):

Corollary 3.3.14 *Let R be a Gorenstein ring, and let I be a Cohen–Macaulay ideal of height g that can be generated by $g+2$ elements. Then I is strongly Cohen–Macaulay.*

Corollary 3.3.15 *Let R be a Gorenstein ring, and let I be an ideal satisfying $\mathcal{SD}$. If R/I satisfies Serre's condition S_2, then I is Cohen-Macaulay.*

Proof. $\mathcal{SD}$ implies that the canonical module of R/I, H_{n-g}, is Cohen–Macaulay. The argument above shows that $R/I = \mathrm{Ext}_R^g(H_{n-g}, R)$ given the condition S_2. □

Sliding depth and strongly Cohen–Macaulay ideals

We give a criterion for an ideal with sliding depth to be strongly Cohen–Macaulay ([124]). It requires a standard result of local duality ([101]), which for convenience is given a short proof.

If R is a Cohen–Macaulay ring with canonical module ω, for any R–module M, we denote $M^\vee = \mathrm{Hom}_R(M, \omega)$.

Proposition 3.3.16 *Let $(R, \mathfrak{m})$ be a Cohen–Macaulay local ring with canonical module ω and let M be a finitely generated R–module with* $\dim M = \dim R$. *The following conditions are equivalent:*

(a) *M is Cohen–Macaulay.*

(b) *The canonical mapping $M \longrightarrow M^{\vee\vee}$ is an isomorphism and for each prime ideal $\wp$ with* $\dim R_\wp \geq 2$

$$\operatorname{depth} M_\wp + \operatorname{depth} (M^\vee)_\wp \geq \dim R_\wp + 2.$$

Proof. We only have to show $(b) \Rightarrow (a)$. We may also assume that M is Cohen–Macaulay on the punctured spectrum of R. If depth $M = r$; we may assume $r < d = \dim R$, and argue by descending induction on r. Note that if $d \leq 2$, then M is already Cohen–Macaulay. Consider an exact sequence

$$0 \to L \longrightarrow F \longrightarrow M \to 0,$$

where F is a finitely generated free module. Dualizing with respect to ω we obtain the exact sequence

$$0 \to M^\vee \longrightarrow F^\vee \longrightarrow L^\vee \longrightarrow \mathrm{Ext}_R^1(M, \omega) \to 0.$$

By assumption, depth $M^\vee \geq d + 2 - r \geq 3$ and $\mathrm{Ext}_R^1(M, \omega)$ is a module of finite length. Counting depths shows that this last module must vanish.

We thus have that the same hypotheses on M hold for L and depth $L = r + 1$. Therefore L is Cohen–Macaulay and so is $L^\vee$ as well. It follows from the second of the sequences that $M^\vee$ is Cohen–Macaulay, and so its dual M is as asserted. □

Theorem 3.3.17 *Let R be a Gorenstein local ring and let I be a Cohen–Macaulay ideal. If I satisfies $\mathcal{SD}$ and*

$$\nu(I_P) \leq \max\{\text{height } I, \text{height } P - 1\}$$

for each prime ideal $P \supset I$, then I is strongly Cohen–Macaulay.

Proof. Since $\mathcal{SD}$ and the other conditions localize, we may assume that I is SCM on the punctured spectrum of R. By adding a set of indeterminates to R and to I, we may assume the height g of I is larger than $n - g + 1$, where n is the minimum number of generators of the new ideal. This clearly leaves the Koszul homology and $\mathcal{SD}$ unchanged. The net effect however is that we have a Koszul complex $\mathbb{K}$ whose exact tail is longer than the rest of the complex.

(i) In the conditions above, H_{n-g-i} is the H_{n-g}–dual of H_i ([142]); to use Proposition 3.3.16 one has to verify that the left hand side of the inequality

$$\text{depth } H_i + \text{depth } H_{n-g-i} \geq (d-n+i) + (d-n+n-g-i) = (d-g) + (d-n),$$

exceeds $(d-g)+1$. If, therefore, $n < d-1$, it will follow that each H_i is Cohen-Macaulay.

(ii) Suppose $n = d-1$; we first examine H_1. Here depth $H_{n-g-1} \geq d-g-1$ and depth $H_1 \geq 2$; we will strengthen the first inequality. Suppose it cannot be done and consider the exact sequence

$$0 \to B_{n-g-1} \longrightarrow Z_{n-g-1} \longrightarrow H_{n-g-1} \to 0.$$

By Proposition 3.3.13 depth $B_{n-g-1} \geq d-g+1$ so that if depth $H_{n-g-1} = d-g-1$ then depth $Z_{n-g-1} = d-g-1$ as well. It will follow that depth $B_{n-g-2} = d-g-2$. A similar sequence for $i = n-g-2$, still by duality, says that depth $H_{n-g-2} = d-g$ or $d-g-2$. In either case we get that depth $Z_{n-g-2} = d-g-2$. We repeat this argument until we get

$$\text{depth } B_1 = \text{depth } B_{n-g-(n-g-1)} = d-g-(n-g-1) = d-n+1 = 2.$$

Since depth $Z_1 = d-g+2 > 2$, we get a contradiction.

(iii) To set up the induction routine, suppose we have shown that H_k and H_{n-g-k} are Cohen–Macaulay; we show that depth $Z_{n-g-k} \geq d-g+2$. The argument is similar to Proposition 3.3.13. We have the exact homology sequence

$$\begin{aligned} 0 \to \operatorname{Ext}_R^{g-1}(Z_{n-g-k}, R) &\longrightarrow \operatorname{Ext}_R^{g-1}(B_{n-g-k}, R) \\ \longrightarrow \operatorname{Ext}_R^{g}(H_{n-g-k}, R) &\longrightarrow \operatorname{Ext}_R^{g}(Z_{n-g-k}, R) \to 0, \end{aligned}$$

since depth $B_{n-g-k} \geq d-g+1$, by induction. But we also have the isomorphisms

$$\begin{aligned} \operatorname{Ext}_R^{g-1}(B_{n-g-k}, R) &= \operatorname{Ext}_R^{g-2}(Z_{n-g-k+1}, R) = \operatorname{Ext}_R^{g-2}(B_{n-g-k+1}, R) \\ &= \cdots = \operatorname{Ext}_R^{g-k-1}(B_{n-g}, R). \end{aligned}$$

This is possible by our 'increase' in g. The last module, from the self duality in the Koszul complex, is nothing but H_k. Since $\operatorname{Ext}_R^g(H_{n-g-k}, R)$ is also a Cohen–Macaulay module, we conclude that depth $Z_{n-g-k} \geq d-g+2$. □

Betti numbers

Let $(R, \mathfrak{m})$ be a local ring and let E be an R–module. Let

$$0 \to L \longrightarrow F \longrightarrow E \to 0$$

be a minimal presentation of E, and set $B = S(F)$. We relate the acyclicity of $\mathcal{Z}(E)$ to the resolution of $S(E)$ as a B–module. Let $\{\mathcal{G}, \partial\}$ be a minimal B–resolution of $S(E)$ as a B–module, i.e. a graded B–projective resolution of $S(E)$ with $\partial(\mathcal{F}) \subset (\mathfrak{m}B + B_+)\mathcal{G}$. Define the filtration $\mathcal{F}_i\mathcal{G}$ on $\mathcal{G}$ by

$$(\mathcal{F}_i\mathcal{G})_j = \bigoplus_{a_{jk} \leq i} B[-a_{jk}].$$

For each i, $\mathcal{F}_{-i+1}\mathcal{G}$ is a subcomplex of $\mathcal{F}_{-i}\mathcal{G}$ and

$$\mathcal{F}_{-i}\mathcal{G}/\mathcal{F}_{-i+1}\mathcal{G} = \mathcal{L}_i \otimes B[-i],$$

where $\mathcal{L}_i$ is a complex of R–modules.

Theorem 3.3.18 *With the notation above, $\mathcal{Z}(E)$ is acyclic if and only if all the complexes $\mathcal{L}_i$ are acyclic. Furthermore, if the equivalencies hold, then $\mathcal{L}_i$ is a minimal R–free resolution of $Z_i(E)$ shifted i steps to the left. In particular, one has the relation of Betti numbers*

$$\beta_i^B(S(E)) = \sum_j \beta_{i-j}^R(Z_j(E)).$$

Proof. For a graded B–module M, put $M^* = M/B_+M$. We have the isomorphism of graded modules

$$H_i(S_+; S(E)) = \operatorname{Tor}_i^B(R, S(E)) = H_i(\mathcal{G})^*,$$

with

$$\mathcal{G}^* = \sum_{i \geq 0}(\mathcal{F}_{-i}\mathcal{G}/\mathcal{F}_{-i+1}\mathcal{G})^* = \sum_{i \geq 0} \mathcal{L}_i,$$

where the $\mathcal{L}_i$ are complexes of R–modules L_{ik} which, considered as B–modules, are concentrated in degree i. It follows that

$$H_i(S_+; S(E))_j = H_i(\mathcal{L}_j).$$

The equivalence of (a) and (b) now follows from Proposition 3.2.9. The additional assertions of the theorem follow trivially. □

Cohen–Macaulay algebras

We shall now begin to feed sliding depth conditions into the coefficient modules of the $\mathcal{Z}$–complex. It will provide a major source of Cohen–Macaulay algebras. In the next chapter we shall see how they may be realized.

Theorem 3.3.19 *Let R be a Cohen–Macaulay local ring and let E be an R–module of rank $E = e$. The following conditions are equivalent:*

(a) *$\mathcal{Z}(E)$ is acyclic and $S(E)$ is Cohen–Macaulay.*

(b) *E satisfies sliding depth and $\mathcal{F}_0$.*

Proof. (b) $\Rightarrow$ (a): We begin with a general observation. Let R be a local ring of dimension d, and let E be a module of rank e, minimally generated by n elements. The conditions $\mathcal{F}_0$ and $\mathcal{SD}_e$ together on the module E place strong requirements on the coefficient modules of $\mathcal{Z}(E)$. Indeed, $\mathcal{F}_0$ implies $d - n + e \geq 0$, while $\mathcal{SD}_e$ requires depth $Z_t(E) \geq (d - n + e) + t$, for all t. These are the conditions that are present in Lemma 3.1.3.

If $\mathcal{Z}(E)$ is not acyclic, let $\wp$ be a minimal prime in the support of the module $\oplus_{i \geq 1} H_i(\mathcal{Z}(E))$. Because the conditions of (b) localize and the homology of $\mathcal{Z}(E)$ does not depend on the generating set, we may assume that $\wp$ is the maximal ideal of R. It is enough to use Lemma 3.1.3.

(a) $\Rightarrow$ (b): Since $S(E)$ is graded Cohen–Macaulay, its dimension can be expressed by $\dim R + \operatorname{rank} E$, and therefore, by Corollary 1.3.6, E satisfies $\mathcal{F}_0$.

We may assume R admits a canonical module ω_R; note that $\omega_B = \omega_R \otimes S$ is a canonical module for B. If M is a B–module, we put $M^\vee = \operatorname{Hom}_B(M, \omega_B)$. We also abbreviate the notation of Theorem 3.3.18 and write $\mathcal{F}_{-i}$ instead of $\mathcal{F}_{-i}\mathcal{G}$.

Claim: $H^j((\mathcal{F}_{-i}/\mathcal{F}_{-i+1})^\vee) = 0$ for $j > \ell = n - e$. Note that ℓ is the codimension of $S(E)$ relative to B.

Let us first see the consequences of this claim. If $\mathcal{L}_i$ is a minimal R–free resolution of $H_i(S_+; S(E))_i = Z_i(E)$, then from Theorem 3.3.18,

$$\mathcal{F}_{-i}/\mathcal{F}_{-i+1} = \mathcal{L}_i \otimes B[-i].$$

Therefore

$$\begin{aligned} H^j(\operatorname{Hom}_B(\mathcal{L}_i \otimes_R B[-i], \omega_B)) &= H^j(\operatorname{Hom}_R(\mathcal{L}_i[-i], \omega_R)) \otimes_R B[i] \quad (3.4) \\ &= H^{j-i}(\operatorname{Hom}_R(\mathcal{L}_i, \omega_R)) \otimes_R B[i] \\ &= \operatorname{Ext}_R^{j-i}(Z_i(E), \omega_R) \otimes_R B[i]. \end{aligned}$$

If the last module vanishes, then depth $Z_i(E) \geq d - (j - i) + 1$; in particular, this holds for $j > n - e$, from which sliding depth follows.

To prove the claim, we show by induction on r that

$$H^j((\mathcal{F}_{-\ell+r}/\mathcal{F}_{-\ell+r+1})^\vee) = H^j((\mathcal{F}_{-\ell+r})^\vee) = 0, \text{ for } j > \ell. \quad (3.5)$$

If $r = 0$, since $\mathcal{F}_{-\ell} = \mathcal{G}$ and $S(E)$ is Cohen–Macaulay, we have

$$H^j((\mathcal{F}_{-\ell})^\vee) = \operatorname{Ext}_B^j(S(E), \omega_B) = 0, \text{ for } j > \ell.$$

On the other hand, the exact sequence

$$0 \to \mathcal{F}_{-\ell+1} \longrightarrow \mathcal{F}_{-\ell} \longrightarrow \mathcal{F}_{-\ell}/\mathcal{F}_{-\ell+1} \to 0,$$

gives rise to the homology exact sequence

$$H^{j-1}((\mathcal{F}_{-\ell+1})^{\vee}) \stackrel{\varphi}{\longrightarrow} H^{j}((\mathcal{F}_{-\ell}/\mathcal{F}_{-\ell+1})^{\vee}) \longrightarrow H^{j}((\mathcal{F}_{-\ell})^{\vee}).$$

Suppose $j > \ell$; then φ is surjective by the case $r = 0$. From equation 3.4,

$$H^{j}((\mathcal{F}_{-\ell}/\mathcal{F}_{-\ell+1})^{\vee})$$

is generated by elements of degree $-\ell$, while

$$H^{j-1}((\mathcal{F}_{-\ell+1})^{\vee})$$

is generated by elements of degree $\geq -\ell + 1$. It follows that

$$H^{j}((\mathcal{F}_{-\ell}/\mathcal{F}_{-\ell+1})^{\vee}) = 0.$$

The induction step is similar. □

Corollary 3.3.20 *Let R be a Cohen–Macaulay domain and let E be a finitely generated module satisfying $\mathcal{SD}_e$ and $\mathcal{F}_1$. Then $S(E)$ is a Cohen–Macaulay integral domain.*

Proof. The previous application of the Lemma 3.1.3 shows that in each localization $R_{\wp}$, height $\wp \geq 1$, the modules $S_t(E)$ have depth at least 1. □

Corollary 3.3.21 *Let R be a Cohen–Macaulay ring and let I be an ideal generated by a d–sequence. The following conditions are equivalent:*

(a) *The Rees algebra $\mathcal{R}(I)$ is Cohen–Macaulay.*

(b) *The associated graded ring $\mathrm{gr}_I(R)$ is Cohen–Macaulay.*

(c) *I satisfies sliding depth.*

Let us spell out a useful special case (see also [24], [148, Theorem 3.1]):

Corollary 3.3.22 *Let R be a Cohen–Macaulay ring and let I be an almost complete intersection. If I is a generic complete intersection, then I is of linear type. In this case, $\mathcal{R}(I)$ is Cohen–Macaulay if and only if* depth $R/I \geq \dim R/I - 1$.

Gorenstein algebras

The following gives a partial description of the canonical module of certain symmetric algebras

Theorem 3.3.23 *Let R be a Cohen–Macaulay local ring and let E be a finitely generated R–module. Suppose $\mathcal{Z}(E)$ is acyclic and $S = S(E)$ is Cohen–Macaulay. Then*

(a) $\omega_S/S_+\omega_S = \bigoplus_{i=0}^{\ell} \operatorname{Ext}_R^{\ell-i}(Z_i(E), \omega_R)$.

(b) *$S(E)$ is Gorenstein if and only if*

$$\text{depth } Z_i(E) \geq d - n + e + i + 1,\ i \leq \ell - 1, \text{ and } \operatorname{Hom}_R(Z_\ell(E), \omega_R) = R.$$

Proof. (b) is a consequence of (a). To prove (a), let $\omega_S = \operatorname{Ext}_B^{\ell}(S, \omega_B)$ be the canonical module of S. We use the result of Theorem 3.3.19: By equation 3.5, we have exact sequences

$$H^{\ell-1}((\mathcal{F}_{-i+1})^\vee) \xrightarrow{\varphi_i} H^\ell((\mathcal{F}_{-i}/\mathcal{F}_{-i+1})^\vee) \xrightarrow{\psi_i} H^\ell((\mathcal{F}_{-i})^\vee) \rightarrow H^\ell((\mathcal{F}_{-i+1})^\vee) \rightarrow 0.$$

Denote again by '*' the reduction $S \rightarrow S/S_+$. We obtain the sequence of R–modules

$$(H^\ell((\mathcal{F}_{-i}/\mathcal{F}_{-i+1})^\vee))^* \xrightarrow{\psi_i^*} (H^\ell((\mathcal{F}_{-i})^\vee))^* \longrightarrow (H^\ell((\mathcal{F}_{-i+1})^\vee))^* \rightarrow 0.$$

Since, by equation 3.4, $H^\ell((\mathcal{F}_{-i}/\mathcal{F}_{-i+1})^\vee)$ is generated by elements of degree $-i$, ψ_i^* equals the $(-i)$–graded part of ψ_i:

$$(\psi_i)_{-i}\colon H^\ell((\mathcal{F}_{-i}/\mathcal{F}_{-i+1})^\vee)_{-i} \longrightarrow H^\ell((\mathcal{F}_{-i})^\vee)_{-i}.$$

On the other hand, since

$$\operatorname{image}(\varphi_i) \subset \bigoplus_{j \geq -i+1} H^\ell((\mathcal{F}_{-i}/\mathcal{F}_{-i+1})^\vee)_j,$$

it follows that $(\psi_i)_{-i} = \psi_i^*$ is injective. Hence we obtain the exact sequence

$$0 \rightarrow (H^\ell((\mathcal{F}_{-i}/\mathcal{F}_{-i+1})^\vee))^* \xrightarrow{\psi_i^*} (H^\ell((\mathcal{F}_{-i})^\vee))^* \rightarrow (H^\ell((\mathcal{F}_{-i+1})^\vee))^* \rightarrow 0.$$

Arguing with degrees, we see that this exact sequence splits, which gives

$$(\omega_S)^* = (\operatorname{Ext}_B^\ell(S, \omega_B))^* = (H^\ell((\mathcal{F}_{-\ell})^\vee))^* = \bigoplus_i (H^\ell((\mathcal{F}_{-i}/\mathcal{F}_{-i+1})^\vee))^*.$$

The assertion now follows from equation 3.4. □

Corollary 3.3.24 *Let R be a Gorenstein ring and let I be an ideal generated by a d–sequence. The following conditions are equivalent:*

(a) *The associated graded ring $\operatorname{gr}_I(R)$ is Gorenstein.*

(b) *I is strongly Cohen–Macaulay.*

That $S(I)$ are Gorenstein only in very circumscribed cases is brought home by ([235]):

Corollary 3.3.25 *Let R be a Cohen–Macaulay ring and let I be an ideal containing regular elements such that $S(I)$ is a Gorenstein ring. Then R is a Gorenstein ring and I has codimension at most two.*

Corollary 3.3.26 *Let R be a Gorenstein local ring and let I be a strongly Cohen–Macaulay ideal, of codimension g, satisfying $\mathcal{F}_0$. Then $S(I)$ is a Cohen–Macaulay ring of type $g-1$.*

Proof. If $n = \nu(I)$, $\ell = n-1$. It suffices to determine the modules $\operatorname{Ext}_R^{\ell-i}(Z_i(I), R)$. Because depth $Z_i(I) = \dim R - g + 2$, we obtain one copy of R, for $i = n-1$, and $g-2$ copies of R/I, for $n-g+1 \leq i \leq n-2$. □

Summary of results

The following statement encapsulates some of the most important aspects of these complexes.

Theorem 3.3.27 *Let R be a local ring with infinite residue field, and let E be a finitely generated R–module.*

1. *The following are equivalent:*
   (a) *$\mathcal{Z}(E)$ is acyclic.*
   (b) *S_+ is generated by a d–sequence of linear forms of $S(E)$.*
2. *If $\mathcal{Z}(E)$ is acyclic, the Betti numbers of $S(E)$ as a module over $B = S(R^n)$ are given by*
$$\beta_i^B(S(E)) = \sum_j \beta_j^R(Z_{i-j}(E)).$$
3. *If R is Cohen–Macaulay and E has rank e, the following conditions are equivalent:*
   (a) *$\mathcal{Z}(E)$ is acyclic and $S(E)$ is Cohen–Macaulay.*
   (b) *E satisfies $\mathcal{F}_0$ and*
$$\operatorname{depth} Z_i(E) \geq d - n + i + e,\ i \geq 0.$$
4. *Moreover, if R is Cohen–Macaulay with canonical module ω_R then*
   (a)
$$\omega_S/S_+\omega_S = \bigoplus_{i=0}^{\ell} \operatorname{Ext}_R^{\ell-i}(Z_i(E), \omega_R).$$

(b) *$S(E)$ is Gorenstein if and only if* $\text{Hom}_R(Z_{n-e}(E), \omega_R) = R$ *and*

$$\text{depth } Z_i(E) \geq d - n + i + e + 1, \ i \leq n - e - 1.$$

Question 3.3.28 Let R be a regular local ring and let I be a Gorenstein ideal of deviation three. Is I strongly Cohen–Macaulay?

3.4 Modules of Projective Dimension Two

We have pointed out some of the difficulties in studying the symmetric algebras of modules of projective dimension greater than 1 from the data expressed in its projective resolution. Two cases are now discussed where those difficulties are circumvented.

Almost complete intersections

It is straightforward to characterize when certain symmetric algebras are almost complete intersections.

Proposition 3.4.1 *Let R be a local ring and let E be a finitely generated R–module. Suppose that either E has finite projective dimension or R is a factorial domain. Then $S(E)$ is an almost complete intersection if and only if E has a projective resolution of the type*

$$0 \rightarrow R \xrightarrow{\psi} R^m \xrightarrow{\varphi} R^n \longrightarrow E \rightarrow 0. \tag{3.6}$$

Proof. We leave it to the reader. □

For this resolution, we denote by I the ideal $I_1(\psi)$. We describe several relationships between I and the ideal $J(\varphi)$ of definition of $S(E)$ (*cf.* [288]).

Proposition 3.4.2 *Let E be a module as in* (3.6)*. Then*

(a) *If E satisfies $\mathcal{F}_1$ then so does I.*

(b) *The converse holds if E is a torsion–free module and E^*, the R–dual of E, is a third syzygy module.*

Proof. We may assume that $(R, \wp)$ is a local ring and the resolution above is minimal.

(a) $\mathcal{F}_1$ for E means that the rank of φ is at most height $\wp - 1$, so that height $\wp \geq m \geq \nu(I)$, as desired.

(b) Let $L = \text{image } (\varphi)$; since E is torsion–free, L is a reflexive module. Moreover, E^* being a third syzygy module is equivalent to $\text{Ext}^1_R(E, R) = 0$, and thus we have the exact sequence

$$0 \rightarrow E^* \longrightarrow R^{n*} \longrightarrow L^* \rightarrow 0.$$

It follows that φ is the composite of the mappings

$$R^m \longrightarrow L \longrightarrow L^{**} \xrightarrow{\varphi^t} R^n.$$

This has the following obvious consequence: If $r = \nu(I) < m$, L would have a free summand of rank $m - r$ and could be written, after a basis change, as

$$\varphi = \begin{bmatrix} 1 & 0 \\ 0 & \varphi_1 \end{bmatrix}$$

where 1 is an identity matrix of size $m-r$ and φ_1 has all of its entries in $\wp$, contradicting the minimality hypothesis. Thus the condition height $\wp \geq \nu(I) = m >$ rank (φ) is verified and $\mathcal{F}_1$ holds for E as well. $\square$

The next result leads to the construction of prime, almost complete intersection ideals. In the sequel, I and E are related by the construction above.

Theorem 3.4.3 *Let R be a Cohen–Macaulay integral domain and let I be a strongly Cohen–Macaulay ideal of height three satisfying $\mathcal{F}_1$, and let E be a torsion–free module such that E^* is a third syzygy module. Then $J = J(\varphi)$ is a Cohen–Macaulay prime ideal.*

Proof. We may assume that R is a local ring of dimension d. To show that each symmetric power $S_t(E)$ is torsion-free, and that $S(E)$ is Cohen-Macaulay we use the approximation complex of the module E:

$$0 \to Z_s(E) \otimes B[-s] \longrightarrow \cdots \longrightarrow Z_1(E) \otimes B[-1] \longrightarrow B \longrightarrow S(E) \to 0,$$

where for each $1 \leq t \leq s =$ rank $L = m - 1$, $Z_t(E) = (\wedge^t L)^{**}$. Since L is the R–dual of the first syzygy module $Z_1(I)$ of I and height $I \geq 2$, by duality $Z_t(E) = Z_{s-t}(I)$.

On the other hand, as I is a strongly Cohen–Macaulay ideal of height 3 the depths of the modules of cycles of its Koszul complex are given by

$$\text{depth } Z_t(I) = \begin{cases} d, & t = 0, s \\ d-1, & 0 < t < s. \end{cases}$$

The assertions follow now from Theorem 3.3.27. $\square$

We now explore some partial converses of this result. Assume that R is a Cohen-Macaulay integral domain.

Proposition 3.4.4 *Let $J = J(\varphi)$ be an almost complete intersection. If J is a prime ideal then* height *I is odd.*

Proof. Since E must be torsion–free, height $I \geq 3$. Localize at a minimal prime $\wp$ of I. Changing notation we may assume that I is generated by a system of parameters.

Suppose $d = \dim R = 2r$, $r > 1$. Consider the rth piece of the approximation complex:

$$0 \to Z_r(E) \to Z_{r-1}(E) \otimes B_1 \to \cdots \to Z_1(E) \otimes B_{r-1} \to B_r \to S_r(E) \to 0.$$

As in the proof of Theorem 3.4.3, the depths of the modules $Z_t(E)$ are given by the formula (derived now from the Koszul complex of a system of parameters of R)

$$\text{depth } Z_t(E) = \text{depth } Z_{s-t}(I) = t.$$

Since E has projective dimension at most 1 on the punctured spectrum, with the depths available it follows directly from the acyclicity lemma that the complex above is acyclic. As, by hypothesis, $S_r(E)$ is torsion–free, chasing depths we conclude that depth $Z_r(E) \geq r + 1$, which is a contradiction. □

Theorem 3.4.5 *Let R be a Gorenstein integral domain, and suppose $J = J(\varphi)$ is an almost complete intersection. If the complex $\mathcal{Z}(E)$ is acyclic and $J(\varphi)$ is a Cohen–Macaulay prime ideal, then I is a strongly Cohen-Macaulay ideal of codimension three.*

Proof. The hypotheses on $\mathcal{Z}(E)$ and $S(E)$ imply according to Theorem 3.3.27 that

$$\text{depth } Z_t(E) \geq \begin{cases} d, & t = 0, s \\ d - s + t, & 0 < t < s. \end{cases}$$

By the afore–mentioned duality, $Z_{s-t}(I) = Z_t(E)$, so that

$$\text{depth } Z_t(I) \geq \begin{cases} d, & t = 0, s \\ d - t, & 0 < t < s. \end{cases}$$

We first observe depth $Z_1(I) \geq d - 1$, so that I is a Cohen–Macaulay ideal of height 3. To show that the Koszul homology of I is Cohen-Macaulay we use local duality [101]; see however [143, Lemma 5.5] for formulation that is appropriate for the use here.

The argument, to be found in [124], proceeds by inductively estimating the depths of the Koszul homology modules $H_t = H_t(\mathbb{K})$. We assume that R is a local domain of dimension d and that I is SCM on the punctured spectrum of R. Since $d \geq m$ and Cohen-Macaulay ideals of height 3 and 5 generators are SCM, we take $d \geq 6$.

The bounds on the depths of $Z_t(I)$ imply that the depth of the module of boundaries B_t of $\mathbb{K}$ is at least $d - t - 2$, so that depth $H_t \geq d - t - 3$. The argument will involve, in order to use duality, estimating the sum

$$\text{depth } H_t + \text{depth } H_{m-3-t},$$

where H_{m-3-t} is the expected dual of H_t with respect to the canonical module, H_{m-3}, of R/I. From the estimate for depth H_t the sum above is at least

$$(d - t - 3) + (d - (m - 3 - t) - 3) = \dim R/I + (d - m),$$

so that if each H_t has depth at least 2 and $d - m \geq 2$, it will follow from [143, Lemma 5.5] that H_t is Cohen–Macaulay. We shall have to argue inductively instead. Let us show that H_1 and H_{m-4} are Cohen–Macaulay. (H_{m-3} is Cohen–Macaulay since it is the canonical module of R/I.) From Proposition 3.3.11 we have that depth $Z_{m-3}(I) = d-1$, so that depth $B_{m-4} = d-2$. To estimate depth $Z_{m-4}(I)$ one uses the assumption that $\mathcal{Z}(E)$ is acyclic and each $S_t(E)$ is torsion–free: Indeed in the exact sequence

$$\begin{aligned} 0 \rightarrow Z_3(E) \quad \simeq \quad & Z_{m-4}(I) \longrightarrow Z_{m-3}(I) \otimes B_1 \longrightarrow \\ & Z_{m-2}(I) \otimes B_2 \longrightarrow B_3 \longrightarrow S_3(E) \rightarrow 0, \end{aligned}$$

as depth $Z_{m-2}(I) =$ depth $Z_{m-3}(I) = d-1$, it follows that depth $Z_{m-4}(I)$ is at least 4, so that depth $H_{m-4} \geq 3$. Thus both H_1 and H_{m-4} have depth at least two and are therefore the H_{m-3}–duals of one another. Since their depths add up to at least $(d-3)+2$ we get that they are both Cohen–Macaulay.

To start the induction, we first estimate the depth of $Z_{m-4}(I)$. Since H_{m-4} is Cohen–Macaulay, the argument of Proposition 3.3.13 again leads to depth $Z_{m-4} = d-1$. In addition, as H_1 is Cohen–Macaulay, depth $B_2 = d-2$ and therefore the depth of H_2 is least 2. Finally, returning to the approximation complex, with $t = 4$, yields depth $Z_{m-5} \geq 5$, and the induction step is completed. □

Codimension two

We remarked earlier on the brevity of the approximation complex of symmetric algebras of codimension two. In the Cohen–Macaulay case the presentation ideal of the algebra has a nice description.

Let R be an integral domain and assume E is a torsion–free module with a resolution

$$0 \rightarrow R^p \xrightarrow{\psi} R^{p+2} \xrightarrow{\varphi} R^n \longrightarrow E \rightarrow 0.$$

Let $L =$ image (ψ), and put $B = S(R^n)$. We have the following commutative diagram of B–modules:

$$\begin{array}{ccccccccc} & & & & 0 & & & & \\ & & & & \downarrow & & & & \\ & & & & B[-1]^p & & & & \\ & & & & \downarrow & & & & \\ & & & & B[-1]^{p+2} & & & & \\ & & & & \downarrow & & & & \\ 0 & \rightarrow & R \otimes B[-2] & \longrightarrow & L \otimes B[-1] & \longrightarrow & B & \longrightarrow & S(E) & \rightarrow & 0 \end{array}$$

The horizontal sequence is the approximation complex $\mathcal{Z}(E)$, and it is exact if E satisfies $\mathcal{F}_0$. The merging of the complexes leads to the free resolution of $S(E)$:

$$0 \rightarrow B[-2] \oplus B[-1]^p \longrightarrow B[-1]^{p+2} \longrightarrow B \longrightarrow S(E) \rightarrow 0.$$

Theorem 3.4.6 *Let R be a Cohen–Macaulay integral domain and let E be a torsion free R–module with a resolution as above. If E satisfies the condition $\mathcal{F}_0$, then the defining ideal $J(\varphi)$ of $S(E)$ can be obtained as the ideal of $p+1$ sized minors of a matrix gotten by adding to ψ a column of linear homogeneous polynomials in the T variables.*

Problem 3.4.7 Let R be a Gorenstein local ring and let I be an ideal generated by a d–sequence. Suppose $S(I)$ satisfies the condition (S_2) of Serre's. Study the extent to which I has sliding depth.

Chapter 4

Linkage and Koszul Homology

Linkage theory deals, *grosso modo*, with the following general problem. Let X be an algebraic variety and Y a closed subvariety of X contained in the closed subvariety Z. What are the properties, with respect to Z and Y, of the closed subvarieties W of Z such that $Z = Y \cup W$?

The aspects of this notion which will be discussed in this chapter have a more specialized character. The first of those has a vivid directness, which we recall. Let R be a Cohen–Macaulay Noetherian ring, and let I and J be two ideals of R. I and J are said to be *directly linked* if there exists a regular sequence $\mathbf{z} = z_1, \ldots, z_n \subset I \cap J$ such that $I = (\mathbf{z})\colon J$ and $J = (\mathbf{z})\colon I$. Furthermore, they are said to be *geometrically linked ideals* if they are unmixed ideals of the same grade n, without common components and $I \cap J = (\mathbf{z})$ for some regular sequence of grade n. In both cases, the schemes $V(I)$ and $V(J)$ together define a complete intersection: $V(I) \cup V(J) = V(\mathbf{z})$. Another notion of linkage, *residual intersection*, replaces the regular sequence $\mathbf{z}$ by more general ideals.

These relationships shall be noted by: $I \sim J$. We will then also say that J is a direct link of I. It induces the more general notion of linkage: I is linked to J if there exists a sequence of direct links

$$I = I_0 \sim I_1 \sim \cdots \sim I_n = J.$$

Questions relating the properties of $V(I)$ to those of $V(J)$, particularly in the case of curves, are found already in the work of M. Noether; more recent instances are [5], [52], [73]. The modern treatment of linkage may be said to have started in [6] and [221]: It depends too heavily on the concept of homological depth that came to fruition in the sixties.

The treatment of linkage here will be limited to Cohen–Macaulay properties of the linkage class of an ideal which are expressed in the behavior of its Koszul homology modules. In this respect, there may be a significant difference between the whole linkage class of an ideal I and the subset of the ideals linked to I through an even number of links. Even in this context only some results needed in the theory of the approximation complexes will be studied.

Linkage theory has evolved into one of the most powerful techniques of commutative algebra. The exposition here will not document its full reach. Hopefully it will provide an entry point to some of the original papers.

4.1 Elementary Properties of Linkage

From the beginning it is clear that linkage is tailored for the study of Cohen–Macaulay ideals, particularly when they lie in Gorenstein rings.

Theorem 4.1.1 *Let R be a local Gorenstein ring and let I be an unmixed ideal of grade g and let $\mathbf{z} = z_1, \ldots, z_g$ be a regular sequence. If $J = (\mathbf{z}): I$ then $I = (\mathbf{z}): J$. Furthermore, if I is a Cohen–Macaulay ideal then J is a Cohen–Macaulay ideal.*

Proof. Since I and $(\mathbf{z}): J$ are unmixed ideals of the same codimension, it suffices to prove equality of their primary components. We may indeed assume that $(R, \mathbf{m})$ is a local ring and $\mathbf{z}$ is a system of parameters. We may pass to $\overline{R} = R/(\mathbf{z})$, so that $\overline{J} = 0: \overline{I}$. But the functor $\mathrm{Hom}_{\overline{R}}(\cdot, \overline{R})$ is self–dual, so that in particular $\mathrm{Hom}_{\overline{R}}(\overline{I}, \overline{R}) \simeq \overline{R}/\overline{J}$ and therefore

$$0: \overline{J} \simeq \mathrm{Hom}_{\overline{R}}(\overline{R}/\overline{J}, \overline{R}) \simeq \mathrm{Hom}_{\overline{R}}(\mathrm{Hom}_{\overline{R}}(\overline{I}, \overline{R}), \overline{R}) \simeq \overline{I}.$$

The proof of the second assertion is similar, using a few additional elements of the theory of the canonical module of a ring (see [113]). First, the reduction to the case $g = 0$ can be carried out as above; we use the notation $\overline{R}$ for R, etc. The canonical module of R/I

$$\omega_{R/I} = \mathrm{Hom}_R(R/I, R) = J$$

is a Cohen–Macaulay module of dimension $\dim R$. We must show that R/J is a Cohen–Macaulay module (of dimension $\dim R$). We use that $\mathrm{Hom}_R(\cdot, R)$ is a self–dualizing functor on the category of finitely generated Cohen–Macaulay modules of maximal dimension. For that consider the exact sequence

$$0 \to J \longrightarrow R \longrightarrow R/J \to 0,$$

to which we apply $\mathrm{Hom}_R(\cdot, R)$ and obtain the exact sequence

$$\begin{aligned} 0 \to \mathrm{Hom}_R(R/J, R) = I \to R \to \mathrm{Hom}_R(\mathrm{Hom}_R(R/I, R), R) = R/I \\ \to \mathrm{Ext}^1_R(R/J, R) \to 0. \end{aligned}$$

It follows that $\mathrm{Ext}^1_R(R/J, R) = 0$, since all the maps are the natural ones. □

Parts of the following result can be traced to Dubreil [52]; in this format and proof they are due to Ferrand and Peskine–Szpiro [221]. It exhibits a projective resolution of $(\mathbf{z}): I$ whenever a resolution of I is known. These resolutions provide means to better control the linkage process, in particular the numerical data of the modules of syzygies.

Theorem 4.1.2 *Let R be a Gorenstein local ring, let I be a perfect ideal of height g and let $\mathbb{F}$ be a minimal free resolution of R/I. Let $\mathbf{z} = z_1, \ldots, z_g \subset I$ be a regular sequence, let $\mathbb{K} = \mathbb{K}(\mathbf{z}; R)$, and let $u: \mathbb{K} \to \mathbb{F}$ be the comparison mapping induced by the inclusion $(\mathbf{z}) \subset I$. Then the dual $\mathbb{C}(u^*)[-g]$ of the mapping cone of u, modulo the subcomplex $u_0: R \to R$, is a free resolution of length g of $R/(\mathbf{z}): I$.*

Proof. The mapping cone $\mathbb{C}(f)$ of a chain mapping

$$f: (\mathbb{M}, \partial) \longrightarrow (\mathbb{N}, \partial')$$

is the complex

$$C(f)_{k+1} = M_k \oplus N_{k+1} \xrightarrow{\varphi_{k+1}} M_{k-1} \oplus N_k = C(f)_k,$$

where

$$\varphi_{k+1} = \begin{bmatrix} -\partial_k & 0 \\ f_k & \partial'_{k+1} \end{bmatrix}.$$

This construction yields the exact sequence of complexes

$$0 \to \mathbb{N} \longrightarrow \mathbb{C}(f) \longrightarrow \mathbb{M}[-1] \to 0.$$

We use it on the dual of the comparison mapping $u: \mathbb{K} \to \mathbb{F}$ induced by the natural surjection $R/(\mathbf{z}) \to R/I$.

We have the exact sequence of complexes

$$0 \to \mathbb{K}^*[-g] \longrightarrow \mathbb{C}(u^*)[-g] \longrightarrow \mathbb{F}^*[-g-1] \to 0.$$

Since $\mathbb{F}^*[-g]$ and $\mathbb{K}^*[-g]$ are acyclic, $H_i(\mathbb{C}(u^*)[-g]) = 0$ for $i \geq 2$, while in lower dimension we have the exact sequence

$$0 \to H_1(\mathbb{C}(u^*)[-g]) \to H_0(\mathbb{F}^*[-g-1]) \xrightarrow{\psi} H_0(\mathbb{K}^*[-g]) \to H_0(\mathbb{C}(u^*)[-g]) \to 0.$$

But the mapping ψ is the inclusion

$$\psi: ((\mathbf{z}): I)/(\mathbf{z}) \hookrightarrow R/(\mathbf{z}),$$

and thus $\mathbb{C}(u^*)[-g]$ is acyclic and therefore a free resolution of $H_0(\mathbb{C}(u^*)[-g]) = R/(\mathbf{z}): I$. □

Corollary 4.1.3 ([5], [73]) *Let R be a Gorenstein local ring and let I be a perfect ideal of codimension two. Then I lies in the linkage class of a complete intersection.*

Proof. Pick a regular sequence $\mathbf{z} \subset I$ formed by elements in a minimal generating set of I, and set $J = (\mathbf{z}): I$. We may assume $\nu(I) = n > 2$; we are going to induct on n. The canonical module of $R/J = I/(\mathbf{z})$ is therefore generated by $n-2$ elements. Since J is perfect, of codimension two, $\nu(J) = n-1$. □

Corollary 4.1.4 ([172]) *Let R be a Gorenstein local ring and let J be an ideal linked to a Gorenstein ideal I. Then J is an almost complete intersection.*

Proof. Let J be linked to the Gorenstein ideal I,

$$J = (\mathbf{z}) \colon I.$$

Since $J/(\mathbf{z})$ is the canonical module of R/I, it is cyclic, and $\nu(J) \leq \text{height}\, I + 1 = \text{height}\, J + 1$. □

The following result of J. Watanabe [306] was soon followed by the full structure theorem of Buchsbaum–Eisenbud [36].

Theorem 4.1.5 *Let R be a Gorenstein local ring and let I be a perfect, Gorenstein ideal of codimension three. Then I lies in the linkage class of a complete intersection.*

Ideals that lie in the linkage class of a complete intersection play such a dominant role that they have been given a specific designation: they are referred to as *licci ideals*.

4.2 Invariants of Linkage

The properties shared by the ideals in certain linkage classes are rich and varied. In this section we shall be concerned with how high depth properties of the Koszul homology modules pass back and forth between ideals linked by an even or an odd number of direct links.

Koszul homology and even linkage

We shall now derive the result that is the main tool in the production of ideals which are strongly Cohen–Macaulay ([142]). The proof is phrased in the slightly more general framework of ideals with sliding depth because of its use in the approximation complexes and later in this chapter.

Theorem 4.2.1 *Let R be a Cohen–Macaulay local ring and suppose L is an ideal of grade n. Let $\mathbf{x} = x_1, \ldots, x_n$ and $\mathbf{y} = y_1, \ldots, y_n$ be two regular sequences in L and set $I = (\mathbf{x})\colon L$ and $J = (\mathbf{y})\colon L$. If I has sliding depth then J has also sliding depth.*

The proof of the theorem displays a number of techniques on dealing with the Koszul homology of modules, which will also be used later.

Lemma 4.2.2 *Let R be a Cohen–Macaulay ring of dimension d and let I be an ideal of height n. If $H_i(I; R) \neq 0$, then its Krull dimension is $d - n$.*

Proof. Let $\mathbf{x} = x_1, \ldots, x_n$ be a maximal regular sequence in I and let $\mathbb{K}(I; R)$ be the Koszul complex on a set of R generators of I. Since the modules $H_i(I; R)$ are annihilated by I, $\dim H_i(I; R) \leq d - n$.

The first non–vanishing homology module is $H_{d-n}(I; R) = ((\mathbf{x})\colon I)/(\mathbf{x})$. Its associated prime ideals are minimal primes of $(\mathbf{x})$, and so $\dim H_{r-n}(I; R) = d - n$. To show that $\dim H_k(I; R) = d - n$, $0 < k < r - n$, assume otherwise and let

$\wp$ be a minimal prime of $H_{r-n}(I;R)$. Localizing at $\wp$ we get $H_k(I;R)_\wp = 0$ and $H_{n-r}(I;R)_\wp \neq 0$, contradicting the rigidity of the Koszul complex. □

We have already examined in Proposition 3.1.1 the dependence of the Koszul homology modules on the generating set of I. Of relevance here is the fact that if $\mathbf{z}'$ is the generating set obtained by adding a superfluous element to $\mathbf{z}$, say, $\mathbf{z}' = z_1, \ldots, z_r, 0$, then

$$H_i(\mathbf{z}';R) = H_i(\mathbf{z};R) \oplus H_{i-1}(\mathbf{z};R).$$

It follows that the sliding depth condition is independent of the generating set. We also note that if the $H_i(\mathbf{z};R)$ are Cohen–Macaulay in the range $0 \leq i \leq m$, so will be the modules $H_i(\mathbf{z}';R)$ in the same range.

Lemma 4.2.3 *Let $\{x_1, \ldots, x_k\}$ be a regular sequence in I. Let "$'$" denote the canonical epimorphism $R \to R' = R/(x_1, \ldots, x_k)$. I satisfies $\mathcal{SD}$ if and only if I' satisfies $\mathcal{SD}$ (in R').*

Proof. It will suffice to prove the assertion for $k = 1$. Let I be an ideal generated by the sequence $\mathbf{x} = x_1, \ldots, x_r$. We have the exact sequence of Koszul complexes

$$0 \to \mathbb{K}(\mathbf{x};R) \xrightarrow{x_1} \mathbb{K}(\mathbf{x};R) \longrightarrow \mathbb{K}(\mathbf{x};R') \to 0.$$

The last complex is the same as $\mathbb{K}(\mathbf{x}';R')$, $\mathbf{x}' = \{0, x_2', \ldots, x_r'\}$.

Taking the homology sequence, we obtain the exact sequence

$$0 \to H_i(\mathbf{x};R) \longrightarrow H_i(\mathbf{x}';R') \longrightarrow H_{i-1}(\mathbf{x};R) \to 0.$$

Reading in the depths of the end modules proves the assertion. □

We begin the proof of the theorem. First, note that the case $n = 0$ is vacuous. For $n = 1$, consider $I = (x) \colon L$ and $J = (x) \colon L$.

Step 1. The condition $I = (x) \colon L$ implies that $I \simeq \mathrm{Hom}_R(L, R)$, and therefore $I \simeq J$. This isomorphism is realized by multiplication by a fixed element of the total ring of fractions Q of R.

We recall Lemma 3.3.11 asserting that I has sliding depth if the modules of cycles, Z_i, of the Koszul complex $\mathbb{K}(I;R)$ satisfy

$$\text{depth } Z_i \geq \min\{d, d - n + i + 1\}.$$

Assume that we have $J = \alpha \cdot I$, α a fixed element of Q, and let $\mathbf{x} = x_1, \ldots, x_r$ be a generating set for I; $\mathbf{y} = \alpha x_1, \ldots, \alpha x_r$ is a generating set of J. It is clear that the modules of cycles of $\mathbb{K}(\mathbf{x};R)$ and $\mathbb{K}(\mathbf{y};R)$ are the same.

Step 2. Assume $n > 1$. To use induction on n, we first establish the following claim. Let $\mathbf{x} = x_1, \ldots, x_n$ and $\mathbf{y} = y_1, \ldots, y_n$ be regular sequences in L. Then there exist elements $z_1, \ldots, z_{n-1}$ satisfying the following conditions:

(a) $z_i = y_i + a_i$, $a_i \in (y_{i+1}, \ldots, y_n)$.

(b) $\{x_0, x_1, \ldots, x_k, z_1, \ldots, z_i\}$ is a regular sequence for $0 \leq k \leq n - i$ (here $x_0 = 0$).

Suppose $z_1, \ldots, z_{i-1}$ have been chosen. By assumption

$$\{x_0, x_1, \ldots, x_k, z_1, \ldots, z_{i-1}\}$$

is a regular sequence for $0 \leq k \leq n - i + 1$. Thus it suffices to choose z_i as in (a) which is not contained in any associated prime of $R/(x_0, x_1, \ldots, x_k, z_1, \ldots, z_{i-1})$ $0 \leq k \leq n - i$. If $\wp$ is such prime, then $(y_i, \ldots, y_n) \not\subset \wp$, as else $(y_1, \ldots, y_n) \subset \wp$ and grade $\wp \geq n$. It follows that z_i can be chosen satisfying (a) and (b) (*cf.* [165, Theorem 124]).

Put

$$D_i = (x_1, \ldots, x_{n-i}, z_1, \ldots, z_i) \colon L \text{ for } 1 \leq i \leq n - 1.$$

Set also $D_0 = (x_1, \ldots, x_n) \colon L$ and $D_n = (y_1, \ldots, y_n) \colon L$. Then $D_0 = I$, and since $(z_1, \ldots, z_n) = (y_1, \ldots, y_n)$, $D_n = J$.

By induction on k, we will show that D_k satisfies $\mathcal{SD}$ for $0 \leq k \leq n$. For $k = 0$ this is our assumption. The elements $x_1, \ldots, x_{n-k}, z_1, \ldots, z_{k-1}$ form a regular sequence in $D_{k-1} \cap D_k$. Denote by "' " the homomorphism

$$R \longrightarrow R/(x_1, \ldots, x_{n-k}, z_1, \ldots, z_{k-1}).$$

From the construction the z_i, both

$$\{x_1, \ldots, x_{n-k}, z_1, \ldots, z_{k-1}, x_{n-k+1}\}$$

and

$$\{x_1, \ldots, x_{n-k}, z_1, \ldots, z_k\}$$

are regular sequences (R = local, so that regular sequences can be permuted). Hence z'_k and x'_{n-k+1} are regular elements in R'. Note that $D'_{k-1} = (x'_{n-k+1}) \colon L'$, while $D'_k = (z'_k) \colon L'$. Now we use the **Step 1** to complete the induction. □

Corollary 4.2.4 *Let I be a Cohen–Macaulay ideal in the linkage class of a complete intersection. Then I is strongly Cohen–Macaulay.*

Corollary 4.2.5 *Let R be a regular local ring and let I be either a Cohen–Macaulay ideal of codimension two or a Gorenstein ideal of codimension three. Then I is strongly Cohen–Macaulay.*

Koszul homology and the canonical module

Let R be a local Gorenstein ring of dimension d, and let I be a Cohen–Macaulay ideal of grade n. The connection between the Koszul homology of I and the canonical module of R/I, $\omega_{R/I}$, could not be more direct, since the latter is the last non-vanishing homology module. We present a result that connects the depths of $H_1(I; R)$ and the symmetric square of the first link of I [275] (see also [289]).

Theorem 4.2.6 *Let R be a local Gorenstein ring, and consider a link $I \sim J$ of Cohen–Macaulay ideals. Then* depth $H_1(I) =$ depth $S_2(\omega_{R/J})$.

Proof. Let $d = \dim R$ and $g =$ height I. After adjoining a variable to I (by passing to, say, $R[[x]]$), we may assume $g > 0$.

Let $\mathbf{x} = x_1, \ldots, x_g$ be a regular sequence defining the link $I \sim J$, and let $L = (\mathbf{x})$. We first show the existence of an exact sequence

$$0 \to M \longrightarrow S_2(I) \longrightarrow S_2(\omega_{R/J}) \to 0, \tag{4.1}$$

where depth $M = d - g + 1$.

We recall that $(L\colon J)/L$ is the canonical module of R/J, and since $L\colon J = I$ we have the exact sequence

$$0 \to L \longrightarrow I \longrightarrow \omega_{R/J} \to 0.$$

It induces the exact sequence

$$0 \to L \circ I \longrightarrow S_2(I) \longrightarrow S_2(\omega_{R/J}) \to 0,$$

where $L \circ I$ denotes the image of $L \otimes_R I$ in $S_2(I)$. Now equation 4.1 follows once it has been shown that depth $L \circ I = d - g + 1$.

Lemma 4.2.7 *There is an isomorphism $L \circ I \simeq L \cdot I$, where $L \cdot I$ is the ordinary product of the ideals L and I.*

Proof. There is a commutative diagram

$$\begin{array}{ccc} L \circ I & \xrightarrow{\varphi} & L \cdot I \\ \psi \uparrow & & \uparrow \\ S_2(L) & \xrightarrow{\epsilon} & L^2 \end{array}$$

where ϵ is an isomorphism since L is a complete intersection and thus syzygetic. To show φ injective, it suffices to verify $\ker \varphi \subset$ image ψ. So let $y \in \ker \varphi$, and write $y = \sum_{i=1}^{g} x_i \circ h_i$ with $h_i \in I$. Then $\sum_{i=1}^{g} x_i h_i = 0$ in R, and since $\mathbf{x}$ is a regular sequence, $h_i \in L$. But then $y \in$ image ψ, which shows that $L \circ I = L \cdot I$. □

We resume the proof of Theorem 4.2.6. To see that depth $L \cdot I = d - g + 1$ it suffices to note that

$$L/L \cdot I \simeq L \otimes_R R/I \simeq (L/L^2) \otimes_{R/L} R/I \simeq (R/L)^g \otimes_{R/L} R/I \simeq (R/I)^g.$$

Now we connect $S_2(I)$ with the 1–dimensional Koszul homology module $H_1(I)$. From a presentation of I

$$0 \to Z_1 \longrightarrow R^n \longrightarrow I \to 0,$$

we get the exact sequence

$$Z_1 \otimes_R I \longrightarrow I \otimes_R R^n \longrightarrow I \otimes_R I \to 0.$$

If we consider the inverse image of $\delta(I)$ in $I \otimes_R R^n$ (see section 2.1) we obtain the exact sequence

$$0 \to B_1 \longrightarrow I \otimes_R R^n \longrightarrow S_2(I) \to 0. \qquad (4.2)$$

The assertion follows from depth–chasing the sequences 4.1 and 4.2. □

The following provides information about the symmetric square of the canonical module of a special class of ideals.

Corollary 4.2.8 *Let R be a Gorenstein local ring and let I be a perfect ideal of type 2. If $\omega_{R/I}$ is the canonical module of R/I, then $S_2(\omega_{R/I})$ is Cohen–Macaulay.*

If the ideal I has codimension three, the Cohen–Macaulayness of $H_1(I)$ is retained across the full linkage class of I, according to the next result (see [289], [300] for more details).

Theorem 4.2.9 *Let R be a local Gorenstein ring, let I be a perfect ideal of codimension three, and denote by $\omega_{R/I}$ the canonical module of R/I. Then $H_1(I)$ is Cohen–Macaulay if and only if $S_2(\omega_{R/I})$ is Cohen–Macaulay.*

Proof. Set $d = \dim R$ and fix a free resolution of I:

$$0 \to R^p \longrightarrow R^m \stackrel{\varphi}{\longrightarrow} R^n \longrightarrow I \to 0.$$

Denote by $Z_1 = \text{image}(\varphi)$ the first–order syzygies of I. The starting point is the identification of the double dual of $\wedge^2 Z_1$ with Z_2, *cf.* the previous chapter:

$$0 \to C \longrightarrow \wedge^2 Z_1 \stackrel{\psi}{\longrightarrow} Z_2 \longrightarrow D \to 0,$$

with C and D to be computed from [9, p.52].

Consider the Lebelt–Weyman complex associated to $\wedge^2 Z_1$:

$$0 \to D_2(R^p) \stackrel{\alpha}{\longrightarrow} R^p \otimes R^m \stackrel{\beta}{\longrightarrow} \wedge^2 R^m \longrightarrow \wedge^2 Z_1 \to 0.$$

This complex is exact by the acyclicity lemma and therefore $\wedge^2 Z_1$ is torsion–free, that is $C = 0$. Dualizing we obtain the complex

$$0 \to (\wedge^2 Z_1)^* \longrightarrow (\wedge^2 R^m)^* \stackrel{\beta^*}{\longrightarrow} (R^p \otimes R^m)^* \stackrel{\alpha^*}{\longrightarrow} (D_2(R^p))^* \to 0.$$

We note that $(\wedge^2 Z_1)^* = Z_{n-3}$, and that $\text{Ext}^3_R(\wedge^2 Z_1, R) = S_2(\omega_{R/I})$.

Set $M = \text{coker}(\beta^*)$, $N = \text{image}(\beta^*)$ and $L = \ker(\alpha^*)$. From [9], we have $D = \text{Ext}^2_R(M, R)$, and with the other modules we set up the exact sequences

$$0 \to N \longrightarrow L \longrightarrow \text{Ext}^1_R(\wedge^2 Z_1, R) \to 0,$$

and

$$0 \to L \longrightarrow (R^p \otimes R^m)^* \longrightarrow (D_2(R^p))^* \longrightarrow S_2(\omega_{R/I}) \to 0.$$

Because D is a module that vanishes in codimension two or less, we obtain

$$\mathrm{Ext}^1_R(\wedge^2 Z_1, R) \simeq \mathrm{Ext}^1_R(Z_2, R) \simeq \mathrm{Ext}^1_R(B_1, R) \simeq \mathrm{Ext}^3_R(H_1(I), R),$$

and also

$$D \simeq \mathrm{Ext}^2_R(M, R) \simeq \mathrm{Ext}^1_R(N, R) \simeq \mathrm{Ext}^1_R(L, R) \simeq \mathrm{Ext}^3_R(S_2(\omega_{R/I}), R).$$

We are now ready to prove the assertion. Clearly $H_1(I)$ is Cohen–Macaulay if and only if Z_2 has depth $d-1$. This implies that D is a Cohen–Macaulay module, of codimension 3 and therefore

$$\mathrm{Ext}^3_R(D, R) \simeq \mathrm{Ext}^2_R(\wedge^2 Z_1, R) = S_2(\omega_{R/I})$$

is Cohen–Macaulay.

Conversely, if $S_2(\omega_{R/I})$ is Cohen–Macaulay, so is D by the identification above, a fact that already implies depth $Z_2 \geq d-3$. We show that

$$\mathrm{Ext}^i_R(Z_2, R) = 0 \text{ for } i = 2, 3.$$

From the exact sequence that defines D we obtain the following exact homology sequence

$$0 \to \mathrm{Ext}^2_R(Z_2, R) \to S_2(\omega_{R/I}) \to \mathrm{Ext}^3_R(\mathrm{Ext}^3_R(S_2(\omega_{R/I}), R), R) \to \mathrm{Ext}^3_R(Z_2, R) \to 0.$$

But if $S_2(\omega_{R/I})$ is Cohen–Macaulay, the middle mapping is the isomorphism derived from local duality. This shows that

$$\mathrm{Ext}^i_R(Z_2, R) = 0 \text{ for } i \geq 2,$$

and therefore depth $Z_2 = d-1$, as we had to establish. □

Corollary 4.2.10 *Let R be a local Gorenstein ring and let I be a perfect ideal of codimension three. Assume that J is an ideal linked to I. If $H_1(I)$ is Cohen–Macaulay, then $H_1(J)$ is also Cohen–Macaulay.*

Corollary 4.2.11 *Let R be a Gorenstein local ring and let I be a perfect ideal of codimension three. If I has type two, then $H_1(I)$ is Cohen–Macaulay.*

The twisted conormal module

Let R be a Gorenstein local ring and let I be a Cohen–Macaulay ideal. The *twisted conormal module* of R/I is $I \otimes_R \omega_{R/I}$.

This module plays a marked role in the deformation theory of analytic algebras [111]; it is particularly useful to ascertain when it is Cohen–Macaulay. In particular, R/I (or, I by abuse of notation) is said to be strongly unobstructed if its twisted conormal module is Cohen–Macaulay.

The following property was discovered by Buchweitz [38]. (See also its multiple extensions in [39]); we follow the treatment of [110].)

Theorem 4.2.12 *Let R be a Gorenstein local ring and assume I is a Cohen–Macaulay ideal geometrically linked to J. Set $S = R/I$ and $T = R/J$. Then $I \otimes \omega_S$ is a Cohen–Macaulay module if and only if $J \otimes \omega_T$ is a Cohen–Macaulay module.*

Proof. Set $g = \text{height}\, I$, and $(\mathbf{z}) = I \cap J$. Changing notation, we may assume $(0) = I \cap J$. The assumption includes that both I and J are generically complete intersections, and therefore the modules $I \otimes \omega_S$ and $J \otimes \omega_T$ have rank g over S and T, respectively. Let us assume that the former is Cohen–Macaulay. To prove the Cohen–Macaulayness of the latter, we use the criterion of [107, Satz 1.1]: If T is a Cohen–Macaulay ring and E is a T–module with rank e, then E is Cohen–Macaulay of dimension $\dim T$ if and only if for a system of parameters $\mathbf{x}$

$$\ell(E/(\mathbf{x})E) = e \cdot \ell(T/(\mathbf{x})).$$

Note the exact sequence (and its twin)

$$0 \to \omega_T \longrightarrow R \longrightarrow S \to 0,$$

of Cohen–Macaulay modules of maximal dimension. We can choose a system of parameters $\mathbf{x}$ for R which is also a system of parameters for S and T. Tensoring this sequence by $\overline{R} = R/(\mathbf{x})$ and denoting $S = S/(\mathbf{x})S$ and $T = T/(\mathbf{x})T$ we have

$$0 \to \omega_T \longrightarrow \overline{R} \longrightarrow S \to 0.$$

We are going to compare the lengths

$$\ell(I \otimes \omega_S) - g \cdot \ell(S)$$

and

$$\ell(J \otimes \omega_T) - g \cdot \ell(T).$$

It suffices to show, by the afore–mentioned criterion, that if one expression vanishes, so will the other. For that, tensor the sequence by T to obtain the homology exact sequence

$$\mathrm{Tor}_2(\overline{R}, T) \xrightarrow{\beta_2} \mathrm{Tor}_2(S, T) \longrightarrow \mathrm{Tor}_1(\omega_T, T) \longrightarrow \mathrm{Tor}_1(\overline{R}, T) \xrightarrow{\beta_1} \mathrm{Tor}_1(T, S).$$

Similarly there will be the exact sequence

$$\mathrm{Tor}_2(\overline{R}, S) \xrightarrow{\alpha_2} \mathrm{Tor}_2(S, T) \longrightarrow \mathrm{Tor}_1(\omega_S, S) \longrightarrow \mathrm{Tor}_1(\overline{R}, S) \xrightarrow{\alpha_1} \mathrm{Tor}_1(T, S).$$

After noting $I \otimes \omega_S = \mathrm{Tor}_1(\omega_S, S)$, $S^g = \mathrm{Tor}_1(\overline{R}, S)$, and similar expressions for T, the assertion is a consequence of the fact that image α_i = image β_i, for all i. This equality, in turn, follows from the commutative diagram, where all mappings are the natural ones:

$$\begin{array}{ccccc} & & \mathrm{Tor}_i(\overline{R}, S) & \xrightarrow{\beta_i} & \mathrm{Tor}_i(S, T) \\ & \overset{\sigma_i}{\nearrow} & & & \\ \mathrm{Tor}_i(\overline{R}, \overline{R}) & & & & \| \\ & \underset{\tau_i}{\searrow} & & & \\ & & \mathrm{Tor}_i(\overline{R}, T) & \xrightarrow{\alpha_i} & \mathrm{Tor}_i(S, T) \end{array}$$

From the Koszul complex $\mathbb{K}(\mathbf{x})$, it follows directly that

$$\begin{aligned}\operatorname{Tor}_i(\overline{R}, \overline{R}) &= \wedge^i \overline{R}^g \\ \operatorname{Tor}_i(\overline{R}, S) &= \wedge^i S^g \\ \operatorname{Tor}_i(\overline{R}, T) &= \wedge^i T^g,\end{aligned}$$

making it easy to see that the mappings σ_i and τ_i are surjective. □

Numerical test

Another useful method to check for licci ideals is the following theorem of Huneke–Ulrich ([156, Corollary 5.13]):

Theorem 4.2.13 *Let I be a homogeneous ideal of $R = k[x_1, \ldots, x_n]$. Suppose I is Cohen–Macaulay, of codimension g, with a minimal graded free resolution:*

$$0 \to \oplus_{j=1}^{b_g} R[-n_{gj}] \longrightarrow \cdots \longrightarrow \oplus_{j=1}^{b_1} R[-n_{1j}] \longrightarrow R \longrightarrow R/I \to 0.$$

If $\max\{n_{gj}\} \leq (g-1)\min\{n_{1j}\}$, *then $I_{(x_1,\ldots,x_n)}$ is not in the linkage class of a complete intersection.*

Example 4.2.14 The following example of edge-ideal associated with a graph was pointed out by Villarreal:

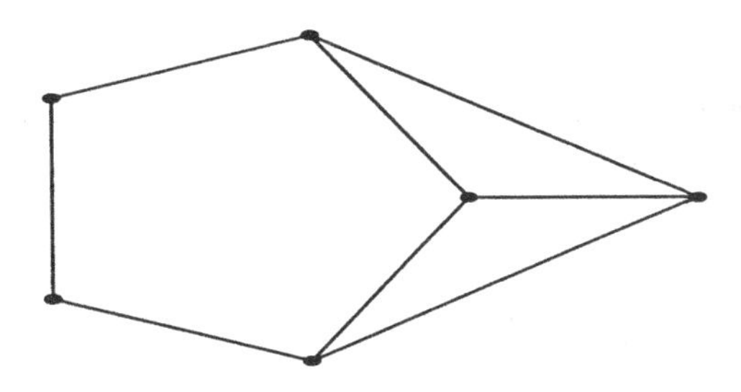

Figure 4.1: Strongly unobstructed but not strongly Cohen–Macaulay graph

Exercise 4.2.15 Let R be a Gorenstein local ring and let I be a perfect ideal of codimension three and type two. If I has deviation at most four, prove that I is strongly Cohen–Macaulay.

4.3 Residual Intersections

The formulation of the notion of residual intersection was made by Artin and Nagata [6], while the recognition of the role of Koszul homology as a controlling element of the process is due to Huneke [142].

Definition 4.3.1 Let R be a Noetherian ring, I be an ideal and s be an integer $s \geq$ height I.

(a) An *s–residual intersection* of I is an ideal J such that height $J \geq s$ and $J = \mathfrak{a}\colon I$, for some s–generated ideal $\mathfrak{a} \subset I$.

(b) A *geometric s–residual intersection* is an s–residual intersection J of I such that height $(I+J) \geq s+1$.

Remark 4.3.2 These definitions mean the following. Let $\mathfrak{a} = (a_1, \ldots, a_s) \subset I$, $J = \mathfrak{a}\colon I$. Then J is an s-residual intersection of I if for all prime ideals $\wp$ with $\dim R_\wp \leq s-1$ we have $I_\wp = \mathfrak{a}_\wp$. A geometric s–residual intersection requires that in addition for all $\wp \in V(I)$ with $\dim R_\wp = s$ the equality $\mathfrak{a}_\wp = I_\wp$ also holds. The case where $s =$ height I is the notion of linkage.

A basic theme of [6] was to find conditions under which a residual intersection of a Cohen–Macaulay ideal is still Cohen–Macaulay. The broad hypotheses of [6] were found by Huneke to be faulty. He then showed how to correct the result. The missing condition turned out to be high depth requirements on the Koszul homology of the ideal I [143]. Shortly after the naturality of sliding depth in this theorem was established [124].

Before we discuss this result, let us describe two examples of residual linkage found in [143].

Example 4.3.3 Let $\mathbf{X}$ be a generic $n \times m$ matrix, $2 \leq n \leq m$, and let $J = I_n(\mathbf{X})$ be the ideal of $k[\mathbf{X}]$ generated by the maximal minors of $\mathbf{X}$. Partition

$$\mathbf{X} = \left[\begin{array}{c|c} \quad \mathbf{Y} \quad & \quad\quad \end{array} \right],$$

where $\mathbf{Y}$ is a $n \times (n-1)$ block. Let $I = I_{n-1}(\mathbf{Y})R$. Then J is a geometric s–residual intersection of I, and $\mathfrak{a}$ is the ideal generated by the $s = m-n+1$ minors of order n obtained by adding to $\mathbf{Y}$ each of the remaining columns of $\mathbf{X}$.

Let I be an ideal and let $T = \operatorname{Spec}(R[It, t^{-1}])$ be the scheme associated with the extended Rees algebra of I. If $G = \operatorname{Spec}(\bigoplus_{n\geq 0} I^n/I^{n+1})$, then $G = V(t^{-1}) \subset T$.

Set also $S = \operatorname{Spec}(S(I))$, and let $W = \operatorname{Spec}(R[T_1, \ldots, T_n, U])$, with one indeterminate T_i for each element a_i of a generating set $(a_1, \ldots, a_n)$ of I. S is the closed subscheme of W defined by the vanishing of the equations U and the syzygies $\sum_{i=1}^{n} b_i T_i$ of the a_i's. Set

$$T' = \operatorname{Spec}(R[T_1, \ldots, T_n, U]/J),$$

where J is generated by the forms derived from the syzygies, together with the equations $UT_i - a_i$.

There is a surjection

$$R[T_1, \ldots, T_n, U]/J \longrightarrow R[It, t^{-1}] \to 0$$

by mapping $T_i \mapsto a_i t$ and $U \mapsto t^{-1}$. It gives a closed immersion of T into T'.

Finally, set $Y = \operatorname{Spec}(R/I)$ and $Y' = \operatorname{Spec}(R[T_1, \ldots, T_n, U]/(I, U))$, the n–affine space over Y.

Proposition 4.3.4 ([143]) *If* $\operatorname{codim} Y \geq 1$ *and I satisfies $\mathcal{F}_1$, then T' is a residual intersection of Y.*

Proof. Recall that $V(T')$ is defined by

$$J = (\sum_{i=1}^{n} c_i T_i \mid \sum_{i=1}^{n} c_i a_i = 0) + (UT_i - a_i).$$

Set $b_i = UT_i - a_i$. It will be shown that

$$(b_1, \ldots, b_n) \colon (I, U) = J.$$

First we prove that J is contained in the ideal on the left–hand side. This follows from the equations

$$\begin{aligned} U(\sum_{i=1}^{n} c_i T_i) &= \sum_{i=1}^{n} c_i (T_i U - a_i) \\ a_j(\sum_{i=1}^{n} c_i T_i) &= \sum_{i=1}^{n} c_i (a_j T_i - a_i T_j). \end{aligned}$$

In addition,

$$a_j T_i - a_i T_j = (UT_i - a_i)T_j - (UT_j - a_j)T_i.$$

Conversely, suppose $F = F(T_1, \ldots, T_n, U) \in (\mathbf{b})\colon (I, U)$. Then there is an equation

$$FU = \sum_{i=1}^{n} G_i b_i = \sum_{i=1}^{n} G_i (UT_i - a_i),$$

and therefore

$$U(F - \sum_{i=1}^{n} G_i T_i) = -\sum_{i=1}^{n} a_i G_i.$$

Write $G_i = G_i' + UG_i''$ where G_i' is free of U. The previous equation becomes

$$U(F - \sum_{i=1}^{n} G_iT_i + \sum_{i=1}^{n} a_iG_i'') = \sum_{i=1}^{n} a_iG_i'.$$

Since the last expression is free of U, it must vanish

$$\sum_{i=1}^{n} a_iG_i' = 0,$$

which means that

$$\begin{aligned} F &= \sum_{i=1}^{n} G_iT_i - \sum_{i=1}^{n} a_iG_i'' \\ &= \sum_{i=1}^{n} (UT_i - a_i)G_i'' + \sum_{i=1}^{n} G_i'T_i. \end{aligned}$$

Therefore, modulo J, $F = \sum_{i=1}^{n} G_i'T_i$. As the mapping $R \mapsto R[T_1, \ldots, T_n]$ is faithfully flat, there are relations

$$\sum_{i=1}^{n} a_ic_{ij} = 0,\ 1 \leq j \leq m$$

in R, and polynomials $H_j \in R[T_1, \ldots, T_n]$ so that

$$G_i' = \sum_{j=1}^{m} c_{ij}H_j.$$

Then

$$\begin{aligned} F &= \sum_{i=1}^{n} T_i(\sum_{j=1}^{m} c_{ij}H_j) \\ &= \sum_{j=1}^{m} H_j(\sum_{i=1}^{n} c_{ij}T_i), \end{aligned}$$

which lies in J since $\sum_{i=1}^{n} c_{ij}a_i = 0$.

It follows that $J = (b_1, \ldots, b_n) : (I, U)$. Therefore T' is a residual intersection of Y' provided codim $T' = n$. This follows from Theorem 1.2.1. □

Throughout the rest of this section $(R, \mathfrak{m})$ is a Cohen–Macaulay local ring of dimension d with infinite residue field.

There are several conjectural settings under which a residual intersection of a Cohen–Macaulay ideal is Cohen–Macaulay. Here we shall require that the ideal A satisfies a condition very close to $\mathcal{F}_1$. Our goal is to connect the notion of residual intersection with sliding depth. This is achieved through the introduction of the notion of a residually Cohen–Macaulay ideal.

Remark 4.3.5 We will be particularly interested in residual intersections that arise in the following fashion. Let I be an ideal of R of height g and let $\mathbf{x} = \{x_1, \ldots, x_s\}$ be a sequence of elements of I satisfying:

(1) height $(\mathbf{x})\colon I \geq s \geq g$.

(2) For all primes $\wp \supset I$ with height $\wp \leq s$, one has

  (i) $(\mathbf{x})_\wp = I_\wp$;
  (ii) $\nu((\mathbf{x})_\wp) \leq$ height $\wp$.

These sequences have the following additional properties:

(a) height $(\mathbf{x}) =$ height I;

(b) $\nu((\mathbf{x})_\wp) \leq$ height $\wp$ for all primes $(\mathbf{x}) \subset \wp$.

To prove (a), let $\wp$ be a minimal prime of $(\mathbf{x})$. Suppose $I \not\subset \wp$; then $((\mathbf{x})\colon I)_\wp = (\mathbf{x})_\wp$. It will follow from (1) that height $\wp \geq s \geq$ height I.

To verify (b), if height $\wp \geq s$, the assertion is trivial; meanwhile, if height $\wp < s$, the proof of (a) shows that $\wp \supset I$ and (2) applies.

Definition 4.3.6 The ideal I is said to be *residually Cohen–Macaulay* if for any sequence $\mathbf{x} \subset I$ with the properties (1) and (2) of the previous remark, it holds that:

(a) $R/((\mathbf{x})\colon I)$ is Cohen–Macaulay of dimension $d - s$;

(b) $((\mathbf{x})\colon I) \cap I = (\mathbf{x})$;

(c) height $((\mathbf{x})\colon I + I) >$ height $(\mathbf{x})\colon I$.

The next two statements spell out the significance of these rather technical definitions.

Theorem 4.3.7 *Let R be a Cohen–Macaulay local ring and I be an ideal. If I has sliding depth then it is a residually Cohen–Macaulay ideal.*

Theorem 4.3.8 *Let R be a Cohen–Macaulay local ring and let I be an ideal satisfying the condition $\mathcal{F}_1$. The following conditions are equivalent:*

(a) *I satisfies the sliding depth condition.*

(b) *I is residually Cohen–Macaulay.*

(c) *I can be generated by a d–sequence $\{x_1, \ldots, x_n\}$ such that*

$$(x_1, \ldots, x_{i+1})/(x_1, \ldots, x_i)$$

is Cohen–Macaulay module of dimension $d - i$, for $i = 0, \ldots, n-1$.

Remark 4.3.9 The ideals occurring in the filtration of Theorem 4.3.8.c have the following homological properties. Assume that R is a regular local ring and that I is a Cohen–Macaulay ideal of height g. Consider the sequences

$$0 \to I_i \longrightarrow I_{i+1} \longrightarrow Q_i \to 0,$$

where $I_i = (x_1, \ldots, x_i)$. We claim that the proj dim $I_i = i - 1$ for $i < n$. Suppose one inequality holds; pick j largest with proj dim $I_j < j - 1$. Note that $j < n - 1$ since $I = I_n$ is assumed Cohen–Macaulay and Q_{n-1} has projective dimension $n - 1$. Localize R at an associated prime of Q_j; this implies that each $Q_{j+k} = 0$ for $k > 0$, and thus $I_{j+1} = \cdots = I_n$. Consider the (localized) sequence

$$0 \to I_j \longrightarrow I_{j+1} \longrightarrow Q_j \to 0$$

since proj dim $Q_j = j$, proj dim $I_{j+1} = 0$ or $g - 1$, and we get proj dim $I_j = j - 1$, which is a contradiction.

The proofs of Theorems 4.3.7 and 4.3.8 will require some technical lemmas on sliding depth. We begin with the connectedness lemma of Abhyankar ([3, (4.9)]) and Hartshorne ([100, Proposition 2.1]), and for completeness we reproduce a proof by Kaplansky given in a lecture several years ago.

Lemma 4.3.10 *Let R be a Noetherian ring with no idempotents other than* 0 *or* 1, *and let I and J be nonzero ideals of R such that $I \cdot J = 0$. Then* grade $(I + J) \leq 1$.

Proof. We may assume that $I \cap J \subset (0\colon J) \cap (0\colon I) = 0$, as otherwise for $0 \neq x \in (0\colon J) \cap (0\colon I)$, $x(I + J) = 0$. Now by the connectedness of Spec(R),

$$I + J \subset [(0\colon J) + (0\colon I)] \neq R.$$

Hence localizing at a prime containing $0\colon J + 0\colon I$, we preserve our assumptions $I \neq 0$, $J \neq 0$, $I \cap J = 0$, and $I + J \neq R$. Thus we may from now on assume that R is local.

Suppose grade $(I + J) > 0$. Let $x = a + b$, $a \in I$, $b \in J$, be a nonzero divisor; it is clear that $a \neq 0$, $b \neq 0$. Moreover, $a \notin R(a + b)$ for an equation $a = r(a + b)$ yields $(1 - r)a = rb$, which is a contradiction, whether r is a unit or not. Since $(I + J)a \subset R(a + b)$ it follows that grade $(I + J) = 1$. □

Lemma 4.3.11 *Suppose $I \neq 0$, and $I_P = 0$ for all minimal prime ideals P that contain I. Then* (a) $(0\colon I) \cap I = 0$, *and* (b) height $((0\colon I) + I) = 1$. *Moreover, if I satisfies $\mathcal{SD}$, then so does I^*, and $R/(0\colon I)$ is Cohen–Macaulay. (Here " * " denotes the canonical epimorphism $R \to R/(0\colon I)$.)*

Proof. (a) and (b) follow directly from the previous lemma. To prove the second assertion, we use the exact sequence of complexes

$$0 \to (0\colon I) \cdot \mathbb{K}(I; R) \longrightarrow \mathbb{K}(I; R) \longrightarrow \mathbb{K}(I'; R') \to 0.$$

The first complex has trivial differential, and since $I \cap 0\colon I = 0$, the exact homology is exactly

$$0 \to L_i \longrightarrow H_i(x_1, \ldots, x_n; R) \longrightarrow H_i(x'_1, \ldots, x'_n; R') \to 0,$$

where $L_i = (0\colon I) \cdot \mathbb{K}(I; R)$.

If I satisfies $\mathcal{SD}$, then depth $(0\colon I = Z_n) = d$. From the sequences we have

$$\operatorname{depth} H_i(x'_1, \ldots, x'_n; R') \geq d - n + i$$

for $i < n$, while by (b) height $I' = 1$, and hence $H_n(x'_1, \ldots, x'_n; R') = 0$.

To see that $R/(0\colon I)$ is Cohen–Macaulay, note that $R/(0\colon I) = B_{n-1}$. The assertion then follows from the exact sequence

$$0 \to B_{n-1} \longrightarrow Z_{n-1} \longrightarrow H_{n-1} \to 0$$

and the fact that Z_{n-1} is Cohen–Macaulay, *cf.* section 3.5. □

Lemma 4.3.12 *Suppose I is generated by a proper sequence $\mathbf{x} = \{x_1, \ldots, x_n\}$. The following conditions are equivalent:*

(a) *I satisfies $\mathcal{SD}$.*

(b) depth $R/(x_1, \ldots, x_i) \geq d - i$, *for* $i = 0, \ldots, n$.

(c) depth $(x_1, \ldots, x_{i+1})/(x_1, \ldots, x_i) \geq d - i$, *for* $i = 0, \ldots, n - 1$.

Proof. Since $\mathbf{x}$ is a proper sequence, we have exact sequences

$$0 \to H_i(x_1, \ldots, x_j) \to H_i(x_1, \ldots, x_{j+1}) \to H_{i-1}(x_1, \ldots, x_j) \to 0$$

for all $i > 1$. By descending induction, if $\mathbf{x}$ satisfies $\mathcal{SD}$ then

$$\operatorname{depth} H_1(x_1, \ldots, x_i) \geq d - i + 1$$

for $i = 0, \ldots, n$. It is also clear that, conversely, this diagonal condition will imply that

$$\operatorname{depth} H_i(x_1, \ldots, x_n) \geq d - i + 1 \text{ for } i \geq 1.$$

We shall use this remark later in the proof.

Denote

$$M_i = ((x_1, \ldots, x_i)\colon x_{i+1})/(x_1, \ldots, x_i)$$

and

$$Q_i = (x_1, \ldots, x_{i+1})/(x_1, \ldots, x_i).$$

We have exact sequences

$$0 \to H_1(x_1, \ldots, x_i) \longrightarrow H_1(x_1, \ldots, x_{i+1}) \longrightarrow M_i \to 0$$

$$0 \to M_i \longrightarrow R/(x_1, \ldots, x_i) \longrightarrow Q_i \to 0$$

and

$$0 \to Q_i \longrightarrow R/(x_1, \dots, x_i) \longrightarrow R/(x_1, \dots, x_{i+1}) \to 0.$$

(b) $\Rightarrow$ (c): Follows from the last exact sequence.

(c) $\Rightarrow$ (a): Using the three exact sequences and the earlier remark the assertion follows by induction on i.

(a) $\Rightarrow$ (b): We show by induction on i that

$$\text{depth } R/(x_1, \dots, x_{n-i}) \geq d - n + i.$$

For $i = 0$ this is our assumption. Suppose the assertion has been proved for $j = n - i \leq n$, and assume that

$$\text{depth } R/(x_1, \dots, x_{j-1}) = k < d - j + 1.$$

Now by the first sequence we have depth $M_{j-1} \geq d - j + 1$; hence the map

$$\alpha: \text{Ext}_R^k(R/\mathfrak{m}, R/(x_1, \dots, x_{j-1})) \to \text{Ext}_R^k(R/\mathfrak{m}, Q_{j-1})$$

induced by the second exact sequence is injective. On the other hand the last sequence gives rise to the mapping

$$\beta: \text{Ext}_R^k(R/\mathfrak{m}, Q_{j-1}) \to \text{Ext}_R^k(R/\mathfrak{m}, R/(x_1, \dots, x_{j-1}))$$

that is injective as well. It follows that the composite $\beta \cdot \alpha$ is injective. But this is a contradiction since $\beta \cdot \alpha$ is induced by multiplication by x_j, and is thus the null mapping. □

Proof of Theorem 4.3.7. Suppose I satisfies sliding depth, height $I = g$ and the sequence $\{x_1, \dots, x_s\}$, $s \geq g$, satisfies (1) and (2) of Definition 4.3.6. All assertions depend solely on the ideal $(x_1, \dots, x_s)$; we may therefore switch to a different set of generators. We use the general position argument of [6] to obtain a system of generators $\{x_1, \dots, x_s\}$ such that for all primes $P \supset I$ with $g \leq$ height $P = k \leq s$ we have

$$(\dagger) \qquad (x_1, \dots, x_s)_P = (x_1, \dots, x_k)_P$$

(see Remark 4.3.5.b).

We proceed by induction on s. Let $s = g$. Since by (4.3.5.a)

$$\text{height } (x_1, \dots, x_g) = \text{height } I = g,$$

it follows that $\{x_1, \dots, x_g\}$ is a regular sequence. Let " ' " denote the epimorphism $R \to R/(x_1, \dots, x_g)$. According to Lemma 4.2.3, I' satisfies sliding depth and therefore $R'/(0: I')$ is Cohen–Macaulay of dimension $d - g$ (*cf.* 4.3.11). But

$$R/(x_1, \dots, x_g): I = R'/(0: I'),$$

and hence condition (a) in (4.3.6) is realized. For the conditions (b) and (c), we have by Lemma 4.3.11 that $(0: I') \cap I' = 0$ and height $((0: I') + I') > 0$, which translate as desired.

We now assume that $s > g$.

1. Case $g > 0$: This is immediate from (†) and the reduction to the ring R'. I' and $\{x'_1, \ldots, x'_s\}$ satisfy all the hypotheses of the theorem. By induction the statements (a), (b) and (c) of (4.3.6) then hold and it is easily lifted to R.

2. Case $g = 0$: Let " * " denote the canonical epimorphism $R \to R/(0\colon I)$. By Lemma 4.3.11 R^* is Cohen–Macaulay of dimension d, I^* and $\{x_1^*, \ldots, x_s^*\}$ satisfy (1) of (4.3.6). As for (2), we only have to check that

$$((x_1^*, \ldots, x_s^*)\colon I^*) = ((x_1, \ldots, x_s)\colon I)^*.$$

One of the inclusions being obvious, let a^* be an element of $(x_1^*, \ldots, x_s^*)\colon I^*$; then $aI \subset (x_1, \ldots, x_s) + 0\colon I$. For $x \in I$ we can therefore write $ax = y + z$, $y \in (x_1, \ldots, x_s)$, $z \in 0\colon I$. It follows that $z = ax - y$ lies in $I \cap 0\colon I = 0$, by Lemma 4.3.11. Furthermore we now have height $(x_1^*, \ldots, x_s^*) =$ height $I^* > 0$ and I^* satisfies $\mathcal{SD}$; we are then back in case 1. Therefore $\{x_1^*, \ldots, x_s^*\}$ and I^* satisfy (a), (b) and (c) of (4.3.6); again it is easy to lift back to R. □

Proof of Theorem 4.3.8. (a) ⇒ (b) is already proved more generally in the previous result.

(b) ⇒ (c): Since $\nu(I_P) \leq$ height P for all primes $P \supset I$, we can find elements $\{x_1, \ldots, x_n\}$ of I such that

(i) $(x_1, \ldots, x_s)_P = I_P$, for all $P \supset I$, height $P \leq s$, and

(ii) height $((x_1, \ldots, x_s)\colon I) \geq s$.

Since I is residually Cohen–Macaulay, we have that for $s \geq g =$ height I, (a), (b) and (c) of (4.3.6) hold.

It is clear that $\{x_1, \ldots, x_g\}$ is a regular sequence. Next we show that x_{s+1} is not a zero–divisor on $R/(x_1, \ldots, x_s)\colon I$ for $g \leq s < n$. It will then follow that $(x_1, \ldots, x_s)\colon I = (x_1, \ldots, x_s)\colon x_{s+1}$. Together with condition (b) this will imply that $\{x_1, \ldots, x_n\}$ is a d–sequence.

Denote by " ' " the canonical epimorphism $R \to R/(x_1, \ldots, x_s)\colon I$. Conditions (a) and (c) imply that I' contains a non-zero divisor z. Suppose x_{s+1} is a zero divisor. Let $y \in (x'_{s+1})\colon I'$; then $zy \in (x'_{s+1})$. This shows that $(x'_{s+1})\colon I'$ consists of zero-divisors. Since R' is Cohen–Macaulay, this implies that height $((x_{s+1})\colon I') = 0$, contradicting (a). Since

$$(x_1, \ldots, x_{s+1})/(x_1, \ldots, x_s) \simeq R/(x_1, \ldots, x_s)\colon x_{s+1} = R/(x_1, \ldots, x_s)\colon I,$$

the implication is proved.

(c) ⇒ (a) : It suffices to apply Lemma 4.3.12. □

There are several other results on residual intersections of Cohen–Macaulay ideals. We quote a fuller theorem from [153] because it has a consequence to be made use of in a later chapter.

Theorem 4.3.13 *Let R be a local Gorenstein ring, and let I be an ideal of codimension g linked in an even number of steps to a strongly Cohen–Macaulay ideal K satisfying $\mathcal{F}_1$. Let $J = A\colon I$ be an s–residual intersection. Then J is a Cohen–Macaulay ideal of codimension s,* depth $R/A = \dim R - s$, *and the canonical module of R/J is the symmetric power $S_{s-g+1}(I/A)$.*

Both the strongly Cohen–Macaulay condition and $\mathcal{F}_1$ are not truly required, as wit Dubreil's Theorem. Another case, $s = g+1$, uses only that $H_1(I; R)$ is Cohen–Macaulay and I is generically a complete intersection [175].

Exercise 4.3.14 Let R be a Cohen–Macaulay local ring and let I be a Cohen–Macaulay ideal satisfying sliding depth. Prove that if $\mathbf{x} = x_1, \ldots, x_g$ is a regular sequence contained in I then $J = (\mathbf{x})\colon I$ is a Cohen–Macaulay ideal.

4.4 Koszul Homology of Trees

In this section we seek graph–theoretic interpretations for the Koszul homology modules and links of ideals attached to graphs.

Let G be a graph on the vertex set $V = \{x_1, \ldots, x_n\}$ as in Theorem 1.2.5. Given a field k we define the ideal $I(G)$ of the polynomial ring $R = k[x_1, \ldots, x_n]$ generated by the set of monomials $x_i x_j$ such that x_i is adjacent to x_j. If all the points of G are isolated we set $I(G) = (0)$. A graph G is said to be Cohen–Macaulay if $I(G)$ is a Cohen–Macaulay ideal. More generally, if $\mathcal{P}$ is a property of certain ideals, the graph G is said to be a $\mathcal{P}$–graph if $\mathcal{P}$ holds for $I(G)$.

Cohen–Macaulay graphs

A construction of a broad family of Cohen–Macaulay ideals is the following. Let G be a graph on the vertex set $V = \{y_1, \ldots, y_n\}$ and let $I = I(G)$. Given a new set of variables $x_1, \ldots, x_n$, we make a copy G_1 of G by putting the label x_i instead of y_i in the graph G. Define $S(G)$ to be the new graph obtained by attaching to each vertex y_i a new vertex x_i and the edge $\{x_i, y_i\}$. We set

$$K = I(S(G)) = (x_1 y_1, \ldots, x_n y_n, I) \subset R = k[\mathbf{x}, \mathbf{y}].$$

The S in the notation is intended for *suspension*, or attaching *whiskers* to the base graph. It shares in the following general property:

Proposition 4.4.1 *Let $R = k[x_1, \ldots, x_n, y_1, \ldots, y_n]$ be a polynomial ring over a field k, and let $M \subset \{(i,j) |\ 1 \leq i < j \leq n\}$. If we set*

$$I = (\{x_i y_i, y_j y_\ell \mid i = 1, \ldots, n \text{ and } (j, \ell) \in M\}),$$

then I is a Cohen–Macaulay ideal.

Proof. We start by effecting the change of variables $z_i = x_i - y_i$ for $i = 1, \ldots, n$. The net result is that $A = k[z_1, \ldots, z_n, y_1, \ldots, y_n]/I$ becomes a graded algebra integral over $k[z_1, \ldots, z_n]$, where $I = (y_i^2 + y_i z_i, y_j y_\ell)$. Observe that the images of $y_{i_1} \cdots y_{i_r}$ $(1 \leq i_1 < \cdots < i_r \leq n)$, not divisible by any of the monomials $y_j y_\ell$, form a generating set for the $k[z_1, \ldots, z_n]$–module A. To complete the proof we now show that they are in fact a free basis. We set $z = (z_1, \ldots, z_n)$ and $y = (y_1, \ldots, y_n)$. Assume the following equality

$$\sum f_i(z) y_i + \sum f_{i_1 i_2}(z) y_{i_1} y_{i_2} + \cdots + \sum f_{i_1 \ldots i_m}(z) y_{i_1} \cdots y_{i_m} =$$

$$\sum g_i(y, z)(y_i^2 + y_i z_i) + \sum h_{j\ell}(y, z) y_j y_\ell$$

Using induction we will prove that $f_i = f_{i_1 i_2} = \cdots = f_{i_1 \ldots i_m} = 0$. First we show $f_i = 0$ for $1 \leq i \leq n$. Making $y_j = 0$ for $j \neq i$ in the equation above we obtain $f_i(z) = g_i(y_i, z)(y_i + z_i)$, which forces $f_i = 0$. We set the next induction step by assuming $f_i = f_{i_1 i_2} = \ldots = f_{i_1 \ldots i_{r-1}} = 0$. The substitutions $y_j = 0$ for $y_j \notin \{y_{i_1}, \ldots, y_{i_r}\}$ and $y_{i_p} = -z_{i_p}$ for $1 < p \leq r$ into the first equality give $\pm f_{i_1 \ldots i_r}(z) z_{i_2} \cdots z_{i_r} = g_{i_1}(y_{i_1}, z)(y_{i_1} + z_{i_1})$, therefore $f_{i_1 \ldots i_r}(z) = 0$. □

The more delicate significance of the suspension construction lies partly in the surprising characterization of Cohen–Macaulay trees [305, Theorem 2.4]:

Theorem 4.4.2 *If G is a Cohen–Macaulay tree, then $G = \mathcal{S}(G_0)$ for some tree G_0.*

Multiplicity

As is to be expected, all the structure of $\mathcal{S}(G)$ lies in the graph G. The following simple computation conforms to this view.

Proposition 4.4.3 *Let G_0 be a graph and let $G = \mathcal{S}(G_0)$. The multiplicity of $R/I(G)$ is*

$$\sum_{i=-1}^{d} f_i(G_0),$$

where $f_i(G_0)$ denotes the number of i–faces of the Stanley–Reisner simplicial complex associated to the ideal $I(G_0)$.

Proof. A system of parameters for $R/I(G)$ are the forms $y_i - x_i$, $1 \leq i \leq n$, according to the proof of Proposition 4.4.1. The reduction is

$$k[y_i]/(y_i^2, y_k y_\ell),$$

whose k–vector space basis is formed by the standard monomials, that is, the faces of the attached simplicial complex. □

Linkage

Let $\Upsilon(G)$ be the family of minimal vertex covers for G. The next proposition describes the first link of $I(\mathcal{S}(G))$ relative to the regular sequence $\mathbf{z} = \{x_1y_1, \ldots, x_ny_n\}$.

Proposition 4.4.4 *Let Υ be the set of minimal vertex covers for G. Let*

$$L = (\{x_{i_1}\cdots x_{i_s} \mid \{x_{i_1}, \ldots, x_{i_s}\} \in \Upsilon(G)\}).$$

Then $(\mathbf{z}) \colon K = (\mathbf{z}, L)$.

Proof. Take $\{x_{i_1}, \ldots, x_{i_r}\} \in \Upsilon$ and $y_ky_\ell \in I$. Since $i_s \in \{k, \ell\}$ for some s we obtain $x_{i_1}\cdots x_{i_r}y_ky_\ell \in (\mathbf{z})$. This shows $(\mathbf{z}, L) \subset (\mathbf{z})\colon K$. Conversely, assume M is a monomial and $M \in ((\mathbf{z})\colon K) \setminus (\mathbf{z})$. Let $\{y_k, y_\ell\}$ be any line in G. Since $My_ky_\ell = x_ty_tM_1$, we have that either x_k divides M or x_ℓ divides M; in both cases, $M \in L$. □

Corollary 4.4.5 *The Cohen–Macaulay type of R/K is equal to $|\Upsilon(G)|$.*

Proof. $\operatorname{Ext}^n_R(R/K, R) \simeq ((\mathbf{z})\colon K)/(\mathbf{z})$. □

Corollary 4.4.6 *Let $G =$ Star(n) be the star on n vertices. Then $I(\mathcal{S}(G))$ lies in the linkage class of a complete intersection.*

Proof. Let x_1 be the center of Star(n); then the only minimal vertex covers are $\{x_1\}$ and $\{x_2, \ldots, x_n\}$, so that, in the notation above, the first link is

$$J = (x_1y_1, \ldots, x_ny_n, x_1, x_2\cdots x_n) = (x_1, x_2y_2, \ldots, x_ny_n, x_2\cdots x_n).$$

Take now its link with respect to the regular sequence formed by the first n elements:

$$(x_1, x_2y_2, \ldots, x_ny_n)\colon J = (x_1, y_2, \ldots, y_n),$$

which is a complete intersection. □

We shall determine the type of Cohen–Macaulay ideals of bipartite graphs. Let G be a bipartite Cohen–Macaulay graph on the vertex set V. Assume G has no isolated points. There are disjoint sets V_1 and V_2 so that $V = V_1 \cup V_2$ and every line of G joins V_1 with V_2. Since V_1 and V_2 are minimal vertex covers for G, which is unmixed, we have

$$n = |V_1| = |V_2| = \text{height } I(G).$$

By Hall's theorem ([18]) there are n independent lines. Therefore we may assume $V_1 = \{y_1, \ldots, y_n\}$, $V_2 = \{w_1, \ldots, w_n\}$, and $y_iw_i \in I(G)$. Notice that $\mathbf{w} = \{y_i - w_i\}_{i=1}^n$ is a regular system of parameters for $R = k[y_i, w_i]$ and that $R/I(G)$ modulo $(\mathbf{w})$ reduces to $A = k[y_i]/(y_i^2, I(\overline{G}))$, where $\overline{G}$ is a graph on the vertex set V_1. Observe that by the proposition above the type of $R/I(G)$ is equal to $|\Upsilon(\overline{G})|$. Also notice that $\operatorname{Soc}(A) = (I(\overline{G})\colon (y_i))/I(\overline{G})$ is generated by the images of all the monomials $y_{i_1}\cdots y_{i_r}$ such that $\{y_{i_1}, \ldots, y_{i_r}\}$ is an independent set of G.

Proposition 4.4.7 *Let* $\gamma_n = \sup\{|\Upsilon(G)| \; G \text{ has } n \text{ vertices}\}$. *Let* F_n *denote the Fibonacci numbers* $F_0 = F_1 = 1$ *and* $F_n = F_{n-1} + F_{n-2}$. *Then* $\gamma_n \leq F_n$ *for* $n \geq 1$.

Proof. The proof is by induction on n. The required inequality is clearly satisfied if $n \leq 4$. Assume $\gamma_k \leq F_k$ for $k < n$ and $n \geq 5$. Let G be a graph with n vertices. We write $G = G_1 \cup G_2$ where G_1 is a connected component of G with at least two vertices. Let x be a vertex in G_1 and let $x_1, \ldots, x_r$ be the vertices of G adjacent to x. Let Υ_1 be the set of minimal covers of G that contains x and let $\Upsilon_2 = \Upsilon(G) \setminus \Upsilon_1$. If $A \in \Upsilon_1$ notice that $A \setminus \{x\}$ is a (possibly empty) minimal vertex cover for $G \setminus \{x\}$, hence $|\Upsilon_1| \leq F_{n-1}$. If $A \in \Upsilon_2$ then $A \setminus \{x_1, \ldots, x_r\}$ is a (possible empty) minimal vertex cover for $G \setminus \{x, x_1, \ldots, x_r\}$, which implies $|\Upsilon_2| \leq F_{n-r-1}$. Therefore $\gamma_n \leq F_n$ as required. □

Remark. Let P_n be the path $\{x_1, \ldots, x_n\}$ then $F_n = |\Upsilon(P_n)|$ is given by $F_0 = F_1 = 1$, $F_2 = 2$, and $F_n = F_{n-2} + F_{n-3}$ for $n \geq 3$. To show it note: (i) The minimal vertex covers of P_n not containing x_1 are in one to one correspondence with the minimal vertex covers of $P_n \setminus \{x_1, x_2\}$. (ii) The minimal vertex covers of P_n containing x_1 are in one to one correspondence with the minimal vertex covers of $P_n \setminus \{x_1, x_2, x_3\}$. We use the terminology, the *vertex covering number* of a graph G, $\alpha_0(G)$, is the smallest number of vertices in any minimal vertex cover. It is the height of the ideal $I(G)$.

Proposition 4.4.8 *If* T *is a forest then* $|\Upsilon(T)| \geq \alpha_0(T) + 1$.

Proof. The proof is by induction on the number of vertices. Assume the inequality for all forests with less than n vertices and assume T is a forest with n vertices. Let z be a point of degree one in T and let x be the vertex in T adjacent to z. Set $T_z = T \setminus \{z\}$. Let $A \in \Upsilon(T_z)$; if $x \in A$ then $A \in \Upsilon(T)$ and if $x \notin A$ then $A \cup \{z\} \in \Upsilon(T)$. Hence $|\Upsilon(T)| \geq |\Upsilon(T_z)|$. If $\alpha_0(T) = \alpha_0(T_z)$ the inequality follows by the induction hypothesis. If $\alpha_0(T_z) < \alpha_0(T)$, then $\alpha_0(T_z) = \alpha_0(T) - 1$. Therefore there is a minimal vertex cover A_1 of T so that $|A_1| = \alpha_0(T)$ and $z \in A_1$. Notice that A_1 is not a minimal vertex cover for T_z because $A_1 \setminus \{z\}$ is a minimal cover for T_z. As a consequence A_1 is not a lift of a cover of T_z. The required inequality now follows using the induction hypothesis. □

Strongly Cohen–Macaulay ideals

The main result of this section is the following surprising fact [262]:

Theorem 4.4.9 *For any tree* G *the ideal* $I(G)$ *has sliding depth.*

An indication of this property occurs already in [305, Theorem 3.5], where it is proved that the 1–dimensional Koszul homology module of such ideals are Cohen–Macaulay.

In the case of a tree G, with n vertices and q edges, the sliding depth requirement would read

$$\text{depth } H_i(\mathbb{K}) \geq n - q + i \geq i + 1, \text{ for all } i.$$

It will be seen to be very appropriate for induction arguments.

Let G be a tree and let $\{y_1, y_2\}$ be one of its edges. We consider two disjoint subtrees G_1, G_2 of G such that y_ℓ is a vertex of G_ℓ, $\ell = 1, 2$; the edges of G are those in G_1, G_2 and $\{y_1, y_2\}$, therefore we can write the ideal $I(G)$ as

$$I = (I(G_1), I(G_2), z), \text{ where } z = y_1 y_2.$$

Partition the indeterminates $Z = \{y_1, \ldots, y_n\}$ into two disjoint sets Z_1 and Z_2, in a manner that $y_1 \in Z_1$ and $y_2 \in Z_2$, and

$$J = I_1 + I_2,$$

where

$$I_1 = I(G_1) \subset R_1 = k[Z_1]$$

and

$$I_2 = I(G_2) \subset R_2 = k[Z_2].$$

This leads to a representation $R = R_1 \otimes_k R_2 = k[Z_1] \otimes_k k[Z_2]$. If we let $\mathbb{K}_1$ and $\mathbb{K}_2$ be the Koszul complexes of I_1 and I_2 over R_1 and R_2, respectively, then the Koszul complex $\mathbb{K}$ of J over R is

$$\mathbb{K} = \mathbb{K}_1 \otimes_k \mathbb{K}_2.$$

By Künneth's exact sequence, we have the following expression for the Koszul homology modules of $\mathbb{K}$:

$$H_s(\mathbb{K}) = \bigoplus_{i=0}^{s} H_i(\mathbb{K}_1) \otimes_k H_{s-i}(\mathbb{K}_2).$$

Note that over the ring R each of these summands of $H_s(\mathbb{K})$ has depth at least

$$(i+1) + (s-i+1) = s+2,$$

an estimate which will be critical to our argument. We are going to feed it into the Koszul homology of $I = (J, z)$. Denote by $\mathbb{L}$ the Koszul complex of the full set of the basic monomials of I.

The basic relationship between the homologies of $\mathbb{K}$ and $\mathbb{L}$ is ([192, Theorem 16.4]):

Proposition 4.4.10 *There exists an exact sequence*

$$0 \to \overline{H_i(\mathbb{K})} \longrightarrow H_i(\mathbb{L}) \longrightarrow {}_zH_{i-1}(\mathbb{K}) \to 0,$$

where for any module E we put $\overline{E} = E/zE$ and ${}_zE = \{e \in E \mid z \cdot e = 0\}$.

There will be preparatory steps to accomplish first. Let $L_\ell = (I_\ell, y_\ell)$, for $\ell = 1, 2$. I_ℓ can be viewed as the ideal attached to a forest with fewer vertices than G and has therefore sliding depth, by the appropriate induction hypothesis. We use it to obtain the following estimates:

Lemma 4.4.11 *Let $\mathbb{K}_\ell$ denote the Koszul complex of I_ℓ. Then*

$$\begin{aligned} \text{depth}\ \overline{H_i(\mathbb{K}_\ell)} &\geq i+1 \\ \text{depth}\ {}_{y_\ell}H_i(\mathbb{K}_\ell) &\geq i+1. \end{aligned}$$

Proof. The defining exact sequence for these modules is

$$0 \to {}_{y_\ell}H_i(\mathbb{K}_\ell) \to H_i(\mathbb{K}_\ell) \stackrel{y_\ell}{\longrightarrow} H_i(\mathbb{K}_\ell) \to \overline{H_i(\mathbb{K}_\ell)} \to 0. \tag{4.3}$$

Inducting on the number of edges of the graph, we may assume that each I_ℓ satisfies the sliding depth condition. We turn the use of Proposition 4.4.10 around, and apply it to L_ℓ, with $z = y_\ell$. Let $\mathbb{A}_\ell$ denote the Koszul complex of L_ℓ. More precisely, the Koszul complex of the generators of I_ℓ and y_ℓ; observe that this will not be a minimal generating set. By induction, since I_ℓ has sliding depth we have:

$$\text{depth}\ H_i(\mathbb{A}_\ell) \geq i+1.$$

We have the exact sequence

$$0 \to \overline{H_i(\mathbb{K}_\ell)} \longrightarrow H_i(\mathbb{A}_\ell) \longrightarrow {}_zH_{i-1}(\mathbb{K}_\ell) \to 0. \tag{4.4}$$

We prove the assertions by induction on i. For $i=0$, $H_i(\mathbb{K}_\ell) = R_\ell/I_\ell$ has depth at least 1. When placed in the sequence 4.3 it implies that ${}_zH_0(\mathbb{K}_\ell)$ has also depth at least 1. We cycle the latter in the sequence 4.4 to obtain that $\overline{H_1(\mathbb{K}_\ell)}$ has depth at least 2. We reenter the sequence (4.3) to get that ${}_zH_1(\mathbb{K}_\ell)$ has depth at least 2 since, by induction, $H_1(\mathbb{K}_\ell)$ has depth at least 2. This process clearly leads to the assertion. □

Proof of Theorem 4.4.9. We are finally ready to show that

$$\overline{H_i(\mathbb{K})} \text{ and } {}_{y_1y_2}H_i(\mathbb{K})$$

have depth at least $i+1$. It is enough to establish this for a component of $H_i(\mathbb{K})$ of the form $H_j(\mathbb{K}_1) \otimes_k H_{i-j}(\mathbb{K}_2)$. For simplicity we denote these homology modules by E_1 and E_2, and set $E = E_1 \otimes_k E_2$. Multiplication of y_1y_2 on E can be factored as

$$E_1 \otimes_k E_2 \stackrel{y_1\otimes 1}{\longrightarrow} E_1 \otimes_k E_2 \stackrel{1\otimes y_2}{\longrightarrow} E_1 \otimes_k E_2.$$

The snake lemma applied to this composition of maps leads to the following exact sequence (tensor products taken over k, and $z = y_1y_2 = (y_1 \otimes 1)\cdot(1 \otimes y_2)$)

$$0 \to {}_{y_1}E_1 \otimes E_2 \to {}_zE \to E_1 \otimes {}_{y_2}E_2 \to (E_1/y_1E_1) \otimes E_2 \stackrel{f}{\to} E/zE \to E_1 \otimes (E_2/y_2E_2) \to 0.$$

Chasing the exact sequence and considering that tensor products are taken over the field k, it is easy to see that the image of f is isomorphic to $E_1/y_1E_1 \otimes y_2E_2$. This module, by the previous lemma, has depth $i+1$. Thus all the modules in the sequence have depth at least $i+1$. □

Example 4.4.12 Sliding depth is not a property of all Cohen–Macaulay graphs. A case in point is that of a square with whiskers attached at two consecutive vertices. In other words, the ideal

$$I = (x_1x_2, x_2x_3, x_3x_4, x_4x_1, x_1x_5, x_2x_6),$$

is Cohen–Macaulay but the depth of $H_1(I)$ is 0 rather than the required ≥ 1.

One main application of Theorem 4.4.9 is:

Corollary 4.4.13 *If G is a Cohen–Macaulay tree, then $I(G)$ is a strongly Cohen–Macaulay ideal.*

Proof. For each prime $\wp$ containing I, it is clear that

$$\nu(I_\wp) \leq \max\{\text{height } I, \text{height } \wp - 1\},$$

as the localization of a tree is a forest! Now use Theorem 3.3.17. □

One reason for seeking ideals that are strongly Cohen–Macaulay is to use them in residual intersections to obtain other classes of Cohen–Macaulay ideals.

Let us give an application of the latter and the mechanism of [143] to generate new Cohen–Macaulay ideals. We use the notation of Proposition 4.4.4. If G is a tree, we denote by t the sum of its edges; we refer to t as the *tree generator*. Set $I = S(G)$ and $A = (\mathbf{z}, t) \subset I$.

Corollary 4.4.14 *$J = A\colon I$ is a Cohen–Macaulay ideal of codimension $|G| + 1$.*

Proof. It suffices to observe that the minimal primes of L (in the notation of Proposition 4.4.4) are $(\mathbf{x}, y_j)$, where y_iy_k is an edge of G. It follows that t cannot belong to any minimal prime of $(\mathbf{z})\colon I = (\mathbf{z}, L)$, so that $A\colon I$ has height at least $\nu(A) = n + 1$. Since $H_1(I; R)$ is Cohen–Macaulay and I is generically a complete intersection, from [175] J must be Cohen–Macaulay. □

Problem 4.4.15 Find a characterization of Cohen–Macaulay graphs; see [305] for some preliminary results.

Chapter 5

Arithmetic of Rees Algebras

This chapter deals with central techniques to study the arithmetic of Rees algebras and certain symmetric algebras. It treats normality and divisor class groups, describes canonical modules and contains several constructions to generate new algebras from old ones. The more novel material are some techniques to effect a bridge between the special theory of ideals of linear type with more general ideals.

A natural limitation of the methods pursued in Chapter 3 is the requirement that one of the $\mathcal{F}_k$–condition holds. To help span the gap it is appropriate to view the Rees algebra of a 'general' ideal as an extension of the Rees algebra of an ideal of linear type. The vehicle for this theory is the notion of reduction of an ideal, an object first isolated by Northcott and Rees ([216]), and which plays a premier role in the study of Rees algebras. An ideal $J \subset I$ is a *reduction* of I if $JI^r = I^{r+1}$ for some integer r; the least such integer, $r_J(I)$, is the *reduction number* of I with respect to J. Phrased otherwise, J being a reduction means that

$$R[Jt] \hookrightarrow R[It]$$

is a finite morphism of graded algebras. We shall introduce several structures where the two algebras interact. One of these is the notion of reduction modules, which while reflecting properties of $R[It]$ are more easily studied over $R[Jt]$. To be truly useful however we must be able to convert homological data on I into similar properties of some of its reductions. Most of the effort will be on Koszul homology so as to be able to use the approximation complexes on the reduction and use mixed structures to derive arithmetical properties of $R[It]$.

The nature of the defining equations of symmetric algebras makes for an easier approach to testing for normality. This chapter contains normality criteria for symmetric algebras of easy applicability, and others, less amenable, for Rees algebras. The divisor class groups of these algebras is expressed as an extension of the divisor class group of the base ring by a finitely generated free abelian group. The rank of this group, in the case of symmetric algebras, is given by a formula read off the presentation of the module. This explicitness makes possible the determination of the canonical module of certain symmetric algebras. In chapter 6, we return to this question with regard to a conjecture on the factoriality of symmetric algebras.

5.1 Reduction Modules

Our exposition in this section and the next one follows closely [298]. Perhaps the more interesting aspects are several new approaches to the study of Hilbert functions and the depth properties of Rees algebras.

Tangent cones

We begin by making more precise the concepts sketched above.

Definition 5.1.1 Let R be a commutative Noetherian ring and let I be an ideal. The *reduction number* of I is the smallest reduction number $r_J(I)$ amongst all reductions J of I.

The reader can easily check that this means that the Rees algebra $R[It]$ is a finitely generated graded module over the algebra $R[Jt]$, and $r_J(I)$ is the infimum of the top degree of any homogeneous set of generators.

Note that I and J share several properties, among which they have the same radical. One of the advantages of J is that it may have a great deal fewer generators. We indicate how this may come about, with the notion of *minimal reduction*.

Definition 5.1.2 Let $(R, \mathfrak{m})$ be a Noetherian local ring and let I be an ideal. The *special fiber* of the Rees algebra $R[It]$ is the ring

$$F(I) = R[It] \otimes_R R/\mathfrak{m}.$$

Its Krull dimension is called the *analytic spread* of I, and is denoted $\ell(I)$.

If $I = \mathfrak{m}$, $F(\mathfrak{m})$ is the Zariski's *tangent cone* of R. $F(I)$ is also called the *fiber cone* of I.

Suppose the residue field of R, $k = R/\mathfrak{m}$, is infinite. We may then pick a Noether normalization of $F(I)$,

$$A = k[z_1, \ldots, z_\ell] \hookrightarrow F(I),$$

where $\ell = \ell(I)$, and the z_j can be chosen in degree 1. Let further $b_1, \ldots, b_s$ be a minimal set of homogeneous module generators of $F(I)$ over the algebra A:

$$F(I) = \sum_{1 \leq i \leq s} Ab_i, \ \deg(b_i) = r_i.$$

Proposition 5.1.3 *Let $a_1, \ldots, a_\ell$ be elements of I that are a lift of $z_1, \ldots, z_\ell$. Then $J = (a_1, \ldots, a_\ell)$ is a reduction of I and*

$$r_J(I) = \sup\{\deg(b_i),\ 1 \leq i \leq s\}.$$

Proof. Both assertions follows easily by lifting the equality $F(I) = \sum_i Ab_i$ to $R[It]$ and using the Nakayama Lemma. □

Corollary 5.1.4 *The following holds*

$$\text{height } I \leq \ell(I) \leq \dim R.$$

Proof. The second inequality arises from the formula for the Krull dimension of $R[It]$. The other assertion follows since any minimal prime of J is also minimal over I, but the former have, by Krull principal ideal theorem, codimension at most ℓ. □

There are several measures of size for ideals in local rings. The following notions will play a role in the sequel. The *deviation* of I is the non–negative integer $\nu(I) -$ height I. Huckaba and Huneke [130] have defined the *analytic deviation* of I as $\ell(I)-$ height I; for ideals of linear type $\nu(I) = \ell(I)$. The ideals of analytic deviation zero are called *equimultiple* (see [103] for a wealth of information about these ideals). There is still room for another definition: the difference $\nu(I) - \ell(I)$ will be called the *second deviation* of the ideal I, the terminology arising because it is the difference between the other two deviations.

Reduction modules

Let R be a local ring, and let I be an ideal with a reduction J. To be able to use the known properties of the Rees algebra of J as a tool to obtain properties of the Rees algebra of I, it is necessary to build structures in which the two algebras intermingle.

We are going to mention only two such structures, one introduced in [298] and another of an older vintage [280].

The Sally module

Consider the exact sequence of finitely generated $R[Jt]$–modules:

$$0 \rightarrow I \cdot R[Jt] \longrightarrow I \cdot R[It] \longrightarrow S_J(I) = \bigoplus_{n=0}^{\infty} I^{n+1}t^n / IJ^n t^n \rightarrow 0. \tag{5.1}$$

Definition 5.1.5 The *Sally module* of I with respect to J is $S_J(I)$ viewed as an $R[Jt]$-module.

A motivation for this definition is the work of Sally, particularly in [237], [238], [239], and [240], where there is an extensive theory of the structure of the components of $S_J(I)$.

To be useful, this sequence requires information about $I \cdot R[Jt]$—which is readily available in many cases—and finer properties of $S_J(I)$. Its main feature is the relationship it bears with the ring $R[Jt]$, a ring that has many well–known properties.

We shall develop some general properties of the Sally module, and make widespread use of them. Let us begin by pointing out a simple but critical property of $S_J(I)$:

Proposition 5.1.6 *Let $(R, \mathfrak{m})$ be a Cohen–Macaulay local ring of dimension d, let I be a $\mathfrak{m}$–primary ideal and let J be a minimal reduction of I. If $S_J(I) \neq 0$ its associated prime ideals have codimension 1, in particular its Krull dimension as an $R[Jt]$–module is d.*

Proof. We first argue that $I \cdot R[Jt]$ is a maximal Cohen–Macaulay $R[Jt]$–module. For this it suffices to consider the exact sequence

$$0 \to I \cdot R[Jt] \longrightarrow R[Jt] \longrightarrow R[Jt] \otimes_R (R/I) \to 0,$$

and observe that the module on the right is a polynomial ring in d variables over R/I, and therefore Cohen–Macaulay of dimension d. Since $R[Jt]$ is a Cohen–Macaulay ring, $I \cdot R[Jt]$ has depth $d+1$.

We may assume $d \geq 1$, and that $S_J(I) \neq 0$. Let $P \subset A = R[Jt]$ be an associated prime of $S_J(I)$; it is clear that $\mathfrak{m} \subset P$. If $\mathfrak{m}A \neq P$, P has grade at least two.

Consider the homology sequence of the functor $\mathrm{Hom}_A(A/P, \cdot)$ on the sequence (5.1.5): we get

$$0 \to \mathrm{Hom}_A(A/P, S_J(I)) \longrightarrow \mathrm{Ext}^1_A(A/P, I \cdot R[Jt]).$$

Note that $I \cdot R[Jt]$ is a maximal Cohen–Macaulay module, hence the Ext module must vanish since P contains a regular sequence of two elements on it. □

The Valabrega–Valla module

Consider the exact sequence of finitely generated $R[Jt]$–modules:

$$0 \to J \cdot R[It] \longrightarrow J \cdot R[t] \bigcap I \cdot R[It] \longrightarrow VV_J(I) = \bigoplus_{n=0}^{\infty} (J \cap I^{n+1})/JI^n \to 0.$$

Definition 5.1.7 The *Valabrega–Valla module* of I with respect to J is $VV_J(I)$ viewed as an $R[Jt]$-module.

Its significance lies in the following Cohen–Macaulay criterion of [280]:

Theorem 5.1.8 *Let $(R, \mathfrak{m})$ be a Cohen–Macaulay local ring and let I be an $\mathfrak{m}$–primary ideal with infinite residue field. Suppose J is a minimal reduction of I. Then $\mathrm{gr}_I(R)$ is Cohen–Macaulay if and only if $VV_J(I) = 0$.*

From Rees algebras to associated graded rings

An elementary but important relationship between Rees algebras and associated graded rings is the following result ([138]):

Theorem 5.1.9 *Let R be a Cohen–Macaulay ring and let I be an ideal containing regular elements such that $R[It]$ is Cohen–Macaulay. Then $\mathrm{gr}_I(R)$ is Cohen–Macaulay.*

Proof. Since I is a regular ideal, that is, I contains regular elements, $\dim R[It] = \dim R + 1$. We may assume that R is a local ring. To show that the depth of $\mathrm{gr}_I(R)$ (relative to its irrelevant maximal ideal) is at least $\dim R$, we make use of the two exact sequences:

$$0 \to I \cdot R[It] \longrightarrow R[It] \longrightarrow \mathrm{gr}_I(R) \to 0 \tag{5.2}$$

$$0 \to It \cdot R[It] \longrightarrow R[It] \longrightarrow R \to 0. \tag{5.3}$$

The second sequence says that $It \cdot R[It]$ is a maximal Cohen–Macaulay module. Since it is isomorphic to $I \cdot R[It]$, the associated graded ring will be Cohen–Macaulay. □

Example 5.1.10 Let $R = k[x, y]/(x^3+y^3)$, and let $I = (x, y) \subset R$. The rings R and $\mathrm{gr}_I(R)$ are isomorphic, so the associated graded ring is Cohen–Macaulay. It is easy to see that the element $xt \in R[It]$ is a maximal regular sequence of the irrelevant maximal ideal (x, y, xt, yt) of the Rees algebra.

Although there is no full converse to Theorem 5.1.9, there exist significant cases where it holds—e.g. ideals generated by d–sequences (Theorem 3.3.19); later we shall return to this question (see Theorem 5.1.22).

In addition to the basic sequences that connect R, $R[It]$ and $\mathrm{gr}_I(R)$, there are other sequences that are useful in induction arguments (cf. [138, Lemma 1.1]).

Proposition 5.1.11 *Let R be a local ring and $a \in I$ be such that its image a^* in I/I^2 is a regular element of $\mathrm{gr}_I(R)$. Denote $R_1 = R/(a)$ and $I_1 = I/(a)$. Then there are exact sequences*

$$\begin{aligned} 0 \to (a, at) \longrightarrow R[It] &\longrightarrow R_1[I_1t] \to 0 \\ 0 \to \mathrm{gr}_I(R)[-1] \longrightarrow R[It]/(a) &\longrightarrow R_1[I_1t] \to 0 \\ 0 \to R \longrightarrow R[It]/(at) &\longrightarrow R_1[I_1t] \to 0. \end{aligned}$$

Proof. The key fact, that (a, at) is the kernel of the natural surjection of Rees algebras, is a consequence of $0 \colon a^* = 0$ in $\mathrm{gr}_I(R)$. It will be left as an exercise for the reader. □

Reduction number one

Several authors have studied, under various guises, what amounts to the vanishing of some $S_J(I)$. In addition to the afore-mentioned research of Sally on $\mathfrak{m}$–primary ideals, recent work by Huckaba and Huneke ([130], [131]) has established the vanishing of the Sally module for large classes of ideals of positive dimension.

We want to see this module as a vehicle for bookkeeping information on algebraic properties of the associated graded ring of an ideal. This occurs as simply as:

Proposition 5.1.12 *Let $(R, \mathfrak{m})$ be a Cohen–Macaulay local ring of dimension d. Suppose I is a Cohen–Macaulay ideal and $J \subset I$ is generated by a regular sequence of s elements. If $I^2 = JI$ then $\mathrm{gr}_I(R)$ is a Cohen–Macaulay ring.*

Proof. In this case $S_J(I) = 0$. On the other hand, the exact sequence

$$0 \to I \cdot R[Jt] \longrightarrow R[Jt] \longrightarrow R/I[T_1, \ldots, T_s] \to 0,$$

means that $I \cdot R[Jt]$ is a Cohen–Macaulay module of dimension $\dim R/I + s + 1 = d + 1$. Plugging this first in equation (5.3), and then in (5.2), yields the assertion. □

Corollary 5.1.13 ([280, Proposition 3.1]) *Let $(R, \mathfrak{m})$ be a Cohen–Macaulay local ring of dimension d and let I be a $\mathfrak{m}$–primary ideal. Suppose J is a reduction of I generated by a regular sequence and $I^2 = JI$. Then $\mathrm{gr}_I(R)$ is a Cohen–Macaulay ring.*

Reduction number two

If the reduction number is 2, the structure of $S_J(I)$ depends on the structure of the module I^2/JI. Note that $S_J(I)$ is then a module over $\mathrm{gr}_J(R)$.

Proposition 5.1.14 *Let $(R, \mathfrak{m})$ be a Cohen–Macaulay local ring of dimension d and let I be a $\mathfrak{m}$–primary ideal. Suppose J is a reduction of I generated by a regular sequence and $I^3 = JI^2$.*

(a) *If $I^2/JI = R/\mathfrak{m}$ then* $\mathrm{depth}\ \mathrm{gr}_I(R) \geq d - 1$.

(b) *If $I^2/JI = (R/\mathfrak{m})^2$ then* $\mathrm{depth}\ \mathrm{gr}_I(R) \geq d - 2$.

Proof. (a) The assumptions imply that $S_J(I)$ is a cyclic $R[Jt]$–module, annihilated by $\mathfrak{m}R[Jt]$. But in the sequence of definition of $S_J(I)$, with $I \cdot R[Jt]$ Cohen–Macaulay, no associated prime of $S_J(I)$ can have height greater than 1. This shows that $S_J(I) \simeq R/\mathfrak{m}[T_1, \ldots, T_d]$. From this we again go to the other two sequences to obtain the desired depth estimate.

(b) The module $S_J(I)$ is either isomorphic to two copies of the polynomial ring $B = R/\mathfrak{m}[T_1, \ldots, T_d]$ or to an ideal of this ring, generated by two elements. In the first case, one proceeds as in (a). If $S_J(I)$ is an ideal of B, it has depth $d - 1$. This means that $I \cdot R[It]$ has also depth $d - 1$ from which we get that the assertion. □

How to tell one case in (b) from the other? It may take place as follows. Suppose $I^3 = JI^2$ and $I^2 = (a_1, \ldots, a_s, JI) \subset J$, and $\mathfrak{m}I^2 \subset JI$. It is clear that $S_J(I)$ will be isomorphic to an ideal of $R/\mathfrak{m}[T_1, \ldots, T_d]$ generated by indeterminates and therefore its depth is available.

Remark 5.1.15 If $I = \mathfrak{m}$ has reduction number 2, Sally ([239]) proved that $\mathrm{gr}_I(R)$ is Cohen–Macaulay.

Example 5.1.16 The ideal $I = (x^3, y^3, x^2y)$ satisfies the condition (a) and

$$\mathrm{depth}\ \mathrm{gr}_I(R) = 1.$$

The Cohen–Macaulayness of the Rees algebra $R[It]$ is more delicate. For $\mathfrak{m}$–primary ideals there is a general condition due to Goto–Shimoda [88, Theorem 3.1] (see also [93, Theorem 4.8], [86], [158]) and Exercise 5.1.26).

Theorem 5.1.17 *The algebra $R[It]$ is Cohen–Macaulay if and only if $\mathrm{gr}_I(R)$ is Cohen–Macaulay and the reduction number of I is at most $d-1$.*

Remark 5.1.18 Craig Huneke has pointed out the following sharpening of this theorem. When R is a regular local ring the Briançon–Skoda theorem holds ([186]), and therefore if I is any $\mathfrak{m}$–primary ideal and J is a minimal reduction of I, then I^d is contained in J, where $d = \dim R$. On the other hand, if $\mathrm{gr}_I(R)$ is Cohen–Macaulay, the images of a minimal generating set of J form a system of parameters for this ring, and therefore $J \cap I^k = JI^{k-1}$ for all k. This shows that the reduction number is at most $d-1$ whenever $\mathrm{gr}_I(R)$ is Cohen–Macaulay, and one can delete this hypothesis from the theorem.

Let us see one application ([151, Theorem 3.2]) :

Theorem 5.1.19 *Let $(R, \mathfrak{m})$ be a two–dimensional regular local ring and let I be a $\mathfrak{m}$–primary ideal. If $R[It]$ is normal then it is a Cohen–Macaulay ring.*

Proof. We may assume that the residue field of R is infinite. By [186, Proposition 5.5] (another proof of which is given in [151]), I has reduction number 1. Thus by Proposition 5.1.12 $\mathrm{gr}_I(R)$ is Cohen–Macaulay and therefore $R[It]$ is Cohen–Macaulay by Theorem 5.1.17. □

The a–invariant

The relationship between the Cohen–Macaulayness of $R[It]$ and $\mathrm{gr}_I(R)$ was shown by Ikeda and Trung ([158]) to depend on the degrees of the minimal generators of the canonical module of $\mathrm{gr}_I(R)$.

Let us recall the notion of the a–invariant of a graded ring introduced by Goto and Watanabe ([89]). We follow the exposition of [31]. To simplify, we are going to assume that the algebras here are homomorphic images of Gorenstein rings. In addition, we speak of canonical modules of rings which are not necessarily Cohen–Macaulay see [113]. This simply means in the local case that if S is a Gorenstein ring and $R = S/I$, where I is an ideal of height g, then we call the R–module $\mathrm{Ext}^g_S(R, S)$ the *canonical module* of R. If it is isomorphic to R we say that R is *quasi–Gorenstein*.

Definition 5.1.20 *Let $R = R_0 + R_1 + \cdots$ be a graded ring of Krull dimension d with irrelevant maximal ideal $M = (\mathfrak{m}, R_+)$. ($(R_0, \mathfrak{m})$ is a local ring.)*

$$a(R) = \sup\{i \mid H^d_M(R)_i \neq 0\}, \tag{5.4}$$

where $H^d_M(R)$ denotes the (graded) d–dimensional local cohomology module of R with respect to M, is the a–invariant of R.

That the module $H^d_M(R)$ is Artinian ensures that $a(R)$ is finite. If ω_R is the canonical module of R, by local duality, it follows that

$$a(R) \quad = \quad -\inf\{\, i \mid (\omega_R/M\cdot\omega_R)_i \neq 0\}. \tag{5.5}$$

The two aspects of this definition complement one another very well, one being useful for its good localization properties, the other because it deals with modules which are Artinian and therefore very useful in induction arguments.

Remark 5.1.21 Let $x \in R$ be a homogeneous, regular element of degree $\deg x = r$. From the exact sequence

$$0 \to R(-r) \xrightarrow{\cdot x} R \longrightarrow R/(x) \to 0,$$

we have the cohomology exact sequence

$$H^{d-1}_M(R) \to H^{d-1}_M(R/(x)) \to H^d_M(R)(-r) \xrightarrow{\cdot x} H^d_M(R) \to 0,$$

from which we get

$$a(R/(x)) \geq r + a(R),$$

with equality if R is Cohen–Macaulay. For a polynomial ring $R = A[x_1,\ldots,x_d]$, A a local ring and the x_i indeterminates of degree 1, one has $a(R) = -n$.

The following version of [158, Theorem 1.1] will serve our needs. Actually the proofs to be given later use the methodology of this result.

Theorem 5.1.22 *Let R be a Noetherian local ring of dimension d and let I be an ideal of positive height. The Rees algebra $R[It]$ is Cohen–Macaulay if and only if $H^r_M(\mathrm{gr}_I(R))_i = 0$ for $i \neq -1$ and $r < d$ and $H^d_M(\mathrm{gr}_I(R))_i = 0$ for $i \geq 0$.*

Putting some of this assertion with Theorem 5.1.9 we have the useful criterion:

Theorem 5.1.23 *Let R be a Cohen–Macaulay local ring and let I be an ideal of positive height. Then $R[It]$ is Cohen–Macaulay if and only if $\mathrm{gr}_I(R)$ is Cohen–Macaulay and $a(\mathrm{gr}_I(R)) < 0$.*

The following elementary observation is useful here. About terminology: For a local ring the punctured spectrum has the usual meaning: the set of non-maximal prime ideals. For a graded algebra we mean the set of homogeneous, non–maximal prime ideals.

Proposition 5.1.24 *Let R be a Noetherian local ring and let I be an ideal of positive grade. Suppose $\mathcal{R}(I)$ is Cohen–Macaulay on the punctured spectrum of R. If $\mathrm{gr}_I(R)$ is Cohen–Macaulay on the punctured spectrum then $\mathcal{R}(I)$ is also Cohen–Macaulay on the punctured spectrum.*

Proof. We may assume that $I = (f_1, \ldots, f_n)$ with each f_i an R–regular element. Set $\mathcal{R}(I) = R[f_1t, \ldots, f_nt]$; it suffices to prove that $\mathcal{R}(I)_{f_it}$ is Cohen–Macaulay for each f_i. Note that f_i is a regular element of $\mathcal{R}(I)_{f_it}$ and there is the canonical isomorphism

$$\mathcal{R}(I)_{f_it}/(f_i)_{f_it} \simeq \mathrm{gr}_I(R)_{f_it}.$$

The assertion follows since the last ring is Cohen–Macaulay. □

To enhance the comparison between the properties of $R[It]$ and of $\mathrm{gr}_I(R)$, we give a result of [133] in a situation where $\mathrm{gr}_I(R)$ is not Cohen–Macaulay (see also [94]).

Theorem 5.1.25 *Let R be a local ring with* depth $R \geq d$ *and let I be an ideal. If* depth $\mathrm{gr}_I(R) < d$*, then*

$$\text{depth } R[It] = \text{depth } \mathrm{gr}_I(R) + 1.$$

Proof. This proof avoids the use of generalized depth employed in [133]. We can compute all depths with respect to the irrelevant maximal ideal $P = (\mathfrak{m}, ItR[It])$ of $R[It]$. For simplicity put $\mathcal{R} = R[It]$, $G = \mathrm{gr}_I(R)$, $L = ItR[It]$ and $L[+1] = IR[It]$.

We are going to determine the depth of $R[It]$ by examining the exact sequences of local cohomology modules derived from the sequences (5.2) and (5.3). For simplicity of notation we denote $H^j_P(\bullet)$ by $H^j(\bullet)$.

Since depth $R = d$ and depth $G = r < d$, we have

$$H^j(R) = 0, \; j < d \tag{5.6}$$
$$H^j(G) = 0, \; j < r. \tag{5.7}$$

The portions of the cohomology sequences that we are interested in are:

$$H^{j-1}(R) \longrightarrow H^j(L) \longrightarrow H^j(\mathcal{R}) \longrightarrow H^j(R)$$
$$H^{j-1}(G) \longrightarrow H^j(L[+1]) \longrightarrow H^j(\mathcal{R}) \longrightarrow H^j(G).$$

Taking into these two sequences the conditions in (5.6) and (5.7) yield the exact sequences

$$H^j(L) \simeq H^j(\mathcal{R}), \; j < d \tag{5.8}$$
$$0 \longrightarrow H^j(L[+1]) \longrightarrow H^j(\mathcal{R}) \longrightarrow H^j(G), \; j \leq r, \tag{5.9}$$

from which we claim that $H^j(\mathcal{R}) = 0$, for $j \leq r$. This will suffice to prove the assertion.

Denote the modules $H^j(L)$ and $H^j(\mathcal{R})$ respectively by M_j and N_j. The sequences (5.8) and (5.9) then give rise to graded isomorphisms (of degree zero)

$$M_j[+1] \simeq N_j, \; j < r$$
$$M_j \simeq N_j, \; j < d$$

and the monomorphism

$$0 \longrightarrow M_r[+1] \xrightarrow{\varphi} N_r. \tag{5.10}$$

Since $N_r \simeq M_r[+1] \simeq M_r$ as (ungraded) modules, φ is a monomorphism of isomorphic Artinian modules, and therefore must be an isomorphism. This means that we have isomorphisms of Artinian graded modules

$$M_j \simeq N_j \simeq M_j[+1],\ j \leq r,$$

with mappings of degree zero. Since the graded components of these modules are zero in all degrees sufficiently high, the modules must vanish. $\square$

Exercise 5.1.26 Let $(R, \mathfrak{m})$ be a Cohen–Macaulay local ring of dimension d and let I be a $\mathfrak{m}$–primary ideal such that the associated graded ring $G = \mathrm{gr}_I(R)$ is Cohen–Macaulay. Let J be a minimal reduction of I and denote $r = r_J(I)$. Prove that $a(G) = r - d$. (*Hint:* Let $J = (a_1, \ldots, a_d)$ and let

$$\mathbb{G} \qquad 0 \to G[-d] \longrightarrow G^d[-d+1] \longrightarrow \cdots \longrightarrow G^d[-1] \longrightarrow G \to 0$$

be the (graded) Koszul complex built on the 1–forms $\overline{a}_i$ of G defined by the elements a_i. Observe that $\mathbb{G}$ is acyclic and that r is the maximum degree of the nonvanishing components of $H_0(\mathbb{G})$. From this result and Theorem 5.1.22, it follows that $R[It]$ is Cohen–Macaulay if and only if G is Cohen–Macaulay and $r < d$ (cf. [88])).

5.2 Hilbert Functions of Local Rings

There will be no attempt here to develop fully a theory of the Hilbert function of primary ideals in local rings. We will however indicate the usefulness of the setting of reduction modules to prove rapidly several known inequalities on the coefficients of the Hilbert polynomial of a primary ideal.

Let $(R, \mathfrak{m})$ be a Noetherian local ring of Krull dimension d and I an $\mathfrak{m}$–primary ideal. The *Hilbert function* of I is the assignment

$$HF : n \rightsquigarrow \lambda(R/I^n).$$

For $n >> 0$ (but often not much greater than zero!) $HF(n)$ is equal to a polynomial $H(n)$ of degree d, the *Hilbert polynomial of* I.

To study either of these functions, without loss of generality, it is convenient to assume that the residue field of R is infinite. Here we limit ourselves mostly to Hilbert polynomials leaving aside the rich area of *irregularities*, to wit the detailed comparison between the two functions.

We shall further assume that R is a Cohen–Macaulay ring. In such cases, if I is generated by a system of parameters the functions are given simply by

$$HF(n) = H(n) = \lambda(R/I) \cdot \binom{n+d-1}{d},\ \forall n.$$

It suggests that the Hilbert function of an arbitrary ideal I be approached through the Hilbert function of one of its minimal reductions J.

We look at two exact sequences as the vehicle for this comparison:

$$0 \to IJ^{n-1}/J^n \longrightarrow I^n/J^n \longrightarrow I^n/IJ^{n-1} \to 0$$

and

$$0 \to IJ^{n-1}/J^n \longrightarrow J^{n-1}/J^n \longrightarrow J^{n-1}/IJ^{n-1} \to 0.$$

The function $\lambda(I^n/J^n)$ is our center of interest since it equals $\lambda(R/J^n) - \lambda(R/I^n)$, the first term being very well behaved. Using the other sequence we have the following expression for the Hilbert function $HF(n)$ of I:

$$\begin{aligned} HF(n) &= \lambda(R/J^n) - \lambda(IJ^{n-1}/J^n) - \lambda(I^n/IJ^{n-1}) \\ &= \lambda(R/J^n) - \lambda(J^{n-1}/J^n) + \lambda(J^{n-1}/IJ^{n-1}) - \lambda(I^n/IJ^{n-1}) \\ &= \lambda(R/J^{n-1}) + \lambda(J^{n-1}/IJ^{n-1}) - \lambda(I^n/IJ^{n-1}), \end{aligned}$$

the first 2 terms of which can be collected since the ring $\oplus(J^{n-1}/IJ^{n-1})$ is a polynomial ring in d variables with coefficients in R/I. We obtain

$$HF(n) = \lambda(R/J) \cdot \binom{n+d-2}{d} + \lambda(R/I) \cdot \binom{n+d-2}{d-1} - \lambda(I^n/IJ^{n-1}).$$

This expression turns the focus on $\lambda(I^n/IJ^{n-1})$. We note that both I^n/J^n and I^n/IJ^n are components of modules over the Rees algebra $R[Jt]$: the first comes from the quotient $R[It]/R[Jt]$, the other being a component of the Sally module $S_J(I)$. The latter has many advantages over the former, a key one being that it vanishes in cases of considerable interest.

Sometimes it will be convenient to write $H_I(\cdot)$ for the Hilbert function of I; the reader is also warned about shifts in the arguments arising from the grading of the modules. For example, the degree 1 component of $S_J(I)$ is I^2/IJ.

Proposition 5.2.1 *Let $(R, \mathfrak{m})$ be a Cohen–Macaulay local ring of dimension d, with infinite residue field, and let I be an $\mathfrak{m}$–primary ideal. Denote by $H_I(n) = \lambda(R/I^n)$ the Hilbert function of I, and let*

$$e_0\binom{n+d-1}{d} - e_1\binom{n+d-2}{d-1} + \cdots + (-1)^{d-1}e_{d-1}\binom{n}{1} + (-1)^d e_d \quad (5.11)$$

be its normalized Hilbert polynomial. Suppose J is a minimal reduction of I and let $S = S_J(I)$ be the corresponding Sally module. Then for $n \gg 0$

$$H_I(n) = e_0\binom{n+d-1}{d} + (\lambda(R/I) - e_0)\binom{n+d-2}{d-1} - \lambda(S_{n-1}). \quad (5.12)$$

Proof. The proof is a straightforward calculation that takes into account the equality $e_0 = \lambda(R/J)$. □

Corollary 5.2.2 *If $S_J(I) \neq 0$, the function $\lambda(S_n)$ has the growth of a polynomial of degree $d-1$.*

Proof. By Proposition 5.1.6, if $S_J(I) \neq 0$ then its Krull dimension is d. □

The following formulas can be used to establish several vanishing results on the coefficients of Hilbert functions.

Theorem 5.2.3 *Let $s_0, s_1, \ldots, s_{d-1}$ be the coefficients of the Hilbert polynomial of $S_J(I)$. Then*

$$\begin{aligned} s_0 &= e_1 - e_0 + \lambda(R/I) \\ s_i &= e_{i+1}, \text{ for } i \geq 1. \end{aligned}$$

Corollary 5.2.4 *The following hold:*

(a) $\lambda(R/I) \geq e_0 - e_1$ ([215]).

(b) *If equality above holds then* $S_J(I) = 0$ ([144, Theorem 2.1], [217]).

Proof. (a) This is clear since e_1 will be obtained by adding to $e_0 - \lambda(R/I)$ the contribution from the leading term in the Hilbert polynomial of $S_J(I)$.

(b) For equality to hold there must be no contribution from $S_J(I)$, which means that $S_J(I)$ is a module of Krull dimension $\leq d-1$. From Proposition 5.1.6 this implies $S_J(I) = 0$. □

Remark 5.2.5 Another result that can be derived from the formulas above is Narita's inequality: $e_2 \geq 0$.

Proposition 5.2.6 *Let $(R, \mathfrak{m})$ be a Cohen–Macaulay local ring of dimension d, with infinite residue field, and let I be an $\mathfrak{m}$–primary ideal with a minimal reduction J. Let e_0 and e_1 be the first two coefficients of the Hilbert polynomial of I. If*

$$e_1 = e_0 - \lambda(R/I) + 1, \tag{5.13}$$

then $S_J(I)$ is isomorphic to an ideal of $B = R/\mathfrak{m}[T_1, \ldots, T_d]$.

Proof. The assumption is that the Hilbert polynomial of $S_J(I)$ has the form

$$\binom{d+n-1}{d-1} - e_2\binom{d+n-2}{d-2} + \cdots.$$

From Proposition 5.1.6 it follows that $\mathfrak{m}R[Jt]$ is the only associated prime of $S_J(I)$. Let $0 \neq a \in S_J(I)_1$ be an element annihilated by $\mathfrak{m}$. There exists a form $f \in R[Jt] \setminus \mathfrak{m}R[Jt]$ such that

$$f \cdot S_J(I) \subset R[Jt]a.$$

Because f is a regular element on $S_J(I)$, the equation implies that $S_J(I)$ is annihilated by $\mathfrak{m}$. □

Corollary 5.2.7 *Suppose further that $R[It]$ is integrally closed. If $e_1 = e_0 - \lambda(R/I) + 1$ then $L = B$.*

Proof. The hypothesis on $R[It]$ says that it satisfies the condition S_2 of Serre's, so that from the exact sequences (5.2) and (5.3) it follows that $I \cdot R[It]$ will inherit the condition. It is now easy to see from the definition of $S_J(I)$ that it has the S_2 property as a B–module. This means that L is a reflexive ideal and it is therefore free. $\square$

The method serves well to prove the following result of Sally ([240, Theorem 1.4]).

Theorem 5.2.8 *Let $(R, \mathfrak{m})$ be a Cohen–Macaulay local ring of dimension d, with infinite residue field, and let I be an $\mathfrak{m}$–primary ideal. Let e_0, e_1 and e_2 be the first three coefficients of the Hilbert polynomial of I. If $e_1 = e_0 - \lambda(R/I) + 1$ and $e_2 \neq 0$, then $L \simeq B$. In particular, I has reduction number 2.*

Proof. In the equality $f \cdot S = L \cdot a$ we take f of least degree. Note that if f is a constant then L must be equal to B. Suppose L has a gcd $= \Delta \neq 1$, $L = \Delta L_0$. There is then an equation $fa = \Delta ca$ that shows $\Delta | f$.

Suppose f has degree $h \geq 1$. From Proposition 5.2.6, the Hilbert polynomial of $f \cdot S$ is then

$$
\begin{aligned}
H_{fS}(n) = H_S(n-h) &= \binom{d+n-h-1}{d-1} - e_2\binom{d+n-h-2}{d-2} + \text{lower terms} \\
&= \binom{d+n-1}{d-1} - (e_2+h)\binom{d+n-2}{d-2} + \text{lower terms},
\end{aligned}
$$

while the Hilbert polynomial of La has the form

$$
H_{La}(n) = H_L(n-1) = \binom{d+n-2}{d-1} - b\binom{d+n-4}{d-3} + \text{lower terms}.
$$

This equation arises from the fact that the Hilbert polynomial of L is the difference between the Hilbert polynomials of B and of B/L, and B/L has dimension at most $d-2$.

A comparison of coefficients shows that $h + e_2 = 1$. By [210] $e_2 \geq 0$, while $h \geq 1$. Thus $e_2 = 0$, which is a contradiction. $\square$

Corollary 5.2.9 *If $e_1 = e_0 - \lambda(R/I) + 1$ and $L \neq B$, then*

$$
\operatorname{depth} \operatorname{gr}_I(R) \geq \operatorname{depth} B/L.
$$

The technique also suggests that one may approach the computation of Hilbert functions by using more general reductions as long as their Rees algebras are nice to handle.

Exercise 5.2.10 Suppose $(R, \mathfrak{m})$ is a local Noetherian ring of dimension d. Let I be an $\mathfrak{m}$–primary ideal and let J be one of its minimal reductions. Prove that

$$r_J(I) \leq a(\mathrm{gr}_I(R)) + d.$$

(It is not assumed that R is Cohen–Macaulay.)

Exercise 5.2.11 Let R be a regular local ring and let $\wp$ be an equimultiple prime ideal. Prove that $\wp$ is generated by a regular sequence (see [48]).

Exercise 5.2.12 Let R be a Cohen–Macaulay local ring of dimension d and let I be an ideal with height $I > 0$. Suppose the algebra $\mathrm{gr}_I(R)$ is Cohen–Macaulay on the punctured spectrum. Let M be the irrelevant maximal ideal of $R[It]$. Prove that the modules

$$H^i_M(R[It]),\ 0 \leq i \leq d$$

are Noetherian.

Conjecture 5.2.13 Let R be a regular local ring and let $\wp$ be a prime ideal. If the associated graded ring $\mathrm{gr}_\wp(R)$ is Cohen–Macaulay then $R[\wp t]$ is Cohen–Macaulay (see [2]).

Problem 5.2.14 Let I be an ideal and J one of its reductions. Define the *second Sally module* of I with respect to J as $S_J^{(2)}(I) = I^2R[It]/I^2R[Jt]$. Carry out a study of this module with emphasis on its dimension and its vanishing.

5.3 Koszul Homology of Reductions

Let R be a Cohen–Macaulay local ring of infinite residue field, with canonical module ω_R, and suppose I is an unmixed ideal which is generically a complete intersection. We examine some homological properties of the reductions of I. It will permit estimates of analytic spreads of several classes of ideals and of depths of associated graded rings. Throughout we assume that I is not generated by analytically independent elements.

Burch's formula

The relationship between analytic spread and asymptotic depth was first pointed out by Burch ([40]):

Theorem 5.3.1 *Let $(R, \mathfrak{m})$ be a local ring and let I be an ideal of height at least one. Then*

$$\ell(I) \leq \dim R - \inf_n \{\mathrm{depth}\ R/I^n\}.$$

Moreover, if $\mathrm{gr}_I(R)$ is Cohen–Macaulay then equality holds.

The inequality itself is due to Burch, but its sharpening in the Cohen–Macaulay case is [60, Proposition 3.3]. If on one hand it shows the difficulty in using the formula to determine $\ell(I)$, on the other hand the relationship that exists between asymptotic depth and Koszul homology in significant cases (e.g. such as those expressed in the approximation complexes) indicates a deeper role for the Koszul homology of I in estimating its analytic spread. This will be borne out for an interesting class of ideals.

We want to state a similar result which emphasizes the module structure of the conormal modules I^m/I^{m+1}. The following is a composite appropriate for some of our main applications later on; some of its hypotheses are unnecessarily restrictive.

Theorem 5.3.2 *Let R be a Cohen–Macaulay local ring and let I be an ideal of R.*

(a) *Suppose that for all integers m, the module I^m/I^{m+1} has depth at least r. Then $\ell(I) \leq \dim R - r$.*

(b) *In addition, suppose R is a regular local ring and assume that I is a Cohen–Macaulay ideal that satisfies sliding depth and $\mathcal{F}_1$. If $r \geq 1$ then I is strongly Cohen–Macaulay and R/I is locally a complete intersection in codimension r.*

Proof. (a): Follows from the fact that (cf. [23])

$$\inf{}_m\{\text{depth } R/I^m\} \quad = \quad \inf{}_m\{\text{depth } I^m/I^{m+1}\}.$$

(b): That such ideals are strongly Cohen–Macaulay is Theorem 3.3.17; the other assertion follows from [260, p. 356]. □

The top Koszul homology of reductions

We begin to derive properties of the Koszul homology of a reduction.

Proposition 5.3.3 *Let I be a height unmixed ideal which is generically a complete intersection and let J be a reduction of I. Then R/I and R/J have the same canonical module, and the same top Koszul homology module.*

Proof. Suppose height $I = g$, and consider the sequence

$$0 \rightarrow I/J \longrightarrow R/J \longrightarrow R/I \rightarrow 0.$$

I/J is a module annihilated by a power of I, whose support contains no minimal prime of I; therefore it has dimension at most $d - g - 1$, $d = \dim R$. In the cohomology sequence

$$\operatorname{Ext}_R^{g-1}(I/J, \omega_R) \rightarrow \operatorname{Ext}_R^g(R/I, \omega_R) \rightarrow \operatorname{Ext}_R^g(R/J, \omega_R) \rightarrow \operatorname{Ext}_R^g(I/J, \omega_R)$$

the modules at the ends vanish. The other assertion has a similar proof. □

The next result reveals several obstructions for a Cohen–Macaulay ideal to have analytic deviation 1. It is the strand of an idea traced to [277, Proposition 4].

Theorem 5.3.4 *Let R be a Gorenstein local ring of dimension d, and infinite residue field. Let I be an ideal of codimension g that is generically a complete intersection and has analytic spread $\ell(I) = g+1$. Let J be a proper minimal reduction of I.*

(a) *If I is Cohen–Macaulay then* depth $R/J = d-g-1$.

(b) *If R/I has the condition S_2 of Serre's then J has no associated prime ideal of codimension $g+2$.*

Proof. (a) Let $J = (a_1, \ldots, a_{g+1})$, and consider the acyclic portion of the Koszul complex $\mathbb{K}$ built on the a_i's:

$$0 \to K_{g+1} \longrightarrow K_g \longrightarrow \cdots \longrightarrow K_2 \longrightarrow B_1 \to 0. \tag{5.14}$$

From Proposition 5.3.3 we have that $H_1(\mathbb{K})$, the canonical module of R/J, is isomorphic to the canonical module of R/I, and therefore is Cohen–Macaulay.

If Z_1 denotes the module of 1–cycles of $\mathbb{K}$, the assertion is that depth $Z_1 = d - g + 1$. Let

$$0 \to B_1 \longrightarrow Z_1 \longrightarrow H_1(\mathbb{K}) \to 0 \tag{5.15}$$

be the sequence that defines $H_1(\mathbb{K})$. Since B_1 has projective dimension $g-1$, it has depth $d-g+1$ and thus the depth of Z_1 is at least $d-g$ since $H_1(\mathbb{K})$ is Cohen–Macaulay of dimension $d-g$. Into the cohomology exact sequence

$$\mathrm{Ext}_R^{g-1}(B_1, R) \longrightarrow \mathrm{Ext}_R^g(H_1(\mathbb{K}), R) \longrightarrow \mathrm{Ext}_R^g(Z_1, R) \longrightarrow \mathrm{Ext}_R^g(B_1, R)$$

we feed

$$\begin{cases} \mathrm{Ext}_R^{g-1}(B_1, R) = R/J \\ \mathrm{Ext}_R^g(B_1, R) = 0 \\ \mathrm{Ext}_R^g(H_1(\mathbb{K}), R) = R/I \end{cases}$$

the first two from the Koszul complex, and the last from the local duality theorem since R/I is Cohen–Macaulay:

$$R/J \longrightarrow \mathrm{Ext}_R^g(H_1(\mathbb{K}), R) = R/I \longrightarrow \mathrm{Ext}_R^g(Z_1, R) \to 0.$$

We claim that $\mathrm{Ext}_R^g(Z_1, R)$ vanishes. By induction on the dimension of R, we may assume that this module has finite length, and $\dim R/I \geq 2$. If the module is different from zero, from the sequence it must have the form $R/(I, a)$ for some a. By Krull's theorem this is not possible.

Finally, to argue that the equality of depth holds, note that if depth $R/J = d-g$, I and J would necessarily have the same associated prime ideals, and therefore would be equal since they agree at such primes.

(b) The equality $\mathrm{Ext}_R^g(H_1(\mathbb{K}), R) = R/I$ still holds and therefore one still gets that $\mathrm{Ext}_R^g(Z_1, R) = 0$. This means that $\mathrm{Ext}_R^{g+2}(R/J, R) = 0$, which is a strong form of the assertion. □

In the following corollaries, I is still assumed to be generically a complete intersection and not generated by analytically independent elements.

Corollary 5.3.5 *Let I be a Cohen–Macaulay ideal of codimension g that is of linear type in codimension $\leq g+1$. Then $\ell(I) \geq g+2$.*

Corollary 5.3.6 *Let I be a Cohen–Macaulay prime ideal all of whose powers I^m are unmixed. Then $g+2 \leq \ell(I) \leq \dim R - 1$. In particular, if $\dim R/I = 2$ no such prime exists, and if $\dim R/I = 3$ then $\ell(I) = g+2$.*

Proof. The upper bound is given by Theorem 5.3.1, while the equality $\ell(I) = g+1$ is ruled out by [48] and the previous corollary. □

The following is a special case of [131, Theorem 2.1]:

Corollary 5.3.7 *Let I be a Cohen–Macaulay ideal of analytic deviation 1 and reduction number 1. Then $\mathrm{gr}_I(R)$ is Cohen–Macaulay.*

Proof. Let J be a minimal reduction of I, of reduction number 1. The previous result saying that J is an ideal with sliding depth, we have by Corollary 3.3.22, that $R[Jt]$ is Cohen–Macaulay. On the other hand, the approximation complex of J

$$0 \to H_1(\mathbb{K}) \otimes B[-1] \to R/J \otimes B \to S(J/J^2) = R[Jt]/(J \cdot R[Jt]) \to 0, \quad (5.16)$$

where $B = R[T_1, \ldots T_{g+1}]$, splits as a complex of R–modules at the minimal primes of I. Since $H_1(\mathbb{K})$ is already a R/I–module of rank 1, tensoring the sequence with R/I we obtain an exact sequence of R/I–modules

$$0 \to H_1(\mathbb{K}) \otimes B[-1] \to R/I \otimes B \to R[Jt]/(I \cdot R[Jt]) \to 0 \quad (5.17)$$

from which it is easy to see that $I \cdot R[Jt]$ is a maximal Cohen–Macaulay module. We can then argue as in Proposition 5.1.12 since by hypothesis $I^2 = JI$. □

The mixed graded algebra and reduction number one

Motivated by the previous proof, we introduce the following algebra.

Definition 5.3.8 Let I be an ideal and J be one of its reductions. The *mixed graded algebra* of I relative to J is the ring $\mathrm{gr}_J(R) \otimes R/I$.

To show its usefulness, let us give another treatment of an interesting fact in [130]. Although not appreciably shorter, this proof will display features that can be used under other hypotheses.

Theorem 5.3.9 ([130, Theorem 2.2]) *Let R be a Cohen–Macaulay local ring and I an ideal of analytic deviation one and height $g \geq 1$. Let r be a positive integer and assume that the minimal primes of R/I have the same height, and the associated primes of R/I have height at most $g+r$. Assume also that I is a complete intersection at each of its minimal primes and there exists a minimal reduction J of I such that $r_J(I_\wp) \leq 1$ for every prime ideal $\wp \supset I$ with* height $\wp/I \leq r$. *Then $r_J(I) = 1$.*

Proof. We may assume that the residue field of R is infinite. Let $J = (a_1, \ldots, a_g, c)$ be a reduction of I where we may assume that $J_0 = (a_1, \ldots, a_g)$ generates I at each of its minimal primes and c is a regular element. Consider the complex defining the conormal module of J

$$0 \to H_1(J) \longrightarrow (R/J)^{g+1} \longrightarrow J/J^2 \to 0, \tag{5.18}$$

where $H_1(J)$ is the 1–dimensional Koszul homology module on the $g+1$ generators of J.

From the exact sequence (5.15), it follows that $H_1(J)$ is an R/J–module that satisfies the condition S_2 of Serre's. In particular its minimal primes are all of codimension g. As a consequence the equality $I \cdot H_1(J) = 0$ results by localization. This means that if we tensor the sequence (5.18) with R/I we obtain the complex

$$0 \to H_1(J) \longrightarrow (R/I)^{g+1} \longrightarrow J/JI \to 0, \tag{5.19}$$

whose exactness follows from the acyclicity lemma. In turn, this implies that the associated prime ideals of J/JI have codimension at most $g+r$. Thus to show the equality $I^2 = JI$ it suffices to prove $I^2 \subset J$, as it leads to an embedding

$$I^2/JI \hookrightarrow J/JI,$$

and the vanishing of I^2/JI would follow from the assumptions.

The point is to show $I^2 \subset J_0 + cI$, as the height of the associated primes of the latter can be estimated. Indeed, from the exact sequence induced by addition of ideals

$$0 \to J_0 \cap cI \longrightarrow J_0 \oplus cI \longrightarrow J_0 + cI \to 0,$$

and the elementary equality $J_0 \cap cI = cJ_0$ ([130, Remark 2.1(iii)]), we observe that the associated primes of $J_0 + cI$ have codimension at most $g+r$; the inclusion $I^2 \subset J_0 + cI$ can then be checked at such primes and the remaining assumption takes care of it. □

Remark 5.3.10 Another known result, but in analytic deviation two [130, Theorem 3.1], also yields to the same analysis.

We shall use the higher degree components of the sequence (5.17) to give one of the main applications of the previous result.

Corollary 5.3.11 ([130, Theorem 2.5]) *Let R be a Cohen–Macaulay local ring and I an analytic deviation one unmixed ideal. Assume* height $I = g \geq 1$, *I is generically a complete intersection. Then the following are equivalent:*

(a) $I^n = I^{(n)}$ *for every n;*

(b) $\ell(I_\wp) <$ height $\wp$ *for every $\wp \supset I$ of codimension $g+1$.*

Proof. (a) $\Rightarrow$ (b): Follows from Burch's formula (Theorem 5.3.1).

(b) $\Rightarrow$ (a): Let J be a reduction of I generated by $g+1$ elements. By Theorem 5.3.9, $I^2 = JI$. As in [130], we argue by induction on n, that is, assume that the equality $I^{(i)} = I^i$ holds for $i \leq n$ and that $I^{(n+1)}$ and I^{n+1} have the same primary components in the punctured spectrum of R. This implies that $I^{(n+1)} \subset I^n$, and we want to argue that $I^{(n+1)} \subset J^n$. To this end, we seek to embed $I^{(n+1)}$ into a subideal of J^n whose associated primes do not include the maximal ideal $\mathfrak{m}$. Using the notation of the previous proof and [130, Remark 2.1(iii)] we have that $\mathfrak{m}$ is not an associated prime of $J_0^n + cI^n$. This means that the inclusion

$$I^{(n+1)} \subset J_0^n + cI^n \subset J^n$$

can be checked on the punctured spectrum only. In the current case this chore is taken care by the hypotheses.

Finally, we may assume that

$$I^{(n+1)}/I^{n+1} \subset J^n/IJ^n.$$

From the degree n component of the approximation complex (5.17), $\mathfrak{m}$ is not an associated prime of J^n/IJ^n, which shows that $I^{(n+1)}/I^{n+1}$ must vanish. □

Cohen–Macaulay type

We give a computation to determine the Cohen–Macaulay type of the algebra $R[Jt]$.

Theorem 5.3.12 *Let I be a Cohen–Macaulay ideal as in* Theorem 5.3.4, *and let J be a minimal reduction. Then $R[Jt]$ is a Cohen–Macaulay ring of type $\nu(I) - 2$.*

Proof. We have only to prove the assertion about the Cohen–Macaulay type. Denote $S = R[Jt]$ and ω_S its canonical module, and use the formula of Theorem 3.3.23

$$\omega_S/S_+\omega_S = \bigoplus_{i=1}^{g} \operatorname{Ext}_R^{g-i}(Z_i, R),$$

where Z_i are the cycles of the Koszul complex on a minimal set of generators of J. A simple calculation shows:

$$\omega_S/S_+\omega_S = R \oplus (R/J)^{g-2} \oplus \operatorname{Ext}_R^{g+1}(R/J, R). \tag{5.20}$$

We must now produce the module $\operatorname{Ext}_R^{g+1}(R/J, R)$, for which we use an exact sequence in the proof of [277, Proposition 4]. The ideal J has obviously I for the intersection of the isolated primary components; a remainder of the intersection of the codimension $g+1$ components can be obtained as follows (cf. [277]). Let $J = (a_1, \ldots, a_g, a_{g+1})$ and assume that $J_0 = (a_1, \ldots, a_g)$ generates J at each minimal prime of I. Let $K = J_0 \colon I$; then a_{g+1} is regular on R/K and $J = I \cap L$, $L = (K, a_{g+1})$. There results the exact sequence of natural maps

$$0 \to R/J \longrightarrow R/I \oplus R/L \longrightarrow R/(I+L) \to 0.$$

Note that $(I, L) = (I, K)$, and that the latter is a Gorenstein ideal of codimension $g+1$ ([221, Remarque 1.4]). Consider now the long exact sequence of Ext's:

$$\begin{aligned}&\mathrm{Ext}^g_R(R/(I+L), R) \to \mathrm{Ext}^g_R(R/I, R) \oplus \mathrm{Ext}^g_R(R/L, R) \to \mathrm{Ext}^g_R(R/J, R) \to \\ &R/(I+L) \to \mathrm{Ext}^{g+1}_R(R/I, R) \oplus \mathrm{Ext}^{g+1}_R(R/L, R) \to \mathrm{Ext}^{g+1}_R(R/J, R) \to 0.\end{aligned}$$

From the proof of Proposition 5.3.3, this sequence yields the short exact sequence

$$0 \to R/(I+L) \to \mathrm{Ext}^{g+1}_R(R/L, R) = I/(J_0, Ia_{g+1}) \to \mathrm{Ext}^{g+1}_R(R/J, R) \to 0.$$

The image of the mapping on the left being the class of a_{g+1}, it follows that

$$\mathrm{Ext}^{g+1}(R/J, R) = I/J.$$

Since J consists of minimal generators of I, the Cohen–Macaulay type of $R[Jt]$ from equation (5.20) is given by the count

$$1 + (g-2) + \nu(I) - (g+1) = \nu(I) - 2,$$

as claimed. □

Koszul homology of reductions

Theorem 5.3.13 *Let R be a Cohen–Macaulay local ring of dimension d, let I be an ideal, let J be a reduction of I with $\nu(J) = s \leq d$, and assume that I satisfies $\mathcal{F}_1$, locally in codimension $\leq s-1$. If I has sliding depth, then J has sliding depth, and in particular J is of linear type and $R[Jt]$ is Cohen–Macaulay.*

Proof. We may assume that $g = \text{height } J < s \leq d$. Further notice that $I_\wp = J_\wp$ for all $\wp \in \mathrm{Spec}(R)$ with $\dim R_\wp \leq s-1$.

At this juncture, it is no restriction to assume that the residue field of R is infinite. Now let $a_1, \ldots, a_s$ be a generating sequence of J, and for $g \leq i \leq s-1$ write $L_i = (a_1, \ldots, a_i)$ and $K_i = L_i \colon J$. Since J satisfies $\mathcal{F}_1$, we may choose $a_1, \ldots, a_s$ in such a way that $J_\wp = (L_i)_\wp$ for all $\wp \in \mathrm{Spec}(R)$ with $\dim R_\wp \leq i-1$ and all $\wp \in V(J)$ with $\dim R_i \leq i$. In other words, height $K_i \geq i$ and height$(J+K_i) \geq i+1$. We claim that for $g \leq i \leq s-1$, R/K_i is Cohen–Macaulay of dimension $d-i$ and $J \cap K_i = L_i$. Once this is shown, the theorem will follow from Theorem 4.3.8.

To prove the claim notice that since $i \leq s-1$, we have $I_\wp = J_\wp = (L_i)_\wp$ for all $\wp \in \mathrm{Spec}(R)$ with $\dim R_\wp \leq i-1$ and all $\wp \in V(I)$ with $\dim R_\wp \leq i$. Thus height $L_i \colon I \geq i$ and height$(I + (L_i \colon I)) \geq i+1$. However, $\nu(I_\wp) \leq \dim R_\wp$ for all $\wp \in V(I)$ with $\dim R_\wp \leq i \leq s-1$, and I is assumed to have sliding depth. In this situation, Theorem 4.3.7 implies that $R/L_i \colon I$ is a Cohen–Macaulay ideal of dimension $d-i$ and $I \cap (L_i \colon I) = L_i$.

Now it suffices to prove that $L_i \colon I = K_i$. The inclusion $L_i \colon I \subset L_i \colon J = K_i$ being trivial, we only need to show the asserted equality at every associated prime $\wp$ of $L_i \colon I$. Since the latter ideal is Cohen–Macaulay of height i, we know that $\dim R_\wp = i \leq s-1$. Thus $I_\wp = J_\wp$, and $(L_i \colon I)_\wp = (K_i)_\wp$.

Finally, that $R[Jt]$ is Cohen–Macaulay follows from Theorem 3.3.19. □

Corollary 5.3.14 *Let I and J be as in* Theorem 5.3.13. *Let $\mathbb{K}$ be the Koszul complex on a minimal set of generators of J. Then I annihilates $H_i(\mathbb{K})$ for $i > 0$.*

Proof. Since $I \cdot H_i(\mathbb{K}) \hookrightarrow H_i(\mathbb{K})$, and this is a module of depth $\geq d - s + i$, it suffices to check the prime ideals $\wp$ of codimension at most $s - i \leq s - 1$. But in this range we have $I_\wp = J_\wp$. □

This result provides the means for a broad generalization of Corollary 5.3.7:

Corollary 5.3.15 *Let R, I and J be as above, and assume that* depth $R/I \geq \dim R - s$. *Then $I \cdot R[Jt]$ is a maximal Cohen–Macaulay module. If the reduction number $r_J(I) = 1$ then $\text{gr}_I(R)$ is Cohen–Macaulay.*

Corollary 5.3.16 *Let R be a Cohen–Macaulay local ring of dimension d, with infinite residue field, and let I be an ideal with sliding depth. If I is of linear type in codimension h then*

$$\ell(I) \geq \inf\{h + 1, \nu(I)\}.$$

Cohen–Macaulay Rees algebras

We shall now dig deeper into the properties of the approximation complexes to lift the assertion of Corollary 5.3.15 to the Rees algebra of the ideal.

Theorem 5.3.17 *Let R be a Cohen–Macaulay local ring and let I and J be ideals as in* Theorem 5.3.13. *Suppose that* depth $R/I \geq \dim R - s$. *If I has codimension at least two and $r_J(I) = 1$ then $R[It]$ is a Cohen–Macaulay algebra.*

Proof. It will be necessary to re-assemble some elements that appeared earlier. First, from Theorem 5.3.13, $R[Jt]$ is a Cohen–Macaulay ring, while by Corollary 5.3.15 $\text{gr}_J(R) \otimes R/I$ is also Cohen–Macaulay. This implies that $I \cdot R[Jt] = I \cdot R[It]$ is a maximal Cohen–Macaulay module.

To show that $R[It]$ is Cohen–Macaulay, we are going to prove the vanishing of certain local cohomology modules. We may assume that $(R, \mathbf{m})$ is a local ring and write $N = (R[It]_+, \mathbf{m})$ for the maximal graded ideal of $R[It]$.

Let $\dim R = d$. Write also $L = It \cdot R[It]$ for the augmentation ideal of $R[It]$. Consider the local cohomology modules attached to the sequence (5.3):

$$0 \to H^d_N(R[It]) \longrightarrow H^d_N(R) \longrightarrow H^{d+1}_N(L) \longrightarrow H^{d+1}_N(R[It]) \to 0.$$

Because $H^d(R)$ is concentrated in degree 0, to show that $H^d_N(R[It])$ vanishes it suffices to show that it does not have any nonzero component in degree 0. On the other hand, using the sequence (5.2) we have the exact sequence

$$0 \to H^d_N(R[It]) \to H^d_N(\text{gr}_I(R)) \to H^{d+1}_N(L[+1]) \to H^{d+1}_N(R[It]) \to 0.$$

We are going to show that $H^d_N(\text{gr}_I(R))$ has trivial components in non–negative degrees.

Proposition 5.3.18 *Let $(R, \mathfrak{m})$ be a Cohen–Macaulay local ring, and let J be an ideal of codimension g, with sliding depth and satisfying the condition $\mathcal{F}_1$. Then $a(\mathrm{gr}_J(R)) \leq -g$.*

Proof. Let $\nu(J) = t+g$. The hypotheses include depth $H_i(J) \geq \dim R-(g+t)+i$, which together with $\mathcal{F}_1$ implies that the approximation complex

$$0 \rightarrow H_t(J) \otimes B[-t] \stackrel{\varphi_t}{\rightarrow} \cdots \rightarrow H_1(J) \otimes B[-1] \rightarrow H_0 \otimes B \rightarrow S(J/J^2) \rightarrow 0,$$

is acyclic and $S(J/J^2) = \mathrm{gr}_J(R)$ is Cohen–Macaulay.

We denote this algebra by G_0 and use the complex to estimate $a(G_0)$. (A similar calculation is carried out in [105] in case J is strongly Cohen–Macaulay.) Let $M_i = H_i \otimes B[-i]$ and let L_i be the image of φ_i. Note that M_i has dimension $d-g+(t+g) = d+t$ and depth at least $d-(t+g)+i+(t+g) = d+i$.

Decompose the approximation complex into short exact sequences

$$0 \rightarrow L_{i+1} \longrightarrow M_i \longrightarrow L_i \rightarrow 0.$$

As in [105], we obtain exact sequences

$$0 \rightarrow H_N^d(M_0)_j \rightarrow H_N^d(G_0)_j \rightarrow H_N^{d+1}(L_1)_j \quad (5.21)$$

$$0 \rightarrow H_N^{d+i}(M_i)_j \rightarrow H_N^{d+i}(L_i)_j \rightarrow H_N^{d+i+1}(L_{i+1})_j, \ 1 \leq i < t \quad (5.22)$$

$$H_N^{d+t}(M_t)_j \simeq H_N^{d+t}(L_t)_j, \quad (5.23)$$

from which we want to argue that $H_N^d(G_0)_j = 0$ for $j > -g$, where $n = t+g$. We first note that $H_N^{d+i}(M_i)_j = 0$ for $j > -n+i$. This a consequence of the duality theorem. Indeed, let $M = H \otimes R[T_1, \ldots, T_n]$, where H is an R–module of depth $\geq d-n+i$. Consider the exact sequence

$$0 \rightarrow M[-1] \longrightarrow M \longrightarrow M/T_1M \rightarrow 0.$$

Assume, by induction, that $H_N^{d+i-1}(M/T_1M)_j = 0$ for $j > -(n-1)+i$. In the cohomology sequence

$$0 \rightarrow H_N^{d+i-1}(M/T_1M)_j \longrightarrow H_N^{d+i}(M)_{j-1} \longrightarrow H_N^{d+i}(M)_j,$$

we obtain the inclusion

$$H_N^{d+i}(M)_{j-1} \hookrightarrow H_N^{d+i}(M)_j,$$

for $j > -(n-1)+i$. Since $H_N^{d+i}(M)_j = 0$ for $j \gg 0$, this implies our claim.

Now use this fact and $j > g$ in the sequences (5.21), (5.22) and (5.23) proves the assertion. □

We come now to the core of the proof of the theorem. The approach is by connecting properties of the Sally module $S_J(I)$ with $\mathrm{gr}_I(R)$. Consider the commutative diagram of exact sequences

$$\begin{array}{ccccccccc} 0 & \rightarrow & R[Jt] & \longrightarrow & R[It] & \longrightarrow & C & \rightarrow & 0 \\ & & \uparrow & & \uparrow & & \uparrow & & \\ 0 & \rightarrow & I \cdot R[Jt] & \longrightarrow & I \cdot R[It] & \longrightarrow & S_J(I) & \rightarrow & 0. \end{array}$$

From the snake lemma there is an acyclic complex

$$0 \to K_S \longrightarrow \mathrm{gr}_J(R) \otimes R/I \longrightarrow \mathrm{gr}_I(R) \longrightarrow K_C \to 0, \tag{5.24}$$

where K_S and K_C are defined through the exact sequence

$$0 \to K_S \longrightarrow S_J(I) \longrightarrow C \longrightarrow K_C \to 0 \tag{5.25}$$

of natural inclusions.

Proposition 5.3.19 *If $r_J(I) = 1$ there exist exact sequences of Cohen–Macaulay modules of dimension d*

$$0 \to \mathrm{gr}_J(R) \otimes R/I \longrightarrow \mathrm{gr}_I(R) \longrightarrow C \to 0, \tag{5.26}$$

and

$$0 \to C[+1] \longrightarrow \mathrm{gr}_J(R) \longrightarrow \mathrm{gr}_J(R) \otimes R/I \to 0. \tag{5.27}$$

Proof. It is clear from the sequences (5.24) and (5.25) and the argument used in the proof of Corollary 5.3.15. □

We now conclude the proof of the theorem. By Proposition 5.3.18,

$$a(\mathrm{gr}_J(R)) \leq -g \leq -2.$$

We take this into the sequence (5.27) and get that both $H^d_N(C[+1])_j$ and $H^d_N(\mathrm{gr}_J(R) \otimes R/I)_j$ vanish for $j > -g$. Going over (5.26), we get that $H^d_N(\mathrm{gr}_I(R))_j = 0$ in non-negative degrees, which by the early part of the argument implies that $H^d(R[It])_0 = 0$, and therefore $R[It]$ is Cohen–Macaulay. □

Reduction number

We make some observations about what is required for the equation $r_J(I) = 1$ to hold.

Theorem 5.3.20 *Let I and J be ideals as in* Theorem 5.3.13, *and suppose every associated prime ideal of I has codimension at most s. If for each prime ideal $\wp$ of codimension s the equality $I^2_\wp = (JI)_\wp$ holds then $I^2 = JI$.*

Proof. It will be enough to show that the associated prime ideals of JI have codimension at most s. From the proofs of Corollaries 5.3.14 and 5.3.15 we use the exact sequence

$$0 \to H_1(J) \longrightarrow (R/I)^t \longrightarrow J/JI \to 0,$$

which will be combined with the sequence

$$0 \to J/JI \longrightarrow R/JI \longrightarrow R/J \to 0.$$

Since depth $H_1(J) \geq d - s + 1$, it follows from the first sequence and the condition that $\dim R_{\wp} \leq s$ for every associated prime $\wp$ of I, that any prime in $\mathrm{Ass}(J/JI)$ has codimension at most s. The claim now follows from the second sequence, since the associated prime ideals of R/J have codimension at most s.

Alternatively, we can argue as follows to show the vanishing of the Sally module $S_J(I)$: In the exact sequence (5.1), as in Proposition 5.1.6, $I \cdot R[Jt]$ being a maximal Cohen–Macaulay module (and therefore an unmixed ideal of codimension one) implies that $S_J(I)$ either vanishes or has Krull dimension d. By induction on $\dim R$ we may assume that $I^2 = JI$ holds on the punctured spectrum of the local ring $(R, \mathbf{m})$. This means that $S_J(I)$ is annihilated by some power of $\mathbf{m}$ so that the dimension of $S_J(I)$ is at most $\nu(J) = s < d$, which is a contradiction unless $r_J(I) = 1$. □

Symbolic powers

We make an application of these techniques to the comparison of the ordinary and the symbolic powers of a prime ideal.

Theorem 5.3.21 *Let I and J be ideals as in* Theorem 5.3.13 *with* $r_J(I) = 1$. *Suppose I is a prime ideal and* depth $R/I \geq d - s$. *If for each prime ideal $\wp$, of codimension at most s, the powers $I_{\wp}^m$ are primary ideals then the ideals I^m are also primary.*

Proof. This follows from the exact sequences (5.26) and (5.27): Together they say that the associated prime ideals of the conormal modules I^m/I^{m+1} have codimension at most s. This suffices, along with the hypothesis, to ensure that each such module is a torsion–free R/I–module. □

Here is a surprising application observed by Bernd Ulrich who proved it by other means in the case of ideals lying in the linkage class of a complete intersection (see also [278] for other developments).

Theorem 5.3.22 *Let $(R, \mathfrak{m})$ be a Gorenstein local ring and let I be a Cohen–Macaulay ideal satisfying sliding depth which is generically a complete intersection. Suppose that for all integers m the conormal module I^m/I^{m+1} is torsion free as a R/I–module. Then I satisfies $\mathcal{F}_1$, it is strongly Cohen–Macaulay (and in particular its Rees algebra $R[It]$ is Cohen–Macaulay), and is generated by at most* $\dim R - 1$ *elements.*

Proof. The main point is to show that for each prime ideal $\wp$, which is not a minimal prime of I, the localization $I_{\wp}$ is generated by at most $\dim R_{\wp} - 1$ elements. We may assume that the assertion holds for the punctured spectrum of R, and thus I is generated by analytically independent elements on the punctured spectrum of R.

Harmlessly we may also assume that the residue field of R is infinite. Let J be a minimal reduction of I; by Theorem 5.3.2.a, $\ell(I) \leq \dim R - 1$, so that $\nu(J) \leq \dim R - 1$. By Theorem 5.3.13 however J satisfies sliding depth so that its associated primes have codimension at most $\nu(J)$ and therefore $\mathfrak{m}$ cannot be one of them. By the induction hypothesis this implies that $I = J$.

Now we use Theorem 3.3.17 to get that I is strongly Cohen–Macaulay. The assertion about the Rees algebra will then follow from Theorem 3.3.19. □

Remark 5.3.23 The argument can be re-arranged in the following manner. We actually want to show the equality between $R[Jt]$ and the symbolic power algebra $\mathcal{R}_s(I)$. Because I and J agree on the punctured spectrum, for any element $z \in \mathcal{R}_s(I)$ there exists a power of $\mathfrak{m}$ conducting z into $R[Jt]$. Since $R[Jt]$ satisfies the condition S_2 of Serre's and $\mathfrak{m}R[Jt]$ has height at least two, z must lie in $R[Jt]$.

Exercise 5.3.24 Prove that if in Proposition 5.1.13 I is an almost complete intersection and

$$0 \to L \longrightarrow R[T_1, \ldots, T_{d+1}] \longrightarrow R[It] \to 0,$$

is a presentation of $R[It]$ then L is generated by the syzygies of J plus one quadratic polynomial. Furthermore the content ideal of the syzygies of I is a Gorenstein ideal.

Problem 5.3.25 Let a, I and R be as in Proposition 5.1.11. If $R_1[I_1t]$ is Cohen–Macaulay, prove that $R[It]$ is also Cohen–Macaulay (cf. [131, Lemma 2.11]). We leave the converse as an open question.

5.4 Jacobian Criteria

The normality tests here are variations on the Krull–Serre's criterion, combined with the re–writing device of the equations of symmetric algebras discussed in Section 1.5. The next section has several constructions geared to checking normality in Rees algebras proper.

Symmetric algebras

We first discuss a very direct test but which still requires that the condition (S_2) of Serre's be detected by other means ([288]).

Suppose that $S = S(E)$ is an integral domain. Let

$$0 \to L \longrightarrow R^n \longrightarrow E \to 0,$$

be a minimal presentation of E. Because of $\mathcal{F}_1$ we have $\ell = \text{rank} L \leq \dim R - 1$.

To apply Serre's normality criterion we assume that S satisfies condition (S_2) and examine the localizations S_Q where Q is a prime of height 1. If $Q \bigcap R = P = (0)$, there is nothing to do since S_P is a polynomial ring over the field of fractions of R. If $P \neq (0)$, localize at P and again let R stand for R_P. We thus have that $Q = PS$. Since $\dim S/PS = \nu(E) = \dim S - 1$, we get $\ell = \text{rank}(L) = \text{height}(P) - 1$. This means that $Q = T(P)$.

Denote by B the polynomial ring $S(R^n) = R[T_1, \ldots, T_n]$ and let J be the ideal of linear forms generated by the images of the elements of L in B. Let $M = PB$. We now convert to B the condition that S_Q be a discrete valuation domain. This is the case if and only if $(M/J)_M$ is a principal ideal, that is, if and only if the image of J in $(M/M^2)_M$ has rank equal to edim $(R) - 1 = \nu(\mathfrak{m}) - 1$.

Now we rephrase this condition into a more visible criterion. Let $f_j = (a_{ij})$, $j = 1, \ldots, m$, be a minimal generating set of L. Choose a minimal set of generators $\{x_1, \ldots, x_r\}$ of the maximal ideal P, we may then write each a_{ij} as a linear combination of the x_k

$$a_{ij} = \sum_{k=1}^{r} a_{ij}^{(k)} x_k.$$

The equations of the presentation ideal $J(E)$ are then written

$$J(E) = [T_1, \ldots, T_n] \cdot \varphi = [x_1, \ldots, x_r] \cdot B(\varphi).$$

The Serre's condition (R_1) at $T(P)$ is thus coded by the rank of the matrix $B(\varphi)$. These remarks show the following:

Proposition 5.4.1 *Let R be an integral domain, and let E be a finitely generated R–module whose symmetric algebra $S(E)$ is an integral domain with the Serre's condition (S_2). $S(E)$ is normal if and only if for each prime ideal P of R such that*

$$\nu(E_P) - \text{rank } E = \text{height } P - 1$$

the rank of the matrix $B(\varphi)$ has, mod $T(P)$, *rank equal to* edim $R_P - 1$.

In the next section, we shall describe the finite set of primes that must be tested. Let us apply this criterion to one of the cases of Example 1.3.4. (They will all be discussed later.) Because E is free on the punctured spectrum, we only have to verify the criterion at the maximal ideal. The Jacobian matrix of $J(E)$ with respect to the x–variables is:

$$B(\varphi) = \begin{bmatrix} 0 & T_1 & T_2 & T_3 & T_4 \\ -T_1 & 0 & -T_3 & -T_4 & T_5 \\ -T_2 & T_3 & 0 & T_5 & T_6 \\ -T_3 & T_4 & -T_5 & 0 & T_7 \\ -T_4 & -T_5 & -T_6 & -T_7 & 0 \end{bmatrix}.$$

Since rank $B(\varphi) = 4$, $S(E)$ is integrally closed.

For another example, consider a complete intersection (cf. [241]):

Example 5.4.2 Let R be a 3-dimensional regular local ring, and let E be the module defined by the matrix

$$\begin{bmatrix} a & 0 \\ b & a \\ 0 & b \\ c & 0 \\ 0 & c \end{bmatrix}$$

where $\{a, b, c\}$ is a regular sequence in R. E is free on the punctured spectrum of R and $S(E)$ is a domain and a complete intersection. By the criterion it follows that $S(E)$ is normal if and only if $\{a, b, c\}$ contains at least two independent minimal generators of the maximal ideal of R.

Example 5.4.3 If $(R, \mathfrak{m})$ is a Cohen–Macaulay local domain, and I is an ideal generated by a system of parameters $\{a_1, \ldots, a_n\}$, $S(I)$ is a Cohen–Macaulay domain. The criterion above means that $S(I)$ is normal if and only if R is a regular local ring, and $\dim((I, \mathfrak{m}^2)/\mathfrak{m}^2) \geq n - 1$.

Rees algebras

Let us recall the notion of integral closure of ideals. Let I be an ideal of a ring R. An element $z \in R$ is integral over the ideal I if it satisfies an equation

$$z^n + a_1 z^{n-1} + \cdots + a_n = 0, \; a_i \in I^i.$$

The *integral closure* of the ideal I is the set of all such $z's$.

This set I_a is an ideal of R, and I is said to be *integrally closed*—or *complete*—if $I = I_a$. The ideal I will be called *normal* if all of its powers are complete. Another way to view this notion is the following. Let $\mathcal{R}(I) = R[It]$ be the Rees algebra of the ideal I. It is not hard to prove that $z \in R$ is integral over I if and only if $zt \in R[t]$ is integral over $\mathcal{R}(I)$.

Coding together the integral closures of the ideal I^n leads to the integral closure

$$\mathcal{R}_a(I) = \sum_{n \geq 0} I_a^n t^n$$

of $\mathcal{R}(I)$. Thus $\mathcal{R}_a(I)$ is the Rees algebra attached to the filtration $\{I_a^n, \; n \geq 0\}$.

References for these concepts are [309, Appendices 4 and 5] and [184, Section II], which have many applications. In particular they prove that if R is a two–dimensional regular local ring—and more generally in [184], if R is a two–dimensional rational singularity—then complete ideals are normal. (For more recent developments, see [49].) This turns out to be no longer valid in higher dimensions and the extent to which this is violated is one of the motivations to study the normality of Rees algebras.

Normality of extended Rees algebras

To phrase Theorem 1.1.14 in a form appropriate for Rees algebras, it is convenient to introduce the extended Rees algebra of the ideal $I \subset R$:

$$A = R[It, u], \; u = t^{-1}.$$

Part of its usefulness comes from the isomorphism $A/uA \simeq \mathrm{gr}_I(R)$, which for the algebra A leads to the representation

$$A = A_u \cap (\bigcap A_{\wp}),$$

where $\wp$ runs over the associated primes of uA. Since $A_u = R[t, t^{-1}]$, we have

Proposition 5.4.4 *If R is a normal domain then A is normal if and only if $A_{\wp}$ is normal for each associated prime $\wp$ of uA.*

This characterization of normality has an immediate application to a case of interest. We first recall:

Definition 5.4.5 An ideal I is said to be *normally torsion–free* if all the powers I^t have the same associated primes.

Corollary 5.4.6 *Let R be a normal domain and let I be a radical ideal which is generically a complete intersection. If I is normally torsion–free then the Rees algebra $\mathcal{R}(I)$ is a normal domain.*

Proof. The hypotheses imply that uA is a radical ideal. □

A key to the understanding of the divisor class group of a normal Rees algebra $\mathcal{R}(I)$ is the identification of the minimal primes of $\text{gr}_I(R)$

When R is a normal domain, then it is clear that the Rees algebra $R[It]$ is normal if and only if the algebra $A = R[It, t^{-1}]$ is normal. This permits a look at the minimal primes of $L = I\mathcal{R}(I)$. Let

$$L = Q_1^{(e_1)} \cap \cdots \cap Q_s^{(e_s)}$$

be the primary decomposition of L. For each Q_i denote by v_i the corresponding discrete valuation of the field of quotients of $R[t]$. Since $L = uA \cap \mathcal{R}(I)$, the integers e_i are given by $e_i = v_i(u)$. From this it is easy to see that

$$I^n = \{x \in R \mid v_i(x) \geq n \cdot e_i,\ 1 \leq i \leq s\}.$$

Rees algebras and the property S_k

Some facts of a very general nature about the normality of the Rees algebra $\mathcal{R}(I)$ are examined next.

A criterion for a Rees algebra to have the (S_2) condition is given in [25]. We give a slight variation of it based on Theorem 5.1.25.

Theorem 5.4.7 *Let R be a Noetherian ring with (S_{k+1}) and let I be an ideal of R such that* height $I \geq k+1$. *Then $R[It]$ has (S_{k+1}) if and only if $\text{gr}_I(R)$ has (S_k).*

Proof. $(\Rightarrow)$ Set $G = \text{gr}_I R$ and let P^* be a prime ideal of G and P be its inverse image in $R[It]$. Localize R at $\wp = P \cap R$ and assume that R is local with the maximal ideal $\mathfrak{m} = \wp$ and $I \subset \mathfrak{m}$. If height $P^* \leq k$, then height $(P) \leq k+1$ and hence $P \not\supset It$ since height $(I, It) \geq k+2$ by the assumption on the height of I. In this case, there exists a nonzero divisor $x \in I$ such that $xt \notin P$. Noting the isomorphism $(R[It]/xR[It])_{xt} = G_{xt}$, we have that

$$\begin{aligned} P^*\text{–depth}(G) &\geq P\text{–depth}(R[It]/xR[It]) \\ &= P\text{–depth}(R[It]) - 1 \\ &\geq \inf\{k, \text{height } P - 1\} \\ &= \inf\{k, \text{height } P^*\} \end{aligned}$$

since $R[It]$ has (S_{k+1}). We may thus assume that height $P^* \geq k+1$ and $P \supseteq (I, It)$. Since P contains $\mathfrak{m}$ it must be the irrelevant maximal ideal of $R[It]$. In this case Theorem 5.1.25 implies that

$$P^*\text{–depth } G \geq P\text{–depth } R[It] - 1 \geq k.$$

($\Leftarrow$) Let P be a prime of $R[It]$ and let $\wp = P \cap R$. Localize R at $\wp$ and assume that R is local with the maximal ideal $\mathfrak{m} = \wp$. If $I \not\subset \mathfrak{m}$, then $R[It] = R[t]$ and hence $R[It]$ has (S_{k+1}) since R has (S_{k+1}). This means that we may take $I \subset \mathfrak{m}$.

We may further assume that $It \subset P$ as otherwise there exists a nonzero divisor $x \in I$ such that $xt \notin P$. Localizing $R[It]$ at xt one gets that x is a regular element of $R[It]$ and $(R[It]/xR[It])_{xt} = G_{xt}$; thus P-depth $(R[It]) \geq k+1$ since G has (S_k).

Finally assume that $(I, It) \subset P$ so that height $P \geq k+2$. Furthermore P is the irrelevant maximal ideal of $R[It]$, and depth $R \geq k+1$. We again use Theorem 5.1.25 to conclude that G has depth at least k (or argue directly from the sequences (5.2) and (5.3)). $\square$

Primary Gorenstein ideals

Our purpose next is to prove the somewhat unexpected:

Theorem 5.4.8 *Let $(R, \mathfrak{m})$ be a Gorenstein local ring of Krull dimension d and let I be an $\mathfrak{m}$–primary, perfect, Gorenstein ideal. If the associated graded ring $\mathrm{gr}_I(R)$ satisfies the condition (S_{d-2}), then I is a complete intersection.*

Proof. We may assume that $d \geq 3$, since as it is well known, perfect Gorenstein ideals of codimension one or two are complete intersections. We may also assume that the residue field of R is infinite. We shall need the following lemma ([283]):

Lemma 5.4.9 *Let $(R, \mathfrak{m})$ be a Noetherian local ring of Krull dimension d and let I be an $\mathfrak{m}$–primary ideal. Let $J = (a_1, \ldots, a_d)$ be a minimal reduction of I. The sequence $\{a_1, a_1t - a_2, a_2t - a_3, \ldots, a_dt\}$ is a system of parameters of the algebra $R[It]$.*

Proof. Since $R[It]$ is a finite $R[Jt]$-algebra, it is enough to show that these elements form a system of parameters of $R[Jt]$. Let Q be the radical of the ideal $(a_1, a_1t - a_2, \ldots, a_dt)$; we show that Q is the irrelevant maximal ideal of $R[Jt]$. For that note the equality

$$a_2^2 = a_2t \cdot a_1 - a_2(a_1t - a_2),$$

that shows $a_2 \in Q$, and consequently $a_1 t \in Q$. In this manner we prove $(J, Jt) \subset Q$. □

To prove the theorem, we use that any two of these parameters generate an ideal of codimension two in $R[It]$. By Theorem 5.4.7, the Rees algebra $R[It]$ satisfies (S_2). We now test this condition with the sequence $\{a, bt\} = \{a_1, a_d t\}$: a must be regular on $R[It]/(bt)$. It means that a is regular on each of the modules I^{n+1}/bI^n. We examine the case $n = 1$. Consider the exact sequence

$$0 \to I^2/bI \longrightarrow I/bI \longrightarrow I/I^2 \to 0.$$

The mid module I/bI contains the submodule

$$R/I \simeq (b)/bI \hookrightarrow I/bI,$$

so that since a is regular on I^2/bI by hypothesis and annihilates $(b)/bI$, we must have

$$I^2/bI \cap (b)/bI = 0.$$

We have shown that I/I^2 contains a submodule isomorphic to R/I, and therefore since R/I is self-injective and I/I^2 is an R/I–module, we have a direct sum decomposition

$$I/I^2 \simeq (R/I) \oplus T.$$

This puts us in the position to use the argument of [285, Lemma 2]: There exists a decomposition

$$I/bI = (b/bI) \oplus I/(b).$$

On the other hand, as I/bI is a module of finite projective dimension over $R^* = R/(b)$, the direct sum shows that $I^* = I/(b)$ is a perfect Gorenstein ideal over R^*.

We argue by induction on the dimension of the ring R. If $d = 3$ the proof is complete as I^* would be a complete intersection. Suppose $d > 3$: as the depth of the ring $\mathrm{gr}_I(R)$ is at least $d - 2$, there exists a regular element

$$b = \sum_{i=1}^{d} c_i a_i, \; b \in J \setminus \mathbf{m} \cdot J$$

such that the form $b^* \in I/I^2$ is a regular element of $\mathrm{gr}_I(R)$. This yields the isomorphism

$$\mathrm{gr}_I(R)/(b^*) \simeq \mathrm{gr}_{I^*}(R^*),$$

so that the latter is a ring with the property (S_{d-3}). This completes the induction step of the proof. □

Corollary 5.4.10 *Let I be a perfect, Gorenstein ideal whose Rees algebra satisfies (S_2). Then I is generically a hypersurface section.*

Corollary 5.4.11 *Let I be a normal Gorenstein ideal of a regular local ring of dimension 3. Then I is a complete intersection.*

The equality of the ordinary and symbolic powers of an ideal

Finally we take up a case of equality between the algebras $\mathcal{R}(I)$ and $\mathcal{R}_a(I)$, when I is a prime ideal. Let R be a Noetherian ring and let P be a prime ideal. Suppose the localization of the algebra $G = \text{gr}_P(R)$ at the prime P is an integral domain (e.g. R_P = regular local ring). Let us see the meaning of this condition as reflected on the Rees algebras of P. Denote by A, B and C, the extended Rees algebras of P corresponding, respectively, to the P–adic, P–integral closure and P–symbolic filtrations. That is, the algebras $\mathcal{R}(P)[t^{-1}]$, $\mathcal{R}(P)_a[t^{-1}]$ and $\mathcal{R}(P)_s[t^{-1}]$.

Remark 5.4.12 If G is an integral domain, then $B = C$. Indeed, from Corollary 5.4.4 we have the equality $A = A_u \cap A_{(u)}$. But both A_u and $A_{(u)}$ contain B and C.

Theorem 5.4.13 *Let R be a Noetherian domain and let P be a prime ideal such that $G_P = \text{gr}_P(R_P)$ is an integral domain. The following are equivalent:*

(a) $B = C$.

(b) *G has a unique minimal prime.*

Proof. The hypothesis on G_P has the following immediate consequences. (i) $B \subset C$: localize at P and use Remark 5.4.12; (ii) uC is a prime ideal: C/uC is a torsion–free R/P–module; localizing at P we obtain an embedding of C/uC into the integral domain G_P.

(a) $\Rightarrow$ (b): It is clear by the lying over theorem, since uB is a prime ideal.

(b) $\Rightarrow$ (a): Since uA has a unique minimal prime Q, and $A_P = B_P = C_P$, there is a unique minimal prime Q^* of B lying over uB. Denote by $\widehat{B}$ the integral closure of B in its field of quotients; $\widehat{B}$ is a Krull domain. The minimal primes $Q_1, \ldots, Q_s$ of $u\widehat{B}$ must each contract to Q^* in B. Localizing at P we conclude that $u\widehat{B}$ is a prime ideal. Let $q \in Q^*$; we have $q = ub$, $b \in \widehat{B}$. Since $q \in uC$, $b \in R[t]$, that is $b \in B$ as desired. □

An extension to radical ideals works as follows. Let

$$I = P_1 \cap \cdots \cap P_n$$

be a radical ideal and suppose that for each minimal P of I the algebra G_P is an integral domain. Denote by $\mathcal{P}$ the corresponding prime ideal of the extended Rees algebra A of I.

Theorem 5.4.14 *The following conditions are equivalent:*

(a) $B = C$.

(b) *$\mathcal{P}_1, \ldots, \mathcal{P}_n$ are the minimal primes of uA.*

Proof. Let us give a slightly different version of the argument above. We prove (b) $\Rightarrow$ (a). Let $\mathcal{Q}_i$ be the unique minimal prime of uB lying over $\mathcal{P}_i$; by assumption we have a decomposition

$$uB = \mathcal{Q}_1{}^{(e_1)} \cap \cdots \cap \mathcal{Q}_n{}^{(e_n)}.$$

Let z be a homogeneous element of C; then z is conducted into B by an element $s \in R$ lying outside of any P_i—since each G_{P_i} is a domain—and by a power of u. Since (s, u) has height at least two in the Krull domain B, we must have $z \in B$. □

Theorem 5.4.15 *Let R be a Cohen–Macaulay ring and let I be an equimultiple ideal of codimension $g \geq 1$. If $R[It]$ has the (S_2) condition then all the powers I^n are unmixed ideals (i.e. I is normally torsionfree).*

Proof. We may assume that $(R, \mathfrak{m})$ is a local ring and $g < \dim R$. Let $J = (x_1, \ldots, x_g)$ be a reduction of I.

Let z be an element of $\mathfrak{m}$ which is regular on R/J; note that z is a regular element on R itself. We claim that z is regular on R/I^n, $n \geq 1$. This is clearly enough to prove the assertion. Since height $(J, z) > g$, there exists an element $w \in J$ such that z, w is a regular sequence in R. For each integer n consider the exact sequence

$$0 \to J^n \longrightarrow R \longrightarrow R/J^n \to 0.$$

Since the associated prime ideals of J^n are the same as those of J, it follows that the (z, w)–depth of R/J is 1 and therefore z, w is a regular sequence on each J^n. This shows that z, w is a regular sequence of the algebra $R[Jt]$. But $R[It]$ is a finite algebra over $R[Jt]$ so that height $(z, w)R[It] = 2$. If $R[It]$ has (S_2) then z, w will be a regular sequence on each power I^n, thus implying that z will be regular on R/I^n since multiplication by w acts nilpotently on this module. □

5.5 Construction of Rees Algebras

We discuss elementary means to build 'larger' Rees algebras from simpler ones. Given two ideals I_1 and I_2, it is generally impossible to control for any property of $\mathcal{R}(I_1+I_2)$ in terms of the same property in $\mathcal{R}(I_1)$ and $\mathcal{R}(I_2)$. We consider special instances of this problem. The case of product varieties is given a fuller examination.

Transversal ideals

Ideals that are particularly amenable for the constructions here have the relationship described by the following:

Definition 5.5.1 Two ideals I and J of a ring R are said to be *transversal* if $I \cap J = I \cdot J$. They are said to be *normally transversal ideals* if for all positive integers m, n the ideals I^m and J^n are transversal.

The first example arises from a product of affine varieties. Let k be a field and let $I_1 \subset R_1$ and $I_2 \subset R_2$ be ideals of affine k–algebras. Set $R = R_1 \otimes_k R_2$. It is easy to see that the ideals I_1R and I_2R are (normally) transversal.

Let I be an ideal of a ring R and let f be a regular element of R which is a nonzero divisor of R/I^m for all m. Then I and (f) are normally transversal.

Detecting (S_r)

Let A be an affine algebra over a field k. We discuss two methods to ascertain whether A satisfies the condition (S_r) of Serre's. When $A = k[x_1, \ldots, x_n]/J$, it can be expressed in terms of projective dimensions.

Proposition 5.5.2 *Let*

$$R = k[x_1, \ldots, x_d] \subset A$$

be a Noether normalization of A, and denote by

$$0 \to F_s \xrightarrow{\varphi_s} F_{s-1} \longrightarrow \cdots \longrightarrow F_1 \xrightarrow{\varphi_1} F_0 \longrightarrow A \to 0$$

a projective resolution of A as an R–module. Denote by $I_j(A)$ the ideal $I_{r_j}(\varphi_j)$, $r_j = \operatorname{rank} \varphi_j$. Then A satisfies (S_r) if and only if

$$\text{height } I_j(A) \geq j + r, \; j \geq 0.$$

Proof. It is a consequence of the Auslander–Buchsbaum formula. □

The other method does not require a Noether normalization and uses the ability of programs such as *Macaulay* to compute Ext's and annihilators of modules.

Proposition 5.5.3 *Let B be the polynomial ring $k[x_1, \ldots, x_n]$ and suppose $A = B/J$ is an equi–dimensional algebra of codimension g. Then A satisfies S_r if and only if the following two conditions hold:*

$$\begin{cases} \text{height}(\text{ann}(\text{Ext}^i_B(A, B))) \geq j + 1, \; g < i \leq j \leq g + r, \\ \text{height}(\text{ann}(\text{Ext}^i_B(A, B))) \geq j + 1, \; g + r < j \text{ and } j - r < i \leq j. \end{cases}$$

Remark 5.5.4 For an example on how this is used, suppose $I = (a_1, \ldots, a_m)$ is an ideal of $R = k[x, y]$ and let $A = R[It] = R[T_1, \ldots, T_m]/J = B/J$ be its Rees algebra. Suppose that I is (x, y)–primary. Set $n = m + 2$. Then we have:

(a) $R[It]$ has (S_2) if and only if

$$\begin{aligned} \text{Ext}^{n-1}_B(A, B) &= 0 \\ \text{height ann}(\text{Ext}^{n-2}_B(A, B)) &\geq n. \end{aligned}$$

(b) By cutting down the number of variables, the following trick is useful: Let $J_0 = (J, x)$ and put $A_0 = B/J_0$. Then $R[It]$ has (S_2) if and only if

$$\begin{aligned} \operatorname{Ext}^n_B(A_0, B) &= 0 \\ \text{height ann}(\operatorname{Ext}^{n-1}_B(A_0, B)) &\geq n. \end{aligned}$$

Let us comment on the proof of (b). Since the localization $R[It]_x$ is a polynomial ring, the Ext modules above are annihilated by a power of x. Consider the canonical exact sequence induced by multiplication by x:

$$0 \to B/J \xrightarrow{x} B/J \longrightarrow B/J_0 \to 0,$$

and the long exact sequence of cohomology

$$\begin{aligned} &\operatorname{Ext}^{n-2}_B(A, B) \xrightarrow{x} \operatorname{Ext}^{n-2}_B(A, B) \longrightarrow \operatorname{Ext}^{n-1}_B(A_0, B) \longrightarrow \\ &\operatorname{Ext}^{n-1}_B(A, B) \xrightarrow{x} \operatorname{Ext}^{n-1}_B(A, B) \longrightarrow \operatorname{Ext}^{n}_B(A_0, B) \longrightarrow 0. \end{aligned}$$

It follows that if $\operatorname{Ext}^n_B(A_0, B) = 0$, then $\operatorname{Ext}^{n-1}_B(A, B) = 0$ since multiplication by x is a nilpotent endomorphism. More interesting is the next step, that $\operatorname{Ext}^{n-1}_B(A_0, B)$ and $\operatorname{Ext}^{n-2}_B(A, B)$ have the same codimension given the vanishing of their higher Ext's. We leave this for the reader.

Theorem 5.5.5 *Let A and B be affine algebras over a field k. If they satisfy (S_r) then $C{=}A \otimes_k B$ will also satisfy (S_r).*

Proof. Let P be a prime ideal of C and set $\wp = P \cap A$. If height $\wp \geq r$, any regular sequence in $\wp$, of length r, will extend by faithful flatness (C is a free A–module!) to a regular sequence in C. If height $\wp < r$, we mod out a regular sequence in $\wp$ of length equal to its height to obtain a new algebra $C' = A' \otimes B$.

Changing notation we may assume that $\wp$ is a minimal prime of A. We must find a regular sequence of the appropriate length in C_P. It will be enough to find it in $D = A_\wp \otimes_k B$. Note that $\wp D$ is a nilpotent ideal so that the prime ideal $P/\wp D \subset D/\wp D$ has the same height as P. Furthermore, because C is faithfully flat over $A_\wp$ then, by [192, 20.E], any regular sequence of D_P lifts to a regular sequence of C_P.

These reductions show that in order to establish the assertion, it suffices to prove that if B is a k–algebra and K is a field extension of k, then if B has S_r, so will $C = K \otimes_k B$. Now we appeal to Proposition 5.5.2, noting that C is equi–dimensional and that the heights of ideals in a polynomial ring $R = k[x_1, \ldots, x_n]$ are maintained under extension of the base field. □

We shall need a simple primality criterion [294].

Proposition 5.5.6 *Let R be a Noetherian domain satisfying the (S_2) condition of Serre's, and let u be an element in its field of fractions. Let D be a divisorial ideal that conducts u into R. Then the ideal $K = (D(U - u))$ is prime if and only if* $\text{grade}(D + Du) \geq 2$.

Proof. Note that K is a divisorial ideal of $R[U]$. Let P be one of its associated primes; P is an ideal of height 1. If $Q = P \cap R = 0$, $K_P = P_P$; otherwise, $K \subset QR[U]$, which is impossible as height $Q = 1$ contradicts the condition on the grade. This shows that K is prime. □

Gluing of Rees algebras

Suppose R_1 and R_2 are affine algebras over a field k and let $I_1 \subset R_1$ and $I_2 \subset R_2$ be ideals. Let

$$I = (I_1, I_2) \subset R = R_1 \otimes_k R_2.$$

We focus on the Cohen–Macaulayness and normality of the Rees algebra $\mathcal{R}(I)$. The difficulty lies in understanding the structure of the presentation of $\mathcal{R}(I)$.

Suppose that

$$0 \to L_1 \longrightarrow R_1[T_1, \ldots, T_m] \longrightarrow R_1[f_1t, \ldots, f_mt] \to 0,$$

and

$$0 \to L_2 \longrightarrow R_2[U_1, \ldots, U_n] \longrightarrow R_2[g_1t, \ldots, g_nt] \to 0,$$

are algebra presentations of the component Rees algebras. We seek the additional generators

$$0 \to (L_1, L_2, J) \longrightarrow R[\mathbf{T}, \mathbf{U}] \longrightarrow \mathcal{R}(I) \to 0,$$

in a presentation of $\mathcal{R}(I)$.

We are going to show that the obvious Koszul elements, $g_iT_j - f_jU_i$, generate J. It will follow from a more efficient way to verify the properties of $\mathcal{R}(I)$.

Consider the tensor product of the two Rees algebras

$$R_1[I_1t_1] \otimes R_2[I_2t_2] \subset R[t_1, t_2],$$

and the natural homomorphism

$$0 \to (t_1 - t_2) \longrightarrow R[t_1, t_2] \longrightarrow R[t] \to 0.$$

Proposition 5.5.7 *The relations counted in J are those that map to*

$$L = (t_1 - t_2)R[t_1, t_2] \cap R[I_1t_1, I_2t_2].$$

Furthermore

$$L = (I_1 \cdot I_2)(t_1 - t_2)R[I_1t_1, I_2t_2].$$

Proof. We only have to prove the last assertion. L is a homogeneous ideal of

$$R[I_1t_1, I_2t_2],$$

which itself is a graded subring of $R[t_1, t_2]$. Let

$$f = \sum_{i=0}^{n} c_i t_1^i t_2^{n-i} \in R[I_1t_1, I_2t_2]$$

be an element of L; f must be a multiple of $t_1 - t_2$ in $R[t_1, t_2]$. Write

$$f = g \cdot (t_1 - t_2) = (\sum_{j=0}^{n-1} a_j t_1^j t_2^{n-1-j})(t_1 - t_2).$$

Comparison of coefficients leads to equations

$$\begin{aligned} c_0 &= -a_0 \\ c_1 &= a_0 - a_1 \\ &\vdots \\ c_{n-1} &= a_{n-2} - a_{n-1} \\ c_n &= a_{n-1}. \end{aligned}$$

Since $c_i \in I_1^i I_2^{n-i} R$, using the first n equations we get successively

$$\begin{aligned} a_0 &\in I_2^n R \\ a_1 &\in I_2^{n-1} R \\ &\vdots \\ a_{n-1} &\in I_2 R. \end{aligned}$$

Now use the last n equations to conclude

$$\begin{aligned} a_{n-1} &\in I_1^n R \\ a_{n-2} &\in I_1^{n-1} R \\ &\vdots \\ a_0 &\in I_1 R, \end{aligned}$$

and therefore

$$a_i \in I_1^{i+1} R \cap I_2^{n-i} R.$$

Since $I_1 R$ and $I_2 R$ are normally transversal, we finally have

$$a_i \in I_1^{i+1} R \cdot I_2^{n-i} R = (I_1 \cdot I_2) I_1^i \cdot I_2^{n-1-i}.$$

These equations establish the claim. □

Corollary 5.5.8 *Suppose* $\mathrm{gr}_{I_1}(R_1)$ *and* $\mathrm{gr}_{I_2}(R_2)$ *satisfy* (S_r). *Then* $\mathrm{gr}_I(R)$ *has the same property.*

Proof. There is an isomorphism

$$\mathrm{gr}_I(R) \simeq \mathrm{gr}_{I_1}(R_1) \otimes_k \mathrm{gr}_{I_2}(R_2),$$

derived from the relations above. We may then use Theorem 5.5.5. □

Theorem 5.5.9 *If the Rees algebras $R_1[I_1t]$ and $R_2[I_2t]$ are Cohen–Macaulay then $R[(I_1, I_2)t]$ is Cohen–Macaulay.*

Proof. From the proof of Theorem 5.1.9, $I_1R_1[I_1t_1]$ and $I_2R_2[I_2t_2]$ are both maximal Cohen–Macaulay modules over the respective rings. It follows that

$$I_1R_1[I_1t_1] \otimes_k I_2R_2[I_2t_2]$$

is a maximal Cohen–Macaulay module over

$$R_1[I_1t_1] \otimes R_2[I_2t_2].$$

This says that L is also a maximal Cohen–Macaulay module, and thus the Rees algebra $R[(I_1, I_2)t]$ is Cohen–Macaulay. □

The next result gives an instance where the technique of product varieties works in other cases as well.

Theorem 5.5.10 *Let R be a regular local ring and let I_1 and I_2 be normally transversal ideals. If the Rees algebras $\mathcal{R}(I_1)$ and $\mathcal{R}(I_2)$ are Cohen–Macaulay then $\mathcal{R}(I_1, I_2)$ is also Cohen–Macaulay.*

Proof. The transversality condition means that $\text{Tor}_1(R/I_1^m, R/I_2^n) = 0$. By the rigidity of Tor in regular local rings [181], we have $\text{Tor}_j(R/I_1^m, R/I_2^n) = 0$, for $j \geq 1$. Taken together, these equations imply

$$\text{Tor}_j^R(R[I_1t], R[I_2t]) = 0, \text{for } j \geq 1.$$

In particular this means that we have an embedding

$$R[I_1t_1] \otimes_R R[I_2t_2] \hookrightarrow R[t_1, t_2].$$

We claim that

$$S = R[I_1t_1] \otimes_R R[I_2t_2] = R[I_1t_1, I_2t_2]$$

is a Cohen–Macaulay ring. First, note that S has Krull dimension $\dim R + 2$. Let $R[\mathbf{T}]/L_1$ and $R[\mathbf{U}]/L_2$ be R–algebra presentations of $R[I_1t_1]$ and $R[I_2t_2]$, respectively. In view of the vanishing of the Tor's, it follows that if $\mathbb{F}$ and $\mathbb{G}$ are projective resolutions of $R[I_1t_1]$ and $R[I_2t_2]$ over $R[\mathbf{T}]$ and $R[\mathbf{U}]$, respectively, then $\mathbb{F} \otimes_R \mathbb{G}$ is a $R[\mathbf{T}, \mathbf{U}]$–projective resolution of S. Furthermore, reading the length of a projective resolution of minimal length, it follows easily that S is indeed Cohen–Macaulay.

For the same reasons, $I_1R[I_1t_1] \otimes_R I_2R[I_2t_2]$ is a maximal Cohen–Macaulay module over $S = R[I_1t_1] \otimes_R R[I_2t_2]$, so that from the exact sequence of Proposition 5.5.7

$$0 \rightarrow (t_1 - t_2)I_1I_2R[I_1t_1, I_2t_2] \longrightarrow R[I_1t_1, I_2t_2] \longrightarrow \mathcal{R}(I_1, I_2) \rightarrow 0$$

we affirm the claim. □

If f is normally transversal to I, then the argument above applies to (I, f) without the requirement of regularity. More precisely

Proposition 5.5.11 *Let I and J be ideals of a Cohen–Macaulay ring. Suppose J is generated by a regular sequence, and it is transversal to all the powers of I. Then I and J are normally transversal. Furthermore, if $\mathcal{R}(I)$ is Cohen–Macaulay then $\mathcal{R}(I, J)$ is also Cohen–Macaulay.*

Its straightforward proof is left as an exercise.

Addition of torsion–freeness

With a small degree of care, Corollary 5.5.8 provides the setting to add two normally torsion–free ideals into another such ideal.

We use the following terminology: An ideal $I \subset R$ is *generically a complete intersection* if for each associated prime $\wp$, $I_\wp$ is generated by a regular sequence. It is not required that the ideal be equi–dimensional, that is, the minimum number of generators of $I_\wp$, denoted $\nu(I_\wp)$, may vary with $\wp$.

Proposition 5.5.12 *Let R_1 and R_2 be polynomial rings over a field k, and let $I_1 \subset R_1$ and $I_2 \subset R_2$ be reduced ideals. Suppose the total rings of fractions of R_1/I_1 and R_2/I_2 are separably generated over k. If I_1 and I_2 are normally torsion–free, then $I = (I_1, I_2)$ is a reduced, normally torsion–free ideal of $R = R_1 \otimes_k R_2$.*

Proof. By assumption each R_i/I_i embeds into a k–direct product of field extensions separably generated over k. Since I_i is normally torsion–free, $\text{gr}_{I_i}(R_i)$ embeds into a direct product of polynomial rings $F[x_1, \ldots, x_r]$, where $F = (R/I_i)_\wp$ for some associated prime $\wp$ of I_i and $r = \nu((I_i)_\wp)$. This means, from Corollary 5.5.8, that $\text{gr}_I(R)$ will embed into a direct product of tensor products of polynomial rings over separably generated extensions of k. In particular $\text{gr}_I(R)$ is a reduced algebra by [192, Proposition 27.G], and by [152, Corollary 1.10] I is normally torsion–free. □

For the purpose of getting the algebra $\text{gr}_I(R)$ to be reduced—which is the main ingredient of normality in some Rees algebras—it is not required that the ideals I_i have finite projective dimension so as to apply [152]. It turns out however that normal torsion–freeness is often easier to control than reducedness.

Corollary 5.5.13 *Let $I = (I_1, I_2)$ be an ideal generated by square–free monomials, listed in I_1 or I_2. Suppose I_1 and I_2 do not share a variable. Then if I_1 and I_2 are normally torsion–free then I is normally torsion–free.*

Proof. Denote by R_i the polynomial subring defined by the variables in I_i. To apply the Proposition it suffices to note that each of the localizations $(R_i/I_i)_\wp$ is a rational function field over k. □

Addition of normal algebras

The following simple example shows that normal Rees algebras, of normally transversal ideals, do not always add:

Example 5.5.14 Let $I_1 = (xy, xz, yz) \subset R_1 = k[x, y, z]$; I_1 is a normal ideal. Make $I_2 = (XY, XZ, YZ) \subset R_2 = k[X, Y, Z]$ be a second copy of I_1. The element $xyzXYZ$ is integral over $(I_1 + I_2)^3$ but does not belong to this ideal.

Problem 5.5.15 Let $R = k[z_1, \ldots, z_d]$ be a polynomial ring over a field k of characteristic zero and let I be an ideal of codimension at least two. Suppose the Rees algebra $\mathcal{R}(I)$ is Cohen–Macaulay and normal. If f is a polynomial that is normally transversal to I, find out under what conditions $\mathcal{R}(I, f)$ is Cohen–Macaulay and normal.

5.6 The f–Number of a Module

This section examines the divisor class group of blowup algebras. It is based on the machinery of divisor theory allied with the notion of **f**–numbers of modules. Together they will made several aspects of the subject amenable to algorithmic methods.

The f –number

Let R be a normal domain and suppose E is a finitely generated module of rank e, such that $S = S(E)_{red}$ is an integrally closed domain. Let

$$R^m \xrightarrow{\varphi} R^n \longrightarrow E \to 0$$

be a presentation of E.

There exists an exact sequence of divisor class groups

$$0 \to F \longrightarrow Cl(S(E)_{red}) \longrightarrow Cl(R) \to 0,$$

where F is a finitely generated free abelian group, whose rank can be read off the matrix φ ([119]).

To establish it we identify some of the divisorial prime ideals of $S(E)$. Assume that $T(P)$ is a prime ideal of height 1. We have

$$\dim S/T(P) = \dim S - 1 = \dim R + \text{rank } E - 1 = d + e - 1.$$

On the other hand, assume that E is minimally generated by $n - t$ elements at P, $t \geq 0$. In view of Theorem 1.2.1,

$$\dim S/T(P) = \dim R/P + \nu(E_P),$$

from which we get

$$\text{height } P = n - t - e + 1.$$

The condition $\nu(E_P) = n - t$ implies that $I_t(\varphi) \not\subset P$, but $I_{t+1}(\varphi) \subset P$. Because of the $\mathcal{F}_1$–condition on E, this implies that

$$\text{height } P \geq (n - e) - (t + 1) + 2,$$

and therefore we must have that P is a minimal prime of $I_{t+1}(\varphi)$, of height $n-e-t+1$.

Of course this argument breaks down when $t = n - e$, but then it will not matter because the localization E_P is a free module.

Definition 5.6.1 Let E be an R-module satisfying $\mathcal{F}_1$ with a given presentation a defined by the matrix φ. For each integer $1 \leq t \leq n - e$, denote by h_t the number of minimal primes of $I_t(\varphi)$ of height $n - e - t + 2$. The sum

$$\sum_{t=1}^{n-e} h_t$$

is the **f**–*number* of E.

We are going to show that **f**–number does not depend on the chosen presentation. (It is more useful, however, to recover it directly in terms of divisorial primes of the corresponding symmetric algebra.) Let E be a module with a rank, satisfying $\mathcal{F}_1$. This means that for each prime ideal $\wp$ of R we have:

$$\nu(E_\wp) \leq \dim R_\wp + \operatorname{rank} E - 1.$$

The **f**–number of E is simply the cardinality of the set of primes where equality holds. The finiteness is a consequence of the identification of these primes with some of the minimal primes of the Fitting ideals of E, as follows from the previous discussion.

The starting point is to identify some distinguished height one primes in $S(E)$. For a prime ideal $\mathbf{p} \subset R$ there is an associated the prime ideal

$$T(\mathbf{p}) := \ker(S_R(E) \longrightarrow S_{R/\mathbf{p}R}(E/\mathbf{p}E)_\mathbf{p}).$$

Note that if, moreover, $S(E)$ is a domain then $T(\mathbf{p})$ is nothing but the contraction $\mathbf{p}S(E)_\mathbf{p} \cap S(E)$, so height $T(\mathbf{p})$ = height $\mathbf{p}S(E)$.

Assume, for the rest of this section, that R is universally catenarian in order to avoid dwelling into technicalities. In this case, if in addition $S(E)$ is equidimensional, one sees that

$$\text{height } T(\mathbf{p}) = 1 \ \text{ if and only if } \ v(E_\mathbf{p}) = \text{height } \mathbf{p} + \operatorname{rank} E - 1.$$

We also observe, for further considerations, that unless $\mathbf{p}$ is a height one prime itself, E is not free if $T(\mathbf{p})$ is to have height one.

Proposition 5.6.2 *Let R be a universally catenarian Noetherian ring and let E be a finitely generated module such that $S(E)$ is a domain. Then the set*

$$\{T(\mathbf{p}) \mid \text{height } \mathbf{p} \geq 2 \textit{ and } \text{height } T(\mathbf{p}) = 1\}$$

is finite. More precisely, for any presentation $G \xrightarrow{\varphi} F \longrightarrow E \rightarrow 0$, this set is in bijection with

$$\{\mathbf{p} \subset R \mid E_\mathbf{p} \textit{ not free}, \ \mathbf{p} \in \operatorname{Min}(R/I_t(\varphi)) \textit{ and } \text{height } \mathbf{p} = \operatorname{rank}(\varphi) - t + 2\},$$

where $1 \leq t \leq \operatorname{rank}(\varphi)$.

Proof. Given a prime $\mathbf{p} \subset R$, set $t = \text{rank } F - \nu(E_{\mathbf{p}})$. If height $\mathbf{p} \geq 2$ and if height $T(\mathbf{p}) = 1$ then, since $\text{rank}(\varphi) = \text{rank } F - \text{rank } E$, by the preceding remarks we have

$$\text{height } \mathbf{p} = \text{rank}(\varphi) - t + 2 \text{ and } \mathbf{p} \supset I_t(\varphi) \setminus I_{t-1}(\varphi).$$

On the other hand, since $S(E)$ is a domain, E satisfies $\mathcal{F}_1$ by Remark 1.3.9. Therefore, one has height $I_t(\varphi) \geq \text{rank}(\varphi) - t + 2$. It follows that $\mathbf{p} \in \text{Min}(R/I_t(\varphi))$. The converse is similar. □

Divisor class group of symmetric algebras

The next result expresses the divisor class group of a normal integral symmetric algebra as an extension of a well prescribed finitely generated free abelian group (see [119] for additional details). If A is a normal domain, we denote by $Cl(A)$ its divisor class group; the class of an ideal I will be denoted by $[I]$.

Theorem 5.6.3 *Let R be a universally catenarian domain and let E be a finitely generated module such that $S(E)$ is a normal domain. Then there is an exact sequence*

$$0 \to F \longrightarrow Cl(S(E)) \longrightarrow Cl(R) \to 0$$

where F is a free (abelian) group of rank equal to the **f**–*number of E.*

Proof. We first observe that R is automatically normal. We use a well–known result on the generation of $Cl(S(E))$ by height one primes whose contraction to R is nonzero [72, Proposition 10.2]. Thus, let $P \subset S(E)$ be such a prime and set $\mathbf{p} = P \cap R$. Then $\mathbf{p}S(E_{\mathbf{p}})$ is a nonzero prime of $S(E_{\mathbf{p}})$ contained in $PS(E_{\mathbf{p}})$, hence $\mathbf{p}S(E_{\mathbf{p}}) = PS(E_{\mathbf{p}})$. By definition, also $T(\mathbf{p})_{\mathbf{p}} = \mathbf{p}S(E_{\mathbf{p}})$. Therefore $P = T(\mathbf{p})$.

To deal with height one primes $\mathbf{p}$ obtained in this process, one defines a convenient map from $Cl(S(E))$ to $Cl(R)$. First we define a map on the level of prime divisors of $S(E)$ whose contraction to R is nonzero. We set, namely:

$$[P] \stackrel{\psi}{\mapsto} \begin{cases} [P \cap R] & \text{if height } P \cap R = 1 \\ 0 & \text{if height } P \cap R \geq 2. \end{cases}$$

Embed E into a free R-module. This will say that the fraction field of $S(E)$ is isomorphic to $K(\mathbf{t})$ for a suitable set of indeterminates $\mathbf{t}$, with K standing for the fraction field of R. Now, let $g \in K(\mathbf{t})$ with corresponding divisor $[g]_{S(E)} = \Sigma_{P \cap R \neq 0} v_P(g)[P]$. One sees that $v_P(g) = v_{P \cap R}(g)$ as $\mathbf{p}S(E_{\mathbf{p}})$ is prime, where $\mathbf{p} := P \cap R$. It follows that $\psi([g]_{S(E)}) = [g]_R$.

Therefore, we have obtained a homomorphism $Cl(S(E)) \longrightarrow Cl(R)$, which is clearly surjective since E is locally free at the height one primes of R. The kernel of this homomorphism is generated by the classes $cl(T(\mathbf{p}))$, where height $\mathbf{p} \geq 2$. It is actually freely generated by the classes $cl(T(\mathbf{p}))$ such that height $\mathbf{p} \geq 2$ and height $T(\mathbf{p}) = 1$. To see this, one argues as in the proof of [256, Theorem 2.1] or [305, Theorem 2.1.4]. By Proposition 5.6.2, the rank of the kernel is the **f**–number of E. □

Remark 5.6.4 Under additional conditions one can say more about the splitting of the sequence of divisor class groups cf. [119, Theorem 3.2.1].

Example 5.6.5 Let I be the ideal generated by the six square–free monomials

$$\{x_1x_2, x_2x_3, x_2x_5, x_3x_4, x_3x_5, x_5x_6\}.$$

This is just the ideal associated to the suspension of a triangle. A calculation with *Macaulay* [14], shows that I is an ideal of linear type, and that the Rees algebra $R[It]$ is normal. From its presentation, we obtain:

$$h_1 = 1, h_2 = h_3 = h_4 = 0, h_5 = 4.$$

Its **f**–number is 5, and therefore the divisor class group of $R[It]$ is $\mathbb{Z}^5$.

Exercise 5.6.6 Let G_0 be a tree, let $G = S(G_0)$ be its suspension and let I be the ideal associated to G. It follows from Theorem 4.4.9 that the algebra $\mathcal{R}(I)$ is normal (Much more generally, the Rees algebra of $I(G)$, where G is a bipartite graph, is always normal ([262]). Prove that its divisor class group is free of rank equal to the type of I.

Submaximal minors

Let X be a $n \times n$ matrix of one of the following kind: a (completely) generic matrix, a generic symmetric matrix, or a generic skew symmetric matrix (all considered over a field). We will round out the picture regarding the normality, the divisor class group and the canonical module of the associated Rees algebras of certain determinantal ideals attached to X, through the mechanism of **f**–numbers.

Example 5.6.7 Generic matrices

Theorem 5.6.8 *Let X be a generic $n \times n$ matrix, $n \geq 2$, and let I be the ideal $I_{n-1}(X)$. I is a normal ideal and the divisor class group of $S = R[It]$ is a free module of rank $n - 1$.*

Proof. We shall make use of certain elements of the proof of Theorem 2.4.4 that I is of linear type. Noteworthy is the following fact: $S/(\Delta)$ is a reduced ring, where Δ is the determinant of X.

Lemma 5.6.9 *I is a normal ideal.*

Proof. We are going to argue by induction on n, the case $n = 2$ being clear. From general facts, we know that $S = \cap_P S_P$, P running over the associated primes of principal ideals. We claim that each such localization is a discrete valuation domain. By the remark above, this is certainly clear for any such prime containing Δ— in particular for the ideal MS generated by the entries of X. Note also that any ideal properly containing MS has grade at least 2. Therefore any other prime of S of grade

1 must intersect R properly inside of M, so that the induction hypothesis takes over (see [140] for details). □

Now we get to the determination of the divisor class group $Cl(S)$ of S; according to Theorem 5.6.3, it is enough to compute the f–number of I, or equivalently, the number of minimal primes of IS.

Lemma 5.6.10 *$Cl(S)$ is free of rank $n-1$.*

Proof. In case $n = 2$, IS itself is prime, since I is a prime ideal generated by a regular sequence. Henceforth we assume $n > 2$. Let P be a minimal prime of IS; put $\mathbf{p} = P \cap R$. Because S is normal, IS is a divisorial ideal and each P has height one. Furthermore, if $\mathbf{p} = M$ = irrelevant maximal ideal of R, then $P = MS$, since by the previous lemma any prime properly containing MS must have grade at least two.

We are now in a position to describe all the minimal primes of IS. Because I is of linear type, the prime ideals we want are those $T(\mathbf{p})$ of height one.

For each integer $1 \leq t \leq n-1$, denote by $\mathbf{p}_t$ the prime ideal generated by the minors of order t of the matrix X. We claim that the $T(\mathbf{p}_t)$'s are the desired primes. Note each of these is a minimal prime of Δ. This is clear for $\mathbf{p}_1 = M$ and $\mathbf{p}_{n-1} = I$. Because $M \not\subset \mathbf{p}_t$ for $t \geq 2$, we can localize at the powers of some entry x_{ij} of X and reduce the matrix to a smaller matrix of indeterminates X' such that $I_t(X)_{x_{ij}} = I_{t-1}(X')_{x_{ij}}$, thus showing that $T(\mathbf{p}_t)$, for $t > 1$, may be defined in terms of the matrix X'. Since Δ belongs to each of the $\mathbf{p}_t$'s, the assertion now follows. □

Remark 5.6.11 In characteristic zero, Bruns [28] has proved the normality and computed the divisor class group of the determinantal ideals of arbitrary size of a generic $m \times n$ matrix.

Example 5.6.12 Symmetric matrices

Until recently a similar theory for generic symmetric matrices was missing. In the characteristic zero case, however, the condition that $I = I_{n-1}(X)$ be of linear type can be bypassed. The normality would follow from the decomposition results of [1], as has been pointed out in [28]. The rank formula could be derived as above, that is, the **f**–number of I is $n-1$.

The situation has become much neater in view of the recent proof of Kotsev ([169]) that such ideals are of linear type. He proves also the normality of the Rees algebras and computes the divisor class group.

Example 5.6.13 Pfaffians

There are two cases to consider, according to whether n is odd or even. The first of these is understood best: I is the ideal generated by the Pfaffians of order $n-1$; it is of linear type, its Rees algebra $S = R[It]$ is normal with free divisor class group, generated by the class of the canonical module $\omega_S \simeq IS$.

For n even, $n = 2p$, I is generated by the Pfaffians of order $2p-2$. It follows directly from [162] that I is locally generated by analytically independent elements and that its **f**–number is $p-1$. It is not known however whether I is of linear type.

Remark 5.6.14 These ideals are rarely generated by d–sequences. It is known that for the generic matrices (and $n \geq 3$) they cannot be generated by d–sequences, for the Pfaffians it is affirmative if n is odd, and, in the symmetric case it is positive if $n = 3$ but not if $n > 3$ according to A. Corso and S. Morey.

5.7 The Canonical Module

We provide a brief description of the theory of the canonical module for Rees algebras and symmetric algebras.

Rees algebras

Throughout $(R, \mathfrak{m})$ is a Cohen–Macaulay local with a canonical module ω_R. Under these conditions, the Rees algebra $S = R[It]$ and associated graded ring $G = \mathrm{gr}_I(R)$ of an ideal I have also canonical modules ω_S and ω_G, respectively. We shall assume that height $I = g > 0$. We discuss here the relationship between these modules. It will be mediated through the examination of the canonical module ω_T of the extended Rees algebra $R[It, t^{-1}]$. Our basic source here is [118], from which the material is lifted.

The guidepost to the canonical module of Rees algebras is the case of a complete intersection. For instance, if $\mathbf{f} = f_1, \ldots, f_g$ is a regular sequence then the Rees algebra of $I = (\mathbf{f})$ is determinantal

$$R[It] \quad \simeq \quad R[Z_1, \ldots, Z_g]/I_2 \begin{pmatrix} Z_1 & \cdots & Z_g \\ f_1 & \cdots & f_g \end{pmatrix}, \tag{5.28}$$

and its canonical is $\omega_R(1, t)^{g-2}$ ([30]).

We are going to use the following notation. Suppose M is an R–module and $J = \oplus J_i t^i$ is a fractionary, graded S–ideal. We set $MJ = \oplus J_i M t^i$ which has a natural structure of S–module. Similar notations will be used with respect to T. For instance, we set

$$\omega_T = \cdots \oplus \omega_R t^{-2} \oplus \omega_R t^{-1} \oplus \omega_R \oplus I\omega_R t \oplus I^2 \omega_R t^2 \oplus \cdots.$$

We begin by introducing an auxiliary notion, the initial type of the ring G. It is closely related to the a–invariant of G that we considered earlier in the chapter. Let ω_G be the (graded) canonical module of G. There exists an integer k such that $(\omega_G)_k \neq 0$ but $(\omega_G)_i = 0$ for $i < k$. Actually we are not going to identify this integer k with $a(G)$, since by simplicity in some arguments we want room to shift at will.

Definition 5.7.1 The integer $r_0(G) = \nu((\omega_G)_k)$ is the initial type of G.

This invariant counts the number of minimal generators of least degree of the canonical module of G. The ordinary type of G will be denoted by $r(G)$. It is a notion that can be extended to S.

Proposition 5.7.2 $r_0(G) \leq r(R) \leq r(G)$.

Proof. From the isomorphism $(\omega_T)_{t^{-1}} \simeq \omega_R[t, t^{-1}]$, we conclude that $\nu(\omega_R) = \nu((\omega_T)_{t^{-1}})$, and up to multiplication by some power of t, one has

$$\omega_T \quad = \quad \cdots \oplus \omega_R t^{-2} \oplus \omega_R t^{-1} \oplus \omega_R \oplus \omega_1 t \oplus \omega_2 t^2 \oplus \cdots, \tag{5.29}$$

where ω_1 is contained properly in ω_R, and $\omega_i \subset \omega_{i-1}$ for all $i > 1$. It follows that $\omega_G \simeq \omega_R/\omega_1 \oplus \omega_1/\omega_2 \oplus \cdots$, from which we have

$$r_0(G) = \nu(\omega_R/\omega_1) \leq \nu(\omega_R) = r(R) = \nu((\omega_T)_{t^{-1}}) \leq \nu(\omega_T) = \nu(\omega_G) = r(G),$$

where we used that $\omega_G \simeq \omega_T/t^{-1}\omega_T$. □

Corollary 5.7.3 *If* $r_0(G) = r(G)$, *then*

(a) $r(R) = r(G)$.

(b) *R is Gorenstein if and only if G is Gorenstein.*

Theorem 5.7.4 ([118, Theorem 2.4]) *Let R be a Cohen–Macaulay local ring with canonical module ω_R, and let I be an ideal of height $g > 0$ such that the Rees algebra $R[It]$ is Cohen–Macaulay. The following conditions are equivalent:*

(a) $r_0(G) = r(G)$.

(b) $\omega_T \simeq \omega_R T$.

(c) $\omega_G \simeq \operatorname{gr}_I(\omega_R)$.

(d) $\omega_S \simeq \omega_R(1, t)^m$ *for some* $m \geq -1$.

If one of equivalent conditions hold we say that the corresponding canonical module has the expected form.

Proof. (a) ⇒ (b): We may assume that ω_T is as in (5.29). Then $(\omega_T/t^{-1}\omega_T)_i = 0$ for $i < 0$ and $(\omega_T/t^{-1}\omega_T)_0 \neq 0$. Identifying ω_G with $\omega_T/t^{-1}\omega_T$, the assumption $r_0(G) = r(G)$ guarantees that $(\omega_G)_0$ generates ω_G as a G–module. This implies that $\omega_i = I^i\omega_R + \omega_{i+1}$ for all $i \geq 0$. It follows that $\omega_i = \bigcap_{j \geq i}(I^i\omega_R + \omega_j)$.

Since ω_T is finitely generated there exists an integer k such that $\omega_{j+1} \subseteq \mathfrak{m}\omega_j$ for all $j \geq k$. By Krull intersection theorem we obtain that $\omega_i = I^i\omega_R$ and therefore $\omega_T = \omega_R T$.

The implications (b) ⇒ (c) and (c) ⇒ (a) are obvious.

To prove the equivalence of (c) and (d) we consider the exact sequences (5.3) and (5.2):

$$0 \to S_+ \longrightarrow S \longrightarrow R \to 0 \tag{5.30}$$

$$0 \to IS \longrightarrow S \longrightarrow G \to 0. \tag{5.31}$$

We fix ω_S such that $(\omega_S)_i = 0$ for $i < 0$ and $(\omega_S)_0 \neq 0$. Dualizing with ω_S we obtain the exact sequences

$$0 \to \omega_S \longrightarrow (\omega_S\colon S_+) \longrightarrow \omega_R[+1] \to 0 \tag{5.32}$$
$$0 \to \omega_S \longrightarrow (\omega_S\colon IS) \longrightarrow \omega_G[-m-1] \to 0 \tag{5.33}$$

for some $m \geq -1$. Here we consider ω_R as a graded S–module concentrated in degree 0, and ω_G is chosen such that $(\omega_G)_i = 0$ for $i < 0$ and $(\omega_G)_0 \neq 0$.

Now assume $\omega_S = \omega_R(1,t)^k$, then (5.32) implies

$$(\omega_S : S_+)_{-1} = \omega_R,$$

and

$$(\omega_S : S_+)_i = (\omega_R(1,t)^k)_i, \text{ for } i \geq 0.$$

Since

$$(\omega_S : IS)_i = (\omega_S\colon S_+)_{i+1}, \text{ for } i \geq 0, \tag{5.34}$$

we see that $\omega\colon IS = \omega_R(1,t)^{k+1}$. Using (5.33) it follows that

$$\omega_G = \mathrm{gr}_I(\omega_R)[-k-1],$$

and hence $\omega_G = \mathrm{gr}_I(R)$ and $k = m$.

Conversely, suppose $\omega_G = \mathrm{gr}_I(\omega_R)$: putting $\omega_S = \omega_0 \oplus \omega_1 t \oplus \omega_2 t^2 \oplus \cdots$, (5.32) implies $(\omega_S\colon S_+)_{-1} = \omega_R$ and $(\omega_S\colon S_+)_i = \omega_i t^i$. Then by (5.34) we get $(\omega_S\colon IS)_0 = \omega_R$ and $(\omega_S\colon IS)_i = \omega_{i-1}t^i$ for $i \geq 1$. Using (5.33) it follows that $\omega_i = \omega_R$ for $i = 0, \ldots, m$. One has also the exact sequences

$$0 \to (\omega_S)_{m+j} \longrightarrow (\omega_S\colon IS)_{m+j} \longrightarrow I^{j-1}\omega_R/I^j\omega_R \to 0.$$

By induction on j it follows that $(\omega_S)_{m+j} = I^j\omega_R$ since the mid term of the exact sequence is $(\omega_S)_{m+j-1}$ as we have seen before. Hence we have

$$\omega_i = \begin{cases} \omega_R & \text{for } i = 0, \ldots, m \\ I^{i-m}\omega_R & \text{for } i \geq m. \end{cases}$$

In other words, $\omega_S = \omega_R(1,t)^m$. □

Corollary 5.7.5 *The following conditions are equivalent:*

(a) *G is Gorenstein.*

(b) $\omega_S \simeq (1,t)^m$ *for some* $m \geq -1$.

Proof. (a) $\Rightarrow$ (b) follows directly from Theorem 5.7.4. For the converse one show as in the proof of the theorem that $(1,t)^m : IS = (1,t)^{m+1}$. Then again using the exact sequence

$$0 \to (1,t)^m \longrightarrow (1,t)^m : IS \longrightarrow \omega_G \to 0$$

we see that $\omega_G = (1,t)^{m+1}/(1,t)^m \simeq G$. □

Remark 5.7.6 We have played loose with the proper degrees of the canonical modules, so we remedy this here.

(a) In the situation of Theorem 5.7.4

$$\begin{aligned}\omega_S &= \omega_R(1,t)^m[-1]\\ \omega_G &= \mathrm{gr}_I(\omega_R)[-m-2]\end{aligned}$$

(b) Furthermore, if I is generically a complete intersection of codimension g, in the formulas above $m = g - 2$.

(c) If $g = 1, \omega_S \simeq S_+$.

Proof. The assertions about the degrees all follow by localization. □

Symbolic powers

Theorem 5.7.7 *Let R be a Gorenstein local ring and let $\wp$ be a prime ideal of codimension $g > 0$, such that $R_\wp$ is a regular local ring and the ordinary and symbolic powers of $\wp$ coincide. If $\mathrm{gr}_\wp(R)$ is Cohen–Macaulay then $R[\wp t]$ is also Cohen–Macaulay.*

Proof. We may assume that height $\wp = g > 0$. Let ω_G be the canonical module of $G = \mathrm{gr}_\wp(R)$:

$$\omega_G = \cdots \oplus \omega_g \oplus \omega_{g+1} \oplus \cdots$$

where ω_g denotes the component of degree g. We claim that there are no components in lower degree. For that, it suffices to observe two facts. First, since G is an integral domain, ω_G is a torsion free $R/\wp$–module. Then, localizing at $\wp$, ω_G localizes onto the canonical module of $\mathrm{gr}_P(R_\wp)$, with $P = \wp R_\wp$. But the associated graded ring of P is a polynomial ring so that $a(G) = -g < 0$. We now use Theorem 5.1.22 to complete the proof. □

Symmetric algebras

We shall now apply the results of the previous section to study the divisor class of the canonical module of $S(E)$, which we denote by $\omega_{S(E)}$. In some significant cases, it will be possible to exhibit $\omega_{S(E)}$ itself.

Theorem 5.7.8 *Let E be a module over $R = k[x_1, \ldots, x_d]$ with a linear presentation* (cf. Section 1.5), *assume that $S(E)$ is normal, let $\{\mathfrak{p}_1, \ldots, \mathfrak{p}_t, \mathfrak{m}\}$ be the set of primes $\mathfrak{q}$ with $\nu(E_\mathfrak{q}) = \dim R_\mathfrak{q} + \mathrm{rank}\, E - 1$, and write*

$$g = \mathrm{height}\, I = \mathrm{codim}(\mathrm{Sing}(S(E)) \bigcap V(\mathfrak{m})) - 1.$$

The following are equivalent:

(a) $\omega_{S(E)} \simeq (\mathfrak{m}S(E))^{s**}$, *for some* s.

(b) $S(E_{\mathbf{p}_i})$ *is quasi–Gorenstein for* $1 \leq i \leq t$.

(c) $S(E_{\mathbf{q}})$ *is quasi–Gorenstein for* $\mathbf{q} \neq \mathfrak{m}$.

(d) *(If* $S(E)$ *is Cohen–Macaulay)* $\mathrm{gr}_I(S)$ *is Gorenstein.*

If any of these conditions is satisfied, then $\omega_{S(E)} \simeq (\mathfrak{m}S(E))^{g-2*}$. *Moreover* I *is equidimensional. If in addition* $S(E)$ *satisfies* (S_3), *then*

$$\omega_{S(E)} = (x_1, f_1)^{g-2} = S_{g-2}((x_1, f_1)),$$

in particular $\nu(\omega_{S(E)}) = g - 1$.

Proof. The equivalence of (a), (b) and (c) is clear since one has (cf. Theorem 5.6.3):

$$Cl(S(E)) = \mathbb{Z}[\mathfrak{m}S(E)] \oplus (\oplus_{i=1}^{t}\mathbb{Z}[T(\mathbf{p}_i)]).$$

We now show that if $[\omega_{S(E)}] = s[\mathfrak{m}S(E)]$, then $s = 2 - g$ and I is equidimensional. However, $[\omega_{S(E)}] = [\omega_{R[It]}]$ and $[\mathfrak{m}S(E)] = [ItR[It]]$. Hence $[\omega_{R[It]}] = s[ItR[It]]$ and the same holds true as we replace S by $S_{\mathbf{p}}$ with $\mathbf{p}$ a minimal prime of I of height h. But then $I = (f_1, \ldots, f_h)$ is a complete intersection (since I is of linear type), and therefore

$$R[It] \simeq S[Z_1, \ldots, Z_h]/I_2\begin{pmatrix} Z_1 & \cdots & Z_h \\ f_1 & \cdots & f_h \end{pmatrix}.$$

Moreover, in $R[It]$, $(z_1, \ldots, z_h) \subset (z_1) \colon (f_1, z_1)$ (z_i denotes the image of Z_i), and hence $(z_1, \ldots, z_h) = (z_1) \colon (f_1, z_1)$ since the former is prime. Furthermore, $[\omega_{R[It]}] = (h - 2)[(f_1, z_1)^{**}]$ ([30]), and therefore $[\omega_{R[It]}] = (2 - h)[(z_1, \ldots, z_h)] = (2 - h)[ItR[It]]$. Since on the other hand, $[\omega_{R[It]}] = s[ItR[It]]$ and $Cl(R[It])$ is torsion–free and $[ItR[It]] \neq 0$, it follows that $s = 2 - h$. Thus $s = 2 - g$ and I is equidimensional.

Now let $f \neq 0$ be an element of I. Then as above, in the ring $R[It]$, $ItR[It] \subset (ft) \colon (f, ft)$, and since $ItR[It]$ is prime, one has equality. Therefore

$$\omega_{S(E)} \simeq (\mathfrak{m}S(E))^{g-2^*} = (ItR[It])^{g-2^*} \simeq (f, ft)^{g-2^{**}}.$$

We are now going to prove that if $S(E) = R[It]$ satisfies (S_3), then

$$(f, ft)^{k^{**}} \simeq S_k(f, ft) \quad \forall \quad k \geq 0.$$

In particular, this will imply that $\omega_{S(E)} = (f, ft)^{g-2} = S_{g-2}(f, ft)$, where we may choose $f = f_1$, $ft = x_1$. To show this, since

$$IR[It] + ItR[It] \subset [(f) \colon (f, ft)] + [(ft) \colon (f, ft)],$$

it follows that $(f, ft)_{\mathbf{p}}$ is principal for all $\mathbf{p} \in \mathrm{Spec}(R[It])$ with $\mathbf{p} \not\supset IR[It] + ItR[It]$, in particular for all $\mathbf{p}$ with height $\mathbf{p} <$ height$(IR[It] + ItR[It]) = g + 1$. Since $g \geq 2$ and $R[It]$ satisfies (S_3) it follows that $(f, ft)_{\mathbf{p}}$ is principal for all $\mathbf{p} \in \mathrm{Spec}(R[It])$

with depth $R[It]_{\mathbf{p}} \leq 2$. Furthermore the equality $ItR[It] = (ft):(f, ft)$ easily implies that (f, ft) is a divisorial ideal. One can then use [289, the proof of Theorem 2.7].

To finish the proof, we only need to show the equivalence of (a) and (d) under the assumption that $S(E)$, and hence $R[It]$, is Cohen–Macaulay. However if $R[It]$ satisfies (S_3), we have seen above that for every integer k, $(\mathfrak{m}S(E))^{k^*} \simeq (f, ft)^k$. Hence condition (a) reads as $\omega_{S(E)} \simeq (f, ft)^k$. Moreover we had seen that if (a) holds then $k = g - 2 \geq 0$. Thus (a) is equivalent to $\omega_{R[It]} = \omega_{S(E)} \simeq (f, ft)^k$, $k \geq -1$. On the other hand by [117, Corollary 2.5], the latter condition is equivalent to (d) (in case $R[It]$ is Cohen–Macaulay). □

Remark 5.7.9 The condition 5.7.8.c is clearly satisfied if for every $\mathbf{q} \neq \mathfrak{m}$, either pd $E_{\mathbf{q}} \leq 1$ or $\nu(E_{\mathbf{q}}) \leq \max\{\text{rank } E, \dim R_{\mathbf{q}} + \text{rank } E - 2\}$.

A well known result of Grothendieck says that a complete intersection is factorial if it is factorial locally in codimension three ([91, Corollaire 3.14]). Certain symmetric algebras exhibit a similar behavior.

Corollary 5.7.10 *Let E be a module over $R = k[x_1, \ldots, x_d]$ with a linear presentation matrix and assume that $E_{\mathbf{q}}$ satisfies $\mathcal{F}_2$ for all $\mathbf{q} \neq \mathfrak{m}$. The following are equivalent.*

(a) *$S(E)$ is factorial.*

(b) *$S(E)$ satisfies (S_2) and is quasi–Gorenstein, and $S(E)_{\wp}$ is factorial for all primes $\wp$ of $S(E)$ with* $\dim S(E)_{\wp} \leq 3$.

Proof. Assume (b) and suppose that $S(E)$ is not factorial. Then $\nu(E) = \dim R + \text{rank } E - 1$ (cf. Theorem 5.6.3) and we may apply Theorem 5.7.8. It follows that $S(E) \simeq \omega_{S(E)} \simeq (\mathfrak{m}S(E))^{g-2^*}$. Therefore $g = 2$, since $[\mathbf{m}S(E)]$ generates the infinite cyclic group $Cl(S(E))$. Hence $S(E) = R[It]$, where I is an S–ideal of linear type and height 2. Now let $\mathbf{p} \in \text{Spec}(S)$ be a minimal prime of I with $\dim S_{\mathbf{p}} = 2$. Then $I_{\mathbf{p}}$ is a complete intersection of height 2. Therefore $R[I_{\mathbf{p}}t]$ has dimension 3 and is not factorial. Write $\wp = (\mathbf{p}, \mathbf{x}) \in \text{Spec}(R[It])$; then $\dim R[It]_{\wp} = 3$ and $R[It]_{\wp}$ is still not factorial. This yields a contradiction to the assumptions of part (b). □

Exercise 5.7.11 (B. Ulrich) Let R be a Cohen–Macaulay local ring and let I be an ideal of positive height. Prove that the following are equivalent:

(a) $\text{gr}_I(R)$ is Cohen–Macaulay;

(b) the Rees algebra of the ideal $(I, X_1, \ldots, X_r)$ in $R[X_1, \ldots, X_r]_{(\mathfrak{m}, X_1, \ldots, X_r)}$ is Cohen–Macaulay for some r.

Exercise 5.7.12 Let R be a Cohen–Macaulay local ring and let I be a Cohen–Macaulay ideal of height at least 2. Prove that if I is equimultiple with reduction number 1 then $R[It]$ is Cohen–Macaulay.

Chapter 6

Factoriality

One of the questions driving this chapter is: When is the symmetric algebra of the R–module E, $S(E)$, factorial? One condition that has been identified is simply $\mathcal{F}_2$ (see below), but it is hardly enough, except for modules of projective dimension one. On the other hand, all known examples are complete intersections.

This puzzle is related to another, more general, phenomenon: It has been observed that symmetric algebras of modules of finite projective dimension are Gorenstein only under certain circumscribed cases. (For modules of infinite projective dimension the phenomenon is rather common.) To help explain this, we discuss a method to derive the canonical module of $S(E)$, and use it to understand some cases of the factorial problem.

While symmetric algebras that are factorial seem rare, there is a straightforward process that produces the *factorial closure* of any symmetric algebra $S(E)$. The setting is a sequence of modifications of the algebra $S(E)$, each more drastic than the preceding. Let R be a Noetherian factorial domain, and let E be a finitely generated R–module. Define:

(i) $D(E) = S(E)/\text{mod } R$–torsion ($D(E)$ is a domain).

(ii) $C(E)$ = integral closure of $D(E)$.

(iii) $B(E)$ = graded bi–dual of $S(E)$, that is if

$$S(E) = \bigoplus_{t \geq 0} S_t(E)$$

then

$$B(E) = \bigoplus_{t \geq 0} S_t(E)^{**},$$

where (**) denotes the bi-dual of an R-module. $B(E)$, for reasons discussed below, is the factorial closure of $S(E)$.

These algebras are connected by a sequence of homomorphisms:

$$S(E) \to D(E) \to C(E) \to B(E).$$

Some comparisons will be made in this chapter, but the emphasis is on constructions leading to $B(E)$.

The algebra $D(E)$ is easy to obtain from $S(E)$; sometimes we refer to it as the Rees algebra of E. The other two algebras, $C(E)$ and $B(E)$, are a different matter. The significance of the algebra $B(E)$ is that it is a factorial domain (*cf.* [46]; see also [117], [241]), although it may fail to be Noetherian ([234]).

6.1 The Factorial Conjecture

In this section we assume that R is a factorial domain. It looks at ways symmetric algebras turn out to be factorial.

The following theorem of Samuel ([241]) exploits the relationship between the factoriality of a graded ring $A = \bigoplus_{n\geq 0} A_n$, and the A_0–module structure of the components A_n.

Theorem 6.1.1 *Let $A = \bigoplus_{n\geq 0} A_n$ be a Noetherian, integral domain. The following conditions are equivalent:*

(a) *A is factorial.*

(b) *A_0 is factorial, each A_n is a reflexive A_0–module and $A \otimes_{A_0} K$ is factorial (K = field of quotients of A_0).*

It justifies the assertion above about the algebra $B(E)$. Let us use it to derive a first necessary condition for a symmetric algebra $S(E)$ to be factorial.

Proposition 6.1.2 *If $S(E)$ is factorial then E satisfies $\mathcal{F}_2$.*

Proof. Let $I_t(\varphi)$ be a Fitting ideal associated to E. We already know that E satisfies $\mathcal{F}_1$, so that each of these ideals has height at least 2. We may then find a prime element $x \in R$ contained in their intersection (possibly after a polynomial change of ring; see [45]). Since each $S_n(E)$ is a reflexive R–module, its reduction modulo x is a torsion-free $R/(x)$–module. Therefore $S_{R/(x)}(E/xE)$ is an integral domain, and from its $\mathcal{F}_1$–condition we get the assertion. □

Theorem 6.1.3 *Let R be a factorial domain, and let E be a module of projective dimension 1. Then $S(E)$ is factorial if and only if E satisfies $\mathcal{F}_2$.*

The result, and its proof, is similar to Theorem 3.1.6.

The nature of a module whose symmetric algebra is factorial remains elusive. Although there are many examples, they all seem to fit a mold. It has led us to formulate:

Conjecture 6.1.4 (Factorial conjecture) *Let R be a regular local ring. If $S(E)$ is factorial domain then proj dim $E \leq 1$, that is, $S(E)$ must be a complete intersection.*

To lend evidence, we give other instances where it holds, and connect it to other conjectures. To inject a word of caution, there are modules whose symmetric algebra is factorial but fail to be complete intersection; the base rings are however not regular and the modules have infinite projective dimension (*cf.* [27]).

We begin with a case where this condition gets somewhat strengthened ([288, Theorem 3.1]):

Theorem 6.1.5 *Let R be a regular local ring containing a field, and suppose E is a finitely generated R–module such that the enveloping algebra of $S(E)$, i.e., $S(E)_e = S(E) \otimes_R S(E)$, is an integral domain. Then proj dim $E \leq 1$.*

Proof. $S(E)_e$ is just the symmetric algebra of the 'double module' $E \oplus E$. We may assume that for each non-maximal prime ideal $\mathbf{p}$, $E_{\mathbf{p}}$ has projective dimension at most 1 over $R_{\mathbf{p}}$. Let

$$0 \rightarrow L \longrightarrow R^n \longrightarrow E \rightarrow 0$$

be a minimal presentation of E. Since $\mathcal{F}_1$ applies to the module $E \oplus E$, we have that ($\ell = \text{rank}(L)$, $d = \dim R$): $2\ell \leq d - 1$.

Let t be the depth of L; because L is free on the punctured spectrum of R, L is a t–syzygy module. We claim that $t \geq \ell + 2$; it will follow from [65] that L must be free. Since $S(E)_e$ is a domain, $E \otimes_R E$ is torsion-free, so that by [7], [181],

$$\text{proj dim } (E \otimes_R E) = 2 \text{ proj dim } E \leq d - 1.$$

But proj dim $E =$ proj dim $L + 1 = d - t + 1$ and thus $t \geq \ell + 2$. □

Proposition 6.1.6 ([117, Proposition 7.2]) *Let R be a Cohen–Macaulay ring and let E be a finitely generated R–module. If E satisfies $\mathcal{F}_2$ then* proj dim $E \neq 2$.

Proof. Suppose otherwise; pick R local with lowest possible dimension, in particular we may assume proj dim$_{R_P}$ $E_P \leq 1$ for each $P \neq \mathfrak{m} =$ maximal ideal of R. Let

$$0 \rightarrow R^r \xrightarrow{\psi} R^m \xrightarrow{\varphi} R^n \longrightarrow E \rightarrow 0$$

be a minimal resolution of E. On account of $\mathcal{F}_2$, we have

$$n = \nu(E) \leq \dim R + \text{rank } (E) - 2,$$

that is

$$n - r = \ell = \text{rank } (\varphi) = n - \text{rank } (E) \leq \dim R - 2.$$

Since $r > 0$, the ideal $I_r(\psi)$ is $\mathfrak{m}$–primary. From Theorem 1.1.16, we have

$$\dim R = \text{height } I_r(\psi) \leq m - r + 1 = \ell - 1,$$

which is a contradiction. □

A few other cases of projective dimension three were resolved in [117], but nothing much beyond is known in the dimension scale.

6.2 Homological Rigidity

There seems to be a connection between the conjecture above and another conjecture on the homological rigidity of the module of differentials. Let k be a field of characteristic zero, and let A be a finitely generated k–algebra whose module of differentials, $\Omega_{A/k}$, has finite projective dimension over A. It is conjectured in [286] that A must necessarily be a complete intersection. We follow [292] closely.

To explore this, assume R is a polynomial ring over k, E is a module of projective dimension r, and that $\Omega_{S(E)/k}$ has finite projective dimension over $S(E)$. The validity of the conjecture would imply that $r \leq 1$.

From the exact sequences of modules of differentials

$$0 \rightarrow \Omega_{R/k} \otimes S(E) \longrightarrow \Omega_{S(E)/k} \longrightarrow \Omega_{S(E)/R} \longrightarrow 0,$$

and the isomorphism $\Omega_{S(E)/R} \simeq E \otimes_R S(E)$ of Proposition 6.3.4, we get that $\Omega_{S(E)/k}$ has finite projective dimension over $S(E)$ if and only if $E \otimes_R S(E)$ does so.

One way to attempt to find a finite projective resolution for $\Omega_{S/R}$ is the following. Let

$$\mathbb{F} : 0 \rightarrow F_r \longrightarrow \cdots \longrightarrow F_1 \longrightarrow F_0 \longrightarrow E \rightarrow 0$$

be a resolution of E. Tensoring it with $S(E)$, over R,

we obtain a complex of free $S(E)$–modules over $E \otimes_R S(E)$. It will be a consequence of Proposition 6.2.2 that if $r \geq 2$, this complex is never acyclic.

We begin with an useful feature of symmetric algebras.

Proposition 6.2.1 *Let R be a regular affine domain over a field k of characteristic zero, and let E be a finitely generated R–module. If $\Omega_{S(E)/k}$ has finite projective dimension over $S(E)$, then $S(E)$ is an integral domain.*

Proof. From the earlier remark, the hypothesis is equivalent to proj $\dim_S E \otimes S < \infty$.

Since S has no nontrivial idempotent, any module of finite projective dimension over it has a well–defined rank. In this case the rank of $E \otimes S$ must be equal to the R–rank of E. Moreover, by the Auslander–Buchsbaum formula ([192, p. 114]) and the ordinary Jacobian criterion for simple points, S is reduced.

Let Q be a minimal prime of S; put $P = Q \cap R$. Localizing at P, and changing the notation, we may assume that P is the maximal ideal of the local domain R. We claim that $P = 0$; for this it suffices to prove that E is a free R–module.

Let

$$R^m \xrightarrow{\varphi} R^n \longrightarrow E \rightarrow 0, \quad L = \varphi(R^m),$$

be a minimal presentation of E. Tensoring over with S_Q we obtain a minimal presentation of $E \otimes S_Q$ since the entries of $\varphi \otimes S_Q$ lie in the maximal ideal of S_Q. Therefore rank $(E \otimes S) = n$, which implies $L = 0$ and E is free as asserted. □

The next result works as a mechanism to shrink the projective dimension of E.

Proposition 6.2.2 *Let R be a regular local ring which is the localization of an affine algebra over a field of characteristic zero and let E be a finitely generated R–module with* proj dim $E = r$. *If*

$$\operatorname{Tor}_1^R(E, S(E)) = 0,$$

then $r \leq 1$.

Proof. If $r > 1$, we may assume that on the punctured spectrum of R the projective dimension of E is at most 1. The hypothesis means

$$\operatorname{Tor}_1^R(E, S_i(E)) = 0, \ i \geq 1,$$

for each of the symmetric powers of E.

We bring in the rigidity of Tor, *cf.* [7], [184], and the complexes of Lebelt–Weyman. We obtain a contradiction by progressively increasing the dimension of R. First, since $\operatorname{Tor}_1^R(E, E) = 0$, by [7], [184], the projective dimension of $E \otimes E$ is $2r$, so that $\dim R \geq 2r$.

We assume that the complex $\mathbb{F}$ is a minimal resolution of E, and recall the complexes $\mathcal{S}_t(\mathbb{F})$ of Section 3.3, lying over the symmetric powers of E. Their length is given by the formula 3.1.10. On the other hand, by Proposition 6.2.1, E satisfies the condition $\mathcal{F}_1$. Since E has projective dimension at most 1 on the punctured spectrum of R, it follows that these complexes have homology concentrated on the maximal ideal of R.

We now repeatedly apply the *lemme d'acyclicité*. For $t = 2$, the length of $\mathcal{S}_2(\mathbb{F})$ is at most $\dim R$, so that the complex is acyclic, and therefore a minimal resolution of $S_2(E)$.

Again, from $\operatorname{Tor}_1^R(E, S_2(E)) = 0$, we increase the dimension of R enough to guarantee the acyclicity of $\mathcal{S}_3(\mathbb{F})$, and so on. □

Corollary 6.2.3 *Let R be a regular affine domain over a field k of characteristic zero, and assume that* proj dim $E = 2$. *Then*

$$\text{proj dim}_{\,S(E)}(\Omega_{S(E)/k}) = \infty.$$

Proof. We replace R by a localization at a prime ideal and assume that E has projective dimension at most 1 on the punctured spectrum of R. Let

$$0 \to F_2 \xrightarrow{\psi} F_1 \xrightarrow{\varphi} F_0 \longrightarrow E \to 0$$

be a minimal free resolution of E.

If proj dim $\Omega_{S/k}$ is finite, S is an integral domain by Proposition 6.2.1 and therefore $\operatorname{Tor}_2^R(E, S) = 0$. We then have

$$T = \operatorname{Tor}_1^R(E, S) = \ker(\varphi \otimes S)/\text{image}(\psi \otimes S).$$

The claim is that this module vanishes.

T is a torsion module of finite projective dimension over S. Furthermore it is graded and annihilated by some power of the maximal ideal of R.

Step 1. To a graded presentation of T:

$$S^p \xrightarrow{\alpha} S^q \longrightarrow T \rightarrow 0,$$

there is an associated divisorial ideal

$$\mathbf{d}(T) = (I_q(\alpha)^{-1})^{-1}$$

where $I_q(\alpha)$ denotes the ideal generated by the q-sized minors of a matrix representation of α. Because T has finite projective dimension, $\mathbf{d}(T)$ is an invertible ideal of S, according to Theorem 1.1.8 . As T is graded and R is a local ring, $\mathbf{d}(T)$ must be generated by a homogeneous element $f \in S$. But Sf contains the annihilator of T, in particular a power of the maximal ideal of R. Thus $f \in R$.

Step 2. If f is not a unit of R, $\dim R = 1$, which is a contradiction. Otherwise the annihilator of T has grade at least 2, according to Theorem 1.1.8. Since both image$(\psi \otimes S)$ and $\ker(\varphi \otimes S)$ are second syzygy modules, this is impossible by standard depth considerations, unless $T = 0$. □

6.3 Regular Primes

This section is concerned with the regular prime ideals of a symmetric algebra $S = S(E)$. It begins with the identification of some of the regular primes of a symmetric algebra $S(E)$, and as an application has the proof of the *Zariski–Lipman conjecture* for symmetric algebras over regular rings. This relative version is much simpler than the other absolute cases of the conjecture that have been established.

There is a set of such primes that is easy to deal with:

Proposition 6.3.1 *Let $Q \supseteq S_+$ be a regular prime of S; put $P = Q \bigcap R$. Then R_P is a regular local ring and E_P is a free R_P–module.*

Proof. After localizing at P, Q becomes the irrelevant maximal ideal of the graded algebra S_P. Since its embedding dimension, $\nu(P) + \nu(E_P)$, must equal the Krull dimension of S_P, it is clear that P and E are as asserted. □

The following syzygy theorem is immediate:

Corollary 6.3.2 *Let R be a Noetherian ring of finite Krull dimension and let E be a finitely generated R–module. Then $S(E)$ has finite global dimension if and only if R is regular and E is a projective R–module.*

Remark 6.3.3 The global dimension of symmetric algebras of non–finitely generated modules is discussed in [42] and the appropriate section of [83].

Modules of differentials

A key to several questions regarding the regularity of $S(E)$ is the close relationship between the modules of differentials of R and of $S(E)$. The next elementary result describes the module $\Omega_{S/R}$ of relative differentials.

Proposition 6.3.4 $\Omega_{S/R} \simeq E \otimes_R S$.

Proof. Write $S = B/J$, where B is a polynomial ring $R[T_1, \ldots, T_n]$. The exact sequence of modules of differentials,

$$J/J^2 \longrightarrow \Omega_{B/R} \otimes_B S \longrightarrow \Omega_{S/R} \to 0$$

is precisely the presentation of E over R tensored by S. □

The following relative version of the *Zariski–Lipman conjecture* is inspired by [182] and [127].

Theorem 6.3.5 *Let R be an affine algebra over a field k of characteristic zero, and let E be a finitely generated R–module. If the module of k-derivations*

$$\mathcal{D} = Der_k(S(E), S(E))$$

is a projective $S(E)$–module, then $S(E)$ is a smooth R–algebra.

Proof. It consists of several steps. To begin we localize R at a maximal ideal; we shall prove that E is a free R–module.

Step 1. $\mathcal{D}$ is a graded S–module ($S = S(E)$). Indeed, if we present

$$S = B/J = R[T_1, \ldots, T_n]/(f_j = \sum a_{ij}T_i,\ j = 1, \ldots, m)$$

the module $\Omega_{S/k}$ is obtained as follows. First, grade the B–module $\Omega_{B/k}$ so that the differentials $d(r)$, $r \in R$, have degree zero, and $d(T_i)$ has degree one. From the fundamental sequence for modules of differentials, $\Omega_{S/k}$ is the quotient of $\Omega_{B/k}$ by the relations

$$\sum d(a_{ij})T_i + \sum a_{ij}d(T_i) = 0,\ j = 1, \ldots, m,$$

and the equations of $J(E)$. Since these are all homogeneous, $\Omega_{S/k}$ is a graded S-module and therefore its dual $\mathcal{D}$ will also be graded.

Step 2. The inclusion $R \hookrightarrow S$ gives rise to the exact sequence

$$\Omega_{R/k} \otimes S \longrightarrow \Omega_{S/k} \longrightarrow \Omega_{S/R} \to 0,$$

which by dualizing and the formula for relative differentials yields

$$0 \to \mathrm{Hom}_S(E \otimes_R S, S) \longrightarrow \mathcal{D} \longrightarrow \mathrm{Hom}_S(\Omega_{R/k} \otimes_R S, S).$$

By adjointness, the module on the left can be written

$$\mathrm{Hom}_R(E,S)=\mathrm{Hom}_R(E,R)\oplus\mathrm{Hom}_R(E,E)\oplus\ldots$$

with the appropriate grading.

From [182] we know that S is a normal domain. Therefore $\dim S=\dim R+\mathrm{rank}(E)=d+e$. Pick, by the previous step, a homogeneous basis $\{D_s,\ s=1,\ldots,d+e\}$ of $\mathcal{D}$. Assume that D_s, $s\leq r$, are the derivations of degree -1. We claim that when restricted to E, they generate $\mathrm{Hom}_R(E,R)$. Suppose otherwise, and let $\varphi\in\mathrm{Hom}_R(E,R)$ but not in the span of the D_s, $s\leq r$:

$$\varphi=\sum_{s\leq r}a_sD_s+\sum_{s>r}a_sD_s.$$

But it is clear that if the degree of D_s is not -1 then $a_s=0$ since S is positively graded.

By dimension counting, $\mathrm{Hom}_R(E,S)$ is a module of rank e over S, containing a free summand of rank equal to the R-rank of $\mathrm{Hom}_R(E,R)$. Thus $\mathrm{Hom}_R(E,S)$ is a free S-module, of rank e.

Step 3. Since $\mathrm{Hom}_R(E,S)$ is free on elements of degree -1, we get that

$$\mathrm{Hom}_R(E,R)\simeq R^e\ \text{ and }\ \mathrm{Hom}_R(E,E)\simeq E^e.$$

We claim that E is R-free. Note that R is a (normal) domain and E is torsion-free ($E\subset S$), and we may assume that E is free on the punctured spectrum of R, with $\dim R>1$.

We recall the canonical mapping

$$E\otimes_R\mathrm{Hom}_R(E,R)\xrightarrow{\ \psi\ }\mathrm{Hom}_R(E,E)$$

defined by

$$\psi(e\otimes f)(x)=f(x)\cdot e,\ \text{ for }\ e,x\in E,\ f\in\mathrm{Hom}_R(E,R).$$

According to [10, Proposition A.1], E is R–projective if and only if ψ is surjective.

With the identifications above, ψ is an endomorphism of E^e:

$$E^e\xrightarrow{\ \psi\ }E^e.$$

Because E is torsion-free and $\mathrm{Hom}_R(E,R)$ is R–free, this map is injective. Denote the cokernel by H. It suffices to prove that $H=0$; this will follow from the next lemma.

Lemma 6.3.6 *Let R be a Noetherian ring and let G be a finitely generated R-module with* $\dim G=\dim R$. *Then any injective endomorphism α of G, which is an isomorphism in codimension at most one, is an isomorphism.*

Proof. Consider the exact sequence

$$0 \to G \xrightarrow{\alpha} G \longrightarrow H \to 0.$$

By induction we may assume that R is a ring of dimension at least 2, and H is a module of finite length. We replace R by the polynomial ring $R[t]$ modulo the characteristic polynomial of α, while preserving all the other hypotheses—that is, we may assume that α is multiplication by an element of the ring. But then if α is not a unit, $\dim H \geq \dim G - 1$ according to Krull principal ideal theorem. □

6.4 Canonical Modules

Thus far, the methods to study factoriality in $S(E)$ focus mostly on quasi–Gorenstein symmetric algebras.

We aim at describing in general terms only the canonical module of the symmetric algebra of a module. It will require restrictions on the nature of the base ring and homological properties of the module.

Let R be a smooth algebra over a field k of characteristic zero. Let E be an R–module whose symmetric algebra, $S = S(E)$, is a normal domain. We may then use the result of [91] that the canonical module of S, ω_S, is the determinantal divisor of the module $\Omega_{S/k}$ of Kähler k–differentials.

It is convenient to work out a fixed presentation of the module E:

$$R^p \xrightarrow{\psi} R^m \xrightarrow{\varphi} R^n \longrightarrow E \longrightarrow 0.$$

We denote by L the image of φ, and $\ell = \operatorname{rank} L$.

Let $B = R[T_1, \ldots, T_n] = S(R^n)$; $S = B/J$, where J is generated by the linear forms

$$\varphi \cdot [T_1, \ldots, T_n].$$

Consider the exact sequences of modules of differentials:

$$J/J^2 \longrightarrow \Omega_{B/k} \otimes S \longrightarrow \Omega_{S/k} \to 0,$$

and

$$0 \to \Omega_{R/k} \otimes S \longrightarrow \Omega_{S/k} \longrightarrow \Omega_{S/R} \to 0.$$

In the second sequence, one has an inclusion on the left, from the smooth condition on R, and the isomorphism $\Omega_{S/R} = E \otimes S$. In addition we have a surjection

$$L \otimes S \longrightarrow J/J^2$$

of S–modules of the same rank.

It follows readily from [91] that under the present conditions, the canonical module of S, $\omega_S \simeq \operatorname{Ext}^{\ell}_B(S, B)$, can be identified to the dual of the determinantal divisor of J/J^2. Together with the observations above we have:

Proposition 6.4.1 $\omega_S \simeq (\wedge^\ell L \otimes S)^*$.

As a matter of fact, this result only requires that S be a domain which is a complete intersection at depth 1 primes.

To get hold of ω_S, first observe that from the presentation

$$R^p \xrightarrow{\psi} R^m \longrightarrow L \rightarrow 0$$

we get first the sequence

$$R^p \otimes \wedge^{\ell-1} R^m \longrightarrow \wedge^\ell R^m \longrightarrow \wedge^\ell L \rightarrow 0,$$

and then an exact sequence,

$$0 \rightarrow R \longrightarrow (\wedge^\ell R^m)^* \longrightarrow (R^p \otimes \wedge^{\ell-1} R^m)^*,$$

since $(\wedge^\ell L)^* \simeq R$. On the other hand, if we tensor the presentation of L with S, and repeat the construction, we get two sequences:

$$\begin{array}{ccccccc} 0 & \longrightarrow & R \otimes S & \longrightarrow & (\wedge^\ell R^m)^* \otimes S & \longrightarrow & (R^p \otimes \wedge^{\ell-1} R^m)^* \otimes S \\ & & \downarrow & & \| & & \| \\ 0 & \longrightarrow & \omega_S & \longrightarrow & (\wedge^\ell R^m \otimes S)^* & \longrightarrow & (R^p \otimes \wedge^{\ell-1} R^m \otimes S)^* \end{array}$$

and therefore a canonical mapping

$$\varphi \colon S \longrightarrow \omega_S. \tag{6.1}$$

A less formal description of ω_S is the following. For each ℓ–vector v in $\wedge^\ell R^m$, denote by $g(v)$ its image in $\wedge^\ell R^n$. The $g(v)$'s define a rank 1 torsion–free module so there is a set of scalars $r(v)$ such that

$$r(u) \cdot g(v) = r(v) \cdot g(u),$$

for all u, v. If we denote by I the ideal of R generated by the $r(v)$'s, we have:

Proposition 6.4.2 $\omega_S \simeq \mathrm{Hom}_S(IS, S)$.

Corollary 6.4.3 *Let E be a module with a resolution*

$$0 \rightarrow R^p \xrightarrow{\psi} R^m \longrightarrow R^n \longrightarrow E \rightarrow 0.$$

Denote by J the ideal generated by the $p \times p$ minors of ψ. If $S = S(E)$ is normal, then

$$\omega_S = \mathrm{Hom}_S(JS, S).$$

Proof. It suffices, in the identification above, to apply the theory of multipliers of [35, Theorem 3.1]. □

Chapter 7

Ideal Transforms

Let I be an ideal of the Noetherian integral domain R, of field of fractions K. The *ideal transform* of I, $T_R(I)$, is the subring of all elements of K that can be transported into R by a high enough power of I:

$$B = T_R(I) = \bigcup_{i \geq 0} I^{-i}.$$

In other words, B is the ring of global sections of the structure sheaf of R on the open set defined by I. The reference for this notion is [208], with its exposition of Nagata's fundamental results on the Hilbert's 14th Problem.

This chapter is devoted to the examination of two kinds of ideal transforms. The first associates a factorial domain to any symmetric algebra over a factorial domain; the second leads to a computational approach to symbolic blowups. The concern is when the process leads to a finitely generated algebra. Examples by Goto, Nishida and Watanabe [87] and by Roberts [234] show that the algebras that result are not always Noetherian.

The second half of the chapter contains a detailed analysis of modules with a linear presentation. Their symmetric algebras are defined by quadrics, which leads to the notion of jacobian duality. It determines the factorial closure of a broad class of modules of projective dimension two, with application to the number of elements that define varieties set–theoretically. Finally a method is discussed to produce prime ideals generated by analytically independent elements.

7.1 Factorial Closure

To begin our discussion of the factorial closure of a symmetric algebra $S(E)$, we first recall a basic notion of commutative algebra.

Let R be Cohen-Macaulay factorial domain and let E be a module with a presentation:

$$R^m \xrightarrow{\varphi} R^n \longrightarrow E \to 0.$$

Denote by $D(E)$ the quotient of $S(E)$ modulo the ideal of torsion elements. To obtain $B(E)$ one might as well apply the bi-dualizing procedure on $D(E)$. In particular we may assume that E is a torsion–free module. According to [241], $B(E)$ can be described in the following manner. First, embed $D(E)$ into a polynomial ring $K[U_1, \ldots, U_e]$, K the field of fractions of R and e the rank of E. Then:

$$B(E) = \bigcap_{\wp \subset R} D(E)_{\wp}, \text{ height } \wp = 1. \tag{7.1}$$

Proposition 7.1.1 *Let J be the ideal $I_{n-e}(\varphi)$ and put $M = JD(E)$. Then $B(E)$ is the M–ideal transform of $D(E)$.*

Proof. Because E is assumed torsion free, J is an ideal of height at least two. If T denotes the ideal transform of $D(E)$ with respect to J, it is clear from the equation above that $T \subset B(E)$. Conversely, if $b \in B(E) \setminus D(E)$, denote by L the conductor ideal $D(E){:}_R\, b$. Let Q be a minimal prime of L; if J is not contained in Q, the localization E_Q is a free R_Q–module and therefore $S(E_Q) = B(E_Q)$, so that $L_Q = R_Q$—which would be a contradiction. This shows that the radical of L contains J. □

Corollary 7.1.2 *There exists an ideal I generated by a regular sequence $\{f, g\}$ such that $B(E) = T_{D(E)}(I)$.*

Proof. Let $\{f, g\}$ be any regular sequence contained in the ideal J. It is clear that $T_{D(E)}(J) \subseteq T_{D(E)}(I)$. On the other hand, by Equation 7.1 any such transform must be contained in $B(E)$. □

The equality of the algebras $B(E)$ and $C(E)$ has a direct formulation ([117, Theorem 2.1]).

Theorem 7.1.3 *Let R be a normal, Cohen–Macaulay, universally Japanese domain.*

(a) *If E satisfies $\mathcal{F}_2$ then $B(E) = C(E)$.*

(b) *Conversely, if $S(E)$ is a domain and $B(E) = C(E)$ then E satisfies $\mathcal{F}_2$.*

Proof. First, we recall the assumption that R is universally Japanese, or a Nagata domain [193, p. 264], means that the integral closure of an affine domain B over R is finitely generated.

We may assume that R is a local ring. On account of $\mathcal{F}_2$, $\dim S(E) = \dim R + \operatorname{rank}(E)$, and therefore $\dim S(E) = \dim D(E) = \dim C(E)$.

Let f be a homogeneous element of $B(E)$. The set

$$I = \{r \in R \mid r \cdot f \in C(E)\}$$

is an ideal of height at least 2. It will follow from Remark 1.3.9 that height $IS(E)$ is at least 2. If $I \neq R$, we shall find this impossible. For simplicity we first argue the case $S(E) = D(E)$. Here we have

$$\text{height } IC(E) = \text{height } (IC(E)\bigcap D(E)) \geq \text{height } ID(E) \geq 2,$$

the equality on the left following from [208, Theorem 34.8]. As $C(E)$ is a Krull domain and f lies in its field of fractions, this is impossible. If $S(E)$ is not a domain, $D(E) = S(E)/P$, where P is a prime of height 0. In this case, height$((IS(E) + P)/P)$ is at least 2, and the argument applies.

For the converse, we verify $\mathcal{F}_2$ in terms of the local number of generators. Let P be the maximal ideal of R. We may assume height $(P) \geq 2$ and E is not free; we must show

$$\nu(E) \leq \text{height } P + \text{rank } E - 2 = \dim S(E) - 2.$$

The ideal I contains a regular sequence $\{a, b\}$ on R, which is also regular on $B(E)$, so we have height $(PB(E)) \geq 2$. By the result of Nagata,

$$\text{height } PC(E) = \text{height } (PC(E)\bigcap S(E)).$$

Since $PS(E)$ is a prime ideal of $S(E)$, $PC(E)\bigcap S(E) = PS(E)$, height $PS(E) \geq 2$. This gives the required condition. □

For later reference we state the following elementary observation:

Proposition 7.1.4 *Let R be a Noetherian domain and let I be an ideal. Let T be the I–ideal transform of R, and let C be a Noetherian ring $R \subset C \subset T$. If* depth $IC \geq 2$ *then $C = T$.*

Proof. This is clear since T is also the IC–ideal transform of C. □

For the remainder of this section, R is either a polynomial ring over a field, or a geometric regular local ring. We shall now look at the Noetherianess of the factorial closure of a symmetric algebra $S(E)$. Since $B(E) = B(E^{**})$, for the bi–dual of E, we may henceforth assume that E is a reflexive module.

Here is a sketch of the method. Since the graded bi–duals of $S(E)$ or $D(E)$ coincide, we consider the embedding

$$D(E) \hookrightarrow B(E) = \sum_{j\geq 0} B_j.$$

Suppose the two algebras first differ in degree $r-1$. We seek to determine the module C_r in the exact sequence

$$0 \to D_r \longrightarrow B_r \longrightarrow C_r \to 0,$$

adding to $D(E)$ the necessary generators in degree r. The algebra obtained will be denoted by $B(r)$. This algebra is checked for equality with $B(E)$; otherwise the preceding step is repeated for the next missing generators.

Here we shall only discuss the extreme case when $B(E)$ is obtainable from $D(E)$ by the addition of a single generator. For simplicity sake, let R be a regular local ring, and let E be a reflexive module of projective dimension one, free on the punctured spectrum of R:

$$0 \to R^m \xrightarrow{\varphi} R^n \longrightarrow E \to 0. \tag{7.2}$$

We shall assume that $S(E)$ is a domain, a fact that is equivalent here to the condition $\mathcal{F}_1$. If $\mathcal{F}_2$ does hold, $S(E)$ is a factorial domain. If $B(E) \neq S(E)$, we must have $m = d - 1, d = \dim R$.

Corollary 3.3.24 is the necessary background for our development of the computation of $B(E)$. A consequence is that the extended Rees algebra of I

$$A = R[It, t^{-1}]$$

is a Gorenstein algebra as well. This formulation provides for a presentation for A once the ideal I has been given by its generators and relations. Specifically, suppose $I = (x_1, \ldots, x_n)$ has a presentation

$$R^m \xrightarrow{\varphi} R^n \longrightarrow I \to 0, \; \varphi = (a_{ij}).$$

We obtain A as the quotient of the polynomial ring $R[T_1, \ldots, T_n, U]$ modulo the ideal J generated by the 1-forms in the T_i's

$$f_j = a_{1j}T_1 + \cdots + a_{nj}T_n, \; j = 1, \ldots, m,$$

together with the linear polynomials

$$UT_i - x_i, \; i = 1, \ldots, n.$$

Since the Krull dimension of A is $\dim R + 1$, J is an ideal of height n. Let us indicate the cases we shall make use of.

(a) If X is a generic $n - 1 \times n$ matrix (x_{ij}), and I is the ideal generated by the $n - 1 \times n - 1$ minors of X, then

$$J = (UT_1 - D_1, \ldots, UT_n - D_n, \; \Sigma_i x_{ji} T_i, \; j = 1, \ldots, m),$$

which by specializing $U = 0$, leads to the ideal whose explicit resolution is given in [109].

(b) Let $X = (x_{ij})$ be the generic, skew–symmetric matrix of order $n+1$, n even. Denote by I the ideal generated by the Pfaffians of X of order n. I is strongly Cohen-Macaulay and satisfies the condition on the local number of generators. J, in turn, is obtained as indicated above.

We require a strong assumption on the ideal $I_1(\varphi)$, that it be generated by a regular sequence. It is automatically satisfied if $S(E)$ is normal by Proposition 5.4.1, but it is present in other cases as well, e.g. in Example 5.4.2. With $I_1(\varphi) = (x_1, \ldots, x_d)$, as in Proposition 5.4.1, we write the presentation ideal $J(E)$ as

$$[x_1, \ldots, x_d] \cdot J(\varphi),$$

where $J(\varphi) = (b_{ij})$ is a $d-1 \times d$ matrix of linear forms in the T_i-variables. Let D be the ideal generated by the $(d-1)$–sized minors of $J(\varphi)$. $D = (D_1, \ldots, D_d)$ is a Cohen-Macaulay ideal of height two. An important role is that of the ideal D *evaluated at* the maximal ideal $\mathbf{p}$ of R, that is, taken mod $\mathbf{p}$; it will be denoted by Δ.

Note that normality requires $\Delta \neq 0$, while height $\Delta \leq 2$ by standard considerations. Δ can be interpreted as a kind of Jacobian ideal, and its height is independent of the choices made. It has a visible role ([290, Theorem 2.2]):

Theorem 7.1.5 *Let E be a module as in the exact sequence* (7.2) *and let $B(d-1)$ be the subalgebra of $B(E)$ obtained by adding to $S(E)$ the component of degree $d-1$ of $B(E)$. Then:*

(a) *$B(d-1)$ is a Gorenstein ring.*

(b) *$B(E) = B(d-1)$ if and only if height $\Delta = 2$.*

Proof. We begin by showing that $S(E)$ and $B(E)$ first differ in degree $d-1$. Since E is free on the punctured spectrum and $S(E) \neq B(E)$, we must have in the resolution of E that $m = d-1$. The projective resolution of a symmetric power $S_t(E)$ is then:

$$0 \to \wedge^m R^m \otimes S_{t-m}(R^n) \to \wedge^{m-1} R^m \otimes S_{t-m+1}(R^n) \to \cdots \to S_t(R^n) \to 0.$$

Because this complex is exact and E is free outside of $\mathbf{p}$, we have that $S_t(E)$ is a reflexive module for $t \leq d-2$, but not reflexive outside this range. Furthermore, a direct calculation shows that

$$\mathrm{Ext}_R^{d-1}(S_{d-1}(E), R) \;=\; R/I_1(\varphi).$$

Note that this measures the difference between B_{d-1} and S_{d-1}. Indeed, from the exact sequence

$$0 \to S_{d-1} \longrightarrow B_{d-1} \longrightarrow C_{d-1} \to 0$$

we obtain that

$$\mathrm{Ext}_R^{d}(C_{d-1}, R) \;=\; \mathrm{Ext}_R^{d-1}(S_{d-1}(E), R),$$

because B_{d-1} is a reflexive R-module. On the other hand, since the socle of $R/I_1(\varphi)$ is principal, we get, by duality, that $C_{d-1} = R/I_1(\varphi)$. B_{d-1} is therefore obtainable from $S_{d-1}(E)$ by the addition of a single generator. We proceed to find this element. Denote by $D_1, \ldots, D_d$, the maximal minors of the matrix $J(\varphi)$. From the equations

$$f_j = 0,\ j = 1, \ldots, d-1$$

in $S(E)$, we obtain

$$x_i D_j \ \equiv \ x_j D_i.$$

Let u be the element D_1/x_1 of the field of fractions of $S(E)$, and by Theorem 6.1.1, $u \in B$. We claim that $S(E)[u] = B(d-1)$, and that all of the asserted properties hold. First define the ideal $J(d-1)$ of the polynomial ring $P(d-1) = R[T_1, \ldots, T_n, U]$, generated by $J(E)$ and the polynomials

$$x_i U - D_i,\ 1 \leq i \leq d.$$

By hypothesis $J(E)$ is a prime ideal of height $d-1$, and as $\Delta \neq 0$, one of the forms D_i has unit content. Thus the ideal $J(d-1)$ has height at least d. It is then a proper specialization of the ideal defined by part (a) of Corollary 3.3.24 and is therefore a perfect Gorenstein ideal. Furthermore, since $\mathbf{p}P(d-1)$ is not an associated prime of $J(d-1)$, and E is free on the punctured spectrum of R, it follows easily that $J(d-1)$ is a prime ideal. This means that

$$P(d-1)/J(d-1) \ \simeq \ S(E)[u] \ \subseteq \ B(d-1).$$

We are now ready to prove the assertions. We begin by showing that u generates $B(d-1)$ over $S(E)$. Let v be the homogeneous generator of $B(d-1)$; we must have $u = bv + f$, where $b \in R$ and f is a form of $S(E)$ of degree $d-1$. We claim that b is a unit. Since the conductor of v, in degree 0, is $I_1(\varphi)$, the equations for v in degree two must be similar to those for u above, that is

$$x_i U - h_i = 0,\ 1 \leq i \leq d.$$

By assumption, one of the D_i does not have all of its coefficients in $\mathbf{p}$; assume D_1 is such a form. $x_1 bU - D_1 + x_1 f$ must be a linear combination of the linear equations for v:

$$x_1 bU - D_1 + x_1 f = \sum_i b_i (x_i U - h_i)$$

with

$$(b_1 - b)x_1 + \sum_{1<i\leq d} b_i x_i = 0.$$

But this is clearly impossible, unless b is a unit.

To prove (b) we use Proposition 7.1.4: As $B(d-1)$ is Cohen-Macaulay, we must have height $\mathbf{p}B(d-1) \geq 2$, which is expressed by height $(\Delta) = 2$. □

Example 7.1.6 Let us return to Example 5.4.2 and compute $B(E)$. The assumption on $I_1(\varphi)$ is realized since $\{a, b, c\}$ is a regular sequence. Here

$$J(\varphi) = \begin{bmatrix} T_1 & T_2 \\ T_2 & T_3 \\ T_4 & T_5 \end{bmatrix}.$$

Because height $(\Delta) = 2$, $B(E) = B(2)$.

For another set of examples, let E be a reflexive module such that its symmetric algebra is an almost complete intersection. This means, according to Proposition 3.4.1, that E has a projective resolution

$$0 \to R \xrightarrow{\psi} R^m \xrightarrow{\varphi} R^n \longrightarrow E \to 0.$$

Whether $S(E)$ is an integral domain depends on the Koszul homology of the ideal $I = I_1(\psi)$. We consider modules that are reflexive and free on the punctured spectrum of a regular local ring R. It implies that (with $\mathcal{F}_1$) $m = \dim R$. According to Proposition 3.4.4 this has the following consequence: If m is even then $S_{m/2}(E)$ is not a torsion–free R–module.

We focus on the case m odd. An application of the complexes (3.1.9) shows that $S_t(E)$ is reflexive for $t \leq (m-3)/2$ and torsion–free but not reflexive if $t = r = (m-1)/2$. We seek the equations for $B(r)$.

Lemma 7.1.7 *The quotient $B(E)_r/S_r(E)$ is a cyclic R-module.*

Proof. As in the proof of Theorem 7.1.5 we consider the exact sequence

$$0 \to S_r(E) \longrightarrow B(E)_r \longrightarrow C_r \to 0,$$

and compute $Ext^{m-1}(S_r(E),\ R) = Ext^m(C_r,\ R)$. A direct calculation of the complexes (3.1.9) shows that this module is isomorphic to $R/I_1(\psi)$. Because $I_1(\psi)$ is generated by a system of parameters, by duality it follows that C_r is cyclic. □

To get the equations for $B(r)$, first observe that the rows of the matrix φ are syzygies of the system of parameters $I_1(\psi) = (x)$. In particular $I_1(\varphi)$ is contained in (x). Denote by P the polynomial ring $R[T_1, \ldots, T_n]$. If $S(E)$ is a normal domain, we can write its defining ideal

$$J(E) = (f) = (x) \cdot J(\varphi).$$

Consider the exact sequence:

$$P^m \xrightarrow{J(\varphi)} P^m \longrightarrow C \to 0.$$

Assume that $J(\varphi)$ has rank $m-1$; thus $\ker J(\varphi)$ is cyclic, generated, say, by the vector of forms $(g) = (g_1, \ldots, g_m)$. Furthermore the associated prime ideals of C as an R-module are trivial, since E is free in codimension at most two, and therefore C is actually an ideal of P. It follows easily that C is a Gorenstein ideal and thus isomorphic to G, the ideal generated by the entries of (g).

Theorem 7.1.8 *Let E be an R-module that is reflexive and free on the punctured spectrum of R, whose symmetric algebra is an almost complete intersection that is a normal domain.*

(a) *$B(r)$ is a Gorenstein ring singly generated over $S(E)$.*

(b) *Denote by $G(0)$ the ideal G evaluated at the origin. Then $B(r) = B$ if and only if height $G(0) \geq 2$.*

Proof. Consider the equations

$$\begin{aligned} (x)\cdot J(\varphi) &= (f) \\ (g)\cdot J(\varphi) &= 0 \end{aligned}$$

Reading them mod (f), that is, in $S(E)$, when $J(\varphi)$ still has rank $m-1$ by the normality hypothesis, we get an element h in the field of fractions of $S(E)$, such that $h\cdot(x) \equiv (g)$. It follows from Theorem 6.1.1 that h actually lies in B.

(a) Let $J(r)$ be the ideal of the polynomial ring $P[U]$

$$((f), (x)\cdot U - (g)).$$

$J(r)$ is by Theorem 3.3.24 a Gorenstein prime ideal of height m.

(b) Because B is the ideal transform of $B(r)$ with respect to the ideal (x), these rings are equal if and only if the grade of $(x)\cdot B(r)$ is at least two. Since $B(r)$ is Cohen–Macaulay this translates into the condition above. □

7.2 Symbolic Power Algebras

In this section we give the formulation of symbolic blowups as ideal transforms, in a manner that is convenient for computation.

Let R be a Noetherian domain and let I be an ideal. The intersection of the minimal primary components of I^n will be denoted by $I^{(n)}$. Because the embedded associated primes of I do not play a role, we can assume that I is unmixed.

We show how the symbolic blowup

$$\mathcal{R}_s(I) = \sum_{n\geq 0} I^{(n)}t^n$$

can be derived from the ordinary Rees algebra of I via an ideal transform, and then discuss a controlling element, the divisor class group of $\mathcal{R}_s(I)$.

Let us indicate how symbolic power algebras came into the scene. When R satisfies the condition S_2 of Serre's, the algebras $\mathcal{R}(I)$ and $\mathcal{R}_s(I)$ are related by the fact that the latter is the ring of regular functions on an open set of $Spec(\mathcal{R}(I))$, defined by the complement of $V(f,g)$, $f,g \in R$. Rees [227] used the following elegant argument in the construction of counterexamples to the Zariski–Hilbert problem, if T_1 is an indeterminate over $\mathcal{R}(I)$ and T_2 is defined by the equation

$$fT_1 + gT_2 = 1,$$

then

$$\mathcal{R}(I)[T_1, T_2] \bigcap K = \mathcal{R}_s(I), \tag{7.3}$$

K being the field of fractions of $\mathcal{R}(I)$. In Rees examples, R is the ring of the affine cone over an elliptic curve and I is the ideal of a non–torsion point. The Noetherianess of $\mathcal{R}_s(I)$, for dim $R = 2$, is expressed already in the divisor class group of R.

Ideal transform

We first recall some facts about asymptotic prime divisors. Let I be an ideal of the Noetherian ring R. Consider the finite sets $\text{Ass}(R/I^n)$. There is an interesting relationship between these sets and the associated prime ideals of $\text{gr}_I(R)$. It is part of a broad theory of asymptotic sequences started out by Ratliff in [224]. The slice that will be offered here is taken from the elegant treatment in [196] and [197]. Among other facts it reveals that these sets have a stable value, that is, $\text{Ass}(R/I^n)$ is constant for $n \gg 0$. There is difficulty in finding the stable value set without using the equations of $\text{gr}_I R$. If, however, the approximation complex of I is acyclic, it is often possible to determine it.

Proposition 7.2.1 ([197]) *Let $A = \sum_{n\geq 0} R_n$ be a Noetherian, graded, homogeneous (i.e. generated by R_1) ring. There exists an integer m such that*

$$\text{Ass}_{R_0}(R_n) = \text{Ass}_{R_0}(R_m)$$

for all $n \geq m$.

We need two lemmas first.

Lemma 7.2.2 *Let $A = \sum_{n\geq 0} R_n$ be a Noetherian, homogeneous, graded ring. Then there exists an integer m such that $(0\colon R_1) \cap R_n = 0$ for $n \geq m$.*

Proof. The ideal $(0\colon R_1)$ is generated by homogeneous elements $a_1, \ldots, a_s$. Let $m \geq 1+\max\{\deg a_i\}$. If $x = \sum r_i a_i$ lies in $(0\colon R_1) \cap R_n$, we may assume r_i homogeneous and therefore lying in (R_1) and thus $x = 0$. □

Lemma 7.2.3 *Let $A = \sum_{n\geq 0} R_n$ be a Noetherian graded ring, and let c be a homogeneous element. Suppose that S is a multiplicative monoid of R_0 such that $(0\colon c) \cap S = \emptyset$. Then there is a homogeneous element d such that $(0\colon cd)$ is a prime ideal and $(0\colon cd) \cap S = \emptyset$.*

Proof. Among homogeneous elements r with $(0\colon rc) \cap S = \emptyset$, pick d so that the colon ideal $P = (0\colon cd)$ is maximal. If x and y are homogeneous elements not in P and such that $x \cdot y \in P$, $(0\colon xdc)$ is larger than P. This means that there exists $s \in S$ such that $x \in (0\colon cds)$, and therefore this ideal must meet S. Thus there is $t \in S$ with $tscd = 0$, which means that $(0\colon cd)$ meets the monoid S. □

Proof of Proposition. Let $P \in B = \bigcup_{n \geq 0} \text{Ass}(R_n)$. By the previous lemma, for $P = (0\colon c)$ there is $Q \in \text{Ass}(A)$ with $P = Q \cap R_0$, which shows B is finite.

To prove stability, pick m as in the first lemma and let $n \geq m$. If $P \in \text{Ass}(R_n)$ write $P = (0\colon c)_{R_0}$, $c \in R_n$. As $n \geq m$, $P = (0\colon cR_1)_{R_0}$. Since $cR_1 \subset R_{n+1}$, $P \in \text{Ass}(R_{n+1})$, that is $\text{Ass}(R_n) \subset \text{Ass}(R_{n+1})$. The conclusion follows since B is finite. □

One of its applications is the result of Brodmann ([23]):

Theorem 7.2.4 *Let R be a Noetherian ring and let I be an ideal. The sets $A(n) = \text{Ass}(R/I^n)$ have a stable value.*

Proof. Apply Proposition 7.2.1 to the algebra $\text{gr}_I(R)$. Let $B(n) = \text{Ass}(I^{n-1}/I^n)$ be the corresponding stable value. From the exact sequence

$$0 \to I^n/I^{n+1} \longrightarrow R/I^{n+1} \longrightarrow R/I^n \to 0,$$

it follows that $A(n+1) \subset A(n) \bigcup B(n+1)$. Since $B(n+1) = B(n) \subset A(n)$, $A(n+1) \subset A(n)$. □

The symbolic powers of an ideal

As an application, we obtain the following description of the symbolic powers of an ideal. Let R be a Noetherian ring and I be an ideal. Let $\mathcal{U} = \cup_{n \geq 0} \text{Ass}(R/I^n)$. Since $\mathcal{U}$ is a finite set of prime ideals, pick f lying in each of the embedded primes but avoiding every minimal prime of I (make $f = 1$ if that set is empty).

Corollary 7.2.5 $I^{(n)} = \bigcup_{t \geq 0} (I^n :_R f^t)$.

For each n, define $t(n)$ to be the least integer t such that

$$I^n : (f, g)^t = I^n : (f, g)^{t+1}.$$

If $t(\cdot)$ is a bounded function, $\mathcal{R}_s(I)$ is an integral extension of $\mathcal{R}(I)$. It is easy to see that if $\mathcal{R}_s(I)$ is Noetherian, then $t(n)$ has the growth of a linear polynomial.

The following formulation is useful when setting up the computation of successive symbolic powers (see Chapter 10).

Proposition 7.2.6 *Let R be a Noetherian ring satisfying Serre's condition S_2. Let $\mathcal{R}_s(I)$ be the symbolic blowup of the ideal I. There exists a regular sequence $\{f, g\}$ in R such that $\mathcal{R}_s(I)$ is the ideal transform of the Rees algebra $\mathcal{R}(I)$ with respect to the ideal (f, g).*

Proof. Choose f as above, and pick $g \in I$ but avoiding all the associated primes of (f). Since (f, g) has grade two, it ensures that the ideal transform of $R(I)$ lives in the polynomial ring $R[t]$. We now apply the formula of the corollary above. □

Divisor class group

Let R be a Noetherian normal domain and let I be an ideal. If the Rees algebra of I is normal, there is an extensive body of results on its divisor class group (see [119], [256] and their references). Here we make a brief analysis of the divisor class group of $\mathcal{R}_s(I)$. The requirement for normality is:

Proposition 7.2.7 *The algebra $R_s(I)$ is normal if and only if for any minimal prime $\wp$ of I the Rees algebra $\mathcal{R}(I_\wp)$ is normal.*

Proof. The equality

$$\mathcal{R}_s(I) = R[t] \cap \Big(\bigcap_{\wp \in \mathrm{Min}(R/I)} R_\wp[I_\wp t]\Big)$$

describes $\mathcal{R}_s(I)$ as an intersection of Krull domains with the same field of fractions. The assertion follows from [165, see discussion of Theorem 114]. The converse is clear. □

Theorem 7.2.8 *Let R be a Noetherian normal domain and let I be an ideal such that $\mathcal{R}_s(I)$ is normal. Then there is an exact sequence*

$$0 \to F \longrightarrow Cl(\mathcal{R}_s(I)) \longrightarrow Cl(R) \to 0$$

where F is a free (abelian) group of rank equal the number of minimal primes of $I\mathcal{R}_s(I)$ whose contraction of R has codimension at least 2. Furthermore, if I is generically a complete intersection, the rank of F is the number of minimal primes of I of codimension at least 2.

Proof. It is similar to that of Theorem 5.6.3, so we only sketch the details. As in [*loc. cit.*], the homomorphism of divisor classes groups is defined first at the level of divisors: For each prime divisor $[P]$ of $R_s(I)$, define $\psi[P] = [P \cap R] \in Div(R)$. The argument is repeated to give the exact sequence.

On the other hand, to find the generators of F, suppose P is a prime ideal of $R_s(I)$, $I \subset P$. If $\wp = P \cap R$ is not a minimal prime of I, by the proof of Proposition 7.2.7, the divisor of P is trivial. The assertion of linear independence follows as in the other proof. Finally, if $\wp$ is a minimal prime of I and $I_\wp$ is generated by a regular sequence of $r \geq 2$ elements, then $\mathcal{R}_s(I_\wp) = \mathcal{R}(I_\wp)$, and its divisor class is $\mathbb{Z}[\wp\mathcal{R}(I_\wp)] \oplus Cl(R_\wp)$. □

If R is a regular local ring, and $\wp$ is a prime ideal, then $\mathcal{R}_s(\wp)$ is normal, since this is the case for the Rees algebra of the maximal ideal of $R_\wp$. There is a single generator for F, the ideal $P = \wp\mathcal{R}(\wp_\wp) \cap \mathcal{R}_s(\wp)$. In this case ([256], [123]):

Corollary 7.2.9 *Let R be a regular local ring and let $\wp$ be a prime ideal of height $g \geq 2$. Then $\mathcal{R}_s(I)$ is a Krull domain of divisor class group $\mathbb{Z}[P]$. The class of the canonical module is $(g-2)[P]$. In particular, if* height $\wp = 2$, *and $\mathcal{R}_s(\wp)$ is Noetherian, then the symbolic blowup algebra is quasi–Gorenstein.*

7.3 Finiteness of Ideal Transforms

There are relatively few criteria that predict the Noetherianess of an ideal transform. We discuss one discovered by Huneke ([144]) that has had wide use.

We need the following framework to state Huneke's criterion.

Proposition 7.3.1 *Let R be a Noetherian normal domain and let $\wp$ be a prime ideal that is generically a complete intersection. Let $\mathcal{R}_s(\wp)$ be the symbolic power algebra of $\wp$, and denote by*

$$A = \mathcal{R}_s^{(k)}(\wp) = \sum_{j \geq 0} I^{(jk)} t^{jk}$$

the kth Veronese subring. Then $\mathcal{R}_s(\wp)$ is Noetherian if and only if A is Noetherian.

It is a consequence of the following general fact (pointed out by an anonymous reviewer):

Proposition 7.3.2 *Let R be a commutative ring and let $\mathcal{F} = \{I_n,\ n \geq 0\}$ be a filtration of ideals. Then the Rees algebra $\mathcal{R}(\mathcal{F})$ is Noetherian if and only if one of its Veronese subrings*

$$A^{(r)} = \oplus_{n \geq 0} I_{nr}$$

is Noetherian.

Proof. It is enough to prove that $\mathcal{R}(\mathcal{F})$ is Noetherian along with $A^{(r)}$. But $\mathcal{R}(\mathcal{F})$ is isomorphic to the direct sum of r

$$\oplus_{n \geq 0} I_{nr+j},\ 0 \leq j < r,$$

each of which is isomorphic to an ideal of $A^{(r)}$. □

Throughout the section, we use $\lambda(\cdot)$ to denote the length, over the appropriate ring, of a module of finite length. Here is the finiteness criterion of [144] (further generalized and simplified in [204]):

Theorem 7.3.3 *Let $(R, \mathfrak{m})$ be a regular local ring of dimension 3 with infinite residue field, and let $\wp$ be a codimension two prime ideal. Then the following are equivalent:*

(i) *$\mathcal{R}_s(\wp)$ is Noetherian.*

(ii) *There exist k, ℓ, two elements $f \in \wp^{(k)}$, $g \in \wp^{(\ell)}$, and an $x \in \mathfrak{m} \setminus \wp$ such that*

$$\lambda(R/(f, g, x)) = k\ell \cdot \lambda(R/(\wp, x)).$$

Before embarking on the proof we note the following elementary notion of intersection multiplicities.

Proposition 7.3.4 *Let $(R, \mathfrak{m})$ be a Noetherian local ring of Krull dimension 1 and let x be a system of parameters. The numerical function on finitely generated R–modules*

$$\chi(M) = \lambda(M/xM) - \lambda({}_xM),$$

where ${}_xM$ denotes the set of elements of M annihilated by x, is additive on short exact sequences. In particular if R has a unique minimal prime $\wp$ then $\chi(M) = \lambda(M_\wp) \cdot \lambda(R/(\wp, x))$.

Proof. The additivity of χ follows from the snake lemma. To prove the other assertion apply χ to a filtration of M

$$M = M_0 \supset M_1 \supset M_2 \supset M_3 \supset \cdots,$$

whose factors M_i/M_{i+1} are either $R/\mathfrak{m}$ or $R/\wp$. The number of the latter is $\lambda(M_\wp)$, while $\chi(R/\mathfrak{m}) = 0$. □

Proof of Theorem. (a) $\Rightarrow$ (b): Let k be an integer such that the Veronese subring $\mathcal{R}_s^{(k)}(\wp)$ is Noetherian, in other words, $\wp^{(kn)} = (\wp^{(k)})^n$ for all n. By Burch's formula, the analytic spread of $\wp^{(k)}$ is two. Let f, g generate a minimal reduction of $\wp^{(k)}$. Choose $x \in \mathfrak{m} \setminus \wp$ and set $S = R/(x)$ and $I = (\wp^{(k)}, x)$. The images $\overline{f}, \overline{g}$ of f, g in S form a reduction of I. Because $I^n = ((\wp^{(k)})^n, x) = (\wp^{(kn)}, x)$, it follows that $\lambda(S/I^n) = \lambda(R/(\wp^{(kn)}, x))$.

Since x is a regular element on any $R/\wp^{(m)}$, it follows by Proposition 7.3.4 applied to the module $R/\wp^{(m)}$ that

$$\begin{aligned} \lambda(R/(\wp^{(m)}, x)) &= \lambda(R/(\wp, x)) \cdot \lambda(R_\wp/(\wp_\wp^m)) \\ &= e\binom{m+1}{2}, \end{aligned} \tag{7.4}$$

where $e = \lambda(R/(\wp, x))$. This means that

$$\lambda(R/I^n) = \lambda(R/(\wp^{(kn)}, x)) = e\binom{kn+1}{2} = ek^2\binom{n+1}{2} - e\binom{k}{2}n,$$

holds for all $n \geq 0$. Therefore the multiplicity of I is ek^2 and $\lambda(R/(f, g, x)) = ek^2$, which is (b).

(b) $\Rightarrow$ (a): We may assume that $k = \ell$. Indeed, if $f \in \wp^{(k)}$, $g \in \wp^{(\ell)}$ and $x \in \mathfrak{m}\setminus\wp$ are such that $\lambda(R/(f, g, x)) = ek\ell$, then $\lambda(R/(f^\ell, g^k, x)) = k\ell ek\ell = e(k\ell)^2$.

Set $I = (\wp^{(k)}, x)/(x) \subset S$, and let

$$P_I(n) = e_0\binom{n+1}{2} - e_1 n + e_2$$

be the Hilbert polynomial of I. Applying equation (7.4) to $J_n = (\wp^{(kn)}, x)$ gives

$$\lambda(S/J_n) \quad = \quad ek^2\binom{n+1}{2} - e\binom{k}{2}n, \qquad (7.5)$$

where $e = \lambda(R/(\wp, x))$.

Let $f, g \in \wp^{(k)}$ be as in the assumption, that is, $\lambda(S/(\overline{f}, \overline{g})) = ek^2$. As $(\overline{f}, \overline{g}) \subset I$, $e_0 = e(I) \leq e((\overline{f}, \overline{g})) = \lambda(S/(\overline{f}, \overline{g})) = ek^2$. This says that

$$e_0 \quad \leq \quad ek^2. \qquad (7.6)$$

On the other hand, since $I^n \subset J_n$ for all $n \geq 1$, one has

$$e_0\binom{n+1}{2} - e_1 n + e_2 \geq ek^2\binom{n+1}{2} - e\binom{k}{2}n$$

for all $n \gg 0$. Hence $e_0 \geq ek^2$, which together with equation (7.6) gives $e_0 = ek^2$. We then return to equation (7.6) to obtain

$$e\binom{k}{2} \geq e_1.$$

These estimates yield the inequality

$$e_0 - e_1 = ek^2 - e_1 \geq ek^2 - e\binom{k}{2} = \lambda(S/J_1) = \lambda(S/I).$$

We have thus shown

$$e_0 - e_1 \geq \lambda(R/I).$$

By Corollary 5.2.4 we have first that $e_0 - e_1 \geq \lambda(R/I)$, and then that $e_2 = 0$.

We have shown that the Hilbert function and Hilbert polynomials of I coincide, and furthermore that

$$\lambda(S/I^n) = \lambda(S/J_n).$$

As $I^n \subset J_n$, we obtain $I^n = J_n$ for all $n \geq 1$. Hence

$$((\wp^{(k)})^n, x) = (\wp^{(kn)}, x), \quad \text{for } n \geq 1,$$

from which, by a direct use of Nakayama's lemma, it follows that

$$\wp^{(kn)} = (\wp^{(k)})^n.$$

We may then apply Proposition 7.3.1 to complete the proof. □

In an intricate and fascinating argument involving this finiteness criterion, Goto, Nishida and Watanabe [87] established the existence of prime ideals of monomial curves in 3–space, whose symbolic Rees algebras are not Noetherian. They proved, for instance, that the ideal

$$\wp = (x^{11} - yz^7, y^3 - x^4z^4, z^{11} - x^7y^2)$$

corresponding to the monomial curve

$$\begin{aligned} x &= t^{25} \\ y &= t^{72} \\ z &= t^{29} \end{aligned}$$

has a Noetherian symbolic power algebra in all characteristics except zero.

Non–Noetherian factorial closures

We shall now give Roberts' example [234] of a module whose factorial closure is not Noetherian. Actually, his aim was twofold: (i) To find a counterexample to Hilbert's 14th Problem, and (ii) to construct a prime ideal in a power series ring whose symbolic blowup is not Noetherian. (His earlier example [233] of the latter was not analytically irreducible.)

There are two parts to his construction. First, establishing a setting whereby the construction permits deciding several features of the factorial closure; then, a delicate analysis of a family of examples.

Let $R = k[x_1, \ldots, x_d]$ be a polynomial ring and let $I = (a_1, \ldots, a_n)$ be an ideal. Attach to I the following derivation of the ring $S = R[T_1, \ldots, T_n]$:

$$\partial = a_1 \frac{\partial}{\partial T_1} + \cdots + a_n \frac{\partial}{\partial T_n}.$$

One of its properties is that it vanishes on any linear form

$$b_1 T_1 + \cdots + b_n T_n,$$

that corresponds to a syzygy of the a_i's. The ring of constants B of ∂ will contain the subring D of S, generated by all such forms, and in addition, all the polynomials that can be conducted into D by nonzero elements of R.

This can be made more precise when k is a field of characteristic zero. Noting that ∂ is a locally finite derivation of S (that is, for any $z \in S$ there exists an integer n such that $\partial^n(z) = 0$), its exponential e^{∂} is an automorphism of S. Using Taylor expansions it is then easy to show:

Proposition 7.3.5 *The ring of constants of ∂ is the ring of invariants of e^{∂}.*

Let us phrase this in the setting of symmetric algebras. Let R be a regular local ring (*resp.* a polynomial ring over a field), and let E be a finitely generated module (*resp.* a graded R–module). Recall that in studying the factorial closure of $S(E)$, we may assume that E is a reflexive module. This allows for an exact sequence

$$0 \rightarrow E \xrightarrow{\psi} F \xrightarrow{\varphi} G \rightarrow 0,$$

where F is a free module and G is a torsion–free module.

If $T_1, \ldots, T_n$ is a basis of F, E is generated by some linear forms

$$f_j = \sum_{i=1}^{n} a_{ij} T_i, \; j = 1, \ldots, m.$$

$D(E)$ is the subring of $R[T_1, \ldots, T_n]$ generated by $f_1, \ldots, f_m$ over the base ring R. At this point we can assume that $E \subset \mathfrak{m}F$, $\mathfrak{m}$ = maximal ideal of R. When E is a free module this is exactly $S(E)$. Let I be any height two ideal in the ideal defining the free locus of E.

The next observation follows from the earlier discussion.

Lemma 7.3.6 *Let $D(E)$ be the Rees algebra of the module E. The factorial closure $B(E)$ consists of the elements of $R[T_1, \ldots, T_n]$ conducted into $D(E)$ by some power of I.*

However explicit, this does not provide the setting to carry out computations. The other leg of the construction is a homomorphism similar to the downgrading homomorphism in the symmetric algebra of a module.

Let $\varphi: F \to R$ be a homomorphism of an R–module. There is a R–module endomorphism of $S(F)$ that extends φ. It is defined as follows. Set $\varphi_0 = 0$ and $\varphi_1 = \varphi$. Then $\varphi_n: S_n(F) \to S_{n-1}(F)$ is given by

$$\varphi_n(e_1 \cdots e_n) = \sum \varphi(e_i)(e_1 \cdots \widehat{e_i} \cdots e_n).$$

Proposition 7.3.7 *Assume that the image of the mapping $\varphi: F = R^r \to R$ is minimally generated by r elements. Then* $\ker(\varphi_n) \subset \mathfrak{m}S_n(F)$.

Actually this is somewhat misleading. If R contains a field of characteristic 0, the construction stands as above. In general however $S(F)$ has to be replaced by the divided power algebra.

Proof. Once a basis $\{e_1, \ldots, e_r\}$ is chosen, $S_n(F)$ is freely generated by the monomials $e_1^{a_1} \cdots e_r^{a_r}$, $\sum a_i = n$. If such monomial occurs, with coefficient 1, in an element of the kernel of φ_n, then $a_1\varphi(e_1)e_1^{a_1-1} \cdots e_r^{a_r}$ (say $a_1 \geq 1$) would be a combination of elements all with coefficients in $\varphi(Re_2 + \cdots + Re_r)$, contradicting the minimality hypothesis. □

One uses this in conjunction with the previous lemma. That is, assume that we have an exact sequence

$$0 \to E \longrightarrow F = R^r \longrightarrow I \to 0,$$

with I an ideal minimally generated by r elements. There is a surjective homomorphism from the symmetric algebra of F onto the Rees algebra of I:

$$S(F) \xrightarrow{\alpha} S(I) \xrightarrow{\beta} \mathcal{R}(I)$$

where α is induced by φ and β is the natural mapping.

The kernel J of α is generated by the 1-forms that define E; that of $\beta \circ \alpha$, J_∞, will be larger, if I is not of linear type.

Corollary 7.3.8 *In the situation above* (K = *field of quotients of* R)*:*

$$\begin{aligned} J_\infty &= \sum_{i\geq 1} \ker\,(\varphi_i \circ \cdots \circ \varphi_1) \\ B_n(E) &= \ker(\varphi_n) = (J_\infty)_n \bigcap (S(E) \otimes K) \end{aligned}$$

This implies that the components $B_n(E)$ lie in $\mathfrak{m}S_n(F)$, so that if $B(E)$ is finitely generated then for all large n, $B_n(E)$ will be contained in $\mathfrak{m}^\ell S_n(F)$ for ℓ large as well.

Here is the critical point of the whole construction.

Theorem 7.3.9 *Let I be the ideal of $k[x, y, z]$ generated by x^{t+1}, y^{t+1}, z^{t+1}, $x^t y^t z^t$, for some integer $t \geq 2$. For each integer n there exists an element in $B_n(E)$ whose coefficient of T_4^n is x.*

Exercise 7.3.10 Establish the validity of equation (7.3).

7.4 Modules with Linear Presentation

This section, and the next two, examines several aspects of the symmetric algebras with modules with linear presentation, and mild generalizations of them. It focuses on their Rees algebras and the corresponding factorial closures. It is essentially the treatment of [257].

Throughout this section $R = k[x_1, \ldots, x_d]$, $d \geq 2$ and $\mathfrak{m} = (x_1, \ldots, x_d)$, and E is a finitely generated R–module of linear type with a linear presentation matrix φ. If $n = \nu(E)$ then write $S = k[\mathbf{T}] = k[T_1, \ldots, T_n]$, and let B be the matrix of linear entries in S such that $\mathbf{T} \cdot \varphi = \mathbf{x} \cdot B$. By the linear type requirement, E necessarily satisfies $\mathcal{F}_1$ and therefore $\nu(E) \leq \dim R + \operatorname{rank} E - 1$. We consider modules for which equality holds. This means, according to Proposition 1.5.1, that the dual module F is a torsion-free S–module of linear type with rank $F = 1$. We may therefore identify F with an S–ideal of linear type and height at least 2. We denote this ideal by I.

Now let $\mathbf{f} = f_1, \ldots, f_d$ be homogeneous generators of I such that in the identification $\mathcal{R}(I) = S(E)$, $f_i t$ corresponds to the image of T_i in $S(E)$. The elements $\mathbf{f}$ appear in the exact sequence

$$0 \to S \xrightarrow{\begin{bmatrix} f_1 \\ \vdots \\ f_d \end{bmatrix}} S^d \xrightarrow{B^t} \cdots$$

and they can be obtained in the following way: choose a $d \times (d-1)$ submatrix B' of B with rank $B' = d-1$, let Δ_i be the minor (with sign) obtained from B' by deleting the ith row, and set $\Delta = \gcd(\Delta_1, \ldots, \Delta_d)$; then $f_i = \Delta_i/\Delta$.

We can now define the algebra

$$C = k[\mathbf{x}, \mathbf{T}, U]/(\mathbf{x} \cdot B, x_iU - f_i \mid 1 \leq i \leq d).$$

It is easy to see that

$$V((I, \mathfrak{m})) = V((I_{d-1}(B), \mathfrak{m})) = \operatorname{Sing}(S(E)) \bigcap V(\mathfrak{m}) \subset \operatorname{Spec}(S(E))$$

(cf. [288, Proposition 1.1] for the second equality), and therefore

$$\text{height } I = \text{codim } (\operatorname{Sing}(S(E)) \bigcap V(\mathfrak{m})) - 1.$$

Normality and factorial closure

Theorem 7.4.1 *Let E be a module over $R = k[x_1, \ldots, x_d]$ of linear type with linear presentation matrix and $\nu(E) = d + \text{rank } E - 1$. The following are equivalent.*

(a) $E_{\mathfrak{q}}$ *satisfies $\mathcal{F}_2$ for all $\mathbf{q} \neq \mathfrak{m}$ and $S(E)$ is normal.*

(b) $\text{gr}_I(S)$ *is a domain.*

Proof. (b) $\Rightarrow$ (a): By assumption, $\text{gr}_I(S)$ is a domain, hence $\mathcal{R}(I) = S(I)$ is normal and $Cl(\mathcal{R}(I)) \simeq \mathbb{Z}$. In particular the **f**–number of E is 1 cf. Theorem 5.6.2.

(a) $\Rightarrow$ (b): It follows that $Cl(S(E)) \simeq \mathbb{Z}$, hence $S(I)$ is normal with $Cl(S(I)) \simeq \mathbb{Z}$. Therefore the **f**–number of I is 1, in particular $\sqrt{I}$ is prime and $\nu(I_{\mathbf{q}}) < \dim S_{\mathbf{q}}$, for all $\sqrt{I} \neq \mathbf{q} \in V(I)$. Write $\wp = \sqrt{I}$.

We will now show that for every $\mathbf{q} \in \text{Ass}(\text{gr}_I(S))$, $\mathbf{q} \bigcap S/I = \wp/I$. Suppose $\mathbf{q} \bigcap S/I \neq \wp/I$, then localizing we may assume that $\mathbf{q} \bigcap S/I = \mathbf{p}/I$ where $\mathbf{p}$ is the maximal ideal of the local ring S, and $\mathbf{p} \neq \wp$; this implies $\nu(I) < \dim S$. Let Q be the preimage of $\mathbf{q}$ under the map $\mathcal{R}(I) \to \text{gr}_I(S)$. Using the exact sequences of Theorem 5.1.9 we obtain

$$\begin{array}{ccccccccc}
0 & \longrightarrow & I \cdot \mathcal{R}(I)_Q & \longrightarrow & \mathcal{R}(I)_Q & \longrightarrow & (\text{gr}_I(S))_{\mathbf{q}} & \longrightarrow & 0 \\
 & & \| \wr & & & & & & \\
0 & \longrightarrow & (\mathcal{R}(I)_+)_Q & \longrightarrow & \mathcal{R}(I)_Q & \longrightarrow & S_Q & \longrightarrow & 0
\end{array}$$

where $S_Q = 0$, or $S_Q = S_{\mathbf{p}} = S$. Now since $\mathbf{p} \subset Q$ and $\nu(I) < \dim S$, it follows that $\dim(\mathcal{R}(I)_Q \geq 2$ and hence $\text{depth}(\mathcal{R}(I)_Q) \geq 2$. In this situation the sequences imply $\text{depth}(\text{gr}_I(R))_{\mathbf{q}} \geq 1$. This contradiction shows that $\mathbf{q} \bigcap S/I = \wp/I$.

To prove that $\text{gr}_I(S)$ is a domain, it therefore suffices to show that $\text{gr}_{I_\wp}(S_\wp)$ is a domain. Thus we only need to show that $I_\wp = \wp_\wp$.

From our assumptions on $S(E)$, $Cl(S(E)) = \mathbb{Z}[\mathfrak{m}S(E)]$ (cf. Theorem 5.6.2), and from the way the equality $S(E) = S(I)$ arose, $[\mathfrak{m}S(E)] = [It\mathcal{R}(I)]$ with $S(I) = \mathcal{R}(I) = S[It]$. Therefore $Cl(\mathcal{R}(I)) = \mathbb{Z}[It\mathcal{R}(I)]$. This property is preserved as we localize at $\wp$ to assume that $\wp$ is the maximal ideal of S. But then $\mathbb{Z}[\wp\mathcal{R}(I)] = Cl(\mathcal{R}(I)) = \mathbb{Z}[It\mathcal{R}(I)]$, hence $[\wp\mathcal{R}(I)] = \pm[It\mathcal{R}(I)]$. We now need to show that $\wp = I$.

However, as above it follows that depth $\mathrm{gr}_I(S) > 0$, hence there exists a non–zerodivisor $f \in I \setminus I^2$ whose leading form in $\mathrm{gr}_I(S)$ is regular (after extending the ground field if necessary). But then $(f, ft) = (ft){:}(It) \simeq (It)^*$ ([138]). In this case, the equation $[\wp\mathcal{R}(I)] = \pm[It\mathcal{R}(I)]$ implies $\wp\mathcal{R}(I) \simeq It\mathcal{R}(I)$ or $\wp\mathcal{R}(I) \simeq (f, ft)\mathcal{R}(I)$. Thus there are nonzero elements a, b in $\mathcal{R}(I)$, such that $a\wp\mathcal{R}(I) = bIt\mathcal{R}(I)$ or $a\wp\mathcal{R}(I) = b(f, ft)\mathcal{R}(I)$. Comparing elements of lowest degree it follows that in S, $\wp \simeq I$ or $\wp \simeq (f)$. This implies $\wp = I$, since height $\wp$ = height $I \geq 2$. □

Corollary 7.4.2 *Let E be a module over $R = k[x_1, \ldots, x_d]$ with linear presentation and $\nu(E) = d + \mathrm{rank}\, E - 1$.*

(a) *If E is of linear type, then $C = S(E)[f_1/x_1]$ in the quotient field of $S(E)$. If in addition $S(E)$ is Cohen–Macaulay, then C is also Cohen–Macaulay.*

(b) *If $E_{\mathbf{q}}$ satisfies $\mathcal{F}_2$ for all $\mathbf{q} \neq \mathfrak{m}$ and $S(E)$ is normal, then C is the factorial closure of $S(E)$.*

Proof. Since $I = (\mathbf{f})$ is an S–ideal of linear type it follows that

$$C = S[\mathbf{x}, U]/(\mathbf{x} \cdot B, x_iU - f_i,\ 1 \leq i \leq d)$$

is the extended Rees algebra $S[It, t^{-1}]$ of I. In particular C is a birational extension of $\mathcal{R}(I) = S(E)$.

We may now prove (a). The polynomial ring $S = k[\mathbf{T}]$ is a factor ring of $S(E) = \mathcal{R}(I)$ and hence $\mathrm{rank}_{\mathcal{R}(I)} B \geq \mathrm{rank}_S B = d - 1$. Therefore the equations $\mathbf{f} \cdot B = 0 = \mathbf{x} \cdot B$ in $\mathcal{R}(I)$ imply that the vectors $\mathbf{f}$ and $\mathbf{x}$ are proportional in $\mathcal{R}(I)$. It follows that C maps onto $S(E)[f_1/x_1]$, and hence $C = S(E)[f_1/x_1]$, since both algebras are birational extensions of $S(E)$.

If in addition $S(E)$ is Cohen–Macaulay, then $\mathcal{R}(I)$ is Cohen–Macaulay and therefore $S[It, t^{-1}] = C$ has the same property by Theorem 5.1.9.

To prove part (b) first recall that the factorial closure $B(E)$ of the domain $S(E)$ is the intersection $\bigcap_{\mathbf{q} \subset R} S(E_{\mathbf{q}})$, height $\mathbf{q} = 1$ cf. (7.1). However, by the definition of C, $C_{\mathbf{q}}$ is a factor ring of $S(E_{\mathbf{q}})$ for all primes $\mathbf{q}$ of R with $\mathbf{q} \neq \mathfrak{m}$. Therefore $C_{\mathbf{q}} = S(E_{\mathbf{q}})$ since $S(E)$ embeds into C by part (a). We conclude that $S(E) \subset C \subset B(E)$.

On the other hand, Theorem 7.4.1 implies that $\mathrm{gr}_I(R)$ is a domain, and hence $S[It, t^{-1}]$ is factorial ([125], see also Theorem 7.2.8). Thus C is factorial and it follows that $C = B(E)$. □

Applications to ideals

Corollary 7.4.3 *Let $J \neq \mathfrak{m}$ be an ideal of $R = k[x_1, \ldots, x_d]$ with a linear presentation matrix. Suppose $\nu(J) = d$, and set*

$$g = \text{height } I = \mathrm{codim}(\mathrm{Sing}(\mathcal{R}(J)) \bigcap V(\mathfrak{m})) - 1.$$

(a) *J is set–theoretically generated by $d - g + 1$ homogeneous polynomials.*

(b) *If $J_{\mathfrak{q}}^{(r)} = J_{\mathfrak{q}}^r$ for all r and all $\mathfrak{q} \neq \mathfrak{m}$ and if $\mathcal{R}(J)$ satisfies (S_2), then $C = \mathcal{R}(J)[f_1/x_1]$ is the full symbolic Rees algebra $\oplus_{r\geq 0} J^{(r)}t^r$ of J.*

Proof. Notice that $\mathfrak{m}f_1/x_1 \subset \mathcal{R}(J) \subset R[t]$ and therefore $f_1/x_1 \in R[t]$. Since moreover

$$\mathfrak{m}f_1/x_1 = Rf_1 + \cdots + Rf_d \subset \mathcal{R}(J)$$

it even follows that $f_1/x_1 \in \oplus_{r\geq 0} J^{(r)}t^r$. Therefore C is a graded subalgebra of the symbolic Rees algebra. Furthermore, height $\mathfrak{m}C = \dim C - \dim C/\mathfrak{m}C = g$.

We first prove part (b). By assumption $\mathcal{R}(J) = \mathcal{R}(I)$ satisfies (S_2). Then a simple depth chase using the exact sequences of Theorem 7.4.1 shows that $S[It, t^{-1}] = C$ satisfies (S_2). In particular grade $\mathfrak{m}C \geq 2$. Also by our assumption on J, for every element y in the symbolic Rees algebra, there exists a power $\mathfrak{m}^s$ such that $\mathfrak{m}^s y \subset \mathcal{R}(J) \subset C$. Since grade $\mathfrak{m}C \geq 2$ we conclude that $y \in C$. This proves part (b).

Part (a) will follow from a more general observation (cf. also [47]):

Proposition 7.4.4 *Let J be an arbitrary homogeneous ideal in $R = k[\mathbf{x}]$, and let C be a finitely generated, graded subalgebra of $R[t]$ containing the Rees algebra and contained in the symbolic algebra of J. If $\dim C/\mathbf{x}C = r$, then the closed variety $V(J)$ is set–theoretically defined by r homogeneous equations.*

Proof. Pick homogeneous elements $u_1, \ldots, u_r$ of C so that $k[u_1, \ldots, u_r]$ is a Noether normalization of $C/\mathbf{x}C$. This means that for large s we have

$$C_s = \sum C_{s-r_i}u_i + \mathbf{x}C_s,$$

where degree $u_i = r_i$. By Nakayama's lemma, we may drop the term $\mathbf{x}C_s$ from the equation, from which it follows that

$$J^s \subset \sum Rh_i,$$

with $h_i = u_i t^{-r_i} \in J^{(r_i)} \subset \sqrt{J}$. □

We next drop the linear type requirement. All the other assumptions and notations are still in effect. This result also follows from [58, Theorem 2] in the case k is algebraically closed.

Proposition 7.4.5 *Let $J \neq \mathfrak{m}$ be an ideal in $R = k[x_1, \ldots, x_d]$ satisfying $\mathcal{F}_0$, with a linear presentation matrix and $\nu(J) = d$. Then J is set–theoretically generated by $d - 1$ homogeneous polynomials.*

Proof. Let $\overline{B}$ be the image of B in $\mathcal{R}(J)$. If rank $\overline{B}$ < rank B, then there exists a nonzero minor of B contained in the defining ideal of $\mathcal{R}(J)$, say D, and hence

$$\dim \mathcal{R}(J) \otimes k[\mathbf{x}]/(\mathbf{x}) \leq \dim S/(D) \leq d - 1.$$

Therefore we may assume rank $\overline{B}$ = rank B. Let F = coker B be the dual module of J. Then

$$d + \text{rank } F \leq \dim S(F) = \dim S(J) = d + 1,$$

hence rank $B \geq d - 1$. As $\mathbf{x}\overline{B} = 0$, it follows that rank B = rank $\overline{B} = d - 1$.

Let $\Delta_1, \ldots, \Delta_d$ be the maximal minors of $d-1$ columns of B such that $\overline{\Delta_1}, \ldots, \overline{\Delta_d}$ are not all zero; let $\Delta = \gcd(\Delta_1, \ldots, \Delta_d)$, and write $f_i = \Delta_i/\Delta$. Then $(\mathbf{f}) = S$ or height $(\mathbf{f}) \geq 2$.

Since rank $\overline{B} = d-1$ it follows that $\overline{\mathbf{f}} \cdot \overline{B} = 0$. Then $\mathbf{x} \cdot \overline{B} = 0$ and rank $\overline{B} = d-1$ imply that the vectors $\mathbf{x}$ and $\overline{\mathbf{f}}$ are proportional over $\mathcal{R}(J)$. Let u be an element of the field of fractions of $\mathcal{R}(J)$ with $u\mathbf{x} = \overline{\mathbf{f}}$. This shows that $(\mathbf{f}) \neq S$. Now $u \in J^{(r)}t^r$ for some $1 \leq r \leq d - 1$.

$\mathcal{R}(J)[u]$ is a homomorphic image of $\mathcal{R}(J)[U]/(x_iU - \overline{f_i},\ 1 \leq i \leq d)$. Therefore

$$\dim \mathcal{R}(J)[u] \otimes R/\mathfrak{m} \leq \dim S/(\mathbf{f}) \leq d + 1 - 2.$$

Now use Proposition 7.4.4. □

7.5 Modules with Second Betti Number One

Since the presentations of the modules treated here are already circumscribed, we use the terminology "relations" for the second–order syzygies.

Throughout most of this section, $(R, \mathfrak{m})$ will be a regular local ring of dimension $d \geq 3$. We want to apply our results to minimal resolutions of the form

$$0 \to R \xrightarrow{\psi} R^d \xrightarrow{\varphi} R^n \longrightarrow E \to 0,$$

$$\psi = \begin{bmatrix} x_1 \\ x_2 \\ \vdots \\ x_d \end{bmatrix}$$

with the x_i's forming an R–regular sequence. These assumptions are equivalent to saying that the second Betti number $\beta_2(E)$ is one and that pd $E_{\mathfrak{q}} \leq 1$ for all $\mathfrak{q} \neq \mathfrak{m}$. Later, we shall return to our assumption that $R = k[x_1, \ldots, x_d]$, $\mathfrak{m} = (x_1, \ldots, x_d)$, and ψ as well as φ are matrices with linear entries.

Write $Q = R[\mathbf{T}] = R[T_1, \ldots, T_n]$. We have seen in Proposition 1.5.5 that there exists a skew symmetric $d \times d$ matrix B with entries in Q such that $\mathbf{T} \cdot \varphi = \mathbf{x} \cdot B$. By F we will denote the Q–module with

$$Q^d \xrightarrow{B} Q^d \longrightarrow F \to 0.$$

For a linear presentation matrix φ we could replace Q by $S = k[\mathbf{T}]$ and then $S_R(E) = S_S(F)$ as in Proposition 1.5.1. Although this cannot be expected in the general case, it nevertheless will turn out that the symmetric algebras of E and F are related by a deformation. Here we say that a Noetherian local ring S is a deformation of a Noetherian local ring R if $R \simeq S/(\mathbf{y})$, $\mathbf{y}$ an S–regular sequence.

Odd–dimensional rings

Let $\mathbf{g} = (g_1, \ldots, g_d)$ denote the ideal generated by the $d-1$ sized Pfaffians of B.

Theorem 7.5.1 *Let E be a module over a regular local ring $(R, \mathfrak{m})$ of odd dimension such that $\beta_2(E) = 1$, proj dim $E_{\mathbf{q}} \leq 1$ for all $\mathbf{q} \neq \mathfrak{m}$, and $\nu(E) = d + \text{rank } E - 1$.*

(a) *If E satisfies $\mathcal{F}_0$, then $S(E)$ is a Cohen–Macaulay ring of type 2.*

(b) *If E satisfies $\mathcal{F}_1$, then E is of linear type. Furthermore, $(x_1, g_1)S(E)$ is a canonical module of $S(E)$.*

Proof. From the definition of B it follows that $S(E) = Q/(\mathbf{x} \cdot B)$. Now if E satisfies $\mathcal{F}_0$ then $\dim S(E) = d + \text{rank } E = n+1$ cf. Corollary 1.3.6, and therefore height $(\mathbf{x} \cdot B) = d + n - (n+1) = d - 1$. Because B is an alternating $d \times d$ matrix with homogeneous entries, we then conclude that $(\mathbf{x} \cdot B)$ defines a Cohen–Macaulay ideal of type 2 ([154], the latter assertion holds because $(\mathbf{x} \cdot B)$ is obtained from a deviation 2 perfect ideal by linkage using minimal generators as a regular sequence). This proves part (a).

Now assume that E satisfies $\mathcal{F}_1$. Then $E_{\mathbf{q}}$ is of linear type for $\mathbf{q} \neq \mathfrak{m}$ because pd $E_{\mathbf{q}} \leq 1$. Therefore any R–torsion element in $S(E)$ is annihilated by a power of $\mathfrak{m}$. On the other hand, height $\mathfrak{m}S(E) = 1$ and $S(E)$ is Cohen–Macaulay. Hence $\mathfrak{m}S(E)$ contains a nonzero divisor on $S(E)$. It follows that $S(E)$ is R–torsion-free and hence a domain.

In order to show the claim concerning the canonical module, we replace $\mathbf{x}$ and B by generic matrices $\widetilde{\mathbf{x}}$ and $\widetilde{B}$. Write $\widetilde{\mathbf{h}} = \widetilde{\mathbf{x}} \cdot \widetilde{B}$, then the ideals $(\widetilde{\mathbf{x}} \cdot \widetilde{B})$ and $(\widetilde{x_1}, \widetilde{g_1}, \widetilde{h_2}, \ldots, \widetilde{h_d})$ are geometrically linked with respect to the regular sequence $\widetilde{h_2}$, ..., $\widetilde{h_d}$ ([153]). It follows that $(\widetilde{x_1}, \widetilde{g_1}, \widetilde{\mathbf{x}} \cdot \widetilde{B})/(\widetilde{\mathbf{x}} \cdot \widetilde{B})$ is a canonical ideal of the Cohen–Macaulay ring defined by $(\widetilde{\mathbf{x}} \cdot \widetilde{B})$. Now we are able to specialize to $(\mathbf{x} \cdot B)$ as long as $(x_1, g_1, \mathbf{x} \cdot B)/(\mathbf{x} \cdot B)$ has height at least one. But the latter condition is obviously satisfied because $(x_1, g_1)S(E) \neq 0$ and $S(E)$ is a domain. □

Theorem 7.5.2 *Let E be as in* Theorem 7.5.1, *let $F =$* coker *B over the ring $Q = R[\mathrm{T}]$, and write M and N for the irrelevant maximal ideals of $S_R(E)$ and $S_Q(F)$ respectively.*

(a) *If E satisfies $\mathcal{F}_0$, then* rank $_Q(F) = 1$ *and the algebra $(S_Q(F))_N$ is a deformation of $(S_R(E))_M$.*

(b) *If E satisfies $\mathcal{F}_1$, then $F \simeq (\mathbf{g})Q$, and $(\mathbf{g})Q$ is a grade 3 Gorenstein ideal of linear type.*

Proof. Let $\mathbf{w} = w_1, \ldots, w_d$ be d new variables over Q, then $S_Q(F) \simeq Q[\mathbf{w}]/(\mathbf{w} \cdot B)$. From this presentation we see that modulo the sequence $w_1 - x_1, \ldots, w_d - x_d$, the algebra $(S_Q(F))_N$ specializes to $(S_R(E))_M$. In particular,

$$\dim S_Q(F) = \dim(S_Q(F))_N \leq \dim(S_R(E))_M + d = n + 1 + d$$

and we conclude that height $(\mathbf{w} \cdot B) \geq d-1$. Hence $(\mathbf{w} \cdot B)$ is a perfect ideal of height $d-1$ (*e.g.* [154]). Therefore

$$\dim(S_Q(F))_N = n+1+d = \dim(S_R(E))_M + d.$$

It follows that $w_1 - x_1, \ldots, w_d - x_d$ form part of a system of parameters of the Cohen–Macaulay ring $(S_Q(F))_N$. Hence

$$(S_Q(F))_N/(w_i - x_i,\ 1 \leq i \leq d) \simeq (S_R(E))_M$$

with $w_1 - x_1, \ldots, w_d - x_d$ a regular sequence.

To show that rank $F = 1$, we first notice that rank $F \geq 1$ by the skew–symmetry of B. On the other hand,

$$\operatorname{rank} F \leq \dim S_Q(F) - \dim Q = n+1+d-(n+d) = 1.$$

This finishes the proof of the first part.

Now assume that E satisfies $\mathcal{F}_1$. Then by Theorem 7.4.1.b, $(S_R(E)_M)$ is a domain, and therefore by part (a), $(S_Q(F))_N$ and hence $S(F)$ is a domain. Since moreover rank $F = 1$, it follows that F is an ideal of linear type. Thus F satisfies $\mathcal{F}_1$ and in particular, height $I_{d-2}(B) \geq 3$. From the skew–symmetry we have that

$$\text{height } I_{d-2}(B) = \text{height } I_{d-1}(B) = \text{height } (\mathbf{g})$$

([36]), so that height $(\mathbf{g}) = 3$. Then $(\mathbf{g}) \simeq$ coker B, and hence $(\mathbf{g}) \simeq F$. □

As an illustration let x_1, x_2, x_3 be a regular sequence in a three–dimensional regular local ring $(R, \mathfrak{m})$, let

$$\Lambda = \left(\begin{array}{c|c} x_1 & \\ x_2 & \star \\ x_3 & \end{array} \right)$$

be a 3×4 matrix with entries in $\mathfrak{m}$ such that height $I_3(\Lambda) \geq 2$, and let J be the ideal generated by all 3×3 minors of Λ fixing the first column of Λ. Then $E = J$ satisfies the assumptions of Theorems 7.5.1.a and 7.5.2.a. If in addition height $I_2(\Lambda) \geq 3$, then Theorems 7.5.1.b and 7.5.2.b apply too.

Now define $C = Q[U]/(\mathbf{x} \cdot B, x_i U - g_i, 1 \leq i \leq d)$. If the module E happens to have a linear resolution over $R = k[\mathbf{x}] = k[x_1, \ldots, x_d]$, then we may choose the matrix B to be linear with entries in $S = k[\mathbf{T}] = k[T_1, \ldots, T_n]$. Assuming that E satisfies $\mathcal{F}_1$ we conclude from Theorem 7.5.2.b that coker $B \simeq (\mathbf{g})S$ and that $(\mathbf{g})S$ is a grade 3 Gorenstein ideal of linear type. In particular, in the notation of section 2, $I = (\mathbf{g})S, \mathbf{f} = \mathbf{g}$, and the algebra C we defined there coincides with the algebra C we introduced just now. Furthermore, the image of $(\mathbf{g})Q$ in $Q \otimes_R R/\mathfrak{m}$ can be identified with $(\mathbf{g})S$, and hence is an ideal of height 3. In particular, two of the assumptions for the next result are automatically satisfied in the linear case.

Corollary 7.5.3 *Let E be as in* Theorem 7.5.1, *and assume that E satisfies $\mathcal{F}_1$.*

(a) *If* $\mathbf{g} \not\subset \mathfrak{m}Q$, *then* $C = S(E)[g_1/x_1]$ *in the quotient field of* $S(E)$. *Furthermore* C *is Gorenstein.*

(b) *The following are equivalent:*

(i) $E_{\mathbf{q}}$ *satisfies* $\mathcal{F}_2$ *for all* $\mathbf{q} \neq \mathfrak{m}$ *and the image of* $\mathbf{g}$ *in* $Q \otimes_R R/\mathfrak{m}$ *generates an ideal of height at least* 2.

(ii) C *is the factorial closure of* $S(E)$.

Proof. It was shown in Theorem 7.5.1.b that $S(E)$ is a domain. To prove part (a) recall that by Theorem 7.5.2.b, $(\mathbf{g})Q$ is a grade 3 Gorenstein ideal of linear type. Therefore the extended Rees algebra $Q[\mathbf{g}t, t^{-1}]$ of this ideal is a Gorenstein domain which, in the notation of the proof of Theorem 7.5.2, can be described as

$$Q[\mathbf{w}, U]/(\mathbf{w} \cdot B, w_i U - g_i, 1 \leq i \leq d)$$

(e.g. [140, Theorem 2.2]). Furthermore, for every maximal ideal N of $Q[\mathbf{g}t, t^{-1}]$ containing $\mathfrak{m}$, $\dim Q[\mathbf{g}t, t^{-1}]_N = d + n + 1$. On the other hand, the defining ideal of C in $Q[U]$ properly contains the prime ideal $(\mathbf{x} \cdot B)Q[U]$ of height $d-1$. Hence for every maximal ideal M of C, the defining ideal of C_M has height at least d. Now it follows as in the proof of Theorem 7.5.2 that for a suitable maximal ideal N of the extended Rees algebra, $Q[\mathbf{g}t, t^{-1}]_N$ is a deformation of C_M. In particular C_M is Gorenstein for every maximal ideal M, and therefore C is Gorenstein. Moreover $\dim C_M = d + n + 1 - d = n + 1$ for every maximal ideal M containing $\mathfrak{m}$. From this we conclude that

$$\text{height}\, \mathfrak{m}C = n + 1 - \dim C/\mathfrak{m}C = \text{height}(\mathbf{g})(Q \otimes_R R/\mathfrak{m}) > 0,$$

and therefore grade $\mathfrak{m}C > 0$.

To show the equality $C = S(E)[g_1/x_1]$, we proceed as in the proof of Corollary 7.4.2.a. Since rank ${}_{S(E)}B \geq$ rank ${}_{Q \otimes R/\mathfrak{m}}B \geq d - 1$, the equations $\mathbf{g} \cdot B = 0 = \mathbf{x} \cdot B$ in $S(E)$ imply that the vectors $\mathbf{g}$ and $\mathbf{x}$ are proportional over $S(E)$. Therefore C maps onto $S(E)[g_1/x_1]$. Let Γ be the kernel of this map. For every prime $\mathbf{q}$ of R with $\mathbf{q} \neq \mathfrak{m}$, $S(E_{\mathbf{q}})$ maps onto $C_{\mathbf{q}}$ and hence $\Gamma_{\mathbf{q}} = 0$. Thus given an element $y \in \Gamma$, we have $\mathfrak{m}^s y = 0$ for some s. Therefore $(\mathbf{m}C)^s y = 0$, and hence $y = 0$ since grade $\mathfrak{m}C > 0$. This finishes the proof of part (a).

To show part (b), one first concludes as in the proof of Corollary 7.4.2.b that C is contained in the factorial closure $B(E)$. Then by Propositions 7.1.1 and 7.1.4 there exists an ideal K of height 2 in R such that for $\mathbf{q} \in \text{Spec}(R)$, $C_{\mathbf{q}} = B(E_{\mathbf{q}})$ if and only if grade $KC_{\mathbf{q}} \geq 2$. Since C is Cohen–Macaulay we may replace "grade" by "height".

So assume that (i) holds and let $\wp \in \text{Spec}(C)$ be a minimal prime of KC. If $\wp \bigcap R = \mathbf{q} \neq \mathfrak{m}$, then $S(E_{\mathbf{q}})$ is factorial because pd $E_{\mathbf{q}} \leq 1$ and $E_{\mathbf{q}}$ satisfies $\mathcal{F}_2$ ([260, Remark, p. 346]); therefore $S(E_{\mathbf{q}}) = C_{\mathbf{q}} = B(E_{\mathbf{q}})$ and hence height $\wp =$ height $\wp_{\mathbf{q}} \geq 2$. If on the other hand $\wp \bigcap R = \mathfrak{m}$, then $\mathfrak{m}C \subset \wp$ and we obtain that

$$\text{height}\, \wp \geq \text{height}\, \mathfrak{m}C = n + 1 - \dim C/\mathfrak{m}C = \text{height}\,(\mathbf{g})(Q \otimes_R R/\mathfrak{m}) \geq 2.$$

Conversely, assume that (ii) holds, then for $\mathbf{q} \neq \mathfrak{m}$, $S(E)_{\mathbf{q}} = C_{\mathbf{q}}$ is factorial and hence $E_{\mathbf{q}}$ satisfies $\mathcal{F}_2$ by Proposition 6.1.2. Furthermore, by the preceding,

$$\text{height}\,(\mathbf{g})(Q \otimes_R R/\mathfrak{m}) = \text{height}\,\mathfrak{m}C \geq \text{height}\,KC \geq 2. \ \square$$

We want to point out that if E has a linear resolution over $k[x_1, \ldots, x_d]$ and satisfies the assumptions of Corollary 7.5.3.b.i, then $S(E)$ is normal, because

$$\text{codim}\,(\text{Sing}(S(E)) \cap V(\mathfrak{m})) = \text{height}\,I + 1 \geq 2.$$

Even–dimensional rings

Using the assumptions of the beginning of the section, we define $\overline{Q} = Q/(Pf(B))$, $\overline{F} = F \otimes_Q \overline{Q}$, and $D = S_R(E) \otimes_Q \overline{Q} = Q/(\mathbf{x} \cdot B, Pf(B))$.

Theorem 7.5.4 *Let E be a module over a regular local ring $(R, \mathfrak{m})$ of even dimension such that $\beta_2(E) = 1$, pd $E_{\mathbf{q}} \leq 1$ for all $\mathbf{q} \neq \mathfrak{m}$, and $\nu(E) = \dim R + \text{rank}\, E - 1$.*

(a) *If E satisfies $\mathcal{F}_0$, then D is a Gorenstein ring.*

(b) *If E satisfies $\mathcal{F}_1$, then D is the Rees algebra of E.*

Proof. The proof proceeds as the proof of Theorem 7.5.1. For part (a), one uses the fact that $(\mathbf{x} \cdot B, Pf(B))$ is a Gorenstein ideal if B is an alternating matrix of even size and if height $(\mathbf{x} \cdot B, Pf(B)) \geq d - 1$ ([153, Proposition 5.9]). To prove part (b), one shows as in the proof of Theorem 7.5.1.b that D is a domain. Furthermore, $(\mathbf{x})$ annihilates $Pf(B)$ in $S(E)$, and therefore $Pf(B)$ is contained in the R–torsion of $S(E)$. It follows that D is the Rees algebra of E. $\square$

Theorem 7.5.5 *Let E be as in* Theorem 7.5.4, *and write M and N for the irrelevant maximal ideals of D and $S_{\overline{Q}}(\overline{F})$ respectively.*

(a) *If E satisfies $\mathcal{F}_0$, then $(S_{\overline{Q}}(\overline{F}))_N$ is a deformation of D_M.*

(b) *If E satisfies $\mathcal{F}_1$, then $\overline{Q}$ is a domain and $\overline{F}$ is a rank 2 torsion–free $\overline{Q}$–module of linear type.*

Proof. Let $\mathbf{w} = w_1, \ldots, w_d$ be d new variables over $\overline{Q}$, then $S_{\overline{Q}}(\overline{F}) = \overline{Q}[\mathbf{w}]/(\mathbf{w} \cdot B)$. Now it follows as in the proof of Theorem 7.5.2.a that D_M can be obtained from $S_{\overline{Q}}(\overline{F}))_N$ by factoring out the regular sequence $w_1 - x_1, \ldots, w_d - x_d$.

Theorem 7.5.4.b and the preceding imply that $S_{\overline{Q}}(\overline{F})$ is a domain. Therefore $\overline{Q}$ is a domain and $\overline{F}$ is a torsion–free $\overline{Q}$–module of linear type having a rank. This rank has to be at least two because $Pf(\overline{B}) = 0$ and the skew–symmetry of $\overline{B}$. On the other hand,

$$\text{rank}_{\,\overline{Q}}(\overline{F}) \leq \dim S_{\overline{Q}}(\overline{F}) - \dim \overline{Q} = n + 1 + d - (n + d - 1) = 2. \qquad \square$$

As an example, let $E = J$ be an ideal in $R = k[x_1, \dots, x_4]$, without $(x_1, \dots, x_4)$–primary component, defining an arithmetically Buchsbaum curve in $\mathbb{P}^3_k$ which is locally Cohen–Macaulay but not arithmetically Cohen–Macaulay. Further assume that $\nu(J) = 4$ and that J satisfies $\mathcal{F}_0$ or $\mathcal{F}_1$ respectively; then Theorems 7.5.1 and 7.5.2 apply to $E = J$. To see this it suffices to show that $\beta_2(J) = 1$. However, the assumption $\nu(J) = 4$ implies that the Buchsbaum invariant $i(R/J)$ of R/J is one ([25, Lemma 2]), and hence $H^1_{\mathfrak{m}}(R/J) \simeq k$ ([268, Proposition 2.6]), and therefore by local duality, $\operatorname{Ext}^3_R(R/J, R) \simeq k$. In particular, $\beta_2(J) = 1$. It should be mentioned that the above class of curves includes any curve in $\mathbb{P}^3_k$ which is geometrically linked to two skew lines ([22, Theorem 1] and [252, §5]).

Corollary 7.5.6 *Let E be as in* Theorem 7.5.4, *and assume that E satisfies $\mathcal{F}_1$. The following are equivalent:*

(i) *$E_{\mathfrak{q}}$ satisfies $\mathcal{F}_2$ for all $\mathfrak{q} \neq \mathfrak{m}$ and $Pf(B) \notin \mathfrak{m}Q$.*

(ii) *D is the factorial closure of $S(E)$.*

Proof. By Theorem 7.5.4.b, D is the Rees algebra of E, and hence D is contained in the factorial closure of $S(E)$. Now the assertion follows as in the proof of Corollary 7.5.3.b, because D is Cohen–Macaulay by Theorem 7.5.4.a, and height $\mathfrak{m}D =$ height $Pf(B)(Q \otimes_R R/\mathfrak{m}) + 1$. □

As before, the assumption $Pf(B) \notin \mathfrak{m}Q$ in the above corollary is automatically satisfied if E has a linear resolution over $R = k[x_1, \dots, x_d]$.

Applications to ideals

We wish to consider the case where $E = J$ is an ideal of R. One may assume that $d \geq 4$ and that height $J \geq 2$, which forces J to be a non–perfect ideal of height 2.

We first point out

Lemma 7.5.7 *If J has a linear resolution over $R = k[x_1, \dots, x_d]$, then J is generated by polynomials of degree $d-2$, and R/J is a Buchsbaum ring of multiplicity $\frac{d(d-3)}{2}$.*

Proof. The generators of J are proportional to the maximal minors of a $d \times (d-1)$ submatrix of φ, and hence they all have the same degree, say s; then the Hilbert series of R/J is

$$\frac{1 - dt^s + dt^{s+1} - t^{s+2}}{(1-t)^d}.$$

Since $\dim R/J = d-2$, it follows that $t = 1$ is a pole of order $d-2$ of this function, and therefore $s = d-2$. The Buchsbaum property of R/J follows from [268, Proposition 2.12]. □

We now have to distinguish between the odd and the even cases. We treat the odd case first.

For odd d, examples of ideals J satisfying the above conditions can be constructed in the following way. One starts with a $d \times d$ skew–symmetric matrix B with linear entries in $k[\mathbf{T}] = k[T_1, \ldots, T_d]$ with the growth condition $\mathcal{F}_1$ on its Fitting ideals to cover the linear type requirement. To find J explicitly, after φ has been determined by the formalism of the dual module, J is the ideal generated by the vector generating the first syzygy module of the transpose of φ. Notice that by the skew–symmetry of B, $0 = \mathbf{x} \cdot B \cdot \mathbf{x}^t = \mathbf{T} \cdot \varphi \cdot \mathbf{x}^t$ and therefore $\varphi \cdot \mathbf{x}^t = 0$. In other words, J satisfies the assumptions from the beginning of the section.

If, for instance, $d = 5$, it suffices to pick a 5×5 matrix B with linear entries whose Pfaffians generate an ideal of height 3 and whose entries generate the irrelevant maximal ideal of R. In this case J is generated by 5 cubics (*cf.* Lemma 7.5.7). For an explicit example, consider the matrix

$$\begin{bmatrix} 0 & -T_1 & -T_2 & -T_3 & -T_4 \\ T_1 & 0 & -T_5 & -T_1 - T_2 & -T_2 - T_3 \\ T_2 & T_5 & 0 & -T_3 - T_4 & -T_4 - T_5 \\ T_3 & T_1 + T_2 & T_3 + T_4 & 0 & -T_1 + T_3 \\ T_4 & T_2 + T_3 & T_4 + T_5 & T_1 - T_3 & 0 \end{bmatrix}$$

The resulting ideal J is prime, and a complete intersection on the punctured spectrum.

We now return to the situation described earlier in the section, when the algebras C and D were defined.

Corollary 7.5.8 *Let J be an ideal in a regular local ring $(R, \mathfrak{m})$ of odd dimension d such that $(R/J)_\mathfrak{q}$ is Cohen–Macaulay for $\mathfrak{q} \neq \mathfrak{m}$, $\beta_2(J) = 1$, $\nu(J) = d$, and J satisfies $\mathcal{F}_1$.*

(a) *If the image of $\mathbf{g}$ in $Q \otimes_R R/\mathfrak{m}$ generates an ideal of height at least 3, then J is set–theoretically generated by $d - 2$ elements.*

(b) *The following are equivalent:*

(i) *$\nu(J_\mathfrak{q}) \leq \max\{2, \dim R_\mathfrak{q} - 1\}$ for $J \subset \mathfrak{q} \neq \mathfrak{m}$, and the image of $\mathbf{g}$ in $Q \otimes_R R/\mathfrak{m}$ generates an ideal of height at least 2.*

(ii) *$C = \mathcal{R}(J)[g_1/x_1]$ is the full symbolic Rees algebra $\oplus_{r \geq 0} J^{(r)} t^r$ of J.*

Proof. First notice that under assumption (b.i), $J_\mathfrak{q}^{(r)} = J_\mathfrak{q}^r$ for all r and $\mathfrak{q} \neq \mathfrak{m}$ ([12, Theorem 3.5], [260, Theorem 5.2]). Furthermore from Corollary 7.5.3.a and its proof we know that

$$\text{grade } \mathfrak{m}C = \text{height } \mathfrak{m}C = \text{height } (\mathbf{g})(Q \otimes_R R/\mathfrak{m}).$$

Now one shows as in the proof of Corollary 7.4.3 that part (a) holds and that (i) implies (ii) in part (b).

In order to prove the reverse implication, we assume that (ii) holds. Then for $J \subset \mathfrak{q} \neq \mathfrak{m}$, $C_\mathfrak{q} = \mathcal{R}(J_\mathfrak{q})[g_1/x_1]$ is the symbolic Rees algebra of $J_\mathfrak{q}$ and hence the analytic spread $\ell(J_{\mathfrak{q}})$ is at most $\max\{2, \dim R_{\mathfrak{q}} - 1\}$ (*e.g.* [40, Corollary]). On the other hand, $\nu(J_{\mathfrak{q}}) = \ell(J_{\mathfrak{q}})$ since $J_{\mathfrak{q}}$ is of linear type (*e.g.* [260, Theorem 3.4]). Again using the assumption (ii) it follows from [290, Proposition 1.2.1 and 1.3] that grade $\mathfrak{m}C \geq 2$. Hence we also obtain height $(\mathbf{g})(Q \otimes_R R/\mathfrak{m}) = \text{grade } \mathfrak{m}C \geq 2$. □

In the next result, $\ell(J)$ will again denote the analytic spread of J.

Corollary 7.5.9 *Let J be an ideal in a regular local ring $(R, \mathfrak{m})$ of even dimension d such that $(R/J)_{\mathfrak{q}}$ is Cohen–Macaulay for $\mathfrak{q} \neq \mathfrak{m}$, $\beta_2(J) = 1$, $\nu(J) = d$, and J satisfies $\mathcal{F}_1$.*

(a) *If $Pf(B) \notin \mathfrak{m}Q$, then J is set–theoretically generated by $d - 1$ elements.*

(b) *The following are equivalent:*

  (i) $\nu(J_{\mathfrak{q}}) \leq \max\{2, \dim R_{\mathfrak{q}} - 1\}$ *for* $J \subset \mathfrak{q} \neq \mathfrak{m}$, *and* $Pf(B) \notin \mathfrak{m}Q$.

  (ii) $\ell(J_{\mathfrak{q}}) \leq \max\{2, \dim R_{\mathfrak{q}} - 1\}$ *for* $J \subset \mathfrak{q}$.

  (iii) *The Rees algebra and the symbolic Rees algebra of J coincide.*

Proof. It follows from the description of the Rees algebra in Theorem 7.5.4.b that $Pf(B) \notin \mathfrak{m}Q$ if and only if $\ell(J) \leq d-1$. This proves part (a) and also the equivalence of (i) and (ii) in part (b) once we keep in mind that $J_{\mathfrak{q}}$ is of linear type for $J \subset \mathfrak{q} \neq \mathfrak{m}$. Moreover, (iii) clearly implies (ii) (*e.g.* [40, Corollary]). The reverse implication follows as in the proof of Corollary 7.5.8 because grade $\mathbf{m}D$ = height $\mathfrak{m}D = 2$, by the proof of Corollary 7.5.6. □

In Corollaries 7.5.8 and 7.5.9 we needed to impose conditions on the heights of the ideals generated by the images of $\mathbf{g}$ or $Pf(B)$ in $Q \otimes_R R/\mathfrak{m}$. These assumptions are automatically satisfied if J has a linear resolution over $R = k[x_1, \ldots, x_d]$. Furthermore in this case, the set–theoretic generation of J can be achieved by homogeneous polynomials.

As an example, consider the ideal

$$J = (x_1^2, x_1x_2, x_2^2, x_1x_4 - x_2x_3)$$

in $R = k[x_1, \ldots, x_4]$ which defines a multiplicity 2 scheme structure on the line $x_1 = x_2 = 0$ in $\mathbb{P}^3_k$ (cf. [252, §4]). Now Theorems 7.5.4, 7.5.5 and Corollary 7.5.9 (parts (a) and (b.iii)) apply to $E = J$.

For another illustration, let J be defined by

$$\begin{bmatrix} x_2 & -x_1 + x_4 & 0 & -x_2 \\ x_3 & x_4 & -x_1 & -x_2 \\ x_4 & 0 & x_4 & -x_1 - x_3 \\ 0 & x_3 & -x_2 + x_4 & -x_3 \end{bmatrix}.$$

Then J is a prime ideal satisfying the assumptions of Theorems 7.5.4, 7.5.5 and Corollary 7.5.9 (parts (a) and (b.iii)).

Finally let J be the defining (prime) ideal of any monomial arithmetically Buchsbaum curve in $\mathbb{P}^3_k$ (k an algebraically closed field) which is not arithmetically Cohen–Macaulay. Then J does not have a linear presentation matrix, however by [25, Theorem 3], J satisfies all the assumptions of Theorems 7.5.4, 7.5.5 and Corollaries 7.5.9.a and 7.5.9.b.iii (cf. also [246] and [130, Example 4.4]).

It would be interesting to know when the bound (if not sharper) in Corollaries 7.5.8.a and 7.5.9.a is attained, at least when the ideal J in $R = k[x_1, \ldots, x_d]$ has a resolution such as

$$0 \to R^p \xrightarrow{\psi} R^d \xrightarrow{\varphi} R^n \longrightarrow J \to 0,$$

with $p \geq 2$, and the rows of φ are linear syzygies of the variables $\mathbf{x} = x_1, \ldots, x_d$.

If the ideal J is of linear type (or maybe just satisfies $\mathcal{F}_1$), the dimension of $S(J)$ is $d+1$, and the rank of the dual matrix B is $d-p$. This means that in the complex

$$0 \to C \longrightarrow k[\mathbf{T}]^d \xrightarrow{B} k[\mathbf{T}]^d \longrightarrow F \to 0,$$

C has rank p.

Finally note that since the cokernel F of B is torsion–free, its minors of order $d-p$ generate an ideal of height at least 2. But F satisfies the condition $\mathcal{F}_1$, so height $I_{d-p-1}(B) \geq 3$. Since B is skew symmetric, these ideals actually have the same radical [36, Corollary 2.6].

Somehow these conditions should be used to improve the dimension of the special fiber of some subring of the symbolic Rees algebra. Here is a simple example to test whether this is possible.

Example 7.5.10 The following ideal ($p = 2$)

$$\begin{array}{l} x_1^2 + x_1x_2 + x_5x_6 - x_6^2 \\ x_1x_2 + x_2^2 - x_5^2 + x_5x_6 \\ x_1x_3 + x_2x_3 + x_4x_5 - x_4x_6 \\ x_1x_4 + x_2x_4 + x_3x_5 - x_3x_6 \\ x_1x_5 + x_2x_6 \end{array}$$

is of linear type, its symmetric algebra is Cohen–Macaulay, but the ideal is not prime. In fact, at this size (*i.e.* $d = 6$, $p = 2$), the ideal J cannot be prime.

The Vetter modules

Given a polynomial ring $R = k[\mathbf{x}] = k[x_1, \ldots, x_d]$ with $d \geq 4$, recall the modules of Example 7.5.11: They have a resolution

$$0 \to R \xrightarrow{\psi} R^d \xrightarrow{\varphi} R^{2d-3} \longrightarrow E \to 0$$

with φ a linear matrix and $\psi = \begin{bmatrix} x_1 \\ \vdots \\ x_d \end{bmatrix}$. Now all earlier results of this section apply to these modules.

Corollary 7.5.11 *Let E be a Vetter module over $k[x_1, \ldots, x_d]$.*

(a) *If d is odd, then $S(E)$ is a normal Cohen–Macaulay domain of type 2 with canonical module $(x_1, g_1)S(E)$, and*

$$C = k[\mathbf{x}, \mathbf{T}, U]/(\mathbf{x} \cdot B, x_iU - g_i, \ 1 \le i \le d)$$

is a Cohen–Macaulay domain and the factorial closure of $S(E)$.

(b) *If d is even, then $k[\mathbf{x}, \mathbf{T}]/(\mathbf{x} \cdot B, Pf(B))$ is a Cohen–Macaulay domain and the factorial closure of $S(E)$.*

7.6 Analytically Independent Elements

In this section we will present yet another attempt to generalize the concept of a dual module to the case where the module E does not have linear presentation and is not necessarily of linear type. It will lead to a negative answer to a question of G. Valla (cf. [104]).

Modules with near linear presentation

Throughout this section, $R = k[\mathbf{x}] = k[x_1, \ldots, x_d]$, $\mathfrak{m} = (\mathbf{x})$, E is a finitely generated graded R–module with all its generators in the same degree, and $n = \nu(E)$. Then the presenting matrix φ of E can be decomposed in the form $\varphi = (\varphi_1 \mid \varphi_2)$ where the entries of φ_1 are linear and the entries of φ_2 are homogeneous of degree at least two. Further write $S = k[\mathbf{T}] = k[T_1, \ldots, T_n]$, and let B_1 be a matrix with linear entries in S such that $\mathbf{T} \cdot \varphi_1 = \mathbf{x} \cdot B_1$. Of course $S_R(\text{coker } \varphi_1) \simeq S_S(\text{coker } B_1)$. Now if $S(E)_{\mathfrak{m}S(E)}$ is a discrete valuation ring, then rank $B_1 = d - 1$, by the Jacobian criterion of Proposition 5.4.1, and hence we have an exact sequence

$$0 \to S \xrightarrow{\begin{bmatrix} f_1 \\ \vdots \\ f_d \end{bmatrix}} S^d \xrightarrow{B_1^t} \cdots .$$

Write $I = (\mathbf{f}) = (f_1, \ldots, f_d)$. Then height $I \ge 2$ and I is isomorphic to the cokernel of B_1 modulo its S–torsion. Observe that, provided $\varphi = \varphi_1$ and E is of linear type, I gives back the ideal defined in section 2. If one takes f_i to be the maximal minors of a suitable $d \times (d-1)$ submatrix of B_1 divided by their greatest common divisor, one sees that the f_i are homogeneous of the same degree $\delta \le d - 1$, where $\delta = d - 1$ if and only if B_1 is a $d \times (d-1)$ matrix and height $I_{d-1}(B_1) = 2$.

Denote by $\varphi_1(\mathbf{f})$ the matrix obtained from φ_1 as we replace x_i by f_i, and let $\ell(I)$ stand for the analytic spread of I.

Lemma 7.6.1 *Let E be a graded module over $R = k[x_1, \ldots, x_d]$ with all its generators in the same degree such that $n = d + \text{rank } E - 1$ and $\text{rank } B_1 = d - 1$ (the latter equality holds if $S(E)_{\mathfrak{m}S(E)}$ is a discrete valuation ring).*

(a) *If $\text{rank } \varphi_1(\mathbf{f}) \leq d - 2$, then $\ell(I) \leq \dim R/I_{d-1}(\varphi_1)$.*

(b) *If $\text{rank } \varphi_1(\mathbf{f}) \geq d - 1$, then $f_1, \ldots, f_d$ are analytically independent.*

Proof. To prove part (a), let X be a new variable over $k[\mathbf{x}, \mathbf{T}]$. Then $\text{rank } \varphi_1(\mathbf{f}) \leq d - 2$ if and only if $\text{rank } \varphi_1(f_1X, \ldots, f_dX) \leq d - 2$. But the latter condition is equivalent to

$$I_{d-1}(\varphi_1) \subset \{(x_i - f_iX \mid 1 \leq i \leq d)k[\mathbf{x}, \mathbf{T}, X]\} \bigcap k[\mathbf{x}].$$

On the other hand

$$\mathcal{R}(I) = S[IX] \simeq S[\mathbf{x}]/\{(x_i - f_iX \mid 1 \leq i \leq d)S[\mathbf{x}, X]\} \bigcap S[\mathbf{x}],$$

and hence

$$\begin{aligned} \mathcal{R}(I) \otimes_S k &\simeq k[\mathbf{x}, \mathbf{T}]/\{(x_i - f_iX \mid 1 \leq i \leq d)S[\mathbf{x}, X]\} \bigcap k[\mathbf{x}, \mathbf{T}] + (\mathbf{T}) \\ &= k[\mathbf{x}, \mathbf{T}]/\{(x_i - f_iX \mid 1 \leq i \leq d)S[\mathbf{x}, X]\} \bigcap k[\mathbf{x}] + (\mathbf{T}) \\ &\simeq k[\mathbf{x}]/\{(x_i - f_iX \mid 1 \leq i \leq d)k[\mathbf{x}, \mathbf{T}, X]\} \bigcap k[\mathbf{x}]. \end{aligned}$$

Thus we have seen that $\text{rank } \varphi_1(\mathbf{f}) \leq d - 2$ if and only if $I_{d-1}(\varphi_1)$ is contained in the defining ideal of $\mathcal{R}(I) \otimes_S k$. Now part (a) follows since $\ell(I) = \dim \mathcal{R}(I) \otimes_S k$.

To prove part (b), we will show that the identity map on $k[\mathbf{T}, \mathbf{x}]$ induces an isomorphism between the Rees algebras $\mathcal{R}_S(I)$ and $\mathcal{R}_R(E)$ of I and E respectively. Then

$$\mathcal{R}_S(I)/(\mathbf{T})\mathcal{R}_S(I) \simeq \mathcal{R}_R(E)/(\mathbf{T})\mathcal{R}_R(E) \simeq R,$$

and therefore $\ell(I) = \dim \mathcal{R}_S(I) \otimes_S k = \dim R = d$.

We first show that there is a natural epimorphism

$$\Phi \colon \mathcal{S}_R(E) \longrightarrow \mathcal{R}_S(I) = S[IX].$$

To do so we have to prove that $\mathbf{T} \cdot \varphi(f_1X, \ldots, f_dX) = 0$, or equivalently that $\mathbf{T} \cdot \varphi(\mathbf{f}) = 0$. Notice that

$$\mathbf{T} \cdot \varphi_1(\mathbf{f}) = \mathbf{f} \cdot B_1(\mathbf{f}) = \mathbf{f} \cdot B_1 = 0,$$

since $B_1^t \cdot \mathbf{f}^t = 0$. On the other hand,

$$d - 1 \leq \text{rank } \varphi_1(\mathbf{f}) \leq \text{rank } \varphi(\mathbf{f}) \leq \text{rank } \varphi = d - 1,$$

where the latter equality follows from the fact that $n = d + \text{rank } E - 1$. Therefore rank $\varphi_1(\mathbf{f}) = \text{rank } \varphi(\mathbf{f})$, and hence $\mathbf{T} \cdot \varphi_1(\mathbf{f}) = 0$ implies that $\mathbf{T} \cdot \varphi(\mathbf{f}) = 0$.

Now that the map Φ is defined we prove that $\Phi(\Gamma) = 0$, where Γ is the R–torsion of $S_R(E)$. Since rank $\varphi = d - 1$, it follows that $(I_{d-1}(\varphi))^r \subset \text{ann}_R(\Gamma)$ for some r. However, we have seen that $I_{d-1}(\varphi_1) \not\subset (\mathbf{T})\mathcal{R}_S(I)$, in particular $I_{d-1}(\varphi) \neq 0$ in the domain $\mathcal{R}_S(I)$. Therefore $I_{d-1}(\varphi)^r \Phi(\Gamma) = 0$ implies $\Phi(\Gamma) = 0$. Now Φ induces an epimorphism from $\mathcal{R}_R(E) \simeq S_R(E)/\Gamma$ onto $\mathcal{R}_S(I)$, which is an isomorphism since both algebras are domains of the same dimension $d + \text{rank } E = n + 1$. □

Theorem 7.6.2 *Let E be a graded module over $R = k[x_1, \ldots, x_d]$ with all its generators in the same degree such that $n = d + \text{rank } E - 1$, $\text{rank } B_1 = d - 1$, and* rank $\varphi_1(\mathbf{f}) \geq d - 1$.

(a) *The identity on $k[\mathbf{x}, \mathbf{T}]$ induces an isomorphism between the Rees algebras $\mathcal{R}_R(E)$ and $\mathcal{R}_S(I)$. Furthermore, I is generated by d analytically independent elements.*

(b) *Assume E to be torsion–free. If φ has any quadratic relations, then I is not syzygetic. If $\varphi \neq \varphi_1$, then I is not of linear type.*

(c) *E is of linear type and $\varphi = \varphi_1$, if and only if, I is of linear type and $I \simeq$* coker B_1.

Proof. Part (a) follows from Lemma 7.6.1.b and its proof. To prove (b), assume that E is torsion–free and that φ has entries of degree two or at least two respectively. Then $\mathcal{R}_R(E)$ has minimal relations of degree two or at least two in the variables $\mathbf{x}$, and hence the same holds for $\mathcal{R}_S(I)$ by part (a). Thus I is not syzygetic or not of linear type respectively.

Part (c) follows from the diagram

$$\begin{array}{ccc} S_R(\text{coker } \varphi_1) & \simeq & S_S(\text{coker } B_1) \\ \downarrow & & \downarrow \\ S_R(E) & & S_S(I) \\ \downarrow & & \downarrow \\ \mathcal{R}_R(E) & \simeq & \mathcal{R}_S(I) \end{array}$$

Remark 7.6.3 Let E be a module with a linear presentation over the polynomial ring $R = k[x_1, \ldots, x_d]$, such that $\nu(E) = d + \text{rank } E - 1$, $E_\wp$ is free whenever $\dim R/\wp \geq 2$, and $S(E)_{\mathfrak{m}S(E)}$ is a discrete valuation ring. Then the assumptions of Theorem 7.6.2 are satisfied. Indeed, we only have to check that rank $\varphi(\mathbf{f}) \geq d - 1$. Suppose that rank $\varphi(\mathbf{f}) \leq d - 2$, then by Lemma 7.6.1.a,

$$2 \leq \text{height } I \leq \ell(I) \leq \dim R/I_{d-1}(\varphi_1).$$

Since $\varphi_1 = \varphi$ it would follow that $\dim R/I_{d-1}(\varphi) \geq 2$, and hence $E_\wp$ would not be free for some $\wp$ with $\dim R/\wp = 2$.

An extended example

G. Valla has conjectured [104] that an ideal in a regular local ring or a homogeneous ideal in a polynomial ring is of linear type provided it is prime and generated by analytically independent elements. Theorem 7.6.2 and Remark 7.6.3 produce an abundance of ideals which are generated by analytically independent elements, but not of linear type. Some of these will indeed yield counterexamples to Valla's conjecture.

We give an example that is monomial, reduced, and Cohen–Macaulay. We obtain this example by applying Theorem 7.6.2 to the case where E is an ideal generated by square free quadratic monomials, that is, edge ideals associated with graphs (see section 4.4). Since an ideal of linear type is $\mathcal{F}_1$, an edge ideal of linear type must be such that each connected component of the corresponding graph has at most one cycle and this cycle must be odd (in fact, the converse holds as well [305, Theorem 3.4]).

A fairly easy step is to make sure that some of the Koszul relations among the generating monomials are not multiples of the reduced relations among these monomials, otherwise one will end up with a dual ideal I of linear type. Also, some care has to be exercised so as to attain the condition rank $\varphi_1(\mathbf{f}) \geq d-1$.

Example 7.6.4 In $R = k[x_1, \ldots, x_{10}]$ let $J = E$ be the ideal generated by the monomials

$$x_1x_6, x_1x_2, x_2x_7, x_2x_3, x_3x_8, x_3x_4, x_4x_9, x_4x_5, x_5x_{10}, x_5x_1.$$

This is the ideal associated to the suspension, $S(Z_5)$, of a pentagon. One easily checks that E satisfies the assumptions of Theorem 7.6.2. Hence by Theorem 7.6.2.a, the dual ideal I in $S = k[T_1, \ldots, T_{10}]$ is generated by 10 analytically independent elements. Since $J = E$ is torsion–free and φ has some quadratic entries, Theorem 7.6.2.b implies that I is not syzygetic and hence not of linear type. In fact, I is generated by the monomials

$$T_1T_4T_8, T_2T_4T_8, T_2T_5T_8, T_2T_6T_8, T_2T_6T_9,$$

$$T_2T_6T_{10}, T_3T_6T_{10}, T_4T_6T_{10}, T_4T_7T_{10}, T_4T_8T_{10}.$$

Using *Macaulay* ([14]) one quickly sees that I is a reduced perfect ideal of grade three having a linear resolution and satisfying $\mathcal{F}_1$.

Starting from the above ideal I, we perform a sufficiently general double link (see Theorem 7.6.6), localize, and homogenize to obtain the following ideal $\wp$ in $R = k[s_1, \ldots, s_9]$, which is prime and hence produces a counterexample to Valla's conjecture.

Proposition 7.6.5 *There exists a homogeneous ideal $\wp \subset R = k[s_1, \ldots, s_9]$, of codimension three and deviation three, satisfying the following conditions:*

(a) *$k[s_1, \ldots, s_9]/\wp$ is a normal Cohen–Macaulay domain, and $\wp$ satisfies $\mathcal{F}_1$.*

(b) *$\wp$ is normally torsionfree, hence $\mathcal{R}(\wp)$ coincides with the symbolic Rees algebra and is normal. Moreover this algebra is Cohen–Macaulay.*

(c) *$\wp$ is a prime ideal generated by analytically independent elements, but is not of linear type. In fact, $\wp$ is not even syzygetic.*

Proof. We simply exhibit a candidate and argue that it has the required properties. Thus, consider the ideal

$$\begin{gathered}\wp = s_5s_7 + s_6s_9,\ s_3s_6s_8 + s_1s_4s_9,\ s_1s_4s_5 - s_3s_5s_8 + s_2s_6s_8 - s_1s_8s_9 \\ s_3s_5s_6 + s_2s_5s_7 + s_3s_5s_9 + s_1s_9^2,\ s_3s_6^2 + s_2s_6s_7 + s_3s_6s_9 - s_1s_7s_9 \\ s_3s_4s_6 + s_2s_4s_7 + s_3s_7s_8 + s_3s_4s_9.\end{gathered}$$

Since the ideal is explicitly given by means of its generators, one may resort to a computer algebra system. A run of *Macaulay* yields a resolution for $\wp$

$$0 \to R[-6]^2 \oplus R[-5] \longrightarrow R[-5]^2 \oplus R[-4]^6 \xrightarrow{\varphi} R[-3]^5 \oplus R[-2] \longrightarrow \wp \to 0,$$

whose details are sufficiently uncomplicated to be checked by hand calculation. For example, it is easy to see that the complete intersection locus of $\wp$, defined by $I_3(\varphi)$, has codimension six, and has the same radical as the ideal

$$\mathbf{K} = (s_5, s_6, s_7, s_9, s_4s_8(s_2s_4 + s_3s_8), s_4(s_1s_4 - s_3s_8), s_8(s_2s_4 + s_3s_8)).$$

In particular, $\wp$ satisfies $\mathcal{F}_1$. Since $\wp$ has codimension three, from its resolution $R/\wp$ must be Cohen–Macaulay. To show it is normal, one uses the ordinary Jacobian criterion which requires the ideal $\wp + I_3(J(\wp))$ to have codimension five at least, where $J(\wp)$ stands for the Jacobian matrix of the generators of $\wp$ with respect to the s-variables. The codimension of the singular locus is indeed five. But a normal homogeneous ring is a domain. This proves (a).

In order to prove (b), one proceeds as follows. A second run of *Macaulay* produces the equations of the Rees algebra of $\wp$: $\mathcal{R}(\wp) \simeq R[\mathbf{u}]/\mathbf{J} \simeq R[u_1, \ldots, u_6]/\mathbf{J}$, with

$$\mathbf{J} = (\mathbf{u} \cdot \varphi, h = s_1s_3u_1u_2 + s_3u_2u_4 + s_2u_2u_5 - s_3u_3u_5 + s_1s_2u_1u_6 - s_1u_4u_6).$$

Observe that since $\wp_{s_5}$ is locally a complete intersection, $\mathbf{J}$ can be computed as

$$\mathbf{J} = \bigcup_{t \geq 1}((\mathbf{u} \cdot \varphi) \colon s_5^t)$$

The computation, using a partial system of parameters, also shows that $\mathbf{J}$ is a Cohen–Macaulay ideal. More concretely, one verifies first that $\{s_5, u_6, u_1 - s_1^2, u_2 - s_9^3\}$ is a regular sequence modulo $\mathbf{J}$; the computer program *Macaulay* will then compute the resolution of the residue ring.

We argue now that the presentation ideal $L = (\wp, \mathbf{J})$ of $\mathrm{gr}_\wp(R)$ is prime. According to [138, Proposition 1.1], L is a Cohen–Macaulay ideal along with $\mathbf{J}$. If Q is one of its associated primes, it cannot contain any of the minimal primes of $\mathbf{K}$, since height $Q = 6$, while $(\mathbf{J}, \wp, \mathbf{K})$ has codimension 7. This means that there is an element $g \in \mathbf{K}$ which is regular modulo L. Since $\wp_g$ is locally a complete intersection, L_g must be a prime ideal, and therefore L is prime.

The content of (c) is submersed in the discussion above. □

Deforming ideals to prime ideals

In the previous section, we faced the problem of converting the radical ideal I of Example 7.6.4 into the prime ideal $\wp$ of Proposition 7.6.5. We want to address this general issue in a more systematic fashion. For two ideals K and I in Noetherian local rings S and R respectively, we say that K is a deformation of I if there exists a sequence $\mathbf{x}$ which is regular on S and S/K such that $R = S/(\mathbf{x})$ and $I = (K, \mathbf{x})/(\mathbf{x})$.

Theorem 7.6.6 *Let $(R, \mathfrak{m}, k)$ be a local Gorenstein domain, and let I be a Cohen–Macaulay ideal of grade g which is generically a complete intersection.*

(a) *If k is infinite then there exists a regular sequence $\alpha_1, \ldots, \alpha_g$ contained in I and an element f contained in the link $J = (\alpha_1, \ldots, \alpha_g)\colon I$, such that grade $(I, f) = g + 1$.*

(b) *Let $\alpha_1, \ldots, \alpha_g, f$ be elements as above. Then $I = (\alpha_1, \ldots, \alpha_g)\colon (f)$. Furthermore let $x_1, \ldots, x_g$ be variables over R, set $S = R[x_1, \ldots, x_g]_{(\mathfrak{m}, x_1, \ldots, x_g)}$, $\beta_i = \alpha_i + x_i f$, and $K = (\beta_1, \ldots, \beta_g)S\colon (f)S$. Then K is a prime ideal, and K is a deformation of I.*

Proof. To show part (a) simply choose $\alpha_1, \ldots, \alpha_g$ to be a regular sequence in I such that $I = (\alpha_1, \ldots, \alpha_g)_P$ for all $P \in \operatorname{Ass} R/I$. Then grade $(I + J) \geq g + 1$, and hence there exists $f \in J$ with grade $(I, f) = g + 1$.

To prove part (b) we first show that $(\alpha_1, \ldots, \alpha_g, f)_P = J_P$ for all prime ideals P of R with $\dim R_P = g$. So let $\dim R_P = g$. If $I \subset P$, then $f \notin P$ and therefore $R_P = (\alpha_1, \ldots, \alpha_g, f)_P \subset J_P$. If $I \not\subset P$, then $(\alpha_1, \ldots, \alpha_g)_P = J_P \supset (\alpha_1, \ldots, \alpha_g, f)_P \supset (\alpha_1, \ldots, \alpha_g)_P$. In any case, $(\alpha_1, \ldots, \alpha_g, f)_P = J_P$ for all P with $\dim R_P = g$, and therefore also $(\beta_1, \ldots, \beta_g, f)_Q = (\alpha_1, \ldots, \alpha_g, f)_Q = J_Q$ for all prime ideals Q of S with $\dim S_Q = g$. It follows that

$$I = (\alpha_1, \ldots, \alpha_g)R\colon JR = (\alpha_1, \ldots, \alpha_g)R\colon (f)R$$

and

$$K = (\beta_1, \ldots, \beta_g)S\colon (f)S = (\beta_1, \ldots, \beta_g)S\colon JS.$$

Here we have used the fact that $\beta_1, \ldots, \beta_g$ form a Q–regular sequence, which is clear since modulo the S–regular sequence $x_1, \ldots, x_g$, the elements $\beta_1, \ldots, \beta_g$ specialize to the R–regular sequence $\alpha_1, \ldots, \alpha_g$. The latter fact also implies that $K = (\beta_1, \ldots, \beta_g)S\colon JS$ is a deformation of $I = (\alpha_1, \ldots, \alpha_g)\colon J$ ([153, Lemma 2.12]). In particular,

$$\operatorname{grade}(K, f)Q \geq \operatorname{grade}(I, f)R = g + 1,$$

and hence f is regular on S/K. Thus in order to show the primeness of K it suffices to prove that $(S/K)_f$ is a domain.

However,

$$(S/K)_f = S_f/(\beta_1, \ldots, \beta_g)\colon (f) = S_f/(\beta_1, \ldots, \beta_g),$$

which is a localization of

$$R_f[x_1, \ldots, x_g]/(\alpha_i + x_i f \mid 1 \leq i \leq g) \simeq R_f.$$

But the latter ring is obviously a domain. (For similar arguments, see [153, proof of Theorem 2.6].) □

A result similar to Theorem 7.6.6 was shown in [153, Theorem 2.6] and [155, Theorem 3.10]. There, however, one needed to adjoin more than g new variables $x_1, \ldots, x_g$ and one had to deal with more than one equation f, rendering the procedure almost useless for machine computations. In contrast, Theorem 7.6.6 seems to yield a rather efficient way of deforming an ideal to a prime ideal. Applied to the ideal I of Example 7.6.4, it would produce another prime ideal $\wp = K$ which is generated by analytically independent elements (K being a deformation of I) but which is not even syzygetic (K being a double link of I, cf. [140, proof of Theorem 1.11].).

Chapter 8

The Equations of Rees Algebras

In this chapter we view the blowup algebras exclusively from the point of view of their presentation as quotients of polynomial algebras. Thus for a Rees algebra $\mathcal{R}(I)$, it means the study of the natural homomorphism associated to a set $(a_1, \ldots, a_n)$ of generators of I

$$B = R[T_1, \ldots, T_n] \xrightarrow{\varphi} \mathcal{R}(I), \ \varphi(T_i) = a_i t,$$

and particularly of how to find $L = \ker(\varphi)$, and analyze its properties. L will be referred to as the *equations* of $R(I)$.

A starting point toward the equations of $\mathcal{R}(I)$ is the observation that L is a graded ideal

$$L = \bigoplus_{\ell \geq 1} L_\ell$$

with L_1 consisting of the linear forms $\sum_{i=1}^n c_i T_i$ such that $\sum_{i=1}^n c_i a_i = 0$, that is, the first order syzygies of I. In general, L_ℓ are the first order syzygies of I^ℓ.

The methods discussed in this chapter are incremental in the sense that we focus almost entirely on the modules

$$\tau_\ell(I) = L_\ell / \sum_{i=1}^{\ell-1} L_i \cdot B_{\ell-i}.$$

Often it will be possible to know enough about these modules to predict the number and degrees of fresh generators.

One approach to get at these equations goes as follows. Let

$$R^m \xrightarrow{\varphi} R^n \longrightarrow I \rightarrow 0$$

be a presentation of the ideal I. L_1 is generated by the 1–forms

$$L_1 = [f_1, \ldots, f_m] = [T_1, \ldots, T_n] \cdot \varphi = \mathbf{T} \cdot \varphi.$$

The ring $R[T_1, \ldots, T_n]/(L_1)$ is the symmetric algebra of the ideal I, and we write $\mathcal{A} = L/(L_1)$ for the kernel of the canonical surjection

$$0 \to \mathcal{A} \longrightarrow S(I) \longrightarrow \mathcal{R}(I) \to 0.$$

If R is an integral domain, $\mathcal{A}$ is the R–torsion submodule of $S(I)$; in fact, there exists an element $0 \neq f \in I$ such that $f\mathcal{A} = 0$. To put to use the underlying torsion phenomenon, the main technical device used is a rewrite of the equations for L_1, the same method of the Jacobian dual that we have used in earlier chapters. Let $\mathbf{x} = [x_1, \ldots, x_r]$ be the ideal generated by the entries of the matrix φ. We may write L_1 now as

$$L_1 = \mathbf{T} \cdot \varphi = \mathbf{x} \cdot B(\varphi),$$

where $B(\varphi)$ is an $r \times m$ matrix of linear forms in the variables T_i's.

The matrix $B(\varphi)$ will often hold the key to a full set of equations for $\mathcal{R}(I)$. For instance, an application of Cramer's rule implies that the $r \times r$ minors of $B(\varphi)$, $I_r(B(\varphi))$, are conducted into the ideal (L_1) by nonzero elements of R, and thus $I_r(B(\varphi)) \subset L$. This places the problem of finding L in a context that has been studied earlier, that of ideals associated to a sequence $\mathbf{x}$ and a matrix Y ([37, Theorem 5.1], [32, Theorem 3.6]):

$$J = (\mathbf{x} \cdot Y, I_r(Y)).$$

In a number of instances, it will be shown that the equations of $\mathcal{R}(I)$ can be described in this manner. In other cases, we are going to see that the matrix $B(\varphi)$ is skew–symmetric and a clear role is played by its Pfaffians.

In the case of $\mathcal{R}_s(I)$ one seeks presentations for its subalgebras

$$\mathcal{R}(I) = R[It] \subset R^{(2)}(I) = R[It, I^{(2)}t^2] \subset \cdots,$$

and looks for means to detect eventual stabilization. This was the game played in [288], [291] with tools from computer algebra. Interestingly enough, the question dealt with in Proposition 7.4.4 may often already be answered in the partial information contained in the details of one of the algebras $R^{(n)}(I)$ (*cf.* [122]). As things have worked out, this approach has provided some of the means to show Noetherianess in $\mathcal{R}_s(I)$. For the opposite goal, it has been necessary to work directly with the ideals $I^{(n)}$.

In order to view $\mathcal{R}_s(I)$ as an extension of $\mathcal{R}(I)$ one introduces modules that reflect the additional generators. Since we are looking at the algebra $\mathcal{R}_s(I)$ from the perspective of its (partial) presentations, it is of interest to consider the modules

$$\sigma_n(I) = I^{(n)} / \sum_{i=1}^{[(n+1)/2]} I^{(i)} I^{(n-i)}.$$

In particular $\sigma_n(I)$ provides a portrait of the fresh generators that must be added in degree n.

In great many instances, the first non–vanishing of the modules $\sigma_n(I)$ and $\tau_n(I)$ will be given by closed expressions, *content formulas*, obtained from the syzygetic data on I. This homological approach is somewhat facilitated if the ideal has a finite free resolution.

8.1 Almost Complete Intersections

The main object here are almost complete intersection prime ideals of dimension one. The result proved is a formula describing the second symbolic power of almost complete intersections of dimension one. In a number of cases, *e.g.* when I is a normal ideal, it gives rise to the presentation ideal of the algebra $R[It, I^{(2)}t^2]$. We then examine when this algebra is the full symbolic power algebra.

There are, through homological filters, several means to determine the symbolic powers of an ideal. Let us outline how this occurs. For simplicity we assume that R is a Gorenstein ring.

Let I be an ideal with a decomposition $I = J \cap Q$, where J is equicodimensional of height g and height $Q > g$, so that we are assuming the I and J have the same minimal primes and primary components. At issue is how to recover J. This follows simply from the equality derived from local duality (see also Proposition 5.3.3).

Proposition 8.1.1 *Let R be a Gorenstein local ring and let J be an ideal whose minimal primes have the same height g. Suppose $J = I \cap Q$, where I is height unmixed and* height $Q > g$. *Then*

$$\mathrm{Ext}_R^g(R/I, R) = \mathrm{Ext}_R^g(R/J, R).$$

Proof. Consider the sequence

$$0 \to I/J \longrightarrow R/J \longrightarrow R/I \to 0$$

and the associated sequence of Ext's:

$$\mathrm{Ext}_R^{g-1}(I/J, R) \longrightarrow \mathrm{Ext}_R^g(R/I, R) \longrightarrow \mathrm{Ext}_R^g(R/J, R) \longrightarrow \mathrm{Ext}_R^g(I/J, R),$$

where the two modules at the ends vanish since I/J has an annihilator of codimension at least $g + 1$. □

Corollary 8.1.2 *Let R be a Gorenstein ring and let I be an equicodimensional ideal of height g. Then the intersection of the primary components of height g of the ideal I^n is the annihilator of the module* $\mathrm{Ext}_R^g(R/I^n, R)$.

It would be of interest to obtain more explicit formulas, particularly not involving the higher syzygies of the ideal. We discuss one of these in the sequel.

The content of syzygies

Let R be a regular local ring of dimension d and let I be a prime ideal of height g. Denote by $I^{(2)}$ the I-primary component of I^2. There is, in general, no explicit formula for the modules $\sigma_n(I)$. In the case of a Cohen–Macaulay almost complete intersection there exists a simple expression for $\sigma(I) = \sigma_2(I)$ (see also [145] and [290]).

To derive it, note that $\sigma(I)$ is the torsion submodule of I/I^2. Assume that $I = (x_1, \ldots, x_n)$ is a Cohen–Macaulay prime ideal of codimension g. There exists an exact sequence

$$H_1 \xrightarrow{\varphi} (R/I)^n \longrightarrow I/I^2 \longrightarrow 0$$

where H_1 is the first Koszul homology module of I. If I is a strongly Cohen–Macaulay ideal, this sequence will imply that proj dim $I^2 \leq g$.

The long exact sequence of the functor $\operatorname{Ext}_R(*, R)$ yields ($S = R/I$):

$$\operatorname{Ext}^g_R(S^n, R) \longrightarrow \operatorname{Ext}^g_R(H_1, R) \longrightarrow \operatorname{Ext}^{g+1}_R(I/I^2, R) \simeq \operatorname{Ext}^g_R(I^2, R) \longrightarrow 0,$$

the isomorphism because I is a Cohen–Macaulay ideal.

On the other hand, from the sequence

$$0 \to I^2 \longrightarrow I^{(2)} \longrightarrow \sigma(I) \to 0$$

we obtain the long exact sequence

$$0 \to \operatorname{Ext}^g_R(I^{(2)}, R) \to \operatorname{Ext}^g_R(I^2, R) \xrightarrow{\psi} \operatorname{Ext}^{g+1}_R(\sigma(I), R) \to \operatorname{Ext}^{g+1}_R(I^{(2)}, R) \to 0,$$

since $\operatorname{Ext}^{g+1}_R(I^2, R) = 0$. Furthermore, it is easy to see that ker ψ and coker ψ have codimension at least $g+2$ and $g+3$, respectively. It follows that

$$\operatorname{Ext}^{g+1}_R(\operatorname{Ext}^g_R(I^2, R), R) \simeq \operatorname{Ext}^{g+1}_R(\operatorname{Ext}^{g+1}_R(\sigma(I), R), R).$$

Finally we observe that since I^2 has projective dimension at most g, $\sigma(I)$ satisfies, locally, the condition

$$\operatorname{depth} \sigma(I) \geq \inf\{2, \dim R - g + 1\},$$

and therefore has the property S_2 of Serre's on its support. From this fact, it follows that the canonical mapping given by local duality

$$\sigma(I) \longrightarrow \operatorname{Ext}^{g+1}_R(\operatorname{Ext}^{g+1}_R(\sigma(I), R), R)$$

is an isomorphism.

To make full use of these observations, assume that I is an almost complete intersection, that is, $n = g+1$. Let ω denote the canonical module of S, $\omega = \operatorname{Ext}^g_R(S, R) \simeq H_1$ (*e.g.* [172], [194]), and use the duality in the Koszul homology ([106], see also [12]) to get the exact sequence

$$\omega^{g+1} = \operatorname{Hom}_S(S^{g+1}, \omega) \xrightarrow{\varphi^*} \operatorname{Hom}_S(\omega, \omega) = S \longrightarrow \operatorname{Ext}^{g+1}_R(I^{(2)}/I^2, R) \to 0.$$

Note that if $[z = \sum z_i e_i]$ is an element of ω, $\varphi([z]) = \sum \overline{z_i} e_i \in S^n$. From this we can read the image of φ^*: For any $\lambda_\omega \in \omega^n$, $\lambda_\omega(s_1, \ldots, s_n) = \sum s_i w_i$, $\varphi^*(\lambda_\omega([z])) = \sum z_i w_i$.

Theorem 8.1.3 *Let R be a regular local ring and let I be a Cohen–Macaulay almost complete intersection of height g. Then $I^{(2)}/I^2 = \mathrm{Ext}_R^{g+1}(R/I_1(\varphi), R)$, where $I_1(\varphi)$ denotes the ideal generated by the entries of the first order syzygies of I. In particular,*

$$I^{(2)} = I^2 \colon I_1(\varphi).$$

Proof. To complete the proof it suffices to pick a generating set for I such that any g elements in it form a regular sequence. To find the image of φ^* in $\mathrm{Hom}(H_1 = \omega, \omega) = S$, it suffices to observe that for any two cycles $\alpha = \sum a_i e_i$ and $\beta = \sum b_i e_i$ we have $a_i\beta = b_i\alpha$ in S^{g+1}. It follows that the image of φ^* is precisely $I_1(\varphi)$. □

Remark 8.1.4 If $I_1(\varphi) = J_1 \cap J_2$ is a decomposition of $I_1(\varphi)$, where J_1 is unmixed of codimension $g+1$ and J_2 has codimension $\geq g+2$, it is not difficult to show that $\mathrm{Ext}_R^{g+1}(R/I_1(\varphi), R) \simeq \mathrm{Ext}_R^{g+1}(R/J_1, R)$. In particular, $I^2 = I^{(2)}$ precisely when J_1 is trivial. In the sequel we assume that these powers differ.

Corollary 8.1.5 *$\sigma(I)$ is cyclic if and only if J_1 is a quasi–Gorenstein ideal.*

Example 8.1.6 Let $I \subset k[x, y, z, w]$ be the ideal of a Cohen–Macaulay monomial curve defined by $x = t^a, y = t^b s^{a-b}, z = t^c s^{a-c}, w = s^a$. According to [22], I is given by the 2×2 minors of

$$\varphi = \begin{bmatrix} x^{a_1} w^{d_1} & x^{c_2} \\ z^{c_1} & y^{b_2} \\ y^{b_1} & x^{a_2} w^{d_2} \end{bmatrix}.$$

The component J_1 of the ideal $I_1(\varphi)$ is $(y^\beta, z^\gamma, x^\alpha w^\delta)$ where

$$\begin{aligned} \alpha &= \gcd(a_1, a_2) \\ \beta &= \gcd(b_1, b_2) \\ \gamma &= \gcd(c_1, c_2) \\ \delta &= \gcd(d_1, d_2) \end{aligned}$$

and therefore $I^{(2)}/I^2$ is a cyclic module, a fact also observed by Herzog, Schenzel and Simis.

Corollary 8.1.7 *If $I_1(\varphi)$ is a Gorenstein ideal then $I^{(2)}$ is a Cohen–Macaulay ideal.*

Proof. Since $I^{(2)}/I^2 \simeq R/I_1(\varphi)$, it is enough to show that

$$\mathrm{Ext}_R^g(I^{(2)}, R) = \mathrm{Ext}_R^{g+1}(I^{(2)}, R) = 0.$$

If $I_1(\varphi)$ is a Gorenstein ideal, both ends of the mapping ψ are $R/I_1(\varphi)$, so the mapping is realized as multiplication by a ring element. But localizing at a prime $\wp$ of codimension $g+2$ kills coker ψ, which shows that ψ is an isomorphism by Krull's theorem. □

Corollary 8.1.8 *Let I be a prime ideal of a regular local ring $(R, \mathfrak{m})$. If I is a normal, almost complete intersection of dimension one, then $I^{(2)}/I^2$ is cyclic. Furthermore if $I \subset \mathfrak{m}^2$, then the Cohen-Macaulay type of I is at least* $\dim R - 1$.

Proof. Because the symmetric and Rees algebras of I coincide [116] and it is Cohen-Macaulay, by the normality criterion of Proposition 5.4.1, we not only must have $I_1(\varphi) = (y_1, \ldots, y_{d-1}, y_d^a)$ for some regular system of parameters (y), which proves the first assertion, but also the rank of the Jacobian matrix at the origin must be $d-1$. Picking a generating set for I such that $\{x_1, \ldots, x_{d-1}\}$ is a regular set, as $x_i \in \mathfrak{m}^2$, this implies that there must be at least $d-1$ syzygies that contribute to the non–vanishing of the Jacobian ideal. □

Remark 8.1.9 The minimum number of generators of the module $I^{(2)}/I^2$, even for a prime ideal generated by 3 elements, may be arbitrarily large, *cf.* [145, Proposition 2.12].

The symbolic square algebra of a prime ideal

We assume from now on that $I_1(\varphi)$ is a complete intersection $(z_1, \ldots, z_d)$ of dimension one. In such case $I^{(2)} = (I^2, u)$ and $I^2 : u = I_1(\varphi)$. We seek ways of finding such element u and the equations it satisfies over the Rees algebra $R[It]$.

Since for each $z \in I_1(\varphi)$, $zu \in I^2$, we may associate a polynomial $zU - f$ where f is a (homogeneous) quadratic in the variables $T_1, \ldots, T_d$ that under the evaluation $T_i \longrightarrow x_i$ is mapped to zu. Consider the ideal J of $\mathcal{R}(I)[U]$ generated by

$$z_i U - f_i, \ i = 1, \ldots, d.$$

We describe when this is the ideal of relations of the homomorphism from the polynomial ring on U onto $R[It, I^{(2)}t^2] = R[It, ut^2]$.

Corollary 8.1.10 *Let I be a normal almost complete intersection and let u be as above. Then $(I_1(\varphi)(U-u))$ is a prime ideal of $\mathcal{R}(I)[U]$.*

Proof. We make use of the primality criterion expressed by Proposition 5.5.6. Since $\mathcal{R}(I)$ is normal, the ideal of relations of u is precisely $(D(U-u))$ (*cf.* [208, (11.13)]). D is a graded ideal, whose component in degree 0 is $I_1(\varphi)$. We must show that this component already generates D. Note that $I_1(\varphi)\mathcal{R}(I)$ is a primary ideal because the coefficients of the syzygies of $\mathcal{R}(I)$ all lie in $I_1(\varphi)$. If M is the radical of $I_1(\varphi)\mathcal{R}(I)$, its primary ideals are the M-primary components of the powers M^s, $s \geq 1$ (since $\mathcal{R}(I)_M$ is a discrete valuation domain) and therefore are all determined by their degree 0 component. □

Denote by Δ the ideal generated by the f_i's evaluated at the origin. That is, if $\mathfrak{m}$ is the maximal ideal of R, Δ is the ideal generated by the f_i in the polynomial ring $R/\mathfrak{m}[T_1, \ldots, T_d]$.

Theorem 8.1.11 *Let I be an almost complete intersection and let $f_1, \ldots, f_d$ be the polynomials as above. Then*

(a) *If $\Delta \neq 0$ then $R[It, ut^2]$ is a Cohen–Macaulay ring.*

(b) *If height $(\Delta) \geq 2$, then $R[It, ut^2]$ is the symbolic power algebra of I.*

Proof. Both conditions imply that the ideal of relations is given as in the previous lemma. (a) follows since $I_1(\varphi)\mathcal{R}(I)$ is a Cohen–Macaulay ideal. As for (b), one now has that the ideal $\mathfrak{m}R[It, ut^2]$ has height at least two. Since any element of the symbolic power algebra is conducted into $R[It, ut^2]$ by a power of $\mathfrak{m}$, it must lie in $R[It, ut^2]$. □

The condition (a) is always satisfied if I is normal. If the ideal Δ has a nontrivial gcd but it is not generated by a square then $R[It, ut^2]$ is also normal.

Example 8.1.12 Let I be the ideal of the monomial curve $(t^7, t^9, t^{12}, t^{15})$:

$$I = (x^3 - yz, y^3 - zw, z^2 - yw, w^2 - x^3y).$$

I is not a normal ideal, however $I_1(\varphi) = (x^3, y, z, w)$, so that $I^{(2)} = (I^2, u)$, in fact

$$u = w^3 - 3x^3yw - 2yz^3 + 3x^3z^2 + y^5$$

will do.

A computation showed that $R[It, ut^2] = \mathcal{R}(I)[U]/(I_1(\varphi)(U - u))$ so that it is Cohen–Macaulay. Furthermore

$$\Delta = (T_1^2, T_2^2 - T_1T_3, T_1T_2),$$

and therefore it is already the full symbolic power algebra.

It then follows by Proposition 7.4.4 that I is set–theoretically a complete intersection.

To construct examples with a more delicate structure—*e.g.* normal ideals—we use a recipe of [157]:

Theorem 8.1.13 *Let $(R, \mathfrak{m})$ be a regular local ring with infinite residue class field, and let I be a radical, almost complete intersection of dimension one. Then the following are equivalent:*

(a) *I is a normal ideal.*

(b) *I^k is integrally closed for some $k > 1$.*

(c) *I is geometrically linked to a regular ideal J (that is, such that R/J is a regular local ring).*

(d) *There exists an $n \times 1$ matrix* $Y = \begin{pmatrix} y_1 \\ \vdots \\ y_n \end{pmatrix}$ *with $y_1, \ldots, y_n, y_{n+1}$ forming a regular system of parameters and an $n \times n$ matrix X with entries in R and* $\det(X) \notin (y_1, \ldots, y_n)R$ *such that*

$$I = I_1(X \cdot Y) + I_n(X).$$

Furthermore, if char $R/\mathfrak{m} \neq 2$, *then the matrix X can be chosen symmetric.*

Example 8.1.14 Let

$$I = (xw + y^3, yw + z^3, zw + x^2, w^3 + xy^2z^2).$$

I is obtained by linking the ideal generated by the first 3 elements with the regular prime $P = (x, y, z)$.

More precisely, let

$$\varphi = \begin{bmatrix} w & 0 & x \\ y^2 & w & 0 \\ 0 & z^2 & w \end{bmatrix}.$$

Then

$$I = ([x, y, z] \cdot \varphi, \det(\varphi)).$$

To prove that I is prime, one observes the existence of normalizations

$$k[z] \hookrightarrow k[y, z]/(z^{10} - y^9) \hookrightarrow R = k[x, y, z, w]/I.$$

It is easy to see that as an $k[z]$–module, R has rank 9. It follows that R is an integral domain since it is torsion–free over $k[z]$. The ideal I is therefore normal according to the quoted result. Since $\Delta = T_4(T_1, T_2, T_3, T_4)$, the symbolic square algebra does not equal $\mathcal{R}_s(I)$. (Mark Johnson has informed us that $\mathcal{R}_s(I)$ is Noetherian.)

Monomial curves

As the previous discussion indicates, it is a simple affair to get hold, computationally, of the element u, and consequently of the polynomials f_i. For $n = 3$, that is, for an ideal with a resolution

$$0 \to R^2 \xrightarrow{\varphi} R^3 \longrightarrow I \to 0,$$

it will be shown that the f_i's are the 2×2 minors of $B(\varphi)$.

Let us examine, with some detail, prime ideals defining monomial curves in affine 3–space. These curves, given parametrically by

$$\begin{cases} x = t^m \\ y = t^n \\ z = t^p \end{cases}$$

where $\gcd(m, n, p) = 1$, have for equations, according to [109], the 2×2 minors of a matrix

$$\varphi = \begin{bmatrix} x^{a_1} & z^{b_1} \\ y^{a_2} & x^{b_2} \\ z^{a_3} & y^{b_3} \end{bmatrix}.$$

The content ideal of this matrix is $I_1(\varphi) = (x^a, y^b, z^c)$, where the exponents are the minimum of the exponents for each of the variables. In particular it implies that $I^{(2)}/I^2$ is cyclic (see also [145]).

We write the equations of $\mathcal{R}(I)$ as

$$J = \varphi \cdot [T_1, T_2, T_3] = [x^a, y^b, z^c] \cdot B(\varphi), \; B(\varphi) = \begin{bmatrix} A_1 & B_1 \\ A_2 & B_2 \\ A_3 & B_3 \end{bmatrix},$$

where the A_i, B_i are linear forms in the T variables.

Proposition 8.1.15 *Let u be the element of $I^{(2)}$ such that $I^{(2)} = (I^2, u)$. u satisfies the equations*

$$[x^a, y^b, z^c] \cdot U - [\Delta_1, \Delta_2, \Delta_3],$$

where the Δ_i are the corresponding 2×2 minors of $B(\varphi)$.

Proof. We set $R = k[x, y, z]$. Let $L = L(\varphi)$ be the ideal of $B = R[T_1, T_2, T_3, U]$ generated by the five polynomials

$$[x^a, y^b, z^c] \cdot \varphi, \; [x^a, y^b, z^c] \cdot U - [\Delta_1, \Delta_2, \Delta_3].$$

There is a mapping $B/L \longrightarrow \mathcal{R}[It, I^{(2)}t^2]$, which is an isomorphism when localized at the prime ideal I. If B is graded by $\deg(T_i) = 1, \deg(U) = 2$, and the remaining variables made of degree 0, note that the homogeneous component of degree 2 of B/L maps into $I^{(2)}$, and our assertion will be established if we show that L_2 is a free R–module. Indeed, it would show the existence of an inclusion of modules of projective dimension one, $B_2/L_2 \hookrightarrow I^{(2)}$, whose cokernel has finite length.

The ideal L is generated by the maximal Pfaffians of the matrix

$$\Phi = \begin{bmatrix} 0 & U & A_1 & A_2 & A_3 \\ -U & 0 & -B_1 & -B_2 & -B_3 \\ -A_1 & B_1 & 0 & z^c & -y^b \\ -A_2 & B_2 & -z^c & 0 & x^a \\ -A_3 & B_3 & y^b & -x^a & 0 \end{bmatrix}.$$

By [36] the resolution of this ideal has the form

$$0 \to B[-4] \longrightarrow B[-3]^2 \oplus B[-2]^3 \longrightarrow B[-2]^3 \oplus B[-1]^2 \longrightarrow L \to 0.$$

The R-projective resolution of the degree 2 component is therefore:

$$0 \to R^3 \xrightarrow{\gamma} R^3 \oplus B_1^2 \longrightarrow L_2 \to 0.$$

The mapping γ, when evaluated at the origin of R, maps k^3 into $k[T_1, T_2, T_3]^2$ and agrees with the matrix $B(\varphi)^t$ evaluated at the origin of R. But the rows of this matrix are clearly k–linearly independent. This shows that

$$\mathrm{Tor}_1^R(L_2, R/(x, y, z)) = 0,$$

and therefore L_2 is R–free. □

The proof does *not* show that L is the presentation ideal J of $\mathcal{R}[It, I^{(2)}t^2]$. It is clear that L is prime if and only if one of the Δ_i is not contained in

$$(x, y, z)R[T_1, T_2, T_3].$$

On the other hand, the argument also shows that there will be no fresh generators, in degree 3, for the presentation ideal. We discuss a mechanism to find more generators of J.

To a set of pure monomials $\{x^a, y^b, z^c\}$ we associate the vector $v = (a, b, c)$. Given two such sets of monomials v_1 and v_2, we say that $v_1 \geq v_2$ if the inequality holds in each component. We attach to each column of the matrix φ two numerical vectors v_1 and v_2.

Theorem 8.1.16 $J = L(\varphi)$ *if and only if one of the following holds:* (i) *the vectors* v_1 *and* v_2 *are not comparable, or* (ii) v_1 *and* v_2 *have at least one component in common.*

The proof is clear from the previous discussion, as such are the conditions that make the R–content of L to be the unity.

If the Rees algebra of the monomial ideal I is normal, then the condition above holds. But there are several other cases where they occur.

Before we go on to find these generators, we record:

Proposition 8.1.17 *Let I be the ideal of the monomial curve* (t^m, t^n, t^p), $m < n < p$. *The Rees algebra of I is not normal in precisely the following cases:*

$$\begin{cases} \min\{a_1, a_2, a_3\} > 1 \text{ or } \min\{b_1, b_2, b_3\} > 1 \\ a_1 = b_2 = 1, \text{ and the other exponents are } > 1. \end{cases}$$

We shall illustrate, with two examples, how the equations of the symbolic square algebra may often be found.

Example 8.1.18 We view the exponents in the matrix φ as two integral vectors, $v_1 = (a_1, a_2, a_3)$ and $v_2 = (b_1, b_2, b_3)$. Note that $\Delta = 0$ means that componentwise, say $v_2 > v_1$. There will be, as will be proved, two basic cases to consider: $v_2 \geq 2v_1$ or otherwise.

We assume throughout that $v_2 > v_1$. This means, in the matrix $B(\varphi)$, the entries B_1, B_2, B_3 lie in the ideal $(x, y, z)B$ and $(A_1, A_2, A_3) = (T_1, T_2, T_3)$. If we let

$$(\alpha, \beta, \gamma) = \inf\{v_2 - v_1, v_1\}$$

then the coefficients of the equations 8.1.15 all lie in the monomial ideal $(x^\alpha, y^\beta, z^\gamma)$.

To obtain additional equations, note that the R–content ideal of L is still monomial, and henceforth the method that yielded the quadratic equations of L can be used again. Eliminating the monomials x^α, y^β, z^γ we get the equation

$$U(x^a y^b z^c U^2 + x^a Y^{b_2} z^{b_3} T_1^3 T_3 + x^{b_1} y^b z^{b_3} T_1 T_2^3 + x^{b_1} y^{b_2} z^c T_2 T_3^3),$$

where $(a, b, c) = v_1 - (\alpha, \beta, \gamma)$ and $(b_1, b_2, b_3) = v_2 - v_1 - (\alpha, \beta, \gamma)$. The quadratic polynomial in U (without the U–factor) will belong to the presentation ideal. It is easy to see that except under the following conditions

$$\begin{cases} a_1 < b_1 < 2a_1 \\ a_2 < b_2 < 2a_2 \\ a_3 < b_3 < 2a_3 \end{cases}$$

and

$$\begin{cases} 2a_1 < b_1 \\ 2a_2 < b_2 \\ a_3 < b_3 < 2a_3 \end{cases}$$

(and symmetric instances of the latter) there will be a unit coefficient in the polynomial.

We provide a template for this equation. Let $w = [X, Y, Z]$ be a vector of indeterminates, and consider the matrix

$$\Phi = \begin{bmatrix} p_1y_1 & Xq_1y_1 & y_1U & -p_3q_1 & p_2q_1 \\ p_2y_2 & Yq_2y_2 & p_3q_2 & y_2U & -p_1q_2 \\ p_3y_3 & Zq_3y_3 & -p_2q_3 & p_1q_3 & y_3U \end{bmatrix}.$$

The ideal L is a specialization of $w \cdot \Phi$. On the other hand, the element

$$G = q_2q_3y_1{p_1}^2 + q_1q_3y_2{p_2}^2 + q_1q_2y_3{p_3}^2 + y_1y_2y_3U^2$$

is obtained by eliminating X, Y, Z from three of these equations. Note that in the specialization w maps to

$$[x^\alpha, y^\beta, z^\gamma].$$

A calculation with *Macaulay* shows that $(w \cdot \Phi, G)$ is a Cohen–Macaulay ideal of codimension 3 and type 3. Denote by K the ideal of $B = R[T_1, T_2, T_3, U]$ obtained by the specialization. If one of the coefficients of G specializes to a unit, then K is a prime ideal and therefore is the presentation ideal of $R[It, I^{(2)}t^2]$.

A projective resolution of K, with the previous grading of B, has for numerical data

$$0 \to B[-4]^3 \xrightarrow{f} B[-4]^3 \oplus B[-2]^3 \oplus B[-3]^2 \xrightarrow{g} B[-4] \oplus B[-2]^3 \oplus B[-1]^2 \to K \to 0.$$

Here is an instance of the case $v_2 \geq 2v_1$. For concreteness, consider the monomial curve defined by $(n^2+n+1, n^2+2n+1, n^2+2n+2), n > 1$. The matrix of relations is

$$\varphi = \begin{bmatrix} x & z^n \\ y & x^{n+1} \\ z & y^n \end{bmatrix}$$

while the matrix $B(\varphi)$ is

$$\begin{bmatrix} T_1 & x^n T_2 \\ T_2 & y^{n-1} T_3 \\ T_3 & z^{n-1} T_1 \end{bmatrix}$$

so that the ideal L is given by

$$\begin{array}{l} xT_1 + yT_2 + zT_3 \\ z^n T_1 + x^{n+1} T_2 + y^n T_3 \\ xU - z^{n-1} T_1 T_2 + y^{n-1} T_3^2 \\ yU - x^n T_2 T_3 + z^{n-1} T_1^2 \\ zU - y^{n-1} T_1 T_3 + x^n T_2^2 \end{array}$$

The last three equations imply that u is integral over I^2. If we eliminate the "coefficients" x, y and z from the first, fourth and fifth equations, we obtain

$$T_1(U^2 + x^{n-1} z^{n-2} T_1 T_2^3 + x^{n-1} y^{n-2} T_2 T_3^3 + y^{n-2} z^{n-2} T_1^3 T_3).$$

When this quadratic monic polynomial in U is added to L, one obtains a Cohen–Macaulay prime ideal, and therefore the presentation ideal of the symbolic square algebra. If $n = 2$, it is easy to see that the algebra is the integral closure of $\mathcal{R}(I)$.

We shall now look at the other case. The curve is (t^{13}, t^{14}, t^{17}), with

$$\varphi = \begin{bmatrix} x^2 & z^3 \\ y & x^3 \\ z & y^3 \end{bmatrix}.$$

While in the previous example the coefficients of the five polynomials belonged to the content ideal $I_1(\varphi)$, that is no longer the case. Here one still obtains the quadratic polynomial

$$xU^2 + xyzT_1^3 T_3 + yT_2 T_3^3 + zT_1 T_2^3.$$

Applying the Cramer's rule to the first, fourth and this quadratic polynomial to eliminate the "new" content ideal (x, y, z), gives the polynomial

$$U^3 + T_2^2 T_3^4 - 3xyT_1^3 T_2^2 T_3 - 2x^3 T_1^4 T_2 T_3 - T_1 T_2^5 + yzT_1^5 T_2 + x^2 z^2 T_1^6$$

These seven elements generate a non–Cohen–Macaulay prime ideal. The symbolic square algebra is the integral closure of $\mathcal{R}(I)$.

8.2 Codimension Two

Let R be a regular local ring of dimension d, and let I be a Cohen–Macaulay prime ideal generated by n elements. For $d = 3$, $n = 2$ and 3, the associated Rees algebra

$$\mathcal{R}(I) = \sum I^s t^s$$

is always Cohen–Macaulay. The intent here is to link the normality and the Cohen-Macaulayness of $\mathcal{R}(I)$, especially when $n \geq 4$.

Normality and Cohen–Macaulayness

We begin with a detailed examination of four–generated prime ideals. Pick a presentation

$$0 \to R^3 \xrightarrow{\varphi} R^4 \longrightarrow I \to 0,$$

and associate to the matrix

$$\varphi = \begin{bmatrix} a_{11} & a_{12} & a_{13} \\ a_{21} & a_{22} & a_{23} \\ a_{31} & a_{32} & a_{33} \\ a_{41} & a_{42} & a_{43} \end{bmatrix}$$

the defining ideal of the symmetric algebra $S(I)$ of I

$$J = [f, g, h] = [T_1, T_2, T_3, T_4] \cdot \varphi.$$

The algebra $S(I)$ is a complete intersection—a fact we shall want to make use of to get the equations of $R[It]$. For that we use one simplifying hypothesis: $I_1(\varphi)$ is a complete intersection. Because the general condition follows closely the examination of any specific example, we start with one.

Let I be the ideal of the space curve of equations:

$$\begin{cases} x &= t^6 \\ y &= t^8 \\ z &= t^{10}(t+1). \end{cases}$$

This ideal is minimally generated by the maximal minors of

$$\varphi = \begin{bmatrix} x & -x & y \\ -z & 2y+z & -2x^2 \\ -2y & -2x+2y & -z \\ 2x^2 & 4z & xz \end{bmatrix}.$$

The ideal J is thus generated by the entries of the product of two matrices:

$$[f, g, h] = [x, y, z] \cdot B(\varphi),$$

where

$$B(\varphi) = \begin{bmatrix} T_1 + 2xT_4 & -T_1 - 2T_3 & -2xT_2 + zT_4 \\ -2T_3 & 2T_2 + 2T_3 & T_1 \\ -T_2 & T_2 + 4T_4 & -T_3 \end{bmatrix}.$$

$F = \det B(\varphi)$ is a polynomial with some unit coefficient (assume characteristic $\neq 2$). The image of F in $S(I)$ is annihilated by the ideal (x, y, z).

Proposition 8.2.1 *Let $(R, \mathfrak{m})$ be a regular local ring of dimension three, and let I be a prime ideal minimally generated by four elements. Using the notation above, assume that $I_1(\varphi) = (x_1, x_2, x_3)$, and that $F = \det B(\varphi) \neq 0$ mod $\mathfrak{m}$. Then $R[It]$ is a Cohen–Macaulay ring whose defining ideal is (f, g, h, F).*

Proof. We know that I is not of linear type, that is, the canonical homomorphism from $S(I)$ onto $R[It]$ has a non-vanishing kernel:

$$0 \to \mathcal{A} \longrightarrow S(I) \longrightarrow R[It] \to 0.$$

On the other hand, I is syzygetic so the two algebras can only begin to differ in degree 3.

Lemma 8.2.2 *Let $\mathcal{A}_3$ be the component of $\mathcal{A}$ in degree 3. Then*

$$\operatorname{Ext}^3_R(\mathcal{A}_3, R) \simeq R/I_1(\varphi).$$

Proof. Applying the functor $\operatorname{Hom}(*, R)$ to the degree three component of the exact sequence above, and taking into account that $\mathcal{A}_3$ is a module of finite length, we get

$$\operatorname{Ext}^3_R(\mathcal{A}_3, R) = \operatorname{Ext}^3_R(S_3(I), R).$$

To obtain this module, it suffices to use the Weyman resolution (cf. [307]; see also Section 3.3). □

By duality, it follows that $\mathcal{A}_3$ is always annihilated by $I_1(\varphi)$, and in case this ideal is a complete intersection (or, more generally, Gorenstein) $\mathcal{A}_3$ is cyclic.

Proof of Proposition. In the symmetric algebra $S(I)$, the ideal Q generated by (x_1, x_2, x_3) is a Gorenstein ideal—$S(I)/Q$ is a polynomial ring over $R/(x_1, x_2, x_3)$—and therefore its annihilator is generated by a single element, necessarily the image of F. By linkage then $\mathcal{A}_3 S(I) = (F)$ is a C–M ideal. On the other hand, $S(I)$ has two minimal primes, $\mathfrak{m}S(I)$ and $\mathcal{A}$, and since F is not contained in the first and the other is the $\mathcal{A}$-primary component of the zero ideal, we must have $(F) = \mathcal{A}$. □

Theorem 8.2.3 *Let R be a regular local ring of dimension three and let I be a prime ideal generated by 4 elements. Suppose that the ideal $\mathcal{A}$ is annihilated by $I_1(\varphi)$. Then I has reduction number 2, $\mathcal{R}(I)$ is Cohen–Macaulay and its ideal of definition is given by 4 equations. Furthermore $I_1(\varphi)$ is a complete intersection.*

Proof. Let $I_1(\varphi) = \mathbf{x} = [x_1, \ldots, x_r]$ and write

$$J = [f, g, h] = \mathbf{x} \cdot B(\varphi).$$

Note that $B(\varphi)$ is a $r \times 3$ matrix.

Let s be the reduction number of I. There exists a polynomial G in L with unit content and degree $s + 1$. By assumption we have

$$G \cdot \mathbf{x} \subset J.$$

Writing in matrix form we have

$$G \cdot \mathbf{x}^t = C \cdot [f, g, h]^t = C \cdot B(\varphi)^t \cdot \mathbf{x}^t$$

with C an $r \times 3$ matrix of s–forms in the T_i's. In other words

$$(C \cdot B(\varphi)^t - G \cdot I_r) \cdot \mathbf{x}^t = 0,$$

where I_r is the $r \times r$ identity matrix.

The rows of $C \cdot B(\varphi)^t - G \cdot I_r$ are syzygies of $\mathbf{x}$ and therefore, by flatness, must be linear combinations of the syzygies of $\mathbf{x}$ over the ring R. This implies that all the coefficients lie in the extension of the maximal ideal $\mathfrak{m}$ of R (r taken minimal). Reducing modulo $\mathfrak{m}$ we obtain

$$C^* \cdot (B(\varphi)^t)^* = G^* \cdot I_r \neq 0,$$

which implies that $r = 3$ (and $I_1(\varphi)$ is a complete intersection) and

$$\det B(\varphi)^* \quad \neq \quad 0. \tag{8.1}$$

By Proposition 8.2.1, $L = (f, g, h, \det B(\varphi))$. □

Corollary 8.2.4 *If I has reduction number* 2 *then the conditions of* Theorem 8.2.3 *hold.*

Proof. In the proof above it suffices that $I_1(\varphi)$ conducts a polynomial of unit content into J. For reduction number 2, the Lemma does just that. □

Corollary 8.2.5 *The canonical ideal of $\mathcal{R}(I)$ is $I_1(\varphi)\mathcal{R}(I)$.*

Proof. The canonical module of $\mathcal{R}(I)$ is isomorphic to the annihilator of $\mathcal{A}$ [113, Korollar 5.14]. □

These considerations lend support to:

Conjecture 8.2.6 Let R be a regular local ring of dimension 3, and assume 2 is invertible. If I is a 4–generated, normal, prime ideal then it has reduction number 2. In particular, the algebra $R[It]$ is Cohen–Macaulay.

Remark 8.2.7 That not all four–generated prime ideals in a 3–dimensional regular local ring have this property has been shown by M. Johnson [160].

If $I_1(\varphi)$ is a complete intersection but the matrix $B(\varphi)$ is singular mod $\mathfrak{m}$, its minors might still be connected to the reduction number of I. If we denote $P = R[T_1, \ldots, T_4]$, $C(I) = B/(f, g, h, \det B(\varphi))$, the R–resolution of the components of C are

$$0 \to P^3_{t-3} \xrightarrow{B(\varphi)\oplus\psi} P^3_{t-2} \oplus P^3_{t-3} \longrightarrow P^3_{t-1} \oplus P_{t-3} \longrightarrow P_t \longrightarrow C_t \to 0,$$

with ψ the matrix of syzygies of $I_1(\varphi)$.

Let A be the matrix obtained by deleting one column of $B(\varphi)$ and assume that the ideal of $P/\mathfrak{m}P$ generated by its maximal minors has height 2. It follows that C_4 is a torsion–free R–module, from which it is easy to see that the reduction number of I cannot be 3.

If I has more than 4 generators, there is still an expected set of defining equations, at least as long as $I_1(\varphi)$ is a complete intersection. Indeed from a resolution

$$0 \to R^{n-1} \xrightarrow{\varphi} R^n \longrightarrow I \to 0,$$

$I_1(\varphi) = (x, y, z)$, one gets first the equations

$$J_1 = (f_1, \ldots, f_{n-1}) = [T_1, \ldots, T_n] \cdot \varphi = [x, y, z] \cdot B(\varphi),$$

and then, as in the earlier argument, those given by the 3×3 minors of $B(\varphi)$.

Proposition 8.2.8 *Let* $J = (J_1, I_3(B(\varphi)))$. *If* height $J = n - 1$ *then* J *defines* $\mathcal{R}(I)$.

Proof. The ideal J is a proper specialization of the ideal described in [154, Example 3.4], [32, Theorem 3.6]. To prove that it is prime—and therefore the ideal of definition of $\mathcal{R}(I)$—it will be enough to show that the maximal ideal of R is not contained in an associated prime of B/J. To see this, we appeal again to [*loc. cit.*], where it is shown that a power of (x, y, z) generates the canonical module of B/J, which would rule that out. □

Denote by Δ the ideal $I_3(B(\varphi))$ evaluated at $R/\mathfrak{m}$. It is an ideal of 3–forms over the residue field $R/\mathfrak{m}$.

Corollary 8.2.9 *The ideal* J *defines* $\mathcal{R}(I)$ *if and only if* height $\Delta = n - 3$.

Proof. If height $J = n - 1$, it follows by the above and [32] that

$$3 + \text{height } \Delta = \text{height}(\mathfrak{m},\ I_3(B(\varphi))) = n.$$

Conversely, let P be a minimal prime of J. If $P \cap R = \mathfrak{m}$, then height $P \geq 3 + (n-3)$. On the other hand, if $P \cap R = \mathfrak{p} \neq \mathfrak{m}$, the component J_1 will generate the whole of the defining ideal of the Rees algebra of the localization $I_{\mathfrak{p}}$, since this ideal is now of linear type, and therefore height $J \geq n - 1$. □

The equations of the symbolic square

Let I be a prime ideal as above. Denote by C the adjoint of $B(\varphi)$. The condition of normality implies that when evaluated at the origin C does not vanish. This is similar to what happens in 3 generators [288]. For the moment we consider the case when $I_1(\varphi)$ is a complete intersection–which we denote by (x, y, z)–along with $C(0) \neq 0$.

If $u_1, \ldots, u_m$ are the new generators of $I^{(2)}$, we denote by $J(2)$ the kernel of the presentation homomorphism:

$$R[T_1, \ldots, T_4, U_1, \ldots, U_m] \longrightarrow R[It, I^{(2)}t^2].$$

We first determine m, the minimal number of generators of the module $I^{(2)}/I^2$.

In the usual manner, one can get $C_2 = \operatorname{Ext}^3(I^{(2)}/I^2, R)$ through the Weyman's complex:

$$R^3 \otimes R^4 \xrightarrow{\psi} \wedge^2(R^3) \longrightarrow C_2 \longrightarrow 0.$$

Let us proceed as above, using the example as guide. The matrix of ψ is schematically:

$$\begin{bmatrix} 0 & h & -g \\ -h & 0 & f \\ g & -f & 0 \end{bmatrix}.$$

In the example we have

$$\begin{bmatrix} -x & 2y+z & 2y-2x & 4z & -x & z & 2y & -2x^2 & 0 & 0 & 0 & 0 \\ y & -2x^2 & -z & xz & 0 & 0 & 0 & 0 & -x & z & 2y & -2x^2 \\ 0 & 0 & 0 & 0 & y & -2x^2 & -z & xz & x & z-2y & 2x-2y & -4z \end{bmatrix}$$

A computation shows that $C_2 = (R/(x, y, z))^3$; thus $I^{(2)}/I^2$ is generated by 3 elements.

We make use of the method of [291] and [290] to get the equations for the generators. First one identifies some candidates. Thus from the equation in $R[It]$:

$$[x, y, z] \cdot B = 0,$$

we get equations

$$\frac{c_{1i}}{x} = \frac{c_{2i}}{y} = \frac{c_{3i}}{z} = u_i \qquad \text{for } i = 1, 2, 3$$

where c_{ij} are the entries of the adjoint matrix C. This gives rise to a set of nine polynomials

$$xu_i - c_{1i},\ yu_i - c_{2i},\ zu_i - c_{3i},\ i = 1, 2, 3.$$

To get $J(2)$, we must add to $(f, g, h, \det B(\varphi))$ these nine polynomials to begin with. There are three additional, natural, equations to be given next. The ideal defined by the entries of $C(0)$ play a decisive role, similar to the earlier condition $\Delta \neq 0$. In this example, it has height 3.

For a slightly different example, let I be the ideal of the curve

$$\begin{cases} x = t^6 \\ y = t^7 + t^{10} \\ z = t^8 \end{cases}$$

One again verifies that the Rees algebra is Cohen-Macaulay and normal. Its symbolic power algebra is proved to be Noetherian by Huneke [140].

For the matrix B we get:

$$B = \begin{bmatrix} T_1 + T_4 & -2xT_2 & T_3 \\ -T_2 & -T_3 & T_4 \\ -2T_3 & T_1 + xT_4 & -T_2 \end{bmatrix}$$

There is a change from the previous case: The ideal of entries of $C(0)$ now has height 4.

The ideal $J(2)$ was given by the expected equations: If we write

$$J = [x, y, z] \cdot B$$

as above, and denote by C the adjoint of B, the equations were those of J, the entries of the matrix

$$[u_1, u_2, u_3] \cdot B^t$$

and the 9 earlier equations:

$$[u_1, u_2, u_3]^t \cdot [x, y, z] - C.$$

($\det B$ can be obtained from the last two sets of equations.)

Ideals associated to two sequences and a matrix

Let us look for the generic description of these ideals. For a given integer n define the following matrices with indeterminate entries:

$$\begin{cases} \mathbf{x} = \begin{bmatrix} x_1 & \cdots & x_n \end{bmatrix} \\ \\ \mathbf{u} = \begin{bmatrix} u_1 & \cdots & u_n \end{bmatrix} \\ \\ B = \begin{bmatrix} b_{11} & \cdots & b_{1n} \\ \vdots & \ddots & \vdots \\ b_{n1} & \cdots & b_{nn} \end{bmatrix} \end{cases}$$

Denote by C the adjoint of B. Let $H(n)$ be the ideal defined by the entries of the following matrices:

$$\begin{cases} \mathbf{x} \cdot B \\ \mathbf{u} \cdot B^t \\ \mathbf{u}^t \cdot \mathbf{x} - C \end{cases}$$

Note that the matrix of the example above is a specialization of $H(3)$.

Conjecture 8.2.10 The ideal $H(n)$ is a prime, Gorenstein ideal of codimension $2n$.

For $n = 3$ and 4, *Macaulay* shows this to be the case. Furthermore, it provides very good guesses of what their projective resolutions should be like.

Proposition 8.2.11 *If $H(n)$ is unmixed of codimension $2n$ then it is a prime ideal.*

Proof. It suffices to show that the localization $H(n)_{x_1}$ is a prime ideal, generated by a regular sequence of $2n$ elements.

Let $I = (\mathbf{x} \cdot B, x_1 u_i - c_{i1},\ i = 1, \ldots, n)$. I_{x_1} is obviously a prime ideal of height $2n$. We claim that $H(n)_{x_1} = I_{x_1}$.

First, multiplying $\mathbf{x} \cdot B$ by $(c_{11}, \cdots, c_{1n})$, we get that $\det B \in I_{x_1}$. Next, if $C_1 = (c_{11}, \cdots, c_{n1})$, we have

$$x_1 \mathrm{u} B^t = (x_1 \mathrm{u} - C_1) B^t + C_1 B^t$$

whose entries lie in I_{x_1}.

For the last set of equations in $H(n)$, it is enough to show that for any pair of indices $(i,\ k)$, $x_1 c_{ik} - x_k c_{i1}$ lies in I_{x_1}. This follows from straightforward elimination in $\mathbf{x} \cdot B$. □

This ideal contains a smaller Gorenstein ideal, of a kind discovered by Huneke and Ulrich [153]:

Proposition 8.2.12 *Let $L(n) = (\mathbf{x} \cdot B, \mathbf{u} \cdot B^t,\ \det(B))$. $L(n)$ is a prime, Gorenstein ideal of codimension $2n - 1$.*

Proof. $L(n)$ has the following description. Let

$$\mathbf{z} = (x_1, \cdots, x_n, u_1, \cdots, u_n),$$

and put

$$A = \begin{bmatrix} 0 & B \\ -B^t & 0 \end{bmatrix}.$$

$L(n)$ is generated by the entries of $\mathbf{z} \cdot A$ and the Pfaffian of A; it is therefore a specialization of [153, Example 5.12] (see also [174]).

To prove that $L(n)$ is Gorenstein, we must show that it has codimension at least $2n-1$. For that, note the following estimates for the sizes of the determinantal ideals of A

$$\begin{aligned} \text{height } I_{2t}(A) &= (n-t+1)^2 \geq \text{rank } A - 2t + 1 \\ \text{height } I_{2t-1}(A) &= (n-t+1)^2 \geq \text{rank } A - 2t + 2 \end{aligned}$$

hold for all $t < n$, and gives a deficit of 1 for $t = n$. According to Theorem 1.3.5 the ideal $I_1(\mathbf{z} \cdot A)$ has codimension $2n-1$.

The primality assertion follows as in the previous proposition. □

We now give conditions under which $H(3)$ specialize properly to the defining ideals of the symbolic square algebra. Because the equations $H(n)$ always define elements of $I^{(2)}$, there is a map

$$A = R[T_1, T_2, T_3, T_4, U_1, U_2, U_3] \longrightarrow R[It, I^{(2)}t^2],\ U_i \to u_i,$$

whose kernel contains the specialization of $H(3)$ ($I_1(\varphi) = (x_1, x_2, x_3)$):

$$J(2) = ((f,g,h)\cdot B, (U_1,U_2,U_3)\cdot B^t, I_1((U_1,U_2,U_3)^t \cdot (x_1,x_2,x_3) - C)).$$

Theorem 8.2.13 *Let I be a four generated prime ideal as in* Theorem 8.2.3. *If the entries of $C(0) = (C_{ij})$ generate an ideal of height* 4, *then*

$$R[It, I^{(2)}t^2] = A/J(2).$$

Moreover, this algebra is never the full symbolic power algebra.

Proof. If height $I_1(C(0)) = 4$, from the free resolution of such ideals [96] it follows that $I_1(C(0))$ is minimally generated by 9 elements and $\det C(0) \neq 0$. Furthermore, height$(x, y, z, J(2)) = 7$, from which it is easy to see that $J(2)$ must be a prime ideal.

Denote by I_2t^2 the degree 2 component of the image of A:

$$0 \to I_2t^2 \longrightarrow I^{(2)}t^2 \longrightarrow C_2 \to 0.$$

We are going to show that $C_2 = 0$; it will be enough to show that I_2 has projective dimension 1.

The degree 2 component of the ideal $J(2)$, obtained from *Macaulay*, is

$$0 \to R^9 \xrightarrow{\psi} L \oplus L \oplus L \oplus R^9 \longrightarrow J(2)_2 \to 0,$$

where $L = RT_1 + \cdots + RT_4$. Put $\psi = \psi_1 + \psi_2$, $\psi_1 \colon R^9 \longrightarrow L^3$, $\psi_2 \colon R^9 \longrightarrow R^9$.

The map ψ_1 is derived from Koszul relations on the rows of the matrix $B(\varphi)$. If, for simplicity, we write this matrix as

$$\begin{bmatrix} a_1 & a_2 & a_3 \\ b_1 & b_2 & b_3 \\ c_1 & c_2 & c_3 \end{bmatrix},$$

the image of ψ are 9 triples of 1–forms such as $A_1 = (0, -a_3, a_2)$, $A_2 = (-a_3, 0, a_1)$, and $A_3 = (a_2, -a_1, 0)$ derived from the first row of $B(\varphi)$, and similarly $B_1, \ldots, C_3$. We claim they define a direct summand; it is enough to show they generate a pure submodule of rank 9. We will work in the residue field of R but still use A_i, B_i, C_i for the residues of the triples above, without danger of confusion. From a linear relation

$$\sum \alpha_i A_i + \sum \beta_i B_i + \sum \gamma_i C_i = 0,$$

multiplying successively by $a_1, -a_2, a_3$, and adding, we obtain the equation of linear dependence on the elements of $C(0)$

$$\sum (\beta_i C_{i2} + \gamma_i C_{i3}) = 0,$$

proving that $J(2)_2$ is R–free.

Because $J(2)$ is Cohen–Macaulay and $\mathcal{R}_s(I)$ is the ideal transform of the algebra $R[It, I^{(2)}t^2]$ with respect to (x, y, z) (cf. Proposition 7.1.4), equality of these two algebras requires that height$(x, y, z, J(2)) \geq 8$, which cannot hold as we observed earlier. □

Remark 8.2.14 The condition height$(I_1(C(0))) = 4$ is not necessary, as examples show. The proof above uses that $\det(B(0)) \neq 0$, and that the cofactors of any of two rows of $B(0)$ be linearly independent.

Corollary 8.2.15 *If I is a prime ideal as above, then $I^{(2)}/I^2$ is generated by 3 elements.*

In case $d = 4$, $n = 5$, with I a complete intersection in codimension 3, the situation is similar.

5 or more generators

We briefly discuss the similarities between the case of $n = 4$ and ideals with more generators.

Rees algebra

Let $(R, \mathfrak{m})$ be a regular local ring of dimension at least 3 and let I be a CM ideal of codimension 2 generated by n elements ($n \geq 5$)

$$0 \rightarrow R^{n-1} \xrightarrow{\varphi} R^n \longrightarrow I \rightarrow 0.$$

Assume further that for each prime ideal $P \neq I$, or $\mathfrak{m}$, I_P can be generated by less that height$(P) - 1$ elements (e.g. I is a complete intersection on the punctured spectrum). This implies that the ideal of the coefficients of its first-order syzygies, $I_1(\varphi)$, is $\mathfrak{m}$–primary. We shall assume that it is a complete intersection $\mathbf{x} = (x_1, \ldots, x_d)$.

Note that if $n \leq d$, then I is of linear type, while $n < d$ implies that the Rees and symbolic power algebras of I coincide (e.g. [116]).

We write the equations of the symmetric algebra of I in its dual form:

$$[f_1, \ldots, f_{n-1}] = [T_1, \ldots, T_n] \cdot \varphi = [x_1, \ldots, x_d] \cdot B(\varphi).$$

The assumption implies that the first symmetric power of I with torsion is $S_d(I)$. It follows that the d–sized minors of the matrix $B(\varphi)$ are elements in the equations J of the Rees algebra of I.

Proposition 8.2.16 *Denote by Δ the ideal obtained by evaluating $I_d(B(\varphi))$ at the origin of R, that is, in $R/\mathfrak{m}[T_1, \ldots, T_n]$. If* height$(\Delta) = n - d$ *(its maximum size), then*

$$J = (\mathbf{x} \cdot B(\varphi), I_d(B(\varphi))).$$

Proof. The assumption implies that height$(\mathfrak{m}, \mathbf{x} \cdot B(\varphi), I_d(B(\varphi))) = n$, from which we get that no associated prime of $(\mathbf{x} \cdot B(\varphi), I_d(B(\varphi)))$, of codimension $n-1$, can contain $\mathfrak{m}$. On the other hand, inverting an element of $\mathfrak{m}$, $(\mathbf{x} \cdot B(\varphi))$ localizes into an ideal of codimension $n-1$, since by assumption I is of linear type on the punctured spectrum.

But with height $(\mathbf{x} \cdot B(\varphi), I_d(B(\varphi))) = n-1$, the ideal is a proper specialization of [153, Example 5.12] and is therefore Cohen–Macaulay. Together with the estimate above on Δ one gets that it is prime as well. $\square$

Symbolic square algebra

There is much variance between the cases $n = 4$ and $n \geq 5$. For simplicity, we assume $d = 3$ and $n = 5$.

Write

$$B(\varphi) = \begin{bmatrix} t_{11} & t_{12} & t_{13} & t_{14} \\ t_{21} & t_{22} & t_{23} & t_{24} \\ t_{31} & t_{32} & t_{33} & t_{34} \end{bmatrix}$$

where the t_{ij} are independent indeterminates. Each pair of equations in the Rees algebra of I

$$\begin{aligned} x_1 t_{1i} + x_2 t_{2i} + x_3 t_{3i} &= 0 \\ x_1 t_{1j} + x_2 t_{2j} + x_3 t_{3j} &= 0 \end{aligned}$$

gives rise by elimination to equations

$$\frac{t_{2i}t_{3j} - t_{2j}t_{3i}}{x_1} = -\frac{t_{1i}t_{3j} - t_{1j}t_{3i}}{x_2} = \frac{t_{1i}t_{2j} - t_{1j}t_{2i}}{x_3}$$

Since this (quadratic) ratio is conducted into the Rees algebra of I by elements outside of I, they define an element, denoted by u_{ij}, in the symbolic square of I. (In this case, $n = 5$, there are 6 such elements; we write $\mathbf{u} = (u_1, \ldots, u_6)$ for them.)

Let us assemble a presumed set of equations for $J(2)$. Denote by C the matrix of 2×2 minors of $B = B(\varphi)$. One starts with

$$\begin{cases} \mathbf{x} \cdot B \\ I_1(\mathbf{x}^t \cdot \mathbf{u} - C) \end{cases}$$

In addition, we have 12 equations arising from cofactor expansions, e.g.

$$u_1 t_{13} + u_2 t_{12} + u_4 t_{11}$$

along with the Plücker relation

$$u_1 u_6 - u_2 u_5 + u_3 u_4.$$

A computation on these 35 polynomials with *Macaulay*, and appealing to [264, Theorem 4.4], showed they define a Gorenstein prime ideal of codimension 10. The computed system of parameters was :

$$\{x_1, x_2, u_1, u_2 - x_3 + t_{31} + t_{12}, t_{11}, t_{21} - u_5 + u_6, t_{23} - u_3 - t_{31} + t_{13}, t_{14} - u_4,$$
$$t_{33} - 3u_3 - 2u_5 + u_6 + 2t_{31} + 3t_{12} + 2t_{32} - t_{13} - 2t_{23}, t_{24} - t_{32}, t_{34} - t_{23}\}$$

It is easy to see however that it does not specialize properly to the ideal defining the symbolic square algebra.

Non–perfect ideals

For comparison, we examine one case of non–perfect prime ideals. Let R be a regular local ring of dimension 4, and let I be a prime ideal of codimension 2 with a resolution

$$0 \rightarrow R^p \xrightarrow{\psi} R^m \xrightarrow{\varphi} R^4 \longrightarrow I \rightarrow 0.$$

Rees algebra

We begin by noting that $I_p(\psi)$ has codimension 4, making it straightforward to show that the Weyman complex for the symmetric square of I is acyclic:

$$0 \rightarrow D_2(R^p) \xrightarrow{\varphi_4} R^p \otimes R^m \xrightarrow{\varphi_3} R^p \otimes R^4 \oplus \wedge^2 R^m \rightarrow R^m \otimes R^4 \rightarrow S_2(R^4) \rightarrow S_2(I) \rightarrow 0.$$

It follows that $S_2(I)$ has torsion. To identify it:

Proposition 8.2.17 *Let*

$$0 \rightarrow \delta(I) \longrightarrow S_2(I) \longrightarrow I^2 \rightarrow 0$$

be the canonical mapping from $S_2(I)$ onto I^2. Then $\delta(I) \simeq \mathrm{Ext}^2(I, R)$. In particular, to obtain the equations of the Rees algebra of I we must add p equations to

$$L_1 = [T_1, T_2, T_3, T_4] \cdot \varphi.$$

Proof. Because I is perfect on the punctured spectrum, $\delta(I)$ is a module of finite length. The cohomology sequence then gives

$$\mathrm{Ext}^4_R(\delta(I), R) \simeq \mathrm{Ext}^4_R(S_2(I), R) = \mathrm{coker}(\varphi_4{}^*) = S_2(C),$$

where we put $C = \mathrm{coker}(\psi^*)$.

There is another way to describe this module. Let Z denote the first–order syzygies of I. From the sequence

$$0 \to Z \longrightarrow R^4 \longrightarrow I \to 0$$

we get an acyclic portion of the approximation complex

$$0 \to (\wedge^2 Z)^{**} \longrightarrow Z \otimes R^4 \longrightarrow S_2(R^4) \longrightarrow S_2(I) \to 0.$$

Moreover as Z has rank 3, the canonical pairing $Z \otimes \wedge^2 Z \longrightarrow R$ identifies $(\wedge^2 Z)^{**}$ with Z^*. We then have

$$\mathrm{Ext}^4_R(S_2(I), R) \simeq \mathrm{Ext}^2_R(Z^*, R).$$

For the computation of this module one starts from the sequence

$$\begin{array}{ccccccccc} 0 & \to & R & \longrightarrow & R^{4*} & \longrightarrow & Z^* & \longrightarrow & \omega & \to & 0 \\ & & & & & \searrow & & \nearrow & & & \\ & & & & & & L & & & & \end{array}$$

where ω is the canonical module of R/I. From the cohomology exact sequence

$$0 \to \mathrm{Ext}^1_R(Z^*, R) \to \mathrm{Ext}^1_R(L, R) = R/I \to \mathrm{Ext}^2_R(\omega, R) \to \mathrm{Ext}^2_R(Z^*, R) \to 0$$

it follows that $\mathrm{Ext}^1_R(Z^*, R)$ has to vanish, since it has finite length and therefore cannot embed in R/I. This yields the exact sequence

$$0 \to R/I \longrightarrow \mathrm{Ext}^2_R(\omega, R) \longrightarrow \mathrm{Ext}^2_R(Z^*, R) \to 0.$$

After noting that the mid module has depth 2, one obtains

$$\mathrm{Ext}^4_R(\mathrm{Ext}^4_R(\delta(I), R), R) \simeq \mathrm{Ext}^4_R(\mathrm{Ext}^2_R(Z^*, R), R) \simeq \mathrm{Ext}^3_R(R/I, R) = C.$$

The assertion now follows by duality. □

It is not clear where we should look for the missing p equations, except when $p = 1$. This case has several singular features, some of which are discussed in Chapter 7. They arise from the fact that $\psi(1)^* = \mathbf{x} = [x_1, x_2, x_3, x_4]$ is a regular sequence and thus the rows of φ are combinations of Koszul relations on $\mathbf{x}$. As a consequence, we can write

$$\mathbf{T} \cdot \varphi = \mathbf{x} \cdot B(\varphi)$$

with $B(\varphi)$ a skew symmetric matrix of linear forms in $\mathbf{T}$. Let G be the Pfaffian of $B(\varphi)$. We seek conditions under which $J = (\mathbf{x} \cdot B(\varphi), G)$ are the equations of $\mathcal{R}(I)$.

Let $C(I)$ denote the algebra $R[\mathbf{T}]/J$. Since J is a proper specialization of [153, Example 5.12], J is a Gorenstein ideal of codimension 3; in particular, $J \neq L_1$, because the latter is an almost complete intersection. On the other hand, the degree 2 component of $C(I)$ is isomorphic to I^2. This is seen by looking at the projective resolution of J: the component of degree 2 in $\mathbf{T}$ has projective dimension 2. That is, we have a sequence

$$0 \to \mathcal{B} \longrightarrow C(I) \longrightarrow \mathcal{R}(I) \to 0,$$

and J is prime if and only if $\mathcal{B} = 0$; note that already $\mathcal{B}_2 = 0$.

Theorem 8.2.18 *The equations of $\mathcal{R}(I)$ are given by J if and only if the Rees algebra is quasi–Gorenstein. This condition will hold if I is generated by 3 elements on the punctured spectrum.*

Proof. Because $C(I)$ is Gorenstein, of the same dimension as $\mathcal{R}(I)$, the module

$$\mathrm{Hom}_{C(I)}(\mathcal{R}(I), C(I)) = 0 \colon \mathcal{B}$$

is the canonical module of $\mathcal{R}(I)$ [113, Korollar 5.14]. It is therefore cyclic, necessarily generated by a homogeneous element of $C(I)$, if and only if $\mathcal{R}(I)$ is quasi–Gorenstein. On the other hand, $\mathcal{B}$, like the kernel of $S(I)$ onto $\mathcal{R}(I)$, is annihilated by a power of I, which is a contradiction unless $\mathcal{B} = 0$.

The other condition means that $\mathcal{B}$ is annihilated by a power of the maximal ideal $\mathfrak{m}$ of R, because I is of linear type on the punctured spectrum. But height $\mathfrak{m}C(I) \geq 1$ as height $J = 3$, and therefore it contains regular elements since $C(I)$ is Cohen–Macaulay. It shows that $\mathcal{B} = 0$. □

Remark 8.2.19 (i) The first condition is realized if $\mathcal{R}(I)$ is already the symbolic power algebra of I, for according to [256], $\mathcal{R}_s(I)$ is then quasi–Gorenstein. (ii) The argument also implies that if the canonical module of $\mathcal{R}(I)$ is minimally generated by two elements, then I is set–theoretically generated by 2 elements.

Symbolic square algebra

We characterize the equality of the two algebras.

Proposition 8.2.20 *The algebras $\mathcal{R}(I)$ and $\mathcal{R}_s(I)$ are isomorphic if and only if*

$$G(0) \neq 0 \text{ and height } I_2(\varphi) = 4.$$

Proof. Equality (for $\dim R = 4$) implies that I is a complete intersection in the punctured spectrum–that is height $I_2(\varphi) = 4$. Grade the ring $B = R[T_1, \ldots, T_4]$ in the $\mathbf{T}$ variables. Since, by the Remark 8.2.19, $\mathcal{R}(I) = B/J$, the graded components of the resolution of J [36] give R–resolutions for the powers of I:

$$0 \to B_{t-3} \longrightarrow B_{t-2}^4 \oplus B_{t-1} \longrightarrow B_{t-1}^4 \oplus B_{t-2} \longrightarrow B_t \longrightarrow I^t \to 0.$$

It says that if all the R–coefficients of the ideal J lie in the maximal ideal of R, then the powers $I^t, t \geq 3$, have projective dimension 3. This is not possible if $I^t = I^{(t)}$.

The equality $C(I) = \mathcal{R}(I)$ follows from Theorem 8.2.18. The resolution says that each I^t has projective dimension at most 2. On the other hand, since the I is a complete intersection on the punctured spectrum, the ideals I^t do not have associated primes of height 3 either. In turn, this clearly implies that I is of linear type on the punctured spectrum. □

The symbolic cube algebra

There are very few general descriptions of the equations of the symbolic cube algebra $R[It, I^{(2)}t^2, I^{(3)}t^3]$ of an ideal. Even for the ideals defining a monomial space curve the picture is incomplete.

One of these cases is the following theorem of Herzog ([112]):

Theorem 8.2.21 *Let $(R, \mathfrak{m})$ be a regular local ring of dimension 3, and let I be a prime ideal generated by the minors $\Delta_1, \Delta_2, \Delta_3$ of the matrix*

$$\varphi = \begin{bmatrix} x_1 & x_2 & x_3 \\ x_3^2 a_1 & x_1^2 a_2 & x_2^2 a_3 \end{bmatrix} \tag{8.2}$$

(a) *If 2 is a unit in R, then there exist u_1, u_2, u_3 in $I^{(3)}$, such that their images generate $I^{(3)}/I^3$, and the u_i satisfy the following equations*

$$\begin{aligned} x_1u_1 &= \frac{1}{2}(a_3\Delta_3^3 + a_1\Delta_2^2\Delta_1) \\ x_2u_2 &= \frac{1}{2}(a_1\Delta_1^3 + a_2\Delta_3^2\Delta_2) \\ x_3u_3 &= \frac{1}{2}(a_2\Delta_2^3 + a_3\Delta_1^2\Delta_3) \\ x_1u_2 + x_2u_1 &= -a_1\Delta_1^2\Delta_2 \\ x_1u_3 + x_3u_1 &= -a_3\Delta_3^2\Delta_1 \\ x_2u_3 + x_3u_2 &= -a_2\Delta_2^2\Delta_3 \end{aligned}$$

(b) *If char $R = 2$, there exists $u \in I^{(3)}$ whose image generates $I^{(3)}/I^3$, and u satisfies the following equations*

$$\begin{aligned} x_1x_2x_3u &= x_1a_2\Delta_2^2\Delta_3 + x_2a_3\Delta_3^2\Delta_1 + x_3a_1\Delta_2^2\Delta_1 \\ x_1^2u &= a_3\Delta_3^3 + a_1\Delta_2^2\Delta_1 \\ x_2^2u &= a_1\Delta_1^3 + a_2\Delta_3^2\Delta_2 \\ x_3^2u &= a_2\Delta_2^3 + a_3\Delta_1^2\Delta_3 \end{aligned}$$

A computer proof

The case (a) is susceptible of the following computer algebra treatment. Suppose the different symbols in the matrix (8.2) are independent indeterminates. Let $I = (\Delta_1, \Delta_2, \Delta_3)$; the symbolic powers of I are given by

$$I^{(n)} = \bigcup_{t \geq 1} (I^n : x_1^t).$$

Theorem 8.2.22 *$\{x_1, x_2\}$ is a regular sequence on $I^{(3)}$, and $I^{(3)}/I \cdot I^{(2)}$ is minimally generated by 3 elements.*

Proof. $I^{(2)}$ and $I^{(3)}$ were defined in the computer (we never actually saw them) and proceeded to verify the assertions. □

Full sets of equations for certain monomial ideals are found in [244]. On the other hand, M. Vaz Pinto has shown that for monomial curves, $I^{(3)}$ needs $1, 2$ or 3 fresh generators, and all cases occur (see also [167]).

Exercise 8.2.23 (Bernd Ulrich) Prove the following abstracted form of the matrix argument of Theorem 8.2.3: Let $I \subset J$ be ideals, and denote by content(J) the ideal generated by the entries of a presentation matrix for J. Suppose

$$I\colon J \not\subset \sqrt{\text{content}(J)}.$$

Then $\nu(J) \leq \nu(I)$.

Exercise 8.2.24 Let I be a four–generated prime ideal in a 3–dimensional regular local ring. Let φ be its 4×3 presentation matrix. Suppose $I_1(\varphi)$ is a complete intersection (a, b, c) and these elements occur in one row of the matrix φ. Show that $\det B(\varphi)^* \neq 0$ (see equation (8.1)), and therefore the Rees algebra of I is Cohen–Macaulay by Theorem 8.2.3.

8.3 Higher Codimension

The ways to get the equations of the Rees algebras of ideals of codimension three or higher are markedly different from what we have seen thus far. In fact, the only systematic approach known to us are the computational methods of Chapter 10. We outline some of the reasons for this state of affairs.

Let R be a regular local ring of dimension d, and let I be a codimension three Cohen–Macaulay ideal with a resolution

$$0 \to R^p \xrightarrow{\ \phi\ } R^m \xrightarrow{\ \varphi\ } R^n \longrightarrow I \to 0.$$

Suppose $\mathbf{x}$ is the vector of coefficients of the matrix φ, and let

$$\mathbf{T} \cdot \varphi = \mathbf{x} \cdot B(\varphi)$$

be the equations of the symmetric algebra of I. Assume that $\mathbf{x}$ defines a complete intersection of codimension ℓ. Then $I_\ell(B(\varphi))$ is contained in the equations of the Rees algebra $\mathcal{R}(I)$, an ideal of codimension $n - 1$. It follows that the ideal $(\mathbf{x} \cdot B(\varphi), I_\ell(B(\varphi)))$ will never have the codimension of the generic case—quite unlike what occurred in codimension two.

Syzygetic torsion

We begin by noting that the complexes of [307] no longer provide projective resolutions of $S_2(I)$, even in the case of a complete intersection, which inhibits the comparison between $S_2(I)$ and I^2, and therefore the identification of $\delta(I)$.

For this purpose we use an approach suggested by [124, Theorem 2.1]. Denote by ω the canonical module of R/I and assume that I is syzygetic on the punctured spectrum. There are exact sequences

$$0 \to \mathrm{Tor}_1(I, R/I) \longrightarrow Z/IZ \longrightarrow (R/I)^n \longrightarrow I/I^2 \to 0$$

and

$$0 \to \mathrm{Tor}_2(I, R/I) \longrightarrow (R/I)^p \longrightarrow (R/I)^m \longrightarrow Z/IZ \to 0,$$

where Z denotes the first–order syzygies of I. If $2R = R$, we can identify

$$\mathrm{Tor}_1(I, R/I) \simeq \wedge^2 I \oplus \delta(I)$$

and

$$\mathrm{Tor}_2(I, R/I) \simeq \mathrm{Hom}(\omega, R/I) = \omega^*.$$

These isomorphisms imply that $\delta(I)$, which by assumption has finite length, is the R/I–torsion submodule of Z/IZ and therefore we have

$$\mathrm{Ext}_R^d(\delta(I), R) \simeq \mathrm{Ext}_R^{d-2}(\omega^*, R).$$

There is a simple way to describe ω^*. First, assume that $u = (a_1, \ldots, a_p)$ is an element in the kernel of $\phi \otimes R/I$, with say $a_1 \neq 0$. The restriction to ω of the projection on the first component of $(R/I)^p$ defines a nonzero mapping from ω^* onto an ideal L of R/I; since these are modules of rank 1, $\omega^* \simeq L$. From each $p-1 \times p$ submatrix of ϕ the vector $v = (\Delta_1, \ldots, \Delta_p)$ of its $p-1$–sized minors, mod I, belongs to ω^*. Let D be the submodule of L generated by the Δ_1's; D is the image of the ideal generated by the maximal minors of the matrix obtained by deleting the first column of ϕ.

Proposition 8.3.1 *Let R be a Gorenstein local ring and let I be an ideal that is Gorenstein on the punctured spectrum, then*

$$\delta(I) = \mathrm{Ext}_R^d(\mathrm{Ext}_R^{d-1}(R/(I, D), R), R).$$

Proof. We claim that $\mathrm{Ext}_R^{d-2}(D, R) = \mathrm{Ext}_R^{d-2}(L, R)$. Indeed, denote by C the cokernel of the embedding

$$0 \to D \longrightarrow L \longrightarrow C \to 0.$$

It suffices to show that C has finite length.

If I is a Gorenstein ideal, then all the modules in the exact sequences above are Cohen–Macaulay. This is clear for all modules, with the possible exception of $\mathrm{Tor}_1(I, R/I) = \wedge^2 I$. But this module has projective dimension 3, as an R–module, according to [179]. This means that ω^* splits off $(R/I)^p$. From this it is easy to see that $D = L$. □

Strongly Cohen–Macaulay ideals

The following content formula is very similar to what we have seen in the previous sections.

Theorem 8.3.2 *Let R be a Gorenstein local ring of dimension d and let I be a strongly Cohen–Macaulay ideal of codimension* $g \geq 2$*. Suppose* $\nu(I) = n = d+1$ *and I satisfies* $\mathcal{F}_1$ *on the punctured spectrum of R. Then the symmetric and Rees algebras of I first differ in degree* $d-g+2$*. Moreover in the canonical homomorphism*

$$0 \to \mathcal{A}_{d-g+2} \longrightarrow S_{d-g+2}(I) \longrightarrow I^{d-g+2} \to 0$$

$$\mathcal{A}_{d-g+2} = \operatorname{Ext}^d_R(R/I_1(\varphi), R),$$

where φ *is the matrix of a minimal presentation of I.*

Proof. The hypothesis says that I satisfies the condition $\mathcal{F}_0$, so that we can use Theorem 3.3.19 which assures that the complex $\mathcal{Z}(I)$ is acyclic (and $S(I)$ is Cohen–Macaulay).

Since I has codimension g, the modules of cycles Z_i of the Koszul complex of I have depth $d-g+2$, for $i = 1, \ldots, d-g+2$. Consider the component $\mathcal{Z}_t$ of $\mathcal{Z}(I)$ lying over some $S_t(I)$. If $t < d-g+2$, the acyclicity of $\mathcal{Z}_t$ implies that $S_t(I)$ has depth at least 1. Since I is of linear type on the punctured spectrum of R, each $\mathcal{A}_t$ is a module of finite length. It follows that $S_{d-g+2}(I)$ is the first component that may have non–trivial torsion.

To determine $\mathcal{A}_{d-g+2}$, we make use of the obvious isomorphism

$$\operatorname{Ext}^d_R(\mathcal{A}_{d-g+2}, R) \simeq \operatorname{Ext}^d_R(S_{d-g+2}(I), R).$$

Chasing the depths in the complex

$$0 \to Z_{d-g+2} \to Z_{d-g+1} \otimes B_1 \to \cdots \to Z_1 \otimes B_{d-g+1} \to B_{d-g+2} \to S_{d-g+2}(I) \to 0$$

we obtain the exact sequence

$$\operatorname{Ext}^{g-2}_R(Z_{d-g+1} \otimes B_1, R) \xrightarrow{\psi} \operatorname{Ext}^{g-2}_R(Z_{d-g+2}, R) \longrightarrow \operatorname{Ext}^d_R(S_{d-g+2}(I), R) \to 0.$$

It is now easy to verify that

$$\operatorname{Ext}^{g-2}_R(Z_{d-g+1}, R) = H_1(I)$$

and

$$\operatorname{Ext}^{g-2}_R(Z_{d-g+2}, R) = R/I.$$

Furthermore, it is not difficult to see that the mapping

$$\psi \colon H_1(I) \otimes B_1^* \to R/I$$

just reads the coefficients of the various syzygies of I. □

It is not clear how to extend this formula to larger sets of ideals, such as those with sliding depth.

Example 8.3.3 The Rees algebras of Gorenstein ideals of codimension three is a rich source of interesting anomalies. For an example, consider the ideal I defined by the Pfaffians of

$$\varphi = \begin{bmatrix} 0 & -a^2 & -b^2 & -c^2 & -d^2 \\ a^2 & 0 & -d^2 & -ab & -c^2 \\ b^2 & d^2 & 0 & -a^2 & -ab \\ c^2 & ab & a^2 & 0 & -b^2 \\ d^2 & c^2 & ab & b^2 & 0 \end{bmatrix}.$$

The content ideal, $I_1(\varphi) = (a^2, b^2, c^2, d^2, ab)$, has type two so that $\mathcal{A}_3$ is generated by 2 elements. A computation showed that the Rees algebra of I is generated by the linear forms coming from the syzygies, these two 3–forms, and a sextic. The algebra $\mathcal{R}(I)$ is neither normal nor Cohen–Macaulay.

Exercise 8.3.4 Let X be a generic $n\times(n+1)$ matrix and let X' be the matrix obtained by deleting two columns of X.

$$I = I_n(X) + I_{n-1}(X').$$

Show that I is a Gorenstein ideal of linear type.

Exercise 8.3.5 Let R be a regular local ring and let E be a reflexive module of rank two such that $S(E)$ is a normal quasi–Gorenstein algebra. If E is free in codimension three, then $S(E)$ is factorial. (*Hint*: Show that $Cl(S(E))$ is generated by the $cl(T(\wp))$ for which $E_\wp$ has projective dimension at most 1. Then use the fact that there is always an isomorphism $E \simeq E^*$ (rank $E = 2$), to show that such prime ideals $\wp$ have codimension at most three.)

8.4 Ideals with Linear Presentation

This section has aims similar to those of section 2 as it examines ideals that fail minimally of being of linear type. We follow [279] closely. Throughout we assume R is either a Cohen–Macaulay local ring of dimension d or a polynomial ring $k[x_1, \ldots, x_d]$. It will be one of the few cases where the structure of the kernel of the natural mapping

$$0 \rightarrow \mathcal{A} \longrightarrow S(I) \longrightarrow \mathcal{R}(I) \rightarrow 0$$

throws some light on the Cohen–Macaulayness of $\mathcal{R}(I)$.

Analytic spread and reduction number

We begin with the determination of the analytic spread of such ideals.

Proposition 8.4.1 *Let $(R, \mathfrak{m})$ be a Cohen–Macaulay local ring of dimension d, and let I be an ideal of codimension at least 1. Suppose that $\nu(I) = d+1$ and that I satisfies sliding depth and $\mathcal{F}_1$ on the punctured spectrum of R. Then $\ell(I) = d$.*

Proof. Let

$$0 \to \mathcal{A} \longrightarrow S(I) \longrightarrow \mathcal{R}(I) \to 0$$

be the canonical isomorphism. By assumption $\mathcal{A}$ is annihilated by a power $\mathbf{m}^r$ of the maximal ideal. The equation $\mathbf{m}^r \cdot \mathcal{A} = 0$ implies, by Lemma 4.3.10, that

$$\text{grade}\,(\mathbf{m}^r S(I), \mathcal{A}) \leq 1.$$

Since $S(I)$ is a Cohen–Macaulay ring by Theorem 3.3.19, the height of this ideal will also be bounded by 1. It follows that the image of $\mathcal{A}$ in $S(I) \otimes_R (R/\mathbf{m})$ has height at most 1, and therefore the analytic spread of I is at least d and thus equals d. □

Expected reduction number

Theorem 8.3.2 shows that $\mathcal{A}_{d-g+2} \neq 0$. It may occur (cf. Example 8.3.3) that this component is contained in $\mathbf{m} \cdot S(I)$, that is, the reduction number of I is greater than $d - g + 1$. Nevertheless, we shall take $d - g + 1$ to represent the *expected* reduction number of the ideal I.

Proposition 8.4.2 *Let R be a Gorenstein local ring with infinite residue field and I be an ideal as* Theorem 8.3.2. *Let*

$$R^m \xrightarrow{\varphi} R^{d+1} \longrightarrow I \to 0$$

be a minimal presentation of I. If I has reduction number $d - g + 1$, then $\nu(I_1(\varphi)) \leq m$.

Proof. Denote by $\mathbf{x} = [x_1, \ldots, x_r]$ a minimal set of generators of the ideal $I_1(\varphi)$. We write the equations of the symmetric algebra of $S(I)$ in its dual form:

$$J = \mathbf{T} \cdot \varphi = \mathbf{x} \cdot B(\varphi).$$

By assumption there exists $f \in \mathcal{A}_{d-g+2}$, $f \notin \mathbf{m}S$. By Theorem 8.3.2, the ideal $(\mathbf{x})$ conducts f into J, which means that we can write

$$f \cdot \mathbf{x}^t \;=\; G \cdot B(\varphi)^t \cdot \mathbf{x}^t,$$

where G is a $r \times m$ matrix. We thus have

$$(f \cdot I_r - G \cdot B(\varphi)^t) \cdot \mathbf{x}^t \;=\; 0,$$

that is,

$$f \cdot I_r - G \cdot B(\varphi)^t \tag{8.3}$$

consists of syzygies of $\mathbf{x}$ and thus has all of its entries in $\mathbf{m}$. This implies that reducing the matrix (8.3) mod $\mathbf{m}$, we must have that the rank of $B(\varphi)$ (mod $\mathbf{m}$) is at least r, and therefore $m \geq r$. □

Corollary 8.4.3 *Let $(R, \mathbf{m})$ be a regular local ring of dimension d, with infinite residue field and let I be a Gorenstein ideal of codimension three. Suppose $\nu(I) = d + 1$, I is $\mathcal{F}_1$ on the punctured spectrum of R, and has reduction number $d - 2$. Then $\nu(I_1(\varphi)) = d$, in particular $I_1(\varphi)$ is generated by a system of parameters.*

Proof. We use the notation of Proposition 8.4.2. Now $m = d+1$, and φ is alternating. From

$$0 = \mathbf{T} \cdot \varphi \cdot \mathbf{T}^t = \mathbf{x} \cdot B(\varphi) \cdot \mathbf{T}^t,$$

we have that $B(\varphi) \cdot \mathbf{T}^t$ is a syzygy of $\mathbf{x}$ and thus has all of its entries in $\mathbf{m}$. This implies that the matrix $B(\varphi)$ mod $\mathbf{m}$ has $\mathbf{T}$ in the kernel, so it has rank at most d. But $I_1(\varphi)$ is a $\mathbf{m}$–primary ideal so $r \geq d$ and thus $r = d$. □

Cohen–Macaulay algebras

The following conjecture will be the focus of our attention from now on.

Conjecture 8.4.4 *Let R be a regular local ring of dimension d and let I be a strongly Cohen–Macaulay ideal of codimension g. Suppose $\nu(I) = d + 1$ and I satisfies $\mathcal{F}_1$ on the punctured spectrum of R. If the reduction number of I is $d - g + 1$ then $\mathcal{R}(I)$ is Cohen–Macaulay.*

The number of equations of the Rees algebra

We give a count for the number of generators of $\mathcal{A}$ for ideals with a linear presentation. Example 8.3.3 will not fit this description but we expect prime ideals to conform to it.

Theorem 8.4.5 *Let $R = k[x_1, \ldots, x_d]$ be a polynomial ring and let I be a strongly Cohen–Macaulay ideal of codimension g having a linear presentation matrix. Suppose $\nu(I) = d + 1$ and I satisfies $\mathcal{F}_1$ on the punctured spectrum of R. Then I has reduction number $d - g + 1$ (in case k is infinite), $\mathcal{A}$ is generated by a single element, and $S(I)$ is reduced.*

We start with a result in the folklore of the theory of Rees algebras. Let I be generated by forms $\{f_1, \ldots, f_q\}$ of the same degree r, and let $k[M]$ denote the k–subalgebra of $\mathcal{R}(I)$ generated by $f_j t$. Set $\mathbf{m} = (x_1, \ldots, x_d)$.

Proposition 8.4.6 *There is a decomposition*

$$\mathcal{R}(I) = k[M] + \mathbf{m} \cdot \mathcal{R}(I),$$

where

$$k[M] \cap \mathbf{m} \cdot \mathcal{R}(I) = 0.$$

In particular, the special fiber of $\mathcal{R}(I)$, $\mathcal{R}(I) \otimes (R/\mathbf{m})$, is an integral domain.

Proof. It is clear. □

Proof of Theorem 8.4.5. The prime ideal $\mathcal{A}$ is a primary component of (0) in $S(I)$. We claim that $\mathbf{m}S(I)$ is the other primary component. Let

$$k[\mathbf{T}]^m \xrightarrow{B(\varphi)} k[\mathbf{T}]^d \longrightarrow F \to 0$$

be the dual Jacobian of I ([257]). We have that

$$S(I) = S_{k[\mathbf{T}]}(F),$$

so that the prime ideal which is the $k[\mathbf{T}]$–torsion submodule of $S(I)$ is also a primary component of (0). But this component contains a power of $\mathbf{m}S(I)$ so must equal this prime ideal.

The equality $0 = \mathbf{m}S(I) \cap \mathcal{A}$ yields the exact sequence

$$0 \to \mathcal{A} \longrightarrow S(I) \otimes (R/\mathbf{m}) \longrightarrow \mathcal{R}(I) \otimes (R/\mathbf{m}) \to 0.$$

From Proposition 8.4.1 we have that $\mathcal{A}$ is a height 1 ideal, while Proposition 8.4.6 says that it is prime. This shows that $\mathcal{A}$ is generated by one single element in degree $d - g + 2$. □

Characteristic polynomial and dual module

The preceding arguments raise the issues of the explicit determinations of the module F and of the polynomial f that generates $\mathcal{A}$. With a slight abuse of terminology we use the following

Definition 8.4.7 In the setting of Theorem 8.4.5, let f be a homogeneous polynomial in $k[\mathbf{T}] = k[T_1, \ldots, T_{d+1}]$ that generates $\mathcal{A}$. We call f the *characteristic polynomial* of the ideal $\mathcal{A}$.

We point out two properties of f and F, the first of which is already embedded in the previous proof.

Proposition 8.4.8 *Let I be an ideal as in* Theorem 8.4.5. *The Rees algebra* $\mathcal{R}(I)$ *has the following representation,*

$$\mathcal{R}(I) \simeq S_{k[\mathbf{T}]/(f)}(F),$$

where f is an irreducible polynomial and F is an ideal of $k[\mathbf{T}]/(f)$, of linear type.

The last assertion implies that viewed as a $Q = k[\mathbf{T}]/(f)$–module, F satisfies the condition $\mathcal{F}_1$—which means that in the ring $k[\mathbf{T}]$ the heights of the determinantal ideals of the matrix $B(\varphi)$ satisfy the estimates

$$\text{height } I_t(B(\varphi)) \geq d - t + 2,\ 1 \leq t \leq d - 1.$$

Proposition 8.4.9 *Let I be an ideal as in* Theorem 8.4.5. *The localization*

$$\mathcal{R}(I)_{\mathbf{m}\mathcal{R}(I)}$$

is a discrete valuation domain, and $\mathcal{R}(I)$ satisfies (S_3).

Proof. By Proposition 8.4.6 we know that $\mathbf{m}\mathcal{R}(I)$ is a prime ideal, and by Proposition 8.4.8,

$$\mathcal{R}(I)_{\mathbf{m}\mathcal{R}(I)} \simeq S_Q(F)_{(\mathbf{x})} \simeq S_K(F \otimes_R K)_{(\mathbf{x})},$$

where K denotes the quotient field of Q. But now $S_K(F \otimes_R K)$ is a polynomial ring over K, and therefore $\mathcal{R}(I)_{(\mathbf{m})\mathcal{R}(I)}$ is a regular local ring, necessarily of dimension one (cf. Proposition 8.4.1).

To prove that $\mathcal{R}(I)$ satisfies (S_3), let $\mathbf{q}$ be a prime ideal of $\mathcal{R}(I)$ such that $\mathcal{R}(I)_\mathbf{q}$ is not Cohen–Macaulay. We want to think of $\mathbf{q}$ as a prime ideal of $B[T_1, \ldots, T_{d+1}]$. By our assumption on I, $\mathbf{m} \subset \mathbf{q}$, hence $\mathbf{q} = (\mathbf{m}, \mathbf{p})B$ with $\mathbf{p} = \mathbf{q} \cap k[\mathbf{T}]$.

We first suppose that $\dim \mathcal{R}(I)_\mathbf{q} \leq 3$, in which case $\dim k[\mathbf{T}]_\mathbf{p} \leq 3$. Now $\mathbf{q}$ contains the characteristic polynomial f, and $\mathcal{R}(I)_\mathbf{q} \simeq (S_Q(F))_\mathbf{q}$ is not Cohen–Macaulay, hence $\mathbf{p} \in V(F) \subset \operatorname{Spec}(Q)$, where $\dim Q_\mathbf{p} \leq 2$. Since by Proposition 8.4.8, F is a Q–ideal of linear type, we conclude that $F_\mathbf{q}$ is an almost complete intersection with $\dim Q_\mathbf{p}/F_\mathbf{p} \leq 1$. Therefore the first Koszul homology of $F_\mathbf{p}$ is Cohen–Macaulay ([4, Lemma]). Now $F_\mathbf{p}$ satisfies sliding depth and $\mathcal{F}_1$, which would force $(S_Q(F))_\mathbf{q}$ to be Cohen–Macaulay.

Finally assume that $\dim \mathcal{R}(I)_\mathbf{q} \geq 4$. From the proof of Theorem 8.4.11, we have an exact sequence

$$0 \to \overline{S} \longrightarrow S \longrightarrow \mathcal{R}(I) \to 0$$

where both algebras $\overline{S}$ and S are Cohen–Macaulay B–modules. By a depth count it follows that $\mathcal{R}(I)_\mathbf{q} \geq 3$. □

Corollary 8.4.10 *Let I be an ideal as in* Theorem 8.4.5. *Then I is normal if and only if it is normal on the punctured spectrum of R.*

There remains to make f and F more explicit. It is not entirely clear how it can be accomplished. If $g = 2$, the module F has a presentation

$$k[\mathbf{T}]^d \xrightarrow{B(\varphi)} k[\mathbf{T}]^d \longrightarrow F \to 0.$$

Since f generates the annihilator of F, since $\det B(\varphi)$ both annihilates F, and since both polynomials have the same degree, we conclude that

$$f = \det B(\varphi),$$

a somewhat curious way of obtaining irreducible polynomials!

Associated Graded Ring and Cohen–Macaulayness

For a given ideal I of a Cohen–Macaulay ring R, the relationship between the Cohen–Macaulay properties of the Rees algebra $\mathcal{R}(I)$ and the associated graded ring $\mathrm{gr}_I(R)$ is very close.

Theorem 8.4.11 *Let R be a Gorenstein local ring of dimension d and let I be a strongly Cohen–Macaulay ideal of codimension $g \geq 2$. Suppose $\nu(I) = d+1$ and I satisfies $\mathcal{F}_1$ on the punctured spectrum of R. If $\mathcal{A}$ is singly generated and $\mathrm{gr}_I(R)$ is Cohen–Macaulay, then $\mathcal{R}(I)$ is Cohen–Macaulay.*

Proof. By assumption and Theorem 8.3.2, we have an exact sequence

$$0 \to \mathcal{A} = Sf \longrightarrow S \longrightarrow \mathcal{R}(I) \to 0,$$

with f a homogeneous element of degree $d - g + 2$. From Theorem 8.3.2 it also follows that if φ is a minimal presentation matrix of I then $I_1(\varphi)$ is a Gorenstein ideal and $I_1(\varphi) \cdot f = 0$. Since $Sf = \mathcal{A} \not\subset \mathbf{m}S$ and $\mathbf{m}S$ is the only associated prime of $I_1(\varphi)S$, it follows that $0 :_S Sf = I_1(\varphi)$. Therefore

$$Sf \simeq \overline{S} = S(I) \otimes_R R/I_1(\varphi) \simeq R/I_1(\varphi)[T_1, \ldots, T_{d+1}].$$

We denote $B = R[T_1, \ldots, T_{d+1}]$ and examine the exact sequence of graded B–modules (graded in the T_j variables alone)

$$0 \to \overline{S} \longrightarrow S \longrightarrow \mathcal{R}(I) \to 0.$$

It shows $\mathcal{R}(I)$ to be nearly Cohen–Macaulay as a B–module since it has depth at least d.

The cohomology sequence of canonical modules is

$$0 \to \omega_{\mathcal{R}(I)} \longrightarrow \omega_S \stackrel{\epsilon}{\longrightarrow} \omega_{\overline{S}} \simeq \overline{S} \longrightarrow \mathrm{Ext}_B^{d+1}(\mathcal{R}(I), B) \to 0.$$

Thus $\mathcal{R}(I)$ is Cohen–Macaulay if and only if $\mathrm{Ext}_B^{d+1}(\mathcal{R}(I), B)$ vanishes.

The canonical module of S, ω_S, has by Corollary 3.3.26, $g - 1$ generators, so that the image of ϵ is a homogeneous ideal of $\overline{S}$ of codimension at most $g - 1$. The proof will be complete once it is shown that $\mathrm{Ext}_B^{d+1}(\mathcal{R}(I), B)$ is a module of Krull dimension zero. In particular this will occur if $\mathcal{R}(I)$ is Cohen–Macaulay on the punctured spectrum. The assertion will now follow from Proposition 5.1.24. □

Degree and Cohen–Macaulayness

Theorem 8.4.12 *Let $R = k[x_1, \ldots, x_d]$ be a polynomial ring and let I be a strongly Cohen–Macaulay ideal of codimension g having a linear presentation matrix. Suppose $\nu(I) = d+1$ and I satisfies $\mathcal{F}_1$ on the punctured spectrum of R. Then I is generated by forms of degree δ with $\delta \leq \frac{d}{g-1}$. Furthermore $\mathcal{R}(I)$ is Cohen–Macaulay if and only if $\delta = \frac{d}{g-1}$.*

Proof. We use the same exact sequence as in the proof of Theorem 8.4.11,

$$0 \to \mathcal{A} = Sf \longrightarrow S \longrightarrow \mathcal{R}(I) \to 0,$$

$S = S(I)$ is Cohen–Macaulay, where f is a homogeneous polynomial of degree $d-g+2$ in $k[T_1, \ldots, T_{d+1}]$, and $0:_S Sf = \mathbf{m}S$. The polynomial ring

$$B = k[x_1, \ldots, x_d, T_1, \ldots, T_{d+1}]$$

has also a natural grading given by $\deg x_i = 1$ and $\deg T_j = \delta$. In this grading, the characteristic polynomial f has degree $(d-g+2)\delta$, and therefore

$$\mathcal{A} = Sf \simeq (S/\mathbf{m}S(-(d-g+2)\delta) = k[\mathbf{T}](-(d-g+2)\delta).$$

Thus

$$0 \to k[\mathbf{T}](-(d-g+2)\delta) \longrightarrow S \longrightarrow \mathcal{R}(I) \to 0$$

is an exact sequence of graded B–modules, which induces the cohomology exact sequence

$$\operatorname{Ext}^d_B(\mathcal{R}(I), B) \longrightarrow \operatorname{Ext}^d_B(S, B) \stackrel{\epsilon}{\longrightarrow} \operatorname{Ext}^d_B(k[\mathbf{T}](-(d-g+2)\delta), B) \quad (8.4)$$

$$\simeq \quad k[\mathbf{T}](d + (d-g+2)\delta) \longrightarrow \operatorname{Ext}^{d+1}_B(\mathcal{R}(I), B) \to 0.$$

Notice that $\epsilon \neq 0$ and that ϵ is surjective if and only if $\mathcal{R}(I)$ is Cohen–Macaulay. We are going to examine the degrees of the generators of $\operatorname{Ext}^d_B(S, B)$.

The approximation complex

$$0 \to Z_d \otimes B \to Z_{d-1} \otimes B \to \cdots \to Z_1 \otimes B \to B \to S \to 0$$

is a homogeneous complex (with respect to the natural grading on B) that is exact. It can be used as in Theorem 3.3.19 to show that

$$\operatorname{Ext}^d_B(S, B) \otimes_B k \simeq \bigoplus_{i=d-g+2}^{d} \operatorname{Ext}^{d-i}_R(Z_i, R) \otimes_R k \simeq k((d+1)\delta)^{g-1},$$

the degrees of the Z_i being given by the acyclic portion of the Koszul complex on the h_j.

Comparing generator degrees in (8.4) and keeping in mind that $\epsilon \neq 0$ it follows that $(d+1)\delta \leq d + (d-g+2)\delta$. Thus $\delta \leq \frac{d}{g-1}$, where equality holds if and only if ϵ is surjective. □

Corollary 8.4.13 *With the assumptions of* Theorem 8.4.12, *the Rees algebra of I is Cohen–Macaulay if $g \leq 4$ or if $d \leq 2g-2$ (in particular if* $\dim R/I \leq 3$).

Proof. If $d \leq 2g-2$ then $\frac{d}{g-1} \leq 2 \leq \delta$, and the claim follows from Theorem 8.4.12.

Next assume that $g \leq 4$ and suppose that $\mathcal{R}(I)$ is not Cohen–Macaulay. We use the notation of the theorem above. As $\nu(\operatorname{Ext}^d_B(S, B)) = g-1 \leq 3$, the exact sequence (8.4) implies that $\dim \operatorname{Ext}^{d+1}_B(\mathcal{R}(I), B) \geq d-2$. Hence there exists $\mathbf{q} \in \operatorname{Spec}(\mathcal{R}(I))$ with $\dim \mathcal{R}(I)_{\mathbf{q}} \leq 3$. Now Proposition 8.4.9 yields a contradiction. □

Chapter 9

Commuting Varieties of Algebras

Let A be a variety defined over a field k, equipped with a structure of k-algebra. Broadly speaking, commuting varieties based on A are subvarieties of A^m defined by commutation conditions. These algebras may be associative algebras with identity, Lie algebras or admit other structures. We key on Lie algebras, as the questions are often reduced to an algebra of inner derivations.

Let W be a subvariety of A. The commuting variety $\mathcal{C}(W)$ is the variety of commuting pairs of points of W. In the case that A and W are linear varieties, a set of defining equations for $C(W)$ is obtained as follows. Let $\{e_1, \ldots, e_n\}$ be a basis of A, $n = \dim_k A$, and let $\{g_1, \ldots, g_m\}$ be a basis of W. Consider two generic elements of W, $x = \sum x_i g_i$ and $y = \sum y_i g_i$. The coordinates of their commutator

$$[x, y] = \sum_{k=1}^{n} f_k e_k$$

gives an ideal of definition

$$J(W) = (f_1, \ldots, f_n) \subset k[x_1, \ldots, x_m, y_1, \ldots, y_m]$$

for $C(W)$. Thus if $W = A$ and if $\{c_{ijk}\}$ are the structure constants for the basis, then

$$f_k = \sum_{i<j} c_{ijk}(x_i y_j - x_j y_i).$$

We refer to these quadrics as the *equations* of $C(A)$.

For an example, let H_n be the Heisenberg algebra of dimension $2n + 1$, and commutation relations $[P_i, Q_i] = I$, for $i = 1, \ldots, n$. The commuting variety $C(H_n)$ is defined by a single equation

$$\sum_{i=1}^{n} x_i y_{n+i} - x_{n+i} y_i.$$

We point out the existence of other kinds of commuting varieties. First there are the more general commuting n–tuples of a Lie algebra $\mathfrak{g}$:

$$C^n(\mathfrak{g}) = \{(x_1, \ldots, x_n) \in \mathfrak{g} \mid [x_i, x_j] = 0, \forall\, 1 \le i, j \le n\}.$$

It differs from the other variety in a fundamental way if $n \geq 3$. Both are defined by an ideal $I = (f_1, \ldots, f_m)$ of a polynomial ring $k[z_1, \ldots, z_d]$, generated by quadrics, all of whose monomials are square–free. Whenever this is the case, there is a graph $\mathcal{G}$ associated to the monomials which actually occur. For the commuting variety proper, $\mathcal{G}$ is bipartite, leading naturally to the naive method of the Jacobian module developed here. For commuting triples, this is no longer always the case.

One interesting commuting variety, not defined by quadrics, is the variety of Lie planes of an algebra $\mathfrak{g}$:

$$\{(x, y) \in \mathfrak{g} \times \mathfrak{g} \mid kx + ky \text{ is a subalgebra of } \mathfrak{g}\}.$$

It can be defined by a set of quartics, obtained from the 3×3 minors of the matrix whose columns are the corresponding coordinates of x, y and $[x, y]$, denoted by $I_3(x, y, [x, y])$. Their significance lie in the fact that they are simple solutions of the classical Yang–Baxter equations.

While a great deal is known about the first of these varieties—which might be labelled ordinary commuting varieties—the others are extremely unwieldy but loom much more interesting. For technical reasons we shall study only the first kind of variety.

There are several questions about the varieties $C(W)$ and the ideals $J(W)$: (i) Whether it gives the radical ideal of definition of $C(W)$; (ii) to describe the ring of regular functions on $C(W)$; (iii) to study depth properties of this variety, in particular whether it is Cohen-Macaulay; and (iv) the nature of its singularities.

The partial answers given here will be shown to depend on an analysis of the fibers of the canonical projection $\mathcal{C}(W) \to W$. For this purpose, we introduce a module, E, over the affine ring of W—to be called the *Jacobian* module of $\mathcal{C}(W)$—that plays an important role in capturing its reduced ideal of definition. It is an ancestor of the module of relative differentials of the morphism.

9.1 The Jacobian Module of an Algebra

This section introduces the Jacobian module attached to commuting variety of a linear subspace of an algebra A. Throughout k is an algebraically closed field, of characteristic zero. (This is not strictly required since we could, often, get away with fields of characteristic different from two. It will make however for an uniform setting.) If $\dim_k A = n$ we denote by $R = k[x_1, \ldots, x_n]$ the ring of polynomial functions on the k–vector space A.

We begin with the technical construction of what will be our main vehicle for analysis. Let $J(A) = (f_1, \ldots, f_n) \subset k[x_1, \ldots, x_n, y_1, \ldots, y_n]$ be the defining equations of $C(A)$. In the manner of section 1.5, we consider the Jacobian of the f_j's with respect to the variables y_i's:

$$\varphi = \frac{\partial(f_1, \ldots, f_n)}{\partial(y_1, \ldots, y_n)}.$$

This is a matrix with entries in R defining an R–module

$$R^n \xrightarrow{\varphi} R^n \longrightarrow E \to 0.$$

Definition 9.1.1 The *Jacobian module* of the commuting variety of A is the R–module $E = \text{coker}\,(\varphi)$.

Its significance lies in the following ([20, Proposition 1.2]) :

Proposition 9.1.2 *Let $S(E)$ denote the symmetric algebra of the R–module E. Then*

$$C(A) \simeq \text{Spec}(S(E))_{red}.$$

There is another description of the Jacobian module. Let $\{e_1, \ldots, e_n\}$ be a basis of A and put $x = \sum x_i e_i$. Dualizing the exact sequence

$$A \otimes_k R \xrightarrow{\varphi} A \otimes_k R \longrightarrow E \to 0,$$

and identifying $A \otimes_k R$ with $(A \otimes_k R)^*$, it is easy to see that

$$\varphi^*(a) = \text{ad}x(a) = [x, a].$$

It follows that the dual of E is the centralizer of the generic element of A.

Example 9.1.3 (a) Let $\mathfrak{g}$ be the 3-dimensional Lie algebra $\{e, f, g\}$ defined by

$$[e, f] = 0,\ [e, g] = e,\ [f, g] = f.$$

The ideal $J(\mathfrak{g})$ is defined by the forms $x_1y_3 - x_3y_1$ and $x_2y_3 - x_3y_2$. The Jacobian module E has a presentation

$$0 \to R^2 \xrightarrow{\varphi} R^3 \longrightarrow E \to 0,$$

$$\varphi = \begin{pmatrix} -x_3 & 0 \\ 0 & -x_3 \\ x_1 & x_2 \end{pmatrix}.$$

It is easy to see that $S(E)$ is reduced, so that $C(\mathfrak{g}) = \text{Spec}(S(E))$. It has two irreducible components.

(b) Denote by DS_n the space of all $n \times n$ matrices with equal line sums, that is, essentially the space of doubly stochastic matrices. If $n = 3$, a calculation with the Bayer and Stillman *Macaulay* program ([14]) shows that the Jacobian module E of $C(DS_3)$ has projective dimension two and $S(E)$ is a Cohen-Macaulay integral domain.

(c) Let V be an n-dimensional vector space over k. There is a natural Lie algebra structure on $L = V \oplus \wedge^2 V$ that makes $\wedge^2 V$ the center of L. The Jacobian module of L is the direct sum of a free module of rank $n(n-1)/2$, corresponding to its center, and the module that has n generators and for relations the forms $x_i y_j - x_j y_i$, $1 \leq i < j \leq n$, that is, the ideal $(x_1, \ldots, x_n)$. The projective dimension of Jacobian modules can thus attain any value.

9.2 Complete Intersections

Theorem 3.1.6 gives the means for testing whether a commuting variety is a complete intersection. We visit two candidates: the space of symmetric matrices and Borel subalgebras of semisimple Lie algebras.

Symmetric matrices

Let k be an algebraically closed field, of characteristic 0, and let $S_n(k)$ be the affine space of all symmetric matrices of order n defined over k. The commuting variety of $S_n(k)$ is defined by the ideal generated by the entries of

$$Z = [X, Y] = X \cdot Y - Y \cdot X, \tag{9.1}$$

where X and Y are generic symmetric $n \times n$ matrices in $n(n+1)$ indeterminates. $Z = [z_{ij}]$ is an alternating matrix of 2-forms.

The main application of this section is ([20]):

Theorem 9.2.1 *The entries of Z form a regular sequence generating a prime ideal.*

The proof will follow from Theorem 3.1.6, once certain details of the structure of the Jacobian module are made clear.

Lemma 9.2.2 *The Jacobian module of $C(S_n(k))$ has projective dimension one.*

Proof. It suffices to show that the presentation matrix φ of E has rank $n(n-1)/2$. If we specialize the matrix X to a generic diagonal matrix, it is easy to see that the forms z_{ij} specialize to

$$z^*_{ij} = (x_{ii} - x_{jj})y_{ij},$$

so that the corresponding matrix has full rank. □

The next result is the required geometric ingredient. It was also pointed out to us, independently, by Robert Guralnick and David Rohrlich. We recall that a square matrix is *non–derogatory* provided its minimal polynomial is its characteristic polynomial.

Proposition 9.2.3 *$\mathcal{C}(S_n(k))$ is an irreducible variety.*

Proof. It suffices to show that if W is a generic symmetric matrix commuting with X then the pair (X, W) is a generic point of $\mathcal{C}(S_n(k))$. This is a formal consequence of the proof of [77, Theorem 1, p. 341–342] once the lemmas below have been established.

Indeed, let A and B be a pair of commuting symmetric matrices, and G be a generic point of the irreducible variety defined by the centralizer of B in $S_n(k)$. By Lemma 9.2.5, G must be a non–derogatory matrix and thus (A, B) is the specialization of a pair $(G, f(G))$ for some polynomial f. But the latter is a specialization of $(X, f(X))$. We have thus shown:

$$(X, W) \longrightarrow (X, f(X)) \longrightarrow (G, f(G)) \longrightarrow (A, B),$$

thus completing the proof. □

Lemma 9.2.4 *Let A be a square matrix. Then the following are equivalent.*

(i) *A is non–derogatory.*

(ii) *If B is a matrix and $[A, B] = 0$, then there is a polynomial $p(t)$ with $p(A) = B$.*

Proof. This is [77, Proposition 4]. □

Lemma 9.2.5 *Let B be a $n \times n$ symmetric matrix. There exists a non-derogatory element of $S_n(k)$ that commutes with B.*

Proof. By [76, Corollary 2, p. 13], and the Jordan decomposition theorem, there exists an orthogonal matrix O such that

$$O^t B O = \bigoplus_{i=1}^{s} (\lambda_i I_i + N_i),$$

with N_i nilpotent, symmetric, and $\lambda_i I_i + N_i$ irreducible. (M^t is the transpose of the matrix M.) The matrix

$$O(\bigoplus_{i=1}^{s} (\mu_i I_i + N_i))O^t,$$

with distinct μ_i's, is non-derogatory and commutes with B. □

Here is an application to more general commutators; for the generic case, see [97], [134] and [211]. For each integer $0 \leq r \leq n$ define

$$M_n^r = \{(A, B) \in S_n(k) \times S_n(k) \mid \text{rank}\,[A, B] \leq r\}.$$

Corollary 9.2.6 *$M_n^{2r} = M_n^{2r+1}$ is an irreducible Gorenstein variety of codimension $(n - 2r - 1)(n - 2r)/2$. Its reduced equations are the Pfaffians of Z of order $2r + 2$.*

Proof. The entries of $Z = [z_{ij} | \ 1 \leq i < j \leq n]$, are homogeneous elements forming a regular sequence in $A = k[x_{pq}, y_{pq} | \ 1 \leq p \leq q \leq n]$, so that the inclusion $R = k[z_{ij}] \subset A$ is a faithfully flat homomorphism; see [192, p. 176]. Furthermore, by Theorem 9.2.1 the irrelevant maximal ideal of R, $\mathbf{p} = (z_{ij})$, extends to a prime ideal of A, $P = \mathbf{p}A$.

Denote by I the ideal of R generated by the Pfaffians of order $2s$ of the matrix Z. According to [166, Theorem 17], I is a prime, Gorenstein ideal of R, with the codimension given by the formula above. In addition, the singular locus of this variety is given by the ideal of Pfaffians of order $2(s-1)$; in particular it is normal.

We now show that the homogeneous ideal $J = IA$ is prime. We claim that the zero divisors of A/J all lie in P. Since the associated primes of A/J are graded, we only have to check that for homogeneous polynomials $f \in A \setminus P$. By the graded version of [192, Theorem 22.5]

$$0 \to A \xrightarrow{f} A \longrightarrow A/(f) \longrightarrow 0$$

is an exact sequence of flat R-modules, since f is regular mod $\mathbf{p}$. Therefore multiplication by f is also regular on any module $M \otimes_R A$.

To verify that J is prime it suffices therefore to check primality in the localization J_P. We show that JA_P is analytically irreducible. By going over to the completion B of A_P at the maximal ideal, the question reduces to verifying a local version of [170, Theorem 17(i)]. Precisely, if $B = K[[z_{ij}]]$ (K is the appropriate coefficient field), as IB is defined over $K[z_{ij}]$, the normality hypothesis sets up the conditions for an application of Zariski's theorem on analytical normality (*cf.* [208, Theorem 37.5]). □

Borel subalgebras

It would be interesting to find out which, amongst distinguished subalgebras of $\mathfrak{g}$, have a nice theory of the Jacobian module. We pick for an incomplete discussion its Borel subalgebras. As it will be seen, not even the dimension of the commuting variety is fully determined.

Let $\mathfrak{g}$ be a semisimple Lie algebra and let B be one of its Borel subalgebras. More precisely, let $\mathfrak{h}$ be a Cartan subalgebra, with a root system Φ and let

$$\mathfrak{g} = \mathfrak{h} + \sum_{\alpha \in \Phi} \mathfrak{g}^{\alpha}$$

be the corresponding root space decomposition. Denote

$$B = \mathfrak{h} + \sum_{\alpha > 0} \mathfrak{g}^{\alpha}.$$

Proposition 9.2.7 *The Jacobian module of B has projective dimension* 1.

Proof. Let h_i, $1 \leq i \leq \ell$, and $e_\alpha \in \mathfrak{g}^\alpha$, $\alpha \in \Phi$ be basis elements and consider the bracket of two generic elements:

$$[\sum_{i=1}^{\ell} x_i h_i + \sum_{\alpha > 0} x_\alpha e_\alpha, \sum_{i=1}^{\ell} y_i h_i + \sum_{\alpha > 0} y_\alpha e_\alpha]$$

$$= \sum (x_i y_\alpha - y_i x_\alpha)[h_i, e_\alpha] + \sum x_\alpha y_\beta [e_\alpha, e_\beta].$$

This gives rise to a set of equations, one for each value of $\alpha > 0$, and therefore the Jacobian module has a presentation

$$R^r \xrightarrow{\varphi} R^s \longrightarrow E \rightarrow 0,$$

where $r = (n-\ell)/2$, $s = (n+\ell)/2$, $n = \dim \mathfrak{g}$. We claim φ is of maximal rank.

To see this, it is enough to specialize to $x_\alpha = 0$ for all α. The αth–equation specializes to

$$f_\alpha = \sum_i x_i y_\alpha \alpha(h_i),$$

and the Jacobian matrix has only the diagonal elements

$$\frac{\partial f_\alpha}{\partial y_\alpha} = \sum_i x_i \alpha(h_i) \neq 0,$$

which proves the assertion. □

The proof shows that E has rank ℓ. It follows that $C(B)$ has a component of dimension $(n+\ell)/2 + \ell = (n+3\ell)/2$. The following argument of Guralnick and Richardson will show that $C(B)$, unlike the case of symmetric matrices, is not always irreducible.

Let $T_r(k)$ denote the algebra of upper triangular matrices over the field k. Assume that $r = 3m$. From the remark above, there is an irreducible component of $C(T_n(k))$ of dimension

$$(9m^2 + 9m)/2.$$

Consider the subalgebra D of $T_n(k)$ of all matrices

$$\begin{bmatrix} 0 & A_1 & A_2 \\ 0 & 0 & A_3 \\ 0 & 0 & 0 \end{bmatrix}.$$

Its commuting variety is contained in affine space of dimension $6m^2$ and is defined by m^2 equations, so that $C(D)$ has dimension at least $5m^2$. But this exceeds the other dimension if $m \geq 10$.

Remark 9.2.8 Recently it has been shown that the variety $C(S_n(k))$ is normal ([219]; see also [19]).

9.3 Semisimple Lie Algebras

Let $\mathfrak{g}$ be a semisimple Lie algebra over a field k of characteristic zero, and let R be the ring of polynomials functions on $\mathfrak{g}$. We aim here at constructing a complex of R-modules that reflects several of the properties of $\mathfrak{g}$. It provides a setting whereby properties of $\mathfrak{g}$ can be piped in, interpreted in algebraic terms and used to clarify several questions.

Applying our definition of the Jacobian module to a semisimple Lie algebra $\mathfrak{g}$ ($\mathbf{A} = \mathfrak{g} \otimes R$):

$$0 \to E^* \longrightarrow \mathbf{A}^* \xrightarrow{\varphi^*} \mathbf{A}^*,$$

identifying $\mathbf{A}$ and $\mathbf{A}^*$, as pointed out earlier,

$$\varphi^*(a) = \mathrm{ad}x(a) = [x, a].$$

We shall refer to the kernel $\mathcal{C}$ of adx, that is the centralizer of x in the Lie algebra $\mathbf{A} = \mathfrak{g} \otimes R$, as the *generic Cartan subalgebra* of $\mathfrak{g}$. Thus $E^* \simeq \mathcal{C}$, and E is a somewhat more primitive object than $\mathcal{C}$.

Once introduced, it becomes compelling to derive the homological properties of E. Thus, for instance, the relative success in dealing with the case of symmetric matrices rested with the fact that E has projective dimension one. When turned to Lie algebras, however, the picture is much different. We have mentioned that any (non–negative) integer can occur as the projective dimension of a Jacobian module. Nevertheless, for semisimple Lie algebras, over fields of characteristic zero, the projective dimension of E is always two.

To put this more precisely, let there be given a semisimple Lie algebra $\mathfrak{g}$ over an algebraically closed field k, of characteristic zero. Denote by R the ring of polynomial functions on $\mathfrak{g}$, $R = k[x_1, \ldots, x_n]$, $n = \dim \mathfrak{g}$. Let $B(,)$ denote the Killing form of $\mathfrak{g}$; we extend it to $\mathbf{A}$. To make phenomena of skew symmetry more visible, choose a base of $\mathfrak{g}$ with $B(e_i, e_j) = \delta_{ij}$. The mapping $\mathrm{ad}x$ is now skew symmetric, so we may identify $\mathcal{C}$ and $\ker(\varphi)$. These definitions and observations when collected yield:

Proposition 9.3.1 *The Jacobian module of $\mathfrak{g}$ gives rise to the exact complex $C_1(\mathfrak{g})$*

$$0 \to \mathcal{C} \xrightarrow{\psi} \mathbf{A} \xrightarrow{\varphi} \mathbf{A} \longrightarrow E \to 0, \tag{9.2}$$

where $\varphi(a) = [x, a]$, x a generic element of $\mathfrak{g}$.

We call $C_1(\mathfrak{g})$ the *Cartan complex* of $\mathfrak{g}$. As will be shown, it gathers a considerable amount of algebro-geometric data about the algebra. This complex could be defined for an arbitrary Lie algebra as the minimal resolution of the Jacobian module. Unlike the semisimple case, we do not know of an interpretation for the second syzygy module. More generally, we denote by $C_n(\mathfrak{g})$ the minimal resolution of the nth symmetric power of E.

Let us briefly describe the information that is known about $C_1(\mathfrak{g})$. (i) Let G the group of inner automorphisms of $\mathfrak{g}$. If $p_1, \ldots, p_\ell$ are the homogeneous polynomials minimally generating the subring of invariants of G, then $\mathcal{C}$ is a free R–module generated by the gradients $\nabla p_1, \ldots, \nabla p_\ell$ (Theorem 9.3.4). (ii) We indicate later how these *vectors* can be computed, with the p_i restituted in the usual manner. They will appear again in the characterization of the ring of regular functions on $C(\mathfrak{g})$ (Corollary 9.3.17). (iii) The module E is torsion-free and its Fitting ideals have strong growth conditions. In addition, $\mathrm{Ext}^2_R(E, R)$ is a perfect module of codimension 3 and its support is the locus of the non-regular elements of the Lie algebra (Corollary 9.3.12).

The other advantages of dealing with E, rather than $J(\mathfrak{g})$, are of two kinds. First, the number of variables gets smaller, which tends to cut down on computational complexity. Another benefit of this change of focus, from the equations to the Jacobian module, is that it permits to uncover some properties the commuting varieties do *not* have.

Remark 9.3.2 It is clear that the definition of the Jacobian module can be given for any representation of an algebra. For a detailed extension to special representations of simple Lie algebras, see [219].

Invariant polynomials and syzygies

If $f(x)$ is a polynomial function on $\mathfrak{g}$, define its *gradient*, $\bigtriangledown f(x)$, by

$$B(\bigtriangledown f(x), y) = df_x(y).$$

Since B is unimodular, $\bigtriangledown f(x)$ is an element of $\mathbf{A}$.

Let G be the group of inner automorphisms of $\mathfrak{g}$. The following result ([84, Lemma 8.1]) was pointed out to us by N. Wallach.

Proposition 9.3.3 *Let $p(x)$ be an invariant polynomial under G. Then*

$$[\bigtriangledown p(x), x] = 0.$$

Proof. If $\mathrm{ad} y$ is nilpotent, then

$$p(x) = p(e^{t\ \mathrm{ad} y} x) = p(x + t[x, y] + O(t^2)).$$

Hence $dp_x([y, x]) = 0$ for y nilpotent. But as $\mathfrak{g}$ is the span of the nilpotent elements, $dp_x([x, \mathfrak{g}]) = 0$.

Now if $y \in \mathfrak{g}$,

$$B([\bigtriangledown p(x), x], y) = B(\bigtriangledown p(x), [x, y]) = dp_x([x, y]) = 0.$$

Thus $[\bigtriangledown p(x), x] = 0$. □

This identifies sufficiently many elements to generate $\mathcal{C}$ ([20], [293]):

Theorem 9.3.4 *Let $\mathfrak{g}$ be a semisimple Lie algebra of rank ℓ, over an algebraically closed field of characteristic zero. Let $p_1, \ldots, p_\ell$ be homogeneous polynomials generating the subring of invariants of $\mathfrak{g}$. The subalgebra $\mathcal{C}$ is generated as an R–module by $\bigtriangledown p_1, \ldots, \bigtriangledown p_\ell$.*

Proof. Let $\mathcal{C}_0$ be the submodule of $\mathcal{C}$ spanned by ∇p_i, $1 \leq i \leq \ell$. According to Kostant [168, Theorem 0.8], the cone $V(p_1, \ldots, p_\ell)$ is a normal, complete intersection of codimension ℓ. This means that the Jacobian matrix of the invariant polynomials p_i has rank ℓ, and the ideal generated by its $\ell \times \ell$ minors is not contained in an ideal of height $\ell + 1$. This implies already that $\mathcal{C}_0$ is a free R–module. Consider the complex

$$0 \to \mathcal{C}_0 \to \mathbf{A} \xrightarrow{\varphi} \mathbf{A} \longrightarrow E \to 0.$$

To show it is acyclic, we use the criterion of [34]. Because $\mathcal{C}_0$ and E have rank ℓ, it suffices to note that the ideal generated by the $\ell \times \ell$ minors of the embedding $\mathcal{C}_0 \hookrightarrow \mathbf{A}$ has codimension at least two, by Kostant's result. The equality $\mathcal{C}_0 = \mathcal{C}$ follows. □

Note that the theorem provides a template to find the Cartan subalgebra attached to a regular element of $\mathfrak{g}$. More precisely, for $z \in \mathfrak{g}$, z is regular if and only if $\text{rank}(\psi(z)) = \ell$ ($\psi(z)$ is the matrix ψ, whose entries are polynomials, evaluated at the point z), in which case $\mathfrak{g}^z$ is generated by the columns of $\psi(z)$. (See Corollary 9.3.11 for additional details.)

What is not fully clear is the relationship between the projective dimension of the Jacobian module and the structure of the Lie algebra. It is not the case that proj dim $E = 2$ characterizes reductive Lie algebras.

Remark 9.3.5 The theorem may provide the means for computing the invariant polynomials, at least when $\mathfrak{g}$ is an algebra of low rank. The point is that programs such as *Macaulay* will determine the syzygies of φ, that is, the $\nabla p(x)$, with $p(x)$ recovered through Euler's formula

$$\deg(p) \cdot p(x) = B(\nabla p(x), x).$$

It is straightforward to set up the computation.

To illustrate, in the case of an algebra of type G_2, using its 'most' natural basis (*cf.* [135])—not orthonormal—one obtains for the generic Cartan subalgebra, in addition to x, another vector with fifth degree components $(b_1, \ldots, b_{14})$. They are however nearly dense polynomials. Here, for instance, is b_1 taken from a *Macaulay* session. The second invariant polynomial p_2 can be found by the method indicated earlier; it is about 30 times longer.

```
-x[2]3x[3]x[5]+1/2x[2]x[3]2x[5]2+x[2]3x[4]x[7]-x[2]2x[3]x[6]x[7]
+1/2x[3]2x[5]x[6]x[7]-1/2x[2]x[4]2x[7]2+1/2x[3]x[4]x[6]x[7]2
-x[2]2x[4]x[5]x[8]+1/2x[3]x[4]x[5]2x[8]-x[2]x[3]x[5]x[6]x[8]
+1/2x[4]2x[5]x[7]x[8]+x[2]x[4]x[6]x[7]x[8]-x[3]x[6]2x[7]x[8]
-x[4]x[5]x[6]x[8]2+6x[2]x[4]x[9]2x[10]-6x[3]x[6]x[9]2x[10]
-3x[4]x[5]x[9]x[10]2+3x[5]x[6]x[10]3+6x[2]x[3]x[9]2x[11]
+6x[4]x[8]x[9]2x[11]+3x[3]x[5]x[9]x[10]x[11]
-3x[4]x[7]x[9]x[10]x[11]-6x[2]x[5]x[10]2x[11]+3x[6]x[7]x[10]2x[11]
+3x[3]x[7]x[9]x[11]2-6x[2]x[7]x[10]x[11]2-3x[5]x[8]x[10]x[11]2
-3x[7]x[8]x[11]3-3x[2]x[3]x[5]x[9]x[12]+3x[2]x[4]x[7]x[9]x[12]
-3x[3]x[6]x[7]x[9]x[12]-3x[4]x[5]x[8]x[9]x[12]
+4x[2]2x[5]x[10]x[12]-1/2x[3]x[5]2x[10]x[12]
```

```
-1/2x[4]x[5]x[7]x[10]x[12]+4x[5]x[6]x[8]x[10]x[12]
+4x[2]2x[7]x[11]x[12]-1/2x[3]x[5]x[7]x[11]x[12]
-1/2x[4]x[7]2x[11]x[12]+4x[6]x[7]x[8]x[11]x[12]
+4x[2]2x[3]x[9]x[13]-1/2x[3]2x[5]x[9]x[13]
-1/2x[3]x[4]x[7]x[9]x[13]+4x[3]x[6]x[8]x[9]x[13]+x[2]3x[10]x[13]
-2x[2]x[3]x[5]x[10]x[13]-5x[2]x[4]x[7]x[10]x[13]
+3/2x[3]x[6]x[7]x[10]x[13]+3/2x[4]x[5]x[8]x[10]x[13]
+x[2]x[6]x[8]x[10]x[13]+x[2]2x[8]x[11]x[13]
-7/2x[3]x[5]x[8]x[11]x[13]-7/2x[4]x[7]x[8]x[11]x[13]
+x[6]x[8]2x[11]x[13]-9x[2]x[9]x[10]x[12]x[13]
+9/2x[5]x[10]2x[12]x[13]-9x[8]x[9]x[11]x[12]x[13]
+9/2x[7]x[10]x[11]x[12]x[13]-6x[2]x[7]x[12]2x[13]
+6x[5]x[8]x[12]2x[13]+9/2x[3]x[9]x[10]x[13]2-9/2x[2]x[10]2x[13]2
-9/2x[8]x[10]x[11]x[13]2+3x[3]x[7]x[12]x[13]2-3x[3]x[8]x[13]3
+4x[2]2x[4]x[9]x[14]-1/2x[3]x[4]x[5]x[9]x[14]
-1/2x[4]2x[7]x[9]x[14]+4x[4]x[6]x[8]x[9]x[14]+x[2]2x[6]x[10]x[14]
-7/2x[3]x[5]x[6]x[10]x[14]-7/2x[4]x[6]x[7]x[10]x[14]
+x[6]2x[8]x[10]x[14]-x[2]3x[11]x[14]+5x[2]x[3]x[5]x[11]x[14]
+2x[2]x[4]x[7]x[11]x[14]+3/2x[3]x[6]x[7]x[11]x[14]
+3/2x[4]x[5]x[8]x[11]x[14]-x[2]x[6]x[8]x[11]x[14]
-9x[6]x[9]x[10]x[12]x[14]+9x[2]x[9]x[11]x[12]x[14]
+9/2x[5]x[10]x[11]x[12]x[14]+9/2x[7]x[11]2x[12]x[14]
-6x[2]x[5]x[12]2x[14]-6x[6]x[7]x[12]2x[14]
+9/2x[4]x[9]x[10]x[13]x[14]-9/2x[6]x[10]2x[13]x[14]
+9/2x[3]x[9]x[11]x[13]x[14]-9/2x[8]x[11]2x[13]x[14]
-3x[3]x[5]x[12]x[13]x[14]+3x[4]x[7]x[12]x[13]x[14]
+6x[2]x[3]x[13]2x[14]-3x[4]x[8]x[13]2x[14]+9/2x[4]x[9]x[11]x[14]2
-9/2x[6]x[10]x[11]x[14]2+9/2x[2]x[11]2x[14]2-3x[4]x[5]x[12]x[14]2
+6x[2]x[4]x[13]x[14]2+3x[3]x[6]x[13]x[14]2+3x[4]x[6]x[14]3
```

In the non–semisimple case there are several questions. First, because there may not exist any underlying skew symmetry in the mapping ad$x$, the second syzygies of E can no longer be interpreted as the generic Cartan subalgebra. What are they? Then, "invariant polynomials" are nowhere visible.

Irreducibility

We shall first focus on the sizes of the determinantal ideals defined by the matrices ψ and φ. For a matrix φ, denote by $I_t(\varphi)$ the ideal generated by the $t \times t$ minors; these are the Fitting ideals of the module defined by φ.

The following result of Richardson [228] is a key geometric ingredient from this point on. (The case of $M_n(k)$, $n \times n$ matrices, was proved by Motzkin and Taussky [207] and Gerstenhaber [77].)

Theorem 9.3.6 *The commuting variety of a semisimple Lie algebra is irreducible.*

The more precise statement he proved is the following. Let $\mathfrak{g}$ be a reductive Lie algebra over an algebraically closed field of characteristic zero. Let $(x, y) \in C(\mathfrak{g})$; for any neighborhood N of (x, y) there exists a Cartan subalgebra $\mathfrak{h}$ such N meets $\mathfrak{h} \times \mathfrak{h}$. Since the Cartan subalgebras are conjugate under the action of the adjoint group G of $\mathfrak{g}$, this means that $C(\mathfrak{g})$ is the closure of the image of the morphism: Fix a Cartan subalgebra $\mathfrak{h}$ and define

$$\eta: G \times \mathfrak{h} \times \mathfrak{h} \longrightarrow \mathfrak{g} \times \mathfrak{g}, \ \eta(a, x, y) = (a(x), a(y)).$$

It follows therefore that the presentation matrix φ of the Jacobian module of $\mathfrak{g}$ satisfies the conditions (i) and (iii) of Theorem 3.1.6.

Its proof is beyond our means here. Nevertheless, we reproduce the following proof of Guralnick ([98]) of the result of [77] and [207].

Theorem 9.3.7 *The commuting variety of $M_n(k)$ is irreducible.*

Proof. First note that every matrix A over k commutes with a regular (i.e. non-derogatory) matrix. Indeed, we can assume that A is in Jordan form; say

$$A = \text{diag}(J_1, \ldots, J_r),$$

where the J_i are the Jordan blocks. Let b_i be the eigenvalue of J_i. Then for any pairwise distinct scalars $a_1, \ldots, a_r$, the matrix $R = \text{diag}((a_1 - b_1)I + J_1, \ldots, (a_r - b_r)I + J_r)$ is regular and commutes with A. So consider a commuting pair (A, B). Now let R be regular matrix commuting with A. Then $(A, B+xR)$ is contained in the commuting variety for each $x \in F$. Moreover except for finitely many values of x, $B + xR$ is also regular. Thus the pair (A, B) is in the closure of the set of commuting pairs where the second term is regular.

Now consider the morphism φ: $M_n(k) \times k^n \longrightarrow C(M_n(k))$ (with k^n viewed as the space of polynomials of degree at most $n - 1$) defined by $\varphi(A, g) = (A, g(A))$. The image is irreducible (since the domain is) and is dense, since it contains all pairs where the first component is regular. Thus its closure is the variety and is irreducible. □

Now we give two examples, the first in positive characteristic.

Example 9.3.8 Let W_1 be the Witt algebra that has for basis

$$\{e_i \mid i \in \mathbb{Z}/(p)\}$$

and multiplication

$$[e_i, e_j] = (j - i)e_{i+j}.$$

For $p = 5$, the Jacobian module of W_1 satisfies the condition $\mathcal{F}_0$ but not $\mathcal{F}_1$ and therefore $C(W_1)$ is not irreducible. ($J(W_1)$ is however Cohen–Macaulay.) There are additional differences between this case and those in characteristic zero. For example, the kernel of φ is generated by an element in degree 3, rather than the generic element x. This implies the well-known fact that the algebra cannot support an invariant form.

Example 9.3.9 This example is to inject a word of caution regarding the primeness of the ideal $J(A)$. Let A be the algebra of *Cayley* numbers. This is the algebra of pairs of quaternions, written $q + re$, with multiplication

$$(q + re)(s + te) = (qs - \bar{t}r) + (tq + r\bar{s})e.$$

The base ring is $R = \mathbb{C}[x_1, \ldots, x_8]$; the Jacobian module $E = R \oplus F$, where F has a resolution

$$0 \to R \longrightarrow R^7 \xrightarrow{\varphi} R^7 \longrightarrow F \to 0,$$

and φ is a skew symmetric matrix. The ideal generated by its 6×6 minors has height 1, implying that F has torsion. A computation showed that $J(A)$ is Cohen–Macaulay.

The free locus

Let I be the determinantal ideal $I_\ell(\psi)$. By the result of Kostant, height $I \geq 2$. Since E is torsion–free, by [34], height $I \geq 3$. We call I the Jacobian ideal of $\mathfrak{g}$, and will see that the radical of I is unmixed, of codimension 3, and exhibits traits of Gorenstein ideals (Corollary 9.3.12).

Let E be a torsion–free module with a resolution as (9.2). We now describe several of its technical features.

Proposition 9.3.10 $\mathrm{Ext}^1_R(E, R) = 0$ *and* $\mathrm{Ext}^2_R(E, R)$ *is a Cohen–Macaulay module of codimension* 3.

Proof. Dualizing the resolution of E we obtain the complex

$$0 \to E^* \longrightarrow (R^n)^* \xrightarrow{-\varphi} (R^n)^* \longrightarrow R^\ell \to 0.$$

Its homology modules are

$$\mathrm{Ext}^1_R(E, R) \text{ and } \mathrm{Ext}^2_R(E, R).$$

We claim that $\mathrm{Ext}^1_R(E,\ R) = 0$ and that

$$W = \mathrm{Ext}^2_R(E,\ R) = \mathrm{coker}(\psi^*)$$

is a Gorenstein module of codimension three. -1z First, note that $E^* \simeq R^\ell$. This implies that the height of the ideal $I_{n-\ell}(\varphi)$ must indeed be three as otherwise E would be reflexive and hence free by the previous observation. The vanishing of $\mathrm{Ext}^1_R(E,\ R)$ follows by depth–chasing. The last assertion now is a consequence of the fact that E is free in codimension two. □

Corollary 9.3.11 *The following equality of radical ideals hold:*

$$\sqrt{I_\ell(\psi)} = \sqrt{I_{n-\ell}(\varphi)}.$$

Corollary 9.3.12 *The radical of the Jacobian ideal of a semisimple Lie algebra is unmixed, of codimension* 3.

Proof. By the Proposition and the previous Corollary, $I_\ell(\psi)$ and the annihilator of $\mathrm{Ext}^2_R(E,\ R)$ have the same radical. The claim follows because $\mathrm{Ext}^2_R(E, R)$ is perfect, of codimension 3. □

The ideal I is a defining ideal for the *non–regular* elements of $\mathfrak{g}$, a result due to Kostant. Note that $I_\ell(\psi)$ and the ideal generated by the Pfaffians of order $n-\ell$ of φ are generated by polynomials of the same degree, $\frac{1}{2}(n-\ell)$. They are likely equal.

The components of the singular variety

Let $\mathfrak{g}$ be semisimple Lie algebra. We have seen above that the sub–variety of singular elements, $\mathfrak{g}_{sing}$ is unmixed, of codimension 3. The following result of N. Wallach sharpens this considerably.

Let $\mathfrak{g}$ be a simple Lie algebra. Fix a Cartan subalgebra $\mathfrak{h} \subset \mathfrak{g}$. Let $\Phi = \Phi(\mathfrak{h}, \mathfrak{g})$ be the root system of $\mathfrak{g}$ with respect to $\mathfrak{h}$, and W the Weyl group of $\mathfrak{h}$. Fix Φ^+ a set of positive roots of Φ.

Theorem 9.3.13 *The number of irreducible components of* $\mathfrak{g}_{sing}$ *is* order(Φ/W).

Let for $\alpha \in \Phi$, $\mathfrak{h}^\alpha = \{H \in \mathfrak{h} \mid \alpha(H) = 0\}$. Set

$$(\mathfrak{h}^\alpha)' = \{H \in \mathfrak{h} \mid \alpha(H) = 0\} \text{ and if } \beta \in \Phi, \beta \neq \pm\alpha \text{ then } \beta(H) \neq 0\}.$$

Lemma 9.3.14 $\mathfrak{g}_{sing} = \bigcup_{\alpha \in \Phi^+} Clos(Ad(G)(\mathfrak{h}^\alpha))$.

Proof. For $x \in \mathfrak{g}_{sing}$, let $x = x_s + x_n$ be its Jordan decomposition; x_n is singular in $\mathfrak{g}^{x_s}$. We may assume $x_s \in \mathfrak{h}$.

Let $\Phi_1{}^+ = \{\alpha \in \Phi^+ \mid \mathfrak{g}_\alpha \subset \mathfrak{g}^{x_s}\}$. Then if $x_n = 0$, $x_s \in \bigcap_{\alpha \in \Phi_1{}^+} \mathfrak{h}^\alpha$, and we are done.

If $x_n \neq 0$, then there is a TDS (3–dimensional simple Lie algebra), $X = x_n, Y, H$ in $\mathfrak{g}^{x_s}$. Up to conjugacy in G^{x_s} we may assume $H \in \mathfrak{h}$ and $\alpha(H) \geq 0$, $\alpha \in \Phi_1^+$. Thus $x_n = \sum_{\alpha \in \Phi_1{}^+} (x_n)_\alpha$. If $(x_n)_\alpha \neq 0$ for all simple $\alpha \in \Phi_1{}^+$, then $\alpha(H) = 2$ for all $\alpha \in \Phi_1{}^+$. But then $\dim(\mathfrak{g}^{x_s})^H = \dim(\mathfrak{g}^{x_s})^{x_n}$. Since $H \in \mathfrak{g}^{x_s}$ is regular semisimple, x_n would be regular in $\mathfrak{g}^{x_s}$. Thus there is $\alpha \in \Phi_1{}^+$, α simple, such that $(x_n)_\alpha = 0$.

Define $\mathfrak{n}^\alpha = \sum_{\beta \in \Phi_1{}^+ - \alpha} \mathfrak{g}_\beta$. Set also

$$\Phi(X, h) = e^{\mathrm{ad} X} h - h \in \mathfrak{n}^\alpha, \; X \in \mathfrak{n}, \; h \in \mathfrak{h}'.$$

Then $d\Phi_{X,h}(Z, 0) = e^{\mathrm{ad} X}([Z, h])$, so $d\Phi_{X,h}(\mathfrak{n}^\alpha) = \mathfrak{n}^\alpha$. Thus $e^{\mathrm{ad}\, \mathfrak{n}^\alpha} \cdot h$ is Zariski dense in $h + \mathfrak{n}^\alpha$, so $e^{\mathrm{ad}\, \mathfrak{n}^\alpha} \cdot \mathfrak{h}^\alpha$ is also Zariski dense in $\mathfrak{h}^\alpha \oplus \mathfrak{n}^\alpha$. Thus $x_n \in Ad(G^{x_s})(\mathfrak{h}^\alpha)$. Since $x_s \in \mathfrak{h}^\alpha$, we have $x = x_s + x_n \in Clos(Ad(G)(\mathfrak{h}^\alpha))$. □

Lemma 9.3.15 $\dim Clos(Ad(G)(\mathfrak{h}^\alpha)) = \dim \mathfrak{g} - 3$.

This is already known, but it is clear by computing the differential.

Proof of Theorem. We note that since G is connected $Clos(Ad(G)(\mathfrak{h}^\alpha))$ is irreducible. If $\beta = s\alpha$, $s \in W$, then clearly $Ad(G)(\mathfrak{h}^\alpha) = Ad(G)(\mathfrak{h}^\beta)$. Thus

$$\mathfrak{g}_{sing} = \bigcup_{\alpha \in \Phi/W} Clos(Ad(G)(\mathfrak{h}^\alpha)).$$

Notice that each $Ad(G)(\mathfrak{h}^\alpha)'$ is Zariski open in $\mathfrak{g}_{sing}$. Thus if

$$Clos(Ad(G)(\mathfrak{h}^\alpha)) \subset Clos(Ad(G)(\mathfrak{h}^\beta)),$$

then

$$Ad(G)(\mathfrak{h}^\alpha)' \bigcap Ad(G)(\mathfrak{h}^\beta)' \neq \emptyset.$$

But then there exist $X \in (\mathfrak{h}^\alpha)'$, $Y \in (\mathfrak{h}^\beta)'$, $g \in G$ such that $Ad(g)X = Y$. Hence one has $s \in W$ such that $s \cdot X = Y$. But

$$\{\pm\alpha\} = \{\gamma \in \Phi \mid \gamma X = 0\}, \quad \{\pm\beta\} = \{\gamma \in \Phi \mid \gamma X = 0\},$$

hence $s \cdot \alpha = \pm\beta$. Thus from $s_\beta \cdot \beta = -\beta$ we see that $Clos(Ad(G)(\mathfrak{h}^\alpha)) \subset Clos(Ad(G)(\mathfrak{h}^\beta))$ if and only if $\alpha \in W\beta$. Thus each $Clos(Ad(G)(\mathfrak{h}^\alpha))$ is an irreducible component. □

Corollary 9.3.16 *$\mathfrak{g}_{sing}$ is irreducible if and only if there is only one root length, i.e. A_n, D_n, E_6, E_7, E_8. Otherwise there are two irreducible components.*

Regular functions

Another application of the details of the resolution of E is the following description of the ring $D(\mathfrak{g})$ of regular functions on $C(\mathfrak{g})$. It is a consequence of the identification $E^* \simeq R^\ell$, and Theorems 9.3.4 and 9.3.6 (see [20, Theorem 4.4]).

We begin with the observation that since x is a semisimple element, $\ker(\varphi) = \ker(\varphi^2)$, so that using the snake lemma on the composition of φ with itself, we get the exact sequence

$$0 \to \ker(\varphi) \to \ker(\varphi^2) \to \ker(\varphi) \to \operatorname{coker}(\varphi) \to \operatorname{coker}(\varphi^2) \to \operatorname{coker}(\varphi) \to 0$$

and therefore

$$0 \to \mathcal{C} \longrightarrow E \longrightarrow \operatorname{coker}(\varphi^2) \longrightarrow E \to 0.$$

This yields a canonical map from the symmetric algebra of $\mathcal{C}$ into $S(E)$.

A set of generators for the affine ring of the commuting variety of $\mathfrak{g}$ can be obtained in the following manner. Let $z_1, \ldots, z_\ell$ be the corresponding basis for the generic Cartan subalgebra of $\mathfrak{g}$; $z_i = \nabla p_i$. Let $T_1, \ldots, T_\ell$ be a set of ℓ new indeterminates. The symmetric algebra $S(E) = k[X,Y]/I_1([XY])$ maps onto the subring $D(\mathfrak{g})$ of $R[T_1, \ldots, T_\ell]$ generated by coordinates of the element

$$z = \sum_{i=1}^{\ell} T_i z_i$$

by specializing $Y \mapsto z$.

More precisely, write each $z_i = \sum_{j=1}^{n} b_{ij} e_j$ and consider the homomorphism

$$\Phi \colon S(\mathfrak{g}) \longrightarrow R[T_1, \ldots, T_\ell], e_j \mapsto \sum_{k=1}^{\ell} b_{kj} T_k.$$

On $[e_i, x] = \sum_{j=1}^{n} f_{ij} e_j$

$$\Phi([e_i, x]) = \sum_{j=1}^{n} f_{ij}\Phi(e_j) = \sum_{k=1}^{\ell}(\sum_{j=1}^{n} f_{ij} b_{kj})T_k.$$

But

$$\sum_{j=1}^{n} f_{ij} b_{kj} = B([e_i, x], x) = 0.$$

Since Φ vanishes on the image of $\mathrm{ad}(x)$, it induces the desired map from $S(E)$ to $D(\mathfrak{g})$.

Theorem 9.3.17 *Let $\mathfrak{g}$ be a semisimple Lie algebra of rank ℓ, with $p_1, \ldots, p_\ell$ its invariants as above. Let $T_1, \ldots, T_\ell$ be a fresh set of indeterminates. Consider the element*

$$z = \sum_{i=1}^{\ell} p_i T_i \in S = R[T_1, \ldots, T_\ell].$$

$D(\mathfrak{g})$ is the R–subalgebra of S generated by the components of the gradient of z.

Proof. $D(\mathfrak{g})$ has obviously transcendence degree $\dim \mathfrak{g} + \ell$ over k. By Richardson's theorem and Proposition 9.1.2, the nilradical of $S(E)$ must be the kernel of the homomorphism. □

Note that Φ restricted to $\mathcal{C}$ induces mappings

$$S(\mathcal{C}) \hookrightarrow D \hookrightarrow R[T_1, \ldots, T_\ell].$$

The composite is the mapping of symmetric algebras derived from the homomorphism of free modules

$$A\colon R^\ell \longrightarrow R^\ell$$

associated to the Cartan matrix $[B(z_i, z_j)]$. None of these algebra homomorphisms is finite.

We rule out one structural property of the ring $D(\mathfrak{g})$.

Corollary 9.3.18 *$C(\mathfrak{g})$ is not a Gorenstein variety.*

Proof. It is enough to show this after localizing at a codimension 3 prime of R. Because of the condition $\mathcal{F}_1$, the localization of E (and we still denote it by E) has a minimal resolution of the form ($p \geq 1$)

$$0 \to R^p \xrightarrow{\alpha} R^{p+2} \longrightarrow R^m \longrightarrow E \to 0.$$

The symmetric algebra of E is a domain, so that $\mathrm{Spec}(S(E)) = C(\mathfrak{g})$, and its ideal of definition is given by the maximal minors of a matrix obtained by augmenting α

with a row of linear forms, *cf.* Theorem 3.4.6. It is an ideal of Cohen–Macaulay type $p+1 \geq 2$, and therefore is not Gorenstein. □

It follows that $D(\mathfrak{g})$ is not the ring of invariants of a connected semisimple group acting on a polynomial ring, since these are, by the theorem of Hochster and Roberts [128], always Gorenstein.

Question 9.3.19 Concerning the commuting variety of a simple Lie algebra $\mathfrak{g}$ there are several outstanding questions:

- Do the natural equations $J(\mathfrak{g})$ define a prime ideal? (In particular, if $\mathfrak{g}$ are the 4×4 matrices of trace 0 is $J(\mathfrak{g})$ prime?)
- Is the variety $C(\mathfrak{g})$ normal?

9.4 Commuting Zoo

There are several other problems regarding commuting varieties (even those of the ordinary kind) that were not raised. The reader is invited to consider hers own.

When the symmetric algebra of the Jacobian module E is written as

$$S(E) = R + E + S_2(E) + \cdots,$$

it might be said that what we have touched thus far is on the structure of E, not of the higher symmetric powers of E. In this section we discuss some issues that show up in calculations.

In the second part of the section we present some comments of Buchweitz on the Yang–Baxter equations, and some clarifications of Guralnick on varieties of Lie planes. This section brings in some facts that intrude into the theory of commuting varieties. They are not well–understood by this author.

Mixed commuting varieties

These would be commuting varieties of pairs in $A \times B$. There would be two Jacobian modules, one over the ring $k[A]$, another over $k[B]$. We leave a single question:

Question 9.4.1 Let g let a simple Lie algebra and let h be a Cartan subalgebra. Determine the projective dimension of the two Jacobian modules.

Vastly more untractable are commuting varieties of non–linear varieties, such as the commuting variety of pairs of nilpotent matrices of a fixed size.

Some trace identities

If $\mathfrak{g}$ is a semisimple Lie algebra (char 0) and R is the ring of polynomial functions on $\mathfrak{g}$, set $\mathbf{F} = \mathfrak{g} \otimes R$. We will view $\mathbf{F}$ both as a free R–module of rank $n = \dim \mathfrak{g}$, or as a Lie R–algebra.

The Cartan complex is best described after picking a basis for $\mathfrak{g}$ (and hence for $\mathbf{F}$) that is orthonormal for the Killing form. The complex $C(\mathfrak{g})$ is

$$0 \to \mathcal{C} = R^{\ell} \xrightarrow{\psi} \mathbf{F}^* \xrightarrow{\varphi} \mathbf{F} \longrightarrow E \to 0.$$

If $e_i,\ \epsilon_i,\ i = 1, \ldots, n$ are dual bases of $\mathbf{F}$ and $\mathbf{F}^*$, then

$$\varphi(\epsilon_i) = [x, e_i].$$

To look at the postulation question (i.e. the Hilbert functions) for the higher symmetric powers of E, one experiments with their projective resolutions. As the case of $S_2(E)$ will indicate, there will not be a canonical method to obtain such resolutions from the matrices ψ and φ alone, that is, they will not be obtained from a master template followed by specialization. For instance, to go after $S_2(E)$, there is the natural complex

$$\mathbf{F}^* \otimes \mathbf{F} \xrightarrow{\alpha} S_2(\mathbf{F}) \longrightarrow S_2(E) \to 0,$$

$\alpha(u \otimes v) = \varphi(u) \cdot v$. After identifying

$$\mathbf{F}^* \otimes_R \mathbf{F} = \mathrm{Hom}_R(\mathbf{F}, \mathbf{F}),$$

note that as an element of the endomorphism ring,

$$\mathrm{ad}(x) = \sum_{i,j,k} x_i c_{ijk} \epsilon_j \otimes e_k,$$

where the $c_{ijk} = B([e_i, e_j], e_k)$ are the structure constants of the basis.

One has the following description of the kernel L of α:

Proposition 9.4.2 $L = \{z \in \mathrm{Hom}(\mathbf{F}, \mathbf{F}) \mid z \cdot \mathrm{ad}(x) = \mathrm{ad}(x) \cdot z^t\}$. *In words, L consists of the endomorphisms z of $\mathbf{F}$ such that $z \cdot \mathrm{ad}(x)$ is skew–symmetric.*

Proof. If $z = \sum_{i,j} z_{ij} \epsilon_i \otimes e_j \in L$, and $\varphi(\epsilon_i) = [x, e_i] = \sum_k a_{ki} e_k$ then

$$\alpha(z)(e_k) = \sum_{j,k} \sum_i z_{ij} a_{ki} e_k \cdot e_j = 0.$$

This means that

$$\begin{cases} \sum_i z_{ij} a_{ji} = 0 \ (k = j) \\ \sum_i z_{ij} a_{ki} + z_{ik} a_{ji} = 0 \ (k \neq j). \end{cases}$$

Together these equations mean that $z \cdot \mathrm{ad}(x)$ is skew–symmetric. □

We would like to describe L. It will not be a free R–module. Three types of generators for it can be found. The first two result from the construction:

$$u \otimes \varphi(v) - v \otimes \varphi(u), \text{ for } u, v \in \mathbf{F}^*,$$

and

$$\psi(w) \otimes v, \text{ for } w \in C, \ v \in \mathbf{F}.$$

The last batch of generators are coming from the invariants $p_1, \ldots, p_\ell$.

Proposition 9.4.3 *For each invariant p, one has that*

$$z = \sum p_{ij} \, \epsilon_i \otimes e_j, \ p_{ij} = \frac{\partial^2 p}{\partial x_i \partial x_j}$$

lies in L.

For a polynomial u, we use the notation $u_r = \frac{\partial u}{\partial x_r}$.

Lemma 9.4.4 *For each invariant polynomial u,*

$$\sum_q u_{\ell,q} \cdot B(x, [e_s, e_q]) = -\sum_q u_{s,q} \cdot B(x, [e_\ell, e_q]).$$

Proof. This will be derived from the identity $[\nabla p, x] = 0$. The kth component of this equation is

$$\sum_{i,j} u_i x_j c_{ijk} = 0;$$

differentiating, first with respect to x_ℓ and then with respect to x_s, we get

$$\sum_{i,j} u_{i,\ell} \frac{\partial x_j}{\partial x_s} c_{ijk} + \sum_{i,j} u_{i,\ell,s} x_j c_{ijk} + \sum_{i,j} u_{i,s} \frac{\partial x_j}{\partial x_\ell} c_{ijk} = 0,$$

which may be rewritten as

$$\sum_i u_{i,\ell} c_{isk} + \sum_{i,j} u_{i,\ell,s} x_j c_{ijk} + \sum_i u_{i,s} c_{i\ell k} = 0.$$

Multiplying this equation by x_k and summing over k, one obtains the stated identity once it is observed that $\sum_{jk} c_{ijk} x_j x_k = 0$. □

We are going to view these endomorphisms as defining a mapping

$$\beta: C \to \mathbf{F}^* \otimes \mathbf{F}.$$

The proof of the proposition is now the straightforward computation that the symmetric matrix $\sum p_{ij} \, \epsilon_i \otimes e_j$ commutes with $\sum_{i,j,k} x_i c_{ijk} \epsilon_j \otimes e_k$. □

The map β will be used to construct the following complex of free R–modules:

$$0 \to \wedge^2\mathcal{C} \xrightarrow{\gamma} \mathcal{C}\otimes\mathbf{F}^* \to \begin{array}{c}\mathcal{C}\\ \oplus \\ \wedge^2\mathbf{F}^* \\ \oplus \\ \mathcal{C}\otimes\mathbf{F}\end{array} \to \mathbf{F}^*\otimes\mathbf{F} \to S_2(\mathbf{F})$$

The mapping γ is defined as follows: Order (by degree) a basis $\{e_1, \ldots, e_\ell\}$ of $\mathcal{C}$ and fix the basis $e_i \wedge e_j,\ i < j$ of $\wedge^2\mathcal{C}$. Set

$$\gamma(e_i \wedge e_j) = e_i \otimes \psi(e_j) + e_j \otimes \psi(e_i).$$

This is not a minimal resolution. The question is whether one (which?) of its subcomplexes is a projective resolution of $S_2(E)$.

Classical Yang–Baxter equations

The following is Belavin-Drinfel'd's description of the Classical Yang–Baxter variety. It was explained to us by R.-O. Buchweitz, who pointed out that it is contained in [16], particularly Proposition 2.4 and Remark 7.3.

If $\mathfrak{g}$ is any Lie algebra, the Yang–Baxter map is defined as a map from $\wedge^2\mathfrak{g} \to \wedge^3\mathfrak{g}$. This map is quadratic and defined on 2–forms as follows:

$$\begin{aligned} q_{YB}(g_1 \wedge g_2 + g_3 \wedge g_4) \;=\; & g_1 \wedge [g_2, g_1] \wedge g_2 + 2(g_1 \wedge [g_3, g_2] \wedge g_4 \\ & - g_2 \wedge [g_3, g_1] \wedge g_4 - g_1 \wedge [g_4, g_2] \wedge g_3 \\ & + g_2 \wedge [g_4, g_1] \wedge g_3) + g_3 \wedge [g_4, g_3] \wedge g_4. \end{aligned}$$

A solution of the Classical Yang–Baxter Equation is any 2–form $\omega \in \wedge^2\mathfrak{g}$ for which $q_{YB}(\omega) = 0$. Now, if we look for solutions given by decomposable 2–forms $g \wedge h$, then this means exactly:

$$q_{YB}(g \wedge h) = g \wedge [h, g] \wedge h = 0,$$

in other words, the bracket $[g, h]$ has to be linearly dependent on g, h. Hence such solutions correspond precisely to Lie planes in $\mathfrak{g}$.

In general, choosing a symplectic basis, one may write any 2–form ω as

$$\omega = \sum_{i=1,\ldots,k} x_i \wedge y_i,$$

where $\{x_1, \ldots, x_k; y_1, \ldots, y_k\}$ is part of a basis of g. Evaluating q_{YB} on this expression, one sees that all the brackets $[x_i, y_j]$, $[x_i, x_j]$ and $[y_i, y_j]$ have to lie in the subspace spanned by $\{x_i; y_i\}$. Furthermore, if one writes the brackets in terms of this basis, and thinks of the coefficients as defining a linear map ψ on $\wedge^3(\{x_i; y_i\})$, one can interpret the solution as a linear map B from $\wedge^2(\{x_i; y_i\})$ into the field over which the Lie algebra $\mathfrak{g}$ is defined, having the following property:

$$\psi(x \wedge y \wedge z) = B([x, y], z) + B([y, z], x) + B([z, x], y) = 0.$$

This means that a solution to the Classical Yang–Baxter Equation (CYBE) can be represented as a Lie subalgebra (the one spanned by $\{x_1, ..., x_k; y_1, ..., y_k\}$), together with a 2–cocycle on it which has maximal rank. Such a representation of a solution of CYBE is not unique: It depends on the choice of a Lie subalgebra on which the solution is non–degenerate. In case of solutions of rank 2 (i.e. $k = 1$), the cocycle condition becomes void, as up to scalars there is only one non–degenerate cocycle.

Lie planes

Let $\mathfrak{g}$ be a simple Lie algebra over $\mathbb{C}$ of rank ℓ and dimension n. Denote by

$$P(\mathfrak{g}) = \{(x, y) \in \mathfrak{g} \times \mathfrak{g} \mid \{x, y\} \text{ is a Lie subalgebra}\}$$

the variety of Lie planes of $\mathfrak{g}$.

Clearly $P(\mathfrak{g})$ contains $C(\mathfrak{g})$, the commuting variety of $\mathfrak{g}$. We recall that the latter, by Richardson's theorem, is an irreducible variety of dimension $n+\ell$. The components of $P(\mathfrak{g})$ will be more difficult to describe, as indicated by the following computations of Guralnick.

Proposition 9.4.5 *Let W be an irreducible component of $P(\mathfrak{g})$ such that the projection π of W into $\mathfrak{g}$ is surjective. Then* $\dim W = n + 2$ *unless* $P(\mathfrak{g}) = C(\mathfrak{g})$.

Proof. Denote still by π the projection of $W' = W \setminus C(\mathfrak{g})$ into $\mathfrak{g}$ (on the first component). We only need to show that the dimension of a generic fiber is 2. So let x be in $\mathfrak{g}$ such that x is a regular semisimple element and that all of the eigenspaces of $\text{ad}(x)$ (other than the zero eigenspace) have dimension 1. Since these are both open conditions, the set of such x is open. Then the fiber is a finite union of $\{y \mid y = ax + bz(i)\}$, where $\{z(i)\}$ are a complete set of independent eigenvectors of $\text{ad}(x)$ for the nonzero eigenvalues. Thus the fiber has dimension two and the result follows. □

Corollary 9.4.6 *If $\ell > 1$, then $C(\mathfrak{g})$ is an irreducible component of $P(\mathfrak{g})$.*

Proof. If W is an irreducible component of $P(\mathfrak{g})$ containing $C(\mathfrak{g})$, and W is not $C(\mathfrak{g})$, then by the preceding $\dim W = n+2$ is at most $n + \ell = \dim C(\mathfrak{g})$. Thus $W = C(\mathfrak{g})$. □

If $\ell = 1$, then $P(\mathfrak{g})$ is itself irreducible of dimension 5 (and so properly contains $C(\mathfrak{g})$).

This still leaves open the question, what are the other components and their dimensions? From the analysis above, it is not too hard to see that the irreducible components of $P(\mathfrak{g})$ which have surjective projections are $C(\mathfrak{g})$ and one or two others depending on whether there are one or two root lengths (if there is a single root length, then all the $z(i)$ above are conjugate under the Weyl group, whence since the irreducible variety is invariant under the Lie group G, all such pairs are in the same component).

A set of equations for $P(\mathfrak{g})$ was indicated in the introduction: Form the $n \times 3$ matrix ψ whose columns are two generic elements x, y and $[x, y]$. A defining ideal for $P(\mathfrak{g})$ is $I_3(\psi)$. By Theorem 1.1.16,

$$\text{height } I_3(\psi) \leq n - 3 + 1 = n - 2.$$

A computation for the algebras of rank 2, with the exception of G_2, showed that one obtains equality. This means that $I_3(\psi)$ is a Cohen–Macaulay ideal, and therefore all components of $P(\mathfrak{g})$ have dimension $n + 2$.

Chapter 10

Computational Methods in Commutative Algebra

This chapter contains computer algebra methods tailored for commutative algebra. Some were specifically developed to study blowup algebras. Most however have a general nature and may be useful in other areas of commutative algebra.

The intrinsic intractability of very large scale computation in algebraic geometry and commutative algebra demand a continued refinement of techniques to ensure that the cost of computation be met largely by theoretical means. How to inform a computation of mathematical knowledge constitutes a most challenging problem.

The overall aim here is a discussion of basic constructions in and how they can effectively be carried out by symbolic computation programs.

The constructions are those for the radical of ideals, with some steps towards a facilitation of primary decomposition, the integral closure, primality testing, and the setting up of ideal transform computations. There is also a varied discussion of approaches to carry out primary decomposition of zero–dimensional ideals.

As a matter of general strategy, the constructions are mediated through the funnel of homological algebra. They can be characterized as successive layers of syzygies computations.

This has an advantage that often one can supply explicit formulas for the objects to be computed. On the other side of the ledger, the reader will not find here estimates for the complexity of the proposed constructions. There are several reasons for the omission, the main one being the near impossibility of estimating the complexity of Gröbner basis of ideals whose generators have many relationships. Nevertheless, close attention was paid to the known optimizations of these computations, and in this we followed [33] and [230].

10.1 Generalized Division Algorithms

This section gives a bird's eye view of the methods that undergird large sectors of computational commutative algebra. They are often used in conjunction with more classical tools, *e.g.* factorization algorithms.

Let $R = k[x_1, \ldots, x_n]$ be a polynomial ring and let I be an ideal given by a set $\{f_1, \ldots, f_m\}$ of generators. In concrete situations, these generators carry along many mutual relationships. Furthermore, they were likely obtained in a 'natural' setting. Among the properties I may have is a great measure of sparsity in its description.

To deal with most questions about I—such as, membership, primality, primary decomposition, homological dimensions, etc—may require an appropriate change in the system of generators. Among the primitive means to produce fresh elements of the ideal we single out:

(a) *Pseudo division.* This views R as a polynomial ring $S[x]$, where one of the variables $x = x_i$ is singled out to play the role of indeterminate. It leads to: If $f(x)$ and $g(x)$ are elements of R, of degrees $r = \deg f(x) \geq \deg g(x) = s$ and if $a \in S$ is the leading coefficient of $g(x)$ then we can write

$$a^{r-s+1} f(x) \quad = \quad q(x) \cdot g(x) + h(x), \ \deg h(x) < \deg g(x), \qquad (10.1)$$

(where $\deg 0 = -\infty$).

There are not many strategies that guide the selection of the elements. At the hands of Ritt and Wu, however, it became the basic operation of their selection mechanisms that from an ideal I obtain collections of short subsets whose zeros conveniently circumscribe the algebraic variety defined by I. With tongue-in-cheek, one might liken their methods as attempts to make rings of polynomials in several variables behave as a ring in a single variable.

(b) *Resultants.* In general they provide a check on the 'compatibility' of a system of polynomial equations. This is a very rich and classical/modern theory extremely useful for predictions in theoretical studies. Some of the best known examples—Sylvester's and Macaulay's—are very direct and may produce entirely fresh elements in the ideal. Being determinantal–based however leads them to produce dense/high degree polynomials. A concerted effort is being pursed by Sturmfels [270] and others to deal with the sparsity issue.

(c) *Syzygies.* One of the most fruitful approaches to the study of an ideal I passes through the examination of its syzygies. The study of the relations amongst the given generators of I—or, more properly, amongst especially obtained generators—will be a key enabling element in questions such as membership, radical, and elements of primary decomposition of I.

Gröbner bases

We are going to see next the division method that has had the greatest impact on computer algebra.

The core problem addressed is the following. Let k be a field, and let R be the polynomial ring $k[x_1, \ldots, x_n]$. If I is an ideal of R, given by a set $\{f_1, \ldots, f_m\}$ of generators, is there a canonical basis for the k–vector space R/I? If $I = (0)$, the monomials $M = x_1^{a_1} \cdots x_n^{a_n}$ provide the obvious basis. More generally, if the f_j are themselves monomials, then the M's which are not multiples of any of the f_j give the sought after basis.

Roughly speaking, a key to the various methods that have been developed lies in associating to the ideal I another ideal, $init(I)$, generated by monomials, so that the vector spaces R/I and $R/init(I)$ be isomorphic. Macaulay first brought up this idea in a classical examination of the Hilbert function of an ideal ([189]).

Monomials, orderings and weight vectors

Denote by $\mathbb{M}$ the monoid of all monomials in the x_i's (including 1). They are given an additional measure of structure by the choice of a total ordering T for $\mathbb{M}$, such that $1 <_T m$, $m \neq 1$, and which is compatible with the monoid multiplication: For $m_1, m_2, m_3 \in \mathbb{M}$, if $m_1 <_T m_2$ then $m_1 m_3 <_T m_2 m_3$. These orderings are referred to as admissible term orders, or simply a term order.

A basic example of a term ordering is the *lexicographic* order (*lex* for short):

$$m_1 = x_1^{a_1} \cdots x_n^{a_n} >_{lex} x_1^{b_1} \cdots x_n^{b_n} = m_2$$

if

$$a_1 = b_1, \cdots, a_{r-1} = b_{r-1}, a_r > b_r, \text{ for } 1 \leq r \leq n.$$

More general term orders arise by combining several admissible partial orders through their lexicographic product of orderings. If $T_1, \ldots, T_s$ are such partial orders, the product order

$$T = T_1 \times_{lex} T_2 \times_{lex} \times \cdots \times_{lex} T_s$$

is defined as above

$$m_1 >_T m_2 \Longleftrightarrow m_1 =_{T_1} m_2, \ldots, m_1 =_{T_{r-1}} m_2, m_1 >_{T_r} m_2, \text{ for } 1 \leq r \leq s.$$

For example, if T_1 is the total degree partial order and T_2 is *lex*, their product is the so–called graded lexicographic ordering:

$$(a_1, \ldots, a_n) < (b_1, \ldots, b_n) \Leftrightarrow$$
$$\text{first nonzero entry of } \left(\sum b_i - \sum a_i, b_1 - a_1, \ldots, b_n - a_n\right) \text{ is positive.}$$

Particularly striking properties are enjoyed by the *reverse lexicographic order*, defined by changing the last requirement above to: the last nonzero entry is negative. Bayer and Stillman [15] have discovered many of its interesting properties and incorporated its efficiencies into their *Macaulay* program.

A result of Robbiano ([232]) asserts that every compatible partial ordering of $\mathbb{M}$ is realized in this manner, and that every term ordering is the lexicographic product of at most n such orderings. The study of all possible orderings is subsumed in the theory of the Gröbner fan of Mora and Robbiano [203].

Initial ideals

We define the homomorphism

$$\log : \mathbb{M} \longrightarrow \mathbb{N}^n,$$

by

$$\log(x_1^{a_1} \cdots x_n^{a_n}) = (a_1, \ldots, a_n),$$

so that orderings are the means to pass back and forth between these monoids.

If $0 \neq f \in R$, $f = C(f) \cdot M(f) + \sum a_i M_i$, where $M(f)$ is the highest monomial that occurs in the representation of f; $0 \neq C(f)$ is its leading coefficient; the product $L(f) = C(f) \cdot M(f)$ is the leading term or initial term $\mathrm{in}(f)$ of f. We define $\log(f) = \log(M(f))$.

For a nonzero ideal I, define $\log(I)$ to be the union of $\log(f)$, $0 \neq f \in I$. $\log(I)$ is a sub–monoid of $\mathbb{N}^n$, stable under the addition of quadrants:

$$\log(I) = \log(I) + \mathbb{N}^n.$$

By the Hilbert basis theorem there are finitely many elements $g_1, \ldots, g_r$ in I so that

$$\log(I) = \bigcup(\log(g_i) + \mathbb{N}^n), \; 1 \leq i \leq r.$$

Definition 10.1.1 Let T be an admissible order on $k[x_1, \ldots, x_n]$. The set

$$\{g_1, \ldots, g_r\}$$

is a Gröbner *basis* of the ideal I if

(a) $g_i \in I, i = 1, \ldots, r$, and

(b) $\log(g_1), \ldots, \log(g_r)$ generate $\log(I)$.

It is easy to see that the images of the monomials

$$x^a \text{ where } a \in \mathbb{N}^n \setminus \log(I),$$

form a basis for the k–vector space R/I. They are the *standard monomials* associated to the Gröbner basis. The ideal of R generated by x^a, for $a \in \log(I)$, is $\mathit{init}(I)$, the *initial ideal* of I.

Definition 10.1.2 For a given $f \in R$, the (unique) polynomial

$$\mathrm{NormalForm}(f) = \sum c_a x^a,$$

where x^a are standard monomials, such that

$$f - \mathrm{NormalForm}(f) \in I$$

is the *normal form* of f.

We have the following fundamental fact

Theorem 10.1.3 (Macaulay)

$$\text{NormalForm}: R/I \longrightarrow R/init(I)$$

is an isomorphism of k–vector spaces.

Remark 10.1.4 If k is a ring, it still makes sense to define the initial ideal of an ideal I of R. Of course this may lead to two non–comparable initial terms $c \cdot \mathbf{x}^{\mathbf{a}}$ and $d \cdot \mathbf{x}^{\mathbf{a}}$ lying in in(I). For reasonable rings, say $k = \mathbb{F}_q[t]$, it is possible to compute initial ideals and Gröbner bases ([164]). This capability will be very important for some of the later algorithms.

Division algorithm

The significant property of these sets is

Proposition 10.1.5 *A set $\{g_1, \ldots, g_r\}$ of elements of the ideal I is a Gröbner basis of I if and only if every nonzero element of I can be written as $f = \sum a_i g_i$, with* $\log(f) \geq \log(a_i g_i)$. *In particular, a Gröbner basis of I is a generating set for I.*

The easy proof is left to the reader. Note the close parallel with the Euclidean algorithm in the ring $k[t]$ and the elements of Gaussian elimination.

Buchberger algorithm

The previous proposition is a basis for solving several general questions about the ideal I, particularly the membership problem. There remains to find such bases. This results from the following analysis due to Buchberger (see [33] for extensive details and related bibliography; see also [201] and [229]).

Let I be defined by a generating set $F = \{f_1, \ldots, f_m\}$. One must have a criterion to decide whether $\log(F) = \{\log(f_1), \ldots, \log(f_m)\}$ generates $\log(I)$, and if not, a device to add new elements to the f_i's. These steps come together in the same argument.

Reduction

We begin with the observation on how to add a possible new generator to F. Let f be a nonzero element of I. If $\log(f)$ is not a multiple of any the $\log(f_i)$'s, one has a new generator. However, even if $\log(f)$ is already a multiple of a $\log(f_i)$ it may still contribute a new generator. This is decided as follows. If the leading monomial $M(f)$ of f is divisible by, say, the leading monomial $M(f_i)$ of f_i, pick $q \in R$ such that $\log(f - qf_i) < \log(f)$. If $M(f)$ is not a multiple of any such $M(f_i)$ we could stop here. (It is also used to effect this operation on the next lower monomial of f

which does not belong to the span of the $M(f_i)$'s.) On iterating we end up with an element

$$g = f - \sum_i a_i f_i,$$

with the property that $g = 0$ or $\log(g)$ is not divisible by $\log(F)$. In either case we say that f *reduces* to g relative to F.

If $g = 0$, f is ignored; otherwise adding $\log(g)$ to the submonoid of $\mathbb{N}^n$ generated by the $\log(f_i)$'s gives rise to a larger submonoid. The Hilbert theorem guarantees that such additions cannot go on forever.

S–resultant

The issue is how to pick appropriate elements of I. The basic step goes to the core of both the Euclidean and Gaussian algorithms. It is embodied in the notion of the (resultant) S–polynomial attached to two polynomials $f, g \in R$. If $M(f)$ and $M(g)$ are their leading monomials, set

$$S(f, g) := a_g \cdot f - (C(f)/C(g)) \cdot b_f \cdot g,$$

where $a_g \cdot M(f) = b_f \cdot M(g)$ is the least common multiple of $M(f)$ and $M(g)$. The collections of such objects have a very natural place in the theory of Taylor resolutions ([271], see also [64], [188]).

The Buchberger algorithm is made up of the following result and the scheme that follows to produce the required elements.

Theorem 10.1.6 *A set of generators $F = \{f_1, \ldots, f_m\}$ of the ideal I is a Gröbner basis of I if and only if the S–polynomial $S(f_i, f_j)$ of each pair (f_i, f_j) of elements of F reduces to* 0 *with respect to F.*

Proof. The proof of the necessity is clear. For the converse, let f be an element of I. We may assume that f is its own normal form with respect to F. Let $f = \sum_j h_j f_j$, and consider the $\log(h_j f_j)$'s (for $h_j \neq 0$). $\log(f)$ cannot be equal to one of the $\log(h_j f_j)$, as it is already in normal form. This means that there must be some cancelling out at the top monomial occurring in the products $h_j f_j$. More precisely, suppose

$$M(h_1) \cdot M(f_1) = M(h_2) \cdot M(f_2) = \cdots = M(h_k) \cdot M(f_k)$$

are the top monomials that occur in the right hand side of the representation of f. Their cancelling out means that the vector of leading terms

$$(L(h_1), \ldots, L(h_k))$$

is a syzygy of

$$(L(f_1), \ldots, L(f_k)).$$

But from the theory of Taylor resolutions, such relations are combinations of the syzygies of pairs $\{C(f_r)M(f_r), C(f_s)M(f_s)\}$. This means that we have a representation

$$f = \sum_j h'_j f_j + \sum a_{rs} S(f_r, f_s),$$

where $\log(h'_j f_j) < \log(h_1 f_1)$. An easy induction completes the proof. □

Algorithm 10.1.7 *Let $F = \{f_1, \ldots, f_m\}$ be a set of generators of the ideal I, and let T be a term order for $\mathbb{M}$.*

$G := F$.

$B := \{\{f_1, f_2\} \mid f_1, f_2 \in F$, and $f_1 \neq f_2\}$.

while $B \neq \emptyset$ do

$\{f_1, f_2\} :=$ a pair in B
$B := B \setminus \{f_1, f_2\}$
$g :=$ normal form of $S(f_1, f_2)$ with respect to G
if $g \neq 0$, then
$B := B \cup \{\{g, h\} \mid h \in G\}$
$G := G \cup \{g\}$

The theoretical cost of these computations can be staggering, doubly exponential in the number of variables, according to [195]. This feature was already present in the classical analysis of the cost of computation in polynomial ideal theory by Grete Hermann [102]. On the other hand, the dynamic behavior of Buchberger's algorithm benefits from the average cost of the computation (linear in the number of variables). Furthermore, unlike the classical methods that had to work out always from a worst case assumption, Gröbner bases algorithms are eminently programmable.

10.2 Basic Methods

We shall point out only the most basic techniques, mostly associated with a single Gröbner basis computation. The more refined methods make use of the theory of syzygies.

Elimination of variables

Let R be the polynomial ring $k[x_1, \ldots, x_n]$, let t be an indeterminate over R, and put $B = R[t]$. Let I be an ideal of B.

Proposition 10.2.1 *Let T be a lexicographic ordering of the variables such that $t >_T x_i$. Let F be a Gröbner basis of I. Then $F \cap R$ is a Gröbner basis of $I \cap R$.*

Proof. Follows immediately from the division algorithm. □

Corollary 10.2.2 *If I and J are two ideals of R, then $I \cap J$ can be computed.*

Proof. Note that if t is a new variable, then $I \cap J = (I \cdot t, J \cdot (1-t)) \cap R$. □

Regular elements and ideal quotients

Proposition 10.2.3 *An element $f \in R$ is regular modulo the ideal I if and only if one of the following conditions hold:*

(a) $((I \cdot t, (1-t) \cdot f) \cap R) \cdot f^{-1} = I$.

(b) $(I, 1 - f \cdot t) \cap R = I$.

The first formula computes $I{:}_R\, f$, the second determines $\bigcup_{n \geq 1}(I{:}_R\, f^n)$.

Given two ideals I and J the ideal quotient $I{:}\, J = \{r \in R \mid r \cdot J \subset I\}$ will figure prominently in our constructions. If $J = (a_1, \ldots, a_n)$, it can be computed as

$$I{:}\, J = \bigcap_{i=1}^{n} (I{:}\, a_i).$$

An alternative is the following construction.

Proposition 10.2.4 *Let x be a variable over the ring R and let*

$$f = a_1 + a_2 x + \cdots + a_n x^{n-1}.$$

Then

$$I{:}\, J = (I \cdot R[x]{:}\, f) \cap R.$$

We shall find that the *saturated* ideal quotient

$$I{:}\, J^{\infty} = \bigcup_{k} (I{:}\, J^k),$$

has many uses. In [15] the following device is given to compute it. Let f be the polynomial defined above and let y be another variable.

Proposition 10.2.5

$$I{:}\, J^{\infty} = ((I, y - f){:}\, y^{\infty}) \cap R.$$

Rings of endomorphisms

There are two related constructions that occur repeatedly. Let D be a Noetherian ring and let Q denote its total ring of fractions. For two ideals K and L, we use the notation

$$K:_Q L = \{x \in Q \mid xL \subset K\}.$$

The significance lies in the following two identifications: If L contains regular elements then

$$\begin{aligned}\mathrm{Hom}_D(L, D) &\simeq D:_Q L\\ \mathrm{Hom}_D(L, L) &\simeq L:_Q L.\end{aligned}$$

The actual determination can be carried out as follows:

Proposition 10.2.6 *Let $L = (a_1, \ldots, a_n)$ be an ideal with each a_i a regular element. Then:*

$$D:_Q L = ((a_1):_D (a_2, \ldots, a_n))a_1^{-1} \quad (10.2)$$

$$L:_Q L = (\bigcap_{i=1}^n L \cdot (a_1 \cdots \widehat{a_i} \cdots a_n))(a_1 \cdots a_n)^{-1}. \quad (10.3)$$

Proof. We only verify the second assertion. Denote $a = \prod_{i=1}^n a_i$. We have

$$\begin{aligned}L:_Q (a_1, \ldots, a_n) &= \bigcap_{i=1}^n L:_Q (a_i)\\ &= \bigcap_{i=1}^n La_i^{-1},\end{aligned}$$

which upon multiplying and dividing by a yields the desired expression. □

Let us sketch out how the ring $I{:}\,I$ can be represented. Suppose A is an integral domain and $I \subset A = R/\wp$, where $R = k[x_1, \ldots, x_n]$. Write $I = L/\wp$, and pick a representation $L = (\wp, a_1, \ldots, a_n)$, with $a_i \notin \wp$.

Proposition 10.2.7 *Let*

$$(\bigcap_{i=1}^n La/a_i)a^{-1} = (b_1, \ldots, b_m)a^{-1}.$$

For each b_j pick a new variable y_j and consider the ideal

$$J = (\wp, ay_1 - b_1, \ldots, ay_m - b_m).$$

Then

$$P = J : a^\infty$$

is the presentation ideal of $I{:}\,I$, that is

$$I : I = k[x_1, \ldots, x_n, y_1, \ldots, y_m]/P.$$

Computing syzygies

Unquestionably, the most useful approach to computation in commutative algebra is through the theory of syzygies. For a general discussion, see [201].

Here we indicate how the first–order syzygies of an ideal $I = (F) = (f_1, \ldots, f_n)$ may be determined. The generators are going to be taken as ordered, and we write one of its syzygies as a column vector $v \in R^n$ such that $F \cdot v = 0$.

Let $G = \{g_1, \ldots, g_m\}$ be a Gröbner basis of I. For each pair of polynomials $\{g_i, g_j\}$ in G the S–polynomial $S(g_i, g_j)$ has a reduction

$$S(g_i, g_j) = a_j g_i - a_i g_j = \sum_{k=1}^{m} h_k g_k,$$

by Theorem 10.1.6. In particular the vector

$$\begin{pmatrix} h_1 \\ \vdots \\ h_i - a_j \\ \vdots \\ h_j + a_i \\ \vdots \\ h_m \end{pmatrix} \tag{10.4}$$

is a syzygy of G. The following elementary fact ([247]) is extremely useful.

Theorem 10.2.8 *The syzygies of G are generated by the vectors* (10.4).

Proof. We leave its proof to the reader. □

To be more useful it is essential to extend the notion of Gröbner bases from ideals of $R = k[x_1, \ldots, x_n]$ to submodules of a free R–module P. One could, for instance, view any submodule of F as ideals of a big polynomial ring $k[x_1, \ldots, x_n, e_1, \ldots, e_m]$, generated by linear forms in the variables e_i and the quadratic polynomials $e_j^2 - e_j$. But this approach is obviously wasteful of resources.

It is better (see [55], [200], [201]) to proceed as follows. Given a free module P, order one of its bases, $\{e_1, \ldots, e_m\}$, and define a monomial of F as an element of the form $\mathbf{x}^A e_i$, where $\mathbf{x}^A$ is a monomial of R. A *monomial submodule* M is generated by such elements. The notion of a Gröbner basis can then be applied to E as well, and the calculus of syzygies above can be repeated for submodules of free modules.

Assembling the syzygies

It is necessary to convert the syzygies of a Gröbner basis G into the syzygies of the basis F it was derived from. It involves the two matrices that convert one set of generators into the other. The following observations were kindly pointed out to us by H.-G. Gräbe.

Denote by F and G the (ordered) given generators and the computed Gröbner basis of a module Q. This gives rise to two transition matrices A and B

$$G = F \cdot A \tag{10.5}$$
$$F = G \cdot B, \tag{10.6}$$

the first of which is obtained in the execution of the Buchberger algorithm, while the other is the representation of the elements of F by the normal form algorithm.

Proposition 10.2.9 *If S_G is a basis for the syzygies of G then the columns of*

$$\left[\; E - A \cdot B \mid A \cdot S_G \;\right]$$

is a basis for the syzygies of F. (E is the $n \times n$ identity matrix.)

Proof. The columns of this matrix are clearly syzygies of F. Conversely, if $F \cdot v = 0$, then $B \cdot v \in S_G$. But

$$\begin{aligned} v &= (E - A \cdot B) \cdot v + A \cdot B \cdot v \\ &\in \text{column space}(E - A \cdot B) + \text{column space}(A \cdot S_G), \end{aligned}$$

which proves the assertion. □

The ability to compute syzygies gives a distinct advantage in carrying out the ideal theoretic operations mentioned earlier. For instance, the computation of the intersection of two ideals $I = (a_1, \ldots, a_m)$ and $J = (b_1, \ldots, b_n)$ is now handled as that of finding the syzygies of the matrix

$$\begin{bmatrix} 1 & a_1 & \cdots \\ 1 & b_1 & \cdots \end{bmatrix},$$

(arranged to accommodate both sets of generators). The desired intersection are the entries at $(1, 1)$.

Noether normalization

A Noether normalization of $A = R/I$ is a polynomial ring $k[\mathbf{z}] \hookrightarrow A$ over which A is finitely generated as a module. Preferably $\mathbf{z}$ should be a subset of the $\mathbf{x}$'s. An algorithm to find $k[\mathbf{z}]$ ([25]) is based on the following observation (see [187] for a detailed analysis).

Let $F = \{f_1, \ldots, f_m\}$ be a set of generators of the ideal I of the polynomial ring $k[x_1, \ldots, x_n]$. Suppose $f_1 \in F$ is monic in one of the variables, say x_n. Let $J = I \cap k[x_1, \ldots, x_{n-1}]$. Then

$$k[x_1, \ldots, x_{n-1}]/J \hookrightarrow k[x_1, \ldots, x_n]/I$$

is an integral extension. In particular, height I = height $J + 1$.

It will provide for systems of parameters and projective dimensions as well. We blur the distinction between an ideal and a set of generators.

Algorithm 10.2.10 *Let $F = \{f_1, \ldots, f_m\}$ be a set of generators of the ideal I.*

$V := \{x_1, \ldots, x_n\}$.

$G := F$.

while $G \neq \emptyset$ *do*

if there is a monic polynomial in x_i in G *then*

$V := \{x_1, \ldots, \widehat{x_i}, \ldots, x_n\}$.
$G := G \cap k[x_1, \ldots, \widehat{x_i}, \ldots, x_n]$.

else effect a change of variables φ so that a monic polynomial occurs in $\varphi(G)$ and $G := \varphi(G)$.

Note that when all the changes of variables are taken into account, we have a sequence of polynomials

$$f_i(y_1, \ldots, y_i) \in k[z][y_1, \ldots, y_{n-d}] = k[x_1, \ldots, x_n],$$

monic in y_i, lying in the ideal I. This leads immediately to a presentation of A as an $k[z]$–module.

It is of great significance that the process be carried out very efficiently. It might be fair to say that almost anything that may be obtained from generic projections can be obtained through Noether normalizations.

Homogenization

Homogeneous ideals have attached numerical data that is useful in controlling a computation. They also benefit from the efficiencies of certain orderings. It is therefore quite common to homogenize a module and later capture information of the original module. This is carried out as follows.

Let $f \in R = k[x_1, \ldots, x_n]$, and denote by T a fresh variable. The homogenization of f is:

$$f^H = T^{\deg f} f(x_1/T, \ldots, x_n/T).$$

For an ideal $I = (f_1, \ldots, f_m)$, we have two ways to define its homogenization: $J = (f_1^H, \ldots, f_m^H)$ or as $L = (f_1^H, \ldots, f_m^H) : T^\infty$. The first depends on the generating set but the other doesn't. The key property of both is that $T - 1$ is a regular element and they deform into the original ideal: $J_{T=1} = L_{T=1} = I$. Whenever we speak of homogenizing we refer to the second process. The same steps can be carried out on a module by acting on its presentation matrix.

The equations of Rees algebras

Let $I = (f_1, \ldots, f_m)$ be an ideal of the ring R. The presentation of the Rees algebra $R[It]$ is obtained as:

Proposition 10.2.11 *In the ring $A = R[T_1, \ldots, T_m, t]$ consider the ideal L generated by the polynomials $T_j - tf_j$, $j = 1, \ldots, m$. Then $R[It] = A/J$, where J is the contraction of L in the subring $B = R[T_1, \ldots, T_m]$.*

Proof. It is clear that $J \supset L \cap B$. Conversely, if $f(T_1, \ldots, T_m)$ is an element of J, we write

$$f(T_1, \ldots, T_m) = f(tf_1 + (T_1 - tf_1), \ldots, tf_m + (T_m - tf_m))$$

and use the Taylor expansion to show $f \in L$. □

Note that J is graded in the T_j variables, with its nth component giving the syzygies of I^n. A variation on the method computes the syzygies of I and forms an appropriate closure:

Proposition 10.2.12 *Let φ be a $m \times n$ matrix whose columns generate the syzygies of I, and let J_0 be the defining ideal*

$$[T_1, \ldots, T_m] \cdot \varphi,$$

of the symmetric algebra of I. Let f be a nonzero element of I. Then

$$J = (J_0, 1 - f \cdot t) \bigcap R[T_1, \ldots, T_m].$$

Remark 10.2.13 Actually, we may replace J_0 by the ideal generated by the polynomials $f_1 T_j - f_j T_1$, $j = 2, \ldots, m$, and use $f = f_1$ in the intersection. An appropriate name for this could be taking the *rational closure* of J_0.

Computer Algebra systems

The division algorithms above, addressing so quickly membership and syzygy problems, gives an early indication of the usefulness of this theory. These algorithms have been programmed into general purposes routines in various computer algebra systems, often taking into account features present in special cases. Among the implementations, we single out the following. (It is of interest to keep in mind that some systems seek to fit simultaneously several roles and may therefore fail to take advantage of the specificity of the computation.)

Macaulay is the most widely used computer algebra system in commutative algebra and algebraic geometry. It lacks a programming language but makes up for it with a rich library of scripts. *CoCoA* is a developing system, with a friendly user interface, and a more powerful version under development. *Maple* and *Reduce* are full–fledged computer algebra systems whose Gröbner basis algorithms are periodically updated; the latter has an extremely useful forum/bulletin board, with a worldwide community of contributors. *Macsyma* has also a Gröbner basis algorithm but it does not incorporate recent optimizations. *Mathematica* has a limited implementation thus far. *Axiom* promises to pick up where *Scratchpad* left off, and its advanced factorization techniques should be very useful. The ideal environment might consist of a system such as *Mathematica* or *Maple*, from which one could call on fast Gröbner *engines*.

Exercise 10.2.14 Let I be and ideal and let E be an R–module. Prove the following statements:

- I is a prime ideal if and only if I^H is a prime ideal.
- $(\sqrt{I^H})_{T=1} = \sqrt{I}$.
- (annihilator(E^H))$_{T=1}$ = annihilator(E).

10.3 Nullstellensätze

Loosely speaking, by *a* Nullstellensatz we shall mean the description of the nil radicals of the ideals in a given class, the quintessential example of which is Hilbert's theorem. For ideals in more narrowly defined classes, it would be desirable to have sharper descriptions of their radicals.

In this chapter a mix of differential and homological approaches are used to produce explicit formulas and algorithms that compute radicals.

Generic socle formulas

The aim here is to set up an environment to develop some closed formulas for the radical of an ideal.

Let $R = k[x_1, \ldots, x_n]$ be a polynomial ring over a field k. Here we prove a result in the form of a closed formula for the radical of an unmixed ideal of R—an ideal all of whose primary components have the same codimension [62]. It is a generalization, to polynomials in several variables, of the venerable formula:

$$\boxed{\sqrt{(f)} = (f) \colon f'} \tag{10.7}$$

Let $(A, \mathfrak{m})$ be a local algebra of dimension zero. Recall that the *socle of an algebra* of A is the annihilator of $\mathfrak{m}$. More generally, let k be a field and let $0 \neq f \in k[x]$ be a polynomial. Suppose that the characteristic of k does not divide the degrees or the multiplicities of the irreducible factors of f. The algebra $A = k[x]/(f(x))$ decomposes into a finite direct product of local algebras

$$A = A_1 \times \cdots \times A_r,$$

and the image of $f'(x)$, the derivative of $f(x)$, in each A_i generates its socle. It is this fact that underlies the explicit radical formula above.

The point will be that we shall want to interpret this formula as a statement about the structure of the socle of certain Gorenstein algebras. For more general algebras, it is not well known how to predict, from the generators and relations of the algebra, which elements will generate its socle. An important exception was a trace formula developed by Scheja and Storch in the mid 70's ([243]).

Definition 10.3.1 Let R be a Noetherian ring. An ideal I is said to be *height unmixed* if all the associated primes of I have the same codimension. When R is a polynomial ring, this means that the dimensions of the irreducible components of the algebraic variety $V(I)$ are the same. For such rings, I will be called *equidimensional* if the minimal associated primes have the same codimension.

The following notion will play a continued role throughout. The reader will observe that we began to use *codimension* as synonymous with *height*, which seems convenient when we must repeatedly keep track of both the height and the dimension of an ideal.

Definition 10.3.2 Let I be an unmixed ideal of codimension m, and let $P_1, \ldots, P_s$ be its associated prime ideals. Denote $B_t = (R/I)_{P_t}$. By a *generic socle formula* we mean an ideal $G \subset R$ such that for each $1 \leq t \leq s$

$$G \cdot B_t = \begin{cases} B_t, & \text{if } B_t \text{ is a domain} \\ \text{otherwise a nonzero submodule of the socle of } B_t. \end{cases}$$

If G is generated by a single element f, we shall call it a *generic socle generator.*

It is clear that generic socle generators always exist. The point is to try and identify them in the data associated with the ideal I. The following seems to justify it.

Proposition 10.3.3 *Let I be a height unmixed ideal and let G be a generic socle formula for I. Then*

$$\boxed{\sqrt{I} = I \colon G} \tag{10.8}$$

Proof. Both sides of the equation are unmixed ideals of the same codimension, so it suffices to prove equality at localizations $R_\wp$, where $\wp$ is a minimal prime of I. But then it is clear from the definition of G. □

Jacobian ideals

Let $R = k[x_1, \ldots, x_n]$, k is a perfect field, and let $I = (f_1, \ldots, f_p)$ be an ideal of R. We shall recall some basic properties of the module of Kähler differentials of the algebra $A = R/I$. Our basic reference will be [173].

The module of k–differentials of the algebra A will be denoted by $\Omega_{A/k}$. Although this module is independent on how it is presented as a quotient of a polynomial ring, it can be conveniently described by the exact sequence of modules of differentials

$$I/I^2 \stackrel{d}{\longrightarrow} \Omega_{R/k} \otimes_R A \longrightarrow \Omega_{A/k} \to 0,$$

where d is the universal derivation: $\mathrm{d}f = \sum_{i=1}^{n} \frac{\partial f}{\partial x_i} \mathrm{d}x_i$.

The Jacobian ideals of I are the determinantal ideals of the matrix

$$\varphi = \left(\frac{\partial(f_1, \ldots, f_p)}{\partial(x_1, \ldots, x_n)} \right) \mod I.$$

We denote them by

$$J_a(I) := I_{n-a}(\varphi).$$

If $m = \operatorname{codim} I$, the *Jacobian ideal* proper is the ideal of $m \times m$ minors of φ; it shall be denoted by J. Because the $J_a(I)$ are the Fitting ideals of the module $\Omega_{A/k}$, they are independent of the generating set of I, and behave well with regard to many processes such as localization and completion.

Another part of the theory of modules of differentials that we shall use is the notion of the Jacobi–Zariski sequence associated to a sequence of algebra homomorphisms. Recall that if $k \to K \to B$ are homomorphisms of algebras, then the natural sequence of modules of differentials

$$\Omega_{K/k} \otimes_K B \longrightarrow \Omega_{B/k} \longrightarrow \Omega_{B/K} \to 0,$$

is exact. It provides a mechanism to define intermediate Jacobian ideals, as we shall do later.

Scheja–Storch's formula

Socle formulas have been very difficult to identify. One that plays a key role in the exposition here is the following. We quote, for ease of reference, the theorem of Scheja–Storch ([243]):

Theorem 10.3.4 *Let k be a field and let $A = k[[x_1, \ldots, x_n]]/I$ be a finite dimensional k–algebra. Assume* $\dim_k A$ *is not divisible by the characteristic of k. Denote by J the Jacobian ideal of A. If A is a complete intersection then J generates the socle of A. Conversely, if k has characteristic zero and A is not a complete intersection then $J = 0$.*

We only comment on the proof of the last assertion. We may assume that I has a generating system $\{f_1, \ldots, f_m\}$ for which each subset of n elements, $\{f_{i_1}, \ldots, f_{i_n}\}$, is a regular sequence. From the first part of the theorem, the corresponding Jacobian determinant will generate the socle of

$$B = k[[x_1, \ldots, x_n]]/(f_{i_1}, \ldots, f_{i_n}),$$

and therefore is contained in I, since the image of I is a proper ideal in B.

It provides the tools to seek some generic socle formulas:

Theorem 10.3.5 *Let $I = (f_1, \ldots, f_p) \subset k[x_1, \ldots, x_n]$ be an unmixed ideal of codimension m, which is generically a complete intersection, and let*

$$k[\mathbf{z}] = k[z_1, \ldots, z_d] \hookrightarrow A = R/I$$

be a Noether normalization of the k–algebra A. Let J be the ideal generated by the $n \times n$ minors of the Jacobian matrix

$$\varphi = \frac{\partial(z_1, \ldots, z_d, f_1, \ldots, f_p)}{\partial(x_1, \ldots, x_n)}.$$

Then J is a generic socle formula for I.

The assertion on J means that at each minimal prime P of A, $J \cdot A_P$ is a nonzero ideal that is contained in the socle of A_P. Observe that if I is a complete intersection, then J is generated by a single element. Furthermore, if the z_i's are linear forms on the x_j's—as it will be usually the case—then a large block of the matrix φ is scalar.

Another point that will be made in the proof will be a restriction on the characteristic of the field k. Loosely speaking, characteristics below the degree of A should be avoided.

Northcott ideals

In the perspective above, one of the earliest examples of generic socle formulas is provided by the following theorem of Northcott [214]. Because it has played a role in our exposition, we outline its proof.

Let $\mathbf{u} = u_1, \ldots, u_n$ be a sequence of elements in the ring R, and let $\varphi = (a_{ij})$ be a $n \times n$ matrix with entries in R. Denote by $\mathbf{v} = v_1, \ldots, v_n$ the sequence

$$[\mathbf{v}]^t = \varphi \cdot [\mathbf{u}]^t.$$

The *Northcott ideal* associated to the sequence $\mathbf{u}$ and the matrix φ is $(\mathbf{v}, \det \varphi)$.

Theorem 10.3.6 *If* grade $(\mathbf{v}) = n$, *then*

$$(\mathbf{v}) : \det \varphi \quad = \quad (\mathbf{u}). \tag{10.9}$$

If $\mathbf{u}$ is a regular system of parameters for a n–dimensional regular local ring, the formula actually describes the socle of the Artin ring $R/(\mathbf{v})$: It is generated by $\det \varphi$. Thus, if $R = k[[x_1, \ldots, x_n]]$ and I is an ideal generated by a regular sequence of homogeneous polynomials $f_1, \ldots, f_n$, then the Euler formulas

$$\deg f_i \cdot f_i = \sum_{j=1}^{n} x_j \frac{\partial f_i}{\partial x_j}$$

will imply that the Jacobian determinant

$$\det \left(\frac{\partial(f_1, \ldots, f_n)}{\partial(x_1, \ldots, x_n)} \right)$$

is a nonzero element of the socle of R/I. (In fact, it generates the socle.)

Proof. It is not difficult to see that we may assume that R is a local ring, and the sequences $\mathbf{u}$ and $\mathbf{v}$ are (proper) regular sequences.

From Cramer's rule, it follows that the right–hand side of the equation is contained in $(\mathbf{v})\colon(\mathbf{u})$. We have to show that any element r that is conducted by $D = \det\varphi$ into $(\mathbf{v})$ must be an element of $(\mathbf{u})$.

To prove the assertion we are going to induct on n, the statement being clear for $n = 1$. Using prime avoidance (see [165, Theorem 124]) we can replace the generators $u_1, \ldots, u_n$ of $(\mathbf{u})$ by another sequence $\mathbf{w} = w_1, \ldots, w_n$ such that $v_1, \ldots, v_{n-1}, w_n$ is a regular sequence. Furthermore, the change of generators matrix ψ,

$$[\mathbf{u}]^t = \psi \cdot [\mathbf{w}]^t,$$

by Nakayama's Lemma, is invertible over R. We may then use $\mathbf{w}$ instead of $\mathbf{u}$.

Let then

$$D \cdot z = \sum_{i=1}^{n} c_i v_i.$$

Since we also have

$$D \cdot w_n = \sum_{i=1}^{n} A_{in} v_i,$$

where the A_{ij} are the entries of the adjoint of φ, we can write

$$(zA_{1n} - w_n c_1)v_1 + \cdots + (zA_{nn} - w_n c_n)v_n = 0.$$

Because the v_i's form a regular sequence, we must have

$$zA_{nn} - w_n c_n \in (v_1, \ldots, v_{n-1}).$$

Passing to the ring $\overline{R} = R/(w_n)$, the conditions of the theorem remain and the proof is easily completed by the induction hypothesis. □

Explicit Nullstellensätze

We will describe variations on a closed formula for the radical of a complete intersection, and its use to find the radical of an ideal. It requires that certain elements be in place. To make up for this, we provide other Jacobian methods to compute radicals.

We begin by stating two consequences of Theorem 10.3.5.

Theorem 10.3.7 *Let $I = (f_1, \ldots, f_m)$ be an ideal of codimension m, and let J be its Jacobian ideal. Then*

$$\boxed{\sqrt{I} = I\colon J} \qquad (10.10)$$

This formula is the analog of (10.7), with J playing the role of the derivative.

Theorem 10.3.8 *Let I be an ideal whose primary components all have codimension m. Let $\{f_1, \ldots, f_m\}$ be a regular sequence in I, and let J_0 be the Jacobian ideal of $I_0 = (f_1, \ldots, f_m)$. Then*

$$\boxed{\sqrt{I} = (I_0 \colon J_0) \colon ((I_0 \colon J_0) \colon I)} \tag{10.11}$$

There will be 'dynamic' constraints on the characteristic of the ground field, connected to the choice of the complete intersection I_0.

Proof of Theorem 10.3.5. There will be a sequence of change of base rings each accompanied by an examination of the Jacobian ideal.

The degree of A over $k[\mathbf{z}]$ is the dimension of the vector space

$$r = \dim_{k(\mathbf{z})} A \otimes_{k[\mathbf{z}]} k(\mathbf{z}).$$

If A is a graded ring and the $z_i's$ are forms of degree 1, then r is the usual degree of the ring, or of the ideal I. In general the z_i are linear forms in the x_j, and we may then assume, without changing the Jacobian ideal, that the z_i's form a subset of the x_i's.

We shall assume that the characteristic of k is larger than r. The condition on the characteristic implies that for any minimal prime P of A the field of fractions of A/P is separably generated over k. This will permit several changes of base fields in such way that the nil radical will change simply by extension.

Consider the module $\Omega_{A/k[\mathbf{z}]}$ of $k[\mathbf{z}]$–differentials of the algebra A. From the Jacobi–Zariski sequence of $k \to k[\mathbf{z}] \to A$, it follows that $\Omega_{A/k[\mathbf{z}]}$ is precisely defined by the matrix φ taken mod I.

Since I is an unmixed ideal, A is a torsion–free $k[\mathbf{z}]$–module, and we pass from $k[\mathbf{z}]$ to its field of fractions K. The algebra $B = A \otimes_{k[\mathbf{z}]} K$ is a direct product

$$B = B_1 \times \cdots \times B_s$$

of local Artinian algebras whose residue fields K_i are separable over K. Each B_t is by hypothesis a complete intersection. Finally we pass from K to a joint Galois closure L of all the K_t/K. This will imply that $C = A \otimes_{k[z]} L$ is a product of local Artinian algebras

$$C = C_1 \times \cdots \times C_q$$

with residue fields isomorphic to L. By Cohen's Theorem, each C_i is an algebra of the form

$$L[[y_1, \ldots, y_m]]/(g_1, \ldots, g_m),$$

where $(g_1, \ldots, g_m) = IC_i$.

Note that the Jacobian ideal of the g_i's with respect to the y_j is given by the image of J, which means that

$$J \cdot C_i = \det\left(\frac{\partial(g_1, \ldots, g_m)}{\partial(y_1, \ldots, y_m)}\right) \cdot C_i.$$

Since the dimension of each C_i relative to L is at most r, which is less than the characteristic of L, we apply [243, Korollar 4.7] (see also [173, Exercise 3, p. 382]) to conclude that J is as stated. □

Proof of Theorem 10.3.7. It will be enough to show that for each minimal prime P of I, if $(R/I)_P$ is not a domain then the image of J lies in the socle of $(R/I)_P$.

Suppose $I \subset k[x_1, \ldots, x_n]$ has codimension m. We make a change of base field k to another large enough to allow for a change of variables such that $k[x_{j_1}, \ldots, x_{j_{n-m}}]$ is a Noether normalization of R/I for any subset of $n - m$ variables. (For example, pass from k to the rational function field $k(u_{ij})$ in n^2 new indeterminates u_{ij} and make a linear change to the original variables, as specified by the matrix (u_{ij}).)

Now we compute the Jacobian ideal with respect to the new variables. Each $m \times m$ minor of the Jacobian matrix, by the argument of the Theorem 10.3.5, ends up in the socle of the localization. □

Proof of Theorem 10.3.8. By Theorem 10.3.7, J_0 is a generic socle formula of I_0, and therefore by Proposition 10.3.3 $\sqrt{I_0} = I_0 \colon J_0$. Every minimal prime of I is a minimal prime of I_0, so that $\sqrt{I_0} = \sqrt{I} \cap L$, where L, if not equal to (1), is the intersection of prime ideals of height m, none of which is an associated prime of I. L is therefore given by $\sqrt{I_0} \colon I$, and I is obtained by $\sqrt{I_0} \colon L$. □

General ideals

Let now I be an ideal, written as $I_1 \cap I_2 \cap I_3$. I_1 is the intersection of the primary components of codimension $m = \text{height } I$, while I_2 is the intersection of the primary components of codimension at least $m + 1$ but still associated to irreducible components. The ideal I_3 is the joint contribution of the embedded primary components.

The preceding supports the following method to compute radicals, by iteratively determining the radical of the higher dimensional components. We shall later consider another instance, that starts at the other end.

Algorithm 10.3.9 *Let I be an ideal of codimension m, and let $(f_1, \ldots, f_m)$ be a regular sequence in I generating the ideal I_0. Let J be the Jacobian ideal of I_0.*

(a) *$\sqrt{I_1}$ can be determined as follows:*

$$\begin{aligned} \sqrt{I_0} &= I_0 \colon J \\ L &= \sqrt{I_0} \colon I \\ \sqrt{I_1} &= \sqrt{I_0} \colon L \end{aligned}$$

(b) *The stable value of*

$$I\colon(\sqrt{I_1})^j = (I\colon(\sqrt{I_1})^{j-1})\colon\sqrt{I_1}$$

is $I_2 \cap I_3'$.

Note that in the ideal $I_2 \cap I_3'$, some of the primary components of I_3' may no longer be embedded. What this amounts to is that a certain amount of unnecessary computation creeps in.

The method of the lower Jacobians

In the same framework of Jacobian ideals, there is another method to compute radicals in [62]. It makes extensive use of the lower Jacobian ideals and often offers significant performance gains. It addition, it is more programmable.

Theorem 10.3.10 *Let k be a perfect field, and let $R = k[x_1, \ldots, x_n]$ be a polynomial ring. Assume $I = (f_1, \ldots, f_m) \subset R$ is an ideal of codimension g. Denote by $d = n - g$ the dimension of I. If the characteristic of k is not zero, suppose that the nil radical of R/I is generated by elements whose index of nilpotency is less than the characteristic of k. If for some integer $a \geq d$ we have*

$$\dim J_{a+1}(I) < d$$

then

$$L = I : J_a(I)$$

has the same equidimensional radical as I. Further, if $a = d$ then L is radical in dimension d; that is, the primary components of L having dimension d are prime.

We only consider the characteristic zero case, and refer to [62] for the full proof.

Proof. Suppose that $\dim J_{a+1}(I) < d$. To check that L has the same equidimensional radical as I, it is enough to verify that for each prime $\wp \supset I$ of dimension d we have

$$I\colon J_a(I) \subset \wp,$$

or equivalently $J_a(I)_\wp \not\subset I_\wp$. If it were otherwise the case, for $A = R/I$, the module $(\Omega_{A/k})_\wp$ has all of its Fitting ideals equal to $A_\wp$. It is therefore a free $A_\wp$–module, and by the ordinary Jacobian criterion, $A_\wp$ is a field of transcendence degree $a + 1$ over k. Since $a + 1 > d$, we have a contradiction.

For the remainder of the assertion, the condition implies that I is a complete intersection at each prime $\wp$ of dimension d, and Theorem 10.3.8 will apply. □

Exercise 10.3.11 Compute the radical of the almost complete intersection

$$\begin{cases} a+b+c+d+e+f \\ ab+bc+cd+de+ef+fa \\ abc+bcd+cde+def+efa+fab \\ abcd+bcde+cdef+defa+efab+fabc \\ abcde+bcdef+cdefa+defab+efabc+fabcd \end{cases}$$

Finding regular sequences

Using Theorem 10.3.8 to compute radicals requires an efficient mechanism to select regular sequences. Furthermore, to be truly effective, the chosen sequence must reflect the naturality of the given generators of the ideal, which may not be a simple task. In contrast, Theorem 10.3.10 has broader validity and the added feature of simpler programmability.

Regular sequences can be found in generic combinations of the generators of an ideal. More precisely, if $I = (\mathbf{f}) = (f_1, \ldots, f_m)$ is an ideal of codimension g and $\varphi = (c_{ij})$ is a sufficiently generic $m \times g$ matrix with entries in k, then the entries of $\mathbf{f} \cdot \varphi$ generate an ideal of codimension g. A drawback in this approach lies in the loss of whatever sparseness is present in the data, which is a resource which must be preserved. Another source of regular sequences in ideals are some of the characteristic sets of Wu and Ritt (see [75]). But this aspect of the subject deserves additional examination. We give here two other approaches to this problem.

Setwise regular sequences

The first is a method of [63], that is an elegant variation of the generic approach but that seeks to contain the loss of sparseness. It is based on:

Theorem 10.3.12 *If $\mathcal{F}$ is a set of polynomials whose initial terms, with respect to some term order on $R = k[x_1, \ldots, x_n]$, generate an ideal of codimension c, then $\mathcal{F}$ can be partitioned into disjoint subsets $\mathcal{F}_1, \ldots, \mathcal{F}_c$ such that for each i, the initial terms of the polynomials in $\mathcal{F}_i$ have a nontrivial common factor. For any such partition, and for almost every choice of coefficients $r_{i,f} \in k$, the linear combinations*

$$f_1 = \textstyle\sum_{f \in \mathcal{F}_1} r_{1,f} f$$
$$\vdots$$
$$f_c = \textstyle\sum_{f \in \mathcal{F}_c} r_{c,f} f$$

generate an ideal of codimension c. Further, each polynomial in $\mathcal{F}$ may be multiplied by any product of the factors of its initial term without spoiling this property.

Partition of the unity

We discuss next a mechanism that calls on Theorem 10.3.7 itself to generate its own regular sequences. Actually it only produces an ideal which is locally a complete intersection; but this is obviously all that is required to use Theorem 10.3.8. It is more of an approach than an algorithm proper. It permits however a great deal of *manual* control. Its shape and usage are based on the following elementary observations (cf. [299]).

Proposition 10.3.13 *Let R be a reduced Noetherian ring with n minimal prime ideals. Let I be an ideal of codimension at least one. Then*

(a) *For $a \in I$ there exists $b \in I \cap (0\colon a)$ such that $a + b$ is a regular element.*

(b) *If* $a \neq 0$ *for each* $0 \neq b \in I \cap (0\colon a)$, $(0\colon a) \neq (0\colon (a+b))$.

(c) *Iteration leads to a regular element contained in* I *using at most* n *steps.*

Proof. (a): Any prime ideal $\wp \supseteq (a, I \cap (0\colon a))$ either contains I, which has codimension at least one, or will contain $(a, (0\colon a))$. Since R is reduced the latter has also codimension at least 1. This means that there exists a regular element of the form $ra + b$, with $b \in I \cap (0\colon a)$. Again using that R is reduced, $a + b$ is regular as well.

(b): This is immediate since any annihilator of $a + b$ must annihilate both a and b.

(c): Let $a_1, a_2, \ldots, a_r$ be obtained by this process: in the notation above, we repeatedly set $a = a_1 + \cdots + a_{j-1}$, $b = a_j$. The descending chain of ideals

$$0\colon a_1 \supset 0\colon (a_1 + a_2) \supset \cdots \supset 0\colon (a_1 + a_2 + \cdots + a_r)$$

gives rise to a similar sequence in the total ring of fractions S of R. Since S has a composition series of length n, the last ideal vanishes if $r \geq n$. □

Remark 10.3.14 In practice this runs as follows. Suppose $f_1, \ldots, f_s \in I$ have been chosen so that they generate an ideal of codimension $s <$ height I. Let

$$L = \sqrt{(f_1, \ldots, f_s)}.$$

We follow the scheme above, but with colon ideals computed relative to L. We note by f_{s+1} the element of I obtained from the lift. The measure of manual control over the sparseness comes in because in selecting the a_j's we may also use reduction modulo the previously chosen a_i for $i < j$.

Wu–Ritt characteristic sets

Another place to look for regular sequences is among *triangular sets* of polynomials.

Definition 10.3.15 Let $R = k[y_1, \ldots, y_n]$ be a polynomial ring and let $\mathbf{f}$ be the set of polynomials $f_1, \ldots, f_m$. $\mathbf{f}$ is a *triangular set* if there is a permutation $x_1, \ldots, x_n$ of the y_i such that

$$\begin{aligned} f_1 &\in k[x_1, x_{m+1}, \ldots, x_n] \\ f_2 &\in k[x_1, x_2, x_{m+1}, \ldots, x_n] \\ &\vdots \\ f_m &\in k[x_1, \ldots, x_m, x_{m+1}, \ldots, x_n]. \end{aligned}$$

It would be of interest to find when such sequences define ideals of codimension m.

Content formulas

The point we want to make here is: Where, in a computation of the syzygies, is part of the radical of an ideal? The explicit Nullstellensätze are instances of this.

A homological Nullstellensatz

It is not likely, through strictly homological means, that one should be able to produce the radical of an ideal. Our aim here is more modest: Given an ideal I, to obtain from its syzygies another ideal L, with the same radical, but which is a generic complete intersection. The Jacobian methods would then be used on L. What might make this possible are various criteria of local complete intersection.

Here we make some observations on this, and begin quoting a result of Gerson Levin [180].

Theorem 10.3.16 *Let R be a local ring, I an ideal in R of finite projective dimension, which is not a complete intersection. Let J be the content of I, i.e. the ideal of all coefficients of relations on a minimal set of generators of I. Let $r = 1+\text{proj dim}_R R/I$; then $(I : J)^r \subset I$.*

The harder part of the proof is the assertion on the index of nilpotency. That $I\colon J$ is contained in the radical of I can be shown as follows. Localize at a minimal prime of I, so that we may assume that I is primary for the maximal ideal of the local ring $(R, \wp)$. If $\nu(I_\wp) < \nu(I)$, $J_\wp = R$ and there is nothing to prove. If $\nu(I_\wp) = \nu(I)$, $I_\wp$ is not a complete intersection and $J_\wp$ is a primary ideal properly containing $I_\wp$, and consequently $I_\wp\colon J_\wp$ is also primary. Of interest is the fact that the quotient is then larger than I and therefore is a better approximation of $\sqrt{I}$ than I itself.

Here is another link between the content of an ideal and radicals. We leave its proof as an exercise for the reader.

Proposition 10.3.17 *Let I and J be ideals, $I \subset J$. Denote by L the content of I. If $I\colon J \not\subset \sqrt{L}$, then $\nu(I) \geq \nu(J)$.*

10.4 Primary Decomposition

Effective primary decomposition of ideals has received systematic study in classical [102], neo–classical [248], [250] and modern treatments [81], [79], [236]. Our aim here will be to show how certain tools from Homological Algebra can be engaged in carrying out part of the task. It is an approach that cannot be up to the full problem. It will only lead the solution up to the point where refined factorization techniques come into play.

Let A be a Noetherian ring and let I be an ideal. The proof by Emmy Noether of primary decomposition of ideals ([212])—in an argument that played a major role in the ushering in of Modern Algebra—is made up of two parts:

(i) I is a finite intersection of irreducible ideals

$$I = L_1 \cap \cdots \cap L_r, \tag{10.12}$$

(ii) and each irreducible ideal L_i is primary. Collecting together those L_i with the same radical yields a primary decomposition of I:

$$I = Q_1 \cap \cdots \cap Q_s. \tag{10.13}$$

There is a measure of uniqueness in some of the Q_i but almost none amongst the L_j, beyond their numbers. The irreducible components may, on occasion, have additional structure that make (10.12) more appealing. For instance, if $R = k[x, y]$, k a field, then every irreducible ideal can be generated by two elements.

Each of these representations present formidable difficulties for explicitly computing them. In this chapter we focus on another representation, which can be enabled along by the radical formulas of the previous chapter. We consider approaches to primary decomposition that at worst require only Gröbner bases calculations over the field of rational functions over a ground field k. However this is not good enough as one would like to perform *all* the arithmetic over k and to be fully deterministic. But, *alas*, we are not there yet.

After carrying out an equidimensional decomposition, we discuss an approach to primary decomposition that reduces the question to zero–dimensional ideals (the subject of the next chapter).

Equidimensional decomposition of an ideal

Bits and pieces of Homological Algebra can be engaged to provide a decomposition of a polynomial ideal I as an intersection of known ideals of the same codimension.

Let I be an ideal of codimension m. Given a primary decomposition of I, we collect into I_i those primary components of a given codimension i.

Definition 10.4.1 In the representation

$$\boxed{I = I_m \cap I_{m+1} \cap \cdots \cap I_r \cap \cdots \cap I_q,} \tag{10.14}$$

the ideal I_i is called the ith *equidimensional component* of I.

Equidimensional factors of an ideal

We attach to an ideal I of a ring R a sequence of ideals J_r, $r \geq 0$, that are useful in obtaining an equidimensional decomposition for I. It is based on the notion of the hull of a module introduced in [62].

Definition 10.4.2 Let I be an ideal. For each non–negative integer r, the rth *equidimensional factor* of I is the ideal:

$$J_r = \text{annihilator}(\text{Ext}_R^r(\text{Ext}_R^r(R/I, R), R)). \tag{10.15}$$

It may well occur that some of these ideals are improper, that is, equal to R. We list some of its elementary properties.

Proposition 10.4.3 *Let R be a Gorenstein ring and let I be an ideal of codimension m. Let J_r denote the rth equidimensional factor of I. Then*

(a) *$J_r = R$ for $0 \leq r < m$.*

(b) *J_r is either R or an equidimensional ideal of codimension r.*

(c) *If $\mathbf{x} = \{x_1, \ldots, x_m\}$ is a regular sequence contained in I, then*

$$J_m = (\mathbf{x}) \colon ((\mathbf{x}) \colon I).$$

Proof. (a) The module $\text{Ext}_R^r(R/I, R)$ vanishes for $r < m$.

(b) First observe that the support of the module $\text{Ext}_R^r(R/I, R)$ has codimension at least r, since R is a Gorenstein ring and therefore for each prime ideal $\wp$, with height $\wp < r$ the injective dimension of $R_\wp < r$. We claim that J_r is the intersection of the primary components of codimension r of the annihilator of $\text{Ext}_R^r(R/I, R)$. On the other hand, if the support of $\text{Ext}_R^r(R/I, R)$ has codimension $> r$ then $J_r = R$, because for the same reasons as above, the double Ext will vanish. More precisely, let E_0 be the subset of all elements of $E = R/I$ whose support has codimension $> r$; E_0 is a submodule of E and $\text{Ass}(F = E/E_0)$ consists of primes of codimension r. Consider the long exact sequence of the functor Ext:

$$\text{Ext}_R^{r-1}(E_0, R) \longrightarrow \text{Ext}_R^r(F, R) \longrightarrow \text{Ext}_R^r(E, R) \longrightarrow \text{Ext}_R^r(E_0, R).$$

Since $\text{Ext}_R^{r-1}(E_0, R) = \text{Ext}_R^r(E_0, R) = 0$, the assertions will be proved once it is shown that F and $\text{Ext}_R^r(F, R)$ have the same annihilator. Let f be an element of R regular on F:

$$0 \to F \xrightarrow{f} F \longrightarrow F/fF \to 0.$$

In the cohomology sequence

$$\text{Ext}_R^r(F/fF, R) \longrightarrow \text{Ext}_R^r(F, R) \xrightarrow{f} \text{Ext}_R^r(F, R) \longrightarrow \text{Ext}_R^{r+1}(F, R),$$

$\text{Ext}_R^r(F/fF, R) = 0$ since f is regular on F and therefore the support of F/fF has codimension $r + 1$. This implies that f is regular on $\text{Ext}_R^r(F, R)$.

If I_0 is the intersection of the (unique) primary components of I of codimension m, as above we have

$$\text{Ext}_R^m(R/I_0, R) = \text{Ext}_R^m(R/I, R).$$

To show (c) we use the standard formula of Rees (cf. [165, p. 155]):

$$\text{Ext}_R^m(R/I, R) \simeq \text{Hom}(R/I, R/(\mathbf{x})) \simeq ((\mathbf{x}) \colon I)/(\mathbf{x}) \simeq ((\mathbf{x}) \colon I_0)/(\mathbf{x}).$$

This is the canonical module ω of R/I_0, so by local duality it follows that the annihilator of $\text{Ext}_R^m(\omega, R)$ is still I_0 since it is grade unmixed. $\square$

Unmixedness test

We single out from Proposition 10.4.3.(c) a test to check whether an ideal is has all of its associated primes of the same codimension.

Corollary 10.4.4 *Let R be a Gorenstein ring and I and ideal of codimension m. Let $\mathbf{x} = \{x_1, \ldots, x_m\} \subset I$ be a regular sequence. Then I is unmixed if and only if $I = (\mathbf{x})\colon((\mathbf{x})\colon I)$.*

Equidimensional decomposition

We are going to show how the capability to compute the equidimensional factors leads to an explicit equidimensional decomposition.

We first illustrate the method with a simple example, an ideal I of a polynomial ring $k[x, y]$. Suppose $I = a \cdot J$, with J an ideal of codimension two. There are only two factors, $J_1 = (a)$ and $J_2 = J$. If a is a unit, I is already equidimensional. In general, set $I_1 = J_1$, and let I_2 be the stable value of $(I + J_2^n) : J_1$. The latter exists because each of these ideals contains J_2 so that the descending chain of ideals corresponds termwise to a descending chain in the Artinian ring R/J_2. I_2 is clearly equidimensional and we claim $I = I_1 \cap I_2$; for this we can localize and apply Krull intersection theorem.

In the general case the features that are automatic in this example, such as the stabilization step, must be made to happen. Let us indicate the steps. Let I be an ideal of codimension m and let J_r, for $r \geq m$, be its equidimensional factors.

Step 1. Set $I_m = J_m$; this is just the intersection of the primary components of I of codimension m.

Step 2. Suppose $I_m, \ldots, I_s$ are equidimensional ideals, height $I_j = j$, with the property that I and $L_s = I_m \cap \cdots \cap I_s$ coincide at any localization $R_\wp$ such that $\dim R_\wp \leq s$. By the previous step this already occurs for $s = m$.

Step 3. Consider the descending sequence of ideals

$$A_i = (I + J_{s+1}^i)\colon L_s.$$

We claim that the sequence of the $(s+1)$th factors of the A_i stabilizes. To prove this we are going to make several calculations.

Let J be any of these factors; note $J_{s+1} \subset J$. Because they are both ideals with pure components of codimension $s + 1$, we may assume that $(R, \wp)$ is a local ring of dimension $s + 1$ and both J and J_{s+1} are $\wp$–primary ideals.

From the previous step, we have that $I = L_s \cap L$, where L is an ideal of pure codimension $s + 1$. Consider the cohomology sequence of the exact sequence

$$0 \rightarrow L_s/I \longrightarrow R/I \longrightarrow R/L_s \rightarrow 0.$$

We have

$$\operatorname{Ext}_R^{s+1}(R/L_s, R) \longrightarrow \operatorname{Ext}_R^{s+1}(R/I, R) \longrightarrow \operatorname{Ext}_R^{s+1}(L_s/I, R) \longrightarrow \operatorname{Ext}_R^{s+2}(R/L_s, R),$$

with the end terms of the sequence vanishing. This means that the annihilator of $\mathrm{Ext}_R^{s+1}(R/I, R)$, which is J_{s+1} by definition (we do not have to consider the double Ext at the top of the dimension), will also annihilate $\mathrm{Ext}_R^{s+1}(L_s/I, R)$ and thus its dual L_s/I. It shows that $J_{s+1} \cdot L_s \subset I$. This implies that the sequence

$$J_{s+1} \subset \cdots \subset J_{s+1}(A_n) \subset \cdots \subset J_{s+1}(A_1)$$

of equidimensional ideals of codimension $s + 1$ will stabilize because it depends exclusively at what happens at the minimal primes of J_{s+1}.

Let q be the maximum of the two numbers: the exponent where the chain stabilizes, and any integer that makes $J_{s+1}^q \cap L_s \subset J_{s+1} \cdot L_s$ to hold (such integers exist by the Artin–Rees Lemma).

Here is the summary up of this construction:

Proposition 10.4.5 *Let I_{s+1} be the $(s+1)$th factor of $I + J_{s+1}^q$. Then I and $L_s \cap I_{s+1}$ agree in codimension $\leq s + 1$.*

Effecting primary decomposition

Arguably the natural approach to primary decomposition of ideals of a polynomial ring $R = k[x_1, \ldots, x_n]$ would involve only the field operations of k. It is not clear to us how this can be achieved. It is however possible to outline methods that require only computation over fields generated by subsets of the x_i but which do not use change of variables, or other methods that carry out only computation over k but in larger polynomial rings.

In this section we are going to follow some ideas in [62] and [79] towards that goal. If the truth be known, much remains to be done!

Reduction to dimension zero

There is no question that the real issue in primary decomposition consists in determining the associated primes of the ideal. This perspective leads one to focus on other ideals with the same associated primes and to consider reductions to other problems.

Given the ideal I we sketch three reductions of the problem to the case of ideals of dimension zero. Another method is briefly discussed in the next section (see Remark 10.4.20).

Equidimensional components

Let J_0 be an equidimensional component of I of dimension d. Since the associated primes of J_0 and of its radical J are the same, we may assume that J is a radical, equidimensional ideal. This means that there exists a Noether normalization

$$k[\mathbf{z}] = k[z_1, \ldots, z_d] \hookrightarrow R/J$$

such that R/J is a torsion free module over $k[\mathbf{z}]$. This means that the associated prime ideals of J correspond to the maximal ideals of the affine $k(\mathbf{z})$–algebra $A \otimes_{k[\mathbf{z}]} k(\mathbf{z})$. More precisely,

$$\{\wp \in \operatorname{Ass}(R/J)\} \Longleftrightarrow \{P \in \operatorname{Ass}(R/J) \otimes k(\mathbf{z})\},$$

where the second set consists of zero–dimensional ideals, with the correspondence being given by $\wp = P \cap R$.

This leads to the localization problem: Given a multiplicative set $S \subset R$ and an ideal $L \subset R$ find $R \cap S^{-1}L$. Some special cases are treated in [79]; see how Proposition 10.4.6 deals with the cases needed here. We expect to come back to the case above using the techniques of [62].

One step at a time

There is no need to look for Noether normalization above by allowing for the computation of superfluous prime ideals that can be filtered out later.

Let $P_1, \ldots, P_s$ be the minimal prime ideals of J. Pick one variable, say $x = x_1$, and consider $(f) = J \cap k[x]$.

- If $f = 0$, the associated prime ideals of J split into two subsets:

$$\{P_1, \ldots, P_s\} = \{P_1, \ldots, P_r\} \cup \{P_{r+1}, \ldots, P_s\},$$

  where $r \geq 1$. The primes in the first subset correspond to those of $J \otimes_{k[x]} k(x)$. After they are isolated, the remaining primes, if any, are obtained by taking the ideal quotient

$$J\colon (P_1 \cap \cdots \cap P_r)$$

  into the next step.

- If $f \neq 0$, f is the product of the irreducible polynomials f_i:

$$f = f_1 \cdots f_r,$$

  and the prime ideals of J correspond to the primes of

$$R/(f_i, J \cap k[x_2, \ldots, x_n]).$$

If J is not zero–dimensional, that $J \cap k[x_j] = 0$ for one of the variables. It would be interesting to benefit from the second case more directly.

Integral closure

The integral closure B of $A = R/J$ is a direct product of integral domains. If we succeed in computing B (see later) as an affine domain $B = D/L$, with $D = k[x_1, \ldots, x_n, y_1, \ldots, y_m]$,

$$B = S/L_1 \times \cdots \times S/L_r,$$

where L_i is generated by L and one idempotent e_i. The associated prime ideals of J are then the ideals $L_i \cap R$.

The case of a graded ring A is rather pleasing: The idempotents must be homogeneous of degree zero and therefore live in the finite dimensional algebra B_0.

Localization

In [79], the following method is proposed to determine some localizations. (See Remark 10.1.4 for definitions.)

Proposition 10.4.6 *Let R be a principal ideal domain of quotient field K. For an ideal $I \subset k[\mathbf{x}]$ it is possible to find an element $s \in R$ such that*

$$I \cdot K[\mathbf{x}] \cap R[\mathbf{x}] = I \cdot R_s[\mathbf{x}] \cap R[\mathbf{x}].$$

Proof. (Sketch) The proposed s is obtained as follows. Let $\{g_1, \ldots, g_r\}$ be the Gröbner basis of I *computed with respect to* R. The initial term of g_i has the form $\text{in}(g_i) = s_i \mathbf{x}^{\mathbf{a_i}}$. s is taken as the product of the s_i's. □

Top radical of an ideal

It would be desirable to have the means to pick apart certain components of the equidimensional decomposition of an ideal. Theorem 10.3.8 is essentially the realization of this goal for the highest dimensional component. Unfortunately the computation of the radical of the lower dimensional components would have to pass through the previous processing of the heavier components.

By bringing in the upper Jacobian ideals into the picture, we discuss an explicit formula for lightest component of the radical of an ideal. Let I be an ideal of codimension m, minimally generated by $m + r$ elements; r is the *deviation* of I. According to Krull's principal ideal theorem, any minimal prime of I has codimension at most $m + r$. We are going to key on those primes.

The decomposition (10.4.1) can be refined by breaking up each I_i into two pieces: I_i', corresponding to minimal primes of I, and I_i'' derived from the embedded primes of I. If one of these is not present, such as I_0'' or I_i', $i > r$, by abuse of notation we set it equal to R.

The radical of I is then given by the expression

$$\sqrt{I} = \sqrt{I_0'} \cap \sqrt{I_1'} \cap \cdots \cap \sqrt{I_r'}.$$

It suggests that one focus on obtaining formulas for the $\sqrt{I_i'}$'s.

Definition 10.4.7 Given an ideal I of deviation r, $\sqrt{I_r'}$ is the *top radical* of I (Notation: topradical(I).)

Conjecture 10.4.8 Let I be an ideal of codimension m and deviation r, and denote by L the ideal generated by the minors of size $m+r$ of the Jacobian matrix of I. Then

$$\text{topradical}(I) = I \colon L.$$

The following provides a measure of support (assume from now on that k is a field of characteristic zero):

Theorem 10.4.9 *This conjecture holds if every embedded prime of I has codimension at most $m + r$. In particular it holds if $v(I) \geq$ proj dim R/I.*

Proof. We have

$$I \colon L = (I_0 \colon L) \cap (I_1' \colon L) \cap (I_1'' \colon L) \cap \cdots \cap (I_r' \colon L) \cap (I_r'' \colon L),$$

since by assumption $I_j'' = R$ for $j > r$. We claim that all the quotients on the right side, with the possible exception of $I_r' \colon L$, are equal to R. This will suffice to establish the claim since at each minimal prime $\wp$ of I_r', we have $I_\wp = (I_r')_\wp$ and $L_\wp$ is nothing but the Jacobian ideal of $(I_r')_\wp$; we may then apply Theorem 10.3.8.

We begin by showing that $I_r'' \colon L = R$. If this is not so, any associated prime of the left–hand side has codimension $m + r$. Let $\wp$ be then one of its primes and localize R at $\wp$ (but keep the notation); denote also $A = R/I$. (We warn the reader about a possible confusion: Sometimes we shall say that an ideal is zero when it would be more appropriate to say it is zero mod I.)

As before we may assume that $A = k[[x_1, \ldots, x_n]]/I$, and $\wp$ is the maximal ideal of A. We consider the Jacobian ideals of the Artin algebras $B_s = A/\wp^s$. Because A has positive dimension, L will map into the Jacobian ideal of B_s, for each s. If B_s is not a complete intersection, by Theorem 10.3.4, its Jacobian ideal vanishes and thus $L \subset \wp^s$. On the other hand, if B_s is a complete intersection its socle must be $\wp^{s-1} B_s$ and $L \subset \wp^{s-1}$. This means that

$$L \subset \bigcap_{s \geq 0} \wp^s,$$

which by Krull's intersection theorem implies $L = 0$.

The case of a component such as $I_j' \colon L$, for $j < r$, or $I_j'' \colon L$, $1 \leq j < r$, is easier to deal with. In fact, if $\wp$ is a minimal prime of say I_j' and $(I_j')_\wp$ is not a complete intersection, then already the $m + j$–sized minors of the Jacobian matrix of $(I_j')_\wp$ vanish by Theorem 10.3.4. On the other hand, if $(I_j')_\wp$ is a complete intersection, the rank of its Jacobian matrix is $m + j < m + r$, so L vanishes at $(I_j')_\wp$ anyway.

The last assertion follows from the Auslander–Buchsbaum formula [192] and standard facts on depth. □

The top of a system of equations

The following retrieves exactly the isolated zeros of a set of polynomials (cf. [299]).

Corollary 10.4.10 *Let $f_1, \ldots, f_n \in k[x_1, \ldots, x_n]$ be a set of n polynomials, and let Δ be its Jacobian determinant. Then*

$$\boxed{(f_1, \ldots, f_n) \colon \Delta}$$

is the radical of the minimal primary components of dimension 0. *In particular this ideal is either* (1) *or a complete intersection.*

To illustrate, suppose $I = (f(x,y)g(x,y), f(x,y)h(x,y))$, where $f(x,y)$ is the gcd of I. Denote by Δ the Jacobian determinant of these two generators of I, and denote by Δ_0 the Jacobian of g, h. A simple calculation shows that

$$I \colon \Delta = ((g,h) \colon \Delta_0) \colon f.$$

By Theorem 10.3.7

$$I \colon \Delta = \sqrt{(g,h)} \colon f,$$

in agreement with the assertion of the theorem.

Let us recall a result of Eisenbud–Evans [58] and Storch [266]:

Theorem 10.4.11 *Let S be a Noetherian ring of dimension $d-1$. Then the radical of any ideal I of $R = S[x]$ is the radical of a d–generated ideal.*

Their proof is fairly constructive already. It is based on pseudo–division of polynomials and an appropriate induction argument on the dimension of S. We will just rewrite it in sketching out some moves to make it more amenable for a Gröbner basis computation.

We assume that S is a reduced ring. Let $I = (f_1, \ldots, f_m)$. In their proof it is argued that there exists a regular element u of S and an element h_1 of I such that

$$u \cdot I \subseteq (h_1).$$

The issue is how u is to be found. We employ the argument of Proposition 10.3.13. Given two elements f, g of I, with $\deg f(x) \geq \deg g(x)$ (as polynomials in x), let α be the leading coefficient of $g(x)$. For some power of α we have

$$\alpha^s \cdot f = q \cdot g + r, \ \deg r < \deg g.$$

Replace then f by r and keep processing the list of generators of I until we have a nonzero element $a_1 \in S$ such that $a_1 \cdot I \subseteq (g_1)$, with $g_1 \in I$.

Repeat this step on the generators of $I \cap (0\colon a_1)$, if the latter is nonzero. This produces a sequence $a_1, \ldots, a_r \in S$, with corresponding elements $g_i \in I$ such that

$$\begin{cases} a_i \cdot a_j = 0 \text{ if } i \neq j \\ a_1 + a_2 + \cdots + a_r \text{ is regular on } S \\ a_i \cdot I \subseteq (g_i) \end{cases}$$

Observe that when the process is applied to an ideal such as $I \cap (0\colon a_1)$, it returns an element $b_1 \in (0\colon a_1)$ such that

$$b_1 \cdot (I \cdot (0\colon a_1)) \subset b_1 \cdot (I \cap (0\colon a_1)) \subset (g_2), g_2 \in I.$$

From this equation we then select some $a_2 \in b_1 \cdot (0\colon a_1)$.

Set now

$$\begin{aligned} u &= a_1 + a_2 + \cdots + a_r \\ h_1 &= a_1 g_1 + a_2 g_2 + \cdots + a_r g_r, \end{aligned}$$

and consider the image I^* of I in $(S/\sqrt{(u)})[x]$. By induction select $h_2, \ldots, h_d \in I$ whose images generate an ideal with the same radical as I^*.

It suffices to verify

$$\sqrt{I} = \sqrt{(h_1, h_2, \ldots, h_d)}.$$

Let $\wp$ be a prime ideal with $(h_1, \ldots, h_d) \subseteq \wp$. If $u \in \wp$, by hypothesis $I \subseteq \wp$. Assume otherwise; then $a_i \notin \wp$ for some i which implies that $a_j \in \wp$ for $j \neq i$. Since $h_1 \in \wp$, this means that $g_i \in \wp$ which from the equation $a_i \cdot I \subseteq (g_i)$ finally implies $I \subseteq \wp$, as desired.

Remark 10.4.12 This arrangement leaves much to be desired: There is a great deal less of control here than that afforded by Proposition 10.3.13.

Theorem 10.4.13 *Let I be an ideal of $R = k[x_1, \ldots, x_n]$. We can find n polynomials $(f_1, \ldots, f_n)$ generating an ideal with the same radical as I. If Δ is the Jacobian determinant of the f_j's, then $(f_1, \ldots, f_n)\colon \Delta$ is the top radical of I.*

Primality testing

Let $R = k[x_1, \ldots, x_n]$ be a polynomial ring over a field k. The question of deciding whether an ideal $I = (f_1, \ldots, f_m)$ of R is prime is simple to approach but often turns out difficult to verify. Because this problem occurs with some regularity, it would be convenient to identify algebraic elements that might see their way into workable algorithms.

To decide primality of an ideal I in a polynomial ring $R = B[t]$, the natural method is first to compute $J = I \cap B$. It reduces the problem to that of the primality of J, and to the determination of irreducibility of polynomials in t over the field of fractions of B/J. The latter is often a formidable task. On the other hand, factorization algorithms over various polynomial rings are reasonably optimized, so that we may rely on them in cases of interest.

Some of the drawbacks and advantages of the method will be obvious. Several Gröbner basis computations will be required, often with one tag variable added. Factorization of polynomials of very large degree may become necessary. It attempts however, the normalization steps notwithstanding, to keep the data sparse. Moreover, it only covers the characteristic zero case; the elements to extend it to finite characteristics exist in the literature.

Computing degrees

One possible intermediate element would be a radical test. It may be used as a screening device. We consider a method of [287].

Let

$$I = (f_1, \ldots, f_m) \subset k[x_1, \ldots, x_n],$$

where k has characteristic zero. Assume height $I = g$, and suppose I is unmixed (e.g. it has passed the test of Corollary 10.4.4). Let J be the Jacobian ideal of I.

Proposition 10.4.14 *Let f be an element of $J \setminus I$. If $I\colon f = I$, then I is a radical ideal. Conversely, if I is radical, such elements exist.*

Proof. This follows from the fact that the localization R_f is now a smooth k–algebra by the Jacobian criterion [192, Theorem 95], while the choice of f assures that the primary decomposition of I is not disturbed. The converse follows because radical ideals are generically smooth. □

The ideal I is then prime if and only if the localization I_f is prime as well. We may then replace I by $L = (I, 1 - Tf) \subset R[T]$. What this means is that we traded I by the smooth ideal L, that is $R[T]/L$ is a regular ring. Since such rings decompose into direct products of integral domains, I is prime if and only if $R[T]/L$ has no non–trivial idempotent elements.

There are several instances of algebras where testing for reducedness already leads to integrality (see [152]).

Cohen–Macaulay ideals

In the affine case, Cohen–Macaulay rings decompose into direct products of equidimensional subrings. The latter, sometimes referred to as *perfect*, have the additional property of being free modules over normalizing subrings.

We sketch a method that has given good results, particularly when R/I is Cohen–Macaulay, and its Krull dimension or the codimension of I is low.

Let

$$B = k[\mathbf{z}] \hookrightarrow A = k[\mathbf{z}, \mathbf{x}]/I = k[z_1, \ldots, z_d, x_1, \ldots, x_m]/I$$

be a Noether normalization of A. The *degree* of A over B is the dimension

$$\deg(A/B) = \dim_{k(\mathbf{z})}(A \otimes_B k(\mathbf{z})),$$

where $k(\mathbf{z})$ is the field of fractions of B. It equals the torsion–free rank of A as a B–module.

Proposition 10.4.15 *Suppose A is a Cohen–Macaulay, equidimensional ring. The degree of A over B is the dimension of the vector space*

$$\ell = \dim_k(k[\mathbf{z}, \mathbf{x}]/(I, \mathbf{z})).$$

A is an integral domain if and only if there exists a subring

$$B \hookrightarrow S = k[\mathbf{z}, U]/(f(\mathbf{z}), U)) \hookrightarrow A,$$

where $f(\mathbf{z}, U)$ is an irreducible polynomial of degree ℓ.

Proof. The assertions are clear from the fact that A is a free module of rank ℓ over B, and therefore lies in the field of fractions of S. □

Example 10.4.16 The ideal

$$I = (xw + y^3, yw + z^3 + xw^3, zw + x^2, w^3 + xy^2z^2 - y^2w^4)$$

is Cohen–Macaulay of codimension 3.

The subalgebra $k[w]$ is a Noether normalization of $k[x, y, z, w]/I$, and the irreducible polynomial

$$f(y, w) = y^{17} - w^{11}y^2 - w^{10}$$

lies in I. Since it is clear that

$$\dim_k(k[x, y, z, w]/(I, z)) = 17,$$

I is a prime ideal.

The computation of the degree by the formula above hides a very powerful criterion ([107, Proposition 1.1]):

Proposition 10.4.17 *Let $B = k[\mathbf{z}] \hookrightarrow A = k[\mathbf{z}, \mathbf{x}]/I$ be a Noether normalization of A. Then*

$$\deg(B/A) \geq \dim_k(k[\mathbf{z}, \mathbf{x}]/(I, \mathbf{z})),$$

with equality holding if and only if $A_{(\mathbf{z})}$ is a free $B_{(\mathbf{z})}$–module.

Because k is infinite, such element u could be found as a linear combination

$$u = \sum_{i=1}^{m} c_i x_i.$$

We pick linear combinations such that for any m–subset $u_1, \ldots, u_m$, the coefficient matrix (c_{ij}) is invertible. For instance, one could take $c_i = c^{i-1}$. Since the number of $k(\mathbf{z})$–subfields of $K = A \otimes k(\mathbf{z})$ is bounded by $\ell\,!$, by Steinitz's theorem ([159, Theorem 4.25]), making sufficiently many choices of u, say

$$(m-1) \times (\ell\,! - 1) + 1,$$

will assure that the question could be ascertained. Indeed, it is clear that any subfield of K that contains m linearly independent such combinations must equal K. The pigeon–hole principle guarantees that a simple extension such as S will occur if A is a domain. In fact, this also says that for most choices of u the expected extension $k(\mathbf{z}, u)$ is maximal, which opens possibilities for probabilistic approaches.

Alternatively, if the computation of Gröbner bases can be carried out over a rational function field, then by choosing the c_i's as independent variables only one computation becomes necessary.

Simple rational extensions

The integer $\deg(A/B)$ is embedded in the details of a presentation of A as a B–module, that is, could be derived from a set of module generators and relations.

View $A = k[\mathbf{z}, x_1, \ldots, x_m]/I$ as the composite of the extensions $A_i = k[\mathbf{z}, x_i]/I_i$, where $I_i = I \cap k[\mathbf{z}, x_i]$. If I is prime, each I_i is generated by an irreducible polynomial $f_i(\mathbf{z}, x_i)$.

For each i let $h_i(\mathbf{z})$ be a nonzero polynomial obtained by eliminating x_i from $f_i(\mathbf{z}, x_i)$ and the derivative $\partial f_i(\mathbf{z}, x_i)/\partial x_i$, for instance the resultant of $f_i(\mathbf{z}, x_i)$ with respect to x_i—that is, any nonzero element in

$$(f_i(\mathbf{z}, x_i), \partial f_i(\mathbf{z}, x_i)/\partial x_i) \bigcap k[\mathbf{z}].$$

Proposition 10.4.18 *Let $a = (a_1, \ldots, a_d) \in k^d$ be such that $\forall\, i$, $h_i(a) \neq 0$. Then the degree of A over B is*

$$\ell = \dim_k(k[\mathbf{z}, \mathbf{x}]/(I, z_1 - a_1, \ldots, z_d - a_d)).$$

Proof. The choice of a implies that each of the extensions A_i is unramified at the maximal ideal $\wp = (z_1 - a_1, \ldots, z_d - a_d)$. Since $k[\mathbf{z}]_\wp$ is regular, it follows from [8, Propositions 4.6 and A.1] that their composite $A_\wp$ is an unramified extension of $k[\mathbf{z}]_\wp$ and is therefore a free $k[\mathbf{z}]_\wp$–module. □

Since such points always exist in profusion, we may also find them by trial and error, without the bother of elimination. For instance, evaluation at points $\mathbf{z} = \mathbf{a}$

followed by computation of gcd works extremely fast in most systems. (For details on the structure of unramified extensions, see [226, Chap. V, Th. 1].)

Once the degree has been found, we have the following general primality test. Observe that we have traded U by one of the x_i–variables.

Proposition 10.4.19 *Let A be an equidimensional reduced algebra and let*

$$B = k[\mathbf{z}] \hookrightarrow A = k[\mathbf{z}, \mathbf{x}]/I,$$

be a normalization as above. Let ℓ be the degree of A over B. Suppose that for some variable U there exists a polynomial

$$f(\mathbf{z}, U) \in I \cap k[\mathbf{z}, U]$$

of degree ℓ. Then A is an integral domain if and only if $f(\mathbf{z}, U)$ is an irreducible polynomial.

Proof. The proof is essentially the same as that of Proposition 10.4.15, but let us highlight the help supplied by the assumptions on A.

First, note that for any variable U, the ideal $J = k[\mathbf{z}, U] \cap I$ is principal. Indeed, since A is a torsion–free $k[\mathbf{z}]$–module (the equidimensional hypothesis), the chain

$$B = k[\mathbf{z}] \hookrightarrow S = k[\mathbf{z}, U]/J \hookrightarrow A = k[\mathbf{z}, \mathbf{x}]/I,$$

says S is also a torsion free as a B–module, which clearly implies that J is equidimensional. Since J has codimension one it must be principal.

If $\deg f(\mathbf{z}, U) = \ell$, the total ring of fractions of S and A are equal, proving the assertion. □

Remark 10.4.20 Observe that if h has degree ℓ, but it is not necessarily irreducible, its prime factors can be used to obtain a primary decomposition of I. This is very close to the classical approach to verify primality, mod the equidimensional and radical reductions that might facilitate the task.

Example 10.4.21 Craig Huneke asked about the primeness of the ideal

$$I \subset R = k[a, b, c, d, e], \quad \text{char } k \neq 2, 5,$$

generated by the following polynomials:

$$\begin{array}{l}
5abcde - a^5 - b^5 - c^5 - d^5 - e^5 \\
ab^3c + bc^3d + a^3be + cd^3e + ade^3 \\
a^2bc^2 + b^2cd^2 + a^2d^2e + ab^2e^2 + c^2de^2 \\
abc^5 - b^4c^2d - 2a^2b^2cde + ac^3d^2e - a^4de^2 + bcd^2e^3 + abe^5 \\
ab^2c^4 - b^5cd - a^2b^3de + 2abc^2d^2e + ad^4e^2 - a^2bce^3 - cde^5 \\
a^3b^2cd - bc^2d^4 + ab^2c^3e - b^5de - d^6e + 3abcd^2e^2 - a^2be^4 - de^6 \\
a^4b^2c - abc^2d^3 - ab^5e - b^3c^2de - ad^5e + 2a^2bcde^2 + cd^2e^4 \\
b^6c + bc^6 + a^2b^4e - 3ab^2c^2de + c^4d^2e - a^3cde^2 - abd^3e^2 + bce^5
\end{array}$$

This ideal has codimension 2 and projective dimension 3. Because it is homogeneous, the rank of $A = R/I$ (relative to any normalizing subring generated by 1–forms) is determined by the Hilbert polynomial; *Macaulay* showed that it is 15. A test promptly affirmed I to be radical and equidimensional.

A Noether normalization of R/I is $B = k[a, b, f]$, $f = d - c$ (if char $k \neq 2$), while $I \cap k[a, b, c, d]$ is generated by the irreducible polynomial

$$a^{10}b^5 + a^5b^{10} + 5a^4b^8c^2d + 10a^3b^6c^4d^2 + 10a^2b^4c^6d^3 + 5a^6b^2c^3d^4 + c^5d^{10} + 5ab^2c^8d^4 + a^{10}d^5 + 2a^5b^5d^5 + 2a^5c^5d^5 + c^{10}d^5 + 5a^4b^3c^2d^6 + a^5d^{10}.$$

10.5 Ideal Transforms

The computation of ideal transforms, being inherently non–terminating, leads to a variety of *ad hoc* schemes to deal with special cases.

In the two classes of ideal transforms that have been considered here, symbolic blowups and factorial closures, the ground ring R is a normal domain that is the degree zero component of a graded domain:

$$A = R \oplus A_1 \oplus A_2 \oplus \cdots,$$

whose components are finitely generated R–modules. The transform

$$B = \bigcup_{t\geq 1}(A : I^t)$$

was taken relative to an ideal I with generators of degree zero. Actually, one could take $I = (f, g)$, with f and g forming a regular sequence on R.

As a result, B is a graded algebra

$$R \oplus B_1 \oplus B_2 \oplus \cdots,$$

with each B_i a finitely generated R-module. B_i is contained in the bi–dual of A_i as an R–module by Proposition 1.1.15. We denote by $B(r)$ the subalgebra of B generated by its homogeneous components up to degree r. B is also the ideal transform, with respect to the same ideal, of any of the $B(r)$'s ($r \geq 1$). We use Proposition 7.1.4 as the termination criterion, which is rephrased as:

Proposition 10.5.1 $B = B(r)$ *if and only if grade* $I \cdot B(r) \geq 2$.

This section consists of a series of observations (*cf.* [290]) on how to use the theory of Gröbner bases to compute ideal transforms of the blowup algebras. It is clear that similar methods could be used to deal with other ideal transforms. It is rather special in that each of the rings is a graded algebra, the ideal is generated by elements of degree zero, and the ideal transform lives naturally in a polynomial ring. We make use of each of these features.

We represent each $B(r)$ as a quotient $P(r)/J(r)$, where $P(r)$ is a graded polynomial ring. Thus $J(E) = J(1)$ is the ideal of 1-forms given by the presentation of the module E. We seek to construct $J(r+1)$ from $J(r)$ and to test whether the equality $B = B(r)$ already holds.

To get started, let us indicate how to obtain the algebra $D(E)$. We assume the module E is given as the cokernel of the matrix $\varphi = (a_{ij})$, an $n \times m$ matrix. Let f be a nonzero element of R chosen so that the localization of E at f is a free R_f–module. If E has rank e, f can be taken in the $(n-e)$th Fitting ideal of E; if E is a prime ideal, f may be an element of E itself. Set

$$L(E) = \bigcup_{t\geq 1} (J(E) : f^t).$$

Then $D(E) = R[T_1, \ldots, T_n]/L(E)$. Note that if E is an ideal, $D(E)$ is the corresponding Rees algebra. In particular, it provides a presentation of the module E_0 obtained by moding out the R-torsion of E.

If we are looking for the factorial closure, the double dual of E must be determined first. Instead of using the method to be discussed later, at this point the following alternative procedure may be applied. From the preceding one may assume that E is a torsion-free R-module. In fact, we suppose that E is given as a submodule of a free module $RT_1 \oplus \cdots \oplus RT_e$, generated by the forms

$$g_k = b_{k1}T_1 + \cdots + b_{ke}T_e, \; k = 1, \ldots, s.$$

$D(E)$ is then the subring of the polynomial ring $R[T_1, \ldots, T_n]$ generated by the forms g_k's over R. To obtain the bi–dual of E we found useful to compute the ideal transform of $\{f, g\}$, chosen earlier, with respect to the subring generated by the g_k–forms and all the products T_iT_j. This is somewhat cumbersome but effective.

There are three parts to obtaining $J(r+1)$ from $J(r)$. Assume that R is a polynomial ring in the variables $x_1, \ldots, x_d$, and that E is a module with a presentation as before.

Step 1. Let $\{f, g\}$ be a regular sequence as in Corollary 7.1.2. To determine whether new generators must be added to $B(r)$ in order to obtain $B(r+1)$, we must compute the component of degree $r+1$ of the ideal transform $T_{B(r)}(f, g)$, that is

$$\bigcup_{t\geq 1} B(r)_{r+1} : (f, g)^t.$$

This module is nothing but B_{r+1}. It is convenient to keep track of degrees and work in the full ring $A = B(r)$. Observe that for any ideal $L = (f_1, \ldots, f_p)$ of A,

$$A : L = \bigcap_{1\leq i\leq p} (Af_1 :_A f_i) f_1^{-1}.$$

When applied to $L = (f, g)^t$, we obtain:

$$A(t) = A : (f, g)^t = \Big(\bigcap_{1\leq i\leq t} (Af^t :_A g^i f^{t-i})\Big) f^{-t} = \Big(\bigcap_{1\leq i\leq t} (Af^i :_A g^i)\Big) f^{-t}.$$

Phrased in terms of the ring $P(r)$ this would be:

$$(\bigcap_{1\leq i\leq t} (L_{r+1}, f^i):_{P(r)} g^i)f^{-t},$$

where L_{r+1} denotes the subideal of $J(r)$ generated by the elements of degree at most $r+1$. (Note: One can also use the full $J(r)$ instead of L_{r+1}, a move that is useful when $B(r)$ turns out to be $B(r+1)$.) Because each $B(r)_j$, $j \leq r$, is a reflexive module, $A(t)$ and $A(t+1)$ can only begin to differ in degree $r+1$. Stability is described by the equality:

$$f(\bigcap_{1\leq i\leq t} Af^i :_A g^i)_{r+1} = (\bigcap_{1\leq i\leq t+1} Af^i :_A g^i)_{r+1}.$$

Suppose $u_i = e_i f^{-t}, i = 1, \ldots, s$, generate $A(t)_{r+1}/A_{r+1}$; we now get to the determination of the equations for the u_i's.

Step 2. The brute force application of the Gröbner basis algorithm permits mapping the polynomial ring $P(r+1)$ onto the new generators to get the ideal $J(r+1)$. It does not work well at all. It is preferable making use of the fact that B is a rational extension of $B(r)$, by seeking the $B(r)$-conductors of each of the u_i. It provides for a number of equations of the form

$$A_{ji}U_i - B_{ji},\ j = 1, \ldots, r_i.$$

This is useful when $B(r)$ is normal and $B(r+1)$ is singly generated over it; the set above would be all that is needed. In general one needs one extra move.

As a matter of fact, it suffices to consider the R-conductors of the elements u_i whenever the next step is going to be used.

Step 3. Denote $P(r+1) = P(r)[U_1, \ldots, U_s]$ and let $L(r+1) = J(r)$ together with all the linear equations above. If $L(r+1)$ is a prime ideal, then it obviously equals $J(r+1)$. Let h be an element such that the localization $S(E_h) = B_h$, e.g. pick $h = f$ above. It is easy to see that

$$J(r+1) = \bigcup_t (L(r+1):_{P(r+1)} h^t).$$

Remark 10.5.2 In view of the complexity of Gröbner basis algorithms, considerable experimentation must be exercised. For instance, (i) the choice of f, g may be changed from one approximation to the next, and (ii) the monomial order has to be played with to permit the computation to go through. It always worked better when f and g were actual variables.

Example 10.5.3 Without intent to assault the reader's goodwill, here is the defining ideal of the symbolic power algebra of the monomial curve (7, 9, 10). (That the algebra had to be Noetherian is proved in [146]). It led to a computation that stretched over nearly ten days. (The characteristic of the field is assumed to be zero; the notation u_{ij} denotes one fresh generator of degree i.)

$(u_{71}^2 + t_1u_{33}u_{51}^2 - t_3^2u_{41}^3,\ -u_{52}u_{71} + t_1u_{33}^2u_{51} + t_3u_{32}u_{41}^2,\ u_{33}u_{71} + u_{51}u_{52} + t_1t_3u_{41}^2,\ -u_{32}u_{71} + t_1^2u_{33}u_{51} - t_1t_2u_{41}^2 +$
$u_{31}u_{33}u_{41},\ u_{31}u_{71} + u_{32}^2u_{41} + t_1u_{33}^3,\ u_{21}u_{71} + t_1t_3^2u_{33}^2 - t_1^3u_{32}u_{33} + u_{32}^3,\ t_3u_{71} - u_{31}u_{51} + 2t_1u_{32}u_{41},\ t_2u_{71} - t_3u_{31}u_{41} -$
$2t_1^2u_{33}^2,\ t_1u_{71} + u_{32}u_{51} - t_3u_{33}u_{41},\ u_{52}^2 + (t_1^3u_{33} + u_{32}^2)u_{41} + t_1u_{33}^3,\ u_{33}u_{52} + u_{31}u_{51} - t_1u_{32}u_{41},\ u_{32}u_{52} + t_3u_{31}u_{41} +$
$t_1^2u_{33}^2,\ u_{31}u_{52} - t_2u_{32}u_{41} - t_1t_3u_{33}^2,\ u_{21}u_{52} + t_1^3t_3u_{32} + t_3u_{31}^2 - t_1^4u_{31},\ t_3u_{52} - t_1t_2u_{41} + u_{31}u_{33},\ t_2u_{52} + t_1^2t_3u_{33} +$
$u_{31}u_{32},\ t_1u_{52} + t_3^2u_{41} - u_{32}u_{33},\ -u_{31}u_{32}u_{51} + (t_3u_{31}u_{33} + t_1u_{32}^2)u_{41} + t_1^2u_{33}^3,\ -u_{21}u_{51} + t_3u_{31}u_{33} - t_1^4u_{33} +$
$t_1u_{32}^2,\ u_{31}^2u_{51} + (t_3u_{32}^2 - 2t_1u_{31}u_{32})u_{41} + t_1t_3u_{33}^3,\ t_3u_{51} - t_1^2u_{41} - u_{33}^2,\ (t_1^3u_{33} + u_{32}^2)u_{51} - t_1^2t_2u_{41}^2 + t_1u_{31}u_{33}u_{41} -$
$t_3u_{32}u_{33}u_{41},\ t_2u_{51} + t_3^2u_{41} - 2u_{32}u_{33},\ t_1^2u_{31}^2u_{41} - t_1t_3u_{31}u_{32}u_{41} + t_1t_3^2u_{33}^3 + (t_1^3u_{32} + u_{31}^2)u_{33}^2 + u_{32}^3u_{33},\ t_1u_{31}u_{51} -$
$t_3^2u_{33}u_{41} + t_1^2u_{32}u_{41} + u_{32}u_{33}^2,\ (t_1^2u_{31}^3u_{33} + t_1t_3u_{32}^4 - t_1^2u_{31}u_{32}^3)u_{41} + t_1t_3^2u_{31}u_{33}^4 + (t_1^2t_3u_{32}^2 + 2t_1^3u_{31}u_{32} + u_{31}^3)u_{33}^3 +$
$u_{31}u_{32}^3u_{33}^2,\ t_1t_2u_{32}u_{41} + t_3^2u_{31}u_{41} + t_1^2t_3u_{33}^2 - u_{31}u_{32}u_{33},\ u_{32}^3 - t_1t_2u_{31}u_{41} + t_1t_3^2u_{33}^2 + (t_1^3u_{32} + u_{31}^2)u_{33},\ t_1t_3^3u_{33} +$
$t_2u_{32}^2 + t_1^3t_3u_{32} + t_3u_{31}^2 - t_1^4u_{31},\ (t_2u_{32}^2 + t_3u_{31}^2)u_{41} + (t_1t_3u_{32} + t_1^2u_{31})u_{33}^2,\ (t_3^3 + t_1^2t_2)u_{41} - t_3u_{32}u_{33} - t_1u_{31}u_{33},\ xu_{41} -$
$t_3u_{32} - t_1u_{31},\ -u_{21}u_{33} + t_1t_2u_{32} + t_3^2u_{31},\ -t_2u_{33} + t_3u_{32} - t_1u_{31},\ zu_{33} + t_3u_{21} + t_1t_2^2,\ yu_{33} - t_1u_{21} + t_2t_3^2,\ -xu_{33} +$
$t_3^3 + t_1^2t_2,\ u_{21}u_{32} - t_2t_3u_{31} + t_1^2t_3^3 + t_1^4t_2,\ t_2^2u_{32} + u_{21}u_{31} + t_1t_3^4 + t_1^3t_2t_3,\ zu_{32} - xt_1^2t_3 + t_2u_{21},\ yu_{32} + xt_1^3 + t_2^2t_3,\ xu_{32} -$
$t_1u_{21} - t_2t_3^2,\ (t_3^2u_{31}u_{33} + t_1t_3u_{32}^2 - t_1^2u_{31}u_{32})u_{41} + t_1^2t_3u_{33}^2 - u_{31}u_{32}u_{33}^2,\ u_{21}u_{41} + t_1^2t_3u_{33} - u_{31}u_{32},\ yu_{51} + t_3^2u_{32} -$
$2t_1t_3u_{31} + 2t_1^5,\ zu_{51} + 2t_1t_2u_{32} + t_3^2u_{31} - 2t_1^4t_3,\ yu_{52} - t_2t_3u_{31} - t_1^2t_3^3 + t_1^4t_2,\ t_1^2u_{31}^4u_{41} + t_1t_3^2u_{31}^2u_{33}^3 + (t_1t_3^2u_{32}^3 +$
$t_1^2t_3u_{31}u_{32}^2 + 2t_1^3u_{31}^2u_{32} + u_{31}^4)u_{33}^2 + (t_1^3u_{32}^4 + 2u_{31}^2u_{32}^3)u_{33} + u_{32}^6,\ zu_{52} + t_2^2u_{32} + t_1t_3^4 - t_1^3t_2t_3,\ t_3^2u_{31}^2u_{41} + t_1t_3^2u_{32}u_{33}^2 +$
$t_1^2t_3u_{31}u_{33}^2 + t_1^3u_{32}^2u_{33} + u_{32}^4,\ (t_3^2u_{32} - t_1t_3u_{31})u_{41} - t_1^3u_{33}^2 - u_{32}^2u_{33},\ -t_2t_3u_{41} + t_1^3u_{33} + u_{32}^2,\ t_2^2u_{41} + t_1t_3^2u_{33} +$
$u_{31}^2,\ t_1t_3u_{31}^3u_{41} + (t_1t_3^2u_{32}^2 + t_1^2t_3u_{31}u_{32} + t_1^3u_{31}^2)u_{33}^2 + t_1^3u_{32}^3u_{33} + u_{31}^2u_{32}^2u_{33} + u_{32}^5,\ -zu_{41} - t_2u_{31} + 2t_1^2t_3^2,\ -zu_{31} +$
$xt_1t_3^2 + t_2^3,\ yu_{41} + t_2u_{32} + 2t_1^3t_3,\ (t_3u_{31}^2u_{33} + t_3u_{32}^3 - t_1u_{31}u_{32}^2)u_{41} + (t_1t_3u_{32} + t_1^2u_{31})u_{33}^3,\ yu_{31} + xt_1^2t_3 + t_2u_{21},\ xu_{31} -$
$t_3u_{21} + t_1t_2^2,\ xt_1t_3^3 + xt_1^3t_2 + u_{21}^2 + t_2^3t_3,\ zu_{21} + yt_2^2 - x^2t_1t_3,\ zt_2t_3 - yu_{21} - x^2t_1^2,\ zt_3^2 - yt_1t_2 + xu_{21},\ z^2t_3 + y^2t_2 +$
$x^3t_1,\ zt_1 + yt_3 + zt_2,\ xu_{51} - t_3^2u_{33} - 2t_1^2u_{32},\ xu_{52} - u_{21}u_{33} + 2t_3^2u_{31},\ xu_{71} - u_{21}u_{51} - t_1^4u_{33} + 3t_1u_{32}^2,\ yu_{71} + (t_2u_{31} +$
$2t_1^2t_3^2)u_{33} - 2t_1^4u_{32} + t_1u_{31}^2,\ zu_{71} + 2t_1^3t_2u_{33} + 2t_1^3t_3u_{32} + t_3u_{31}^2).$

This ideal is Cohen–Macaulay. On the other hand, according to Corollary 7.2.9, the symbolic blowup of a codimension two prime ideal is always quasi–Gorenstein, so that it must be Gorenstein.

10.6 Integral Closure

The problem of describing constructively the integral closure of an affine domain has been dealt with, previously, in [265] and particularly in [249]. In [295] this question was visited again, and a method was proposed that makes use of resources that have since become available. Amongst these is a better recognition of the algebraic methods which have a more smooth merging with computational techniques based on the Buchberger algorithm. (See [273] for another discussion of this problem.)

The more general problem of finding the integral closure of a ring A in one of its overrings B is also treated here [21]. The motivation for such methods comes from the presence of such constructions in the Zariski's Main Theorem, to create means to effect Stein factorizations (see [99]), and in the computation of rings of invariants.

S_2–ification of an Affine Ring

Set $S = k[x_1, \ldots, x_n]$, and let $A = S/P$ be an affine domain of dimension d. We seek to express the integral closure of A as an affine domain $k[y_1, \ldots, y_m]/Q$. The setting used is that provided by the normality criterion of Krull–Serre's (Theorem 1.1.14).

The task here is, given an affine domain A, find an integral rational extension B that satisfies both conditions (S_2) and R_1. The integral closure B will be determined by two types of extensions. First, through a preprocessing, that enables (S_2); this is the so-called S_2-ification of A. It is followed by a sequence of extensions of a second kind, each a step in the desingularization of A in codimension one; it could be labelled the R_1-ification of A. If A already satisfies R_1 then B can be computed in a single step.

S_2-ification and the canonical module

We shall now explain several issues connected to the construction of the S_2-ification of A. The ring B will be exhibited in terms of local duality, from which it will be clear that it does not depend on the Noether normalization.

One aims, for a given Noetherian ring A with total ring of fractions Q, to find integral extensions

$$A \hookrightarrow B \hookrightarrow Q,$$

with the property that if $\wp \subset A$ has height at least 2 then $\wp B$ has grade at least 2 (see [213]).

If A is an integral domain, there is a natural candidate for B: Let

$$A^{(1)} = \bigcap_{\text{height } \wp=1} A_{\wp}.$$

Since A is Noetherian, it is easy to see that $A^{(1)}$ is an integral extension of A with the property that, if x is an element of the field of quotients of A and its conductor into A is an ideal of height at least two, then $x \in A^{(1)}$.

The construction of $A^{(1)}$ presents obvious difficulties, the most serious of which is that it may not be Noetherian. We aim at a more computational approach to its determination in some cases of interest.

Canonical module

We shall need some elementary properties of the canonical module of a ring. The basic reference used is [113], or the forthcoming [31]. In particular we shall use:

Theorem 10.6.1 *Let $A = S/P$, where S is a Gorenstein ring and let P be an ideal of codimension g. Then $\omega_A = \operatorname{Ext}^g_S(A, S)$ is a canonical module for A.*

This module has the following properties:

(a) ω_A satisfies the (S_2)-condition.

(b) $\operatorname{Hom}_A(*, \omega_A)$ is a self-dualizing functor on the category of A-modules with the condition (S_2).

This immediately leads to the following description of the S_2–ification of A. As a matter of terminology, we use *extension* to denote an injective, finite extension of A.

Theorem 10.6.2 *Let A be a Noetherian ring with canonical module ω_A, and suppose that A is generically a Gorenstein ring. Then $B = \text{Hom}_A(\omega_A, \omega_A)$ is the minimal extension of A with the property (S_2).*

The condition that A is generically a Gorenstein ring is equivalent to saying that its canonical module may be identified to an ideal J containing regular elements.

Proof. For a given extension $A \subset C$ consider the diagram of natural maps

$$\begin{array}{ccc} A & \longrightarrow & C \\ \downarrow & & \downarrow \\ \text{Hom}_A(\text{Hom}_A(A,\omega_A),\omega_A) & \longrightarrow & \text{Hom}_A(\text{Hom}_A(C,\omega_A),\omega_A). \end{array}$$

If C is an extension, satisfying (S_2), there results an embedding

$$B = \text{Hom}_A(\omega_A, \omega_A) \hookrightarrow C,$$

that proves the assertion. □

Another way to get to B is through a subring $C \subset A$, A finite over C, for which the canonical module ω_C is known, as is the case for the Noether normalization of an affine ring A. One then has $\omega_A = \text{Hom}_C(A, \omega_C)$, cf. [113].

Theorem 10.6.3 *The S_2–ificiation of A is given by* $\text{Hom}_C(\text{Hom}_C(A, \omega_C), \omega_C)$.

Proof. From above we obtain the identifications

$$\begin{aligned} B = \text{Hom}_A(\omega_A, \omega_A) &= \text{Hom}_A(\text{Hom}_C(A,\omega_C), \text{Hom}_C(A,\omega_C)) \\ &= \text{Hom}_C(\text{Hom}_C(A,\omega_C) \otimes_A A, \omega_C) \\ &= \text{Hom}_C(\text{Hom}_C(A,\omega_C), \omega_C) \end{aligned}$$

by adjointness. □

Corollary 10.6.4 *If C has (S_2) and is Gorenstein in codimension* 1, *then*

$$B = \text{Hom}_C(\text{Hom}_C(A, C), C).$$

Proof. Note that $D = \text{Hom}_C(\text{Hom}_C(A, C), C)$ satisfies (S_2), being a C–dual. By Theorem 10.6.2 we have an inclusion of extensions of A

$$B \hookrightarrow D,$$

which by assumption is an isomorphism (as C–modules) in codimension at most 1. A depth count shows that its cokernel must be trivial. □

Computation of the canonical ideal

We describe a procedure to represent the canonical module of an affine domain A as one of its ideals. This facilitates the computation of the S_2–ification of A.

Let

$$A = R/P \quad = \quad k[x_1, \ldots, x_n]/(f_1, \ldots, f_m) \tag{10.16}$$

where P is a prime ideal of codimension g. Pick a regular sequence

$$\mathbf{x} = \{x_1, \ldots, x_g\} \subset P.$$

The canonical module ω_A has the following representation

$$\omega_A = \mathrm{Ext}^g_R(R/P, R) \simeq ((\mathbf{x})\colon P)/(\mathbf{x}).$$

Theorem 10.6.5 *Let P and $(\mathbf{x})$ be as above. There exist*

$$\begin{aligned} a &\in (\mathbf{x})\colon P \setminus (\mathbf{x}) \\ b &\in (a, \mathbf{x})\colon ((\mathbf{x})\colon P) \setminus P. \end{aligned} \tag{10.17}$$

Then if set

$$L \quad = \quad (b((\mathbf{x})\colon P) + (\mathbf{x}))\colon a$$

then

$$((\mathbf{x})\colon P)/(\mathbf{x}) \simeq L/P \subset R/P.$$

Proof. That a can be so chosen is clear. To show the existence of b, we localize at P. In the local ring R_P, PR_P is generated by a regular sequence $\mathbf{y} = \{y_1, \ldots, y_g\}$ so that we can write

$$\mathbf{x} = \mathbf{y} \cdot \varphi$$

for some $g \times g$ matrix φ. By Theorem 10.3.6,

$$((\mathbf{x})\colon P)_P = (\mathbf{x}, \det \varphi)$$

and the image of $\det \varphi$ in $(R/(\mathbf{x}))_P$ generates its socle.

On the other hand, ω_A is a torsion–free A–module, so that the image of a in $((\mathbf{x})\colon P/(\mathbf{x}))_P$ does not vanish. This means that $(a, \mathbf{x})_P$ must contain $(\mathbf{x}, \det \varphi)$, which establishes (10.17).

Let L be defined as above. For $r \in (\mathbf{x})\colon P$ we can write

$$rb = ta \bmod (\mathbf{x}).$$

Another such representation

$$rb = sa \bmod (\mathbf{x}),$$

would lead to an equation

$$(t - s)a \in (\mathbf{x}).$$

But a already conducts P into $(\mathbf{x})$, so that $t - s$ must be contained in P since the associated prime ideals of $(\mathbf{x})$ have codimension g. This defines a mapping

$$(\mathbf{x}): P/(\mathbf{x}) \longrightarrow L/P,$$

in which the class of a is mapped to the class of $b \in R/P$, and therefore completes the proof as both torsion–free modules have rank one. □

The double dual

There are other approaches to realize the S_2–ification of the ring A. Our method of choice involves the computation of the canonical ideal as above. There will be times when certain conditions are present that may be worthwhile to take advantage of. We are going to discuss one of them.

We assume that we have a Noether normalization of A

$$R = k[\mathbf{z}] = k[z_1, \ldots, z_d] \hookrightarrow A.$$

A is a torsion free R–module, and the process that gives the Noether normalization also yields a presentation of A as a R–module:

$$R^p \xrightarrow{\varphi} R^q \longrightarrow A \rightarrow 0.$$

The double dual of A is the module

$$A^{**} = \mathrm{Hom}_R(\mathrm{Hom}_R(A, R), R).$$

Proposition 10.6.6 *A^{**} is a subring of B.*

Proof. Follows from Proposition 1.1.15. □

Note that as a consequence of Theorem 1.1.14 and Proposition 1.1.15 B is a reflexive R–module. We must now find a finite formula for A^{**}, for which purpose we make use of the presentation of A.

For an ideal I of an integral domain A, $I^{-1} = A\colon I$ has the usual meaning:

$$I^{-1} = \{x \in \text{field of fractions of } A \mid x \cdot I \subset A\}.$$

I^{-1} is an A–module, canonically isomorphic to $\mathrm{Hom}_A(I, A)$. As an R–module I^{-1} is reflexive if A is a reflexive R-module. To see this, it is enough to show that I^{-1} is the R–dual of some R–module. From $A = \mathrm{Hom}_R(A^*, R)$, we have by adjointness

$$\mathrm{Hom}_A(I, A) = \mathrm{Hom}_A(I, \mathrm{Hom}_R(A^*, R)) = \mathrm{Hom}_R(I \otimes_A A^*, R).$$

Let J be the first non–vanishing Fitting ideal of A as an R–module. Since A is a torsion–free R–module, height $J \geq 2$. Pick f, g, two relatively prime polynomials in J. Define

$$C = \bigcup_{t\geq 1}(A\colon (f,g)^t). \tag{10.18}$$

Proposition 10.6.7 $A^{**} = C$.

Proof. From Proposition 1.1.15, it follows that the right–hand side of this formula is contained in A^{**}. (At the same time, this shows that the right–hand side has a stable value, since B is a finitely generated R–module.) Conversely, let $x \in A^{**}$ and denote by L the R–conductor of A^{**} into C, that is

$$L = \{r \in R \mid rx \in C\}.$$

L is an ideal of R whose radical contains J. Indeed, if $\wp$ is a prime containing L but not J, $A_\wp$ is a free $R_\wp$–module and therefore coincides with $A^{**}_\wp$. But $A_\wp \subset C_\wp$ contradicts the definition of L. This means that for some integer t, $(f,g)^t \cdot x \subset A$ and therefore $x \in C$. □

The point of this construction is that A^{**} will satisfy condition (S_2). The other overrings to be constructed will also preserve this property. We have the following:

Theorem 10.6.8 *If A satisfies the condition R_1 then $B = C$.*

Proof. C is a finite rational extension of A, and therefore inherits the Serre's condition R_1 from it. □

In many cases it is unnecessary to find the Noether normalization, since it is only used to provide the elements f and g. It is clearly enough that they define part of a normalizing system of parameters and lie in the ideal that defines the Cohen–Macaulay locus of A.

Proposition 10.6.9 *Let $A = k[\mathbf{x}]/P$. The formula of* Proposition 10.6.7 *holds if f, g are chosen such that* height$(P, f, g) = 2 +$ height P *and lying in the ideal*

$$L = \bigcap_{j\geq r+1} \text{annihilator}(\text{Ext}^j_{k[\mathbf{x}]}(A, k[\mathbf{x}])) \tag{10.19}$$

that defines the Cohen–Macaulay locus of A.

Generators and relations

There remains to obtain a presentation of $A^{**} = k[y_1, \ldots, y_m]/Q$. Once that is done, we replace A by A^{**}.

We use a method developed in [290], exploiting the fact that the transform is computed relative to a two–generated ideal. There will be two steps to consider:

Algorithm 10.6.10 *Let A be an affine domain and let f, g be a regular sequence as above.*

(a) *The formula*

$$A:(f,g)^t = (\bigcap_{1\leq i\leq t}(Af^t:_A g^i f^{t-i}))f^{-t} = (\bigcap_{1\leq i\leq t}(Af^i:_A g^i))f^{-t}$$

yields $a_1 f^{-r}, \ldots, a_s f^{-r}$, such that $A^{**} = A[a_1 f^{-r}, \ldots, a_s f^{-r}]$, where r is the stable value of t.

(b) *For each $a_i f^{-r}$ consider the linear polynomial $f^r U_i - A_i$ where the U_i are indeterminates and A_i is a representative in $k[x_1, \ldots, x_n]$ of a_i. Let $K = (P, f^r U_1 - A_1, \ldots, f^r U_s - A_s)$. Let Q be the "rational closure" of K:*

$$Q = \bigcup_{t\geq 1}(K:_{k[x_1,\ldots,x_n,U_1,\ldots,U_s]} f^t).$$

Then

$$A^{**} = k[x_1, \ldots, x_n, U_1, \ldots, U_s]/Q.$$

The proof of correctness is clear from the discussion. An advantage of computing the transforms $A(t)$ lies in the convenience of testing for stabilization.

In the sequel, when we replace A by a larger ring C, the latter will always be a reflexive A–module and therefore (S_2) passes from A to C.

Example 10.6.11 The affine domain defined by the prime ideal of Example 10.4.21 fits nicely here. $A = k[a, b, c, d, e]/P$ satisfies the condition of Theorem 10.6.8. Its integral closure is $B = k[a, b, c, d, e, u, v]/Q$, with Q generated by the polynomials (where P is the radical of the ideal generated by the first three polynomials):

$$
\begin{array}{l}
ab^3c + bc^3d + a^3be + cd^3e + ade^3 \\
a^2bc^2 + b^2cd^2 + a^2d^2e + ab^2e^2 + c^2de^2 \\
a^5 + b^5 + c^5 + d^5 - 5abcde + e^5 \\
a^4bc - ab^4e - 2b^2c^2de + a^2cde^2 + bd^3e^2 + du \\
a^3b^2c - bc^2d^3 - b^5e - d^5e + 2abcde^2 - ev \\
a^2b^2cd + b^2c^3e + a^4de - 2bcd^2e^2 - abe^4 - cu \\
bc^5 - 2ab^2cde + c^3d^2e - a^3de^2 + be^5 - bv \\
b^5c + a^2b^3e - abc^2de - ad^3e^2 + cv \\
ab^2c^3 + abcd^2e - a^2be^3 - de^5 + dv \\
abc^2d^2 - b^3c^2e + ad^4e - a^2bce^2 + b^2d^2e^2 - cde^4 + bu \\
b^3c^2d - cd^2e^3 - av \\
bcd^4 - c^4de - au \\
abc^4 - b^4cd - a^2b^2de + ac^2d^2e + b^2c^2e^2 - bd^2e^3 - eu \\
ab^4c^2e^3 - a^4b^3c^2d + a^2b^3c^4e + a^6bcde + ab^6cde + 3b^4c^3d^2e\backslash \\
-a^3b^4de^2 - 3a^2b^2c^2d^2e^2 - b^3cd^4e^2 + a^2bd^4e^3 - b^5c^5 - a^3b^2ce^4\backslash \\
+bc^2d^3e^4 - abcde^6 - b^2cd^2u + a^2d^2eu + c^2de^2u - abcdev + e^5v - v^2 \\
b^4cde^4 + 2a^3d^3e^4 + 4a^2b^2de^5 - ac^2d^2e^5 + 2bd^2e^7 + abcdeu\backslash \\
+ac^2d^2v - b^2c^2ev - 4a^2ce^2v - bd^2e^2v + uv \\
ab^3c^6 + a^5bc^3d - a^2b^4c^2de - 2ab^2c^4d^2e + a^5cd^3e\backslash \\
-a^2b^3d^3e^2 - 2abc^2d^4e^2 - a^2b^2c^3e^3 + ad^6e^3 - a^2bcd^2e^4 - cd^3e^6\backslash \\
+b^2c^2eu - bd^2e^2u - u^2 + bc^3dv - cd^3ev - ade^3v
\end{array}
$$

(\ denotes continuation.)

Desingularization in codimension one

We must find a way to further enlarge A, if necessary, into its integral closure B, which brings us to the problem of desingularization in codimension one. The approach here will fail to work in many cases of interest as it depends on Jacobian ideals, and we restrict ourselves to mostly to characteristic zero, or at least to cases where certain separability conditions are met.

Computational test of normality

It may be advisable to check for normality in advance, so for completeness we discuss the standard normality test and a variant mechanism.

Theorem 10.6.12 *Let A be an affine domain and let J denote its Jacobian ideal. Then A is normal if and only if*

$$\boxed{A = A\colon J.} \tag{10.20}$$

Proof. This is just the assertion that J is an ideal of grade at least two, so that at each prime $\wp$ such that depth $A_\wp = 1$ the localization is a discrete valuation domain. Since A is the intersection of all such integrally closed domains ([165, Theorem 53]), it is also normal. □

There will be occasions when it is not necessary to compute the full Jacobian ideal J, as any of its subideals that satisfy the equation above suffices for certification of normality. One of these may be found as follows ([185]):

Theorem 10.6.13 *Let A be an affine domain and let R be one of its Noether normalizations. Suppose the field of fractions of A is a separable extension of the field of fractions of R. Let $A = R[x_1, \ldots, x_n]/P$ be a presentation of A and denote by J_0 the Jacobian ideal of A relative to the variables $x_1, \ldots, x_n$. If B is the integral closure of A then $J_0 \cdot B \subset A$. In particular if $A = A\colon J_0$ then A is integrally closed.*

Unlike the singular locus, through the Jacobian criterion, there is no explicit description of the non–normal locus. This opens the door to *ad hoc* methods, each suited to a limited situation. In the case of Rees algebras, there are a few procedures that have been proved useful. Some are discussed in [25]; here we present two additional methods.

The proof of the first of these will be left as an exercise. Its practice only requires that part of the Jacobian ideal be known with the stated property.

Proposition 10.6.14 *Let R be a normal affine domain over a field of characteristic zero, and let I be an ideal. Let*

$$0 \to L \longrightarrow R[T_1, \ldots, T_n] \longrightarrow \mathcal{R}(I) \to 0$$

be the presentation of the Rees algebra of I. Denote by J the Jacobian ideal of $\mathcal{R}(I)$. Then I is a normal ideal if and only if

$$(L, I)\colon J = (L, I).$$

The second method is less general but it is modeled on the situation of a Rees algebra $\mathcal{R}(I)$ of a local ring R that is normal on the punctured spectrum of R.

Proposition 10.6.15 *Let R be a Noetherian domain with a finite integral extension S. Let J be the conductor of the extension. Suppose there is a prime ideal $\wp$ and an element $f \notin \wp$ with the following properties:*

(i) *J contains a power of $\wp$;*

(ii) *$\wp R_f$ is a principal ideal.*

If grade $(\wp, f) \geq 2$ *then $R = S$.*

Proof. Suppose that grade $(\wp, f) \geq 2$, but $R \neq S$. The latter says that J is a proper ideal of grade 1. Let P be one of its associated prime ideal; it follows that the local ring R_P has depth one, and therefore $\wp \subset P$, but $f \notin P$. If $\wp = P$, R_P is a discrete valuation domain and thus $J_P = R_P$, which contradicts the choice of P. So we must have $\wp \neq P$, which in turn implies that PR_P contains a regular sequence with 2 elements since $\wp_P$ is a principal ideal. □

Augmenting the ring

If A fails the normality test we embark on the task of enlarging it. Perhaps the most natural approach is the following: Let I be an ideal of A; then $B = \operatorname{Hom}_A(I, I)$ is an integral extension of A contained in the field of fractions of A. There are two requirements one would like to place on the choice of I: (i) B is a proper extension of A if the latter is not integrally closed; (ii) the computation of the endomorphisms is given, often, through a direct syzygy computation.

We are going to use (cf. [295]) an opening provided by the following regularity criterion of Lipman [183]:

Theorem 10.6.16 *If J is an invertible ideal then A is regular.*

While the usual Jacobian criterion requires $J = A$, this version permits one to work with much less. It will provide the required approach.

We introduce the following notation: $J^{-1} = \operatorname{Hom}_A(J, A)$, and $C = (J \cdot J^{-1})^{-1}$. Both J^{-1} and C are A–modules, contained in the field of fractions of A. We note that C is a ring. Indeed, if $x \in C$, the equation

$$x \cdot J \cdot J^{-1} = x \cdot J^{-1} \cdot J \subset A$$

implies that $x \cdot J^{-1} \subset J^{-1}$, and therefore x belongs to $\operatorname{Hom}_A(J^{-1}, J^{-1})$. On the other hand, every element of this endomorphism ring is realized by multiplication by an element in the field of fractions of A, so that if $z \cdot J^{-1} \subset J^{-1}$ we have $z \cdot J^{-1} \cdot J \subset J^{-1} \cdot J \subset A$, and thus $z \in C$.

Algorithm 10.6.17 *Let A be an affine domain satisfying Serre's condition (S_2), and let J be its Jacobian ideal.*

(a) *If* height $J \geq 2$, *then $B = A$.*

(b) *If* height $J = 1$, *then C is a subring of B properly containing A.*

(c) *Replace A by C until* (a) *is realized.*

Proof. If height $J \geq 2$, A satisfies R_1 and is therefore normal by Theorem 1.1.14. If height $J = 1$, we claim that $A \neq C \subset B$. Indeed, the ideal $J \cdot J^{-1}$ must have depth 1 as otherwise $J_\wp$ would be invertible for each prime $\wp$ of height 1; by Theorem 10.6.16 $A_\wp$ would be regular. But an ideal I of a Noetherian domain A has depth 1 if and only if $I^{-1} \neq A$ (see [165, Exercise 2, p. 102]), and therefore C is a proper extension of A. Moreover, since $C = \operatorname{Hom}_A(J \cdot J^{-1}, A)$, it inherits (S_2) from A. □

Remark 10.6.18 If A is known to be a Gorenstein ring in codimension one then almost any ideal can be used instead of J^{-1}: It can then be proved that for any ideal I which is not principal in codimension one the endomorphism ring of I is strictly larger that A.

Trace rings

The ideal $I \cdot I^{-1}$ is the so–called *trace* ideal of I. If computation of syzygies is possible over the ring A, it can be obtained as follows. Let

$$A^m \xrightarrow{\varphi} A^n \longrightarrow I \to 0$$

be a presentation of I. Let ψ be the $n \times p$ matrix, whose columns generate the relations of φ^* (the transpose of φ).

Proposition 10.6.19 $I \cdot I^{-1} = I_1(\psi)$.

We leave the straightforward proof to the reader.

Hypersurface rings

We illustrate with the important case of a hypersurface. Let

$$f \in R = k[x_1, \ldots, x_n]$$

be an irreducible polynomial. (To simplify matters, assume that the characteristic of k is large enough.)

Set $L = (f, \frac{\partial f}{\partial x_1}, \ldots, \frac{\partial f}{\partial x_n})$; then $J = L/(f) \subset A = R/(f)$ is the Jacobian ideal of A. We assume that A is not normal but that L is a Cohen–Macaulay ideal, necessarily of codimension two. (This will be the case for a plane curve.)

Let

$$0 \to R^n \xrightarrow{\alpha} R^{n+1} \longrightarrow L \to 0,$$

be a resolution of L. The matrix β obtained from α by deleting the first row (the row that corresponds to the generator f of L) yields a projective resolution of J as an R–module:

$$0 \to R^n \xrightarrow{\beta} R^n \longrightarrow J \to 0.$$

Tensoring with A we get the exact sequence of Cohen–Macaulay A–modules

$$0 \to J \longrightarrow A^n \xrightarrow{\varphi} A^n \longrightarrow J \to 0.$$

Because A is a Gorenstein ring, dualizing the sequence we get another exact sequence

$$0 \to \mathrm{Hom}(J, A) \longrightarrow A^n \xrightarrow{\varphi^*} A^n \longrightarrow \mathrm{Hom}(J, A) \to 0.$$

If we now use the result of Eisenbud [56] on the relationships between maximal Cohen–Macaulay modules over hypersurface rings and matrix decompositions (succinctly here: ker φ^* is given by the image of its matrix adjoint) we obtain:

Proposition 10.6.20 *In the situation above,*

$$J \cdot J^{-1} = \Delta/(f),$$

where Δ is the ideal generated by the minors of size $n-1$ of the matrix β. In addition, if $g \in \Delta$ is not divisible by f, then

$$\boxed{C = \operatorname{Hom}(J \cdot J^{-1}, A) = (((f, g) \colon \Delta)/(f))g^{-1}.} \tag{10.21}$$

For curves, and some low dimensional rings, this method gives a reasonable performance.

The reader will note that we have not tapped the rich vein of methods of desingularization of curves that bring in very early the underlying geometry. The stakes tend to be higher here since "curves" and one–dimensional rings are not equivalent objects.

There is a large bibliography on this topic, that is periodically visited and enlarged by geometers, number theorists and code theorists ([301]).

Problem 10.6.21 In the case of a curve, find reasonable conditions for the ring of the proposition above to be the integral closure of A.

Integral closure of a morphism

Here we address the question of the effective construction of the integral closure of an affine domain in another affine domain. It is the affine version of the question of how to carry out the Stein factorization of a morphism ([21]).

The procedures that are presented here are closely modeled on the proof of Zariski's Main Theorem ([220]). It is particularly instructive in the manner in which the algebraic techniques are realized computationally through the utilization of Gröbner basis algorithms.

Birational morphisms

We are first going to isolate the elements which will play the key roles in the constructions.

Let $A \subset B$ be affine integral domains. Consider the diagram of inclusions:

$$\begin{array}{ccccc} A & \longrightarrow & A' = \overline{A} \cap B & \longrightarrow & B \\ & & \downarrow & & \downarrow \\ & & \overline{A} & \longrightarrow & \overline{B} \end{array}$$

where $\overline{A}$ and $\overline{B}$ are the integral closures of A and B respectively in their (common) quotient field. A' is the desired integral closure of A in B.

Since A and B have a common quotient field, and B is a k–algebra of finite type (and hence an A–algebra of finite type) we have an expression of B in the form:

$$k[f_1, \ldots, f_n, \frac{a_{n+1}}{b}, \ldots, \frac{a_m}{b}], \tag{10.22}$$

where A is the subalgebra generated by the set $\{f_1, \ldots, f_n\}$ and b is an element of A.

We actually taken B as being given by

$$k[x_1, \ldots, x_m]/\mathfrak{P},$$

with A the k–subalgebra generated by the set $\{x_1, \ldots, x_n\}$, this new representation can be obtained in the following manner. For each x_i, $i = n+1, \ldots, m$ consider the k–subalgebra generated by the set $\{x_1, \ldots, x_n, x_i\}$. Choose an elimination term ordering on the indeterminates $\{x_1, \ldots, x_m\}$ such that $x_1, \ldots, x_n$ are the lowest indeterminates and x_i is the smallest of the remaining indeterminates (see [251] for details). There will exist an element of the form $\beta_i x_i - \alpha_i$ in the Gröbner basis found by this process. Thus $x_i = \dfrac{\alpha_i}{\beta_i}$, and if one sets:

$$a_i = \alpha_i \prod\nolimits_{j \neq i} \beta_i$$

and

$$b = \prod\nolimits_{j=n+1}^{m} \beta_i,$$

one will then have the representation (10.22).

We recall two elementary properties of the integral closure $\overline{I}$ of an ideal I.

Proposition 10.6.22 *Let R be an integral domain. Then*

(a) *Let $b \in R$. If $\overline{R}$ is the integral closure of R then $\overline{(b)} = \overline{R}b \cap R$.*

(b) *If R is an affine domain and I is an ideal of R then $\overline{I^{s+1}} = I\overline{I^s}$ for $s \gg 0$.*

Proof. The first assertion being clear, the second follows from the fact that the integral closure of the Rees algebra $R[It]$ is a finitely generated graded module over $R[It]$. □

We shall now organize the two approaches to the determination of A'. They shall use some of the same elements. The first method, less general, takes place entirely in A, in a manner of speaking, while the other involves B.

From $A' = \overline{A} \cap B$, every element z of A' has an representation of the form:

$$z = \frac{w}{b^r}, \text{ where } w \text{ is in } \overline{(b^r)} \cap (a_{n+1}, \ldots, a_m, b)^r.$$

This representation need not however be a minimal one in terms of the degree of the denominator.

For $i \in \mathbb{N}_0$, let us define S_i to be the A-submodule of B given by:

$$S_i = b^{-i}[\overline{(b^i)} \cap (a_{n+1}, \ldots, a_m, b)^i]. \tag{10.23}$$

These remarks yield the following expression of A' as an ascending union of submodules.

Proposition 10.6.23 *The integral closure A' of A in B is given by:*

$$A' = \bigcup_{i \in \mathbb{N}_0} S_i. \tag{10.24}$$

This gives us the necessary vehicle for the computation of A'. Recall that an element b of an ideal I is *superficial* (of order 1) with *defect* c ([208]) if

$$(I^i : b) \cap I^c = I^{i-1} \ \forall\, i > c. \tag{10.25}$$

It is a consequence of the Artin–Rees lemma that if b is a regular element by enlarging c we may remove I^c from this equation (see [192, Proposition 11.E]). We shall refer to such c as the *strong* defect of b.

Theorem 10.6.24 *If b is a superficial element of the ideal $(a_{n+1}, \ldots, a_m, b)$ with defect c then:*

(a) *There exists an $r \in \mathbb{N}_0$ such that $S_r = S_{r+k}$ for all $k \in \mathbb{N}_0$.*

(b) *If the strong defect is c and $r = \max\{c, \min\{i \mid \overline{(b^{i+1})} = b\overline{(b^i)}\}\}$ then $S_r = S_{r+k}$ for all $k \in \mathbb{N}_0$.*

Proof. (a) follows since A' is a Noetherian A–module, while (b) is a direct verification. □

This means that if b is a superficial element of the ideal $(a_{n+1}, \ldots, a_m, b)$, the integral closure A' of A in B can be found by computing the submodule S_r where r is as described in the theorem. The value of r can be found c and by the successive computation of the ideals

$$b^i \overline{A} \cap A = \overline{(b^i)}$$

and testing at each stage the condition

$$\overline{(b^i)} = b\overline{(b^{i-1})},$$

or equivalently

$$\overline{(b^i)} \subset (b).$$

We note that there are limits for this value which are independent of b. This is a consequence of the uniform Briançon–Skoda theorem (cf. [149, Theorem 4.13]).

The defect of the superficial element b can be determined by computing the bound on the upper degree of an element in the annihilator of the image of b in the associated graded ring of A with respect to the $(a_{n+1}, \ldots, a_m, b)$-adic filtration. Hence if an oracle indicates to us that the element b is a superficial element of the ideal $(a_{n+1}, \ldots, a_m, b)$, the computation of the A–submodule S_r of B yields the integral closure of A in B. In the event that b is not superficial we must find an alternative path to the solution.

Theorem 10.6.25 *Let $A \subset B$ be finitely generated affine domains over a computable field k. Then the integral closure of A in B can be determined as follows. Let $\overline{A}$ be the integral closure of A in its field of fractions, and suppose B is generated over A by fractions having powers of b as denominators. Let r be an integer such that $\overline{A}b^{r+1} \cap A \subset (b)$. Then the integral closure of A in B is*

$$A' = b^{-r}[b^r B \cap b^r \overline{A} \cap A].$$

Now consider the elements of B which have representations of the form

$$z = \frac{w}{b^r}$$

then w is an element of the ideal $b^r \overline{A} \cap A$ of A. By representing the k–algebra B as:

$$k[x_1, \ldots, x_m]/P$$

with A the subalgebra generated by the set $\{x_1, \ldots, x_n\}$ of indeterminates, a generating set for the ideal $b^r B \cap A$ is obtained through a Gröbner basis computation on the earlier elimination order of the ideal generated by the elements of the ideal P and b^r.

To represent the algebra A', in addition to $x_1, \ldots, x_n$, we choose new variables $h_1, \ldots, h_s$ corresponding to the new elements in intersection formula of the previous theorem, thus yielding:

$$\begin{aligned} B &= k[x_1, \ldots, x_m, h_1, \ldots, h_s]/Q = (P, h_i - g_i(x_1, \ldots, x_m),\ i = 1, \ldots, m) \\ A' &= k[x_1, \ldots, x_n, h_1, \ldots, h_s]/Q \cap k[x_1, \ldots, x_n, h_1, \ldots, h_s]. \end{aligned}$$

Reduction to the birational case

The computation of the integral closure in general relies on the reduction of the problem to the case where the morphism is birational. Again there are several approaches.

In a nutshell:

Proposition 10.6.26 *Suppose $A \hookrightarrow B$ is a morphism of affine domains and that $B = A[f_1, \ldots, f_r]$. Let D be the subring of the polynomial ring $B[t]$ given by $D = A[t, f_1 t, \ldots, f_r t]$. If $B[t]$ is graded so that $\deg t = 1$ and the elements of B have degree zero, then D is a graded subalgebra of $B[t]$.*

(a) *$B[t]$ is a birational extension of D.*

(b) *The integral closure of D in $B[t]$ is a graded algebra whose degree zero component is the integral closure of A in B.*

Consider the k–algebras A and B and the morphism $\varphi: A \to B$ to be represented as before. Then the k–algebra $B[t]$ has a representation

$$k[x_1, \ldots, x_n, t, u_{n+1}, \ldots, u_m, x_{n+1}, \ldots, x_m]/\mathrm{P}.$$

The ideal P in this representation is the ideal generated by the ideal P together with the polynomials $u_j - tx_j$ for $j = n+1, \ldots m$. As for the subalgebra D, it is represented by the set $\{x_1, \ldots, x_n, t, u_{n+1}, \ldots, u_m\}$; D is birationally equivalent to $B[t]$.

Since the morphism $\psi: D \to B[t]$ is birational, the integral closure D' of D in $B[t]$ can be computed as in the previous section, with representation for B[t] in the form

$$k[x_1, \ldots, x_m, t, u_{n+1}, \ldots, u_m, h_1, \ldots, h_r]/\Omega,$$

where D' is the k–subalgebra generated by the set

$$\{x_1, \ldots, x_n, t, u_{n+1}, \ldots, u_m, h_1, \ldots, h_r\}.$$

Moreover as D is a graded k–subalgebra of $B[t]$ so is D'.

The k–algebra B therefore has a representation of the form:

$$k[x_1, \ldots, x_m, h_1, \ldots, h_r, t, u_{n+1}, \ldots, u_m]/\mathrm{P},$$

where P is generated by Ω and the element t.

The k–subalgebra of B generated by the set:

$$\{x_1, \ldots, x_n, h_1, \ldots, h_r, t, u_{n+1}, \ldots, u_m\}$$

is $D' \cap B$.

Proposition 10.6.27 *The k–algebra $D' \cap B$ is the integral closure A' of A in B.*

Proof. If x lies in A', then x lies in B and hence x is in $B[t]$ and is integral over A, and therefore x is integral over D so x lies in $D' \cap B$. Conversely if x is in $D' \cap B$, then x is integral over D, and of degree 0. But x satisfies a monic homogeneous equation, which must therefore be of degree 0. Hence x is integral over A, and so x lies in A'. □

An alternative offers itself if B is algebraic over A. The device that found the linear forms $\alpha_i x_i - \beta_i$ would now return a polynomial

$$c_s x_i^s + c_{s-1} x_i^{s-1} + \cdots + c_0 = 0,$$

of least degree s. The $c_s x_i$ would be added to A to obtain A^*, an integral extension of A, with B a rational extension of A^*.

Bibliography

[1] S. Abeasis, Gli ideali GL(V)-invarianti in $S(S^2(V))$, *Rendiconti di Matematica* **13** (1980), 235–262.

[2] I. M. Aberbach and C. Huneke, An improved Briançon–Skoda theorem with applications to the Cohen–Macaulayness of Rees algebras, *Math. Annalen*, to appear.

[3] S. Abhyankar, Concepts of order and rank on a complex space, and a condition for normality, *Math. Annalen* **141** (1960), 171–192.

[4] Y. Aoyama, A remark on almost complete intersections, *Manuscripta Math.* **22** (1977), 225–228.

[5] R. Apéry, Sur les courbes de première espèce de l'espace de trois dimensions, *C. R. Acad. Sci. Paris* **220** (1945), 271–272.

[6] M. Artin and M. Nagata, Residual intersections in Cohen–Macaulay rings, *J. Math. Kyoto Univ.* **12** (1972), 307–323.

[7] M. Auslander, Modules over unramified regular local rings, *Illinois J. Math.* **5** (1961), 631–647.

[8] M. Auslander and D. Buchsbaum, On ramification theory in Noetherian rings, *American J. Math.* **81** (1959), 749–765.

[9] M. Auslander and M. Bridger, *Stable Module Theory*, *Memoirs Amer. Math. Soc.* **94**, 1969.

[10] M. Auslander and O. Goldman, Maximal orders, *Trans. Amer. Math. Soc.* **97** (1960), 1–24.

[11] L. Avramov, Complete intersections and symmetric algebras, *J. Algebra* **73** (1980), 249–280.

[12] L. Avramov and J. Herzog, The Koszul algebra of a codimension 2 embedding, *Math. Z.* **175** (1980), 249–280.

[13] J. Barshay, Graded algebras of powers of ideals generated by A–sequences, *J. Algebra* **25** (1973), 90–99.

[14] D. Bayer and M. Stillman, *Macaulay*, A computer algebra system for computing in algebraic geometry and commutative algebra, 1990.

[15] D. Bayer and M. Stillman, A criterion for detecting m–regularity, *Invent. Math.* **87** (1987), 1–11.

[16] A. A. Belavin and V. G. Drinfel'd, Solutions of the Classical Yang–Baxter Equation for simple Lie algebras, *Functional Analysis and its Applications* (English Translation) **16** (1982), 159–180.

[17] E. R. Berlekamp, Factoring polynomials over large finite fields, *Math. Comp.* **24** (1970), 713–715.

[18] B. Bollobás, *Graph Theory*, Springer–Verlag, Berlin–Heidelberg–New York, 1979.

[19] J. Brennan, On the normality of commuting varieties of symmetric matrices, Preprint, 1993.

[20] J. Brennan, M. Vaz Pinto and W. V. Vasconcelos, The Jacobian module of a Lie algebra, *Trans. Amer. Math. Soc.* **321** (1990), 183–196.

[21] J. Brennan and W. V. Vasconcelos, Effective computation of the integral closure of a morphism, *J. Pure & Applied Algebra* **86** (1993), 125–134.

[22] H. Bresinsky, P. Schenzel and W. Vogel, On liaison, arithmetical Buchsbaum curves and monomial curves in $\mathbb{P}^3$, *J. Algebra* **86** (1984), 283–301.

[23] M. Brodmann, Asymptotic stability of $\text{Ass}(M/I^n M)$, *Proc. Amer. Math. Soc.* **74** (1979), 16–18.

[24] M. Brodmann, Rees rings and form rings of almost complete intersections, *Nagoya Math. J.* **88** (1982), 1–16.

[25] P. Brumatti, A. Simis and W. V. Vasconcelos, Normal Rees algebras, *J. Algebra* **112** (1988), 26–48.

[26] W. Bruns, The Eisenbud–Evans principal ideal theorem and determinantal ideals, *Proc. Amer. Math. Soc.* **83** (1981), 19–24.

[27] W. Bruns, Additions to the theory of algebras with straightening law, in *Commutative Algebra* (M. Hochster, C. Huneke and J. D. Sally, Eds.), MSRI Publications **15**, Springer–Verlag, Berlin–Heidelberg–New York, 1989, 111–138.

[28] W. Bruns, Algebras defined by powers of determinantal ideals, *J. Algebra* **142** (1991), 150–163.

[29] W. Bruns, Zur Konstruktion basischer Elemente, *Math. Z.* **172** (1980), 63–75.

[30] W. Bruns, The canonical module of a determinantal ring, in *Commutative Algebra*: Durham 1981 (R. Sharp, Ed.), London Math. Soc. Lecture Notes Series **72**, Cambridge University Press, Cambridge, 1982, 109–120.

[31] W. Bruns and J. Herzog, *Cohen–Macaulay Rings*, Cambridge University Press, to appear.

[32] W. Bruns, A. Kustin and M. Miller, The resolution of the generic residual intersection of a complete intersection, *J. Algebra* **128** (1990), 214–239.

[33] B. Buchberger, Gröbner bases: An algorithmic method in polynomial ideal theory, in *Recent Trends in Mathematical Systems Theory* (N.K. Bose, Ed.), D. Reidel, Dordrecht, 1985, 184–232.

[34] D. Buchsbaum and D. Eisenbud, What makes a complex exact?, *J. Algebra* **25** (1973), 259–268.

[35] D. Buchsbaum and D. Eisenbud, Some structure theorems for finite free resolutions, *Advances in Math.* **12** (1974), 84–139.

[36] D. Buchsbaum and D. Eisenbud, Algebraic structures for finite free resolutions, and some structure theorems for ideals of codimension 3, *American J. Math.* **99** (1977), 447–485.

[37] D. Buchsbaum and D. Eisenbud, Generic free resolutions and a family of generically perfect ideals, *Advances in Math.* **18** (1975), 245–301.

[38] R.-O. Buchweitz, *Contributions à la théorie des singularités*, Thèse, Université de Paris, 1981.

[39] R.-O. Buchweitz and B. Ulrich, Homological properties which are invariant under linkage, preprint.

[40] L. Burch, Codimension and analytic spread, *Proc. Camb. Phil. Soc.* **72** (1972), 369–373.

[41] L. Caniglia, *Complejidad de Algoritmos en Geometria Algebraica Computacional*, Ph.D. Thesis, Universidad de Buenos Aires, 1989.

[42] J. Carrig, The homological dimensions of symmetric algebras, *Trans. Amer. Math. Soc.* **236** (1978), 275–285.

[43] R. Chibloum, A. Micali and J.-P. Olivier, Sur la dimension des algèbres symétriques, Preprint, 1991.

[44] D. Costa, Sequences of linear type, *J. Algebra* **94** (1985), 256–263.

[45] D. Costa, L. Gallardo and J. Querré, On the distribution of prime elements in polynomial Krull domains, *Proc. Amer. Math. Soc.* **87** (1983), 41–43.

[46] D. Costa and J.L. Johnson, Inert extensions for Krull domains, *Proc. Amer. Math. Soc.* **59** (1976), 189–194.

[47] R. C. Cowsik, Symbolic powers and number of defining equations, in *Algebra and its Applications*, Lecture Notes in Pure & Applied Mathematics **91**, Marcel Dekker, New York, 1984, 13–14.

[48] R. C. Cowsik and M. V. Nori, On the fibers of blowing up, *J. Indian Math. Soc.* **40** (1976), 217–222.

[49] S. Cutkosky, A new characterization of rational surface singularities, *Invent. Math.* **102** (1990), 157–177.

[50] E. Davis, Ideals of the principal class, R–sequences and a certain monoidal transformation, *Pacific J. Math.* **20** (1967), 197–205.

[51] A. M. Dickenstein and C. Sessa, Duality methods for the membership problem, in *Effective Methods in Algebraic Geometry*, Progress in Mathematics **94**, Birkhäuser, Boston–Basel–Berlin, 1991, 89–103.

[52] P. Dubreil, *Quelques propriétés des variétés algèbriques*, Actualités Scientifiques et Industrielles **210**, Hermann, Paris, 1935.

[53] J. Eagon and D. G. Northcott, Ideals defined by matrices and a certain complex associated with them, *Proc. Royal Soc.* **269** (1962), 188–204.

[54] J. Eagon and M. Hochster, Cohen–Macaulay rings, invariant theory and the generic perfection of determinantal loci, *American J. Math.* **93** (1972), 1020–1058.

[55] D. Eisenbud, *Commutative Algebra with a view toward Algebraic Geometry*, in Preparation.

[56] D. Eisenbud, Homological algebra on a complete intersection with an application to group representations, *Trans. Amer. Math. Soc.* **260** (1980), 35–64.

[57] D. Eisenbud, Open Problems in Computational Algebraic Geometry, in *Computational Algebraic Geometry and Commutative Algebra*, Proceedings, Cortona 1991 (D. Eisenbud and L. Robbiano, Eds.), Cambridge University Press, to appear.

[58] D. Eisenbud and E. G. Evans, Every algebraic set in n–space is the intersection of n hypersurfaces, *Invent. Math.* **19** (1973), 107–112.

[59] D. Eisenbud and E. G. Evans, A generalized principal ideal theorem, *Nagoya Math. J.* **62** (1976), 41–53.

[60] D. Eisenbud and C. Huneke, Cohen–Macaulay Rees algebras and their specializations, *J. Algebra* **81** (1983), 202–224.

[61] D. Eisenbud and C. Huneke, Ideals with a regular sequence as syzygy, in *Appendix* to: A. Beauville, Sur les hypersurfaces dont les sections hyperplanes sont à module constant, in The Grothendieck Festschrift, vol. I (P. Cartier *et al*, Eds.), Birkhäuser, Boston-Basel-Berlin, 1990, 121–132.

[62] D. Eisenbud, C. Huneke and W. V. Vasconcelos, Direct methods for primary decomposition, *Invent. Math.* **110** (1992), 207–235.

[63] D. Eisenbud and B. Sturmfels, Finding sparse systems of parameters, *J. Pure & Applied Algebra*, to appear.

[64] S. Eliahou, A problem about polynomial ideals, *Contemporary Mathematics* **58** (1986), 107–120.

[65] E. G. Evans and P. Griffith, The syzygy problem, *Annals of Math.* **114** (1981), 323–333.

[66] E. G. Evans and P. Griffith, *Syzygies*, London Math. Soc. Lecture Notes **106**, Cambridge University Press, Cambridge, 1985.

[67] G. Faltings, Ein Kriterium für Vollständige Durchschnitte, *Invent. Math.* **62** (1981), 393–401.

[68] M. Fiorentini, On relative regular sequences, *J. Algebra* **18** (1971), 384–389.

[69] H. Fitting, Die Determinantenideale Moduls, *Jahresbericht DMV* **46** (1936), 192–228.

[70] H. Flenner, Die Sätze von Bertini für lokale Ringe, *Math. Annalen* **229** (1977), 97–111.

[71] O. Forster, Über die Anzahl der Erzeugenden eines Ideals in einem Noetherschen Ring, *Math. Z.* **84** (1964), 80–87.

[72] R. M. Fossum, *The divisor class group of a Krull domain*, Ergebnisse der Mathematik und ihrer Grenzgebiete, B. **74**, Springer–Verlag, Berlin–Heidelberg–New York, 1973.

[73] F. Gaeta, Détermination de la chaine syzygétique des idéaux matriciels parfaits et son application à la postulation de leurs variétés algébriques associées, *C. R. Acad. Sci. Paris* **234** (1952), 1833–1835.

[74] A. Galligo and C. Traverso, Practical determination of the dimension of an algebraic variety, in *Computers in Mathematics* (E. Kaltofen and S. M. Watt, Eds.), Springer–Verlag, Berlin–Heidelberg–New York, 1989, 46–52.

[75] G. Gallo and B. Mishra, Efficient algorithms and bounds for Wu–Ritt characteristic sets, in *Effective Methods in Algebraic Geometry*, Progress in Mathematics **94**, Birkhäuser, Boston–Basel–Berlin, 1991, 119–142.

[76] F. R. Gantmacher, *Matrix Theory*, vol. II, Chelsea, New York, 1959.

[77] M. Gerstenhaber, On dominance and varieties of commuting matrices, *Annals of Math.* **73** (1961), 324–348.

[78] P. Gianni and T. Mora, Algebraic solution of systems of polynomial equations using Gröbner bases, Proceedings 5th AAEEC, Lecture Notes in Computer Science **356**, Springer–Verlag, Berlin–Heidelberg–New York, 1989, 247–257.

[79] P. Gianni, B. Trager and G. Zacharias, Gröbner bases and primary decomposition of polynomial ideals, *J. Symbolic Computation* **6** (1988), 149–167.

[80] R. Gilmer, *Multiplicative Ideal Theory*, Marcel Dekker, New York, 1972.

[81] M. Giusti and J. Heintz, Algorithmes - disons rapides - pour la décomposition d'une variété algébrique en composantes irréductibles et équidimensionelles, in *Effective Methods in Algebraic Geometry*, Progress in Mathematics **94**, Birkhäuser, Boston–Basel–Berlin, 1991, 169–194.

[82] S. Glaz, Regular symmetric algebras, *J. Algebra* **112** (1988), 129–138.

[83] S. Glaz, *Commutative Coherent Rings*, Lecture Notes in Mathematics **1371**, Springer–Verlag, Berlin–Heidelberg–New York, 1989.

[84] R. Goodman and N. R. Wallach, Classical and quantum–mechanical systems of Toda lattice type.I, *Commun. Math. Phys.* **83** (1982), 355–386.

[85] S. Goto, The divisor class group of a certain Krull domain, *J. Math. Kyoto Univ.* **17** (1977), 47–50.

[86] S. Goto and S. Huckaba, On graded rings associated to analytic deviation one ideals, *American J. Math.*, to appear.

[87] S. Goto, K. Nishida and K. Watanabe, Non–Noetherian symbolic blow–ups for space monomial curves, Preprint, 1991.

[88] S. Goto and Y. Shimoda, On the Rees algebras of Cohen–Macaulay local rings, Lecture Notes in Pure & Applied Mathematics **68**, Marcel Dekker, New York, 1979, 201–231.

[89] S. Goto and K. Watanabe, On graded rings I, *J. Math. Soc. Japan* **30** (1978), 179–213.

[90] H.-G. Gräbe, Moduln Über Streckungsringen, *Results in Mathematics* **15** (1989), 202–220.

[91] A. Grothendieck, Théorèmes de dualité pour les modules cohérents, Séminaire Bourbaki, **149**, Secr. Math. I.H.P., Paris, 1957.

[92] A. Grothendieck, *Local Cohomology*, (Notes by R. Hartshorne) Lecture Notes in Mathematics **41**, Springer–Verlag, Berlin–Heidelberg–New York, 1967.

[93] U. Grothe, M. Herrmann and U. Orbanz, Graded rings associated to equimultiple ideals, *Math. Z.* **186** (1984), 531–556.

[94] A. Guerrieri, On the depth of the associated graded ring, *Proceedings Amer. Math. Soc.* , to appear.

[95] T. H. Gulliksen and G. Levin, *Homology of Local Rings*, Queen's Papers in Pure and Applied Math. **20**, Queen's University, Kingston, 1969.

[96] T. H. Gulliksen and O. Negård, Un complexe résolvant pour certains idéaux déterminantiels, *C. R. Acad. Sci. Paris* **274** (1972), 16–18.

[97] R. M. Guralnick, A note on pairs of matrices with rank one commutators, *Linear and Multilinear Algebra* **8** (1979), 97-99.

[98] R. M. Guralnick, A note on commuting pairs of matrices, *Linear and Multilinear Algebra* **31** (1992), 1–4.

[99] R. Hartshorne, *Algebraic Geometry*, Springer–Verlag, Berlin–Heidelberg–New York, 1977.

[100] R. Hartshorne, Complete intersections and connectedness, *American J. Math.* **84** (1962), 497–508.

[101] R. Hartshorne and A. Ogus, On the factoriality of local rings of small embedding codimension, *Comm. Algebra* **1** (1974), 415–437.

[102] G. Hermann, Die Frage der endlich vielen Schritte in der Theorie der Polynomideale, *Math. Annalen* **95** (1926), 736–788.

[103] M. Herrmann, S. Ikeda and U. Orbanz, *Equimultiplicity and Blowing up*, Springer–Verlag, Berlin–Heidelberg–New York, 1988.

[104] M. Herrmann, B. Moonen and O. Villamayor, Ideals of linear type and some variants, in *The Curves Seminar* at Queen's University, Vol. VI, Queen's Papers in Pure and Applied Math. **83** (1989).

[105] M. Herrmann, J. Ribbe and S. Zarzuela, On the Gorenstein property of Rees and form rings of powers of ideals, *Trans. Amer. Math. Soc.*, to appear.

[106] J. Herzog, *Komplexe, Auflösungen und Dualität in der lokalen Algebra*, Habilitationsschrift, Regensburg Universität, 1974.

[107] J. Herzog, Ein Cohen–Macaulay Kriterium mit Anwendungen auf den Konormalenmodul und Differentialmodul, *Math. Z.* **163** (1978), 149–162.

[108] J. Herzog, Certain complexes associated to a sequence and a matrix, *Manuscripta Math.* **12** (1974), 217–247.

[109] J. Herzog, Generators and relations of abelian semigroups and semigroup rings, *Manuscripta Math.* **3** (1970), 153–193.

[110] J. Herzog, Homological properties of the module of differentials, Atas VI Escola de Algebra, Soc. Brasileira de Matemática, 1981, 33–64.

[111] J. Herzog, Deformationen von Cohen–Macaulay Algebren, *J. reine angew. Math.* **318** (1980), 83–105.

[112] J. Herzog, A homological approach to symbolic powers, in *Commutative Algebra*, Proceedings, Salvador 1988 (W. Bruns and A. Simis, Eds.), Lecture Notes in Mathematics **1430**, Springer–Verlag, Berlin–Heidelberg–New York, 1990, 32–46.

[113] J. Herzog and E. Kunz, *Der kanonische Modul eines Cohen–Macaulay Rings*, Lecture Notes in Mathematics **238**, Springer–Verlag, Berlin–Heidelberg–New York, 1971.

[114] J. Herzog, A. Simis and W. V. Vasconcelos, Approximation complexes of blowing–up rings, *J. Algebra* **74** (1982), 466–493.

[115] J. Herzog, A. Simis and W. V. Vasconcelos, Approximation complexes of blowing-up rings.II, *J. Algebra* **82** (1983), 53–83.

[116] J. Herzog, A. Simis and W. V. Vasconcelos, Koszul homology and blowing–up rings, in *Commutative Algebra*, Proceedings: Trento 1981 (S. Greco and G. Valla, Eds.), Lecture Notes in Pure and Applied Math. **84**, Marcel Dekker, New York, 1983, 79–169.

[117] J. Herzog, A. Simis and W. V. Vasconcelos, On the arithmetic and homology of algebras of linear type, *Trans. Amer. Math. Soc.* **283** (1984), 661–683.

[118] J. Herzog, A. Simis and W. V. Vasconcelos, On the canonical module of the Rees algebra and the associated graded ring of an ideal, *J. Algebra* **105** (1987), 285–302.

[119] J. Herzog, A. Simis and W. V. Vasconcelos, Arithmetic of normal Rees algebras, *J. Algebra* **143** (1991), 269–294.

[120] J. Herzog and N. V. Trung, Gröbner bases and multiplicity of determinantal and Pfaffian ideals, *Advances in Math.* **96** (1992), 1–37.

[121] J. Herzog, N. V. Trung and B. Ulrich, On the multiplicity of blow-up rings of ideals generated by d–sequences, Preprint, 1991.

[122] J. Herzog and B. Ulrich, Self–linked curve singularities, *Nagoya Math. J.* **120** (1990), 129–153.

[123] J. Herzog and W. V. Vasconcelos, On the divisor class group of Rees algebras, *J. Algebra* **93** (1985), 182–188.

[124] J. Herzog, W. V. Vasconcelos and R. Villarreal, Ideals with sliding depth, *Nagoya Math. J.* **99** (1985), 159–172.

[125] M. Hochster, Criteria for the equality of ordinary and symbolic powers of primes, *Math. Z.* **133** (1973), 53–65.

[126] M. Hochster, Properties of noetherian rings stable under general grade reduction, *Arch. Math.* **24** (1973), 393–396.

[127] M. Hochster, The Zariski–Lipman conjecture in the graded case, *J. Algebra* **47** (1977), 411–424.

[128] M. Hochster and J. Roberts, Rings of invariants are Cohen–Macaulay, *Advances in Math.* **13** (1974), 115–175.

[129] S. Huckaba, On complete d-sequences and the defining ideals of Rees algebras, *Math. Proc. Cambridge Phil. Soc.* **106** (1989), 445–458.

[130] S. Huckaba and C. Huneke, Powers of ideals having small analytic deviation, *American J. Math.* **114** (1992), 367–403.

[131] S. Huckaba and C. Huneke, Rees algebras of ideals having small analytic deviation, *Trans. Amer. Math. Soc.*, to appear.

[132] S. Huckaba and T. Marley, Depth properties of Rees algebras and associated graded rings, *J. Algebra* **156** (1993), 259–271.

[133] S. Huckaba and T. Marley, Depth formulas for certain graded rings associated to an ideal, *Nagoya Math. J.*, to appear.

[134] K. Hulek, A remark on certain matrix varieties, *Linear and Multilinear Algebra* **10** (1981), 169–172.

[135] J. E. Humphreys, *Introduction to Lie Algebras and Representation Theory*, Springer–Verlag, Berlin–Heidelberg–New York, 1972.

[136] C. Huneke, Almost complete intersections and factorial rings, *J. Algebra* **71** (1981), 179–188.

[137] C. Huneke, The theory of d–sequences and powers of ideals, *Advances in Math.* **46** (1982), 249–279.

[138] C. Huneke, On the associated graded ring of an ideal, *Illinois J. Math.* **26** (1982), 121-137.

[139] C. Huneke, On the symmetric algebra of a module, *J. Algebra* **69** (1981), 113–119.

[140] C. Huneke, Determinantal ideals of linear type, *Arch. Math.* **47** (1986), 324–329.

[141] C. Huneke, On the symmetric and Rees algebras of an ideal generated by a d–sequence, *J. Algebra* **62** (1980), 268–275.

[142] C. Huneke, Linkage and Koszul homology of ideals, *American J. Math.* **104** (1982), 1043–1062.

[143] C. Huneke, Strongly Cohen–Macaulay schemes and residual intersections, *Trans. Amer. Math. Soc.* **277** (1983), 739–763.

[144] C. Huneke, Hilbert functions and symbolic powers, *Michigan Math. J.* **34** (1987), 293–318.

[145] C. Huneke, The primary components of and integral closures of ideals in 3–dimensional regular local rings, *Math. Annalen* **275** (1986), 617–635.

[146] C. Huneke, On the finite generation of symbolic blow–ups, *Math. Z.* **179** (1982), 465–472.

[147] C. Huneke, Criteria for complete intersections, *J. London Math. Soc.* **32** (1985), 19–30.

[148] C. Huneke, Symbolic powers of prime ideals and special graded algebras, *Comm. Algebra* **9** (1981), 339–366.

[149] C. Huneke, Uniform bounds in Noetherian rings, *Invent. Math.* **107** (1992), 203–223.

[150] C. Huneke and M. E. Rossi, The dimension and components of symmetric algebras, *J. Algebra* **98** (1986), 200–210.

[151] C. Huneke and J. D. Sally, Birational extensions in dimension two and integrally closed ideals, *J. Algebra* **115** (1988), 481–500.

[152] C. Huneke, A. Simis and W. V. Vasconcelos, Reduced normal cones are domains, Amer. Math. Soc., Contemporary Mathematics **88** (1989), 95–101.

[153] C. Huneke and B. Ulrich, Residual intersections, J. reine angew. Math. **390** (1988), 1–20.

[154] C. Huneke and B. Ulrich, Divisor class groups and deformations, American J. Math. **107** (1985), 1265–1303.

[155] C. Huneke and B. Ulrich, Algebraic linkage, Duke Math. J. **56** (1988), 415–429.

[156] C. Huneke and B. Ulrich, The structure of linkage, *Annals Math.* **126** (1987), 277–334.

[157] C. Huneke, B. Ulrich and W. V. Vasconcelos, On the structure of certain normal ideals, *Compositio Math.* **84** (1992), 25–42.

[158] S. Ikeda and N. V. Trung, When is the Rees algebra Cohen–Macaulay?, *Comm. Algebra* **17** (1989), 2893–2922.

[159] N. Jacobson, *Basic Algebra* I, W. H. Freeman, San Francisco, 1974.

[160] M. Johnson, Ph.D. Thesis, Michigan State University, 1994.

[161] T. Józefiak, Ideals generated by the minors of a symmetric matrix, *Comment. Math. Helvetici* **53** (1978), 595–607.

[162] T. Józefiak and P. Pragacz, Ideals generated by Pfaffians, *J. Algebra* **61** (1979), 189–198.

[163] E. Kaltofen, Polynomial factorization 1987–1991, Proceedings Latin'92, Lecture Notes in Computer Science **583**, Springer–Verlag, Berlin–Heidelberg–New York, 1992, 294–313.

[164] A. Kandri-Rody and D. Kapur, Computing a Gröbner basis of a polynomial ideal over an Euclidean domain, *J. Symbolic Computation* **6** (1988), 37–57.

[165] I. Kaplansky, *Commutative Rings*, University of Chicago Press, Chicago, 1974.

[166] H. Kleppe and D. Laksov, The algebraic structure and deformation of Pfaffian schemes, *J. Algebra* **64** (1980), 167–189.

[167] G. Knödel, P. Schenzel and R. Zonsarow, Explicit computations on symbolic powers of monomial curves in affine space, *Comm. Algebra* **20** (1992), 2113–2126.

[168] B. Kostant, Lie group representations on polynomial rings, *American J. Math.* **85** (1963), 327–404.

[169] B. V. Kotsev, Determinantal ideals of linear type of a generic symmetric matrix, *J. Algebra* **139** (1991), 488–504.

[170] T. Krick and A. Logar, An algorithm for the computation of the radical of an ideal in the ring of polynomials, Proceedings 9th AAEEC, Lecture Notes in Computer Science **539**, Springer–Verlag, Berlin–Heidelberg–New York, 1991, 195–205.

[171] M. Kühl, On the symmetric algebra of an ideal, *Manuscripta Math.* **37** (1982), 49–60.

[172] E. Kunz, The conormal module of an almost complete intersection, *Proc. Amer. Math. Soc.* **73** (1979), 15–21.

[173] E. Kunz, *Kähler Differentials*, Vieweg, Wiesbaden, 1986.

[174] A. Kustin, The minimal free resolution of the Huneke–Ulrich deviation two Gorenstein ideals, *J. Algebra* **100** (1986), 265–304.

[175] A. Kustin, M. Miller and B. Ulrich, Generating a residual intersection, *J. Algebra* **146** (1992), 335–384.

[176] Y. N. Lakshman and D. Lazard, On the complexity of zero–dimensional algebraic systems, in *Effective Methods in Algebraic Geometry*, Progress in Mathematics **94**, Birkhäuser, Boston–Basel–Berlin, 1991, 217–225.

[177] D. Lazard, Résolution des systèmes d'equations algébriques, *Theoretical Comp. Sci.* **15** (1981), 77-110.

[178] K. Lebelt, Freie Auflösungen äu*β*erer Potenzen, *Manuscripta Math.* **21** (1977), 341–355.

[179] K. Lebelt, Zur homologischen Dimension äußerer Potenzen von Moduln, *Arch. Math.* **26** (1975), 596–501.

[180] G. Levin, Personal Communication, 1990.

[181] S. Lichtenbaum, On the vanishing of Tor in regular local rings, *Illinois J. Math.* **10** (1966), 220–226.

[182] J. Lipman, Free derivation modules on algebraic varieties, *American J. Math.* **87** (1965), 874–898.

[183] J. Lipman, On the Jacobian ideal of the module of differentials, *Proc. Amer. Math. Soc.* **21** (1969), 422–426.

[184] J. Lipman, Rational singularities with applications to algebraic surfaces and unique factorization, *Inst. Hautes Études Sci. Publ. Math.* **36** (1969), 195–280.

[185] J. Lipman and A. Sathaye, Jacobian ideals and a theorem of Briançon–Skoda, *Michigan Math. J.* **28** (1981), 199–222.

[186] J. Lipman and B. Teissier, Pseudo–rational local rings and a theorem of Briançon–Skoda about integral closures of ideals, *Michigan Math. J.* **28** (1981), 97–116.

[187] A. Logar, A computational proof of the Noether normalization lemma, Proceedings 6th AAEEC, Lecture Notes in Computer Science **357**, Springer–Verlag, Berlin–Heidelberg–New York, 1989, 259–273.

[188] G. Lyubeznik, A new explicit finite free resolution of ideals generated by monomials in an R–sequence, *J. Pure & Applied Algebra* **51** (1988), 193–195.

[189] F. S. Macaulay, Some properties of enumeration in the theory of modular systems, *Proc. London Math. Soc.* **26** (1927), 531–555.

[190] R. MacRae, On an application of the Fitting invariants, *J. Algebra* **2** (1965), 153–169.

[191] T. Marley, The coefficients of the Hilbert polynomial and the reduction number of an ideal, *J. London Math. Soc.* **40** (1989), 1–8.

[192] H. Matsumura, *Commutative Algebra*, Benjamin/Cummings, Reading, Massachusetts, 1980.

[193] H. Matsumura, *Commutative Ring Theory*, Cambridge University Press, Cambridge, 1986.

[194] T. Matsuoka, On almost complete intersections, *Manuscripta Math.* **21** (1977), 329–340.

[195] E. Mayr and A. Meyer, The complexity of the word problem for commutative semigroups and polynomial ideals, *Advances in Math.* **46** (1982), 305–329.

[196] S. McAdam, *Asymptotic Prime Divisors*, Lecture Notes in Mathematics **1023**, Springer–Verlag, Berlin–Heidelberg–New York, 1983.

[197] S. McAdam and P. Eakin, The asymptotic ass, *J. Algebra* **61** (1979), 71–81.

[198] A. Micali, Sur les algèbres universalles, *Annales Inst. Fourier* **14** (1964), 33–88.

[199] A. Micali, P. Salmon and P. Samuel, Integrité et factorialité des algèbres symétriques, Atas do IV Colóquio Brasileiro de Matemática, SBM, (1965), 61–76.

[200] H. M. Möller, Computing syzygies à la Gauss–Jordan, in *Effective Methods in Algebraic Geometry*, Progress in Mathematics **94**, Birkhäuser, Boston–Basel–Berlin, 1991, 335–345.

[201] H. M. Möller and F. Mora, New constructive methods in classical ideal theory, *J. Algebra* **100** (1986), 138–178.

[202] H. M. Möller and F. Mora, The computation of Hilbert functions, Lecture Notes Computer Science **162**, Springer–Verlag, Berlin–Heidelberg–New York, 1983, 157–167.

[203] T. Mora and L. Robbiano, The Gröbner fan of an ideal, *J. Symbolic Computation* **6** (1988), 183–208.

[204] M. Morales, Noetherian symbolic blow-ups, J. Algebra **140** (1991), 12–25.

[205] M. Morales and A. Simis, The symbolic powers of monomial curves in $\mathbb{P}^3$ lying on a quadric surface, *Comm. Algebra* **20** (1992), 1109–1121.

[206] M. Morales and A. Simis, Arithmetically Cohen–Macaulay monomial curves in $\mathbb{P}^3$, *Comm. Algebra* **21** (1993), 951–961.

[207] T. Motzkin and O. Taussky, Pairs of matrices with property L II, *Trans. Amer. Math. Soc.* **80** (1955), 387–401.

[208] M. Nagata, *Local Rings*, Interscience, New York, 1962.

[209] M. Nagata, *Lectures on the Fourteenth Problem of Hilbert*, Tata Institute, Bombay, 1964.

[210] M. Narita, A note on the coefficients of Hilbert characteristic functions in semi–regular local rings, *Proc. Camb. Phil. Soc.* **59** (1963), 269–275.

[211] M. G. Neubauer, The variety of pairs of matrices with rank(AB-BA) ≤ 1, *Proc. Amer. Math. Soc.* **105** (1989), 787–792.

[212] E. Noether, Idealtheorie in Ringbereichen, *Math. Annalen* **83** (1921), 24–66.

[213] S. Noh and W. V. Vasconcelos, The S_2–closure of a Rees algebra, *Results in Mathematics* **23** (1993), 149–162.

[214] D. G. Northcott, A homological investigation of a certain residual ideal, *Math. Annalen* **150** (1963), 99–110.

[215] D. G. Northcott, A note on the coefficients of the abstract Hilbert function, *J. London Math. Soc.* **35** (1960), 209–214.

[216] D. G. Northcott and D. Rees, Reductions of ideals in local rings, *Proc. Camb. Phil. Soc.* **50** (1954), 145–158.

[217] A. Ooishi, δ-genera and sectional genera of commutative rings, *Hiroshima Math. J.* **17** (1987), 361–372.

[218] E. Platte, Ein elementarer Beweis des Zariski–Lipman–Problems für graduierte analytische Algebren, *Arch. Math.* **31** (1978), 143–145.

[219] D. I. Panyushev, The Jacobian modules of a representation of a Lie algebra and geometry of commuting varieties, Preprint, 1993.

[220] C. Peskine, Une généralisation du main theorem de Zariski, *Bull. Sc. Math.* (2) **90** (1966), 119–127.

[221] C. Peskine and L. Szpiro, Liaison des variétés algébriques, *Invent. Math.* **26** (1974), 271–302.

[222] K. N. Raghavan, *Uniform Annhilation of Local Cohomology and Powers of Ideals Generated by Quadratic Sequences*, Ph.D. Thesis, Purdue University, 1991.

[223] K. N. Raghavan, A simple proof that ideals generated by d–sequences are of linear type, *Comm. Algebra* **19** (1991), 2827–2831.

[224] L. J. Ratliff, Jr., On prime divisors of I^n, n large, *Michigan Math. J.* **23** (1976), 337–352.

[225] L. J. Ratliff, Jr. and D. E. Rush, Two notes on reductions of ideals, *Indiana Univ. Math. J.* **27** (1978), 929–934.

[226] M. Raynaud, *Anneaux Locaux Henséliens*, Lecture Notes in Mathematics **169**, Springer–Verlag, Berlin–Heidelberg–New York, 1970.

[227] D. Rees, On a problem of Zariski, *Illinois J. Math.* **2** (1958), 145–149.

[228] R. W. Richardson, Commuting varieties of semisimple Lie algebras and algebraic groups, *Compositio Math.* **38** (1979), 311–327.

[229] L. Robbiano, On the theory of graded structures, *J. Symbolic Computation* **2** (1986), 139–170.

[230] L. Robbiano, Introduction to the theory of Gröbner bases, Queen's Papers in Pure and Applied Mathematics, 1988.

[231] L. Robbiano, Bounds for degrees and number of elements in Groöbner bases, Proceedings 8th AAEEC, Lecture Notes in Computer Science **508**, Springer–Verlag, Berlin–Heidelberg–New York, 1991, 292–303.

[232] L. Robbiano, Term orderings on the polynomial ring, Proceedings EUROCAL 85, Lecture Notes in Computer Science **204**, Springer–Verlag, Berlin–Heidelberg–New York, 1985, 513–517.

[233] P. Roberts, A prime ideal in a polynomial ring whose symbolic blow-up is not Noetherian, *Proc. Amer. Math. Soc.* **94** (1985), 589–592.

[234] P. Roberts, An infinitely generated symbolic blow–up in a power series ring and a new counterexample to Hilbert's Fourteenth Problem, *J. Algebra* **132** (1990), 461–473.

[235] M. E. Rossi, A note on symmetric algebras which are Gorenstein, *Comm. Algebra* **11** (1983), 2575–2591.

[236] E. Rutman, Gröbner bases and primary decompositions of modules, *J. Symbolic Computation*, **14** (1992), 483–503.

[237] J. D. Sally, On the associated graded ring of a local Cohen–Macaulay ring, *J. Math. Kyoto U.* **17** (1977), 19–21.

[238] J. D. Sally, Cohen–Macaulay local rings of maximal embedding dimension, *J. Algebra* **56** (1979), 168–183.

[239] J. D. Sally, Tangent cones at Gorenstein singularities, *Compositio Math.* **40** (1980), 167–175.

[240] J. D. Sally, Hilbert coefficients and reductionn number 2, *J. Alg. Geo. and Sing.* **1** (1992), 325–333.

[241] P. Samuel, Anneaux gradués factoriels et modules réflexifs, *Bull. Soc. Math. France* **92** (1964), 237–249.

[242] H. Sanders, *Approximationskomplexe und Koszulhomologie*, Ph.D. Thesis, Essen Universität, 1985.

[243] G. Scheja and U. Storch, Über Spurfunktionen bei vollständigen Durchschnitten, *J. reine angew. Math.* **278** (1975), 174–190.

[244] P. Schenzel, Examples of Noetherian symbolic blowup rings, *Rev. Roumaine Math. Pures Appl.* **33** (1988), 375–383.

[245] P. Schenzel, Symbolic powers of prime ideals and their topology, *Proc. Amer. Math. Soc.* **93** (1985), 15–20.

[246] P. Schenzel, Examples of Gorenstein domains and symbolic powers of monomial space curves, *J. Pure & Applied Algebra* **71** (1991), 297–311.

[247] F. Schreyer, *Die Berechnung von Syzygien mit dem verallgemeinerten Weierstrasschen Divisionsatz*, Diplomarbeit, U. Hamburg, 1980.

[248] A. Seidenberg, Constructions in algebra, *Trans. Amer. Math. Soc.* **197** (1974), 273–313.

[249] A. Seidenberg, Construction of the integral closure of a finite integral domain II, *Proc. Amer. Math. Soc.* **52** (1975), 368–372.

[250] A. Seidenberg, On the Lasker–Noether decomposition theorem, *American J. Math.* **106** (1984), 611–638.

[251] D. Shannon and M. Sweedler, Using Gröbner bases to determine algebra membership, split surjective algebra homomorphisms determine birational equivalence, *J. Symbolic Computation* **6** (1986), 267–273.

[252] A. Simis, Multiplicities and Betti numbers of homogeneous ideals, in *Space Curves*, Proceedings, Rocca di Papa (Roma 1985), Lecture Notes in Mathematics **1266**, Springer–Verlag, Berlin–Heidelberg–New York, 1987, 232–250.

[253] A. Simis, Topics in Rees algebras of special ideals, in *Commutative Algebra*, Proceedings, Salvador 1988 (W. Bruns and A. Simis, Eds.), Lecture Notes in Mathematics **1430**, Springer–Verlag, Berlin–Heidelberg–New York, 1990, 98–114.

[254] A. Simis, *Selected Topics in Commutative Algebra*, Lecture Notes, IX Escuela Latinoamericana de Matematica, Santiago, Chile, 1988.

[255] A. Simis, Koszul homology and its syzygy–theoretic part, *J. Algebra* **55** (1978), 28–42.

[256] A. Simis and N. V. Trung, Divisor class group of ordinary and symbolic blow–ups, *Math. Z.* **198** (1988), 479–491.

[257] A. Simis, B. Ulrich and W. V. Vasconcelos, Jacobian dual fibrations, *American J. Math.* **115** (1993), 47–75.

[258] A. Simis, B. Ulrich and W. V. Vasconcelos, Tangent star cones, in Preparation.

[259] A. Simis and W. V. Vasconcelos, The syzygies of the conormal module, *American J. Math.* **103** (1981), 203–224.

[260] A. Simis and W. V. Vasconcelos, On the dimension and integrality of symmetric algebras, *Math. Z.* **177** (1981), 341–358.

[261] A. Simis and W. V. Vasconcelos, The Krull dimension and integrality of symmetric algebras, *Manuscripta Math.* **61** (1988), 63–78.

[262] A. Simis, W. V. Vasconcelos and R. Villarreal, On the ideal theory of graphs, *J. Algebra*, to appear.

[263] D. Spear, A constructive approach to commutative ring theory, Proc. 1977 MACSYMA User's Conference, 369–376.

[264] R. P. Stanley, Hilbert functions of graded algebras, *Advances in Math.* **28** (1978), 57–83.

[265] G. Stolzenberg, Constructive normalization of an algebraic variety, *Bull. Amer. Math. Soc.* **74** (1968), 595–599.

[266] U. Storch, Bemerkung zu einem Satz von M. Kneser, *Arch. Math.* **23** (1972), 403–404.

[267] E. Strickland, On the conormal bundle of the determinantal variety, *J. Algebra* **75** (1982), 523–537.

[268] J. Stückrad and W. Vogel, *Buchsbaum Rings and Applications*, Springer–Verlag, Berlin–Heidelberg–New York, 1986.

[269] B. Sturmfels, Gröbner bases and Stanley decompositions of determinantal rings, *Math. Z.* **205** (1990), 137–144.

[270] B. Sturmfels, Sparse elimination theory, in *Computational Algebraic Geometry and Commutative Algebra*, Proceedings, Cortona 1991 (D. Eisenbud and L. Robbiano, Eds.), Cambridge University Press, to appear.

[271] D. Taylor, *Ideals generated by monomials in an R–sequence*, Ph.D. Thesis, University of Chicago, 1966.

[272] M. Tomari and K. Watanabe, Filtered rings, filtered blowing–ups and normal two–dimensional singularities with "star–shaped" resolution, Publ. RIMS. Kyoto Univ. **25** (1989), 681–740.

[273] C. Traverso, A study on algebraic algorithms: the normalization, Rend. Semin. Mat. Univ. Politec. Torino (fasc. spec.) *Algebraic Varieties of Small Dimension*, 1986, 111–130.

[274] N. V. Trung, Reduction exponent and degree bound for the defining equations of graded rings, *Proc. Amer. Math. Soc.* **101** (1987), 229–236.

[275] B. Ulrich, Sums of linked ideals, *Trans. Amer. Math. Soc.* **318** (1990), 1–42.

[276] B. Ulrich, *Linkage Theory and the Homology of Noetherian Rings*, Lecture notes, 1992.

[277] B. Ulrich, Remarks on residual intersections, in *Free Resolutions in Commutative Algebra and Algebraic Geometry*, Proceedings, Sundance 1990 (D. Eisenbud and C. Huneke, Eds.), Research Notes in Mathematics **2**, Jones and Bartlett Publishers, Boston–London, 1992, 133–138.

[278] B. Ulrich, Artin–Nagata properties and reductions of ideals, *Contemporary Mathematics*, to appear.

[279] B. Ulrich and W. V. Vasconcelos, The equations of Rees algebras of ideals with linear presentation, *Math. Z.*, to appear.

[280] P. Valabrega and G. Valla, Form rings and regular sequences, *Nagoya Math. J.* **72** (1978), 91–101.

[281] G. Valla, On the symmetric and Rees algebras of an ideal, *Manuscripta Math.* **30** (1980), 239–255.

[282] G. Valla, On set-theoretic complete intersections, in *Complete Intersections* (S. Greco and R. Strano, Eds.), Lecture Notes in Mathematics **1092**, Springer–Verlag, Berlin–Heidelberg–New York, 1984, 85–101.

[283] G. Valla, Certain graded algebras are always Cohen–Macaulay, *J. Algebra* **42** (1976), 537–548.

[284] G. Valla, On form rings which are Cohen–Macaulay, *J. Algebra* **58** (1979), 247–250.

[285] W. V. Vasconcelos, Ideals generated by R–sequences, *J. Algebra* **6** (1967), 309–316.

[286] W. V. Vasconcelos, The complete intersection locus of certain ideals, *J. Pure & Applied Algebra* **38** (1985), 367–378.

[287] W. V. Vasconcelos, What is a prime ideal?, Atas IX Escola de Algebra, IMPA, Rio de Janeiro, 1986, 141–149.

[288] W. V. Vasconcelos, On linear complete intersections, *J. Algebra* **111** (1987), 306–315.

[289] W. V. Vasconcelos, Koszul homology and the structure of low codimension Cohen–Macaulay ideals, *Trans. Amer. Math. Soc.* **301** (1987), 591–613.

[290] W. V. Vasconcelos, Symmetric algebras and factoriality, in *Commutative Algebra* (M. Hochster, C. Huneke and J. D. Sally, Eds.), MSRI Publications **15**, Springer–Verlag, Berlin–Heidelberg–New York, 1989, 467–496.

[291] W. V. Vasconcelos, On the structure of certain ideal transforms, *Math. Z.* **198** (1988), 435–448.

[292] W. V. Vasconcelos, Modules of differentials of symmetric algebras, *Arch. Math.* **56** (1990), 436–442.

[293] W. V. Vasconcelos, Symmetric Algebras, in *Commutative Algebra*, Proceedings, Salvador 1988 (W. Bruns and A. Simis, Eds.), Lecture Notes in Mathematics **1430**, Springer–Verlag, Berlin–Heidelberg–New York, 1990, 115–160.

[294] W. V. Vasconcelos, On the equations of Rees algebras, *J. reine angew. Math.* **418** (1991), 189–218.

[295] W. V. Vasconcelos, Computing the integral closure of an affine domain, *Proc. Amer. Math. Soc.* **113** (1991), 633–638.

[296] W. V. Vasconcelos, Constructions in commutative algebra, in *Computational Algebraic Geometry and Commutative Algebra*, Proceedings, Cortona 1991 (D. Eisenbud and L. Robbiano, Eds.), Cambridge University Press, to appear.

[297] W. V. Vasconcelos, Jacobian matrices and constructions in algebra, Proceedings 9th AAEEC, Lecture Notes in Computer Science **539**, Springer–Verlag, Berlin–Heidelberg–New York, 1991, 48–64.

[298] W. V. Vasconcelos, Hilbert functions, analytic spread and Koszul homology, *Contemporary Mathematics*, to appear.

[299] W. V. Vasconcelos, The top of a system of equations, *Bol. Soc. Mat. Mex.* (issue dedicated to Jose Adem), to appear.

[300] W. V. Vasconcelos and R. Villarreal, On Gorenstein ideals of codimension four, *Proc. Amer. Math. Soc.* **98** (1986), 205–210.

[301] A.T. Vazquez, Rational desingularization of a curve defined over a finite field, in *Number Theory, New York Seminar 1989–1990* (D. V. Chudnovsky *et al.* Eds.), Springer–Verlag, Berlin–Heidelberg–New York, 1991, 229–250.

[302] U. Vetter, Zu einem Satz von G. Trautmann über den Rang gewisser kohärenter analytischer Moduln, *Arch. Math.* **24** (1973), 158–161.

[303] O. Villamayor (h), On class groups and normality of Rees rings, *Comm. Algebra* **17** (1989), 1607–1625.

[304] R. Villarreal, Rees algebras and Koszul homology, *J. Algebra* **119** (1988), 83–104.

[305] R. Villarreal, Cohen–Macaulay graphs, *Manuscripta Math.* **66** (1990), 277–293.

[306] J. Watanabe, A note on Gorenstein rings of embedding codimension three, *Nagoya Math. J.* **50** (1973), 227–232.

[307] J. Weyman, Resolutions of the exterior and symmetric powers of a module, *J. Algebra* **58** (1979), 333–341.

[308] H. Wiebe, Über homologische Invarianten lokaler Ringe, *Math. Annalen* **179** (1969), 257–274.

[309] O. Zariski and P. Samuel, *Commutative Algebra*, Vol. II, Van Nostrand, Princeton, 1960.

Contents

PART 3: REAL ESTATE CONTRACTS

PART 4: REAL ESTATE ANALYSIS

PART 5: FEDERAL LAW AFFECTING REAL ESTATE

PART 6: REAL ESTATE LICENSING EXAMINATIONS

LIST OF ILLUSTRATIONS

LIST OF TABLES

Preface

In recent years, legal, social, economic, and financial matters affecting real estate have become more complex, creating greater educational needs for those who are engaged in the real estate business. In today's environment, a salesperson, broker, or appraiser cannot survive solely on a pleasant disposition and a neat appearance. Real estate professionals must be aware of and obey federal, state, and local regulations. They need in-depth knowledge of each property being brokered, including its physical surroundings and its economic, legal, social, and political environment.

In recognition of these needs, most states have adopted more stringent standards for the licensure of real estate brokers and salespersons. These standards include more education, more rigorous examinations, or both. Population mobility, the need for licensure reciprocity, increased federal regulation, and greater communication and cooperation between state license law officials have all led to greater uniformity in real estate licensing standards. Also, since 1993, federal law has required that real estate appraisers be state certified. Appraisers (both experienced professionals and new entrants) now must take qualifying examinations that can lead to at least two levels of certification: residential and general. The appraisal sections of this book can now serve the needs of those desiring preparation for appraiser certification as well as those who seek sales or broker licenses.

This book was written as an introduction to real estate, taking into consideration real estate complexities, the need for more knowledge, and the trend toward uniformity among states. It is primarily intended to serve those seeking to meet licensure requirements. However, it may also satisfy the needs of people who want to know more about real property law, contracts, finance, brokerage, ethics, leasing, appraisal, mathematics, fair housing, truth in lending, and other such topics. For the person just wanting to buy or sell a home, the glossary may serve as a reference for greater understanding of the transaction. It alone can prevent misunderstandings that are costly in terms of time and money.

A careful reading of any document is crucial to understanding it. Many people have unfortunately lost money by failing to read and understand contracts before signing them. Similarly, a careful reading of this book will help you understand this complex subject—real estate.

At the beginning of each chapter of this book is a two- or three-page list of bullet points summarizing essential matters in the chapter. Each of these lists provides much of the information you need to pass your exam.

Each chapter also includes at least one example of a trick question. These are often asked on licensing exams. The explanation explains why it is a trick question and distinguishes between the correct answer and an incorrect one that might be tempting.

Questions are included at the end of each chapter, and comprehensive exams are presented at the end of the book. These will allow you to test your understanding of the material presented, both chapter by chapter and as a whole. Although these questions can be of valuable assistance, the knowledge that you gain will approximate the effort that you expend. In other words, the more effort you put in, the more information you will retain.

Should you become actively engaged in the real estate business, we wish you success. We hope that all your dealings will be honest and fair. Only in this way can the real estate business become recognized as a profession.

We are grateful to the many people who have aided us in preparing this book. This edition was made possible by Suzanne S. Barnhill and the staff at Barron's.

Jack P. Friedman

PART ONE
Introduction to Real Estate

1 How to Use This Book

If you are reading this book, you are probably interested in employment in the real estate field. Real estate is a fast-growing, exciting field that offers significant income potential to a person who is willing to work hard and to learn the special features of this interesting business.

If you want to enter the real estate business as a salesperson, broker, or appraiser, you will have to take and pass an examination and satisfy other requirements. Potential real estate salespeople and real estate brokers take licensing examinations in order to get sales or brokerage licenses. A person whose ambition is real estate appraisal takes a certification examination in order to become a certified appraiser. In either case, taking the exam is only one step (you also need educational preparation, for example), but it is a very important one. This book is designed to help you to succeed in taking that important step. To fulfill this purpose, this book has a number of important features that enhance its usefulness to you:

1. This book can be used to study for any licensing or certification examination offered in the real estate field by any state. It covers the material in all nationally distributed examinations as well as those given in states that prepare their own examinations. No matter where you live, this is the book you need.
2. Chapters 3–22 each begin with "Essential Matters to Learn from this Chapter." Please use these bullet point lists as both an introduction to and a review of the particular chapters.
3. Special emphasis is given to the topics that license applicants find most troublesome on examinations:
   a. Definitions
   b. Contracts
   c. Arithmetic and real estate math
   d. Federal laws and regulations
   e. Closing statements (for broker applicants)
4. Hundreds of sample questions are included at the end of the chapters, covering all the materials discussed.
5. Nine model examinations are included. Use these to practice. The first exam is entirely about federal law and regulation. Exams 2–8 are for brokers and salesperson applicants. The ninth is for appraiser applicants.
6. A full and complete glossary is included, along with suggestions as to the best study strategy for getting the most out of it.
7. Chapter 23, the final chapter before the model tests, describes the *strategy* and *techniques* of taking examinations so that you can be sure to get every point possible.

Professional Licensing

The purpose of professional licensing is public protection. Business success as a real estate broker or salesperson depends largely on sales ability, personality, integrity, and product knowledge. However, a winning personality and well-coordinated interior colors do not protect the public, so don't expect exam topics to be about housing styles or floor plans. Instead, exam topics will include definitions of terms, real estate and contract law, appraisal and finance, and government regulations affecting real estate, including fair housing laws and land use. Some math is required to assist buyers and sellers. Unless you demonstrate adequate knowledge of these topics, you cannot become licensed. Without a license, you cannot work in real estate. Doing so would be a disservice to the general public, including your clients. In fact, acts by unlicensed individuals are punishable by fines and/or prison terms.

Chapter Topics

This book is divided into 23 chapters of text and the model examinations (Chapter 24), with answers and explanations to math problems. Each chapter covers a specific topic. Some are long, others short. Remember, though, that the material in most of the chapters *complements* the material in the others. This is especially true in Part 3, "Real Estate Contracts." The first part of Part 3 (Chapter 6, "Introduction to Contracts") discusses everything you should know about contracts in general. Each of the other chapters in Part 3 is devoted to one or more types or parts of real estate contracts. For each of these chapters, all the general material in Chapter 6 applies to what you are learning. Chapter 7 discusses land description. Chapters 8–12 point out the specific characteristics of each type of real estate contract. Chapter 13 is on mortgages and finance, including lending institutions. Chapters 14 and 15 cover appraisals. No matter what exam you're studying for, you should study Chapter 14. However, if you aren't preparing for the general appraisal certification exam, you can skip Chapter 15. Chapters 16–20 contain a variety of problems that are principally solved using mathematics. You will need to learn about percentages, ratios, and financial statements that are used in real estate.

Just as our society has become more aware of civil rights and consumer protection, so have real estate licensing exams. Pay very close attention to Part 5, which includes chapters on fair housing law (Chapter 21); subdividing, building, and development; land use controls; and environmental matters (Chapter 22).

Most states have their examinations prepared for them by a testing company, which often administers the examination as well. The three largest testing companies for real estate exams are AMP, PSI, and Pearson-Vue. You can get test information from each of these companies. Some of the largest states prepare their own examinations. It doesn't really matter much to you how the examinations are prepared, because from state to state most of them follow the same format as described in Chapter 23. Your state's examining authority will provide you with a lot of material about the exam you will have to take, including sample questions, information about testing sites and dates, and information about the content of the exam. Inquire whether your state uses only a hard copy paper test or offers an electronic version. Some states have introduced computer-driven exams at remote locations, which you may find more convenient. You should also obtain a copy of your state's real estate license laws (available from the same source), because a portion of the examination will cover those laws. Your copy of them will come in handy as you study Chapter 5 ("State Real Estate License Law"); space is provided in that chapter for you to write in specific features of your state's laws.

Sample Questions and Examinations

Most chapters are followed by sample questions covering the material contained in the chapter and written in the format most frequently encountered on licensing examinations. You can use these questions to test yourself on what you have read.

When you have finished studying the material in the book, you can take the model examinations in Chapter 24. These will allow you to check your retention of the material you have read and will give you some experience in taking real estate examinations in a format similar to the licensing examination. Each model examination has a 3-hour time limit.

You should not take all the examinations at once. Rather, take one and then correct it, using the answers that follows that particular exam. The questions you got wrong will pinpoint the areas in which you need further study. Do this additional studying, and then take the second examination. Once again, use the questions you missed to guide you to further study. For questions involving mathematics, the answer explanations should help you to understand why you missed each question you got wrong and to avoid making the same mistake again. Then take the third examination, once again using the few questions you miss as guides for final study before taking the licensing examination itself.

Taking Examinations

Chapter 23 discusses *techniques* for taking examinations. This chapter will tell you all you must know about how to prepare for an examination, how to approach it, and how to get the highest score possible. *Be sure to read this chapter!* It provides valuable guidelines for taking the licensing examination and will introduce you to the kind of examination you can expect to encounter.

Study Strategy

Each state offers multiple versions of a test and revises the tests and questions frequently. No one can accurately predict the content of a test. Therefore, understanding the general principles of real estate and specifically how to apply those principles is essential to both passing an exam and practicing in a professional manner.

Many states and some exam preparation companies offer exam content outlines on their websites. Search your state's website for its specific real estate exam content outline. The outline will provide a list of topic areas tested by your state. It will also indicate the approximate percentage of the exam devoted to each topic area. Knowing those percentages can help you decide how much time to devote to studying each topic area. For example in Florida, the outline states that approximately 1% of the test content is about planning and zoning while approximately 12% is about real estate contracts. Obviously if you are taking the Florida exam, you will want to devote more time and effort studying information about contracts than information about planning and zoning.

You should approach your study of the materials you must know in a systematic and sensible way in order to get the most out of the information. The following study strategy will be helpful:

1. First, find out everything you can about the licensing examination offered by your state, including where and when it is given, how much it costs, and how to apply. See Chapter 23, "Examination Preparation," for a full checklist of the things you should find out. When you need information, go to your state's real estate regulatory website. There you can find out about the exam, when and where it's offered, how much it costs, how to apply, and what education requirements you have to meet—in short, everything you need to know. You also can download a copy of your state's license law and regulations. If you can't get on the Internet or if the information you want isn't on the website, don't write a letter—telephone. Even if it's long distance, a phone call costs only a few dollars, and you get the information quickly. Getting a response to a letter can take a week or longer.
2. Next, assemble all the study materials you will need. They include:
   - This book.
   - Your state's Real Estate License Law, Rules, and Regulations; disclosure forms; sample contracts; and any other materials provided by the state agency.
   - Any study materials *specifically* recommended or required by your state, most particularly, *all* study materials provided by your state's real estate licensing agency; these will tell you what you must know about the special features of your state's laws and practices.
3. Now begin studying. Start with Chapter 2 in this book, "Definitions of Real Estate Terms." Read it carefully, making sure you take note of all the terms. Don't try to understand them all at this point; just become familiar with them.
4. Proceed through the chapters in the order they appear in the book. When you get to Chapter 5 ("State Real Estate License Law"), have your copy of your state's license law handy so you can refer to it.
5. When you have read through Chapter 23 and have answered all the questions at the end of each chapter, take one model examination and grade it. Note your weak areas, and concentrate on them in later study.
6. At this point, read all the study materials provided by your state, and compare your state's practices and laws with those outlined in this book. You will have noted a number of places in this book where you should fill in your state's practice on a particular point—this is the time to do that.
7. Now that you have covered general real estate principles and your state's special materials, take another model examination.
8. Go over this book again, as well as the materials your state has provided and its license law. Pay particular attention to the problem areas pinpointed by the questions you missed on the first two model examinations. Take a third examination.

Now you should be ready to take, and PASS, your chosen real estate examination. Be sure to schedule your studying so that you finish up only a day or two before the actual examination; in this way, all the material you have learned will be fresh in your mind.

Follow-up

Approximately 50% of those who take a given real estate exam receive a failure notice. This rate varies by exam purpose (sales, broker) and by state.

If you are one of the many who receive unwelcome news, you probably need more preparation. Read/review this book, and consider taking a refresher course. Some states provide grading of tests by topic areas. If your state gives you this information, you can use it to identify your weak areas. Concentrate on those weak areas when preparing to take the exam again.

Most exams are prepared by people called psychometricians. They are experts in occupational exam content and in selecting questions from a pool of questions. However, they are not experts in real estate. So they can unintentionally identify right answers as wrong.

States don't like to admit fault. However, if you think there are errors in exam composition or grading, you can write to the state Real Estate Commission to seek an explanation of how your exam was graded.

For example, if you excel in math and solved math questions where your answers were not among the choices, there could be a test problem. Questions and answer choices may have typographical errors, or the answer key may identify incorrect solutions.

In a question about the deductibility of expense on an income tax return, if the answer choices don't reflect recent tax changes, do inquire.

You should not receive a failure notice because the exam questions have not been updated. One improperly graded question can make the difference between passing and failing. Situations have occurred where applicants brought successful lawsuits based on improper grading. However, litigation should be used only as a last resort, as it is both expensive and unfriendly.

Definitions of Real Estate Terms

2

On the following pages is a glossary of real estate terminology. It is quite extensive and probably covers all the terms you might be expected to be familiar with when you take a licensing examination. We suggest that you begin your preparation for the examination by carefully reading over this glossary. This may seem like a boring and tedious thing to do, but it will have many advantages. The main benefit is that it will acquaint you with some important terminology used in the real estate business. Once you have studied these terms, the remaining material in this book will be much easier for you to absorb; you will already have some familiarity with most of what you read. In fact, the purpose of the rest of the book is to show you how all the many items defined in the glossary relate to one another in the real estate business.

The best way to study the glossary is to go over it once fairly quickly. Many real estate terms may look familiar because they are composed of familiar words. Often, however, you will find that the *real estate meaning* is *not* the one you are used to. This is one of the more serious problems that many people have with the real estate licensing examinations. You must learn real estate terminology. In other words, you must be able to understand and use the special meanings that many words have in a real estate context. These are called "terms of art."

Once you have spent adequate time on the terminology, go through the rest of the book, using the glossary for reference as needed. As you answer questions at the end of each chapter, you will encounter terms that you don't recall. This is a good opportunity to flip back to the definitions in this chapter to master the terms. Doing so will help you understand the question and increase your chances of answering it correctly. Recognizing and understanding terms on an exam will build your confidence and help you pass. Later, as the time for your real estate examination approaches, go over the glossary again. Reading the terms once again will be a sort of thumbnail review for you after you have studied the rest of the book. This will be of considerable help in your final preparation for the examination, especially since you will often encounter strictly definitional questions such as "The term *fee simple* means most nearly _____."

Sometimes in a definition, words or phrases appear in italics or in small capital letters. That means these words or phrases are themselves entries in the glossary. Sometimes they're also defined elsewhere in this book.

Trick Question

The suffix "-or" on a word indicates that this is the party that gives something. The suffix "-ee" on a word indicates that this is the party that receives something. A grantor gives the deed; the grantee receives it. A trick question may try to confuse the terms *mortgagor/mortgagee*. The *mortgagor* is the borrower. The borrower gives the *mortgage*, which is the pledge of property to secure the loan. Many think the mortgage is the loan, that they went to a lender and got a mortgage. That is not so. The borrower gives a mortgage, which pledges the property as collateral. The borrower is the mortgagor; the lender is the mortgagee. The borrower got a mortgage *loan*, not a mortgage.

AAA TENANT / *See* TRIPLE-A TENANT.

ABSTRACT OF TITLE / A summary of all of the recorded instruments and proceedings that affect the title to property.

Importance: A person going to buy real estate or lend money on it wants assurance of ownership. Usually an attorney or title insurance company prepares an abstract of title, based on documents in the county courthouse, to be certain of ownership; and only if title is good or marketable does that person buy or make the loan. Typically, *title insurance* is also acquired.

ABUTTING / Adjoining or meeting. *See also* ADJACENT.

ACCELERATED DEPRECIATION / Depreciation methods, chosen for income tax or accounting purposes, that offer greater deductions in the early years. *Straight-line depreciation,* rather than accelerated depreciation, generally applies to buildings bought after 1986.

ACCELERATION CLAUSE / A provision in a loan giving the lender the right to declare the entire amount immediately due and payable upon the violation of a different loan provision, such as failure to make payments on time.

Importance: Without an acceleration clause, a missed payment is just that—one overdue payment. The acceleration clause means the entire loan is due, and the real estate may be foreclosed on.

ACCEPTANCE / The act of agreeing to accept an offer.

Importance: To have a valid contract there must be an offer and an acceptance. The acceptance may be from either party to the other: for example, buyer/seller; landlord/tenant.

ACCESSION / Additions to property as a result of annexing fixtures or alluvial deposits.

ACCESS RIGHT / The right of an owner to get to and from his/her property.

Importance: Property may be or may become encircled by other property. The access right gives an easement by necessity, which allows the owner of landlocked property to cross over adjacent property to reach a street.

ACCRETION / Addition to land through processes of nature, such as deposits of soil carried by streams. *See* ALLUVIUM.

ACRE / A measure of land: 43,560 square feet. There are 640 acres in a square mile.

ACTION TO QUIET TITLE / *See* QUIET TITLE SUIT.

ACTUAL EVICTION / Expulsion of a tenant from the property.

ADA / *See* AMERICANS WITH DISABILITIES ACT (1990).

ADJACENT / Lying near but not necessarily in actual contact with.

Importance: When describing the physical proximity of property, the word includes touching property and other nearby parcels, for example, property across the street.

ADJOINING / Contiguous; attaching; in actual contact with.

Importance: In describing the proximity of land, adjoining property is actually touching.

ADJUSTABLE-RATE MORTGAGE (ARM) / A mortgage loan with an interest rate that is allowed to vary during the life of the loan; usually there are *caps* on the amounts by which the interest rate can change annually and over the life of the loan.

Importance: ARMs are often a viable alternative to fixed-rate mortgages, and are especially desirable to the borrower who expects that rates will decline. Often the initial rate is lower than that on a fixed-rate mortgage, and there are caps on the ceiling rate that can be charged if interest rates rise.

ADMINISTRATOR / A person appointed by a court to administer the estate of a deceased individual who left no will.

Importance: If a person dies leaving a will, an executor is usually named to carry out the provisions of the will. When there is no *executor,* an administrator is appointed by the court. The administrator receives a fee.

ADMINISTRATOR'S DEED / A deed conveying the property of a person who died without a will (intestate).

Importance: If a person owned real estate and died without leaving a will, the administrator will give a deed. However, the administrator does not want the potential liability associated with a general warranty deed, so an administrator's deed is used.

ADULT / A person who has attained the age of majority.

Importance: In most states, a person who is 18 years of age is considered an adult. Also, some persons who are married or are in military service may be considered adults though not yet 18. A contract with a minor can be *disaffirmed* by the minor or by that person shortly after he/she becomes an adult.

AD VALOREM / According to valuation.

Importance: Used in describing a property tax rate. *See also* AD VALOREM TAX.

AD VALOREM TAX / A tax based on the value of the thing being taxed. **Example:** If the effective tax rate is 1 percent, the tax will be $1 per $100 of property value.

Importance: In some states, there are several taxing jurisdictions that levy ad valorem taxes. Tax districts include the city, county, and school district.

ADVERSE POSSESSION / A means of acquiring title to real estate when an occupant has been in actual, open, notorious, exclusive, hostile, and continuous occupancy of the property for the period required by state law.

Importance: A person can gain title to real estate (or lose title) by acting in a certain way for long enough. In some states, this process may take 3, 5, 10, or 25 years, depending on the circumstances.

AFFIDAVIT / A statement or declaration in writing, sworn to or affirmed before some officer who is authorized to administer an oath or affirmation.

Importance: Some statements must be in the form of an affidavit before they can be recorded. For example, a contractor's lien must be in affidavit form to be perfected.

AFFIRM / To confirm; to ratify; to verify.

AGENCY / The legal relationship between a principal and his/her *agent* arising from a contract in which the principal employs the agent to perform certain acts on the principal's behalf.

Importance: The law of agency governs the rights and obligations of a broker to the principal and of a licensed real estate salesperson to the broker.

AGENCY DISCLOSURE / A written explanation, to be signed by a prospective buyer or seller of real estate, explaining to the client the role that the broker plays in the transaction. The purpose of disclosure is to explain whether the broker represents the buyer or seller or is a dual agent (representing both) or a subagent (an agent of the seller's broker). This allows the customer to understand to which party the broker owes loyalty.

AGENT / A person who undertakes to transact some business or to manage some affair for another, with the authority of the latter. Frequently the term "real estate agent" refers to a licensed salesperson, while a BROKER is called a "broker." Generally, a broker's license requires more experience and education. A salesperson must be sponsored as an active licensee by a licensed broker. **Example:** An owner employs a broker to act as his agent to sell real property; the broker in turn employs salespersons to act as her agents to sell the same property.

Importance: An agent has certain duties to the principal, including loyalty. The agent is to act in the best interest of the principal, even when such action is not in the agent's best interest.

AGREEMENT OF SALE / A written agreement between seller and purchaser in which the purchaser agrees to buy certain real estate, and the seller agrees to sell, upon the terms and conditions of the agreement. Also called *offer and acceptance; contract of sale.*

Importance: When you are buying or selling property, your rights and obligations are described in this document. Offer this document or accept it only when you are satisfied that it contains what you will agree to in a final transaction.

AIR RIGHTS / The right to use, control, or occupy the space above a designated property. **Example:** The MetLife Building in New York City is built on air rights above the Grand Central Railroad Station.

Importance: Most property ownership rights include air rights up to the skies; however, air rights may be limited to a certain height, or, conversely, air rights may be the only property owned, having a floor that begins a stated number of feet above the ground.

ALIENATION / Transferring property to another, as the transfer of property and possession of lands, by gift or by sale, from one person to another.

Importance: In some states, property may be transferred by voluntary alienation, as in a sale for cash, or involuntarily, as in a condemnation.

ALLUVIUM (ALLUVION) / Soil deposited by accretion; usually considered to belong to the owner of the land to which it is added. *See also* ACCESSION.

ALT-A MORTGAGES / Residential property–backed loans made to borrowers who have better CREDIT SCORES than SUBPRIME borrowers but provide less documentation than normally required for a loan application.

Importance: *Alt-A mortgages* typically carry interest rates somewhere between the best rates available and those offered on subprime loans.

ALTERNATIVE MINIMUM TAX / A different way to compute income taxes due, based on a rate of either 26% or 28% of total taxable income plus certain deductions or tax-favored income that increase the amount subject to tax. The taxpayer must pay the greater of either the regular income tax amount or the alternative minimum tax.

AMENITIES / In appraising, the nonmonetary benefits derived from property ownership. **Examples:** pride of home ownership; accessibility to good public schools, parks, and cultural facilities.

Importance: Ownership of real estate may add to one's self-esteem and community involvement.

AMERICANS WITH DISABILITIES ACT (1990) / Requires owners and tenants of "places of public accommodation" to modify practices that discriminate against the disabled, to provide auxiliary aids to communication, and to remove architectural barriers (if removal can be readily achieved). **Examples:** installing ramps, making curb cuts in sidewalks and entrances, repositioning shelves, installing raised toilet seats and grab bars in toilet stalls, and rearranging partitions to increase maneuvering space.

AMORTIZATION / A gradual paying off of a debt by periodic installments.

Importance: Most mortgages require the payment of interest plus at least some amortization of principal so that the loan is eventually retired by the amortization payments.

AMORTIZATION TERM / The time required to retire a debt through periodic payments; also known as the *full amortization term.* Many mortgage loans have an amortization term of 15, 20, 25, or 30 years. Some have an amortization schedule as a 30-year loan but require a *balloon payment* in 5, 10, or 15 years.

Importance: Both the lender and the borrower are assured that the loan will be retired through regular payments without the need to refinance.

ANCHOR TENANT / The main tenant in a shopping center, usually a department store in a regional shopping center or a grocery store in a neighborhood shopping center. Large shopping centers may have more than one anchor tenant. The anchor tenant attracts both customers and other tenants.

ANNEXATION

(1) The process by which an incorporated city expands its boundaries to include a specified area. The rules of annexation are established by state law.

Importance: *Annexation* is generally sought by a city to expand its boundaries by taking in an area to which it may already be providing services. Many unincorporated suburban areas, however, resist efforts to annex them into the city because of possibly higher tax rates and loss of local control over schools and other services.

(2) Permanent attachment to property.

Importance: Personal property becomes a FIXTURE (part of the real estate) depending, in part, on the method of *annexation.*

ANNUAL PERCENTAGE RATE (APR) / The cost of credit, expressed as an annual interest rate, which must be shown in consumer loan documents, as specified by the Federal Reserve's *Regulation Z* that implements the federal TRUTH IN LENDING ACT.

ANNUITY / A series of equal or nearly equal periodic payments or receipts. **Example:** The receipt of $100 per year for the next five years constitutes a $100 five-year annuity.

ANTICIPATION, PRINCIPLE OF / In real estate appraisal, the principle that the value of a property today is the present value of the sum of anticipated future benefits.

Importance: The anticipation principle serves as the basis for the *income approach* to appraisal.

ANTITRUST LAWS / Federal and state acts to protect trade and commerce from monopolies and restrictions.

APPORTIONMENT / (1) *Prorating* property expenses, such as taxes and insurance, between buyer and seller. (2) The partitioning of property into individual parcels by tenants in common.

APPRAISAL / A professional opinion or estimate of the value of a property.

Importance: If a buyer is unfamiliar with the value of real estate in an area, it is prudent to require that, as a condition of the agreement of sale, the property be appraised for an amount at least equal to the price being paid. Also, lenders require an appraisal of property as a requirement for making a loan. The borrower will pay the appraisal fee.

APPRAISAL APPROACH / One of three methods used in estimating the value of property: *income approach, sales comparison approach,* and *cost approach.* See each.

Importance: Having three separate methods to estimate the value of property will add confidence to an appraiser's value estimate. However, not all approaches are applicable for all properties.

APPRAISAL BY SUMMATION / *See* COST APPROACH.

APPRAISAL DATE / In an *appraisal report*, the date to which the value applies; distinguished from *report date.*

Importance: Because market conditions change, the appraisal date establishes the conditions for which the value is estimated.

APPRAISAL FOUNDATION / An organization that came into existence in 1987 in an effort to encourage uniform requirements for appraisal qualifications and reporting standards. *See also* UNIFORM STANDARDS OF PROFESSIONAL APPRAISAL PRACTICE.

APPRAISAL INSTITUTE / An organization of professional appraisers that was formed in 1991 by the merger of the American Institute of Real Estate Appraisers and the Society of Real Estate Appraisers. It offers *MAI* and *SRA* designations.

APPRAISAL MANAGEMENT COMPANY (AMC) / A firm that acts as an intermediary between certified or licensed real estate appraisers and clients (often mortgage lenders) seeking appraisal services.

Importance: *Appraisal management companies* became firmly entrenched in the appraisal market when the Home Valuation Code of Conduct was adopted by FREDDIE MAC in 2009. The code required lenders to go through an intermediary to order appraisals so as to prevent lenders from overly influencing the appraisal process. The Code was superseded by the DODD-FRANK WALL STREET REFORM AND CONSUMER PROTECTION ACT in 2010, but AMCs continue as a prominent part of the appraisal market.

APPRAISAL REPORT / A document that describes the findings of an appraisal engagement. *See* UNIFORM STANDARDS OF PROFESSIONAL APPRAISAL PRACTICE STANDARD 2.

APPRAISER / One who is qualified to provide a professional opinion on the value of property. All states in the United States offer at least two certifications for real estate appraisers: CERTIFIED GENERAL APPRAISER, awarded to one who is qualified to appraise any property, and CERTIFIED RESIDENTIAL APPRAISER, awarded to one who is qualified to appraise residences with up to four units and vacant lots. State certification requires education, experience, and passing a written examination. Some states offer additional categories for licensees and trainees. Several appraisal organizations offer professional designations. The Appraisal Institute is best known in real estate; it offers the MAI (for income property) and SRA (residential) designations. Others include American Society of Appraisers, International Association of Assessing Officers, International Right of Way Association, and American Society of Farm Managers and Rural Appraisers.

APPRECIATION / An increase in the value of property.

Importance: Appreciation is one of the most significant benefits from real estate ownership. Gains from inflation as well as real income are included.

APPURTENANCE / Something that is outside the property itself but is considered a part of the property and adds to its greater enjoyment, such as the right to cross another's land (i.e., a *right-of-way* or *easement*).

Importance: The right to use another's land for a special purpose can materially affect the value of the subject property.

APR / *See* ANNUAL PERCENTAGE RATE.

ARBITRATION / Process of settling disputes through a neutral third party agreed upon by each side in the dispute. In most cases, arbitration is decided by private firms established to provide this service and is an alternative to filing suit in the public courts.

ARM / *See* ADJUSTABLE-RATE MORTGAGE.

ARM'S-LENGTH TRANSACTION / The "typical, normal transaction," in which all parties have reasonable knowledge of the facts pertinent to the transaction, are under no pressure or duress to transact, and are ready, willing, and able to transact.

ASA / A senior professional designation offered by the American Society of Appraisers. The ASA designation is awarded upon meeting rigorous requirements that include extensive experience, education, and approved sample reports.

AS IS / Without guarantees as to condition, as in a sale.

ASSESSED VALUE OR VALUATION / The value against which a property tax is imposed. The assessed value is often lower than the MARKET VALUE due to state law, conservative tax district appraisals, and infrequent reassessment. It may also exceed market value.

Importance: The assessed value of a property is typically a reasonable estimate by the tax assessor, who must periodically review all property in the jurisdiction. By contrast, an appraisal of a single property is more likely to approximate that property's market value.

ASSESSMENT / A charge against real estate made by a government to cover the cost of an improvement, such as a street or sewer line. *See also* ASSESSED VALUATION.

Importance: Property buyers and owners need to recognize that they may have to pay assessments for municipal improvements that affect their property. Frequently, the owner has little or no influence on the decision.

ASSESSMENT RATIO / The ratio of assessed value to market value. **Example:** A county requires a 40% assessment ratio on all property to be taxed. Property with a $100,000 value is therefore assessed at $40,000 (40 percent of $100,000), and the tax rate is applied to $40,000.

ASSESSOR / A government official who is responsible for establishing the value of property for the purpose of *ad valorem taxation.*

Importance: The assessor estimates the value of each piece of real estate in the tax jurisdiction but does not fix the tax rate or amount.

ASSIGNEE / The person to whom an agreement or contract is sold or transferred.

Importance: This legal term refers to the person who receives the contract.

ASSIGNMENT / The method or manner by which a right or contract is transferred from one person to another.

Importance: Contracts for the sale of real estate are generally assignable, except when financing must be arranged or certain conditions are imposed on a party. In a lease, an assignment gives all the rights of the original tenant to the new tenant.

ASSIGNOR / A party who assigns or transfers an agreement or a contract to another.

Importance: Although many contracts can be assigned, the assignor is not necessarily relieved of the obligation. For example, debts cannot be assigned. Another party can assume a debt, but the original borrower remains liable.

ASSUMABLE LOAN / A mortgage loan that allows a new home purchaser to undertake the obligation of the existing loan with no change in loan terms. Loans without due-on-sale clauses, including most FHA and VA mortgages, are generally assumable.

ASSUMPTION OF MORTGAGE / The purchase of mortgaged property whereby the buyer accepts liability for the debt that continues to exist. The seller remains liable to the lender unless the lender agrees to release him/her.

Importance: When entering an agreement for the purchase or sale of real estate, the financing should be checked to determine whether the mortgage is assumable. If the mortgage carries favorable terms and is assumable, this fact will add value to the transaction.

ATTACHMENT / Legal seizure of property to force payment of a debt.

Importance: Property can be taken as security for a debt provided that such action has court approval.

ATTEST / To witness to; to witness by observation and signature.

Importance: Certain documents must be attested to as a condition for recording at the county courthouse.

ATTORNEY-IN-FACT / A person who is authorized to act for another under a power of attorney, which may be general or limited in scope.

Importance: A person may give another the right to act for him/her in some or all matters (*power of attorney*). The designated person, called an attorney-in-fact, need not be an attorney-at-law.

AUTOMATED VALUATION MODEL (AVM) / Computerized method for estimating the value of a property. Often used for mass *appraisal* purposes, such as reassessment of a city's property tax base. **Example:** An appraisal firm collected extensive data on home sales and the distinguishing characteristics of each home. By applying an automated valuation model to the data, the firm could quickly estimate a preliminary value of any home in the local market.

AVERAGE TAX RATE / The total income tax paid compared to one's total income. Contrast *marginal tax rate,* the tax rate on the next additional dollar of income.

AVULSION / The sudden removal of land from one owner to another that occurs when a river abruptly changes its channel.

Importance: If a stream or river is a boundary line, the boundary may change with the addition or removal of land.

BABY BOOMERS / Individuals who were born during the years following World War II, generally defined as 1946–1964. This group represents a sizable portion of the consuming public, and their spending habits and lifestyle have a powerful influence on the economy.

BAILOUT / An effort by the government to provide sufficient financial assistance to prevent the failure of a specific private or quasi-private entity. The program may consist of loans or grants to satisfy outstanding debts or may involve government purchase of an equity position in the firm.

BALANCE, PRINCIPLE OF / In real estate appraisal, the principle that there is an optimal mix of inputs that, when combined with land, will result in the greatest land value. Inputs, or *factors of production,* include labor, capital, and entrepreneurship.

BALLOON PAYMENT / The final payment on a loan when that payment is greater than the preceding installment payments and satisfies the note in full. **Example:** A debt requires interest-only payments annually for 5 years, at the end of which time the balance of the principal (a balloon payment) is due.

Importance: In arranging financing for real estate, if a balloon is involved, the borrower will need to plan for the large balance that will come due.

BAND OF INVESTMENT / A weighted average of debt and equity rates.

BANKRUPTCY / The financial inability to pay one's debts when due. The debtor seeks relief through court action that may work out or erase debts. There are a number of chapters in bankruptcy law; a certain chapter requires liquidation of all assets to pay off debts, and another chapter allows the debtor to continue operations. Certain property is free from attachment by creditors, depending on state law.

BARGAIN AND SALE DEED / A deed that conveys real estate, generally lacking a warranty. The *grantor* will thus claim to have ownership but will not defend against all claims.

Importance: Sometimes a property owner does not wish to offer a *warranty deed* but will give more assurance than is offered by a *quitclaim deed.* A bargain and sale deed is a compromise between those two.

BASE AND MERIDIAN / Imaginary lines used by surveyors to find and describe the location of land.

Importance: In states that use a rectangular or government survey method, these lines are similar to latitude and longitude.

BASE LINE / Part of the *government rectangular survey* method of land description. The base line is the major east-west line to which all north-south measurements refer.

BASIC INDUSTRY MULTIPLIER / In economic base analysis, the ratio of total population in a local area to employment in basic industry. *Basic industry* is considered to be any concern that attracts income from outside the local area. The jobs added in basic industry may also contribute to the need for local service jobs (telephone operator, nurse, supermarket clerk, etc.); and, if the workers have families, additional new people may be brought to the area.

BASIS POINT / One-100th of 1%.

BEFORE-AND-AFTER RULE / In an *eminent domain* award, practice followed by many jurisdictions of appraising the property value both before and after the taking, considering enhancement or injury to the property that was the result of *condemnation.*

BENEFICIARY / The person who receives or is to receive the benefits resulting from certain acts.

BEQUEATH / To give or hand down personal property by a will.

BEQUEST / Personal property that is given by the terms of a will.

BILATERAL CONTRACT / A contract under which each party promises performance. *Contrast with* UNILATERAL CONTRACT.

BILL OF ASSURANCE / Recorded restrictions affecting a subdivision and a part of all deeds to lots therein.

BILL OF SALE / A written instrument that passes title of personal property from a seller to a buyer.

Importance: A real estate sales agreement is prepared on an *agreement of sale.* A bill of sale is used when furniture and portable appliances are sold.

BINDER / An agreement, accompanied by a deposit, for the purchase of real estate, as evidence of good faith on the part of the purchaser.

Importance: Binders are seldom used in some states because of general agreement and requirements on brokers to use forms specially promulgated for use in their states.

BIWEEKLY LOAN / A mortgage that requires principal and interest payments at two-week intervals. The payment is exactly half of what a monthly payment would be. Over a year's time, the 26 payments are equivalent to 13 monthly payments on a comparable monthly payment mortgage. As a result, the loan will be amortized much faster than a loan with monthly payments.

BLANKET MORTGAGE / A single *mortgage* that includes more than one parcel of real estate as security.

Importance: When one piece of property is insufficient collateral for a loan, the borrower may be able to satisfy the requirement by putting up more property as security. It is a good idea to negotiate release clauses whereby each parcel can be released from the mortgage when a portion of the loan is paid off, without having to pay off the entire loan.

BLOCKBUSTING / A racially discriminatory and illegal practice of coercing a party to sell a home to someone of a minority race or an ethnic background and then using scare tactics to cause others in the neighborhood to sell at depressed prices. **Example:** A sales agent arranges a sale in which a minority family enters a previously all-white neighborhood. The agent then engages in *blockbusting* by contacting other owners in the neighborhood and informing them that their property's value will fall if they don't sell right away at a depressed offered price.

BOARD OF DIRECTORS / People selected by stockholders to control the business of the corporation.

BONA FIDE / In good faith; without fraud. **Example:** A purchaser pays for property without knowledge of any title defects. In a bona fide sale, the seller accepts consideration without notice of any reason against the sale.

Importance: An act in good faith is open, sincere, and honest.

BOND / A certificate that serves as evidence of a debt. *See also* MORTGAGE.

Importance: There are many types of bonds that may be used in real estate. A *mortgage bond* is secured by a mortgage on property. A *completion bond* (also called *performance bond*) is usually issued by a bonding company to assure completion of construction if a contractor fails.

BOOT / In an exchange of like property (i.e., real estate for real estate), boot is unlike property included to balance the value of the exchange. *See* SECTION 1031.

BOY / Abbreviation for "beginning of year."

Importance: In most leases, rent is due in the beginning of each period. If a commercial lease requires annual payments, rent is due in the beginning of the year (BOY).

BRANCH OFFICE / A place of operation for a real estate firm that is located apart from the main office. The office is owned by the firm owner but is managed by another licensed broker, who may be called an *associate broker* or *broker associate*, and may have other agents working out of the location.

BRIDGE LOAN / Loan to cover a shortage of cash from the sale of one property for the purchase of another property. **Example:** A relocated seller nets $100,000 equity from a home sale but needs $140,000 of cash to purchase his next home. The seller requires a $40,000 bridge loan on his personal signature; he expects to repay the $40,000 bridge loan after receiving a signing bonus at his new job.

BROKER / A person who is licensed by a state to act for property owners in real estate transactions, within the scope of state law.

Importance: In most states, a person must be licensed as a broker to act in a real estate transaction for another. A licensed real estate salesperson must be sponsored by a broker who accepts responsibility for the salesperson's acts. A broker is regulated by the law of agency, which requires the broker to act in the best interest of the principal.

BROKERAGE / The business of being a *broker*.

Importance: Parties are brought together in a real estate transaction, such as a sale, lease, rental, or exchange.

BROKER'S OPINION OF VALUE (BOV) / Also called Broker's Price Opinion (BPO). An analysis provided by a real estate broker to assist a buyer or seller in making decisions about the listing price of real estate or a suitable bid for purchase; similar to *comparative market analysis*. A fee may or may not be charged. This analysis may not be represented as an APPRAISAL. For mortgage lending purposes, the lender requires an appraisal performed by an APPRAISER who is certified or licensed by the state.

BUBBLE / Situation in which the prices of stock, real estate, or some other asset rise to levels well beyond reasonable VALUATION before "the bubble bursts" and prices eventually return to normal.

BUILDING CAPITALIZATION RATE / In appraisal, the *capitalization rate* is used to convert an *income* *stream* into one lump-sum value. The rate for the building may differ from that for the land because the building is a wasting asset.

BUILDING CODES / Regulations established by local governments and describing the minimum structural requirements for buildings, including foundation, roofing, plumbing, electrical, and other specifications for safety and sanitation.

Importance: Work on real estate must be in compliance with the building codes, or it will not pass the required inspections.

BUILDING INSPECTION / Inspection of property as it proceeds under construction to ensure that it meets building codes: foundation, plumbing, electrical wiring, roofing, materials. Also, periodic inspection of existing public buildings for health and safety considerations.

BUILDING LINE / A line fixed at a certain distance from the front and/or sides of a lot, beyond which the building may not project.

Importance: When situating improvements on a lot, it is important to observe the building line. Often, one must check with the city to ascertain exactly where this line is.

BUILDING LOAN AGREEMENT / *See* CONSTRUCTION LOAN.

BUILDING PERMIT / Permission granted by a local government to build a specific structure at a particular site.

BULLET LOAN / Typically a loan with a 5- to 10-year term and no amortization. At the end of the term, the full amount is due.

BUNDLE OF RIGHTS THEORY / The theory that ownership of realty implies rights, such as occupancy, use and enjoyment, and the right to sell, bequeath, give, or lease all or part of these rights.

Importance: In a real estate transaction, knowing what one has or is acquiring is vital, because some of the bundle of rights may be missing.

BUYDOWN / Payment of discount points at loan origination in order to secure a lower interest rate; the rate may be bought down for only a few years or for the life of the loan.

Importance: When offered low-rate financing on property being bought, one should determine whether the buydown applies to just the first few years or the full term of the loan.

BUYER'S AGENCY AGREEMENT / Contract or verbal understanding that creates the relationship between a potential buyer and a BUYER'S BROKER. Essentially, the broker who is the buyer's agent agrees to represent the interests of the buyer, and the buyer agrees to use the agent in any sales transaction occurring under the terms of the agreement. Although the buyer's broker represents the buyer, the COMMISSION is typically paid by the seller.

BUYER'S BROKER / An agent hired by a prospective purchaser to find an acceptable property for purchase. The broker then represents the buyer and negotiates with the seller in the purchaser's best interest. **Example:** Abel wishes to purchase a small apartment complex. Abel hires Baker, a licensed real estate broker, to find such a property. Baker is acting as a buyer's broker and earns a commission from Abel when an acceptable property is found and the owner is willing to sell.

CANCELLATION CLAUSE / A provision in a contract that gives the right to terminate obligations upon the occurrence of certain specified conditions or events.

Importance: A cancellation clause in a lease may allow the landlord to break the lease upon sale of the building. A lessor may need a cancellation clause if he/she plans to sell the building; the buyer may have another use for it.

CAP / A limit on the amount by which the interest rate on an *adjustable-rate mortgage* may be changed; usually there are annual caps and lifetime caps.

Importance: Without caps, interest rates of adjustable-rate mortgages can increase without limit. Also, abbreviation for income *capitalization.*

CAPITALIZATION / A process whereby anticipated future income is converted to one lump sum capital value.

Importance: Rental property evaluation is enhanced by the capitalization process. Net operating income is divided by a *capitalization rate* to estimate value, using the following formula:

$$\text{Property Value} = \frac{\text{Rental Income less Operating Expenses}}{\text{Capitalization Rate}}$$

CAPITALIZATION RATE (CAP RATE) / A rate of return used to convert anticipated future income into a capital value. The capitalization rate includes interest and principal recovery.

Importance: *See also* CAPITALIZATION.

CAPITALIZED INCOME / The value estimated by the process of converting an *income stream* into a lump-sum amount; also called *capitalized value.*

CAPTURE RATE / The portion of total sales in the real estate market that is sold by one entity or one project.

CARRYING CHARGES / Expenses necessary for holding property, such as taxes and interest on idle property or property under construction.

Importance: When considering nonrental real estate as an investment, carrying charges should be taken into account.

CASH FLOW / Periodic amounts accruing from an investment after all cash expenses, payments, and taxes have been deducted.

CASH-ON-CASH / *See* EQUITY DIVIDEND.

CAVEAT EMPTOR / "Let the buyer beware." The buyer must examine the goods or property and buy at his/her own risk. In recent years, the doctrine has been eroded, especially by the disclosure requirements (regarding property conditions) introduced in many states.

Importance: This was once an accepted condition of sale. Homebuyers in some states have much more protection now, as sellers and brokers are required to disclose problems or face possible penalties.

CAVEATS / Warnings, often in writing to a potential buyer, to be careful.

CCIM / *See* CERTIFIED COMMERCIAL INVESTMENT MEMBER.

CC&Rs / Covenants, conditions, and restrictions. These are limitations on land use, usually in a deed, imposed in a subdivision. Protects homeowners by preventing certain uses and assuring uniformity.

CEASE AND DESIST / Order by a court or administrative agency prohibiting a person or business from continuing an activity. Used in real estate brokerage to prevent antitrust behavior among firms or to prevent illegal discrimination.

CENSUS TRACT / Geographical area mapped by the U.S. government for which demographic information is available. This information may be used by retailers, real estate developers, and brokers to estimate consumer purchasing power in a market area.

CERCLA / *See* COMPREHENSIVE ENVIRONMENTAL RESPONSE COMPENSATION AND LIABILITY ACT.

CERTIFICATE OF ELIGIBILITY / Document issued by the DEPARTMENT OF VETERANS AFFAIRS to indicate the recipient's ability to get a home loan GUARANTEED by the agency. The certificate may be obtained by submitting the proper forms to the agency or from a lender participating in the agency's automated certificate program.

CERTIFICATE OF NO DEFENSE / *See* ESTOPPEL CERTIFICATE.

CERTIFICATE OF OCCUPANCY / A document issued by a local government to a developer permitting the structure to be occupied by members of the public. Issuance of the certificate generally indicates that the building is in compliance with public health and *building codes.*

CERTIFICATE OF REASONABLE VALUE (CRV) / Required for a VA-guaranteed home loan, the CRV is based on an appraiser's estimate of value of the property to be purchased. Because the loan amount may not exceed the CRV, the first step in getting a VA loan is to request an appraisal. Although anyone can request a VA appraisal, it is customary for the lender or the real estate agent to do so. The form used is VA Form 26-1805, Request for Determination of Reasonable Value.

CERTIFICATE OF REDEMPTION / Document provided by county tax collectors showing that all past-due property taxes have been paid.

CERTIFIED COMMERCIAL INVESTMENT MEMBER (CCIM) / A designation awarded by the CCIM Institute, which is affiliated with the *National Association of Realtors®.*

CERTIFIED GENERAL APPRAISER / A person qualified to appraise any property, under appraiser certification law adopted by all states.

CERTIFIED PROPERTY MANAGER (CPM) / A member of the Institute of Real Estate Management, an organization affiliated with the *National Association of Realtors®.*

CERTIFIED RESIDENTIAL APPRAISER / A person qualified to appraise residences and up to four units of housing, under appraiser certification law. Standards call for less education, less experience, and a less comprehensive examination than are required for a certified general appraiser.

CERTIFIED RESIDENTIAL BROKER (CRB) / A designation awarded by the Realtors National

Marketing Institute, which is affiliated with the *National Association of Realtors®*.

CFPB / *See* CONSUMER FINANCIAL PROTECTION BUREAU.

CHAIN / A unit of land measurement, 66 feet in length.

Importance: In surveying, land descriptions sometimes use the chain.

CHAIN OF TITLE / A history of *conveyances* and *encumbrances* affecting a title from the time the original patent was granted, or as far back as records are available. *See also* ABSTRACT OF TITLE.

CHATTEL / Personal property, including autos, household goods, and clothing.

Importance: Many laws have different applications, depending on whether a property is real estate or chattel. It can be important to know the type of property.

CHATTEL MORTGAGE / A pledge of personal property as security for a debt.

CLAWBACK / Provision in a law or contract that limits or reverses a payment or distribution for specified reasons.

CLEAR TITLE / A title that is free and clear of all *encumbrances*.

Importance: When buying or selling real estate, it is essential to know whether the title is encumbered or clear. If it is not clear, the effect of the encumbrances on the value or use of the real estate must be checked.

CLIENT / The person who employs a broker, lawyer, accountant, appraiser, and so on.

Importance: The law describes certain relationships that a professional has with a client. In a real estate transaction, it is essential to know exactly what that relationship is. *See also* AGENCY.

CLOSING / (1) The act of transferring ownership of a property from seller to buyer in accordance with a sales contract. (2) The time when a closing takes place.

CLOSING COSTS / Various fees and expenses payable by the seller and buyer at the time of a real estate CLOSING (also called *settlement* or *transaction costs*).

The following are some closing costs:

- BROKERAGE commissions
- lender DISCOUNT POINTS/other fees
- TITLE INSURANCE premium
- deed RECORDING fees
- loan PREPAYMENT PENALTY
- INSPECTION and APPRAISAL fees
- attorney's fees

A rough approximation of a homebuyer's *closing costs* is 2%–4% of the purchase price. Similarly, a seller's closing costs may be 3%–10% of the sale price. A brokerage commission of 3%–7% is included, often paid by the seller.

CLOSING DATE / The date on which the seller delivers the deed and the buyer pays for the property; also called *settlement date*.

Importance: The *agreement of sale* will reflect when closing is to take place. A buyer or seller may be considered to have defaulted on a contract if either one is unable to close by the agreed upon date.

CLOSING DISCLOSURE / A completed form required by *TRID* (TILA-RESPA Integrated Disclosure) that details loan features and financial requirements. Lenders must give a filled-in *Closing Disclosure* form to borrowers at least three days before closing (six days if mailed). Follows the *Loan Estimate* form.

CLOSING STATEMENT / An accounting of funds from a real estate sale, made to both the seller and the buyer separately. Most states require the broker to furnish accurate closing statements to all parties to any transaction in which he/she is an agent.

Importance: This statement shows the accounting, which should be consistent with the agreement of sale. If an error is suspected in the closing statement, the closing should not take place.

CLOUD ON THE TITLE / An outstanding claim or *encumbrance* that, if valid, will affect or impair the owner's title.

Importance: A cloud on title can restrict use and affect ownership. This should be cleared before closing. An attorney or title company should be consulted for assurance of title.

COLLATERAL / Property pledged as security for a debt.

COLOR OF TITLE / That which appears to be good title but is not.

Importance: A person can be fooled by believing he/she is receiving good title. An attorney or title company's input is essential.

COMMERCIAL PROPERTY / Property designed for use by retail, wholesale, office, hotel, and service users.

Importance: Nearly all cities and towns have zoning that restricts the location of commercial property.

COMMINGLE / To mingle or mix, as by the deposit of another's money in a broker's personal account.

Importance: Brokers generally must maintain all *earnest money* in an account that is separate from their personal funds.

COMMISSION / (1) The amount earned by real estate brokers for their services; (2) the state agency that enforces real estate license laws.

Importance: (1) Commissions are the way real estate brokers earn money. (2) The Real Estate Commission, in most states, licenses brokers and salespersons and may suspend a license for certain behavior.

COMMITMENT / A pledge or promise; a firm agreement.

Importance: When financing is needed to buy property, a commitment from a lender must be obtained. The loan terms are noted in the commitment, which may include an interest rate lock-in.

COMMON ELEMENTS / Generally, in a *condominium* development, land or a tract of land considered to be the property of the development in which all owners can enjoy use.

Importance: Owners of condominiums share in the use of and payment for common areas, such as walkways, recreational facilities, and ponds. A homeowner's association typically manages the common area.

COMMON LAW / The body of law that has grown out of legal customs and practices that developed in England. Common law prevails unless superseded by other law.

COMMUNITY PROPERTY / Property accumulated through joint efforts of husband and wife and owned by them in equal shares. The doctrine now exists in Arizona, California, Idaho, Louisiana, Nevada, New Mexico, Texas, Wisconsin, and the state of Washington.

Importance: Husband and wife must agree to all real estate transactions involving community property.

COMPARATIVE (OR COMPETITIVE) MARKET ANALYSIS (CMA) / An estimate of the value of property using only a few indicators taken from sales of comparable properties, such as price per square foot. A CMA is not an appraisal.

Importance: Real estate brokers and agents, because they are not state-certified appraisers, may not perform appraisals. So they estimate the value of a subject property using a CMA in order to serve their clients with value information.

COMPARATIVE SALES APPROACH / *See* SALES COMPARISON APPROACH.

COMPARATIVE UNIT METHOD / An appraisal technique to establish relevant units as a guide to appraising the subject property. **Examples:** (1) Parking garages are compared per space. (2) Bowling centers are compared per lane. (3) Land may be sold per square foot or per front foot.

COMPLETION BOND / A legal instrument used to guarantee the completion of a development according to specifications. More encompassing than a performance bond, which assures that one party will perform under a contract under condition that the other party performs. The completion bond assures production of the development without reference to any contract and without the requirement of payment to the contractor.

COMPOUND INTEREST / Interest paid on the original principal and also on the unpaid interest that has accumulated. **Example:** $100 deposited in a 5 percent savings account earns $5 interest the first year. If this interest is not withdrawn, the account's second-year earnings are 5 percent of $105, or $5.25.

Importance: Compound interest is the cornerstone of all financial computations, including monthly mortgage payments and remaining balances.

COMPREHENSIVE ENVIRONMENTAL RESPONSE COMPENSATION AND LIABILITY ACT (CERCLA) / Federal law, known as SUPERFUND, passed in 1980 and reauthorized by SARA in 1986. The law imposes strict joint and several liability for cleaning up environmentally contaminated land. Potentially responsible parties include any current or previous owner, generator, transporter, disposer, or party who treated hazardous waste at the site. Strict liability means that each and every party is liable for the full cost of remediation, even parties who were not contaminators.

COMPS / An appraisal term, short for "comparables," that is, comparable properties.

COMPUTER-AIDED DESIGN (CAD) / The use of a computer for design work in fields such as engineering and architecture; the computer's graphics capabilities replacing work that would traditionally have been done with pencil and paper.

COMPUTERIZED LOAN ORIGINATION (CLO) / Origination of a mortgage loan, usually by someone who is not a loan officer, with the assistance of specialized computer software that ties the originator to one or more mortgage lenders. CLO systems allow real estate brokers to provide a wider array of services.

CONCESSION

(1) A business that operates within the property of another business and provides goods or services.

Example: The gift shop in the lobby of a hotel, where a businessman rents space from the hotel to sell newspapers, magazines, and sundries, is a *concession.*

(2) Reduction in price, RENT, or other benefit provided to a TENANT or buyer as an inducement to buy or LEASE.

Example: Baker is converting apartments to CONDOMINIUM units. In order to induce tenants to purchase their units, Baker offers a *concession* of a 20% discount on the sales price, available only to current tenants.

CONDEMNATION / (1) The taking of private property for public use, with just compensation to the owner, under *eminent domain*; used by governments to acquire land for streets, parks, schools, and by utilities to acquire necessary property; (2) declaring a structure unfit for use.

Importance: All property is subject to condemnation, though the government must show need. The amount of compensation can be disputed.

CONDITION(S) / Provision(s) in a contract that some or all terms of the contract must be met or the contract need not be consummated. **Examples:**

- Buyer must obtain certain financing.
- House must be appraised at a certain amount.
- City must give occupancy permit.
- Seller must pay for certain repairs.

CONDITIONAL SALES CONTRACT / A contract for the sale of property stating that the seller retains title until the conditions of the contract have been fulfilled.

Importance: Generally, buyers have less of an interest under this type of contract than is conveyed by the receipt of a deed at closing.

CONDITIONAL USE PERMIT (CUP) / A VARIANCE granted to a property owner that allows a use otherwise prevented by zoning.

CONDO / See CONDOMINIUM.

CONDOMINIUM / A system of ownership of individual units in a multiunit structure, combined with joint ownership of commonly used property (sidewalks, hallways, stairs). *See also* COMMON ELEMENTS.

Importance: A condominium can be mortgaged by its individual owner, who must pay assessments for common area maintenance.

CONFORMING LOAN / A mortgage loan that is eligible for purchase by FNMA or FHLMC.

Example: The maximum size of a loan that could be purchased by FNMA or FHLMC, or *conforming loan*, in 2019 was $484,350 when secured by a single-family home (up to $726,525 in certain high-cost areas). The limit is revised periodically according to the change in average sales price of conventionally financed single-family homes.

CONFORMITY PRINCIPLE / An appraisal principle that holds that property values tend to be maximized when the neighborhood is reasonably homogeneous in social and economic activity.

CONSENT DECREE / A judgment whereby the defendant agrees to stop the activity that was asserted to be illegal, without admitting wrongdoing or guilt.

CONSIDERATION / Anything of value given to induce entering into a contract; it may be money, personal services, love and affection, etc.

Importance: A contract must have some consideration to be legally binding.

CONSTANT / *See* CONSTANT PAYMENT LOAN.

CONSTANT PAYMENT LOAN / A loan on which equal payments are made periodically so that the debt is paid off when the last payment is made.

Importance: Although each periodic payment is the same, the portion that is interest declines over time, whereas the principal portion increases.

CONSTRUCTION LOAN / A loan used to build on real estate.

Importance: Many construction lenders require, among other things, that the builder obtain a commitment for a permanent loan before they will issue a construction loan. Commercial banks are the most common source of construction loans. The rate is often the prime rate plus 2 percent plus 1 or more discount points. The loan is advanced in stages as the project is completed.

CONSTRUCTIVE EVICTION / An eviction existing when, through the fault of the landlord, physical conditions of the property render it unfit for the purpose for which it was leased.

CONSTRUCTIVE NOTICE / The legal presumption that everyone has knowledge of a fact when that fact is a matter of public record. **Example:** A buys land from B, believing that B is the owner. However, B was a huckster; C owned the property. Because C's deed had been properly recorded, A had constructive notice of C's ownership and cannot claim ownership against C.

CONSUMER FINANCIAL PROTECTION BUREAU (CFPB) / U.S. government agency that enforces TRID (TILA-RESPA Integrated Disclosure), governing lending and closing of home mortgage loans.

CONSUMER PRICE INDEX (CPI) / A historical record of the price of a typical basket of products purchased by a household. Published monthly by the Bureau of Labor Statistics, U.S. Department of Labor.

CONTIGUOUS / Actually touching; contiguous properties have a common boundary. *See also* ADJOINING.

CONTINGENCY CLAUSE / *See* CONDITION(S).

CONTRACT / An agreement between competent parties to do or not to do certain things for a consideration.

Importance: A valid contract is enforceable in a court of law. All contracts for real estate must be in writing to be enforceable, except leases for less than 1 year.

CONTRACT FOR DEED / *See* LAND CONTRACT.

CONTRACT OF SALE / *See* AGREEMENT OF SALE.

CONVENTIONAL LOAN / A mortgage loan other than one guaranteed by the Veterans Administration or insured by the *Federal Housing Administration.*

Importance: Conventional loans generally require a larger down payment than others, although the required down payment for conventional loans can be decreased with private mortgage insurance.

CONVERTIBLE ARM / An *adjustable rate mortgage* that offers the borrower the option to convert payments to a fixed-rate schedule at a specified point within the term of the loan. Conversion is made for a nominal fee, and the interest rate on the fixed-rate loan is determined by a rule specified in the ARM loan agreement.

CONVEY / To deed or transfer title to another.

Importance: This term is used to imply a sale or transfer of ownership.

CONVEYANCE / (1) The transfer of the title of real estate from one to another; (2) the means or medium by which title of real estate is transferred.

Importance: This term usually refers to use of a deed but can also be used for a lease, mortgage, assignment, or encumbrance.

COOPERATIVE / A type of corporate ownership of real property whereby stockholders of the corporation are entitled to use a certain dwelling unit or other units of space.

Importance: Special income tax laws allow the tenant stockholders to deduct on their tax returns the housing interest and property taxes paid by the corporation.

CORPOREAL / Visible or tangible. Corporeal rights in real estate include such things as the right of occupancy under a lease.

CO-SIGNER / A person who signs a credit application with another person, agreeing to be equally responsible for the repayment of the loan. *Same as* ACCOMMODATION PARTY.

COST APPROACH / One of the three appraisal methods of estimating value. The estimated current cost of reproducing the existing improvements, less the estimated depreciation, added to the value of the land, gives the appraised value; also called *appraisal by summation.*

Importance: Most properties sell for market value. The cost approach is useful for proposed construction or for estimating the amount of insurance needed.

CO-TENANCY AGREEMENT / Provision in a shopping center lease that requires the operations of one or more other named tenants. If the specified tenant(s) GO DARK, the tenant can significantly reduce rent, reduce payments for common area maintenance, and/or terminate its lease, according to the provision.

COTERMINOUS / Having the same period or termination point (leases).

COUNSELORS OF REAL ESTATE / A professional organization of real estate investment counselors and consultants. Affiliated with the NATIONAL ASSOCIATION OF REALTORS®. Awards the designation of Counselor of Real Estate (CRE). Publishes the journal *Real Estate Issues.*

COUNTEROFFER / Rejection of an offer to buy or sell, responding with a simultaneous substitute offer.

COVENANTS / Promises written into deeds and other instruments and agreeing to do or not to do certain acts, or requiring or preventing certain uses of the property.

Importance: When buying real estate, it is prudent to determine whether any of the covenants would inhibit or prevent a proposed use of the property.

CPI / *See* CONSUMER PRICE INDEX.

CPM / *See* CERTIFIED PROPERTY MANAGER.

CRB / *See* CERTIFIED RESIDENTIAL BROKER.

CRE / *See* COUNSELORS OF REAL ESTATE.

CREDIT RATING (REPORT) / An evaluation of a person's capacity for (or history of) debt repayment. Generally available for individuals from a local retail credit association; for businesses from companies such as Dun & Bradstreet; and for publicly held bonds from Moody's, Standard & Poor's, and Fitch's. Individuals have access to their own files.

CREDIT SCORE / A number that purports to predict the probability that a person will default on a loan. Generally, the higher the number, the better risk the individual is considered to be. The score may determine whether the person gets the loan and how favorable the terms will be. The score is estimated from information contained in the individual's CREDIT REPORT. *See also* FICO.

CREDIT TENANT / A shopping center or office building tenant that is large enough, in business long enough, and financially strong enough to be rated at least investment grade by one of the major credit rating services. A property leased to such a tenant may obtain low-interest rate, long-term mortgage financing underwritten on the basis of the tenant's likelihood of honoring its lease.

CREDIT UNION / A nonprofit financial institution federally regulated and owned by the members or people who use its services. Credit unions serve groups with a common interest, and only members may use the available services.

CREDITOR / One to whom money is owed by the debtor; one to whom an obligation exists. In its strict legal sense, a creditor is one who voluntarily gives credit to another for money or other property. In its more general sense, a creditor is one who has a right by law to demand and recover from another a sum of money on any account.

CREDITWORTHINESS / A measure of the ability of a person to qualify for and repay a loan.

CUL-DE-SAC / A street with an intersection at one end and a closed turning area at the other. Often valued in the design of residential subdivisions for the privacy provided to homes on the street.

CURABLE DEPRECIATION / *Depreciation* or deterioration that can be corrected at a cost less than the value that will be added.

Importance: It is economically profitable to correct curable depreciation.

CURTESY / The right of a husband to all or part of his deceased wife's realty regardless of the provisions of her will. Curtesy exists in only a few states.

Importance: Where curtesy exists, it is the husband's counterpart to *dower.*

DAMAGES / The amount recoverable by a person who has been injured in any manner, including physical harm, property damage, or violated rights, through the act or default of another.

Importance: Damages may be awarded for compensation and as a punitive measure to punish someone for certain acts.

DATE OF APPRAISAL / *See* APPRAISAL DATE.

DEBT/EQUITY RATIO / The relationship of these two components of *purchase capital.* **Example:** Mortgage = $75,000; equity = $25,000. Therefore, the debt/equity ratio is 3 : 1. This is equivalent to a 75% *loan-to-value ratio* loan.

DECREE / An order issued by a person in authority; a court order or decision.

Importance: A decree may be final or interlocutory (preliminary).

DEDICATION / The gift of land by its owner for a public use and the acceptance of it by a unit of government. **Example:** Streets in a subdivision, land for a park, or a site for a school.

DEDICATION BY DEED / *See* DEDICATION; DEED.

DEDUCTIBLE / The amount of cash that must be paid first by an insured homeowner to cover a portion of a damage or loss. The insurance company will pay the balance of losses exceeding this amount. *See also* TAX DEDUCTIBLE.

Importance: Policies with high deductible amounts are often available at lower premiums than low-deductible policies.

DEED / A written document, properly signed and delivered, that conveys title to real property. *See also* BARGAIN AND SALE DEED; GENERAL WARRANTY DEED; QUITCLAIM DEED; SPECIAL WARRANTY DEED.

DEED IN LIEU OF FORECLOSURE / A deed that conveys a defaulting borrower's realty to a lender, thus avoiding foreclosure proceedings.

DEED OF TRUST / *See* TRUST DEED.

DEED RESTRICTION / A clause in a deed that limits the use of land. **Example:** A deed may stipulate that alcoholic beverages are not to be sold on the land for 20 years.

Importance: A buyer should check that deed restrictions will not inhibit an intended use of property; a seller should consider whether he/she wants to restrict the use of land.

DEFAULT / (1) Failure to fulfill a duty or promise, or to discharge an obligation; (2) omission or failure to perform any acts.

Importance: Upon default, the defaulting party may be liable to the other party(ies).

DEFEASANCE / A clause in a mortgage that gives the borrower the right to redeem his/her property after he/she has defaulted, usually by paying the full indebtedness and fees incurred. **Example:** A late payment on a mortgage or other *default* doesn't necessarily cause the borrower to lose the property. Defeasance may allow redemption, though the loan and fees may have to be paid.

DEFENDANT / The party sued in an action at law.

DEFERRED GAIN / Any gain not subject to tax in the year realized but postponed until a later year.

DEFERRED PAYMENTS / Money payments to be made at some future date.

DEFICIENCY JUDGMENT / A court order stating that the borrower still owes money when the security for a loan does not entirely satisfy a defaulted debt.

Importance: When property is foreclosed on, the amount realized by the sale of the collateral may not satisfy the debt. The lender may be able to get a deficiency judgment to recover the balance owed.

DELAYED EXCHANGE / *See* SECTION 1031; TAX-FREE EXCHANGE, DELAYED.

DELEVERAGE / Become less reliant on debt. Traditionally, the term usually referred to financial leverage in corporations. Since the 2007 Great Recession, usage has broadened to include any entity that suffers from too much debt, including the U.S. government, economic sectors such as real estate, and individuals. Reduces financial risk.

DELIVERY / Transfer of the possession of a thing from one person to another.

Importance: In a real estate transaction, there should be delivery of a deed. For example, if an owner dies without giving a deed to a relative while alive, a verbal promise to give the property is inadequate.

DEMOGRAPHY / The study of population characteristics of people in an area, including age, sex, and income.

DEPARTMENT OF VETERANS AFFAIRS / *See* VA.

DEPRECIATION / (1) In appraisal, a loss of value in real property due to age, physical deterioration, or functional or economic obsolescence; (2) in accounting, the allocation of the cost of an asset over its economic useful life.

Importance: It should be recognized that the value of real estate may decline; also, that one can reduce income taxes by claiming depreciation as a tax expense.

DEPRECIATION RECAPTURE / When real property is sold at a gain and ACCELERATED DEPRECIATION had been claimed, the owner may be required to pay a tax at ordinary rates to the extent of the excess accelerated depreciation. Excess depreciation on residential real estate after 1980 is recaptured; all depreciation on commercial property after 1980 is recaptured when an accelerated method had been used, under SECTION 1250 of the INTERNAL REVENUE CODE. (**NOTE:** The importance of *depreciation recapture* for real estate was minimized by the Tax Reform Act of 1986, which mandated straight-line depreciation for buildings bought after 1986.)

DEPRESSION / Economic conditions causing a severe decline in business activity, reflecting high unemployment, excess supply, and public fear.

DEVISE / A gift of real estate by will or last testament.

DEVISEE / A person who inherits real estate through a will.

DIRECT CAPITALIZATION / The value estimated by dividing *net operating income* by an overall *capitalization rate* to estimate value. *See also* CAPITALIZATION. **Example:**

Gross income	$100,000
Operating expenses	−$40,000
Net operating income	$ 60,000
Capitalization rate	÷ 0.12
Value estimate	$500,000

DIRECT SALES COMPARISON APPROACH / *See* SALES COMPARISON APPROACH.

DIRECTIONAL GROWTH / The location or direction toward which a city is growing.

Importance: An investor can often make a profit by purchasing land in the path of a city's growth.

DISCHARGE IN BANKRUPTCY / The release of a bankrupt party from the obligation to repay debts that were, or might have been, proved in bankruptcy proceedings.

DISCOUNT / The difference between the FACE AMOUNT of an obligation and the amount advanced or received. **Example:** Abel sells land for $100,000 and receives a $60,000 mortgage at 7 percent interest as part of the payment. Abel then sells the mortgage (the right to collect payments on the mortgage) at a $15,000 *discount,* thereby receiving $45,000.

DISCOUNT BROKER / A licensed broker who provides brokerage services for a lower commission than that typical in the market. Generally, the services provided are less extensive than those of a full-service broker or may be unbundled so that a client may contract for specific services. Many discount brokers charge a flat fee rather than a percentage of the selling price.

DISCOUNTED CASH FLOW / *See* DISCOUNTING.

DISCOUNTED LOAN / One that is offered or traded for less than its face value. *See* DISCOUNT; DISCOUNT POINTS.

DISCOUNTED PRESENT VALUE / *See* DISCOUNTING.

DISCOUNTING / The process of estimating the present value of an *income stream* by reducing expected *cash flow* to reflect the *time value of money.* Discounting is the opposite of compounding; mathematically they are reciprocals.

DISCOUNT POINTS / Amounts paid to the lender (usually by the seller) at the time of origination of a loan, to account for the difference between the market interest rate and the lower face rate of the note (often required when FHA or VA financing is used).

Importance: Discount points must be paid in cash at closing. Buyer and seller should agree, at the time they prepare an agreement of sale, as to who is to pay the points or what the limits are on the number of points to be paid.

DISCRIMINATION / Applying special treatment (generally unfavorable) to an individual solely on the basis of the person's race, religion, sex, color, national origin, handicap, or familial status. *See* FAIR HOUSING LAW. **Example:** Abel is accused of *discrimination* under the Fair Housing Law because he refused to rent apartments to non-white families.

DISPOSSESS PROCEEDINGS / The legal process by a landlord to remove a tenant and regain possession of property.

Importance: If a tenant breaches a lease, the landlord will want to regain possession of the property.

DISTINGUISHED REAL ESTATE INSTRUCTOR (DREI) / A *real estate* teacher, typically of licensing preparation courses, who has been designated by the *Real Estate Educators Association.* **Example:** To qualify as a DREI, Don had to demonstrate classroom teaching techniques to a panel of teachers, who judged his effectiveness. Don also offered experience and expertise in real estate law.

DISTRIBUTEE / A person receiving or entitled to receive land as the representative of the former owner; an heir.

DOCUMENTARY EVIDENCE / Evidence in the form of written or printed papers.

Importance: Documentary evidence generally carries more weight than oral evidence.

DODD-FRANK WALL STREET REFORM AND CONSUMER PROTECTION ACT / Massive overhaul implementing financial regulatory reform, signed into law on July 21, 2010, in response to the financial crisis of 2007–2010. Named for Senator Chris Dodd and Representative Barney Frank and known as the *Dodd-Frank Act.*

DOMINANT TENEMENT / Land that benefits from an EASEMENT on another property. The other property, which is usually ADJACENT, is the SERVIENT TENEMENT.

DOWER / Under common law, the legal right of a wife or child to part of a deceased husband's or father's property.

Importance: Dower has been generally abolished or severely altered in most states. *Community property* applies in some states.

DREI / *See* DISTINGUISHED REAL ESTATE INSTRUCTOR.

DUAL AGENT / An agent who represents more than one party to a transaction.

DUE DILIGENCE / (1) Making a reasonable effort to perform under a contract. (2) Making a reasonable effort to provide accurate, complete information. A study that often precedes the purchase of property or the underwriting of a loan or investment; considers the physical, financial, legal, and social characteristics of the property and expected investment performance.

DURESS / Unlawful constraint exercised upon a person whereby he/she is forced to do some act against his/her will. **Example:** "Your signature or your brains will be on this contract."

Importance: A person who signs a contract or performs another act under duress need not go through with the agreement.

EARNEST MONEY / A deposit made by a purchaser of real estate as evidence of his/her good faith.

Importance: Earnest money should accompany an offer to buy property. Generally, a broker, attorney, or title company deposits the money in a separate account, beyond the control of the principals, until the contract is completed.

EARNEST MONEY CONTRACT / The term used in some states for an *agreement of sale* contract. In some states preprinted forms, promulgated by the state Real Estate Commission, are available for several types of transactions.

EASEMENT / The right, privilege, or interest that one party has in the land of another. **Example:** The right of public utility companies to lay their lines across others' property.

Importance: Easements allow utility service without requiring the public utility to buy the land. However, a potential real estate purchaser should determine exact locations of all easements to be sure they won't interfere with planned land uses.

ECCR / Easements, covenants, conditions, restrictions. Adds EASEMENTS to CONDITIONS, COVENANTS, and RESTRICTIONS.

ECHO BOOMERS / Generation representing the children of BABY BOOMERS who were born immediately following World War II.

ECONOMIC DEPRECIATION / Loss of value from all causes outside the property itself. **Example:** An expensive private home may drop in value when a sanitary landfill is placed nearby.

ECONOMIC LIFE / The remaining period for which real estate improvements are expected to generate more income than operating expenses cost.

Importance: As land improvements age, they tend to command less rent (in real terms) while maintenance costs rise. The economic life is the expected life of positive contributions to value by buildings or other land improvements.

ECONOMIC OBSOLESCENCE / *See* ECONOMIC DEPRECIATION.

EFFECTIVE GROSS INCOME / For income-producing property, *potential gross income*, less a vacancy and collection allowance, plus miscellaneous income. *See also* GROSS INCOME. **Example:** An office building rents for $12 per square foot and contains 100,000 leasable square feet. A 5% vacancy and collection allowance is expected. A small concession stand provides $1,000 of annual revenue.

Potential gross income: $12 × 100,000	= $1,200,000
Less: Vacancy and collection allowance @ 5%	−$60,000
Add: Miscellaneous income	+$1,000
Effective gross income	$1,141,000

EFFICIENCY RATIO / The ratio of usable square footage in a building to total square footage. **Example:** An office building's exterior is 100 feet by 100 feet, or 10,000 total square feet. However, its walls, elevator shafts, utility closets, broom closets, restrooms, and hallways represent 20% of the total area. The building's efficiency ratio is 80%.

EJECTMENT / An action to regain possession of *real property* and to obtain damages for unlawful possession.

Importance: Ejectment allows a rightful owner to remove a squatter or trespasser.

EMINENT DOMAIN / The right of the government or a public utility to acquire property for necessary public use by *condemnation*; the owner must be fairly compensated.

Importance: Governments or those with governmental authority can acquire property they need under the Federal Fifth Amendment and state constitutions. As to the amount paid, the parties are entitled to a "day in court."

ENCROACHMENT / A building, a part of a building, or an obstruction that physically intrudes upon, overlaps, or trespasses upon the property of another.

Importance: A *survey* is often required as part of a real estate contract, to determine whether there are any encroachments on the property.

ENCUMBRANCE / Any right to or interest in land that diminishes its value. Included are outstanding mortgage loans, unpaid taxes, easements, deed restrictions, mechanics' liens, and leases.

ENDORSEMENT / The act of signing one's name on the back of a check or note, with or without further qualification; also, the signature itself.

ENTITLEMENT / (1) The right to develop land with government approvals for ZONING density, utility installations, occupancy permits, use permits, streets. (2) In a VA loan, the dollar amount of loan guarantee that the Department of Veterans Affairs (VA) provides to each eligible veteran. VA loan guarantees are available to military veterans who served in the armed services during specified war periods. The guarantee allows the veteran to borrow money to buy a home without the need for a cash down payment. Generally, LENDERS will lend up to four times the amount of the entitlement with no cash investment by the borrower.

ENVIRONMENTAL ASSESSMENT / A study of the property and area to determine health hazards:

Phase I. To identify the presence of hazards (e.g., asbestos, radon, PCBs, leaking underground storage tanks).

Phase II. To estimate the cost of remediation or cleanup.

Phase III. To remediate the environmental contamination.

ENVIRONMENTAL IMPACT REPORT (or STATEMENT) / Describes probable effects on the environment of a proposed development; may be required by a local government to assure the absence of damage to the environment.

EOY / Abbreviation for "end of year."

Importance: Most mortgage loans call for payments at the end of each period. If the loan requires annual payments, a payment would be due at the end of the year (EOY).

EQUAL AND UNIFORM TAXATION / Principle that all persons of the same class must be treated equally. The same rate and value should apply to property being taxed.

EQUITABLE TITLE / The interest held by one who has agreed to purchase but has not yet closed the transaction.

EQUITY / The interest or value that the owner has in real estate over and above the liens against it. **Example:** If a property has a market value of $10,000, but there is a $7,500 mortgage loan on it, the owner's equity is $2,500.

EQUITY DIVIDEND / The annual *cash flow* that an equity investor receives; same as *cash-on-cash* return.

EQUITY OF REDEMPTION / The right of a real estate owner to reclaim property after default, but before foreclosure proceedings, by payment of the debt, interest, and costs. *See also* DEFEASANCE.

EQUITY YIELD RATE / The rate of return on the *equity* portion of an investment, taking into account periodic *cash flow* and the proceeds from resale. The timing and amounts of *cash flow* after debt service are considered, but are not income taxes.

EROSION / The gradual wearing away of land through process of nature, as by streams and winds.

Importance: Planting vegetation can often prevent soil erosion.

ERRORS AND OMISSIONS INSURANCE / Liability protection against professional malpractice, mistakes in business dealings by insured, etc.

Importance: Real estate BROKERS usually have *errors and omissions insurance* to protect against claims arising from mistakes they and their agents make in their representation of clients.

ESCALATOR MORTGAGE / *See* ADJUSTABLE RATE MORTGAGE.

ESCAPE CLAUSE / A provision in a contract that allows one or more of the parties to cancel all or part of the contract if certain events or situations do or do not occur.

ESCHEAT / The reversion of property to the state in the event that the owner dies without leaving a will and has no legal heirs.

Importance: In the absence of a will, the assets of a person who has no legal heirs will escheat to the state.

ESCROW / An agreement between two or more parties providing that certain instruments or property be placed with a third party for safekeeping, pending the fulfillment or performance of some act or condition.

Importance: It is prudent to place money or property in escrow rather than to give it to the other principal of a pending transaction.

ESCROW ACCOUNT / Account where funds are held for safekeeping until needed, usually without interest; in real estate, an account that receives the funds needed for expenses such as property taxes, HOMEOWNER'S INSURANCE, and MORTGAGE INSURANCE when they come due. *See also* TRUST ACCOUNT.

ESCROW AGENT / A neutral third party (such as a lawyer, broker, or title company) who is trusted by buyer and seller to close a transaction. The buyer and seller give instructions with conditions that must be met in order to close.

ESTATE / The degree, quantity, nature, and extent of interest a person has in real or personal property.

ESTATE AT SUFFERANCE / The wrongful occupancy of property by a tenant after his/her lease has expired.

ESTATE AT WILL / The occupation of real estate by a tenant for an indefinite period, terminable by one or both parties at will.

ESTATE FOR LIFE / An interest in property that terminates upon the death of a specified person. *See also* LIFE ESTATE.

ESTATE FOR YEARS / An interest in land that allows possession for a definite and limited time.

ESTATE IN REVERSION / An estate left by a *grantor* for himself/herself, to begin after the termination of some particular estate granted by him/her. **Example:** A landlord's estate in reversion becomes his/hers to possess when the lease expires.

ESTOPPEL CERTIFICATE / A document by which, for example, the mortgagor (borrower) certifies that the mortgage debt is a lien for the amount stated. He/she is thereafter prevented from claiming that the balance due differed from the amount stated. Estoppels apply also to leases.

Importance: The buyer of property on which there is a mortgage or lease should get estoppels to be sure the terms of those agreements are as expected. The right to get estoppel certificates may be written into the lease or mortgage.

ET AL. / Abbreviation of *et alii* ("and others").

ET UX. / Abbreviation of *et uxor* ("and wife").

ETHICS / Principles by which one treats colleagues, clients, and the public in a fair, just, and truthful manner. Adherence to such standards is considered one of the requisites for recognition as a profession.

EVALUATION / A study of the potential uses of a property but not to determine its present value.

Importance: Evaluations include studies as to the market for and the marketability of a property, feasibility, *highest and best use*, land use, and supply and demand.

EVICTION / A legal proceeding by a lessor (landlord) to recover possession of property.

Importance: Eviction allows a landlord to regain property when a tenant does not uphold the lease. A legal process must be followed.

EVICTION, PARTIAL / A legal proceeding whereby the possessor of the property is deprived of a portion thereof.

EXCESS (ACCELERATED) DEPRECIATION / The accumulated difference between ACCELERATED DEPRECIATION claimed for tax purposes and what STRAIGHT-LINE DEPRECIATION would have been. Generally, *excess accelerated depreciation* is recaptured (taxed) as ORDINARY INCOME upon a sale, instead of receiving more favorable CAPITAL GAINS treatment. *See* DEPRECIATION RECAPTURE.

EXCESS RENT / The amount by which the *rent* under an existing *lease* exceeds the rental rate on comparable existing space.

Importance: Should the lease expire or the tenant break the lease, the new rental rate will probably be at (lower) market rates.

EXCHANGE / Under Section 1031 of the Internal Revenue Code, like-kind property used in a trade or business or held as an investment can be exchanged tax free. *See also* BOOT.

EXCLUSIONARY ZONING / Zoning laws of a community that would serve to prohibit low- and moderate-income housing; considered to be illegal.

EXCLUSIVE AGENCY LISTING / An employment contract giving only one broker the right to sell the property for a specified time and also allowing the owner to sell the property himself/herself without paying a commission.

Importance: This is sometimes arranged when an owner wants to continue personal selling efforts while employing a broker. Acceptable to some brokers for some properties.

EXCLUSIVE RIGHT TO SELL LISTING / An employment contract giving the broker the right to collect a commission if the property is sold by anyone, including the owner, during the term of the agreement, and often beyond the term to someone the broker introduced. *Compare* EXCLUSIVE AGENCY LISTING; *contrast with* OPEN LISTING.

EXECUTE / (1) To make out a contract; (2) to perform a contract fully.

Importance: In real estate, unsigned contracts are generally meaningless.

EXECUTED CONTRACT / A contract where all terms and conditions of which have been fulfilled.

Importance: When executed, a contract has been completed.

EXECUTOR / A person designated in a will to carry out its provisions concerning the disposition of the estate.

Importance: The person making a will should name an executor or co-executor who is trustworthy and capable of carrying out the terms of the will. If there is no will and no executor, the court appoints an *administrator.*

EXECUTORY / Contracts that are incomplete, not final, or not yet fully performed or executed. **Example:** A real estate sales contract is executory until the sale is closed.

EXECUTRIX / A woman who performs the duties of an executor.

EXPENSE RATIO / A comparison of *operating expenses* to *potential gross income.* This ratio can be compared over time and with that of other properties to determine the relative operating efficiency of the property considered. **Example:** An apartment complex generates potential gross income of $1,000,000 annually and incurs operating expenses of $400,000 over the same time period. The expense ratio of 40% can be compared to its historical rate and the same ratios competitive properties. The comparison may disclose reasons for differences that can be used to bolster the property's efficiency.

EXPRESS AGENCY / A clear and explicit relationship between an AGENT and a PRINCIPAL, in the form of either a written contract or an oral understanding. *Contrast with* IMPLIED AGENCY.

EXPRESS CONTRACT / Written agreement that includes elements needed for a valid CONTRACT and specific terms of the contract. *Contrast with* IMPLIED CONTRACT.

FACTORS OF PRODUCTION / In economics, land, labor, capital, and entrepreneurship.

FAIR HOUSING LAW / A federal law that forbids discrimination on the basis of race, color, sex, religion, handicap, familial status, or national origin in the selling or renting of homes and apartments.

FAMILY LIMITED PARTNERSHIP / A limited partnership whose interests are owned by members of the same family. By this arrangement, gift and estate taxes may be reduced. However, owners will not enjoy the freedom of complete ownership or free transferability of interest provided by other ownership vehicles. *See* MINORITY DISCOUNT.

FANNIE MAE / Nickname for Federal National Mortgage Association (FNMA), one of two companies (the other is FHLMC or FREDDIE MAC) that purchase in the secondary market a large share of the residential mortgage loans originated each year. Most of these loans are used to support special bond-type securities sold to investors. *Fannie Mae,* now a semiprivate corporation (a GOVERNMENT-SPONSORED ENTERPRISE) was once an agency of the federal government. On September 8, 2008, in the wake of the financial crisis, Fannie Mae was placed under conservatorship by the FEDERAL HOUSING FINANCE AGENCY. The U.S. government announced the right to 80% ownership of its stock.

FARMER MAC / *See* FEDERAL AGRICULTURAL MORTGAGE CORPORATION.

FARMERS HOME ADMINISTRATION (FmHA) / Former agency of the U.S. Department of Agriculture that administered assistance programs for purchasers of homes and farms in small towns and rural areas. In 1994, the USDA was reorganized, and the functions of the FmHA were transferred to the Farm Service Agency.

FDIC / *See* FEDERAL DEPOSIT INSURANCE CORPORATION.

FEDERAL AGRICULTURAL MORTGAGE CORPORATION / Federal agency established in 1988 to provide a secondary market for farm mortgage loans. Informally called *Farmer Mac.*

FEDERAL DEPOSIT INSURANCE CORPORATION (FDIC) / A U.S. government agency that insures depositors' accounts in commercial banks and savings and loan associations.

Importance: Federal insurance, up to $250,000 per depositor for each insured bank, provides depositor confidence in the banking system and in the individual bank. Banks and savings and loan associations add liquidity to the real estate market.

FEDERAL EMERGENCY MANAGEMENT AGENCY (FEMA) / Federal agency that administers the national FLOOD INSURANCE program. More information at *www.fema.gov.*

FEDERAL HOUSING ADMINISTRATION (FHA) / A U.S. government agency that insures to lenders the repayment of real estate loans.

Importance: The FHA is instrumental in assuring that financing is available for housing for low- and moderate-income levels. Programs include single-family homes, condos, apartments, nursing homes, even new towns.

FEDERAL HOUSING FINANCE AGENCY (FHFA) / U.S. government agency created in 2008 under the Housing and Economic Recovery Act to replace the Federal Housing Finance Board and assume oversight of the FEDERAL HOME LOAN BANK SYSTEM that includes the housing-related GSEs Fannie Mae, Freddie Mac, and the Federal Home Loan Banks.

Importance: Shortly after it was created, the *Federal Housing Finance Agency* was given conservatorship of FANNIE MAE and FREDDIE MAC, providing the agency a strong hand in the ongoing business of the two GSEs. The FHFA was granted enhanced powers to establish standards, restrict asset growth, increase enforcement, and put entities into receivership.

FEDERAL HOUSING FINANCE BOARD / A federal agency created under the FINANCIAL INSTITUTIONS REFORM, RECOVERY, AND ENFORCEMENT ACT (FIRREA) to regulate and supervise the 12 district Federal Home Loan Banks. The Board was abolished in 2008, and its functions were transferred to the newly created FEDERAL HOUSING FINANCE AGENCY (FHFA).

FEDERALLY RELATED MORTGAGE / A MORTGAGE LOAN that is, in some way, subject to federal law because it is guaranteed, insured, or otherwise regulated by a government agency such as the FEDERAL DEPOSIT INSURANCE CORPORATION (FDIC), FEDERAL HOME LOAN MORTGAGE CORPORATION (FHLMC), FEDERAL HOUSING ADMINISTRATION (FHA), FEDERAL NATIONAL MORTGAGE ASSOCIATION (FNMA), the board of governors of the FEDERAL RESERVE SYSTEM, National Credit Union Association (NCUA), Office of the Comptroller of the Currency (OCC), Office of Thrift Supervision (OTS), or Department of Veterans Affairs (VA).

FEDERALLY RELATED TRANSACTION / Real estate transaction that is overseen by a federal agency, including: Federal Reserve Board; FEDERAL DEPOSIT INSURANCE CORPORATION (FDIC); Office of the Comptroller of the Currency (OCC); Office of Thrift Supervision (OTS); National Credit Union Association (NCUA); FANNIE MAE; FREDDIE MAC; Federal Housing Administration (FHA); and Department of Veterans Affairs (VA).

FEDERAL NATIONAL MORTGAGE ASSOCIATION (FNMA) / *See* FANNIE MAE.

FEE SIMPLE OR FEE ABSOLUTE / Absolute ownership of real property; the owner is entitled to the entire property with unconditional power of disposition during his/her life, and the property descends to his/her heirs and legal representatives upon his/her death intestate.

Importance: When buying real estate, the seller can give only the rights he/she has. A buyer should determine whether complete ownership in fee simple will be received.

FF&E / FURNITURE, FIXTURES, AND EQUIPMENT.

FHA / FEDERAL HOUSING ADMINISTRATION.

FHA LOAN / A mortgage loan insured by the FHA.

Importance: FHA loans generally reduce the required down payment to 3 percent (sometimes less) but require FHA mortgage insurance, in addition to interest, at 0.5 percent annually.

FHFA HOUSE PRICE INDEX (HPI) / A home price index compiled by the FEDERAL HOUSING FINANCE AGENCY. The index is based on data taken from loans held by the home mortgage GSEs. Values are available for each state and metropolitan area in the country.

FICO (FAIR ISAAC COMPANY) SCORE / A measure of borrower credit risk commonly used by MORTGAGE underwriters when originating loans on owner-occupied homes. The score is based on the applicants' credit history and the frequency with which they use credit. Expressed as a number between 300 and 850. **Example:** When the mortgage company processed the loan application, it ordered a *FICO score* for the applicant. If the score is high enough, the applicant will be offered a loan with the lowest interest rate and lowest required down payment. If not, the borrower must settle for a SUBPRIME loan with less-favorable terms.

FIDUCIARY / (1) A person who, on behalf of or for the benefit of another, transacts business or handles money or property not his/her own. (2) Founded on trust; the nature of trust.

Importance: A fiduciary, or a person in such capacity, must act in the best interest of the party who has placed trust.

FILTERING DOWN / The process whereby, over time, a housing unit or neighborhood is occupied by progressively lower-income residents.

FINAL VALUE ESTIMATE / In an *appraisal* of *real estate*, the appraiser's value conclusion. **Example:** An appraiser has determined these three amounts of value based on each appraisal approach:

Cost approach	$600,000
Sales comparison approach	$575,000
Income approach	$560,000

She then reconciles these amounts, decides which is most relevant, and offers a final value estimate. If the income approach is considered most indicative of purchaser behavior, the amount may be $560,000 to $570,000, depending on the relative weight assigned to each approach.

FINANCIAL FEASIBILITY / The ability of a proposed land use or change of land use to justify itself from an economic point of view.

Importance: Financial feasibility is one test of the *highest and best use* of land, but not the only test. Nor does the financial feasibility of a project necessarily make it the most rewarding use of land.

FINANCING LEASE / A lease wherein lessee becomes lessor to the operating tenant.

FIRREA / Financial Institutions Reform, Recovery, and Enforcement Act.

FIRST MORTGAGE / A mortgage that has priority as a *lien* over all other mortgages.

Importance: Generally, the first mortgage is the one recorded first. When a first mortgage is retired, existing mortgages of lower priority will move up. In case of foreclosure, the first mortgage will be satisfied before other mortgages.

FISCAL POLICY / Decisions by the federal government regarding spending and public finance. Includes decisions on the federal budget, the budget deficit and public debt, and taxation. *Contrast with* MONETARY POLICY.

FISCAL YEAR / A continuous 12-month time interval used for financial reporting; the period starts on any date after January 1 and ends one year later.

FIXED EXPENSES / In the operation of *real estate*, expenses that remain the same regardless of occupancy. *Contrast* VARIABLE EXPENSES. **Examples:** Insurance and interest expenses are expected to be the same, whether or not a building is occupied, so they are fixed expenses. By contrast, the costs of utilities and office cleaning will vary with occupancy.

FIXTURES / Personal property or improvements attached to the land so as to become part of the real estate.

Importance: When buying or selling real estate, it is best to specifically identify which appliances remain and which do not. Otherwise, fixtures remain with the property.

FLIP / Purchase and immediate resale of property (within hours or days) at a quick profit. Often has a negative connotation, attributed to shysters who profit illegally or at the expense of an innocent party.

FLOOD INSURANCE / An insurance policy that covers property damage due to natural flooding. Flood insurance is offered by private insurers but is subsidized by the federal government. *See* FEDERAL EMERGENCY MANAGEMENT AGENCY (FEMA). **Example:** Abel owns a home within an area defined as a 100-year FLOODPLAIN. If the local community is participating in a special federal program, Abel may purchase *flood insurance* to cover damage from periodic flooding.

FLOOD PLAIN / A level land area subject to periodic flooding from a CONTIGUOUS body of water. *Flood plains* are delineated by the expected frequency of flooding. For example, an annual flood plain is expected to flood once each year.

FLOOR LOAN / The minimum that a lender is willing to advance on a *permanent mortgage* on INCOME PROPERTY. An additional principal amount will be loaned upon attainment of a certain occupancy rate.

FNMA / *See* FEDERAL NATIONAL MORTGAGE ASSOCIATION.

FORBEARANCE / A policy of restraint in taking legal action to remedy a *default* or other breach of contract, generally in the hope that the default will be cured, given additional time.

FORCE MAJEURE / An unavoidable cause of delay or of failure to perform a *contract* obligation on time.

FORECLOSURE / A legal procedure whereby property pledged as security for a debt is sold to pay a defaulted debt.

Importance: Foreclosure gives a lender the right to sell property that was pledged for a debt. All parties to a mortgage contract should recognize its consequences.

FORFEITURE / Loss of money or anything else of value because of failure to perform under contract.

FORM 1099 / Federal income tax form used to report interest, dividends, non-employee compensation, and other income.

FRACTIONAL INTEREST / Ownership of some but not all of the rights in *real estate.* Examples are *easement,* hunting rights, and *leasehold.*

FRACTURED CONDOMINIUM / Housing development containing part rental apartments and part owner-occupied condominiums.

FRAUD / The intentional use of deception to purposely cheat or deceive another person, causing him/her to suffer loss.

Importance: Fraud, intentionally deceiving another, is far worse than misrepresentation, which is an incorrect or untrue statement. Consequently, punishment for fraud is more severe.

FRAUD AND FLIPPING / Illegal practice of purchase and immediate resale, often to defraud an innocent lender.

FREDDIE MAC / Nickname for Federal Home Loan Mortgage Corporation (FHLMC), counterpart to FANNIE MAE as a major purchaser of residential mortgage loans in the secondary market. This company was originally created to buy loans from federally chartered savings and loan associations. Today, there is little difference in the operations of Fannie Mae and Freddie Mac. Together, these companies wield significant influence on the standards required of mortgage borrowers and the types of loans offered. Both companies were placed in conservatorship of the FHFA in 2008 and delisted from the New York Stock Exchange in June 2010.

FREEHOLD / An interest in real estate without a predetermined time span. **Example:** A fee simple or a life estate.

FREE RENT / Space that is occupied but generates no income. **Example:** An apartment may be occupied by a resident manager who is partially compensated by free rent. Apartment units may be occupied during an initial period of free rent offered as a lease CONCESSION. Units may be occupied by firms that supply service to the property owner in exchange for free rent.

FRONT FOOT / A standard measurement of land, applied at the frontage of its street line. The first dimension is the frontage along a street or body of water. **Example:** a rectangular commercial lot described as 50 feet by 100 feet has 50 feet of frontage on the street and a 100-foot depth.

Importance: This measure is used for city lots of generally uniform depth. Prices are often quoted as the number of dollars per front foot.

FULL AMORTIZATION TERM / *See* AMORTIZATION TERM.

FULLY AMORTIZED LOAN / A loan having payments of *interest* and *principal* that are sufficient to liquidate the loan over its term; self-liquidating. *See also* AMORTIZATION TERM.

FUNCTIONAL DEPRECIATION / Loss of value from all causes within the property, except those due to physical deterioration. **Example:** A poor floor plan or outdated plumbing fixtures.

FUNCTIONAL OBSOLESCENCE / *See* FUNCTIONAL DEPRECIATION.

FUNDS FROM OPERATIONS (FFO) / A measure of the profitability of a real estate investment trust (REIT). *FFO* begins with net income as derived using generally accepted accounting principles (GAAP). To that it adds depreciation deductions and deductions for amortization of deferred charges, which are noncash deductions. FFO does not consider extraordinary items and gains (losses) on the sale of real estate.

FURNITURE, FIXTURES, AND EQUIPMENT (FF&E) / A term frequently found in the ownership of a hotel or motel. This type of property wears out much more rapidly than other components of a hotel or motel, so an owner or prospective buyer needs to establish the condition, cost, and frequency of replacement of FF&E.

FUTURE VALUE OF ONE / *See* COMPOUND INTEREST.

GAAP / Generally accepted accounting principles. The set of rules considered standard and acceptable by certified public accountants. Accounting deductions are required for real estate depreciation, even for assets that appreciate in value.

GABLE ROOF / A pitched roof with sloping sides.

GAMBREL ROOF / A double pitched roof having a steep lower slope with a flatter slope above.

GAP MORTGAGE / A loan that fills the difference between the *floor loan* and the full amount of the *permanent mortgage.*

GENERAL AGENT / One with powers to act for a specific person or business in all matters. Has broader authority than a *special agent.* However, the general agent's authority is not as comprehensive to all businesses owned by the principal as is the authority of a *universal agent.*

GENERAL WARRANTY DEED / A deed in which the *grantor* agrees to protect the *grantee* against any other claim to title of the property and also provides other promises. *See also* WARRANTY DEED.

Importance: This is the best type of deed to receive.

GEOGRAPHIC INFORMATION SYSTEMS (GIS) / A computer mapping program whereby land characteristics and/or demographic information are color coded and often overlaid. The purpose is to determine locations of various business activities and demographics.

GIFT DEED / A deed for which the consideration is love and affection, and no material consideration is involved.

Importance: A gift deed is frequently used to transfer real estate to a relative.

GI LOAN / *See* VA LOAN.

GLA / *See* GROSS LEASABLE AREA.

GLOBAL POSITIONING SYSTEM (GPS) / A network of satellites allowing users with portable GPS devices to determine precise locations on the surface of the Earth. The portable GPS device measures the exact time taken for signals to reach it from at least four different satellites; from this, the instrument can compute its location. Some GPS devices merely report the position in the form of geographic coordinates; others can provide sophisticated maps and driving directions, including turn-by-turn voice narration.

GNMA / Government National Mortgage Association (Ginnie Mae).

GO DARK / Terminate operations; said of an anchor retail tenant in a shopping center. Even though the tenant may continue to pay rent, its absence from the shopping center may cause the shopping center to suffer or close. *See* CO-TENANCY AGREEMENT.

GOOGLE EARTH / Application provided by Google that allows the user to browse satellite images of the world by street address or geographic coordinates (latitude and longitude).

GOVERNMENT RECTANGULAR SURVEY / A rectangular system of land survey that divides a district into 24-mile-square tracts from the *meridian* (north-south line) and the *base line* (east-west line). The tracts are divided into 6-mile-square parts called townships, which are in turn divided into 36 tracts, each 1 mile square, called sections.

Importance: This system is still used in several western states. For urban or suburban purposes, the *lot and block number* and/or *metes and bounds* methods predominate.

GOVERNMENT-SPONSORED ENTERPRISE (GSE) / Quasi-governmental organization that is privately owned but was created by the government and retains certain privileges not afforded totally private entities.

Both FANNIE MAE and FREDDIE MAC are *government-sponsored enterprises* that were created by the federal government but transferred to private owners. After 40 years of operation, on September 6, 2008 and in response to the financial crisis, 80% of their stock was deemed owned by the U.S. government. Both organizations have the implicit backing of the U.S. Treasury, which allows them to raise funds at lower interest rates than other private concerns and assist housing finance.

GRACE PERIOD / Additional time allowed to perform an act or make a payment before a default occurs.

Importance: Many mortgage contracts have a grace period before a late payment is considered a default. It is usually wise to solve the problem before the grace period expires.

GRADE / (1) Ground level at the foundation of a building; (2) the degree of slope on land (e.g., a 2 percent grade means that the elevation rises 2 feet for every 100 linear feet).

Importance: The grade of land should be checked to determine whether it suits a planned use of the land.

GRADED LEASE / *See* GRADUATED LEASE.

GRADIENT / The slope, or rate of increase or decrease in elevation, of a surface; usually expressed as a percentage. *See also* GRADE.

GRADUATED LEASE / A lease that provides for graduated changes in the amount of rent at stated intervals; seldom used in short-term leases.

Importance: Graduated leases allow rent changes automatically so that there is no need to revise the entire lease just to change the rent. These are often long-term leases that suit both landlord and tenant.

GRANT / A technical term used in deeds of conveyance of property to indicate a transfer.

GRANTEE / The party to whom the title to real property is conveyed; the buyer.

GRANTOR / The person who conveys real estate by deed; the seller or donor.

GRAYWATER / Household wastewater that is reused for purposes that do not require high sanitation, such as irrigation of household plants and other vegetation. May include waste from bathtubs, washbasins, and laundry appliances but not kitchen sink or dishwasher drainage.

Importance: Programs to encourage the use of *graywater* are used by municipalities as part of an effort to conserve potable water.

GREEN BUILDING / Building that is built or developed specifically to minimize utility costs or to maximize positive environmental considerations (reduce damage to the environment).

Importance: Many environmentally conscious people seek to buy or rent space in *green buildings* to feel they are part of the solution to the world's energy and environmental problems.

GRI / A graduate of the Realtors® Institute, which is affiliated with the *National Association of Realtors®*.

Importance: The GRI designation indicates that a real estate salesperson or broker has gone beyond the minimum educational requirements.

GRM / *See* GROSS RENT MULTIPLIER.

GROSS BUILDING AREA / The total floor area of a building, usually measured from its outside walls.

GROSS INCOME / Total income from property before any expenses are deducted. Gross income may be further described as *potential*, which assumes neither vacancy nor collection losses, or *effective*, which is net of vacancy and collection losses.

GROSS LEASABLE AREA (GLA) / The floor area that can be used by a *tenant*; generally measured from the center of joint partitions to outside wall surfaces. *Contrast with* NET LEASABLE AREA.

GROSS LEASE / A lease of property whereby the landlord (lessor) is responsible for paying all property expenses, such as taxes, insurance, utilities, and repairs. *Contrast with* NET LEASE.

Importance: Landlord and tenant agree in writing as to who pays each operating expense. Otherwise there is strong likelihood for disagreement and litigation.

GROSS POSSIBLE RENT / *See* POTENTIAL GROSS INCOME.

GROSS RENT MULTIPLIER (GRM) / The sales price divided by the rental rate. **Example:** The sales price is $40,000; the gross monthly rent is $400; the GRM = $40,000/$400 = 100. It may also be expressed as an annual figure (8.333), that is, the number of years of rent equaling the purchase price.

Importance: In many investment situations, the price is set based on a multiple of the rent level.

GROUND LEASE / An agreement for the rent of land only, often for a long term, at the expiration of which all of the real estate belongs to the landowner.

Importance: Sometimes land can be purchased or leased separately from buildings, thus splitting ownership into components that are more desirable. A property buyer or lessee must be mindful of the lease terms and the effect of such a lease on using or financing the property.

GROUND RENT / The rent earned by leased land.

Importance: Ground leases may be net or gross. In a net lease, the *tenant* pays expenses, such as insurance and real estate taxes.

GUARANTOR / A third party to a lease, mortgage, or other legal instrument who guarantees performance under the contract.

GUARDIAN / A person appointed by a court to administer the affairs of an individual who is not capable of administering his/her own affairs.

Importance: An *incompetent* cannot enter a valid contract. It is important to deal with the person's guardian.

HABENDUM CLAUSE / The "to have and to hold" clause that defines or limits the quantity of the estate granted in the deed. **Example:** "To have and to hold for one's lifetime" creates a life estate.

HAMP / *See* HOME AFFORDABLE MODIFICATION PROGRAM.

HANDYMAN SPECIAL / In real estate brokerage jargon, a property that is in need of repair, a *fixer upper*. The implication is that the property is a bargain for someone who can accomplish the repairs economically.

HARP / *See* HOME AFFORDABLE REFINANCE PROGRAM.

HASP / *See* HOMEOWNERS AFFORDABILITY AND STABILITY PLAN.

HEIRS AND ASSIGNS / Terminology used in deeds and wills to provide that the recipient receive a *fee simple* estate in lands rather than a lesser interest.

Importance: These words give the recipient complete ownership, not just an estate for a limited duration of time.

HELOC / *See* HOME EQUITY LINE OF CREDIT.

HEREDITAMENTS / Any property that may be inherited, whether real or personal, tangible or intangible.

HIGHEST AND BEST USE / The legally and physically possible use that, at the time of *appraisal*, is most likely to produce the greatest net return to the land and/or buildings over a given time period.

Importance: To realize the full value of land, the improvements built on it must represent its highest and best use.

HIP ROOF / A pitched roof formed by four walls sloped in different directions. The two longer sides of the roof form a ridge at the top.

HISTORIC DISTRICT / A designated area where the buildings are considered to have some significant historic character. Such designation makes the area eligible for certain federal assistance programs and protects the area from clearance in conjunction with federally sponsored programs.

HOLDER IN DUE COURSE / A person who has taken a note, check, or similar asset (1) before it was overdue, (2) in good faith and for value, and (3) without knowledge that it had been previously dishonored and without notice of any defect at the time it was negotiated to him/her.

Importance: A holder in due course is an innocent buyer of paper (a debt).

HOLDOVER TENANT / A tenant who remains in possession of leased property after the expiration of the lease term.

Importance: A holdover tenant has a *tenancy at sufferance*. The landlord may dictate the terms of occupancy.

HOME AFFORDABLE MODIFICATION PROGRAM (HAMP) / Program of the U.S. Departments of the Treasury and Housing and Urban Development intended to reduce mortgage payments to as little as 31% of verified monthly gross pretax income for homeowners who are not unemployed. *See also* HOMEOWNERS AFFORDABILITY AND STABILITY PLAN (HASP).

More information: Find further information at *http://www.makinghomeaffordable.gov/programs/lower-payments/Pages/hamp.aspx*. Call 888-995-4673 to discuss this with a government representative.

HOME AFFORDABLE REFINANCE PROGRAM (HARP) / Program set up by the FEDERAL HOUSING FINANCE AGENCY in March 2009 to help UNDERWATER and near-underwater homeowners refinance their mortgages. Unlike the HOME AFFORDABLE MODIFICATION PROGRAM (HAMP), which aims to assist homeowners who are in danger of foreclosure, this program targets homeowners who are current on their monthly mortgage payments but are unable to refinance because of dropping home prices in the wake of the U.S. housing market price correction during 2006–2011.

HOME EQUITY CONVERSION MORTGAGE (HECM) / A reverse mortgage insured by the FEDERAL HOUSING ADMINISTRATION (FHA). The program enables older homeowners to withdraw some of the equity in their homes in the form of monthly payments for life or a fixed term, or in a lump sum, or through a line of credit.

Importance: LENDERS making FHA *home equity conversion mortgages* are insured against loss when the home is sold. FHA insurance often requires that an upfront insurance premium be paid by the borrower, typically out of the proceeds of the loan.

HOME EQUITY LOAN / A loan secured by a second mortgage on one's principal residence, generally to be used for some nonhousing expenditure.

HOMEOWNERS AFFORDABILITY AND STABILITY PLAN (HASP) / U.S. program, announced on February 18, 2009, to help up to nine million homeowners avoid foreclosure. Initial funding of $75 billion was supplemented by $200 billion in additional funding for FANNIE MAE and FREDDIE MAC to purchase and more easily refinance mortgages.

More information: The *Homeowners Affordability and Stability Plan* is funded mostly by the HOUSING AND ECONOMIC RECOVERY ACT OF 2008. It uses cost sharing and incentives to encourage lenders to reduce homeowners' monthly payments to 31% of their gross monthly income. Under the program, a lender would be responsible for reducing total monthly mortgage payments (PITI) to no more than 38% of the borrower's income, with the government sharing the cost to further reduce the payment to 31%. The plan also involves potentially forgiving or deferring a portion of the borrower's mortgage balance.

HOMEOWNER'S INSURANCE POLICY / An insurance policy designed especially for homeowners. Usually protects the owner from losses caused by most common disasters, hazards, theft, and LIABILITY. Coverage and costs vary widely. Flood insurance is generally not included in standard policies and must be purchased separately.

HOMESTEAD / The status provided to a homeowner's principal residence by some state statutes; protects the home against judgments up to specified amounts.

Importance: In states that honor homesteads, the owner can continue possession and enjoyment of a home against the wishes of creditors.

HOMESTEAD EXEMPTION / In some jurisdictions, a reduction in the assessed value allowed for a person's principal residence.

Importance: In some states, the homestead exemption reduces assessed values for owners; additional exemptions may be claimed by persons past age 65.

HOUSING AND ECONOMIC RECOVERY ACT OF 2008 / Legislation passed to address the subprime housing crisis. The Act created a new regulator for the housing-related GSEs: FANNIE MAE, FREDDIE MAC, and the FEDERAL HOME LOAN BANKS.

More information: Under the *Housing and Economic Recovery Act of 2008*, the FEDERAL HOUSING FINANCE AGENCY (FHFA), which was created by merging the Office of Federal Housing Oversight and the Federal Housing Finance Board, was granted enhanced powers to establish standards, restrict asset growth, increase enforcement, and put entities into receivership.

Importance: Under the Act, the FHFA was given authority to provide capital infusions to Fannie Mae and Freddie Mac to avoid a failure on their obligations, thus increasing confidence in government-sponsored housing finance.

HOUSING STOCK / The total number of DWELLING UNITS in an area.

Example: In 2017, the *housing stock* in the United States comprised approximately 137 million units, of which 87% were occupied, with an owner-occupied rate of 64%. The housing stock includes mobile homes but excludes GROUP QUARTERS such as prisons, college dormitories, hospitals, hotels, and CONGREGATE HOUSING. The stock also includes approximately 17 million vacant housing units.

HUD / U.S. Department of Housing and Urban Development.

HVAC / An acronym that refers to the climate control system in buildings (heating, ventilation, and air-conditioning).

HYPOTHECATE / To pledge a thing as security without having to give up possession of it.

Importance: The word *hypothecate* comes (through Late Latin and French) from a Greek word meaning "to put down as a deposit" and has the same meaning as the French-derived word *mortgage.*

ILLIQUIDITY / Inadequate cash to meet obligations. Real estate is considered an illiquid investment because of the time and effort required to convert it to cash.

IMPACT FEE / An expense charged against private developers by the county or city as a condition for granting permission to develop a specific project. The purpose of the fee is to defray the cost to the city of expanding and extending public services to the development. Though most impact fees are collected from residential property, there can be no discrimination in favor of commercial or industrial property. By law, impact fees cannot be imposed to build new schools.

IMPLIED AGENCY / Occurs when the words and actions of the parties indicate that there is an agency relationship.

IMPROVEMENT RATIO / The value of *improvements* relative to the value of unimproved land. **Example:** Land worth $250,000 was improved with a $750,000 building. The improvement ratio is $750,000/$250,000 or 3 : 1.

IMPROVEMENTS / Additions to raw land that tend to increase value, such as buildings, streets, sewers, etc.

Importance: An improvement is anything except the raw land. *See also* HIGHEST AND BEST USE.

INCHOATE / (1) Recently or just begun; (2) unfinished, begun but not completed. **Examples:** In real estate, this term can apply to *dower* or *curtesy* rights prior to the death of a spouse, instruments that are supposed to be recorded, and interests that can ripen into a vested estate.

INCOME / The money or other benefit coming from the use of something. Gross sales or income is the full amount received; net income is the remainder after subtracting expenses. Many persons in real estate prefer to use *cash flow* as the measure of income, whereas those in accounting prefer *net income.*

INCOME APPROACH / One of the three appraisal methods used in arriving at an estimate of the market value of property; the value of the property is the present worth of the income it is expected to produce during its remaining life.

Importance: Annual income for rental property can be capitalized into value to estimate the property's worth.

INCOME MULTIPLIER / The relationship of price to income. *See also* GROSS RENT MULTIPLIER.

INCOME PROPERTY / Property whose ownership appeal is that it produces income. **Examples:** Office buildings, shopping centers, rental apartments, hotels.

INCOME STREAM / A regular flow of money generated by a business or an investment.

INCOMPETENT / A person who is unable to manage his/her own affairs by reason of insanity, imbecility, or feeblemindedness.

Importance: When conducting business with an incompetent, his/her guardian's consent is required.

INCURABLE DEPRECIATION / (1) A defect that cannot be cured or is not financially practical to cure; (2) a defect in the "bone structure" of a building.

Importance: When appraising real estate using the *cost approach,* incurable depreciation is separated from curable to indicate the actual loss in value sustained by the property.

INDENTURE / A written agreement made between two or more persons having different interests.

Importance: Indentures are used in mortgages, deeds, and bonds. They describe the terms of the agreement.

INDEPENDENT CONTRACTOR / A contractor who is self-employed for tax purposes. When real estate salespeople are self-employed, the broker is not required to withhold taxes.

INDEPENDENT FEE APPRAISER / A person who estimates the value of property but has no interest in the property and is not associated with a lending association or other investor.

INDEX / (1) A statistic that indicates some current economic or financial condition. Indexes are often used to make adjustments in wage rates, rental rates, loan interest rates, and pension benefits set by long-term contracts. (2) To adjust contract terms according to an index.

INDEX LEASE / A lease in which rentals are tied to an agreed upon index of costs. **Example:** Rentals are to increase along with the Consumer Price Index.

Importance: An index lease can provide fairness to both parties in a long-term leasing arrangement.

INDUSTRIAL PROPERTY / Property used for industrial purposes, such as factories and power plants.

Importance: Land to be used for industrial purposes must be zoned for that purpose.

INFLATION / A loss in the purchasing power of money; an increase in the general price level. Inflation is generally measured by the Consumer Price Index, published by the Bureau of Labor Statistics.

Importance: Real estate is considered a hedge against inflation because it tends to be long lasting and holds its value in real terms. As the value of the dollar drops, real estate tends to command more dollars. For example, a home purchased in 1967 for $50,000 was resold in 1976 for $100,000 and in 2019 for $300,000. The home did nothing to cause its price to change; inflation caused the house to command more dollars.

INFRASTRUCTURE / The basic public works of a city or subdivision, including roads, bridges, sewer and water systems, drainage systems, and essential public utilities.

INJUNCTION / A writ or order issued by a court to restrain one or more parties to a suit or a proceeding from performing an act that is deemed inequitable or unjust in regard to the rights of some other party or parties in the suit or proceeding.

Importance: An injunction can prevent a wrongdoing while the legal process continues, that is, before a final judgment is made about the rights of the parties.

INNOCENT PURCHASER / One who buys an asset without knowing of a flaw in the title or property. *See* BONA FIDE.

IN REM / Latin for "against the thing." A proceeding against *realty* directly, as distinguished from a proceeding against a person (used in taking land for nonpayment of taxes, and so on). By contrast, *in personam* means "against the person."

INSTALLMENTS / Parts of the same debt, payable at successive periods as agreed; payments made to reduce a mortgage.

Importance: Many debts are paid in installments that include interest for a recent period plus some amount for *amortization.*

INSTALLMENT TO AMORTIZE ONE DOLLAR / A mathematically computed factor, derived from *compound interest* functions, that offers the level periodic payment required to retire a $1.00 loan within a certain time frame. The periodic installment must exceed the periodic interest rate. *See also* AMORTIZATION; AMORTIZATION TERM.

INSTRUMENT / A written legal document, created to effect the rights and liabilities of the parties to it. **Examples:** Deed, will, lease.

INSURABLE TITLE / A title that can be insured by a title insurance company.

Importance: When acquiring real estate, a buyer should determine whether the title is insurable. If not, there are probably valid claims that will affect his/her use or ownership.

INSURANCE / Protection against loss resulting from hazards, such as fire and wind, over a period of time; the property owner's risk is assumed by the insurer in return for the payment of a policy premium.

Examples:

- FLOOD INSURANCE
- HAZARD INSURANCE
- HOMEOWNER'S (INSURANCE) POLICY
- TITLE INSURANCE

INSURANCE COVERAGE / Total amount and type of insurance carried.

Importance: It is conservative to retain insurance coverage based on the replacement value or cost of one's valuables.

INTANGIBLE VALUE / Value that cannot be seen or touched. **Example:** The goodwill of an established business.

INTEREST / (1) Money paid for the use of money; (2) the type and extent of ownership.

INTEREST RATE / (1) The percentage of a sum of money charged for its use; (2) the rate of return on an investment.

Importance: The loan interest rate is an important ingredient in determining the periodic installment payment.

INTERIM FINANCING / Loan, including a construction loan, with a term of generally three years or less; often used when a buyer expects interest rates to decline so as not to be locked into a long-term, *permanent mortgage* at a higher rate.

INTERPLEADER / A proceeding initiated by a neutral third party to determine the rights of rival claimants to property or a transaction.

Importance: An escrow agent can call for an interpleader when there is a dispute between the buyer and seller.

INTER VIVOS TRUST / A trust set up during one's lifetime.

INTESTACY / *See* INTESTATE.

INTESTATE / A person who dies leaving no will or a defective will. His/her property goes to his/her legal heirs.

Importance: State law determines inheritance rules for intestates. This is known as intestate succession. If there are no heirs, the property *escheats* to the state.

INVESTMENT ANALYSIS / A study of the likely return from a proposed investment with the objective of evaluating the amount an investor may pay for it, the investment's suitability to that investor, or the feasibility of a proposed real estate development.

INVESTMENT PROPERTY / Property that is owned for its income-generating capacity or expected resale value. **Examples:** Apartments, office buildings, undeveloped land.

INVESTMENT VALUE / The estimated value of a certain real estate investment to a particular individual or institutional investor; may be greater or less than *market value,* depending on the investor's particular situation.

INVOLUNTARY ALIENATION / A loss of property for nonpayment of debts such as taxes or mortgage foreclosure.

INVOLUNTARY LIEN / A lien imposed against property without the consent of the owner (unpaid taxes, special assessments, etc.).

Importance: A lien can be created without any action by the landowner.

IRREVOCABLE / Incapable of being recalled or revoked; unchangeable, unalterable.

JEOPARDY / Peril, danger, risk. **Example:** Property pledged as security for a delinquent loan is in jeopardy of *foreclosure.*

JOINT TENANCY / Ownership of realty by two or more persons, each of whom has an undivided interest with *right of survivorship.* **Example:** A and B own land in joint tenancy. Each owns half of the entire (undivided) property. Upon A's death, B will own the entire property, or vice versa.

JOINT VENTURE / An agreement between two or more parties who invest in a single business or property.

JUDGMENT / A court decree stating that one individual is indebted to another and fixing the amount of the indebtedness.

Importance: A judgment is a final determination of the matter, decided by a court.

JUDGMENT CREDITOR / A person who has received a court decree or judgment for money due to him/her.

JUDGMENT DEBTOR / A person against whom a judgment has been issued by a court for money owed.

JUDGMENT LIEN / The claim upon the property of a debtor resulting from a judgment. **Example:** A won't pay his debt to B. After establishing the debt in court, B may be allowed by the court to put a lien on A's real estate.

JUDICIAL FORECLOSURE / Procedure used when a trustee or mortgagee requests court supervision of a foreclosure action.

JUMBO LOAN / A loan whose principal is too large to be purchased by FNMA or FHLMC, a loan of $484,350 or more in 2019.

JUNIOR LIEN / *See* JUNIOR MORTGAGE.

JUNIOR MORTGAGE / A *mortgage* whose claim against the property will be satisfied only after prior mortgages have been sold; also called *junior lien.*

Importance: A junior (second, third) mortgage has value as long as the borrower continues payments or the property's value is in excess of the mortgage debts.

LACHES / Delay or negligence in asserting one's legal rights.

Importance: If a person does not act in a reasonable time to assert his/her rights, he/she may be barred from doing so because of the delay.

LAND / The surface of the earth; any part of the surface of the earth. (**Note:** Legal definitions often distinguish land from water.)

LAND CONTRACT / A real estate installment selling arrangement whereby the buyer may use, occupy, and enjoy land but no *deed* is given by the seller (so no title passes) until all or a specified part of the sale price has been paid.

Importance: In comparison to a deed, a land contract is easier to foreclose should the buyer fail to make the payments.

LAND LEASE / Only the ground is covered by the lease. *See also* GROUND LEASE.

LANDLORD / A person who rents property to another; a *lessor.*

LANDMARK / A fixed object serving as a boundary mark for a tract of land.

Importance: In surveying, a landmark serves as a reference point.

LAND PATENT / *See* PATENT.

LAND, TENEMENTS, AND HEREDITAMENTS / A phrase used in early English law to express all sorts of *real estate.*

Importance: This is the most comprehensive description of real estate.

LEAD-BASED PAINT / Considered a hazardous material. It is potentially poisonous, and its existence in property is to be disclosed. Its presence is often difficult to determine because applications of lead-based paint may have been covered over by more recent applications of paint that is free of lead.

LEASE / A contract in which, for a consideration called *rent,* one who is entitled to the possession of real property (the *lessor*) transfers those rights to another (the *lessee*) for a specified period of time.

Importance: A lease is an essential agreement, allowing an owner to transfer possession to a user for a limited amount of time.

LEASE ABSTRACT / A synopsis of the major points of a lease.

LEASED FEE / The landlord's ownership interest in a *property* that is under *lease. Contrast with* LEASEHOLD.

LEASEHOLD / The interest or estate on which a *lessee* (tenant) of real estate has his/her lease.

Importance: A leasehold can be quite valuable when the tenant's rent is below the market rate and the lease is for a long term.

LEASEHOLD IMPROVEMENTS / Fixtures, attached to *real estate*, that are generally acquired or installed by the tenant. Upon expiration of the *lease*, the tenant can generally remove them, provided that removal does not damage the property and is not in conflict with the lease. **Examples:** Cabinets, light fixtures, window treatments of a retail store in a leased building.

LEASEHOLD VALUE / The value of a tenant's interest in a *lease*, especially when the *rent* is below market level and the lease has a long remaining term.

LEASE WITH OPTION TO PURCHASE / A lease that gives the lessee (tenant) the right to purchase the property at an agreed-upon price under certain conditions.

Importance: Because the option allows but does not compel the purchase, it gives the tenant time to consider acquisition.

LEGAL DESCRIPTION / Legally acceptable identification of real estate by the (1) *government rectangular survey*, (2) *metes and bounds*, or (3) *lot and block number* method.

LESSEE / A person to whom property is rented under a *lease*; a tenant.

LESSOR / A person who rents property to another under a *lease*; a landlord.

LETTER OF INTENT (LOI) / The expression of a desire to enter into a CONTRACT without actually doing so.

LEVEL ANNUITY / *See* ANNUITY.

LEVERAGE / The use of borrowed funds to increase purchasing power and, ideally, to increase the profitability of an investment.

Importance: If the property increases in value or yields financial benefits at a rate above the borrowed money interest rate, leverage is favorable, also called positive. However, if the rate of property benefits is less than the interest rate, the investor's losses are increased.

LIAR LOAN / Loan in which the proposed borrower provided credit information that was not intended to be verified. *See also* NO-DOCUMENTATION LOAN.

Importance: During the housing boom of 2002–2006, *liar loans* were provided to borrowers who could have asserted an income level that was five times their actual income because their income and employment were not verified before the loan was originated.

LICENSE / (1) Permission; (2) a privilege or right granted by a state to an individual to operate as a real estate broker or salesperson.

Importance: (1) License allows a person to use property for a limited time. (2) In most states, a person must be licensed as a broker (or salesperson) to receive payment for a sale, lease, or other transaction.

LICENSED APPRAISER / Generally, an *appraiser* who meets certain state requirements but lacks the experience or expertise of a certified appraiser. *See also* CERTIFIED GENERAL APPRAISER; CERTIFIED RESIDENTIAL APPRAISER.

LICENSEE / A person who holds a real estate license.

Importance: In most states, only licensees are entitled to receive compensation for assisting with a real estate transaction. Education and the passing of examinations are requirements of licensing.

LIEN / A charge against property making it security for the payment of a debt, judgment, mortgage, or taxes; a lien is a type of *encumbrance*.

Importance: A lien makes the property collateral for a debt. Some liens may allow the property to be sold to satisfy the debt.

LIFE ESTATE / A freehold interest in land that expires upon the death of the owner or some other specified person.

Importance: A person with a life estate may use the property, but not abuse it, for as long as he/she lives. Then it reverts to the *remainderman*.

LIFE TENANT / A person who is allowed to use property for his/her lifetime or for the lifetime of another designated person. *See also* LIFE ESTATE.

LIKE-KIND PROPERTY / Property having the same nature. *See* SECTION 1031.

LIMITED LIABILITY COMPANY (LLC) / Organization form recognized in most states that may be treated as a partnership for federal tax purposes and has limited liability protection for the owners at the state level. The entity may be subject to the state franchise tax as a corporation. Most states also recognize *limited partnerships,* in which the individual partners are protected from the liabilities of the other partners. These entities are considered *partnerships* for both federal and state tax purposes. **Example:** A limited liability company may be an excellent way to own real estate, because it may provide many of the legal advantages of a corporation and the tax advantages of a partnership. States may impose restrictions, for example, by limiting the number of owners.

LIMITED PARTNERSHIP / PARTNERSHIP in which there is at least one partner who is passive and whose LIABILITY is limited to the amount invested and at least one partner whose liability extends beyond monetary investment. *See* GENERAL PARTNER; FAMILY LIMITED PARTNERSHIP. **Example:** Abel, a syndicator, forms a *limited partnership* with Price, Stone, and Wise. Abel invests his time and talent, is the general partner, and owns 10% of the partnership. Price, Stone, and Wise have each invested $30,000 cash and are limited partners. They buy property with a $90,000 down payment and a $500,000 mortgage. The property drops in value by $250,000. Price, Stone, and Wise lose their equity, and Abel, the general partner, is responsible for additional losses.

LIQUIDATED DAMAGES / An amount agreed upon in a contract that one party will pay the other in the event of a breach of the contract.

LIS PENDENS / Latin for "suit pending"; recorded notice that a suit has been filed, the outcome of which may affect title to a certain land.

Importance: Title to the property under consideration may be in jeopardy.

LISTING / (1) A written employment contract between a *principal* and an *agent* authorizing the agent to perform services for the principal involving the latter's property; (2) a record of property for sale by a broker who has been authorized by the owner to sell; (3) the property so listed. In some states, a listing must be in writing. *See also* EXCLUSIVE AGENCY LISTING; EXCLUSIVE RIGHT TO SELL LISTING; NET LISTING; OPEN LISTING.

LISTING PRICE / The price a seller puts on a home when it is placed on the market. Listing price, or ASKING PRICE, is generally considered a starting place for negotiations between the seller and a prospective buyer.

LITIGATION / The act of carrying on a lawsuit.

LITTORAL / Part of the shoreline of a static body of water, such as an ocean, bay, sea, lake, or pond.

Importance: Littoral rights differ from *riparian rights,* which pertain to a river or stream.

LOAN CLOSING / *See* CLOSING.

LOAN COMMITMENT / An agreement to lend money, generally of a specified amount, at specified terms at some time in the future. A locked-in interest rate may be included. *See* INTEREST RATE; LOCK-IN.

LOAN CONSTANT / *See* MORTGAGE CONSTANT.

LOAN ESTIMATE / A completed three-page form required by TRID to be given to residential mortgage applicants within three days of applying. The form provides a list of charges for mortgage fees and expenses. The borrower can compare this to the *Closing Disclosure* form that must be provided at least three days before closing.

LOAN EXPENSES / Fees associated with arranging a mortgage loan, including discount points, recording fees, appraisals, credit check, and legal fees.

LOAN FRAUD / Purposely giving incorrect information on a loan application to better qualify for a loan; may result in civil liability or criminal penalties.

LOAN-TO-VALUE RATIO (LTV) / The ratio obtained by dividing the mortgage principal by the property value.

Importance: Lenders typically provide loans with a stated maximum loan-to-value ratio. For conventional home loans, it is typically 80%. In many cases, it can be increased to 95% if mortgage insurance is purchased. VA and FHA loans may offer higher ratios.

LOCK-IN / An agreement to maintain a certain price or rate for a certain period of time.

Importance: In many *mortgage* commitments, the lender agrees to lock-in the interest rate for a certain period, such as 60 days. Sometimes a lock-in is provided only upon payment of a *commitment* fee.

LOT AND BLOCK NUMBER / A land description method that refers to a recorded plat. **Example:** Lot 6, Block F of the Sunnybrook Estates, District 2 of Rover County, Rhode Island.

LOT LINE / A line bounding a lot as described in a survey of the property.

Importance: Lot lines mark boundaries. There may also be building *setback* requirements, or *building lines*, within a lot.

LTV / *See* LOAN-TO-VALUE RATIO.

MAI / A professionally designated member of the Appraisal Institute.

Importance: The MAI designation is one of the most coveted in real estate. Many appraisals of large commercial properties are done by MAIs.

MAJORITY / The age at which a person is no longer a minor and is fully able to conduct his/her own affairs; in some states majority is 18.

Importance: A contract with a minor is voidable by the minor.

MANAGEMENT AGREEMENT / A CONTRACT between the owner of property and someone who agrees to manage it. Fees are generally 2–6% of the rental income. *See* PROPERTY MANAGEMENT.

MARGIN / A constant amount added to the value of the *index* for the purpose of adjusting the interest rate on an *adjustable rate mortgage.*

MARGINAL INCOME TAX RATE / The additional income tax paid on the next dollar earned.

MARGINAL PROPERTY / Property that is barely profitable to use. **Example:** The sale of cotton that has been efficiently raised yields $100, but the cotton cost $99.99 to raise. The land is therefore considered marginal land.

MARKETABILITY STUDY / An analysis, for a specific client, of the probable sales of a specific type of real estate product.

MARKETABLE TITLE / A title that a court will consider so free from defect that it will enforce its acceptance by a purchaser; similar to *insurable title.*

MARKET ANALYSIS / A study of the supply and demand conditions in a specific area for a specific type of property or service. A market analysis report is generally prepared by someone with experience in real estate, economics, or marketing. It serves to help decide what type of project to develop and is also helpful in arranging permanent and construction financing for proposed development.

MARKET APPROACH / *See* SALES COMPARISON APPROACH.

MARKET DATA APPROACH / *See* SALES COMPARISON APPROACH.

MARKET PRICE / The actual price paid in a market transaction; a historical fact.

Importance: *Market value* is a theoretical concept, whereas market price has actually occurred.

MARKET STUDY / *See* MARKET ANALYSIS.

MARKET VALUE / The highest price a buyer, willing but not compelled to buy, will pay, and the lowest price a seller, willing but not compelled to sell, will accept. Many conditions are assumed to exist.

Importance: In theory, property would sell for its market value.

MASS APPRAISING / An effort, typically used by tax *assessors,* to determine the salient characteristics of properties in a given submarket, to allow an approximation of value for each. Sophisticated statistical techniques are used frequently in mass appraising.

MATERIAL FACT / A fact that is germane to a particular situation; one that participants in the situation may reasonably be expected to consider.

Importance: In a contract, a material fact is one without which the contract would not have been made.

MECHANIC'S LIEN / A lien given by law upon a building or other improvement upon land, and upon the land itself, as security for the payment for labor done upon, and materials furnished for, the improvement.

Importance: A mechanic's lien protects persons who helped build or supply materials.

MEDIAN PRICE / The price of the house that falls in the middle of the total number of homes sold in that area. The median is generally less than the average because the average is increased by a few high-priced sales.

MEETING OF THE MINDS / Agreement by all parties to a contract to its terms and substance.

Importance: When there is a meeting of the minds, the contract is not based on secret intentions of one party that were withheld from another.

MERIDIAN / North-south line used in government rectangular survey.

METES AND BOUNDS / A land description method that relates the boundary lines of land, setting forth all the boundary lines together with their terminal points and angles.

Importance: A person can follow a metes and bounds description on a plat or on the ground.

MILL / One-tenth of a cent; used in expressing tax rates on a per dollar basis. **Example:** A tax rate of 60 mills means that taxes are 6 cents per dollar of assessed valuation.

MINERAL LEASE / An agreement that provides the lessee the right to excavate and sell minerals on the property of the lessor or to remove and sell petroleum and natural gas from the pool underlying the property of the lessor. In return, the lessor receives a royalty payment based on the value of the minerals removed.

MINERAL RIGHTS / The privilege of gaining income from the sale of oil, gas, and other valuable resources found on land.

MINOR / A person under an age specified by law (18 in most states).

Importance: Real estate contracts entered into with minors are voidable by the minor.

MINORITY DISCOUNT / A reduction from the MARKET VALUE of an asset because the MINORITY INTEREST owner(s) cannot direct the business operations and the interest lacks marketability.

MISREPRESENTATION / An untrue statement, whether deliberate or unintentional. It may be a form of nondisclosure where there is a duty to disclose or the planned creation of a false appearance. Where there is misrepresentation of *material fact,* the person injured may sue for *damages* or rescind the contract.

MONETARY POLICY / Actions of the Federal Reserve Bank that affect the value of the nation's currency. Monetary policy has a strong influence on market interest rates and the rate of monetary inflation (declining purchasing power of the dollar). *Contrast with* FISCAL POLICY.

MONUMENT / A fixed object and point designated by surveyors to establish land locations. **Examples:** Posts, pillars, stone markers, unique trees, stones, pipes, watercourses.

MORATORIUM / A time period during which a certain activity is not allowed.

MORTGAGE / A written instrument that creates a lien upon real estate as security for the payment of a specified debt.

Importance: The mortgage allows a defaulted debt to be satisfied by forcing a sale of the property.

MORTGAGE BANKER / One who originates, sells, and services *mortgage* loans. Most loans are insured or guaranteed by a government agency or private mortgage insurer.

MORTGAGE BROKER / One who, for a fee, places loans with investors but does not service such loans. On residential loans, the fee is commonly paid by the lender so that the broker represents the lender rather than the borrower.

MORTGAGE COMMITMENT / An agreement between a lender and a borrower to lend money at a future date, subject to the conditions described in the agreement.

Importance: The terms of the commitment are important, especially the interest rate *lock-in,* if there is one.

MORTGAGE CONSTANT / The percentage ratio between the annual mortgage payment and the original amount of the debt.

MORTGAGEE / A person who holds a *lien* on or *title* to property as security for a debt.

Importance: The mortgagee receives the lien; the *mortgagor* receives the loan.

MORTGAGE LIFE INSURANCE / A policy that guarantees repayment of a mortgage loan in the event of death or, possibly, disability of the MORTGAGOR. Sometimes called MORTGAGE INSURANCE, not to be confused with PRIVATE MORTGAGE INSURANCE.

MORTGAGE MODIFICATION / Legislation and Treasury Department actions aimed at providing lenders with incentives to work with borrowers to avoid foreclosure.

Importance: To avoid foreclosure on homes they held loans on, some banks encouraged *mortgage modifications* to reduce or postpone a homeowner's mortgage payments. Some programs were limited to borrowers who were not in default. *See also* HOME AFFORDABLE MODIFICATION PLAN; HOMEOWNERS AFFORDABILITY AND STABILITY PLAN.

MORTGAGOR / A person who pledges his/her property as security for a loan. *See also* MORTGAGEE.

MOST PROBABLE SELLING PRICE / A property's most likely selling price when not all the conditions required for a *market value* estimate are relevant. **Example:** An appraiser estimated a property's most likely sales price at $100,000, assuming a sale within 20 days, whereas its market value of $120,000 might require up to 6 months to realize.

MULTIPLE LISTING / An arrangement among a group of real estate *brokers*; they agree in advance to provide information about some or all of their listings to the others and also agree that commissions on sales of such listings will be split between listing and selling brokers.

MULTIPLIER / A factor, used as a guide, applied by multiplication to derive or estimate an important value. **Examples:** (1) A *gross rent multiplier* of 6 means that property renting for $12,000 per year can be sold for six times that amount, or $72,000. (2) A population multiplier of 2 means that, for each job added, two people will be added to a city's population.

NAR / *See* NATIONAL ASSOCIATION OF REALTORS®.

NATIONAL ASSOCIATION OF REALTORS® (NAR) / An organization devoted to encouraging professionalism in real estate activities.

Importance: The NAR has strong lobbyists who protect the interest of the real estate community, especially of homeowners. It also has affiliates related to appraising, counseling, and managing real estate.

NEGATIVE AMORTIZATION / An increase in the outstanding balance of a loan resulting from the inadequacy of periodic debt service payments to cover required interest charged on the loan. Generally occurs under indexed loans for which the applicable interest rate may be increased without increasing the monthly payments. Negative amortization will occur if the indexed interest rate is increased.

NET INCOME / In real estate, this term is now *net operating income*. In accounting, net income is the actual earnings, after deducting all expenses, including interest and depreciation, from gross sales.

NET LEASABLE AREA (NLA) / For office and retail properties, the portion used exclusively by the tenant; generally excludes hallways, restrooms, and other common areas.

Importance: The rent per square foot may be judged consistently among buildings when it is based on this space measurement.

NET LEASE / A lease whereby, in addition to the rent stipulated, the lessee (tenant) pays such things as taxes, insurance, and maintenance. The landlord's rent receipt is thereby net of those expenses.

Importance: The responsibility for maintenance costs is shifted to the lessee; the lessor is a passive investor.

NET LISTING / A listing in which the broker's *commission* is the excess of the sale price over an agreed-upon (net) price to the seller; illegal in some states. **Example:** A house is listed for sale at $100,000 net. The broker's commission is $1 if it sells for $100,001. But if it sells for $150,000, the broker receives $50,000.

NET OPERATING INCOME (NOI) / Income from property or business after operating expenses have been deducted but before deducting income taxes and financing expenses (interest and principal payments).

NLA / *See* NET LEASABLE AREA.

NO-DOCUMENTATION LOAN / Mortgage loan, sometimes called a "no doc," for which the borrower is not required to provide proof of income, employment, or assets. Such loans evolved from *low-documentation loans*, which were originally designed for applicants in the Alt-A loan category, who could satisfy credit standards with high credit scores or low loan-to-value ratios. *See also* ALT-A MORTGAGES; LIAR LOAN.

NOI / *See* NET OPERATING INCOME.

NONCONFORMING LOAN / Home mortgage loan that does not meet the standards of, or is too large to be purchased by, FNMA or FHLMC. Typically, the interest rate is at least half a percentage point higher than for a conforming loan.

NONCONFORMING USE / A use that violates *zoning ordinances* or codes but is allowed to continue because it began before the zoning restriction was enacted.

Importance: This allows a prior use to continue but puts restrictions on future uses of the property.

NONRECOURSE / Carrying no personal liability. Lenders may take the property pledged as collateral to satisfy a debt but have no *recourse* to other assets of the borrower.

NOTARY PUBLIC / An officer who is authorized to take acknowledgments to certain types of documents, such as *deeds, contracts,* and *mortgages,* and before whom affidavits may be sworn.

Importance: Most documents must be notarized as a condition of being recorded.

NOTE / A written instrument that acknowledges a debt and promises to pay.

Importance: A note is enforceable in a court of law. Collateral for the note may be sold to satisfy the debt.

NOTICE TO QUIT / A notice to a tenant to vacate rented property.

Importance: The tenant is permitted to complete the term of the lease, except in cases of *tenancy at will* or *tenancy at sufferance.*

NOVATION / Substitution of a revised agreement for an existing one, with the consent of all parties involved. Technically, any time parties to an existing contract agree to change it, the revised document is a *novation.*

NULL AND VOID / Having no legal validity.

OBLIGEE / The person in whose favor an obligation is entered into.

OBLIGOR / The person who binds himself/herself to another; one who has engaged to perform some obligation; one who makes a bond.

OBSOLESCENCE / (1) A loss in value due to reduced desirability and usefulness of a structure because its design and construction have become obsolete; (2) loss due to a structure's becoming old-fashioned, not in keeping with modern needs, with consequent loss of income; and (3) changes outside the property.

Importance: Obsolescence can cause a loss in value just as *physical deterioration* does.

OFFER AND ACCEPTANCE / *See* AGREEMENT OF SALE.

OFFERING PRICE / The amount a prospective buyer offers for a property on the market. When an offer is submitted via a sales contract, it constitutes a bona fide offer for the property.

OFFEROR / One who extends an offer to another.

OFFICE BUILDING / A structure primarily used for conducting business, such as administration, clerical services, and consultation with clients and associates. Such buildings can be large or small and may house one or more business concerns.

OFFICE OF FEDERAL HOUSING ENTERPRISE OVERSIGHT (OFHEO) / *See* FEDERAL HOUSING FINANCE AGENCY.

OPEN-END MORTGAGE / A mortgage under which the mortgagor (borrower) may secure additional funds from the mortgagee (lender), usually stipulating a ceiling amount that can be borrowed.

Importance: A development in real estate finance is a line-of-credit home equity loan. This works the same as an open-end mortgage.

OPEN LISTING / A listing given to any number of brokers without liability to compensate any except the one who first secures a buyer ready, willing, and able to meet the terms of the listing or secures the seller's acceptance of another offer. The sale of the property automatically terminates all open listings.

OPEN MORTGAGE / A mortgage that has matured or is overdue and is therefore open to foreclosure at any time.

OPERATING EXPENSE RATIO / A mathematical relationship derived by dividing *operating expenses* by *potential gross income.*

Importance: A comparison of rents for properties would be incomplete without also comparing operating expenses. Apartments generally have operating expense ratios between 30 and 50 percent; this may be exceeded when the lessor pays utilities or the apartments are in low-rent areas. Office buildings often have higher operating expense ratios (40% and 60%) because more intensive management and maintenance, such as cleaning services, are provided.

OPERATING EXPENSES / Amounts paid to maintain property, such as repairs, insurance, property taxes, but not including financing costs or depreciation.

OPERATING LEASE / A lease between the lessee and a sublessee who actually occupies and uses the property.

Importance: In an operating lease, the lessee runs the property; by contrast in a *financing lease,* the lessee becomes lessor to the operating tenant.

OPERATING STATEMENTS / Financial reports on the cash flow of a business or property. *See* CASH FLOW; RENT ROLL.

OPTION / The right, but not the obligation, to purchase or lease a property upon specified terms within a specified period. **Example:** Moore purchases an *option* on a piece of land. The option cost is $2,500. The option runs 90 days, and the land will cost $1,000 per acre. If Moore wishes to purchase the land, she has 90 days to exercise her right and may buy the land for $1,000 per acre. If she decides not to purchase, she FORFEITS the $2,500 option cost.

OPTIONEE / One who receives or purchases an option.

OPTIONOR / One who gives or sells an option.

OPTION TO PURCHASE / A contract that gives one the right (but *not* the obligation) to buy a property within a certain time, for a specified amount, and subject to specified conditions.

ORAL CONTRACT / An unwritten agreement. With few exceptions, oral agreements for the sale or use of real estate are unenforceable. However, an oral lease for less than 1 year is valid in many states.

Importance: When dealing with real estate, agreements should be in writing.

ORIGINAL EQUITY / The amount of cash initially invested by the underlying *real estate* owner; distinguished from sweat equity or payments made after loan is made.

OUTPARCEL / *See* PAD SITE.

OVERAGE / *See* PERCENTAGE RENT.

OVERALL CAPITALIZATION RATE (OVERALL RATE OF RETURN) / The rate obtained by dividing *net operating income* by the purchase price of the property.

Importance: Rates of return from properties may be compared to each other, or the rate may be divided into income to estimate property value.

OWNERSHIP RIGHTS TO REALTY / Possession, enjoyment, control, and disposition.

PACKAGE MORTGAGE / A mortgage arrangement whereby the principal amount loaned is increased to include *personalty* (e.g., appliances) as well as *realty*; both realty and personalty serve as collateral.

PAD SITE / An individual freestanding site for a retailer, often adjacent to a larger shopping center. The site is generally more than ½ acre but less than 2 acres. Also referred to as an *outparcel.*

P&I / Principal and interest (payment).

PARCEL / A piece of property under one ownership; a lot in a subdivision.

PARTIALLY AMORTIZED LOAN / A loan that requires some payments toward *principal* but does not fully retire the debt, thereby requiring a *balloon payment. See also* AMORTIZATION.

PARTIAL OR FRACTIONAL INTEREST / The ownership of some, but not all, the rights in *real estate.* **Examples:** (1) *Leasehold;* (2) *easement;* (3) *hunting rights.*

PARTITION / The division of real property between those who own it in undivided shares. **Example:** A and B own land as tenants in common until they partition it. Thereafter, each owns a particular tract of land.

PARTNERSHIP / An agreement between two or more entities to go into business or invest. Either partner may bind the other, within the scope of the partnership. Each partner is liable for all the partnership's debts. A partnership normally pays no taxes but merely files an information return. The individual partners pay personal income tax on their share of income.

PARTY WALL / A wall built along the line separating two properties, lying partly on each. Either owner has the right to use the wall and has an *easement* over that part of the adjoining owner's land covered by the wall.

PATENT / Conveyance of title to government land; also called a *land patent.*

PAYMENT CAP / A contractual limit on the percentage amount of adjustment allowed in the monthly payment for an *adjustable rate mortgage* at any one adjustment period. Generally, it does not affect the interest rate charged. If the allowable payment does not cover interest due on the principal at the adjusted rate of interest, *negative amortization* will occur.

PERCENTAGE LEASE / A lease of property in which the rental is based on a percentage of the volume of sales made upon the leased premises. It usually provides for minimum rental and is regularly used for retailers who are tenants.

Importance: The retailer pays additional rent only if sales are high; the shopping center owner has an incentive to make an attractive shopping area.

PERCENTAGE RENT / The rent payable under a *percentage lease*; also called *overage.* Typically, the percentage applies to sales in excess of a preestablished base amount of the dollar sales volume.

Importance: Percentage rent provides incentive to a landlord for making a store or shopping area appeal to the market.

PERIODIC ESTATE / A lease, such as from month to month or year to year. Also known as periodic tenancy.

PERMANENT MORTGAGE / A mortgage for a long period of time (more than 10 years).

Importance: A permanent mortgage usually replaces construction or interim financing, and provides steady interest income plus *amortization* payments to the lender. For the borrower, it means that there is no need to seek new financing.

PERSONAL LIABILITY / An individual's responsibility for a debt. Most mortgage loans on real estate are *recourse* (i.e., the lender can look to the property and the borrower for repayment). *Contrast with* NONRECOURSE.

PERSONALTY / Personal property, that is, all property that is not *realty.*

Importance: Many laws and terms that apply to real property are not the same as those for personalty. When dealing with both types, appropriate law and terminology must be applied.

PHYSICAL DEPRECIATION (DETERIORATION) / The loss of value from all causes of age and action of the elements. **Examples:** Faded paint, broken window, hole in plaster, collapsed porch railing, sagging frame.

PLANNED UNIT DEVELOPMENT (PUD) / A zoning or land-use category for large tracts that allows several different densities and forms of land use, planned as a single, well-integrated unit.

PLAT / A plan or map of a certain piece or certain pieces of land. **Examples:** A subdivision plat or a plat of one lot.

PLAT BOOK / A public record containing maps of land showing the division of the land into streets, blocks, and lots and indicating the measurements of the individual parcels.

Importance: The tax assessors office, usually in a city or county, maintains a plat book that is open for public inspection.

PLOTTAGE / Increment in the value of a plot of land that has been enlarged by assembling smaller plots into one ownership.

Importance: Combining several small tracts into one ownership can provide a large enough land area for a more profitable use than would be possible otherwise. However, it is often difficult to get several owners to sell as some hold out for a high price.

PMI / Abbreviation for private mortgage insurance.

POACHER / Someone who trespasses upon another's land for the purpose of illegal hunting, trapping, or fishing. Unlike a squatter, a poacher does not build or occupy a structure on the land but removes property (wildlife) that belongs to the owner.

POCKET CARD / Identification required for *salespersons* and *brokers* in most states.

Importance: Issued by the state licensing agency, it identifies its holder as a licensee and must be carried at all times.

POINTS / Fees paid to induce lenders to make a *mortgage* loan. Each point equals 1% of the loan principal. Points have the effect of reducing the amount of money advanced by the lender, thus increasing the effective interest rate.

POLICE POWER / The right of any political body to enact laws and enforce them for the order, safety, health, morals, and general welfare of the public.

Importance: Government authorities get the power of *eminent domain* through their police power.

POTENTIAL GROSS INCOME / The theoretical amount of money that would be collected in a year if all units in a rental building were fully occupied all year; also called *gross possible rent.*

POWER CENTER / A shopping center with few TENANTS, most of them, junior anchor, or ANCHOR TENANTS. Generally a *power center*'s anchor tenants are "category killers," that is, the dominant retailers in the markets they serve.

POWER OF ATTORNEY / An instrument authorizing a person to act as the agent of the person granting it.

Importance: By using a power of attorney, one can designate a specific person to do everything or just certain limited activities.

PREAPPROVAL / A bank or mortgage banker's determination of the amount to be lent to a borrower based on verification of the borrower's financial information.

Importance: Preapproval is stronger than *prequalification,* which is a preliminary estimate of a loan amount, typically without income verification or assurance of offering a loan.

PREMISES / Land and tenements; an estate; the subject matter of a *conveyance.*

PREPAYMENT CLAUSE / A clause in a mortgage that gives a mortgagor (borrower) the privilege of paying the mortgage indebtedness before it becomes due.

Importance: Sometimes a penalty must be paid if prepayment is made, but payment of *interest* that is not yet due is waived.

PREQUALIFICATION / A real estate salesperson's or broker's approximation of housing price affordability based on a buyer's estimated future income. Typically, there is no verification or assurance of a loan.

Importance: Prequalification provides an estimate to allow a prospective buyer to consider only housing choices that are within his/her means. It does not assure a loan.

PRESERVATION DISTRICT / A zoning designation to protect and maintain wildlife, park land, scenic areas, or historic districts.

PRICE FIXING / Illegal effort by competing businesses to maintain the same price, such as the commission rate on the sale of real estate.

PRIMA FACIE EVIDENCE / Evidence that on its face will establish a given fact.

PRIMARY LEASE / A lease between the owner and a tenant who, in turn, has sublet all or part of his/her interest.

Importance: The tenant in the primary lease is still responsible to the landlord, even though the subtenant(s) occupy the space.

PRINCIPAL / (1) The employer of an *agent* or *broker*; the broker's or agent's client; (2) the amount of money raised by a mortgage or other loan, as distinct from the interest paid on it.

PRIVATE MORTGAGE INSURANCE (PMI) / Protection for the lender in the event of DEFAULT, usually covering a portion of the amount borrowed. *See also* MORTGAGE LIFE INSURANCE.

PROBATE (PROVE) / To establish the validity of the will of a deceased person.

Importance: Probate relates not only to the validity of a will but also to matters and proceedings of estate administration.

PROBATE COURT / *See* SURROGATE'S COURT.

PROCURING CAUSE / A legal term that means the cause resulting in accomplishing a goal. Used in real estate to determine whether a broker is entitled to a commission.

PRODUCTION BUILDER, PRODUCTION HOME / One who builds homes with plans and specifications that have been replicated in an efficient or assembly-line process. *See also* TRACT HOME; *contrast with* CUSTOM BUILDER.

PROGRESSIVE TAX / Tax whose burden falls more heavily on the wealthy or high-income groups than on poor or low-income groups. *Contrast* PROPORTIONAL TAX; REGRESSIVE TAX. **Example:** The federal estate tax is considered a *progressive tax* because it taxes only high-value estates. It is paid by estates of those who die with significant assets.

PROJECTION PERIOD / The time duration for estimating future *cash flows* and the resale proceeds from a proposed real estate investment.

PROMISSORY NOTE / A promise to pay a specified sum to a specified person under specified terms.

PROPERTY / (1) The rights that one individual has in lands or goods to the exclusion of all others; (2) rights gained from the ownership of wealth. *See also* PERSONALTY; REAL PROPERTY.

PROPERTY LINE / The recorded boundary of a plot of land.

Importance: A *survey* is performed in order to establish property lines and describe them on a *plat.*

PROPERTY MANAGEMENT / The operation of property as a business, including rental, rent collection, maintenance.

Importance: Property managers remove the daily burden from real estate investors, thus allowing investors to be free from daily business operations.

PROPORTIONAL TAX / Tax whose burden is applied at the same rate to the poor as to the wealthy. *Contrast* PROGRESSIVE TAX; REGRESSIVE TAX.

PROPRIETORSHIP / Ownership of a business, including income-producing real estate, by an individual, as contrasted with a partnership or corporation.

PRORATE / To allocate between seller and buyer their proportionate shares of an obligation paid or due; for example, to prorate real property taxes or insurance.

Importance: Many items of expense are prorated between buyer and seller to the date of closing.

PROTECTED CLASS / Identified minority subgroup of the population that cannot be legally discriminated against under federal law.

Importance: A landlord or home seller cannot legally refuse to lease or sell to members of *protected classes.* The seven protected classes under the *Fair Housing Law* are race, color, religion, sex, handicap, familial status, and national origin.

PROTECTION CLAUSE / Clause in a LISTING contract that protects the BROKER who shows the listed property. If the owner waits for the listing to expire before selling to a potential buyer to whom the property was shown, the broker is still entitled to a commission.

PUD / PLANNED UNIT DEVELOPMENT.

PUR AUTRE VIE / For the life of another. A life estate pur autre vie grants a life estate to one person that expires on the death of another person.

Importance: Allows an estate to terminate upon the death of someone other than the life tenant.

PURCHASE AGREEMENT / *See* AGREEMENT OF SALE.

PURCHASE AND SALE AGREEMENT / *See* AGREEMENT OF SALE.

PURCHASE CAPITAL / The amount of money used to purchase real estate, regardless of the source.

PURCHASE CONTRACT / *See* AGREEMENT OF SALE; CONTRACT OF SALE.

PURCHASE MONEY MORTGAGE / A mortgage given by a grantee (buyer) to a grantor (seller) in part payment of the purchase price of real estate.

Importance: Institutional lenders are often unable or unwilling to finance certain types of property, so the seller must accept a purchase money mortgage to facilitate a sale.

QUALIFIED / *See* PREQUALIFICATION.

QUIET ENJOYMENT / The right of an owner, or any other person legally entitled to possession, to the use of property without interference.

Importance: No interference should be caused by a landlord to a tenant who is in compliance with a lease.

QUIET TITLE SUIT / A suit in court to remove a defect, cloud, or suspicion regarding the legal rights of an owner to a certain parcel of *real property.*

Importance: A potential claimant is told to bring forward his/her claim so its validity can be judged. If it is not valid, the claimant must stop interference with the owner.

QUITCLAIM DEED / A deed that conveys only the *grantor's* rights or interest in real estate without stating their nature and with no warranties of ownership.

Importance: This deed is often used to remove a possible cloud from the title.

RANGE LINES / In the *government rectangular survey* method of land description, lines parallel to the principal *meridian,* marking off the land into 6-mile strips known as ranges; they are numbered east or west of the principal meridian. *See also* BASE AND MERIDIAN.

RATE OF INTEREST / *See* INTEREST RATE.

REAL ESTATE / Land and all attachments that are of a permanent nature.

Importance: Real estate is distinguished from personal property. At one time real estate was the sole source of wealth and achieved a special place in the law because of its importance.

REAL ESTATE AGENT / *See* AGENT.

REAL ESTATE EDUCATORS ASSOCIATION / A professional organization composed primarily of persons who teach *real estate* in junior colleges and proprietary license preparation schools.

REAL ESTATE INVESTMENT TRUST (REIT) / A *real estate* mutual fund, allowed by income tax laws to avoid the corporate tax if 95% of its income is distributed. It sells shares of ownership and must invest in real estate or mortgages.

Importance: A REIT allows small investors to participate in the ownership of large, potentially profitable real estate projects.

REAL ESTATE SETTLEMENT PROCEDURES ACT (RESPA) / A law that states how *mortgage* lenders must treat those who apply for federally related real estate loans on property with 1–4 dwelling units. Intended to provide borrowers with more knowledge when they comparison shop for mortgage money. Generally replaced by TRID in October 2015. *See* TRID.

REAL PROPERTY / The right to use real estate as (1) *fee simple* estate, (2) *life estate,* or (3) *leasehold* estate; sometimes also defined as *real estate.*

REALTOR® / A professional in real estate who subscribes to a strict code of ethics as a member of the local and state boards and of the *National Association of Realtors®*.

Importance: Fewer than half of those licensed to sell real estate are REALTORS®. In many areas, a person must be a REALTOR® to participate in the predominant *multiple listing* service.

REALTY / The property rights to real estate.

REAPPRAISAL LEASE / A lease whereby the rental level is periodically reviewed and reset by independent appraisers.

Importance: Landlord and tenant can agree on a long-term lease knowing that the rent will be fair throughout the term because of the reappraisal clauses in the lease.

RECAPTURE CLAUSE / In a contract, a clause permitting the party who grants an *interest* or right to take it back under certain conditions.

RECESSION / Economic slowdown; officially declared after two consecutive quarters of reduced gross domestic product.

RECIPROCAL EASEMENT AGREEMENT (REA) / Agreement between owners of adjacent parcels allowing mutual use of both parcels for certain purposes.

RECISION / *See* RESCISSION.

RECORDING / The act of entering *instruments* affecting the title to real property in a book of public record.

Importance: Recording in this manner gives public notice of the facts recorded.

RECOURSE / The ability of a lender to claim money from a borrower in *default*, in addition to the property pledged as *collateral*.

REDFIN / Popular website of homes for sale.

REDLINING / An illegal practice of a lender refusing to make home loans in certain areas. The term is derived from circling, with red pencil on a map, areas where the institution will not lend.

Importance: If home loans are not made in a certain area, property values will plummet and neighborhoods will deteriorate rapidly. Redlining is an illegal, discriminatory practice.

REDUCTION CERTIFICATE / A document in which the mortgagee (lender) acknowledges the sum due on the mortgage loan.

Importance: This is used when mortgaged property is sold and the buyer assumes the debt.

REGRESSIVE TAX / Tax that takes a higher percentage of the earnings of a low-income family than of those of a high-income family. A sales tax on food is considered regressive because low-income people pay the same amount of tax on a loaf of bread, for example, as do high-income people.

REGULATION Z / Implementation by the Federal Reserve of the federal Truth in Lending Act; it specifies how the *annual percentage rate* of a loan is calculated and expressed in consumer loan documents.

REGULATORY TAKING / A series of government limits to property use that constitutes a condemnation of property.

REIT / *See* REAL ESTATE INVESTMENT TRUST.

RELEASE / The act by which some claim or interest is surrendered.

RELEASE CLAUSE / A clause in a mortgage that gives the owner of the property the privilege of paying off a portion of the indebtedness, thus freeing a part of the property from the mortgage.

Importance: Release clauses are frequently used when a mortgage covers more than one property (*blanket mortgage*) so that a particular parcel can be released upon some payment.

RELICTION / Gradual subsidence of waters, leaving dry land.

Importance: Ownership of land beneath a lake, for example, can become more important as reliction occurs.

RELOCATION SERVICE / A company that contracts with other firms to arrange the relocation of employees from one city to another. The service generally handles the sale of the employee's home and purchase of a new home. Furniture-moving services may also be included.

REMAINDER / An estate that takes effect after the termination of a prior estate, such as a *life estate*.

Importance: The *remainderman* owns the property outright upon the death of the *life tenant*.

REMAINDERMAN / The person who is to receive possession of a property after the death of a *life tenant*.

Importance: Many people wish to allow a surviving spouse to occupy property for the rest of his/her life, with a child as the remainderman.

REMEDIATION / The cleanup of an environmentally contaminated site. *See* CERCLA.

RENT / The compensation paid for the use of real estate.

Importance: Rent is the most important portion of a lease and may be paid in money, services, or other valuable.

RENT MULTIPLIER / *See* GROSS RENT MULTIPLIER.

RENT ROLL / A list of tenants, generally with the lease rent and expiration date for each tenant.

REPLACEMENT COST / The cost of erecting a building to take the place of or serve the functions of a previous structure.

Importance: Replacement cost often sets the upper limit on value; it is often used for insurance purposes.

REPORT DATE / In an appraisal, usually the date of the last property inspection.

REPRODUCTION COST / The normal cost of exact duplication of a property as of a certain date. (**Note:** Replacement requires the same functional utility for a property, whereas a reproduction is an exact duplicate, using the same materials and craftsmanship.)

RESALE PRICE / In a projection of real estate investment performance, the selling price that it is assumed a property could fetch at the end of the projection period. *See also* RESALE PROCEEDS.

RESALE PROCEEDS / Net cash remaining to investor after sale of investment property and paying mortgage payoff and selling costs.

RESCISSION / The act of canceling or terminating a contract. Rescission is allowed when the contract was induced by fraud, duress, misrepresentation, or mistake. *Regulation Z* allows one to rescind certain credit transactions within three business days (not applicable to first mortgages on a home); purchasers of certain land that must be registered by the Department of Housing and Urban Development may rescind within three business days.

RESERVATION PRICE / The highest price a buyer can pay and still achieve his or her primary objectives, such as keeping the monthly payments affordable or paying no more than MARKET VALUE for the property. A buyer will negotiate in hopes of keeping the sales price at or below his or her reservation price.

RESIDENTIAL SALES COUNCIL / An affiliate of the Realtors National Marketing Institute of the *National Association of Realtors®* that provides educational and promotional materials for members, most of whom are involved in residential real estate sales or brokerage.

RESPA / *See* REAL ESTATE SETTLEMENT PROCEDURES ACT.

RESTRAINT ON ALIENATION / A legal situation that would, if allowed to be enforced, prevent property from being sold easily. Restraints on alienation are against public policy, so they cannot be enforced by law.

RESTRICTION / A limitation placed upon the use of property, contained in the deed or other written *instrument* in the chain of title.

Importance: If buying property with restrictions, the buyer should determine its suitability for the uses he/she requires.

RESTRICTIVE COVENANT / *See* RESTRICTION.

REVERSION / The right of a lessor to possess leased property upon the termination of a lease.

Importance: A lease is valid for an established term, after which the lessor receives the reversion.

REVERSIONARY INTEREST / The interest a person has in property upon the termination of the preceding estate.

Importance: A lessor's interest in leased property is a reversionary interest.

REVOCATION / The recalling of a power of authority conferred, as a revocation of a power of attorney, a license, an agency, etc.

Importance: A person with the authority to convey may also have authority to revoke, with reason.

RIDER / An amendment or attachment to a contract.

RIGHT OF FIRST REFUSAL / The opportunity of a party to match the TERMS of a proposed contract before the contract is EXECUTED. **Example:** In some states, tenants whose apartments are converted to CONDOMINIUMS are given the *right of first refusal* for the unit when it is sold. This means the tenant may purchase the unit at the same price and terms offered to any outside purchaser.

RIGHT OF SURVIVORSHIP / The right of a surviving joint tenant to acquire the interest of a deceased joint owner; the distinguishing feature of *joint tenancy* and *tenancy by the entireties.*

Importance: The right of survivorship is often used where the joint tenants are closely related.

RIGHT-OF-WAY / (1) The right to use a particular path for access or passage; a type of easement; (2) the areas of subdivisions dedicated to government for use as streets, roads, and other public access to lots.

RIPARIAN OWNER / A person who owns land bounding a river or stream.

RIPARIAN RIGHTS / Rights pertaining to the use of water on, under, or adjacent to one's land. Riparian rights apply only to moving water. Frontage on a body of water that is relatively static, such as a bay, sea, ocean, lake, or pond, confers *littoral rights.*

Importance: In most states, riparian rights do not permit property owners to alter the flow of water to their downstream neighbors.

RISK PREMIUM / The difference between the required interest rate on an investment and the rate on risk-free investments such as U.S. Treasury securities.

ROBO-SIGNER / Epithet applied to persons at banks who signed hundreds of forms each day to proceed with FORECLOSURE without investigating the validity of information on the forms, as was required by law.

Importance: When it was disclosed that thousands of foreclosures were based on documents signed by *robo-signers,* an uproar occurred that embarrassed lenders who were responsible for WRONGFUL FORECLOSURES.

SALE-LEASEBACK / The simultaneous purchase of property and lease back to the seller. The lease portion of the transaction is generally long term. The seller-lessee in the transaction is converted from an owner to a tenant.

SALES ASSOCIATE / In some states, notably Florida, a licensed real estate AGENT who is employed by the designated or employing BROKER. *See* SALESPERSON.

SALES COMPARISON APPROACH / One of three appraisal approaches; also called *market approach* and *market data approach.* Value is estimated by analyzing sales prices of similar properties (comparables) recently sold.

Importance: Virtually all appraisals of homes, and many appraisals of other properties, rely most heavily on the sales comparison approach. Two other approaches are *cost* and *income.*

SALES CONTRACT / A contract by which the buyer and seller agree to the terms of sale.

Importance: This document (in some states called an *agreement of sale, a contract of sale,* or an *earnest money contract*) stipulates the rights and responsibilities of buyers and sellers.

SALESPERSON / A person who is licensed to deal in real estate or perform any other act enumerated by state real estate license law while in the employ of a *broker* licensed by the state.

Importance: A salesperson's license is required for anyone to sell another's property. The salesperson must have a sponsoring broker.

SALES PRICE / The amount of money required to be paid for real estate according to a contract, or previously paid.

SALVAGE VALUE / The estimated value that an asset will have at the end of its useful life.

Importance: Real estate improvements, though long lasting, have a limited useful life at the end of which there may be salvage or scrap value.

SANDWICH LEASE / A lease held by a lessee who sublets all or part of his/her interest, thereby becoming a lessor. Typically, the sandwich leaseholder is neither the owner nor the user of the property.

Importance: The sandwich lessee tries to profit from income tax advantages or the rent differential between the other leases.

S&P CORELOGIC CASE-SHILLER HOME PRICE INDEX / A home price index developed by economists Carl Case and Robert Shiller for the Standard & Poor's Financial Services, LLC. The index is based on a constant quality, repeat sale methodology that accounts for variations in the characteristics and locations of the single-family homes used. Data on home sales are drawn from local deed records.

Importance: The S&P CoreLogic Case-Shiller Home Price Index is used to calculate the average price appreciation or decline in each of the 20 metropolitan districts covered by the index.

SARA / *See* SUPERFUND AMENDMENTS AND REAUTHORIZATION ACT.

SATISFACTION OF MORTGAGE / *See* SATISFACTION PIECE.

SATISFACTION PIECE / An *instrument* for recording and acknowledging final payment of a mortgage loan.

Importance: After a loan has been paid off, the borrower should record a satisfaction of mortgage or satisfaction piece.

SECTION (of Land) / One square mile in the *government rectangular survey.* There are 36 sections in a 6-mile-square township.

SECTION 1031 / The section of the INTERNAL REVENUE CODE that deals with tax-deferred exchanges of certain property. General rules for a tax-deferred exchange of real estate are as follows.

The properties must be

1. exchanged or qualify as a DELAYED EXCHANGE,
2. LIKE-KIND PROPERTY (real estate for real estate), and
3. held for use in a trade or business or held as an investment. *See also* BOOT.

Example: Under *Section 1031,* Lowell trades her APPRECIATED land for Baker's shopping center. EQUITIES are the same. Lowell's adjusted tax basis in the land becomes her adjusted tax basis in the shopping center.

SECURITY INSTRUMENT / An interest in real estate that allows the property to be sold upon a default on the obligation for which the security interest was created. The security interest is more specifically described as a *mortgage* or a *trust deed.*

SEIZIN / The possession of realty by a person who claims to own a *fee simple estate* or a *life estate* or other salable interest.

Importance: Seizin is a covenant needed to transfer ownership to another.

SEPARATE PROPERTY / Property acquired by either spouse prior to marriage or by gift or devise after marriage, as distinct from *community property.*

Importance: In community property states, property that is separate before marriage can remain that way; property acquired during marriage by joint effort is community property.

SERVIENT TENEMENT / *Contrast with* DOMINANT TENEMENT.

SETBACK / The distance from the curb or other established line within which no buildings may be erected. *Compare with* BUILDING LINE.

Importance: Setbacks must be observed; if they are violated during construction, the property may have to be razed.

SETTLEMENT DATE / *See* CLOSING DATE.

SETTLEMENT STATEMENT / *See* CLOSING STATEMENT.

SEVERALTY / The ownership of *real property* by an individual as an individual.

Importance: Severalty is distinguished from joint ownership, whereby two or more persons are owners.

SHADOW ANCHOR / An anchor store appearing to be part of a shopping center but situated on land that is owned separately and so not actually a tenant of the shopping center. The shopping center enjoys the traffic generated by the shadow anchor but not the rental income.

SHADOW BANKING SYSTEM / The aggregate system of finance that occurs outside of banks and has therefore not been subject to bank regulation. *See also* DODD-FRANK WALL STREET REFORM AND CONSUMER PROTECTION ACT.

Importance: Components of the *shadow banking system* include HEDGE FUNDS, private equity firms, securitized investment vehicles, and other nonbank financial institutions. An estimated 25–30% of financial transactions occur in the shadow banking system.

SHORT SALE / Arrangement between a MORTGAGOR (borrower) and MORTGAGEE (lender) through which the mortgagor retires the mortgage obligation with a payment of something less than the total outstanding PRINCIPAL BALANCE. Any principal forgiven in the transaction is considered by the IRS to be taxable income to the borrower. At this time, Congress is considering a change in tax law to facilitate short sales or WORKOUTS of subprime residential mortgages. (**Note:** This is entirely different from a short sale of stock, which occurs when the seller doesn't own the stock. The broker finds stock to borrow and deliver for the sale, while the seller hopes the value will decline so he can purchase at a lower price than he sold at.)

SINGLE AGENT / A broker who represents, as a FIDUCIARY, either the buyer or the seller but not both in the same transaction.

SITUS / The economic attributes of location, including the relationship between the property and surrounding properties, as well as distant points of interest and the linkages to those points. *Situs* is considered to be the aspect of location that contributes to the market value of a real property.

Importance: The *situs* of a commercial property includes surrounding commercial establishments and the location of suppliers' warehouses, competing establishments, and the homes of employees and customers.

SPECIAL AGENT / One with authority to act in a particular aspect of a business. Compare this to a *general agent,* who has authority to act for all matters of the business, and to a *universal agent,* who has authority to act for all matters of all the principal's businesses and personal issues. A real estate salesperson or broker is a special agent under a listing contract.

SPECIAL ASSESSMENT / An *assessment* made against a property to pay for a public improvement by which the assessed property is supposed to be especially benefited.

Importance: A municipality may install a new sewer line or sidewalk; each owner along the path may be charged a special assessment in addition to a regular tax.

SPECIAL-PURPOSE PROPERTY / A building with limited uses and marketability, such as a church, theater, school, or public utility.

SPECIAL WARRANTY DEED / A deed in which the *grantor* limits the title warranty given to the *grantee* to anyone claiming by, from, through, or under him/her; the grantor. The grantor does not warrant against title defects arising from conditions that existed before he/she owned the property.

Importance: The seller does not guarantee title against all claims—just those while he/she was the owner.

SPECIFIC LIEN / Loan secured by a specific property as COLLATERAL. **Example:** The mortgage on Sam's house was a *specific lien* against the property. The lender may be allowed to foreclose on and sell the property to satisfy the debt.

SPECIFIC PERFORMANCE / A legal action in which the court requires a party to a contract to perform the terms of the contract when he/she has refused to fulfill his/her obligations.

Importance: This action is used in real estate because each parcel of land is unique; consequently, a contract concerning one parcel cannot be transferred or applied to another.

SPOT ZONING / The act of rezoning a parcel of land for a different use from all surrounding parcels, in particular where the rezoning creates a use that is incompatible with surrounding land uses.

SQUATTER / One who occupies property without permission or legal authority.

SRA / A designation awarded by the *Appraisal Institute.*

Importance: Many of the most qualified residential appraisers have this designation.

STAGFLATION / A term coined in the 1970s to describe an economic situation of stagnant economic condition with inflation.

STAGING / Remodeling and interior decoration to enhance the marketability of a house prior to sale.

STANDARD DEDUCTION / An allowance provided as an income tax deduction in lieu of itemizing each tax-deductible expense on an income tax return.

STATE-CERTIFIED APPRAISER / *See* CERTIFIED GENERAL APPRAISER; CERTIFIED RESIDENTIAL APPRAISER.

STATUTE / A law established by an act of a legislature.

Importance: Statutes are written laws; laws are also made through judicial interpretation and government administration.

STATUTE OF FRAUDS / A state law that provides that certain contracts must be in writing in order to be enforceable; applies to deeds, mortgages, and other real estate contracts, with the exception of leases for periods shorter than 1 year.

Importance: The statute of frauds requires that contracts involving real estate be in writing.

STATUTE OF LIMITATIONS / A certain statutory period after which a claimant is barred from enforcing his/her claim by suit.

Importance: If a practice continues beyond the statute of limitations, the person who is adversely affected may be barred from trying to prevent it.

STATUTORY DEDICATION / The owners of a subdivision or other property file a plat that results in a grant of public property, such as the streets in a development.

STEERING / An illegal practice of limiting the housing shown to a certain ethnic group. *See* DISCRIMINATION; FAIR HOUSING LAW. **Example:** Although the minority couple wanted to be shown all available rental housing that they could afford, the unlicensed person *steered* them to or from certain neighborhoods.

STEP-UP LEASE / *See* GRADUATED LEASE.

STIPULATIONS / The terms within a written *contract.*

STRAIGHT-LINE DEPRECIATION / Equal annual reductions in the book value of property; used in accounting for replacement and tax purposes.

Importance: Straight-line provides less depreciation in the early years of an asset than does an accelerated method. Most taxpayers prefer accelerated depreciation because it minimizes current taxes. However, for financial reporting purposes, most companies prefer straight-line because it provides a higher net income.

SUBAGENCY / The relationship under which a sales agent tries to sell a property listed with another agent. This situation is common under a *multiple listing service* (MLS). A listing contract is taken by a listing broker and entered into the MLS, from which any member broker may sell the property. The listing broker and the selling broker split the commission.

SUBDIVIDING / The division of a tract of land into smaller tracts.

Importance: Subdividing allows raw acreage to be developed with streets, utilities, and other amenities added, resulting in lots ready for houses to be built.

SUBDIVISION / A tract of land divided into lots or plots suitable for home building purposes. Some states and localities require that a subdivision *plat* be recorded.

SUBJECT TO MORTGAGE / A method of taking title to mortgaged real property without being personally responsible for the payment of any portion of the amount due. The buyer must make payments in order to keep the property; however, if he/she fails to do so, only his/her equity in that property is lost.

SUBLEASE / A lease from a lessee to another lessee. The new lessee is a sublessee or subtenant. *See also* SANDWICH LEASE.

SUBLET / *See* SUBLEASE.

SUBORDINATED GROUND LEASE / A lease used when the *mortgage* has priority over the *ground lease.*

Importance: In case of a *default,* the unsubordinated interest has a prior claim to the subordinated interest.

SUBORDINATE MORTGAGE / One having a lower priority to another; the subordinate mortgage has a claim in foreclosure only after satisfaction of mortgage(s) with priority.

SUBORDINATION CLAUSE / A clause or document that permits a mortgage recorded at a later date to take priority over an existing mortgage.

Importance: Ordinarily a second mortgage automatically moves up to become a first *lien* when the first mortgage is retired. If the second mortgage has a subordination clause, it will remain a second mortgage when a first mortgage is refinanced.

SUBPRIME / A mortgage loan issued to a party who does not meet the standards set by FNMA and FHLMC for a CONFORMING LOAN.

SUBSURFACE RIGHTS / *Same as* MINERAL RIGHTS.

SUPERFUND / The commonly used name for CERCLA, the federal environmental cleanup law. If a site is on the Superfund list, it is required to be cleaned up by any and all previous owners, operators, transporters, and disposers of waste to the site. The federal government will clean such sites, requiring the responsible parties to pay the cleanup costs. Imposes strict liability.

SUPERFUND AMENDMENTS AND REAUTHORIZATION ACT (SARA) / Law that confirmed the continued existence of SUPERFUND. SARA put more teeth into CERCLA, though SARA provides an innocent landowner defense for a buyer who conducted a Phase I environmental study before the acquisition, with negative results.

SURETY / A person who guarantees the performance of another; a guarantor.

Importance: The surety becomes liable for the contract, just like the original principal. The surety is called when the principal fails to perform some duty.

SURRENDER / The cancellation of a lease before its expiration by mutual consent of the lessor and the lessee.

Importance: Surrender occurs only when both parties agree to it.

SURROGATE'S COURT (PROBATE COURT) / A court having jurisdiction over the proof of wills and the settling of estates and of citations.

SURVEY / (1) The process by which a parcel of land is measured and its area ascertained; (2) the blueprint showing the measurements, boundaries, and area.

Importance: A survey is needed to determine exact boundaries and any *easements* or *encroachments.*

SWING LOAN / Type of *bridge loan,* sometimes *interim financing* to cover the difference between funds realized from the sale of one property and the amount required to purchase another.

TAG SALE / Sale where individuals mark used household items with tags displaying the price. Tag sales are commonly held on weekends at the vendor's home, often having the merchandise displayed in the front yard or driveway. Tag sales, also called *yard sales* or *garage sales*, are an American tradition.

TALF / *See* TERM ASSET-BACKED SECURITIES LOAN FACILITY.

TARP / *See* TROUBLED ASSETS RELIEF PROGRAM.

TAX / A charge levied upon persons or things by a government.

Importance: Many different types of taxes affect real estate. Most local governments levy an *ad valorem tax* based on the property value. The federal government and many states have an income tax, but rental property owners may deduct operating expense, interest, and depreciation expense, thereby reducing their taxable income.

TAX DEDUCTIBLE / A type of expense that is allowed under tax law. Tax-deductible expenses can be used to reduce taxable income. *See* TAX DEDUCTION.

Importance: INTEREST and AD VALOREM TAXES are generally tax deductible for all types of property. DEPRECIATION, repairs, MAINTENANCE, UTILITIES, and other ordinary and necessary business expenses are tax deductible for INCOME PROPERTY.

TAX-FREE EXCHANGE / *See* SECTION 1031.

TAX SALE / The sale of property after a period of nonpayment of taxes.

Importance: Unpaid taxes become a *lien*. Property may be sold for the nonpayment of taxes.

TEASER RATE / An unusually low interest rate offered for the first few months or year of a mortgage loan; used as an enticement to potential borrowers.

Importance: When comparing interest rates on loans offered, a buyer should determine future rate adjustments as the initial rate may be discovered to be a teaser.

TENANCY AT SUFFERANCE / Tenancy established when a lawful tenant remains in possession of property after expiration of a lease. *See* HOLDOVER TENANT.

Importance: The tenant at sufferance has no estate or title; the landlord may oust the tenant at any time.

TENANCY AT WILL / A license to use or occupy lands and tenements at the will of the owner.

Importance: There is no fixed length of possession. The tenant may leave or may be put out at any time. Some states require notice, such as 30 days.

TENANCY BY THE ENTIRETIES / An estate that exists only between husband and wife, with equal right of possession and enjoyment during their joint lives and with the *right of survivorship*; that is, when one dies, the property goes to the surviving tenant.

TENANCY IN COMMON / An ownership of realty by two or more persons, each of whom has an undivided interest, without the *right of survivorship*. Upon the death of one of the owners, his/her ownership share is inherited by the party or parties designated in his/her will.

TENANCY IN SEVERALTY / Ownership of real property by an individual as an individual; ownership by one person or by a legal entity.

Importance: Tenancy in severalty is distinguished from *joint tenancy* and/or *tenancy in common*, whereby two or more persons are owners.

TENANT / A person who is given possession of real estate for a fixed period or at will. *See also* LEASE.

TENANT IMPROVEMENTS (TIs) / Those changes, typically to office, retail, or industrial property, to accommodate specific needs of a tenant. TIs include installation or relocation of interior walls or partitions, carpeting or other floor covering, shelves, windows, toilets, and so on. The cost of these is negotiated in the lease.

TENEMENTS / (1) Everything of a permanent nature; (2) anything attached to the soil. In common usage, a tenement is a run-down apartment building.

TERM ASSET-BACKED SECURITIES LOAN FACILITY (TALF) / Funding facility under which the Federal Reserve Bank of New York will lend up to $200 billion on a nonrecourse basis to holders of certain AAA-rated asset-based securities (ABS) backed by newly and recently issued consumer and small business loans. On August 17, 2009, the Federal Reserve Board and the Treasury announced that, to promote the flow of credit to businesses and households and to facilitate the financing of commercial properties, the *TALF* program would continue to provide loans against newly issued ABS and legacy commercial mortgage–backed securities (CMBS) through June 30, 2010. TARP guarantees $20 billion of TALF loans.

TERMITES / Insects that bore into wood and destroy it.

Importance: Termite inspections by reputable pest control companies are often required in a real estate transaction. In some places, it is customary to require a seller to post a termite bond as assurance that the foundation has been properly treated.

TERMS / Conditions and arrangements specified in a contract.

Importance: Anything lawful may be included in a contract and becomes part of its terms.

TESTAMENT / A will.

TESTAMENTARY TRUST / Created by a will that comes into effect only after the testator's death.

TESTATE / Having made a valid will. *Contrast with* INTESTATE.

TESTATOR / A man who makes a will.

TESTATRIX / A woman who makes a will.

TILA / *See* TRUTH IN LENDING ACT.

TILA-RESPA INTEGRATED DISCLOSURE / *See* TRID.

"TIME IS OF THE ESSENCE" / A phrase that, when inserted in a contract, requires that all references to specific dates and times of day concerning performance be interpreted exactly.

TIMESHARING / A form of property ownership under which a property is held by a number of people, each with the right of possession for a specified time interval. *Timesharing* is most commonly applied to resort and vacation properties. Often there is a network where vacation dates and *timeshare* locations can be exchanged. **Example:** Ingram is an owner in a *timesharing* arrangement for a lakefront cottage. Ingram is entitled to use the cottage each year from July 1 to July 15. Use of the property during the remainder of the year is divided among other owners. All property expenses are paid by an owners' association to which Ingram pays an annual fee.

TIME VALUE OF MONEY / The concept that money available now is worth more than the same amount in the future because of its potential earning capacity.

TITLE / Evidence that the owner of land is in lawful possession thereof; evidence of ownership. The word is often clarified or qualified by an adjective, such as *absolute, good, clear, marketable, defective, legal.*

TITLE ABSTRACT / *See* ABSTRACT OF TITLE.

TITLE INSURANCE / An insurance policy that protects the holder from any loss sustained by reason of defects in the title.

Importance: The premium is paid once and is good only until ownership changes.

TITLE SEARCH / An examination of the public records to determine the ownership and *encumbrances* affecting real property.

Importance: A title search is typically performed before *title insurance* is issued. If the search shows a title risk, the policy may contain an exception or may not be issued.

TOPOGRAPHY / The state of the surface of the land; may be rolling, rough, flat, etc.

Importance: The topography may affect the way land can be developed, including potential uses.

TORRENS SYSTEM / A title registration system used in some states; the condition of the title can easily be discovered without performing a title search.

TORT / A wrongful act that is not a crime but that renders the perpetrator liable to the victim for damages.

TOWNSHIP / A 6-mile-square tract delineated by *government rectangular survey.*

TOXIC ASSETS / Nontechnical term that supplanted the less dramatic and urgent designation "troubled assets" used during 2007 and 2008 to describe mortgage-related loans and investments that had lost most if not all value in a market that was virtually not functioning. Carried on the books of banks and other financial institutions on a market to market basis, they were "poisoning" balance sheets and causing a wave of insolvencies.

TRACT / A *parcel* of land, generally held for *subdividing;* a *subdivision.*

TRADE FIXTURES / Articles placed in rented buildings by the tenant to help carry out a trade or business. The tenant may remove the fixtures before the expiration of the lease, but if the tenant fails to do so shortly after the lease expires, the fixtures become the landlord's property.

TRADING UP / Buying a larger, more expensive property.

TRAFFIC COUNTS / Generally, vehicular traffic tallies performed by the state department of transportation or a local government. These are helpful to determine how to prevent or resolve traffic jams, to plan new roads, or to widen existing streets. Retailers use them to determine potential support for a new store.

TRANSACTION BROKER / A broker who provides limited representation to a buyer, seller, or both but does not represent either in a FIDUCIARY capacity or as a SINGLE AGENT. Does not provide undivided loyalty to either party. Facilitates transaction through assistance but not representation. In some states, this term is applied to a broker in an arrangement similar to a no-brokerage agency relationship.

TRANSFER DEVELOPMENT RIGHTS / A type of *zoning ordinance* that allows owners of property zoned for low-density development or conservation use to sell development rights to other property owners. The development rights purchased permit the landowners to develop their parcels at higher density than otherwise. The system is designed to provide for low-density uses, such as historic preservation, without unduly penalizing some landowners.

TRID / TILA-RESPA Integrated Disclosure. Federal law applying to most home mortgages beginning October 2015. Combines provisions of the Truth in Lending Act (TILA) and the Real Estate Settlement Procedures Act (RESPA). Requires the *Loan Estimate* and *Closing Disclosure* forms be provided to borrowers.

TRIPLE-A TENANT / A tenant with an excellent credit record; also called *AAA tenant.* **Example:** The U.S. Postal Service and the American Telephone and Telegraph Company are examples of triple-A tenants because they are unlikely to *default* on a lease.

TRIPLE NET LEASE / A lease whereby the tenant pays all expenses of operations, including property taxes, insurance, utilities, maintenance, and repair.

TRULIA / Popular website of homes for sale.

TRUST ACCOUNT / A bank account separate, apart, and physically segregated from a broker's own funds, in which the broker is required by state law to deposit all money collected for clients; called an *escrow account* in some states.

TRUST DEED / A *conveyance* of real estate to a third person to be held for the benefit of another; commonly used in some states in place of mortgages that conditionally convey title to the lender.

TRUSTEE / A person who holds property in trust for another to secure performance of an obligation; the neutral party in a *trust deed* transaction.

TRUSTEE'S DEED / The deed received by the purchaser at a foreclosure sale; issued by the *trustee* acting under a *trust deed.*

TRUSTOR / The person who conveys property to a trustee, to be held in behalf of a *beneficiary;* in a *trust deed* arrangement, the *trustor* is the owner of real estate and the *beneficiary* is the lender.

TRUTH IN LENDING ACT (TILA) / Federal law that requires disclosure of financial terms and conditions for common loans and certain real estate mortgage transactions. Known as *Regulation Z. See* TRID.

TURNKEY PROJECT / A development in which a developer completes the entire project on behalf of a buyer; the developer turns over the keys to the buyer at completion, and the buyer need only turn the key in the door.

Importance: Many government-owned public housing projects are *turnkey projects.* A private developer undertakes all activities necessary to produce the project, including land purchases, permits, plans, and construction, and then sells the project to the housing authority.

UNDERWATER / Slang term used to refer to a property whose debt exceeds its value.

UNDIVIDED INTEREST / An ownership right to use and possession of a property that is shared among co-owners, with no one co-owner having exclusive rights to any portion of the property. *Compare with* PARTITION.

UNEARNED INCREMENT / An increase in the value of real estate due to no effort on the part of the owner; often due to an increase in population.

Importance: *Appreciation* of land in the path of growth is considered an unearned increment because the landowners did nothing to cause it.

UNIFORM COMMERCIAL CODE (UCC) / A group of laws to standardize the state laws that are applicable to commercial transactions. Few of the laws have relevance to real estate.

UNIFORM RESIDENTIAL APPRAISAL REPORT (URAR) / A standard form for reporting the *appraisal* of a dwelling.

Importance: This form is required for use by the major secondary mortgage purchasers. It provides numerous checklists and appropriate definitions and certifications that are preprinted on the form.

UNIFORM RESIDENTIAL LANDLORD AND TENANT ACT (URLTA) / A model law governing residential leasing practice and leases, adopted wholly or in part by about 15 states.

UNIFORM STANDARDS OF PROFESSIONAL APPRAISAL PRACTICE (USPAP) / Standards promulgated by the *Appraisal Foundation* that set forth the requirements for research and reporting with which a professional appraiser is to comply.

UNILATERAL CONTRACT / An obligation given by one party contingent on the performance of another party but without obligating the second party to perform. *Compare with* BILATERAL CONTRACT.

UNITY / Four unities are required to create a joint tenancy: interest, possession, time, and title. In other words, the joint tenants must have an equal interest arising from the same conveyance, the same undivided possession, and the same use over the same time.

UNIVERSAL AGENT / An agent with authority to act for the principal in all businesses and in all matters rather than in only one particular business (GENERAL AGENT) or in only one aspect (SPECIAL AGENT).

UPFRONT PAYMENT / In a mortgage origination transaction, any payment that must be made at the time of origination. Examples of upfront payments are FEDERAL HOUSING ADMINISTRATION (FHA) insurance prepayments and origination fees.

UPGRADERS / People who currently own a home but are seeking to buy what they consider a better home; also called "move-up" buyers. The "upgrade" may involve getting a larger home, a more conveniently located home, or one with special amenities. In almost all cases, the upgrader is looking to spend more for the new home than the proceeds received from selling the old home.

UPLANDS / Land within the mitigation parcel adjacent to surrounding wetlands, often having exotic plant species like the wetlands.

Example: *Uplands* include rangeland, improved pastures, pine flatwoods, and tropical hardwood hammocks all with varying degrees of melaleuca infestation.

UPSIDE-DOWN MORTGAGE / A mortgage loan whose balance is greater than the value of the property providing the security for the loan. A homeowner with such a loan cannot sell the home or refinance the loan without surrendering cash. The homeowner has negative equity in the home. *See also* SHORT SALE; UNDERWATER.

URAR / *See* UNIFORM RESIDENTIAL APPRAISAL REPORT.

URBAN PROPERTY / City property; closely settled property.

Importance: Urban property is more valuable than rural or suburban land because of its greater business activity.

URLTA / *See* UNIFORM RESIDENTIAL LANDLORD AND TENANT ACT.

U.S. GREEN BUILDING COUNCIL / Nonprofit organization that offers LEED certification of GREEN BUILDINGS, educational resources, and a network of chapters and affiliates to support and encourage environmentally sensitive construction of buildings, including homes, and communities.

USPAP / *See* UNIFORM STANDARDS OF PROFESSIONAL APPRAISAL PRACTICE.

USURY / A rate of interest higher than that permitted by law.

Importance: Each state has its own usury laws, with different ceiling interest rates applying to each type of loan. Penalties are severe for usury rates.

VA / Department of Veterans Affairs (formerly Veterans Administration), a government agency that provides certain services to discharged servicemen and women. See also VA LOAN OR MORTGAGE.

VA LOAN OR MORTGAGE / Home loan guaranteed by the U.S. Department of Veterans Affairs (VA) under the Servicemen's Readjustment Act of 1944 and later. Discharged servicemen and women with more than 120 days of active duty are generally eligible for a VA loan, which typically does not require a down payment.

Importance: The VA guarantees restitution to the lender in the event of default, up to a stated amount.

VALID / (1) Having force, or binding force; (2) legally sufficient and authorized by law.

Importance: A valid contract can be enforced in court.

VALUATION / (1) Estimated worth or price; (2) valuing by appraisal.

Importance: Valuation is the process of estimating the worth of an object.

VALUE / (1) The worth of all the rights arising from ownership; (2) the quantity of one thing that will be given in exchange for another.

Importance: Price is the historic amount that was paid; value is an estimate of what something is worth. Often value is qualified as to a specific type: market, user, assessed, insurable, speculative.

VARIABLE EXPENSES / Property operating costs that increase with occupancy.

VARIANCE / Permission granted by a ZONING authority to a property owner to allow for a specified violation of the zoning requirements. *Variances* are generally granted when compliance is impossible without rendering the property virtually unusable. **Example:** Sherman owns a lot that is zoned for low-density housing. Because of the peculiar shape and TOPOGRAPHY of the lot, Sherman cannot build a home with the required minimum floor area without violating the SETBACK requirements for the area. Sherman may be granted a *variance* to either construct a smaller dwelling or encroach on the setback line.

VENDEE / A purchaser; a buyer.

VENDEE'S LIEN / A *lien* against property under a *contract of sale*, to secure the deposit paid by a purchaser.

VENDOR / A seller.

VERIFICATION / Sworn statements before a duly qualified officer as to the correctness of the contents of an *instrument*.

VICARIOUS LIABILITY / The responsibility of one person for the acts of another.

VIOLATION / An act, a deed, or conditions contrary to law or the permissible use of real property.

Importance: When there is a violation of a law or contract, the perpetrator may be liable for damages and/or penalties.

VOID / Having no force or effect; unenforceable.

VOIDABLE / Capable of being voided but not void unless action is taken to void. **Example:** Contracts to real estate entered into by minors are voidable only by the minors.

VOLUNTARY ALIENATION / Legal term describing a sale or gift made by the seller or donor of his/her own free will.

WAIVER / The voluntary renunciation, abandonment, or surrender of some claim, right, or privilege.

Importance: A person may waive a right when that right is not especially important to him/her in the overall transaction.

WARRANTY / A promise or representation contained in a *contract*.

Importance: Usually a seller's warranty pertains to the quality, character, or title of goods that are sold.

WARRANTY DEED / A deed that contains a covenant that the *grantor* will protect the *grantee* against any and all claims; usually contains covenants assuring good title, freedom from *encumbrances*, and *quiet enjoyment*. *See also* GENERAL WARRANTY DEED; SPECIAL WARRANTY DEED.

WASTE / Often found in a mortgage or lease contract, this term refers to property abuse, destruction, or damage (beyond normal wear and tear). The possessor causes unreasonable injury to the holders of other interests in the land, house, garden, or other property. The injured party may attempt to terminate the contract or sue for damages.

WATER TABLE / The distance from the surface of the ground to a depth at which natural groundwater is found.

Importance: The water table may affect the type of buildings that are possible on a parcel of land and the ability to get well water.

WETLANDS / Land, such as swamps, marshes, and bogs, normally saturated with water. Development may be prohibited, because it could disturb the environment. The U.S. Army Corps of Engineers (COE) and the U.S. Environmental Protection Agency (EPA) have adopted a regulatory definition for administering the Section 404 permit program of the Clean Water Act (CWA) as follows: "Those areas that are inundated or saturated by surface or groundwater at a frequency and duration sufficient to support, and that under normal circumstances do support, a prevalence of vegetation typically adapted for life in saturated soil conditions."

WILL / The disposition of one's property to take effect after death.

Importance: If a person dies without a will (*intestate*), the property goes to his/her heirs at law. If a person dies intestate without heirs, the property *escheats* to the state.

WITHOUT RECOURSE / Words used in endorsing a note or bill to denote that the future holder is not to look to the debtor personally in the event of nonpayment: the creditor has recourse only to the property and the borrower is held harmless after foreclosure.

Importance: In *default,* the borrower can lose the property mortgaged but no other property. There can be no *deficiency judgment.*

WRAPAROUND MORTGAGE / A mortgage that includes in its balance an underlying mortgage. Instead of having distinct and separate first and second mortgages, a wraparound mortgage includes both. For example, suppose that there is an existing first mortgage of $100,000 at 4 percent interest. A second mortgage can be arranged for $50,000 at 8 percent interest. Instead of getting that second mortgage, the borrower arranges a wraparound, subject to the first mortgage, for $150,000 at 6 percent. The first mortgage of $100,000 stays intact (and is included in the $150,000 wraparound). The borrower pays the wraparound lender one payment on the $150,000 wraparound, and the wrap lender remits the payment on the first mortgage to the first mortgage lender.

WRONGFUL FORECLOSURE / A FORECLOSURE that was allegedly performed illegally or incorrectly. *See also* ROBO-SIGNER.

ZILLOW / Popular website that offers data on residential housing prices and sales.

ZONE / An area set off by the proper authorities for specific use, subject to certain restrictions or restraints. Changing the zoning of property usually requires approval by the city council.

ZONING ORDINANCE / An act of city, county, or other authorities specifying the type of use to which property may be put in specific areas. **Examples:** residential, commercial, industrial.

Fundamentals of Real Estate Law

3

ESSENTIAL MATTERS TO LEARN FROM THIS CHAPTER

- *Property* refers to the legal rights to use and enjoy things or, in layperson's terms, the things themselves.
- *Real estate* (also called *realty*) refers to land and everything permanently affixed to land.
- *Personal property* (also called *personalty*) refers to things other than real estate.
- *Chattels* are items of personal property, including autos, clothing, and movable furniture.
- A *fixture* is an object that was personal property before being affixed to real estate in a manner that caused it to be part of the real estate. A business may remove a trade fixture (signs, counters, bar stools) when its lease expires.
- *Annexation* is the manner of attachment of personal property to real estate. It is an important determinant of whether the object attached becomes a fixture (part of the real estate). Annexation is also the process applied by a city to extend its boundaries by including adjacent land.
- *Corporeal* refers to property that is tangible (i.e., can be seen or touched). *Incorporeal* refers to intangible property.
- An *easement* is the right to use the property of another for a specific purpose.
- An *estate* is the degree, quantity, nature, and extent of interest a person has in a property. There are many types of estate.
- A *life estate* expires on the death of the *life tenant*. Then the *remainderman* acquires ownership of the property (called the *reversion*). *Pur autre vie* ("for another's life") is a type of life estate that ends on the death of someone other than the life tenant.
- *Fee simple* (also called *fee simple absolute*) is the most complete form of ownership one can have. It is subject to eminent domain and land use controls.
- A *freehold estate* is an interest in real estate without a known termination date. When there is an interest in real estate with a preset termination date, it is called *nonfreehold* or *less than freehold*.
- An *interest in real estate* allows a party certain use but is less than fee simple.
- *Tenancy in severalty* is ownership by one person or one entity.
- *Tenancy in common* is a form of co-ownership by two or more people without survivorship. Each tenant in common owns an undivided interest in the whole and may dispose of his/her interest any way he/she pleases.
- *Joint tenancy* is a form of ownership characterized by survivorship. When one tenant dies, the remaining joint tenants inherit the portion owned by the decedent.

- *Tenancy by entireties* is a form of joint tenancy (having survivorship) that exists only in certain states and is available only to husband and wife. It protects one spouse's rights to the home from certain actions of the other spouse.
- *Community property* is a form of ownership recognized in seven states, most of Spanish origin. When a husband and wife acquire property by joint effort during their marriage, each spouse owns one-half of that property. Property either owned separately before marriage or inherited during marriage remains *separate property.*
- A *condominium (condo)* is a form of ownership whereby a person owns a specific unit, can mortgage it, and has rights to use the common property jointly.
- A *cooperative (co-op)* is a form of ownership whereby a person owns stock in a corporation, receiving exclusive use of a specific unit and rights to use the common property jointly. Co-op and condo owners must pay assessments.
- *Homestead* is the protection of a family's home from creditors in certain instances, such as bankruptcy. The dollar amount or other limits vary by state.
- *Foreclosure* is the process to force the sale of a property for nonpayment of a claim on it.
- *Riparian* refers to land bounding on a stream or river.
- *Littoral* refers to the shoreline of property on a slow-moving or static body of water, such as a pond, lake, bay, sea, or ocean. Generally, private ownership stops at the mean high-water mark.
- *Intestate* refers to a person who died without a will. State law determines heirs. *Intestate succession* is the state's law for distributing the property of someone who died without a will. If the decedent left no heirs, the decedent's property *escheats* to the state.
- *Survivorship* is when a co-owner automatically inherits ownership of the portion that belonged to another upon death of that other owner.
- *Adverse possession* allows a person who uses another's property to gain ownership when the use is actual, open, notorious, exclusive, hostile, continuous, and long enough (set by state law).
- *Devise* is a transfer of real estate by will or last testament. *Bequest* is a transfer of personal property by will or last testament.
- *Dower rights,* in states which recognize them, give the wife (or child) an interest in part of the deceased husband's (or father's) estate. A few states also recognize a husband's rights, called *curtesy.*
- *Constructive notice* occurs when an instrument, such as a deed, is recorded in the county courthouse to give notice to the world of its existence.
- *Color of title* is what appears to be good title but is not.
- *Prescription* refers to an easement acquired by adverse possession.
- *Erosion* is the gradual removal of land by nature. *Avulsion* is the sudden removal of land, as by a hurricane. *Accretion* is the natural addition to land.
- A *license* gives a person permission to go onto another's land for a special purpose.
- *License* is also the privilege given by the state to a person to operate as a real estate salesperson or broker.
- *Eminent domain* is the right of government, or an entity with governmental authority, to take private property for public use. The process is called *condemnation,* which requires payment of *just compensation* for the property taken.

This is the first chapter of subject matter material. Much of what is written in this chapter will be discussed in greater detail in later chapters. This chapter summarizes real estate law. Contract law and professional licensing are discussed in later chapters.

The subject matter of this chapter is a very fertile source of license examination questions. This chapter is particularly important because it concerns things that the average layperson knows little or nothing about. In this manner, the examining authorities can make sure that successful licensing applicants have studied the laws and customs of the real estate business and know enough about it to be worthy of public trust.

PROPERTY

Property refers to the legal rights to use and enjoy any thing. Strictly speaking, the term *property* does not refer to the things themselves but to the *legal rights* that a society allows someone with regard to the *use* of and *enjoyment* of these things. In practice, however, we tend to use the word *property* to mean the things themselves, so we don't think it odd for someone to say, "That car is my property." To be absolutely proper, however, he should be saying, "The *rights* to use and enjoy that car are my property."

A *bundle of rights* is a concept of ownership used to describe various things that may be legally owned, enjoyed, bought, and sold. The bundle of rights includes the right to use, sell, lease, mortgage, give, or devise to heirs. The bundle includes occupancy, farming, exploration, and construction. Upon the purchase of real estate, the buyer typically receives the rights the seller had upon the sale. If the seller has previously sold or reserved certain rights, they are not transferred to the buyer. A buyer cannot acquire more rights or property than the seller had.

Property is divided into two kinds: *real property* and *personal property*. These are also known as *realty* and *personalty*. Realty is the property rights to real estate; personalty is the property rights to everything else. Personal property may be called *chattel*.

Real estate is defined as "land and everything permanently attached to it." *Horizontal boundaries* of land ownership are defined by distances and markers on the surface. *Vertical boundaries* of land ownership are downward to the center of the earth and upward to the heavens. However, the rights to use this space can be subdivided. An owner can retain surface rights while selling or leasing air rights (above the land) and/or mineral rights, which are the rights to extract minerals from beneath the surface. Mineral rights can extend from a depth of 50 or 100 feet beneath the surface down to a much lower level. A "permanent attachment" is (a) anything that grows on the land, (b) anything that is built upon the land (including roads, fences, etc., as well as buildings), and (c) fixtures. *Fixtures* are items that may appear to be personal property but are considered to be part of the real estate. Fixtures may include window air conditioners, built-in appliances, draperies, and other items of that nature. There is no cut-and-dry definition of a fixture, so many of the court disputes concerning real estate contracts involve misunderstandings concerning what are and what aren't fixtures.

An *improvement* is something that generally adds value to the raw land. Improvements *of* land include sidewalks, streets, sewers and other utilities, grading the land surface, etc. Improvements *on* the land include fences, garage additions, patios, and buildings.

Many objects can be either realty or personalty, depending on how they are used. A brick is personalty until it is mortared into place in the wall of a building, at which time it becomes realty. A tree is realty when it is growing, but it becomes personalty when it is cut down. When it is used as a building material, the wood becomes real estate again. Fixtures present a "gray area" where an argument can be made either way as to the nature of the item. When a contract isn't clear about what is and what is not included, a court often has to decide such a problem based upon the answers to these questions:

1. How is the item attached to the property? The manner of attachment is called *annexation.* The more "tightly" it is attached, the more likely it is that the item will be considered part of the real estate.
2. What was the intention of the person who attached it? If he intended it to be a permanent part of the real estate, then it probably will be a fixture.
3. What is the prevailing custom in that particular jurisdiction? Depending on the location, some things may or may not be considered fixtures. For example, in some places, built-in appliances, air conditioners, draperies, and the like are considered part of the real estate, whereas elsewhere the seller of real estate would be considered perfectly within her rights to take them with her. A *trade fixture,* such as a business sign or a bar stool, may be removed by a tenant upon lease expiration.

When referring to legal matters, you must be precise and clear. This is essential. If a contract is written vaguely or if the description of a piece of land is too vague, the preparer can be held responsible for problems caused by his/her negligence.

Both realty and personalty can be divided into the *tangible* and the *intangible.* Tangibles can be seen and touched; intangibles lack physical existence but may represent valuable rights. When we speak of realty, these property types are referred to as *corporeal* and *incorporeal.* Corporeal property can be seen and touched. Incorporeal property, which is intangible, includes the right to use property that is owned by another. Examples of incorporeal property include *easements, rights-of-way,* and the mere permission to use someone else's property.

Corporeal property involving real estate is almost always real property. Incorporeal property can be either real or personal property. Easements usually are real property; short-term leases (in most states leases of one year or less) are personal property.

ESTATES IN LAND

An *estate* is the degree, quantity, nature, and extent of interest a person has in real estate. Estates are divided into two groups: *freehold* and *nonfreehold.* The basic difference is that a freehold estate is of uncertain duration because it extends until the owner chooses to dispose of it or, if he doesn't, until he dies. We don't know how long that will be. Freehold estates may be *inheritable* or *noninheritable.* Nonfreehold estates include various kinds of leases. Leases are contracted for specific periods of time, so it is known exactly when these kinds of estate will cease to exist, causing them to be nonfreehold.

Fee simple estates are inheritable. Basically, these give their owners absolute rights to do whatever they want with the land involved, subject only to general law, such as eminent domain, zoning, and building codes. Most important, as far as the law is concerned, fee simple provides the right to *dispose* of the estate in any legal manner: by selling it, by giving it away, or by devising it to someone (transferring it on death by will).

Life estates are noninheritable. In a life estate arrangement the owner, called the *life tenant,* has the right to use, occupy, and enjoy the property so long as he lives. Upon his death, the ownership of the property goes to a predetermined person (or persons) referred to as the *remainderman.* Since the life tenant has rights to the property only so long as he lives, he can transfer to others only those rights. He may sell his life estate to someone, but that person must relinquish the property upon the original life tenant's death. The same would happen if the life tenant leased the property to someone and died before the lease expired. Furthermore, the life tenant is considered to be the custodian of the property for the remainderman: he cannot allow the property to deteriorate beyond ordinary wear and tear, because he must protect the remainderman's interest in it.

There is a special form of life estate that is *pur autre vie*. This term is French, meaning "for another life." In such a life estate, the duration is not for the life tenant's life but for the life of another person. For example, I purchase a specially built home to accommodate my severely disabled daughter. The purpose is to provide her with a place to reside, with or without my son, for the rest of her life, after which time I want the home to become the property of a charity. My daughter is legally incompetent and cannot own real estate in her own name. Another thought that I have would be to grant the home to my son pur autre vie; the "other life" would be that of my disabled daughter. The charity would be the remainderman. My son would be the life tenant so long as my daughter lives; upon her death, the property would revert to the charity.

Nonfreehold estates include leases for specific terms. The lessee, or tenant, has the right (unless she contracts to give it up) to assign or otherwise dispose of her leasehold rights if she has a *leasehold estate*. Furthermore, she has the right to use, occupy, and enjoy the real estate during the period of her lease.

An estate in land requires that its owner have the following rights: the right to possess the land, the right to *use* the land, and the right to *dispose* of his estate to someone else. Possession means, basically, just that: the right to occupy the land and make use of it, and to exclude everyone else from the land. A person who has *some* rights to use land, but not *all* possessory rights, is said to have an *interest* in land. An interest occurs when a person has a month-to-month lease that either party can terminate at any time (*tenancy at will*) because no specific duration is mentioned. An interest also occurs when *license to use* is involved; here there is mere permission and not necessarily a written arrangement. This might occur if someone asked permission to come onto your land one afternoon to go fishing.

Trick Question

A trick question defines an *estate* as a lavish house on a large tract of land or farm. Although this may be what laypeople think an estate is, it is the wrong answer. An *estate* is the interest one has in a piece of real estate.

CREATION OF ESTATES IN LAND

Estates in land are created in the following ways: by *will, descent, voluntary alienation, involuntary alienation*, and *adverse possession*. One receives an estate *by will* if he inherits it by being *so designated in someone's will*. This is also called *devise*. One receives an estate *by descent* if he is *designated by law* as the recipient of property of a deceased person. When a property owner dies, his/her will typically states who will inherit the property. *Probate* court, called *surrogate's court* in some jurisdictions, is the court where the deceased's will is proven to be valid.

If someone dies without a will, he/she died *intestate*. His/her property passes to heirs according to the state law of *intestate succession*. If the deceased had no heirs, his/her property escheats (reverts by default) to the state.

Some states also have *dower* laws. Dower laws state that a person's spouse and sometimes his/her minor children are entitled to receive a certain interest in his/her property when he/she dies. Most states have changed their laws (or are in the process of changing them) to give husbands and wives equal rights to one another's estates.

Voluntary alienation is the most common means whereby a person receives estates in land. This term refers to voluntary exchanges, sales, and gifts wherein the one who gives up the property does so willingly. *Involuntary alienation* occurs where the owner of an estate is forced, in a legal manner, to give up some or all of her rights by the action of law. This most commonly occurs in the case of bankruptcy or of having some other kind of legal judgment entered against one, such as a foreclosure or failure to pay property taxes. Involuntary alienation also occurs by *foreclosure, escheat, eminent domain* (see page 73) and *adverse possession.*

Adverse possession is a special means of acquiring ownership. All states have statutes permitting adverse possession, which is the right of a person who has used another's land to receive a legal claim to that land, in fee simple. The historical reason behind this is that, if the true owner is so uninterested in his land that he does nothing during this period of time to prevent someone else from putting it to use, the community is better off by letting this other person actually have the full rights to the land. Adverse possession cannot be acquired overnight; depending on the state, it takes from 7 to 40 years to establish these rights. The adverse possession must be *open, notorious, exclusive, continuous,* and *hostile.* This means that the claimant, during the required time period, must have acted completely as if the property were his own, including defending his "rights" against encroachment by others and being open in his actions. The adverse possession must continue throughout the statutory period; a person can't use someone else's land for the summer of 1989 and later for the summer of 2019 and then claim that he has been using it for 30 years. Further, no tenant can claim against his landlord, because every time a tenant pays rent he legally acknowledges that the person receiving payment is the true owner. Nor may one claim against the government, on the theory that one is already the nominal "owner" of public property. *Easements* (see page 71) also can be acquired by adverse possession; they are called *easements by prescription.*

Trick Questions

Questions about inheritance can be confusing. Here are two sample trick questions and explanations of the correct answers.

Question: Mr. Jones died without a will. His property escheats to the state. Is this true or false?

Answer: This is false because Mr. Jones's property is inherited by his heirs at law.

Question: Mr. Smith died without heirs at law. His property escheats to the state. Is this true or false?

Answer: This is false because his property is inherited according to his will.

Remember that property escheats to the state *only* when a person leaves no will (dies intestate) *and* has no heirs at law.

TENANCIES

Estates in land may be owned in a variety of ways, depending on the number of people involved and their relationship. These various forms of ownership are called *tenancies. Tenancy in severalty* is ownership by one person or one entity. *Tenancy in common* is a form of ownership by two or more persons. Each one owns an *undivided interest,* which means that he owns a fraction of each part of the realty. In this arrangement, each is entitled to his share of the profits and is responsible to the other owners for his share of the costs and expenses. No part of the land owned together may be disposed of without the consent of all, unless they have a specific arrangement whereby less than unanimity can force a decision. Tenants in common need not own equal shares. They may dispose of their shares in any way they choose, unless they have specifically agreed upon some limitation. This means that A and B may buy real estate as tenants in common and later on B may sell his share to C, making A and C tenants in common. Further, a tenant in common may devise his share to someone else or give it away.

Joint tenancy is similar to tenancy in common except for one important difference: joint tenants have the *right of survivorship.* This means that if a joint tenant dies, his/her share is divided proportionally among the surviving joint tenants. Consequently, a joint tenant cannot devise (will) her share to someone else. However, she can sell or give her interest away; the new owner will then have the status of a tenant in common. Joint tenancy requires four unities: interest, possession, time, and title.

Tenancy by the entireties is a special form of joint ownership allowed only to married couples. In states where it is used, it protects the rights of a spouse to jointly owned property by providing the same basic rights and responsibilities as joint tenancy while also protecting the property against any foreclosure due to judgment or debt against one of the parties.

Community property is a system that exists in a few states, including the populous ones of California and Texas.* Where community property laws exist, half of all property acquired by either husband or wife during a marriage is considered to belong to each partner. These ownership rights transcend death and divorce. Community property states do not have dower or curtesy laws.

All states have passed laws that permit the establishment of *condominium* interests in land. *Condominium* is a legal term, describing a certain kind of ownership of land; it does *not* refer to any particular architectural style. In theory, any kind of property, serving any use, can be owned in condominium. In this form of ownership, one has title to a part of a larger piece of property, with the right to use his part exclusively and the right to share some of the rest with the other owners. Typically in a residential condominium, one would have the right to use his own dwelling unit and no right to enter the units owned by others. However, one would share the right, with the others, to use the *common property,* which might include walkways within the project, recreational facilities, common hallways, and elevators. Most important, the law allows an owner of a condominium unit to mortgage and otherwise encumber his unit. If he defaults, a lienholder may force the sale of his unit, but the other owners in the condominium project cannot be held responsible for the defaulting owner.

A residential condominium development usually has a *homeowners' association* or some similar group that is elected by the owners of the units in the project. The function of the association is to ensure that the common property in the development is taken care of and to make and enforce whatever rules the owners as a group want within their private community. Many people who have bought condominium units have found later that the rules and regulations of the association do not permit them to do everything they want to. For example, the association may have a rule limiting overnight guests to no more than two per unit or requiring that only certain kinds of flowers be planted in the front yards. An owner who may wish to have more guests or to plant different flowers can find herself in conflict with the association and can be required to conform to the rules.

---

*The others are Arizona, Idaho, Louisiana, Nevada, New Mexico, Washington, and Wisconsin.

Many condominium projects have quite a lot of common property; swimming pools, park areas, parking lots, bicycle paths, tennis courts, clubhouses, and gatehouses are examples. The owners of units in the project own these common elements together, as a group. Their association looks after the common property and assesses each unit owner a fee (called a *condominium fee*), usually monthly, to get the funds necessary to maintain it. An owner who does not pay this fee will usually find that the association has the legal power to file a lien on his unit in order to collect the money.

The *cooperative* form of ownership is common in only a few cities, including New York, Miami, and Washington, D.C. In a cooperative, the entire real estate is owned by a *cooperative corporation.* Instead of owning a unit outright, the cooperative owner owns a proportion of the *shares of stock* of the cooperative corporation along with the right to exclusive use of a particular unit using what is called a *proprietary* lease. The owners elect a board of directors that is responsible for the operation of the corporation and, therefore, of the property. Each owner pays monthly assessments to the corporation, which, in turn, pays its property taxes, mortgage payments, maintenance bills, and so forth. A big disadvantage of cooperative ownership is that, if one or more of the owners does not pay, the others will have to make up the difference or face the risk of having the entire property foreclosed.

Tenancy at will occurs when there is a lease arrangement but no specific time period is agreed upon. Essentially, then, either party can terminate the arrangement whenever he wants (i.e., at will), so its duration is uncertain.

Tenancy at sufferance occurs when a tenant remains on the property after the expiration of a lease. The occupant is called a *holdover tenant.* This differs from tenancy at will in that often some aspects of the original lease contract may be considered to be in force (such as those requiring notice before termination or prohibitions against certain kinds of uses or activities by the tenant). Many states have enacted laws requiring that, even in cases of tenancy at sufferance or tenancy at will, the parties give one another certain minimum notice before termination of the arrangement.

HOMESTEAD LAWS

Many states have what are known as *homestead* laws. These have a number of effects, some of which are similar to those provided by dower and curtesy, and are referred to as *probate homestead.* In many states, the homestead laws also protect one's home against certain kinds of judgments and claims. Usually, this protection extends only to a specific dollar amount; the intent of these laws is to ensure that even under the most dire financial circumstances, a homeowner will not lose her entire investment in her home.

LIMITATIONS ON ESTATES AND INTERESTS

Very few estates or interests are completely free of restrictions or encumbrances; one or more of the following will affect most of them: encumbrances, easements, restrictive covenants, and liens.

Encumbrances

An *encumbrance* is a claim, liability, or charge against real estate that may affect the real estate's value, generally downward. An encumbrance may affect the title (examples include a mortgage, lien, or judgment) or the physical use of property (examples include an easement or restriction on use).

Easements

An *easement* is the right of one landowner to use land belonging to another for a specific purpose.

The two classes of easements are *appurtenant* and *in gross*. Easements appurtenant benefit a certain tract of land and transfers with the land. An example is an access easement; it benefits a tract of land called the *dominant estate*. The property used (the one allowing the easement) is the *servient estate*. If title to the dominant estate is transferred, the easement will benefit the new owner.

An easement in gross benefits a specific person or entity. It does not pass with the land; rather it remains with the specific owner. An example is a utility easement. It allows the utility company to install, maintain, and repair its utility lines upon the premises of the servient estate. It benefits the utility company. When the servient estate is transferred, the easement remains the property of its owner, i.e., the utility company.

The most common kinds of easements are access easements, utility easements, and drainage easements. All of these allow someone who is *not* an owner of the affected property to make use of that property for some purpose.

An *access easement* allows someone to cross a property to reach (i.e., obtain access to) another property. These rights exist where one parcel of land is completely blocked by others from access to a public right-of-way.

A *utility easement* is a right held by a utility company to put its utility lines on private property, usually for the purpose of providing service to that property.

A *drainage easement* requires that a property owner not disturb a natural or man-made drainage pattern that crosses his land.

Easements are created in all the ways that estates are; most commonly, however, they are created by voluntary alienation or adverse possession. When created by adverse possession, easements are called *prescriptions*. In some states, a person who sells a part of his land that does not have direct access to a public right-of-way is legally bound to provide the purchaser with an access easement.

Restrictive Covenants

A *restrictive covenant* is a contract whereby a group of neighboring landowners agrees to do (or *not* to do) a certain thing or things so as to mutually benefit all. Covenants most commonly occur in residential subdivisions and are created by the developer or a homeowners' association as a means of assuring prospective buyers that the development will be required to meet certain standards. Examples of covenants are prohibitions against using a lot for more than one single-family dwelling, limitations on the minimum size of structures to be built, and specification of architectural style. In some states, the entire collection of restrictions applying to a subdivision is called a *bill of assurance*. Also, in some states, restrictive covenants are called *deed restrictions*. Some properties refer to them as Covenants, Conditions, and Restrictions (CCRs). A seller who wants to control or limit the future use of property could insert a deed restriction when transferring ownership. Deed restrictions that are against public policy are not enforceable. For example, discriminatory deed restrictions cannot be enforced.

Liens

Liens are claims against a property, with that property being usable as security for the claim. Mortgages are liens. Claims for unpaid property taxes are liens. Also, anyone who has done work or provided materials so as to enhance the value of a property may, if not paid on time, secure a *mechanic's lien* against the property. If liens are not paid in the legal manner, the holder of the lien can foreclose against the property. *Foreclosure* is a legal process whereby a person holding a legal claim against a property may have the court order the property to be sold so as to provide funds to pay the claim.

Liens, restrictive covenants, and easements are, legally speaking, *a part of the property* and cannot be separated from it without the consent of all parties involved. A person who is considering buying property must be aware that a purchase does not remove these. All encumbrances on the property will remain with it after she acquires title unless something is done to remove them.

Water Rights

Riparian refers to property that borders on a moving body of water such as a river or stream. The owner of such property has *riparian rights.* In most states, a riparian property owner cannot accelerate or detain the natural flow of water.

Many western states allow *usufructory rights,* whereby an owner may divert a stream for a beneficial purpose, such as to water a growing crop.

Littoral rights are held by owners of a static or slow-moving body of water: lake, pond, bay, ocean, or sea. Generally, private ownership ends at the mean high-water level.

Accretion and Erosion

Accretion is an increase in land through the processes of nature. A sudden change in a water-course can result in a loss of soil, called *avulsion,* or an increase in alluvial deposits, called *alluvium* or *alluvion.*

Erosion is the gradual wearing away of land through the processes of nature. Soil erosion may cause *subsidence.* This could result in a loss of lateral support of a building, diminishing its structural integrity.

TITLE AND RECORDATION

A person who owns real estate is said to hold *title* to the land. There is no such thing as "title papers" to real estate, as there are with automobiles and certain other chattels. The term *title* refers to the validity of the available evidence that backs up one's claim to ownership of land. Normally, all deeds, liens, restrictions, and easements affecting land are recorded, usually in the courthouse of the county in which the land is located. Some leases, contracts of sale, and other documents may be recorded as well. The object of *recordation* is to provide a means whereby anyone may check the validity of any claim to land. In a legal sense, therefore, the ownership of all land and all rights to land should be a matter of record. These records are public, which means that anyone may examine them. It also means that anything that is recorded gives everyone *constructive notice* of the information recorded; therefore, in a lawsuit concerning ownership, one is assumed to be aware of all information in the public records.

The total of the evidence in the records provides the validity of the title an owner holds. Good title is title that cannot be impeached by the records. Title that appears to be good but is not is said to be *color of title.*

Attorneys, especially those who serve title companies (also called title insurance companies), including both attorney employees and independent attorneys, search titles to assure buyers that title is good and marketable. *Marketable title* is defined as being adequate for the property to be sold. Title companies have records and facilities giving them the best ability to research titles. They provide insurance for titles to lenders (mortgagees) and/or borrowers (owners or mortgagors). Sometimes title companies make a mistake and must pay a claim for the insurance they provide. Ultimately, the buyer assumes responsibility for receiving good title and for insuring it. Title insurance is paid only once by each owner, unlike fire or other home insurance that must be renewed each year.

EMINENT DOMAIN AND ESCHEAT

There are two special ways in which the government may acquire title to real estate. The most important is *eminent domain.* This is the right of the government to acquire privately owned property to be put to a public use, even if the private owner is unwilling to part with it. Government needs this power in order to operate efficiently for the benefit of everyone. Imagine the problems if, when the government decided to build a road, it had to look all over to find willing sellers for land needed for the project—the road might wander all around, and quite possibly it couldn't be built at all.

In an eminent domain situation, the government can't just seize the land. It must pay the owner just compensation, generally considered the "fair market value" of the land. However, once the local government has established that it needs the land for a legitimate public use, there is no way to stop it. All a landowner can do is to dispute the price being offered; if he thinks it is unfair, he may sue for a higher award and have the court make the final decision. Usually, the government will approach the landowner to work out an amicable arrangement, and quite often the landowner agrees to the offered price and there is no dispute. If, however, the landowner is reluctant or refuses to negotiate, the government will *condemn* the affected land. *Condemnation* is a legal process whereby the owner is dispossessed of the property and must leave. He retains the right to dispute the price being offered. Once condemnation has begun, a time will come when he must leave the land, which will be transferred to the government whether or not the payment question has been settled. All levels of government have power of eminent domain. States have also granted limited eminent domain power to private enterprises such as utility companies and railroads so these enterprises can place their lines. However, power plants and railroad terminals do not qualify for eminent domain.

Escheat is a process whereby land for which no legal owner exists reverts to government. The government can either dispose of or use the land as it sees fit. Escheat most commonly occurs when someone dies intestate and no legal heirs can be found.

YOUR STATE'S REAL ESTATE LAW

The laws of each state vary considerably. Most state real estate licensing authorities make available literature that explains some of the special provisions of law that affect real estate in that state. To find the answers to the questions on this page, you should study Table 3-1 and the materials from your state. Then check the appropriate boxes, and write your other answers in the spaces provided.

1. Does your state have *dower laws*?

   ☐ No   ☐ Yes; they apply to husbands *and* wives

   ☐ Yes, for wives only

   What are the dower rights of wives? ______________________

   ____

   ____

   What are the dower rights of children? ______________________

   ____

   ____

2. Is your state a *community property* state?

   ☐ No   ☐ Yes

3. Which tenancies does your state allow?

Joint tenancy	☐ No	☐ Yes
Tenancy by the entirety	☐ No	☐ Yes

   (All states allow tenancy in common, at sufferance, at will.)

4. Does your state have homestead laws?

   ☐ No   ☐ Yes

   If yes, what are the major provisions?

   ____

   ____

5. What is the statutory period for adverse possession in your state?

   For fee simple ownership ________ years

   For easements ________ years

   Are there any special situations that would shorten or lengthen the statutory period?

   ☐ No   ☐ Yes

   If yes, what are they?

   ____

   ____

   ____

Table 3-1

OWNERSHIP AND INTERESTS IN REAL ESTATE:
THE FIFTY STATES AND THE DISTRICT OF COLUMBIA*

State	Joint Tenancy	Tenancy by the Entirety	Community Property	Dower	Homestead
Alabama	•			•	•
Alaska		•			•
Arizona	•		•		•
Arkansas	•	•		•	•
California	•		•		•
Colorado	•				•
Connecticut	•				
Delaware	•	•		•	
District of Columbia	•	•		•	
Florida	•	•			•
Georgia				•	•
Hawaii	•	•		•	•
Idaho	•		•		•
Illinois	•	•			•
Indiana	•	•			
Iowa	•			•	•
Kansas	•			•	•
Kentucky	•	•		•	•
Louisiana	See below		•	See below	•
Maine	•				•
Maryland	•	•			
Massachusetts	•			•	•
Michigan	•	•		•	•
Minnesota	•				•
Mississippi	•	•			•

State	Joint Tenancy	Tenancy by the Entirety	Community Property	Dower	Homestead
Missouri	•	•			•
Montana	•			•	•
Nebraska	•				•
Nevada	•		•		•
New Hampshire	•				•
New Jersey	•	•		•	•
New Mexico	•		•		•
New York	•	•			•
North Carolina	•	•		•	•
North Dakota	•				•
Ohio		•		•	•
Oklahoma	•	•			•
Oregon		•		•	•
Pennsylvania	•	•			
Rhode Island	•	•		•	
South Carolina	•			•	•
South Dakota	•				•
Tennessee	•	•		•	•
Texas	•		•		•
Utah	•	•			•
Vermont	•	•		•	•
Virginia	•	•		•	•
Washington	•		•		•
West Virginia	•	•		•	•
Wisconsin	•		•	•	•
Wyoming	•	•			•

*The states are arranged alphabetically; each column represents one of five estates or ownership interests. A bullet (•) in a column means that the state *permits* that feature. If there is no bullet, the state *does not permit* it.

Louisiana law comes from French civil law, whereas the law in the other 49 states has evolved from English common law. Therefore, the law in Louisiana is different, and Louisiana license applicants should study that state's law especially diligently. Of the other 49 states:

All permit fee simple estates.
All permit life estates.
All permit individual ownership (tenancy in severalty).
All permit tenancy in common.
All permit condominium.

QUESTIONS ON CHAPTER 3

1. Which of the following is most accurately described as personal property?
   (A) a fixture
   (B) a chattel
   (C) an improvement
   (D) realty

2. Which of the following is *not* corporeal property?
   (A) a building
   (B) growing crops
   (C) monthly rent
   (D) a fixture

Questions 3–5 concern the following situation:

Mr. Jones died; his will left to Mrs. Jones the right to use, occupy, and enjoy Mr. Jones's real estate until her death. At that time, the real estate is to become the property of their son Willis.

3. Mrs. Jones is a
   (A) remainderman
   (B) life tenant
   (C) joint tenant
   (D) tenant in common

4. Willis is a
   (A) remainderman
   (B) life tenant
   (C) joint tenant
   (D) tenant in common

5. Mrs. Jones has a
   (A) fee simple estate in joint tenancy with Willis
   (B) fee simple estate as tenant in common with Willis
   (C) life estate
   (D) life estate in joint tenancy with Willis

6. Which of the following is *not* realty?
   (A) a fee simple estate
   (B) a leasehold for indefinite duration
   (C) lumber
   (D) a life estate

7. Real estate is best defined as
   (A) land and buildings
   (B) land and all permanent attachments
   (C) land and everything growing on it
   (D) land only

8. A freehold estate is
   (A) one acquired without paying anything
   (B) any leasehold
   (C) any estate wherein one may use the property as he wishes
   (D) an estate of uncertain duration

9. An item of personalty that is affixed to realty so as to be used as a part of it is
   (A) a fixture
   (B) a chattel
   (C) personal property
   (D) an encumbrance

10. Which is *not* considered a permanent attachment to land?
    (A) anything growing on it
    (B) fixtures
    (C) chattels
    (D) anything built upon the land

11. A person who has some rights to use land, but not all possessory rights, is said to have
    (A) an interest in land
    (B) an estate in land
    (C) a life estate in land
    (D) a tenancy in common

12. A person who has permission to use land but has no other rights has
    (A) tenancy at sufferance
    (B) tenancy in common
    (C) license to use
    (D) a fee simple estate

13. A person receives title to land by virtue of having used and occupied it for a certain period of time, without actually paying the previous owner for it, receives title by
    (A) will
    (B) descent
    (C) alienation
    (D) adverse possession

14. A person who dies leaving no will is said to have died
    (A) intestate
    (B) without heirs
    (C) unbequeathed
    (D) unwillingly

15. A person who owns an undivided interest in land with at least one other and who has the right of survivorship is said to be a
    (A) tenant in common
    (B) tenant at will
    (C) joint tenant
    (D) tenant at sufferance

16. Tenancy in severalty refers to
    (A) ownership by one person or one entity only
    (B) ownership by two persons only
    (C) ownership by at least three persons
    (D) a special form of joint ownership available only to married couples

17. Dower rights are rights that assure that
    (A) a spouse receives a certain portion of his/her deceased spouse's estate
    (B) dowagers may have property rights
    (C) a homeowner cannot lose his entire investment in his home
    (D) husbands and wives share equally in property acquired during marriage

18. An easement is
    (A) the right to use the property of another for any purpose
    (B) the right to use the property of another for a specific purpose
    (C) a private contract and does not permanently affect the realty
    (D) the right to keep another from using one's land illegally

19. The process whereby a person holding a claim against property can have the property sold to pay the claim is
    (A) a lien
    (B) a covenant
    (C) a mechanic's lien
    (D) a foreclosure

20. A person who appears to own property but does not is said to have
    (A) good title
    (B) recorded evidence of title
    (C) constructive notice
    (D) color of title

21. Two people who own undivided interests in the same realty, with right of survivorship, have
    (A) tenancy in severalty
    (B) tenancy in common
    (C) tenancy at sufferance
    (D) joint tenancy

22. A fee simple estate is
    (A) a freehold estate
    (B) a life estate
    (C) an estate for years
    (D) an estate in abeyance

23. Which of the following is *not* considered a permanent attachment to land?
    (A) improvements to land
    (B) a fixture
    (C) trees growing on the land
    (D) a load of bricks

24. A person who has some rights to land, but not all the possessory rights, has
    (A) a life estate
    (B) an interest in land
    (C) an estate at will
    (D) a fixture

25. A person who has color of title to land has
    (A) the appearance of title
    (B) forgeries of state title papers
    (C) a right to one-half the income from the land
    (D) foreign ownership

26. A life estate *pur autre vie* is one in which
    (A) the life tenant must occupy the premises and may not lease them to anyone else
    (B) the duration of the life estate is based upon the duration of the life of someone other than the life tenant
    (C) the life tenant has died
    (D) the life tenant may not occupy the property himself/herself

27. Which of the following would not be real property?
    (A) a lease for one month
    (B) an easement
    (C) permanent ownership of mineral rights on someone else's land
    (D) a life estate

28. *Adverse possession* may establish
    (A) use of property
    (B) a claim to title
    (C) who owes back taxes
    (D) homestead rights

29. An easement by prescription may establish
    (A) use of property
    (B) a claim to title
    (C) who owes back taxes
    (D) homestead rights

30. Fifty feet of land was added to Apple's farm because a river changed its course. This is an example of
    (A) accretion
    (B) avulsion
    (C) riparian rights
    (D) usufructory

31. Fifty feet of beachfront land was removed by a flood. This is an example of
    (A) riparian rights
    (B) estovers
    (C) avulsion
    (D) emblements

32. As used in real estate practices, the land of a *riparian* owner borders on
    (A) a river
    (B) a stream
    (C) a moving watercourse
    (D) any of the above

33. A person holding title to real property in severalty would most likely have
    (A) a life estate
    (B) an estate for years
    (C) ownership in common with others
    (D) sole ownership

34. Joint ownership of real property by two or more persons, each of whom has an undivided interest (not necessarily equal) without right of survivorship, is
    (A) a tenancy in partnership
    (B) a tenancy by the entireties
    (C) a tenancy in common
    (D) a leasehold tenancy

35. Generally, the taking of private land by governmental bodies for public use is governed by due process of law and is accomplished through
    (A) exercise of police power
    (B) eminent domain
    (C) reverter
    (D) escheat

36. Governmental land use planning and zoning are important examples of
    (A) exercise of eminent domain
    (B) use of police power
    (C) deed restrictions
    (D) encumbrance

37. After Harry bought his apartment, he started receiving his own property tax bills. This indicates that he has bought into a
    (A) condominium
    (B) cooperative
    (C) leasehold
    (D) syndicate

38. Helen and David, who are not married, want to buy a house together. To ensure that if one dies the other automatically becomes the full owner, their deed must state that they are
    (A) tenants in common
    (B) tenants in severalty
    (C) tenants by the entirety
    (D) joint tenants

39. Real property is
    (A) the land
    (B) the land, the improvements, and the bundle of rights
    (C) the land and the right to possess it
    (D) the land, buildings, and easements

40. A house has a built-in bookcase and elaborate chandelier. The buyer is also getting a dinette set and a table lamp. Which is correct?
    (A) The built-in bookcase and chandelier are included in the real estate contract. The lamp and dinette set are chattels and are sold separately.
    (B) All the items must be included in the real estate contract.
    (C) The chandelier is real estate; the other items must be sold on a bill of sale.
    (D) All of the items are real estate because they are sold with the house.

41. Man-made additions to land are known as
    (A) chattels
    (B) trade fixtures
    (C) easements
    (D) improvements

42. Land can be owned apart from the buildings on it under the arrangement known as a
    (A) net lease
    (B) life estate
    (C) remainder
    (D) ground lease

43. No one seems to own the vacant land next to Mary's house, so she has used it for a pasture for many years. She may have a good chance of obtaining ownership by
    (A) poaching
    (B) foreclosure
    (C) adverse possession
    (D) eminent domain

44. An abstract of title contains
    (A) the summary of a title search
    (B) an attorney's opinion of title
    (C) a registrar's certificate of title
    (D) a quiet title lawsuit

45. In order to reach the beach, the Smiths have a permanent right-of-way across their neighbor's beachfront property. The Smiths own
    (A) an easement
    (B) a license
    (C) a deed restriction
    (D) a lien

46. Which of the following is real estate?
    (A) fence, driveway, shrubs
    (B) a dishwasher in a box that needs to be installed
    (C) a cord of timber that has been cut
    (D) a piece of paper on which is a written and signed deed

47. Littoral rights for properties abutting the ocean give ownership
    (A) to the mean high-water mark
    (B) to a point 100 feet from the shore
    (C) to the state ownership line
    (D) to a point designated by the U.S. Army Corps of Engineers

48. Personal property is best defined as
    (A) anything that can be owned aside from real estate
    (B) intangibles
    (C) fixtures attached to real estate
    (D) furniture and built-in appliances

49. The lumber for a deck that has not yet been built is
    (A) a trade fixture
    (B) real estate
    (C) personal property
    (D) a fixture

50. The vertical boundaries of land ownership are best described as
    (A) from the center of the earth to 1,000 feet above ground
    (B) from the surface of the earth to 1,000 feet below ground
    (C) from the center of the earth upward to the heavens above
    (D) from the surface of the earth to 1,000 feet below ground and 1,000 feet above ground

51. The type of co-ownership that allows an owner's property to be left to his or her heirs is
    (A) tenancy by the entirety
    (B) tenancy in severalty
    (C) tenancy in common
    (D) joint tenancy

52. One who owns a life estate cannot
    (A) sell his interest
    (B) mortgage his interest
    (C) devise his interest
    (D) lease his interest

53. A condominium owner can avoid payment of his or her share of the common expenses by doing which of the following?
    (A) not using certain common elements
    (B) abandoning her apartment
    (C) defaulting on mortgage payments
    (D) none of the above: payment cannot be avoided

54. Which type of property can become real estate by the manner of physical attachment?
    (A) fixtures
    (B) emblements
    (C) portable appliances
    (D) deeds

55. Owners of cooperative apartments and condominiums both
    (A) own real estate
    (B) own shares in a corporation
    (C) must live full-time in their units
    (D) must pay assessments

56. The right of _________ gives the state ownership of property left by one who died without heirs and without a will.
    (A) estovers
    (B) intestate
    (C) emblements
    (D) escheat

57. A fee simple estate is best described as
    (A) ownership for one's lifetime
    (B) complete ownership, subject to eminent domain, zoning, taxation, and other government land use limitations
    (C) interest in a property as limited by a deed
    (D) ownership as long as one pays a fee for title

58. Which of the following is true about eminent domain?
    (A) The process used to acquire the property is called condemnation.
    (B) Governments can use it whenever and wherever they want.
    (C) Governments can pay whatever amount they like.
    (D) Constitutionally, the right of eminent domain is guaranteed by the First Amendment.

59. Three people bought property with the understanding that if one of them died, his heirs would inherit that interest. Most likely they took title under which type of estate?
    (A) joint tenancy
    (B) tenancy in common
    (C) estate in common
    (D) tenancy by entireties

60. A city wants to buy land to build a convention center. Which of the following is true regarding eminent domain?
    (A) It is not allowed because a convention center is not a necessity.
    (B) It can be applied with just compensation paid for the land.
    (C) It will be too costly.
    (D) Only federal and state governments have that power.

61. A person who inherits real estate under the terms of a will is a(n)
    (A) bequest
    (B) life estate
    (C) devisee
    (D) executor

62. When title to property is taken as tenants in common,
    (A) all parties have an equal interest
    (B) each tenant in common may sell his interest
    (C) each tenant in common must pay rent to the landlord
    (D) when a tenant dies, the share is distributed equally to the surviving tenants in common

63. Your grandfather died intestate, and you received his house. Title was acquired
    (A) by will
    (B) by estovers
    (C) by descent
    (D) by mitigation

64. The most extensive form of ownership in real estate is
    (A) a paid-up fee
    (B) life estate
    (C) fee tail
    (D) fee simple

65. What type of rights does not refer to the use of water?
    (A) riparian
    (B) usufructory
    (C) dower
    (D) littoral

66. Allen has given Betty the right to park in his driveway for three weeks. This right is best called
    (A) license
    (B) easement
    (C) right of way
    (D) lease

67. Unmarried domestic partners wish to buy a house using an ownership agreement whereby if one dies, that party's heirs become owners of the share. Which form of ownership should they use?
   (A) joint tenancy
   (B) tenancy in severalty
   (C) limited partnership
   (D) tenancy in common

68. Which is *not* true of a joint tenancy?
   (A) Its use is limited to a husband and wife.
   (B) It offers the right of survivorship.
   (C) It is a form of co-ownership.
   (D) It differs from tenancy in common.

69. A proprietary lease is a unique characteristic of what form of ownership?
   (A) condominium
   (B) cooperative
   (C) timeshare
   (D) PUD

70. Approximately 41 states do *not* recognize
   (A) partnerships
   (B) joint tenancy
   (C) community property
   (D) tenancy in common

71. The exercise of eminent domain results in
   (A) seizure
   (B) unjust compensation
   (C) condemnation
   (D) government authority

72. Susan lived rent free in a remote cabin, with the owner's permission, for 40 years. Susan
   (A) owns the property under adverse possession
   (B) owns the property under prescription
   (C) must pay rent to the owner
   (D) has no ownership interest

73. I give my brother permission to live in a house I own until the death of his daughter. This is
   (A) an ordinary life estate
   (B) a life estate pur autre vie
   (C) a leasehold interest
   (D) a tenancy for life

74. Upon expiration of a life estate, the property passes to the
   (A) life tenant
   (B) beneficiary
   (C) remainderman
   (D) trustee

75. Upon marriage in community property states, real property previously owned individually by each party
   (A) becomes jointly owned
   (B) doesn't change
   (C) becomes community property
   (D) is held by a tenancy in common

76. A widower dies without making a will. His surviving children inherit the property by
   (A) escheat
   (B) law of intestate succession
   (C) estover
   (D) emblements

77. Which of the following provides the most complete form of ownership?
   (A) fee tail
   (B) fee simple absolute
   (C) fee on condition
   (D) leasehold

78. Which of the following is a property improvement?
   (A) a pool table in a basement
   (B) an in-ground swimming pool
   (C) an above-ground plastic swimming pool
   (D) a merger of two REITs treated as a pooling of interests

79. The *bundle of rights* refers to
    (A) the U.S. Constitution's Bill of Rights
    (B) sticks carried to a property as firewood
    (C) the right to sell your property
    (D) a set of legal rights to property that an owner enjoys

80. Riparian rights apply to land along
    (A) any body of water
    (B) ocean waters
    (C) ripe vegetables and fruits
    (D) rivers and streams

81. Littoral rights apply to land adjacent to
    (A) a city's reservoir
    (B) man-made lakes and ponds
    (C) oceans or seas
    (D) all of the above

82. How is a fixture best defined?
    (A) personal property included in the sale of real estate
    (B) personal property more or less permanently attached to real estate
    (C) personal property left with real estate when sold
    (D) any chandelier or ceiling light

83. Two unmarried individuals own a house with equal interests. What form of ownership is most likely?
    (A) tenants by the entireties
    (B) limited partnership
    (C) tenancy in common
    (D) condominium

84. The right to use, occupy, and enjoy real estate for a limited amount of time specified in a lease is
    (A) fee simple
    (B) cooperative
    (C) leased fee estate
    (D) leasehold estate

85. Morris grants the use of a property to his daughter, Susie, for her lifetime. When Susie dies, the property will go to Morris's son, Herman. Susie has a
    (A) remainder interest
    (B) life estate pur autre vie
    (C) life estate
    (D) riparian rights

86. A tenant who stays without permission after a lease expires is a
    (A) holdover tenant
    (B) tenant at will
    (C) squatter
    (D) eviction

87. Which states recognize community property?
    (A) California and Texas
    (B) New York and New Jersey
    (C) Ohio and Pennsylvania
    (D) Virginia and West Virginia

88. When a joint tenant dies, his or her share
    (A) is owned by the surviving joint tenants in proportion to their ownership share
    (B) has its share of mortgage debt canceled
    (C) escheats to the state
    (D) becomes community property

89. In a cooperative apartment arrangement, the building and land are owned by
    (A) the state
    (B) a corporation
    (C) husband and wife
    (D) a condominium

90. When a tenant in common dies, his or her interest
    (A) is inherited by the surviving tenants in common
    (B) is inherited by his or her heirs in accordance with his or her will
    (C) is inherited by the remainderman
    (D) escheats to the state

91. Larry, as a tenant, rents a house for one year. Larry has the right to
(A) sell the house
(B) assign or sublet the house
(C) borrow against the house
(D) use and possess the property for one year upon payment of rent

92. A tenancy at will expires
(A) when either the owner or the tenant decides, or upon the death of either
(B) on the termination date stated in the lease
(C) every year unless renewed
(D) on the date written in the tenant's will

93. Which of the following is an encumbrance on real estate?
(A) a restriction on use
(B) a mortgage loan
(C) an easement
(D) all of the above

94. An easement can apply to use by
(A) people only
(B) property only
(C) government only
(D) people or property

95. The property that benefits from use of an easement is the _________ estate.
(A) servient
(B) life
(C) credit
(D) dominant

96. An easement by prescription is needed by a person whose property is
(A) on a body of water
(B) landlocked
(C) fenced
(D) zoned

97. A seller who wants to control the future use of property could insert a(n) _________ when transferring ownership.
(A) deed restriction
(B) zoning clause
(C) zip lock
(D) exculpatory clause

98. A lien is a
(A) form of property tax
(B) type of lease
(C) claim against a property with the property being the security for the claim
(D) form of foreclosure

99. Recording a deed at a county courthouse gives
(A) actual notice to the world of the existence of a document
(B) constructive notice to the world of the existence of a document
(C) notice to title companies
(D) no notice at all

100. The court where an estate is settled is called the _________ court in most jurisdictions.
(A) claims
(B) appeals
(C) probate or surrogate's
(D) tax

ANSWERS

1. B
2. C
3. B
4. A
5. C
6. C
7. B
8. D
9. A
10. C
11. A
12. C
13. D
14. A
15. C
16. A
17. A
18. B
19. D
20. D
21. D
22. A
23. D
24. B
25. A
26. B
27. A
28. B
29. A
30. A
31. C
32. D
33. D
34. C
35. B
36. B
37. A
38. D
39. B
40. A
41. D
42. D
43. C
44. A
45. A
46. A
47. A
48. A
49. C
50. C
51. C
52. C
53. D
54. A
55. D
56. D
57. B
58. A
59. B
60. B
61. C
62. B
63. C
64. D
65. C
66. A
67. D
68. A
69. B
70. C
71. C
72. D
73. B
74. C
75. B
76. B
77. B
78. B
79. D
80. D
81. D
82. B
83. C
84. D
85. C
86. A
87. A
88. A
89. B
90. B
91. D
92. A
93. D
94. D
95. D
96. B
97. A
98. C
99. B
100. C

PART TWO
Real Estate Brokerage

Agency Law

4

ESSENTIAL MATTERS TO LEARN FROM THIS CHAPTER

- An *agency* relationship involves two parties: the *principal* and the *agent.*
- The *principal* is the party who owns property and employs the agent to locate a buyer or lessee, or is the party who employs the agent to find property to buy or lease.
- An agency relationship may be *express* or *implied.*
- In an *express* agency, the principal and agent have an agreement, which is either oral or written.
- In an *implied* agency, the parties have an agreement based on the nature of the principal's business and his/her relationship to the agent's business. This type of agency implies that the agent has permission to perform certain activities.
- A person can earn a *license* to be a real estate broker by meeting educational, testing, and experience requirements that are set by each state. A real estate broker legally serves as an agent for the principal.
- One who holds a real estate license is a licensee. In most states, he/she may be a broker or salesperson.
- A real estate broker is a *special agent* for the principal because the agent's powers are limited. This contrasts with a *general agent,* who is entrusted by the principal with broader powers to act for him/her in the business, or with a *universal agent,* who has full authority for all of the principal's businesses.
- A real estate broker may employ licensed *salespeople* to assist in performing his role. Typically, these salespeople must meet licensure requirements of education and testing. The broker is responsible for the acts of salespeople he employs. The salespeople are *subagents* of the broker and are paid by the broker.
- A *REALTOR®* is a broker or salesperson who is a member of the National Association of Realtors® (NAR), which is a trade association. NAR provides educational training, economic data, and political lobbying. It is not a state agency, and it does not offer a license to sell real estate belonging to others.
- A broker typically represents a single party, either the buyer or seller, in a transaction. Then the broker owes *fiduciary* duties to that party. Although a fiduciary of one party in a single agency relationship, the broker must deal honestly with all parties in the transaction.
- As a fiduciary, the broker must deal honestly and fairly, be loyal to and obey his principal's instructions, protect confidential matters of his principal, provide full disclosure to his principal, properly account for funds, observe skill and care, present all offers and counteroffers in a timely manner, and disclose all known material facts about the property and its value, except those that are readily observable.
- The broker may not obey a principal's request that is illegal. Of particular interest is the requirement to follow fair housing laws and not violate anyone's civil rights.

- A single agency broker for an owner must not tell a prospective buyer that the owner is desperate to sell at a low price because he is getting a divorce. That is a personal matter not affecting the real estate. However, the broker must honestly answer the buyer's questions about the condition of the property to the extent of his knowledge because the buyer is interested in buying the property. This is different from answering questions about personal matters of the owner, which the buyer does not acquire.
- *Dual agency* occurs when a broker or the brokerage firm represents opposing parties (both the buyer and seller) in the same transaction. In states that allow dual agency, both principals must understand the situation and receive disclosure, often required to be in writing. Professional behavior is expected of a dual agent, including honesty, skill, care, accounting for funds, and presentation of offers.
- *Transaction brokerage* occurs when the broker does not represent either party but, instead, provides services and general information to either party, without looking out for the interests of a particular party.
- *No-brokerage agency* may occur when parties to a transaction have reached agreement but seek help, such as how to fill out a contract for sale. Written disclosure should be provided to both parties so all can understand and agree to the broker's role.
- A *listing* is a contract between a principal and a broker to offer real estate for sale. Many states require a listing to be in writing for a broker to collect a commission, and many states require a listing to have a fixed expiration date. Listings are explained in more detail in Chapter 11.
- A broker earns a commission when he or she brings a ready, willing, and able buyer at the seller's terms as set forth in the listing. If the owner decides not to sell, the owner still owes the broker the full commission.
- An *open listing,* often used for high-value commercial property, gives the owner the right to list the property with more than one broker. The broker who sells the property earns the commission.
- An *exclusive-agency listing* allows only one broker to sell, but the owner may sell through his own efforts without paying a commission.
- An *exclusive right to sell listing* provides a commission to only one broker, even if the owner himself makes the sale.
- A *multiple listing service* (MLS) is a local organization in which many brokers participate. Typically, the listing broker brings the listing into a pool, through which all listings are known to members. The listing broker usually shares the commission with the selling broker.
- A *net listing* is one in which a minimum selling price is specified. The broker's commission is the amount above the net price specified. Net listings are frowned upon or illegal in most states because in some contracts for sale, the broker's interest may be opposed to the owner's.
- Ways to terminate an agency relationship include:

  1. Death or incompetence of either party
  2. Bankruptcy of either party
  3. Destruction of the property
  4. Performance of the contract
  5. Expiration of the contract
  6. Mutual agreement of the parties

7. Resignation of the agent (but may require payment for damages)
8. Discharge of the agent by the principal (principal may be liable for damages)

- Commission rates are set by negotiation between the principal and broker.
- The *Sherman Antitrust Act* prevents brokers from *price fixing.* It can apply monetary penalties and a prison term to brokers who agree to fix commission rates.
- Be sure to study Chapter 11 on Listings/Property Management and Chapter 21 on Fair Housing Law.

PARTIES IN REAL ESTATE AGENCIES

A licensed real estate broker or salesperson is an *agent* of the principal; therefore, their relationship comes under the *laws of agency.* He/she is also affected by the real estate license law of the particular state, a topic that will be discussed in Chapter 5.

An agency relationship involves two parties: the *principal* and the *agent.* The agency relationship is contractual, but it also is covered by the requirements of agency law. In the real estate business, a broker acts as an agent for his *employer,* or principal, who is the owner of the property for which the broker is seeking a buyer or tenant or who is the potential buyer or lessee. A licensed salesperson usually is treated as the agent of the *broker,* because license law does not permit salespersons to act without the supervision of a broker.

In an *express agency,* the principal and agent have an agreement that is either written or oral. In an *implied agency,* the relationship is implied by the business positions of the principal and agent. For example, when the principal is the property owner and the agent is a broker used by sellers, the relationship is an implied agency.

A real estate broker is a *special agent.* He is called this because a special agent's powers are limited, usually to finding someone with whom the principal can deal. In contrast, a *general agent* has authority to act in all aspects of the business (sales, rentals, property management). A *universal agent* is a person whose powers are broader and may extend to all aspects of the principal's multiple businesses. Note that a real estate broker usually has little or no power actually to commit the principal to anything. The principal may refuse to deal with anyone the broker brings to her, although the principal may still be liable to pay the broker a commission if it has been earned.

DUTIES OF PARTIES

An agent is employed to deal with third parties on behalf of the principal. The agent is an employee of the principal and must follow his principal's instructions implicitly, unless doing so would result in violating law. Furthermore, he has a duty to act in his employer's best interests, even when doing so means that he cannot act in his own favor. More specifically, agency law requires that the agent abide by the following obligations to his principal:

- He must *obey* his principal's instructions (except, of course, when he is instructed to do something illegal).
- He must be *loyal* to his principal's interests.
- He must act in *good faith.*
- He is expected to use *professional judgment, skill,* and *ability* in his actions. This is particularly true in the case of licensed agents such as real estate brokers, who are assumed to have met certain standards by virtue of being licensed.
- He must be able to *account* for all money belonging to others that comes into his possession. Real estate brokers often collect rents, earnest money deposits, and other money on behalf of their principals.

- He must perform his duties *in person*. Normally, this would mean that he could not delegate his duties to anyone else unless the principal approved of the arrangement. Because of the nature of the real estate brokerage business and the establishment of the broker-salesperson relationship by state licensing laws, a real estate broker has the implied right to delegate some of his function to duly licensed salespersons and other brokers in his employ.
- He must keep his principal *fully informed* of developments affecting their relationship. A real estate broker must report all offers made on the principal's property as well as all other information he acquires that may affect the principal's property, price, or other relevant factors.

The principal has obligations to the agent as well. Most of these have to do with money: the principal is obliged to *compensate* the agent for his services, to *reimburse* the agent for expenses paid on behalf of the principal, and to *protect* the agent against any loss due to proper performance of his duties. The principal also has the implied duty to make the agent fully aware of his duties and to inform him fully about anything that will affect his performance.

REAL ESTATE AGENCY CONTRACTS

Agency contracts are *employment* contracts. They do not come under the broad heading of real estate contracts and so are not necessarily required by the Statute of Frauds to be in writing (see Chapter 6). Many states' real estate licensing laws do require listing contracts to be in writing. In real estate, an agency contract usually is called a *listing*. The details of preparing these contracts are discussed in Chapter 11. Here, however, we can briefly distinguish among the various kinds of listings. Among listings of property for *sale* are the following.

An *open* listing is one in which the principal agrees to compensate the agent only if the agent actually finds the buyer with whom the principal finally deals. A single owner may have open listings with many brokers since she is obliging herself to pay only upon performance and the broker makes his effort at his own risk.

An *exclusive agency* listing is one in which the principal agrees to employ no other broker. If the property is sold by any licensed agent, a commission will be paid to the listing broker. State license law usually requires that any other agent must work *through* a broker having an exclusive listing; the participating brokers must then agree on a means by which they will share in the commission. The commission will always be paid to the listing broker, who then may pay a share to the other broker(s) who cooperated in the sales transaction. The brokers may then pay a share to licensed salespersons in their employ. Under an exclusive agency, the principal may find a buyer and owe no commission.

An *exclusive right to sell* listing guarantees that the principal will pay the broker a commission regardless of who actually sells the property. This applies even if the principal herself finds the buyer, with no aid at all from the broker. (Note that under an exclusive agency listing, the principal does not have to pay a commission if she herself finds the buyer.)

Usage of the term *exclusive listing* varies from one place to another. In some localities the term refers to an exclusive agency; in others it refers to the exclusive right to sell; in still others it refers to both types interchangeably.

A *net* listing is one in which the principal agrees to receive a given net amount. The excess by which the sale price exceeds the net amount goes to the broker as his commission. In many states, net listings are illegal; in many others, they are not illegal but are "frowned upon" or otherwise disapproved of in some manner. The big problem with net listings is that the agent is strongly tempted to act in his own interest rather than that of his employer.

To illustrate, suppose a net listing contract with a proposed net to the seller of $180,000 is agreed upon. By law, anytime the broker solicits an offer he must report it to the principal. However, let us assume that an offer of exactly $180,000 is made; if the broker transmits it, the seller probably will accept and the broker will receive no compensation. Clearly, there is the temptation to illegally "hide" the offer, and any subsequent offers, until one yielding a satisfactory excess over the listed price is received to provide an adequate commission. Such action is contrary to the agent's legal responsibilities to his employer.

A further source of trouble with net listings arises when a knowledgeable agent deals with a naive seller. There may exist an opportunity for the agent to convince the seller to sign a net listing at a low price, thereby guaranteeing the agent a large commission when he finds a buyer who will pay market value.

THE AGENCY RELATIONSHIP IN REAL ESTATE BROKERAGE

In a *listing contract* (see Chapter 11), the agency relationship is between the owner of the listed real estate and the broker or the brokerage firm whom the owner hires to sell or lease the real estate. This makes sense; however, in the real estate brokerage business it is very common for other brokers to be involved in the transaction. Brokers very often *cooperate* with one another: they agree to allow one another to seek buyers (or tenants) for each others' listings. If another broker finds a buyer (tenant), then the listing broker agrees to share with that other broker the commission paid by the property owner. This situation leads to the question of whom all these agents actually represent. In *traditional agency*, all agents involved in a cooperated transaction represent *only the seller:* buyers have no agent looking out for their interests. It's called traditional agency because that's the way it was always done until about 30 years ago. At that time the states began to allow *buyer agency.* Buyer agency occurs when the licensee specifically represents the buyer(s) and a contract to that effect is signed. This way, buyers get agency representation: a professional who is looking out for *their* interests. (Buyer agents are compensated with a share of the commission paid by the seller, just as in traditional agency.) More recently, many states have created a "non-agency" brokerage. In this arrangement the licensee isn't really an agent at all; buyer and seller have to look out for themselves and the broker just assists in showing property and moving the transaction along.

Most states now allow *buyer agency.* This is a situation in which a licensed broker is the agent of the buyer, rather than of the seller. In such a situation, the loyalty of the licensee is to the buyer rather than to the seller. The states now allow this form of agency in order to provide a means whereby buyers receive the agency loyalty and service that has always been enjoyed by sellers. Buyer agency also allows licensees to engage in the "selling process" more effectively, because they are not restricted as to the information and advice that they can give to buying prospects.

Many states now require *agency disclosure* by real estate brokers. Upon first contact with a prospective buyer, the broker must provide notification that he actually represents the seller and does not look out for the specific needs and interests of the buyer. Usually, the law specifies a certain form that must be given to the buyer. The broker cannot proceed with any dealings with that buyer until the broker has received from the buyer a signed form acknowledging that the buyer understands the agency relationship the broker has with the seller.

Trick Question

Trick questions often overstate or understate the role of a real estate broker by implying that the broker can make decisions for the homeowner. The broker cannot do this. Instead, the broker must obey the principal's instructions, provided those instructions are legal. However, the principal need not accept the broker's suggestions.

Single Agent Relationship

In a single agent relationship, the agent may represent either the buyer or the seller. The broker (agent) owes fiduciary duties to the party represented and must disclose these in writing in a residential transaction. The fiduciary duties of a single agent to the principal include:

- Dealing honestly and fairly
- Loyalty
- Confidentiality
- Obedience
- Full disclosure
- Accounting for all funds
- Skill, care, and diligence in the transaction
- Presenting all offers and counteroffers in a timely manner, unless a party has previously directed the licensee otherwise in writing
- Disclosing all known facts that materially affect the value of residential real property and are not readily observable

Dual Agency

Suppose a real estate licensee has contracted to be the agent of a prospective buyer and then begins to show property to that buyer. Suppose, also, that this agent wants to show the buyer property that the agent himself/herself, or the agent's brokerage firm, has listed. In this case, the agent would become the agent of both parties to the transaction. It is clearly impossible to have absolute loyalty to the interests of both parties, so a means has to be devised that allows the licensee to proceed in such a situation. Most states now allow *dual agency* in such cases, though some states prohibit this practice. In order to solve the problem of representing both sides to the transaction, a *dual agency disclosure* has to be presented to, and agreed to, by both the prospective buyer and the owner of the property to be shown. In it, the parties all agree that the licensee's agency relationship with both buyer and seller will be limited so as not to compromise the interests of either party.

Typically, the dual agency disclosure says, at the very least, that the agent cannot reveal to the seller the highest price the buyer will pay and cannot reveal to the buyer the lowest price the seller will accept. While dual agency is a solution that allows the licensee to show the property in question, it does so by limiting the agency obligation to both parties. Therefore, some states are trying variations on dual agency, such as requiring that a different licensee step in to represent one of the parties fully while the original licensee continues to fully represent the other party.

Transaction Brokerage

Many states have adopted a form of brokerage called *transaction brokerage* (also called *facilitative brokerage* in a few states). A transaction broker signs an agreement with either the buyer or seller or both, stating that the broker does not represent the interests of either party but merely provides brokerage service and general information to both parties. Essentially, the buyer and seller are on their own with respect to looking out for their interests.

In some states, the duties of a transaction broker, which are to be disclosed in a residential transaction, are the following:

- Dealing honestly and fairly
- Accounting for all funds
- Using skill, care, and diligence in the transaction
- Disclosing all known facts that materially affect the value of residential real property and are not readily observable to the buyer
- Presenting all offers and counteroffers in a timely manner, unless a party has previously directed the licensee otherwise in writing.
- Limited confidentiality, unless waived in writing by a party. This limited confidentiality will prevent disclosure that the seller will accept a price less than the asking or listed price, that the buyer will pay a price greater than the price submitted in a written offer, or the motivation of any party for selling or buying property, that a seller or buyer will agree to financing terms other than those offered, or of any other information requested by a party to remain confidential.
- Any additional duties that are entered into by separate written agreement

Note that loyalty, confidentiality, obedience, and full disclosure are not included. In a transaction broker relationship, the seller (or buyer) is considered to be a customer of the broker, not a principal. The customer is not responsible for the acts of the broker-licensee.

In other states, a transaction brokerage relationship is similar to a no brokerage agency relationship, as described below.

No Brokerage Agency Relationship

When buyer and seller have tentatively reached an agreement without a broker but seek some help, they may engage a broker with *no brokerage* relationship. The broker's assistance may be needed to help complete a contract, and payment is by a fixed fee or hourly rate rather than by commission. There is no agency relationship.

The duties of a licensee who has no brokerage agency relationship with a buyer or seller must be fully described and disclosed in writing to the buyer or seller. The disclosure must be made before the showing of property. When incorporated into other documents, the required notice must be of the same size type, or larger, as other provisions of the document and must be conspicuous in its placement so as to advise customers of the duties of a licensee who has no brokerage relationship with a buyer or seller. Those duties include:

- Dealing honestly and fairly
- Disclosing all known facts that materially affect the value of residential real property that are not readily observable to the buyer
- Accounting for all funds entrusted to the licensee

Buyer Brokerage

Buyer brokerage occurs when a broker enters into a contract to represent a buyer. The typical arrangement is that the broker owes the buyer full agency representation and does not represent the seller at all. This is in spite of the fact that in the typical buyer's broker arrangement, the seller will pay the entire commission to the listing broker, who will share the commission as agreed with the buyer's broker. In effect, then, the seller ends up paying the other party's (buyer's) representative. Typical listing contracts contain clauses in which the listing owner agrees to these provisions.

Most states' license laws mandate a *default relationship,* either explicitly (it is spelled out in the law) or implicitly (the law does not specify a default relationship, but it is implied). This means that upon the "first contact" between a licensee and a buyer (or a seller), a certain form of agency relationship exists by law, unless and until it is replaced by another form. The nature of the default relationship varies from state to state. In some, the default is transaction brokerage. This means that no agency relationship exists until the parties involved agree to one. In some states, the default is seller agency; others have buyer agency as the default.

A key point is that everyone involved in a transaction must be aware of what the agency relationships are before the negotiations for the transaction begin. Exactly when disclosure of buyer agency must be made varies among the states. Some states require that buyer agency be disclosed at first contact with the seller and/or the seller's broker before an offer actually might be made. Others allow the broker's relationship to be disclosed only when an offer actually is made. This usually is accomplished by including language in the offer (or contract of sale: see Chapter 8) that specifically states that the broker is representing the buyer even though the buyer's broker will share in the commission paid by the seller.

A 21st-century innovation that is beginning to appear in some states is *designated agency.* This is supposed to solve a problem brought on by buyer agency. Suppose a licensee has a buyer agency with someone who is looking for a house. The licensee thinks that the prospect will be interested in a certain home for which this same licensee also has a listing contract. Now the broker is under agency contract to both parties. One solution is to have both buyer and seller agree to a dual agency with the licensee, but, as we have seen, dual agency has its drawbacks. With designated agency, the licensee's principal broker appoints another licensee (the *designated agent*) from the same office to represent one of the parties (usually the buyer), while the original licensee represents *only* the other party in this one situation.

EARNING THE COMMISSION

Under common law, a broker earns his commission when he produces a ready, willing, and able buyer (or tenant). From then on, the risk that the deal actually will go through rests with the principal. However, many listing contracts today specify that the commission is not payable until the sale closes.

As mentioned previously, the principal is not obliged to deal with anyone whom the broker brings; that is, the principal, by hiring the broker, incurs no legal responsibility to the *third party.* Consequently, a buyer who is willing to pay the asking price and meet all the conditions set down by the principal in the listing agreement normally will have no legal recourse if the principal refuses to deal with him, unless the buyer's civil rights have been violated.

Nevertheless, the principal's obligations to the broker remain. In the listing agreement an acceptable price and terms of sale are specified. If the broker produces a bona fide offer that meets all these conditions, then the principal is responsible for paying the broker a commission even if she chooses not to accept the offer. This is required, because the broker has satisfied the terms of his employment contract and there *is* a contract between him and the principal. There is *no* contract, however, between the principal and third parties (such as potential buyers), so the principal is not liable to them.

If the broker's principal refuses to pay him a commission that the broker feels he has earned, he may sue in court to receive it. To assure success, the broker must prove three things:

1. That he was *licensed* throughout the time, beginning with the solicitation of the listing until the closing of the transaction and the passing of title (or the notification by the principal that she will not accept an offer that meets the terms of the listing agreement).
2. That he had a contract of employment with the principal. The best evidence here is a written contract, and some state licensing laws require that listings be written to be enforceable.
3. That he was the "efficient and procuring cause" of the sale; that is, that he actually brought about the sale within the terms of the listing contract. In an open listing, this would mean that he actually found the eventual buyer. In an exclusive agency listing, he must have found the buyer or the buyer must have been found by a licensed agent. An exclusive right to sell listing effectively defines the broker as having earned a commission when and if the property is sold because only one specific broker has that right.

TERMINATION OF AN AGENCY CONTRACT

Agency contracts can be terminated by a variety of events, though some of them are relatively uncommon. In a broad sense, contracts can be terminated in two ways: by the *actions of the parties* to them or *by law* when certain events occur.

Termination by the actions of the parties includes the following:

The contract is terminated by *performance* when both parties perform their duties as prescribed and the event for which the agency is created ends. In the real estate business, a listing agency contract is terminated by performance when there is a "meeting of the minds" between the principal and the third party found by the agent. Sometimes, however, the contract will specify some other event (usually title closing) as the actual termination of the contractual relationship.

The parties may *mutually agree* to terminate the relationship before it has been terminated by performance.

The agent may *resign*. In this case, the agent may be liable to the principal for damages due to his breach of the contract, but he cannot be held to perform under the contract.

The principal may *discharge* the agent. Once again, the principal, too, can be liable for damages due to her breach of the contract, but she cannot be forced to continue the employment of the agent.

The agent may resign, or the principal may discharge the agent without penalty if it can be proved that the other party was not properly discharging his or her duties under the contract. An agent would be justified in resigning if, for example, his principal did not provide him with enough information to do his job well or required that he perform some illegal act in the execution of his duties. An agent could be discharged justifiably if it could be shown that he was not faithful to his duties or was acting contrary to the interests of the principal.

Termination of the contractual relationship also occurs automatically, *by law*, with the occurrence of certain events, such as the following:

- The *death of either party* terminates an agency relationship.
- If either party becomes *legally incompetent*, the agency relationship ceases.
- *Bankruptcy* of either party, so as to make continuation of the relationship impossible.
- *Destruction of the subject matter* terminates any agency relationship. In real estate this would include such events as the burning down of a house or the discovery of another claim on the title that would make it impossible for the owner of the property to pass good and marketable title.

QUESTIONS ON CHAPTER 4

1. The typical relationship of a real estate broker to the principal is that of
   (A) general agent
   (B) special agent
   (C) secret agent
   (D) travel agent

2. Which of the following is *not* required of a single agent with respect to his principal?
   (A) to be loyal
   (B) to provide full disclosure
   (C) to account for his own personal finances
   (D) to deal honestly and fairly

3. A listing contract that says that the broker will receive a commission no matter who sells the property is called
   (A) an open listing
   (B) a net listing
   (C) an exclusive agency listing
   (D) an exclusive right to sell listing

4. Which of the following does *not* terminate an agency relationship?
   (A) making an offer
   (B) the death of either party
   (C) the resignation of the agent
   (D) the destruction of the subject matter

5. To prove her right to a commission, a broker must show
   (A) that she was licensed throughout the transaction
   (B) that she had a contract of employment
   (C) that she was the "efficient and procuring cause" of the sale
   (D) all of the above

6. A net listing is one
   (A) that requires the broker to seek a net price for the property
   (B) that is legal in all states
   (C) that most ethical brokers would prefer to use
   (D) in which the broker's commission is the amount by which the sale price exceeds the agreed-upon net price the seller specifies

7. Among other things, the principal is obligated to
   (A) compensate the agent for his services
   (B) give the agent full authority to accept or turn down offers
   (C) give the agency the principal's financial statement
   (D) none of the above

8. A real estate agent
   (A) must always represent the seller
   (B) may represent the buyer if the buyer-broker agency is disclosed to all parties
   (C) may not take listings if she represents buyers
   (D) need not disclose a buyer-broker agency relationship

9. A special agent is one
   (A) whose powers are limited to certain specific functions
   (B) who must report to a licensing agency
   (C) who has less than a certain amount of experience as an agent
   (D) who is not a licensed agent

10. Among other things, an agent is obligated to
    (A) act in the principal's best interests unless these conflict with the agent's own interests
    (B) keep an accurate account of all money
    (C) never disclose the agency relationship to anyone
    (D) represent only one principal at a time

11. A *no agency* brokerage relationship means that
    (A) there is no agency representation
    (B) the broker represents the person (buyer or seller) he/she meets first
    (C) the broker makes the transaction easy and problem free
    (D) there is no broker involved in the transaction

12. In a *dual agency*,
    (A) the broker owes complete loyalty to the interests of all parties he/she represents
    (B) the broker is also either the buyer or the seller in the transaction
    (C) there is more than one broker involved in the transaction
    (D) none of the above

13. The rate of commission on a real estate sale is
    (A) typically 6% of the selling price
    (B) mandated at various rates by law
    (C) paid to the salesperson, who gives 50% to the broker
    (D) negotiable between the broker and (typically) seller

14. As a general rule, escrow money received by the broker
    (A) is given to the seller on signing the contract
    (B) may be spent by the broker for repairs to the property
    (C) is deposited in the broker's trust account soon after receipt
    (D) is shared equally by the broker and salesperson

15. An unlicensed employee of a broker who prepares an advertisement to sell a house
    (A) is free to do so
    (B) must have the broker's prior written approval
    (C) must become licensed
    (D) must accurately describe the house

16. An owner who sells property that he developed
    (A) may engage an unlicensed sales staff who receive commissions
    (B) may engage a licensed sales staff who receive salaries only
    (C) must acquire a broker's license
    (D) must acquire a salesperson's license

17. In a transaction broker relationship, the seller or buyer
    (A) is responsible for the broker's acts
    (B) is not responsible for the broker's acts
    (C) is not considered a customer of the broker
    (D) is an agent of the broker

18. In a transaction broker arrangement, parties give up their rights to a broker's
    (A) undivided loyalty
    (B) honest dealings
    (C) use of skill and care
    (D) accounting for funds

19. A way to terminate a brokerage relationship is
    (A) agreement between the parties
    (B) fulfillment of the contract
    (C) expiration of the term
    (D) all of the above

20. A broker with no agency relationship may assist
    (A) the buyer
    (B) the seller
    (C) both buyer and seller
    (D) all of the above

21. A salesperson is _____________ of the broker.
    (A) an agent
    (B) an employee
    (C) a paid contractor
    (D) a REALTOR®

22. *Prima facie evidence* refers to evidence that is good and sufficient on its face to
    (A) sell real estate
    (B) buy real estate
    (C) establish a given fact
    (D) file a lawsuit

23. Brokerage representation may legally include
   (A) a broker working for the seller
   (B) a broker working as a single agent for buyer or seller (but not both in the same transaction)
   (C) a broker representing neither buyer nor seller but facilitating the transaction
   (D) all of the above

24. Termination of a brokerage relationship may occur as a result of
   (A) bankruptcy of the principal or customer
   (B) death of either party
   (C) destruction of the property
   (D) all of the above

25. When a broker is an agent of the owner, the salesperson is
   (A) a REALTOR®
   (B) an independent contractor
   (C) a subagent
   (D) an employee

26. A single agent must
   (A) deal honestly and fairly
   (B) be loyal
   (C) provide full disclosure
   (D) all of the above

27. A person who wishes to sell his or her own house
   (A) must employ a broker
   (B) must employ a salesperson
   (C) must employ a REALTOR®
   (D) may employ any of the above, but is not required to do so

28. Which of the following, when performed for others, are included as real estate services for the purposes of determining licensure requirements?
   (A) buying and selling property
   (B) renting and leasing property
   (C) auction or exchange of property
   (D) all of the above

29. The legal relationship of a single-agent listing broker to the owner is that of
   (A) trustee
   (B) general agent
   (C) dummy agent
   (D) fiduciary

30. *Dual agency* is defined as an arrangement in which
   (A) the same broker represents both the buyer and the seller
   (B) the broker is the seller
   (C) the salesperson and the broker split the commission
   (D) the listing broker and selling broker cooperate with each other

31. If a broker uses a title company to hold escrow deposits, the broker must turn over the funds
   (A) on the same day
   (B) in the same time frame as would be required if the broker maintained the account
   (C) when convenient
   (D) at the end of each month

32. If a buyer and seller reach oral agreement over the price and terms of a contract but seek a broker's help to complete written contract forms, the relationship with the broker is likely to be
   (A) single agent
   (B) dual agent
   (C) no brokerage relationship
   (D) any of the above

33. A single agent relationship provides a ______________ with a seller.
   (A) fiduciary relationship
   (B) transaction agency
   (C) no business relationship
   (D) subagency

34. When a real estate attorney wants to earn a commission for performing real estate services, he or she must
   (A) become a REALTOR®
   (B) obtain a real estate license
   (C) join a franchise
   (D) be a member of the bar association

35. It is a good practice for a broker to prequalify
   (A) a prospective buyer
   (B) a seller
   (C) a sales agent
   (D) all clients

36. A person with broad powers to act for another may be called
   (A) a general agent
   (B) a universal agent
   (C) both A and B
   (D) neither A nor B

37. A potential buyer visits a real estate agency asking to view a certain house. The salesperson doesn't know the buyer. He or she
   (A) must show the property
   (B) should take a firearm for self-defense purposes
   (C) may refuse to show the house
   (D) should take whatever safety precautions are necessary

38. In most transactions where there is a broker, the broker represents
   (A) the brokerage
   (B) the seller
   (C) the buyer
   (D) all of the above

39. A broker is killed in a plane crash. His listings are
   (A) binding on the owners
   (B) transferable to his heirs
   (C) terminated
   (D) transferred to his salespersons

40. A broker is representing a buyer and showed him a lot in a new subdivision. The broker did not ask the developer but, knowing that the developer had put in streets in other subdivisions, told the buyer that the developer would do the same in this one. The buyer relied on the broker's statement. The developer did not put in streets in this subdivision. Who is liable for damages?
   (A) no one because nothing was put in writing
   (B) the developer for not doing here what he had done elsewhere
   (C) the broker because he misrepresented what would happen
   (D) both the broker and the developer

41. A salesperson may accept compensation from
   (A) employing broker only
   (B) co-broker
   (C) principal
   (D) mortgage company

42. To be a dual agent (in states that allow this), the licensee must
   (A) receive compensation only from the seller
   (B) receive compensation only from the buyer
   (C) disclose the relationship and gain consent from both parties
   (D) be certain that both husband and wife approve the transaction

43. The buyer gives a $5,000 earnest money check to the seller's broker with instructions not to deposit the check for a week and not to tell the seller that the check has not been deposited. The seller's broker should
   (A) follow the buyer's instructions
   (B) disclose the conditions to the seller
   (C) refuse to accept the check
   (D) notify the Board of Realtors

44. A listing is given to the broker by the
   (A) seller
   (B) buyer
   (C) lawyer
   (D) lender

45. The seller has accepted an offer. An hour later, the broker receives another offer. The broker should
    (A) tell the seller of the second offer
    (B) reject the second offer on the seller's behalf
    (C) accept the second offer if it is for more money
    (D) seek a better offer from the first buyer

46. Most of the work in a real estate sale is carried out by a licensed salesperson. Who is the legal agent?
    (A) salesperson
    (B) broker
    (C) lawyer
    (D) principal

47. To be entitled to a commission, a broker with an open listing must prove that he or she
    (A) was the procuring cause of sale
    (B) was licensed
    (C) brought a ready, willing, and able buyer
    (D) all of the above

48. Seller Agent Orange receives three offers on Green's house. He should
    (A) advise Green of the best one
    (B) advise Green of the best two, with a comparison of pros and cons
    (C) advise Green of all three offers
    (D) wait for more offers to come in

49. At a cocktail party, Henry meets a man looking to buy a windmill. Henry has a friend who is looking to sell his windmill, and Henry brokers the deal. To collect a commission, Henry needs a
    (A) business broker's license
    (B) real estate broker's license
    (C) mortgage banker's license
    (D) no license since this is a private arrangement outside Henry's usual business duties

50. A broker's principal role when representing a seller is to
    (A) bring a ready, willing, and able buyer
    (B) advise the buyer as to the suitability of a purchase
    (C) advise the seller as to price and terms
    (D) sponsor salespersons

51. When a broker is a selling agent, his fiduciary obligations to his principal include
    (A) acting in the principal's best interest
    (B) keeping the principal's interest above his own
    (C) keeping the principal informed
    (D) all of the above

52. Commission splits between brokers and salespersons are
    (A) prescribed by law
    (B) 50/50
    (C) negotiated
    (D) suggested by NAR

53. Appraising, brokerage, consulting, development, economics, finance, and property management are
    (A) activities requiring a state license or certification
    (B) specializations within real estate
    (C) unregulated activities
    (D) economic opportunities

54. An agency relationship can be established by
    (A) an oral agreement
    (B) a written agreement
    (C) an agreement implied by law
    (D) all of the above

55. If the owner informs the broker that he wants single agent representation, what should the broker do?
(A) decline the assignment
(B) ask NAR for permission
(C) ask MLS for permission
(D) provide the single agency notice at or before taking the listing

56. Mr. and Mrs. Buyer asked broker Bob about the best way to take title to the property they had agreed to buy. What should Bob do?
(A) advise the Buyers to ask their parents
(B) advise the Buyers to contact an attorney
(C) have his accountant advise the Buyers of the best way to take title
(D) advise the Buyers to contact the county recorder

57. Broker Martin holds $10,000 as earnest money on the purchase made by Will. When can Martin remove the earnest money from the escrow account?
(A) when the purchase closes escrow
(B) when the buyer gives him permission
(C) when the seller gives him permission
(D) when he needs the money for business expenses, provided it is replaced within five days

58. A fiduciary relationship could exist between a principal and all of the following *except*
(A) an appraiser
(B) a receiver
(C) a trustee
(D) an administrator

59. When a salesperson shows a house that is listed with her firm to a prospective buyer whom she represents, there is a(n)
(A) dual agency
(B) general agency
(C) multiple listing
(D) exclusive right to sell

60. You are representing seller Sam and bring buyer Bill, your customer, to see the house. Bill makes an offer. Bill has confidentially told you that he went through bankruptcy two years ago, but he thinks he will be financially approved to buy the house. What, if anything, should you do with the information about the bankruptcy?
(A) Tell Sam.
(B) Keep the information confidential as Bill expects you to.
(C) Keep the information confidential unless the deal doesn't go through.
(D) Contact the bank to verify Bill's financial status.

61. A licensed real estate salesperson can lawfully accept an extra commission in a difficult sale from
(A) a broker-employer
(B) the mortgage lender
(C) an appreciative seller
(D) a thankful buyer

62. A broker with a listing brings a contract from a ready, willing, and able buyer, at the listed price, which the seller is considering. Generally, the broker has earned his commission
(A) now
(B) when the seller accepts
(C) when contract contingencies are met
(D) after closing

63. A listing agent knows that the house has a roof leak, the water heater is not up to city code, and the seller must sell quickly because of financial difficulties. The agent is showing a potential buyer the property. What can or must the agent legally say?
(A) "The roof leaks, and the water heater does not meet city code."
(B) "The seller will take less than the asking price."
(C) "The seller is in financial difficulty and must sell quickly."
(D) "The roof leaks, the water heater does not meet city code, and the seller needs a quick sale."

64. Clients may be liable for errors made by brokers or subagents through
(A) vicarious liability
(B) casual liability
(C) absolute liability
(D) estovers

65. When money is deposited in a broker's trust account, part of which will be used to pay the broker's commission,
(A) the broker can withdraw his or her share of the money before the real estate transaction is consummated or terminated
(B) accurate records must be kept on the account
(C) interest on the account is by law the property of the broker
(D) the salespeople can withdraw money prior to closing

66. The rate of brokerage commissions for the sale of a house is determined by
(A) standard practice in the community
(B) realtor guidelines in the state
(C) national norms
(D) negotiations between the owner and the broker

67. When answering questions a buyer raises about home defects, the salesperson should
(A) not disclose any defects
(B) answer honestly to the best of his knowledge
(C) plead ignorance of any defects
(D) tell only the inspector

68. A broker listed a house for sale. A potential buyer is owed what duty?
(A) loyalty
(B) honesty
(C) full disclosure
(D) none

69. The owner of a shopping center signed a management agreement with a real estate broker. Is there an agency relationship?
(A) yes
(B) no
(C) maybe
(D) a good management question

70. The fiduciary responsibilities of a real estate broker should be explained to all parties *except*
(A) broker's salespersons
(B) co-brokers
(C) clients
(D) mortgage lenders

71. A real estate broker should not deposit earnest money in his personal account to avoid
(A) IRS regulations
(B) reconciliation
(C) refunds
(D) commingling

72. A homeowner sent a letter to six local real estate brokerage firms offering to pay a 10% commission to any firm that produced a buyer within 30 days. This created
    (A) an exclusive right to sell with the six firms
    (B) an open listing with the six firms
    (C) an intense competition
    (D) an invitation to talk with the owner

73. An owner tells the broker of serious hidden defects in the property. The broker advises the buyer to have the property inspected. The inspection does not reveal these defects. The sale is closed, and three months later the buyer learns of the problem. Which of the following is true?
    (A) Tough luck for the buyer.
    (B) The broker and seller are guilty of misrepresentation.
    (C) The inspector is incompetent.
    (D) The inspector is liable for repairs.

74. A broker has a listing on a house. A buyer tells the broker that he will pay more than the listing price if the owner requests it. The broker
    (A) must tell the owner
    (B) must not tell the owner
    (C) must ask the buyer his top price
    (D) must step aside and let the owner negotiate

75. Under what circumstances is a salesperson permitted to spend a client's money for personal purposes?
    (A) never
    (B) an advance deposit
    (C) when approved by the broker
    (D) when offered by the client

76. A prospective buyer is interested in investment property. The broker he contacts has no experience in investment property. The broker should
    (A) conceal his lack of experience
    (B) disclose his lack of experience
    (C) reach out to other brokers for their listings
    (D) ask an appraiser for assistance

77. A real estate salesperson resigned from one broker and was soon employed by another. The listings the salesperson brought to the first broker
    (A) remain with the first broker
    (B) can be transferred to the new broker after 10 days
    (C) can be terminated by the client
    (D) can be kept by a salesperson if he or she becomes a broker

78. A fiduciary must act in the best interest of
    (A) all parties
    (B) his principal
    (C) law enforcement
    (D) his family

79. A state-licensed real estate salesperson can collect a commission from
    (A) everyone
    (B) his sponsoring broker
    (C) any broker
    (D) the closing company

80. A buyer's broker in a residential house sale typically collects a commission paid by the
    (A) buyer
    (B) seller
    (C) lessee
    (D) lessor

81. A dual agent represents
    (A) both parties in the same transaction
    (B) either the buyer or the seller but not both
    (C) the buyer and the mortgage lender
    (D) a modern approach to doing business

82. A broker acting as a dual agent may not
    (A) operate without a written listing
    (B) act in one party's interest to the detriment of the other
    (C) tell jokes
    (D) make suggestions to improve the property's appearance

83. Buyer and seller agree to a transaction but wait until a week after the listing expires before they close. Broker Bob brought the buyer and seller together during the term of the listing. However, the seller does not want to pay the commission. Which of the following is true?
    (A) The listing expired, so no commission is due.
    (B) Only if the listing has an extender clause can the broker claim a commission.
    (C) The broker can likely collect his commission in court.
    (D) The buyer and seller are ethical principals permitted to save expenses.

84. A real estate broker is engaged by the owner with a listing. The broker is a(n) _________ agent.
    (A) general
    (B) universal
    (C) special
    (D) limited

85. A transaction broker
    (A) exercises no professional standards
    (B) takes directions from the title company
    (C) does what the buyer directs
    (D) does not represent the interests of either party

86. In order to collect a commission from the sale of real estate, a broker must be licensed
    (A) before closing
    (B) before the title is recorded
    (C) before a contract is accepted
    (D) at all times from listing to closing

ANSWERS

1. **B**	12. **D**	23. **D**	34. **B**	45. **A**	56. **B**	67. **B**	78. **B**
2. **C**	13. **D**	24. **D**	35. **A**	46. **B**	57. **A**	68. **B**	79. **B**
3. **D**	14. **C**	25. **C**	36. **C**	47. **D**	58. **A**	69. **A**	80. **B**
4. **A**	15. **B**	26. **D**	37. **D**	48. **C**	59. **A**	70. **D**	81. **A**
5. **D**	16. **B**	27. **D**	38. **B**	49. **B**	60. **A**	71. **D**	82. **B**
6. **D**	17. **B**	28. **D**	39. **C**	50. **A**	61. **A**	72. **D**	83. **C**
7. **A**	18. **A**	29. **D**	40. **C**	51. **D**	62. **A**	73. **B**	84. **C**
8. **B**	19. **D**	30. **A**	41. **A**	52. **C**	63. **A**	74. **A**	85. **D**
9. **A**	20. **D**	31. **B**	42. **C**	53. **B**	64. **A**	75. **A**	86. **D**
10. **B**	21. **A**	32. **C**	43. **B**	54. **D**	65. **B**	76. **B**	
11. **A**	22. **C**	33. **A**	44. **A**	55. **D**	66. **D**	77. **A**	

State Real Estate License Law

5

ESSENTIAL MATTERS TO LEARN FROM THIS CHAPTER

- Each state has its own license laws to regulate real estate brokers and salespersons. This material should be readily available online at the state real estate commission's website. Please download a copy.
- Each state has a real estate commission or other agency that regulates licensee qualifications and practices. The state real estate commission has a board of commissioners that sets policy as well as a staff of employees headed by a full-time executive called the commissioner.
- Please determine for your state:

  1. How many board members there are.
  2. What occupations board members must be drawn from.
  3. Who appoints commissioners (often the state governor) and confirms them (typically the state senate).
  4. How often the commission must meet.

- Each real estate commission in each state regulates:

  1. What constitutes *real estate* that requires a license to sell or lease for others.
  2. What types of licenses it issues—typically broker and salesperson. Appraisers are considered in Chapters 14 and 15 of this book.
  3. Qualifications to become a broker or salesperson: age, education, testing, experience.
  4. Agency relationships permitted: buyer, seller, dual, transaction, no agency.
  5. Amounts of fees for applying for and maintaining each license or service.
  6. Schools that offer real estate education.
  7. Discipline. Accepting and investigating complaints, holding hearings, punishment, penalties.
  8. Nonresident brokers: reciprocity, examinations.
  9. Real estate recovery fund disbursements.

- Each state will have a list of do's and don'ts. Items covered include:

  1. Mishandling money belonging to others. Paramount is the deposit of money belonging to others promptly into escrow accounts; commingling others' funds with a broker's own money is prohibited.
  2. Practices that are illegal include misrepresentation and fraud. This includes false or misleading advertising, dishonesty, and misleading someone.
  3. Business practices, including a requirement to submit all offers, that prohibit a net listing.

4. Violating civil rights, blockbusting, steering, and discrimination in housing and real estate sales.
5. Imparting all required information, such as leaving the principal a copy of all signed documents.
6. Improper payment of commissions, such as to unlicensed individuals.
7. Disclosure that describes agency relationship to principal.
8. Being convicted of certain crimes or being negligent in performing duties.
9. A requirement for brokers and salespersons to display their licenses in a public area of the office and for each licensee to keep a pocket card with him/her.

- Please obtain a copy of the real estate license law in your state and highlight provisions that are related to licensees.

STATE REAL ESTATE LICENSE LAW

Every state requires that real estate brokers and salespersons be licensed by that state before they can do business. Therefore, each state has a set of laws, referred to as the Real Estate License Law, that describes the process of becoming licensed, the responsibilities of licensees, the administration of the license law, the means of removing licenses from those unfit to keep them, and miscellaneous items. The details of these laws vary from state to state, but many of their general provisions are the same.

Understanding Your State's Laws

Every state has at least two kinds of licenses for real estate transaction professionals: the *broker's license* and the *salesperson's license.* (Note that in a very few states, the titles for these licenses may be somewhat different.) The main difference between the two is that a licensed broker may own, operate, and manage a real estate brokerage business; a licensed salesperson *cannot* do any of these things. This means that every state requires licensed salespeople to work for and be under the supervision of a licensed broker. Usually this is referred to as the broker's "holding" the salesperson's license. The license is not actually issued to the salesperson; rather, it is issued to that salesperson's broker. The supervision responsibility of the broker is a serious matter because the broker is liable for anything and everything the salesperson does in the course of the brokerage business. In fact, all agency contracts negotiated by a salesperson actually are contracts between the seller (or buyer) and the *broker,* not between the salesperson and the principal.

Some states distinguish between two types of broker: the *principal broker* and the *associate broker.* The principal broker is the managing broker, who takes the ultimate responsibility for the actions of all agents in the firm or office. An associate broker may be an assistant manager or merely someone with a broker's license who has chosen to act as a salesperson and have another broker hold his/her license. In many states, staff at a broker's branch office must include an associate broker. Some states may have different examination and education requirements for principal and associate brokers.

To study the material in this chapter effectively, please get a copy of your state's law; every state licensing agency will provide a copy of the state's licensing law to applicants for the licenses. Be sure to study this law carefully! Every state devotes a considerable portion

(20–40%) of the licensing examination to a test of the applicant's knowledge of the license law; if you have not studied your state's law, you will reduce your ability to pass the examination.

This chapter will serve as a guide to the study of your state's licensing laws. It would be confusing to try to cover the details of every state's variations in this chapter. A better way is to provide this study guide, which you will supplement with a copy of your state's law. The second section of sample questions at the end of the chapter (pages 131–142) is designed to show you the *kinds* of questions you will encounter as well as to give you an opportunity to demonstrate your knowledge of your own state's law.

Licensing law can be subdivided into several general areas, such as those used below. Of course, your state may not cover these areas in the same order as shown below. However, nearly everything in your state's law will be reviewed here. Do not be concerned if you have to skip around in your state's law to follow the study pattern in this book; test graders don't care whether you know the sequence in which something appears. They want to make sure that you know the *content* of the law; if you follow the study pattern, you *will* know it.

In each section, plenty of blank space is provided for you to note the particular features of your state's law. Sometimes this can be accomplished by actually taping or pasting the relevant part of your state's law in the available space. Other times it will be easier just to make handwritten notes. Feel free to *cross out* items that *do not apply* to your state!

In any event, the study pattern below breaks down the license law into logical segments for the most effective studying.

Before beginning the study pattern, please *carefully read your state's law.* This overview will familiarize you with the basic provisions and will make it easier for you to look up items as they are discussed in the study pattern. As you read through the first time:

- Look for notes referring to specific legal cases in which a certain provision was upheld or interpreted.
- Check for explanatory notes providing examples that illustrate important points.
- Be on the watch for portions of the law that have since been repealed or amended! Some states print only the law as it currently stands. Others include all enacted laws relevant to licensing since the original one, even though some portions may have been repealed or amended later on. In such cases, there usually will be a notation to point out when the repeal or amendment took place. For study purposes, cross out any repealed or otherwise invalid parts of the printed license law you have.
- Look also for provisions that have been enacted but are not yet effective. Often a state law will allow a considerable time after it has passed until it takes effect. This is particularly evident in the case of laws that raise or tighten standards (especially educational) for licensure. Look also for laws containing schedules of provisions that take effect in the future. Many states have adopted schedules by which educational standards for licensure will be raised every so often until a particular goal is reached.
- Be sure to note the way in which the various provisions are numbered or otherwise identified. This will make it easier to understand parts of the law where reference is made to another part of the law.

COMMON PRACTICES NOT REQUIRED BY STATE LAW

Some common brokerage practices and terms are included in exam questions or answer choices even though they are not part of state law.

- **INDEPENDENT CONTRACTOR.** This is a federal income tax status that allows salespersons and brokers to be taxed as having their own business instead of being employees. Most brokers prefer this arrangement because it doesn't require brokers to pay minimum wage, social security contributions, or other payroll taxes on the earnings of salespeople or associate brokers. Federal tax rules must be followed to ensure that a salesperson is not an employee.
- **REALTOR®.** The National Association of Realtors (NAR) is a private trade organization of which about 30% of state-licensed brokers and salespersons are members. Those who are members are encouraged by NAR to refer to themselves as REALTORS®. They subscribe to a code of ethics and claim that their title of REALTOR® bestows special professional recognition. However, the state license as broker or salesperson, not membership in the NAR, allows one person to sell real estate for another person legally.

Trick Question

A trick question commonly substitutes the word "REALTOR®" for "broker" or "salesperson." These terms are not interchangeable. Being a REALTOR® doesn't allow a person to sell real estate. A REALTOR® is a member of a professional trade association. A broker or salesperson possesses a state license that is needed to sell real estate to the public.

STUDY PATTERN FOR THE LICENSE LAW

I. Definitions

A. *Real Estate.* The definition of real estate usually includes most interests in land, buildings, and other improvements. Some states also require licensure for handling investment interests involving real estate (such as syndicates), the sale of businesses, and so on.

B. *Broker.* Note the functions that are defined as those of a real estate broker. Usually there will be many. If your state has associate brokers, how are their functions defined?

C. *Salesperson.* Some states may use a different name, such as sales associate, for this job. Usually most of the functions will be the same as those for brokers, with the provision that the salesperson is employed by the broker.

D. *Broker employed by another broker.* Most states provide for this situation; the specific name given to it (associate broker, affiliate broker, employee broker, etc.) varies.

E. *Principal broker and associate broker.* If your state distinguishes between principal brokers and associate brokers, describe the differences between them.

____

____

____

____

F. *Corporate broker.* Some states provide that a corporation may have a license if it satisfies certain requirements.

____

____

____

____

____

G. *Branch offices.* Under what conditions (if any) does your state allow a brokerage firm to have branch offices?

____

____

____

____

____

II. Agency Relationships

A. Which of the following agency relationships is allowed in your state?

☐ buyer agency

List restrictions, if any: ____

____

____

☐ dual agency

List restrictions, if any: ____

____

____

☐ no agency brokerage

List restrictions, if any: ____

____

____

☐ transaction brokerage

List restrictions, if any: ______________________________

____

____

☐ designated agency

List restrictions, if any: ______________________________

____

____

B. Is there a default agency relationship in your state? In other words, if no agency relationship is stated in a contract, the default will apply. If so, describe it.

____

____

____

____

III. Exemptions to the License Law

List here all those who are exempt from your state's license law. Exemptions usually include owners of property for sale or lease, those holding an owner's power of attorney, attorneys acting in the exercise of their duties, trustees, certain employees of owners, government, and some others.

____

____

____

____

IV. Administration of the Law

Most states have a Real Estate Commission that administers the license law. However, some states do not; in these states, the law is administered by a state agency such as the Department of State, or a State Licensing Board or Department. In any case, you will be asked how this administration is set up and what the organizational pattern is.

If your state has a Real Estate Commission, note the following information. How many commissioners are there? What are their required qualifications (i.e., residence in state, experience, licensed broker, etc.)? What is their compensation? How are they chosen and confirmed, and by whom? Is there a required geographical distribution among them? How is the chairperson chosen, and by whom? How are vacancies filled? How long do commissioners serve? Is there a date or time limit for the naming of new commissioners? How frequently does the commission meet? Where does it meet? How many commissioners constitute a quorum?

If your state does not have a Real Estate Commission, look for answers to these questions. Which state agency administers the Real Estate License Law? Where is it located? What, if any, representation is permitted to members of the real estate industry? How is such representation effected? Who appoints whom? Are any qualifications required?

Regardless of whether or not your state has a commission, the following questions will be pertinent. What are the general powers of the agency or commission in the enforcement of the law? How is the *staff* of the agency or commission organized (e.g., does the head have the title of Secretary, Executive Secretary, Commissioner, etc.; are there limits on the size of the staff; who hires them; who determines their pay)? Does the commission or agency have a seal? Where are the offices located?

V. The Licensing Process

You must know *all* the steps to take to get *any* license: broker, associate broker (if applicable), or salesperson. Also, you must know the necessary qualifications to be met by anyone seeking a license. Section VI is devoted to cataloging the fees that licensees pay.

A. Must an applicant for either license be sponsored by anyone (either by a person who already has a license or, if a salesperson applicant, by the future employing broker)?

B. What is the minimum age for the following licenses?
1. salesperson:
2. broker:
3. associate broker:

C. How much *general* education must applicants have (e.g., high school or the equivalent)?
1. salesperson:
2. broker:
3. associate broker:

D. What *special* education requirements must licensees meet? Distinguish between education required *before* getting a license and continuing education required *after* obtaining a license in order to retain the license. Also, are any substitutions (such as experience) permitted for any of these requirements?
1. salesperson:
2. broker:

3. associate broker: ____

____

____

E. Will changes in the educational requirements that the present law dictates go into effect in the future? What are they, and when do they become effective?

____

____

____

____

____

F. Are there any specific *experience* requirements for either license? (Often a broker applicant is required to have been licensed as a salesperson for a certain minimum time.)

____

____

____

____

____

G. Are licensees required to be bonded or to contribute to a "real estate recovery fund" or similar fund?

____

____

____

____

____

H. Does your state have any other specific requirements?

____

____

____

____

____

I. On what grounds can your state refuse to issue a real estate license to an applicant?

____

____

____

____

____

J. Where must a sales license or broker license be displayed? Often this is on the wall of a public area of the broker's office. A *pocket card* should be held on the person of every licensee.

____

____

____

____

____

VI. Fees

Fill in the blank spaces in the following form to show the fees, and other pertinent information concerning them, that your state requires. The information should be broken down by broker/salesperson; there is a column for each. *Cross out* the fees that your state does not charge.

Fee	Broker	Salesperson	Associate Broker
APPLICATION FEE			
EXAMINATION FEE			
LICENSE FEE:			
(a) Amount of fee			
(b) How often is it payable?			
(c) When is it due?			
RENEWAL FEE:			
(a) Amount of fee if different from license fee			
(b) Penalty for late renewal? How much?			
(c) When is license renewal due?			
TRANSFER FEES:			
(a) To new employing broker			
(b) Change of business address			
(c) Change to different branch of same firm			
(d) Any penalty for frequent transfers?			
DUPLICATE (REPLACEMENT) LICENSE FEE			

Fee	Broker	Salesperson	Associate Broker
REAL ESTATE RECOVERY FUND:			
(a) Payment when becoming licensed			
(b) Later periodic payments			
(c) When are payments due?			
OTHER FEES:			
(a) Reactivating inactive license			
(b) License fee while inactive			
(c) ____________			
(d) ____________			

VII. How to Lose a Real Estate License

The license law of your state will spell out a number of practices for which the regulatory authorities are allowed to revoke or suspend a license. These practices will be found in the license law itself, in a "code of ethics" or a set of "rules and regulations," or in both places. We have grouped and listed the more common practices and have provided space for you to note any special provisions of your own law. *Cross out* the ones that your state's law does not mention, but *make sure* that your law does not specify the item in somewhat different wording! When in doubt, leave it in. All the practices listed below are poor business practices if not specifically illegal. So even if you aren't sure your state law prohibits one of them, you ought to avoid it anyway. There is space for you to write in comments, if any.

The practices are grouped according to type.

A. MISHANDLING MONEY BELONGING TO OTHERS

1. Commingling trust (or escrow) money with one's own funds.

____

____

____

2. Not depositing or remitting funds quickly, once they are received.

____

____

____

3. Salesperson not remitting funds quickly to broker.

____

____

____

4. Accepting noncash payments on behalf of principal without principal's agreement. (Checks are considered to be cash.)

B. MISREPRESENTATION AND FRAUD

All of these practices are *misrepresentation or fraud.* Some state laws list a number of such practices; others simply say, ". . . for any misrepresentation or fraud . . . ," or some words to that effect.

1. False advertising.

2. Intentionally misleading someone.

3. Acting for more than one party to a transaction without the knowledge of all parties.

4. Using a trademark or other identifying insignia of a firm or organization (such as the National Association of Realtors®) of which one is not a member.

5. Salesperson pretending to be a broker or to be employed by a broker not his/her own.

6. Not identifying oneself as a licensee in any transaction in which one is also a party.

7. Taking kickbacks, referral fees, commissions, placement fees, and so on in association with one's duties from persons who are not one's principal and without the principal's knowledge.

8. Guaranteeing future profits from resale.

9. Offering property on terms other than those authorized by the principal.

10. Pretending to represent a principal by whom one is not actually employed.

11. Failing to identify the broker in all advertising.

12. Others (specify).

C. IMPROPER BUSINESS PRACTICE

1. Failing to submit all offers to the principal.

2. Attempting to thwart another broker's exclusive listing.

3. Inducing someone to break a contract.

4. Accepting a net listing.

5. Failing to put an expiration date in a listing.

6. Putting one's sign on a property without the owner's permission.

7. Failing to post a required bond; failing to keep the bond up to date.

8. Blockbusting.

9. Discriminating.

10. Steering.

11. Other violation of state, federal, or local fair housing or civil rights laws.

D. FAILURE TO IMPART REQUIRED INFORMATION

1. Failure to leave copies of contracts with all parties involved.

2. Failure to deliver a closing statement to all parties entitled to one.

3. Failure to inform some or all parties of closing costs.

E. IMPROPER HANDLING OR PAYMENT OF COMMISSIONS

1. Paying a commission to an unlicensed person.

2. Paying a commission to a licensee not in one's employ.

3. Salesperson receiving a commission from anyone other than his/her employing broker.

4. Unlicensed person receiving or sharing in a commission.

F. AGENCY DISCLOSURE

1. Failure to disclose to prospective buyers that broker is agent of the seller.

2. Failure to disclose buyer-broker agency to seller or seller's broker (include when a broker must do so).

____

____

____

3. Failure to disclose facilitative (transaction) brokerage requirements and limitations.

____

____

____

4. Required form to use for agency disclosure.

____

____

____

G. OTHER

1. Being convicted of certain crimes (specify).

____

____

____

2. Violating any part of the license law.

____

____

____

3. Making false statements on the license application.

____

____

____

4. Showing any evidence of "incompetence," "unworthiness," "poor character," and so on. This is a catch-all provision usually found at the end of the list of specific offenses to drive home the point that *any* improper action can cost a licensee his/her license. How does your state phrase this provision?

____

____

____

5. Other provisions not covered above (specify).

VIII. Revocation Procedure

Every state has a legal process whereby a license can be revoked or suspended. Your state law will spell this out very clearly. Look for the following:

A. Who can make a complaint against a licensee?

B. What form must the complaint take?

C. To whom must the complaint be submitted?

D. What discretion does the licensing authority have with regard to acting on complaints?

E. How are licensees notified of complaints against them?

F. Is there a time limit with regard to the time between the act and the complaint or between receipt of the complaint and notification of the licensee?

G. What rights does the licensee have to respond to administrative acts?

____

____

____

H. How much time may elapse between a hearing and a decision?

____

____

____

I. May any (or only some) of the parties to a hearing appeal to the courts? If so, which courts? How soon must an appeal be filed?

____

____

____

J. Does a licensee retain his/her license if an appeal is filed?

____

____

____

K. What are the penalties that the licensing authority may impose directly?

____

____

____

L. What happens to the license of a salesperson or associate broker whose principal broker loses his/her license?

____

____

____

IX. Handling of Trust Money

Brokers usually keep their principals' money in trust, in the form of rents, earnest money deposits, and so on. The law usually contains provisions regulating such money and its handling. As a rule, it has to be put into a special account called a *trust account* or an *escrow account*. Things to look for are: (1) How many different kinds of accounts must be maintained? (2) Can they be interest bearing; if so, who gets the interest? (3) Must they be deposited in particular types of banks or with particular agents?

____

____

____

____

____

X. Nonresident Licenses

Every state has some regulation regarding out-of-state residents who desire to pursue the real estate business within the state. There are three ways to handle this situation; find the one that your state uses and make the appropriate notes in the space provided below. A state may choose to:

A. Issue nonresident licenses by requiring nonresidents to take the state's licensing examination. What are the fees such applicants must pay? Do they have to meet all the state's other qualifications? Must they maintain a license in their home state?

B. Prohibit out-of-state residents from operating within the state. The few states that have this provision usually allow the out-of-state licensee to operate within the state provided that all business is funneled through a *resident* licensee.

C. Allow some or all out-of-state licensees to operate on the basis of *reciprocity*; that is, the state will recognize the nonresident's qualification as having been determined by his/her home state and will issue the applicant an in-state nonresident license on the basis of licensure in a reciprocating state.

____

____

____

____

____

____

____

____

XI. Real Estate Recovery Fund

Not all states have real estate recovery funds or similar funds with different names. If yours does not, then omit this section.

A real estate recovery fund is a fund set up and administered by an agency of the state for the purpose of providing redress for damages caused to members of the public by improper actions by real estate licensees. Usually the money comes from an "assessment" made upon licensees at the time the fund is set up and continuing periodically until the fund reaches a certain level. After that time if the fund is depleted below a specified point, the assessments begin again and continue until the fund is back to the desired level. Look for the following with regard to the real estate recovery fund.

How does one establish a claim? What process must be gone through before a claim can be paid? What happens to a licensee if a claim is paid because of an improper action on his/her part? Are there any exceptions to the kinds of claims that may be paid? Are there any time limits within which claims must be made? Are there limits to the dollar amount of any single award?

____

____

____

____

XII. Regulation of Real Estate Schools

More and more states are requiring some kind of real estate training of applicants and licensees. To assure that the training is proper and effective, some states have amended their licensing laws to give the state real estate licensing agency the power to regulate real estate schools that teach courses used to meet the educational requirements. Look for the following:

Does the law specify any particular curriculum that any or all of the courses must follow? Does the law permit the licensing agency to specify the curriculum? Does the law require that teachers have any particular qualifications? Is the licensing agency given the power to determine the qualifications of teachers?

____

____

____

____

XIII. Penalties for Violating the License Law

Violation of the license law is a crime. Is it a misdemeanor or a felony or other type of crime? What is the maximum penalty—imprisonment, fine, or both? Is there any provision for repeat offenders? What happens to the license of a licensee who is convicted of violating the law? How does a conviction of violating the law affect a person's future right to get a real estate license?

____

____

____

XIV. Other Provisions

Use this space to note other important provisions of your state's license law that have not been covered so far.

____

____

____

QUESTIONS ON CHAPTER 5

You will note that there are two sections of questions for this chapter. The first section is composed of 28 general questions, such as would apply in all states. The answers to these questions appear after them.

The second section is sort of a do-it-yourself question set. A great many questions are provided, but the answers to them depend on the specifics of your state's laws. Therefore, no key to them is given; you must look up the answers in your copy of the law or your notes. More specific instructions appear with the questions in the second set.

GENERAL QUESTIONS

1. A licensee's license must be
   (A) carried in his wallet at all times
   (B) kept on the wall in his home
   (C) posted in a public place in the broker's office
   (D) kept on the wall at the Real Estate Commission

2. A person must be licensed if she is to sell
   (A) her own home
   (B) property belonging to an estate for which she is executor
   (C) property belonging to other clients who pay her a commission
   (D) property she has inherited

3. A broker must place funds belonging to others in
   (A) his office safe, to which only he knows the combination
   (B) a safety deposit box
   (C) an account maintained by the Real Estate Commission
   (D) a trust, or an escrow, account

4. Which of the following would *not* be deposited in a broker's escrow account?
   (A) earnest money check of a buyer
   (B) commissions earned, received after closing
   (C) tenant security deposit
   (D) sufficient funds to open an escrow account at the given bank

5. A broker's license can be revoked for
   (A) closing a deal
   (B) intentionally misleading someone into signing a contract that he/she ordinarily would not sign
   (C) submitting a ridiculous offer to a seller
   (D) all of the above

6. Real estate licenses, once received,
   (A) remain in effect indefinitely
   (B) are good for a limited period of time
   (C) must be filed in the county records
   (D) may be inherited by the licensee's spouse

7. A broker's unlicensed secretary may
   (A) sell property provided that he does it under the broker's direct supervision
   (B) sell or negotiate deals so long as he does not leave the office
   (C) refer interested clients to the broker or his employed licensees
   (D) all of the above

8. A broker may use escrow moneys held on behalf of others
   (A) for collateral for business loans
   (B) for collateral for personal loans
   (C) to make salary advances to licensees in his employ
   (D) none of the above

9. In order to do business, a licensee must
   (A) make proper application for a license
   (B) pass a licensing examination
   (C) have a license issued by the appropriate state agency
   (D) all of the above

10. A broker can lose his license for
    (A) paying a commission to an unlicensed person
    (B) using moneys received as commissions to pay office help
    (C) returning earnest money to a buyer after an offer is refused
    (D) placing his license on the wall of his office lobby

11. A salesperson can lose her license for
    (A) selling properties quickly, at low prices
    (B) buying property for her own use from her principal
    (C) ethical behavior
    (D) misrepresenting facts

12. Which of the following is not required to have a real estate license?
    (A) a resident manager of an apartment project
    (B) a resident manager of an apartment project who, for a fee, sells a house across the street
    (C) a student who sells houses as part of a research project
    (D) all of the above

13. A licensed salesperson
    (A) must be under the supervision of a broker
    (B) can collect commission payments only from his broker
    (C) must have his license held by his employing broker
    (D) all of the above

14. In order to sell property belonging to a trust for which he is trustee, the trustee must have a
    (A) broker's license
    (B) salesperson's license
    (C) trustee's license
    (D) none of the above

15. A person who is employed to sell real estate
    (A) must be licensed
    (B) is an agent
    (C) may be paid a salary or commission
    (D) all of the above

16. A real estate salesperson's compensation is
    (A) 50% of the broker's share
    (B) whatever the seller agreed to
    (C) paid by the broker
    (D) set by state law

17. After one becomes licensed as a broker, the most appropriate term for that person is
    (A) REALTOR®
    (B) realtist
    (C) associate
    (D) broker

18. A salesperson may
    (A) open his or her own office
    (B) work out of a broker's branch office
    (C) advertise in his or her personal name
    (D) all of the above

19. Which type of listing is discouraged because it creates a conflict of interest?
    (A) net listing
    (B) open listing
    (C) exclusive listing
    (D) exclusive right to sell

20. Those exempt from licensure include
    (A) people acting for their own ownership interest
    (B) part owners in proportion to their ownership interest
    (C) businesses that sell their own property
    (D) all of the above

21. In a transaction in which the licensed salesperson performed nearly all the work, who is the legal agent?
    (A) salesperson
    (B) broker
    (C) both salesperson and broker
    (D) property owner

22. A cause of action against a real estate licensee arises when
    (A) the licensee has failed to disclose a material fact
    (B) the licensee has knowingly made an inaccurate statement
    (C) the buyer relied on information provided by the licensee and was damaged
    (D) any of the above conditions are present

23. The best way for a broker acting as a dual agent to minimize his or her risk is to
    (A) hire an attorney
    (B) obtain written permission from both parties to act for both sides
    (C) read the state's real estate law
    (D) provide full disclosure to all parties

24. When representing a buyer in a real estate transaction, a broker must disclose to the seller all of the following *except*
    (A) the buyer's motivation
    (B) the broker's relationship with the buyer
    (C) any agreement to compensate the broker out of the listing broker's commission
    (D) the broker's agreement with the buyer

25. Broker Herman must have an escrow account
    (A) before the state licensing agency will issue his broker's license
    (B) when he holds earnest money
    (C) before his agents can work
    (D) when his personal account is overdrawn

26. Which of the following best describes the role of a broker who seeks a buyer?
    (A) general contractor
    (B) independent contractor
    (C) special agent
    (D) attorney-in-fact

27. How is a buyer's broker typically paid?
    (A) by the buyer at closing
    (B) by a finder's fee paid outside of closing
    (C) by the buyer prior to closing
    (D) by the listing broker through a commission split

28. A salesperson working for a broker may
    (A) be classified as an independent contractor for income tax purposes
    (B) be paid the minimum wage rate
    (C) receive a commission from another broker
    (D) advertise in her own name, without including the broker, if she pays for the ad herself

ANSWERS

1. **C**	5. **B**	9. **D**	13. **D**	17. **D**	21. **B**	25. **B**
2. **C**	6. **B**	10. **A**	14. **D**	18. **B**	22. **D**	26. **C**
3. **D**	7. **C**	11. **D**	15. **D**	19. **A**	23. **B**	27. **D**
4. **B**	8. **D**	12. **A**	16. **C**	20. **D**	24. **A**	28. **A**

SPECIFIC QUESTIONS CONCERNING YOUR STATE'S LICENSE LAW

No key is provided to the following questions because the answers can differ from one state to another. Try to answer the questions; then check your answers by referring to your state's law and the notes you have made. A good practice is to write your answers lightly in *pencil.* When you check them over after answering the questions, you will be able to erase the incorrect answers and to write in the correct answers in dark pencil or pen. That way, the questions you missed will stand out, and in your future study you will be sure to take note of them as possible weak spots.

Fill-Ins

Fill in the blanks with the words or phrases that reflect the license law of *your* state.

1. The examination fee for a salesperson's license is $ ______________.

2. My state will reciprocally issue a real estate license to a licensee from any of the following states: ____
____.

3. An applicant for a broker's license must be at least ______________ years old.

4. A broker must deposit all earnest money payments in an account called a(n) _______
________________ .

5. To get a salesperson's license, one must have the following general education achievement (high school, college, etc.): ______________________________.

6. The maximum imprisonment for violating license law is ______________________.

7. Three kinds of people who are exempt from license law are __________________, ____________________, and ____________________.

8. The license law is administered by ______________________________________.

9. Before taking a licensing examination, a salesperson applicant must have at least ______________ classroom hours of instruction in Real Estate Principles.

10. After passing his examination and becoming licensed, a salesperson has ______________ months/years to take ________________ classroom hours of (specify required subject matter) ______________________________ in order to maintain his licensed status.

11. If a salesperson transfers her employment to a new broker, a transfer fee of $__________ must be paid.

12. The examination fee for the broker's license examination is $____________.

13. Violation of license law is a (misdemeanor, felony, etc.) ______________________.

14. A salesperson's license fee is $____________ every ____________ year(s).

15. License renewal fees must be paid by (date due) ____________.

16. A salesperson may collect commission payments from ____________.

17. A broker may pay a "referral" fee of up to $____________ to unlicensed persons.

18. If a salesperson's license is renewed late, an additional fee of $____________ must be paid.

19. A broker's license fee is $____________ every ____________ year(s).

20. A broker who is employed by another broker pays a license fee of $____________.

21. To take the broker's license examination, a person must have completed the following *real estate* education: ____
____
____.

22. Nonresident license fees are ____.

23. Only the following have the authority to revoke a real estate license: ____________
____.

24. Earnest money deposits paid to brokers must be deposited in the appropriate account within (time limit) ____.

25. A salesperson's license renewal fee is $______________.

26. A broker's license renewal fee is $______________.

27. In the case of a broker who transfers from employment by another broker to independent status, a fee of $____________ must be paid.

28. A salesperson whose license is inactive must reactivate it within ______________ years or it will expire.

29. The fee for reactivating an inactive broker's license is $______________.

30. A salesperson may not transfer his employment to a new broker more often than ____________________ times per year.

31. When a complaint is lodged against a licensee, she must be given at least __________ days' notice of any hearings that will be held.

32. The real estate salesperson's license examination is given ____________ times per year.

33. [If your state has associate broker licenses] The education and experience requirements for an associate broker are ______________________________.

34. When a broker's license is renewed late, a late fee of $__________ is charged.

35. If a broker moves his place of business, a fee of $__________ is charged for reissuing new licenses showing the new address.

36. The fee for replacing a lost, stolen, or damaged license is $__________.

37. Real estate licenses must be displayed (where?) ______________________.

38. Before a person may take the broker's license examination, she must have been a licensed salesperson for at least __________ years.

39. The license fee for a broker's branch office is $__________.

40. After a hearing of a complaint against a licensee, a decision must be rendered within __________ days/months.

True/False

Write *T* for true, *F* for false.

____ 41. Net listings are illegal in my state.

____ 42. The Real Estate Commission has the power to specify the curriculum of the courses that licensees and prospective licensees must take to qualify for, or to keep, real estate licenses.

____ 43. It is illegal to use sellers' powers-of-attorney in order to circumvent the license law.

____ 44. My state has a real estate recovery fund.

____ 45. Violation of license law is a misdemeanor.

____ 46. A person who has lost his real estate license as a result of a violation of license law may never be issued another license.

____ 47. A convicted felon may not be issued a real estate license.

____ 48. Brokers may deposit trust and escrow moneys in interest-bearing accounts.

____ 49. A nonresident of this state may not get any kind of license to do real estate business in this state.

____ 50. There is no requirement for special real estate education for salespersons.

____ 51. An out-of-state resident may be issued a license to do real estate business in my state provided that she passes the required examinations, meets the educational and age requirements, and pays the specified fees.

____ 52. When a hearing is held on a complaint against a licensee, the licensee may be represented at the hearing by an attorney.

____ 53. Complaints against licensees must be submitted to the Real Estate Commission in writing.

____ 54. Every time a complaint is filed against a licensee, a hearing must be held.

____ 55. License fees must be paid every 2 years.

____ 56. The license law requires that, at a specified date in the future, the real estate education required of licensees will be greater than it is now.

____ 57. In order to take the broker's license examination, a person must satisfy a number of requirements, including at least 3 years' experience as an actively licensed salesperson.

____ 58. To hold a broker's license a person must be at least 21 years of age.

____ 59. All licensees must have a high school education or the equivalent.

____ 60. An applicant for a salesperson's license must be sponsored by a licensed broker.

The following acts concern violations of license law. Write *T* for those that your state's license law specifies as a violation. Write *F* for those that are *not* specified.

____ 61. Accepting a net listing

____ 62. Making false statements on license applications

____ 63. Specifying a closing to occur within 15 days of the signing of a contract of sale

____ 64. Refusing to convey a ridiculous offer to a principal

____ 65. A broker paying a commission directly to another broker's licensed salesperson

____ 66. Paying a small referral fee to a person who refers a prospect but takes no other action in the eventual transaction

____ 67. Inducing a seller to break a listing contract with another broker

____ 68. Advertising the broker's commission rates

____ 69. Negotiating a commission rate on a sale

____ 70. Failing to put an expiration date on a listing

____ 71. Posting "for sale" signs in violation of local ordinances

____ 72. Selling a home to a black family in a white neighborhood

____ 73. Refusing to show a home in a white neighborhood to a black family

____ 74. Failing to leave copies of contracts with all parties involved

____ 75. Allowing a person other than a licensed real estate broker to prepare a closing statement

____ 76. Failing to inform a homebuyer of all closing costs that he must pay

____ 77. Refusing to negotiate a loan for a buyer

____ 78. Putting one's sign on property without the owner's permission

____ 79. Allowing a licensed salesperson to show property without the broker being present

____ 80. Allowing a salesperson to handle a closing, even if the broker is present and supervising the salesperson

____ 81. Accepting a placement fee from a loan company without the buyer's or seller's knowledge and consent

____ 82. Guaranteeing to someone that she will be able to sell a property within 2 years at a profit of 50% or more

____ 83. Intentionally lying in order to induce someone to enter into a contract

____ 84. Buying property on which one collects a commission without identifying oneself as a licensee

____ 85. Offering property at a lower price than the seller has authorized in the listing or in a later communication

____ 86. Failing to identify the broker in all advertising

____ 87. Placing a blind ad (one in which the fact that the property is being offered by an agent is not stated)

____ 88. Collecting a commission from more than one party to a transaction

____ 89. False advertising

____ 90. Offering for sale property belonging to others, for which one has no listing

____ 91. A salesperson is not remitting earnest money quickly to his broker

____ 92. Accepting a postdated check as earnest money after the seller has agreed to this

____ 93. Commingling trust or escrow money with personal funds

____ 94. Commingling trust or escrow money with the brokerage firm's general funds

____ 95. Putting trust or escrow moneys belonging to different people into the same trust or escrow account

____ 96. Paying license fees for someone else

____ 97. Doing business while one's license is inactive or is being held by the Real Estate Commission

____ 98. Secretly collecting a commission from more than one party to a transaction

____ 99. Attempting to deal directly with a seller who has signed an exclusive listing agreement with another licensee

____ 100. Selling and renting property from the same office

Multiple-Choice Questions

It is difficult to provide examples of such questions concerning specifics of state law. General examples were given in the preceding section. For reference purposes, we include the following to give you an idea of the *kinds* of multiple-choice questions you may encounter. Note that one answer in some questions is left blank. If none of the other three answers is correct for your state, write the correct answer in the blank. Otherwise, ignore it.

101. License renewal fees are paid
    (A) every year
    (B) every 2 years
    (C) ____________________
    (D) every 3 years

102. The license fees for brokers/salespersons are
    (A) $50/$20
    (B) the same for both
    (C) $40/$10
    (D) ____________________

103. In my state, it is illegal to
    (A) accept a net listing
    (B) post "SOLD" signs on property for which a contract has been accepted
    (C) be a buyer's broker
    (D) negotiate commission rate

104. In my state
   (A) buyer-broker agency must be disclosed (when?) ________
   (B) disclosure of the broker-seller agency must be made to buyers
   (C) no agency disclosure is required
   (D) ________________________

105. Violating license law is punishable by a fine of up to
   (A) $500, imprisonment for up to 6 months, or both
   (B) $500, imprisonment for up to 1 year, or both
   (C) $1,000, imprisonment for up to 1 year, or both
   (D) ________________________

106. Licensees must pay fees for the following:
   (A) transferring employ to another broker
   (B) replacing a lost pocket card
   (C) reactivating an inactive license
   (D) all of the above
   (E) one or two of the above (specify which) ________________________
   (F) none of the above

107. When a complaint is made against a licensee and a hearing is scheduled, the licensee must be given the following advance notice of the hearing:
   (A) 1 week
   (B) 10 days
   (C) 2 weeks
   (D) 1 month
   (E) ________________________

108. A licensee whose license is inactive or is being held by the Real Estate Commission must pay the following fees:
   (A) ________________________
   (B) the same as active licensees
   (C) half those of active licenses
   (D) $40, broker; $20, salesperson

109. A licensee may appeal a decision made at a hearing for a complaint against him directly to
   (A) a circuit court
   (B) the state supreme court
   (C) a federal court
   (D) ________________________

110. A broker must keep all trust and escrow moneys in
   (A) a separate checking account
   (B) a non-interest-bearing account
   (C) an account that can bear interest
   (D) ________________________

Questions 111–140 concern your state's Real Estate Commission.

Fill-Ins

Fill in the blanks with the appropriate words or phrases.

111. The Real Estate Commission has ____________ members.

112. They serve terms of ____________ years.

113. They are appointed by ____.

114. Persons may be nominated for consideration for the post of Real Estate Commissioner by ____.

115. The Chairman of the Real Estate Commission is chosen by ____________________ ____.

116. The Real Estate Commission must meet (how often?) ________________________.

117. The Real Estate Commissioners are paid $ _______________.

118. The Real Estate Commission meets (where?) ______________________________.

119. The geographical distribution of the homes of the Real Estate Commissioners must be ____.

120. A Real Estate Commissioner must have been a resident of this state for ____________ years prior to assuming her post.

121. A Real Estate Commissioner must hold a _____________ license and must have held it for at least _____________ years prior to assuming his post.

122. When a vacancy is created on the commission, _________________ appoints a new commissioner to fill the unexpired portion of the term.

123. The head of the *staff* of the Real Estate Commission is called _________________ ____.

124. The offices of the Real Estate Commission are located at ______________________ ____.

125. For the commission to transact any business, at least ____________ members must be present.

True/False

Write *T* for true, *F* for false.

____ 126. The Real Estate Commission sets the salaries of all people working for it.

____ 127. The Real Estate Commission's budget is composed of all fees and charges it receives from licensees.

____ 128. Real Estate Commission members cannot be reappointed to additional terms.

____ 129. All Real Estate Commission members' terms expire at the same time.

____ 130. Real Estate Commission members must be licensed brokers or salespersons.

____ 131. The Real Estate Commission has no power to revoke or suspend licenses.

____ 132. The Real Estate Commission may deny a license to anyone for any reason.

____ 133. The Real Estate Commission has five members.

____ 134. The Real Estate Commission cannot hold a hearing on a complaint against a licensee unless a quorum is present.

____ 135. The Real Estate Commission has three members.

____ 136. At least one member of the Real Estate Commission must be a member of the general public served by real estate licensees.

Multiple-Choice Questions

Fill in the blanks where necessary.

137. The members of the Real Estate Commission, when they assume their posts,
    (A) must have been licensed brokers for at least ____________ years
    (B) must have been residents of the state for at least ____________ years
    (C) need not be licensed if they represent particular groups (e.g., consumers)
    (D) ________________________

138. The members of the Real Estate Commission
    (A) must be REALTORS®
    (B) may not participate in the real estate business while holding their posts
    (C) must resign if they are sued for improper business practice
    (D) ________________________

139. The Real Estate Commission is responsible for
    (A) issuing real estate licenses
    (B) conducting hearings on complaints against licensees
    (C) setting fees for licenses
    (D) more than one of the above (specify which) ________________________

140. The term of a Real Estate Commissioner is
   (A) 2 years
   (B) 3 years
   (C) 4 years
   (D) ____________________

If your state has a real estate recovery fund (or similar fund), answer questions 141–160. The real estate recovery fund is referred to as the "fund" in these questions, regardless of its specific name in your state.

Fill-Ins

Fill in the blanks with the appropriate words or phrases.

141. When a salesperson first becomes licensed, she must pay $ ____________ into the fund.

142. The maximum amount that can be paid on a single claim from the fund is $ _______.

143. The fund must be maintained at a minimum amount of $____________.

144. When the fund falls below $____________, assessments will be made against all active licensees in the amounts of $____________ for brokers and $____________ for salespersons.

145. Claims must be made against the fund within (time limit) ____________________ of the time the alleged improper action took place.

146. No more than $ ____________ may be paid out of the fund as a result of a single transaction.

147. In my state, the name of the fund is ____________________________________.

148. If the fund is too low to pay a claim in full, the claim plus ____________ percent interest on the unpaid balance will be paid when the fund has risen.

149. A licensee on whose behalf a payment from the fund has been made must pay it back, along with ____________ percent interest.

150. When a broker first becomes licensed, he must pay $ ______________ into the fund.

True/False

Write *T* for true, *F* for false.

____ 151. When a claim against the fund has been paid on behalf of a licensee, her license is suspended or revoked and cannot be reissued until she has fully reimbursed the fund.

____ 152. A claim cannot be made against the fund unless a court judgment has been issued against the affected licensee and it is obvious that he cannot pay the amount in full.

____ 153. A discharge in bankruptcy does not relieve a licensee of any obligation she may have to the fund.

____ 154. A licensee may be exempted from fund assessments if he posts a satisfactory bond with the Real Estate Commission.

____ 155. Money in the fund may be invested in state or federal obligations.

____ 156. The Real Estate Commission has custody of the fund.

____ 157. If the fund becomes dangerously depleted, the Real Estate Commission can levy a special assessment on all licensees for the purpose of augmenting the fund.

____ 158. Claims cannot be paid from the fund unless they are due to fraudulent actions by licensees.

Multiple-Choice Questions

Fill in the blanks where necessary.

159. To make a claim against the fund, one must
    (A) secure a court judgment against a licensee
    (B) submit proof that the licensee cannot pay the claim in full
    (C) file the claim within __________ years of the infraction
    (D) ____________________

160. The assessment to be paid into the fund by new licensees is $ __________ for new salespersons and $ __________ for new brokers.
    (A) $10/$20
    (B) $20/$40
    (C) $50/$100
    (D) ____________________

If your state permits the Real Estate Licensing Authority to regulate real estate schools, answer questions 161–170.

Fill-Ins

Fill in the blanks with the appropriate words or phrases.

161. The curriculum required of an applicant for a salesperson's license must include __________ hours, distributed as follows: ______________________________
____.

162. The curriculum required of an applicant for a broker's license must include __________ hours, distributed as follows: ______________________________
____.

163. Teachers of authorized courses must have the following qualifications: __________
____.

164. Required courses cannot meet for more than __________ hours per day.

165. A real estate school must be given __________ days' notice of the licensing authority's intention to withdraw the school's approval.

True/False

Write *T* for true, *F* for false.

____ 166. The licensing authority may determine the qualifications of teachers of approved courses.

____ 167. A course cannot meet for more than 3 hours per day.

____ 168. A broker's license applicant must have passed a course providing at least 30 hours of study in the area of valuation and appraisal.

____ 169. The licensing authority must approve all real estate schools, even if their courses do not fulfill the educational requirement for licensure.

____ 170. The licensing authority forbids the offering of courses designed specifically to enable people to pass the licensing examination.

PART THREE
Real Estate Contracts

Introduction to Contracts

6

ESSENTIAL MATTERS TO LEARN FROM THIS CHAPTER

- A *contract* is an agreement between two or more parties in which each party agrees to do something or not do something.
- Common contracts used to transfer rights in real estate matters include a *Contract for Sale* (also called by other names such as *Agreement of Sale*), mortgage, lease, deed, and listing.
- A contract involving real estate matters must meet the same requirements as any other contract with one exception. That is, contracts for the sale or use of real estate must be in writing to be enforceable. Every state has a provision called the *Statute of Frauds* that requires contracts in real estate to be in writing to be enforceable. One exception is that leases of one year or less need not be in writing.
- To distinguish parties, the one who *gives* the contract has the suffix "–or." The one who *receives* the contract has the suffix "–ee":

  1. The *vendor* is the seller. The *vendee* is the buyer.
  2. The one who grants a deed is the *grantor*. The one who receives the deed is the *grantee.*
  3. The one who gives property for another's use is the *lessor*. The user is the *lessee.*
  4. The one who pledges property as collateral for a debt (the borrower) is the *mortgagor*. The lender receives the mortgage, which is the collateral for a debt, and so the lender is the *mortgagee.*

- All contracts must have four elements:

  1. Mutual agreement (offer and acceptance).
  2. Consideration.
  3. Legally competent parties.
  4. Lawful purpose.

- Some contracts must have a legal form that includes certain elements. For example, a deed requires a property description.
- *Mutual agreement* means that the parties must agree. There is an offer and acceptance of that offer. Any change that a party makes to an offer received constitutes a rejection of that offer and, if presented, a substitution of a new offer is called a counteroffer.
- *Consideration* is something of value that is provided. Most often in real estate, *valuable consideration* is money given in exchange for property. Another common form of consideration, in a gift, is love and affection, called *good consideration.*

- An *executed* contract is complete. An *executory* contract is missing something, such as signatures. A contract for sale is executory until a closing occurs.
- *Legally competent parties* in real estate are adults in age, 18 or 21 in most states (not minors), who are mentally competent. An illiterate person may enter a contract when the written words have been read or explained.
- A contract with a *minor* is *voidable* by the minor (the other party is bound) while still a minor or shortly after attaining majority age.
- A contract with one who is *mentally incompetent,* such as a person suffering from dementia, is *void.* A legal guardian may contract for that party.
- *Lawful purpose* means that the object of the contract is a lawful one. A contract to do something illegal, such as to pay someone to rob a store or to murder another person, is void.
- Most of the time a contract may be prepared in any form. A few contracts, however, such as deeds, require a *legal form* that describes the property.
- *Duress* is pressure. If a certain type of duress is present, such as a threat against the life of a party by the one who stands to benefit, the contract is void.
- *Misrepresentation* is the misstatement or the nondisclosure of a *material fact.* The party injured may sue for damages or rescind the contract.
- *Fraud* is intentional, willful deceit. Fraud may be a crime. The party injured may be able to recover losses and punitive damages. The perpetrator may be prosecuted as a criminal.
- *Discharge* is reached when no one must perform any longer. Often a contract is discharged when all parties perform under the contract or all parties mutually agree to cancel the contract.
- *Breach* occurs when a party violates a provision or makes it impossible for the other party to perform.
- When a real estate seller breaches a contract, the buyer may go to court to require *specific performance,* that is, make the seller do what the contract requires. In transactions involving real estate, specific performance is often enforced by the court because each parcel of real estate is unique and cannot be exactly replaced by a substitute property.
- *Liquidated damages* is enforcement of the amount specified in a contract that must be paid by the party who breached the contract.

GENERAL CONTRACT INFORMATION

Most people have some experience with contracts. A popular misconception is that a contract must be filled with obscure, legalistic language and printed in nearly invisible type. While it is true that contracts must conform to certain standards, usually the only requirement of the *language* is that it *say clearly what it is supposed to say.*

A little study of contracts will reveal the basic logic behind the laws and customs that surround them and will eliminate the mystery that usually clouds the layperson's view of these documents. This will give you more confidence in dealing with contracts.

To start at the beginning, a contract is an agreement between two or more *parties* in which each of the parties pledges to do or not do something in return for *consideration.* Consideration can take a multitude of forms. A common one is a payment of money, though contracts involving no money payment can be perfectly valid.

REAL ESTATE CONTRACTS

The real estate business is dominated by contracts. Buying, selling, mortgaging, leasing, and listing real estate all involve particular kinds of contracts. To become familiar with the real estate business, we must devote a considerable part of our study to contracts *in general* as well as to the *specific kinds* that concern real estate. The contracts associated with the kinds of real estate transactions that most frequently occur are *listings, contracts of sale, deeds, mortgages,* and *leases.*

Figure 6-1 displays contracts in the order in which they might be encountered in a typical series of transactions. First an owner (a seller) enters into a *listing contract* with a broker. In this contract, the seller agrees to pay a commission when the broker has found a ready, willing, and able buyer for the listed property. In many states, when a buyer is contacted, the law requires that an *agency disclosure* be made to the buyer by the seller's broker at that time (see Chapter 4 for discussion of agency disclosure). When a buyer becomes interested in the listed property, negotiations begin. If there is a buyer-broker agency relationship (see Chapter 4), disclosure of it must be made to the seller and/or his broker during the negotiations. If the buyer and the seller agree to transact, they will sign a *contract of sale* in which they agree that at some time in the future the seller will convey the property to the buyer. This conveyance will occur when the *deed* is executed by the seller and delivered to the buyer. In some states, a *binder* is used to tie up property until a more formal contract of sale is prepared. The binder includes the essential terms of sale. It is expected to be followed in a few days by a contract of sale, which is typically prepared by an attorney.

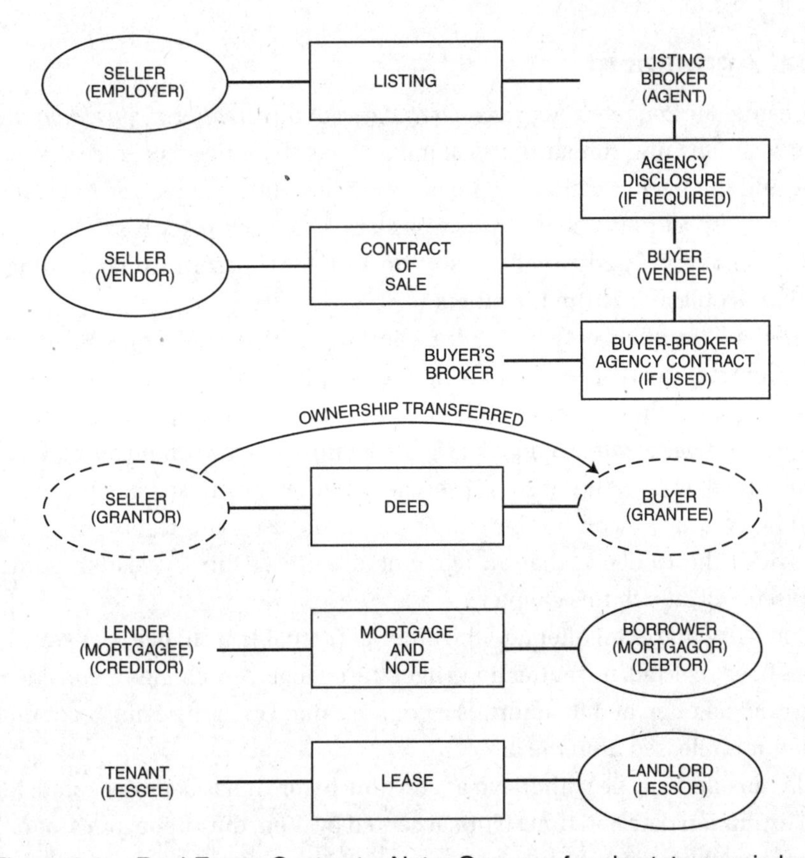

Figure 6-1. Real Estate Contracts. Note: Owners of real estate are circled.

A real estate transaction often triggers the sale of personal property, such as furniture, artwork, tools, and portable appliances. A *bill of sale* is prepared as a separate document used to sell personal property.

A real estate owner may give *power of attorney* to a trusted friend, relative, or business associate. The power of attorney, which can be either general or limited to certain property, allows the person to act for the owner. The person holding the power of attorney is called an *attorney-in-fact.* Such a person is not (necessarily) an attorney-at-law. The attorney-in-fact may act for the property owner. The buyer must be confident that the power of attorney is properly written to delegate the appropriate authority to the attorney-in-fact.

The buyer may seek to borrow part of the cost of the property, so he offers a lender a *mortgage* as collateral for the *note,* which serves as evidence of the debt he owes. Also, if he decides to rent part of the premises to someone else, he enters into a *lease* contract.

REQUIREMENTS OF VALID CONTRACTS

A contract cannot be considered by a court unless it is valid. While certain kinds of contracts may have additional requirements for validity, *all* contracts must have the following attributes:

1. Mutual agreement
2. Consideration
3. Legally competent parties
4. Lawful purpose
5. Legal form (for some contracts)

1. Mutual Agreement

Mutual agreement (often referred to as *offer and acceptance, reality of consent, mutual assent, meeting of the minds,* and similar phrases) means that all parties recognize that an offer has been made and has been accepted by the other parties. All are agreed as to what each party has done, with respect to the contract, as well as to what each party is still required to do in the future. Generally, the parties will agree to do, or not to do, some thing or things in return for some kind of obligation from the others.

Provided that the other essentials of a contract are present, once an offer has been accepted, a contract exists. The one making the offer is the *offeror,* and the one to whom it is made is the *offeree.* The actions of both of them in creating the contract must be shown to be intentional and deliberate. Generally, their signatures to a written contract are evidence of this, provided that no fraud, misrepresentation, duress, or mistake is present. However, many binding contracts need not be written. In such cases, the testimony of the affected parties provides the evidence that an agreement exists. (Duress, misrepresentation, and fraud are discussed later in this chapter.)

When one party makes an offer and the other party makes a change, signs, and returns it, the offer has been rejected, no matter how slight the change. Any change is considered a rejection of the original offer, and its return is a counteroffer. Upon receiving a counteroffer, the original offeror is released from the offer.

Generally, an offer may be withdrawn at any time before it has been accepted. Many offers include an expiration date and time. If not accepted by then, the offer expires, and the offeror is relieved of any obligation.

Once an offer has been terminated by rejection or expiration, the other party cannot go back and accept the terminated or expired offer.

> **Trick Question**
>
> A trick question is often based on the "common knowledge" that as parties exchange offers they are getting closer to a contract. For contracts, this is not true. In law, any change to a proposed contract is actually a rejection of the contract presented. A counteroffer is a rejection of a proposed contract and is a substitution of a new offer.
>
> Another trick question may state that an offer may not be withdrawn until the other party has either accepted or rejected it. This, too, is not correct. An offer may be withdrawn any time before it has been accepted.

2. Consideration

Legal consideration refers to the "benefit" received by the parties. Each party must be shown to be obligated to do (or not to do) something in return for some action or nonaction by the other parties to the contract. The benefit can take many forms. The only requirement is to show that the party receiving the benefit agreed that it was of use to him and that he was willing to obligate himself in some way in order to get it. Payment of money is always consideration, as is the transfer of ownership of anything that can be shown to have a market value. In addition, the transfer of many *rights* is consideration; for example, a lease transfers rights to use property but not ownership of the property.

Examples. Here are some examples of consideration in a contract.

1. A pays $50,000 to B. B transfers ownership of the house to A.
2. A pays B $1,200 per month. B gives A the right to use an apartment for each month for which A pays.
3. A pays B $7.50. B lets A watch a movie in B's theater.
4. A agrees not to hang her laundry outside where B and C can see it. B agrees not to hang his laundry where A and C can see it. C agrees not to hang her laundry where A and B can see it.
5. A agrees to use her best efforts to try to find a buyer for B's house. B agrees to pay A a commission of 6% of the sale price if A finds a suitable buyer.
6. A agrees to pay B $392,000 in 60 days, at which time B will transfer to A ownership of a particular house.
7. A pays B $1,000. B gives A the right to purchase B's land for $25,000 at any time of A's choosing during the next 6 months.
8. A transfers ownership of her house to B. B transfers to A the ownership of his parking lot, four albums of rare stamps, two sets of dishes, and a pedigreed dog.
9. A pays B $10,000. B transfers to A the ownership of a producing diamond mine in South Africa.
10. A transfers to her son, C, the ownership of the family estate in return for the "love and affection" that C has given to A in the past.
11. A pays B $10,000. B agrees to pay A $126.70 per month for the next 10 years.
12. A pays B $10,000. B gives A the right to have B's house sold if B defaults on his debt to A.

In all of these instances, there exists legal consideration; these examples show but a few of the many forms it can take. In many cases, it includes the payment of money. In (4), (8), and (10), however, no money changes hands. In most, the consideration involves the parties doing something or performing some act. In (4), however, the parties agree *not* to do something. Examples (2), (3), (4), (7), and (12) include the exchange of rights; (5) includes the exchange of services; and the others involve the exchange of ownership of things.

These are examples of real estate contracts we will examine in more detail later. Example (2) is a *lease*; (4) is a *restrictive covenant*; (5) is a *listing*; (6) is a *contract of sale*; (7) is an *option*; (12) is a *mortgage*; and (1), (8), (9), and (10) are *deeds*. Many of these contracts are purchases of one kind or another. Examine (11) closely and you will discover that it is a *loan* or a *note*: A lends B $10,000, which B agrees to repay (at a 9% interest rate) in equal monthly installments of $126.70 for the next 10 years. Example (5) is a *listing*; it is also an *employment contract*, wherein the services of one of the parties are purchased by the other party.

GOOD CONSIDERATION, VALUABLE CONSIDERATION

In all except (10) we have *"valuable" consideration*. Valuable consideration is anything that is of value to practically everyone and thus can be sold for money. Money, of course is "valuable," as is ownership of things, receiving someone's services, receiving various kinds of rights, and the like. However, things such as friendship, love, and affection are not valuable consideration, because their value is entirely subjective. You may set a very high value upon the attention and friendship you receive from your best friend, but to most of the rest of us the attention is valueless since we don't even know him. Furthermore, a person's affection is not a transferable benefit. Rights, ownership, money, services, and such things can be passed on (or sold) to practically anyone, but you can't sell your best friend's affection to someone else. Consequently, such things as friendship and love are called *"good" consideration*. They are of value to the one who receives them, and if she agrees to transfer benefit in return for them, *she* can be held to her bargain.

The law does *not* require that *equal consideration* accrue to all parties. So long as consideration is there, the law is satisfied, provided of course that no fraud or duress is involved. There is only one exception: the courts will not enforce a contract requiring the simultaneous exchange of different sums of money.

In some cases, a contract doesn't have to spell out the exact amount of consideration. Of course, a deed contract ought to describe the real estate very carefully. Often in deeds, the price paid may be described, for example, as ". . . ten dollars and other good and valuable consideration." Because deeds are recorded and therefore available to the public to see, many buyers don't want nosy people to find out exactly what they paid for their real estate. This kind of wording, especially for money consideration, is perfectly all right in most states. It says that one has received consideration, and that is all that is necessary. In some states, the amount paid must be stated on the deed.

3. Legally Competent Parties

Competent parties are persons who have the capacity to enter into contractual arrangements. Not all legal persons are human beings: corporations, partnerships, trusts, and some other organizations can also be parties to a contract. You can contract to purchase an automobile from ABZ Auto Sales, Inc., even though this company is not a living, breathing being.

Furthermore, not all humans are competent to contract. Many are legally unable to do so, and many more may contract only under carefully defined circumstances.

NONHUMAN PARTIES

Usually, each party to a contract must assure himself of the other's legal competence. Although the law places few restrictions upon the possible contracting ability of nonhuman parties, often these organizations will operate under self-imposed restrictions. A corporation's charter, for example, can permit or prevent certain kinds of contractual agreements on the part of the organization. In addition, it can provide that some or all of the permitted functions be carried out in certain ways. For this reason, a person who contracts with an organization must assure himself, first, that the organization is permitted to engage in such a contract and, second, that the contract is executed and carried out consistent with the organization's permitted limitations and prescribed methods.

PEOPLE

Among people, there are two major categories of incompetent parties: infants and mentally incompetent persons.

a. Infants (or Minors). These are people who have not yet attained the age of majority, which is 18 to 21 in various states. While you may not think that a hulking 17-year-old football star could be incompetent at *anything*, the law says he is. Furthermore, a person who may be immature, ignorant, or naive in business judgment still may be considered by the law to be perfectly competent to take responsibility for her actions if she has attained the age of majority. Minors are not prohibited from contracting, so no law is broken if a minor signs a contract. However, the contract is viewed in a special way by the law: the minor can *void* the contract later on, if he wishes, but adult parties to the contract do *not* have that privilege. The minor may not, however, void (or *disaffirm*) only *part* of the contract; he must abandon it entirely. The contract is characterized as *voidable* by the minor. If he contracts to buy property from an adult, he will have the right to declare the contract void, if he wishes, any time before it is carried out. The adult, however, must honor the contract unless and until the minor actually disaffirms it. If the adult does not keep his bargain, the minor can sue. The adult will receive no help from the courts if the minor disaffirms. However, the minor cannot, say, disaffirm the part of the contract that specifies he must pay for the property while at the same time requiring the adult to transfer it to him anyway; he must abandon the *entire* contract. Furthermore, if he disaffirms it after the adult has provided him some benefit under it, the minor may have to pay for (or return) that which he had already received by the time he decided to void the contract.

The whole idea behind the legal specification of some people as incompetent is a matter of protection. An *infant* is considered too young to bind herself to agreements; the law gives her protection by allowing her to void most agreements that she gets herself into. Once an infant attains the age of majority, she becomes an adult and loses this. However, she *is* given a "reasonable" time beyond the attainment of majority to decide whether she will disaffirm any contracts that she entered into as a minor.

b. Mentally Incompetent Persons. The law also extends this "protection" to other categories of people, the largest being mentally incompetent persons. Unlike infancy, however, mental incompetency is a state that can come and go. One who is competent (sane) can become incompetent (insane), and one who is incompetent can be cured and so become competent again. Also, it is possible for a person who is under severe stress or suffering from some other transitory disturbance to become so disoriented that the law may extend the protection of incompetence to him during that episode.

Legal protection of a mentally incompetent person takes two forms. First, if he has had a guardian appointed by the court, then all contracts he enters into are *void* and he had no capacity to contract at all. Notice that the infant's contracts are *voidable*, whereas the contracts of an incompetent person under guardianship are *void*. If the incompetent person has no guardian, his contracts, too, are voidable rather than void. He has the choice to require that his bargains with others be carried out or to disaffirm the contracts.

Neither the infant nor the mentally incompetent person may disaffirm *all* kinds of contracts; generally, courts will enforce against them contracts for *necessaries*. Examples of necessaries are food, clothing, and shelter; sometimes other things are included as well. For example, in some cases contracts for employment or for education may be considered necessaries. Once again, the law is extending another form of protection: an adult usually will not care to contract with an infant or someone else she knows may later plead incompetence and void the contract. Consequently, a truthful incompetent may be at a serious disadvantage if his personal circumstances *require* him to contract. Therefore, the law allows some incompetents what amounts to a limited competency. For things that are "necessary" to them, they are allowed to create binding contracts, which they cannot disaffirm and which a court can enforce against them. As a result, competent parties receive the protection *they* seek in a contractual arrangement and any reluctance they may have to deal with an incompetent is eliminated.

c. Other Incompetency. Certain other forms of incapacity "earn" one the legal protection of incompetency. A contract made by a person who is severely intoxicated with alcohol or is under the dominating influence of drugs known to affect judgment may later be considered voidable. Many states consider felon prisoners to be without the right to contract; this situation is described by the rather grim term of *civilly dead.* Usually, when the sentence has been served and sometimes at the time parole is secured, most rights will be restored.

4. Lawful Purpose

A contract that requires any of its parties to violate the law usually is void. However, if it is possible to do so without impairing the contract's basic purpose, a court will uphold the contract but strike out that part requiring an illegal act. A very good example concerns the federal Fair Housing Act of 1968. Among other things, this law declared illegal certain restrictive covenants that required buyers of some real estate to agree they would resell only to people of certain racial or cultural backgrounds. Such covenants are against public policy. At the time the law was enacted, vast numbers of deeds to homes contained such clauses, but that fact did not void those deeds in their entirety. The outlawing of these clauses had no great effect upon the main purpose of the deeds, which was to transfer title to realty in return for consideration. However, if someone had created a special agreement for the sole purpose of restricting the racial or cultural background of purchasers of certain real estate, then that entire contract would have been voided by the enactment of the law. If this kind of agreement came into being after the law was passed, it would be illegal from the beginning and would *never* be a contract.

The essential rule of lawful purpose is that a contract that exists substantially for the purpose of requiring an illegal act is void. So also is a contract in which an illegal act is required, the elimination of which would materially alter the purpose or effect of the contract. When an illegal act is included in a contract but is not essential to the basic intent or purpose of the agreement, a court often will void only the illegal part while letting the rest stand.

5. Legal Form

Some contracts are required to follow a certain form or to be prepared in a certain manner. For example, deeds and contracts of sale require that "legal" descriptions of the property be included. If the legal description is missing or is defective, the entire contract may be invalid. It should be pointed out, though, that much of the wordy, archaic language that frequently appears in contracts is *not* required by any law and appears only as a result of tradition.

STATUTE OF FRAUDS: REAL ESTATE CONTRACTS MUST BE WRITTEN TO BE ENFORCEABLE

For most real estate contracts there is one critical requirement with respect to form. Generally, all real estate contracts, except (1) listings (in some states) and (2) leases for 1 year or less, *must be written to be enforceable.* This requirement is an outgrowth of the fact that each state has enacted a law called the Statute of Frauds, which requires that all contracts of certain types be written if legal means are to be used to enforce them. Included in these categories are all contracts in which land or an interest in land is sold and contracts that cannot be performed within 1 year. This latter category, then, includes lease contracts extending for more than 1 year.

As noted in Chapter 4, a listing contract does not require the sale of land or an interest in land among its parties. It is a contract of *employment,* which is a category not covered by the Statute of Frauds. However, most states, in their real estate licensing laws, require that residential listings be in writing.

The definition of "land or an interest in land" is very specific and also quite broad. It includes future interests, so, in addition to deeds, contracts of sale must be written. It includes partial interests, so mortgages must be written. It includes improvements and attachments to land. So contracts involving the purchase of buildings or growing plants must be written, even if the actual land beneath them is not part of the transaction. However, once an attachment to land is severed from it (e.g., trees are cut down or crops are reaped), the attachment becomes personal property and is no longer subject to the part of the Statute of Frauds that covers realty. This law covers more than just realty contracts, so many other nonrealty contracts can be required to be written to be enforceable.

DURESS, MISREPRESENTATION, AND FRAUD

All parties to a valid contract must have entered into it willingly and without having been misled as to pertinent facts concerning the arrangement. Consequently, you cannot expect a court to enforce a contract against someone if you pointed a gun at his head to force him to sign or if you deliberately or unintentionally misled him so that he agreed to its terms on the basis of this false information. The former kind of situation is described as one of *duress*; the latter is either *misrepresentation* or *fraud.*

Duress

Duress can take any number of forms. When circumstances force a person to do something against her will, one could say that duress exists, but the law does not always see the situation in that way. If the duress is caused by a party who benefits under the contract, then a court usually will recognize it as such; this covers situations such as forcing someone's assent by threatening her illegally. But other circumstances can also create duress: Jones discovers that he has a rare, expensive disease and must sell his home in order to raise money to pay for the cure. A person who buys from him at this time might make an advantageous deal legally if Jones is in a hurry to sell. If the duress is caused by one's own action, the court is reluctant to recognize it. If you buy a new house and therefore are desperate to sell your old home, that situation is pretty much your own fault. If you create your own duress, you usually have to suffer with it.

Misrepresentation and Fraud

Misrepresentation and fraud go a few steps further: here someone is led into a bad bargain on the basis of false information. Misrepresentation is any untrue statement, whether deliberate or unintentional. It may be a form of nondisclosure where there is a duty to disclose or the planned creation of a false appearance. Where there is misrepresentation of *material fact*, the person injured may sue for *damages* or rescind the contract.

Fraud, on the other hand, is always deliberate, with an intent to deceive. There is another difference, too: fraud is a *crime* as well as a civil wrong. Whereas a victim of either fraud or misrepresentation is entitled to compensation for what he was misled into losing, a victim of fraud may also sue for punitive damages, which are a form of extra payment required of the culprit as a kind of punishment. Moreover, the perpetrator of a fraud may also be prosecuted in criminal court and may end up being convicted and imprisoned.

PROVING MISREPRESENTATION OR FRAUD

A victim must prove three things to show that misrepresentation or fraud exists: (1) that the incorrect information was relevant to the contract she is disputing—that is, it was *material*; (2) that it was reasonable for her to rely on this false information; and (3) that this reliance led her to suffer some loss as a result.

Let's consider an example. Brown considers buying Green's house. Brown asks Green whether the house is free of termites, and Green says that it is. Actually, the house is infested. If Brown buys the house based, in part, upon Green's information and subsequently discovers termite infestation, he may have a case. The point at issue is material: it is reasonable for Brown to base his decision to buy at least partly upon the assurance that no termites inhabit the premises. When he finds them, it is obvious that he has suffered a loss; therefore the first and third conditions are met. Now he must show that it was reasonable to rely upon Green (the owner at the time) to give him accurate information on this matter. If the court agrees with Brown, he has proved his case. If Green knew about the termites and deliberately lied, then he committed fraud. If he truly thought there were no termites, then he misrepresented.

If one or more of the three conditions are missing, fraud or misrepresentation cannot be claimed successfully. If Brown had not bought Green's house, he could not claim relief, because he suffered no loss due to the falsehood. Also, if he had asked a neighbor a few houses away about the termite problem in Green's house and had relied on the neighbor's incorrect information, it is unlikely that he could get help from the court. Brown would have great difficulty showing that it was reasonable to rely on a neighbor for this kind of information.

SOME INFORMATION MUST BE DIVULGED

For some kinds of contracts, the law requires that certain information be divulged by at least some of the parties. If it is not, then misrepresentation or fraud may exist. For example, most licensing laws require that brokers inform other parties if the brokers are acting as principals in the contract as well as agents. Also, most states require a broker to disclose which party she is acting for, especially if more than one party to a transaction is paying her a commission.

DISCHARGE AND BREACH OF CONTRACTS

Contractual arrangements can be terminated in two general ways. *Discharge* of a contract occurs when no one is required to perform under it anymore. *Breach of contract* occurs when one party violates a provision or makes performance impossible, even though other parties are willing, by refusing to do his part or by otherwise preventing discharge of the contract.

Discharge

Discharge can be considered the "amicable" situation in which the parties to a contract agree that the arrangement is terminated. Most often, discharge occurs by *performance*; that is, everyone has done what he has promised to do and nothing more is required of anyone. Other forms of discharge include agreement to terminate a contract, for one reason or another, before it would have been discharged by performance. Such a reason might be the substitution of a new party for one of the original parties; this original party would have had her duties discharged. Sometimes the parties may decide to terminate one agreement by substituting another one for it. Discharge can also occur when the parties simply agree to abandon the contract without substituting a new party or another agreement for it. Some contracts, such as listings, are discharged by death. A contract for sale is not discharged by death.

A party to a contract may *assign* the contract to another unless prohibited by the contract. An assignment does not relieve the party of obligation. For that, a *novation* is used. This agreement between all parties substitutes a new party and releases the original party to the contract.

STATUTE OF LIMITATIONS

The law limits the time during which contracting parties can take a dispute concerning a contract to court; this is the Statute of Limitations. Different states set different kinds of limits on different kinds of contracts. Once the limit has passed, a dispute will not be heard in court. In that sense, the contract may be said to be discharged, because it can no longer be enforced. It is greatly advisable that any agreement discharging a contract for reasons other than performance or limitation be in writing. Under some situations (particularly when some, but not all, parties have performed or begun to perform under the contract), an agreement discharging an existing contract *must* be in writing.

Breach

Breach occurs when a party violates the provisions of a contract; often this action can be serious enough to terminate the contract by making it impossible to continue the arrangement. The injured parties must seek remedy from the court; usually it takes the form of a judgment for *damages* against the party causing the breach. Sometimes it may be feasible to secure a judgment for *specific performance*, which requires the breaching party to perform as he had agreed. A buyer of real estate may gain specific performance because each parcel of real estate is unique. To be successful in court, the injured parties must be able to show that a breach has indeed occurred. Failure by a party to perform entirely as specified in the contract constitutes a breach, as does a declaration that he does not intend to perform. Finally, there is a breach if a party creates a situation wherein it is impossible for every party to the contract to perform as specified. Once any of these events has occurred, there is a breach. One attribute of a breaching of a contract is that it terminates the obligations of the other parties as well. The person who has breached a contract cannot have that agreement held good against any of the other parties. When a monetary amount of damages for nonperformance is stated in a contract, it is called *liquidated damages.*

Injustice

The court will not enforce a contract if doing so will result in an obvious injustice. While the law is precise, its enforcement is allowed to be compassionate. Consider a case where Smith and Jones come to an *oral* agreement, wherein Smith pays money to Jones and Jones transfers title to a piece of land to Smith. Smith then builds a house upon the land, whereupon Jones claims title to the land (and to the attachments, including the house) on the grounds that the contract transferring title was oral and therefore invalid. In a case as cut and dry as this, the court would award title to Smith, despite the absence of a written contract, for two reasons. First, there is a principle of law that recognizes an oral contract of this nature provided that the purchaser has "substantially" improved the property; otherwise he clearly would suffer unjustly. Second, if Jones had engineered the entire scheme specifically in order to take advantage of Smith, then Jones's case would be dismissed. A person can't use the law to obtain an unfair advantage over someone else who is acting in good faith.

Outside Circumstances

Finally, a contract can be terminated because of some outside circumstance. For example, new legislation may invalidate some kinds of contracts. The death of one of the parties will void a listing but not a contract of sale. If a party becomes ill or injured to the point where she cannot perform, she may sometimes void a contract without penalty. If it turns out that the parties have made a mistake as to what their agreement constitutes, then the contract can be nullified. For example, if Emmett thinks a contract of sale involves his purchase of Joan's property on *First* Street and Joan thinks she has agreed to sell Emmett the property she owns on *Twenty-first* Street, there is no contract, because there hasn't been a meeting of the minds. Finally, if the subject matter of the contract is destroyed, the contract ceases to exist.

QUESTIONS ON CHAPTER 6

1. A contract in which an owner of real estate employs a broker for the purpose of finding a buyer for the real estate is a
   (A) deed
   (B) contract of sale
   (C) listing
   (D) lease

2. A contract in which ownership of real property is transferred from one person to another is a
   (A) deed
   (B) contract of sale
   (C) listing
   (D) lease

3. The two parties to a lease contract are the
   (A) landlord and the serf
   (B) rentor and the rentee
   (C) lessor and the lessee
   (D) grantor and the grantee

4. The requirement that all parties to the contract have an understanding of the conditions and stipulations of the agreement is
   (A) reality of consent
   (B) meeting of the minds
   (C) offer and acceptance
   (D) all of the above

5. Consideration in the form of "love and affection" is
   (A) illegal in any transaction
   (B) good consideration
   (C) valuable consideration
   (D) contingent consideration

6. A person who is too young to be held to a contractual arrangement is called
   (A) youthful
   (B) incompetent
   (C) an emancipated minor
   (D) civilly dead

7. One who is otherwise incompetent to make a contract may be bound to contracts for
   (A) anything but real estate
   (B) real estate only
   (C) necessaries
   (D) food, clothing, and shelter not to exceed $100 per week

8. Which of the following contractual arrangements would be unenforceable?
   (A) A agrees to buy B's house.
   (B) A agrees with B that B shall steal money from C.
   (C) A agrees to find a buyer for B's car.
   (D) A agrees with B that B shall make restitution to C for money stolen by B.

9. The law that requires that most real estate contracts be written to be enforceable is the
   (A) Statute of Limitations
   (B) Statute of Frauds
   (C) Statute of Written Real Estate Agreements
   (D) Statute of Liberty

10. A contract in which A agrees to allow B to use A's real estate in return for periodic payments of money by B is a
    (A) deed
    (B) contract of sale
    (C) lease
    (D) mortgage

11. When a person deliberately lies in order to mislead a fellow party to a contract, this action is
    (A) fraud
    (B) misrepresentation
    (C) legal if no third parties are hurt
    (D) all right if it is not written into the contract

12. Holding a gun to someone's head to force him to sign a contract is
    (A) attempted murder
    (B) permissible only in exceptional circumstances
    (C) duress
    (D) rude and inconsiderate but not illegal so long as the gun doesn't go off

13. When a party to a contract makes performance under it impossible, the result is
    (A) breach of contract
    (B) discharge of contract
    (C) performance of contract
    (D) abandonment of contract

14. A contract in which A agrees to purchase B's real estate at a later date is
    (A) a deed
    (B) an option
    (C) a contract of sale
    (D) a lease

15. Contracts made by a minor are
    (A) enforceable at all times
    (B) void
    (C) voidable by either party
    (D) voidable only by the minor

16. Incompetent parties include
    (A) minors
    (B) insane persons
    (C) people with court-appointed guardians
    (D) all of the above

17. The parties to a deed are the
    (A) vendor and vendee
    (B) grantor and grantee
    (C) offeror and offeree
    (D) acceptor and acceptee

18. If A and B have a contractual arrangement and B violates the contract, A may
    (A) do nothing but suffer the consequences
    (B) sue in court for damages and/or specific performance
    (C) call the police and have B arrested unless B agrees to cooperate
    (D) damage B to the extent that B has damaged A

19. A agrees to sell his brand-new limousine to B in return for one dollar. Later A wishes to back out of the deal.
    (A) A may not do so.
    (B) A may do so because he is not getting the true value of the limousine.
    (C) A may only require that B pay a fair price for the limousine.
    (D) Both actions are clear evidence of insanity, so the contract is void.

20. A agrees to trade her car to B in exchange for a vacant lot that B owns.
    (A) This is a valid contractual arrangement.
    (B) This is not a contract, because no money changes hands.
    (C) This is not a contract, because unlike items can't be traded.
    (D) This is not a valid contract, because the car is titled in A's name.

21. Two offers are received by the broker at the same time. The broker must
    (A) present both offers
    (B) present only the higher offer
    (C) present only the offer that provides a greater commission
    (D) use his or her best judgment as to which offer to present

22. In a legal sale contract, the seller is often referred to as the
    (A) trustor
    (B) divisor
    (C) donor
    (D) vendor

23. A contract based on an illegal consideration is
(A) valid
(B) void
(C) legal
(D) enforceable

24. A valid purchase and sale agreement must contain
(A) listing price
(B) broker's commission
(C) financing requirements
(D) purchase price, terms, and conditions of sale

25. The terms of a purchase and sale agreement can be changed only
(A) in writing and signed by all parties to the transaction
(B) by the broker in a dual agency
(C) by the title company
(D) by the attorneys for both parties

26. A person who has the right to act for another is
(A) a broker
(B) a witness
(C) an attorney-at-law
(D) an attorney-in-fact

27. Which is *not* necessary for a valid contract?
(A) offer
(B) legal object
(C) acceptance
(D) due date

28. A contract is accepted with minor modifications. It is
(A) effective
(B) binding
(C) not binding
(D) voidable

29. A contract that requires a party *not* to do something is
(A) negative
(B) restraint on alienation
(C) acceptable
(D) unjust

30. Which can bind a principal to a contract?
(A) general agent
(B) special agent
(C) subagent
(D) attorney-in-fact

31. Broker Bill was given oral instructions to submit an offer for $47,000, but he erroneously prepared an offer for $49,000, which all parties signed. As a result,
(A) the contract is voidable
(B) the buyer is exempt
(C) Broker Bill must forfeit any commission
(D) the contract is binding

32. A bill of sale is used to convey
(A) retail store inventory
(B) leasehold
(C) leased fee
(D) all of the above

33. A contract that states the price and terms in writing is
(A) express
(B) implied
(C) boilerplate
(D) unenforceable

34. In a contract where there is a dispute because of a contradiction, which holds more weight?
(A) preprinted form
(B) words written in
(C) legalese
(D) boilerplate

35. A counteroffer is
(A) good negotiating technique
(B) rejection of the offer with substitution of a new offer
(C) an implied contract
(D) a selling strategy

36. Which of the following is *not* an essential element of a contract?
(A) offer and acceptance
(B) consideration
(C) competent parties
(D) earnest money

37. Which of the following would *not* have the legal capacity to enter into a valid and enforceable contract for the sale of real estate?
    (A) an illiterate person
    (B) a minor
    (C) a woman in a nursing home
    (D) an authorized officer of a partnership

38. A binder given by a buyer in a real estate transaction
    (A) may be withdrawn by the buyer at any time before the seller signs his or her acceptance
    (B) draws interest in favor of the broker
    (C) must exceed 2% of the sales price
    (D) must be monetary

39. In case of a breach of contract, a party may enforce *specific performance*. What is specific performance?
    (A) a lawsuit for actual and treble damages
    (B) a lawsuit by the nondefaulting party to specifically collect damages agreed to in the contract
    (C) a lawsuit to compel action called for in the agreement
    (D) all of the above

40. A written and signed real estate contract is voidable by only one party when
    (A) the market value was less than the sales price
    (B) the market value declines
    (C) one of the parties failed to read the instrument before signing it
    (D) one of the parties was an unmarried minor at the time the contract was signed

41. An enforceable contract may *not* include
    (A) affection
    (B) money
    (C) duress
    (D) love

42. The law requiring that contracts for the sale of real estate be in writing to be enforceable in a court of law is the
    (A) Statute of Conveyance
    (B) Statute of Frauds
    (C) Statute of Enforceability
    (D) Statute of Writing

43. A contract is orally explained to an adult who is illiterate. The buyer signs with an X. The contract is
    (A) void
    (B) voidable by the illiterate
    (C) unenforceable
    (D) valid

44. A buyer offers $100,000 for a house listed at $100,000 and requires that the refrigerator and washer-dryer be included. The buyer is told that the seller has refused. Two days later, the seller changes his mind and will include these appliances. Which of the following is true?
    (A) A contract has been reached.
    (B) The buyer is relieved of his offer.
    (C) The buyer is bound by his first offer because it has been accepted.
    (D) The seller must wait to hear what the lawyer says.

45. A buyer who personally signed a contract of sale is diagnosed with dementia. His legal caretaker seeks to set the contract aside and declare it void. Most likely
    (A) it can be declared void because of the buyer's incompetence
    (B) the deal is done
    (C) the seller may keep the earnest money
    (D) the broker has earned a commission

46. For a contract to be binding, there must be
    (A) a meeting of the minds
    (B) consideration
    (C) legal purpose
    (D) all of the above

47. A 17-year-old contracts to purchase a house and closes on it. The contract may be voided by
    (A) the buyer while still a minor or soon after reaching majority age
    (B) neither party because the purchase was closed
    (C) the seller at any time
    (D) the minor during his lifetime

48. A contract calls for a specific amount of damages to be paid if a party defaults. The damages are called
    (A) severance
    (B) liquidated
    (C) actual
    (D) compensatory

49. If title and possession do not occur simultaneously, there should be an interim
    (A) lease (interim occupancy agreement)
    (B) option
    (C) exchange
    (D) novation

50. To be valid, a contract must include
    (A) money
    (B) consideration
    (C) property
    (D) signature

51. Which of the following is an executory contract?
    (A) a recorded deed
    (B) one year into a three-year lease
    (C) an agreement of sale that is unsigned
    (D) an option that has been exercised

52. A contract requiring performance of something that is illegal is
    (A) enforceable
    (B) void
    (C) valid
    (D) illegal

53. A contract with a minor is
    (A) voidable by either party
    (B) voidable by the minor
    (C) enforceable in court
    (D) enforceable if notarized

54. A valid contract requires
    (A) clear writing
    (B) review by an attorney
    (C) a broker
    (D) an offer and an acceptance

55. A breach of contract occurs when
    (A) both parties agree to terminate it
    (B) one party fails to perform as required
    (C) there is an oral misunderstanding
    (D) one party files a lawsuit

ANSWERS

1. **C**	11. **A**	21. **A**	31. **D**	41. **C**	51. **C**
2. **A**	12. **C**	22. **D**	32. **A**	42. **B**	52. **B**
3. **C**	13. **A**	23. **B**	33. **A**	43. **D**	53. **B**
4. **D**	14. **C**	24. **D**	34. **B**	44. **B**	54. **D**
5. **B**	15. **D**	25. **A**	35. **B**	45. **A**	55. **B**
6. **B**	16. **D**	26. **D**	36. **D**	46. **D**	
7. **C**	17. **B**	27. **D**	37. **B**	47. **A**	
8. **B**	18. **B**	28. **C**	38. **A**	48. **B**	
9. **B**	19. **A**	29. **C**	39. **C**	49. **A**	
10. **C**	20. **A**	30. **D**	40. **D**	50. **B**	

Description of Land

7

ESSENTIAL MATTERS TO LEARN FROM THIS CHAPTER

- The purpose of describing land is to determine its boundaries so an owner or buyer can determine what he owns. A survey is used to do this, with the professional called a surveyor; this person may be qualified by the state, i.e., a *registered surveyor.* A survey is often used to determine whether a structure is encroaching on (overlapping) someone else's property.
- A survey often contains a map with a text description.
- There are four methods to describe land:

1. Metes and bounds.
2. Lot and block numbers. (See Figure 7-2)
3. Monuments or occupancy.
4. Rectangular survey (in certain states).

- A *metes and bounds* description begins at a place identified as the point of beginning (POB) and ends at the same place. In between the point of beginning and its end to the same place, the survey calls distances and angles, enclosing a tract of land.
- A *lot and block* description is used after a subdivider or developer has filed a subdivision *plat* at the courthouse. The subdivision is given a name. Blocks are parcels of land within the subdivision that have been carved into individual lots suitable to build upon. The lots are numbered.
- *Monuments or occupancy* uses certain objects to help establish boundaries. A certain stream, large tree, rock outcropping, or other natural or man-made object is referenced in the description. Occupancy might be stated as ownership, e.g., "Smith's farm."
- The *rectangular survey* method is used in certain states. It is based on *principal meridians,* which are imaginary lines that run north-south, and on *base lines,* which are imaginary east-west lines:

1. Land is divided into *townships* having 6 miles on each side, comprising 36 square miles.
2. Townships are divided into 36 *sections,* each 1 mile square.
3. Townships are identified by the distance in miles from the base line and meridian.
4. Sections are divided into fractional parts, e.g., ½ or ¼.
5. Fractional parts are identified by the portion of the section they occupy, for example, NW or SE.
6. A *section* is 1 square mile, therefore it includes *640 acres.*
7. There are *43,560 square feet* in 1 acre.

- A legal description of property is prepared by a surveyor.
- To learn the intricacies of surveying, especially rectangular surveying, requires many hours of effort. Typically, there are only two or three questions on land description in most 100-question salesperson or broker exams. There may be no questions or only one on the rectangular survey system. Therefore, when you study for the exam, your time will be better spent on other matters.
- Please memorize the size of an acre (43,560 square feet), number of acres in a square mile (640), size of a township (36 square miles), and size of a section (1 square mile).

RECOGNIZED METHODS OF LAND DESCRIPTION

Recognized, uniform methods of land description are essential to proper contractual transactions involving land. If the land in a transaction cannot be identified, the legal system will not recognize the transaction as binding. Therefore, proper description of the land involved is an essential part of contracts of sale, deeds, leases, options, mortgages, listings, and virtually all other kinds of real estate contracts.

There are four generally accepted methods of land description in the United States:

1. Metes and bounds descriptions
2. Lot and block number descriptions
3. Monument or occupancy descriptions
4. Rectangular survey

The purpose of all of these is to *identify* real estate. Often, in the terminology of the real estate business, descriptions are called "legal descriptions," giving the impression that there is some particular formula that must be followed. Actually, the only legal point of any importance is that the description be *sufficient to identify the property.* If the property can be identified from the description given, then the description is good. Otherwise, no matter how elaborate it may appear, the description is faulty.

Most description methods are designed to demonstrate to a surveyor a means of marking the outline of the land on the ground. Most descriptions do not include buildings. Remember that real estate is land *and all attachments to it.* So a description of the land automatically includes all improvements unless they are *specifically excluded.*

Metes and Bounds Descriptions

A metes and bounds description has a point of beginning, which is where the perimeter of the property line begins and ends. Metes and bounds descriptions indicate to the surveyor how to locate the corners of a tract. By locating the corners, one also locates the sides since they run from corner to corner. The following is a metes and bounds description of Lot 15 in Figure 7-1. This lot is shown enlarged in Figure 7-2.

BEGINNING at a point being the southeast corner of the intersection of Vicious Circle and Alongthe Avenue, thence one hundred and six and forty-hundredths feet (106.40′) north 90°0′0″ east, thence one hundred and twenty feet (120.00′) south 0°0′0″ east, thence seventy-three and fourteen-hundredths feet (73.14′) south 90°0′0″ west, thence one hundred and twenty-four and fifty-three hundredths feet (124.53′) north 15°29′26″ west to the point of beginning; said tract lying and being in Gotham City, Copperwire County, Montana.

Note that a metes and bounds description must have the following components:

1. A properly identified *point of beginning.*
2. *Distance* and *direction* given for each side.

The description above contains all these. The point of beginning is identified as the intersection of two public streets; this is permissible since a person may always consult the public records to find out exactly where these streets are. From that point, the surveyor is directed to proceed a certain distance and in a certain direction. By following this description, he can arrive *only* at the specific points described, and these points are the corners of the lot. The surveyor will have outlined the particular plot of land in question by following this description; thus it is a good description.

The accuracy of the description is the only point of any consequence in a court of law. The description can progress clockwise or counterclockwise. Directions (north, south, etc.) may be capitalized or not. The description should include the name of the city (or county) and the state in which the real estate is located.

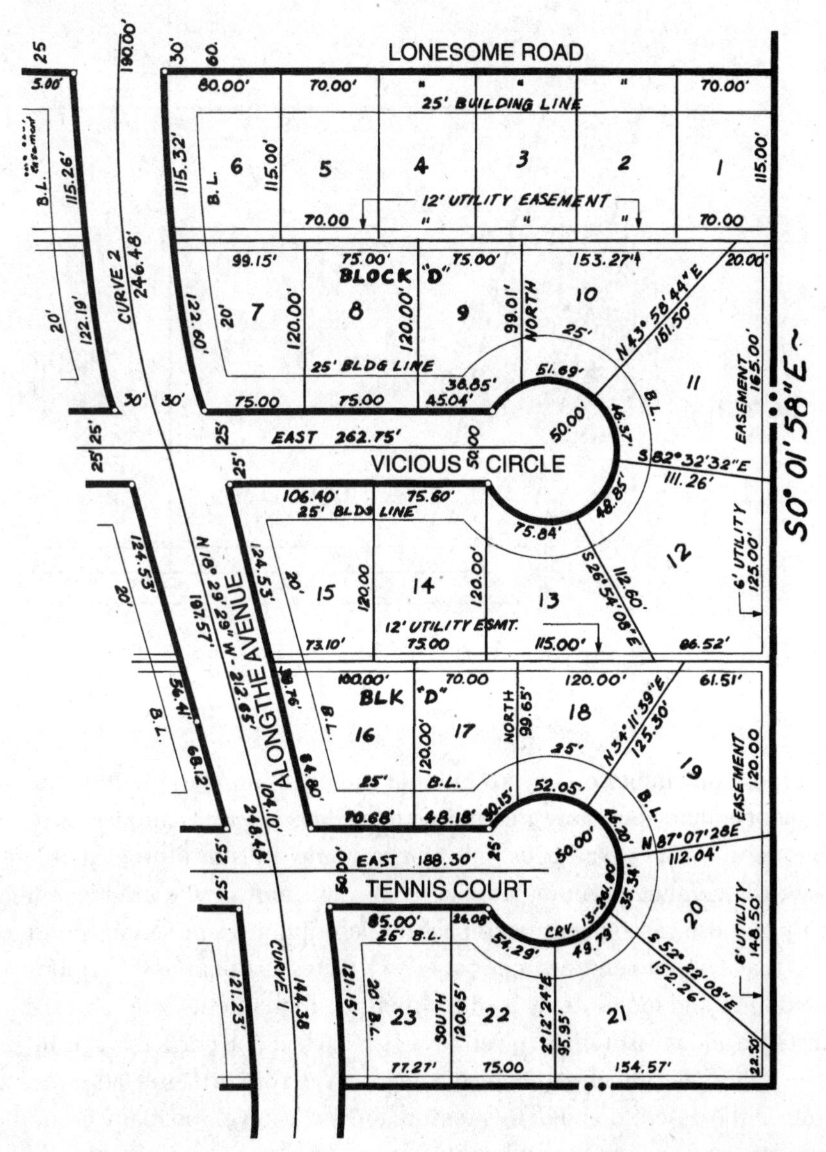

Figure 7-1. Block D, Stripmine Estates, Gotham City, Copperwire County, Montana

Often a metes and bounds description may include more language than in the fairly minimal example provided here. Common additional elements are as follows:

1. To define the street rights-of-way, lot lines of adjacent lots, and so on, along which directed lines travel. Example: ". . . thence one hundred feet (100.00′) north 63°30′0″ west along the southern right-of-way boundary of Thirtieth Street, thence one hundred feet (100.00′) north 88°0′0″ west along the southern boundary of Lot 44, ..."
2. To end each direction with the words ". . . to an iron pin . . ." or ". . . to a point . . ." or some other language suggesting that there is a mark on the ground to refer to. More often than not, there is no marker now in existence, though at the time the lot originally was laid out, many years ago, there indeed was. However, over the years these things disappear, erode, and otherwise become difficult to locate.

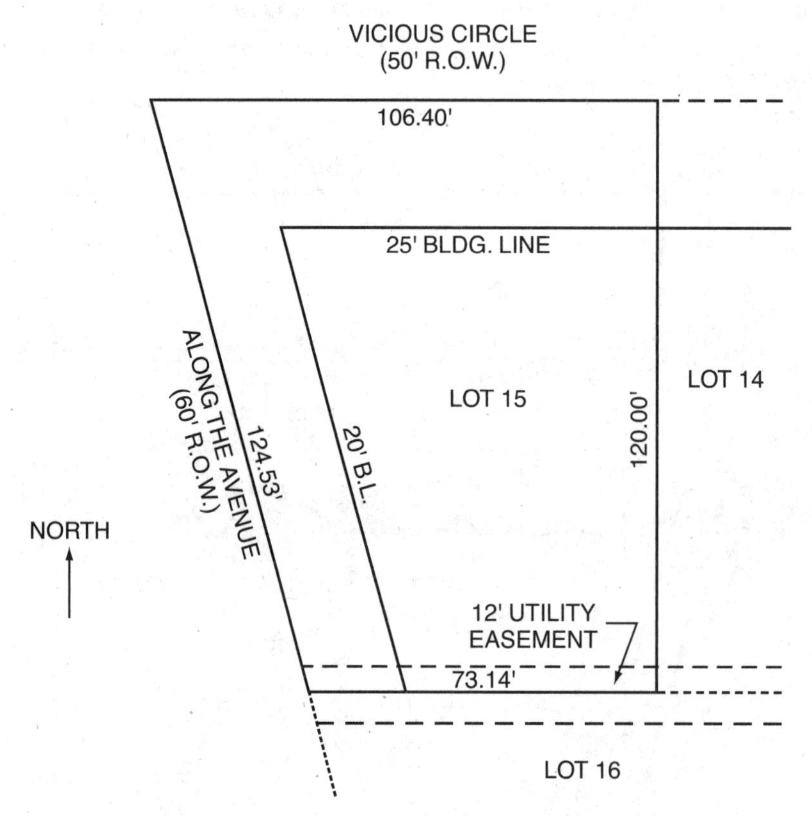

Figure 7-2. Lot 15, Block D, Stripmine Estates

3. To refer to actual things on the ground that can be found. If these things are relatively permanent and obvious, they are referred to as *monuments*. Examples: large and identifiable trees, visible streambeds, milestones, stone walls or other constructions, and occasionally genuine monuments placed at a location by the original surveyors for the very purpose of serving as references for descriptions. One serious point to keep in mind is that if monuments are mentioned in a metes and bounds description and if the directed lines and monuments conflict, then the monuments rule. This means that if the distance and measurement given say to go 200 feet due north to a certain milestone and the milestone actually is 193 feet, 6 degrees west of north, the latter measurement will rule in the description. For this reason, the inclusion of too many monuments in a description can sometimes cause confusion or error.

4. To add the statement "... containing x number of acres (square feet, etc.), more or less." This is convenient but unnecessary. Given the measurements and directions supplied by the description, it is always possible to calculate the area of the parcel if need be. The major reason for including the acreage is to protect the grantor against an error in the description that might indicate the transfer of much more (or much less) land. For example, if the description mentions a tract of 100 acres and the tract actually is only 10 acres, the inclusion of a statement of area would be an immediate indication that the description contained a mistake.

The point of beginning of a metes and bounds description must use a point of reference outside the description of the property itself. In urban areas, it is very common to use the nearest intersection of public roads or streets, as in the sample description given on page 164.

Distance may be expressed in any known unit of measurement, although the most common by far is feet. Directions are given according to the compass. However, in most parts of the United States there are two "sets" of compass directions: magnetic bearings and true bearings. Magnetic bearings refer to the magnetic poles; true bearings refer to the geographic poles. If magnetic bearings are used, then, in order to achieve true bearings, they must be corrected for the location at which the survey was made and for the date. The reason is that the magnetic pole is not at the same location as the true pole; furthermore, the magnetic pole "drifts" and so is located slightly differently at different times.

By convention, compass bearings, as shown in Figure 7-3, are described as numbers of degrees east or west (or south). There are 360 degrees in a full circle, 90 degrees in a right (square) angle. The symbol for *degree* is °, so 90° means 90 degrees. Each degree is divided into 60 *minutes* (symbol: ′), and each minute is composed of 60 *seconds* (symbol: ″). The statement N 44°31′56″ E is read as "North 44 degrees, 31 minutes, 56 seconds east."

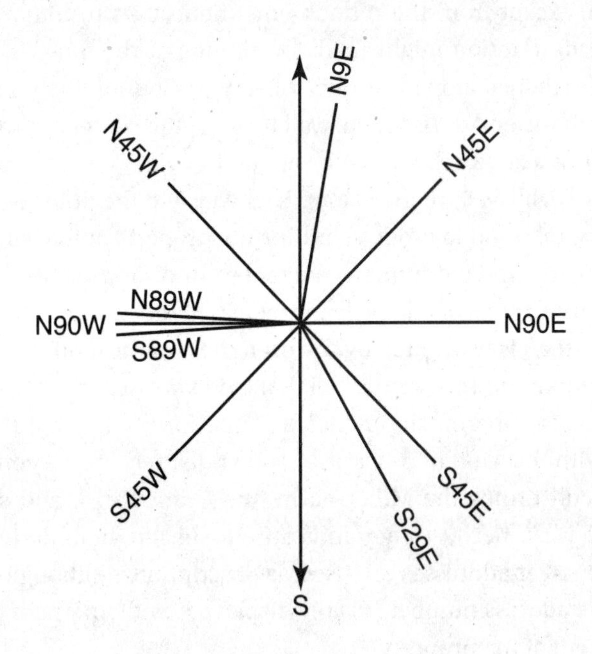

Figure 7-3. Compass Bearings

Lot and Block Number

Lot and block number descriptions are common in urban areas where *plats* of subdivisions must be recorded. A *plat* is a map of the manner in which land is divided into lots; Figure 7-1 is an example of a plat. If the plat is a matter of public record (i.e., has been recorded), it is quite sufficient to refer to it for the legal description of property it illustrates. For example, it would be enough to describe Lot 15 on that plat as "Lot 15, Block D, Stripmine Estates, Gotham City, Copperwire County, Montana." Given this description, a surveyor could look up this subdivision in the records, find the plat of Block D, and from it get all the information he needed to do an accurate survey of the lot.

Lot and block number descriptions are very simple to do and take little space. They are applied after an official plat of a subdivision has been recorded at the county courthouse. Therefore, in areas where the records are such that they can be used, lot and block number descriptions are very popular. It should be pointed out also that reference can be made to *any* recorded document in the preparation of a description. Thus, if lot and block descriptions are impossible, it may be possible simply to refer to a previous deed or other recorded instrument in which a description of the property appears. This should be done, however, only if the previous description is exactly the same as the one to be used at present. Even so, it is potentially inconvenient in that, whereas copies of plats are easily obtained and often can be received directly from developers and dealers, old deeds and other documents have to be dug out of the records.

Monument and Occupancy Descriptions

Monument descriptions are similar superficially to metes and bounds descriptions. The difference is that monument descriptions rely entirely upon monuments and specify no distance or direction, except from the monuments themselves to important points of the survey. A monument description might read: "Beginning at the intersection of Highway 9 and County Road 445, thence along the right-of-way of County Road 445 to the end of a stone wall; thence from the end of the stone wall through the center of a large, free-standing oak tree to the center of a stream bed; thence along the stream bed to the point where the stream bed intersects Highway 9; thence along Highway 9 to the point of beginning." Since this description can be followed in order to outline the property, it is a valid description, but there is always the risk that the monuments may move or disappear. These descriptions are used mainly for large rural tracts of relatively inexpensive land, where a full survey may entail much greater expense than is warranted by the usefulness of the land.

Occupancy descriptions are very vague: "All that land known as the Miller Farm"; "All that land bordered by the Grant Farm, Happy Acres Home for the Elderly, the Ellis Farm, and State Highway 42." With this kind of description, one must rely on the community's general impression of what constitutes the Miller Farm, the Grant Farm, and so on. Occupancy descriptions are very weak because they indicate no specific boundaries at all. A similar situation arises when street addresses are used as descriptions. Although houses and other buildings usually have address numbers visibly displayed on them, these give no indication at all of the actual extent of the property.

Rectangular Survey Descriptions

The U.S. government rectangular survey is used in the states indicated in Figure 7-4. This survey divides the land into squares 6 miles on a side; these are called *townships.* Each township is identified with respect to its distance from the *base line* and the *principal meridian.* There are several sets of principal meridians and base lines throughout the country, so each principal meridian carries a name or number to distinguish it from all the others. Each principal meridian has a single base line paired with it.

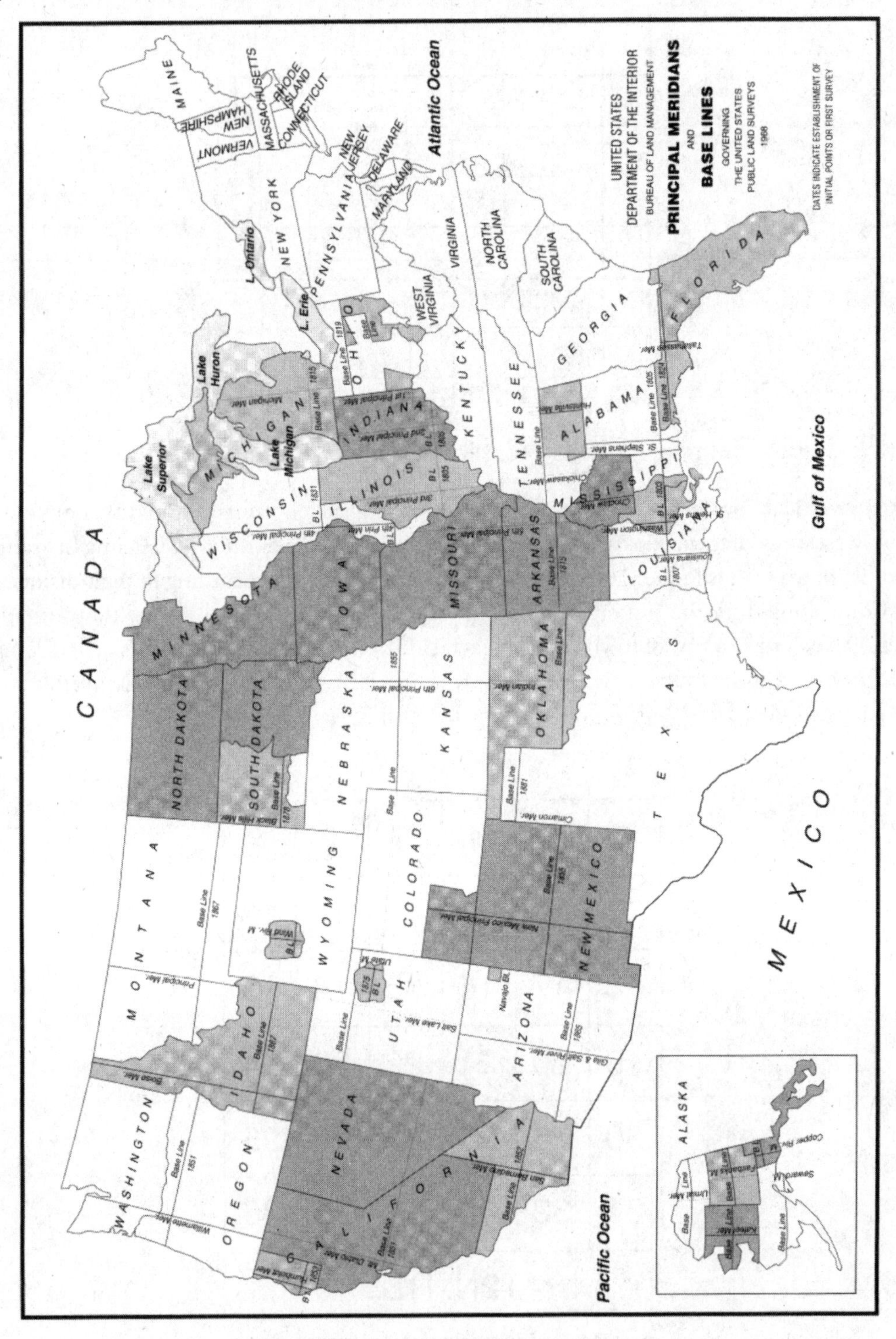

Figure 7-4. Government Rectangular Survey

Principal meridians are imaginary lines that run north-south. Base lines are imaginary lines that run east-west. Parallel to the principal meridians and 6 miles apart from each other are other *meridians.* Parallel to the base line are other lines called *parallels*; these are also 6 miles apart. Altogether, this system of north-south meridians and east-west parallels cuts the map up into a grid of 6-mile squares, as in Figure 7-5.

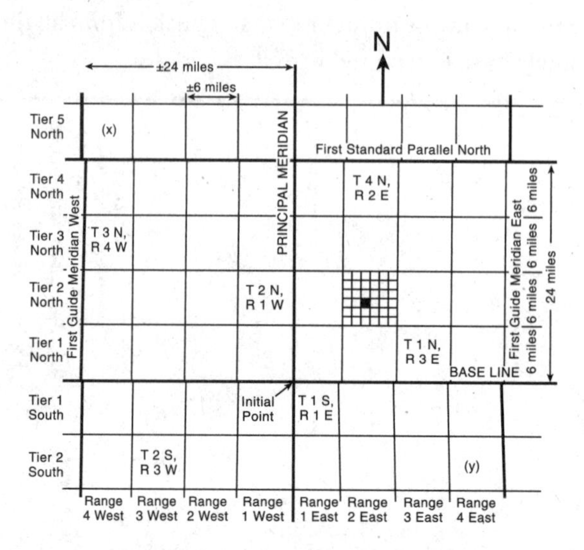

Figure 7-5. Six-Mile Square Grid

The vertical (north-south) rows of townships are called *ranges,* and the horizontal (east-west) rows of townships are *tiers.* Each township can be identified by labeling the range and tier in which it is found. The ranges and tiers are numbered according to their distances and directions from the principal meridian, or the base line. For example, the principal meridian will have a row of townships on each side of it. The one to the east is called *Range 1 East,* while the one to the west is *Range 1 West.* Similarly, the tier immediately north of the base line is called *Tier 1 North,* while the one below it is *Tier 1 South.*

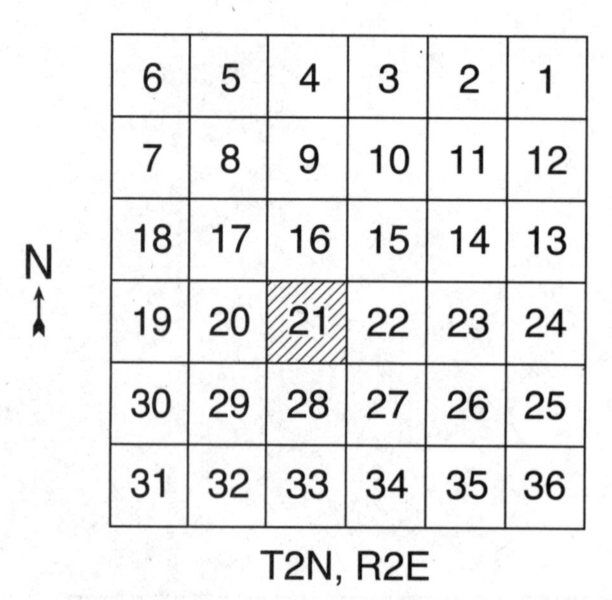

Figure 7-6. Numbering of Township Grids

LIMITS OF DESCRIPTIONS

It is important to understand that descriptions, while they may accurately describe land, do not provide any guarantee of *title* to that land. A deed that contains a perfectly usable description of a tract of land may be worthless if the grantor has little or no right to convey title to that particular tract. The mere fact that a deed or contract of sale contains a particular description does not guarantee to the buyer that his claim to the described land will hold up in court. It can only do so if the seller actually had the legal right to transfer the land she described. One cannot sell, give, or otherwise bargain away what one does not own.

Trick Question

A trick question refers to the use of the house number, street, city, and state on a deed to identify a property. This is inadequate information. Instead, the deed should have an accurate property description. A postal address is not adequate for a deed.

QUESTIONS ON CHAPTER 7

1. The number of acres in 1 square mile is
   (A) 640
   (B) 5,280
   (C) 1,000
   (D) 160

2. The best way to determine whether there is a boundary line encroachment is
   (A) appraisal
   (B) survey
   (C) title search
   (D) environmental investigation

3. "All that land known as the Jones Farm" is which kind of description?
   (A) invalid
   (B) occupancy
   (C) metes only
   (D) rural

Questions 4–7 refer to the lot shown in Figure A.

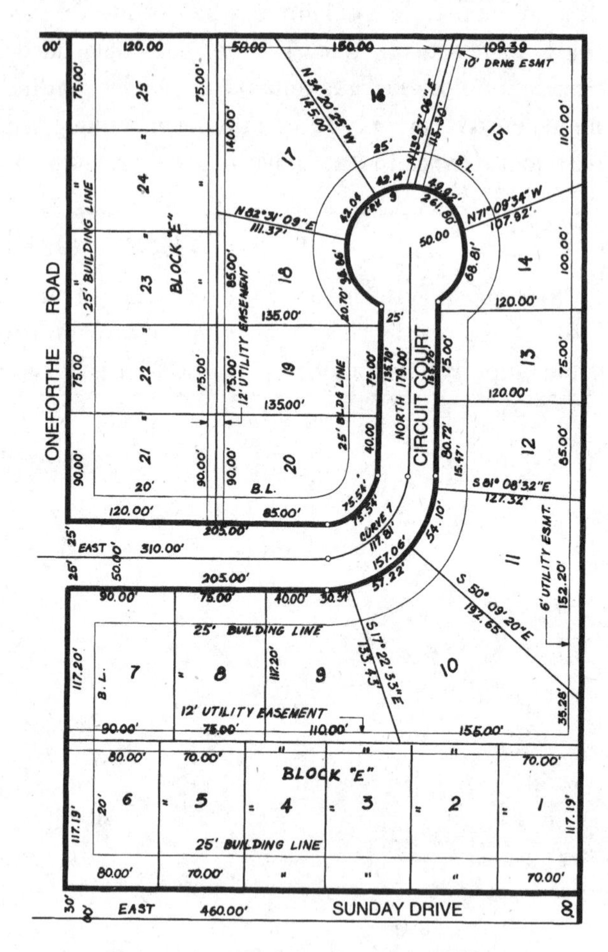

Figure *A.* Block E, Quagmire Village, Metropolis, Elko County, Nevada

4. Which lot has 80 feet of frontage on Sunday Drive?
   (A) Lot 6
   (B) Lot 5
   (C) Lot 4
   (D) Lot 3

5. Which lot has the most street frontage?
   (A) Lot 7
   (B) Lot 18
   (C) Lot 20
   (D) Lot 21

6. How many lots border on Circuit Court?
   (A) 9
   (B) 15
   (C) 16
   (D) 21

7. How many lots front on more than one street?
   (A) 1
   (B) 2
   (C) 3
   (D) 4

8. A parcel of land is square, ½ mile by ½ mile. How many acres does it contain?
   (A) 40
   (B) 160
   (C) 320
   (D) 640

9. A township is _____ square mile(s).
   (A) 1
   (B) 4
   (C) 36
   (D) 64

10. Base lines in the rectangular survey method run
    (A) north and south
    (B) northeast
    (C) northwest
    (D) east and west

11. The POB in a metes and bounds survey is
    (A) period of basis
    (B) position of boundary
    (C) point of beginning
    (D) pier of boarding

12. The perimeter is 6 miles for each side. The number of townships is
    (A) 1
    (B) 2
    (C) 3
    (D) 4

13. Principal meridians are imaginary lines that run
    (A) north-south
    (B) east-west
    (C) northeast
    (D) southwest

14. Developers subdivide land and assign lot and block numbers to identify lots in the subdivision. This is a legal description known as
    (A) a block party
    (B) recorded plat description
    (C) developer plat
    (D) lot survey method

15. In the rectangular survey method, a township is
    (A) 36 square miles
    (B) one square mile
    (C) a small city
    (D) 640 acres

16. What is the type of legal description that uses imaginary lines of meridians and base lines?
    (A) metes and bounds
    (B) recorded plat
    (C) lot and block
    (D) rectangular survey

17. A metes and bounds description
    (A) provides a point of beginning, which is also the end
    (B) is not used in deeds
    (C) must include a monument
    (D) is less precise than a description based mostly on monuments

18. Why do property owners sometimes ask surveyors to plant stakes in the ground?
    (A) helps to locate boundaries
    (B) makes property marketable
    (C) provides exact dimensions
    (D) assists tax assessor

19. What is the best method to determine whether there is an encroachment?
    (A) get a survey
    (B) read a title policy
    (C) check the deed
    (D) get an appraisal

20. A legal description of property is prepared by a(n)
    (A) appraiser
    (B) attorney
    (C) broker
    (D) surveyor

21. The metes and bounds description has
    (A) a base line
    (B) monuments
    (C) lot and block numbers
    (D) a point of beginning

22. The rectangular survey system includes
    (A) base lines
    (B) meridians
    (C) townships
    (D) all of the above

23. The lot and block method is used in
    (A) farm and ranch land
    (B) life estates
    (C) blockchain computations
    (D) subdivisions

24. A section of land contains ________ acres.
    (A) 10
    (B) 100
    (C) 460
    (D) 640

25. The number of square feet in an acre is
    (A) 34,560
    (B) 43,560
    (C) 54,650
    (D) 56,560

26. An encroachment is a(n)
    (A) leased fee
    (B) variance
    (C) nonconforming use
    (D) unauthorized use of another's property

ANSWERS

1. **A**	8. **B**	15. **A**	22. **D**
2. **B**	9. **C**	16. **D**	23. **D**
3. **B**	10. **D**	17. **A**	24. **D**
4. **A**	11. **C**	18. **A**	25. **B**
5. **D**	12. **A**	19. **A**	26. **D**
6. **B**	13. **A**	20. **D**	
7. **C**	14. **B**	21. **D**	

Contracts of Sale

8

ESSENTIAL MATTERS TO LEARN FROM THIS CHAPTER

- A contract for the sale of real estate must contain all the elements required of any other contract to be legally valid:

  1. Mutual agreement (offer and acceptance).
  2. Consideration.
  3. Legally competent parties.
  4. Lawful purpose.
  5. Description adequate to identify property.

- In addition, the *Statute of Frauds* is a law that **requires a contract of sale for real estate to be in writing to be enforceable**.
- Various names are used by local custom for a contract to sell real estate. These names include "contract of sale," "agreement of sale," "purchase and sale agreement," "offer and acceptance," and "earnest money contract."
- A *Contract of Sale* precedes the settlement (or closing), at which the buyer typically provides payment in direct exchange for a deed to the property delivered by the seller. This time interval is necessary because of time required for the buyer to research title, arrange financing, and perform inspections.
- In many states, a specific form is required to be used for a sale by a licensed broker or salesperson. If that form is not used or if the specific form requires more than minor changes, the broker may not make the changes but, instead, must have a form drafted by an attorney.
- A valid contract of sale will typically include:

  1. Interests and identities of the parties.
  2. Description of the real estate.
  3. Price and any financing required.
  4. Contingencies to be resolved and other terms of sale.
  5. Settlement date.
  6. Signatures of buyers and sellers.

- To avoid disputes, a *legal description* of the property (see Chapter 7) should be included in the contract. However, the minimum description is one sufficient to locate the property.
- The price should state the total to be paid and, if financing is to be arranged, the amount, terms, and type of loan that will be sought for the contract to be acceptable.

- *Contingencies* include such things as inspections (property condition including structural, mechanical, electrical, plumbing, termites, environmental) and the buyer's need to do something else, such as arrange financing or sell another property. The contract should specify who is to repair or pay for repairs and what may happen if the cost of repairs exceeds specified amounts.
- Unlike prices set for merchandise in a retail department store, prices and terms of real estate are *negotiable.*
- The negotiation process typically starts with a prospective buyer who presents an *offer,* often through a broker, to an owner. The offer may be withdrawn anytime before it is accepted. The owner may sign the offer as it is presented, making no changes, called *acceptance,* which creates a contract.
- The owner may decide not to respond to an offer, in which case the offer is *void* on its expiration date or, if earlier, when withdrawn by the prospective buyer.
- The owner may change something in the offer. The slightest change constitutes a *rejection* of the offer and substitution of a new offer to the other party, called a *counteroffer.*
- This process may continue back and forth until an acceptance occurs, at which point a *contract* exists, or until a party decides to break off negotiations and no longer pursues a contract.
- *Due diligence* is the process a buyer performs to be certain that the property's condition is as described, with no surprises as to its physical, financial, or legal attributes.
- A *settlement* (closing) date is usually stated in offers, often with the phrase *time is of the essence* so that the deadlines are interpreted exactly. The deadline (or other requirements in a contract) may be amended by mutual agreement of the parties.
- A *power of attorney* (POA) is used to give another person authority to act. The person with such authority is called an *attorney-in-fact.* The party need not be an attorney-at-law. If the intent is to transfer real estate, the POA should be in writing to be enforceable.
- *Earnest money,* also called *hand money,* is the deposit given by a buyer to a seller (or a broker if one is involved) to show good faith. In most situations, the seller keeps the earnest money if the buyer defaults. However, if the contingencies stated in the contract are not met, the buyer is entitled to a full refund of the earnest money.
- *Liquidated damages* are sometimes stated in a contract to define the penalty for failure to complete the contract as specified.
- *Specific performance* is a legal remedy sought by a buyer to compel the seller to complete the contract, often enforced in court, and allowed because each piece of real estate is unique.
- Contracts for the sale of real estate remain valid on the estate of an owner who died (called a *decedent*), as are leases, deeds, and mortgages. Listings normally terminate on the death of either party, the owner, or the broker.
- Contracts for real estate are normally *assignable,* that is, transferrable to someone else. The *assignor* (party who assigned the contract) is not relieved of the obligation unless the opposing party agrees to such a release. The document used to release a party to a contract and substitute another is called a *novation.*

CONTRACTS OF SALE OF REAL ESTATE

A contract for the sale of real estate describes the rights and obligations of the purchaser(s) and seller(s) of real estate. Days, weeks, or months later, at a ceremony called a *closing* or *settle-*

ment, the seller(s) will deliver a deed that transfers legal title to the property; the purchaser(s) will provide payment for the property as described in the contract of sale.

A contract of sale for real estate is "*the deal.*" This means that all the provisions that the parties want, all the "ifs, ands, or buts," must be set forth in this contract. All the negotiation, "wheeling and dealing," "give and take," and such should occur and be settled *before* the contract of sale exists. Once there is a contract of sale, it is too late for any more negotiation. All parties to the agreement must abide by it, and none can change anything in it unless all parties agree to the change. Therefore, it's important that the parties to this contract make sure that all the conditions and requirements that they want with regard to the sale and purchase of the real estate are contained in the sale contract.

Preparing Contracts of Sale of Real Estate

It is possible to sell property merely by delivery of a deed in exchange for payment and thus bypass the contract of sale. However, the parties, especially the buyer, face a lot of risk. A title search must be performed to satisfy the buyer that the seller can give ownership, and time is required for such a search. The buyer may also need time to arrange financing and inspections. Most lenders will consider real estate loans only after they have studied the contract of sale. In short, it is rare for a transfer of real estate to be completed without such a contract. Furthermore, it is of paramount importance that all parties involved in a transaction understand their contractual obligations and privileges.

Upon completion of the contract of sale, the purchaser (*vendee*) receives an "equitable" title. Equity, in law, attempts to treat things as they should be to be fair and considers the purchaser to have "equitable title," although legal title remains with the seller (*vendor*) until closing. Therefore, if there is a casualty loss—a fire, for example—before closing, the risk may be upon the purchaser. However, considering that equity is intended to be fair, the risk of loss before closing should not pass automatically to the purchaser in all instances. There are situations in which the seller should be responsible and courts will hold her to be so. To avoid potential problems concerning risk of loss, the contract of sale should specify who has the burden of loss prior to closing and which party is obligated to complete the transaction and to receive the proceeds from insurance.

The *Statute of Frauds* requires that contracts for the sale of real estate be in writing to be enforceable. The *parol evidence* rule prevents oral testimony from being introduced when it conflicts with better (written) evidence. Consequently, care should be exercised in the preparation of a contract of sale.

A contract of sale should contain any *contingency clauses,* or *contingencies,* that the parties want. These "what if?" provisions often allow a buyer or seller to cancel the deal without penalty if something does or doesn't happen (the *if* part). For example, the seller may have her home on the market because her employer is transferring her, but wants to be able to cancel the deal *if* the transfer falls through. A buyer may need to sell his own home in order to be able to buy the seller's property, so he wants to be able to cancel the deal *if* he can't close the sale of his own property within a reasonable time. A great many homebuying contracts of sale are *contingent upon financing*; this means that the buyer can cancel the deal *if* he is unable to get a suitable mortgage loan. Many preprinted contracts of sale have blank spaces for common contingencies already printed in them: The buyer can cancel the deal *if* the seller can't provide clear title, *if* the property is severely damaged before closing, *if* termite infestation is found. The seller can cancel the deal *if* the buyer doesn't get financing approval within a certain time.

Many states require that the seller provide the buyer with a *disclosure* of certain conditions, defects, and so on, of the property; if there aren't any, then the seller must provide the buyer with a statement that no such conditions exist. Usually the state mandates a certain form upon which the disclosure is made. If possible, the disclosure is made available to a prospective buyer before an offer is made, sometimes when the first contact between prospect and seller and/or listing broker occurs. In a brokered transaction, such disclosure normally will be provided to the broker when the property is listed. The broker then will provide a copy of it to all potential buyers who examine the property. If the transaction is not brokered, it is the seller's responsibility to provide the disclosure to all potential buyers.

A deposit, or *earnest money*, is not necessarily a legal requirement of a contract of sale (the parties' promises to perform in the future usually are sufficient as the required consideration). It is very common for the buyer to put down a cash deposit at the time the contract of sale is agreed upon. The function of earnest money is to serve as evidence of his good-faith intentions to perform under the contract. Some states, however, are fussier about the term *earnest money* and specify that it has certain meaning. For example, in some states, if the deposit is called "earnest money," the buyer may abandon the contract at no penalty, except forfeiture of the deposit, and/or the seller can abandon the contract by returning the buyer's deposit and, usually, an additional sum (such as an amount equal to the buyer's earnest money). In these states, therefore, in order to require that buyer and seller go through with the contract, any deposit must be specified to be something (such as a prepayment toward the purchase price) that is *not* what the state defines as earnest money. A *binder* is used in some states as a preliminary agreement. It contains minimal features of the deal and is expected to be followed in a few days by a contract of sale. Check the law in your state.

REQUIREMENTS OF A VALID CONTRACT OF SALE

To assist you in learning about contracts of sale, get a copy of a preprinted contract from a local real estate brokerage office or buy a generic one for your state. They are usually available online. You might also be able to download one from the state real estate commission's website. Note that in some states, these contracts may be called something different from "contract of sale," such as "agreement of sale." In Texas, it is an "earnest money contract." Don't worry about the name, though. Whatever they are called where you live, they still are generically contracts of sale.

Like any contract, contracts of sale must conform to the general requirements of contracts described in Chapter 6. Within contracts of sale, there are six essentials that also must be observed:

1. The interests and identities of the parties
2. Description sufficient to locate the real estate
3. The price and how it is to be paid
4. Contingencies and other terms of the sale
5. Settlement date
6. Signatures of the parties

1. Interests and Identities of the Parties

Who is (are) the buyer(s)? Who is (are) the seller(s)? These are the parties to the contract. The name(s) of the seller(s) should be as they are on their recorded deed for the property. The name(s) of the buyer(s) should be as they will be on the deed to be delivered.

2. Legal Description of the Real Estate

The real estate has to be described; methods of describing real estate were presented in Chapter 7. Simple descriptions (lot and block number descriptions; uncomplicated rectangular survey descriptions) should be written into the contract, but whoever is writing it should take special care to make sure the description is *accurate.*

If the description is complicated, the chances of error are high, especially with metes and bounds descriptions with all their numbers, directions, and so on. A complicated description often won't be written into the contract but will be *incorporated by reference.* Sometimes an *existing* accurate description will be provided as an attachment. Most commonly, the referred document will be one that already is in the public records; the obvious one is the existing recorded deed of the sellers. (If theirs also incorporates a description by reference, then you refer to whatever document that deed refers to.) Because the public records are open to anyone, it is easy to find the referred document and get the description from it. An example of a reference description could be ". . . all that land described in deed from X to Y, dated June 17, 1941, page 133, book 122 of the Coppermine County, Montana, public records." This tells anyone who reads the contract exactly where to find the description. The description by reference can be used only if the real estate currently being transferred is *exactly the same* as described in the referenced description. If any changes have been made to boundaries of the land, then a new description must be written.

Note that land descriptions don't contain any information about improvements to the land (house, drive, etc.). That's okay; remember, real estate is *land and all permanent attachments to it.* So if you acquire the described land, you're also getting whatever is on it.

When personal property is sold, a *bill of sale* is the document used. Often the buyer purchases furniture or portable appliances using a bill of sale.

3. The Price and How It Is to Be Paid

The exact price must be described. Usually a buyer doesn't have all the cash to pay for the property and must secure a mortgage loan to be able to complete the purchase. The contract should describe the actual cash amount to be provided by the buyer(s) and the amount of borrowed money that they will provide. If the seller is willing to accept an IOU for all or part of the purchase price (a *purchase money mortgage,* or what we call *seller financing*), that should be included, too. It would be wise for the buyer(s) to insist on including the desired terms of the mortgage loan they need: the maximum interest rate they will pay, the term (15 years, 30 years, etc.), and any other loan terms they consider important—as well as a stipulation that if they can't find the desired loan within a reasonable time, they have the right to cancel the purchase with no penalty.

4. Contingencies and Other Terms of the Sale

These are the "what-ifs" and other specific requirements that the parties agree to. We described one contingency above: "what if" the buyers can't get the mortgage loan they want? Then they can back out of the deal without losing anything. Many of these are written into standard contract forms: the seller must provide "good and marketable" title; the buyer must apply for a loan within ___ days. What if the buyer wants to move in *before* the settlement date (or the seller wants to stay on for a while *after* the settlement date)? Should rent be paid? How much? Common requirements include an appraisal and inspections for property conditions (electrical, mechanical, plumbing, and the like), structural condition, and absence of wood-destroying insects. A *walk-through* inspection is often provided within 24–48 hours before closing.

5. Settlement Date

A *settlement date* (in some places it's called the *closing date*) must be in a contract of sale. This is the deadline by which the contract requirements should be completed, the deadline by which the buyer must pay for the property and the seller must provide the deed. Without a settlement date, there isn't a requirement that the contract be completed; either party could dawdle all they want. Most contracts include the phrase "time is of the essence." This means that references to deadlines are to be interpreted exactly. If it turns out that a date in the contract is too optimistic (the parties don't have time to complete everything they have to do), the settlement date can always be amended to a later one, by mutual agreement.

6. Signatures of the Parties

All parties to a contract must sign it; this is the legal way of showing that they all agree to it. Signatures to a contract of sale don't have to be notarized.

If you look at a contract of sale, you may see a section in which the commission to the broker, and when it is to be paid, are described. (Some generic forms include this part, too.) This section actually is a separate contract between seller(s) and broker. It isn't really necessary because the listing contract already obligates the seller to pay the broker, but most brokerage firms include it in their contract-of-sale forms anyway.

Preprinted contracts of sale are almost always used for sales of homes. Suppose there is something already written into the preprinted contract, but the parties *don't* want that part to apply to them? Then they simply cross out that part and have the parties initial or sign near that part of the contract form to acknowledge that they agree to the alteration.

In some states, a form promulgated by the real estate commission must be used by licensees, especially in residential transactions. If anything more than a minor change to a form is used, the contract must be prepared by an attorney. Otherwise, the broker might be accused of practicing law without a license.

A contract of sale that is not yet signed is referred to as *executory*. After the sale has closed, it is called *executed*. A principal (buyer or seller) may give signature authority to someone else using a power of attorney. Such person is called an attorney-in-fact.

Trick Question

A trick question asks whether a person must be an attorney-at-law to have a power of attorney. The answer is NO! You can give any competent person the authority to act for you by using a form called *power of attorney*. That party becomes an attorney-in-fact but need not have a license to practice law. This person can act for you in the situations you agreed to and listed on the power of attorney form.

NEGOTIATION (OFFER, COUNTEROFFER, ACCEPTANCE)

Contracts of sale (or any other kind) just don't pop out of the air, ready to go. They are the result of *negotiation*. Negotiation of a contract of sale for real estate usually begins with an *offer* from an interested potential buyer. (The person making the offer is the *offeror*; the person receiving it is the *offeree*.) The buyer probably knows what the seller wants (price, terms, etc.). Very often, though, the buyer won't agree to everything, so the offer will be somewhat different from what the seller desires. The seller will consider the buyer's offer and can accept it, turn it down, or turn it down and make a "counteroffer" back to the buyer. Now it's the buyer's turn; she may accept the seller's offer, turn it down, or make another offer back to the seller. This can go on for several cycles. Eventually either the parties will agree on a final deal, and a contract will be made, or they may decide that it's hopeless and will stop negotiating and go their own ways.

A key element of the negotiation process is that *no one involved is obligated to anyone until a contract exists*. No one is *required* to negotiate; no one is *required* to keep on negotiating once the process has begun. Anyone involved may terminate the process simply by refusing to go on with it. The law lays down some simple rules for the negotiation process.

Once an offer is *accepted*, it isn't an offer anymore: it has become a *contract*. Once a contract exists, the negotiation process ends; because the parties have finally come to an agreement, there's nothing more to discuss. If a party later on decides that he wants some more conditions, it's too late. The contract already exists, and he must abide by it. Of course, he can always approach the other parties and ask to amend the existing contract, but the other parties have no obligation to revise the contract.

During negotiation, only one offer can exist at any given moment in time. If A makes an offer to buy B's real estate, an offer exists. It is up to B to decide what to do.

1. B can accept A's offer and thus create a contract.
2. B can reject A's offer, in which case the offer no longer exists.
3. B can make a counteroffer back to A, in which case A's offer no longer exists. Now there is *another* offer on the table, from B to A. Making a counteroffer therefore means that two things are being done simultaneously: the offer that was "on the table" is *rejected* and a new offer is made "the other way."

There's one other thing going on. An offer may be withdrawn *at any time until it has been accepted*. While B is pondering A's offer, A can, at any time, notify B that the offer is withdrawn. If that happens, that's the end of the offer: it no longer exists. But if, before then, B notifies A that he accepts the offer, the offer doesn't exist anymore *because it has become a contract*. You can't "withdraw" a contract.

If an offer is rejected, it doesn't exist anymore and can't be "resurrected" later on. Consider the following conversation:

JANE: Mike, I'll give you $150,000 for your house.
MIKE: No, I need at least $175,000.
JANE: Forget it. That's too much.
MIKE: Okay, I'll take the $150,000.

Is there now a contract for Jane to buy Mike's house for $150,000?

NO! Let's take it step by step. First, Jane made an offer to buy Mike's house for $150,000. Mike then counteroffered the house to Jane for $175,000. By doing this, he *rejected* Jane's offer for $150,000 and then replaced her offer to him with *his offer to her* for $175,000. Jane then rejected Mike's offer to her. But she didn't make any other offer back to him. At that point there was no offer on the table at all. When Mike said he'd "take the $150,000," no contract came into being because there wasn't an offer from Jane anymore for him to accept. He had already *rejected* Jane's original offer, so it didn't exist anymore. His statement "Okay, I'll take the $150,000" actually creates a *new* offer *from him to Jane* to sell her the house for $150,000. She doesn't have to accept it. If she does, then Mike and Jane will have a contract. If she says something like "I've changed my mind, I don't want your house," then she has, in turn, rejected Mike's new offer, and there's no offer on the table.

This is a very important aspect of the law of offer and acceptance: *until a contract exists no one has any obligation to anyone else!* If you look at a certain preprinted contract of sale, you may see language on the order of ". . . this offer will remain open until ________." You're supposed to put a date, or a time of day and a date, in the blank space. What does this mean? Must whoever made the offer sit and wait until the deadline to see if the other party accepts the offer? Is that other party guaranteed that length of time to ponder the offer? *NO!* Because this is an offer, no one is obligated to anything. The offeror can withdraw the offer at any time until the offeree accepts it. The only interpretation is that if the deadline arrives and the offeree hasn't responded to the offer, then the offer will automatically be *withdrawn* at that time. In effect, this provision is putting a limited lifetime on the offer and nothing more.

Many real estate agents are apparently unaware of this part of the law. There are frequent situations when someone wanted to withdraw an offer (maybe they got a better one in the meantime) and the licensee told them, "You can't do that; you've agreed to leave the offer open until that time." This is wrong and could expose the licensee to a lawsuit.

If an offer is accepted by the offeree, it doesn't become a contract until the offeror (or the offeror's agent) has been *notified* of the acceptance. The law says you can't be held to a contract if you don't know that it exists. Because notifying an agent is the same as notifying that agent's principal, if a buyer tells the listing agent that he or she will accept a counteroffer from the seller, the contract exists at that point. Or, if a buyer's agent is notified that the seller accepts an offer from the buyer, there is a contract. (In either case, a professional licensee will immediately call his or her principals to notify them of the acceptance.) Notification doesn't have to be in writing; it can be oral, such as by phone.

Finally, it is common practice, especially in brokered transactions, for offers between a potential buyer and seller to be made using contract-of-sale forms. Typically, a buyer wants to make an offer on a home. The buyer (or agent involved) will fill out a contract-of-sale form, with all the terms that the buyer wants written into it. The buyer signs it, and it is submitted to the seller. If the seller accepts the offer, he signs it, and a contract exists once the buyer is notified. The seller presumably is happy because he got an offer he was willing to accept. The buyer is happy because the seller has agreed to all the terms, conditions, price, and so on, that she wanted. Most of the time, though, the offeree does *not* agree with all the terms of the offer. If a counteroffer is made, it is common practice to cross out the parts the offeree doesn't like, write the desired changes in the margins of the form, and add anything else that is desired the same way. The offeree initials or signs by all those changes, signs the document, and submits it to the offeror. (Because the counteroffer rejects the original offer, the original offeror and offeree now have changed places.)

Of course, the new offeree may not like all the changes and might then cross out more stuff, write in more stuff, initial or sign *those* changes, and send it back as a new offer. This process can go back and forth again and again; the cumulative marginal changes and cross-outs can result in a contract form that is becoming an unreadable mess. When it gets to that point, it certainly is good practice to prepare a fresh form with the terms and conditions written into it reflecting the current state of the negotiations.

PARTS OF A REAL ESTATE CONTRACT OF SALE

A contract of sale for real estate will normally contain certain parts. These parts and their purposes are as follows:

PART	PURPOSE
Parties	States the names of the seller (vendor) and buyer (vendee).
Property description	Provides a legal or an adequate description of the subject property and the fixtures included.
Sales price	States the amount to be paid at closing in cash and from the proceeds of financing.
Assignment	Describes whether a party can assign the contract to another party. Assignment does not relieve the assignor of obligation to perform.
Financing conditions	Purchaser stipulates acceptable conditions of financing, including the type of loan, loan-to-value ratio, interest rate, monthly payments.
Earnest money	Notes the amount of the good-faith deposit presented to the seller or his agent. Also called *hand money* or *escrow money.*
Type of deed	Describes the type of deed to be given at closing (see Chapter 9).
Title evidence	Describes the type of evidence to be considered satisfactory assurance that the seller can give title.
Title approval	Describes the rights of purchaser and seller in the event that the purchaser objects to a flaw in the title.
Property condition	States whatever changes (if any) are required to be made to the property, inspections, maximum cost of repairs a party must pay.
Prorations	Calls for the proration of prepaid or unpaid taxes, insurance, rents, etc., between buyer and seller (see Chapter 18).
Loss	Affixes responsibility for casualty losses until closing.
Possession	Establishes a date for possession.
Closing	Establishes a date for closing.
Broker	Acknowledges the existence of a real estate broker, specifies the commission rate, and identifies the principal who will pay the commission.
Default	Describes the obligations of parties in the event one fails to perform.

PART	PURPOSE
Contingencies	Describes any "what if" provisions of the contract and the conditions under which the parties may cancel the contract without penalty.
Miscellaneous provisions	States whatever other provisions buyer and seller agree upon.
Signatures	Principals, brokers, escrow agents.

A real estate owner may give another person a power of attorney (POA) to sell real estate. To be valid for the sale of real estate, the POA must be in writing. The POA expires upon the death of the principal. However, if the property was under a contract of sale before the principal's death, the contract remains valid.

Contracts for the sale, lease, or mortgage of real estate remain binding on heirs after a person's death. A listing expires with the owner's death.

QUESTIONS ON CHAPTER 8

1. Joe offers to buy Susie's house for $225,000. Susie examines Joe's offer and then offers to sell her house to Joe for $250,000. Which of the following is true?
   (A) An offer from Susie to Joe now exists.
   (B) No offer exists.
   (C) Two offers exist: one from Joe to Susie and one from Susie to Joe.
   (D) A contract between Joe and Susie now exists.

2. The law that requires contracts for the sale of real estate to be in writing to be enforceable is the
   (A) Statute of Frauds
   (B) Statute of Limitations
   (C) Statute of Liberty
   (D) parol evidence rule

3. A person who has the power to sign the name of his principal to a contract of sale is a(n)
   (A) special agent
   (B) optionee
   (C) tenant in common
   (D) attorney-in-fact

4. A valid contract of purchase or sale of real property must be signed by the
   (A) broker
   (B) agent and seller
   (C) seller only
   (D) buyer and seller

5. Hand money paid upon the signing of an agreement of sale is called
   (A) an option
   (B) a recognizance
   (C) earnest money
   (D) a freehold estate

6. Joe made an offer to buy Susie's house. The offer says, "This offer will remain open until noon, October 22, 2019." It is now 3 P.M., October 21, 2019. Which of the following is true?
   (A) If Susie does not respond by the deadline, the offer will become a contract.
   (B) Susie must wait until after noon on October 22, 2019, before she can submit a counteroffer.
   (C) Joe must give Susie until noon, October 22, 2019, to consider his offer.
   (D) If Joe has changed his mind, he may now withdraw the offer.

7. A seller of real estate is also known as the
   (A) vendee
   (B) grantor
   (C) vendor
   (D) grantee

8. In any real estate contract of sale there must be
   (A) an offer and acceptance
   (B) a leasing arrangement
   (C) a mortgage loan
   (D) prepayment of taxes

9. If, upon receipt of an offer to purchase under certain terms, the seller makes a counteroffer, the prospective purchaser is
   (A) bound by his original offer
   (B) bound to accept the counteroffer
   (C) bound by the agent's decision
   (D) relieved of his original offer

10. A contract that has no force or effect is said to be
   (A) voidable
   (B) void
   (C) inconsequential
   (D) a contract of uncertain sale

11. To be enforceable, a contract of sale must have
    (A) the signature of the wife of a married seller
    (B) an earnest money deposit
    (C) competent parties
    (D) witnesses

12. A party to a contract of sale normally has the right to
    (A) change the provisions in the contract if it is convenient to her
    (B) change the provisions in the contract if it will save her money
    (C) back out of the contract if she finds a better deal
    (D) none of the above

13. Which of the following would *not* be a contingency clause in a sale contract?
    (A) The settlement will be within 90 days.
    (B) The seller can cancel the deal if the buyer cannot get suitable financing.
    (C) The buyer can cancel the deal if Aunt Minnie won't lend him $10,000.
    (D) The deal is off if the property is infested with termites.

14. Which of the following is *not* true?
    (A) A contract of sale must contain a legal description of the land being sold.
    (B) An unwritten contract of sale is not enforceable in court.
    (C) Once a contract of sale is signed, there is no way to avoid going through with the deal.
    (D) A contract of sale should specify a time limit by which the sale will be closed.

15. A contract of sale contract included a financing contingency. The purchaser could not obtain the loan required. The purchaser
    (A) forfeits the earnest money
    (B) pays the broker
    (C) is refunded all earnest money
    (D) should not have agreed to buy

16. Simon submits a contract of sale through the seller's broker. It gives the seller 48 hours to accept. After 24 hours, Simon changes his mind and decides to withdraw the agreement. Simon
    (A) can withdraw anytime before he is notified of its acceptance
    (B) must give the seller the 48 hours, as promised
    (C) will lose his earnest money
    (D) must pay the broker's commission

17. Buyer and seller shake hands on a transaction for a house at $300,000. The contract is
    (A) illegal
    (B) voidable
    (C) unenforceable
    (D) express

18. A buyer's final inspection, usually performed before closing, is called
    (A) due diligence
    (B) due on sale
    (C) a walk-through
    (D) a punch list

19. The buyer makes an offer in writing. The seller accepts the offer but makes a small, unimportant change before returning it. There is
    (A) no contract
    (B) a valid contract
    (C) further negotiation
    (D) offer and acceptance

20. A valid contract to purchase real estate must include
    (A) a credit report
    (B) an appraisal report
    (C) a title report
    (D) a description sufficient to locate the property

21. There is a clause in a contract of sale allowing an inspection for wood-destroying insects. Termites are discovered. Who pays to correct the problem?
    (A) the buyer
    (B) the seller
    (C) the buyer and seller each pay half
    (D) the person named in the contract as the responsible party

22. To be effective, a contract of sale must include
    (A) the purchase price and terms
    (B) the list price
    (C) the closing date
    (D) boilerplate language

23. Brokers may submit
    (A) listing agreements
    (B) purchase and sale contracts
    (C) options
    (D) all of the above

24. In a contract of sale, the closing date should be
    (A) within 10–15 days
    (B) within 16–20 days
    (C) within 21–30 days
    (D) clearly stated

25. __________ is used to substitute a new borrower, with the lender's approval.
    (A) Assignment
    (B) Assumption
    (C) Novation
    (D) Subject to mortgage

26. A property condition inspection was permitted by a contingency clause to a contract of purchase and sale. The inspection report indicates that multiple repairs are needed, and the buyer doesn't want to purchase the property. Which of the following is true?
    (A) The buyer may get a refund of earnest money if repair costs exceed the limit stated in the contract.
    (B) The buyer must proceed with the purchase if the seller promises to pay for all repairs.
    (C) The buyer may apply the earnest money to the repairs.
    (D) The broker may pay for the repairs out of his or her commission.

27. Which parties determine the amount of earnest money in a real estate contract?
    (A) broker and buyer
    (B) local Board of Realtors
    (C) buyer and seller
    (D) seller and broker

28. A contract for the purchase of real estate entered into by a 17-year-old is
    (A) enforceable
    (B) voidable by the 17-year-old
    (C) void
    (D) unenforceable

29. A signed purchase and sale agreement is
    (A) voidable by the buyer
    (B) voidable by the buyer with forfeit of any earnest money
    (C) an option
    (D) a binding contract

30. Two weeks after purchase, the buyer discovers oil on the property. The mineral rights are most likely
    (A) property of the state
    (B) owned by the buyer unless reserved by the seller in the deed
    (C) owned by the seller
    (D) subject to litigation

31. If the buyer defaults, the earnest money is usually used as
    (A) a commission to the broker
    (B) a return to the buyer
    (C) advance rent
    (D) liquidated damages

32. Herb is 17 when he signs a contract to lease a small house. After the property manager accepts his offer, the lease contract is
    (A) void
    (B) voidable by Herb
    (C) voidable by either party
    (D) illegal

33. Carol is buying a house and has 30 days in which to obtain a financing commitment as specified in the sales contract. The contract states that Carol must notify the seller within that period if she cannot arrange financing or risk default if she does not close. The contract states that "time is of the essence" applies to financing approval. Which is true?
    (A) Carol can arrange financing at her convenience.
    (B) The financing approval is automatically extended.
    (C) The lender must check Carol's credit references.
    (D) Carol must adhere to the stated time frame, which will end at 11:59 P.M. on the 30th day.

34. Harry gives Sam a $5,000 payment and a written agreement stating that, on July 1 of the following year, Harry will purchase the property from Sam for $150,000 cash, or Sam may keep the money. This transaction is
    (A) a lease with an option to purchase
    (B) a purchase contract with a delayed settlement
    (C) an option agreement
    (D) a limited partnership

35. Sally has shown Rebecca 11 homes, and Rebecca has selected one and made an offer. An hour later Rebecca changes her mind. Which is true?
    (A) Rebecca has already committed to the first seller and cannot withdraw.
    (B) Rebecca may withdraw the offer but will forfeit the earnest money.
    (C) The broker earned a commission.
    (D) Rebecca may withdraw before the seller accepts the offer.

36. A fixture
    (A) is normally assumed to accompany a property that is sold
    (B) is always assumed to belong to the seller and may be removed
    (C) is usually sold separately from real estate with a bill of sale
    (D) is not considered real estate

37. One who has a right to sign the name of a principal to a contract of sale is
    (A) a special agent
    (B) an attorney-at-law
    (C) an attorney-in-fact
    (D) a broker with a listing

38. Which of the following is true under a contract of sale in which the date of occupancy is later than the settlement date?
    (A) The buyer does not acquire legal title upon settlement.
    (B) The contract should specify whether the seller is to pay rent to the buyer.
    (C) The buyer cannot obtain hazard insurance.
    (D) The seller is the legal owner until occupancy is ended.

39. A prospective purchaser has a legal right to demand which of the following?
    (A) a copy of the broker's employment contract with the seller
    (B) the return of the earnest money deposit prior to seller's acceptance of the offer to purchase
    (C) a copy of the plans and specifications of the home
    (D) a copy of the seller's utility bills for the past year

40. Which of the following is true regarding the assignment of a sales contract?
    (A) An assignment of the sales contract by the buyer is generally valid.
    (B) Only the original seller can assign rights.
    (C) Assignments relieve the assignor of all obligations.
    (D) Assignment automatically requires the consent of the seller.

41. The legal remedy to force the seller to close the sale is called
    (A) quitclaim
    (B) distraint
    (C) specific performance
    (D) delivery of deed

42. A valid real estate contract of sale must
    (A) include a title search
    (B) include a survey
    (C) include an appraisal
    (D) be signed by the buyer and seller

43. Due diligence is which party's obligation to discover flaws in the property?
    (A) buyer's
    (B) seller's
    (C) broker's
    (D) lender's

44. When a contract of sale includes a due diligence contingency for the buyer's inspection,
    (A) the buyer should provide a punch list for the seller to fix
    (B) the buyer may terminate the contract if the results of an inspection are negative
    (C) the broker should get repairs done
    (D) the seller must repair all problems regardless of the cost

45. A contract form for the sale of real estate that is not yet signed is
    (A) executory
    (B) executed
    (C) oral
    (D) voidable

46. A licensed salesperson presented to the owner an offer for $150,000. The owner counteroffered at $160,000. While the potential buyer is considering the counteroffer, the owner sends an e-mail rescinding the counteroffer. Which of the following is true?
    (A) The buyer gets 48 hours to consider the counteroffer.
    (B) There is no contract.
    (C) The buyer must provide a reply.
    (D) The broker has earned a commission.

47. In the owner's absence, a relative with the owner's written general power of attorney
    (A) may sell the property
    (B) can sell anything except real estate
    (C) needs a court order to sell real estate
    (D) needs a specific power of attorney

48. A contract of sale is signed by the buyer and seller. Before closing, the buyer has
    (A) good title
    (B) marketable title
    (C) no title
    (D) equitable title

49. A buyer defaulted on a contract of sale for which the buyer had paid a $5,000 earnest money deposit. As to the deposit,
(A) the seller may keep it
(B) it belongs to the broker when a broker is involved
(C) the buyer can stop payment on the check if it has not been cashed
(D) the seller and buyer split it

50. When buyer and seller have orally come to an agreement for the sale of real estate, there is
(A) binding contract of sale
(B) unenforceable agreement
(C) binding contract if there is adequate earnest money
(D) contract enforceable by the MLS

51. Which of the following is true about a contract of sale?
(A) It conveys legal title when signed.
(B) It binds both parties to complete the sale when conditions are met.
(C) It requires specific performance of the buyer.
(D) Upon signing, the broker has the right to collect a commission.

52. The owner, Marvin, has given his friend, Kevin, power of attorney. Which of the following is correct?
(A) The power of attorney must be in writing if it provides Kevin with authority to sell Marvin's house.
(B) If Marvin dies, Kevin can exercise the power of attorney to sell the house.
(C) Only an attorney, licensed to practice law, can get a power of attorney.
(D) Power of attorney cannot be terminated.

53. Neil enters into an oral contract to purchase real estate and provides a $1,500 check as earnest money. The oral contract is
(A) unenforceable
(B) binding on the buyer
(C) voidable by the seller
(D) negotiable

54. When a real estate buyer purchases a seller's piano, what document is usually prepared?
(A) bill of sale
(B) chattel mortgage
(C) furnishings purchase
(D) fixture transfer

55. A homebuyer signs a contract with a seller that is contingent on specific financing. The buyer changes his mind and informs the seller that she will never buy. Which of the following is correct?
(A) The seller may sue for specific performance.
(B) The seller may sue for breach of contract.
(C) The lender can sue the buyer.
(D) The broker has not earned a commission.

56. __________ means that something is left to be done to a contract.
(A) Executed
(B) Executory
(C) Unfulfilled
(D) Unlisted

57. An offer to buy real estate may be withdrawn
(A) by a broker
(B) by the seller before closing
(C) by the buyer before acceptance by the seller
(D) by the seller upon return of the deposit received

58. The Statute of Frauds requires
   (A) all contracts to be in writing to be enforceable
   (B) contracts for the sale of real estate to be in writing to be enforceable
   (C) a valid contract to have two witnesses
   (D) all deeds for real estate to be recorded

59. A real estate contract must be recorded within __________ of its signing.
   (A) three days
   (B) one month
   (C) one year
   (D) none of the above; there is no requirement that it be recorded

60. A real estate sales contract is executory until it is
   (A) signed by all parties
   (B) fully negotiated
   (C) approved by an attorney
   (D) closed

61. To be enforceable, a real estate sales contract must be
   (A) witnessed by two honorable citizens
   (B) in writing
   (C) prepared by an attorney-at-law
   (D) recorded in a courthouse

62. A buyer signs a sales contract in the broker's office but changes his mind before he leaves the office. The offer
   (A) can be withdrawn because it has not been accepted
   (B) must be presented to the seller
   (C) forfeits the deposit money
   (D) may then be presented to the seller

63. If a contingency in a real estate sales contract is not met,
   (A) the buyer forfeits any deposit
   (B) the agreement is void
   (C) the seller must pay the commission
   (D) the attorney is at fault

64. A contingency in a real estate sales contract may require
   (A) inspection
   (B) financing
   (C) approval of the buyer's wife
   (D) any of the above

65. To be valid, a contract for the sale of real estate must have a(n)
   (A) broker
   (B) broker and agent
   (C) clear title opinion
   (D) offer and acceptance

ANSWERS

1. **A**	11. **C**	21. **D**	31. **D**	41. **C**	51. **B**	61. **B**
2. **A**	12. **D**	22. **A**	32. **B**	42. **D**	52. **A**	62. **A**
3. **D**	13. **A**	23. **D**	33. **D**	43. **A**	53. **A**	63. **B**
4. **D**	14. **C**	24. **D**	34. **C**	44. **B**	54. **A**	64. **D**
5. **C**	15. **C**	25. **C**	35. **D**	45. **A**	55. **B**	65. **D**
6. **D**	16. **A**	26. **A**	36. **A**	46. **B**	56. **B**	
7. **C**	17. **C**	27. **C**	37. **C**	47. **A**	57. **C**	
8. **A**	18. **C**	28. **B**	38. **B**	48. **D**	58. **B**	
9. **D**	19. **A**	29. **D**	39. **B**	49. **A**	59. **D**	
10. **B**	20. **D**	30. **B**	40. **A**	50. **B**	60. **D**	

Deeds

9

ESSENTIAL MATTERS TO LEARN FROM THIS CHAPTER

- A *deed* is a contract by which a property owner, called the *grantor*, transfers his (or her) rights to property to another, called the *grantee*. A grantor cannot transfer more rights than he or she has.
- Deeds are classified by the type of warranty the grantor provides. There are three main classifications:

  1. General warranty.
  2. Special or limited warranty.
  3. Quitclaim.

- In a *general warranty deed*, the grantor assures that he has title, can convey the property, and will forever defend the grantee's rights. This gives the grantee the strongest assurance of ownership.
- In a *special* or *limited-warranty deed*, the grantor warranties title for the period of his ownership but not for any time prior to his ownership. The grantor may or may not assure the buyer that he has title or whether he has encumbered the property. Deeds from trustees, sheriffs, administrators, and referees in bankruptcy are in this category and may contain little or no guarantees of ownership.
- A *quitclaim deed* offers no warranty. The grantor provides no assurance of ownership. Whatever ownership interest the grantor had, if any, is transferred. A quitclaim deed is often used to remove a *cloud on title*.
- To be valid, a deed must include the signature of the grantor (who must be legally competent), the name of a grantee, and a description of the property. Consideration must be stated (even if a token amount), and the deed must contain words of conveyance and the interest being conveyed. It must be delivered to a grantee who accepts it.
- A deed has five parts:

  1. Premises—names of parties, consideration, property description.
  2. Habendum—"to have and to hold" clause that limits the estate granted.
  3. Testimonium—warranties if any.
  4. Execution—signature of grantor, date, seal.
  5. Acknowledgment—notarization required for recording.

- A deed may be recorded in the county courthouse. Recording is not a requirement for a deed to be valid. Recording gives *constructive notice* (notice to the world) of the existence of the deed, although it does not assure the deed's validity. A deed must be *acknowledged* to be recorded, meaning that the signature of the grantor must be *notarized*.

- If a grantee fails to record a deed, then an unscrupulous seller can sell the property again to someone else. If the second buyer bought in good faith and recorded his deed first, courts will award ownership to the second buyer.
- Sellers often move away or can refuse to honor the warranty. So no matter what type of deed, it is important that a qualified party conduct a *title search* to be certain that the seller is, in fact, the owner of the property and can convey title.
- Lenders almost always require *title insurance* for a real estate loan, and the equity holder can also purchase that type of insurance. Before issuing a mortgage loan, a lender will have a qualified party perform a title search, which further assures the buyer of good title. Either the title company's lawyers or independent lawyers examine the title.
- *Alienation* is a term meaning the transfer of property. It can be a voluntary act of the owner, as would be needed to get a mortgage loan, or a sale. *Involuntary alienation* occurs when ownership is transferred as demanded by someone else, such as a foreclosure or condemnation.
- A *land contract,* also called a *contract for deed* or *installment land sale contract,* does not provide a deed at closing. It is a seller-financed transaction without the seller providing a deed at closing. Often used in the sale of low–down payment resort property, this type of contract transfers the deed after the buyer makes final payment, typically after many years of installment payments. This makes it easy for the lender to continue ownership if the buyer defaults on payments.

USE OF DEEDS

To transfer an interest in real estate during his lifetime, the owner (*grantor*) surrenders a *deed of conveyance* to another (the *grantee*). The deed of conveyance, also simply called a *deed,* is covered by the Statute of Frauds; consequently, it must be in writing to be enforceable. Deeds are contracts and nothing more. They are *not* "title papers" similar to the official documents you get from the state showing that you are the owner of a particular car. Deeds are not "guaranteed" or "backed up" by any government entity. The deed transfers property rights, but those rights are only as "good" as the public records show that they are. Therefore, a deed should be recorded to give the public *constructive notice* of its existence. Any document recorded in the public records must be *acknowledged*; generally, that means it must be *notarized.* Notarization is done by a *notary public,* someone who is licensed by the state to be a witness to the authenticity of signatures. In some counties, a few public officials (mayor, county manager, alderman, etc.) may also be authorized to acknowledge documents. The only guarantee provided by the recording entity (usually a county government) is that the recorded document is a true copy of the original; there is no government guarantee of the validity of the document.

TYPES OF DEEDS

Deeds are classified into three principal types based on the warranty they provide to the grantee:

1. General warranty.
2. Special or limited warranty.
3. Quitclaim.

Any of the three types will convey whatever ownership the grantor has in the property.

The *general warranty* deed contains assurances by the grantor:

1. That he has title to the property and the power to convey it (covenant of seizin).
2. That his title is good against claims by others (covenant of quiet enjoyment).
3. That the property is free of encumbrances except as specified (covenant against encumbrances).
4. That he will perform whatever else is necessary to make the title good (covenant of further assurances).
5. That he will forever defend the rights of the grantee (covenant of warranty forever).

The general warranty deed is the best type of deed for a grantee to receive.

A deed that contains some but not all of the covenants of a warranty deed is a *special* or *limited warranty* deed. Special warranty deeds typically contain assurances only against claims that may have originated while that grantor held title. It does not protect against claims that originated earlier.

Bargain and sale deeds contain no warranties, only an assurance that the grantor has an interest in the property. A grantor need not state precisely what his interest is or whether the property is unencumbered. It is up to the grantee to ascertain, prior to closing, the grantor's interest in the property.

There are many other types of deeds, including a *trustee's deed* (used to transfer property from a trust); a *referee's deed* in foreclosure, a *sheriff's deed,* and an *administrator's deed.* Most of these are used by persons who must transfer the title of someone else's property. In these situations, the authorized transferor does not wish to personally warranty title from all claims, so the deed given contains a limited warranty. Most of these deeds are grouped into a type of *special warranty* deeds.

In contrast, the *quitclaim* deed carries no warranties whatsoever. The grantor gives up all of his rights to the property without even implying that he has or has ever had any such rights. Should the grantor be possessed of the property in *fee simple absolute,* then a quitclaim deed serves to transfer complete ownership. However, if a grantor owns nothing, that is exactly what a grantee will receive from the quitclaim deed. The quitclaim deed is usually used to clear up an imperfection in the title, called a *cloud on title.*

A cloud on title (also called a *title defect*) is a circumstance in which the public records are not perfectly clear about who owns or has rights to certain property rights. For example, Elrod wants to buy Sally Smith's real estate. The public records show that Sally, along with her brothers Melvin and Paul, inherited the property from their father some time ago. Sally insists that she is the sole owner, and there is an executor's deed that names only Sally as the grantee from the estate. So, according to the records, the "cloud" is that there is nothing to show that Melvin and Paul ever gave up their rights to inherit a share of the real estate. What actually happened is that when their father died, the three siblings agreed that Sally would get the house and Sally would give up to them her share of their father's stocks and bonds. But the executor failed to get quitclaim deeds for the house from each of the brothers to Sally. Now Sally must get, and record, quitclaim deeds from them to "clear" the title.

DEED REQUIREMENTS

The following elements must be present in a deed for it to be considered valid:

1. A legally competent grantor.
2. A designated grantee.
3. Consideration—even though only a token amount is shown.
4. Words of conveyance.
5. The interest being conveyed.
6. A description of the property.
7. Grantor's proper signature.
8. Delivery of the deed and acceptance.

PARTS OF A DEED

The traditional deed has the following five parts:

PART OF DEED	PURPOSE
1. Premises	Gives names of parties, exploratory facts, words of conveyance, consideration, legal description of the property.
2. Habendum	Is the "to have and to hold" clause, which may describe a limit to the quantity of the estate being granted, such as "to have and to hold forever" or "to have and to hold for life."
3. Testimonium	Contains warranties (if any).
4. Execution	Contains date, signatures, and seal if applicable.
5. Acknowledgment	Contains attestation by a public officer, usually a notary public, as to the genuineness of the grantor's signature. (Although acknowledgment is not required to transfer title, it is usually a requirement for the deed to be recorded.)

Figures 9-1 and 9-2 identify the relevant parts of a warranty deed and a quitclaim deed. Note that the quitclaim contains nothing that could technically be called a testimonium. This is so because the quitclaim contains no warranty. The grantee receives whatever interest the grantor may have in the property, but the nature of that interest usually is not stated. Since the grantor will defend against no claim but his own, the document itself is adequate evidence against him. Consequently, no warranty is included.

EXECUTION OF THE DEED

To be considered a valid contract, a deed must be signed by the grantor and delivered to the grantee or delivered in escrow (see below). Occasionally, problems arise when an undelivered deed is found in the grantor's home after his death. The grantee named may claim ownership; but since the grantor never delivered the deed, courts will not agree that transfer was intended.

The requirement for delivery of the deed generally is considered very literally; that is, the grantee actually must have come into physical possession of the document in order for title to be considered transferred to her. The grantee may arrange for her agent to accept delivery for her. A common form of such delivery is *delivery in escrow*. In this situation, the grantor delivers the deed to a trustee or escrow agent, who holds it pending completion of some

WARRANTY DEED

STATE OF GEORGIA,

.... County.

THIS INDENTURE, made this ....day of ....
in the year of our Lord Two Thousand and ....
Between ....
of the State of.................................and County of ....of the first part
and ....
of the State of ....and County of.......................................of the second part.

WITNESSETH: That the said part ....of the first part, for and in consideration of the sum of
....DOLLARS
in hand paid at and before the sealing and delivery of these presents, the receipt whereof is hereby acknowledged, ha....
granted, bargained, sold and conveyed and by these presents do..............grant, bargain, sell and convey unto the said
part ....of the second part, ...heirs and assigns, all that tract and parcel of land
lying and being in..

PREMISES

TO HAVE AND TO HOLD the said bargained premises, together with all and singular the rights, members and appurtenances thereof, to the same being, belonging or in any wise appertaining, to the only proper use, benefit and behoof of ....the said part ....of the second part, ...heirs and assigns forever, IN FEE SIMPLE.

HABENDUM

And the said part ....of the first part, for ....heirs, executors and administrators will warrant and forever defend the right and title to the above described property unto the said part ....of the second part,..........................heirs and assigns, against the lawful claims of all persons whomsoever.

TESTIMONIUM

IN WITNESS WHEREOF, That the said part ....of the first part ha ....hereunto set ....hand........ and affixed ....seal ...., the day and year above written.

Signed, sealed and delivered in the presence of:

....(Seal)
....(Seal)
....(Seal)
....(Seal)
....(Seal)

EXECUTION

....
....
....
....

ACKNOWL-
EDGMENT

Figure 9-1. Warranty Deed

QUIT-CLAIM DEED

STATE OF GEORGIA, ...County.

THIS INDENTURE, made this ....day of...in the year of our Lord Two Thousand and...

between ....of the first part,

and ....of the second part.

WITNESSETH: That the said part ....of the first part for and in consideration of the sum of ....Dollars, cash in hand paid, the receipt of which is hereby acknowledged, ...bargained, sold and do ...... by these presents bargain, sell, remise, release, and forever quit-claim to the said part ....of the second part, ... heirs and assigns, all the right, title, interest, claim or demand which the said part..................of the first part ha........or may have had in and to ....

with all the rights, members and appurtenances to the said described premises in anywise appertaining or belonging.

TO HAVE AND TO HOLD the said described premises, unto the said part ....of the second part heirs and assigns, so that neither the said part ....of the first part nor .... heirs, nor any other person or persons claiming under ....shall at any time, claim or demand any right, title or interest to the aforesaid described premises or its appurtenances.

IN WITNESS WHEREOF, the said part..of the first part ha........hereunto set hand ......and affixed ....seal, the day and year above written.

Signed, sealed and delivered in the presence of

	....(Seal)
....	....(Seal)
....	....(Seal)
....	....(Seal)

PREMISES

HABENDUM

(NO TESTIMONIUM)

EXECUTION

ACKNOWLEDGMENT

Figure 9-2. Quitclaim Deed

required action by the grantee (such as payment in full of a note given in payment for all or part of the price). Delivery is considered to have occurred as soon as the escrow agent or trustee received the deed, so title is transferred. However, because the grantee has no deed in her possession, she will have difficulty disposing of the title until she fulfills the necessary requirements to get the deed released to her by the agent.

ALIENATION

The transfer of real property is called *alienation. Voluntary alienation* occurs when the transfer is not forced: when the property is sold, given as a gift, or used to provide a trust deed to a lender to secure a mortgage.

Involuntary alienation occurs when the transfer is forced, as in a foreclosure or condemnation or when the real estate escheats to the state.

CONTRACT FOR DEED

In a *contract for deed,* the seller finances the sale and does not transfer title until the final payment is made. This plan is frequently used to sell resort property. The agreement may also be called an *installment sale contract* or *land contract.*

Trick Question

A trick question uses the term *contract for deed* as though it is a contract to provide a deed in coming days. However, a contract for deed is completed after a much longer timespan. Most contracts for deed provide the deed after all installment payments have been made for years, after those payments fully retire the total cost.

Another trick question considers deeds as though they are cards in the game Monopoly—that is, existing documents that are passed from seller to buyer. That is not correct; deeds are not transferable documents. Instead, deeds are newly written contracts.

QUESTIONS ON CHAPTER 9

1. X hands Y a deed with the intent to pass title and orally requests Y not to record the deed until X dies. When is the deed valid?
   (A) immediately
   (B) when Y records the deed
   (C) when X dies
   (D) never

2. A quitclaim deed conveys only the interest of the
   (A) guarantee (C) claimant
   (B) property (D) grantor

3. Which of the following forms of deeds has one or more guarantees of title?
   (A) quitclaim deed (C) warranty deed
   (B) executor's deed (D) special-form deed

4. The part of conveyance that defines or limits the quantity of the estate granted is
   (A) habendum (C) equity
   (B) premises (D) consideration

5. The Statute of Frauds
   (A) requires certain contracts to be in writing to be enforceable
   (B) requires a license to operate as a broker or a salesperson
   (C) regulates escrow accounts
   (D) regulates fraud conveyance

6. For a deed to be recorded, it must be in writing and must
   (A) be signed by the grantee
   (B) state the actual purchase price
   (C) be acknowledged
   (D) be free of all liens

7. The most comprehensive ownership of land at law is known as
   (A) an estate for years
   (B) a life estate
   (C) a fee simple
   (D) a defeasible title

8. From the standpoint of the grantor, which of the following types of deeds creates the least liability?
   (A) special warranty
   (B) general warranty
   (C) bargain and sale
   (D) quitclaim

9. The recording of a deed
   (A) passes the title
   (B) insures the title
   (C) guarantees the title
   (D) gives constructive notice of ownership

10. Title to real property may pass by
    (A) deed (C) both A and B
    (B) bill of sale (D) neither A nor B

11. Real property title may be conveyed by
    (A) adverse possession (C) deed
    (B) inheritance (D) all of the above

12. A warranty deed protects the grantee against a loss by
    (A) casualty (C) both A and B
    (B) defective title (D) neither A nor B

13. Recordation of a deed is the responsibility of the
    (A) grantor (C) both A and B
    (B) grantee (D) neither A nor B

14. Deeds should be recorded
    (A) as soon as possible after delivery
    (B) within 30 days of delivery
    (C) within one year
    (D) none of the above

15. A deed of conveyance must be signed by
    (A) the grantee and the grantor
    (B) only the grantor
    (C) only the grantee
    (D) none of the above

16. A deed to be valid need not necessarily be
   (A) signed
   (B) written
   (C) sealed
   (D) delivered

17. The party to whom a deed conveys real estate is the
   (A) grantee
   (B) grantor
   (C) beneficiary
   (D) recipient

18. Deeds are recorded in the
   (A) county courthouse
   (B) city hall
   (C) Federal Land Book
   (D) state capital

19. A deed must
   (A) contain the street address of the property
   (B) state the nature of the improvement on the land (dwelling)
   (C) contain an adequate description to identify the property sold
   (D) state the total area in the tract

20. What is the maximum number of grantees that can be named in a deed?
   (A) two
   (B) four
   (C) ten
   (D) there is no limit

21. Property is identified in a deed by the
   (A) habendum
   (B) consideration
   (C) description
   (D) acknowledgment

22. Attestation, by a notary public or other qualified official, of the signature on a deed or mortgage is called an
   (A) authorization
   (B) acknowledgment
   (C) execution
   (D) authentication

23. For which reason or reasons is a deed recorded?
   (A) to insure certain title
   (B) to give notice of a transaction to the world
   (C) to satisfy a requirement of state law
   (D) to save title insurance cost

24. Ownership of property is transferred when
   (A) the grantor signs the deed
   (B) the grantor's signature has been notarized
   (C) delivery of the deed is made
   (D) the correct documentary stamps are put on the deed and canceled

25. From the point of view of the grantee, the safest kind of deed that can be received is a
   (A) general warranty deed
   (B) special warranty deed
   (C) quitclaim or release deed
   (D) trustee's deed

26. Which of the following statement(s) is (are) true?
   (A) A quitclaim deed transfers whatever interest a grantee has in property.
   (B) A quitclaim deed carries a warranty of good title.
   (C) both A and B
   (D) neither A nor B

27. A notary public
   (A) makes the statement
   (B) requires a graduate-level education
   (C) provides acknowledgment
   (D) assures a genuine deed

28. In real estate parlance, *alienation* means
   (A) walking away from a transaction
   (B) offending the seller
   (C) breaking a lease
   (D) transferring title

29. Of the following, the best type of deed to receive is a
   (A) special warranty deed
   (B) trustee deed
   (C) general warranty deed
   (D) quitclaim deed

30. A quitclaim deed will transfer
   (A) with a special warranty
   (B) whatever interest the grantor has
   (C) with a limited warranty
   (D) the leasehold estate

31. To provide security for a loan, the borrower gives a deed to a disinterested third party. This is called a
   (A) junior lien
   (B) general warranty deed
   (C) trust deed
   (D) contract for deed

32. A quitclaim deed is given by the
   (A) grantee
   (B) trustee
   (C) trustor
   (D) grantor

33. One who receives real estate under a will is a(n)
   (A) escheatee
   (B) intestate
   (C) devisee
   (D) bequeathee

34. Which instrument is typically used to convey ownership of real estate?
   (A) devise
   (B) lease
   (C) deed
   (D) mortgage loan

35. To be valid, a deed must include
   (A) grantee's signature
   (B) grantor warranty
   (C) property survey
   (D) legal description

36. Herman prepares a deed for his cattle farm to his nephew. Herman dies before delivering the deed. His nephew
   (A) receives nothing
   (B) inherits the farm
   (C) receives the farm when the deed was signed
   (D) receives the farmland but not the cattle

37. A bank acquired property through foreclosure. Upon sale of the property, the bank covenants and warrants the property's title only against defects occurring during the bank's ownership. This conveyance is a(n)
   (A) administrator's deed
   (B) general warranty deed
   (C) special warranty deed
   (D) covenant deed

38. A deed is the instrument used to convey ownership of
   (A) personal property
   (B) short-term use of real estate
   (C) real property
   (D) any of the above

39. Which of the following is a voluntary alienation of real property?
   (A) escheat
   (B) condemnation
   (C) foreclosure
   (D) a deed of trust

40. Which of the following statements regarding deeds is true?
   (A) A general warranty deed gives the least liability to the grantor.
   (B) A quitclaim deed gives the least protection to the grantee.
   (C) A special warranty deed gives the greatest protection to the grantor.
   (D) A bargain and sale deed is voidable in most states.

41. The clause that defines or limits the quantity of the estate being conveyed is the
   (A) habendum clause
   (B) reversion clause
   (C) partition clause
   (D) revocation clause

42. As far as its validity between grantor and grantee is concerned, a signed deed that is delivered but not dated, acknowledged, or recorded is
   (A) invalid because of these omissions
   (B) void
   (C) revocable by the grantor
   (D) valid despite these omissions

43. The recording system performs which of the following functions?
   (A) insures title against loss due to third-party claims
   (B) cures major defects in title
   (C) gives to the world constructive notice of documents
   (D) handles the closing of real estate transactions

44. Which of the following parties is in the weakest position against a claim of title by a stranger?
   (A) a nonoccupant holder of a warranty deed
   (B) a nonoccupant holder of an unrecorded quitclaim deed
   (C) one who holds an unrecorded deed
   (D) one who holds a recorded quitclaim deed to the property

45. A deed made and delivered but *not* recorded is
   (A) valid between the parties and valid as to third parties with notice
   (B) valid between the parties and valid as to subsequent recorded interests
   (C) valid between the parties and invalid as to subsequent donees of the property
   (D) invalid between the parties

46. You can be most assured that you are getting fee simple ownership if
   (A) the owner will give a general warranty deed
   (B) the seller can furnish title insurance
   (C) you retain an attorney
   (D) you use an escrow company

47. *Alienate title* means
   (A) inherit property
   (B) sell or convey property
   (C) lease property
   (D) sell to a tenant

48. An arrangement whereby the seller retains title until final payment has been made is
   (A) an installment sale contract
   (B) a general warranty deed
   (C) an amortizing mortgage
   (D) a bullet loan

49. A seller wants to convey his interest, if any, in property without providing a guarantee. The deed he would like to use is a __________ deed.
   (A) general warranty
   (B) special warrant
   (C) quitclaim
   (D) bargain and sale

50. A deed to real estate is
   (A) a card used in the game Monopoly
   (B) a title search
   (C) a recorded document
   (D) a title to a specific property that may be recorded in the county courthouse

ANSWERS

1. **A**	11. **D**	21. **C**	31. **C**	41. **A**
2. **D**	12. **B**	22. **B**	32. **D**	42. **D**
3. **C**	13. **B**	23. **B**	33. **C**	43. **C**
4. **A**	14. **A**	24. **C**	34. **C**	44. **B**
5. **A**	15. **B**	25. **A**	35. **D**	45. **A**
6. **C**	16. **C**	26. **D**	36. **A**	46. **B**
7. **C**	17. **A**	27. **C**	37. **C**	47. **B**
8. **D**	18. **A**	28. **D**	38. **C**	48. **A**
9. **D**	19. **C**	29. **C**	39. **D**	49. **C**
10. **A**	20. **D**	30. **B**	40. **B**	50. **D**

Leases and Property Management 10

ESSENTIAL MATTERS TO LEARN FROM THIS CHAPTER

- A *lease* is a contract that transfers possession of real estate in exchange for *rent.*
- The property owner is the *landlord,* also called the *lessor.* The user is the *tenant,* also called the *lessee.*
- The landlord's interest is called the *leased fee* or *leased fee estate.*
- The tenant's interest is called the *leasehold,* or *leasehold estate.*
- Leases for one year or longer must be in writing to be enforceable under the *Statute of Frauds.*
- Types of leasehold estates include:

  1. *Estate for years.* This has a stated expiration date, which could be any period.
  2. *Estate from year to year.* It is automatically renewed unless one party gives notice of termination to the other.
  3. *Tenancy at will.* Either party may terminate the agreement.
  4. *Tenancy at sufferance.* A tenant has remained on the property after lease expiration.

- Common lease provisions include who pays operating expenses:

  1. *Gross lease.* The landlord pays operating expenses.
  2. *Net lease.* The tenant pays operating expenses. Operating expenses generally include property taxes, maintenance, and insurance. *Net-net* and *triple-net* are terms used to emphasize the obligations of the tenant. Operating expenses do not include debt service (interest and principal payments) or income taxes, because they are financing expenses or personal obligations of the lessor.

- *Percentage leases* are often used for retail properties. The tenant pays a *base* or *basic rent* plus a *percentage of gross business sales* above a certain amount.
- *Escalation* provisions, also called *pass-throughs,* allow for increases in expenses to be paid by the lessee.
- *Graduated* leases call for rent increases at preset intervals.
- *Index* or *indexed* leases allow for increases based on changes in an economic index, such as the Consumer Price Index (CPI).
- A lease may or may not include the lessee's right to *assign* (give to another all the rights to the lease) or *sublet* (give part of the lease rights to another by creating a new lease). The original lessee remains liable to fulfill lease obligations unless released by the lessor.
- The sale of a property does not change a lease. The new property owners must continue the tenant's rights as described in the lease.

- A lease is typically terminated by *performance*, which occurs when both parties observe the full term of the lease.
- Upon expiration, a tenant may remove and keep *trade fixtures.* In a restaurant, this often includes kitchen equipment, signage, and barstools.
- A *breach* of a lease occurs when one party violates a lease provision.
- When a tenant vacates without paying rent, that tenant may be liable for *damages*, typically to continue paying rent to the end of the lease, enforced by a court. The landlord must make an effort to *mitigate* damages by seeking a substitute tenant.
- A tenant may withhold rent upon certain *lessor* violations.
- The *death* of either party normally does not terminate a lease. The estate of the decedent is responsible for honoring the remaining term of the lease.
- Some cities have adopted *rent controls* to prevent sharply increasing rents from being passed on, especially to residential tenants.
- Real estate license law in most states requires a property manager to have a broker's license. This requirement does not apply to an onsite resident manager.

WHAT A LEASE IS

A lease is a contract that transfers possession of property in exchange for rent. The real estate owner is known as the *landlord* or *lessor*; the user is the *tenant* or *lessee.* The landlord retains a *reversionary right*; that is, he retains the right to use the property upon expiration of the lease. The tenant holds a *leasehold estate*, that is, a personal property interest that gives the tenant the right to use the property for the lease term and under the conditions stipulated in the lease.

With respect to leases, the Statute of Frauds usually requires that only leases for more than one year have to be in writing. This means that in some states it is possible for shorter-term leases (one year or less) to be oral. (It always is the best practice for any contract to be written; disputing oral contracts in court can be uncertain.) You should check your state's laws on leases. Some do require that shorter-term leases be written.

TYPES OF LEASEHOLD ESTATES

1. Estate for years
2. Estate from year to year
3. Tenancy at will
4. Tenancy at sufferance

An *estate for years* has a stated expiration date. The period included may be only 1 day or more than 50 years. When the lease term ends, the tenant is to vacate the property.

An *estate from year to year* (also called *periodic estate or estate from period to period*) does not have a fixed expiration date. It is automatically renewed for another period (week, month, year) unless adequate notice for termination is given. One period of notice is generally adequate; less than 1 year is acceptable in the case of a year-to-year lease.

A *tenancy at will* allows the tenant to remain with the consent of the landlord. At any time, either the tenant or the landlord may terminate the agreement. "Emblements" arise by operation of law to give tenants-at-will the right to harvest growing crops.

A tenant who holds property beyond the lease term and without the landlord's consent is a "holdover tenant." The form of estate is called *tenancy at sufferance.* A holdover tenancy has the fewest rights of any leasehold estate.

SOME COMMON LEASE PROVISIONS

Rent

Rent may be payable in any form that the parties agree upon. Typically, rent is payable in money, but it may also be payable in crops, labor, and so on—whatever the landlord and tenant agree upon. *Straight* or *flat* leases call for level amounts of rent. *Step-up* or *step-down, graduated,* or *reappraisal* leases provide for fluctuating amounts of rent.

Percentage Lease

A *percentage lease* requires the tenant to pay, as rent, a percentage of sales. Usually a basic rent is stated to guarantee a minimum amount. For example, a lease may require that a retailer-tenant pay $10,000 rent annually plus 2% of gross sales in excess of $500,000. Should sales be under $500,000, the rent is $10,000; for sales of $600,000, the rent is $12,000, that is, $10,000 basic plus 2% of the $100,000 excess sales.

Gross vs. Net Leases

A lease that requires the landlord to pay operating expenses, such as utilities, repairs, maintenance, insurance, and property taxes, is called a *gross lease.* The lessor considers the rent to be gross income and receives net operating income after he pays operating expenses out of his own pocket. A *net lease* requires the tenant to pay operating expenses. The landlord receives the rent as a net return.

The terms *net* and *gross,* though often used, may be inadequate to describe a lease. *Hybrid, semi-net, net-net,* and *triple-net* are used to describe the degree to which a party may be responsible for operating expenses. Interest expenses and income taxes are the lessor's obligations. They are not operating expenses. Careful reading of a lease by interested parties is necessary to avoid misunderstandings.

Trick Question

A trick question confuses a gross lease with a percentage lease. Although the percentage lease uses the word *gross,* a percentage lease refers to gross sales. A *percentage lease* is defined as a lease that requires the tenant to pay a percentage of gross sales as rent. A *gross lease* is defined as a lease where the landlord pays operating expenses. A gross lease does not refer to tenant sales.

Escalation and Graduated Leases

Escalation or *step* provisions are found in long-term gross leases. Should certain operating expenses exceed a designated amount, the increase is passed on to the tenant. *Graduated leases* call for increases at certain intervals, such as every year or at two- or five-year intervals.

Index Lease

An *index* or *indexed lease* is one in which the rent is changed periodically in accordance with a published economic index, such as the *Consumer Price Index* published by the Bureau of Labor Statistics.

Assignment and Subletting

Unless specifically prohibited in a lease, a tenant has the right to *assign* or *sublet*. In either case, the original tenant remains liable to fulfill the lease contract.

An *assignment* occurs when the tenant transfers all the rights and obligations stipulated in the lease to a third party (the assignee).

Should the original tenant wish to keep a portion of the property for his own use or to rent the premises to a third party for a term shorter than his lease, he may sublet. *Subletting* involves a second lease; the original tenant becomes a lessor for part of the property or for the entire property for part of the lease term.

Sale of the Property

When leased property is sold, the lease survives the sale. The tenant is allowed to continue use under the lease terms.

Termination of a Lease

The normal termination of a lease is the result of *performance*. When both landlord and tenant have fulfilled their lease obligations, there is said to be performance.

Another method of terminating a lease is by *agreement*. Landlord and tenant may agree to terminate the lease before its stated expiration date. Compensation may be included in such an agreement.

A *breach* may also cause termination. When either party fails to perform under a lease, there is said to be a breach. Although a breach can terminate a lease, it does not cancel all provisions. The violating party may be held liable for damages. When a tenant has breached a lease, the landlord's remedy is in court. Using a court order, a landlord may eject or *evict* a tenant.

A tenant who vacates property before the fixed expiration date of a lease is responsible for the remaining rent. A landlord must attempt to *mitigate* (reduce) damages. In other words, he must attempt to lease the property to another tenant to offset the lost rent from a tenant who has vacated. If the property is eventually leased, the tenant owes only the difference in rent plus expenses incurred in obtaining a substitute tenant. However, vacating property fails to terminate a lease except by *constructive eviction*. In this form of breach, the landlord makes it very difficult or impossible for the tenant to "enjoy" the leased premises as contracted. In effect, the landlord's actions force the tenant to leave. Examples are very noisy repair work that goes on night after night in an apartment building; turning off utilities for long periods of time; closing the parking lot in a retail shopping center. In most places and situations, death of a lessor or lessee does not terminate a lease. The lessee remains liable to the end of the current lease, and the lease is binding on the lessor's estate and heirs.

Ground Lease

A *ground lease* is a long-term lease of land or air space. The tenant usually builds whatever improvements are desired. The landlord receives what is called *triple-net* rent; this means that the tenant pays not only operating expenses but also property taxes and mortgage payments. Ground leases are for very long periods of time, usually 25 to as many as 99 years.

Sale and Leaseback

In a *sale and leaseback*, a party who owns and uses property sells it to an investor and leases it back, becoming the tenant. Often the sale generates a significant amount of cash for the seller-occupant, which may be profitably used in his business.

Lease provisions frequently encountered are described below. (Not all of the provisions listed will be found in every lease.)

PROVISION	PURPOSE
Premises	Describes the property being leased
Term	States the term of the lease
Rent	Describes rental amount, date, and place payable
Construction of building and improvements	Describes improvements for proposed property or property to be renovated to suit tenant
Common area	Describes maintenance of common-use areas, such as hallways in shopping centers
Taxes	States who will pay ad valorem taxes and tax escalation amounts
Repairs	States which party is responsible for repairs and maintenance
Alterations	States responsibilities and rights of tenant to alter improvements
Liability insurance	Designates party who is responsible to carry liability insurance and the extent
Other insurance	Designates party who is responsible to carry hazard insurance and the extent
Damage by fire and casualty	Describes operation of lease in the event of a fire or other casualty
Payment for utilities	States which party must pay charges for water, heat, gas, electricity, sewage disposal, and other utilities
Subordination	Describes rights between landlord and tenant in the priority of encumbrances
Certificate of lease status	Obligates tenant to inform mortgage lender of occupancy and rent status
Condemnation and eminent domain	Describes operation of lease in the event of a taking or partial taking of the property
Compliance of law	Requires landlord and tenant to comply with laws and ordinances

PROVISION	PURPOSE
Access by landlord	Allows landlord to enter to inspect property
Mechanic's liens	Describes the discharge of mechanic's liens
Assignment and subletting	Limits or provides the tenant's rights to assign or sublet
Nonliability of landlord	Frees landlord from liability due to injury or damage within premises
Right to cure defaults	Allows landlord to cure tenant defaults and require reimbursement
Use	Permits tenant the use of property for specific purposes
Notices	Describes method to serve notice to a party
Definitions	Defines how terms are used in the lease
Partial invalidity	Describes operation of lease when parts of lease are held to be invalid
Waiver of redemption	Permits tenant to waive all rights upon lease expiration
Bankruptcy or insolvency	Describes required notice and operation of lease in the event of bankruptcy
Default	Defines lease default
Surrender	Describes condition tenant is to leave property in upon expiration of the lease
Arbitration	Provides a method to settle disputes outside of court
Brokers	Notes whether a broker was involved in lease negotiation
Encumbrances	Describes rights and obligations of parties to deliver and return property (with respect to liens, easements, etc.)
Quiet enjoyment	Agreement by landlord that tenant, while he performs, may have, hold, and enjoy property without interference
Other restrictions	May limit the leasing of space in a shopping center to a competitor
Entire agreement	States that the written lease is the entire agreement
Percentage clause	Stipulates conditions affecting a percentage lease
Recapture clause	Stipulates that in a percentage lease a lessor may terminate lease if lessee's revenues do not attain a specified level within a certain length of time

LANDLORD-TENANT LAW

Most states and some localities have collected many provisions of lease law and some special enactments into a body of law called *landlord-tenant law*. Most often this body of law applies to residential leasing (apartments, homes, etc.) and specifically defines certain areas of the relationship between the landlord and the tenant. Lease law itself is very old and has been created largely to benefit the owner of the leased property (the landlord). Landlord-tenant law, therefore, tends to concentrate upon *tenants'* rights, although it often also includes specific tenant responsibilities.

Over a dozen states have adopted (some with modifications) a model law called the *Uniform Residential Landlord and Tenant Act*, also known as *URLTA*. This law applies only to residential leasing, with some exceptions such as hotels, motels, and mobile home lots. URLTA concerns such subjects as security deposits, rights of landlord and tenant, and condition of the property. States that have not adopted URLTA as a statute usually have laws covering the same things; however, these laws may, in general, be more favorable to landlords than is URLTA. With or without URLTA, many states and localities have enacted more stringent laws governing tenants' rights. You should check whether your state has laws considering either of the following. (Note that if there are only local laws on either subject, they probably won't be covered on state licensing examinations.)

Rent Control

Rent control allows government to limit the amount of rents charged. They exist in New York City and in a few other major metropolitan areas where market rents are unusually high.

Withholding Rent

Some states and localities allow tenants to withhold rent if the property doesn't meet certain conditions. These usually involve habitability, safety, and the like or conditions that do not meet building or health codes. In some places the law allows the tenant simply to refuse to pay until the landlord remedies the problem. In others the withheld rent must be used to remedy the illegal condition, if possible, or must be paid into some sort of escrow account pending resolution of the problem. These laws prohibit the landlord from evicting tenants for nonpayment of rent when rent is withheld for reasons allowed by the law. In some states that require an implied *warranty of habitability*, tenants may withhold rent, if premises are "nonhabitable," until the landlord remedies the situation. The actual conditions that allow tenants to withhold rent may vary considerably from state to state, although some permit withholding in the case of clear violation of building, safety, sanitation, and health codes.

Trade Fixtures

A *trade fixture* is attached to a rented facility by a tenant who uses that fixture in conducting a business. No matter how trade fixtures are attached to real estate, they are personal property. Some examples are booths, counters, and barstools in a restaurant, tanks and pumps in a service station, and shelves in a retail store. The tenant may remove these at the expiration of the lease.

In determining whether an item is a trade fixture, courts will consider:

1. *The manner of attachment.* Personal property can generally be removed without damage.
2. *The adaptation of the article to the real estate.* Items that are built specifically to be installed in a particular building and are unlikely to be used in another are probably real estate.
3. *The intent of the parties when the property was attached.*

PROPERTY MANAGEMENT

Property management is the business of seeking and negotiating with tenants for other people's real estate. It is a real estate brokerage function, and property managers in all states are required to hold real estate sales or brokerage licenses. (*Resident managers* of apartment buildings or projects are allowed by most states to function without a license, provided that they do not manage any other properties and are employed either by the property owner or a properly licensed property manager.)

Some brokers who claim to be "property managers" may do little more than find tenants for vacant properties and collect payment for doing so. Actually, a *professional* property manager does a lot more. In fact, he is able to do everything associated with the proper management of a property so that the owner need do nothing at all. Property management functions include leasing, negotiating with potential tenants, advertising, developing leasing plans, arranging maintenance and repairs, and paying all bills (out of rent receipts). Of course, property managers are paid for their services. Normally, their payment is a percentage of rent payments collected; however, they may instead (or also) receive a flat fee.

A property manager may have a written contract with the owner, or at least an oral agreement, to explain duties and responsibilities. The agreement should include the manager's name, responsibilities, and fee arrangement, which is often a percentage of the gross rental income received.

The manager's duties typically include maintaining the property and enhancing (or at least preserving) its value in matters under his control and trying to meet the owner's goals. Examples of these goals are maximizing cash flow or minimizing tenant turnover.

Certain special kinds of lease arrangements often are found in managed properties. Retail property (shopping centers, etc.) often use *percentage leases.* In this type of lease, all or part of the tenant's rent is a percentage of her gross business revenues. The theory here is that a well-managed and popular shopping center will generate good business for tenants, and so they will be willing to pay part of their income as rent. Also, the property manager is encouraged to promote and advertise the shopping center because by doing so he gets more people to come to it, spend more money, and thus generate more rent.

Many percentage leases contain *recapture clauses.* These allow the property manager to void the lease if the tenant does not do at least a certain volume of business within a certain time. In this way the management avoids getting stuck with a poor tenant or one who does not do much business.

QUESTIONS ON CHAPTER 10

1. Which of the following statements is correct?
   (A) The tenant is known as the lessor.
   (B) The owner leasing property is the lessee.
   (C) The tenant is the lessee.
   (D) none of the above

2. Under a net lease, who is liable for payment of insurance and repairs?
   (A) lessor
   (B) lessee
   (C) sublessee
   (D) both A and B

3. A gross lease requires the tenant to pay rent based on
   (A) gross sales
   (B) net sales
   (C) gross profit
   (D) none of the above

4. A percentage lease requires the tenant to pay a
   (A) percentage of taxes and insurance
   (B) percentage of net income as rent
   (C) percentage of gross sales as rent
   (D) none of the above

5. A tenant is delinquent in paying rent. The landlord should
   (A) call the sheriff to evict him
   (B) bring a court action
   (C) give him 30 days' notice
   (D) turn off water, lights, etc.

6. A lease that requires the landlord to pay the operating expenses of the property is called
   (A) a gross lease
   (B) an assigned lease
   (C) a percentage lease
   (D) a net lease

7. A lease for less than 1 year
   (A) may be oral
   (B) must be in writing
   (C) may be oral but must be reduced to writing within 1 year
   (D) is always invalid

8. Transfer, by a tenant, of the rights and obligations of an existing lease to another tenant is called
   (A) assignment of lease
   (B) a release
   (C) subletting
   (D) an eviction

9. A contract that transfers possession but not ownership of property is
   (A) a special warranty deed
   (B) an option
   (C) an easement
   (D) a lease

10. When an individual holds property past the expiration of a lease without the landlord's consent, the leasehold estate he has is a
    (A) tenancy at sufferance
    (B) freehold estate
    (C) common of pasturage
    (D) holdover tenant

11. A sublease is a
    (A) lease made by a lessor
    (B) lease made by a lessee and a third party
    (C) lease for basement space
    (D) condition of property

12. A landlord rents a store to a furniture retailer on a percentage lease. On which of the following is the percentage usually based?
    (A) market value
    (B) sales price
    (C) tenant's gross sales
    (D) tenant's net income

13. Which of the following will *not* terminate a lease?
    (A) performance
    (B) breach
    (C) surrender
    (D) vacancy

14. When a leased property is sold, what effect does the sale have upon the tenant?
    (A) The tenant must record the lease in the recorder's office.
    (B) The tenant must obtain an assignment from the purchaser.
    (C) The tenant must move out after reasonable notice.
    (D) There is no effect.

15. A lease for more than 1 year *must* be in writing to be enforceable because the
    (A) landlord or tenant might forget the terms
    (B) tenant must sign the agreement to pay rent
    (C) Statute of Frauds requires it
    (D) lease can then be assigned to another person

16. A lease can state that the rent is to be paid in
    (A) labor
    (B) crops
    (C) cash
    (D) any of the above

17. An estate at will is
    (A) a limited partnership
    (B) a tenancy of uncertain duration
    (C) an inheritance by will
    (D) a life tenancy

18. A reversionary interest
    (A) expires when a lease is signed
    (B) allows land to escheat at the termination of a lease
    (C) is a landlord's right to use of property upon expiration of a lease
    (D) constitutes a breach of a lease

19. In a sale and leaseback, the existing owner sells the property and becomes
    (A) the landlord
    (B) the tenant
    (C) burdened by debt
    (D) free to choose

20. Escalation clauses in a lease provide for
    (A) term extensions
    (B) elevator and escalator maintenance
    (C) increased rentals because of higher operating expenses
    (D) purchase options

21. Which of the following statements is true?
    (A) Property managers are required to be licensed in most states.
    (B) Percentage leases are illegal in many states.
    (C) An unlicensed person may solicit tenants for several properties provided that she works for a licensed property manager.
    (D) Property managers do not negotiate leases.

22. Which of the following is true?
    (A) Property management is a real estate brokerage function.
    (B) Resident managers of apartment complexes don't always have to be licensed.
    (C) Property managers usually are paid a percentage of rents collected as their fee.
    (D) all of the above

23. A person who remains on leased property, without the landlord's consent, after the expiration of a lease is
    (A) a tenant at will
    (B) a tenant *par duration*
    (C) a tenant in common
    (D) a tenant at suffrance

24. A lease that requires that all or part of the tenant's rent be based on the tenant's revenues is
    (A) a lease at will
    (B) a flat lease
    (C) a percentage lease
    (D) a gross lease

25. A triple-net lease is one where the tenant pays
    (A) property taxes
    (B) insurance
    (C) maintenance
    (D) all of the above

26. An economic statistic frequently selected as a basis for periodically adjusting rent is the
    (A) unemployment rate
    (B) gross domestic product
    (C) Bureau of National Affairs
    (D) Consumer Price Index

27. A commercial property investor is typically most interested in
    (A) land value
    (B) current net income
    (C) tax shelter
    (D) appreciation

28. Rent on a two-year commercial lease is to be increased or decreased each month based on certain economic indicators. This is called
    (A) a percentage lease
    (B) an index(ed) lease
    (C) an economic lease
    (D) a convoluted lease

29. To represent an owner in an effort to lease space in a shopping center requires a
    (A) professional corporation
    (B) attorney
    (C) listing
    (D) real estate broker's license

30. In a typical percentage lease, rent is calculated as a percentage of
    (A) assets of the lessee's business
    (B) net sales of the lessee's business
    (C) gross sales of the lessee's business
    (D) net taxable income of the lessee's business

31. In a tenancy at will, how much notice of intent to terminate must the tenant give the landlord?
    (A) forty-five days if the tenancy is month-to-month
    (B) seven days if the tenancy is week-to-week
    (C) both A and B
    (D) neither A nor B

32. A lease in which the tenant pays maintenance expenses, utilities, and property taxes is a __________ lease.
    (A) gross
    (B) percentage
    (C) leasehold
    (D) net

33. An advantage of being a tenant in a home under a month-to-month lease is
    (A) inflation hedge
    (B) mobility
    (C) income tax deductions
    (D) financial leverage

34. A businessman wants to rent a retail space for use as a barbershop. He is willing to pay rent, property taxes, insurance, and utilities. The lease is called
    (A) gross
    (B) net or triple-net
    (C) percentage
    (D) stepped

35. A retail tenant in a mall pays monthly rent and, in addition, pays all property maintenance, taxes, utilities, and cleaning services. This is known as a
    (A) net lease
    (B) percentage lease
    (C) ground lease
    (D) gross lease

36. Michael, a life tenant, decided to remove a backyard deck and tear down the garage so he could park his car behind the house. Can Ellen, the remainderman, prevent Michael from doing this?
   (A) No. Michael owns a life estate and may use the property during his lifetime as he wishes.
   (B) Yes. A life tenant may not commit waste of any kind.
   (C) No. A life tenant is permitted to remove any improvement on the property.
   (D) Yes, but Michael must pay Ellen an amount equal to the value of the deck and garage.

37. The right of a tenant to take possession of leased premises and not be evicted by anyone who claims superior rights is called the
   (A) covenant of possession
   (B) occupancy guarantee
   (C) covenant of protection
   (D) covenant of quiet enjoyment

38. The sale of a property that is under a long-term lease
   (A) cannot be made unless the present tenant is given an opportunity to terminate the lease
   (B) terminates the lease, and the tenant must negotiate a new lease with the new owner
   (C) terminates the lease upon 45 days' notice by the new owner
   (D) has no effect on the lease as far as the tenant is concerned

39. A tenant's rights under a lease are
   (A) terminated when the property is sold
   (B) usually terminated when the lessor dies
   (C) terminated when the property is mortgaged
   (D) terminated upon a surrender

40. The interest of a tenant in a rented building is called a
   (A) leasehold
   (B) leased fee
   (C) tenant's rights
   (D) tenancy discount

41. Which of the following statements about shopping center leases is true?
   (A) A store's lease in a shopping center is often a percentage lease.
   (B) An index lease requires constant payments over the term of the lease.
   (C) A gross lease is one where the rent is based on an agreed percentage of the gross income.
   (D) A net lease is one where the rent is based on a fixed percentage of the net income.

42. Under a net lease, a tenant would pay a set amount of rent plus a portion of all the following *except*
   (A) taxes
   (B) insurance
   (C) maintenance
   (D) debt service

43. When leased premises reach a physical condition whereby the tenant is unable to occupy them for the purpose intended, the situation is legally recognized as
   (A) an action in ejection
   (B) constructive eviction
   (C) passive eviction
   (D) dispossess eviction

44. If a tenant defaults on a lease and abandons the property in good condition, for which of the following could the tenant be held liable?
   (A) balance of the rent plus forfeiture of the security deposit
   (B) balance of the rent plus the cost to find a new tenant
   (C) decrease in the market value of the property
   (D) balance of the rent over the full remaining lease term

45. The term that describes a tenant's interest in a property is
    (A) a fee simple interest
    (B) a reversionary interest
    (C) a remainder interest
    (D) a leasehold estate

46. A leasehold estate that has a definite beginning and ending date and does not automatically renew is known as
    (A) an estate for years
    (B) a periodic tenancy
    (C) an estate at will
    (D) an estate at sufferance

47. Which of the following parties to a long-term ground lease is the holder of the leased fee?
    (A) grantee
    (B) grantor
    (C) lessor
    (D) lessee

48. A tenant has installed a bar and barstools. Upon the expiration of the lease, the tenant
    (A) may remove the trade fixtures
    (B) must leave these fixtures
    (C) can return within a year after lease expiration to take these trade fixtures
    (D) may make the next tenant pay to buy or use these trade fixtures

49. In a gross lease, who pays for operating expenses?
    (A) lessor
    (B) lessee
    (C) tenant
    (D) mortgagee

50. When a lease calls for future rental increases in steps at periodic intervals, it is called
    (A) percentage
    (B) stair stepped
    (C) graduated
    (D) grossed up

51. The lease on a restaurant has expired. The tenant can remove and move everything he has installed except
    (A) hanging light fixtures
    (B) ovens and dishwashers
    (C) bar stools
    (D) underground sprinkler system

52. A lease has two years left. The lessee may withhold rent if
    (A) the property is sold
    (B) the lessee moves out of town
    (C) the property is condemned for health reasons as a result of the lessor's neglect
    (D) the lessee's family outgrows the space

53. A property manager's compensation is most often based on
    (A) a percentage of rental payments collected
    (B) a percentage of net operating income
    (C) a percentage of operating expenses
    (D) a percentage of expenses saved

54. A property management agreement likely includes
    (A) a current rent roll
    (B) a title search
    (C) a description of the manager's responsibilities
    (D) a list of favored repairmen

55. The duties of a property manager include all of the following *except*
    (A) maintenance of the property
    (B) preserving the value of the property
    (C) investing the income stream generated by the property
    (D) meeting the goals set by the property owner

56. A properly drafted property management agreement should contain all of the following *except*
    (A) names of owner and manager
    (B) requirement that the manager provide periodic reports to the owner
    (C) fee schedule
    (D) list of approved janitorial companies

57. A leasehold estate has a
    (A) single owner
    (B) bundle of rights
    (C) limited term
    (D) rent control

58. The lease of a house for ten years is a(n)
    (A) encroachment
    (B) mortgage
    (C) transfer of rights to use in exchange for rent
    (D) life estate pur autre vie

59. Two medical students both sign a one-year lease on a single-family house. Three months later, one moves out. The remaining tenant
    (A) is responsible for all rent for the full remaining term
    (B) must pay half the rent for the remaining lease term
    (C) may move out with no further obligation
    (D) can get the medical school to pay half the remaining rent

60. A commercial tenant has 10,000 square feet under lease. He needs only 6,000 square feet for his use. He has found another business to lease the 4,000 square feet he doesn't need. What document is likely to be used to accommodate his need?
    (A) novation
    (B) assignment of lease
    (C) sublease
    (D) abandonment of lease

61. A landlord agrees to receive $10.00 per square foot rental income. He will pay for taxes, insurance, and utilities. The lease is described as
    (A) percentage
    (B) gross
    (C) net
    (D) proprietary

62. An owner who leases property to a tenant gives up the right of ________ for the duration of the lease.
    (A) occupancy
    (B) survivorship
    (C) estovers
    (D) residence for voting purposes

63. Leases for more than one year must be ________ to be enforceable.
    (A) verbal
    (B) written
    (C) acknowledged
    (D) recorded

64. When property is leased, the landlord owns
    (A) a lease
    (B) a mortgage
    (C) a fee simple estate
    (D) a leased fee estate

65. In a sublease, the original tenant
    (A) is released from all lease obligations
    (B) is denied use of the property
    (C) continues to be liable under the original lease
    (D) must sign a novation

66. An oral lease for less than one year is
    (A) unenforceable
    (B) binding
    (C) terminable on 30 days' notice
    (D) automatically renewable for one year

67. When leased property is sold,
   (A) the lease is canceled
   (B) a new lease must be arranged
   (C) the lease is binding on the new owner and existing tenant
   (D) the tenant may vacate at any time by terminating the lease

68. The difference between a lease assignment and a sublease is
   (A) assignment is for the entire lease
   (B) sublease is always shorter
   (C) sublease is for less space
   (D) assignment frees the first tenant from the lease

ANSWERS

1. **C**
2. **B**
3. **D**
4. **C**
5. **B**
6. **A**
7. **A**
8. **A**
9. **D**
10. **A**
11. **B**
12. **C**
13. **D**
14. **D**
15. **C**
16. **D**
17. **B**
18. **C**
19. **B**
20. **C**
21. **A**
22. **D**
23. **D**
24. **C**
25. **D**
26. **D**
27. **B**
28. **B**
29. **D**
30. **C**
31. **B**
32. **D**
33. **B**
34. **B**
35. **A**
36. **B**
37. **D**
38. **D**
39. **D**
40. **A**
41. **A**
42. **D**
43. **B**
44. **D**
45. **D**
46. **A**
47. **C**
48. **A**
49. **A**
50. **C**
51. **D**
52. **C**
53. **A**
54. **C**
55. **C**
56. **D**
57. **C**
58. **C**
59. **A**
60. **C**
61. **B**
62. **A**
63. **B**
64. **D**
65. **C**
66. **B**
67. **C**
68. **A**

Listing and Buyer Agency Agreements

11

ESSENTIAL MATTERS TO LEARN FROM THIS CHAPTER

- A *listing* is a contract of employment between a broker and a principal who is the property owner. Listings come under the Law of Agency. See Chapter 4 for fiduciary duties of a broker.
- The *listing agreement* obligates the owner to pay the broker a commission when the broker brings a ready, willing, and able buyer who is willing to buy at the listing terms.
- In most states, a listing must be in writing and have a fixed expiration date. Generally the term is between 60 and 120 days. The broker must leave a copy of the listing with the owner immediately after it is signed.
- Listings belong to the broker, even when they are procured by a salesperson. Although a particular salesperson may have procured the listing, when that salesperson leaves the employ of the broker, the listing remains the property of the broker.
- Major types of listings include:

  1. **OPEN.** The owner reserves the right to engage additional brokers. The one who sells the property earns the commission. This is often used with commercial property.
  2. **EXCLUSIVE AGENCY.** Only one broker may earn the commission. The owner can sell by his own efforts without paying a commission.
  3. **EXCLUSIVE RIGHT TO SELL.** The broker gets a commission from a sale, no matter who sells, including the owner.
  4. **NET.** The owner receives a stated net amount from a sale, and the broker keeps the excess price over the stated net amount as a commission. Net listings are illegal or unethical, depending on the state.
  5. **MULTIPLE LISTINGS.** The broker who has the listing shares the commission in a sale with a cooperating broker who brings the purchaser.
  6. **LIMITED SERVICE.** The broker provides limited services for the owner or buyer, usually for a fixed fee. Such services may include helping the seller or buyer complete a contract-of-sale form.

- The listing price is established by the property owner, often with market information provided by the broker.
- The rate of commission is specified in the listing. It is negotiated with the owner.
- Fixed commissions set by different brokers in the local market are prohibited by the Sherman Antitrust Act because they are anticompetitive.
- The listing broker collects the commission at closing. That broker *shares* (or *splits*) *the commission* with the salesperson who procured the listing and with the cooperating broker who brought the buyer.

- *Buyer agency agreements* occur when the buyer agrees to be represented by a broker who seeks a suitable property and owes fiduciary duties to the buyer.
- *Dual agency* occurs when the same broker represents both parties. All parties must be aware that this arrangement exists and understand that the broker cannot provide loyalty to both parties. A broker may require that different salespersons within the same firm represent buyer and seller, respectively.
- A *designated agency* brings in a separate broker to assist a party in a sale instead of having both buyer and seller represented within the same firm.
- *Transaction brokerage* occurs when a buyer or seller asks a broker to provide certain services even though the newly introduced broker did not bring either the buyer or seller.
- *Termination* of a listing occurs in two distinct categories: by the *parties* or by *operation of law.*
- Termination *by parties*:

  1. *Performance.* The property is sold.
  2. *Mutual consent.* Both parties agree to terminate.
  3. *Expiration of time.* Listing period expires.
  4. *Revocation by principal.* Principal may revoke but may owe commission.
  5. *Revocation by broker.* Broker may revoke but may be liable for damages.

- Termination *by law*:

  1. *Death* of either party.
  2. *Bankruptcy* of either party.
  3. *Insanity* of either party.
  4. *Destruction* of the property.

- Be sure to study Chapter 4 on Agency Law and Chapter 21 on Fair Housing law.

HIGHLIGHTS OF LISTINGS

A *listing* is a contract of employment between a principal (property owner) and an agent (broker) to sell or lease real estate. Under the terms of the listing contract, the principal agrees to pay the broker a commission when the broker locates a party who is ready, willing, and able to buy or lease the listed property at its list price. Should an offer be tendered for less than the list price, the broker is entitled to a commission only if the transaction is consummated. The listing is "owned" by the broker, not the salesperson who procured it.

A different kind of contract between a real estate broker and his/her client is a *buyer agency agreement.* We usually don't call them listings, though, because the term *listing* has always referred to property for sale.

Normally, the seller pays the commission at closing. It is possible for a buyer to agree to pay a commission in lieu of the seller or in addition to the seller, so long as all parties are aware of the circumstances.

TYPES OF LISTINGS

Listings can be categorized into six broad types as follows:

1. Open listings
2. Exclusive agency listings
3. Exclusive right to sell listings

4. Net listings
5. Multiple listing service (MLS)
6. Limited-service listing agreement

Each type is described below.

1. Open Listings

Using an *open listing,* the principal offers the broker a commission provided that the broker secures a purchaser. The owner reserves the right to list the property with other brokers or to sell the property himself. Thus, the open listing is open to other real estate brokers. All open listings are automatically canceled upon a sale of the property so that the seller pays no more than one commission to one broker. However, if the owner gave an exclusive listing to a broker and also gave an open listing to a different broker who produced a buyer, the owner is liable for commissions to both brokers.

2. Exclusive Agency Listings

In an *exclusive agency listing,* one broker is employed to the exclusion of all other brokers. The owner retains the right to sell the property herself without paying a commission. However, if the property is sold by a broker other than the listing broker, the listing broker is still entitled to a commission.

3. Exclusive Right to Sell Listings

The *exclusive right to sell listing* allows the employed broker to collect a commission upon the sale, no matter who sells the property. Whether the employed broker, another broker, or the owner procures a purchaser, a broker with an exclusive right to sell receives his fee upon the sale.

At first glance, such a contract may seem to be to a seller's disadvantage. Even if the seller procures a buyer on her own or another broker finds one, the listing broker receives a commission. However, this rigid arrangement gives the listing broker the assurance that the property is available for sale by him only. Brokers employed by other types of listings may be dismayed to learn that property has been sold by the owner or another broker. Since this does not occur under an exclusive right to sell, the employed broker has more incentive to devote his efforts and to spend money to promote the sale of the listed property. In recognition of this fact, most real estate boards encourage the exclusive right to sell listing, particularly for single-family dwellings.

4. Net Listings

In a *net listing,* the property owner fixes the amount that he wishes to receive from a sale. He informs the broker, for example, that he wants a $175,000 check from the sale. The broker may keep any amount as a commission so long as the seller "nets" $175,000. Thus, if the sale price is $175,001, the commission is $1; a sale at $190,000 yields a $15,000 commission.

Generally, net listings are held in poor repute in the real estate brokerage business. In many states they are illegal; in the others they are "frowned upon" or otherwise discouraged by licensing officials. The reason is that they offer too many opportunities for unethical brokers to violate their agency relationships with their principals (the owners who list their properties with them). An owner who isn't knowledgeable about the actual value of her property could be talked into net listing the property at a low price, thereby all but guaranteeing the broker an unusually high payment for selling the property. Even if the listing price is reasonable, brokers may try to "hide" or may illegally turn down offers that the owner probably would accept but that don't provide enough excess for what the broker thinks is an adequate commission. For example, a property is net listed for $200,000. Obviously, the owner will take any offer of $200,000 or more since the net listing assures her of receiving the full $200,000 (but no more) from any offer of at least $200,000. A broker who receives an offer of $210,000 may consider his $10,000 share too little and be tempted to try on his own to negotiate a higher price, turn the offer down, delay transmitting the offer, or otherwise violate his obligation under license law to *immediately* transmit the offer to the owner.

In states that permit net listings, authorities and professional real estate brokerage associations suggest that, when offered a net listing, the broker add her customary charge to the net amount and then inform the seller of the resulting list price. For example, a 6% commission rate added to a $160,000 net figure results in a $170,212.76 offering price. Six percent of $170,212.76 is $10,212.76, leaving $160,000 for the seller. The arithmetic works as follows:

$$\frac{\text{Net Amount to Seller}}{\text{Fraction of Sales Price to Seller*}} = \text{List Price with Commission}$$

(*100% less commission)

When the $160,000 net figure and the 6% rate of commission are used, the list price becomes:

$$\frac{\$160{,}000}{0.94} = \$170{,}212.76$$

Note that you can't simply add the commission as the percentage of dollars above the net price. The list price with commission is calculated by dividing the net amount to the seller by 1 minus the commission rate expressed as a decimal.

5. Multiple Listing Service (MLS)

In many metropolitan areas, brokers who are members of an MLS agree to pool all of their listings. They form a *multiple listing service* to coordinate activities. Through that sevice, brokers agree to share their exclusive right to sell listings; any cooperating broker may sell property that another has listed. Upon the sale of multiple-listed property, the commission is split between the listing and the selling broker in accordance with the agreement established by the multiple listing service. Members pay fees to the service to cover its expenses of operation, which include the periodic publication, in writing and/or by computer, of the features of all listed property of the brokers belonging to that multiple listing service.

6. Limited-Service Listing Agreement

In a *limited-service listing agreement*, the seller agrees to pay the broker. Often payment is a flat fee in advance that is nonrefundable. The broker agrees to perform certain services, such as inclusion of the house in the multiple listing service (MLS), with all inquiries directly to the seller (rather than the broker). The seller may agree to pay a commission (typically 2–4%) to the buyer's licensed representative. The seller may, for an additional fee typically paid at closing, have the listing broker provide seller representation services. Generally, under the agreement, the owner may sell to a buyer he or she procures, without paying a commission.

SELECTING A FORM

No special form is required for a listing. However, it is important to read the form used to gain an understanding of the type of listing.

AGENCY LAW

A broker is an agent for his principal. The broker's duties, obligations, and responsibilities to his principal are covered by the Law of Agency. Chapter 4 of this book describes this body of law as it pertains to real estate brokerage. Many states require that, upon first contact with a prospective buyer, a broker or salesperson must disclose to the prospect the agency relationship with the seller, if there is one. Ordinarily, this is accomplished by having the buyer sign and return to the agent a form (usually designed and prescribed by law) that describes the agency relationship so that the prospect understands to which party the agent owes loyalty. Brokers can represent buyers in negotiations and transactions. In such cases, they are agents of the buyer, not the seller, and there has to be an agency contract between broker and buyer. Disclosure is required here, too; when negotiations begin with a seller (or seller's broker) of a property the buyer is interested in, the seller and/or listing broker must be notified of the buyer-broker agency relationship.

RATE OF COMMISSION

As a matter of law, commission rates for the sale of real estate are negotiable between the principal and his broker. The rate or amount should be fixed in the listing contract to avoid misunderstandings. A "standard" commission rate in a local area violates the Sherman Antitrust Act since it is noncompetitive.

Trick Question

A trick question involves use of the word *fixed.* Commission fixing among brokers in a city or state applies a uniform rate to all listings. This practice is anticompetitive and is outlawed by the Sherman Antitrust Act. A commission rate in a given listing, however, *should be* fixed, meaning set or specified in the listing. In that sense, a *fixed* commission rate is a perfectly sound and legal business arrangement.

LISTING PERIOD

Contracts for the listing of real estate should have a fixed expiration date. Most contracts provide 60 to 120 days for the sale of single-family homes and 6 months to a year for commercial property and apartment complexes. The law does not establish limits, so the actual period agreed upon is a matter of negotiation.

Occasionally, after having been shown a particular piece of property through a broker, a prospect and the property owner may arrange a sale "behind the broker's back." Together they wait for the listing to expire and then contract for the sale without advising the broker in an attempt to elude the payment of a commission. Despite the fact that the listing has expired, courts are likely to require payment to a licensed broker whose listing expired if she can show that she was the procuring cause of the sale. Many listings contain a provision that protects the broker in these situations.

A *broker protection clause* specifies that the broker is entitled to a commission when a prospect he introduced buys the property after the listing has expired.

An *extender clause* has been used to continue the entire listing automatically after its expiration date. However, this use is considered to violate the requirement that a listing have a fixed termination date. In many practices in recent years, the extender clause has functioned in a similar manner to a broker protection clause.

BUYER AGENCY AGREEMENTS

As agents, real estate brokers don't just represent sellers (in listing contracts); they can also represent buyers. When they do, the broker and buyer may execute a *buyer agency agreement.* This contract obligates a broker to represent the buyer and to look out for the buyer's interests. As the buyer's agent, the broker or salesperson can point out other properties that might better suit the buyer's needs, can state that a given property may be overpriced, and can otherwise advise and assist the buyer to get the best possible property to suit the buyer's needs, and at the best possible price. A listing broker or salesperson shouldn't offer this kind of assistance to a potential buyer because it would violate the agency obligation to the seller that was created by the listing contract.

Most buyer agency agreements are exclusive; that is, the buyer agrees not to employ another broker during the term of the buyer agency contract. Of particular concern is how the buyer's agent will be paid. In "traditional" real estate brokerage, all agents are paid out of the proceeds of the commission paid by the seller to his/her listing broker. Almost everywhere in the United States, buyer agency works the same way. Today's listing contracts typically stipulate that the seller will allow a buyer's agent to be paid from the proceeds of the commission that the seller will pay, and the buyer agency contract usually says that this is how the buyer's broker will be paid. However, the buyer agency contract also describes payment that the buyer must make if the buyer receives out-of-the-ordinary kinds of services from the broker or if the buyer decides to buy a property that is under a listing contract that does *not* allow commission sharing with a buyer's broker.

Buyer agency agreements have become popular; the reason is obvious. They give the buyer access to the same level of advice and expertise that sellers have traditionally always enjoyed from their brokers.

Buyer agency agreements are subject to the same agency law provisions as listing contracts and may be terminated the same ways as described previously for listing contracts. Buyer agency agreements also have limited terms (60–120 days is typical).

You may read your state's license law carefully to find out exactly how buyer agency is treated and what rules and regulations apply to it.

DUAL AGENCY, TRANSACTION BROKERAGE

Dual agency and *transaction brokerage* are special cases of agency. In dual agency, the licensee represents both buyer and seller at the same time. In transaction brokerage (some states call it *facilitative brokerage*), the licensee works with both seller and buyer and the licensee's responsibilities to either party are limited. In some states, transaction brokerage is treated as no brokerage agency.

Dual Agency

Dual agency creates potential conflict because it is impossible for a dual agent to provide full agency responsibility to both parties at the same time. Therefore, the dual agency contract severely restricts the actual agency responsibility that the dual agent has. Typically, the dual agent cannot tell the seller the maximum price the buyer is willing to pay and cannot tell the buyer the minimum price the seller will accept. Nor can the dual agent tell either party about any particular urgency to transact (or lack of it) that the other party may have. Essentially, the dual agent is pretty much prohibited from giving either party an advantage.

Transaction Brokerage

A transaction broker is generally engaged to provide limited representation for a buyer or seller. This allows a licensee to assist both the buyer and the seller. However, a licensee will not work to represent one party to the detriment of the other party when acting as a transaction broker to both parties.

The duties of a transaction broker, which in some states are to be disclosed in a residential transaction, are as follows:

- Dealing honestly and fairly.
- Accounting for all funds.
- Using skill, care, and diligence in the transaction.
- Disclosing all known facts that materially affect the value of residential real property and are not readily observable to the buyer.
- Presenting all offers and counteroffers in a timely manner, unless a party has previously directed the licensee otherwise in writing.
- Limited confidentiality, unless waived in writing by a party. This limited confidentiality will prevent disclosure that the seller will accept a price less than the asking or listed price, that the buyer will pay a price greater than the price submitted in a written offer, of the motivation of any party for selling or buying property, that a seller or buyer will agree to financing terms other than those offered, or of any other information requested by a party to remain confidential.
- Any additional duties that are entered into by separate written agreement.

It should be noted that loyalty, confidentiality, obedience, and full disclosure are not included. In a transaction broker relationship, the seller (or buyer) is considered to be a customer of the broker, not a principal. The customer is not responsible for the acts of the broker-licensee.

Transaction brokerage in some states doesn't establish any agency relationship at all. The transaction broker is subject to the same restrictions as the dual agent regarding information for either seller or buyer. The only responsibility the transaction broker has to either party is not to make mistakes in what he does for them and to exercise due diligence in what he does. Essentially, the transaction broker handles the paperwork, provides the buyer access to listed properties, introduces buyer and seller, supervises negotiations (but doesn't offer advice about price or terms of sale)—that sort of thing. It's up to the buyer and seller themselves to get the other information they need and to conduct the negotiations.

In your practice, check your state's license law very carefully to see if it allows either dual agency (most do) or transaction brokerage (many don't, but it is increasing in popularity).

For whichever (or both) is allowed, make sure you understand your state's restrictions and rules regarding them!

Designated Agency

Finally, see if your state allows *designated agency.* This is a new, 21st-century "innovation." It is supposed to "solve" the problem that occurs when a licensee has a buyer agency agreement with a buying prospect and wants to show that prospect a house for which she also has a listing contract (seller agency). This would create a dual agency situation in which both parties lose a lot of agency representation benefits. Designated agency, in the states that allow it, lets the licensee's managing broker assign one "side" of the transaction to a different agent within the firm. This means that one of the parties suddenly finds himself working with a different agency so long as he is interested in this particular property. However, that agent can give full agency representation, and the other party can get full agency representation from the original agent. Typically, it is the buyer who is assigned to the designated agent; this can't happen, though, without the assigned party's consent. This consent may be written into the original agency contract (listing or buyer-agency agreement), or permission may be obtained at the time the designated agency occurs.

TERMINATION OF LISTINGS

Listing agreements may be terminated either by the parties or by operation of law.

Termination by the parties can occur through:

1. **PERFORMANCE.** The broker procures a purchaser; the listing contract is terminated when the sale is completed.
2. **MUTUAL CONSENT.** Both parties agree to terminate the listing.
3. **EXPIRATION OF AGREED TIME.** A listing generally has a definite expiration date. If it doesn't, the listing ends after a reasonable period. In many states, a listing must have a fixed expiration date.
4. **REVOCATION BY THE PRINCIPAL.** At any time, the principal may revoke authority given to the broker. However, the principal may be liable for damages resulting from the breach of contract. The principal is not liable for damages if he can show the agent to have been negligent in her duties, as by disloyalty, dishonesty, or incompetence.
5. **REVOCATION BY THE BROKER.** An agent may terminate the listing contract or abandon it. However, he may be held liable for damages to the principal because of failure to complete the object of the listing contract.

Termination by law can occur through:

1. **DEATH OF EITHER PARTY.** The death of either the principal or the other party (broker or firm) generally causes termination of a listing contract.
2. **BANKRUPTCY OF EITHER PARTY.** Bankruptcy of a principal or an agent normally terminates a contract.
3. **INSANITY OF EITHER PARTY.** If either party is judged to be insane, he is considered incapable of completing the contract; therefore, the listing contract will terminate.
4. **DESTRUCTION OF THE PROPERTY.** If the property upon which the agency had been created is destroyed, the agency is terminated.

QUESTIONS ON CHAPTER 11

1. When more than one broker is employed by an owner to sell real estate, there exists
   (A) an exclusive agency listing
   (B) an open listing
   (C) an exclusive right to sell
   (D) a unilateral listing

2. To collect a commission in court, a broker must show that
   (A) he is licensed
   (B) he had a contract of employment
   (C) he is the cause of the sale
   (D) all of the above

3. The broker's responsibility to the owner of realty is regulated by
   (A) the law of agency
   (B) the law of equity
   (C) rendition superior
   (D) investiture

4. What is a real estate listing?
   (A) a list of brokers and salespersons
   (B) a list of property held by an owner
   (C) the employment of a broker by an owner to sell or lease
   (D) a written list of improvements on land

5. Under an open listing
   (A) the seller is not legally required to notify other agents in the case of sale by one of the listing brokers
   (B) the owner may sell the property without paying a commission
   (C) both A and B
   (D) neither A nor B

6. If the listing owner sells her property while the listing agreement is valid, she is liable for a commission under a(n) ________________ listing agreement.
   (A) net
   (B) exclusive right to sell
   (C) exclusive agency listing
   (D) open

7. Under an exclusive agency listing,
   (A) the seller may sell by his own effort without obligation to pay a commission
   (B) the broker receives a commission regardless of whether the property is sold
   (C) both A and B
   (D) neither A nor B

8. The listing agreement may not be terminated
   (A) without compensation if the broker has found a prospect ready, able, and willing to buy on the seller's terms
   (B) because of incompetence of the prospect
   (C) if the seller files for bankruptcy
   (D) if the listing agreement is for a definite time

9. A multiple listing
   (A) is illegal in certain states
   (B) is a service organized by a group of brokers
   (C) requires that members turn over eligible listings within 12 days
   (D) causes lost commission

10. Under which type of listing are brokers reluctant to actively pursue a sale because of the risk of losing a sale to competing brokers?
    (A) an exclusive right to sell listing
    (B) a net listing
    (C) an open listing
    (D) a multiple listing

11. A dual agent
    (A) may perform the same duties for both buyer and seller that she would if she were agent for only one of them
    (B) is severely restricted in the agency duties that she can perform for either party
    (C) may represent either party but only one of them at a time
    (D) is someone who has more than one property listed

12. Broker Jones is working with buyer Smith and seller Brown in a real estate transaction. Although Jones provides assistance, he is not an agent for either Smith or Brown. Jones is
    (A) a dual agent
    (B) a buyer's agent
    (C) a seller's agent
    (D) a transaction broker

13. A principal may sell her property without paying a commission under
    (A) an open listing
    (B) an exclusive agency listing
    (C) both A and B
    (D) neither A nor B

14. The rate of commission is normally fixed under
    (A) a net listing
    (B) an open listing
    (C) both A and B
    (D) neither A nor B

15. Buyer and seller both may pay a commission so long as
    (A) all parties are aware of the situation
    (B) payments are made to the same broker
    (C) both A and B
    (D) neither A nor B

16. It is possible to terminate a listing by
    (A) revocation by the principal
    (B) revocation by the agent
    (C) both A and B
    (D) neither A nor B

17. An exclusive right to sell listing enables
    (A) the seller to sell the property himself without paying a commission
    (B) an unlicensed broker to sell and earn a commission
    (C) both A and B
    (D) neither A nor B

18. Broker Morris has a written agreement with Sarah Lee as the only broker to find a one-story house with three bedrooms and two baths in Lancaster. What type of agreement does Broker Morris have?
    (A) an open buyer's agent agreement
    (B) a dual agency agreement
    (C) an exclusive buyer's agent agreement
    (D) an agency agreement

19. Which of the following would terminate a listing agreement?
    (A) death of the broker
    (B) revocation of the salesperson's license
    (C) an offer to purchase
    (D) expiration of the salesperson's license

20. The Floyds' home was listed with a broker, but they located a buyer and sold the property themselves. However, they were still obligated to pay a brokerage fee. What kind of listing had they agreed to?
    (A) a net listing
    (B) an open listing
    (C) an exclusive right to sell
    (D) a multiple listing

21. If the agent has not located a buyer by the end of the listing period, what happens?
    (A) The listing is automatically extended for an additional 30 days.
    (B) The listing price is reduced by 10%.
    (C) The agency relationship terminates.
    (D) The broker can charge for advertising.

22. A contract in which an owner promises to pay compensation to a broker for providing a buyer but may give listings to other brokers is known as
    (A) an exclusive right to sell
    (B) a net listing
    (C) an open listing
    (D) a multiple listing

23. Stephen, a salesperson associated with Broker Bob, lists a house for sale for $320,000 with 6% commission due at closing. Three weeks later, the owner accepts an offer of $300,000, brought in by Stephen. Bob's practice is that 45% of commissions goes to the office and the remainder to the salesperson. How much will Stephen make on the sale?
    (A) $18,000
    (B) $10,560
    (C) $9,500
    (D) $9,900

24. Before Mary listed her house, she reserved the right to sell the property to anyone she found without paying a broker. What kind of listing is this?
    (A) exclusive right to sell
    (B) exclusive agency
    (C) multiple
    (D) open

25. Roger has contracted with three brokers to sell his ranch. He has agreed to pay a commission to the broker who brings a ready, willing, and able buyer and who meets his terms. This is a(n)
    (A) exclusive listing
    (B) exclusive right to sell
    (C) open listing
    (D) multiple listing

26. The listing broker will receive a commission fee of 6% on the first $250,000 of the sale price and 4% on any amount above $250,000. If the property sells for $450,000, how much is the total commission?
    (A) $15,000
    (B) $18,000
    (C) $27,000
    (D) $23,000

27. Which of the following is included in an agent's responsibilities to the party the agent is *not* representing?
    (A) The agent must disclose all information to the other party.
    (B) The agent owes no loyalty to the other party.
    (C) The agent must provide all information about the principal.
    (D) The agent must advise the other party.

28. Broker Barb listed Tony's house and subsequently showed it to her sister, who made an offer to purchase. What is Barb's responsibility of disclosure?
    (A) No disclosure is required.
    (B) Barb must disclose her family relationship to the seller.
    (C) No disclosure is required if the property is sold at the list price.
    (D) It is a violation of fiduciary duty for Barb to show the property to a relative.

29. Seller Cynthia agrees to let Broker Bob keep any amount over the $175,000 that she needs if he sells her house. This is called
    (A) a buyer's listing
    (B) an open listing
    (C) a gross listing
    (D) a net listing

30. A real estate listing usually creates
    (A) a general agency
    (B) a private agency
    (C) a universal agency
    (D) a special agency

31. An exclusive listing agreement should include
    (A) an advertising allowance
    (B) a defined market area
    (C) a definite termination date
    (D) a scaled commission percentage

32. A listing agreement states that the seller will pay the broker any amount over the final sales price plus the seller's closing costs. This is
    (A) a sound business practice
    (B) a net listing
    (C) a triple-net lease
    (D) a highly ethical brokerage practice

33. Broker Michele represented a buyer as a buyer's agent in the purchase of a home listed by Broker Susan. How may Michele be paid?
    (A) only by her client, the buyer
    (B) directly by the seller
    (C) by Broker Susan
    (D) by Broker Susan, but a subagency relationship is created

34. The broker's commission rate in a listing is set by a
    (A) Board of Realtors standard commission rates
    (B) broker's firm's policy
    (C) negotiation with the owner-seller
    (D) multiple listing service fixed commission

35. A broker should provide a copy of the listing to the owner
    (A) at closing of the sale
    (B) at presentation of a contract of sale
    (C) at the first open house
    (D) as soon as the owner signs the listing contract

36. Which type of listing gives the most protection for the broker?
    (A) open
    (B) multiple
    (C) net
    (D) exclusive right to sell

37. An owner wants to give a listing to a broker while retaining the owner's right to sell by his own efforts without paying a commission. The listing agreement that is most suitable is
    (A) an exclusive right to sell
    (B) an exclusive agency listing
    (C) a multiple listing
    (D) a net listing

38. A listing must include
    (A) applicable zoning requirements
    (B) the type of deed to be given
    (C) the owner's motivation to sell
    (D) a description of the property sufficient to identify it

39. When listing property, which of the following should a broker describe to a property owner?
    (A) possible buyers
    (B) tax assessment amounts
    (C) environmental risks
    (D) market conditions

40. A broker has an exclusive right to sell listing on a new subdivision. Potential buyers who are individuals without agents want to make offers. The broker prepares contracts for these buyers to sign. The broker
    (A) can represent both parties
    (B) must deal honestly with all parties
    (C) owes loyalty to the buyers
    (D) cannot serve two masters in the same contract

41. When a licensed salesperson prepares a listing that is signed by the property owner, the listing is a contract between
    (A) the property owner and the salesperson's broker
    (B) the property owner and the salesperson
    (C) the salesperson and the buyer
    (D) the salesperson and the broker

42. A salesperson resigns from a broker's firm. The listings the salesperson brought to the firm are the property of
   (A) the salesperson
   (B) the salesperson's broker
   (C) the salesperson but only after he or she becomes licensed as a broker
   (D) no one; the listings expire when the salesperson resigns

43. When stating the seller's listing price and terms to a prospective buyer, the seller's broker can disclose
   (A) the lowest amount the seller will accept
   (B) the price and terms included in the listing
   (C) the lowest amount the broker thinks the property will appraise for
   (D) the highest amount the broker thinks the property will appraise for

44. Which party establishes the listing price?
   (A) buyer
   (B) seller
   (C) broker
   (D) salesperson

45. Which of the following is true of both an exclusive right to sell listing and an exclusive agency listing?
   (A) Only one broker may represent the seller.
   (B) The owner may sell without paying a commission.
   (C) Both are open listings.
   (D) Neither can be entered in the MLS.

46. A salesperson listed a home for sale and then changed brokerage firms. What happened to the listing?
   (A) It stayed with the first brokerage firm.
   (B) It moved to the salesperson's new broker.
   (C) It was terminated.
   (D) The decision is up to the owner/seller who listed the house.

47. A home is listed for sale and shown by four cooperating brokers. How many total owner-agency relationships were created?
   (A) none
   (B) one
   (C) four
   (D) five

48. A licensed real estate salesperson can receive compensation from
   (A) the seller
   (B) any licensed broker
   (C) his sponsoring broker
   (D) all of the above

49. A group of brokers in Small Town, USA, agree to pool their listings. The selling broker will share the commission equally with the listing brokers. This is known as a(n)
   (A) illegal act
   (B) multiple listing service
   (C) open listing
   (D) lockbox listing

50. A broker agrees to accept as a commission whatever amount of sales price is above $100,000. This is a(n)
   (A) open listing
   (B) exclusive listing
   (C) exclusive right to sell
   (D) net listing

51. An owner agrees to pay one broker a commission for procuring a buyer unless the owner finds the buyer. This is a(n)
   (A) implied listing
   (B) open listing
   (C) exclusive listing
   (D) exclusive right to sell

52. Which of the following may a broker with an exclusive right to sell not delegate to other licensees?
(A) authority to advertise the property
(B) authority to show the property
(C) authority to create a flyer
(D) authority to receive or distribute commissions

53. To encourage a broker to market property actively, an owner may provide a(n)
(A) exclusive listing
(B) open listing
(C) exclusive right to sell
(D) implied listing

54. A residential listing contract with an owner generally gives the broker the power to
(A) accept or reject all offers
(B) market and show the property
(C) negotiate the price and terms
(D) all of the above

55. Any licensed broker may be called a
(A) broker
(B) REALTOR®
(C) realtist
(D) agent

56. The term REALTOR® is the same as
(A) salesperson
(B) broker
(C) agent
(D) none of the above

57. All brokers agreeing to charge a standard commission for residential sales throughout a market area is
(A) an ethical business practice
(B) illegal price fixing
(C) fair to all customers
(D) a house rule among brokers

58. Commingling is a broker's practice of
(A) mixing others' money with his own
(B) mixing races in a housing development
(C) selling units in a condominium project
(D) renting to unmarried tenants

59. One party to a listing defaults. What happens?
(A) The listing is terminated.
(B) The defaulting party may owe damages.
(C) The defaulting party may release the injured party.
(D) Both parties must get a lawyer for advice.

60. The law of agency governs
(A) listing contracts
(B) sales contracts
(C) mortgage contracts
(D) leases

61. A salesperson has just received a listing signed by the seller. The next step is for the salesperson to
(A) find a buyer
(B) get his broker to sign the listing
(C) prepare an advertising flyer
(D) enter the listing in the multiple listing service

62. A listing should terminate
(A) in 90 days
(B) in one year
(C) at a fixed date stated in the listing
(D) when the broker gives up efforts to sell

63. A homeowner signs a listing for 90 days but cancels it within 36 hours after signing. What happens?
(A) The seller may owe damages to the broker.
(B) By canceling quickly, the seller has no obligations.
(C) The seller owes a full commission.
(D) The seller owes a partial commission.

64. A prospective buyer tells a broker that he has won a $1 million lottery. The broker should tell
(A) the seller
(B) the title company
(C) the mortgage company
(D) no one

ANSWERS

1. **B**	11. **B**	21. **C**	31. **C**	41. **A**	51. **C**	61. **B**
2. **D**	12. **D**	22. **C**	32. **B**	42. **B**	52. **D**	62. **C**
3. **A**	13. **C**	23. **D**	33. **C**	43. **B**	53. **C**	63. **A**
4. **C**	14. **B**	24. **B**	34. **C**	44. **B**	54. **B**	64. **D**
5. **C**	15. **A**	25. **C**	35. **D**	45. **A**	55. **A**	
6. **B**	16. **C**	26. **D**	36. **D**	46. **A**	56. **D**	
7. **A**	17. **D**	27. **B**	37. **B**	47. **B**	57. **B**	
8. **A**	18. **C**	28. **B**	38. **D**	48. **C**	58. **A**	
9. **B**	19. **A**	29. **D**	39. **D**	49. **B**	59. **B**	
10. **C**	20. **C**	30. **D**	40. **B**	50. **D**	60. **A**	

Other Real Estate Contracts 12

ESSENTIAL MATTERS TO LEARN FROM THIS CHAPTER

- A *contract for exchange* is used when property owners agree to swap property. Properties of the same nature that are exchanged, such as real estate for real estate, are called "like-kind."
- Because it would be unusual for two properties to have exactly the same value, most likely one owner will give something of value to equalize the value of the exchange. That other "unlike" property is called *boot.* Boot may be cash, furniture, a car, a diamond, or anything else of value. Relief of mortgage debt is also boot received.
- An *exchange* of business or investment real estate may be *tax deferred* under *Section 1031* of the *Internal Revenue Code.* Tax deferment under Section 1031 does not apply to one's personal residence, which gets preferential treatment under a different section of the Internal Revenue Code.
- *Tax deferment* means the gain realized will not be recognized (taxable) on the exchange at that time. The potential gain may be taxable in a later transaction. Tax deferment does not apply to unlike property received, the *boot.* Tax must be paid on boot received to the extent that there is gain.
- When one or both properties is mortgaged, the fairness of the exchange is based on equalizing the equity value (market value minus debt) of the properties.
- One need not exchange property simultaneously for the gain to be deferred. A *tax-deferred exchange* is allowed whereby after a sale, the seller identifies a limited number of possible substitute like-kind properties within 45 days and closes within 180 days of the sale.
- A *contract for deed* is used when the seller finances a sale of real estate but doesn't give the buyer title until the final payment is made. This is a common method of selling resort or vacation property.
- Another name for a contract for deed is *installment land sale contract* or simply *land contract.*
- An *option* is a contract that gives a party the right to buy within a certain time but not the obligation to buy. The landowner is caller the *optionor.* The possible buyer is the *optionee.*
- The optionee may buy before the option time period expires but is not obligated to buy. An option is considered a *unilateral* contract (not bilateral) because only one party, the optionee, can act.
- If the optionee decides not to buy, the amount he paid the property owner for the option is forfeited to the owner. If the optionee does buy, the option price may or may not be applied to the purchase price, depending on what is written in the option contract.

- The purchase price should be stated in the option, set at initial negotiation.
- An *assignment* is a contract whereby another person accepts the obligation of a contract.
- An assignment does not release the *assignor* (the one who assigns to another) from the original contract.
- To release the original party to a contract and substitute another person requires a *novation.* All parties must sign the novation, which releases one party from a contract and substitutes a different person.

People dealing in real estate may want to accomplish transactions that do not involve the categories of contracts discussed in Chapters 8 through 11. In this chapter, you will learn about other types of contracts including contracts for the exchange of real estate, contracts for deed, options, assignments, and novations.

CONTRACTS FOR EXCHANGE

A *contract for exchange* is used when principals want to exchange property. You may wonder why people would want to swap their real estate, but there are peculiarities in the U.S. Internal Revenue Code that confer tax advantages on some kinds of exchanges. However, these advantages are available only for investment property (property held for rental and income) and property used in one's business. Therefore, people who exchange homes that they live in (their residences) can't use these tax breaks (those who sell their own homes are offered different tax breaks). So a residential broker or salesperson (one who focuses on housing) is unlikely to encounter an exchange transaction.

If both properties are agreed by the owners to be of equal value and both are unencumbered, they may be swapped without additional consideration. This transaction may be beneficial to the principals. For example, one person may own land and want income-producing property, whereas the other party may have the opposite holdings and needs. A swap accomplishes the objectives of both parties, even though the properties exchanged are worth much more now than they cost originally.

More often than not, the properties exchanged are of different values. Then the party who gives up the lower-valued property must also give other consideration so that the trade becomes a fair one, acceptable to both parties. Personal property (cash, cars, diamonds, etc.) put into an exchange to equalize the total value of all property exchanged is called *boot.* The party who receives boot may be taxed to the extent of the unlike property (boot)received or, if less, to the gain realized. "Unlike" property in a real estate exchange transaction is anything that isn't real estate. "Like-kind" property is real estate. The tax advantage only applies to the like-kind property that is exchanged.

Mortgaged property can be exchanged. Suppose Mr. Jones owns a $10,000 vacant lot free and clear; Mr. Smith owns a $50,000 building with a $40,000 mortgage against it. Since the equity of both is $10,000, the exchange is a fair trade. Smith's *boot* received is relief from a $40,000 debt. Deferred or delayed exchanges are allowed under IRS Section 1031 (see pages 362 and 366), which requires identification of a limited number of possible replacement properties within 45 days of a sale and closing on the one(s) selected within 180 days of a sale. There are some other rigid requirements.

Years ago, when a simultaneous exchange was required for income tax purposes, the same broker might have been involved with the buyer and seller, possibly presenting a conflicting situation (which is why such *dual agency* is now illegal in many states). In a deferred exchange, which is the current norm, the broker of the sale and the broker of the replacement property are likely to be entirely different people. It is unlikely that the same broker would be called on as an agent for parties with conflicting interests.

CONTRACTS FOR DEED

In a *contract for deed* (also known as *land contract* and as *installment land sale contract*), the property seller finances the sale but does not surrender the deed until all payments have been made. This type of contract typically allows the purchaser to gain possession, but not ownership, of the land while paying for it. Because the seller retains the deed until the final payment, he is sufficiently protected in the event of default. In many states, because the land is still legally his, he doesn't have to go through foreclosure; he can have the defaulting buyer evicted quickly and regain full possession and ownership of the land. Even in states where foreclosure is required under a contract for deed, the procedure often is less time-consuming and expensive than it would be if the seller were foreclosing a mortgage.

A contract for deed is frequently used for the sale of subdivided land, especially in resort communities.

OPTIONS

An *option* gives its holder the right, but *not* the obligation, to buy specific property within a certain time at a specified price. The option need not be exercised. An *optionee* can elect not to purchase the property, in which case amounts paid to acquire the option belong to the property owner, who is the *optionor.*

Consideration must be paid with an option. Upon exercise of the option, the amount paid for the option may or may not apply to the sales price, depending upon the agreement.

An option may be used by a speculator who is uncertain whether a significant increase in the value of the property is forthcoming. Through purchasing an option, she can be assured of a fixed price for the property in a set timespan. If the value increase materializes, she exercises the option. Otherwise, she lets it lapse and loses only the cost of the option. The holder of an option may sell the option itself (at a gain or a loss) if she wishes to do so and finds a buyer.

ASSIGNMENTS

Assignments refer to the transfer of contracts. Through an assignment, a person may convey to someone else his rights under a contract. A lease may be assigned, as may a contract of sale, a deed, an option, and so on. The person who gives or assigns the contract is the assignor; the receiver is the assignee. It should be noted that the assignor is not automatically relieved of his duties to other parties under the earlier contract. If a lease contract exists and is assigned, for example, the original lessee is still bound to pay rent if the assignee fails to do so.

Trick Question

A trick question suggests that a person can assign a contract to another and be relieved of the obligation. This is not true. Although the assignee becomes liable for the contract, the assignor is not relieved of any obligations unless the original party on the other side agrees, often through a novation. As a matter of law, one party can't be relieved of obligations simply by assigning them to another party.

NOVATIONS

A *novation* is a contract used to substitute a new contract for an existing contract between the same parties. It may also be used to substitute or replace one party to a contract with a different party. Thus, a novation is used to amend an agreement or substitute the parties involved.

A novation must be bilateral: both parties must agree to it. Suppose a home is to be sold, with the buyer assuming the seller's mortgage. Since the seller made a contract to repay the debt, she cannot automatically substitute the new buyer for herself on the debt. Acceptance of the substitution must come from the lender. The process used for this substitution is called a *novation*. In such a case, the original contract has been assigned, while at the same time the assignor has been relieved of her obligations to other parties to the contract.

QUESTIONS ON CHAPTER 12

1. Property, unlike real estate, that is used to equalize the value of all property exchanged is called
   (A) listing
   (B) gain
   (C) boot
   (D) equalizers

2. Mr. Hill has a 6-month option on 10 acres at $1,000 per acre. Mr. Hill may
   (A) buy the property for $10,000
   (B) sell the option to another
   (C) not buy the land
   (D) all of the above

3. A contract that gives someone the right but not the obligation to buy at a specified price within a specified time is
   (A) a contract for sale
   (B) an option
   (C) an agreement of sale
   (D) none of the above

4. A contract for deed also is known as
   (A) an installment land sales contract
   (B) a mortgage
   (C) land contract
   (D) both A and C

5. Which of the following best describes an installment or land contract?
   (A) a contract to buy land only
   (B) a mortgage on land
   (C) a means of conveying title immediately, while the purchaser pays for the property
   (D) a method of selling real estate whereby the purchaser pays for the property in regular installments while the seller retains title to the property until final payment

6. Mr. Beans owns land worth $5,000; Mr. Pork owns a house worth $30,000, subject to a $20,000 mortgage that Beans will assume. For a fair trade,
   (A) Pork should pay $5,000 cash in addition
   (B) Beans should pay $5,000 cash or boot in addition
   (C) Beans and Pork may trade properties evenly
   (D) none of the above

7. A speculator can tie up land to await value enhancement, without obligating himself to buy, through the use of
   (A) an installment land contract
   (B) an option
   (C) a contract for deed
   (D) A and C only

8. Ms. Smith wants to buy Ms. Jones's house and assume the 6% mortgage. To be released from debt, Jones should
   (A) assign the debt
   (B) grant an option on the debt
   (C) ask the lender for a novation to substitute Smith
   (D) none of the above

9. Larry assigned his rights to use leased property to Allen. Allen stopped paying rent. Larry may
   (A) sue Allen for the rent
   (B) not pay the property owner since he hasn't received payment from Allen
   (C) both A and B
   (D) neither A nor B

10. Treasure Homes sold a lot to Mr. Kay under an installment land sales contract. Upon making the down payment, Kay is entitled to
    (A) a warranty deed
    (B) use of the property
    (C) a fee simple estate
    (D) all of the above

11. A like-kind exchange or deferred like-kind exchange is allowed under Section __________ of the Internal Revenue Code.
    (A) 8
    (B) 1031
    (C) 1231
    (D) 1250

12. The sale of land with the deed delivered upon final payment is known as a(n)
    (A) installment land sale contract
    (B) land contract
    (C) contract for deed
    (D) all of the above

13. In a typical land contract, the seller transfers the deed
    (A) at closing
    (B) upon receipt of disclosures
    (C) after the rescission period expires
    (D) when final payment is made

14. The fairness of like-kind exchanges of mortgaged property is based on
    (A) appraised property value
    (B) equity value
    (C) book value
    (D) adjusted tax basis

15. Before the expiration of an option, the optionee may
    (A) buy the property
    (B) choose to do nothing
    (C) sell the option
    (D) do any of the above

16. If an option to purchase is exercised, which of the following is true?
    (A) The option money is automatically applied to the purchase price.
    (B) The optionor is forced to sell the property.
    (C) The notice to exercise must be in writing.
    (D) Closing takes place before the option is exercised.

17. In an option contract, which of the following is true?
    (A) The seller must sell, but the buyer need not buy.
    (B) It is specifically enforceable by both parties.
    (C) The seller may sell or not at his or her option.
    (D) The buyer must buy.

18. Which of the following concerning a *land contract* is *not* true?
    (A) The seller retains possession of the property.
    (B) The seller retains legal title to the property.
    (C) The seller finances the sale.
    (D) The buyer is protected from loss in value.

19. Someone who receives an option is called the
    (A) optionee
    (B) optionor
    (C) lessee
    (D) grantor

20. An owner sold a lakefront lot using a *contract for deed*, taking 10% down, with payments to be made over the next 10 years. When does the buyer get a deed?
    (A) upon final payment
    (B) upon closing
    (C) when the contract expires
    (D) when the seller dies

21. Before it is exercised, an option is a
    (A) contract for deed
    (B) bilateral contract
    (C) unilateral contract
    (D) trilateral contract

22. A seller provided an option to the buyer for 90 days with an option fee of $5,000 and a property price of $100,000. During the option period, the buyer decides not to buy. What does the buyer need to do?
    (A) nothing
    (B) notify the seller of his intentions
    (C) reconsider the decision once more
    (D) buy as required by the option

23. An option is granted giving the optionee the right to buy during the next 90 days. When should the purchase price be set?
    (A) when the option is granted
    (B) on the ninetieth day
    (C) anytime during the 90 days
    (D) when the broker can help negotiate the amount

24. Which of the following is correct about an option?
    (A) The landowner-optionor can void it during its term.
    (B) The landowner-optionor keeps the option sales price if the option is not exercised.
    (C) The optionee must buy the property before the option expires.
    (D) The optionee must buy the property after the option expires.

25. Mr. X, a property owner, sells an option to buy land for $1,000 per acre for 1½ years. Mr. X dies a month later. What happens?
    (A) The option remains valid.
    (B) The option is void upon his death.
    (C) The option expires one year after his death.
    (D) The option is void but can be renewed by Mr. X's heirs.

26. An option contract for the purchase of real estate
    (A) is a bilateral contract
    (B) gives the buyer time to arrange a purchase
    (C) favors the seller
    (D) includes minerals or royalties

27. Albert sold an option to Baker to buy his (Albert's) land. Which of the following is true?
    (A) Albert is the optionor.
    (B) Baker is the optionor.
    (C) Baker must buy the land.
    (D) The option is a bilateral contract.

28. An option was granted for 90 days. One day after it expired, the option holder offered to close on the purchase. The option holder
    (A) can get his option money back because the option has expired
    (B) can exercise the option because there is a three-day grace period after expiration
    (C) can exercise the option if his lawyer assists
    (D) cannot exercise the option because its rights have expired

29. Purchase of an option
    (A) requires the optionee to buy upon expiration
    (B) gives the optionee the right but not the requirement to perform
    (C) can be applied to the purchase price
    (D) is nonbinding on either party

ANSWERS

1. **C**	6. **B**	11. **B**	16. **B**	21. **C**	26. **B**
2. **D**	7. **B**	12. **D**	17. **A**	22. **A**	27. **A**
3. **B**	8. **C**	13. **D**	18. **A**	23. **A**	28. **D**
4. **D**	9. **A**	14. **B**	19. **A**	24. **B**	29. **B**
5. **D**	10. **B**	15. **D**	20. **A**	25. **A**	

PART FOUR
Real Estate Analysis

Mortgages and Finance

13

ESSENTIAL MATTERS TO LEARN FROM THIS CHAPTER

- A *mortgage loan* is the pledge of property as security for a debt. A mortgage loan combines two contracts: (1) a *note,* which is the debt, and (2) the *mortgage* (derived from Latin as *dead pledge*), which pledges property to secure the debt.
- The borrower *gives* the mortgage (pledge of property as security for the debt) to the lender. The borrower is the *mortgagor.* The lender *receives* the mortgage and is the *mortgagee.*
- In *title theory* states, the lender holds title to the property. In *lien theory* states, the lender has a lien on the property.
- Types of mortgage loans include (alphabetically):

1. **ADJUSTABLE RATE.** The interest rate changes at regular intervals, such as every year. The total rate is the sum of the *index rate* plus a *margin. Caps* are the maximum change allowed in the rate. Caps can provide annual limits and life-of-loan limits.
2. **BIWEEKLY.** Payments are made every two weeks. Two payments equal the same amount as one monthly payment would be. Because there are 26 two-week periods in a year, the total annual payment amount is similar to 13 monthly payments. The extra payment means that the loan is amortized more quickly than a monthly payment loan.
3. **BLANKET.** Includes more than one parcel of real estate.
4. **BRIDGE LOAN.** Used to cover a deficit during the period between a construction loan and a permanent loan or to cover the difference between funds realized from the sale of an existing home and the funds needed to buy a new home.
5. **BUDGET.** Payments include principal and interest, (property) taxes, and insurance (PITI).
6. **CHATTEL MORTGAGE.** Personal property (not real estate) serves as collateral.
7. **CONFORMING.** Loan specifications that would allow purchase by FNMA.
8. **CONSTRUCTION LOAN.** Use of the money is to build. These are high risk because of the uncertainty of completion of construction and sale or occupancy of the property. They carry high interest rates.
9. **CONVENTIONAL LOAN.** Not FHA insured or VA guaranteed. When loan-to-value ratio exceeds 80%, *private mortgage insurance* (PMI) may be required by the lender.
10. **FHA.** Insured by the Federal Housing Authority, a part of the U.S. Department of Housing and Urban Development (HUD).
11. **FLEXIBLE PAYMENT.** Interest rate changes bring about a change in the remaining term, instead of changing the payment amount.

12. **GAP LOAN.** Used to fund the shortfall between a construction loan and permanent loan amount or used to purchase a house while waiting for the sale of a former house.
13. **GRADUATED PAYMENT.** Payments are low in early years and increase to a level amount sufficient to pay interest fully and to amortize principal over the remaining term.
14. **HOME EQUITY LOAN.** Junior mortgage, limited by the amount of equity in a home.
15. **LAND CONTRACT, CONTRACT FOR DEED.** Loan from seller who doesn't transfer title until final payment is received.
16. **OPEN OR OPEN END.** More principal can be borrowed at a later date within the loan agreement.
17. **PACKAGE.** Collateral pledged includes real estate and personal property, such as furniture and appliances.
18. **PURCHASE-MONEY MORTGAGE.** A loan from the seller to the buyer. Seller provides title at closing.
19. **REVERSE ANNUITY.** Lender pays borrower a lump sum and/or monthly amount. Borrower must be over age 62 and have substantial equity in house.
20. **SWING LOAN.** Interim loan to fill a temporary need for funds to buy another property. See BRIDGE LOAN; GAP LOAN.
21. **VA GUARANTEED.** Lender is guaranteed against loss, up to a preset amount, by the U.S. Department of Veterans Affairs.
22. **WRAPAROUND LOAN.** A second mortgage that overstates its balance by including an existing first mortgage.

- Frequently used provisions or mortgage-related terms include (alphabetically):

1. **AMORTIZING.** Payments exceed interest, with the difference reducing principal. A fully amortized loan will be paid to a zero balance at the end of the full amortization term.
2. **ASSUMPTION.** Agreement by a purchaser to become liable to repay the existing loan owed by a seller of a property. Does not relieve the original borrower, who is the property seller, of any obligations.
3. **BALLOON PAYMENT.** Large final payment on a loan.
4. **CLOSING DISCLOSURE.** Form provided by a mortgage lender to a borrower that describes the financial terms of a loan. Borrowers can compare the Loan Estimate to the Closing Disclosure to assure that they get the promised loan.
5. **CONSUMER FINANCIAL PROTECTION BUREAU (CFPB).** U.S. government agency that administers TRID (TILA-RESPA Integrated Disclosure).
6. **DEBT COVERAGE RATIO.** Comparison of net operating income from the property to required debt service. For most income-producing properties, a ratio of at least 1.2 times is sought by a prudent lender.
7. **DEFICIENCY JUDGMENT.** Following a foreclosure sale, the unpaid amount due from the borrower when the foreclosure sale does not bring enough money to fully repay the mortgage balance plus expenses of foreclosure.
8. **FICO SCORE.** Credit score used by mortgage lenders to evaluate applicants, ranging from 300 to 850. A score above about 680 qualifies for a prime loan.
9. **FIRST MORTGAGE.** The one recorded first against the property, recorded at the county courthouse.

10. **FORECLOSURE.** Involuntary sale by a lender after certain borrower default. The foreclosure process must comply with state law.
11. **JUMBO.** Loan for an amount that exceeds FNMA guidelines.
12. **JUNIOR.** Any mortgage lower in priority to the first mortgage.
13. **LOAN ESTIMATE.** Form to be provided by a mortgage lender to a proposed borrower within three days of loan application and that details loan terms. The purpose is to allow borrowers to comparison shop for loans.
14. **LOAN-TO-VALUE RATIO.** Percentage ratio of the mortgage principal compared to the property value.
15. **NEGATIVE AMORTIZATION.** Monthly payment is less than interest. Principal balance increases.
16. **NOVATION.** A three-party agreement whereby one party is substituted for another.
17. **ORIGINATION.** Creation of a mortgage loan. Mortgage originators include mortgage bankers, brokers, and lending institutions.
18. **PERMANENT MORTGAGE.** A mortgage that is not scheduled to be repaid within the next 10 years.
19. **PREPAYMENT PRIVILEGE OR PENALTY.** Right to pay off the mortgage prior to maturity. Mortgage may provide for a penalty. VA and FHA loans cannot have a penalty; these may be prepaid without penalty.
20. **PRIMARY MORTGAGE MARKET.** Participants who originate mortgage loans. These include mortgage bankers, brokers, and lending institutions.
21. **PRIORITY.** Order of payment to satisfy mortgages and other liens in the event of a foreclosure sale.
22. **PRIVATE MORTGAGE INSURANCE (PMI).** Insures partial repayment to lenders of defaulted conventional (non-VA or non-FHA) mortgages.
23. **REFINANCING.** Repayment of a mortgage loan by substituting another mortgage loan. Refinancing is often desired to reduce the interest rate or to extract cash by increasing the principal.
24. **SECOND MORTGAGE.** The one recorded at the county courthouse after the first mortgage. It is below the first mortgage in priority of claim. The amount owed on a second mortgage may be greater though usually is less than the amount owed on a first mortgage.
25. **SECONDARY MORTGAGE MARKET.** Refers to the purchase of existing mortgages. Fannie Mae and Freddie Mac are the largest loan purchasers.
26. **SHORT SALE.** Upon a defaulted mortgage, the lender may agree to a sale at a price that is "short" (less than the full amount owed) to avoid the foreclosure process.
27. **SUBJECT TO.** Agreement by a purchaser to become liable to make payments on the existing loan of a property that he purchased but not become liable for the entire principal balance. Does not relieve the original borrower of any obligations.
28. **SUBORDINATION.** Agreement in a loan that the loan will not move up in priority, even after a mortgage having higher priority is paid off.
29. **SUBPRIME.** Loan that is underwritten with less than prime qualities, such as a low borrower FICO score. Borrowers with a low FICO score, often below 620, may still be eligible for a loan, likely required to be a subprime loan.
30. **TERM.** Period of time until loan maturity. The term may represent the full amortization time or be earlier with a balloon mortgage.

31. **TRID.** TILA-RESPA Integrated Disclosure, which are federal regulations, based on the Truth in Lending Act (TILA) and Real Estate Settlement Procedures Act (RESPA), covering required Loan Estimate and Closing Disclosure forms for mortgages.

- Mortgage participants include (alphabetically):

1. **COMMERCIAL BANKS.** Institutions that accept deposits. They originate mortgages and construction loans in addition to business and consumer loans.
2. **FEDERAL HOME LOAN MORTGAGE CORPORATION (FREDDIE MAC OR FHLMC).** A government-sponsored enterprise (GSE) that performs the same functions as FNMA.
3. **FEDERAL HOUSING ADMINISTRATION (FHA).** Branch of HUD that insures qualified mortgages. Down payments may be as small as 3%.
4. **FEDERAL NATIONAL MORTGAGE ASSOCIATION (FANNIE MAE OR FNMA).** A government-sponsored enterprise (GSE) that buys mortgage loans on the secondary mortgage market. It does not originate mortgage loans.
5. **GOVERNMENT NATIONAL MORTGAGE ASSOCIATION (GINNIE MAE OR GNMA).** A government agency that subsidizes mortgages to serve low- and moderate-income homebuyers.
6. **MORTGAGE BANKER.** A company that originates, services, and sells mortgage loans.
7. **MORTGAGE BROKER.** A company that originates and sells mortgage loans.
8. **SAVINGS AND LOAN ASSOCIATIONS (S&Ls).** Institutions that accept deposits and originate mortgages and construction loans.
9. **VETERANS AFFAIRS, DEPARTMENT OF.** U.S. government agency that guarantees loans made to eligible veterans up to a certain amount. Down payments may be zero.

IMPORTANCE OF BORROWED MONEY

Most real estate professionals learn very early in their careers that borrowed money is the lifeblood of the real estate business, because nearly all real estate purchases are made with borrowed funds. In addition, most real estate development and construction require borrowed money. Without access to such funds, these aspects of the real estate business dry up quickly. Such cessation of activity has occurred several times during periods of "tight" money, when interest rates are high and loanable funds scarce.

There are several reasons for this dependence on borrowed money in the real estate business. Perhaps the most significant reason is that real estate is *expensive*. Even a modest home costs more than $100,000; apartment projects, office buildings, shopping complexes, and the like often cost in the millions. Very few of the approximately 64% of American families that own their own homes could do so if they had not been able to borrow a significant portion of the price at the time of purchase. Quite simply, very few families have the kind of ready cash necessary to buy a home without borrowing. Similarly, very few real estate developers or builders have enough cash on hand to finance completely the ventures they engage in.

Another fact that contributes to the importance of borrowed money in the real estate business is that real estate is *very good collateral* for loans. It lasts a long time and so can be used to back up loans that have very long repayment terms, such as the 15 to 30 years typical of most home mortgage loans. This enables real estate owners to spread the cost of an asset over a long payment term so that they can afford more expensive units as well as spread the payment over the time they expect to use the asset.

Real estate investors like to borrow money to buy real estate because it gives them *leverage.* As a simple example, suppose someone buys an apartment complex costing $5 million by putting up $1 million of his own money and getting a mortgage loan for the other $4 million. In effect, he has gained "control" of a $5 million property but has used only $1 million to do it. By borrowing, he is able to buy considerably more than if he used only his own money; if he is wise in his choices of the real estate to buy, he may improve his investment profits considerably by borrowing.

TRID: Loan Estimate and Closing Disclosure Requirements

Borrowing to purchase real estate can be costly due to the initial fees and charges as well as the ongoing expenses, including interest and other requirements. The U.S. government has played an important role in requiring disclosures by institutional lenders so that potential borrowers can shop for the most suitable loan.

Requirements for institutional lenders changed significantly as of October 3, 2015. The Consumer Financial Protection Bureau (CFPB) then began to require two new forms, Loan Estimate and Closing Disclosure, as part of TRID. For most mortgage loans, these two forms replaced the HUD-1 Settlement Form and the good-faith estimate. The original Truth in Lending Act (TILA), as a standalone law, was reduced to apply to just a few types of real estate loans.

The purpose of these new forms is to allow consumers to shop more effectively for a mortgage loan. Please see Chapter 19 for the Loan Estimate form and Chapter 20 for closing statements and the Closing Disclosure form.

MORTGAGES AND NOTES

In the financing of real estate, two important documents are employed: (1) the *loan,* the evidence for which is a *note* or *bond,* and (2) the *mortgage,* which is a pledge of property as security for a debt. A simplified sample of a note and of a mortgage is shown below.

Since the borrower gives the note and mortgage to the lender, the borrower is the *mortgagor.* The lender is the *mortgagee.* (The suffix *-or* always signifies the giver, whereas *-ee* describes the receiver.)

NOTE	MORTGAGE
Date I owe you $100,000, payable at the rate of $900 per month for 20 years, beginning next month. Signed, T. Borrower	Date If I don't pay on the attached note, you can take 34 Spring Street, Hometown, U.S.A. Signed, T. Borrower

A mortgage and a note are both contracts. Any provisions that are agreed upon may be written into either contract, so long as such provisions are legal. The essential elements of a note are (a) a promise to pay at (b) a fixed or determinable time (c) a certain sum of money (d) to a payee or to bearer, (e) signed by the borrower.

TITLE AND LIEN THEORY

Mortgages (also called *trust deeds* in several states) are viewed either as transfers of property to the lender (title theory) or as the creation of a lien (lien theory). *Title theory* holds that the property is transferred to the lender subject to return of the property to the borrower upon payment of the loan. *Lien theory* holds that a mortgage gives the lender a lien on the property, but not title to it.

MORTGAGE LOAN CLASSIFICATION

Mortgages are often classified or described according to particular provisions. Several types of mortgages and their provisions are described below.

VA Mortgages (Also Known as GI Mortgages)

VA loans are made by private lenders to eligible veterans. The VA guaranty on the loan protects the lender against loss if the payments are not made. A VA-guaranteed loan can be used to buy a home, either existing or preconstruction, as a primary residence, or to refinance an existing loan.

BENEFITS OF A VA-GUARANTEED LOAN

- Equal opportunity for all qualified veterans to obtain a VA loan.
- Reusable, when paid off, on the next home purchase.
- No down payment (unless required by the lender or the purchase price is more than the Certificate of Reasonable Value of the property).
- No mortgage insurance.
- One-time VA funding fee that can be included in the amount borrowed.
- Veterans receiving VA disability compensation are exempt from the VA funding fee.
- VA limits certain closing costs a veteran can pay.
- Can be assumed by qualified persons.
- Minimum property requirements to ensure the property is safe, sanitary, and sound.
- VA staff dedicated to assisting veterans who become delinquent on their loan.

WHO IS ELIGIBLE?

Generally, the following people are eligible:

- Veterans who meet length-of-service requirements.
- Service members on active duty who have served a minimum period.
- Certain Reservists and National Guard members.
- Certain surviving spouses of deceased veterans.

KEY UNDERWRITING CRITERIA

- No maximum debt ratio; however, lender must provide compensating factors if total debt ratio is over 41%.
- No maximum loan amount; however, VA does limit its guaranty. Veterans can borrow up to $453,100 without a down payment in most of the county. Find out the limit in any county by visiting *www.benefits.va.gov/homeloans/*.

- Published residual income guidelines to ensure veterans have the capacity to repay their obligations while accounting for all living expenses.
- No minimum credit score requirement; instead, VA requires a lender to review the entire loan profile to make a lending decision.
- Complete VA credit guidelines are published at *www.benefits.va.gov/warms/pam26_7.asp.*

FHA Mortgages

The Federal Housing Administration (FHA), which is part of the Department of Housing and Urban Development (HUD), has many programs that help to provide housing. They insure single-family and multifamily housing loans, nursing home mortgage loans, and loans on mobile homes and other properties.

Section 203(b) is one of the most popular FHA programs. The purpose of 203(b) is to provide mortgage insurance for a person to purchase or refinance a principal residence. The mortgage loan is funded by a lending institution, such as a mortgage company, bank, or savings and loan association, and the mortgage is insured by HUD. To be eligible for a 203(b) mortgage, the borrower must meet standard FHA credit qualifications. Eligible properties are one-to-four-unit structures. The borrower may borrow up to 97% of the purchase price and may include the upfront mortgage insurance premium in the amount of mortgage principal borrowed. The borrower must have his own cash for the down payment and closing costs. He will also be responsible for paying an annual premium. More information on this program is available at *www.hud.gov/offices/hsg/sfh/ins/sfh203b.cfm.*

Conventional Mortgages: Conforming and Jumbo

Conventional mortgages are loans that are neither guaranteed by the VA nor insured by the FHA. Generally, they are the fixed-rate, fixed-principal-and-interest-payment type. Many of these loans will be purchased in the secondary mortgage market by Fannie Mae (FNMA) or Freddie Mac (FHLMC). The secondary market is where existing first mortgages are bought and sold. In 2019, the maximum amount for a single-family house to conform to FHLMC or FNMA purchase guidelines was $484,350. Loans above that amount are called "jumbo" loans. Jumbo mortgages may be insured privately, with *private mortgage insurance* (called PMI in the mortgage business). Normally, no PMI is required on conventional loans with 80% loan-to-value ratio or less. However, if the loan is for more than 80% of the lower of either the appraised value of the mortgaged property or its sale price, lenders will require PMI, which is paid for by the borrower. Conventional loans may be for as much as 95% of the purchase price provided that PMI can be purchased that covers the loan amount in excess of 80% loan-to-value. As a practical matter, however, the cost of PMI makes loans of much more than 90% loan-to-value so expensive as not to be worthwhile. The loan limit is $726,525 in certain high-cost areas, which is 50% more than $484,350.

Trick Question

A common trick question confuses the secondary mortgage market with a second mortgage. The secondary mortgage market is where mortgages that have been originated are traded. Nearly all of those mortgages traded in the secondary mortgage market are actually first mortgages.

Subprime Mortgages

Loans that do not conform to FHLMC or FNMA guidelines may still be available, though typically at higher interest rates. For example, many people without an adequate historical credit record may arrange a loan, called "subprime." A special case applies to those who have recently become self-employed and cannot meet FNMA or FHLMC for historical self-employment earnings. Lenders may offer a loan. Because it is not a conforming loan and therefore not marketable to FNMA or FHLMC, it carries higher risk. The loan will therefore have a higher interest rate.

FICO Score

Fair Isaac Company (FICO) has developed a credit rating system that most lenders weigh heavily in deciding whether to offer a loan. The system is called FICO. You can get a better understanding of it from *www.fico.com.* It scores one's creditworthiness on a scale of 300 (the worst) to 850 (the best). Major credit bureaus (TransUnion, Experian, and Equifax) apply FICO. Some lenders will relax typical underwriting criteria for those with high FICO scores.

Purchase Money Mortgages

Sometimes a *seller* provides financing for a purchaser by accepting a mortgage loan in lieu of cash for all or part of the purchase price. A mortgage of this type, where the seller accepts a mortgage loan, is called a *purchase money mortgage.* The seller effectively gives purchase money to the buyer.

Construction Loans and Permanent Mortgages

Construction loans are used by builders or developers to improve land. Construction loan advances are permitted as construction progresses. When commercial property is completed, a *permanent loan* is used to pay off completely the construction loan. Construction loans on single-family homes usually continue until a residence under construction is sold and the occupant arranges permanent financing.

The word *permanent* as it is used in real estate financing is a misnomer. Permanent mortgage loans seldom have terms beyond 30 years, and most will be repaid before their full terms. Although the conventional wisdom is that the typical mortgage loan is outstanding for about 12 years, recent experience has shown a shorter term, especially when declining interest rates encourage more frequent refinancing.

Term Loans, Amortizing Loans, and Balloon Payment Loans

A *term loan* requires only interest payments until maturity. At the end of the maturity term, the *entire* principal balance is due.

Amortizing loans require regular, periodic payments. These are called P&I (principal and interest) payments. Each payment is greater than the interest charged for that period so that the principal amount of the loan is reduced at least slightly with each payment. The amount of the monthly payment for an amortizing loan is calculated to be adequate to retire the *entire* debt over the amortization period.

A *balloon payment loan* requires interest and some principal to be repaid during its term, but the debt is not fully liquidated. Upon its maturity, there is still a balance to be repaid, called a balloon payment.

First and Junior Mortgages

Mortgages may be recorded at the county courthouse for the property to give notice of their existence. The mortgage loan that is recorded first in the courthouse against a property remains a *first mortgage* until it is fully paid off. In the event of default and foreclosure, the lender who holds the first mortgage receives payment in full *before* other mortgage lenders receive anything.

Any mortgage recorded after the first mortgage is called a *junior mortgage.* Junior mortgages are further described as *second, third, fourth,* and so on, depending upon the time they were recorded in relation to other remaining mortgage loans on the same property. The earlier a mortgage is recorded, the earlier the mortgagee's claim in a foreclosure action. As mortgages with higher priority are repaid, lower-priority mortgages automatically move up in priority. An exception is that a *subordination* clause in a mortgage may prevent that mortgage from moving to a higher priority, thereby giving priority to liens that were arranged later.

Trick Question

Don't make the mistake of thinking that the word *junior* in *junior mortgage* implies a smaller amount. A junior mortgage is one that is younger. In other words, it was recorded in the county courthouse later than the first mortgage. Therefore a junior mortgage has lower priority as security. The junior mortgage may be for a greater or lesser amount than the first mortgage.

Home Equity Lines of Credit

The *home equity line of credit* is like a second mortgage loan, except that the borrower does not have to draw down all of the money at one time. This type of loan is well suited for borrowers who anticipate that they will need more money in the near future but do not need it immediately.

Setting up a home equity line of credit can be like applying for a credit card. However, because a second mortgage is involved, there is processing, including an appraisal of the property. The application fee may be up to 2% of the line of credit, though some lenders may reduce or even waive this fee. Some plans also charge an annual fee to encourage the borrower to use the entire line of credit amount once it has been granted. In addition, many plans require the borrower to take out a minimum amount when the loan is granted.

Home equity lines of credit offer a flexible way to access home equity, thereby financing periodic needs. Borrowers may tailor the plan to the way they want to handle the payments and can draw upon the line using checks (good for infrequent, large withdrawals) or credit cards (for frequent, smaller withdrawals).

Budget Mortgages

Budget mortgages require a homeowner to pay, in addition to monthly interest and principal payments, one-twelfth of the estimated taxes and insurance into an *escrow account,* also called an *impound account.* These mortgages reduce a lender's risk, for the lender is thus assured that adequate cash will be available when an annual tax or insurance bill comes due. The payment is abbreviated PITI, for **P**rincipal, **I**nterest, (property) **T**axes, and **I**nsurance.

Package Mortgage

Package mortgages, in addition to real estate, may include appliances, drapes, and other items of personal property in the amount loaned. The lender tends to exercise more control over a borrower's monthly obligations, and the borrower is able to spread the payment for such items over the term of the mortgage, which is a lengthy period.

Chattel Mortgage

A *chattel mortgage* is a mortgage on personal property. The property may be a car, boat, furniture, or jewelry.

Blanket Mortgage

A *blanket mortgage* covers more than one parcel of real estate. *Release provisions* are usually included to allow individual parcels to be released from the mortgage upon payment of part of the mortgage principal.

Open End Mortgage

Using this type of mortgage, the initial balance can be increased, that is the funding remains open for an additional amount. A homeowner may buy a house needing extensive repair work. By initially paying, say, $125,000 and initially getting a $100,000 loan, the mortgage balance may be increased to, say, $200,000 as repair work progresses.

Flexible-Payment Mortgage

With an adjustable-rate mortgage (ARM), the payment changes periodically based on changes in interest rates. With a flexible-payment loan, though, changes in the interest rate may not change the payment immediately. Payment caps may limit the change in payments. In a flexible-payment mortgage (FPM), instead of the payment rising to pay for the higher interest requirement, the loan principal balance increases. This is called *negative amortization.*

Wraparound Loan

This is really a second mortgage that includes an existing first mortgage in its balance. Suppose a commercial property is worth $1 million, and it has a $600,000 first mortgage at a low interest rate, say 4%. A lender might be willing to advance $150,000 at 8% in the form of a second mortgage, but there are negative characteristics associated with the second mortgage. So the potential second mortgage lender advances $150,000 in cash and calls the loan a $750,000 wraparound ($600,000 plus $150,000 = $750,000). The interest rate on the wraparound is 6%. The borrower pays the wraparound lender the payment on $750,000, and the wraparound lender pays the first mortgage payment. If the first mortgage prohibits a second mortgage from encumbering the property, then the wraparound may also be prohibited.

Swing Loan

A *swing* loan allows a borrower to "swing a deal." Suppose a homeowner is buying a new house but hasn't sold the old one. A swing loan is provided by a bank for a short term to use as the down payment for the new house; the loan must be repaid when the old house is sold. Variations include the *bridge loan* and *gap loan*, described below.

Bridge Loan

A *bridge loan* is a short-term loan to cover the period between the construction loan and the permanent loan. There may be a time lag during which a short-term loan is needed.

Gap Loan

A *gap loan* is used to fill in for a shortfall until certain conditions have been met, such as an occupancy percentage rate for an office building. The permanent financing may have a floor amount, payable when construction is complete, and a higher amount, payable when occupancy reaches a certain level. A gap loan funds the difference until occupancy reaches the specified level.

Construction Loan

A *construction loan* funds construction costs by advancing money in steps as the project is developed. Generally, it is considered to be a high risk because the lender relies on project completion and on sale or leasing. A construction loan is accompanied by a high interest rate, discount points, and fees. Underwriting is more difficult for a proposed project than for one that is fully occupied. The construction lender does not want to advance more than is put in the project, so the loan is made in increments: when the land is purchased, streets and utilities installed, foundations, framing, roofing, and so on. Before providing a construction loan, a lender generally requires the borrower to have a commitment for permanent financing to be substituted when an income property is built.

For a *subdivision loan, release provisions* allow parcels to be removed as collateral so individual lots may be sold. To release a lot requires a payment against the loan.

Home Equity Loan

For a household that is house rich and cash poor, a *home equity loan* may be a tempting way to finance a child's college education, medical bills, or another purpose. A home equity loan is likely to be simply a second mortgage, but it carries a lower rate than other types of consumer finance. A home equity line of credit, which is a type of home equity loan, was described previously.

Land Contract

In a *land contract,* also called a *contract for deed* or an *installment sales contract,* title does not pass until the final payment is made. This is often used to finance recreational or resort property or for buyers with poor credit ratings. Lenders keep title until full payment has been made.

Sale-Leaseback

A lease and a loan share many of the same characteristics, so a lease can be structured to be like a loan. Instead of borrowing to raise money and repaying the loan, one can buy, simultaneously resell the property, and then lease it back for a long term (25 or 30 years). The sale generates cash (as does borrowing a mortgage loan), and the rental payments are like mortgage payments. At the end of the lease term there may be a purchase option at a small amount for the property user. This is sometimes a method to arrange 100% financing and to control property without taking title as the owner.

MORTGAGE REPAYMENT

For decades only one mortgage repayment method was available for home loans: *fixed-interest, fixed-payment amortization.* On these loans, the interest rate was always the same and the monthly payment was always the same. When all the scheduled payments had been made, the loan balance was paid in full. Beginning in the 1970s, however, a variety of other methods entered the mortgage market. Today, the most popular loans are fixed-rate mortgages (FRMs) and adjustable rate mortgages (ARMs), each of which has several variations.

Fixed-Rate Mortgages

The *fixed-rate mortgage* (*FRM*) is the most popular modern mortgage instrument. It features an interest rate that does not change; it is "fixed" over the life of the loan. Fixed-rate loans can have any term up to 30 years (even longer in certain restricted government-subsidized programs). The most popular term, however, is 30 years, followed by 15 years and 20 years. In a typical FRM, the monthly principal and interest payment remains the same throughout the loan term. By the end of the term, the loan has been completely amortized, that is, repaid to the lender. However, there are some plans in which the monthly payment is not always the same; the most common of these are *graduated-payment mortgages* and *balloon mortgages.*

GRADUATED-PAYMENT MORTGAGE (GPM)

The GPM is designed to allow a relatively low monthly payment in the first years of the loan; the payment is increased by some amount (or percentage) every year until it becomes high enough to fully amortize (pay off) the loan over the remaining term. Typically, this final payment level will be reached after 5 to as many as 10 years. One unpleasant feature of most GPMs is *negative amortization* in the early years. The loan principal amount actually goes up instead of down at first, because the payments during the first 3 to 5 years aren't even enough to pay the interest on the loan. The unpaid interest is added to the loan amount, which keeps rising until the payments eventually become high enough to cover the interest.

BALLOON MORTGAGE

The balloon mortgage features reasonable monthly payments but a term too short to allow for the loan to be fully paid. Therefore, at the end of the term a large final payment (the "balloon" payment) is due. A typical arrangement is for the borrower to make payments as though the loan had a 30-year term; however, after 10 or 15 years the loan comes due and the entire unpaid balance must be paid. Some balloon payment loans call for interest payments only, so there is no principal reduction before maturity.

Adjustable-Rate Mortgages

Adjustable rate mortgages (*ARMs*) feature interest rates that change every so often. A change in the interest rate usually means a change in the amount of the monthly payment. ARMs were first introduced in the 1970s but did not become popular with borrowers until the 1980s.

By that time, several modifications had been introduced that helped overcome consumer reluctance. When ARMs were first offered, a wide variety of terms and conditions were available. Nowadays the ARM has been fairly well standardized into a few popular versions that have many features in common. All have *periodic adjustment*, which is an interest rate based upon some *index* of interest rates, and both periodic and lifetime *caps* (ceilings) on interest rates.

The *period* of the loan is the frequency at which adjustments occur; common periods are 6 months and 1 year, but longer periods (3 to 5 years) are gaining in popularity. Today most ARMs have the interest rate adjusted every year. Interest rates aren't adjusted arbitrarily; they are based upon an *index* of current interest rates. The ARM contract specifies the index used and the *margin*. Each time the rate is adjusted, this margin will be added to the index rate to get the new rate for the ARM loan for the following period.

The *caps* are limits on changes in the interest rate. The *periodic cap* limits the amount of change for any one period; the *lifetime cap* provides upper and lower interest rate limits that apply for the entire life of the loan. The most popular cap arrangements are "1 and 5" (limited largely to FHA and VA loans) and "2 and 6." The first number is the periodic cap; the second is the lifetime cap. An "annual ARM with 2 and 6 caps" will be adjusted every year (annually), by no more than 2 percentage points (the periodic cap) above or below the previous year's rate, and over the life of the loan the interest rate can't rise by more than a total of 6 percentage points (the lifetime cap) above the original rate. ARMs are popular with lenders, who don't want to get stuck with low-interest loans in a period of rising rates. However, ARMs have been difficult to "sell" to borrowers, most of whom are unwilling to risk a rise in interest rates, which would increase their loan payments. The mortgage industry is addressing this concern with a variety of features and concessions tied to ARM loans. The most popular and effective of these are *convertibility* and *teaser rates*.

The convertibility feature, which is not an automatic part of all ARM loans, allows the borrower to convert the ARM loan to an FRM loan at some time in the future, usually 3, 4, or 5 years after the loan is originated. Typically a fee, usually 0.5% to 1% of the loan amount, is charged for conversion. The appeal of convertibility is that the loan doesn't have to be an ARM forever, and the borrower feels "safer" with the ARM because of the conversion privilege.

Teaser rates are unusually low *initial* interest rates. They are very common in ARMs with annual (or shorter) adjustment periods. An interest rate that is attractively low for the first year or two of the ARM translates into low initial payments and thus into genuine money savings for the borrower. Teaser rates can have another useful effect: if the lifetime cap applies to the initial loan rate, then the maximum interest rate on the loan will be relatively low as well. Teaser rates are typically 3 or more percentage points below current FRM interest rates. For example, if FRM loans are available at 4% interest for 30-year mortgages, the teaser rate for a 30-year annual ARM with 2 and 6 caps probably will be around 1%. This would result in a lifetime cap of 1% + 6%, or 7%. Even if, a year after origination, the interest rate were to rise by the full 2% annual cap, the second year's rate would be 3%, still a full point less than the original FRM rate.

To illustrate, for a 30-year FRM for $500,000 at 4% interest, the monthly payment would be $2,387.08. An ARM of $500,000, with a 1% teaser rate, would require a monthly payment of $1,608.20 for the first year and a maximum possible payment (at 3% interest) of $2,091.12 the second year. The ARM borrower would be more than $9,346 ahead the first year and another $3,552 ahead the second year, for a total cash savings of $12,898 over the first two years. This "money in the pocket" provides many borrowers with a strong incentive to choose the ARM, particularly if it also has the convertibility feature

ARMs with longer adjustment periods (3, 5, and even 7 years) are becoming more popular. If they have teaser rates, these rates are much closer to the FRM rate (1% to 1.5% less, at the most); periodic caps on these loans are higher, although a lifetime cap of about 6% is common.

Others

REVERSE MORTGAGE (RM)

Here the lender makes the payments to the borrower, gradually advancing money to a (typically elderly) homeowner. The amounts advanced, plus interest, become the principal of the loan. Payments to the lender are deferred until a sale of the property, the death of the homeowner, or a time when the balance owed approaches the market value of the home. RMs also can be arranged whereby money, borrowed against the home, is used to purchase an annuity contract that will pay the homeowner for life. The homeowner then pays the mortgage interest with income from the annuity. The primary borrower must be at least age 62 to qualify for the FHA program. These are likely to be used by people who are house rich but cash poor.

BIWEEKLY MORTGAGES

Here the lender figures the monthly payment on a 30-year amortizing loan and then requires half to be paid every 2 weeks. Because there are twenty-six 2-week periods in a year, this is the equivalent of paying for 13 months per year. As a result, the loan is retired in 18 to 22 years. For borrowers who are paid by their employers every 2 weeks, such a mortgage may fit the budget well. However, it may not appeal to those who are paid monthly.

PRIMARY MORTGAGE MARKET

The primary mortgage market consists of institutions that originate mortgages. These institutions investigate the borrower's creditworthiness and the property value (see "Lender Criteria" on page 262). Primary mortgages can then be sold on what is called the secondary mortgage market, which often provides a profit for the originator.

Savings and Loan Associations

Savings and loan associations actively originate residential mortgages in the United States. Most sell the loans in the secondary market.

Commercial Banks

Commercial banks originate many residential mortgages; they are more active, however, in the construction loan field and in other types of lending. They have a great deal of latitude in the types of activity in which they engage, but strict limitations are imposed for each type of loan.

Mutual Savings Banks

Mutual savings banks have no stockholders; they operate for the benefit of depositors. Some originate or invest in residential mortgages.

Life Insurance Companies

Life insurance companies invest a relatively small amount of their total assets in residential mortgage loans. They are quite active in permanent mortgage loans on income-producing property, including apartments, shopping centers, and office buildings.

Mortgage Companies

Mortgage companies, also known as *mortgage bankers* and *mortgage brokers*, are the most common sources of home mortgages. They originate almost half of all residential mortgages. However, these businesses do not lend their own money; they are not financial institutions. Rather, they originate and service home mortgage loans on behalf of large investors (insurance companies, banks, pension funds, and the like) in mortgage loans. *Mortgage bankers* receive a fee for originating loans and fees for servicing them: collecting payments, holding escrows, dunning late payers, and so on. *Mortgage brokers* originate loans, unlike mortgage bankers, but do not service them.

SECONDARY MORTGAGE MARKET

Federal National Mortgage Association

The *Federal National Mortgage Association* (also known as *FNMA* or *Fannie Mae*) is a federally chartered corporation that was previously owned by stockholders. In August 2008, it was placed in government conservatorship because of the mortgage crisis. Its future is unknown at this time. It does not make direct loans to homeowners. Rather, it purchases first mortgage loans, many originated by mortgage bankers, in the secondary mortgage market. FNMA thereby aids in the liquidity of mortgage investments, making mortgages more attractive investments. FNMA presently owns a significant percentage of all residential mortgages. FNMA uses the mortgages as collateral for their loans, which are called mortgage-backed securities (MBS).

Federal Home Loan Mortgage Corporation

The *FHLMC*, or *Freddie Mac*, is similar to FNMA. It buys mortgages on the secondary market. It buys much of its inventory from savings and loan associations and commercial banks. It also sells participations in pools of mortgages called mortgage-backed securities (MBS). Like Fannie Mae, it was placed under government conservatorship in August 2008, and its future is uncertain.

Government National Mortgage Association

The *Government National Mortgage Association* (*GNMA* or *Ginnie Mae*) is a U.S. government entity that encourages low-income housing. It guarantees packages of mortgage-backed securities consisting of FHA or VA loans.

Farmers Home Administration

This agency, *FmHA*, which has regional branches, provides financing for the purchase and operations of farms and rural homes and also guarantees loans on such properties made by others.

LENDER CRITERIA

Institutional lenders perform a process called *underwriting*. The purpose of this is to assess the risks of making the loan. They look at the property and the borrower. As to the property, the lender typically arranges for an *appraisal* (cost is paid by the borrower), which will provide information about the property's value, its location, and market conditions, including comparable sales.

The lender will get a *credit report* on the borrower, financial statements, and letters from employers concerning salary, term of employment, and prospects for the continuation of employment. Lenders will verify bank accounts and other assets of the borrower as well as liabilities.

For a conventional loan on a single-family home, the lender will normally require a 20% down payment. This would be a *loan-to-value ratio* of 80%. If the borrower seeks a greater loan-to-value ratio, he will be required to get private mortgage insurance (PMI). Up to 90% and 95% loan-to-value ratios may be obtained. For a duplex or four-unit property, a larger down payment (25%) is normally required.

Lenders have certain guidelines to follow if they want to have a truly marketable loan—that is, a loan that can be sold to FNMA or FHLMC. This is called a *conforming* loan. The lender will provide a *qualifying* ratio of housing payments to gross income of 28% and a ratio of total fixed payments to gross income of 36%. For example, suppose the Andersons earn $10,000 per month. The maximum housing payment they would be allowed is $2,800 per month, including principal, interest, taxes, and insurance. When payments on other long-term debt, such as car loans, are considered, the maximum allowed would be $3,600 (36%). These guidelines may be exceeded by applicants with high FICO scores.

Many lenders will offer *nonconforming* loans to those who don't qualify for a conforming loan. These loans won't conform to FNMA/FHLMC guidelines, and a higher rate of interest is charged. But they may allow a marginal borrower to purchase the home desired. Nonconforming loans include jumbo, subprime, and Alt-A.

With income-producing property, a lender will also look to the debt coverage ratio and insist that the net operating income from the property exceed debt service by a comfortable margin that is by 25% or more. So a debt service coverage ratio of 1.25 times net operating income may be required.

When considering a loan on the stock in a cooperative apartment, lenders will consider the risk. If it is an existing, sold-out development, it is less risky than one that is to be built or converted from rental apartments. For those riskier developments, the developer or sponsor's filings of stock with the attorney general may offer clues as to the likelihood of success.

Qualification Ratios

Generally, ratios of housing expenses to borrower income provide the essential limits for qualifying for a mortgage loan. These are usually determined on a monthly basis. Keep in mind that there are not exactly four weeks in a month; the actual number is $4^1/_3$. If a home-buyer earns $1,000 per week, he earns approximately $4,333 per month. (Be careful of trick math questions on exams that will cause you to choose a wrong answer if you assume there are four weeks in a month.)

PREQUALIFICATION

Real estate salespeople and brokers don't want to waste anyone's time by showing unafford-able houses to prospective buyers. However, a salesperson cannot assure a loan and initially doesn't know the buyer's creditworthiness or FICO score. (A FICO score of 680 or more is considered excellent credit; a score less than 500 probably won't be enough to qualify for a prime loan.) Using estimates of the prospective buyer's income, the salesperson will try to qualify the customer's limit to be able to target a house price range. Typically, the salesperson does not verify the income or creditworthiness. No loan of any amount is assured by prequalification.

PREAPPROVAL

A potential buyer may approach a lending institution and ask to be preapproved. The lender will then underwrite an amount to lend based on the borrower's verified income and creditworthiness. The house must "appraise out," which occurs at a later step in the process.

The salesperson and seller will likely be more willing to deal with a buyer who is preapproved. They are assured that the transaction won't fall through for financial reasons.

MORTGAGE LOAN DISCOUNTING

Points

One *point* is 1% of the amount of the loan and is charged at origination. Some mortgage loan charges are expressed in terms of points. These charges are either *fees* or *discounts*. Typical fees expressed in points are origination fees, conversion fees (ARM to FRM; see page 259), prepayment penalties, and so on. The actual dollar amount of such a fee depends upon the dollar amount of the loan involved. For example, most new mortgage loans have origination fees. Usually, the fee is about one point, or 1% of the loan amount. For a loan of $100,000, then, the fee will be $1,000 (1% of $100,000); for a loan of $75,000 the fee will be $750, and so on.

Discount points are expressed as percentages of the loan amount. Discounts are paid at origination in order to reduce the interest rate on the loan. (You can think of discounts as somewhat like interest paid in advance.) Depending upon the amount of the discount required, a lender will offer a variety of interest rates on new mortgage loans. For example, a "menu" of interest rates might be: 6.5 + 0, 6.25 + 1.25, 6.0 + 2.5. These mean that the lender will offer an interest rate of 6.5% with no discount (usually called "par"); 6.25% in return for a payment of 1.25% of the loan amount (a 1.25-point discount); and 6% in return for a 2.5-point discount. For a $100,000 loan from this menu, the borrower's interest rate will be 6.5% if no discount is paid, 6.25% if $1,250 is paid in advance, and 6% for an advance payment of $2,500. Why would anyone pay discounts? Actually, many borrowers don't. However, *sellers* often do for a variety of reasons, which are usually concerned with making a sale more attractive to a buyer.

Discounts also are used as *buydowns*, often in association with *teaser* interest rates. Buydowns substitute an initial charge for interest later in the loan. Most often they are used to provide an unusually low interest rate during the early period of a loan. For example, if the "normal" rate of interest is 7%, a lender may offer a buydown (for a fee of about 5.5% of the loan amount) that will give the loan an interest rate of 5% the first year, 6% the second, and 7% the third. After the third year, the interest rate will be 7% for the rest of the loan period.

Prepayment Penalties and Privileges

Nearly all home mortgage loans allow the borrower to pay off the principal balance without penalty at any time. Many commercial property mortgages require a penalty, such as 3% of the outstanding balance, for early debt retirement. The penalty diminishes over time.

Acceleration Clauses

An *acceleration clause* states that the full principal balance becomes due upon certain default. If one or more payments are past due (depending on the agreement), the entire loan becomes due. It is common for residential mortgage loans to have a "due on sale" clause. A due on sale clause will cause the full, unpaid amount of the loan to become payable when the property is sold, even when the loan is not in default.

Loan Assumption

Property offered for sale may be mortgaged with a loan bearing a low interest rate, which makes it attractive to the real estate purchaser. The buyer may pay the seller for his equity (the difference between the property value if free and clear and the amount of debt) and then *assume* the loan. The purchaser then becomes liable for the loan and its required payments. The original borrower, however, is not automatically freed from his obligation. He may ask the lender for a *novation* (see pages 240 and 249), which serves to release him from responsibility, and then substitute the new property owner as the sole debtor. In the absence of a novation, the lender may look to the property and to both the original borrower and the subsequent purchaser for repayment.

Rather than assume a loan, a purchaser may agree to take the property *subject* to a mortgage loan. In doing so, the buyer acknowledges the debt, but it does not become her personal liability. In the event of default, she stands to lose no more than her investment.

Usury

Every state has laws prescribing maximum permitted interest rates. The maximum rate varies with different types of loans. Charging a higher interest rate than that permitted by law is considered *usury*. Penalties for usury differ widely from state to state but are severe in most states. In fact, some penalties require the lender to forfeit all interest.

Estoppel Certificate

An *estoppel certificate* is a statement that prevents its issuer from asserting different facts at a later date. For example, a mortgage originator may wish to sell the mortgage loan (the right to collect payments) to another. The loan purchaser may ask for an estoppel certificate, whereby the borrower states the amount he owes. Later, the borrower is stopped from asserting that he owed less at the time he signed the estoppel certificate.

Releasing from Property Mortgages

Property may be released from mortgages using four methods:

1. **SATISFACTION PIECE.** This document from the lender states that the loan is paid off and that the lender releases property from the lien.
2. **PARCEL OR LAND RELEASE.** This releases part of mortgaged property from the mortgage, usually upon payment of part of the debt.
3. **POSTPONEMENT OF LIEN.** The lien is not satisfied but is subordinated, meaning that it assumes a lower priority.
4. **FORECLOSURE.** This is described in detail below.

Foreclosure, Deficiency Judgments, and Equity of Redemption

Upon certain default, a lender may be allowed to go through an authorized procedure, known as *foreclosure,* to have the property applied to enforce the payment of the debt. Upon foreclosure, real estate tax liens are generally paid first. Other tax liens and mechanics' liens are generally next in line. Should the property sell at foreclosure for more than the tax and mechanics' liens, the unpaid first mortgage loan, and the expenses of the foreclosure action, the excess belongs to the junior lenders, if any. After all lenders have received full payment, any excess funds go to the borrower.

It is more likely that the proceeds from a foreclosure will be inadequate to pay mortgage indebtedness. In this case, the lender may attempt to get a *deficiency judgment* for the amount of the loan that remains unpaid. This requires the borrower to pay the deficiency personally. Deficiency judgments are difficult to establish in many jurisdictions. Commercial property mortgagors are frequently able to negotiate *exculpatory* clauses in their loans. Exculpatory clauses cause loans to be *nonrecourse.* When such a clause exists, a defaulting borrower may lose the real estate that he has pledged, but he is not personally liable for the debt. The lender may look only to the property as collateral. A borrower who is behind on payments but doesn't want to go through the foreclosure process may offer the lender a deed in lieu of foreclosure.

Should property be foreclosed, the borrower has the right to redeem the property by repaying in full the principal owed. This is known as *equity of redemption* and must be exercised within the time period prescribed by state law. In some states, the right must be exercised before the actual foreclosure sale; in others, the right remains for up to 2 years after such a sale.

QUESTIONS ON CHAPTER 13

1. A type of mortgage in which the lender makes periodic payments to the borrower, who is required to be age 62 or older in the FHA program, is called
   (A) opposite
   (B) accelerate
   (C) reverse
   (D) deficit

2. A conventional mortgage is
   (A) amortizing
   (B) guaranteed by the FHA
   (C) not guaranteed by any government agency
   (D) approved by the VA

3. A chattel mortgage is usually given in connection with
   (A) realty
   (B) farms
   (C) personal property
   (D) commercial property

4. The lending of money at a rate of interest above the legal rate is
   (A) speculation
   (B) usury
   (C) both A and B
   (D) neither A nor B

5. One discount point is equal to
   (A) 1% of the sales price
   (B) 1% of the interest rate
   (C) 1% of the loan amount
   (D) none of the above

6. CFPB stands for
   (A) Certain Finance Problems Board
   (B) Certified Financial Public Basis
   (C) Consumer Financial Protection Bureau
   (D) Common Focus on Public Behavior

7. The main appeal of VA mortgages to borrowers lies in
   (A) low interest rates
   (B) minimum down payments
   (C) an unlimited mortgage ceiling
   (D) easy availability

8. When a loan is assumed on property that is sold,
   (A) the original borrower is relieved of further responsibility
   (B) the purchaser becomes liable for the debt
   (C) the purchaser must obtain a certificate of eligibility
   (D) all of the above

9. An estoppel certificate is required when the
   (A) mortgage is sold to an investor
   (B) property is sold
   (C) property is being foreclosed
   (D) mortgage is assumed

10. An owner who seeks a mortgage loan and offers three properties as security will give
    (A) a blanket mortgage
    (B) an FHA mortgage
    (C) a conventional mortgage
    (D) a chattel mortgage

11. A clause in a mortgage or accompanying note that permits the creditor to declare the entire principal balance due upon certain default of the debtor is
    (A) an acceleration clause
    (B) an escalation clause
    (C) a forfeiture clause
    (D) an excelerator clause

12. A loan estimate is
    (A) an estimate of approximate total payments on a mortgage
    (B) an estimate of the approximate amount of the monthly payment required
    (C) a form given to mortgage loan applicants that details expected costs to get the mortgage loan
    (D) an estimate of the approximate down payment needed

13. A second mortgage is
    (A) a lien on real estate that has a prior mortgage on it
    (B) the first mortgage recorded
    (C) always made by the seller
    (D) smaller in amount than a first mortgage

14. A large final payment on a mortgage loan is
    (A) an escalator
    (B) a balloon
    (C) an amortization
    (D) a package

15. A requirement for a borrower under an FHA-insured loan is that he
    (A) not be eligible for a VA or conventional loan
    (B) have cash for the down payment and closing costs
    (C) have his wife sign as coborrower
    (D) certify that he is receiving welfare payments

16. A mortgaged property can
    (A) be sold without the consent of the mortgagee
    (B) be conveyed by the grantor's making a deed to the grantee
    (C) both A and B
    (D) neither A nor B

17. In the absence of an agreement to the contrary, the mortgage normally having priority will be the one
    (A) for the greatest amount
    (B) that is a permanent mortgage
    (C) that was recorded first
    (D) that is a construction loan mortgage

18. The mortgagor's right to reestablish ownership after delinquency is known as
    (A) reestablishment
    (B) satisfaction
    (C) equity of redemption
    (D) acceleration

19. The Federal National Mortgage Association is active in the
    (A) principal mortgage market
    (B) secondary mortgage market
    (C) term mortgage market
    (D) second mortgage market

20. The money for making FHA loans is provided by
    (A) qualified lending institutions
    (B) the Department of Housing and Urban Development
    (C) the Federal Housing Administration
    (D) the Federal Savings and Loan Insurance Corporation

21. Amortization is best defined as
    (A) liquidation of a debt
    (B) depreciation of a tangible asset
    (C) winding up a business
    (D) payment of interest

22. A mortgage is usually released of record by a
    (A) general warranty deed
    (B) quitclaim deed
    (C) satisfaction piece
    (D) court decree

23. Loans from savings and loan associations may be secured by mortgages on
    (A) real estate
    (B) mobile homes
    (C) both A and B
    (D) neither A nor B

24. The borrower is the
    (A) mortgagee
    (B) creditor
    (C) mortgagor
    (D) both A and B

25. If a property is foreclosed that has all the liens listed below, which lien would ordinarily be paid first?
    (A) real estate tax lien
    (B) federal income tax lien
    (C) mortgage lien
    (D) mechanic's lien

26. A term *mortgage* is characterized by
    (A) level payments toward principal
    (B) interest-only payments until maturity
    (C) variable payments
    (D) fixed payments including both principal and interest

27. Mortgage bankers
    (A) are subject to regulations of the Federal Reserve System
    (B) are regulated by federal, not state, corporation laws
    (C) act as secondary lenders
    (D) earn fees paid by new borrowers and lenders

28. The seller of realty takes, as partial payment, a mortgage called
    (A) sales financing
    (B) note toting
    (C) primary mortgage
    (D) purchase money mortgage

29. A state in which a borrower retains title to the real property pledged as security for a debt is a
    (A) lien theory state
    (B) title theory state
    (C) creditor state
    (D) community property state

30. A state in which a mortgage conveys title to the lender is known as a
    (A) lien theory state
    (B) title theory state
    (C) conveyance state
    (D) community property state

31. A biweekly mortgage requires
    (A) monthly payments to increase by predetermined steps each year
    (B) payments every two weeks
    (C) that the lender share profits from resale
    (D) none of the above

32. The use of reverse annuity mortgages
    (A) is widespread because of inflation
    (B) helps elderly people who are house rich but cash poor
    (C) is losing importance with inflation
    (D) none of the above

33. The Federal Housing Administration's role in financing the purchase of real property is to
    (A) act as the lender of funds
    (B) insure loans made by approved lenders
    (C) purchase specific trust deeds
    (D) do all of the above

34. The instrument used to remove the lien of a trust deed from record is called a
    (A) satisfaction lien
    (B) release deed
    (C) deed of conveyance
    (D) certificate of redemption

35. The type of mortgage loan that permits borrowing additional funds at a later date is called
    (A) an equitable mortgage
    (B) a junior mortgage
    (C) an open-end mortgage
    (D) an extensible mortgage

36. A loan to be completely repaid, principal and interest, by a series of regular, equal installment payments is a
    (A) straight loan
    (B) balloon-payment loan
    (C) fully amortized loan
    (D) variable-rate mortgage loan

37. Foreclosure is the forced sale of property to
    (A) delay mortgage payments
    (B) accelerate payments
    (C) enforce an unpaid lien
    (D) pay taxes

38. A secured real property loan usually consists of
    (A) financing statement and trust deed
    (B) the debt (note) and the lien (deed of trust)
    (C) FHA or PMI insurance
    (D) security agreement and financing statement

39. When a loan is fully amortized by equal monthly payments of principal and interest, the amount applied to principal
    (A) remains constant
    (B) decreases while the interest payment increases
    (C) increases while the interest payment decreases
    (D) increases by a constant amount

40. A *subordination clause* in a trust deed may
    (A) permit the obligation to be paid off ahead of schedule
    (B) prohibit the trustor from making an additional loan against the property before the trust deed is paid off
    (C) allow for periodic renegotiation and adjustment in the terms of the obligation
    (D) give priority to liens subsequently recorded against the property

41. *PMI* stands for
    (A) public money interest
    (B) payments made on investments
    (C) principal and mortgage interest
    (D) private mortgage insurance

42. High loan-to-value ratio loans generally are accompanied by
    (A) FHA insurance
    (B) VA guarantees
    (C) PMI
    (D) any of the above

43. Low interest rates cause
    (A) higher inflation rates
    (B) housing to be more affordable
    (C) higher loan-to-value ratios
    (D) owner financing to be more readily available

44. A borrower who is behind on payments and doesn't want to go through the foreclosure process may offer the lender a
    (A) second mortgage
    (B) deed in lieu of foreclosure
    (C) sale and leaseback
    (D) note indicating that the keys are in the mailbox

45. Which of the following types of first mortgage loan is typically described as "conventional"?
(A) 80% loan from mortgage banker
(B) 97% FHA loan
(C) 100% VA loan
(D) contract for deed

46. Virtually all deeds of trust will have a clause that states that the full principal is due upon certain default. This is called a(n)
(A) release clause
(B) acceleration clause
(C) novation clause
(D) assignability clause

47. A VA loan amount is based on
(A) Certificate of Reasonable Value
(B) term of military service
(C) secondary mortgage market
(D) FNMA loan limits

48. In a fully amortizing loan,
(A) the payment toward principal increases as the loan is amortized
(B) the payment toward interest decreases as the loan is amortized
(C) the principal and interest payment remains constant
(D) all of the above

49. Prequalification is
(A) assurance that the buyer can get a loan
(B) permission to negotiate as a buyer
(C) a good-faith contract
(D) an estimated amount the purchaser can borrow

50. Because of its illiquidity, large size, and lack of portability, investors expect commercial real estate to provide __________ return, compared to marketable securities.
(A) a higher
(B) a lower
(C) the same
(D) an unknown

51. When the buyer pays the cost with a lump-sum check and the seller delivers the deed, the arrangement is a
(A) purchase money mortgage
(B) contract for deed
(C) contract for purchase and sale
(D) closing

52. After a mortgage is paid off, most lenders are expected to
(A) keep all records for five years
(B) send a recorded satisfaction piece
(C) file a lien
(D) refinance the loan upon request

53. The bank agreed to lend Ira $300,000, with his house as collateral. Of this amount, Ira received $200,000 immediately and could borrow the $100,000 balance in installments. This is called
(A) an open-end mortgage
(B) a blanket mortgage
(C) an amortizing loan
(D) a P&I mortgage

54. A loan that is approved quickly and has a low down payment usually has
(A) higher costs and interest rate
(B) lower costs and interest rate
(C) a longer amortization period
(D) an adjustable rate

55. Two months in a row a homeowner failed to make payments on his trust deed, so the lender recorded a notice of default. The borrower
(A) cannot stop foreclosure
(B) must repair his credit report
(C) may walk away
(D) may request reinstatement

56. A loan from the seller to the buyer is called a
(A) bullet loan
(B) home equity loan
(C) junior mortgage
(D) purchase money mortgage

57. Concerning discount points on a conventional loan, which statement is correct?
(A) Points must be paid by the buyer.
(B) Points must be paid by the seller.
(C) Points increase the effective interest rate the lender receives on the loan.
(D) Points have no effect on the interest rate charged, whether the face rate or the effective rate.

58. A home equity loan is
(A) secured
(B) unsecured
(C) for a 15-, 20-, or 30-year term
(D) less than a first mortgage

59. Leverage is
(A) a broker's employment of a salesperson
(B) using cash for all purchases
(C) using credit cards
(D) using borrowed money to complete an investment purchase

60. A mortgage loan that includes two or more properties is
(A) an acquisition and development mortgage
(B) a total lien mortgage
(C) a blanket mortgage
(D) a participation mortgage

61. Closing Disclosure refers to
(A) a completed form given by a lender to a proposed borrower detailing costs and payments required for a proposed loan
(B) a set time to settle on a contract
(C) defects and repairs needed on a house
(D) permission to place a lockbox

62. When property is sold "subject to" a mortgage, the person primarily responsible for repayment is
(A) the buyer
(B) the lender
(C) the one who assumes the loan
(D) the seller

63. Fannie Mae is
(A) an originator of mortgage loans
(B) a purchaser of mortgage loans
(C) an agency of the Federal Housing Administration
(D) a branch of the Federal Reserve

64. The seller would have the least financing exposure in
(A) new financing by buyer
(B) purchase-money mortgage
(C) assumption of loan by buyer
(D) purchase "subject to" mortgage

65. A first-time buyer's down payment source may be
(A) savings
(B) a gift from a relative
(C) a personal loan
(D) any of the above

66. A lender should consider which of these factors before issuing a loan?
(A) property value
(B) borrower's credit rating
(C) down payment
(D) all of the above

67. A seller took 10% down in cash and provided 90% of the sales price as a first mortgage. The best way to describe the mortgage is
(A) home equity loan
(B) fixed interest rate
(C) purchase-money mortgage
(D) buyer financing

68. When a lender verifies a borrower's employment, income, and creditworthiness and then issues a letter agreeing to a maximum loan, it is called
(A) verification
(B) preapproval
(C) assignment
(D) subordination

69. Which of the following participate(s) in the secondary loan market?
(A) Fannie Mae
(B) Freddie Mac
(C) Ginnie Mae
(D) all of the above

70. Private mortgage insurance protects the
(A) buyer's heirs
(B) lender
(C) buyer's income
(D) investment value

71. *P&I* in real estate finance stands for
(A) property and investments
(B) property and inventory
(C) principal and interest
(D) principal and inventory

72. The number of weeks in a month is most nearly
(A) 4
(B) 5
(C) 2
(D) $4^1/_3$

73. The seller has agreed to pay two points to the lending institution to help the buyers obtain a mortgage loan. The house was listed for $320,000 and is being sold for $300,000. The buyers will pay 20% in cash and borrow the rest. How much will the seller owe to the lender for points?
(A) $6,400
(B) $4,800
(C) $5,120
(D) $6,000

74. The lender's underwriting criteria specify a maximum housing expense to income ratio of 28%. If the applicant proves annual earnings of $75,000 in the previous year, and that salary is continuing, the maximum monthly PITI would be
(A) $2,100
(B) $1,750
(C) $2,800
(D) $2,700

75. Which of the following is an advantage of a biweekly mortgage payment plan?
(A) The borrower pays less per month in return for a longer term.
(B) It is equivalent to 12 monthly payments each year.
(C) There are lower interest rates.
(D) The loan is paid off sooner than it would be with 12 monthly payments.

76. Lender *A* has a first mortgage but allows Lender *B*'s loan to have priority. This is an example of
(A) home equity swap
(B) upside-down loan
(C) subordination
(D) deficiency judgment

77. The function of the Federal Housing Administration (FHA) is to
(A) lend money
(B) insure loans
(C) guarantee loans
(D) buy loans

78. The buyer who agrees to pay an existing loan but does not take personal responsibility
(A) assumes the loan
(B) takes the property subject to the loan
(C) subordinates the loan
(D) forecloses on the loan

79. In general, the lien with first claim on the real estate is the
(A) largest one
(B) one that is signed
(C) one that was recorded first
(D) one that is foreclosed

80. When a borrower is required to maintain an escrow account with the lending institution, money in that account may be used to pay the homeowner's
(A) REALTOR®
(B) utility bills
(C) property taxes or insurance
(D) life insurance

81. Which of the following describes a package mortgage?
(A) a loan that is packaged with other loans to sell to the secondary market
(B) a standardized loan package to be used in financing the purchase of a single-family home
(C) a loan with more than one parcel of real estate as collateral
(D) a loan with both real and personal property as collateral

82. Which one of the following does not affect the interest rate adjustment in an adjustable-rate mortgage loan?
(A) index
(B) equity value
(C) margin
(D) cap

83. To secure a loan in some states, the borrower conveys title to the property to a disinterested third party for safekeeping. In this situation, the document required is a
(A) general warranty deed
(B) deed of trust
(C) involuntary conveyance
(D) mortgage

84. What is the definition of a mortgage?
(A) a promise to repay a loan under specific terms and conditions
(B) a pledge of personal liability in case of delinquency
(C) a description of the loan and the collateral on that loan
(D) a pledge of property as collateral for a debt

85. In a fixed-rate fully amortized loan, which one of the following does not vary from month to month?
(A) monthly principal and interest payment
(B) interest paid
(C) principal paid
(D) remaining principal balance

86. A property is foreclosed and sold for nonpayment of the mortgage loan. The sale does not cover the remaining balance of the mortgage loan. The bank goes to court to collect the shortfall, which is called a
(A) judicial foreclosure
(B) statutory redemption
(C) equitable redemption
(D) deficiency judgment

87. The secondary mortgage market engages in the sale of
(A) first mortgages
(B) second mortgages
(C) junior mortgages
(D) conventional mortgages

88. Which of the following is not an obligation of the mortgagor?
(A) maintain the property
(B) keep hazard insurance in force
(C) get permission for alterations
(D) advise the mortgagee of a change in the escrow account balance

89. Which of the following is the least likely source of originating a home mortgage?
(A) mortgage banker
(B) commercial bank
(C) life insurance company
(D) savings and loan association

90. A person buys a furnished house and wishes to finance the purchase of both the house and the furniture through a mortgage loan. He tries to get
(A) a blanket mortgage
(B) a purchase-money mortgage
(C) a temporary loan
(D) a package mortgage

91. Buying real property "subject to the mortgage" is
   (A) a type of condition loan
   (B) a mortgage bought by FHA and sold to Ginnie Mae
   (C) taking title to property with no personal responsibility for paying the mortgage loan
   (D) the right to foreclose without going to court

92. TRID is an abbreviation for
   (A) third
   (B) trans-identification
   (C) troubled identification
   (D) TILA-RESPA Integrated Disclosure

93. A mortgage has a rate that may change every six months depending on changes in the rate of Treasury bills. This is
   (A) an adjustable-rate mortgage
   (B) a capped loan
   (C) an adjustable-payment mortgage
   (D) an indexed loan

94. Which of the following is *not* a lien?
   (A) mortgage loan
   (B) mechanic's lien
   (C) judgment
   (D) easement

95. The borrower of a mortgage is the
   (A) mortgagee
   (B) mortgagor
   (C) trustor
   (D) trustee

96. In an adjustable-rate mortgage, the rate is composed of the index plus
   (A) points or discount points
   (B) service fees
   (C) closing costs
   (D) margin

97. After paying off a mortgage loan, the property owner should be sure that which of the following instruments is recorded?
   (A) relief of debt
   (B) novation
   (C) satisfaction of mortgage
   (D) burned mortgage

98. Which agency does not buy mortgage loans?
   (A) Fannie Mae
   (B) Freddie Mac
   (C) commercial banks
   (D) Federal Emergency Management Agency

99. Which agency is not active in the secondary mortgage market?
   (A) VA
   (B) FNMA
   (C) GNMA
   (D) FHLMC

100. A house is sold at foreclosure. Unpaid liens include property taxes, mechanic's lien, and first mortgage. Which is paid first?
   (A) taxes
   (B) mechanic's lien
   (C) first mortgage
   (D) whichever gets to the closing first

ANSWERS

1. **C**	21. **A**	41. **D**	61. **A**	81. **D**
2. **C**	22. **C**	42. **D**	62. **D**	82. **B**
3. **C**	23. **C**	43. **B**	63. **B**	83. **B**
4. **B**	24. **C**	44. **B**	64. **A**	84. **D**
5. **C**	25. **A**	45. **A**	65. **D**	85. **A**
6. **C**	26. **B**	46. **B**	66. **D**	86. **D**
7. **B**	27. **D**	47. **A**	67. **C**	87. **A**
8. **B**	28. **D**	48. **D**	68. **B**	88. **D**
9. **A**	29. **A**	49. **D**	69. **D**	89. **C**
10. **A**	30. **B**	50. **A**	70. **B**	90. **D**
11. **A**	31. **B**	51. **D**	71. **C**	91. **C**
12. **C**	32. **B**	52. **B**	72. **D**	92. **D**
13. **A**	33. **B**	53. **A**	73. **B**	93. **A**
14. **B**	34. **B**	54. **A**	74. **B**	94. **D**
15. **B**	35. **C**	55. **D**	75. **D**	95. **B**
16. **C**	36. **C**	56. **D**	76. **C**	96. **D**
17. **C**	37. **C**	57. **C**	77. **B**	97. **C**
18. **C**	38. **B**	58. **A**	78. **B**	98. **D**
19. **B**	39. **C**	59. **D**	79. **C**	99. **A**
20. **A**	40. **D**	60. **C**	80. **C**	100. **A**

14 Appraisals

ESSENTIAL MATTERS TO LEARN FROM THIS CHAPTER

- An *appraisal* is a professionally derived *estimate* or *opinion* of the value of real estate.
- Four principal appraisal classifications are recognized in most states:
  1. **CERTIFIED RESIDENTIAL APPRAISER.** Qualified to appraise a residence having one to four units of any value.
  2. **CERTIFIED GENERAL APPRAISER.** Qualified to appraise any real property of any value.
  3. **LICENSED APPRAISER.** Qualified to appraise a noncomplex residence having one to four units of up to $1,000,000 of value or a complex residential unit up to $250,000 in value.
  4. **REGISTERED TRAINEE APPRAISER.** An apprentice appraiser who must work under another's supervision.
- To be certified or licensed as an appraiser requires education and experience established for all states by the *Appraisal Foundation.* Certified appraisal requirements include a four-year college degree.
- Licensed or certified appraisers, and ones designated by certain professional organizations, must conform to the requirements of the *Uniform Standards of Professional Appraisal Practice* (USPAP), which is published by the Appraisal Foundation.
- *Market value* is the objective of most appraisals. Other types of value that may be sought include *insurance value, investment value, value in use, loan value,* and *book value.*
- A real estate broker or salesperson is not a qualified appraiser and cannot perform an appraisal. A broker or salesperson may prepare a *Comparative Market Analysis* (CMA) or a *Broker's Opinion of Value* (BOV).
- Several *economic principles* underlie appraisal. Two of the most important ones are the following:
  1. **SUBSTITUTION.** What does it cost to acquire a substitute property with the same salient characteristics?
  2. **ANTICIPATION.** What are the anticipated future benefits from the subject property and how much are they worth?
- *Highest and best use of a property* is the use that will provide the highest return. To qualify, the use must be *legal, physically possible, financially feasible,* and from all possible uses, the one that provides the highest return on investment.

- There are three appraisal *approaches*:

  1. **COST.** The appraiser is to answer the question: How much would it cost to *reproduce* an exact replica of the subject, less *depreciation.*

     Alternatively, depending on the property, How much would it cost to *replace* the subject with a modern equivalent, less *depreciation.*

     The market value of land is added to either amount.

     Depreciation is a loss in value. The three forms of depreciation are:

     - Physical deterioration—wear and tear.
     - Functional obsolescence—function is outmoded, out of style.
     - Economic or external obsolescence—loss in value caused by sources outside the property.

  2. **SALES COMPARISON.** How much did comparable sales sell for? Adjust the actual price paid for the comparable sale to the subject for differences. Units of comparison include square footage or other items common to the type of property, such as the number of seats in a movie theater.
  3. **INCOME OR CAPITALIZATION.** Determine the property's net operating income (NOI). This is effective gross income less operating expenses, divided by the *capitalization rate expressed in decimal form.* A capitalization (or cap) rate comprises a return *on* and *of* investment.

     There are two forms of written appraisal report:

     1. Appraisal report.
     2. Restricted appraisal report.

- An *appraisal report* must contain certain elements, stated in USPAP, so that the reader can understand the data the appraiser used and how the appraiser reached the value conclusion.
- A restricted *appraisal report* must contain certain elements that are stated in USPAP but can be abbreviated when the client is the only user and is knowledgeable about the property.

REQUIREMENTS FOR APPRAISAL CERTIFICATION AND LICENSING

Most states have at least two appraiser classifications; some have as many as four.

- **CERTIFIED RESIDENTIAL APPRAISER.** This designation allows its holder to appraise residential properties of one to four units and specifies no limit on the value.

  Education. Requires 200 classroom hours of precertification education, including coverage of the Uniform Standards of Professional Appraisal Practice, and a bachelor's degree or equivalent.

  Experience. 1,500 hours in no less than 12 months.
- **CERTIFIED GENERAL APPRAISER.** This designation allows its holder to appraise any property of any value.

  Education. Requires 300 classroom hours of precertification education, including coverage of the Uniform Standards of Professional Appraisal Practice, and a bachelor's degree or equivalent.

Experience. 3,000 hours in no less than 18 months, where a minimum of 1,500 hours is obtained in nonresidential appraisal work.

- **LICENSED RESIDENTIAL APPRAISER.** This designation allows its holder to appraise noncomplex residential properties of one to four units valued up to $1 million and complex one-to-four units up to $250,000.
  Education. Requires 150 classroom hours of precertification education, including coverage of the Uniform Standards of Professional Appraisal Practice.
  Experience. 1,000 hours in no less than 6 months.
- **REGISTERED TRAINEE APPRAISER.** This is an "apprentice" or "trainee" designation that does not require an examination.
  Education. Requires 75 classroom hours of precertification education, including coverage of the Uniform Standards of Professional Appraisal Practice.
  Experience. None

Appraisal Examinations

No examination is required to be a registered appraiser in the states that offer that designation. All three of the examinations for the other designations consist of 100 questions, and candidates are given four hours for completion. A score of 75% is passing for all exams.

To apply for an appraisal license in most states, an applicant must:

1. Be 18 years old or older.
2. Have a high school diploma or equivalent.
3. Provide evidence of successful completion of the educational requirements.
4. Make it possible for the Appraisal Board to begin an inquiry as to competence and qualifications by disclosing records of crimes or proceedings, mental disabilities, and disclosures of suspensions or revocations of other professional licenses.

Government Regulation of Appraisers

At one time, anyone could claim to be an appraiser since there was no mandatory test of the skills required to estimate property value. The higher qualifications of some individuals were (and still are) recognized by professional designations conferred by appraisal associations, but such designation was not a requirement for practice as an appraiser. Since the early 1990s, however, a real estate appraiser must be licensed or certified in his or her state in order to appraise property in a federally related transaction. Since federally related transactions include any in which an FDIC-insured bank, FNMA, FHLMC, or GNMA is involved, probably 90% or more of real estate appraisals are now prepared by a state-licensed or state-certified appraiser. A licensed real estate broker or salesperson may appraise real estate in situations where there is not a federally related transaction.

Government regulation was imposed as a result of the savings and loan debacle of the 1980s, when numerous savings institutions collapsed, largely as a result of defective or nonexistent real estate appraisals. An appraisal serves as evidence that a loan is properly supported by collateral and thus serves to protect our nation's financial institutions.

WHAT IS AN APPRAISAL?

An *appraisal* is a professionally derived opinion or estimate of value. Appraisals are used in real estate when a professional's opinion of value is needed. Because each property is unique and is not traded in a centralized, organized market, value estimates require the collection and analysis of market data.

To understand what appraisals are, please examine the preceding definition more closely. By *a professional,* we mean a person with the competence and experience to do the type of analysis required. You can get opinions of value from the sales agent, the owner, the tenant, or anyone else familiar with the property. However, these opinions may not be very useful and probably won't be convincing as evidence of value. Usually, an appraisal expert has attained some type of designation through formal study and examination by a recognized body such as the state or a professional organization.

An *opinion* is a judgment supported by facts and logical analysis. The appraiser considers all available information that reflects on the value of the property and then follows a logical process to arrive at an opinion. The result is not merely a guess but a careful reading of the facts in the case.

As shown in Figure 14-1, The Appraisal Process, the preparation of an appraisal report is an eight-step problem-solving exercise that begins with *defining the problem*—for example, "to estimate market value." The second step is to *plan the appraisal.* The third is *data collection and verification.* Fourth, the *highest and best use* of the site is considered, and fifth, the *land value* is estimated. Three *approaches to appraisal—cost, sales comparison,* and *income* (or capitalization) are applied as a sixth step. Each approach provides an independent value estimate. The seventh step is *reconciliation,* in which the value estimates from the three approaches are considered and a *final value conclusion* is reached. The last step is a written *report.* Each of these steps is covered in more detail later in the chapter.

WHY APPRAISALS ARE NEEDED

An appraisal may be sought for any of a number of purposes, including the following:

1. To help buyers and/or sellers determine how much to offer or accept.
2. To assist lenders in determining the maximum prudent amount to lend on real estate.
3. To arrive at fair compensation when private property is taken for public use.
4. To determine the amount to insure or the value loss caused by a natural disaster.
5. To determine the viability of a proposed building, renovation, or rehabilitation program.
6. To assist in corporate mergers, reorganizations, and liquidations.
7. To determine taxes due, including income, gift, estate, and ad valorem taxes.

The following sections present some other specialized uses of appraisal.

Trick Question

A trick question suggests that an appraiser *determines* value. That is not an appraiser's job! The market determines value. An appraiser provides an *estimate or opinion* of value.

Another trick question is that a broker can provide an appraisal. That is not correct. A broker can offer a *Comparative Market Analysis* (CMA) or a *Broker's Opinion of Value* (BOV), but only a certified or licensed appraiser can provide an appraisal.

STEP ONE: DEFINITION OF THE PROBLEM

a. Purpose of assignment
b. Type of value sought
c. Identification of property and legal interests
d. Date of appraisal

↓

STEP TWO: PLAN OF APPRAISAL

a. Determine data requirements
b. Identify appropriate methodology
c. Estimate time and personnel needs
d. Provide fee and assignment proposal

↓

STEP THREE: COLLECTION AND VERIFICATION OF DATA

a. Area and neighborhood
b. Site and off-site
c. Improvement analysis
d. Law and government
e. Economic activity data—income, costs, sales

↓

STEP FOUR: ANALYSIS OF HIGHEST AND BEST USE OF LAND

a. As if vacant
b. As improved

↓

STEP FIVE: ESTIMATE LAND VALUE

↓

STEP SIX: APPLICATION OF RELEVANT VALUATION APPROACHES

a. Income approach
b. Sales comparison approach
c. Cost approach
d. Analysis

↓

STEP SEVEN: RECONCILIATION

a. Review of facts as related to valuation principles
b. Statistical and probability indications
c. Logic and judgment
d. Final indicated value conclusions

↓

STEP EIGHT: REPORT OF FINAL VALUE ESTIMATE

Figure 14-1. The Appraisal Process

Source: INCOME APPRAISAL ANALYSIS, Jack P. Friedman and Nicholas Ordway, © 1988. Adapted by permission of Prentice Hall, Inc., Upper Saddle River, New Jersey.

INFLUENCES ON REAL ESTATE VALUE

Forces affecting real estate values operate on the international and national, regional, and local community levels. These forces can be categorized as follows:

1. Physical and environmental
2. Economic-financial
3. Political-governmental-legal
4. Sociological

Physical and Environmental

Physical and environmental forces include such factors as dimensions, shape, area, topography, drainage, and soil conditions as well as utilities, streets, curbs, gutters, sidewalks, landscaping, and the effect of legal restraints (zoning, deed restrictions) on physical development.

Nuisances and hazards are also to be considered, including contaminated air and water and environmentally hazardous building materials such as asbestos, PCBs, certain types of mold, and urea formaldehyde.

Economic-Financial

The fixed location of land and its immobility distinguish land from other assets. Land is dependent on where it is and what surrounds it. Urban land derives its value from its location. The most significant determinants of value are the type of industry in the area, employment and unemployment rates and types, interest rates, per capita and household income, and stability. These matters also affect the individual property, including real estate prices and mortgage payments.

Political-Governmental-Legal

Political-governmental-legal factors focus on the services provided, such as utilities, spending and taxation policies, police and fire protection, recreation, schools, garbage collection, and planning, zoning, and subdivision regulations. Building codes and the level of taxes and assessments are also important considerations.

Sociological

Sociological factors are concerned with the characteristics of people living in the area, the population density and homogeneity, and compatibility of the land uses with the needs of residents.

LEGAL CONSIDERATIONS IN APPRAISAL

Appraisers are expected to know and consider legal matters affecting real estate in making their appraisals. These matters are covered elsewhere in this book. Specific topics and coverage are as follows:

A. Real estate vs. real property (Chapters 2 and 3)
B. Real property vs. personal property (Chapters 2 and 3)
   1. Fixtures
   2. Trade fixtures
   3. Machinery and equipment
C. Limitations on real estate ownership (Chapters 2 and 3)
   1. Private
      a. Deed restrictions
      b. Leases (Chapter 10)
      c. Mortgages (Chapter 13)
      d. Easements
      e. Liens
      f. Encroachments
   2. Public (Chapter 3)
      a. Police power
         (1) Zoning
         (2) Building and fire codes
         (3) Environmental regulations
      b. Taxation (Chapter 16)
         (1) Property tax
         (2) Special assessments
      c. Eminent domain
      d. Escheat
D. Legal rights and interests (Chapters 3 and 10)
   1. Fee simple estate
   2. Life estate
   3. Leasehold interest
   4. Leased fee interest
   5. Other legal interests
      a. Easement
      b. Encroachment
E. Forms of property ownership (Chapter 3)
   1. Individual
   2. Tenancies and undivided interest
   3. Special ownership forms
      a. Condominiums
      b. Cooperative
      c. Timesharing
F. Legal descriptions (Chapter 7)
   1. Metes and bounds
   2. Government survey
   3. Lot and block
G. Transfer of title (Chapter 9)
   1. Basic types of deeds
   2. Recordation

TYPES OF VALUE APPRAISED

There are numerous types of value that the appraiser may be asked to estimate.

Market Value

Market value is the major focus of most real property appraisal assignments. Both economic and legal definitions of market value have been developed and defined. A current economic definition agreed upon by agencies that regulate federal financial institutions in the United States of America is:[1]

The most probable price that a property should bring in a competitive and open market under all conditions requisite to a fair sale, the buyer and seller each acting prudently and knowledgeably, and assuming the price is not affected by undue stimulus. Implicit in this definition is the consummation of a sale as of a specified date and the passing of title from seller to buyer under conditions whereby:

- Buyer and seller are typically motivated;
- Both parties are well informed or well advised, and acting in what they consider their best interests;
- A reasonable time is allowed for exposure in the open market;
- Payment is made in terms of cash in U.S. dollars or in terms of financial arrangements comparable thereto; and
- The price represents the normal consideration for the property sold unaffected by special or creative financing or sales concessions granted by anyone associated with the sale.

Value Contrasted with Cost and Price

Value, cost, and price are not the same. *Value* is a measure of how much a purchaser would likely be willing to pay for the property being appraised. *Cost* is a measure of the expenditures necessary to produce a similar property. Depending on several factors, this cost could be higher or lower than the current value. *Price* is the historic fact of how much was spent on similar properties in past transactions. Neither past prices nor cost necessarily represent a fair measure of current value.

Other Types of Value

Although market value is the most common value sought, an appraiser may be asked to estimate some other value needed for a specific purpose. Included are such values as *loan value, insurable value, book value, rental value, fair value, salvage value, investment value,* and many others.

LOAN VALUE OR MORTGAGE VALUE

Loan or *mortgage value* is the same as market value. Property serving as collateral for a loan or mortgage is normally valued at market value.

[1] ***Dictionary of Real Estate Appraisal,*** 6th ed. (Chicago: Appraisal Institute, 2015), pp. 141–142; definition 3.

INVESTMENT VALUE

Investment value is the estimated value of a certain real estate investment to a particular individual or institutional investor. It may be greater or less than market value depending on the investor's particular situation. For example, the investment value of vacant land in the path of growth would be greater for a young, aggressive investor who has time to wait for fruition than for an elderly widow who needs available cash for living expenses. Similarly, the investment value of a tax shelter is greater for a high-tax-bracket investor than for a tax-exempt pension plan.

VALUE IN USE AND VALUE IN EXCHANGE

Value in use, which tends to be subjective because it is not set in the market, is the worth of a property to a specific user or set of users. It considers the value of the property when put to a specific use as part of a going concern. For example, the use of a factory where automobiles are assembled may have a high value to the manufacturer even though the market value of the property is difficult to measure and may be low because there are few interested buyers. Value in use is distinguished from *value in exchange,* which is the value of a commodity, in terms of money, to people in general rather than to a specific person. Value in exchange is a more objective measure and is commonly identified with market value.

ASSESSED VALUE

The *assessed value* (or *assessed valuation*) is the value of property established for property tax purposes. Although tax assessors try to value most property at market value (or some specified fraction of market value), the fact that they must assess a great number of properties periodically on a limited budget often means that assessed values are quite different from market values.

INSURABLE VALUE

The *insurable value* of a property is the cost of total replacement of its destructible improvements. For example, suppose a home that sold for $100,000 ten years ago would cost $200,000 to rebuild today, not including land cost. Its insurable value would be $200,000 even if its current market value, including the land, was only $175,000.

GOING-CONCERN VALUE

Going-concern value is the worth of a business, such as a hotel, based on its operations. The replacement cost of the property may be much more or much less than the business is worth.

Comparative Market Analysis

Very often, property owners want to get an idea of what their real estate is worth, without paying for a full appraisal. For example, a homeowner may contact a broker or salesperson to list her house and will accept the broker's or salesperson's judgment of its value or amount for a house to be listed and put on the market. A broker may not wish to spend the time to prepare a full-blown appraisal just to solicit a listing, even if she is perfectly qualified to do so. Further, unless the broker is licensed or certified by the state, the appraisal could not be used to support a loan by a federally chartered bank or savings association.

A Comparative Market Analysis (CMA) will be prepared by the broker or salesperson for this purpose. Unlike the standard preprinted forms required for Fannie Mae/Freddie Mac residential appraisals, there is no standard form for a CMA. Brokers are free to include whatever information they wish to use and present it in any manner.

A CMA is likely to show recent sales of other houses in the neighborhood and salient characteristics. It may also show properties currently for sale on the market. Consider the data in Table 14-1.

Table 14-1
Comparative Market Analysis

Property	Price	Date	Sq. Ft.	# Bedrooms	# Baths	Stores	Per SF
Subject	?	Now	2,000	4	2	1	?
Comp 1	$420,000	Last year	2,100	4	2	2	$200.00
Comp 2	$380,000	Last year	1,950	3	1.5	1	$194.88
Comp 3	$470,000	Six months	2,250	4	2.5	1	$202.88
Asking 1	$430,000	Now	2,050	3	2	2	$209.74
Asking 2	$450,000	Now	2,100	3	2	1	$214.28

This particular CMA might be used to judge a market range (highest and lowest prices per square foot). However, it offers no information on special features, the age or condition of the houses, how long it took to sell them, or how long the existing listings have been on the market. So a CMA is not as detailed or thorough as an appraisal, though a CMA may provide a reasonable indication of property value.

Comparative (Residential) Market Analysis

A *comparative (residential) market analysis* (CMA or RMA) provides information on the current situation of the housing market. It tries to answer the question, "How is the market?" A response might be, "There are currently 102 homes for sale in northeast Tallahassee. Twenty homes were newly listed last month, and 27 sold last month. The average time on the market for those that sold was 62 days. Of those that sold, 67% were in the $400,000–$500,000 price range, with 10% in the $500,000–$600,000 range and 5% above $600,000. Eight percent were below $300,000, and 10% were between $300,000 and $400,000. Houses with fewer than two full bathrooms are not doing well. About half of the buyers are local 'move-ups,' with the other half transferees. Prices seem to be rising slightly."

As is apparent, the purpose of an RMA is to describe the current condition of the real estate market. This may help answer the question of whether this is a good time to buy or sell and whether prices are rising or falling.

Automated Valuation Models

Automated Valuation Models (AVMs) apply complex statistical techniques to data on comparable sales to offer an estimate of value. AVMs are often used by those offering home equity loans who don't want to incur a $350–$400 appraisal fee. For a typical fee of $50, a value estimate is provided, which may consider the size, age, location, and number of rooms in the house. If a lender wants greater assurance of value than an AVM offers, he may ask for a site visit or a full appraisal.

Evaluations

An *evaluation* of real estate is a study that does not lead to an estimate of market value. An evaluation may be performed to consider the feasibility of a proposed use (feasibility study) or the highest and best use of the property. An evaluation could also be a market study that considers the supply and demand in the current market for a certain type of land use.

ECONOMIC PRINCIPLES

Economics involves combining the *factors of production* to result in a product that is worth more than the cost of the individual factors. The factors of production are land, labor, capital, and entrepreneurship. For example, to produce a bushel of wheat may require $1 for land rent, $1 for labor, $1 for equipment, and $1 for the business management and risk-taking entrepreneur. This is a total of $4, so the farming operation would not occur unless the expected selling price was $4.01 or more per bushel. Many economists recognize only three factors of production. These economists consider profit as the owners' compensation for risk taking and for inputs of management and entrepreneurship. Certain economic principles are associated with real estate valuation. The major principles are explained below.

Anticipation

Anticipation is determination of the present worth of income or other benefits one expects to receive in the future from the ownership of property. For income-producing properties, value is based on the anticipated net receipts from the operation of the assets plus any amounts to be received upon resale.

Balance (Proportionality)

For any type of land use, there are optimal amounts of different factors of production that can be combined to create maximum land values. Land, labor, capital, and entrepreneurship can be combined in different proportions, as is demonstrated by the number of houses that may be erected on a parcel of land. Values are maximized when factors are in proportion, or *balance* is achieved.

Change

Real estate values tend not to remain constant but vary over time. New technology and social patterns create new demands for real estate. Demographic *changes* create needs for different kinds of housing. People's desires and tastes undergo transitions. Neighborhoods go through a life cycle of growth, maturity, decline, and renewal. Any of these factors and many others can change the utility of real estate at a given location.

Objects wear out. New businesses are started, and others end. The land use pattern is modified by private and public actions. Money supply and interest rates fluctuate. Economic conditions create opportunities or stifle growth.

Competition

When profits exceed the income necessary to pay for the factor of production, *competition* will tend to enter the market, causing average net incomes to fall. This principle is important to an analyst attempting to estimate the value of property that is selling above the cost of its replacement. Its high cost attracts builders and developers, who can earn a large profit from new construction.

Conformity

Conformity is a measure of how well the architectural style and levels of amenities and services offered by a real estate development meet market needs and expectations. A project that fails to conform to market standards is likely to suffer a financial penalty. This does not mean, for example, that all buildings in a particular location must be of the same architectural style. However, the architectural styles and the land use must be compatible. Consider the consequences of building a brightly painted Victorian house replete with architectural gingerbread in a neighborhood of split-level houses. Consider also what would happen if a municipal incinerator were constructed next to a nursing home. The nursing home might have to close because of the health hazard posed by the smoke and other pollution.

Contribution

Contribution is the amount by which the value or net income of an economic good is increased or decreased by the presence or absence of some additional factor of production. Contribution is the value increment or detriment to an economic good by the addition of some new factor, rather than the actual cost of the new factor itself. In real estate, some things add more than their cost of production, whereas others may actually detract from value. For example, a new exterior paint job may improve the appearance of a house and make it more salable. On the other hand, a potential buyer may regard a swimming pool as a liability rather than an asset.

An example of the principle of contribution would be a builder's deliberation over whether to add a tennis court to an apartment complex. The cost of this feature is $25,000. With the added tennis court, the complex is worth $1,100,000. Without it, the apartment complex is worth only $1,000,000. Thus, the tennis court would add $75,000 to the overall value. Since the cost of adding this amenity is less than its contribution, a prudent builder would construct the tennis court.

Increasing and Decreasing Returns

As resources are added to fixed agents in production, net returns will tend to increase at an increasing rate up to a certain point. Thereafter, total returns will increase at a decreasing rate until the increment to value is less than the cost of resource unit input. A common problem faced by the owners of land is the determination of how intensively their land should be developed. Development should become more intensive as long as profit increases. For example, the profit on a downtown office building, to be sold above cost, may increase with the height of the building, up to a point. Above that height, the profit may decline.

Opportunity Cost

Opportunity cost is the return forfeited by not choosing an alternative. A person with $25,000 to invest may choose to buy equity in a rental residence rather than to purchase a certificate of deposit (CD). The opportunity cost is the interest not received on the CD.

Substitution

The maximum value of a property is set by the lowest price or cost at which another property of equivalent utility may be acquired. This *substitution* principle underlies each of the three traditional approaches to value used in the appraisal process: (1) direct sales comparison, (2) income, and (3) cost. A rational purchaser will pay no more for a property than the lowest price being charged for another property of equivalent utility. Likewise, it is not prudent to pay more for an existing project if another one of equivalent utility can be produced, without unreasonable delay, for a lower cost. If an investor is analyzing an income stream, the maximum price is set by examining the prices at which other income streams of similar risk and quality can be purchased.

Supply and Demand

Supply is the quantity of goods, such as real estate, available at a given price schedule; *demand* is the quantity of goods desired at this price schedule. Demand is based on the desire of potential purchasers to acquire real estate, provided that they also have sufficient sources of financing to act on their desires. Together, supply and demand interact to establish prices.

In the long run, supply and demand are relatively effective forces in determining the direction of price changes. An excessive supply or lack of demand tends to depress price levels. A contrary pressure, which serves to raise prices, occurs when there is either an inadequate supply or a high demand.

Surplus Productivity

Surplus productivity is the net income attributed to land after the costs of labor, capital, and entrepreneurship have been paid. Because land is physically immobile, the factors of labor, capital, and entrepreneurship must be attracted to it; therefore, these factors are compensated first. If any money is left over, this residual amount is paid as rent to the owner. In economic theory, land is said to have "residual value" and has worth only if a surplus remains after paying for the other factors of production.

REAL ESTATE MARKETS AND ANALYSIS

This topic is concerned with the characteristics of the real estate market, the analysis of use of additional space (absorption), the role of money and capital markets, and the financing terms available for a property. Market analysis and absorption are discussed below. Mortgages and finance were considered in Chapter 13.

Real Estate Markets

Real estate is not a single market but instead consists of a series of submarkets with different desires and needs that can change independently of one another. Consequently, real estate markets are said to be *segmented.* The market for retail space in north Atlanta may be quite different from that in east, west, or south Atlanta. Even within the directional quadrants, there are significant differences, depending on the distance from downtown and from the nearest major artery or intersection. Within a narrow geographic area, there may be a saturation of one type of retail space but not enough of another. For example, neighborhood shopping centers, which typically have a supermarket and a drugstore as anchor tenants, may be abundant whereas there is no convenient regional mall.

Availability of Information

One key to successful appraisal is the ability to locate reliable, consistent data. Sources of market data for rental income and expense are described below.

NATIONALLY DISSEMINATED RENTAL AND OPERATING EXPENSE DATA

Certain national organizations collect information from owners and managers in major cities regarding local rents and operating expenses. Such information can be used judiciously to determine whether the data for a particular property appear consistent with experience reported nationally. Because the real estate market is fragmented, a data source appropriate for the specific property must be used.

Office Buildings. The Building Owners and Managers Association (BOMA) International provides information on office building rental rates and operating expenses experienced by its members in major U.S. cities. BOMA's address is

Building Owners and Managers International
1101 15th Street NW, Suite 800
Washington, DC 20005
www.boma.org

Shopping Centers. The Urban Land Institute (ULI) releases a new edition of *Dollars and Cents of Shopping Centers* every two years. ULI's address is

Urban Land Institute
2001 L Street NW, Suite 200
Washington, DC 20036
www.uli.org

Apartments, Condominiums, and Cooperatives. The Institute of Real Estate Management (IREM) of the National Association of REALTORS® periodically provides the *Income/Expense Analysis* for various types of buildings (elevator, walk-up, others) in different cities. IREM's address is

Institute of Real Estate Management
430 N. Michigan Avenue
Chicago, IL 60611
www.irem.org

Hotels and Motels. One source of information on national and local trends is STR Global. Its address is

STR Global
735 E. Main Street
Hendersonville, TN 37075
Phone: (615) 824-8664
Fax: (615) 824-3848
www.strglobal.com

LOCAL SOURCES OF REAL ESTATE DATA

Local organizations often collect real estate data, usually for membership use or for sale to interested parties. These include the following:

LOCAL BOARDS OF REALTORS. Most metropolitan areas have a Board of REALTORS®, or other broker group, that sponsors a multiple listing service (MLS). Upon the sale of property listed through the MLS, the broker must supply information about the completed transaction. Each property sold and its terms of sale are therefore available on computer or in a published book. Some REALTOR® boards provide information to members only; some share with other real estate organizations.

LOCAL TAX ASSESSING OFFICES. Assessing offices usually keep a file card on every property in the jurisdiction, noting property characteristics, value estimate, and data from which the estimate was derived. Many jurisdictions are notified of every sale or building permit, for immediate update of affected properties.

LOCAL CREDIT BUREAUS AND TAX MAP COMPANIES. These may have data on certain parcels. Since their main business is not evaluation, however, they are not regular sources.

UNIVERSITY RESEARCH CENTERS. University research centers, many of which are supported by state broker and salesperson license fees, may have aggregated data on real estate transactions collected from other sources throughout the state. These data are often helpful in identifying trends established over time and by city for various types of property. Additional research and educational information may also be available from such centers.

PRIVATE DATA SOURCES. Many real estate investors retain files on their property. They will often share information, usually for reciprocity rather than payment.

PROPERTY OWNERS. Property owners' records include permanent records such as deeds, leases, and copies of mortgages (lenders have the original mortgage documents) as well as periodic accounting and tax return information about a property's recent past. An owner's report may be of limited immediate use, however, because it may be disorganized, contain extraneous information, or be arranged poorly for appraisal purposes. Also, data from a single property cannot offer a broad perspective on the market.

Demand for Property

An appraiser must always be concerned about the demand for a property. People must want to use the property enough to pay the rent asked. If the demand to use the property is high, the demand to buy the property will also be high. High demand means top rents, low vacancies, and good resale prospects. Poor demand means rent reductions, high vacancies, and a property that is hard to sell.

The following are the key items that produce demand for real estate.

Economic growth in the local area increases demand for all properties. New jobs and more residents increase the need for developed real estate. Rising incomes mean higher rents and prices as well as more retail sales.

Good-quality property raises demand. A property should have all the standard features expected in the market plus something extra that the competition doesn't have. Appearance, features, size, and services are valued in the market.

A good location improves demand. Location can make a poor-quality property profitable, while a good property in the wrong place can suffer.

A competitive price can increase demand. If a property is less than ideal, it may be able to compete on the basis of price. It is important to know what segment of the market the property is intended to serve and to price accordingly.

The cost of alternatives also determines demand. Apartments are more popular when house prices are high. Houses sell better when interest rates are low.

The demand for a type of property can be estimated by a market analysis. This will indicate whether there is room for more of that type of property in the market. The demand for a specific property can be determined by comparing its features, location, and price to those of similar properties.

MARKET ANALYSIS

Market analysis is the identification and study of current supply and demand conditions in a particular area for a specific type of property. Such a study is used to indicate how well a particular piece of real estate will be supported by the market. It identifies the most likely users of the project and determines how well they are being served by the existing supply of properties. In essence, the study shows whether there is a need for a new project or whether an existing project has a good future.

For example, suppose a developer is considering construction of new luxury apartments in a certain town. A market study will first examine the sources of demand for the units. It will identify a target market: the type of tenants most likely to be attracted to the property. Description of the target market may include family income, typical family structure, and the features that potential buyers will desire in a residence. The analysis will then survey the market area and use all relevant available data sources to see how many people of this type exist and where they live. A good study will project growth trends in the target market since a likely source of tenants will be new arrivals.

Next, the study will examine supply conditions. The number and the location of similar properties are identified. A survey of vacancies indicates how well supply matches demand. Features and characteristics of competing properties should be described and some indication of market rents found. In addition, any new projects that will come along should be identified.

An appraiser should determine whether the market is unbalanced. When the supply of a certain type of real estate is short, rents and prices may be high but only temporarily. New competition will add to the supply and drive prices down. By contrast, when a market is oversupplied, the price must be low enough to offer an attractive investment.

Market analysis is used to estimate the pace of rent-up or sales for a new project. This absorption rate may be expressed as an overall rate ("The market needs 1,000 new apartment units per year") or as a specific rate for the project ("Given current competition, the project should capture 200 new rentals per year"). An absorption rate estimate is important in projecting the revenue production of a property.

A market analysis may indicate that there is little demand for the type of project envisioned. Then a change in plans is needed; the project can be redirected to a different target market. The study may also be used to help in the design of the project. It may identify some feature lacking in the existing supply that, if included in the proposed project, will offer a competitive advantage. At the same time, it may be necessary to offer the standard features of the competition. The market study will also help in pricing the product for the indicated market target.

FEASIBILITY ANALYSIS

A *feasibility study* tests not only market conditions but also the financial viability of a proposed project. It is not enough to describe the market for rental units; one must also explore whether the cost of production can be maintained at a price that allows a profit or meets some investment objective.

THE VALUATION PROCESS

As noted earlier in this chapter, the steps in the appraisal or valuation process are as follows:

1. Define the problem to be solved.
2. Plan the appraisal.
3. Collect and verify data.
4. Analyze the highest and best use of the land.
5. Estimate the land value.
6. Apply relevant valuation approaches.
7. Reconcile the value indications and make a final value estimate.
8. Report the final value estimate.

Define the Problem

In defining the problem to be solved, the appraiser must ascertain a number of items. A definition of value should be formulated at the outset of the valuation process. The property to be valued and the property rights to be appraised must be identified. Also, the date of the estimate of value must be determined. The use of the appraisal should be addressed in order to determine the needs and requirements of the client. Finally, the appraiser must consider and set forth any special limiting conditions that will be part of the appraisal.

Plan the Appraisal

The second step is planning the appraisal. This includes determining the data needed and identifying potential data sources, identifying the methodology to be applied, estimating the time and personnel needed, estimating the fee, and scheduling the work to be performed.

Collect and Verify Data

Part three of the valuation process involves gathering data related to the subject property and its environs (region and immediate neighborhood) and comparable market data. A narrative report often starts with national trends and then proceeds to the region and local area. General data gathering relates to the region in which the property is located and the interaction of social, economic, governmental, and environmental forces. A neighborhood description is included. A *neighborhood* is an area of complementary land uses that may be delineated by natural barriers, political boundaries, income levels of inhabitants, or streets.

The appraiser, in gathering specific data and completing research into general data, should seek to identify any factors that directly or indirectly influence the subject property so that these factors can be considered in the appraisal. Specific data are those related directly to the subject property. These include site data, improvement data, zoning data, ad valorem tax data, and market data. Market data is a broad term referring to all real estate sales, listings, leases, and offers that are used in developing the three approaches to value (see pages 299–309). Thus, data requirements are set by the appraisal problem to be addressed. Not only must data be collected, but also the accuracy of the collected information must be verified.

Analyze the Highest and Best Use

After describing and analyzing the subject environs, site, and improvements, the appraiser analyzes the *highest and best use* of the property. The highest and best use analysis for improved properties has a twofold purpose: (1) to determine the highest and best use of the site as if vacant and (2) to analyze the highest and best use of the property as improved. The highest and best use of the site as if vacant is required so that land with similar highest and best uses can be used for comparison purposes in the land value estimate section of the appraisal. Also, the existing improvements are evaluated in regard to the highest and best use of the site as if vacant. If the existing improvements do not conform to this use of the site, changes or modifications should be considered. The highest and best use of a property is a very crucial conclusion, as the three approaches to value are developed on the basis of this determination.

Estimate the Land Value

The next step in the valuation process involves estimating the value of the site as if vacant. In determining this value, consideration must be given to the highest and best use of the site. One or more of the following methods can be used to arrive at an indication of the value of the site as if vacant:

1. Sales comparison
2. Allocation
3. Extraction
4. Subdivision development
5. Land residual technique
6. Ground rent capitalization

Apply Valuation Approaches

After the site value estimate, the three approaches to value are developed: the cost approach, the sales comparison (market) approach, and the income approach. Sometimes, one or more of the three approaches is not applicable to a given property and is not developed in the appraisal. In this case, the appraiser should state why the approach is not applicable. This can be determined, however, only by careful analysis of the property being appraised in relation to the analytical tools available to the appraiser. A short explanation of each approach follows; the three approaches are considered in detail later in this chapter.

COST APPROACH

The cost approach is based on the principle of substitution. It recognizes that buyers often judge the value of an existing structure by comparing it with the cost to construct a new structure of the same type. In developing the cost approach, the appraiser estimates the cost to replace or reproduce the improvements, deducts depreciation (if any), and then adds to this figure the estimated value of the site or land as if vacant. The major limitation of this approach is the difficulty in estimating accrued depreciation, particularly when depreciation is present in more than one form.

SALES COMPARISON APPROACH

The sales comparison approach (formerly called *market approach*) recognizes that no informed and prudent person will pay more for a property than it would cost to purchase an equally desirable substitute property (the principle of substitution). In developing this approach, the appraiser locates sales of properties similar to the property being appraised (called *comparable sales*) and provides details describing them in the appraisal report. These sale properties are then compared to the property being appraised, and adjustments are made for differences between each such property and the property under appraisal. The reliability of this approach is directly related to the quality and quantity of the sales data.

INCOME CAPITALIZATION APPROACH

The income capitalization approach (see pages 341–343) is based on the principle of anticipation, which states that value is created by the anticipation of future benefits to be derived from ownership of a given property. The appraiser is primarily concerned with the future benefits to be derived from operation of the subject property—the net operating income (NOI) or cash flow. The steps in this approach include estimating potential gross income by comparison with competing properties, deducting a market derived vacancy and collection loss allowance, and estimating expenses (derived from historical and/or market experience) to determine a projected stream of net operating income. The income stream is then capitalized into an indication of value by using capitalization rates extracted from competitive properties in the market, or by using other techniques when applicable.

Reconciliation

It should be noted that all three approaches are closely interrelated and that all three use market data: the cost approach utilizes data from the market on labor and materials costs, the direct sales comparison approach analyzes sales of similar properties, and the income capitalization approach determines the market investment return rates. Also, data and conclusions from one approach are often used in one or more of the other approaches. If good data are available for application of all three approaches, the resulting indications should fall within a narrow range.

Among the factors considered in the final estimate of value are the accuracy, reliability, and pertinence of the information available for each approach. Major considerations in the reconciliation of value indications are which approach has the best data, which is the most likely to be free of error, and which is the most reliable for the type of property being appraised. A final value estimate should be rounded so as not to give a false impression of the precision associated with an opinion.

Report

Two types of written reports for real estate appraisals prepared under USPAP are:

1. Appraisal report
2. Restricted appraisal report

An *appraisal report* can summarize many of the details, which are included in work papers. A *restricted appraisal report* is abbreviated so much that its use is restricted to one party, such as the property owner, who knows all about the property and doesn't need extensive explanations in a report.

PROPERTY DESCRIPTION

The site, improvements, and construction design must be described in an appraisal to allow the reader to understand the physical qualities of the property.

Site Description

The size and shape of the site are described, and any impediment to development or any superior qualities are noted. If a parcel is narrow or awkwardly shaped, this fact is mentioned. Road frontage (or lack of access) and topography are noted. Zoning is described, detailing allowable uses and the current use. The appraiser should note, with reference to maps, whether the property is in a flood-prone area. The availability of utilities at the site, including water supply and sewers, gas and electricity, and, if appropriate, cable television, is noted.

Improvements Description

The size, condition, and utility of the improvements must be described in the appraisal report, accompanied by photographs of the property, both inside and out. Photographs give the reader a visual impression in addition to the written description of the information to be conveyed. Anything relevant about the usefulness (or lack thereof) and condition of the property is described.

Basic Construction and Design

When plans and specifications are available, the appraiser can refer to them in describing the design and construction. Otherwise, he must rely on visual impressions and answers to questions provided by owners, lessees, or neighbors. At the least, the report should describe the foundation (concrete slab, pier and beam), framing, finish (inside and out), and electrical and mechanical systems.

The usefulness of the building is important. Functionality is affected by such factors as ceiling height for warehouse space, pedestrian traffic flow for retail locations, and layout of office space for office buildings.

HIGHEST AND BEST USE ANALYSIS

Highest and best use is defined as follows:

> *The reasonably probable and legal use of property that results in the highest value. The four criteria that the highest and best use must meet are legal permissibility, physical possibility, financial feasibility, and maximum profitability.*[2]

The four criteria are applied in sequential order. Potential uses are narrowed through the consideration of each criterion so that, by the time the last criterion is applied, only a single use is indicated. A property often will have numerous uses that are physically possible; a lesser number that are both physically possible and legally permissible; fewer still that are physically possible, legally permissible, and financially feasible; and only a single use that meets all four criteria.

Highest and Best Use of the Land as if Vacant

This type of use forms the basis of land value in the cost approach. Four tests are employed in the effort to determine the highest and best use of land. The first three are physical, legal, and financial. Uses that meet the first three tests are then considered to determine which is the maximally productive. Highest and best use analysis assumes that any existing building can be demolished.

PHYSICALLY POSSIBLE

The appraiser describes the uses surrounding the site and in the neighborhood, along with the traffic flow on the subject's street and nearby thoroughfares. Visibility and access are described as they affect potential uses. The size and shape are considered, and the question of whether these attributes lend themselves to certain uses or are impediments is discussed. Nearby transportation systems (bus, subway, rail) are also considered. Availability of utilities is a physical factor that can limit or impede development.

LEGALLY PERMISSIBLE

Legal restrictions include private restrictions and existing public land use regulations—in most cities, zoning. Often, only a title search by a competent attorney will uncover deed restrictions. It is therefore recommended that a title search be made if any question regarding deed restrictions arises. If common restrictions (e.g., utility and drainage easements) exist, the appraiser can state whether they appear to adversely affect the site's development potential. The appraiser should specify what is allowed within the zoning classification.

FINANCIALLY FEASIBLE

The financial feasibility of the physically possible and legally permissible uses of the site is considered. In this analysis, consideration is given to the supply and demand levels for residential, retail, and industrial properties within the area.

[2] *The Dictionary of Real Estate Appraisal*, 6th ed. (Chicago: Appraisal Institute 2015), p. 109.

MAXIMALLY PRODUCTIVE

On the basis of surrounding existing uses, most probable type of use and, within this range, possible alternatives (e.g., office/warehouse or office/laboratory for industrial property) are stated. Although a great deal of judgment is involved, the appraiser provides a conclusion as to the highest and best use. This use need not be one that can be implemented immediately but may be something to consider for the future.

Highest and Best Use of the Property as Improved

When the property is already improved with a substantial building, there are only two practical alternatives for the subject: razing the improvements or continued operation as improved. When the improvements have lost much of their utility because of their effective age, obsolescence, and condition, demolition could be considered. However, it is often not economically feasible to demolish when there is sufficient income and utility to justify continued operation of the building for an interim period.

APPRAISAL MATHEMATICS AND STATISTICS

Math concepts in appraisal fall into two categories:

1. The math of real estate finance, which is covered in Chapter 15.
2. Statistical concepts of mean, median, mode, range, and standard deviation.

- *Mean*—the average of a set of numbers. It is found by adding the amounts and dividing by the number of items.
- *Median*—the middle item in a ranking.
- *Mode*—the most popular amount, that is, the one that occurs most frequently in a list.
- *Range*—the difference between the largest and the smallest numbers in a list.
  For example, consider seven homes that sold for the following amounts:

Number	Price
1	$ 300,000
2	340,000
3	400,000
4	420,000
5	480,000
6	480,000
7	540,000
Total	$2,960,000

  The *mean*, or average, is found by totaling all the prices and then dividing by the number of sales (seven). In this example, the mean is $422,857. The *median* is $420,000, as three houses sold for more and three for less. The *mode* is the most popular price—in this case, $480,000, as two houses sold for that price and only one for every other price. The *range* is from $300,000 to $540,000, or $240,000.
- *Standard deviation*—a term used to describe the variance of the observations from the mean. It is important to remember that 66% of observations fall within one standard deviation from the mean and 95% fall within two standard deviations.

SALES COMPARISON APPROACH

The sales comparison approach is based primarily on the principle of substitution, which holds that a prudent individual will pay no more for a property than it would cost to purchase a comparable substitute property. The approach recognizes that a typical buyer will compare asking prices and seek to purchase the property that meets his or her wants and needs for the lowest cost. In developing the sales comparison approach, the appraiser attempts to interpret and measure the actions of parties involved in the marketplace, including buyers, sellers, and investors.

Collection of Data

Data are collected on recent sales of properties similar to the subject, called *comparables* (*comps*). Sources of comparable data include REALTOR® publications, public records, buyers and sellers, brokers and salespersons, and other appraisers. Salient details for each comp are described in the appraisal report. Because comps will not be identical to the subject, some price adjustment is necessary. The idea is to simulate the price that would have been paid if the comp were actually identical to the subject. Differences that do not affect value are not adjusted for. If the comp is *superior* to the subject, an amount is *subtracted* from the known sales price of the comp. *Inferior* features of the comp require that an amount be *added* to the comp's known sales price. From the group of adjusted sales prices, the appraiser selects an indicator of value that is representative of the subject.

Selection of Comparables

No set number of comparables is required, but the greater the number, the more reliable the result. How are comps selected? To minimize the amount of adjustment required, comps should be closely similar to the subject. How current and how similar the comps are depends on the availability of data. Comps should be verified by calling a party to the transaction. The appraiser must ascertain that comps were sold in an *arm's length* market transaction; that is, that the sales price or terms were not distorted by, for example, the fact that buyer and seller were relatives.

The following are common shortcomings in the selection process:

- The sale occurred too long ago. (The market may have changed since the sale took place.)
- The location is too different. (Location and surroundings have important effects on value.)
- Special financing was used in a comparable sale. (The price may have been based in part on favorable financing terms.)
- Too few comparables can be found. (One or two sales may not represent the market.)
- The comparable was not an open market transaction. (A sale to a relative or an intra-company sale, foreclosure, or the like is not considered an open market or arm's length transaction.)

When the subject is unusual or when market activity is slow, one or more of these problems may be unavoidable. However, if this is the case, the problem(s) should be acknowledged and taken into consideration when the final value opinion is rendered.

Steps in Approach

The steps involved in developing the sales comparison approach are as follows:

1. Research the market to obtain information pertaining to sales, listings, and sometimes offerings of properties similar to the property being appraised.
2. Investigate the market data to determine whether they are factually accurate and whether each sale represents an arm's length transaction.
3. Determine relevant units of comparison (e.g., sales price per square foot), and develop a comparative analysis for each.
4. Compare the subject and comparable sales according to the elements of comparison, and then adjust each sale as appropriate.
5. Reconcile the multiple value indications that result from the adjustment of the comparables into a single value indication.

Units of Comparison

Units of comparison (see Step 3 above) are defined as "the components into which a property may be divided for purposes of comparison."[3] For example, apartment complexes are typically analyzed on the basis of one or more of the following three units of comparison: sales price per square foot of rentable area, sales price per apartment unit, and gross income multiplier. Land is sold by the acre, front foot (along a road or street), square foot, and developable unit. All appropriate units of comparison should be analyzed for the property type being appraised and the resulting value indications reconciled to a single indicated value or value range. The best unit of comparison for any property is the unit that is considered important in the market for that property type.

Adjustments are usually required to compare the comparable properties selected to the subject property. The adjustments may be applied either to the total sales price or to the unit or units of comparison. Income multipliers are not normally adjusted, however. The comparable sales are adjusted for differences in elements of comparison, which are defined as "the characteristics or attributes of properties and transactions that cause the prices paid for real estate to vary."[4] The adjustment process is an effort to isolate the amount paid for different features or sizes; thus it rests on the principle of contribution.

Use of Units of Comparison in Appraisals

Real estate is more easily appraised using units of comparison. Just as meat is sold by the pound, fabric by the yard, precious metals by the troy ounce, and liquids by the fluid ounce, quart, or gallon, real properties are compared by a unit of comparison after adjustments are made to comparables to obtain a uniform quality level. The unit of comparison used may vary depending on the property type or use. Units of comparison may be used in all approaches to value.

Land

Farm land and large vacant tracts are normally sold by the *acre* (1 acre equals 43,560 square feet). Valuable tracts are sold by the *square foot,* although land along a highway suitable for

[3]Ibid, p. 240.
[4]Ibid, p. 75.

commercial development or waterfront property may be sold by the number of *front feet* (length along the highway or shoreline). A certain minimum or typical depth is associated with the frontage. Another way that land may be sold is per unit of allowable development. For example, when zoning permits 12 apartment units per acre, the price may be $120,000 per acre (or stated as $10,000 per developable unit).

Improved Property

Improved properties may be sold by varying units of measurement, depending on their use. For example, office or retail space may be sold on a per-square-foot, motels on a per-room, and theaters on a per-seat basis. However, quality, condition, and other elements are also considered, and an adjusted price is estimated before the units are compared so that the units of comparison are truly comparable.

Elements of Comparison

The following *elements of comparison* should be considered in the sales comparison approach:

1. Real property rights conveyed
2. Financing terms
3. Conditions of sale
4. Market conditions on date of sale
5. Location
6. Physical characteristics

Adjustments for these are made to the actual selling price of the comparable property.

Real Property Rights Conveyed

In most situations, fee simple title will be conveyed. However, when the comparables differ from the subject in the extent of rights conveyed, adjustment is required. The appraiser must also be aware of the effects of easements and leaseholds on the price and must apply adjustments accordingly.

Financing Terms

Generally, the seller will receive cash from the sale, even when the buyer arranges a loan from a third party. When the property is sold with special financing, however, the price will be affected. Therefore, when the seller provides financing or pays points or excessive fees for the buyer's loan, the price received should be adjusted to a cash equivalent to result in the market value. Terms of financing, whether good or poor, affect the price but not the value.

Conditions of Sale

In some transactions, sellers or buyers may be unusually motivated so that the price agreed upon does not necessarily reflect market value. This is true in cases of foreclosure and bankruptcy as well as transactions in which a buyer requires a certain adjacent parcel for business expansion. A parent may sell property to a child at a bargain price, or a stockholder may sell to his wholly owned corporation at a high price, both in relation to the market. An appraiser must adjust these non-arm's-length transactions to the market or may not use them as comparables.

Market Conditions on Date of Sale

Especially during periods of inflation or local economy changes, it is possible to determine shifts in price levels over time. In a rising market, a comparable property sold just a few months ago would fetch a higher price if sold now. For this reason, it is important to identify the date of sale of each comparable. Unless a comparable sale is very recent, an adjustment may be needed for market conditions.

Location

Since real estate prices are greatly affected by location, an adjustment is warranted for this characteristic when a comparable's location differs significantly from the subject's. Because each location is unique, the adjustment is often subjective. For retail properties, traffic count is often a key variable.

Physical Characteristics

Physical characteristics of the property warrant adjustment. These characteristics include age, condition, construction, quality, maintenance, access, visibility, and utility.

The Adjustment Process

Adjustments may be made in terms of percentages or in dollar amounts. Either the total sales price may be adjusted, or the adjustments can be applied to one or more units of comparison. The adjustments should be made in sequential order, with the adjustment for real property rights always made first, the adjustment for financing terms then made to the sales price adjusted for real property rights conveyed, and so on. Adjustments for location and physical characteristics can be grouped together as a single cumulative amount or percentage. If characteristics are interrelated or interdependent, then cumulative percentage adjustments may be used. If they are independent, however, each should be applied to the actual price of the comparable property.

The paired sale technique is useful in adjusting for just one feature. If other sales are found for properties that are similar except for just one feature, the value of that feature can be captured. It is then applied to the comparable properties.

After adjustment, the sales will indicate a range in total value or unit value for the property being appraised. If the sales data are highly comparable and the sales occurred in a reasonably efficient market, the range in indicated values should be tight. When market conditions are imperfect and comparable sales data are limited, however, the range in indicated values may be wider.

In reconciling the value indications provided by the comparable sales, the appraiser should consider the amount of adjustment required for each sale. Sales requiring lesser degrees of adjustment are typically more comparable and are given greater weight than sales requiring greater adjustment. Other factors must be considered, however, including the reliability of the sales data and the degree of support of the required adjustments. After consideration of these factors, a final point value or value range is set forth.

COST APPROACH

The cost approach to appraising is predicated on the assumption that the value of a structure does not exceed the current cost of producing a replica of it. This tends to hold true when a site is improved with a new structure that represents the highest and best use of the site.

Typically, an appraiser estimates the reproduction cost of the subject property and then subtracts for depreciation. The market value of the site (land prepared for use) is then added to the depreciated reproduction cost to arrive at a final estimate of value.

Steps in the cost approach, also called the summation approach, are as follows:

1. Estimate the value of the site as if vacant and available to be put to its highest and best use as of the date of the appraisal.
2. Estimate the reproduction or replacement cost new of the improvements.
3. Estimate all elements of accrued depreciation, including physical, functional, and economic obsolescence.
4. Subtract the total accrued depreciation from the replacement cost new of the improvements to determine the present worth of the improvements.
5. Add the estimated depreciated worth of all site improvements.
6. Add the total present worth of all improvements to the estimated site value to arrive at the value of the property as indicated by the cost approach.

The procedure for estimating land value was described in the sales comparison approach.

Site Value

A *site* is land that has been cleared, graded, and improved with utility connections so that it is ready to be built upon. In a real estate appraisal, the site value is typically estimated even though the purpose of the appraisal is to estimate the value of the entire property, the improved real estate. The site value is estimated as part of the cost approach.

Site value may also give information about the entire property. For example, suppose that the site value is almost equal to the improved property value; this fact may provide a clue about the highest and best use of the property. It may help also in analyzing the possibility of replacing the existing building or give information on the useful life of the present improvements.

There are several ways to estimate site value; sales comparison, land residual, allocation, extraction, and plottage and assemblage are discussed below.

SALES COMPARISON

The sales comparison technique for land is similar to that described above for improved properties, but it is easier to apply to land because there are fewer factors to consider. It is the most commonly used technique for land valuation.

LAND RESIDUAL

Improved properties may generate annual income. If the building value is known, the income attributable to the building may be subtracted from the total income; the difference is the income to the land. That amount may be converted to a land value by dividing by a capitalization rate (rate of return).

ALLOCATION

Allocation is a method of estimating land value as a fraction of total value. Suppose that, for the type of property in question, data from numerous comparables indicate that land is typically 25% of total value. The land value for the subject property is then estimated at 25% of total value unless there is some reason not to apply this technique (such as the presence of excess land on the subject).

EXTRACTION

The *extraction* method of estimating land value involves subtracting the depreciated cost of improvements from the total value of the property. It is used to best advantage when the improvements have little value.

PLOTTAGE AND ASSEMBLAGE

These terms relate to combining two or more parcels of land. The combination is called *assemblage*. When the combined parcel is worth more, because of increased utility, than the parcels would be if sold separately, the increment of value added is called *plottage*.

Reproduction Cost Versus Replacement Cost

Reproduction cost is the cost of building a replica; *replacement cost* is the cost of replacing the subject with one that has equivalent utility but is built with modern materials and to current standards of design and function. Replacement cost new tends to set the upper limit on current value.

Measuring Reproduction Cost or Replacement Cost

Four methods of measuring reproduction or replacement cost are as follows:

1. **QUANTITY SURVEY**, whereby each type of material in a structure is itemized and costed out (the price of each nail, brick, etc., is reflected). Overhead, insurance, and profit are added as a lump sum.
2. **UNIT-IN-PLACE**, whereby the installed unit cost of each component is measured (e.g., exterior walls, including gypsum board, plaster, paint, wallpaper, and labor). Equipment and fixtures are added as a lump sum.
3. **SEGREGATED COST OR TRADE BREAKDOWN**, which is similar to the unit-in-place method except that the units considered are major functional parts of the structure, such as foundation, floor, ceiling, roof, and heating system. Fixtures and equipment are added as a lump sum.
4. **COMPARATIVE UNIT**, whereby all components of a structure are lumped together on a unit basis to obtain the total cost per square foot or cubic foot. Costs include materials, installation, and builder's overhead and profit.

Cost Services

Companies that provide cost data, either through online computer services or as hard copy, include Marshall & Swift/Boeckh, R.S. Means, and Dodge (McGraw-Hill).

Cost Indexes

Cost indexes are prepared, using a construction inflation index, to update costs from the latest price schedules to the present time. Another index, a local multiplier, is printed for selected cities to show the variation from the benchmark region or city.

Depreciation

Depreciation is a reduction in value, due to all sources, from cost new. *Accrued depreciation* is the total loss of value from all causes, measured from reproduction cost new. Depreciation is broken down into three elements—physical, functional, and external or economic—and each is further classified as curable or incurable.

Physical Deterioration

Physical deterioration is a reduction in the utility of improvements resulting from an impairment of physical condition and is commonly divided into curable and incurable components. *Curable physical deterioration* considers items (referred to as "deferred maintenance") that a prudent purchaser would anticipate correcting immediately upon acquisition of the property. It is assumed that the cost of effecting the correction will not be greater than the anticipated gain in value accrued by virtue of correcting the problem. This estimate is usually computed as a cost to cure. *Incurable physical deterioration* refers to items that cannot be physically or economically corrected.

Functional Obsolescence

Functional obsolescence is a loss of value due to characteristics inherent in the structures themselves. It results in a decreased capacity of the improvements to perform the functions for which they were intended, in accordance with current market tastes and standards of acceptability. Something becomes outmoded. As mentioned above, functional obsolescence consists of curable and incurable items.

CURABLE FUNCTIONAL OBSOLESCENCE

Curable functional obsolescence may be the result of either a deficiency or an excess. The measure of a deficiency is the excess cost to cure. The measure of an excess is the reproduction cost of the superadequacy, less physical deterioration already charged, plus the cost to cure. As noted, in order for an item to be considered curable, the necessary cost to cure must not exceed the anticipated increase in value due to the cure.

INCURABLE FUNCTIONAL OBSOLESCENCE

Incurable functional obsolescence is caused by either a deficiency or an excess that, if cured, would not warrant the expenditure. In other words, the owner/purchaser would not be justified in replacing items whose cost would exceed the anticipated increase in value. If caused by a deficiency, obsolescence is estimated as the capitalized value of the rent loss due to the condition. An excess is measured by the reproduction cost of the item, less physical deterioration already charged, plus the present worth of the added cost of ownership due to the subject property and other factors relating to this property.

External or Economic Obsolescence

External or economic obsolescence is a diminished utility of the structure due to negative influences from outside the site. It is almost always *incurable* on the part of the owner, landlord, or tenant. External obsolescence can be caused by a variety of factors, such as neighborhood decline; the property's location in a community, state, or region; market conditions; or government regulations.

Depreciation Calculations

There are a number of ways to calculate the value loss from depreciation. These include the economic age-life method, the breakdown method, and market extraction. Definitions of two important terms are offered below as an aid to understanding depreciation calculations.

Economic life is the period over which improvements to real estate contribute to the value of the property. Economic life establishes the capital recovery period for improvements in the traditional residual techniques of income capitalization. It is also used in the estimation of accrued depreciation (diminished utility) in the cost approach to value estimation.

Effective age, as applied to a structure, is the age of a similar structure of equivalent utility, condition, and remaining life expectancy, as distinct from chronological age; that is, effective age is the age indicated by the condition and utility of the structure. If a building has had better than average maintenance, its effective age may be less than the actual age. If there has been inadequate maintenance, it may be greater. A 40-year-old building, for example, may have an effective age of 20 years because of rehabilitation and modernization.

ECONOMIC AGE-LIFE METHOD

The ratio between the effective age of a building and its total economic life is its *economic age-life*. To estimate the building's incurable physical depreciation, this ratio is applied to the cost of the building or components after deferred maintenance has been subtracted.

BREAKDOWN METHOD

With this method, the loss in value is separated into physical, functional, and external causes.

Cost Approach Example

Replacement Cost New

Office/warehouse building:		
(24,208 SF @ $41.56/SF)		$1,006,084
Site improvements and equipment:		
Parking lot	$150,000	
Signage, landscaping, misc.	10,000	
Subtotal, site improvements		160,000
Subtotal, hard costs		$1,166,084
Developer/contractor profit:		
Building cost new @ 15%		174,912
Ad valorem taxes @ 2.6% × $140,000		3,600
Marketing, misc.		2,000
Replacement cost new		1,346,596

Less: Accrued depreciation:		
Physical depreciation (based on direct cost)		
Curable	$120,000	
Incurable	802,686	
Subtotal	$922,686	
Obsolescence		
Functional	100,000	
External	220,000	
Subtotal	$320,000	
Total accrued depreciation		−1,242,686
Estimated present value of improvements		103,910
Add: Estimated land value		140,000
Indicated land value via cost approach		$243,910
	Rounded to:	$240,000

Cost Approach Estimate of Value $240,000

INCOME APPROACH

Most of the income approach methodology will be found in the general certification section of Chapter 15. All appraisal candidates, however, are expected to understand gross rent multipliers (or gross income multipliers) and to be able to estimate income and expense.

Gross Income Multiplier

The *gross income multiplier* is simply the sales price divided by the rental income, generally on an annual basis. For example, if a building sold for $1,200,000 and earns $200,000 a year in rent, the multiplier is 6 ($1,200,000/$200,000 = 6). This multiplier is often used on a monthly rent basis, especially for houses. If a home sold for $120,000 and rents for $1,000 per month, the monthly multiplier is 120. The multiplier is applied to the subject's rent to estimate its value.

Generally, the gross rent multiplier is used as a sales comparison measure for income property. It may be used as the income approach for single-family housing.

Estimate of Income and Expenses

The income most commonly used is called *net operating income (NOI)*, which is found for rental properties as follows:

1. Estimate *potential gross income* (the rental collected if all units are rented for an entire year).
2. Subtract a vacancy and collection allowance.
3. Subtract *operating expenses.* These include insurance, maintenance and repairs, real estate taxes, utilities, and other expenses essential to the operation of the property.

   Operating expenses are categorized as fixed or variable. *Fixed expenses* remain constant regardless of occupancy, whereas *variable expenses* go up and down depending on occupancy.

Another category of operating expenses consists of reserves for the replacement of appliances, carpets, and other short-lived assets. Since these may last only 5–10 years, an amount is set aside annually as a provision to prevent distorting the income in the year these assets are replaced.

Interest and principal payments and depreciation are *not* operating expenses.

The result—(1) minus (2) and (3)—is net operating income. NOI is then divided by a capitalization rate, which is the rate used to convert an income stream into a lump-sum capital value. The capitalization rate must be adequate to provide for a return on the entire investment, as well as a recovery of the portion of the property that is subject to depreciation.

The rate of return on the entire investment can be estimated as in the following example:

Basic Rate:	
Safe, liquid rate of return on investment	
(U.S. government bonds, insured savings)	*1%*
Provision for illiquidity	*1%*
Provision for investment management	*1%*
Provision for risk	*2%*
Rate of return on investment	*5%*
Plus capital recovery: 80% of cost represented by improvements, which are subject to a 40-year life = 80% × 2½% (straight-line annual depreciation)	*2%*
Capitalization rate	*7%*

If net operating income is estimated at \$100,000, the resulting value is \$100,000 ÷ 0.07 = \$1,428,571, rounded to \$1,429,000. The formula employed is as follows:

$$\text{Value} = \frac{\text{Income}}{\text{Capitalization Rate}}$$

PARTIAL INTERESTS

The residential appraiser is expected to know the effects of certain partial interests, namely:

- Leaseholds and leased fees
- Life estates
- Easements
- Timeshares
- Cooperatives
- Undivided interests in common areas

Leaseholds and Leased Fees

When property is leased, the tenant has a *leasehold.* This may be a valuable interest, particularly when the lease rent (called the *contract rent*) is below market rent and the lease is for a long term. The landlord's position is called *leased fee.* An appraiser should determine the effect of a lease on the market value of the fee simple and ascertain the definition of value being sought. Is it a leased fee value, or fee simple unencumbered?

Life Estates

A *life estate* allows a person to use property for his or her own lifetime or for the life of another person (*pur autre vie*). Upon death, the remainderman becomes the sole owner. Thus, the market value of the property is divided between two persons: one with the right of use for life and another whose use commences upon the life tenant's death.

Easements

An *easement* is the right to use property owned by another for a specific purpose. For example, billboards may be erected on easements, and power company transmission line paths and other land needed for utilities are often provided by the use of easements. Easements may enhance the value of real estate when they generate adequate revenue (billboard rent), attract desired services (utilities to the property), or provide access to other properties. They may detract from value when they represent nuisances (unsightly utility wires), indicate flood plains, or prevent alternative profitable property uses. Unless easements are specifically excluded by the appraisal report, an appraiser should ascertain the effect of easements on property being considered as well as easements on others' properties that the subject property owner can enjoy.

Timeshares

A *timeshare* is the right to use property for a given period of time each year. Often used for resort property, a timeshare may be ownership of a specific apartment unit for a specific period, such as the first week in April of each year, or it may represent the right to use any of a number of identical units in a condominium development for any week during a certain period (with prices varying for peak and off-season periods). The price of a weekly timeshare, when multiplied by 52 weeks, may exceed two to three times the value of the unit if sold in its entirety. However, timeshare ownership often includes amenities (such as utilities, laundry service, and periodic redecoration) that would not be included in ownership of the unit itself.

Cooperatives

Some apartment buildings, especially in New York, are owned as cooperatives (co-ops). A corporation owns the building and can mortgage it. Stockholders own certain shares, which give each stockholder the right to occupy a certain unit. Stockholders are generally tenants and must pay their pro rata shares of maintenance expenses. Special tax laws allow the tenants to deduct interest on the mortgage and real estate taxes. An appraiser must recognize the valuation process for this type of property and understand how the stock is valued.

Undivided Interests In Common Areas

Condominiums provide for separate ownership of individual units with the right to mortgage each unit. Common areas such as walkways, recreational facilities, and exterior walls are owned jointly. Each condo owner owns a share of the common area and may use it, as may all other condo owners.

APPRAISAL STANDARDS AND ETHICS

The Uniform Standards of Professional Appraisal Practice (USPAP) have been developed by the Appraisal Foundation. Appraisers belonging to an organization that is a member of the Appraisal Foundation adhere to these standards. The following are highlights of certain parts of the USPAP.

Preamble

- The appraisal must be meaningful and not misleading in the marketplace.
- The appraiser is to observe all the ethical standards set forth below.
- Certain competency provisions are to be observed.
- If departure provisions are utilized, the rules concerning their use must be observed. A departure provision is a provision for the appraiser to depart from the standards. For example, a departure provision allows an approach to be waived provided that the appraiser explains why that approach was not used.
- Users of an appraiser's services are encouraged to demand work in conformance with the standards.

Ethics

To promote and preserve the public trust inherent in professional appraisal practice, an appraiser must observe the highest standards of professional ethics. The Ethics Rule of USPAP is divided into four sections: Conduct, Management, Confidentiality, and Record Keeping.[5]

CONDUCT

- An appraiser must perform assignments ethically and competently in accordance with USPAP and any supplemental standards agreed to by the appraiser in accepting the assignment. An appraiser may not engage in criminal conduct. An appraiser must perform assignments with impartiality, objectivity, and independence, and without accommodation of personal interests.
- In appraisal practice, an appraiser must not perform as an advocate for any party or issue.
- An appraiser must not accept an assignment that includes the reporting of predetermined opinions and conclusions.
- An appraiser must not communicate assignment results in a misleading or fraudulent manner. An appraiser must not use or communicate a misleading or fraudulent report or knowingly permit an employee or other person to communicate a misleading or fraudulent report.
- An appraiser must not use or rely on unsupported conclusions relating to characteristics such as race, color, religion, national origin, gender, marital status, familial status, age, receipt of public assistance income, handicap, or an unsupported conclusion that homogeneity of such characteristics is necessary to maximize value.

[5]The sections that follow are taken from Appraisal Standards Board, ***Uniform Standards of Professional Appraisal Practice,*** 2018–2019 edition (Washington, DC: Appraisal Foundation). Comments and footnotes have been omitted.

MANAGEMENT

- The payment of undisclosed fees, commissions, or things or value in connection with the procurement of an assignment is unethical.
- It is unethical for an appraiser to accept an assignment, or to have a compensation arrangement for an assignment, that is contingent on any of the following:

  1. The reporting of a predetermined result (e.g., opinion of value);
  2. A direction in assignment results that favors the cause of the client;
  3. The amount of a value opinion;
  4. The attainment of a stipulated result; or
  5. The occurrence of a subsequent event directly related to the appraiser's opinions and specific to the assignment's purpose.

- Advertising for or soliciting assignments in a manner that is false, misleading, or exaggerated is unethical.

CONFIDENTIALITY

- An appraiser must protect the confidential nature of the appraiser-client relationship.
- An appraiser must act in good faith with regard to the legitimate interests of the client in the use of confidential information and in the communication of assignment results.
- An appraiser must be aware of, and comply with, all confidentiality and privacy laws and regulations applicable in an assignment.
- An appraiser must not disclose confidential information or assignment results prepared for a client to anyone other than the client and persons specifically authorized by the client, state enforcement agencies and such third parties as may be authorized by due process of law, and a duly authorized professional peer review committee except when such disclosure to a committee would violate applicable law or regulation. It is unethical for a member of a duly authorized professional peer review committee to disclose confidential information presented to the committee.

RECORD KEEPING

- An appraiser must prepare a workfile for each appraisal, appraisal review, or appraisal consulting assignment. The workfile must include:

  1. The name of the client and the identity, by name or type, of any other intended users;
  2. True copies of any written reports, documented on any type of media;
  3. Summaries of any oral reports or testimony, or a transcript of testimony, including the appraiser's signed and dated certification; and
  4. All other data, information, and documentation necessary to support the appraiser's opinions and conclusions and to show compliance with this Rule and all other applicable Standards, or references to the location(s) of such other documentation.

- An appraiser must retain the workfile for a period of at least five (5) years after preparation or at least two (2) years after final disposition of any legal proceeding in which the appraiser provided testimony related to the assignment, whichever period expires last.
- An appraiser must have custody of his or her workfile or make appropriate workfile retention, access, and retrieval arrangements with the party having custody of the workfile.

Competency

The appraiser must properly identify the problem to be addressed and have the knowledge and experience to complete the assignment competently. If the appraiser does not have the required knowledge or experience, she must

1. Disclose this fact to the client before accepting the engagement.
2. If engaged, take all steps necessary to complete the assignment competently.
3. Describe the deficiency and the steps taken to complete the assignment competently.

Standards

The following are some standards to be followed in appraising real estate.

- **STANDARD 1.** In developing a real property appraisal, an appraiser must identify the problem to be solved and the scope of work necessary to solve the problem and must correctly complete research and analysis necessary to produce a credible appraisal. An appraiser must be aware of, understand, and correctly employ the recognized methods and techniques necessary to produce a credible appraisal.
- **STANDARD 2.** In reporting the results of a real property appraisal, an appraiser must communicate each analysis, opinion, and conclusion in a manner that is not misleading.
- **STANDARD 3.** In performing a review assignment, an appraiser acting as a reviewer must develop and report a credible opinion as to the quality of another appraiser's work and must clearly disclose the scope of work performed.

QUESTIONS ON CHAPTER 14

1. All of the following types of broad forces affect value *except*
   (A) physical
   (B) intellectual
   (C) political
   (D) social
   (E) economic

2. Social, economic, governmental, and environmental influences that affect property value are called
   (A) laws
   (B) forces
   (C) factors
   (D) consequences
   (E) impacts

3. Compared with other assets, real estate is
   (A) immobile
   (B) expensive
   (C) long-lived
   (D) mortgageable
   (E) all of the above

4. The truly distinguishing characteristic of real estate as compared with other assets is
   (A) large size
   (B) long life
   (C) high price
   (D) uniqueness
   (E) fixed location

5. The primary physical distinguishing characteristic of land is
   (A) homogeneity
   (B) high cost
   (C) immobility
   (D) slow depreciation
   (E) no depreciation

6. Urban land derives its value primarily from its
   (A) size
   (B) natural beauty
   (C) concentration of population
   (D) location
   (E) buildings

7. Which of the following is a basic component of value?
   (A) utility
   (B) scarcity
   (C) demand coupled with purchasing power
   (D) transferability
   (E) all of the above

8. The value of real estate is determined in the market mainly by its
   (A) price
   (B) productivity
   (C) mortgage
   (D) size
   (E) height

9. Which of the following is an important governmental influence on neighborhood values?
   (A) zoning code
   (B) income level
   (C) owner occupancy
   (D) flood plain mapping
   (E) ethnic concentration

10. *Realty* is best defined as rights in
    (A) land
    (B) improvements on and to land
    (C) buildings
    (D) fixtures
    (E) real estate

11. Built-in appliances are legally classified as
    (A) real estate
    (B) trade fixtures
    (C) chattels
    (D) personal property
    (E) amenities

12. The term improvements to real estate refers to
    (A) buildings only
    (B) buildings, fences, walkways, etc.
    (C) trees and buildings only
    (D) all of the above
    (E) none of the above

13. Rights of condemnation are held by a
    (A) state government agency
    (B) local agency or authority representing a federal government agency
    (C) utility company
    (D) small municipality
    (E) all of the above

14. Eminent domain requires
    (A) appraisal processes for the estimation of real property value
    (B) just compensation for specific types of financial loss
    (C) due process for the taking of property through condemnation
    (D) both A and B
    (E) both B and C

15. Zoning and environmental protection regulations represent the exercise of
    (A) land use politics
    (B) public use
    (C) escheat
    (D) eminent domain
    (E) police power

16. All of the following are considered legal encumbrances *except*
    (A) leases
    (B) utilities
    (C) zoning
    (D) mortgages
    (E) liens

17. A two-story house has all the bedrooms upstairs, and the only bathroom is on the first level. This is an example of
    (A) physical obsolescence
    (B) functional obsolescence
    (C) environmental obsolescence
    (D) physical depreciation
    (E) external obsolescence

18. Which of the following expires upon the owner's death?
    (A) fee tail
    (B) remainder interest
    (C) simple interest
    (D) life estate
    (E) estate for years

19. The most complete interest in real property is
    (A) fee tail
    (B) joint tenancy
    (C) tenancy in common
    (D) fee simple
    (E) life estate pur autre vie

20. A deed should be recorded in order to
    (A) give constructive notice of the transaction
    (B) comply with federal and state law
    (C) pass title
    (D) make the deed a legal document
    (E) establish the priority of claim of the purchaser

21. Property description is necessary to
    (A) identify and locate the subject property
    (B) identify the property rights being appraised
    (C) find the appropriate section in the report
    (D) find the official map coding of the subject property
    (E) tell where the property deed is recorded

22. One way in which land may be legally described is by
    (A) highest and best use
    (B) acreage
    (C) mailing address
    (D) metes and bounds
    (E) front footage

23. "Block 17, lot 5 of the Woodcreek Subdivision" is an example of a land description using which of the following systems?
    (A) lot and block
    (B) rectangular method
    (C) metes and bounds
    (D) government survey
    (E) monuments survey

24. Which of the following is not external obsolescence for a house?
    (A) odd floor plan
    (B) nearby landfill
    (C) nearby interstate highway
    (D) nearby leaking underground storage tank
    (E) gas station next door

25. *Loan value* is the same as
    (A) investment value
    (B) market price
    (C) replacement value
    (D) book value
    (E) market value

26. *Mortgage value* is synonymous with
    (A) market price
    (B) liquidation value
    (C) market value
    (D) assessed value
    (E) insurable value

27. An objective kind of value that can be estimated for property bought and sold in the market is called
    (A) value in use
    (B) value in exchange
    (C) economic value
    (D) potential value
    (E) insurable value

28. In appraisal, "market value" is most commonly identified with
    (A) value in use
    (B) value in exchange
    (C) assessed value
    (D) listing value
    (E) none of the above

29. Price and value are
    (A) not necessarily the same
    (B) synonymous
    (C) different, depending on financing terms
    (D) close together in an inactive market
    (E) used interchangeably in a report

30. Market price is the amount for which a property
    (A) should sell
    (B) was sold
    (C) will sell
    (D) could sell
    (E) would be appraised

31. All of the following statements are true *except*
    (A) *real property* refers to items that are not permanently fixed to a part of the real estate
    (B) appraising is the art and science of estimating the value of an asset
    (C) assets typically requiring appraisal include real and personal property
    (D) asset values change with time
    (E) markets change with supply and demand

32. *Investment value* is best described as
    (A) market price
    (B) market value
    (C) the cost of acquiring a competitive substitute property with the same utility
    (D) the present worth of anticipated future benefits to a certain individual or institutional investor
    (E) value in exchange

33. Assessed value is usually based on
    (A) cost value
    (B) book value
    (C) market value
    (D) insurable value
    (E) replacement cost

34. The principle of anticipation
    (A) is future oriented
    (B) is past oriented
    (C) involves the "as of" date for an appraisal
    (D) predicts the loan-to-value ratio for the subject property
    (E) is substitution oriented

35. Which principle of value best affirms that value is the present worth of expected future benefits?
    (A) supply and demand
    (B) balance
    (C) substitution
    (D) anticipation
    (E) conformity

36. What principle states that value levels are sustained when the various elements in an economic or environmental mix are in equilibrium?
    (A) anticipation
    (B) equivalence
    (C) substitution
    (D) balance
    (E) highest and best use

37. The fundamental valuation principle underlying the sales comparison process is
    (A) contribution
    (B) substitution
    (C) conformity
    (D) change
    (E) anticipation

38. Which principle of value best affirms that the maximum value of a property generally cannot exceed its replacement cost new?
    (A) increasing and decreasing returns
    (B) supply and demand
    (C) substitution
    (D) balance
    (E) anticipation

39. The fact that the value of a property tends to equal the cost of an equally desirable substitute is an example of the principle of
    (A) balance
    (B) substitution
    (C) contribution
    (D) diminishing returns
    (E) supply and demand

40. The principle of substitution holds that a purchaser will pay no more for a property than
    (A) the maximum he can afford
    (B) the cost of acquiring an equally desirable substitute
    (C) the price of a previously owned property
    (D) the price of a property with greater utility
    (E) none of the above

41. The function(s) of a real estate market is (are) to
    (A) facilitate exchanges
    (B) set prices
    (C) allocate resources
    (D) adjust supply to demand
    (E) all of the above

42. In estimating current market value, an appraiser assumes all of the following *except*
    (A) the buyer is typically motivated
    (B) the parties are knowledgeable
    (C) a reasonable time will be allowed for market exposure
    (D) the property will be marketed by a professional expert
    (E) the property will be sold for cash or equivalent terms

43. In which market are there many potential buyers but few properties available?
    (A) demand
    (B) buyer's
    (C) seller's
    (D) low-priced
    (E) normal

44. A perfect market occurs when
(A) there are numerous buyers and sellers who are knowledgeable and free to trade
(B) all products are interchangeable and can be transported to better markets
(C) the government allocates supply and demand perfectly
(D) all of the above are true
(E) A and B only are true

45. The identification and study of a pertinent market is called
(A) market analysis
(B) neighborhood review
(C) property research
(D) market segmentation
(E) market interaction

46. If the typical occupancy rate in an area is 95%, what conclusion would you most likely draw about a subject property that has 100% occupancy?
(A) Advertising is average.
(B) The rents are high.
(C) The rents are low.
(D) Management is incompetent.
(E) New construction will occur soon.

47. Which type of studies test the ability of various proposed improvements to meet investment objectives?
(A) market
(B) feasibility
(C) marketability
(D) cost-benefit
(E) prospectus

48. A decrease in land value is the result of
(A) functional obsolescence
(B) physical deterioration
(C) market forces
(D) wear and tear
(E) obsolescence

49. To estimate market value, an appraiser follows the
(A) appraisal report
(B) valuation process
(C) evaluation methodology
(D) appraisal guidelines
(E) report evolution technique

50. In appraising, the most important judgment is probably called for in
(A) projecting selling rates
(B) reconciling value indications
(C) selecting approaches to use
(D) performing neighborhood analysis
(E) making rental market estimates

51. An "as of" date is specified in appraisals to
(A) show when the appraiser inspected the property
(B) indicate the prevailing price level
(C) indicate the market conditions on which the value is estimated
(D) indicate when the buyer agreed to purchase the property
(E) indicate the value as of the closing date

52. An appraisal of real estate
(A) guarantees its value
(B) assures its value
(C) determines its value
(D) estimates its value
(E) segments its value

53. Residential appraisers generally provide an estimate of
(A) market price
(B) mortgage loan value
(C) cash value
(D) assessed value
(E) market value

54. Commercial real estate appraisers are most frequently asked to estimate
(A) assessed value
(B) liquidation value
(C) insurable value
(D) market value
(E) intrinsic value

55. Common transactions that frequently require appraisals are
   (A) income-tax-deductible charitable contribution of property, estate tax, and property settlement upon divorce
   (B) damage lawsuits and security for bail bonds
   (C) company merger, tax basis, and loan assumptions
   (D) all of the above
   (E) none of the above

56. Knowledge of land value is required for all of the following *except*
   (A) condemnation actions
   (B) fire insurance
   (C) property taxation
   (D) a ground lease
   (E) a grazing lease

57. A market value estimate provided in an appraisal
   (A) changes with the use to which it is put
   (B) changes with the function of the appraisal
   (C) remains the same regardless of whom the appraisal is prepared for
   (D) depends upon the use or function of the appraisal
   (E) always reflects market value

58. In preparing an appraisal, definition of the problem identifies all of the following *except*
   (A) the real estate being appraised
   (B) the highest and best use for the property
   (C) the real property rights
   (D) the date of the value estimate
   (E) the use to which the appraisal will be put

59. USPAP, in appraisal, stands for
   (A) United States Preferred Appraisal Path
   (B) United States Principles of Appraisal Practice
   (C) Useful Statistical Principles Applied Practically
   (D) Uniform Standards of Professional Appraisal Practice
   (E) unless stated, prepare another proposal

60. The requirement to verify data in an appraisal report varies according to
   (A) the appraiser's available time
   (B) the purpose and intended use of the appraisal report
   (C) legal restrictions on the property
   (D) the type of value being sought
   (E) the use to which the property has been put

61. A property's immediate environment is
   (A) its pivotal point
   (B) a "comp" grid
   (C) a neighborhood
   (D) the adjacent uses
   (E) a natural boundary

62. The life cycle of a neighborhood illustrates the principle of
   (A) substitution
   (B) highest and best use
   (C) change
   (D) conformity
   (E) anticipation

63. The data required for an appraisal assignment depend on the
   (A) lending association
   (B) appraiser
   (C) buyer
   (D) Office of Thrift Supervision
   (E) nature of the appraisal problem

64. Which of the following exhibits is *least* frequently used in an appraisal?
   (A) a photograph of the subject property
   (B) an aerial photograph of the surrounding area
   (C) a plot plan
   (D) a floor plan
   (E) a photograph of a comparable property

65. In which market is the sales comparison approach most applicable?
(A) seller's
(B) buyer's
(C) reasonable
(D) active
(E) calm

66. The criteria for determining highest and best use include all of the following *except*
(A) physical possibility
(B) financial feasibility
(C) legal permissibility
(D) maximal productivity
(E) effect on community welfare

67. The three approaches to estimating value are
(A) cost, income, and replacement
(B) replacement, income, and reproduction
(C) cost, sales comparison, and income capitalization
(D) reproduction, cost, and income
(E) market, building residual, and multiplier

68. In appraising a dwelling built 50 years ago, the appraiser usually relies on the
(A) current reproduction cost
(B) original construction cost
(C) current replacement cost
(D) sales comparison approach
(E) superior construction technique

69. Which approach would be best when appraising a 15- to 20-year-old house?
(A) cost
(B) feasibility study
(C) sales comparison
(D) income capitalization
(E) replacement cost new less accrued depreciation

70. The appraiser's final value estimate should be based on
(A) an average of the three value indications given by the three approaches to value
(B) a weighing of the reliability of the information analyzed in each of the three approaches to value
(C) an average of the values in the three closest comparable sales
(D) the correlation technique
(E) adjustments for the most recent indicators in the local market

71. Before signing the appraisal report, the appraiser should
(A) reinspect the subject property
(B) evaluate the reliability of each approach to value
(C) review the overall appraisal process and check for technical accuracy
(D) seek the property owner's opinion
(E) average the results from the approaches used

72. Since each value approach has its own strengths and weaknesses, an appraiser should
(A) choose the approach that is the most popular with lenders
(B) choose the approach that is the most popular with buyers
(C) weigh the strengths and weaknesses of each approach and decide which is the most reliable for the subject property
(D) weight each approach equally
(E) use the approach that the client feels is most appropriate

73. The cost approach is often given more weight in the appraisal of
(A) old or obsolete buildings
(B) single-family houses more than 10 years old
(C) new buildings or special-use buildings
(D) commercial and industrial properties
(E) vacant land

74. In reconciliation and conclusion of value, the appraiser should
    (A) describe the relevance of each value approach explored
    (B) discuss the reliability of the data used
    (C) provide arguments to justify his or her final conclusion of value
    (D) explain his or her judgments and reasoning
    (E) do all of the above

75. Value indications are reconciled into a final value estimate
    (A) throughout the appraisal
    (B) after the report of defined value
    (C) after each approach is completed
    (D) at the preliminary stage to let the client know what result to expect
    (E) after all three approaches have been completed

76. During the reconciliation process, an appraiser should ask:
    (A) How appropriate is each approach?
    (B) How adequate are the data?
    (C) What range of values do the approaches suggest?
    (D) all of the above
    (E) A and C only

77. Which of the following is the shortest and least detailed type of appraisal report?
    (A) self-contained
    (B) summary
    (C) restricted
    (D) engagement letter
    (E) detailed

78. All of the following are main sections in the typical narrative appraisal report *except*
    (A) introduction
    (B) engagement letter
    (C) description, analyses, and conclusion
    (D) addenda
    (E) valuation

79. An oral appraisal report
    (A) is not worth the paper it is written on
    (B) must be supported by a work paper file
    (C) is unethical
    (D) is impossible
    (E) is a legitimate substitute for a written report

80. Standardized residential form reports
    (A) are adequate for all appraisals
    (B) may not be as detailed or complete as written reports
    (C) are seldom adequate for appraisals
    (D) allow little space for neighborhood analysis
    (E) are difficult to computerize

81. The highest and best use of a site is its
    (A) existing use
    (B) most probable use that is legally and physically possible and provides the highest financial returns
    (C) immediate next use
    (D) ordinary and necessary use
    (E) least expensive use

82. The important reason(s) for inspecting and analyzing the subject site is (are) to
    (A) ascertain its highest and best use
    (B) note any unusual characteristics
    (C) find comparable sales
    (D) be certain the property exists
    (E) all of the above

83. The length of a tract of land along a street is called the land's
    (A) depth
    (B) width
    (C) frontage
    (D) abutment
    (E) lineage

84. The number of square feet in 1 acre is
    (A) 64,000
    (B) 460
    (C) 440
    (D) 43,560
    (E) 34,560

85. The livable square footage of a single-family residence is usually measured from the
    (A) total interior (including interior walls)
    (B) inside room dimensions
    (C) interior plus basement
    (D) exterior
    (E) exterior, with adjustment for wall width

86. The most important item an appraiser needs for recording data during a building inspection is a
    (A) plat map
    (B) tape measure
    (C) checklist
    (D) clipboard
    (E) blank paper

87. A large home built in an area of small cottages is an example of
    (A) overimprovement
    (B) underimprovement
    (C) land regression
    (D) functional obsolescence
    (E) environmental aesthetics

88. The period over which a building may be profitably used is its
    (A) actual life
    (B) physical life
    (C) useful life
    (D) normal life
    (E) effective age

89. An improvement's remaining economic life is
    (A) its chronological age
    (B) its effective age
    (C) the future time span over which the improvement is expected to generate benefits
    (D) its effective age minus its chronological age
    (E) its effective age plus its chronological age

90. To be considered as a possible alternative for highest and best use, a use must be
    (A) physically and legally possible and financially feasible
    (B) physically and legally possible
    (C) already in existence and legal
    (D) physically possible and appropriate
    (E) legal and profitable

91. When applying the concept of highest and best use of land as if vacant, the assumption is made that
    (A) any existing building can be demolished
    (B) zoning cannot be changed
    (C) the basic characteristics of a site can be changed
    (D) vacant land produces maximum income
    (E) land is not subject to erosion

92. Real estate values always
    (A) increase
    (B) outperform the stock market
    (C) remain stable or rise
    (D) fluctuate
    (E) inflate

93. The best assurance of appraisal accuracy results from the use of
    (A) the greatest number of adjustments
    (B) a gross dollar amount of adjustments
    (C) a large number of truly comparable properties
    (D) a net dollar amount of adjustments
    (E) market segmentation

94. The direct sales comparison approach to appraising derives an estimate of value from
    (A) sales of comparable properties
    (B) comparison of the architecture of properties
    (C) comparison of loans made on properties
    (D) original costs of comparable properties
    (E) checking prices in the *Wall Street Journal* stock market report

95. A gross rent multiplier is often used in the
   (A) direct sales comparison approach for single-family housing
   (B) income approach for income property
   (C) income approach for single-family property
   (D) cost approach
   (E) back-door approach to valuation

96. Gross rent multiplier analysis is
   (A) required in almost every residential appraisal
   (B) applicable to every residential appraisal
   (C) applicable to residential property valuation only
   (D) part of the direct sales comparison approach to valuation of commercial properties
   (E) seldom used for houses

97. In estimating site value, the direct sales comparison approach
   (A) does not apply to older residences
   (B) is the most reliable method available
   (C) is considered inferior to other methods
   (D) is used only when the subject property is unimproved
   (E) is used for real estate but not for personal property appraisals

98. In an open market transaction, the subject property sold would not be
   (A) listed for at least 30 days
   (B) listed on a multiple listing service
   (C) advertised in local newspapers
   (D) sold to a relative
   (E) closed by a title company

99. The minimum number of comparable sales needed to apply direct sales comparison is
   (A) 3
   (B) 4
   (C) 10
   (D) 30
   (E) no set number

100. Physical units of comparison include
   (A) seats in a theater
   (B) rooms in a hotel
   (C) square feet of a house
   (D) cubic feet of a warehouse
   (E) all of the above

101. A nine-story office building without an elevator would suffer from which type of depreciation?
   (A) locational obsolescence
   (B) functional obsolescence
   (C) external obsolescence
   (D) physical deterioration
   (E) accounting depreciation

102. A comparable sale has four bedrooms compared to the subject's three bedrooms. The fourth bedroom is worth $30,000. What do you do with that number?
   (A) Add it to the comparable's sale price
   (B) Subtract it from the comparable's sale price
   (C) Add it to the subject's value
   (D) Subtract it from the subject's value
   (E) Nothing—do not use this sale as a comparable

103. The sales comparison approach requires an appraiser to make adjustments to value. How are the adjustments posted on a grid?
   (A) If the comparable is better, subtract value; if worse, add value.
   (B) If the comparable is better, add value; if worse, subtract value.
   (C) If the subject is better, subtract value; if worse, add value.
   (D) If the subject is better, add value; if worse, subtract value.
   (E) A grid is not the place to post adjustments.

104. The method of estimating replacement or reproduction cost that requires you to know the original cost of construction of the building is
(A) square foot
(B) quantity survey
(C) unit-in-place
(D) index
(E) exact measurement

105. An appraiser uses all of the following approaches in developing an opinion of value for real estate *except*
(A) cost approach
(B) replacement approach
(C) income approach
(D) sales comparison approach
(E) capitalization of income approach

106. "The total number of years a building is considered to contribute to the value of the property" is a good definition of
(A) chronological age
(B) effective age
(C) economic life
(D) depreciable term
(E) usable age

107. A rate or ratio that relates net operating income to the value of income property is
(A) the gross rent multiplier
(B) the capitalization rate
(C) the depreciated cost factor
(D) the effective gross income factor
(E) equity yield rate

108. To estimate the appraised value of an income-producing property, the appraiser will
(A) divide the NOI by the cap rate
(B) divide the cap rate by the NOI
(C) multiply the NOI by the cap rate
(D) divide the equity dividend rate by NOI
(E) none of the above

109. The subject property of an appraisal has a one-car garage and no fireplace. Two similar nearby properties sold in the last month, and each one has a two-car garage and a fireplace. In making adjustments to use the sold properties as comparables, the appraiser will
(A) subtract value from the sales price of comparables
(B) use two properties with one-car garages and no fireplace that sold last year
(C) add value to the sales price of comparables
(D) use the recorded sales prices of each property
(E) consult with active brokers in the local market

110. The economic life of a building has come to an end when
(A) the rent produced is valued at less than a similar amount of money invested
(B) the reserve for depreciation equals the cost to replace the building
(C) the building ceases to represent the highest and best use of the land
(D) the value of the land and the building equals the value of the land alone
(E) the building is vacant and cannot be rented

111. A capitalization rate incorporates return on
(A) land and building and recapture of building
(B) land and building and recapture of land
(C) land and recapture of land and building
(D) building and recapture of land and building
(E) recapture of building only

112. All of the following are examples of external obsolescence *except*
(A) population density
(B) wear and tear
(C) zoning
(D) special assessments
(E) rent control

113. The gross income multiplier is derived by dividing the sales price by
(A) the monthly net income
(B) the monthly effective income
(C) the annual net income
(D) the annual gross income
(E) vacancy and collection allowance

114. In using the direct sales comparison approach to appraisal, the appraiser considers
(A) the sales price of comparable properties
(B) the acquisition cost to the present owner
(C) income tax rates
(D) property tax assessments
(E) the listing price of the subject

115. Which of the following is *not* a physical characteristic of land?
(A) nonhomogeneity
(B) immobility
(C) indestructibility
(D) situs
(E) permeability

116. Which type of depreciation is nearly always incurable?
(A) physical depreciation
(B) functional depreciation
(C) functional obsolescence
(D) external obsolescence
(E) book depreciation

117. An appraiser is trying to estimate the value of a property by capitalizing the income stream. Which of the following will the appraiser use?
(A) equity value
(B) mortgage balance
(C) replacement cost
(D) net operating income
(E) calculated value

118. A real estate speculator purchases three adjacent lots for $10,000 each. When he combines them into one large lot worth $50,000, this is an example of
(A) combinatory increment
(B) sum of parts
(C) accretion
(D) plottage
(E) rational value

119. A miniature golf course opened and enjoyed $75,000 of profits in its first year of operation. Then another miniature golf course opened across the street, and the first one lost $15,000 in that year. This is an example of
(A) regressive profits
(B) competition
(C) market saturation
(D) conformity
(E) substitution

120. A plastics manufacturer opened across the highway from a housing subdivision. The factory emits strong odors that cause nearby houses to lose value. This is an example of
(A) physical deterioration
(B) functional obsolescence
(C) external obsolescence
(D) regression
(E) economic useful life

121. When appraising a vacant building that was used as a church, the most applicable approach would most likely be
(A) cost approach
(B) income approach
(C) capitalization
(D) faith based
(E) special-purpose approach

122. Air-conditioning in a house in Alaska is an example of
(A) functional obsolescence
(B) economic obsolescence
(C) ventilation overimprovement
(D) architectural enlightenment
(E) HVAC supersufficiency

123. Which of the following is a step in the appraisal process?
(A) include the tax assessment
(B) consider a 1031 exchange
(C) collect data
(D) prepare a feasibility study
(E) rationalize value

124. When applying the cost approach, the appraiser considers all of the following *except*
(A) land at market value
(B) replacement cost or reproduction cost
(C) depreciation
(D) gross rental income
(E) renovation cost

125. A competitive market analysis for a house depends most heavily on
(A) replacement cost
(B) reproduction cost
(C) sales comparison approach
(D) highest and best use
(E) broker's negotiation skills

126. A competitive market analysis considers
(A) comparable sales
(B) highest and best use
(C) gross income
(D) leased sales
(E) broker's salesmanship

127. Which of the following is an example of functional obsolescence?
(A) faded paint
(B) nearby highway
(C) awkward floor plan
(D) one-story home
(E) leaky skylight

128. The gross income of an office building is $18,000. Operating expenses are $6,000. The capitalization rate is 12%. What is the value estimate?
(A) $200,000
(B) $150,000
(C) $144,000
(D) $100,000
(E) $12,000

129. Faded and peeling paint is referred to in an appraisal as
(A) deferred maintenance
(B) preventive maintenance
(C) lack of care
(D) property waste
(E) lead-based paint

130. The economic analysis that considers basic employment and service employment is
(A) market analysis
(B) feasibility analysis
(C) investment analysis
(D) economic base analysis
(E) broker analysis

131. Because of a recent hailstorm in the area, the subject house has a roof that needs replacement at a cost of about $8,000. Comparable sales have new roofs. In appraising the subject, the price of comparable sales should be adjusted
(A) upward because of their new roofs
(B) downward to match the subject's characteristics
(C) neither upward nor downward
(D) upward or downward depending on insurance coverage
(E) upward or downward depending on the quality of their roofs compared to the subject's proposed new roof

132. An appraisal approach that places emphasis on market rental income minus operating expenses is
   (A) market approach
   (B) direct sales comparison
   (C) cost approach
   (D) income capitalization
   (E) rental analysis

133. Which of the following tends to increase the value of residential real estate?
   (A) good local services such as schools, police, fire, parks, land use controls, and nearby employment opportunities
   (B) high interest rates
   (C) high housing turnover
   (D) abundant apartment development
   (E) nonconforming use

134. When applying the cost approach to appraisal, the appraiser has estimated land value and replacement cost of the structure. What step has not yet been done to complete the cost approach?
   (A) determine the capitalization rate
   (B) locate comparable properties
   (C) measure the building
   (D) estimate accrued depreciation
   (E) determine cost index

135. From all potential possible uses of a property, the highest and best use of land if vacant
   (A) attracts the greatest gross income
   (B) attracts the greatest net income
   (C) provides the highest return to the land
   (D) has the greatest floor area coverage ratio
   (E) is not allowed under current zoning

136. The highest and best use of land is
   (A) legally permissible
   (B) physically possible
   (C) economically feasible
   (D) all of the above
   (E) none of the above

137. When all three appraisal approaches are relevant, what should the appraiser do to the three indicator values?
   (A) average them
   (B) reconcile them
   (C) select the middle value as the market value
   (D) select the lowest amount for conservatism
   (E) select the highest amount to please the client

138. Which appraisal approach would be most useful when applying for a mortgage loan on a fully occupied shopping center?
   (A) tax assessment
   (B) cost approach
   (C) income approach
   (D) cost segregation
   (E) feasibility study

139. The direct sales comparison approach would be most applicable for a
   (A) retail shopping center
   (B) warehouse
   (C) church
   (D) ten-year-old single-family house in an active housing market
   (E) office building

140. Negative value influences from outside a property create
   (A) physical depreciation
   (B) functional obsolescence
   (C) economic obsolescence
   (D) internalization
   (E) hardship

141. The income approach to value
   (A) is the best approach for appraising a single-family home
   (B) is the best way to determine the highest and best use of vacant land
   (C) is used to estimate the value of investment property
   (D) is one of a dozen acceptable approaches to appraising real estate
   (E) is applicable to all forms of real estate

142. Which of the following is *not* typically used in the cost approach to appraisal?
   (A) replacement cost
   (B) reproduction cost
   (C) accrued depreciation
   (D) capitalization rate
   (E) market value of land

143. The __________ rate is used by appraisers to convert net operating income into a value estimate.
   (A) mortgage
   (B) equity
   (C) capitalization
   (D) return of investment
   (E) interest

144. An office building generates $50,000 effective gross income and $20,000 operating expenses. To achieve an 8% capitalization rate, the buyer would pay up to
   (A) $625,000
   (B) $500,000
   (C) $375,000
   (D) $250,000
   (E) $24,000

ANSWERS

1. **B**	21. **A**	41. **E**	61. **C**	81. **B**	101. **B**	121. **A**	141. **C**
2. **B**	22. **D**	42. **D**	62. **C**	82. **E**	102. **B**	122. **A**	142. **D**
3. **E**	23. **A**	43. **C**	63. **E**	83. **C**	103. **A**	123. **C**	143. **C**
4. **E**	24. **A**	44. **E**	64. **B**	84. **D**	104. **D**	124. **D**	144. **C**
5. **C**	25. **E**	45. **A**	65. **D**	85. **D**	105. **B**	125. **C**	
6. **D**	26. **C**	46. **C**	66. **E**	86. **C**	106. **C**	126. **A**	
7. **E**	27. **B**	47. **B**	67. **C**	87. **A**	107. **B**	127. **C**	
8. **B**	28. **B**	48. **C**	68. **D**	88. **C**	108. **A**	128. **D**	
9. **A**	29. **A**	49. **B**	69. **C**	89. **C**	109. **A**	129. **A**	
10. **E**	30. **B**	50. **B**	70. **B**	90. **A**	110. **D**	130. **D**	
11. **A**	31. **A**	51. **C**	71. **C**	91. **A**	111. **A**	131. **B**	
12. **B**	32. **D**	52. **D**	72. **C**	92. **D**	112. **B**	132. **D**	
13. **E**	33. **C**	53. **E**	73. **C**	93. **C**	113. **D**	133. **A**	
14. **E**	34. **A**	54. **D**	74. **E**	94. **A**	114. **A**	134. **D**	
15. **E**	35. **D**	55. **A**	75. **E**	95. **C**	115. **D**	135. **C**	
16. **B**	36. **D**	56. **B**	76. **D**	96. **D**	116. **D**	136. **D**	
17. **B**	37. **B**	57. **C**	77. **C**	97. **B**	117. **D**	137. **B**	
18. **D**	38. **C**	58. **B**	78. **B**	98. **D**	118. **D**	138. **C**	
19. **D**	39. **B**	59. **D**	79. **B**	99. **E**	119. **B**	139. **D**	
20. **A**	40. **B**	60. **B**	80. **B**	100. **E**	120. **C**	140. **C**	

Appraising Income Property

15

ESSENTIAL MATTERS TO LEARN FROM THIS CHAPTER

- This chapter is about appraising income property. It is principally for those seeking to be certified general appraisers. Those seeking broker or sales licenses or who want to be residential appraisers need not study this chapter.
- The mathematical conversion of a stream of expected future income into a lump-sum present value is called *capitalization of income*. This is the cornerstone of appraising income property.
- Fundamental to capitalizing income properly is an understanding of compound interest and present value.
- *Principal* is a lump-sum amount. It may be the starting amount or a dollar amount to be received at a later date. *Principal* is distinguished from *interest*, which is the earnings on principal.
- *Compound interest* means that the principal grows by interest earned. Interest is earned on both the initial principal and the interest earned in prior periods. A *period* may be a month, a year, or another time interval.
- *Simple interest* is not compounded. This is not considered an appropriate method for evaluating a stream of income.
- A *reciprocal* is the number 1 divided by the amount. For example, the reciprocal of 5 is equal to $^1/_5 = 0.20$.
- An *annuity* is a series of equal or near-equal payments. An *ordinary annuity* provides payments at the end of each period. An *annuity in advance* provides payments at the beginning of each period.
- *A rate of return on investment* is an interest rate that is earned.
- *A rate of return of investment* is the periodic return of principal.
- There are six functions of compound interest:

1. **FUTURE VALUE OF $1.** This is the growth of $1 at compound interest. The $1 is deposited at the *beginning* of the first period. It answers the question: If I deposit $1 now at a fixed rate of compound interest, how much will the total be after *n* periods?
2. **FUTURE VALUE OF AN ANNUITY OF $1 PER PERIOD.** This is the accumulated growth of $1 deposited at the *end* of each period for *n* periods.
3. **SINKING FUND FACTOR.** This is the reciprocal of the future value of an annuity of $1 per period. It answers the question: How much must I deposit each period to reach $1 at the end of *n* periods?
4. **PRESENT VALUE OF $1.** This answers the question: If I expect to receive $1 after *n* periods, how much is that money worth today? It is the reciprocal of the *future value of $1*.

5. **PRESENT VALUE OF AN ANNUITY OF $1 PER PERIOD.** This answers the question: If I expect to receive $1 at the end of each period for *n* periods, how much is the sum worth today, assuming a fixed interest rate?
6. **INSTALLMENT TO AMORTIZE $1.** This is the reciprocal of the *present value of an annuity of $1 per period.* This factor provides the periodic amount necessary to retire (called *amortize*) a $1 loan at a fixed rate of interest over *n* periods.

- In the *income approach* to appraisal, the appraiser follows a systematic pattern to derive an estimate of annual cash flow from the property. The cash flow is considered income. Annual cash flow is *capitalized* into a lump sum to derive the present value of the cash flow stream.
- The process to evaluate the income of property begins with *potential gross income*, which is the total amount that can be collected if all units are rented at market rates all year. If it is a shopping center with percentage leases, that income is included.
- *Effective gross income* provides subtractions for vacancy and collection losses and provides additions for miscellaneous income such as from vending or laundry machines in an apartment complex.
- *Operating expenses* are then subtracted to derive *net operating income.* Operating expenses are needed to operate the property and include such things as real estate taxes, hazard insurance, utilities, maintenance and repairs, painting, trash removal, and often a replacement reserve. Interest expense is not an operating expense. It is a financing expense.
- Often leases must be studied to determine the degree of "net-ness," that is, which expenses or escalations are passed back to tenants.
- Net operating income (NOI) is *capitalized*, that is, divided by an *overall capitalization rate* to derive a property value indication. This will be reconciled with value indications from the sales comparison and/or cost approach to derive a value opinion.
- The capitalization rate is the sum of two components. First is the rate of return *on* investment. To that a rate of recovery *of* investment is added.
- For example, if 10% is the required annual return *on* investment and 2% is the required annual return *of* investment, the capitalization rate is 12%. In decimal form, that is 0.12.
- If the net operating income is $10,000, it is divided by 0.12 to derive a present value of $83,333 with a capitalization rate of 12%.
- Three ways to derive the capitalization rate are the following:

1. **BUILD-UP RATE:** Total of various components. Start with a safe, liquid, riskless rate. Add for illiquidity, investment management, and risk to derive an appropriate return *on* investment. Add an annual return *of* investment.
2. **BAND OF INVESTMENT:** Mortgage and equity rates, weighted by the proportion of purchase capital contributed.
3. **MARKET EXTRACTION.** For sales comparables, divide NOI by selling price.

- *A discounted cash flow* (DCF) analysis is applied by multiplying periodic cash flow each year by the present value factor for each corresponding year. Then add a resale amount (reversion) at its present value to derive a property value estimate.
- When appraising income property, the appraiser must carefully consider the quality, quantity, and duration of the expected future income. This includes the type of lease, creditworthiness of tenant(s), and amounts of expected operating expenses.

This chapter is of interest primarily to those who wish to become licensed as certified general appraisers. Much of the material will be of interest also to brokers, salespersons, and other appraisers but is not presently required by licensing examinations for those groups.

APPRAISAL MATH FOR INCOME PROPERTY

There are six functions of compound interest; each is described below. All six functions are variations on the compound interest formula, which holds that interest in each period is based on the initial principal plus accrued but unpaid interest earned in a prior period. The six functions are as follows:

- Future value of $1
- Future value of an annuity of $1 per period
- Sinking fund factor
- Present value of $1
- Present value of an annuity of $1 per period
- Installment to amortize $1

Future Value of $1

This function is compound interest. Interest that has been earned and left on deposit becomes principal. In the next period, that principal will earn interest along with the initial principal. The formula for compound interest is as follows:

$$S^n = (1 + i)^n$$

where S^n = sum after n periods; i = periodic rate of interest; n = number of periods.

Table 15-1 indicates the growth of a $1.00 deposit that earns 10% compound interest for 5 years. Figure 15-1 illustrates the growth of $1.00 at compound interest.

Although Table 15-1 illustrates a compounding interval of 1 year, compounding may occur daily, monthly, quarterly, or semiannually. Interest rates are usually stated at a nominal annual figure, such as 10%, but more frequent compounding increases the effective rate. The general formula is the same: $S^n = (1 + i)^n$.

Table 15-1
Future Value of $1
$1.00 Deposit at 10% Interest Rate for 5 Years

Year	Compound Interest and Balance	
0	Deposit	$1.00
1	Interest earned	0.10
1	Balance, end of year	1.10
2	Interest earned	0.11
2	Balance, end of year	1.21
3	Interest earned	0.121
3	Balance, end of year	1.331
4	Interest earned	0.1331
4	Balance, end of year	1.4641
5	Interest earned	0.1464
5	Balance, end of year	$1.6105

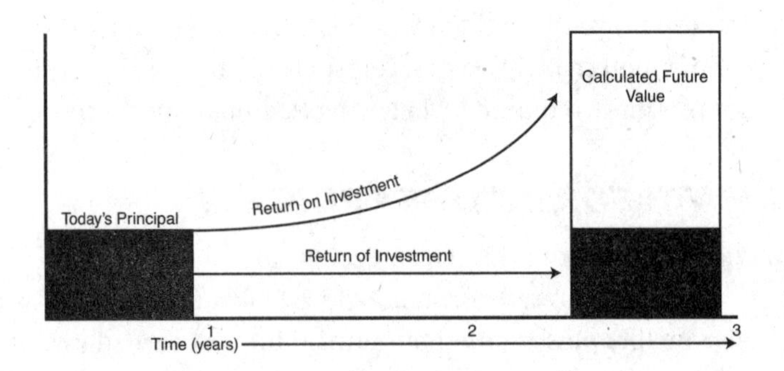

Figure 15-1. Growth of Principal at Compound Interest

Future Value of an Annuity of $1 per Period

This function offers the future value of a series of equal amounts deposited at the ends of periodic intervals. It is the sum of all these individual amounts, each deposited at the end of an interval (period), plus appropriate interest. Note that this differs from the future value of $1 factor in two ways. First, the future annuity factor is based on a series of deposits, whereas the future value of $1 involves a single deposit. Second, the accumulation is based on a deposit at the end of each interval, whereas compound interest is based on a deposit made at the beginning of the total period.

The formula for the future value of $1 per period is as follows:

$$S_n = \frac{s^n - 1}{i} \text{ or } \frac{(1+i)^n - 1}{i}$$

where S_n = future value of annuity of $1 per period; i = periodic interest rate; n = number of periods considered.

Table 15-2 shows the annual interest earned and end-of-period balances of the future value of an annuity of $1 per period at 10% interest. Figure 15-2 is an illustration of the function for four periods.

Table 15-2

Future Value of an Annuity of $1 per Period

Illustrated at 10% Interest Rate for Four Periods

End of period 1, initial deposit	$1.00
Interest, period 1	0.00
Balance, end of period 1	1.00
Interest, end of period 2	0.10
Deposit, end of period 2	1.00
Balance, end of period 2	2.10
Interest, end of period 3	0.21
Deposit, end of period 3	1.00
Balance, end of period 3	3.31
Interest, end of period 4	0.331
Deposit, end of period 4	1.00
Balance, end of period 4	$4.641

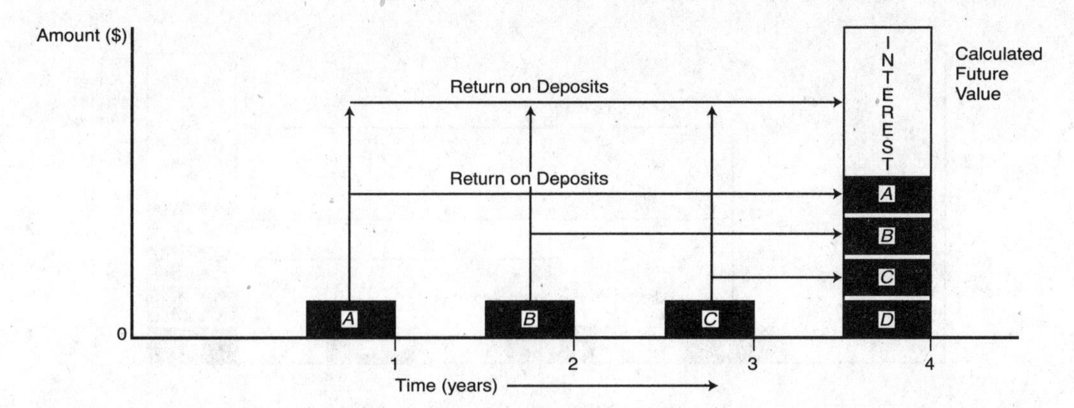

Figure 15-2. Accumulation of $1 per Period (Future Value of an Annuity). *A, B, C,* and *D* each represent $1 deposited at the end of a year. Each deposit earns compound interest from the date deposited until the date on which a terminal amount is sought. Thus, all deposits and interest are allowed to accumulate. The terminal value is the sum of all deposits plus compound interest.

Sinking Fund Factor

This factor shows the deposit that is required at the end of each period to reach $1.00 after a certain number of periods, considering interest to be earned on the deposits. It is the reciprocal of the future value of an annuity of $1 per period.

For example, at a 0% rate of interest, $0.25 must be deposited at the end of each year for 4 years to accumulate $1.00 at the end of 4 years. If, however, the deposits earn compound interest at 10%, only $0.215471 need be deposited at the end of each of the 4 years.

Table 15-3 indicates how four periodic deposits grow to $1.00 with interest. Figure 15-3 offers a graphic presentation of the same data.

Table 15-3
Sinking Fund Factor
Illustrated to Reach $1.00 in Four Periods at 10% Interest Rate

Deposit, end of period 1	$0.215471
Interest for period 1	0.000000
Balance, end of period 1	0.215471
Interest for period 2	0.021547
Deposit, end of period 2	0.215471
Balance, end of period 2	0.452489
Interest for period 3	0.045249
Deposit, end of period 3	0.215471
Balance, end of period 3	0.713209
Interest for period 4	0.071321
Deposit, end of period 4	0.215471
Balance, end of period 4	$1.000000

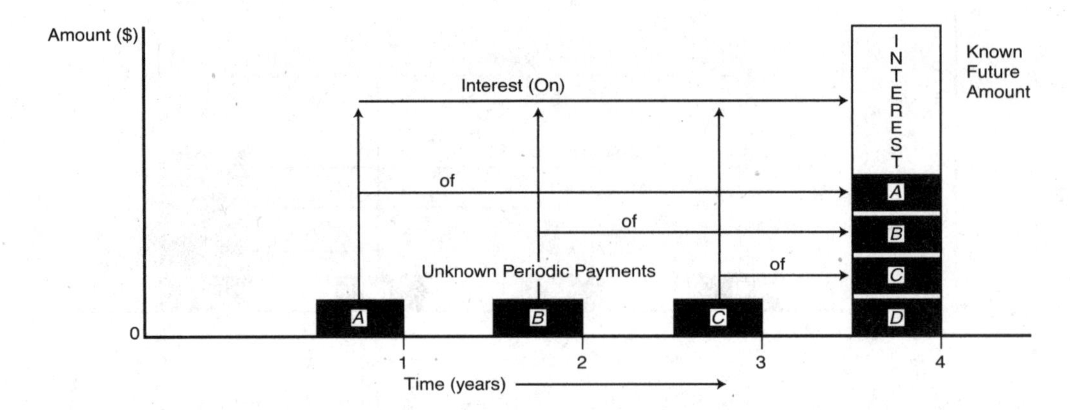

Figure 15-3. Sinking Fund Factor. *A, B, C,* and *D* are equal amounts deposited at the end of each year. Each deposit earns compound interest for the period of time it remains on deposit. At the end of the time period considered, the depositor can withdraw the terminal value. The sinking fund factor is computed in such a way that the terminal value will always equal $1.

Present Value of $1

This factor offers the value now of $1 to be received in the future. Money has a time value; a dollar to be received in the future is worth less than a dollar now. The amount of discount depends on the time between the cash outflow and inflow and on the necessary rate of interest or discount rate.

Since the purpose of investing is to receive returns in the future, applying the present value of $1 factor to anticipated future income is a crucial step in valuing an investment. When applying a present value factor, the terms *discounting* and *discount rate* are used. These are contrasted with *compounding* and *interest rate,* which are used in computing future values. Mathematically, the present value of a reversion is the reciprocal of the future value of $1.

For example, at a 10% discount rate, $100.00 that is expected 1 year from now has a present value of $90.91. As an arithmetic check, consider that, if an investor has $90.91 now and earns 10% during the year, the interest will amount to $9.09, making the principal in 1 year $100.00 ($90.91 original principal + $9.09 interest).

An investor who will get $100.00 in 2 years and pays $82.64 now receives a 10% annual rate of interest. As a check, consider that after 1 year, $82.64 will grow to $90.91 with 10% interest and then to $100.00 in 2 years.

The formula for the present value of $1 is as follows:

$$V^n = \frac{1}{(1+i)^n} \text{ or } \frac{1}{S^n}$$

where V = present value of $1; i = periodic discount rate; n = number of periods.

Table 15-4 shows the present value of a reversion at 10% interest for 4 years. Figure 15-4 illustrates the present value of $1.

Table 15-4
Present Value of $1
Illustrated at 10% Interest Rate for 4 Years

Year	Compound Amount	Reciprocal		Present Value of $1.00 Reversion
1	1.1	1/1.1	=	$0.909091
2	1.21	1/1.21	=	0.826446
3	1.331	1/1.331	=	0.751315
4	1.4641	1/1.4641	=	0.683013

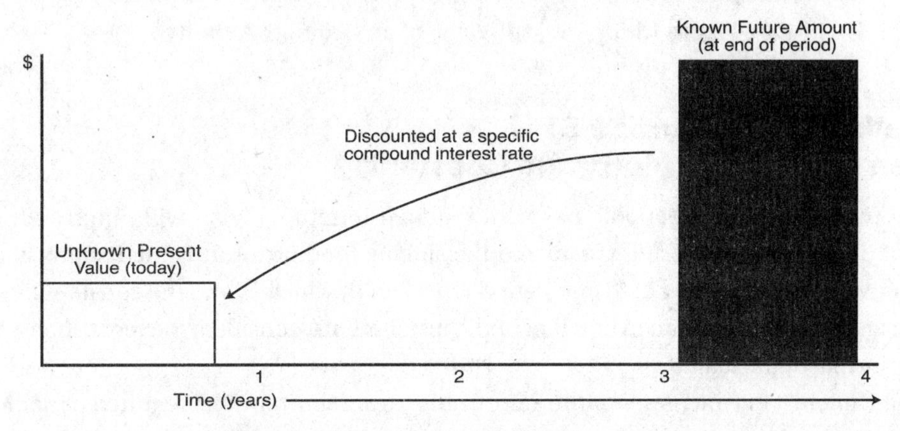

Figure 15-4. Present Value of a Reversion

Present Value of an Annuity of $1 per Period

An ordinary annuity is a series of equal payments beginning one period from the present. It is also defined as a series of receipts. For example, the right to receive $1.00 at the end of each year for the next 4 years creates an ordinary annuity.

Table 15-5 shows the present value of an annuity of $1 each year for 4 years at a 10% interest rate. Figure 15-5 provides an illustration.

Table 15-5
Present Value of an Annuity of $1 per Period
Illustrated for 4 Years at 10% Interest Rate

Year	Present Value of Reversion	Present Value of Annuity
1	$0.9091	$0.9091
2	0.8264	1.7355
3	0.7513	2.4868
4	0.6830	3.1698

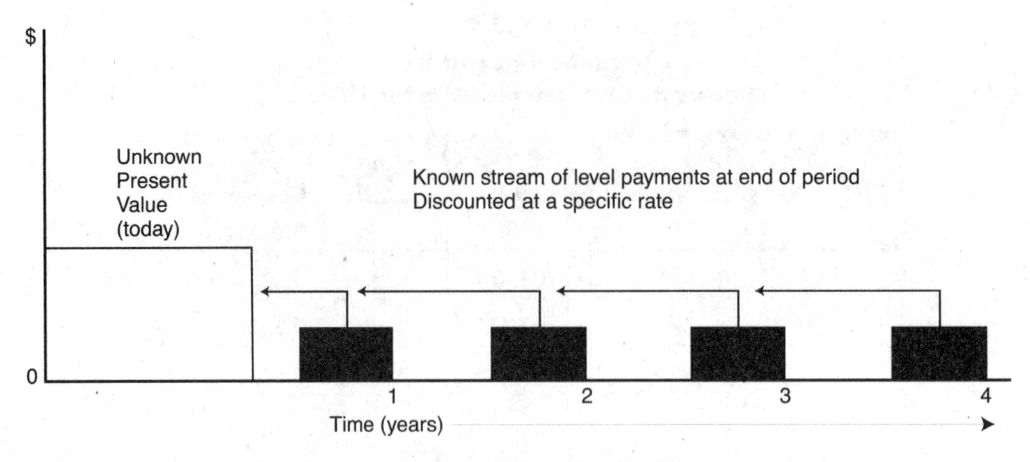

Figure 15-5. Present Value of an Ordinary Annuity

Installment to Amortize $1 (Loan Constant or Mortgage Constant)

This function shows the periodic payment required to retire a loan, with interest. It is the sum of the interest (or discount) rate and the sinking fund factor. It is the reciprocal of the present value of an annuity of $1 per period. This factor, which shows the constant payment necessary to amortize a loan in equal periodic installments, considers the term, interest rate, and principal of the loan.

As the interest rate increases or the amortization term shortens, the required periodic payment is increased. Conversely, lower interest rates and longer repayment periods reduce the required periodic payment. Each level payment in the installment to amortize $1 factor is a blend of interest and a reduction of the original principal.

The formula for the installment to amortize $1 is as follows:

$$\frac{1}{a_n} = \frac{i}{1-V^n} \text{ or } \frac{1}{a_n} = \frac{i}{1-\frac{1}{(1+i)^n}}$$

where a_n = present value of annuity of $1 per period; i = interest rate; n = number of periods; V^n = present value of $1.

Table 15-6 shows that the installment to amortize $1 is the reciprocal of the present value of an annuity of $1 factor. Figure 15-6 illustrates the distribution of periodic payments to interest and principal retirement.

Table 15-6
Installments to Amortize $1 as Reciprocal of Present Value of Annuity
Illustrated for 4 Years at 10% Interest Rate

Year	Present Value of Annuity	Installment to Amortize $1
1	$0.9091	$1.10
2	1.7355	0.5762
3	2.4868	0.4021
4	3.1698	0.3155

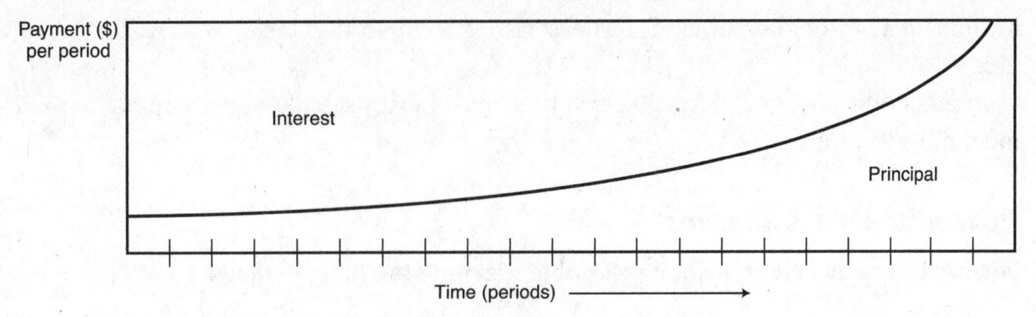

Figure 15-6. Mortgage Payment Application (Installment to Amortize $1)

As an example of use, suppose that a $100 loan requires four equal annual payments at 10% interest. The required payment will be $31.55. As a check, one can construct an amortization schedule as shown in Table 15-7.

Table 15-7
Amortization Schedule for a $100 Loan
Level Payments at 10% for 4 Years

Year	Loan Balance at Beginning of Year	Add Interest at 10%	Less Interest and Principal Payment	Balance at End of Year
1	$100.00	$10.00	$31.55	$78.45
2	78.45	7.85	31.55	54.75
3	54.75	5.48	31.55	28.68
4	28.68	2.87	31.55	-0-

Trick Question

A trick question asks about the growth of an investment at a given rate of return. The problem should state whether the investment is increasing at a compounded or a noncompounded rate. Be cautious. If the investment is growing at a compounded rate, interest for each year is based on the sum of principal plus interest earned during previous years. With compounding, the principal is increased by previously earned interest. However, without compounding, no interest is paid on interest previously earned. The interest is based solely on the original principal. As a result, the amount earned as interest each year does not vary for a noncompounded rate.

INCOME APPROACH

A certified general appraiser is expected to know how to appraise income-producing property on the basis of the revenue it is expected to generate. A number of techniques used for this purpose are described below.

Estimation of Income and Expenses

Income property valuation begins with an estimate of income and expenses. The tiers of income generated by a given property range from a top tier, which is potential gross income, to a bottom tier, which is owner's cash flow. Table 15-8 below is a chart of income tiers.

In addition to income, operating expenses must be considered and an operating statement must be prepared.

Potential Gross Income

Potential gross income is the rent collectible if all units are fully occupied.

TYPES OF RENT

In estimating potential gross income, the appraiser distinguishes between *market rent* (economic rent) and *contract rent.* Market rent is the rate prevailing in the market for comparable properties and is used in calculating market value by the income approach. Contract rent is the actual amount agreed to by landlord and tenant. If the leases are long-term, contract rent is important in calculating investment value.

Market rents represent what a given property should be renting for, based on analysis of recently negotiated contract rents for comparable space—but only those that are considered bona fide "market" rents. The market rent should be the amount that would result from a lease negotiated between a willing lessor and a willing lessee, both free from influence from outside sources.

Table 15-8
Tiers of Real Estate Income

	Potential gross income
Less:	Vacancy and collection allowance
Plus:	Miscellaneous income
Equals:	**Effective gross income**
Less:	Allowable expenses
	Maintenance and operating
	Administrative
	Utilities
	Real estate taxes
	Replacement reserves
	Etc.
Equals:	**Net operating income**
Less:	Debt service
	Mortgage principal
	Mortgage interest
Equals:	**Before-tax cash flow**
	(Cash throw-off or equity dividend)

RENTAL UNITS OF COMPARISON

Rents must be compared in terms of a common denominator. Typical units of comparison are square foot, room, apartment, space, and percentage of gross business income of the tenant, as shown in Table 15-9.

Table 15-9
Rental Units of Comparison for
Types of Income-Producing Property

Rental Unit of Comparison	Type of Income-Producing Property
Square foot	Shopping centers, retail stores, office buildings, warehouses, apartments, leased land
Room	Motels, hotels, apartment buildings
Apartment	Apartment buildings
Space	Mobile home parks, travel trailer parks, parking garages
Percentages of gross business income	Retail stores, shopping centers, restaurants, gas stations

When the square foot unit is used for an office building or shopping center, the appraiser should note that some leases are based on *net leasable area* (NLA) and others on *gross leasable area* (GLA). NLA is the floor area occupied by the tenant. GLA includes NLA plus common areas such as halls, restrooms, and vestibules. An office building with 100,000 square feet of GLA may have only 80,000 square feet of NLA, a difference of 20%. When leases are compared, it must be known whether rent is based on GLA or NLA.

Effective Gross Income

Effective gross income is the amount remaining after the vacancy rate and collection (bad debt) allowances are subtracted from potential gross income and after miscellaneous income is added. Calculating effective gross income is an intermediate step in deriving the cash flow.

VACANCY AND COLLECTION ALLOWANCE

The losses expected from vacancies and bad debts must be subtracted from potential gross income. These losses are calculated at the rate expected of local ownership and management.

MISCELLANEOUS INCOME

Miscellaneous income is received from concessions, laundry rooms, parking space or storage bin rentals, and other associated services integral to operating the project.

Operating Expenses

The potential list of *operating expenses* for gross leased real estate includes

- Property taxes: city, county, school district, special assessments
- Insurance: hazard, liability
- Utilities: telephone, water, sewer or sewage disposal, electricity, gas
- Administrative costs: management fees, clerical staff, maintenance staff, payroll taxes, legal and accounting expenses
- Repairs and maintenance: paint, decoration, carpet replacement, appliance repairs and replacement, supplies, gardening/landscaping, paving, roofing
- Advertising and promotion: media, business organizations

FIXED VERSUS VARIABLE EXPENSES

Operating expenses may be divided into two categories. *Fixed expenses* do not change as the rate of occupancy changes. *Variable expenses*, by contrast, are directly related to the occupancy rate: as more people occupy and use a building, variable expenses increase.

Fixed expenses include property taxes, license and permit fees, and property insurance. Variable expenses generally include utilities (such as heat, water, sewer), management fees, payroll and payroll taxes, security, landscaping, advertising, and supplies and fees for various services provided by local government or private contractors. There may be a fixed component in expenses normally classified as variable—for example, a basic fixed payroll cost regardless of the occupancy rate.

REPLACEMENT RESERVE

Provision must be made for the replacement of short-lived items (e.g., carpeting, appliances, and some mechanical equipment) that wear out. Such expenditures usually occur in large lump sums; a portion of the expected cost can be set aside every year to stabilize the expenses. A *replacement reserve* is necessary because the wearing out of short-lived assets causes a hidden loss of income. If this economic fact is not reflected in financial statements, the net operating income will be overstated. An appraiser should provide for a replacement reserve even though most owners do not set aside money for this purpose.

Net Operating Income

Net operating income (NOI) is estimated by subtracting operating expenses and the replacement reserve from effective gross income. Any interest or principal payments are not considered in computing NOI.

NOI for each comparable sale should be estimated in the same way that the subject property's NOI is considered. All relevant income and operating expenses are included in consistent amounts or rates to uniformly apply a capitalization rate. Each dollar of error in stating NOI will be multiplied, perhaps by 10 or more, in the resulting value estimate.

Operating Statement Ratios

Operating statement ratios are informative comparisons of expenses to income. The *operating expense ratio* is the percentage relationship of operating expenses to potential or effective gross income. There are typical ratios for each type of property in a market area, and much higher or lower ratios may be a clue to the effectiveness of management policies, whether efficient or inefficient.

The *net income ratio* is the reciprocal of the operating expense ratio. It tells what the fraction of property net operating income is compared to potential gross income. The *break-even ratio* is the occupancy rate needed to have zero cash flow. It is the sum of operating expenses plus debt service, divided by potential gross income.

Income Capitalization

Income capitalization is a process whereby a stream of future income receipts is translated into a single present-value sum. It considers

1. Amount of future income (cash flow)
2. Time when the income (cash flow) is to be received
3. Duration of the income (cash flow) stream

A *capitalization rate* is used to convert an income (cash flow) stream into a lump-sum value. The formula used is as follows:

$$V = \frac{I}{R}$$

where V = present value; I = periodic income; R = capitalization rate.

Direct Capitalization

Direct capitalization involves simple arithmetic division of two components to result in a capitalized value, as shown below:

$$\text{Property Value} = \frac{\text{Net Operating Income}}{\text{Overall Rate of Return}}$$

The appraiser considers other properties that have been sold and the net operating income generated by each to derive an *overall rate of return* (OAR). A market-derived OAR is then divided into the subject's NOI to estimate the latter's value.

OVERALL CAPITALIZATION RATE

To derive an overall capitalization rate from market data, the appraiser must select similar properties that have sold recently and are characterized by income streams of risk and duration similar to those of the subject property. If sufficient comparables are available, an OAR can be derived directly from the market and the appraiser can use direct capitalization to convert forecast NOI into a market value estimate.

To illustrate, suppose an appraiser has found the following data:

Comparable	NOI	Sales Price	Indicated OAR
#1	$240,000	$2,000,000	0.120
#2	176,000	1,600,000	0.110
#3	210,000	2,000,000	0.105

If adequate data of this nature were consistently available, appropriate overall rates could be developed for all types of properties, and other methods of capitalization would be of little importance. Unfortunately, reliable, verified sales data are not plentiful, and caution should be exercised in using this capitalization method. The comparability of the sale properties to the subject property must be analyzed carefully. Appropriate application requires that the comparables considered be consistent in these respects:

- Income and expense ratios
- NOI calculations
- Improvement ratio
- Remaining economic life
- Date and terms of sale

The simplicity of direct capitalization makes it attractive, even though complex relationships of income, expenses, and rates of return may underlie the computation. Various ways to derive rates are described in this chapter.

The *overall capitalization rate* must include a return *on* and a return *of* investment. Return *on* investment is the compensation necessary to pay an investor for the time value of the money, risk, and other factors associated with a particular investment. It is the interest paid for the use of money, also referred to as *yield.* Return *of* investment deals with repayment of the original principal, referred to as *capital recovery.* The capitalization rate taken from the market presumably incorporates other factors, such as expected inflation, income tax effect on the property, and interest rates and loan terms. However, this is a presumption, without actual numbers being tested to validate the expectation.

The yield rate is a rate of return on investment. Since real estate tends to wear out over time, though its economic useful life may be 50 years or longer, an investor must receive a recovery *of* the investment in addition to a return on it. There are three ways to provide a return of investment:

1. Straight line
2. Annuity
3. Sinking fund at safe rate

Straight Line. The straight-line technique is useful when building income is forecast to decline. Under a straight-line method, an annual depreciation rate is computed as an equal percentage for each year of the asset's life. When the life is 50 years, the rate is 2% per year, totaling 100% over 50 years. Thus, 2% would be added to the 10% return on investment to provide a 12% capitalization rate. That rate is divided into the estimated building income to derive an estimate of building value.

Annuity. An annuity method is useful when the income is projected to be level annually for a finite life. The appraiser determines the capital recovery rate at the sinking fund factor at the investment's rate of return. At 10% return on investment for 50 years, this factor is 0.000859, offering a total capitalization rate of 0.100859 (0.10 + 0.000859).

Sinking Fund at Safe Rate. The sinking fund factor at a safe rate is used when reinvestment assumptions are conservative. The 5% sinking fund factor for 50 years is 0.004777. This provides a capitalization rate of 0.104777 (0.10 + 0.004777) when given a 10% rate of return on investment.

Band of Investment

Net operating income from real estate is typically divided between mortgage lenders and equity investors. Rates of return can be weighted by the portions of purchase capital contributed, to result in a rate of return applicable to the entire property. This method is called the *band of investment.* The yield or interest or discount rate is then used to capitalize all of the NOI into one lump-sum property value. This value would assume that a capital recovery provision is unnecessary since neither amortization nor recovery of equity invested is considered.

For example, suppose that lenders typically contribute 75% of the required purchase capital and seek a 6% rate of return on investment. Equity investors want a 9% rate of return on investment. The yield or discount rate for the entire property would be 6.75%, as shown in Table 15-10. This rate consists purely of interest and assumes no principal retirement of the loan and no capital appreciation or depreciation of the equity investment. Further, all of the NOI is paid each year to each investor, mortgage and equity. Repayment of all purchase capital will be made upon a resale of the property at an indefinite time in the future.

Table 15-10
Rates of Return for a Mortgaged Property

Source of Funds	Portion of Purchase Capital		Rate of Return on Investment		Weighted Rate
Mortgage	0.75	×	0.06	=	0.0450
Equity	0.25	×	0.09	=	0.0225
Yield or discount rate for entire property					0.0675

Gross Income Multipliers

In the case of multipliers, gross rent (income) estimates are frequently used by investors. The estimated annual gross income is multiplied by a factor that has been derived as an accurate representation of market behavior for that type of property. A monthly gross income multiplier is typically used for residential housing, whereas annual multipliers are used for other types of property.

Ease of computation makes a rent multiplier a deceptively attractive technique. For appropriate application, however, everything about the subject being appraised and the comparables used must be alike, including the lease terms, tenant strength, operating expenses, and physical setting. The rent multiplier is, at best, a crude measuring device.

Residual Techniques

Residual techniques can be applied to income-producing property. The term *residual* implies that, after income requirements are satisfied for the known parts of the property, the residual, or remaining, income is processed into an indication of value for the unknown part. The three most commonly used types are (1) land, (2) building, and (3) equity residual techniques.

LAND RESIDUAL TECHNIQUE

The land residual technique is useful when improvements are relatively new or have not yet been built. Their values (or proposed costs) can be closely approximated, as can their estimated useful lives. For example, suppose recently constructed real estate improvements cost $1 million and have a 50-year economic useful life. The appropriate rate of return on investment is 10% because that rate is competitive with other investment opportunities perceived to have equal risk. Straight-line capital recovery is considered appropriate for the improvements. The annual recovery rate is 2% (100% ÷ 50 years = 2% per year), so the total required rate of return for the improvements is 12%. Annual net operating income for the first year is estimated to be $150,000. The improvements require a 12% return on $1 million of cost, which equals $120,000 of income. Subtracting this amount from the $150,000 net operating income leaves a $30,000 residual for the land. This amount, capitalized in perpetuity at 10%, offers a land value of $300,000. Capitalization of land income in perpetuity is appropriate because land is thought to last forever. This technique is summarized in Table 15-11.

Table 15-11
Land Residual Technique—Straight-Line Capital Recovery for Buildings

Net operating income (first year)	$ 150,000
Less: Income attributable to improvements:	
10% (on) + 2% (of) 5 12%; 0.12 × $1,000,000	– $ 120,000
Residual income to land	$ 30,000
Capitalized at 10% return on investment in perpetuity	$300,000
The total property value is then estimated at $1.3 million ($1 million for improvements plus $300,000 for land).	

The land residual technique is used also in estimating the highest and best use of land. Costs of different proposed improvements can be estimated, as can the net operating income for each type of improvement. A capitalization rate is multiplied by the estimated improvement cost, and the result subtracted from forecast net operating income. The result is forecast land income. Whichever type or level of improvement offers the highest residual income to land is the highest and best use of the land, subject to legal and physical constraints.

BUILDING RESIDUAL TECHNIQUE

The building residual technique works in reverse of the land residual technique. When the land value can be estimated with a high degree of confidence (as when there are many recent sales of comparables), income attributable to the land is subtracted from net operating income. The residual income is attributable to the improvements and can be capitalized to estimate their value. Then the capitalized value of the improvements is added to the land value to obtain a total property value estimate.

As an example of the building residual technique, suppose that the land value is estimated at $200,000 through a careful analysis of several recent sales of comparable vacant tracts of land. A 10% discount rate is considered appropriate. Of the estimated $150,000 of annual NOI, $20,000 is attributable to the land. Capital recovery for land income is unnecessary since the land will last forever. The $130,000 balance of NOI is attributable to the improvements, which would be worth $1,083,333 using a straight-line capital recovery rate and 50-year life. The building residual technique is summarized in Table 15-12.

Table 15-12
Building Residual Technique

Net operating income	$ 150,000
Less: Income to land: 10% of $200,000	– $ 20,000
Income to improvements	$ 130,000
Improvement value at straight-line capital recovery: (over 50 years) $130,000 ÷ 0.12	$ 1,083,333

EQUITY RESIDUAL TECHNIQUE

The equity residual technique is an effort to estimate the value of the property when the mortgage amount and terms are known. Net operating income is calculated as in other techniques, and debt service is subtracted to result in cash flow. The cash flow is capitalized at an appropriate equity dividend rate to obtain equity value, which is added to the mortgage principal to provide a value estimate.

Owner's Cash Flow from Operations

In this technique, net operating income is forecast for a number of years to some period in the future when a property sale is expected to occur. The cash flows from operations and resale are discounted by an appropriate discount rate. Then they are totaled to offer a property value estimate.

Adjustments to Capitalization Rates

Depending on lease and market information, net operating income may be forecast to be level, to grow, or to decline. An appraiser must use known information and provide market evidence for the projections.

Financial results from real estate depend on lease terms and economic conditions. If the tenant's credit is triple A and the lease is long term at a fixed net rental rate, income should be assured. If the tenant's credit rating is not high or if the lease is not net or its term is short, value will depend on such factors as inflation, competition, interest rates, and economic conditions.

Current values may be strongly affected by expectations of future resale prices. An appraiser must therefore consider resale expectations when selecting a capitalization rate. This "cap rate" is raised when value decline is expected in the future, lowered when appreciation is anticipated.

Estimating Resale Price

REVERSION

It is sometimes helpful to estimate a resale price based on expectations for the future. The resale price, less selling costs and outstanding debts, has a present value that, added to the present value of net operating income for each year, may be used to estimate today's value. The estimated resale price is diminished by selling expenses, including commissions and legal expenses, to result in proceeds from sale. Table 15-13 shows how a $1,000,000 resale price is diminished by transaction costs.

Table 15-13
Resale Price Diminished by Transaction Costs

Resale price		$1,000,000
Less seller's estimated expenses of sale:		
Title insurance policy	$ 3,000	
Attorney's fees	4,500	
Release of lien	200	
Survey of property	600	
Escrow fees	500	
Recording fees	200	
Broker's professional service fee (commission)	40,000	
Subtotal		− 49,000
Estimated receipts to seller		$ 951,000

OTHER METHODS OF RESALE PRICE ESTIMATION

The resale price can be estimated in several other ways. First, today's market value with a percentage annual growth or depreciation rate is applied. This method assumes that future values are related to today's value, often adjusted by expected inflation. Second, income forecast for the year after the sale is capitalized to obtain a resale price. Third, the sales price of encumbered property may be the cash flow after debt service, capitalized at an equity dividend rate, and added to the mortgage balance.

Measuring Rate of Cash Flow

Cash flow is usually defined as the amount left after debt service. The rate of cash flow can be derived in two ways: dividend and yield.

Equity Dividend Rate

The *equity dividend rate* (also called *cash-on-cash return*) is cash flow divided by the equity paid for the property. The result is a measure of cash return on investment. When cash flow can be measured accurately, it may be divided by a market-derived equity dividend rate to obtain the equity value. Equity value, plus mortgage balance, offers a property value estimate.

For example, an investor pays $400,000 for a building using $100,000 of her own cash and a $300,000 mortgage loan. The annual rental income is $50,000, annual operating expenses are $8,000, and debt service is $34,000 annually. The cash-on-cash return is 8%, as shown below.

Expenses = $8,000 + $34,000 = $42,000
Annual cash return = $50,000 – $42,000 = $8,000
Cash-on-cash return = $8,000 ÷ $100,000 = 0.08 = 8%

Debt Coverage Ratio

The *debt coverage ratio* indicates a lender's margin of safety. It is computed by dividing net operating income by debt service. Lenders typically want a ratio ranging from at least 1.1 for the safest property, secured by a net lease from a tenant with triple-A credit, up to 1.2 to 1.5 for apartments or office buildings and a minimum of 2.0 for recreational property.

Discounted Cash Flow (Yield) Capitalization

By contrast to first-year capitalization techniques, yield capitalization rates employ compound interest/present value factors. These rates are typically found in six-function compound interest tables, described on pages 331–337, which offer a precise value for periodic cash flow and reversion based on a specific rate of return on investment.

Discounted Cash Flow Application

Discounted cash flow capitalization is best applied when income from the property and a resale price can be forecast with reasonable certainty. This requires income forecasts of rents, operating expenses, and tenant improvements, as previously discussed.

The key to discounted cash flow is the *discounting process*, which is based on the fact that money to be received at some time in the future is worth *less* than money received today. The value that we assign to such *future incomes* is called *present value*, or *present discounted value*. Exactly what the present value of a future income is will depend on the answers to two questions: (1) when will the income accrue and (2) what is the *discount rate*? The discount rate is a rate of return, similar to an interest rate, that measures the risk and time value of the income to be received. How a discount rate is arrived at is discussed below. However, there are a couple of basic rules that are useful: (1) the longer we have to wait for a future income, the less it is worth to us *today* (i.e., the lower is its present value) and (2) the higher the discount rate applied to a future income, the lower its present value (and vice versa).

Appraisers value an income property by examining the net income that the property can be expected to provide in the future and then finding a present value for that income. This can be done fairly easily using six-function tables or a financial calculator. However, most states don't allow applicants for licenses to use financial calculators (not yet, anyway), so we will take a brief look at the math involved.

Essentially, we derive a present value for each part of the future income stream and then add all the values. For example, suppose we are appraising a property that will produce a net income of $10,000 per year for 5 years; at the end of 5 years, it can be sold for $50,000. (This one-time income at the end of the holding period, either sale price or salvage value, is called a *reversion* by real estate appraisers.) Let's assume that we're using a discount rate of 12%. This means that we need a return of 12% a year (compounded) for this investment. To simplify, let's assume also that the incomes accrue at the *end* of each year. Then our income stream will be $10,000 a year for 4 years, with a fifth-year income of $10,000 *plus* the sale proceeds of $50,000, for a total fifth-year income of $60,000. Our income stream looks like the column below headed "Income."

Time	Income	Factor	Present Value
Year 1	$10,000	$1/1.12^1 = 0.892857$	$ 8,928.57
Year 2	$10,000	$1/(1.12)^2 = 0.797194$	7,971.94
Year 3	$10,000	$1/(1.12)^3 = 0.711780$	7,117.80
Year 4	$10,000	$1/(1.12)^4 = 0.635518$	6,355.18
Year 5	$60,000	$1/(1.12)^5 = 0.567427$	34,045.61
		Present value total	**$64,419.10**

We arrived at a present value of about $64,400. What we're saying, with this number is that a future income of $10,000 a year for 5 years, plus a reversion of $50,000 at the end of 5 years, when discounted at 12% a year is worth about $64,400 *today*. In other words, $64,400 invested today, at 12% annual return, will yield the income we project for the property.

How exactly did we do this? Let's look at the first year's income of $10,000. We don't get it for a year, but we have to pay for it *now*. Basically, we ask, "How much money, invested today, will grow to $10,000 in 1 year, given a 12% annual return?" This first year's income, then, will have a present value of $10,000/1.12 because we are looking for an amount that, once it has earned a 12% return, will *become* $10,000. What we did in the table was to figure out the *factor* for 12% for 1 year, which is the decimal equivalent of 1/1.12, and then multiply it by $10,000. The advantage of this method is that once we figure out what 1/1.12 is, we can use that number to get the present value of *any* amount to be received 1 year from now at a rate of 12%.

For the 12% *factors* for periods beyond 1 year, we just divide by 1.12 *again* for *each* year longer that we have to wait for the income. For the present value of the second year's income, we multiply $10,000 by 1/1.12 *twice* (once for each year): $\$10{,}000 \times 1/1.12 \times 1/1.12$, which is the same as $\$10{,}000 \times 1/(1.12)^2$. For the third year we use $1/(1.12)^3$, and so on. Once we've figured out the present value of each of the five annual incomes, we just add them. You can check these numbers out yourself. For each year's income, start with the present value. Multiply it by 1.12; do this for each year that you have to wait, and you should end up with $10,000 for each of the first 4 years and $60,000 for the fifth year.

YIELD RATE SELECTION

An appraiser can attempt to construct a yield rate using the *buildup method* by identifying the components of a required rate of return. Although there is general agreement in identifying the components (see below), the proportion of each component in the overall rate is arbitrary. Furthermore, the buildup is unnecessary because an investor acquires the entire property, which provides one rate. Although the buildup method is theoretical, it helps us to understand capitalization rates and to compare them with non–real estate investments.

To construct a built-up rate, begin with a liquid, risk-free rate, such as the rate on passbook accounts offered by financial institutions whose depositors' accounts are insured by a U.S. government agency. The rate is the minimum that must be paid as compensation for the time value of money.

Add to this risk-free rate a provision for the risk associated with the specific type of real estate being appraised. For example, risk is low where there is a strong tenant with captive customers but high for property leased to weak tenants who have fickle customers.

Provisions must be made for illiquidity. *Liquidity* is a measure of how quickly an asset can be converted into cash. Compared to stocks and bonds, real estate is illiquid, particularly in a weak market or tight-money economy.

The burden of investment management must also be included. This is the decision making required of the investor and is distinguished from the everyday process of property management.

The components of the buildup method and their proportions in the overall built-up rate are summarized as follows:

Risk-free rate	2%
Risk of subject property	2%
Illiquidity of subject property	2%
Investment management	1%
Rate of return on investment	7%

Selecting the appropriate yield rate depends on numerous factors, especially the availability of accurate market data. An appraiser must consider the duration, quality, and amount of income from the property forecast for each future year. Other relevant factors include the earnings or yield rate required by the market and the portion of investment that is subject to capital recovery.

Ratio capitalization requires a realistic estimate of net operating income, which is divided by a supportable capitalization rate. The mathematical composition of the cap rate is derived from the market, tempered by property performance expectations.

VALUATION OF PARTIAL INTERESTS

The certified general appraiser is expected to have an extensive knowledge of partial interests in real estate. These include interests in a lease, lease provisions, and the identification of the various ownership positions within a property having different risk features.

Lease Interests

A *subleasehold* may be carved out of a lease interest. A subleasehold is the interest of the tenant's tenant; that is, a tenant may sublease his property, becoming a landlord to another tenant. The subtenant's interest may be valuable.

Renewal options in a lease give the tenant the right, but not the obligation, to extend the lease at whatever rent and terms are described in the option. This encumbrance can affect the value of the property either positively or negatively.

Tenant improvements must be analyzed to determine how they affect the property value. Are they to remain after the lease expires? Would a substitute tenant find them useful and pay a premium for them, or would they be expensive to demolish on demand?

Concessions granted in the market or by the property to attract tenants must be assessed. For example, suppose an office building rents for $25 per square foot on a 5-year lease, with the first year free. The property is appraised in the second year of the lease. Is the correct rent $25/per square foot or much less? How much rent can be capitalized? The appraiser needs to understand the market.

Lease Provisions

NET VERSUS GROSS LEASES

A lease is typically referred to as *net* or *gross*, although most leases are neither absolutely net nor absolutely gross. In an absolute net lease, the tenant (lessee) pays taxes, insurance, and operating expenses: maintenance, including repairs, alterations, replacements, and improvements, ordinary or extraordinary repairs whether interior or exterior, and so on. A net lease is the most straightforward approach for a single-tenant structure. The terms *triple-net* and *net-net-net* lease are often used to indicate a greater degree of net-ness than that for other leases. As to specific taxes, tenants pay taxes, special assessments, and the like. However, tenants do not pay obligations of the landlord (lessor), such as gift, income, inheritance, franchise, or corporate taxes, or a tax on the rental receipts of the landlord.

The real estate owner is a passive investor when a net lease is used, relying on net rent to provide a return on investment. The property is the owner's investment, not a business.

With a gross lease, the landlord pays all operating expenses. The owner must become a businessperson, keeping firm control over operating expenses and assuring that money spent to comply with the lease is spent prudently.

With a *stop* (or *escalation*) clause in a gross lease, the landlord pays a base-year amount for the expense and the tenant pays the increase in expense each year. A net lease or stop clauses reduce the need for frequent rent adjustments, cost of living indexes, or short-term leases.

OPERATING EXPENSES

A properly written lease should specify whether landlord or tenant pays for each operating expense. It should also specify that the responsible party must present proof of payment, such as a paid property tax bill or insurance policy in conformity with requirements of the lease, to provide assurance of protection from risks covered by the lease. If the lessee in a net lease does not supply the information, the lessor should request it.

Percentage Leases

A *percentage lease*, commonly used for a retail tenant, provides incentive to the landlord to make the property attractive and thus encourage retail sales. The landlord must be prepared to audit the tenant's retail sales records and to enforce provisions of the lease that protect the lessor's rights to a percentage of sales.

Most percentage leases require a fixed *minimum base rent* plus a percentage rent based on gross sales in excess of a certain amount. The base rent is often set as an anticipated average amount of sales per square foot for the particular type of business. For example, suppose industry norms for a ladies' fashion shop indicate that sales of $200 per square foot (SF) are typical. Rent for a 10,000-SF shop could be $100,000 per year plus 5% of sales above $2 million (10,000 SF × $200 typical sales = $2,000,000). Percentage rents would be imposed only when sales exceed the norm of $200/SF of floor area.

Different types of businesses have different typical percentage rents. For example, a jewelry store or other retailer selling luxury items tends to be relatively small. It has a high markup and low inventory turnover in comparison to a grocery or discount store. The percentage rental rate thus tends to be higher for the luxury or specialty store.

Certain lease provisions are necessary to assure fairness to both parties in administering a percentage lease. Monthly or annual sales reports may be required. The landlord may have the right to hire an auditor, whose fee will be paid by the tenant if sales are understated or by the landlord if no discrepancies are found.

Cost of Living Index

Leases for multiple years without rent adjustments result in hardships to landlord or tenant during highly inflationary or deflationary periods. A long-term lease may have a *cost of living adjustment* that calls for changes in rent based on the change in some published index.

The index used and the frequency and amount of adjustment are negotiable. The Consumer Price Index (CPI), published by the Bureau of Labor Statistics of the U.S. Department of Labor, is usually chosen because of its frequency of computation and publication. A lesser known local index tied to rental rates, operating expenses, or real estate values may be satisfactory but can introduce elements of instability or potential manipulation, possibly resulting in litigation.

The adjustment period may be a single year or a multiple of years. This period should be short for a gross lease but may be longer for a net lease. With a net lease, the landlord has less financial exposure to the risk of inflation because the tenant pays operating expenses.

The degree of adjustment may be only part of the full change in the index selected. For example, if 50% of the index was selected as the multiplier and the index rose by 20%, rent would increase by 10%. This provides some protection to the landlord (who perhaps bought the property with preinflation dollars) but does not inflict the full effect of inflation on the tenant.

Excess Rent

Excess rent is the difference between contract and market rents; the term implies that the tenant is paying more than would be paid for comparable property. If a property has several leases, the appraiser should prepare a schedule with the expiration date of each lease to assist in estimating the quality and duration of the income stream.

SITE VALUE

Two methods of estimating site value, in addition to those given in Chapter 14, are *ground rent capitalization* and *subdivision analysis*. These methods apply only in certain situations.

Ground Rent Capitalization

In a few places, notably Hawaii, Baltimore, and Manhattan, there is a high concentration of ground lease ownership; that is, the building owner does not own the land but, as a tenant, leases land from the landowner. An appraiser can estimate land value by *capitalization,* that is, by dividing the annual rent by an appropriate capitalization rate. As a simple example, suppose that the rent on a lot in Baltimore is $120 per year, renewable forever. The appropriate rate of return on investment is 6%. Then the land value is $120 ÷ 0.06 = $2,000.

Subdivision Analysis

Subdivision analysis is a method of estimating the maximum that a subdivider can pay for a tract of land. This method involves a sales estimate (usually monthly or quarterly) and a cost estimate over time. Cash flows are then discounted to a present value.

The example shown in Table 15-14 was prepared on an annual basis. As can be seen from the table, $137,285 is the maximum that can be paid for land and still allow the subdivision to be developed, under the assumptions shown.

Table 15-14
Subdivision Development Analysis

	Year			
	1	2	3	4
Lost sales units	0	10	15	20
Price per lot		$ 20,000	$ 22,000	$ 25,000
Total sales	-0-	200,000	330,000	500,000
Costs:				
Engineering	$ 40,000			
Utility installation	70,000	100,000		
Clearing and grading	50,000			
Advertising	25,000	25,000	25,000	25,000
Streets and sidewalks	50,000	50,000	60,000	-0-
Permits	10,000			
Contractor's overhead and profit	10,000	10,000	10,000	10,000
Legal and accounting	15,000			
Interest	20,000	15,000	15,000	15,000
Entrepreneurial profit			50,000	50,000
Total costs	$290,000	$200,000	$160,000	$100,000
Cash flow (Total sales less Total costs)	(290,000)	-0-	170,000	$400,000
Discount rate @ 10%	× .90909	× .8264	× .7513	× .683
Present value	(263,636)	-0-	$127,721	$273,200
Sum of present value	$137,285			

Summary

The certified general appraiser is expected to have knowledge of income property valuation techniques and an understanding of their applications to specialized properties and interests that are beyond the expectations for a residential appraiser, broker, or salesperson.

QUESTIONS ON CHAPTER 15

1. The factor that indicates the future amount of $1 at compound interest is called
   (A) future value of an annuity of $1 per period
   (B) future value of $1
   (C) present value of $1
   (D) present value of an annuity of $1 per period
   (E) installment to amortize $1

2. The factor that offers the future value of $1 deposited at the end of each period at compound interest is
   (A) present value of an annuity of $1 per period
   (B) present value of $1
   (C) sinking fund factor
   (D) future value of an annuity of $1 per period
   (E) capitalization rate

3. The factor that indicates the periodic deposit at the end of each period needed to reach $1 is
   (A) future value of $1 per period
   (B) present value of $1 per period
   (C) installment to amortize $1
   (D) sinking fund factor
   (E) mortgage constant

4. The factor that indicates the value of $1 to be received in the future is
   (A) mortgage constant
   (B) present value of $1
   (C) equity yield rate
   (D) cap rate
   (E) internal rate of return

5. The factor that indicates the present value of $1 per period is
   (A) present value of $1
   (B) present value of an annuity of $1 per period
   (C) accumulation of $1 per period
   (D) sinking fund factor
   (E) perpetuity factor

6. The installment to amortize $1 is also known as
   (A) internal rate of return
   (B) loan or mortgage constant
   (C) discount rate
   (D) payback period
   (E) present worth of $1

7. The installment to amortize $1 is the reciprocal of the
   (A) future value of $1
   (B) sinking fund factor
   (C) capitalization rate
   (D) equity yield rate
   (E) present value of an annuity

8. The appropriate time adjustment for a property is concluded to be an increase of 7% per year compounded. The time adjustment for a comparable sales property that sold for $40,000 two years ago is
   (A) −$5,796
   (B) −$5,600
   (C) −$2,800
   (D) +$5,600
   (E) +$5,796

9. How much must be deposited today, in a bank that pays 10% rate of interest with annual compounding, in order to realize $10,000 in 4 years?
   (A) $40,000
   (B) $9,077
   (C) $6,830
   (D) $4,000
   (E) $2,500

10. Ms. Brown has just paid $5 for a purchase option on some land. The option gives her the right to buy the property for $10,000 at the end of 2 years. The $5 paid for the option will not be applied to the purchase price. How much must Ms. Brown put aside today, in a bank that pays a 10% rate of interest with monthly compounding, to achieve a balance of $10,000 in 2 years?
    (A) $12,100
    (B) $10,000
    (C) $8,194
    (D) $6,720
    (E) $5,000

Questions 11–16 are based on the following information:

You are appraising a 20-acre tract of unimproved land. The site is zoned for single-family residential use. All utilities are available along the street on which the land fronts. From the engineers who will plat the proposed subdivision, you learn that 20% of the land area will be used for streets and sidewalks. You find that zoning will permit four lots per acre of net developable land after deducting streets. Research indicates that lots similar to those that will be available on the subject land sell for $18,000 and that the entire tract can be developed and sold in 1 year. You find that 40% of the sale price of each lot must be allocated to selling costs, overhead, contingencies, carrying cost, and developer's profit. Finally, your research discloses that 2,000 feet of streets (including water, storm sewer, and sanitary sewer lines) must be installed at a cost of $80 per foot.

11. How many lots can be developed?
    (A) 20
    (B) 64
    (C) 80
    (D) 88
    (E) 16

12. What is the gross sales price from the sale of all lots?
    (A) $288,000
    (B) $360,000
    (C) $1,152,000
    (D) $1,440,000
    (E) $1,584,000

13. What is the cost of installing streets and water and sewer lines?
    (A) $16,000
    (B) $32,000
    (C) $64,000
    (D) $160,000
    (E) none of the above

14. What is the total of selling cost, overhead, contingencies, carrying costs, and developer's profit?
    (A) $144,000
    (B) $460,800
    (C) $576,000
    (D) $633,600
    (E) none of the above

15. What is the total cost of development and overhead, including developer's profit?
    (A) $304,000
    (B) $320,000
    (C) $620,800
    (D) $733,600
    (E) none of the above

16. What is the developer's potential profit on the 20-acre tract?
    (A) less than $160,000
    (B) $160,000
    (C) $531,200
    (D) $1,152,000
    (E) $1,440,000

17. The annual net operating income from an apartment house is $22,000. With a capitalization rate of 11%, the indicated market value is
    (A) $2,420
    (B) $126,000
    (C) $176,000
    (D) $200,000
    (E) $242,000

18. A small office building sold for $120,000. The monthly net operating income is $1,300 per month. What was the overall capitalization rate?
    (A) 1.08%
    (B) 9.2%
    (C) 10.8%
    (D) 12%
    (E) 13%

19. For which of the following would the lowest ratio of operating expenses to gross income be incurred by the landlord?
    (A) resort hotel
    (B) net-leased retail store
    (C) office building
    (D) apartment building
    (E) nursing home for the elderly

20. The band-of-investment technique is most useful when equity investors are primarily concerned with
    (A) land appreciation rates
    (B) building tax shelters
    (C) gross rent multipliers
    (D) overall capitalization rates
    (E) noise abatement

21. Which of the following is the preferred method of deriving a capitalization rate?
    (A) summation
    (B) band of investment
    (C) direct comparison
    (D) bank of investment
    (E) monetary policy

22. Capitalization is employed in the
    (A) cost approach
    (B) direct sales comparison approach
    (C) income approach for income properties
    (D) income approach for residential properties
    (E) none of the above

23. Capitalization is the process whereby
    (A) income is converted to an indication of value
    (B) syndicates are formed
    (C) an asset is removed from accounting records
    (D) both A and B
    (E) none of the above

24. The operating expense ratio for income property is typically
    (A) between 4 and 10
    (B) under 100%
    (C) under 10%
    (D) under 2%
    (E) more than 10

25. In determining income and expenses of an existing office building, the first step is
    (A) a lease and rent analysis
    (B) an effective gross income estimate
    (C) an operating expense estimate
    (D) a reconstructed income tax return
    (E) a feasibility study

26. A forecast using discounted cash flow analysis would include
    (A) income, vacancy, and operating expenses
    (B) an economic analysis
    (C) reversion at the end of the holding period
    (D) discounting expected future cash flows to a present value
    (E) all of the above

27. If the overall capitalization rate for income property were to increase while estimated net operating income remained the same, the resulting value estimate would
    (A) increase
    (B) decrease
    (C) remain the same
    (D) any of the above could occur
    (E) none of the above would occur

28. Estimated income property values will decline as a result of
    (A) increased net cash flows
    (B) lower capitalization rates
    (C) lower vacancy rates
    (D) increased discount rates
    (E) higher standard of living

29. *Income capitalization* is the term used to describe the process of estimating the value of income property by studying expected future income. This process
    (A) converts the net operating income of a property into its equivalent capital value
    (B) reflects the time value of money by reducing or discounting future income to its present worth
    (C) focuses on the present worth of future benefits
    (D) uses market interest rates
    (E) all of the above

30. When estimating the value of an income-producing property, the appraiser will *not* consider
    (A) income taxes attributable to the property
    (B) the remaining economic life of the property
    (C) potential future income
    (D) net operating income
    (E) expected future income patterns

31. In income capitalization, value is measured as the present worth of the
    (A) forecast reversion with a growth factor
    (B) forecast cash flow capitalized in perpetuity
    (C) forecast effective gross income (EGI) plus the reversion
    (D) forecast net operating income plus the reversion
    (E) cost of production

32. Income capitalization techniques are typically *not* used in valuing
    (A) retail properties
    (B) apartment buildings
    (C) office buildings
    (D) motels
    (E) single-family residences

33. The lump sum that an investor receives upon resale of an investment is called
    (A) net income
    (B) gross income
    (C) equity dividend
    (D) reversion
    (E) residual

34. The most commonly used capitalization rate is the
    (A) income rate
    (B) composite capitalization rate
    (C) interest rate
    (D) overall rate
    (E) underall rate

35. An ordinary annuity is
    (A) level in amount and timing
    (B) different from a variable annuity
    (C) received at the end of each period
    (D) a stream of income
    (E) all of the above

36. The procedure used to convert expected future benefits into present value is
(A) residual analysis
(B) capitalization
(C) market capitalization
(D) equity capitalization
(E) compounding

37. A reconstructed operating statement for an owner-operated property should include
(A) income tax
(B) book depreciation
(C) management charges
(D) wages earned outside the property
(E) imputed interest

38. *Yield* is defined as
(A) overall capitalization
(B) stopping to review work
(C) letting another speak first
(D) rate of return on investment
(E) rate of return of investment

39. Which of the following is a specific expense item rather than a category?
(A) escalation
(B) property taxes
(C) operating expenses
(D) fixed charges or expenses
(E) variable expenses

40. In yield capitalization, investor assumptions are
(A) accrued
(B) argued
(C) regulated
(D) disregarded
(E) simulated

41. The basic formula for property valuation via income capitalization is
(A) $V = IR$
(B) $V = I/R$
(C) $V = R/I$
(D) $V = SP/GR$
(E) $V = I/F$

42. All of the following consider the time value of money *except*
(A) net present value
(B) discounted cash flow
(C) internal rate of return
(D) payback period
(E) compound interest

43. A rent survey of apartment buildings reveals that one-bedroom units have a considerably higher occupancy factor than two-bedroom units. If the subject property contains only two-bedroom units, the appraisal should probably project
(A) an average of the vacancy factors for all units surveyed
(B) a higher vacancy factor than was found for one-bedroom units
(C) a lower vacancy factor than was found for one-bedroom units
(D) the same vacancy factor as was found for one-bedroom units
(E) Projection from the information provided is impossible.

44. The quality of a forecast future income stream is indicated by its
(A) amount
(B) length
(C) timing
(D) mortgage
(E) risk

45. The total anticipated revenue from income property operations after vacancy and collection losses are deducted is
(A) net operating income
(B) before-tax cash flow
(C) effective gross income
(D) potential gross income
(E) property residual income

46. Which of the following is used in direct capitalization?
(A) internal rate of return
(B) overall capitalization rate
(C) building capitalization rate
(D) mortgage capitalization rate
(E) equity yield rate

47. Which of the following statements is true?
(A) If the overall yield on the property is less than the mortgage rate, favorable leverage occurs.
(B) When negative leverage occurs, the owner receives a cash yield on the property that is greater than if the property was not financed.
(C) If the overall property yield equals the mortgage rate, the owner receives an additional yield on his investment by trading on the equity.
(D) Financing the property at higher rates of interest than the overall rate may not be beneficial to the owner in terms of yield on the equity investment.
(E) When a property generates cash flow, its owner pays tax on the money received.

48. Which of the following statements is true?
(A) An investor usually considers the return from a property in the light of the risk being assumed.
(B) An investor who assumes higher risk usually expects lower investment yield.
(C) Higher-risk investments are generally associated with lower investment returns.
(D) To maximize the overall risk position, the investor may wish to diversify real and/or personal property investments.
(E) The lowest-risk investment is real estate.

49. Capitalization rate is
(A) the annual rent divided by the purchase price
(B) annual income and all gains or losses prorated to an effective annual amount—an internal rate of return
(C) the percentage or decimal rate that, when divided into a periodic income amount, offers a lump-sum capital value for the income
(D) unchanging with time
(E) none of the above

Questions 50–54 are based on the following information:

A $12 million office building is purchased with an 80% loan-to-value ratio mortgage, payable over 30 years at 11% interest with monthly payments of $91,423. At 100% occupancy, rents would be $2 million. The vacancy trend in the area is 2.5%. Operating expenses amount to $650,000, including $50,000 placed in a replacement reserve. The land is considered to be worth $2 million.

50. What ratio, expressed as a percentage, is between 80 and 85?
(A) improvement
(B) overall rate
(C) mortgage constant
(D) vacancy
(E) operating expense

51. What ratio, expressed as a percentage, is between 10 and 11?
(A) improvement
(B) overall rate
(C) mortgage constant
(D) vacancy
(E) operating expense

52. What ratio, expressed as a percentage, is between 11 and 12?
(A) improvement
(B) overall rate
(C) mortgage constant
(D) vacancy
(E) operating expense

53. What ratio, expressed as a percentage, is between 30 and 40?
(A) improvement
(B) overall rate
(C) mortgage constant
(D) vacancy
(E) operating expense

54. What ratio, expressed as a percentage, is between 0 and 5?
(A) improvement
(B) overall rate
(C) mortgage constant
(D) vacancy
(E) operating expense

---

55. Investors for apartments are seeking a 10% cash-on-cash return in the current market. The current interest rate on a 30-year mortgage for this type of property is 12% with monthly payments. Lenders will fund up to a 75% loan-to-value ratio. What is the overall rate of return using the band of investment technique?
(A) 10%–10.5%
(B) 10.51%–11%
(C) 11.01%–11.25%
(D) 11.26%–12%
(E) none of the above

Questions 56–60 are based on the following information:

A certain older office building in the heart of a downtown metropolitan area currently generates $40,000 net operating income. Income is expected to decline systematically over the estimated 10 remaining years of the building's useful life. Therefore, straight-line capital recovery (declining income to the building) is considered appropriate. The land is currently valued at $250,000, based on numerous recent sales of comparable properties. For this type of property, investors want a 10% interest (yield to maturity) rate.

56. How much of the first year's $40,000 net operating income is attributable to the land?
(A) $15,000
(B) $20,000
(C) $25,000
(D) $30,000
(E) $40,000

57. How much of the net operating income is attributable to the building?
(A) $15,000
(B) $20,000
(C) $25,000
(D) $30,000
(E) $40,000

58. What is the building capitalization rate, assuming straight-line capital recovery?
(A) 5%
(B) 10%
(C) 15%
(D) 20%
(E) 25%

59. What is the value of the building, using the building residual technique?
(A) $75,000
(B) $100,000
(C) $125,000
(D) $130,000
(E) $140,000

60. What is the combined value of the building and land?
(A) $275,000
(B) $300,000
(C) $325,000
(D) $330,000
(E) $340,000

61. The most common type of partial-interest appraisal involves
    (A) a leased property
    (B) a timeshare residence
    (C) a condominium
    (D) rezoning
    (E) condemnation

62. When contract rent exceeds market rent, the leasehold interest
    (A) is subleased
    (B) must be appraised
    (C) has no positive value
    (D) should be mortgaged
    (E) is the same as the leased fee

63. Lease provisions may describe all of the following *except*
    (A) rental payments and the term of use and occupancy
    (B) whether landlord or tenant pays operating expenses
    (C) rent escalations and renewal or purchase options
    (D) the credit rating of the tenant
    (E) assignment or subletting rights

64. *Overage rent* is the
    (A) actual rent over the entire lease term
    (B) percentage rent above guaranteed minimum rent
    (C) amount by which contract rent exceeds market rent
    (D) rent that is past due
    (E) rent the property could command in the open market if not subject to its lease

65. A lease normally states all of the following *except*
    (A) agreed terms of occupancy
    (B) rental payments
    (C) tenant's responsibilities and obligations
    (D) financing of the property
    (E) actions causing default

66. All of the following lease provisions are advantageous to the lessee *except*
    (A) an escape clause
    (B) a renewal option
    (C) a purchase option
    (D) an escalation clause
    (E) All of the above are advantageous.

67. If a $54,000 investment in real estate generates gross earnings of 15%, the gross monthly return most nearly is
    (A) $819
    (B) $705
    (C) $675
    (D) $637
    (E) $8,100

68. In arriving at an effective gross income figure, an appraiser of rental property makes a deduction for
    (A) real property taxes
    (B) repairs
    (C) vacancy
    (D) depreciation
    (E) replacement reserves

ANSWERS

1. **B**
2. **D**
3. **D**
4. **B**
5. **B**
6. **B**
7. **E**
8. **E**
9. **C**
10. **C**
11. **B**
12. **C**
13. **D**
14. **B**
15. **C**
16. **C**
17. **D**
18. **E**
19. **B**
20. **D**
21. **C**
22. **C**
23. **A**
24. **B**
25. **A**
26. **E**
27. **B**
28. **D**
29. **E**
30. **A**
31. **D**
32. **E**
33. **D**
34. **D**
35. **E**
36. **B**
37. **C**
38. **D**
39. **B**
40. **E**
41. **B**
42. **D**
43. **B**
44. **E**
45. **C**
46. **B**
47. **D**
48. **A**
49. **C**
50. **A**
51. **B**
52. **C**
53. **E**
54. **D**
55. **D**
56. **C**
57. **A**
58. **D**
59. **A**
60. **C**
61. **A**
62. **C**
63. **D**
64. **B**
65. **D**
66. **D**
67. **C**
68. **C**

Taxation and Assessment

16

ESSENTIAL MATTERS TO LEARN FROM THIS CHAPTER

- *Property taxes* and *income taxes* on real estate are discussed in this chapter.
- Property taxes are called *ad valorem* taxes, meaning they are applied against the *value* of the property.
- Generally, local government units assess property taxes against the value of real estate. These government units include the city, county, and school district.
- Some property owners or users are exempt from property tax, notably government property and facilities used for religious purposes.
- Some properties or owners are allowed a discount, notably a homestead exemption for an owner-occupant. Other exemptions may be provided for those over a certain age, military veterans, and people having low income.
- Tax districts reappraise real estate periodically at market value. They may change the assessment upon a sale or make a general reevaluation.
- Tax districts provide a method for owners to appeal the assessment.
- If ad valorem taxes are unpaid, they become a lien on property that may be sold to satisfy the lien. Tax liens have priority over other liens, even liens that predate the tax lien.
- *Mills.* Property taxes are generally expressed as mills. A mill is a thousandth of a dollar. There are 10 mills in a penny. If the tax rate is 20 mills, the annual tax is 2 cents per dollar of value, which is 2% of the value. This is $2,000 for a $100,000 property.
- In some jurisdictions, the tax rate is applied to a fraction of market value, such as 40%. In that case, a $100,000 market value property is stated as *assessed* at $40,000 and the tax rate is applied to the assessment amount.
- Income taxes are applied by the federal government and most states.
- Homeowners who itemize tax deductions (rather than take the standard deduction) may deduct mortgage interest on new purchases of up to two homes, up to $750,000 of debt. Previous laws apply to debts incurred before the 2018 law limits. Itemized deductions on federal income tax returns for state and local taxes, including real estate taxes, are limited to a total of $10,000 beginning in 2018.
- Discount points on the purchase of a home may be tax deductible.
- If certain occupancy and ownership holding periods are met, up to $500,000 of the gain on the sale of a house by a married couple is not taxed; up to $250,000 of gain for an unmarried individual is not taxed.
- *Tax basis* is the original cost plus capital improvements. Adjusted tax basis provides for the subtraction of depreciation deductions claimed for income tax purposes on business or investment property. *Adjusted tax basis* is the starting point for determining gain or loss on the sale of real estate.

- For business or investment property, rental income is included in gross income; then ordinary and reasonable business expenses are deducted to determine annual taxable income or loss. Generally, interest expense is tax deductible in addition to tax depreciation expense.
- *Section 1031* is a portion of the Internal Revenue Code that provides an opportunity to defer taxes on the gain from the sale of property used in trade or business or held as an investment. A replacement property, which also must be used in trade or business or held as an investment, must be identified within 45 days and closed within 180 days of the sale of old property. If unlike property, such as cash, is received in the exchange, it is called *boot*. Tax must be paid to the extent of boot received but not more than gain.
- Most real estate brokers arrange for their salespersons to be independent contractors rather than employees. As independent contractors, the salespersons set their own work schedules and pay their own income taxes and social security (self-employment taxes apply).
- Economists classify taxes as *regressive, progressive,* or *proportional.* A regressive tax is one in which people having low income pay a higher proportion of their income than more affluent people. A sales tax on food is regressive because it requires those of low income to pay the same tax on a loaf of bread as a wealthy person pays. A *progressive* tax is one where those of high income pay a higher proportion of their income as tax. Generally, the income tax has a progressive structure. A *proportional* tax is one in which the tax burden is applied equally to high-income and low-income taxpayers.

Real estate, because it is valuable and also hard to hide, has been subject to tax for nearly all of human history. Consequently, it is easy for officials to assess taxes on real estate. The property tax is an ad valorem tax; that is, it is based on the *value* of the thing being taxed. The tax bill on a large, valuable property should be more than the bill on a small, relatively low-valued one in the same tax district.

The property tax usually is the major source of income for city and county governments. In a sense, this may be just, because the real estate that provides the bulk of the taxes collected also benefits greatly from many of the services local government pays for with the property taxes it collects. Most important of these is fire protection, which accrues almost exclusively to real estate. Police protection and the provision and maintenance of local streets and roads also provide obvious benefits. Nearby, locally provided public amenities such as schools and parks make many kinds of real estate more valuable. The control of real estate development through zoning and building codes also tends to preserve the values of existing real estate improvements, as does the provision of planning services. In addition, local courthouses store the records that are necessary for the documentation of title to real estate.

TAX ASSESSMENT

Real estate taxation is relatively straightforward, with some elements that should be studied carefully.

Because the tax is based upon the value of the real estate, some method must exist for determining the values of all parcels of real estate in the taxing jurisdiction. For this purpose, local governments usually have a *tax assessor's* office, although it often goes by a different name. The function of the office is to set a value on each parcel of taxable property in the jurisdiction. Sometimes this is done by actually appraising each property, but this method is very expensive except for the most valuable properties. More often, the assessment is arrived at by simpler means. Frequently the indication of value made when a property is sold will be used to determine the sale price, which, it is assumed, is very close to the true value. (In states that use documentary stamps or deed transfer taxes, the approximate sale price often can be deduced from the value of the stamps affixed to the deed or from the amount of transfer tax paid.) Often all assessments will be adjusted by some specific formula each year to take account of inflation or other reasons for overall value changes. Building permits serve to identify new construction and additions to existing buildings; aerial photography accomplishes the same objective.

More often than not, a single property will be in *more* than one taxing jurisdiction. Most commonly, a property may be in a particular county as well as in a specific city or town, and each may levy taxes. Many areas have independent school districts as well. Furthermore, there can be a multiplicity of other kinds of taxing districts, depending upon what state and local laws will allow and provide for. Usually, however, a single assessing agency will provide a valuation for all taxing jurisdictions that affect a given property; this is done because assessment is an expensive process and duplication of effort is thereby avoided.

Every jurisdiction provides some means whereby property owners can appeal assessments they consider unjust. The exact method of appeal varies considerably across the country. You should find out how it is done in your state.

EXEMPTION FROM PROPERTY TAXES

Not all real estate is subject to the real property tax; some of it is *exempt.* Exactly what is exempt and what is not is a matter decided by both state and local laws. Some kinds of property, however, are exempt almost everywhere. Primary among these is any property owned by government. This exemption seems to make sense: it would be redundant for the government to pay taxes to itself. Of course, there are several governments. Although it is conceivable that they could pay taxes to one another, they don't. Therefore, *all* government-owned property, no matter which government owns it, is exempt from property taxes. In some areas, where a very large proportion of the property is government owned, this situation can work a hardship on local taxpayer-owners of property, especially when the government-owned property receives many of the services paid for by local taxes. Examples of such "problem" areas are state capitals, Washington, D.C., and many small cities and towns that provide school and other services for large government (usually military) establishments. In such cases, the benefiting branch of government often provides some direct grant of money to the affected local government to make up for the unpaid property tax.

Some privately owned property may also be exempt from tax. Most areas exempt certain kinds of land uses that seem to provide other benefits to the community, making them desirable even when exempt from property tax. These uses usually include churches, nonprofit private schools, and some other charitable organizations. In some localities, *all* property owned by such organizations is exempt from tax; however, the current trend is toward granting the exemption only to property that actually is used by the organization for its basic purposes. In such an area, then, a church building used for religious purposes would be exempt but an apartment project owned by the church would not be—even if all its profits went to the church.

Tax increment districts are places where local governments reduce or abate taxes on certain improvements. If a certain tenant gains approval for a proposed improvement, taxes on the improvement are reduced. The improvement is expected to rejuvenate the area and bring in other investors with improvements that generate more tax revenues for the local government.

PAYMENT OF TAXES

Past-due property taxes automatically become a lien on the property so affected. If the taxes remain unpaid, the taxing government eventually can foreclose and have the property sold in order to pay the taxes. A tax lien takes priority over all other liens, including mortgages, even ones that have been entered on the records before the tax lien.

Normally, a property owner has a considerable period of time between the time the taxes are levied and the date payment is due. The specific schedule followed varies from state to state and sometimes even among localities within a state. As a matter of fact, a property owner whose land falls into more than one taxing district may find that the various districts have differing collection schedules.

A homeowner often is spared the chore of remembering when taxes are due because his mortgage lender requires him to make a supplemental payment into an escrow account with each monthly principal and interest payment. This money mounts up to create a fund from which the lender pays taxes (and usually insurance) when they are due. Generally the lender is contractually responsible to the borrower for paying the taxes out of these funds; if a mistake is made or a penalty is required for late payment, the lender can be held liable to the borrower.

CALCULATION OF PROPERTY TAXES

To calculate property taxes, three figures are necessary: the *appraised valuation* of the property, the *assessment ratio*, and the *tax rate*. The assessment ratio is applied to the appraised valuation to obtain the *assessed value*. Then, the tax rate is applied to the assessed value in order to find the tax due. In effect, the tax rate is a certain proportion, such as 7.66% of the assessed valuation, or $76.60 per $1,000 of valuation, or 76.6 mills per dollar of valuation. All three of these represent the same tax rate, expressed in different ways.

To determine the tax rate, it usually is easiest to turn the rate into a percentage figure, because that can be easily applied to the assessed valuation through simple arithmetic procedures. Rarely, however, is the tax rate expressed as a percentage; instead, it usually is given as a *millage rate*. A millage rate specifies how many *mills* must be paid for each *dollar* of assessed valuation. A *mill* is $^1/_{10}$ of a cent; therefore, there are 1,000 mills in 1 dollar. The millage rate shows how many *thousandths* of the assessed valuation must be paid in tax.

Millage rates are easy to convert to percents. Percents measure hundredths. To change a millage rate to a percentage rate, *the decimal point is moved one place to the left*. This is the same as dividing by 10 since there are 10 thousandths in each hundredth. Consequently, a millage rate of 77.6 is the same as 7.76%.

Once the tax rate is expressed as a percent, the next step is to determine the assessed valuation to which it must be applied. In some states and localities, the assessed valuation is the same as the market value of the property as estimated by the tax assessor. In others, however, only part of the market value is subject to tax. Sometimes the taxable portion is fixed by state law; in other localities, the local government determines the taxable portion of the estimated value. Often the figure so arrived at will be altered even further. Many states have a *homestead exemption,* which exempts a certain amount of the assessed valuation from taxation. Often older homeowners get special exemptions, as do low-income homeowners in a few areas. Once these exemptions are determined, they are subtracted from the assessed valuation to derive the net figure upon which the tax is based. However, a further element of confusion can enter the picture. Many properties are subject to tax by more than one jurisdiction. The method of calculating assessed valuation, and the exemptions that may be allowed, can vary among jurisdictions.

When the assessed valuation subject to tax has been determined, the tax rate is applied to it to determine the tax to be paid. For example, assume that a property has a taxable assessed valuation of $100,000 and the millage rate is 77.6, or 7.76%; then 7.76% of $100,000 will be the tax to be paid. This is found by changing the percentage into a decimal and multiplying: $0.0776 \times \$100{,}000 = \$7{,}760$. The tax, then, is $7,760. Typically, the effective annual tax rate is 1% to 3% of the property value.

INCOME TAXATION

Income tax laws are numerous and complex. Some highlights affecting real estate are indicated below, distinguished between those that affect personal residences and those that apply to income property.

Personal Residence Income Taxation

It is important to maintain permanent records concerning the acquisition of a personal residence. The initial cost serves as the basis for determining gain or loss, and that can stay with the owner through sales and purchases of subsequent homes.

HOME ACQUISITION DEDUCTIONS

The acquisition of a home gives rise to a tax deduction for discount points on a mortgage loan. Provided the discount points are for interest (not for services) and are customary in the area, they may be taken as an interest deduction. Since 1994, the Internal Revenue Service has allowed the deduction for discount points to a buyer even when the points have been paid by the seller. Points incurred to refinance a personal residence are treated differently from original debt. Points in a refinancing must be amortized (spread out equally each period over the life of the loan).

HOMEOWNERSHIP TAX DEDUCTIONS

Tax legislation passed in 2017 significantly affects homeownership. First, the *standard deduction* was almost doubled, becoming $12,000 for single individuals and $24,000 for married couples filing jointly. Taxpayers must have more than the standard deduction amount to benefit from itemization.

The itemized deduction for state and local taxes (SALT) is now limited to $10,000. Homeowners with combined property taxes and state income taxes of more than $10,000 will get no income tax deduction for the excess of SALT paid above $10,000.

Deductible mortgage interest is limited. New homebuyers may deduct interest expenses on up to $750,000 of debt. Prior buyers may continue to deduct interest expense up to $1 million of debt on two residences.

Deductions for interest expense on home equity loans are allowed when the loans are used to make substantial home improvements and are within the $750,000 total debt limit.

HOME SALE

A married couple may exclude up to $500,000 of gain on the sale of a principal residence. A single individual may exclude up to $250,000 of gain. There are limits as to the frequency of exclusion and requirements for a principal residence.

Income Taxation of Income Property

The purchase of income-producing real estate ordinarily does not generate immediate income tax deductions. A buyer should indicate how much was paid for land, how much for real estate improvements, and how much for personal property. Except for land that is not depreciable, depreciation for real estate improvements and personal property generate tax deductions while not requiring cash payments.

OPERATIONS OF INCOME PROPERTY

Rental income is included as business income. Expenses of business operations are tax deductible. These include not only interest and ad valorem taxes but also maintenance and repairs, management, insurance, and all other reasonable and necessary expenses to operate the property. Depreciation is a noncash expense that serves to reduce taxable income. Because depreciation provides a deduction without a cash payment, it is considered very favorable for owners. Under present tax law, apartment owners may depreciate their property (not land) over a 27½-year term. Owners of other property types, such as office buildings, warehouses, shopping centers, and so on, must use a 39-year depreciable life. Depreciation deductions reduce the tax basis for property, which could cause a taxable gain upon resale.

RESALE

The adjusted tax basis is subtracted from the resale price of real estate to derive the taxable gain. The adjusted tax basis is original cost plus capital improvements less accumulated tax depreciation. This gain is generally taxable at the same rate as a capital gain. However, certain types of depreciation tax deductions may be taxable as ordinary income upon a sale.

EXCHANGE

Under Section 1031 of the Internal Revenue Code, there are opportunities to postpone or defer the tax on a gain. One must identify potential replacement property within 45 days of the sale date of the previously owned property and close within 180 days of the sale. The old and new properties must be held for use in a trade or business, or held as an investment. The realized gain will be taxable to the extent that the seller receives unlike property, including cash, which is called *boot.*

Trick Question

A trick question tries to confuse property taxation with income taxation. Property taxation is typically imposed by local governments. It is used to fund police and fire protection, schools, libraries, construction and maintenance of public roads and parks, and other community benefits, including government administration. These taxes are called ad valorem taxes because they are based on the value of real property. In contrast, income taxation is applied by the federal government and 45 states. It is based on income earned by individuals and businesses.

OTHER TAXATION ISSUES

Economic Impact of Taxation

Economists classify taxes in relationship to the impact on ability to pay. A tax is "regressive" when those with low income pay a higher proportion of their means than others. For example, poor people pay the same tax on a loaf of bread as the wealthy. A tax is "proportional" when it taxes proportionally to taxpayer means. Ad valorem property taxes are considered proportional because most people buy housing according to affordability. A tax is "progressive" when the wealthy pay a higher rate. Income tax rates increase with higher income, so the income tax is progressive.

Independent Contractor vs. Employee

An independent contractor works for himself/herself, pays self-employment taxes, and pays estimated income taxes.

By contrast, an employee pays social security tax. From his or her paycheck, the employer deducts income taxes, social security tax, and other withholdings. The employer must also contribute to social security and other programs.

Most real estate brokers prefer to treat salespersons as independent contractors. Real estate brokers may do so provided they don't exercise certain controls, such as what hours the person works, ownership of tools and supplies the person uses (vehicles, computer), or payment of dues.

Marginal and Average Income Tax Rates

An *average income tax rate* is the amount of income tax divided by total income. For example, someone who pays $25,000 in tax and has a total income of $100,000 has a 25% average tax rate. However, if that person earns $1,000 more and as a result owes $300 more tax, the *marginal tax rate* is 30%. The 30% marginal tax rate in this example is the tax rate on the next dollar earned.

Alternative Minimum Tax

A 26% or 28% tax is imposed on total income that adds back certain tax-favored income plus certain tax-deductible expenses. The taxpayer must figure taxes two ways and pay the tax that is greater. The two ways are called *regular* and *alternative*.

CHECKLIST FOR YOUR STATE'S AND LOCALITY'S PROPERTY TAXATION PRACTICES

Obtain the answers to the following questions from the literature supplied by your state's licensing agency or from the local tax assessor's office in your area.

What proportion of the estimated market value is subject to tax? ________________%

Are any of the following exemptions applicable to the tax base?

- ☐ Homestead exemption
- ☐ Old-age exemption
- ☐ Exemption for the poor
- ☐ Military veterans
- ☐ Other(s): ________________________________

Is it possible for a property to be subject to taxes by more than one governmental body?

- ☐ No; taxes are levied only by ________________________
- ☐ Yes; these include:
  - ☐ State government
  - ☐ County (parish) government
  - ☐ City (township) government
  - ☐ School district
  - ☐ Special assessment district
  - ☐ Improvement district
  - ☐ Other(s): ________________________________

QUESTIONS ON CHAPTER 16

1. An ad valorem tax is based on
   (A) income earned
   (B) the value of the thing being taxed
   (C) the size of the thing being taxed
   (D) something other than A, B, or C

2. The property tax is a form of
   (A) sales tax
   (B) ad valorem tax
   (C) income tax
   (D) excise tax

3. The major source of income for local government usually is
   (A) income taxes
   (B) licenses and fees
   (C) property taxes
   (D) parking meters

4. Which of the following is (are) always exempt from real property taxes?
   (A) income-producing property owned by a church
   (B) government-owned property
   (C) most restaurants
   (D) a private, for-profit school

5. A millage rate of 84.5 mills is the same as
   (A) $84.50 per $1,000
   (B) 8.45%
   (C) both A and B
   (D) none of the above

6. If a property is assessed at $60,000 and the millage rate is 32.5, the tax is
   (A) $19.50
   (B) $195.00
   (C) $1,950
   (D) $19,500

7. The appraised valuation of a property is $85,000. The property tax is based on 20% of appraised valuation. The city tax is 50 mills; the county tax is 40 mills. Which of the following is true?
   (A) The city tax is $1,850.00.
   (B) The county tax is $850.00.
   (C) The city and county taxes combined add up to $1,530.00.
   (D) The city and county taxes combined add up to $7,650.00.

8. A lien for unpaid property taxes
   (A) can be sold at auction
   (B) takes priority over all other liens
   (C) cannot exist unless taxes are at least 36 months overdue
   (D) is a form of adverse possession

9. Which of the following is *not* a good reason why real property taxes are so effective?
   (A) Real estate ownership is easy to hide.
   (B) Real estate is valuable.
   (C) Real estate is easy to find.
   (D) Real estate can be foreclosed to provide payment of unpaid tax levies.

10. Real property taxes are justifiable because
    (A) real property benefits from many services provided by property taxes
    (B) real property owners are wealthy and can afford to pay taxes
    (C) real property taxes are legal everywhere, whereas many other kinds of taxes are not
    (D) none of the above

11. A tax-deductible front-end interest cost on a mortgage loan or deed of trust for a newly purchased home is called
    (A) interest abatement
    (B) discount points
    (C) interest in advance
    (D) prepaid mortgage insurance

12. A tax deduction for business property that requires no cash outlay is
    (A) interest expense
    (B) vacancy loss expense
    (C) depreciation expense
    (D) management expense

13. A Section 1031 exchange is best described as
    (A) tax exempt
    (B) taxable
    (C) tax deferred
    (D) tax free

14. The initial cost of a business asset, increased by capital expenditures and reduced by tax depreciation expense, from which gain or loss is measured, is
    (A) basic adjustment
    (B) basal metabolism
    (C) adjustment of base
    (D) adjusted basis

15. The federal income tax exemption on the gain from the sale of a personal residence held long enough by a married couple is up to
    (A) $25,000
    (B) $125,000
    (C) $250,000
    (D) $500,000

16. The federal income tax exemption on the gain from the sale of a personal residence held long enough by a single individual is up to
    (A) $25,000
    (B) $125,000
    (C) $250,000
    (D) $500,000

17. Cash or other unlike property included to equalize values in a tax-free Section 1031 exchange is called
    (A) slipper
    (B) equity
    (C) parity
    (D) boot

18. To qualify as a Section 1031 exchange, the property must be
    (A) a personal residence
    (B) used in a trade or business or held as an investment
    (C) land
    (D) mortgaged

19. Federal income tax rates are generally
    (A) regressive
    (B) neutral
    (C) progressive
    (D) none of the above

20. Federal income tax deductions for a principal residence include
    (A) mortgage interest on up to $750,000 of debt
    (B) real estate taxes up to $10,000 annually
    (C) uninsured casualty losses, over a certain amount
    (D) all of the above

21. A homeowner, on his or her federal income tax return, may not deduct
    (A) fixing-up expenses
    (B) cost of a new porch
    (C) repainting
    (D) any of the above

22. The report of a sale to the IRS is normally made on Form 1099 by
    (A) buyer
    (B) seller
    (C) broker
    (D) closing agent

23. In a tax-free exchange under Section 1031 of the Internal Revenue Code, *boot* is
    (A) "unlike" property that may be taxable to the recipient
    (B) like-kind property
    (C) given by a TIC
    (D) an equity kicker

24. Harold bought 20 acres for $125,000 and sold the land three years later for $230,000. On his income tax return he reports
    (A) short-term capital gain of $105,000 in the year of sale
    (B) long-term capital gain of $105,000 in the year of sale
    (C) ordinary income of $105,000 in the year of sale
    (D) annual income of $35,000

25. A principal difference between an independent contractor and an employee is
    (A) dress code
    (B) income taxation
    (C) fiduciary relationship
    (D) public service

26. Indications of independent contractor status include
    (A) freedom of hours to work
    (B) use of own vehicle
    (C) payment of own professional dues
    (D) all of the above

27. The business relationship between broker and salesperson for income tax purposes is usually
    (A) employer-employee
    (B) independent contractor
    (C) single agent
    (D) self-employment

28. The term most closely associated with real estate taxes is
    (A) ad valorem
    (B) appraised value
    (C) market value
    (D) net income tax

29. The assessment ratio can be used to
    (A) determine the tax rate
    (B) convert market value to value in use
    (C) convert market value to assessed value
    (D) determine property tax exemptions

30. Under Internal Revenue Code Section 1031, tax on the profit of an exchange is
    (A) reduced
    (B) deferred
    (C) eliminated
    (D) mitigated

31. Mr. and Mrs. Martin jointly bought a principal residence in 2014 for $325,000 and sold it in 2020 for $500,000. What portion of their $175,000 gain is subject to capital gains tax?
    (A) $0
    (B) $325,000
    (C) $500,000
    (D) $175,000

32. Which of the following is *not* tax deductible for homeowners who itemize tax deductions?
    (A) interest expense
    (B) real estate taxes
    (C) discount points
    (D) utilities

33. When a salesperson's agreement with a broker is to be an independent contractor, the broker
    (A) withholds federal taxes and social security tax
    (B) withholds only social security
    (C) withholds taxes only when they exceed $600 in a year
    (D) does not withhold federal taxes or social security

34. Under the income tax changes passed in 2017, the annual limit of deductible state and local taxes is
    (A) $0
    (B) $10,000
    (C) $24,000
    (D) no limit

35. Under the income tax changes passed in 2017, the maximum amount of debt that can generate tax-deductible interest on new housing purchases is
    (A) $500,000
    (B) $750,000
    (C) $1,000,000
    (D) no limit

36. As a result of the income tax changes passed in 2017, the increased standard deduction has what effect on homeownership for income tax purposes?
    (A) makes it more attractive
    (B) makes it less attractive
    (C) has no effect
    (D) makes it more attractive for married couples only

ANSWERS

1. **B**	7. **C**	13. **C**	19. **C**	25. **B**	31. **A**
2. **B**	8. **B**	14. **D**	20. **D**	26. **D**	32. **D**
3. **C**	9. **A**	15. **D**	21. **D**	27. **B**	33. **D**
4. **B**	10. **A**	16. **C**	22. **D**	28. **A**	34. **B**
5. **C**	11. **B**	17. **D**	23. **A**	29. **C**	35. **B**
6. **C**	12. **C**	18. **B**	24. **B**	30. **B**	36. **B**

Real Estate Investment

17

ESSENTIAL MATTERS TO LEARN FROM THIS CHAPTER

- To be a successful real estate investor, an individual needs to understand how to select property, when to buy and sell, and how to finance and manage real estate.
- Benefits of investing in real estate include:

  1. Cash returns or *cash flow.*
  2. *Appreciation* in value.
  3. *Income tax advantages,* called *tax shelter.*

- Benefits of investing in real estate can be enhanced by borrowing, called *financial leverage.* Leverage adds *risk.*
- Drawbacks of investing in real estate include:

  1. *Large size* of each investment.
  2. *Illiquidity* or a lengthy period needed to sell.
  3. *Management effort* required.
  4. *High transaction costs.*
  5. *Immobility.*
  6. *Government intervention.*
  7. Various *risks.*

- *Investment analysis* is a process of deciding whether a particular investment opportunity is likely to be a good choice.
- A *pro forma* statement of operations is prepared to estimate future performance of a real estate investment.
- Investment criteria are calculated to evaluate real estate investment opportunities.
- The *gross rent multiplier* (GRM) is the selling price divided by the gross income. A selling price of $1 million divided by gross income of $200,000 provides a GRM of 5.0. The lower the GRM, the less a buyer is paying per dollar of gross income. However, the GRM doesn't account for operating expenses. Operating expenses vary by property.
- The *overall rate of return* (OAR) is the percentage relationship between *net operating income* (NOI) and selling price. NOI considers rental income minus operating expenses and thus is more refined than GRM. Real estate investors seek a high OAR.
- The *cash-on-cash return,* known as the *equity dividend rate,* deducts financing expenses from NOI. It compares the cash flow to be received from the property to the amount of equity that must be invested. Therefore, it considers the effects of borrowed money on an owner's rate of return.
- A real estate owner needs to recognize that there are three distinct phases in the life cycle of the investment. It begins with the purchase, continues with the operations, and ends with a sale.

- Financing real estate is crucial to achieving a high rate of return. However, it entails risk.
- Important financing issues include:

    1. Interest rate.
    2. Extent of *leverage*, or *loan-to-value ratio.*
    3. *Amortization* term.
    4. *Prepayment privileges* or penalties.
    5. *Exculpation* or *personal liability.*
    6. *Assumability.*
    7. *Call* or *acceleration.*
    8. *Subordination* to other financing.

- Various property types have different investment characteristics, attractive features, and risks. Major property types include:

    1. Apartments.
    2. Office buildings.
    3. Industrial property.
    4. Shopping centers (retail properties).

- Location is an important attribute. Characteristics that make a location favorable will differ according to the property type.
- Sources of data for real estate information include:

    1. Board of Realtors.
    2. Tax assessor offices.
    3. University real estate research centers.
    4. Private (proprietary) data sources.
    5. Property owners.

- It is important to receive both a return *on* investment and a recovery *of* investment.
- A *return on investment* refers to earnings. The rate should be adequate to account for risk, illiquidity, and investment management required by real estate.
- A *recovery of investment* is considered to ensure that the money received has accounted for a recapture of the cost of a wasting asset.
- Income tax considerations include the tax deduction allowed for *tax depreciation.* This is a noncash expense that is tax deductible, which is quite favorable during the *operations* phase of real estate ownership.
- Depreciation tax deductions reduce the owner's *adjusted tax basis.* The *selling price* minus the *adjusted tax basis* is the measurement of gain or loss on a sale. A tax-free exchange may be used to defer taxes on the gain of a sale.

FUNDAMENTALS OF REAL ESTATE INVESTMENT

Real estate has been the path to riches for many investors. In addition, it has provided security and income to those who had the foresight to build an estate. Success in real estate investing requires knowledge, capital, and the willingness to take risks. An investor must know how to select properties, when to buy and sell, and how to finance and operate a rental property.

Real estate has proven itself to be a lucrative investment through the years. There are many ways to invest in real estate, and each offers distinct benefits. Almost two-thirds of Americans own their own homes. This basic level of real estate investment frees them from dependence on a landlord and hopefully provides financial rewards in the form of value appreciation. In addition to their homes, many people have acquired small holdings of income property. Often self-managed, these properties provide periodic income, some of which is tax sheltered, and opportunity for appreciation gains.

Investors may put their money into a real estate investment trust (REIT). This is like a mutual fund for real estate. Through such entities, investors can own a part of large properties or diversified holdings. This form also provides the advantages of professional property and investment management.

Investment Advantages

Regardless of the form of investment, investors look to real estate for several advantages. Real estate often offers higher *cash returns* than alternative investments. In part, this can be attributed to greater risk and difficulty of resale (illiquidity) associated with the property. Returns may be increased by using borrowed money, called *leverage.* Real property tends to be a good hedge against inflation since rents may be increased as costs rise. Well-located properties may even beat inflation and offer resale profits from *appreciation* in value. Finally, *tax shelter* has traditionally been an attraction of real estate investment. Opportunities for tax shelter have been diminished by tax law changes. However, in some cases, tax benefits still exist. Each of these attractions are described in greater detail in this chapter. Real estate investments cover a wide range, from the relatively secure to the highly speculative. In general, the most secure investments emphasize current income production from highly creditworthy tenants. Speculative investments emphasize appreciation potential from future developments.

Investment Drawbacks

There are serious drawbacks to real estate investing. One factor that limits the field of real estate investors is the *size* of the required investment. A down payment on even a modest rental home requires several thousand dollars. Purchasing a sizable commercial property or assembling a diversified portfolio of properties means a much larger investment, certainly beyond the means of the typical investor. *Syndicators* overcome this problem by gathering a group of investors, each of whom purchases shares of a large property or inventory of properties. Partnership interests may be acquired for as little as $5,000. Furthermore, the investor may buy into several partnerships to gain diversification.

Real estate is an *illiquid* investment. This means that once the investment is made, it is difficult to get money out quickly. It takes at least several weeks, and sometimes months, to sell a property. Market conditions may prevent a sale at a reasonable price. If liquidity is desired, you may purchase shares in a publicly traded syndication, or a REIT. Many of these are traded on the major stock exchanges and offer immediate conversion to cash.

Since real estate is a physical asset, it requires ongoing *management.* Purchase of a property is like buying a small business. The owner is responsible for maintaining the property and keeping it productive. Professional property managers or resident managers can be hired to take on many of these responsibilities. The owner still must decide on tenant selection, rental rates, marketing, alterations, and when to sell.

Real estate markets are local in nature. There is a shortage of available information about current prices, rental, and vacancy rates. The typical purchaser is in the market infrequently and cannot stay attuned to current conditions. Therefore, it is difficult to decide on the best offering price for a property or rental rate to charge.

Aside from the size of the investment, the costs of purchasing a property are increased by transaction expenses. These include sales commissions, attorney fees, title insurance, appraisal and surveying costs, loan fees, and transfer taxes. A seller may pay as much as ten percent of the price of the property in transaction costs. In addition, the buyer may pay five or more percent. These costs tend to make real estate a long-term investment. Short-term profits would have to be substantial to offset transaction costs.

Real estate is *immobile*. For the investor, this means the performance of the property is determined by local market conditions. The demand for the property may be depressed by an economic downturn or an abundant supply of similar properties. The property cannot be moved to another location where conditions are better. In addition to local market conditions, a property is affected by the quality of the neighborhood and what happens to surrounding properties. Therefore, location is a major factor in a property's performance. Furthermore, these conditions may change over time. A big part of investment analysis involves evaluating trends in the local market and surrounding area.

What you can do with your property is subject to government policies. On a broad scale, government influence on interest rates and economic trends will affect demand for the property. On a local scale, government can limit your use of the property through zoning ordinances, building codes, health codes, and environmental laws. In some places, a residential project may come under rent controls. Decisions to extend streets and utilities can change the attractiveness of your property. For these reasons, real estate investors often take an active interest in the affairs of their local government.

INVESTMENT ANALYSIS

Investment analysis is deciding if a particular investment opportunity is a good choice. A projection of expected investment returns is helpful in making the decision. Since the return from real estate investments depends on future rent, operating expenses, financing, and tax considerations, an estimation of what these will be will help you estimate how the property will perform. Projections may be the main determinant of the decision or may supplement other sources of information. The important thing to remember is that projections are only as good as the assumptions used to make them.

Pro Forma Statements

Pro forma statements show expected returns from operating a property. One of the most important results is *cash flow*, the money produced by an investment that you get to keep. Basically, this amounts to the cash, if any, left over after all expenses have been paid. Expenses include those to operate the property and to meet all loan requirements.

Cash flow represents the money produced by operating the investment. It does not reflect any appreciation gains that may eventually be received at resale nor equity buildup due to mortgage amortization. Since these types of return are realized only in the future, cash flow is a measure of the current performance of the investment. For that reason, some investors base their decision to acquire a property on a simple *cash-on-cash* return. This measure is cash flow divided by the required equity investment. For example, a property that is expected to produce a cash flow of $10,000 per year and requires a cash investment of $100,000 offers a cash-on-cash return of 10%.

The following is an example of how to calculate the cash flow from a real estate investment.

STATEMENT OF OPERATIONS

PRO FORMA

Total rental income if building is full (Potential Gross Income)		$200,000
Less, allowance for vacancies and bad debts		– 10,000
Effective Gross income		$190,000
Operating expenses:		
Management	$ 10,000	
Maintenance	25,000	
Utilities	7,000	
Property taxes	15,000	
Insurance	3,000	
Repairs	10,000	
Total operating expenses		– 70,000
Net operating income		$120,000
Debt service (mortgage payments):		
Principal	$ 5,000	
Interest	80,000	
Total debt service		– 85,000
Before-tax cash flow		$ 35,000

A property seller or broker may offer a *pro forma statement of operations,* such as the one above, as part of a sales presentation. It is a projection of what may be expected to occur. The rental income may be attractive estimates rather than actual amounts earned by the property. Alternatively, operating expenses may be based on last year's rates while property taxes, insurance, or utilities have increased in the current year. You may have to recast the amounts to determine what you are likely to incur in expenses or receive in cash flow. Get the owner's rent roll, which is a list of the tenants, space occupied (i.e., apartment number), and rent, and compare the figures to the pro forma statement. You can also look at actual operating statements, not forecasts, to be assured you are not deceived.

It is often useful to separate expenses that are fixed from those that are variable. This can inform you of the effect on cash flow caused by a change in occupancy. A *fixed expense* is the same amount regardless of the vacancy rate or rental rate. Generally, casualty insurance and real estate taxes are fixed in the short run. Maintenance, repairs, and utilities vary with occupancy, so they are *variable expenses.* An investor should always aim for rental income to cover at least the fixed expenses because a drop in occupancy won't result in a corresponding drop in fixed expenses.

Expected cash flow can be an indication of how speculative an investment is. When appreciation rates are high, it is not uncommon for properties with negative cash flow to be purchased. The investor expects to "feed" the property (pay the cash flow deficit out of other income or capital reserves) until the property is sold. An expected large profit at resale offsets the period of negative cash flow. In other cases, investors may purchase properties with turnaround potential—those that are currently losing money but that may become productive given renovation or marketing effort. These properties are expected to have negative cash flow for a time but eventually produce positive income and resale profits. In any case, where the cash-on-cash return is low or negative, the investment is speculative in nature. Eventual returns are based on anticipated improvements in market conditions or the appeal of the property, and this introduces risk that the improvement will not materialize.

Short-term analysis can be used to screen properties. The following indicators may be calculated quickly with a few bits of information that are based on current performance.

GROSS RENT MULTIPLIER

This is the price of the property divided by the gross rent. The lower the number, the less you are paying for the gross income. However, the gross rent multiplier doesn't consider operating expenses, which vary for each property.

OVERALL RATE OF RETURN OR CAPITALIZATION RATE

This is the net operating income (rent less operating expenses) divided by the value of the property. This measure is preferred to the gross rent multiplier because it accounts for operating expenses. Also, it offers a percentage rate, which is a customary way to express rate of return.

CASH-ON-CASH RETURN OR EQUITY DIVIDEND RATE

This is the expected cash flow divided by the required equity investment and is especially useful when you know how the property will be financed. You may use before-tax or after-tax cash flow. If after-tax cash flow is used, you should adjust the returns on alternative types of investment for income taxes to make a meaningful comparison.

Long-term analysis requires more information and assumptions since you are projecting several years into the future. If you are considering a property whose value is likely to increase or decrease in the near future, or if inflation is significant, long-term analysis is worthwhile. The analysis requires that you project growth rates for rents, operating expenses, and property value. Changes in debt service and taxes are also considered. The result of the analysis can be a measure of yield to maturity called the *internal rate of return*. This is the projected annual rate of return over the entire holding period. It can be used to compare other investments with the same holding period. Alternatively, you may calculate the *net present value* of the investment based on your required rate of return. This is often useful when comparing investments that require different levels of cash investment or when putting together a portfolio of several investments. Fortunately, there are computer programs available to perform long-term investment analysis.

Trick Question

A trick question confuses a return *on* investment with a return *of* investment. When you purchase any kind of investment, you expect some kind of profit. This is the return *on* investment. It is represented by a growth in value or a payment to investors without a loss in value. However, you also expect a return *of* investment. In other words, you expect to get back your initial investment intact. If you remove some of the asset value over time (a partial return of investment), you also likely reduce the return *on* investment. This is the case with an amortizing mortgage loan, for example. In such a loan, each total monthly payment includes a return *of* a portion of the principal, called *amortization*, plus the interest due each period, which provides a return on investment to the mortgage investor. Think of a return *on* investment as the fruit from a tree and a return *of* investment as the tree being reduced in size or ultimately being sold.

INVESTMENT STRATEGY

Investment Life Cycle

Like most things, a real estate investment has a beginning and an end. You purchase a property (or share in a partnership) and later sell it, hopefully for a profit. In between, the property may provide income if well managed. This middle period of operation may be long or short, depending on market conditions and the reasons for buying the property. Because of high transaction costs, most investors end up holding real estate for several years.

There are important considerations at each stage of the life cycle of the investment.

Purchase

The investor is concerned with getting the right property at a good price and with favorable financing terms. The property should fit the investment objectives. If cash flow is desired, pick a property with proven ability to produce cash flow. If you want appreciation, pick a property with potential for increased value. The price you pay will have a lot to do with appreciation. If you pay above market value, the property will have to appreciate in value for you to break even. Financing may determine whether the property makes money or even whether you can keep the property. You must decide how much leverage to use, knowing that high leverage increases possible returns on equity as a percentage rate but also reduces current cash flow amounts.

Operation

During this phase, the property should produce some cash flow. This is the income to you as an investor. Some properties have negative cash flow, which means you must pay some expenses out of your pocket. A part of cash flow may be due to income tax benefits. Depending on your tax situation, these benefits allow you to shelter some nonproperty income from taxes. Remember that cash flow may vary from year to year, and the amount depends on your ability to retain rent-paying tenants and control operating expenses.

Resale

The main benefit from a resale is capital gain, the profit on the resale. To make a profitable resale, you will want to maximize the sales price and minimize income taxes on the sale. The highest price depends on the timing of the sale. You want to sell when the market is strong. Sometimes, a better price can be had if you are willing to help finance the purchase. To limit your exposure to taxes, you may wish to structure the sale as a Section 1031 exchange. If properly done, an exchange can defer payment of taxes.

The key to a successful investment is to consider all of the income expected through the cycle, and its timing, to determine whether it is worth the cost.

AVAILABLE FINANCING

Arranging the best financing can be as important as negotiating the best price for a property. Before you talk to a lender, however, you should understand how the various terms of a loan can affect your investment.

Interest Rate

Interest is a charge for the use of money. The lower the rate, the more attractive the loan. The degree of leverage is also important. Many investors will pay a significantly higher interest rate in order to enjoy a lower cash down payment and therefore be able to exercise more leverage. Some equity investors will pay higher prices for property to achieve more leverage. Sellers sometimes accommodate by providing financing to buyers or by paying *discount points* on a buyer's loan, but they will try to raise their price to compensate. Consequently, prices are generally increased by favorable financing.

Degree of Leverage

When interest rates are low, the loan amount is usually limited by the *loan-to-value* (L/V) ratio that a lender offers. When rates are high, the *debt coverage ratio* (DCR) usually sets the maximum loan amount. For conservative lenders, the L/V ratio is usually in the range of 65% to 70%. Other lenders may use L/V ratio criteria of 75% to 80%. A conservative DCR is around 1.25.

Amortization Term

The longer the amortization term of a mortgage loan, the lower the annual debt service and therefore the greater the annual cash flow. However, once loan terms are stretched beyond 30 years, the annual mortgage payment requirement declines but the change is quite small. Annual debt service can never become less than the interest due on the principal balance. Many loans on income property become due years before being fully amortized. These are called *balloon payment* loans.

Prepayment Privileges

In this circumstance, the borrower has the right to pay off the remaining debt before it is due. Some loans include penalties for prepayment, although usually the penalty declines with time. For example, the penalty may be 3% if the loan is prepaid during the first five years, 2% for the next five years, and then declining by 0.5% for each of the following years until there is no penalty. Some mortgage loans are locked in for the first five to ten years; prepayment without the lender's consent is prohibited.

Exculpation

Exculpation is freedom from liability. When a mortgage loan includes an exculpatory clause, the property is the sole collateral for the loan. Should the property be foreclosed, the lender can look only to the property for full satisfaction of the debt, not to other property you own. In the absence of exculpation, you are usually personally liable for the debt. *Nonrecourse* is another term used to describe mortgages with exculpatory clauses; the lender's only recourse in the event of borrower default is to the property itself, not to your personal assets.

Assumability

This right allows the borrower to transfer the mortgage loan with the property to another party. Since the new owner may not be as good a credit risk as the original owner, lenders often reserve the right to approve a loan assumption. Some lenders are willing to approve changes routinely, provided that they can *escalate* (increase) the interest rate on the debt. Such provisions in mortgage loans make assumption privileges less attractive to a buyer. Despite lender approval of changes in assumption, original borrowers remain liable for the mortgage debt. So if there is a default on an assumed loan, sellers are liable, but they can try to collect from their buyer. By contrast, if the buyer took the property "subject to" the loan but did not assume it, the sellers (who were the original borrowers) remain solely liable for default.

Call or Acceleration Provisions

These provisions effectively permit balloon payments on lender demand. Although the loan has a 25- or 30-year term, lenders, at their option, can accelerate payment of the principal after 10 or 15 years, regardless of whether the loan is in default. This obviously puts the lender in a strong position. Upon reaching the "call" date, the lender can force repayment or escalate the interest rate.

Subordination

Subordination is moving a mortgage loan to a lower priority. The priority of a lender's claim is where the lender stands in line for collection after a foreclosure. Undeveloped land financed by a seller with a first mortgage and now ready for development is not acceptable collateral for a construction lender. Construction lenders want a first lien. If the holder of an existing mortgage is willing to subordinate, the existing mortgage will be reduced to a second lien position to allow a first lien for the new mortgage. If a subordination clause will be needed, you should attempt to insert it initially; lenders will not reduce collateral later. Subordinated loans are more favorable to borrowers because these loans increase the flexibility of first-mortgage financing.

Property Types and Investment Characteristics

Real estate comes in many different forms and can be leased or financed in so many different ways that the lease or mortgage will alter its investment characteristics. These characteristics vary from very passive investments to management-intensive ones and from the lowest to the highest levels of risk. Principal property types include apartments, office buildings, industrial properties, and shopping centers.

Classes of Property

Commercial property is often classified in the market by quality, age, and/or size.

Shopping centers, from smallest to largest, are as follows:

- Strip
- Neighborhood
- Community
- Regional
- Superregional

Most successful shopping centers require *anchor tenants* because they attract smaller tenants and customers. An anchor tenant for a neighborhood shopping center is typically a supermarket with a pharmacy attached. A community center may be anchored by a junior department store, and a regional mall may have a department store as its anchor.

Office buildings, from largest and most prestigious to oldest and smallest, are as follows:

- Trophy
- Class A
- Class B
- Class C
- Class D

Large law firms, investment brokers, and banks often occupy trophy buildings. Smaller, older buildings are often occupied by businesses that depend on affordability and services.

Apartments are generally divided in the market as:

- Class A
- Class B
- Class C
- Class D

The classification is generally based on age, size, and amenities provided.

Industrial properties are often classified by the letters A, B, C. Most important, however, is their utility. Ceiling height (16, 18, 22, or 30 feet) may be very important; floor load capacity (weight), loading docks, heavy utility capacity, and truck turnaround opportunities are also important.

Market classifications are generally based on physical attractiveness and amenities. This type of classification differs from the same system—A, B, C, D—used for insurance/fire ratings. Don't confuse fireproof ratings with market attractiveness ratings.

Location

An expression in real estate is that the three key aspects of success are "location, location, location." Very similar buildings can have very different values depending on where they are located. In fact, the value of undeveloped land is almost totally determined by where it is located.

The reason location is so important is that real estate is more than just space in an economic sense. Its surroundings have a lot to do with how useful the property is. One big reason to buy or rent a piece of property is to be near other activities. The type of location desired varies with the type of property.

For *residential properties*, people will want to be relatively close to their job and shopping. But not so close that there is a lot of traffic in their neighborhood. Being in a good school district is important to many home buyers. The neighborhood should be pleasant and well-cared for. A prestigious address can add a great deal of value. If the property has apartments, being on a bus route might be important.

Stores and shopping centers need visibility and traffic flow. The property should be near areas where shoppers live or work, depending on the type of stores. Big stores should not be too close to their competitors. However, small stores often like to be near big stores, hoping to attract shoppers coming and going to the larger store.

Office buildings should be close to business activities. Lawyers want to be near courthouses or government offices; consultants want to be near corporate headquarters or banks; doctors need to be near hospitals or nursing homes; and other businesses need to be near customers. Because of the necessity to be close together, downtown areas used to be the best place for offices. Today, suburban locations may be just as valuable because of improvements in communications.

Industrial properties should be close to the source of their materials and labor. They have great need for good transportation, with heavy industry often locating on rail lines or near seaports. A good highway is almost a necessity.

In real estate, the best locations get the top dollar. As an investor, you must look at more than the property alone. Remember, however, that the quality of a location can change over time. Sometimes, a lesser location can become the best place to be.

SOURCES OF MARKET DATA FOR INCOME AND EXPENSES

In addition to the sources listed in Chapter 14, the following may be helpful.

LOCAL SOURCES OF REAL ESTATE DATA

Local organizations often collect real estate data, usually for membership use or sale to interested parties. These include the following:

- Local Board of Realtors/multiple listing service
- Local tax assessor's office
- Local credit bureaus and tax map reproduction firms
- University research centers
- Private brokerage or appraisal firms

LOCAL BOARDS OF REALTORS

Most metropolitan areas have a Board of Realtors, or other broker group, that sponsors a multiple listing service (MLS). Upon the sale of property listed through MLS, the broker must supply information about the completed transaction. Each property sold and its terms of sale are therefore available on computer or in a published book. Some REALTOR® boards provide information to members only; some share with other real estate organizations.

LOCAL TAX ASSESSOR'S OFFICES

Assessor's offices usually keep a file on every property in the jurisdiction, with property characteristics, value estimate, and how the estimate was derived. Many jurisdictions are notified of every sale or building permit for immediate updating of affected properties.

UNIVERSITY RESEARCH CENTERS

University research centers, many of which are sponsored by broker and salesperson state license fees, may have aggregated data on real estate transactions collected from other sources throughout the state. The data are often helpful in identifying trends established over time, and by city, for various types of property. Additional research and educational information may be available from such centers.

PRIVATE DATA SOURCES

Many real estate appraisers and brokers retain file data for property owned by their clients and others. They will often share information, usually for reciprocity rather than payment.

PROPERTY OWNER

Property owner's records include permanent records such as deeds, leases, and copies of mortgages (lenders have the original mortgage document) as well as periodic accounting and tax return information about the property's recent past. An owner's report may be of limited immediate use, however, because it may be disorganized, have extraneous information, or be arranged poorly for appraisals. Also, data from a single property cannot offer a broad perspective on the market.

RETURN ON INVESTMENT

Return on investment is the actual earnings from the investment; this is apart from any returns that represent repayment (called amortization) of the principal invested. The difference is like distinguishing the fruit from the tree on which it grows. Both can be sold for money, but selling the tree provides a one-time lump sum, a different type of income than would picking the fruit each year.

In real estate, it is sometimes difficult to determine how much of income is return *on* investment and how much is a return *of* investment. For example, if you rent a property that depreciates in value, some of the rental income must go toward significant repairs or the eventual possibility of replacing the building when it becomes useless from age or obsolescence.

Why is this important? It makes a difference in evaluating the performance of the investment. The return on investment determines how well your money is invested. Return of investment affects risk of capital. The sooner your investment is recovered, the less risk there is of losing it.

Measures of return on investment include the *equity dividend rate,* based on a one-year analysis, and the *internal rate of return,* based on a multiyear projection (see DISCOUNTED CASH FLOW) analysis. The rate of return you get includes the following.

A SAFE RATE

This is the rate you could get if you put your money into a perfectly safe, liquid investment, such as a federally insured passbook savings account.

A LIQUIDITY PREMIUM

This compensates you for the difficulty of and time required to sell your property. Stocks and bonds may be sold at market value within a moment's notice, whereas selling real estate may take months or years.

AN INVESTMENT MANAGEMENT PREMIUM

This is for the burden of monitoring and making decisions about the investment. The expense of managing the property is separate and additional.

A RISK PREMIUM

This accounts for the chance that you may not get all your money back or that the return will be lower than expected.

You may use a buildup approach to evaluating the return on a particular investment. Consider a real estate investment that promises an annual return of 10% before income taxes. You would hold the property for at least ten years. The safe liquid rate, as measured by the current yield on passbook savings accounts, is 2%. The real estate looks like a superior investment. But you must remember to account for the premiums included in the rate. The following increased rates are estimates that will vary with the type of property, economic situation, and tenant or user.

The real estate is difficult to sell, so add another 3% for illiquidity. The real estate requires more investment decisions, so add 2%. Finally, there is much more risk that the real estate will deviate from the promised return. Add 4% more. The built-up rate is now 11%. Using these estimates, the real estate should offer at least an 11% rate of return on investment. If it doesn't, you might be better off with the savings account.

Return of Investment

Hopefully, any investment will provide a stream of future income. A portion of this income represents the *return of* the investor's original invested cash. Any excess is a *return on investment.* The return of investment may appear in a lump sum at the end of the investment term. For example, if you deposit money in a savings account or buy a certificate of deposit, your money is refunded when you close the account. In other cases, the return may be a part of the periodic income stream. If you were to borrow a self-amortizing mortgage loan, the principal is paid down with each payment. At maturity there is no lump-sum payment.

The return from a real estate investment is less clear-cut. Whether periodic income includes return of investment depends on how much is realized at resale. Consider an investment of $100,000 that provides an annual cash flow of $15,000. If the property is sold later for $100,000, the resale proceeds provide all of the return *of* investment so that all of the annual cash flow is return *on* investment. If the sale is made for less than $100,000, a portion of the annual cash flow is actually return *of* investment.

In summary, consider the entire cycle of an investment, from purchase to resale, all of the investment returns and contributions, and their timing.

Risk

Risk is the chance that things will not turn out as planned. Since no one knows what will occur in the future, risk is an unavoidable fact of life. However, it is possible to anticipate or forecast the future and project a reasonable range of possible outcomes. This is what is meant by the term *calculated risk.* It does not imply that risk can be precisely measured but that sources of risk can be identified. In this way, it is possible to rank investment opportunities according to their riskiness.

There are different types of risk. Possibly the most familiar type is *business risk.* This is the chance that the investment will not perform as expected. It may turn out that more capital investment is needed to keep the property going. For example, a major component of the building may fail unexpectedly and need replacing. Or a storm may damage the building. Revenues may drop due to a rise in vacancies or the need to lower rental rates. Operating expenses may increase. Expected appreciation may not be realized when the property is sold.

It is even possible that changes in tax laws may affect the return over time. Some of these risks can be reduced by using *hazard insurance.* Others may be minimized by diversifying investment among different locations and types of property.

A second type of risk is *financial risk.* This is the chance that you will be unable to make the debt service payments on the property and thus become insolvent. Financial risk is increased by the use of *leverage.* The more money borrowed to purchase the property, the greater the chance that a declining net operating income will fail to cover debt service. This is especially a problem if you are poorly capitalized or have most of your capital tied up in illiquid assets. This is why any investment is made more risky by using financial leverage. (Likewise, leverage increases the potential return.) The use of short-term financing also exposes you to *interest rate risk.* This is the chance that interest rates will increase and adversely affect return. Such financing as balloon loans or adjustable rate mortgages present the possibility that debt service costs will increase if interest rates rise.

TAX CONSIDERATIONS

One of the traditional attractions of real estate investment is its ability to generate tax losses. These losses arise from certain deductions and credits that reduce taxable income but do not require cash outlays. In other words, a property may show a loss for tax purposes yet produce positive *cash flow.* When this occurs, the loss is said to shelter income from taxation. Reduced income taxes are a form of investment income.

Most of the expenses associated with rental property are deductible from taxable income. These include operating expenses and interest payments on the mortgage loan. Mortgage principal repayment is never deductible. (If the property is your personal residence, you may deduct some or all of the property taxes and interest payments.)

DEPRECIATION/INCOME TAX DEDUCTIONS

A key to real estate investments is depreciation claimed for federal income tax purposes. The Internal Revenue Service allows an owner of business or investment property to claim depreciation as a business expense, even when the property increases in value. Land is not considered to wear out, so depreciation cannot be claimed on it. Appliances, carpets, and furniture can be depreciated over a relatively short life and buildings over a longer period. The exact life to be used has changed frequently. Almost every time the tax law has changed in recent years, the depreciable life allowed for real estate has been changed. Usually tax changes affect new owners, whereas present owners continue the depreciable life they started with.

Depreciation allows a tax deduction without the owner paying for it in cash. This gives the owner the best possible benefit—a tax deduction that doesn't reduce cash flow. For example, suppose an investor buys a small building for $100,000 and is allowed a 27.5-year depreciable life using the straight-line method of depreciation. The owner may claim a tax deduction of $3,640 annually (the first year can be less depending on the month the property was acquired). This deduction reduces the owner's taxable income, which will in turn save income taxes. The deduction was created merely by a bookkeeping entry (not a cash payment) and can be claimed even though the property rises in value. Thus the owner gains by reducing taxes on property that provides cash flow from rents and would otherwise generate taxable income.

However, the taxpayer must be prepared to return the benefits received from depreciation deductions when the property is sold. There is a concept called *basis*—the point from which gain or loss is measured. In general, basis begins as the price paid for the property. As you claim depreciation as a tax deduction, you reduce the basis. This increases the gain when property is sold. If the investment remains the same in value, or rises, all of the depreciation accumulated while holding becomes taxable as a gain when sold. In this way, depreciation that was claimed while owning the property generates taxable gain on a sale, unless the property actually loses value more quickly than the investor depreciates it on his/her income tax return.

Tax depreciation should not be confused with an actual reduction or loss in value. Tax depreciation does not measure the actual loss in value of property sustained in a year. It is an arbitrary deduction allowed by tax law.

To summarize, depreciation claimed for tax purposes allows the investor to save on taxes, perhaps having to pay them back later. It is not a measure of a property's loss in value.

TAX-FREE EXCHANGES

In most real estate investment programs, there comes a time when you need to change properties. Your property may have grown in value and now it is time to move up to a larger property. You may want to get into a different type of property, such as moving from rental housing to an office building. If you change your personal residence, you may want to keep your investments close at hand. You may have several small properties and want to consolidate into one large property.

Many have found that the best way to change properties is through exchanging. They merely trade properties with another investor. This may be simpler than selling the property and reinvesting in a new one. It means you need only one transaction instead of two (or more). Financing is often exchanged along with the property, so you don't necessarily need to get a new loan. There is no need to get out of the market and then get back in, taking the chance that the market changes during that time.

Another big advantage of exchanging is that you may save taxes. If your property has appreciated, you will owe taxes on capital gains when you sell. An exchange can let you defer paying those taxes until some time in the future when you sell the new property. To do this, you must trade for other investment real estate (the property traded or received can't be your home). If you end up with some cash from the exchange, you may have to pay some taxes. In most cases, you will need to trade up in the exchange. If you end up with a smaller debt, you may have to pay some taxes. However, in cases where you pay some taxes, those taxes may be less than if you sold your property for cash.

The rules of a tax-free exchange are described in Section 1031 of the Internal Revenue Code. They allow a sale to qualify as a tax-free exchange, provided that, within 45 days, replacement property is identified and is purchased within 180 days. Be sure to enlist professional tax advice when considering a trade.

QUESTIONS ON CHAPTER 17

1. Investment real estate may offer
   (A) cash flow
   (B) appreciation in value
   (C) income tax shelter
   (D) all of the above

2. REIT is a
   (A) really energized income tax
   (B) real estate investment trust
   (C) real estate income target
   (D) none of the above

3. A REIT is like
   (A) tenancy in common
   (B) a joint tenancy
   (C) community property
   (D) a mutual fund

4. Refinancing is
   (A) arranging a new loan to replace an existing loan
   (B) a wraparound mortgage
   (C) done to postpone income taxes
   (D) reduces operating expenses

5. Depreciation claimed for income tax purposes
   (A) reduces cash flow
   (B) reduces taxable income
   (C) increases potential gross income
   (D) increases operating expenses

6. A list of tenants, their space identified, and rent amounts, is called
   (A) an operating statement
   (B) a tenant list
   (C) a rent roll
   (D) an operating role

7. One of the important drawbacks to direct real estate investing is
   (A) illiquidity
   (B) location
   (C) small size
   (D) disclosure requirements

8. Instead of selling property and paying a tax on the gain, one may prefer
   (A) a capital gain
   (B) Section 1231
   (C) a tax-free exchange
   (D) ordinary income

9. A projection of what is expected to occur from operating a property is a __________ statement.
   (A) conforming
   (B) nonconforming
   (C) consistent
   (D) pro forma

10. An apartment complex currently has 12 vacancies among its 200 units. The current vacancy rate is
    (A) 94%
    (B) 0.5%
    (C) 12%
    (D) 6%

11. Casualty insurance is
    (A) financing expense
    (B) capital expenditure
    (C) fixed expense
    (D) variable expense

12. Taxable gain on the sale of commercial real estate is calculated by
    (A) sales price minus the basis
    (B) adjusted sales price minus the basis
    (C) realized selling price minus the adjusted tax basis
    (D) contract sales price minus the adjusted tax basis

ANSWERS

1. **D**	4. **A**	7. **A**	10. **D**
2. **B**	5. **B**	8. **C**	11. **C**
3. **D**	6. **C**	9. **D**	12. **C**

Real Estate Mathematics and Problem Solving

18

The amount of time you need to spend on this mathematics chapter depends on your experience and comfort working with mathematics. If you excel in math and feel that working out problems is fun, a quick review may be all you need; so spend time on other chapters. However, if you don't recall how to convert decimals to fractions or percentages—and the reverse—then you'll need to concentrate on this chapter.

ESSENTIAL MATTERS TO LEARN FROM THIS CHAPTER

- Most real estate questions on exams are word problems. You will be expected to interpret the question and calculate the answer. Generally, calculators are permitted.
- Frequently, finding the proper solution to a math question requires a disproportionate amount of time. If you can't solve a question in a few minutes, mark it to return to it after you've completed the rest of the test. Each question counts equally, so don't shortchange the time you spend on 20 test questions just to solve one math problem. Return to the time-consuming math questions after you've completed other questions.
- For example, if you have a 100-question test and a 2-hour time limit, the average time to budget for each question is 1 minute. That would leave 20 minutes to check your answers or return to difficult questions that you flagged. If you must spend 15 minutes to solve one math question, you'll have to shortchange the time to answer something else. It is better to miss one math question than not to finish the test by leaving 15 other questions unanswered or rushing through them.
- The same is true while studying in this book. If math is your strong suit, you'll breeze through this chapter. However, if you are weak in math, it may take so much time to learn here that you'll miss material in other chapters. So be mindful of how you spend your time.
- When addressing a math problem, we suggest that you:

  1. Read the question carefully.
  2. Determine what is sought.
  3. Determine how to derive the answer.
  4. Calculate the answer.
  5. If your answer is among the choices, you are probably right.
  6. If time permits, check your work by inputting your answer into the question to determine whether your solution fits the problem.

- Most measurements for lengths and areas are well known. For real estate, you must memorize these:

  1. There are 43,560 square feet in an acre.
  2. There are 640 acres in a square mile.

- You will also need to understand that *frontage* refers to the number of lineal feet along the front of a lot. *Depth* refers to how far back the lot goes.
- *Lineal feet* is the distance along a straight line. It is one-dimensional.
- *Square feet* is an area. It is two-dimensional.
- *Cubic feet* is three-dimensional. The size of a box may be measured in cubic feet.
- You may see these types of math problems:

  1. The *area* of a lot or a building. Area or building measurement questions may ask how many lots can be carved out of a given number of acres, how many square feet are in a building or a lot, or how deep a lot is.
  2. *Percentage* problems:
     - *Commission* problems may ask the dollar amount of commissions when given a real estate sales price and a commission split arrangement.
     - *Interest* problems may require finding the interest rate or the amount of interest to be paid. Typically, interest problems state a given amount of time in days or months.
     - *Depreciation* problems ask you to calculate the property value loss, given its cost, age, and useful life.
     - *Profit or loss* problems ask for an amount or percentage of gain or loss based on an initial investment.
     - *Return on investment* problems are similar to interest problems. They seek the rate of return earned, usually annually, on an investment.
  3. *Proration* problems require you to calculate a daily amount of interest, insurance, or real estate tax expense for part of a year. In many cases, you must determine who pays or who receives credit, based on due dates of payments, and when the closing date will be for the sale.

The mathematics problems you will encounter on the licensing examination are most likely to be in the form of *word problems*, that is, problems that describe a situation for which you have to find a mathematical answer. Here is an example of a simple word problem:

> *I have three apples. John has four apples. How many apples do John and I have together?*

Obviously, some word problems are more complicated. However, all math questions can be solved by performing these steps:

1. *Read* the question.
2. Determine *what* is being asked for.
3. Determine *how* to arrive at the answer.
4. *Calculate* the answer.
5. *Check* the answer.

The last step is particularly important, especially when the problem is a complicated one; you have to make sure that you did all the calculations correctly, or your answer will be wrong. On most examinations, knowing *how* to do a problem will get you no credit; you must provide the *correct answer* as evidence that you know how to do that particular problem. You should be aware of one more fact: on mathematics questions offered in a multiple-choice format, you have both an advantage and a possible disadvantage.

The advantage, which is always present, is that *you are given the correct answer to the problem*! Obviously, it *must* be among the ones you are to choose from; all you have to do is to figure out *which* one it is.

A possible disadvantage is that some of the *incorrect* answers may be answers you would arrive at by making common mistakes. You may be lulled into choosing the incorrect answer by assuming that, since your calculations gave you an answer that appears among the ones you must choose from, it *must* be correct. You should especially be on your guard if one of the choices given is something like "none of the above" or "none of the other answers is correct." This kind of choice *may* mean that none of the answers is correct, but it may also be included as a red herring.

Mistakes on mathematics problems may result from:

1. Incorrect reading of the problem. *Solution*: Read carefully!
2. Misunderstanding what is being asked for. *Solution*: Read carefully and slowly; you have plenty of time, so don't rush!
3. Not knowing how to do the problem correctly. *Solution*: Study this chapter carefully so that you know how to approach all kinds of problems.
4. Incorrect calculation of the answer. *Solution*: Study the methods of calculation for all kinds of real estate problems, as outlined in this chapter.
5. Mistakes in arithmetic. *Solution*: Check your work to make sure that you did your arithmetic correctly.

Note that these five common sources of errors can be controlled by applying the five steps in problem solving offered at the beginning of this chapter.

In discussing the various kinds of real estate mathematics problems, we are going to assume in this chapter that you have an adequate knowledge of basic arithmetic. Have your calculator ready before you tackle the sample problems; use this opportunity to become familiar with the calculator if you aren't already. You must get used to doing these kinds of problems with a calculator because you'll be using one for the exam.

MEASUREMENTS AND TERMS

Measurements of Lengths and Areas

Many real estate problems involve *measurements* of lengths and areas. You will be assumed to know at least the measurements listed below.

Linear measures:

12 inches	=	1 foot
3 feet	=	1 yard
5,280 feet	=	1 mile

Area measures:

144 square inches	=	1 square foot
9 square feet	=	1 square yard
640 acres	=	1 square mile

Most of these measures you already know, but here is one that *you absolutely must remember:*

1 acre = 43,560 square feet

Burn this into your memory! You can bet the farm that you'll be asked questions that *require* you to know the number of square feet in an acre.

Special Terms You Should Know

Also, you must know the following special *terms* used in real estate measurement; these are illustrated in Figure 18-1.

Depth refers to the straight-line distance from the front lot line to the rear lot line. If these lot lines aren't parallel, the term *depth* will refer to the *longest* straight-line distance between them. The lot shown in Figure 18-1 has a depth of 272 feet.

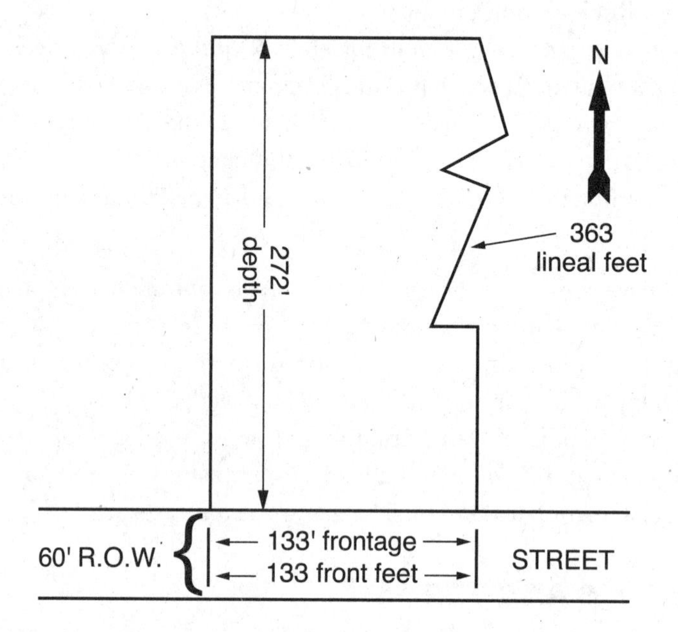

Figure 18-1. Real Estate Measurement

Frontage is the *lineal distance* (see *lineal foot* below) of the lot line that is also a part of a public street or other right-of-way. Frontage of corner lots usually is expressed as separate figures for the distance fronting on each of the streets the lot bounds. This would also apply to very deep lots that border streets at both ends, and so on. The frontage of the lot in Figure 18-1 is 133 feet.

Front foot refers to the distance that a lot borders on a street. For the lot illustrated in Figure 18-1, we can say that it has *133 feet of frontage* or that it has *133 front feet.* (To say that the lot has 133 front feet of frontage is redundant.)

The amount of frontage is usually the most important statistic of the dimensions of a property. For example, when someone buys or sells lakefront property, the frontage along the lake provides the most value, not the depth of the lot.

When the dimensions of a lot are provided, the frontage is stated first. For example, a commercial property lot that has 50 feet of street frontage and is 200 feet deep would be described as 50′ × 200′.

Lineal foot refers to the distance between two particular spots. Note that lineal feet need not be a *straight* measure; rather, the term refers to the number of feet traveled while following a *particular path connecting the particular points involved.* In real estate, this term is typically used to describe irregular lot sides caused by irregular paths along roads. For example, the east side of the lot in Figure 18-1 is 363 lineal feet on a lot that has only 272 feet of depth.

Right-of-way (R.O.W.) refers to the area owned by the government within which a road is located. On maps, plats, and so on, the term R.O.W. usually refers to the width of the right-of-way area (which usually exceeds *pavement* width).

Approaches to Solving Problems

You are assumed to be familiar with the basic methods of calculating distance, area, and volume; here we will confine ourselves to particular kinds of real estate problems that you will encounter. Most of them can be characterized as specific *forms* of problems that are easy to recognize.

Before showing how to deal with them, however, we should pass on one very crucial hint for solving these kinds of problems: *draw a picture of the problem if you possibly can.* Then, with the picture as your guide, go through the problem step by step.

We cannot stress this process too much. By drawing a picture of the problem, you can *look* at it. A picture really is valuable, especially if it helps you to visualize a problem. (Sometimes, of course, the problem on the exam will include a picture or diagram. But if it doesn't, supply your own!) Then when you do the problem, work it very carefully, step by step. Even if you feel you can take some shortcuts or do part of it in your head, DON'T! Solving problems this way may be boring or tedious, but it helps to eliminate mistakes.

Percentage Problems

A great many real estate problems involve percents; these include commissions, interest, mortgage payments, loan discounts (points), depreciation, profit and loss, and return on investment. We will discuss each of these. However, you should remember that the mathematics involved is the same for all of them: you will be looking for one of three things—a base, a rate, or a result.

Every percent problem can be stated as one of three variations of the following statement:

A is *B* percent of *C*.

Numerical examples of such a statement are "14 is 50% of 28" and "63 is 90% of 70." Of course, if you know all three numbers, you don't have a problem.

All percent problems are like the statement above, except that one of the numbers is missing: the problem is to find that missing number.

Type 1:	*A is B percent of ?.*	*(C is missing.)*
Type 2:	*A is ? percent of C.*	*(B is missing.)*
Type 3:	*? is B percent of C.*	*(A is missing.)*

The solutions to these problems are as follows:

$$A = B \times C$$
$$B = A \div C$$
$$C = A \div B$$

An easy way to remember these is that *A* is *B* multiplied by *C*. To find either *B* or *C*, you divide *A* by the other. Here are some examples illustrating the three variations of the percent problem:

1. What is 88.5% of 326? (? is *B* percent of *C*.)
2. 16 is what percent of 440? (*A* is ? percent of *C*.)
3. 43 is 86% of what? (*A* is *B* percent of ?.)

The solutions are as follows:

1. The missing item is *A*.	$A = B \times C$	$A = 0.885 \times 326$	$A = 288.51$
2. The missing item is *B*.	$B = A \div C$	$B = 16 \div 440$	$B = 0.0363636 = 3.63\%$
3. The missing item is *C*.	$C = A \div B$	$C = 43 \div 0.86$	$C = 50$

Ratio Problems

Ratios are fractions that relate one amount to another. Ratio problems are usually a type of percentage problem. For example, a building's operating expenses are $4,500 per year, and its rental income is $10,000 per year. What is its *operating expense ratio*?

Answer: $4,500 divided by $10,000 = 45%

A building's *efficiency ratio* is its usable square footage compared to its total square footage. If 20% of square footage is "lost" to utility closets, restrooms, stairwells, etc., then the building's efficiency ratio is 80%.

Commission Problems

Commission problems naturally are very popular on real estate examinations because most licensees are paid in commissions. A real estate commission is simply what the employer pays the agent for performing his function. Most often, the commission is a portion of the sale price received by a seller because of the efforts of the agent to secure a buyer. Typical commission rates vary from as low as 1% or less to over 10%, depending on the kind of property, the prevailing custom in the area, and the success of the agent, or the employer, in negotiating the listing contract.

The simplest commission problems are like this one:

Jones sells Smith's house for $311,800. Jones's commission is 7% of the sale price. What is her commission, in dollars?

Here the solution is to find 7% of $311,800: 0.07 × $311,800 = $21,826.

This *same* problem, which is a type 1 problem described above, can also be stated as a type 2 or type 3 problem:

Type 2: *Jones sells Smith's house for $311,800 and receives a commission of $21,826. What is her commission rate?*

Type 3: *Jones sells Smith's house and receives a commission of $21,826, which is 7% of the sale price. What was the sale price?*

These problems are simple. To make them more "interesting," they are often expanded in one of two ways. The first way is to make them more complicated, and there are two methods of doing that. The second way is to develop the question *backward*. This often traps many unwary examinees. However, if you pay attention, it won't fool you.

First, let's look at the more elaborate kinds of commission problems. The two methods used are (a) to have you calculate the salesperson's share of a total commission and (b) to make the means of determining the commission more complicated.

Here (a) is illustrated:

> *O'Reilly sells Marino's house for $180,000. The commission is 6½% of the sale price. O'Reilly's broker receives 45% of this amount from the listing broker as the selling broker's share. O'Reilly herself is entitled to 60% of all commissions that she is responsible for bringing in. How much does O'Reilly get?*

This may look complicated, but it is just three consecutive simple percent problems.

(1) How much was the total commission? A is 6½% of $180,000; A = $11,700.
(2) O'Reilly's broker received 45% of this amount. A is 45% of $11,700 = $5,265.
(3) O'Reilly gets 60% of that amount. A is 60% of $5,265 = $3,159.

This is the second kind (b):

> *Albertson sold Harding's house for $115,000. The commission rate is 7% of the first $50,000 of the sale price, 4% of the next $50,000, and 1.5% of everything over that. What was Albertson's total commission?*

Once again, we have several simple percentages to calculate: 7% of the first $50,000 is $3,500; 4% of the next $50,000 is $2,000; adding $2,000 to $3,500 yields a commission of $5,500 on the first $100,000 (or $50,000 + $50,000). This leaves a balance of $15,000, to which we apply the commission rate of 1½% to get another $225, for a total commission of $5,725.

This question could have been complicated further by going an additional step and asking, What was the effective commission rate on the sale? Here the question is "What percent of the sale price is the actual commission?" First we say $5,725 is ? percent of $115,000. Then we take the additional step: $B = 5{,}725 \div 115{,}000 = 0.0498$, so the percentage = 4.98%.

Now let's turn to the other general category of complicated commission problems—those in which the question is developed *backward.* Here the question is phrased like this:

> *Baker sold her house; after paying a 7% commission, she had $260,400 left. What was the sale price of the house?*

Many people will find 7% of $260,400, add it on, and GET THE WRONG ANSWER!! Their mistake is that the commission should be figured on the *sale price,* and they are figuring it on what the seller had left *after* the commission was paid. The way to solve these problems is simple: If Baker paid a 7% commission, then *what she had left* was 93% (or 100% − 7%) of the sale price. Now the problem becomes a familiar variation: $260,400 is 93% of ? $C = \$260{,}400 \div 0.93 = \$280{,}000$.

As you can see, this question is "backward" in that you don't actually have to figure the dollar amount of the commission. On your examination, be sure to look out for these kinds of problems; the best way to spot them is to follow the rule of *reading* the question and *then* trying to determine what you are asked to do.

You should be aware of the terminology used in commission figuring. The *commission* is the dollar amount arrived at by applying the *commission rate* to the *sale price.* If you are asked for the "commission," you are being asked to tell *how many dollars and cents* the commission payment was. If you are asked for the "commission rate" or the "rate," you are being asked to provide the *percent rate* of the commission.

Interest Problems

Interest is money paid to "rent" money from others. It is typically expressed as an annual percentage rate; that is, for each year that the money is borrowed, a certain percentage of the loan amount is charged as interest. Interest will most often be stated in annual percentage terms even for loans that have a duration of much less than 1 year. As an example, the "interest rate" on charge accounts, credit card accounts, and the like may be charged monthly, and the bill may be due within a month, but the interest rate still is expressed as an annual rate. This practice is becoming even more prevalent as a result of federal truth in lending laws, which *require* that the annual percentage rate (APR) be stated.

Usually for short-term arrangements, a monthly or quarterly rate will also be specified. If you are given only an annual rate and are required to calculate based on a monthly or quarterly rate, simply divide the annual rate by 12 to get the monthly rate and then divide by 4 to get the quarterly rate. If a problem gives a monthly or quarterly rate, multiply a monthly rate by 12 or a quarterly rate by 4 to get the annual rate. For example:

> *A rate of 12% annually is 1% monthly or 3% quarterly.*
> *A rate of 9½% annually is 2.375% quarterly or 0.791666 . . . % monthly.*
> *A rate of 0.6% monthly is 1.8% quarterly or 7.2% annually.*

Interest is calculated in two ways, *simple* and *compound.*

Simple interest is calculated quite straightforwardly. To determine the interest charge for any period of time, we calculate the proportional interest, based on the annual rate. If the annual rate is 6%, interest for 2 years is 12%, interest for 6 months is 3%, and so on.

When *compound interest* is used, interest is charged on unpaid interest as well as on the unpaid debt. This adds a complicating element to the calculation since some sort of adjustment may be necessary to allow for the fact that interest is not being paid periodically but is being allowed to accumulate over all or part of the term of the debt. Since in real life most debts are paid in monthly installments, one does not frequently encounter interest compounding in the payment of debts. One *does* encounter it in savings accounts, where a well-advertised feature is that if a depositor does not withdraw the interest his principal earns, that interest is added to his balance and begins to earn interest too.

Before compound interest can be calculated, the compounding frequency must be given. This is expressed in units of time. Compounding can be daily, quarterly, semiannually, annually—indeed it can be over any period of time, even "continuously." Given a compounding term, interest is figured by first deriving an interest rate for the term involved. Then interest is calculated for *each* term, taking into account the fact that some or all of the interest earned so far may not have been paid out. Here is an example:

> *Berger borrows $1,000 at 8% annual interest. The interest is compounded quarterly. How much interest will Berger owe after 9 months?*

First we must change the annual rate to a quarterly rate: 8 ÷ 4 = 2, so the quarterly interest rate is 2%.

For the first quarter, interest is 2% of $1,000, or $20.

To calculate interest for the second quarter, we add the first quarter's interest to the balance to get $1,020, upon which interest for the second quarter is due. Therefore, the second quarter's interest is 2% of $1,020.00, or $20.40. In effect, for the second quarter there is $20 interest, again, on the initial balance of $1,000 plus another 40¢ interest on the unpaid $20 of interest earned in the first quarter.

Interest for the third quarter of the 9-month period will be based on a balance of $1,040.40, which is the original balance plus the interest accumulated in the first two quarters. Therefore, interest for the third quarter is 2% of $1,040.40, or $20.81. The total interest for the 9 months (three quarters) will be $20.00 + $20.40 + $20.81, or $61.21.

Notice that compound interest will amount to *more* than simple interest if the interest period extends over more than one compounding term. We can show this with the same problem, this time assuming that simple interest is paid. In this case, the interest is $20 each quarter, for a total of only $60, compared to the $61.21 charged under compound interest. The reason is that with simple interest, interest is calculated only on the original loan balance. Compound interest includes interest on earned but unpaid interest as well as on the original principal, so a larger amount becomes subject to interest once the first compounding term has passed.

In real estate problems, interest questions usually involve *mortgage loans* and *loan discounts*, or *points.*

MORTGAGE LOANS

Mortgage loans are like any other loans in that interest is charged for the use of someone else's money. Usually these loans require monthly payments, with part of the payment used to pay the interest due since the last payment and the remainder are used to reduce the amount of the loan (the *principal*). Eventually, in this manner the loan will be fully paid off.

A very common question on this type of loan concerns the interest payable in any given month. This requires you to calculate the loan balance for that month before you can find the interest payment due. Here is an example:

> *Smith borrows $30,000 on a 30-year mortgage loan, payable at 9% interest per year. Payments are $241.55 per month. How much interest is to be charged in each of the first 3 months?*

The first step in the solution is to determine the *monthly* interest rate since mortgage interest payments are made monthly. Therefore, we divide the annual rate of 9% by 12 and get an interest rate of ¾% or 0.75% per month.

Now we can find the interest payable in the first month, which is 0.0075 × $30,000.00 = $225.00. The total monthly payment is $241.55, so after the $225.00 interest is paid, there remains $241.55 − $225.00 = $16.55 to be used to reduce the loan. This means that the loan balance for the second month is $30,000.00 − $16.55 = $29,983.45.

Now we calculate interest for the second month: 0.0075 × $29,983.45 = $224.88. This means that in the second month $241.55 − $224.88 = $16.67 is used to reduce the loan, leaving a balance of $29,983.45 − $16.67 = $29,966.78 for the third month.

The third month's interest, then, will be 0.0075 × $29,966.78 = $224.75. We can tabulate these results as follows:

Month	Loan Balance	Loan Payment	Interest	Loan Payoff
1	$30,000.00	$241.55	$225.00	$16.55
2	29,983.45	241.55	224.88	16.67
3	29,966.78	241.55	224.75	16.80
4	29,949.98	......	......	......

This kind of loan is called *self-amortizing, equal payment.* The monthly payments are equal. However, as the loan balance is reduced a little each month, the portion of the payment representing interest decreases monthly while the portion going to reduce the loan increases.

In some loans, called *level principal payment loans,* the loan amount is reduced the *same* amount with each payment. This means that the payments themselves get smaller each time as the interest due decreases because the loan balance is decreasing. Here is an example:

> *Smith borrows $24,000 for 10 years at an annual rate of interest of 12%. He pays it back in the level principal payment manner each month. What will be his payments for the first, fourth, and tenth months?*

The first thing we must do here is to calculate the amount of the monthly level principal payment. Since 10 years is 120 months, the monthly principal payment is $24,000 ÷ 120 = $200. The monthly interest rate is 12% ÷ 12 = 1%.

In the first month, the entire $24,000 is on loan, so the payment is the level $200 payment plus 1% of $24,000, which is another $240. Therefore, the first month's total payment is $200 + $240 = $440.

For the fourth month, Smith will owe the original balance of $24,000 *less* the $200 payments that were made for each of the first 3 months. That amount totals $600, so for the fourth month he owes $24,000 − $600 = $23,400. Then, 1% of $23,400 is $234; this added to the level payment of $200 gives a payment of $434 for the fourth month.

The payment for the tenth month is calculated in the same way. At that time, nine payments will have been made. So the loan will have been reduced by 9 × $200 = $1800, leaving a loan amount of $22,200. Since 1% of that amount is $222, the full payment for the tenth month will be $422.

LOAN DISCOUNTS

The purpose of loan discounts is briefly explained in Chapter 13 ("Mortgages and Finance"). A loan discount is very simple to calculate since it is merely a given percentage of the loan amount. You should be aware of the terminology, though, since these discounts often are referred to as *points.* One discount point is the same as a charge of 1% of the loan amount. Therefore, a discount of three points (often just referred to as "three points") is 3% of the loan amount. Frequently, a problem involving discounts will require you first to determine the loan amount; this can trap the unwary examinee, who makes the mistake of calculating the discount based on the purchase price instead of the *loan amount.* Here is an example:

> *Martha bought a home costing $154,900. She got a 95% loan, on which there was a charge of 3¾ points. What was the discount in dollars and cents?*

First we calculate the loan amount, which is 95% of $154,900: 0.95 × $154,900 = $147,155. Now we can calculate the discount, which is 3¾ points, or 3¾%, of the loan amount:

0.0375 × $147,155 = $5,518.31

Depreciation

Although the tax laws allow all sorts of complicated ways to calculate depreciation (see Chapter 16, page 366, for depreciation expense deductions on business or investment real estate), the only one you have to worry about on licensing examinations is the simplest one, which is called *straight-line depreciation.* Straight-line depreciation assumes that the property depreciates an *equal* amount each year.

We will not argue here whether real estate does or does not, in fact, depreciate. Some does and some does not seem to. However, there are many reasons why one should be aware of the possibility of depreciation, not the least of which is that in the long run a real estate asset *will* wear out. The 50-year-old house that is being offered for ten times the original cost of construction also has a new roof, modernized heating and air-conditioning, new wiring, carpeting, kitchen installations, and plumbing. Nearly every visible surface has been repainted and remodeled.

When calculating depreciation, you must know the *useful life* of the property (sometimes referred to as *economic life,* or just plain *life*). Once you know this, you can easily calculate an annual *depreciation rate*: divide the useful life, in years, *into* 100% to get the percent rate of depreciation per year. For example:

10-year life: 100% ÷ 10 = 10% depreciation per year
50-year life: 100% ÷ 50 = 2% depreciation per year
35-year life: 100% ÷ 35 = 2.857% depreciation per year

Once this has been done, most depreciation problems become only slightly elaborated versions of the standard three types of percentage *problems.* For example:

(a) *Madison owns a building for which he paid $550,000. If it has a total useful life of 30 years, what is its value after 5 years?*
(b) *Madison owns a building for which he paid $550,000. If it has a 40-year life, in how many years will it be worth $426,250?*
(c) *Madison owns a building worth $550,000. If it depreciates to $440,000 in 9 years, what is the total useful life?*
(d) *If a building depreciates at a 2½% per year, in how many years will it be worth 85% of its original value?*

Here is how the answer to each of these is found:

(a) Here you are asked to determine the value after 5 years, so you must determine how much the building depreciates each year. A 30-year life yields 3⅓% depreciation per year. Five years' depreciation, then, is 16⅔%. Therefore, in 5 years the building will be worth 83⅓% (100% − 16⅔%) of its original value: 0.8333 × $550,000 = $458,333.33, its value after 5 years.
(b) Here you want to know how long it takes for a certain depreciation to take place. A 40-year life is depreciation at 2½% per year. $426,250 is 77½% of $550,000 (426,250 ÷ 550,000). 100% − 77½% = 22½% total depreciation. At 2½% per year, that would take 9 years to accumulate (22.5 ÷ 2.5 = 9).
(c) Here you have to determine the useful life. $440,000 is 80% of $550,000 (440,000 ÷ 550,000 = 0.80). Therefore, the building depreciates 20% in 9 years, or 2.2222 . . . % per year. Divide this figure into 100% to get the total number of years required for the building to depreciate fully: 100 ÷ 2.2222 = 45. Thus, the useful life is 45 years.
(d) In this problem, you don't have to calculate the depreciation rate since it is given to you (2½% per year). If the building depreciates to 85% of its original value, it will have depreciated 15%. 15% ÷ 2½% = 6 years' worth of depreciation.

Profit and Loss

Calculation of profit and loss is another slightly different version of the three types of percentage problems. The important thing to remember is that profit and loss are always expressed as percentages of *cost.* Cost is the original price that the seller paid for the property when he acquired it. If *his* selling price is higher than cost, he has a profit. If it is lower, he has a loss.

The *dollar value* of profit or loss is the difference between purchase price and sale price.

The *rate* of profit or loss is the percentage relationship between the purchase price and the dollar value of profit or loss.

Here are some examples:

(a) *Martin bought her house for $449,500 and sold it later for $559,000. What was her rate of profit?*

(b) *Samson Wrecking Company mistakenly tore down part of Habib's home. Before the home was damaged, it was worth $300,000. Afterward, it had sustained 28% loss. What was its value after the wrecking?*

(c) *Harrison sold his home for $420,500 and made a 45% profit. What had he paid for the home?*

Here are the solutions:

(a) First, determine the dollar value of the profit: $559,000 − $449,500 = $109,500. Then determine what percentage proportion the dollar value is of the *purchase price* ($109,500 is ? percent of $449,500). 109,500 ÷ 449,500 = 24.36%.

(b) The loss was 28% of the original value of $300,000: 0.28 × $300,000 = $84,000. $300,000 − $84,000 = $216,000 value afterward.

(c) The original price of the house, plus 45%, is now equal to $420,500. Therefore, $420,500 is 145% of the original price of the home (? is 145% of $420,500). 420,500 ÷ 1.45 = $290,000. As a check, Harrison's profit is $130,500 ($420,500 less $290,000), which is 45% of $290,000 ($130,500 ÷ $290,000 = 0.45).

Return on Investment

The concept of return on investment is very similar to the concept of interest payments on loans. In the investment case, an investor spends money to buy an income-producing asset. She wants to make money from it; otherwise there is no point to the investment. She calculates her return and expresses it as a *percentage* of her investment being paid back to her each year. In a sense, she can be thought of as "lending" her money to the investment and having it "pay" her "interest" on her money.

It is possible to make deceptively simple-sounding investment questions so complicated that they can best be answered with the aid of a computer or a very sophisticated calculator. This fact needn't concern you, though. On licensing examinations, the questions are kept simple enough that they can be calculated quickly by hand or, in states that allow their use, with simple, handheld calculators that do no more than add, subtract, multiply, and divide.

If you think of return on investment problems as similar to interest problems, you should have no trouble with them. Here are some examples:

(a) *Bennett owns a building that cost him $650,000. How much income should the building produce annually to give Bennett a 15% return on his money?*

(b) *Maximilian paid $435,000 for a triplex apartment building. All units are identical. He lives in one unit and rents the other two. The total net income per month from rental is $3,550. What annual rate of return is he getting on his investment in the rental units?*
(c) *What monthly income should a building costing $385,000 produce if the annual return is to be 15%?*

Here are the solutions:

(a) The building should produce an income each year of 15% of $650,000: $0.15 \times \$650,000 = \$97,500$.
(b) This problem involves a lot of steps, but they are simple ones. First we must determine how much of the purchase price should be allocated to the two rental units. Since all three are the same, one-third of the purchase price ought to be allocated to each. This means that two-thirds of the price should be allocated to the two rental units: $2/3 \times \$435,000 = \$290,000$. Next we must determine the annual dollar amount of income received. The monthly income is $3,550, so the annual income is $\$3,550 \times 12 = \$42,600$. Now we must find out what percent $42,600 is of $290,000: $\$42,600 \div \$290,000 = 14.69\%$, the annual rate of return.
(c) Here we must calculate a monthly dollar income. The annual income must be 15% of $385,000, or $57,750. The monthly income is one-twelfth of that amount, or $4,812.50.

Trick Question

A trick question lists monthly amounts for some elements but requires an answer based on an annual amount. For example, the question may provide a monthly rental income and then ask about the annual rate of return. To answer correctly, you must convert everything to the same time period. Since the question asks for an annual amount, convert the given information about monthly values to annual values.

PRORATION

Proration, which is a necessary input into the calculation of closing statements, normally will come up only on examinations for broker's licenses. If you are seeking a brokerage license, you should cover the material in this section. Appraisal or salesperson candidates can skip it.

Two kinds of prorating methods are in common use. The most widely used one employs the *statutory year*; the other, the *actual year*.

The function of proration is to distribute equitably the costs of a particular charge that two or more people must share. In real estate, these costs usually are created by transactions associated with a title closing, in which allocations are made for charges that apply over the period in which both buyer and seller own the property. There is no legal requirement that any proration of any items occur at settlement. However, four items can be prorated:

1. **PROPERTY TAXES** are almost always prorated. Many preprinted sales contracts even contain language requiring proration (although buyer and seller can agree to strike that provision).
2. **INTEREST ON ASSUMED LOANS** is applicable only if there is an assumed loan in the transaction.
3. **PREPAID OR LATER-PAID RENTALS** apply only to rental properties.
4. **PROPERTY INSURANCE** used to be prorated decades ago when multiyear policies were common but is rarely prorated in modern times.

Property taxes are usually paid by the year. When a transaction occurs in the middle of a year, part of that year's property tax usually will be deemed, by the parties to the sale, to be payable by each one. Note that the government that collects the taxes does *not* prorate them; it collects the *full* amount of the taxes from whoever owns the property at the time the taxes are due. If, when the sale occurs, the year's property taxes have *already* been paid, the seller will have paid them and the buyer must recompense the seller for the portion of the taxes that apply to the part of the year when the buyer will own the property. On the other hand, if the taxes are due *after* the close of the sale, the buyer will have to pay them in full, for the entire year, when they come due. At the closing, the seller pays the buyer for the share that applies during the period of the tax year when the seller owned the property.

Insurance usually is paid in advance. Often insurance policies are multiyear policies, although the premiums generally are payable annually. On licensing examinations, however, problems that involve closing often assume that the entire multiyear premium was paid at once when the policy was purchased; this is a device to see whether examinees read the questions completely. Since insurance usually is paid in advance, the buyer, if he assumes the existing insurance policy, recompenses the seller for the prepaid unused portion of the policy that the buyer gets from her.

When the buyer assumes the seller's mortgage loan and the closing date is not the day after the loan payment is due and paid, the month's interest must be prorated between buyer and seller. Loan payments are due at the end of the monthly payment periods. When the next payment is due, the buyer will have to pay it in full, including the interest for an entire month during which he actually owned the property only part of the time. Therefore, the seller will have to pay the buyer for her share of the mortgage interest.

Rents work the other way since they are payments *to* the owner rather than *by* the owner. Rent usually is paid in advance, but it is possible to have a situation where rent is paid at the *end* of the month or lease period. If rent is paid in advance, the seller will have been paid a full month's rent for the month in which the closing occurs; she must pay the buyer his prorated share of that month's rent already received. If the rent is paid at the end of the month, the buyer will get a full month's rent covering the month of sale, and he must recompense the seller for the period during which she owned the property. Because of these two types of situations, examinees must read rent questions carefully.

Computing Prorations

To compute prorations an examinee must know three things:

1. How much—in money—is the item to be prorated?
2. To whom is payment to be made?
3. How much time is involved?

The discussion above covers the question of who pays whom what charges, but in summary we can state two easily remembered rules:

1. If the item was paid *before* closing, the buyer recompenses the seller.
2. If the item will be paid *after* closing, the seller recompenses the buyer.

These rules are fair because the seller had to pay all items due before closing, and the buyer will have to pay everything that comes due after closing.

Calculating the amount of money due in the payment will vary, depending on the complexity of the problem. Sometimes the amount is specified exactly; other times it will have to be calculated. It is quite popular in licensing examinations to require the examinee to calculate the property tax bill before he can begin to prorate it (see Chapter 16). Usually the insurance premium and rentals will be given since it is difficult to incorporate a sensible calculation method for them into a problem. Proration of the interest on assumed mortgages is another favorite candidate for calculation.

Calculating the proration itself includes apportioning the time involved among the parties. Rent and interest usually are apportioned over 1 month, taxes over 1 year, and insurance over 1 or more years.

To prorate correctly, you must have two important pieces of information:

1. Who (buyer or seller) is supposed to "have" the day of closing?
2. Which calculation year (statutory or calendar) is being used?

You must be given this information somewhere in the examination. If you can't find it, ask one of the examination supervisors.

THE CLOSING DAY

The day of closing can be counted either as the buyer's first day or as the seller's last day. Which it is to be usually reflects local custom, so we can't state a general rule. In the East and the South, the closing day most often is considered the seller's, while it tends to be the buyer's in the West.

THE STATUTORY YEAR

Prorating items over a full year can involve very messy calculations since everything has to be divided by 365 (or 366), the number of days in a year. Many banks and other financial institutions therefore have substituted the 360-day "statutory year" as a means of simplifying calculation. This system assumes the year to be made up of twelve 30-day months. Numbers such as 360 and 30, while large, can be divided easily by many other numbers so that calculations involving them are less cumbersome.

For items paid monthly, such as interest and rent, the statutory year rarely is used, since its application can create visible distortion in 31-day months and, especially, in February. However, when spread over a year or more, the error introduced with the statutory year becomes very small, so it is sometimes used for calculating prorated taxes and insurance.

To prorate an item using the statutory year, first find the payment for the full period. Next, divide this payment by the number of months in the period to get the monthly cost, and divide the monthly cost by 30 to find the daily cost. After determining who pays whom and for what length of time, you can calculate the prorated payment.

Using the statutory year was helpful when calculations were made by hand or with adding machines; now that calculators and computers are in common use, dividing by 365 (or 366) instead of 360 isn't such a big deal. As a result, the recent trend has been to use the actual year instead of the statutory year.

THE ACTUAL YEAR

If using the statutory year is forbidden, the actual year must be used. In this case, you need to calculate how many days of the period must be paid for. You then multiply this sum by the *daily* cost of the item to get the prorated payment. The difficulty comes in the fact that to get the daily rate you must divide the annual rate by 365, or 366 in leap years. Further, it often is confusing to try to count days elapsed in a significant part of a year, especially when dealing with odd beginning and ending dates. To try to figure how many days elapse between January 1 and July 26 of a given year is bad enough; to calculate the number of days that elapsed from, say, October 23 of one year to March 14 of the following year is even worse. If you must use the actual year, you have no alternative; you need to remember the number of days in each month and whether or not you're dealing with a leap year. Be sure to keep your calculator in good working order, and to be familiar with its use, to help you make actual-year computations.

ROUNDING IN PRORATION

One very important warning is necessary here. When you are calculating prorations, always carry your intermediate results (i.e., daily and monthly charges) to *at least two decimal places beyond the pennies*. Do *not* round to even cents until you have arrived at your *final* answer, because when you round you introduce a tiny error. This is acceptable with respect to your final answer, but intermediate answers will later be operated on, including multiplication by fairly large numbers. Each time a rounded number is multiplied, the rounding error is multiplied also! To prevent such errors from affecting your final result, always carry intermediate steps to at least two extra decimal places beyond what your final answer will have.

Proration Examples

Here is an example of proration for each of the four commonly prorated items: rent, interest, taxes, insurance.

RENT

Rent usually is paid in advance.

> *The closing date is May 19, 2020, and rent is payable for the calendar month on the first of each month. Who pays whom what in the proration of rent of $795 per month?*

Since the seller received the May rent payment on May 1, he should pay the buyer the portion of rent covering the part of the month *after* the closing date. May has 31 days; the closing date is May 19. Therefore, the buyer will own the property for 31 − 19 = 12 days during May. The daily rent for May is $^1/_{31}$ of the monthly rent since there are 31 days: $795.00 ÷ 31 = $25.6452 (remember to carry to two extra decimal places). The buyer's share of the rent, then, is 12 × $25.6452 = $307.7424, which rounds to $304.74. (Note that if we had rounded the daily rental to $25.65 and then multiplied by the 12 days, the result would have been $307.80, or 5¢ too much.)

INTEREST

Interest is paid after it has accrued. Normally, the interest period ends the day *before* a payment is due—that is, a payment due on the first of the month covers the preceding month; one due on the 18th of the month covers the period from the 18th of the preceding month through the 17th of the current one.

> *The closing date is September 22, 2020. Interest is payable, with the payment on the 16th of the month. The loan amount as of September 16, 2020, is $448,100. The interest rate is 6¼% per year. If the loan is assumed, who pays whom what for prorated interest?*

We have to calculate the monthly interest due and determine how much of that amount is paid by whom. First, the dollar amount of interest—this is $^1/_{12}$ of $6^1/_4$% of the loan balance of \$448,100: $^1/_{12} \times 0.0625 \times \$448{,}100 = \$2{,}333.85$. Note that we rounded off the monthly interest, because the lender does that, too, each month.

The period involved here is 30 days since September has that length. The seller will have to pay the buyer for all of the interest between September 16 and September 22 since, on October 16, the buyer is going to have to pay the full month's interest. The seller will own the property for 7 days of that time. (At a glance, this appears incorrect since $22 - 16 = 6$. However, the seller owns the property on both the 16th and the 22nd, so we must add a day. Another way is to count the days on our fingers, starting with the 16th and ending with the 22nd.)

Now we must determine the daily interest charge, which is $\$2{,}333.85 \div 30 = \$77.7950$. We multiply this by 7 to get the seller's share for 7 days: $\$77.7950 \times 7 = \$544.565$, which rounds to \$544.57.

TAXES

Taxes are usually assessed for a full year, so we would use the statutory year for calculation, if permitted.

> *The closing date is April 27, 2020. Taxes are \$1,188.54 per year. The tax year is March 1 to February 28 (or 29) of the following year. Who pays whom how much in prorated taxes? Use both the statutory and the actual year, assuming (a) that taxes are paid on June 1 and (b) that taxes are paid on March 15 of the tax year.*

This sample problem is probably more complex than any you will encounter on a licensing examination, but it demonstrates all the possibilities.

In situation (a), the seller pays his share to the buyer since the taxes are due after the closing date and so must be paid by the buyer. In situation (b), the buyer pays the seller her share, because the taxes for the full tax year were paid by the seller before the closing date.

Statutory Year Calculations. The closing date is April 27, 2020, and the tax year begins March 1, 2018. Therefore the seller owns the property for 1 month and 27 days:

Note that we did not use April 27, the closing date, in this calculation, because we consider the closing date as belonging to the seller. April 28, then, is the first day that the seller does *not* own the property, and so we must count from that day.

The seller's share is for 1 month and 27 days. The total tax payment is \$1,188.54 per year. Dividing by 12 yields a monthly tax charge of \$99.0450. Dividing this amount by 30 yields a daily tax charge of \$3.3015.

1 month @ \$99.0450	=	\$99.0450	
+ 27 days @ \$3.3015	=	+ \$89.1405	
Total seller's charge		\$188.1855	= \$188.19

The seller's share is \$188.19, so the buyer's share is the rest, or

$$\$1{,}188.54 - \$188.19 = \$1{,}000.35$$

In situation (a), then, the seller pays the buyer \$188.19.

In situation (b), the buyer pays the seller \$1,000.35.

Actual Year Calculations. When calculating using the actual year, we reduce everything to days. The year has 365 days. When using the same problem, the daily tax charge is $1,188.54 ÷ 365 = $3.25627. The seller owns the property for a total of 58 days: 31 days of March and 27 days of April of the tax year. Therefore the seller's share is

58 × $3.25627 = $188.8654 = $188.87

The buyer's share would be $1,188.54 − $188.87 = $999.67.

In situation (a), taxes are paid on June 1 and the seller pays the buyer $188.87.

In situation (b), taxes are paid on March 15 and the buyer pays the seller $999.67.

Note two things about this calculation. First, the results are 68¢ different from those obtained with the statutory year method; this discrepancy occurs because of the different calculation technique. Second, when the daily charge was calculated, it was carried to *three* extra decimal places. In the actual year calculation, we might have to multiply by a number as large as 365, and the rounding error would be multiplied by that much. Carrying to three extra decimal places reduces this error.

INSURANCE

> *A 3-year insurance policy, dated October 22, 2020, is assumed on the closing date of August 11, 2022. The full 3-year premium of $559.75 was paid at the time the policy was bought. Who pays whom what if this policy is prorated?*

First we calculate the monthly insurance charge, which is 1/36 (3 years, remember) of the premium of $559.75: $559.75 ÷ 36 = $15.5486. The daily charge is 1/30 of that amount: $15.5486 ÷ 30 = $0.5183.

Now we must calculate how long the buyer will use the policy. If the policy is dated October 22, 2020, then it expires on October 21 (at midnight) of 2023.

The buyer will own the policy for 1 year, 2 months, and 10 days. He will pay the seller the prorated share since the seller paid for the full 3 years when she bought the policy. (Now we just go ahead and figure 1 year and 2 months to be 14 months to save the problem of calculating the annual premium.)

14 months @ $15.5486	=	$217.6804		
10 days @ $0.5183	=	+$ 5.1830		
Total payable to seller		$222.8634	=	$222.86

Calculating insurance according to the actual year can be cumbersome. To do this we ought to calculate first the annual premium, which is 1/3 of $559.75 = $186.58333. The daily premium charge is this amount divided by 365: $186.58333 ÷ 365 = $0.51119. Note that once again we are carrying actual year calculations to an extra *three* decimal places.

HINTS ON HANDLING MATHEMATICAL PROBLEMS ON EXAMINATIONS

Now that the different kinds of mathematics problems that appear on licensing examinations have been discussed, a few remarks dealing with the proper ways to approach them are in order.

You should remember that your objective on the examination is to get a passing grade. It is *not* necessary that you get 100%—only enough to pass. Mathematical problems can be terrible time consumers; therefore you should devote your time at the outset to the problems that do not take a lot of time. Save the complicated ones for later, when you have had the chance to answer all the "easy" questions. Many examinees determinedly tackle the mathematics first or spend tremendous amounts of time on a very few problems, only to find later that they have to rush just to have a chance of getting to every question on the examination.

Try to determine just how long a problem will take *before* you tackle it. If it's going to take a lot of time, postpone it. Then when you get back to the time-consuming questions, do first those that are worth the most points. This may mean that you will turn in your examination without finishing one or two of the really long arithmetic problems. That doesn't mean you're stupid—it means you're smart. Instead of slaving away over the few points these unfinished problems represented, you used your time to build up a good score on the other parts of the examination that could be answered quickly. However, before turning in your test paper, try to mark an answer for *every* question, even math problems not attempted. You might just guess the right answer!

Another point to remember concerning arithmetic questions is that you may not need all the information given. After you read the question and determine just what you are being asked to do, begin to search for the information you need to provide the answer. Do not assume that just because some information is included you must find some way of using it in your solution. Fairly often extra information has been included just to sidetrack or confuse examinees whose arithmetic skills make them unsure of themselves.

QUESTIONS ON CHAPTER 18

1. A 2,000-square-foot house is on an acre of land, with an additional 4½ adjacent acres. The price is $50 per square foot for the home plus $5,000 for each and every acre of land. The total price is
   (A) $100,000
   (B) $127,500
   (C) $122,500
   (D) $117,500

2. Two adjacent lots, each 50 feet wide by 125 feet deep, cost a total of $50,000. What is the approximate price per square foot?
   (A) $1
   (B) $2
   (C) $3
   (D) $4

3. A property management agreement requires the owner to pay a fee of 6% of the first $100,000 of rental income received and 5% of the excess. Gross income of $140,000 was received. What is the fee?
   (A) $5,000
   (B) $6,000
   (C) $7,000
   (D) $8,000

4. A property manager's fee is set at 6% of rental income collected. The rent roll is $5,500. However, one tenant skipped out, owing $500. Operating expenses were $2,000. How much is the management fee for that month?
   (A) $330
   (B) $300
   (C) $180
   (D) $120

5. A salesperson is trying to determine how much to suggest for a listing price. The lot is worth $20,000, the house is 1,500 square feet at $50 per square foot, and a 6% sales commission is to be included. The price that will meet this is
   (A) $95,000
   (B) $100,700
   (C) $111,170
   (D) $101,064

6. A 20-year-old house is 30 × 40 feet. Replacement cost is $45 per square foot. Depreciation is estimated at 2% per year. The lot is valued at $10,000. The estimated value using the cost approach is
   (A) $10,000
   (B) $32,400
   (C) $64,000
   (D) $42,400

7. The listing broker is to split the commission 50-50 with the selling broker. The land is a 200-acre farm that sold for $2,500 per acre with a 5% commission. Each broker may receive a commission of
   (A) $12,500
   (B) $25,000
   (C) $50,000
   (D) $500,000

8. A mortgage provides a 2% prepayment penalty for the first five years. The mortgage began at $34,800 and requires principal and interest payments of $249.32 each month for thirty years. The interest rate is 7¾%. If the mortgage is prepaid immediately after the first payment, the prepayment penalty is
   (A) $699.88
   (B) $89,505.98
   (C) $695.50
   (D) $34,775.50

9. A property is listed for $340,000 and sells for $330,000. The listing broker splits the 6% commission with the selling broker, each getting 50%. The selling broker's commission from the sale is
   (A) $20,400
   (B) $19,800
   (C) $10,200
   (D) $9,900

10. Property with a market value of $85,000 is assessed at 50% of value, less a $5,000 homestead exemption. The tax rate is $35 per $1,000 taxable. The amount of tax is
    (A) $2,975
    (B) $1,487
    (C) $2,800
    (D) $1,312.50

11. A typical lot sells for $20,000 per acre. The subject lot is 1¼ acres. What is its most likely price?
    (A) $20,000
    (B) $25,000
    (C) $15,000
    (D) to be determined by market analysis

12. A married couple has $78,000 of combined yearly income. A lender will provide a loan of 2.5 times yearly income. What is the maximum house they can afford if they make a 10% down payment?
    (A) $195,000
    (B) $214,500
    (C) $216,666
    (D) $156,000

13. A property sells for $125,000. The seller's closing costs are 2%, and the broker's commission is 6%. The mortgage payoff is $68,000. How much will the seller receive in cash at closing?
    (A) $52,000
    (B) $53,000
    (C) $47,000
    (D) $62,000

14. A vacant lot was bought for $12,000 and resold for $15,000. What is the percentage of gross profit?
    (A) 3%
    (B) 20%
    (C) 25%
    (D) 125%

15. A house was sold with no mortgage. The seller received a check for $60,000 after paying $800 of closing expenses and a 6% brokerage fee. What was the gross sales price?
    (A) $63,600
    (B) $64,400
    (C) $64,681
    (D) $66,000

16. An apartment rents for $1,200 a month; the first month is payable before move-in. A one-month security deposit is required plus half a month's rent for a pet deposit, with deposits payable before move-in. How much must Veronica pay before she moves in with her dog?
    (A) $1,200
    (B) $2,400
    (C) $3,000
    (D) $3,600

17. Three rectangular lots contain a total of 21,000 square feet. All are 100 feet deep. How much road frontage does each have?
    (A) 700 feet
    (B) 70 feet
    (C) 210 feet
    (D) 2,100 feet

18. The monthly gross rent multiplier was 100. The property rented for $1,200 per month. The value was
    (A) $120,000
    (B) $100,000
    (C) $1,200,000
    (D) not enough information to determine

19. Interest for the most recent month was $450. The annual interest rate is 5%. How much is the current principal balance?
    (A) $108,000
    (B) $90,000
    (C) $900,000
    (D) $1,108,000

20. A salesperson pays $900 per month to the broker for use of a desk and telephone and receives 60% of commissions brought in. Last month, her net was $3,600. How much were the commissions the salesperson earned on her sales?
    (A) $2,700
    (B) $3,600
    (C) $4,500
    (D) $9,000

21. A house sold for $275,000 with a 6% commission. The listing broker paid the selling broker 50%. The listing salesperson and the selling salesperson each received 40% of what their broker received. How much did the listing salesperson receive?
    (A) $16,500
    (B) $8,250
    (C) $4,950
    (D) $3,300

22. Houses in the area have been appreciating by 6% per year. A typical home that sold two years ago for $150,000 would now be worth
    (A) $168,000
    (B) $168,540
    (C) $159,000
    (D) $177,000

23. A 10,000-square-foot warehouse is offered to lease at $4 per square foot per year. The vacancy allowance is 5%. Debt service is $12,000 per year. Depreciation is $6,000 per year. The tenant will pay $12,000 in annual taxes. If the capitalization rate is 8%, the estimated value will be
    (A) $475,000
    (B) $500,000
    (C) $333,000
    (D) $200,000

24. Property has an assessed value of $100,000. The tax rate is 2½% of assessed value. Taxes are payable on December 1. Closing takes place June 1. At closing
    (A) no proration of taxes is necessary
    (B) the buyer is credited with $2,500
    (C) the buyer is debited $1,250
    (D) the buyer is credited $1,250 and the seller is debited $1,250

25. Half of the property taxes are due to the county on June 1 and the other half on December 1 of each year. The penalty is 1% per month of the amount that is late. Harold's tax bill is $1,200, which he pays in full on December 31. How much is his penalty?
    (A) $12
    (B) $42
    (C) $48
    (D) $84

26. Harold agreed to pay $500 per front foot for a 100′ × 200′ lot with frontage on a commercial street. What was the total price?
    (A) $50,000
    (B) $100,000
    (C) both A and B
    (D) neither A nor B

ANSWERS

1. **B**	6. **D**	11. **B**	16. **C**	21. **D**	26. **A**
2. **D**	7. **A**	12. **C**	17. **B**	22. **B**	
3. **D**	8. **C**	13. **C**	18. **A**	23. **A**	
4. **B**	9. **D**	14. **C**	19. **A**	24. **D**	
5. **D**	10. **D**	15. **C**	20. **C**	25. **C**	

TILA-RESPA Integrated Disclosure (TRID)

19

ESSENTIAL MATTERS TO LEARN FROM THIS CHAPTER

- The Consumer Financial Protection Bureau (CFPB) is a U.S. government agency that regulates mortgage loan disclosures by banks and other lenders.
- The CFPB is responsible for implementing the TILA-RESPA Integrated Disclosure (TRID) rule.
- TILA is an acronym for the Truth in Lending Act.
- RESPA is an acronym for the Real Estate Settlement Procedures Act.
- The purpose of TRID is to allow the borrower to "Know Before You Owe."
- TRID was promulgated by the CFPB to ensure that residential mortgage borrowers are provided adequate disclosures and have an opportunity to shop for loans.
- TRID's primary requirement on lenders is that they disclose costs and fees of loans to prospective borrowers.
- The CFPB uses TRID to mandate disclosures, not to limit or regulate fee amounts charged.
- The CFPB's purpose is to assure prospective buyers that they are provided with full disclosure.
- The disclosures are an effort to allow borrowers to shop for the best or most suitable mortgage loan.
- On October 3, 2015, two new forms were mandated for use with most residential mortgages and replace prior forms.
- The *Loan Estimate* is a new form. The lender must provide a Loan Estimate within three days of a borrower's application.
- The Loan Estimate form replaces the Good Faith Estimate and Truth in Lending Act disclosure booklet.
- The Loan Estimate describes initial fees and costs if the borrower goes forward with the loan.
- The Loan Estimate describes the interest rate and required payments on the loan.
- The Loan Estimate can be compared to the *Closing Disclosure* (the second new form), which is provided three days before closing to ensure the borrower is getting the loan he/she was promised.
- The new requirements are called the TILA-RESPA Integrated Disclosure (TRID) rule.

GOVERNMENT AGENCY OVERSIGHT

The Consumer Financial Protection Bureau (CFPB), formed in 2011, was entrusted with the responsibility of regulating banks and other lenders who offer residential mortgage loans. Previously, such regulation was handled by the Department of Housing and Urban Development (HUD).

The CFPB replaced HUD as the agency that enforces the Real Estate Settlement Procedures Act (RESPA). The purpose of government regulation is to assure disclosure of loan terms and fairness to borrowers so they can shop effectively for the most suitable mortgage loan.

New Forms

Beginning on October 3, 2015, CFPB required banks and other lenders to provide a three-page form to mortgage applicants. The form is called the *Loan Estimate.* Most of the three-page form is applicable to all residential mortgages. Some variations provide additional provisions to describe conditions for additional payments (which are often voluntary) on the loan or for interest rate changes. The principal purpose of this form is to allow borrowers to shop for the most suitable loan and to compare loan terms offered by different lenders.

The second new form required by the CFPB is the *Closing Disclosure* (discussed in Chapter 20), which must be provided three days before closing. By comparing the Loan Estimate with the Closing Disclosure, the borrower can easily compare features of the proposed loan with the loan that will actually be provided.

Replacements

The Loan Estimate and Closing Disclosure forms generally replace the Good Faith Estimate (GFE) and Truth in Lending Act (TILA) disclosure booklet. The HUD-1 closing statement is no longer used in most residential transactions.

Exceptions

Borrowers who apply for a reverse mortgage will receive a GFE and TILA disclosure instead of the Loan Estimate and Closing Disclosure forms. Those applying for a home equity line of credit (HELOC), a manufactured housing loan not secured by real estate, or a loan through certain housing assistance programs will receive a TILA disclosure but not a GFE or a Loan Estimate.

Trick Question

A trick question suggests that either the Loan Estimate or Closing Disclosure places limits on charges. This is not so! These forms require the disclosure of loan charges but do not stipulate or regulate the charges themselves.

SAMPLE LOAN ESTIMATE

Figure 19-1 is a three-page Loan Estimate form. This sample has been filled in as an example.

FICUS BANK

4321 Random Boulevard • Somecity, ST 12340

Save this Loan Estimate to compare with your Closing Disclosure.

Loan Estimate

DATE ISSUED	2/15/2013
APPLICANTS	Michael Jones and Mary Stone 123 Anywhere Street Anytown, ST 12345
PROPERTY	456 Somewhere Avenue Anytown, ST 12345
SALE PRICE	$180,000

LOAN TERM	30 years
PURPOSE	Purchase
PRODUCT	Fixed Rate
LOAN TYPE	☒ Conventional ☐ FHA ☐ VA ☐ ____________
LOAN ID #	123456789
RATE LOCK	☐ NO ☒ YES, until 4/16/2013 at 5:00 p.m. EDT *Before closing, your interest rate, points, and lender credits can change unless you lock the interest rate. All other estimated closing costs expire on **3/4/2013** at 5:00 p.m. EDT*

Loan Terms

Loan Terms		Can this amount increase after closing?
Loan Amount	$162,000	**NO**
Interest Rate	3.875%	**NO**
Monthly Principal & Interest *See Projected Payments below for your Estimated Total Monthly Payment*	$761.78	**NO**
		Does the loan have these features?
Prepayment Penalty		**YES** • **As high as $3,240** if you pay off the loan during the first 2 years
Balloon Payment		**NO**

Projected Payments

Payment Calculation	Years 1-7	Years 8-30
Principal & Interest	$761.78	$761.78
Mortgage Insurance	+ 82	+ —
Estimated Escrow *Amount can increase over time*	+ 206	+ 206
Estimated Total Monthly Payment	$1,050	$968

Estimated Taxes, Insurance & Assessments *Amount can increase over time*	$206 a month	**This estimate includes** ☒ Property Taxes ☒ Homeowner's Insurance ☐ Other: *See Section G on page 2 for escrowed property costs. You must pay for other property costs separately.*	**In escrow?** YES YES

Costs at Closing

Estimated Closing Costs	$8,054	Includes $5,672 in Loan Costs + $2,382 in Other Costs – $0 in Lender Credits. *See page 2 for details.*
Estimated Cash to Close	$16,054	Includes Closing Costs. *See Calculating Cash to Close on page 2 for details.*

Visit **www.consumerfinance.gov/mortgage-estimate** for general information and tools.

LOAN ESTIMATE PAGE 1 OF 3 • LOAN ID # 123456789

Figure 19-1.

Closing Cost Details

Loan Costs

A. Origination Charges	$1,802
.25 % of Loan Amount (Points)	$405
Application Fee	$300
Underwriting Fee	$1,097

B. Services You Cannot Shop For	$672
Appraisal Fee	$405
Credit Report Fee	$30
Flood Determination Fee	$20
Flood Monitoring Fee	$32
Tax Monitoring Fee	$75
Tax Status Research Fee	$110

C. Services You Can Shop For	$3,198
Pest Inspection Fee	$135
Survey Fee	$65
Title – Insurance Binder	$700
Title – Lender's Title Policy	$535
Title – Settlement Agent Fee	$502
Title – Title Search	$1,261

D. TOTAL LOAN COSTS (A + B + C)	$5,672

Other Costs

E. Taxes and Other Government Fees	$85
Recording Fees and Other Taxes	$85
Transfer Taxes	

F. Prepaids	$867
Homeowner's Insurance Premium (6 months)	$605
Mortgage Insurance Premium (months)	
Prepaid Interest ($17.44 per day for 15 days @ 3.875%)	$262
Property Taxes (months)	

G. Initial Escrow Payment at Closing		$413
Homeowner's Insurance	$100.83 per month for 2 mo.	$202
Mortgage Insurance	per month for mo.	
Property Taxes	$105.30 per month for 2 mo.	$211

H. Other	$1,017
Title – Owner's Title Policy (optional)	$1,017

I. TOTAL OTHER COSTS (E + F + G + H)	$2,382

J. TOTAL CLOSING COSTS	$8,054
D + I	$8,054
Lender Credits	

Calculating Cash to Close

Total Closing Costs (J)	$8,054
Closing Costs Financed (Paid from your Loan Amount)	$0
Down Payment/Funds from Borrower	$18,000
Deposit	– $10,000
Funds for Borrower	$0
Seller Credits	$0
Adjustments and Other Credits	$0
Estimated Cash to Close	$16,054

LOAN ESTIMATE PAGE 2 OF 3 • LOAN ID # 123456789

Figure 19-1. (continued)

Additional Information About This Loan

LENDER	Ficus Bank	**MORTGAGE BROKER**	
NMLS/__ LICENSE ID		**NMLS/__ LICENSE ID**	
LOAN OFFICER	Joe Smith	**LOAN OFFICER**	
NMLS/__ LICENSE ID	12345	**NMLS/__ LICENSE ID**	
EMAIL	joesmith@ficusbank.com	**EMAIL**	
PHONE	123-456-7890	**PHONE**	

Comparisons	Use these measures to compare this loan with other loans.
In 5 Years	$56,582 Total you will have paid in principal, interest, mortgage insurance, and loan costs. $15,773 Principal you will have paid off.
Annual Percentage Rate (APR)	4.274% Your costs over the loan term expressed as a rate. This is not your interest rate.
Total Interest Percentage (TIP)	69.45% The total amount of interest that you will pay over the loan term as a percentage of your loan amount.

Other Considerations

Appraisal	We may order an appraisal to determine the property's value and charge you for this appraisal. We will promptly give you a copy of any appraisal, even if your loan does not close. You can pay for an additional appraisal for your own use at your own cost.
Assumption	If you sell or transfer this property to another person, we ☐ will allow, under certain conditions, this person to assume this loan on the original terms. ☒ will not allow assumption of this loan on the original terms.
Homeowner's Insurance	This loan requires homeowner's insurance on the property, which you may obtain from a company of your choice that we find acceptable.
Late Payment	If your payment is more than *15* days late, we will charge a late fee of *5% of the monthly principal and interest payment.*
Refinance	Refinancing this loan will depend on your future financial situation, the property value, and market conditions. You may not be able to refinance this loan.
Servicing	We intend ☐ to service your loan. If so, you will make your payments to us. ☒ to transfer servicing of your loan.

Confirm Receipt

By signing, you are only confirming that you have received this form. You do not have to accept this loan because you have signed or received this form.

Applicant Signature Date Co-Applicant Signature Date

LOAN ESTIMATE PAGE 3 OF 3 • LOAN ID #123456789

Figure 19-1. (continued)

QUESTIONS ON CHAPTER 19

1. TRID stands for TILA-RESPA Integrated Disclosure. What does TILA stand for?
   (A) Truth in Lending Act
   (B) Trust in Life Award
   (C) Tricks in Lending Avoidance
   (D) Telling Lies Act

2. Which U.S. government agency administers TRID?
   (A) Department of Housing and Urban Development
   (B) Federal Housing Authority
   (C) Consumer Protection Agency
   (D) Consumer Financial Protection Bureau

3. In what year were the Loan Estimate and Closing Disclosure forms first required for most residential loans?
   (A) 1929
   (B) 1968
   (C) 2011
   (D) 2015

4. A principal purpose of the Loan Estimate form is to
   (A) limit interest rates
   (B) limit closing costs
   (C) restrict prepayment fees
   (D) allow borrowers to shop for loans

5. The Loan Estimate form has how many pages?
   (A) 1
   (B) 2
   (C) 3
   (D) 4 or more depending on the type of loan

6. After applying for a mortgage loan, the lender has __________ days to provide a Loan Estimate to the applicant.
   (A) 1
   (B) 2
   (C) 3
   (D) 4 or more depending on the complexity of the loan

7. A Loan Estimate must be provided to all mortgage applicants *except* those seeking a(n)
   (A) fixed-rate mortgage
   (B) fixed-term mortgage
   (C) adjustable-rate mortgage
   (D) reverse mortgage

8. Which federal form showed debits and credits to the buyer and seller but is no longer provided after a Loan Estimate?
   (A) HUD-1
   (B) Closing Disclosure
   (C) signed loan application
   (D) credit report

9. The TRID program is associated with what phrase?
   (A) "Know Before You Owe"
   (B) "Think Before You Borrow"
   (C) Title Response Procedures Act
   (D) Treasury Interest Reduction Act

ANSWERS

1. **A**	4. **D**	7. **D**
2. **D**	5. **C**	8. **A**
3. **D**	6. **C**	9. **A**

Closing a Real Estate Transaction

20

ESSENTIAL MATTERS TO LEARN FROM THIS CHAPTER

- *Closing* of a real estate transaction is the time when money and property change hands. Closing is also called *settlement.*
- In years past, the seller handed a deed to the buyer; in the same motion, the buyer handed payment to the seller. One party didn't release the paper to be given until the paper to be received was firmly in the other hand.
- In past years in some local customs, brokers prepared closing statements. This is seldom done these days. Accordingly, there are few questions about settlement, mostly fundamental ones, on salesperson exams. More detailed questions, such as those computing prorations, are on broker examinations.
- Nowadays, almost all the effort occurs before closing. Lawyers work for the buyer or for the seller but not for both. Lawyers may also work for the lenders. The seller's lender wants to ensure that money received is sufficient to pay off the loan and will be applied to do so. The buyer's lawyer wants to be certain that the seller can give good title.
- A *closing agent,* often a title insurance company, frequently serves as the intermediary, trusted by both parties. The closing agent receives the buyer's payment and the seller's deed. Only after everything is in acceptable order does the closing proceed.
- The closing agent prepares an accounting for both parties, called a *closing* (or *settlement*) *statement.* In the sale of real estate, numerous fees and charges must be accounted for. The closing agent ensures that the charges are accurately applied and that the proper papers are to be recorded at the county courthouse. The closing agent writes checks to parties who are paid at closing.
- Sometimes a fee may be *paid out* (that is, *outside*) *of closing* (POC). For example, a buyer may pay a fee to a home inspector by writing a separate check. POCs are stated on the closing statement, but the amounts of POC fees are not included in the accounting of closing statements.
- Debits and credits are used in the closing statement as general titles for charges and credits allowed. *Debits* are charges that must be paid by a party, whereas credits recognize that a party has paid something. For example, the seller is credited with the property's price. The buyer is debited for the property's price. The seller receives cash and is debited for that money. The buyer pays cash and is credited with the amount of cash paid.
- Total debits and credits for the seller must equal each other. Total debits and credits for the buyer must equal each other. However, a seller's total debits or credits will not necessarily equal the buyer's total debits or credits.

- If the seller is responsible for paying the broker's sales commission, the seller is debited. However, there is not a corresponding entry for the buyer because the buyer neither pays nor receives the sales commission.
- Certain items, usually expenses, are shared within the year by the buyer and seller. Sharing these items within the year is called *proration.* When the seller has paid taxes for the full year and the buyer will enjoy a partial year without paying the tax collector for the expense, the buyer must pay the seller at closing for the portion of the item he will use. For example, the seller paid $365 of property taxes on April 1, covering the entire year. Closing is on July 1. The buyer must pay the seller for the half year of his (the buyer's) ownership, July 1 through December 31, which is approximately $182. The buyer is debited $182. The seller is credited $182 because the seller paid the tax collector for the full year and deserves to get half a year back from the buyer.
- Here is a different example. Property taxes for the year are due December 1. Closing is March 13. The taxes will be paid by the buyer to the tax collector on December 1. The seller must pay the buyer for the approximately 72 days when he (the seller) owned the property without paying taxes. The estimated annual taxes are prorated at closing, with the seller being debited and the buyer being credited for the same amount. The prorated estimated amount is paid by the seller, to the buyer, at closing.
- Provisions of the Truth in Lending Act (TILA) and Real Estate Settlement Procedures Act (RESPA) have been implemented in a new rule called TILA-RESPA Integrated Disclosure (TRID), which applies to loan applications of a buyer of a personal residence, including a house (or summer house), condominium, co-op, mobile home, or lot on which a house will be built or a mobile home can be placed. (Mobile homes are now called *manufactured* homes.)
- The intent of TRID is to provide disclosure to prospective homebuyers about closing costs so they may shop for loans. TRID does not apply to investors with plans to remodel a house for resale, an apartment complex, or farmland with or without a tenant's dwelling. TRID does apply to a farm with the new owner's residence.
- The Consumer Financial Protection Bureau (CFPB) is the U.S. government agency that administers TRID. RESPA was previously administered by the Department of Housing and Urban Development (HUD).
- TRID does not regulate the amount of fees or charges. Under TRID, within three business days after loan application, the lender is to give the proposed buyer/borrower a Loan Estimate. (A sample Loan Estimate form is provided in Chapter 19.) Then the borrower may shop elsewhere.
- At least three days before the consummation of the loan, a Closing Disclosure form must be given to the borrower. The borrower can compare this to the Loan Estimate to be certain they are in agreement—that the borrower will get what was promised. If a revision is needed to the proposed Closing Disclosure, three business days must be provided before the closing.

CLOSING STATEMENT PREPARATION

Quite some time ago, it was common practice for real estate brokers (but not salespeople) to prepare the *closing statements* (also called *settlement statements, escrow statements*) for the transactions they and their salespeople effected. Therefore, it is traditional for several questions about closing statements to appear on *broker's* license examinations. Salesperson's examinations rarely have anything more than an occasional very simple question about closing statements. Nowadays, although licensing laws usually require brokers to make sure that a closing statement is provided to both buyer and seller, common practice is for someone other than the broker actually to prepare the statement. Often the lender providing a new mortgage loan for the buyer will do so; otherwise closings can be handled by *title companies, abstract companies*, or *escrow agents.*

Usually the title company or other closing agent will provide the deed and mortgage documents to the county clerk to be recorded. Recording gives constructive notice of their existence.

Despite the verification of ownership and recording, sometimes a false document slips through or a valid document is not recorded. When a valid claim appears, some party may lose its ownership interest. Title insurance exists for that reason. Mortgage lenders will demand that a title insurance policy cover their loan amount, the cost paid by someone else. Buyers can also purchase title insurance to cover their equity investment.

Title insurance is paid only once, at closing, unlike hazard and most other types of insurance that must be paid annually. The title insurance policy is issued based on a title company's or an attorney's title search of records kept by the county clerk at the county courthouse.

Internal Revenue Service Reporting Requirements

The Federal Tax Reform Act of 1986 requires that brokers and/or settlement agents report the details of some real estate transactions to the U.S. Internal Revenue Service. The primary responsibility lies with the settlement agent, who is the person actually handling the closing.

If there is no settlement agent, the responsibility rests with the buyer's broker and the seller's broker, in that order. Not all transactions need to be reported.

Purpose of Closing Statements

Closing statements provide an accounting of all funds involved in a real estate transaction. These statements show the amount that the buyer must pay and that the seller will receive from the transaction. Buyer and seller each are given an accounting of all items they must pay for or are credited with in the transaction. A title company may prepare a reconciliation for each sale as well, to "button up" the statements.

Debits and Credits

You do not have to be a bookkeeper to understand closing statements. Two columns are shown for the buyer and two for the seller. The two columns for each party are a debit column and a credit column. Totals of debit and credit columns for the same party must agree with each other.

Listed in the debit column are amounts that the party being considered, buyer or seller, is charged for. Listed in the credit column are those that the party will receive credit for. Cash is needed to balance.

As a simple example, assume the sale of a $520,000 house on July 30, 2020. Assume that the seller has already paid taxes of $6,859.24 for the entire 2020–2021 fiscal year ended June 30, 2021. The seller should be credited with the payment of $520,000 for the house and $6,306.68 of prepaid taxes for the remainder of the 2020 year (July 31, 2020, to June 30, 2021, during which time the buyer will own the house). Cash will be the offsetting debit.

The buyer will be debited for the house and paid-up taxes; cash is the offsetting credit. The closing statements will appear as follows:

	Seller		Buyer	
	Debit	**Credit**	**Debit**	**Credit**
Real property		$520,000.00	$520,000.00	
Prepaid taxes		6,306.68	6,306.68	
Cash due from buyer				$526,306.68
Cash due to seller	$526,306.68			
Totals	**$526,306.68**	**$526,306.68**	**$526,306.68**	**$526,306.68**

(**NOTE:** If you cannot easily tell how the $6,306.68 of prorated property tax was calculated, you should review the "Proration" section in Chapter 18.)

Now let's consider the same transaction except that we will add two more items. A broker was involved, who earned a commission of $31,200 for making the sale; also, the buyer must pay $300 for a survey. The seller is to pay the broker's commission, so the seller's statement will show a debit of $31,200 for that. The buyer's statement shows a debit of $300 for the survey. The new closing statement will appear as follows:

	Seller		Buyer	
	Debit	**Credit**	**Debit**	**Credit**
Real property		$520,000.00	$520,000.00	
Prepaid taxes		6,306.68	6,306.68	
Sales commission	$31,200.00			
Cash due from buyer			300.00	$526,606.68
Cash due to seller	$495,106.68			
Totals	**$526,306.68**	**$526,306.68**	**$526,606.68**	**$526,606.68**

An item that affects only one party is shown as a debit or a credit *only on the statement of the party affected.* For example, if the sales commission is paid by the seller, the amount appears only on the seller's closing statement. If the buyer pays for a survey, it appears only as a debit to him.

Items of value that are sold, exchanged, or transferred between buyer and seller are shown in *opposite* columns of *both* parties. For example, the transferred property is shown as a credit to the seller and as a debit to the buyer. If an item is transferred, sold, or exchanged, it is a debit to one party and a credit to the other.

Trick Question

A trick question suggests impossible answers as choices. Eliminate these impossible choices. An item that is transferred, exchanged, or taken over by the opposite party should never appear as either a debit to both parties or as a credit to both parties. When both parties are affected by the same matter, such as prorated taxes, the item is a credit to one and a debit to the other.

Items usually *debited* (charged) to the buyer that are likely to be encountered include the following:

1. Purchase price of the real property
2. Purchase price of personal property
3. Deed-recording fees
4. Title examination
5. Title insurance
6. Hazard insurance
7. Survey
8. Appraisal fee (sometimes charged to seller)
9. Prepaid taxes
10. Loan assumption fees

Items that are likely to be credited to the buyer are:

1. Earnest money deposits
2. Proceeds of a loan he borrows
3. Assumption of a loan
4. Mortgage to the seller (purchase money)
5. Current taxes unpaid to closing
6. Tenant rents paid in advance (the seller collected these)
7. Balance due to close (paid by buyer to close)

Items likely to be *debited* to a seller are:

1. Sales commission
2. Current but unpaid taxes
3. Existing debt, whether assumed or to be paid off
4. Loan prepayment penalties
5. Discount points for buyer's VA or FHA loan. Discount points on *conventional* loans may be charged to buyer or to seller depending on contractual arrangements, custom in the area, local law, and so on.
6. Rent received in advance
7. Deed preparation

Items usually *credited* to a seller are:

1. Sales price of real property
2. Sales price of personal property
3. Prepaid taxes
4. Prepaid insurance (only if policy is assumed by buyer)
5. Escrow balance held by lender

TILA-RESPA INTEGRATED DISCLOSURE (TRID)

TILA-RESPA Integrated Disclosure (TRID) is a federal law that covers most residential mortgage loans used to finance the purchase of one- to four-family properties. Included are a house, a condominium or cooperative apartment unit, a lot with a mobile home, and a lot on which a house will be built or a mobile home placed using the proceeds of a loan.

Purpose of TRID

The purpose of TRID is to provide potential borrowers with information concerning the settlement (closing) process so that they can shop intelligently for settlement services and make informed decisions. TRID does not set the prices for services; its purpose is merely to provide information about settlement (closing) and costs.

SAMPLE CLOSING DISCLOSURE

Beginning in October 2015, the lender must supply the borrower-applicant with a three-page Loan Estimate within three days of application.

Assuming that the application goes forward, the lender must provide a Closing Disclosure and then wait three days (six days if the form was mailed) before closing so that all revisions to the proposed contract are included. Figure 20-1 shows a five-page Closing Disclosure form. This sample has been filled in as an example.

Previous Form Usage

The disclosure form formerly used for the Truth in Lending Act (TILA) and the Good Faith Estimate (GFE) will have less frequent use in the future. Both will be used for a reverse mortgage. The TILA disclosure form (but not the GFE) will be used for a home equity line of credit (HELOC), a manufactured housing loan that is not secured by real estate, or a loan through certain types of homebuyer assistance programs. The HUD-1 Uniform Settlement Statement form will no longer be used in most residential real estate closings.

Closing Disclosure

This form is a statement of final loan terms and closing costs. Compare this document with your Loan Estimate.

Closing Information

Date Issued	4/15/2013
Closing Date	4/15/2013
Disbursement Date	4/15/2013
Settlement Agent	Epsilon Title Co.
File #	12-3456
Property	456 Somewhere Ave Anytown, ST 12345
Sale Price	$180,000

Transaction Information

Borrower	Michael Jones and Mary Stone 123 Anywhere Street Anytown, ST 12345
Seller	Steve Cole and Amy Doe 321 Somewhere Drive Anytown, ST 12345
Lender	Ficus Bank

Loan Information

Loan Term	30 years
Purpose	Purchase
Product	Fixed Rate
Loan Type	☒ Conventional ☐ FHA ☐ VA ☐ ______
Loan ID #	123456789
MIC #	000654321

Loan Terms		Can this amount increase after closing?
Loan Amount	$162,000	**NO**
Interest Rate	3.875%	**NO**
Monthly Principal & Interest *See Projected Payments below for your Estimated Total Monthly Payment*	$761.78	**NO**
		Does the loan have these features?
Prepayment Penalty		**YES** • **As high as $3,240** if you pay off the loan during the first 2 years
Balloon Payment		**NO**

Projected Payments

Payment Calculation	Years 1-7	Years 8-30
Principal & Interest	$761.78	$761.78
Mortgage Insurance	+ 82.35	+ —
Estimated Escrow *Amount can increase over time*	+ 206.13	+ 206.13
Estimated Total Monthly Payment	$1,050.26	$967.91

		This estimate includes	In escrow?
Estimated Taxes, Insurance & Assessments *Amount can increase over time* *See page 4 for details*	$356.13 a month	☒ Property Taxes	**YES**
		☒ Homeowner's Insurance	**YES**
		☒ Other: Homeowner's Association Dues	**NO**
		See Escrow Account on page 4 for details. You must pay for other property costs separately.	

Costs at Closing

Closing Costs	$9,712.10	Includes $4,694.05 in Loan Costs + $5,018.05 in Other Costs – $0 in Lender Credits. *See page 2 for details.*
Cash to Close	$14,147.26	Includes Closing Costs. *See Calculating Cash to Close on page 3 for details.*

CLOSING DISCLOSURE PAGE 1 OF 5 • LOAN ID # 123456789

Figure 20-1.

Closing Cost Details

Loan Costs	Borrower-Paid		Seller-Paid		Paid by Others
	At Closing	Before Closing	At Closing	Before Closing	
A. Origination Charges	**$1,802.00**				
01 0.25 % of Loan Amount (Points)	$405.00				
02 Application Fee	$300.00				
03 Underwriting Fee	$1,097.00				
04					
05					
06					
07					
08					
B. Services Borrower Did Not Shop For	**$236.55**				
01 Appraisal Fee to John Smith Appraisers Inc.					$405.00
02 Credit Report Fee to Information Inc.		$29.80			
03 Flood Determination Fee to Info Co.	$20.00				
04 Flood Monitoring Fee to Info Co.	$31.75				
05 Tax Monitoring Fee to Info Co.	$75.00				
06 Tax Status Research Fee to Info Co.	$80.00				
07					
08					
09					
10					
C. Services Borrower Did Shop For	**$2,655.50**				
01 Pest Inspection Fee to Pests Co.	$120.50				
02 Survey Fee to Surveys Co.	$85.00				
03 Title – Insurance Binder to Epsilon Title Co.	$650.00				
04 Title – Lender's Title Insurance to Epsilon Title Co.	$500.00				
05 Title – Settlement Agent Fee to Epsilon Title Co.	$500.00				
06 Title – Title Search to Epsilon Title Co.	$800.00				
07					
08					
D. TOTAL LOAN COSTS (Borrower-Paid)	**$4,694.05**				
Loan Costs Subtotals (A + B + C)	$4,664.25	$29.80			

Other Costs	Borrower-Paid At Closing	Borrower-Paid Before Closing	Seller-Paid At Closing	Seller-Paid Before Closing	Paid by Others
E. Taxes and Other Government Fees	**$85.00**				
01 Recording Fees Deed: $40.00 Mortgage: $45.00	$85.00				
02 Transfer Tax to Any State			$950.00		
F. Prepaids	**$2,120.80**				
01 Homeowner's Insurance Premium (12 mo.) to Insurance Co.	$1,209.96				
02 Mortgage Insurance Premium (mo.)					
03 Prepaid Interest ($17.44 per day from 4/15/13 to 5/1/13)	$279.04				
04 Property Taxes (6 mo.) to Any County USA	$631.80				
05					
G. Initial Escrow Payment at Closing	**$412.25**				
01 Homeowner's Insurance $100.83 per month for 2 mo.	$201.66				
02 Mortgage Insurance per month for mo.					
03 Property Taxes $105.30 per month for 2 mo.	$210.60				
04					
05					
06					
07					
08 Aggregate Adjustment	– 0.01				
H. Other	**$2,400.00**				
01 HOA Capital Contribution to HOA Acre Inc.	$500.00				
02 HOA Processing Fee to HOA Acre Inc.	$150.00				
03 Home Inspection Fee to Engineers Inc.	$750.00			$750.00	
04 Home Warranty Fee to XYZ Warranty Inc.			$450.00		
05 Real Estate Commission to Alpha Real Estate Broker			$5,700.00		
06 Real Estate Commission to Omega Real Estate Broker			$5,700.00		
07 Title – Owner's Title Insurance (optional) to Epsilon Title Co.	$1,000.00				
08					
I. TOTAL OTHER COSTS (Borrower-Paid)	**$5,018.05**				
Other Costs Subtotals (E + F + G + H)	$5,018.05				
J. TOTAL CLOSING COSTS (Borrower-Paid)	**$9,712.10**				
Closing Costs Subtotals (D + I)	$9,682.30	$29.80	$12,800.00	$750.00	$405.00
Lender Credits					

CLOSING DISCLOSURE PAGE 2 OF 5 • LOAN ID # 123456789

Figure 20-1. (continued)

Calculating Cash to Close

Use this table to see what has changed from your Loan Estimate.

	Loan Estimate	Final	Did this change?
Total Closing Costs (J)	$8,054.00	$9,712.10	**YES** • See **Total Loan Costs (D)** and **Total Other Costs (I)**
Closing Costs Paid Before Closing	$0	– $29.80	**YES** • You paid these Closing Costs **before closing**
Closing Costs Financed (Paid from your Loan Amount)	$0	$0	**NO**
Down Payment/Funds from Borrower	$18,000.00	$18,000.00	**NO**
Deposit	– $10,000.00	– $10,000.00	**NO**
Funds for Borrower	$0	$0	**NO**
Seller Credits	$0	– $2,500.00	**YES** • See Seller Credits in **Section L**
Adjustments and Other Credits	$0	– $1,035.04	**YES** • See details in **Sections K and L**
Cash to Close	$16,054.00	$14,147.26	

Summaries of Transactions

Use this table to see a summary of your transaction.

BORROWER'S TRANSACTION

K. Due from Borrower at Closing	**$189,762.30**
01 Sale Price of Property	$180,000.00
02 Sale Price of Any Personal Property Included in Sale	
03 Closing Costs Paid at Closing (J)	$9,682.30
04	
Adjustments	
05	
06	
07	
Adjustments for Items Paid by Seller in Advance	
08 City/Town Taxes to	
09 County Taxes to	
10 Assessments to	
11 HOA Dues 4/15/13 to 4/30/13	$80.00
12	
13	
14	
15	
L. Paid Already by or on Behalf of Borrower at Closing	**$175,615.04**
01 Deposit	$10,000.00
02 Loan Amount	$162,000.00
03 Existing Loan(s) Assumed or Taken Subject to	
04	
05 Seller Credit	$2,500.00
Other Credits	
06 Rebate from Epsilon Title Co.	$750.00
07	
Adjustments	
08	
09	
10	
11	
Adjustments for Items Unpaid by Seller	
12 City/Town Taxes 1/1/13 to 4/14/13	$365.04
13 County Taxes to	
14 Assessments to	
15	
16	
17	
CALCULATION	
Total Due from Borrower at Closing (K)	$189,762.30
Total Paid Already by or on Behalf of Borrower at Closing (L)	– $175,615.04
Cash to Close ☒ From ☐ To Borrower	**$14,147.26**

SELLER'S TRANSACTION

M. Due to Seller at Closing	**$180,080.00**
01 Sale Price of Property	$180,000.00
02 Sale Price of Any Personal Property Included in Sale	
03	
04	
05	
06	
07	
08	
Adjustments for Items Paid by Seller in Advance	
09 City/Town Taxes to	
10 County Taxes to	
11 Assessments to	
12 HOA Dues 4/15/13 to 4/30/13	$80.00
13	
14	
15	
16	
N. Due from Seller at Closing	**$115,665.04**
01 Excess Deposit	
02 Closing Costs Paid at Closing (J)	$12,800.00
03 Existing Loan(s) Assumed or Taken Subject to	
04 Payoff of First Mortgage Loan	$100,000.00
05 Payoff of Second Mortgage Loan	
06	
07	
08 Seller Credit	$2,500.00
09	
10	
11	
12	
13	
Adjustments for Items Unpaid by Seller	
14 City/Town Taxes 1/1/13 to 4/14/13	$365.04
15 County Taxes to	
16 Assessments to	
17	
18	
19	
CALCULATION	
Total Due to Seller at Closing (M)	$180,080.00
Total Due from Seller at Closing (N)	– $115,665.04
Cash ☐ From ☒ To Seller	**$64,414.96**

CLOSING DISCLOSURE

PAGE 3 OF 5 • LOAN ID # 123456789

Figure 20-1. (continued)

Additional Information About This Loan

Loan Disclosures

Assumption
If you sell or transfer this property to another person, your lender

☐ will allow, under certain conditions, this person to assume this loan on the original terms.

☒ will not allow assumption of this loan on the original terms.

Demand Feature
Your loan

☐ has a demand feature, which permits your lender to require early repayment of the loan. You should review your note for details.

☒ does not have a demand feature.

Late Payment
If your payment is more than *15* days late, your lender will charge a late fee of *5% of the monthly principal and interest payment.*

Negative Amortization (Increase in Loan Amount)
Under your loan terms, you

☐ are scheduled to make monthly payments that do not pay all of the interest due that month. As a result, your loan amount will increase (negatively amortize), and your loan amount will likely become larger than your original loan amount. Increases in your loan amount lower the equity you have in this property.

☐ may have monthly payments that do not pay all of the interest due that month. If you do, your loan amount will increase (negatively amortize), and, as a result, your loan amount may become larger than your original loan amount. Increases in your loan amount lower the equity you have in this property.

☒ do not have a negative amortization feature.

Partial Payments
Your lender

☒ may accept payments that are less than the full amount due (partial payments) and apply them to your loan.

☐ may hold them in a separate account until you pay the rest of the payment, and then apply the full payment to your loan.

☐ does not accept any partial payments.

If this loan is sold, your new lender may have a different policy.

Security Interest
You are granting a security interest in
456 Somewhere Ave., Anytown, ST 12345

You may lose this property if you do not make your payments or satisfy other obligations for this loan.

Escrow Account
For now, your loan

☒ will have an escrow account (also called an "impound" or "trust" account) to pay the property costs listed below. Without an escrow account, you would pay them directly, possibly in one or two large payments a year. Your lender may be liable for penalties and interest for failing to make a payment.

Escrow		
Escrowed Property Costs over Year 1	$2,473.56	Estimated total amount over year 1 for your escrowed property costs: *Homeowner's Insurance* *Property Taxes*
Non-Escrowed Property Costs over Year 1	$1,800.00	Estimated total amount over year 1 for your non-escrowed property costs: *Homeowner's Association Dues* You may have other property costs.
Initial Escrow Payment	$412.25	A cushion for the escrow account you pay at closing. See Section G on page 2.
Monthly Escrow Payment	$206.13	The amount included in your total monthly payment.

☐ will not have an escrow account because ☐ you declined it ☐ your lender does not offer one. You must directly pay your property costs, such as taxes and homeowner's insurance. Contact your lender to ask if your loan can have an escrow account.

No Escrow		
Estimated Property Costs over Year 1		Estimated total amount over year 1. You must pay these costs directly, possibly in one or two large payments a year.
Escrow Waiver Fee		

In the future,
Your property costs may change and, as a result, your escrow payment may change. You may be able to cancel your escrow account, but if you do, you must pay your property costs directly. If you fail to pay your property taxes, your state or local government may (1) impose fines and penalties or (2) place a tax lien on this property. If you fail to pay any of your property costs, your lender may (1) add the amounts to your loan balance, (2) add an escrow account to your loan, or (3) require you to pay for property insurance that the lender buys on your behalf, which likely would cost more and provide fewer benefits than what you could buy on your own.

CLOSING DISCLOSURE   PAGE 4 OF 5 • LOAN ID # 123456789

Figure 20-1. (continued)

Loan Calculations

Total of Payments. Total you will have paid after you make all payments of principal, interest, mortgage insurance, and loan costs, as scheduled.	$285,803.36
Finance Charge. The dollar amount the loan will cost you.	$118,830.27
Amount Financed. The loan amount available after paying your upfront finance charge.	$162,000.00
Annual Percentage Rate (APR). Your costs over the loan term expressed as a rate. This is not your interest rate.	4.174%
Total Interest Percentage (TIP). The total amount of interest that you will pay over the loan term as a percentage of your loan amount.	69.46%

Questions? If you have questions about the loan terms or costs on this form, use the contact information below. To get more information or make a complaint, contact the Consumer Financial Protection Bureau at **www.consumerfinance.gov/mortgage-closing**

Other Disclosures

Appraisal
If the property was appraised for your loan, your lender is required to give you a copy at no additional cost at least 3 days before closing. If you have not yet received it, please contact your lender at the information listed below.

Contract Details
See your note and security instrument for information about
- what happens if you fail to make your payments,
- what is a default on the loan,
- situations in which your lender can require early repayment of the loan, and
- the rules for making payments before they are due.

Liability after Foreclosure
If your lender forecloses on this property and the foreclosure does not cover the amount of unpaid balance on this loan,

☒ state law may protect you from liability for the unpaid balance. If you refinance or take on any additional debt on this property, you may lose this protection and have to pay any debt remaining even after foreclosure. You may want to consult a lawyer for more information.

☐ state law does not protect you from liability for the unpaid balance.

Refinance
Refinancing this loan will depend on your future financial situation, the property value, and market conditions. You may not be able to refinance this loan.

Tax Deductions
If you borrow more than this property is worth, the interest on the loan amount above this property's fair market value is not deductible from your federal income taxes. You should consult a tax advisor for more information.

Contact Information

	Lender	Mortgage Broker	Real Estate Broker (B)	Real Estate Broker (S)	Settlement Agent
Name	Ficus Bank		Omega Real Estate Broker Inc.	Alpha Real Estate Broker Co.	Epsilon Title Co.
Address	4321 Random Blvd. Somecity, ST 12340		789 Local Lane Sometown, ST 12345	987 Suburb Ct. Someplace, ST 12340	123 Commerce Pl. Somecity, ST 12344
NMLS ID					
ST License ID			Z765416	Z61456	Z61616
Contact	Joe Smith		Samuel Green	Joseph Cain	Sarah Arnold
Contact NMLS ID	12345				
Contact ST License ID			P16415	P51461	PT1234
Email	joesmith@ficusbank.com		sam@omegare.biz	joe@alphare.biz	sarah@epsilontitle.com
Phone	123-456-7890		123-555-1717	321-555-7171	987-555-4321

Confirm Receipt

By signing, you are only confirming that you have received this form. You do not have to accept this loan because you have signed or received this form.

Applicant Signature　　Date　　Co-Applicant Signature　　Date

CLOSING DISCLOSURE　　PAGE 5 OF 5 • LOAN ID # 123456789

Figure 20-1. (continued)

QUESTIONS ON CHAPTER 20

1. The HUD-1 Uniform Settlement Statement
   (A) includes buyer's closing costs only
   (B) includes seller's closing costs only
   (C) is no longer used in most residential real estate closings
   (D) cites mortgage delinquency data

2. If the following transactions involve a new first mortgage, to which would TRID apply?
   (A) summer house to be occupied by purchaser
   (B) house purchased by an investor to remodel and resell
   (C) 50-unit apartment building
   (D) 50-acre farm with home for use by tenant farmer

3. A home seller agrees to take back a $25,000 second mortgage note on the sale. On the closing statement, this is a
   (A) credit to the buyer, debit to the seller
   (B) credit to the buyer, credit to the seller
   (C) debit to the buyer, credit to the seller
   (D) debit to the buyer, debit to the seller

4. TRID applies to
   (A) first mortgage loan on one's proposed home
   (B) seller financing
   (C) investor purchase
   (D) all real estate loans

5. Under TRID, lenders must
   (A) provide a Loan Estimate and a Closing Disclosure form to proposed borrowers for most residential transactions
   (B) limit the number of discount points to no more than 2
   (C) let the buyer void the transaction up to ten days after closing
   (D) not charge more than 1% for lender fees

6. Which of the following is not required to be supplied by residential lenders?
   (A) Loan Estimate
   (B) Closing Disclosure
   (C) accurate accounting
   (D) commissions of brokers

7. A sale will close on March 13. Real estate taxes for the year are $500, payable December 1. At closing,
   (A) credit the buyer $100
   (B) debit the buyer $100
   (C) credit the buyer $400
   (D) debit the buyer $400

8. Three business days before closing, the loan applicant has the right to inspect
   (A) the seller's credit report
   (B) a final Uniform Settlement Statement
   (C) the broker's financial statement
   (D) a Closing Disclosure form

9. The title company may check records of the
   (A) county clerk
   (B) county recorder
   (C) federal court
   (D) all of the above

10. On a closing statement, which is a debit to the buyer?
    (A) purchase price
    (B) new mortgage loan
    (C) down payment cash
    (D) all of the above

11. Who performs a title search prior to a purchase?
    (A) buyer
    (B) seller
    (C) broker
    (D) title company and closing attorney

12. TRID requires the lender to provide a Loan Estimate of closing costs within _____ business day(s) after loan application.
    (A) 1
    (B) 2
    (C) 3
    (D) 5

13. Authority to enforce TRID rests with
    (A) CFPB
    (B) HUD
    (C) FHA
    (D) FNMA

14. Which party is most likely to demand title insurance as a prerequisite to participating in the purchase?
    (A) buyer
    (B) seller
    (C) mortgage lender
    (D) attorney

15. The closing date in a purchase and sale agreement should be
    (A) filled in by the broker
    (B) within 30 days following the contract
    (C) set for a date after loan approval
    (D) clearly stated

16. At closing, which is expected to be prorated?
    (A) taxes
    (B) rental income
    (C) both A and B
    (D) neither A nor B

17. TRID requires the proposed lender to give the borrower
    (A) Loan Estimate
    (B) Closing Disclosure
    (C) both A and B
    (D) neither A nor B

18. TRID applies to
    (A) all real estate financing
    (B) contracts for deed
    (C) second mortgages only
    (D) financing of one- to four-family residences

ANSWERS

1. **C**	6. **D**	11. **D**	16. **C**
2. **A**	7. **A**	12. **C**	17. **C**
3. **A**	8. **D**	13. **A**	18. **D**
4. **A**	9. **D**	14. **C**	
5. **A**	10. **A**	15. **D**	

PART FIVE
Federal Law Affecting Real Estate

Fair Housing Law

21

ESSENTIAL MATTERS TO LEARN FROM THIS CHAPTER

- Increasing a real estate salesperson's or broker's knowledge and observance of *civil rights laws* is important to create equality in housing opportunity. Accordingly, the proportion of questions on this topic has increased significantly on real estate licensing examinations in recent years. Often the questions do not merely ask what the law is but, instead, how it is to be applied in a given fact pattern.
- The importance of equal opportunity in housing is compelling. It leads to equal opportunity in education, jobs, and other important matters.
- *The federal Fair Housing Act of 1968* is part of the Civil Rights Act of 1968. The Fair Housing Act was amended significantly in 1988 and has been modified since then.
- *Protected classes* are based on race, color, religion, sex, national origin, familial status (including pregnant women and families with children under 18), or handicap.
- *Fair Housing law* prohibits discriminatory practices by all real estate brokers, builders, and mortgage lenders when serving a protected class. A real estate appraiser may not describe the racial composition of the neighborhood within which the subject house exists.
- For example, if a real estate broker is asked by an owner to take a listing on a house but not to show the house to members of a certain race, the broker should refuse the listing. Similarly, a broker should treat everyone the same as to purchase and lease opportunities.
- It is unlawful for organizations who offer memberships to real estate service providers to decline to offer membership to those of a protected class.
- Today, nearly all dwelling units are covered by Fair Housing laws. A *dwelling unit* is defined as a structure that is used as a residence by one or more families: a condominium, co-op, mobile home park, trailer court, and timesharing units.
- Refusal to rent or sell, discrimination in terms of rental or sales, advertising preferences, false representation of the unavailability of a unit, and inducing a housing change because of the entry by a protected class are all unlawful.
- *Handicap* includes persons who are physically impaired. Alcoholics are covered. However, drug addicts and persons convicted of drug-related felonies are not.
- A handicapped person who seeks to lease a dwelling must pay the lessor the cost of a change needed (a wheelchair ramp, grab bars in the bathroom) and must restore the property when vacated. A person who requires a service dog may keep it even though dogs are otherwise not allowed in the complex and must repair animal damage when vacating.

- Essential terms to understand include:
    1. **STEERING.** Directing a certain class toward (or away from) a certain racial or ethnic area.
    2. **BLOCKBUSTING.** Attempting to profit by announcing to a neighborhood the entry of a certain class.
    3. **REDLINING.** Refusing to finance housing in a certain neighborhood.
- There are notable exceptions to the Fair Housing laws:
    1. Housing for the elderly only, where 100% of the units are occupied by residents age 62 or older or where 80% of the units are occupied by at least one person age 55 or older and the housing publishes and adheres to policies to demonstrate an intent to serve people age 55 or over.
    2. Apartment complexes with four or fewer units where the owner occupies at least one unit and single-family houses sold or rented by an owner who owns not more than three properties. This exemption is not available when the owner advertises using discriminatory language (such as "exclusive" or "whites only") or when the services of a real estate broker are used.
    3. A church that owns rental housing may require that tenants be members as long as the church doesn't discriminate based on race, color, sex, national origin, familial status, or handicap.
    4. Private clubs that provide lodging as incident to their purpose may give preferential lodging treatment to members.
- In a new housing development of multiple stories, only the ground-floor units must accommodate the handicapped.
- Owners or lessors may refuse to sell or lease to someone who is not economically qualified to purchase or rent.
- An aggrieved person may file a complaint with the Department of Housing and Urban Development within one year of the alleged discriminatory practice.
- A person may file a civil action in the appropriate district court within one year of the alleged violation.
- *The Americans with Disabilities Act*, passed in 1992, does not apply to single-family houses. It is intended to eliminate physical barriers that prevent disabled workers or customers from functioning in the marketplace.

The federal Fair Housing Act, Public Law 90–284, was enacted into law on April 11, 1968, as Title VIII of the Civil Rights Act of 1968. A significant amendment occurred in 1988, and other changes have been passed since then.

PURPOSE OF FAIR HOUSING ACT

The purpose of the Fair Housing Act is expressed by Section 801 of the law, which states:

> *It is the policy of the United States to provide, within constitutional limitations, for fair housing throughout the United States.*

The following explanation of the need for fair housing is quoted directly from *Understanding Fair Housing,* U.S. Commission on Civil Rights.*

> *Housing is a key to improvement in a family's economic condition. Homeownership is one of the important ways in which Americans have traditionally acquired financial capital. Tax advantages, the accumulation of equity, and the increased value of real estate property enable homeowners to build economic assets. These assets can be used to educate one's children, to take advantage of business opportunities, to meet financial emergencies, and to provide for retirement. Nearly two of every three majority group families are homeowners, but less than two of every five nonwhite families own their homes. Consequently, the majority of nonwhite families are deprived of this advantage.*
>
> *Housing is essential to securing civil rights in other areas. Segregated residential patterns in metropolitan areas undermine efforts to assure equal opportunity in employment and education. While centers of employment have moved from the central cities to suburbs and outlying parts of metropolitan areas, minority group families remain confined to the central cities, and because they are confined, they are separated from employment opportunities. Despite a variety of laws against job discrimination, lack of access to housing in close proximity to available jobs is an effective barrier to equal employment.*
>
> *In addition, lack of equal housing opportunity decreases prospects for equal educational opportunity. The controversy over school busing is closely tied to the residential patterns of our cities and metropolitan areas. If schools in large urban centers are to be desegregated, transportation must be provided to convey children from segregated neighborhoods to integrated schools.*
>
> *Finally, if racial divisions are to be bridged, equal housing is an essential element. Our cities and metropolitan areas consist of separate societies increasingly hostile and distrustful of one another. Because minority and majority group families live apart, they are strangers to each other. By living as neighbors they would have an opportunity to learn to understand each other and to redeem the promise of America: that of "one Nation indivisible."*

DISCRIMINATION IN SALE OR RENTAL

The Fair Housing Act recognizes seven *protected classes*: race, color, religion, sex, national origin, familial status, and handicap. Section 804 of the Fair Housing Act makes it unlawful to do any of the following:

> *(a) To refuse to sell or rent after the making of a bona fide offer, or to refuse to negotiate for the sale or rental of, or otherwise make unavailable or deny, a dwelling to any person because of race, color, religion, sex, national origin, familial status, or handicap.*
>
> *(b) To discriminate against any person in the terms, conditions, or privileges of sale or rental of a dwelling, or in the provision of services or facilities in connection therewith, because of race, color, religion, sex, national origin, familial status, or handicap.*

---

*From *Understanding Fair Housing,* U.S. Commission on Civil Rights, Clearinghouse Publication 42, February 1973, p. 1.

(c) *To make, print, or publish, or cause to be made, printed, or published any notice, statement, or advertisement, with respect to the sale or rental of a dwelling that indicates any preference, limitation, or discrimination based on race, color, religion, sex, national origin, familial status, or handicap, or an intention to make any such preference, limitation, or discrimination.*

(d) *To represent to any person because of race, color, religion, sex, national origin, familial status, or handicap that any dwelling is not available for inspection, sale, or rental when such dwelling is in fact so available.*

(e) *For profit, to induce or attempt to induce any person to sell or rent any dwelling by representations regarding the entry or prospective entry into the neighborhood of a person or persons of a particular race, color, religion, sex, national origin, familial status, or handicap.*

1988 AMENDMENTS

Today, nearly all dwelling units are subject to the Fair Housing Act.

Amendments to the federal Fair Housing Act that became law in 1988 extended protection to familial status and handicap. Familial status refers to members of a family. The amendment prohibits discrimination against people with children, adults living with, or in the process of acquiring legal custody of, anyone under age 18, and pregnant women.

The physically and mentally handicapped are also covered. *Handicap* is defined as a mental or physical impairment that substantially limits a person's major life activities; included are persons with records of impairment. Current drug addicts, persons convicted of drug-related felonies, and transvestites are not protected as handicapped under the law, but alcoholics are covered.

The law also appears to prohibit restrictions that would prevent the handicapped from using necessary aids. For example, a person who requires a seeing-eye dog should be allowed to have one even if pets are otherwise prohibited in an apartment building. Similarly, that person would be allowed to take the dog through the hallway even if dogs are generally prohibited in public parts of the building. The pet owner can be required to repair damage by the animal when vacating the property.

DISCRIMINATION IN NEW CONSTRUCTION

Under the 1988 amendments, newly constructed multifamily facilities (four or more units) must be accessible to the handicapped. In buildings with elevators, the handicapped must have access to 100% of the units, but only ground-floor units of garden-type apartments are required to be accessible.

Common areas in buildings must be accessible to all handicapped persons, and doors and hallways must be wide enough to allow passage of wheelchairs.

New living units must be constructed in a way that allows access for the handicapped, including the appropriate location of light switches, plugs, and environmental controls. Bathroom walls must be reinforced to allow future installation of grab rails, and the occupant must be allowed to install them. When the handicapped person vacates, he may be required to restore the property to its original condition.

DISCRIMINATION IN FINANCING

Section 805 of the Fair Housing Act applies to transactions after December 31, 1968. It states that it is unlawful for

> *any bank, building and loan association, insurance company or other corporation, association, firm or enterprise whose business consists in whole or in part in the making of commercial real estate loans, to deny a loan or other financial assistance to a person applying therefore for the purpose of purchasing, constructing, improving, repairing, or maintaining a dwelling, or to discriminate against him in the fixing of the amount, interest rate, duration, or other terms or conditions of such loan or other financial assistance, because of the race, color, religion, sex, or national origin of such person or of any person associated with him in connection with such loan or other financial assistance or the purposes of such loan or other financial assistance, or of the present or prospective owners, lessees, tenants, or occupants of the dwelling or dwellings in relation to which such loan or other financial assistance is to be made or given: Provided, that nothing contained in this section shall impair the scope or effectiveness of the exception contained in Section 803(b).*

Section 803(b) exempts a single-family house sale or lease by owner, if certain provisions are met. This exemption is described on page 438.

Federal law also prohibits lending institutions from *redlining.* Redlining is a practice whereby lenders designate certain areas as "too risky" and refuse to make loans on property in those areas. (The term *redlining* refers to outlining those areas in red on a map.) Before it was outlawed, redlining by lenders usually affected low-income areas and, particularly, areas whose populations were composed largely of minorities.

DISCRIMINATION IN BROKERAGE SERVICES

Section 806 of the Fair Housing Act states:

> *After December 31, 1968, it shall be unlawful to deny any person access to or membership or participation in any multiple-listing service, real estate brokers' organization or other service, organization, or facility relating to the business of selling or renting dwellings, or to discriminate against him in the terms or conditions of such access, membership, or participation, on account of race, color, religion, sex, national origin, familial status, or handicap.*

BLOCKBUSTING

Blockbusting, that is, the soliciting of homeowners by unscrupulous real estate agents, brokers, or speculators who feed upon fears of homeowners, is prohibited by the Fair Housing Act. Blockbusters attempt to buy properties at very low prices from whites who flee racially transitional neighborhoods and then broker or sell them to blacks at high prices. Some blockbusters deliberately incite panic and white flight to achieve their greedy, unlawful goal.

STEERING

Steering is an illegal practice whereby real estate brokers or salespersons show (or don't show) clients certain housing because of the clients' race or ethnicity. It takes two principal forms:

1. Intentionally failing to show houses in certain neighborhoods because of race, color, religion, sex, national origin, familial status, or handicap.
2. Showing clients only certain neighborhoods where their race, color, religion, sex, national origin, familial status, or handicap is common.

The practice of steering is a violation of the federal Fair Housing Act. Some brokers or salespersons act as "gatekeepers" for "exclusive" neighborhoods (a term to be avoided in discussion and advertising), but the practice is absolutely illegal steering.

Other brokers may think they are providing a valuable service when they direct a client to certain neighborhoods where they believe the client will feel "comfortable."

The broker or salesperson should show clients properties that are suitable for their needs, without regard to race, color, or other protected categories.

Trick Question

A common trick question suggests that a broker may show properties only in areas that are predominantly of the client's racial, religious, or ethnic background. However, this is a wrong answer. The broker should show all the properties that meet the client's housing needs and budget. Not doing this is an illegal act called *steering*.

EXEMPTIONS

The federal Fair Housing Act covers all single or multifamily dwelling units, with the few exceptions noted below. A *dwelling* is defined as any building or structure designed as a residence to be occupied by one or more families; included are mobile-home parks, trailer courts, condominiums, cooperatives, and time-sharing units. Community associations and "adult-only" communities are clearly included unless they qualify under an exception.

Housing-for-the-elderly-only communities are built and operated for older persons: specifically, 100% of the units must be occupied by residents 62 years of age or older, or 80% of the units must be occupied by at least one person age 55 or older. The apartments must publish and adhere to policies that demonstrate an intent to serve persons aged 55 and over.

Apartment complexes with four or fewer units are exempt from the federal Fair Housing Act when the owner occupies at least one unit, and single-family homes sold or rented by an owner are exempt. However, the exemption applies only to persons owning no more than three properties at one time and is subject to other restrictions. In selling, the owner may not use a real estate salesperson or broker and may not print, publish, or otherwise make any reference to preference, limitation, or discrimination on the basis of race, color, religion, sex, national origin, familial status, or handicap.

Another exemption allows a religious organization to discriminate with respect to its noncommercial property. It does not, however, allow this exemption if the religion discriminates on the basis of race, color, sex, national origin, familial status, or handicap with respect to its membership. Another exemption allows private clubs that provide lodging as an incident to their main purpose to give preferential treatment to club members.

ENFORCEMENT BY THE FEDERAL GOVERNMENT

Any person who claims to have been injured by a discriminatory housing practice or who believes that he will be irrevocably injured by a discriminatory housing practice that is about to occur (hereafter "person aggrieved") may file a complaint with the Secretary of the Department of Housing and Urban Development (HUD). Complaints must be in writing, must state the facts, and must be filed within 1 year after the alleged discriminatory housing practice occurred. The Attorney General conducts all litigation in which the Secretary of HUD participates as a party pursuant to the Fair Housing Act.

ENFORCEMENT BY PRIVATE PERSONS

The rights granted to private persons by the Fair Housing Act may be enforced by civil action in appropriate U.S. district courts without regard to the amount in controversy and in appropriate state or local courts of general jurisdiction. A civil action must be commenced within 1 year after the alleged discriminatory housing practice occurred.

Upon application by the plaintiff and in such circumstances as the court may deem just, a court of the United States in which a civil action under this section has been brought may appoint an attorney for the plaintiff and may, upon proper showing, authorize the commencement of a civil action without the payment of fees, costs, or security. A court of a state or subdivision thereof may do likewise to an extent not inconsistent with the law or procedures of the state or subdivision.

The court may grant as relief, as it deems appropriate, any permanent or temporary injunction, temporary restraining order, or other order and may award to the plaintiff actual damages, injunctive or other equitable relief, and civil penalties up to $50,000 ($100,000 for repeating violators), together with court costs and reasonable attorney fees in the case of a prevailing plaintiff, provided that said plaintiff, in the opinion of the court, is not financially able to assume said attorney's fees.

PENALTY FOR INJURY, INTIMIDATION, DISCOURAGEMENT

Under Section 901 of the Civil Rights Act of 1968 (Title IX) whoever

- **A.** injures or threatens to injure or interfere with any person because of his race, religion, color, sex, national origin, familial status, or handicap, and who is selling, leasing, occupying, financing, etc., dwelling, or
- **B.** intimidates persons who deal with others in housing on account of race, religion, color, sex, national origin, familial status, or handicap, or
- **C.** discourages others from dealing with others in housing on account of race, religion, color, sex, national origin, familial status, or handicap

shall be fined up to $1,000 or imprisoned for up to 1 year, or both. If bodily injury results, the penalty is a fine of $10,000 maximum or up to 10 years in prison, or both. If death results, the wrongdoer shall be imprisoned for any term of years or for life.

OTHER FAIR HOUSING LAWS

The Fair Housing Act was not the first law intended to prevent discriminatory practice in housing. An 1866 civil rights law barred all racial discrimination in public and private housing. The Supreme Court of the United States, in the 1917 *Buchanan* case, prohibited, on constitutional grounds, local governments from requiring residential segregation. This ruling is noteworthy because in 1896, the Supreme Court established the doctrine that legally compelled segregation in such areas as public transportation and public education was constitutionally permissible. The Buchanan decision destroyed the doctrine as it applied to housing. In 1948, in *Shelley* v. *Kraemer*, the Supreme Court struck down as unconstitutional the legal enforcement of racially restrictive covenants.

The executive branch of the government took fair housing action for the first time in 1962 when President Kennedy issued an executive order on equal opportunity in housing. Although it represented a significant legal step forward, this executive order was limited. Its guarantee of nondiscrimination was restricted largely to housing provided through the insurance and guaranty programs administered by FHA and its sister agency, the Veterans Administration (VA), after the date of the order's issuance (November 20, 1962). Housing financed through conventional loans was not covered by the President's order, which also left hundreds of thousands of existing housing units receiving FHA and VA assistance immune from the nondiscrimination mandate. In fact, barely 1% of the nation's housing was covered by President Kennedy's executive order.

In 1964, Congress enacted Title VI of the Civil Rights Act of 1964, prohibiting discrimination in any program or activity receiving federal financial assistance. Among the principal programs affected by this law were low-rent public housing, a program directed to providing housing for the poor, and urban renewal. Like President Kennedy's executive order, Title VI excluded conventionally financed housing. It also excluded most FHA and VA housing that the executive order covered. Less than half of 1% of the nation's housing inventory was subject to the nondiscrimination requirement through Title VI.

In 1968, Congress enacted Title VIII of the Civil Rights Act of 1968, the federal Fair Housing Law. This law, which is the one described at the beginning of the chapter, prohibits discriminatory practices by all real estate brokers, builders, and mortgage lenders.

In June 1968, two months after enactment of Title VIII, the Supreme Court of the United States, in the landmark case of *Jones* v. *Mayer*, ruled that an 1866 civil rights law passed under the authority of the Thirteenth Amendment (which outlawed slavery) bars all racial discrimination in housing, private as well as public.

Today, over 90% of all U.S. housing is subject to the Fair Housing Law.

THE AMERICANS WITH DISABILITIES ACT

The Americans with Disabilities Act (ADA), passed in 1992, considerably broadens the scope to which society in general must accommodate persons with disabilities. Many provisions of the act address discrimination in employment and access to public services. However, some of the most pervasive problems facing disabled persons have necessarily been involved with real estate: buildings are the most "barrier-prone" part of the physical environment. Narrow doors and hallways, variations in floor levels that require the negotiation of stairs, built-in facilities too high for the wheelchair-bound—the list of problems goes on and on.

Most new construction that may serve the public in general (almost anything except single-family housing) must meet barrier-free standards. Also, existing businesses with 15 or more employees must eliminate physical barriers that prevent disabled workers or customers from "functioning in the marketplace"; this requirement obviously requires considerable "retrofitting" of existing real estate.

QUESTIONS ON CHAPTER 21

1. When discussing the possibility of listing their house, the owners tell the salesperson, "We want to sell only to buyers of a certain national origin." The salesperson should
   (A) accept that condition and write it into the listing to comply with the principal's request
   (B) report the owners to the Department of Housing and Urban Development
   (C) because it is only a listing, accept the condition but be prepared to show the property to others
   (D) explain that this would be a violation of the federal Fair Housing Act; if the owners, persist, refuse the listing

2. A blind man with a service dog seeks to rent an apartment in a building that does not allow pets. Under the federal Fair Housing Act, the apartment building owner
   (A) can reject the applicant as a tenant
   (B) can increase the security deposit compared with what is charged other tenants
   (C) must lease, though the lease can require the tenant to repair any damage caused by the animal
   (D) cannot treat the tenant any differently than any other prospective tenant

3. Which of the following phrases does HUD consider discriminatory advertising?
   (A) "senior housing"
   (B) "friendly neighbors"
   (C) "exclusive neighborhood"
   (D) "no one excluded"

4. An ad reads: "Home for sale only to a family." This ad violates
   (A) Truth in Lending and Fair Housing Acts
   (B) TRID and Fair Housing Act
   (C) Fair Housing Act and Americans with Disabilities Act
   (D) HUD and EPA

5. A minority family asks a salesperson to show a certain listed house. The licensee suggests they seek a home in an area with a greater population of minority residents. This is known as
   (A) illegal steering
   (B) illegal redlining
   (C) legal steering
   (D) helpful sales discussion

6. A homeowner wishes to lease his house for a year while he takes a sabbatical. He instructs the licensed agent to lease to a white family only. The agent must
   (A) be loyal and comply
   (B) ask for the request in writing
   (C) refuse to follow the order because it violates the Fair Housing Act
   (D) ask the broker what the firm's policy is

7. An apartment complex is intended for the elderly only. What percentage of units must have at least one person over age 55?
   (A) 100%
   (B) 80%
   (C) 50%
   (D) 0%

8. Under the federal Fair Housing Act, which of the following is protected as a "handicap"?
   (A) inadequate income to pay rent
   (B) use of a controlled substance for more than a year
   (C) inability to get an FHA mortgage
   (D) a mental handicap that limits a major life activity

9. A real estate broker may not
   (A) identify local schools, whether public or private
   (B) mention ethnic festivities held in the city
   (C) state the racial composition of the neighborhood
   (D) map religious facilities in the area

10. A builder wishes to erect a 150-unit two-story building without an elevator. Under the federal Fair Housing Act, which units must be accessible to the handicapped?
    (A) all
    (B) ground-floor only
    (C) second-floor only
    (D) 20% set-aside

11. Allen, not a real estate licensee, wants to sell his brother's house to a member of their same religious affiliation. Allen can
    (A) do anything provided he has his brother's oral permission
    (B) charge a commission for the sale of the home
    (C) advertise that the house is for sale only to members of that religion
    (D) discuss the matter with his brother

12. Houses located near a college offer rooms or apartments for rent. Which of thc following is violating the federal Fair Housing Act?
    (A) owners who rent out two rooms in their house but not to a Greek couple
    (B) owners who live in their house, renting out three rooms, and require a $500 greater deposit from college students
    (C) owners who live elsewhere and refuse to rent their house to single individuals
    (D) owners who rent a room in the house they live in, as well as two apartments in the house they own next door, but won't rent to a family with 2 young children

13. A broker advertises a home for rent as follows: "4BR, 2BA, nr. schools & shopping. Elderly married couples only." Which law has been broken?
    (A) Uniform Residential Appraisal Report
    (B) Americans with Disabilities
    (C) Truth in Lending
    (D) Federal Fair Housing Act

14. An FDIC-insured lender is considering a loan application to remodel an existing house. The applicant is a member of a minority. Under the federal Fair Housing Act, the lender must make the loan unless
    (A) the home is in a minority neighborhood
    (B) the home is more than 50 years old
    (C) the applicant is not economically qualified
    (D) property has lead-based paint

15. A lawyer wishes to buy a 40-year-old office building. To comply with the Americans with Disabilities Act, the lawyer must
    (A) widen hallways
    (B) install elevators
    (C) install a wheelchair ramp to at least one entrance
    (D) remove carpet from hallways

16. The owner of a house lists it for sale with ABC Realty, asking the firm not to sell to a minority prospect. ABC Realty, to comply with the federal Fair Housing Act, should
    (A) refuse to accept the listing
    (B) disregard the request
    (C) agree to the request but show the house to minority prospects if requested
    (D) report the owner to HUD

17. A prospective tenant in a 50-unit apartment complex requires a wheelchair. The tenant wants to rent a unit, install grab bars in the bathroom, and lower door handles. Under the federal Fair Housing Act, the landlord
    (A) must allow and pay for these changes
    (B) must allow the tenant to make these changes at the tenant's expense and can require the tenant to restore the property at the end of the lease
    (C) can refuse to rent the unit because the unit is not handicapped accessible
    (D) must allow the tenant's modifications and accept the property as modified when the lease ends

18. Under the federal Fair Housing Act, which of the following can a landlord legally refuse to rent to?
   (A) an unmarried couple
   (B) a married couple with children
   (C) unemployed retirees
   (D) a married couple unable to afford the rent

19. Steering is
   (A) helping a buyer find a house in a neighborhood of people of similar ethnicity
   (B) helping a buyer find a house with a suitable floor plan
   (C) helping a buyer to qualify for a mortgage
   (D) encouraging a buyer to look at homes in neighborhoods with specific ethnic characteristics and discouraging the buyer from looking in other neighborhoods

20. A salesperson was taking a listing on a million-dollar-plus house. The owner told the salesperson that she would be unwilling to sell to a minority buyer. The salesperson should
   (A) honor the principal's request
   (B) smile and take the listing but be prepared to disobey
   (C) decline the listing because following the seller's instruction would violate the federal Fair Housing Act
   (D) report the owner to HUD

21. A church owns a retirement facility and requires that tenants be members of the church. This is
   (A) a violation of Fair Housing laws
   (B) acceptable under all tenant requirements
   (C) acceptable provided there is no discrimination based on race, color, sex, national origin, familial status, or handicap
   (D) not permitted as an exception to Fair Housing laws

22. The Civil Rights Act of 1968 prohibits discrimination based on
   (A) race
   (B) sexual preference
   (C) age
   (D) athletic ability

23. A broker shows an Asian family houses only in Asian neighborhoods. This is
   (A) legal steering
   (B) blockbusting
   (C) effective marketing
   (D) illegal steering

24. Which of the following is not protected under the federal Fair Housing Act?
   (A) race
   (B) religion
   (C) national origin
   (D) sexual preference

25. Which of the following is not discriminatory under Fair Housing laws?
   (A) steering
   (B) blockbusting
   (C) refusing to rent to a single woman
   (D) owning two units, living in one, and refusing to rent the other to families with children

26. A lender refuses to make mortgage loans on properties in a certain area of a city regardless of the qualifications of the borrower. This is
   (A) discrimination
   (B) price fixing
   (C) blockbusting
   (D) redlining

27. Which of the following neighborhood characteristics may an appraiser not describe in an appraisal?
   (A) economic characteristics
   (B) homogeneity of property
   (C) racial composition
   (D) proximity to transportation

28. Federal Fair Housing laws protect all except
    (A) racial minorities
    (B) families with children
    (C) pregnant women
    (D) graduate students

29. Which government agency investigates alleged violations of Fair Housing laws?
    (A) FHA
    (B) EEOC
    (C) HUD
    (D) FNMA

30. An American citizen of Mexican ancestry asks to be shown a house in a certain neighborhood. How should the broker respond?
    (A) "I'd be pleased to show you a house for sale anywhere."
    (B) "You'll be happier in a house elsewhere."
    (C) "Neighbors there will not be friendly."
    (D) "The seller does not want me to show that house."

31. The 1968 Fair Housing Act, as amended, prohibits housing discrimination on the basis of
    (A) race and color
    (B) race, color, and religion
    (C) race, color, religion, and national origin
    (D) race, color, religion, national origin, sex, familial status, or handicap

32. The prohibitions of the 1968 Fair Housing Act, as amended, apply to privately owned housing when
    (A) a broker or other person engaged in selling or renting dwellings is used
    (B) discriminatory advertising is used
    (C) both A and B
    (D) neither A nor B

33. The broker's obligation in complying with the requirements for equal opportunity in housing is
    (A) to replace white residents with minority homeowners
    (B) to avoid any acts that would make housing unavailable to someone on account of color
    (C) both A and B
    (D) neither A nor B

34. The Civil Rights Act of 1968
    (A) makes it illegal to intimidate, threaten, or interfere with a person buying, renting, or selling housing
    (B) provides criminal penalties and criminal prosecution if violence is threatened or used
    (C) both A and B
    (D) neither A nor B

35. Court action may be taken by an individual under the Fair Housing Act
    (A) only if a complaint is filed with HUD
    (B) if action is taken within 1 year of the alleged discriminatory act
    (C) if the alleged discriminatory act occurred on public property
    (D) if the alleged discriminatory act caused damage of at least $500

36. Which is exempt from the federal Fair Housing Act of 1968?
    (A) salesperson selling lots in a subdivision
    (B) salesperson renting apartments in a building with 10 units
    (C) owner renting vacant single-family unit
    (D) none of the above

37. If a landlord refuses to rent to a minority prospective tenant with a poor credit record,
    (A) federal fair housing laws are violated
    (B) the minority applicant should report this violation
    (C) there is no violation
    (D) the landlord may face a significant fine

38. Sullivan wants to rent an apartment but is confined to a wheelchair. At Sullivan's request, the landlord must
   (A) build a ramp at the landlord's expense
   (B) build a ramp at Sullivan's expense
   (C) do nothing
   (D) reduce the rent to Sullivan's affordability level

39. A seller asks his broker about the race of a potential buyer. The broker
   (A) must answer the question if a single agent
   (B) violates fair housing laws if he answers the question
   (C) may answer the question because of the Constitutional right to free speech
   (D) may tell the seller but no one else

40. Which constitutional amendment provides support by banning racial discrimination?
   (A) Second
   (B) Third
   (C) Twelfth
   (D) Thirteenth

41. Broker Bob is showing houses to Joe Chen, who is of Chinese ancestry. Bob shows Joe houses in a neighborhood of predominantly Chinese-American residences because Bob thinks Joe would feel more at home there. This is
   (A) illegal steering under fair housing laws
   (B) superior and tactful service
   (C) done for Joe's benefit
   (D) effective sales technique

42. Where the dominant population base of an area consists of minorities,
   (A) brokers need not be concerned with fair housing laws
   (B) brokers must still observe fair housing laws
   (C) there are no reporting requirements
   (D) violation of fair housing laws is not possible

43. The Civil Rights Act of 1968 prohibits
   (A) integration
   (B) greenlining
   (C) discrimination in housing
   (D) school busing

44. Familial status refers to
   (A) families with two or more children
   (B) families with children under age 18
   (C) the earning capacity of the family
   (D) TRID

45. Broker Barb is acting as a buyer's agent for Joe. Joe asks whether any people of a certain race live in a specific neighborhood. How should Barb respond?
   (A) She should advise Joe that, under fair housing laws, she cannot provide that information.
   (B) She should advise Joe of the race of all the neighbors she is aware of.
   (C) She should advise Joe that he should contact the demographic agency to obtain the information.
   (D) She should advise Joe to contact the school district to obtain the information.

46. A broker has contacted owners in a certain subdivision and advised them that several members of a particular race have bought homes in the area. The broker has offered to list properties at a reduced rate and encouraged the owners to sell quickly. This is
   (A) steering
   (B) profiling
   (C) blockbusting
   (D) scaring

47. Ellen owns a two-family house and lives in one side. She wants to advertise the other side as nonsmoking. Can she legally do so?
   (A) Yes, because the right to smoke is not protected by law.
   (B) Yes, because she owns the house.
   (C) No, because a property owner cannot advertise discriminatory practices.
   (D) No, because smoking is perfectly legal.

48. A tenant applicant confined to a wheelchair is interested in renting a patio home. A request is made to the landlord to allow the tenant to build an access ramp. Which of the following is true?
    (A) The tenant's application may be rejected.
    (B) The landlord must allow the tenant to make the modification at the tenant's expense.
    (C) The landlord is required to pay for modifications to the property to accommodate the access.
    (D) The landlord may collect an additional deposit to assure compliance.

49. No federal Fair Housing laws are violated if a landlord refuses to rent to
    (A) families with children
    (B) tenants of Indian descent
    (C) deaf persons
    (D) college students

50. The federal ban on discrimination based on familial status is intended to provide equal access to rentals for
    (A) unmarried couples under 25
    (B) people with children
    (C) single tenants
    (D) the elderly

51. The Civil Rights Act of 1866 prohibits
    (A) zoning
    (B) redlining
    (C) open housing
    (D) racial discrimination

52. A minority group is moving into an area immediately adjacent to an old subdivision. ABC Realty offers to list homes in the subdivision at a lower-than-usual commission rate if the owners list within 45 days. There is no mention of race, and the broker acts in good faith. Which of the following is true?
    (A) Such practice is not illegal.
    (B) Brokers cannot lower their standard rate of commission.
    (C) The broker's license can be revoked for this action.
    (D) This is blockbusting.

53. A couple of Greek ancestry looking for a house asks a real estate agent to show them houses in Greek neighborhoods. Which of the following is correct?
    (A) The agent must accommodate this request because the buyers requested it.
    (B) The agent should tell the buyers that he will show them houses in a number of different neighborhoods.
    (C) The agent should refuse to show any houses to the buyers.
    (D) The agent should agree and show the buyers houses in other neighborhoods.

54. Advising people on which neighborhoods they would be happy in may be construed as an illegal activity called
    (A) steering
    (B) canvassing
    (C) redlining
    (D) blockbusting

55. "Whites only" is the advertisement for a house that's for sale. This is
    (A) legal if the house is owner occupied
    (B) illegal unless a real estate agent is used
    (C) legal because it's a one-family house
    (D) illegal

56. The Aggie Oaks Apartments advertising states that families with children are welcome. The management has allocated a certain section of the complex for residents with children, and those tenants must pay an additional security deposit for each child under twelve. Is this discriminatory?
(A) No, because families with children are welcome to the complex.
(B) Yes, based on marital status.
(C) Yes, based on familial status.
(D) No, because children are not a named protected class.

57. Armando and Maria Rios agreed to sign a buyer's agreement with a real estate agent who showed them homes in predominantly Hispanic areas only. This is an example of
(A) boycotting certain neighborhoods
(B) thoughtful customer service
(C) cultural sensitivity
(D) steering

58. Agent Curt has been asked by his seller not to advertise in a Spanish-language newspaper. Curt and his office would not violate the law if
(A) the paper does not serve the market area of his office
(B) he doesn't like the paper's editorial positions on many issues
(C) his seller directed him in writing
(D) his office has never used the paper because of the high rates charged by the paper

ANSWERS

1. **D**	11. **D**	21. **C**	31. **D**	41. **A**	51. **D**
2. **C**	12. **D**	22. **A**	32. **C**	42. **B**	52. **A**
3. **C**	13. **D**	23. **D**	33. **B**	43. **C**	53. **B**
4. **C**	14. **C**	24. **D**	34. **C**	44. **B**	54. **A**
5. **A**	15. **C**	25. **D**	35. **B**	45. **A**	55. **D**
6. **C**	16. **A**	26. **D**	36. **C**	46. **C**	56. **C**
7. **B**	17. **B**	27. **C**	37. **C**	47. **A**	57. **D**
8. **D**	18. **D**	28. **D**	38. **B**	48. **B**	58. **D**
9. **C**	19. **D**	29. **C**	39. **B**	49. **D**	
10. **B**	20. **C**	30. **A**	40. **D**	50. **B**	

Subdividing, Building and Development, Land Use Controls, Environmental Matters 22

ESSENTIAL MATTERS TO LEARN FROM THIS CHAPTER

- This chapter covers three different, though related, topics:

  1. Subdividing, building, and development.
  2. Land use controls.
  3. Environmental matters.

- *Subdividing* is dividing raw land into usable pieces. It typically begins with acquiring control of the land, rezoning if the land is not zoned for the desired use, putting in infrastructure such as utilities and streets, and then selling the lots in pieces. *Building* is erecting a structure; *development* entails both subdividing and building.
- *Permits* from the municipality are required. The building process starts with a *building permit* and ends with a *certificate of occupancy.*
- *Zoning* is a municipality's right to control land use. It comes under the *police power* of government, which is the right to regulate for the health, safety, morals, and welfare of its citizens.
- Often, a city's zoning designation of a tract of land is preceded by a master plan. Land that is newly annexed by a city may first be brought in with agricultural zoning.
- When the owner wants to use the land, he may request, through the city's administration, the zoning category that he seeks. The city's staff will review the request and make recommendations to the planning and zoning board and then to the elected city council for final decision. Houston, Texas, is the only major city in the United States that does not have zoning.
- *Zoning categories* include agricultural, residential, commercial, and industrial. There are subcategories within each category. Zoning of a parcel of land may change upon application and a hearing.
- *Zoning information* in a municipality includes a *zoning map* showing each parcel and a *zoning ordinance* to explain details of what is allowed within each category and the zoning process.
- A zoning *variance* may be used to waive a minor violation.
- When a structure is built that later becomes in violation of zoning, typically because of a change in zoning law, it is called *nonconforming.* Prior use of the property is allowed to continue unless the building is damaged. Rebuilding may not be allowed unless the property will conform to the current standards. At that point, a replacement structure must conform to the current (not original) zoning ordinance.

- A subdivider or land developer may file a *preliminary plat* to show how he would like to lay out the land in pieces, where the streets and utilities will be. The project may be one in which lots will be sold or where a completed structure will be leased.
- The subdivider may also include in plans a *Declaration of Covenants* that provides restrictions beyond zoning. When building a condominium, *Conditions, Covenants, and Restrictions* (CCRs) are used in a *Declaration of Restrictions.*
- Typically, the real estate entrepreneur requires financing. *Debt* is common, as is investor *equity.* The investor often contributes his own effort and expertise, called "sweat equity."
- Lenders usually fund the project as it progresses, making sure that the improvements are properly built before advancing additional funds.
- A developer may arrange to have a general contractor build the project, using a cost-plus-profit or cost-plus-percentage, fixed-fee, or guaranteed-maximum contract.
- *Acquisition due diligence* is a term used for the process of ensuring that land or a building to be purchased is suitable for the expected purchase. Engineers and inspectors are engaged to study the property.
- *Environmental considerations* are a major item considered in the due diligence process.
- *The Comprehensive Environmental Response, Compensation, and Liability Act* (CERCLA) is commonly known as *Superfund.* Originally passed in 1980, it was reauthorized in 1986 as the Superfund Amendments and Reauthorization Act (SARA).
- *The Environmental Protection Agency* (EPA) is the federal agency that enforces CERCLA. State agencies are also actively involved.
- Superfund imposes strict, joint and several, and retroactive liability on past, present, and future owners, tenants, and transporters of materials to property that requires cleanup. Possibly innocent land buyers may be exempt from paying the full cost of cleanup but only if they meet certain requirements.
- When property under a purchase contract is found to have contamination, the potential buyer may void the purchase. There are generally three phases to an environmental study. Phase I determines whether there is any contamination. Phase II determines the extent and cost of cleanup. Phase III is when the cleanup work is performed.
- Some common environmental problems include leaking petroleum storage tanks, asbestos-containing materials, certain solvents including ones used in dry cleaning plants, electronic equipment such as PCBs, and other materials found in industrial facilities.
- Disturbing *wetlands* can be done only with permission from the Army Corps of Engineers. Wetlands may host wildlife that it is important to preserve.
- *Radon* is a naturally occurring radioactive substance found in air and water that has been linked to lung cancer.
- Homes built prior to 1978 are sold with warnings that they may contain *lead-based paint.* Children may ingest paint chips with lead that are harmful.
- Although strong chemicals may be used in a home, such as drain cleaners, hair coloring, laundry detergent, disinfectants, pesticides, and fungicides, homes are generally exempt from EPA requirements. Removal of asbestos insulation and floor tiles from houses requires permit authorization.

LAND

Subdividing is the process of acquiring raw land and improving it to the point where it is prepared to be built on. At that point, it is called a *site* and it can be sold to a builder or a prospective homeowner who will build on it. *Building* is putting a structure on the site. *Development* is both activities combined, usually including financing.

The value of land depends on what is around it. Land is a prisoner of its environment, so its location is crucial. Attributes of land may be described in four categories: physical, financial-economic, legal-political, and social. Physical characteristics include size, shape, topography, eye appeal, access to utilities, climate, and soil quality. Financial-economic issues include employment and unemployment in the area, wage rates, industry type and growth, interest rates, and inflation rates. Legal-political matters include zoning, building codes, and taxation. Social forces include school quality, hospitals, houses of worship, and availability of services. A key determinant of land value is the rate of appreciation and expectations. This is because tangibles such as land are considered a hedge against inflation.

The value of land is most often appraised using a *sales comparison* (market) approach. Another method is by land development, whereby the final aggregate selling price of the lots are estimated and, working backward, one subtracts the cost of development and adjusts for time and risk.

The cost of the finished lots for a homesite is typically four to five times the cost of the raw land. For example, if the raw land is purchased for $40,000 per acre, a finished lot of one-quarter acre can be expected to sell for $40,000 to $50,000.

The subdivider must have appropriate zoning for end use. For example, the land may be zoned R-1, which in that jurisdiction allows four houses per acre; or R-2, which may allow six houses. The subdivider must engage an architect and engineer to examine the subdivision and file a plat to be approved (map of the subdivision) by the planning and zoning and the engineering departments of the city.

A *building permit* must be obtained from the city before ground can be broken. The municipality's engineer checks the suitability of the land for the planned building. Later, assuming construction progresses, the city's inspector will inspect the foundation, heat and ventilation system, wiring, and plumbing and roofing for compliance with the *building code.* Upon completion of construction, a *certificate of occupancy* will be issued.

ACQUISITION AND DEVELOPMENT LOANS

Often a commercial bank is the likely source of an acquisition and development loan. Such loans generally carry interest at the prime rate plus three to five percentage points plus discount points. The loan is funded in stages as the development proceeds. A subdivision or any other construction loan is considered high risk. Until sales occur in sufficient quantity, money is poured in without any coming out. It is the financial equivalent of swimming underwater. Economics in the area could change, as could the national economy. Home building is sensitive to interest rates and local growth, especially employment. Buyers are fickle—buying a new home is easily put off.

The first funds loaned are to purchase the land. Then, as the land is graded and potential drainage problems are cured, another installment is due. Connecting utilities to each lot—electricity, natural gas, water, sewer, and perhaps cable television—are costly. Another construction draw is due. Paving streets and putting in sidewalks deserve another draw.

Advertising and selling expenses must be funded. So the construction loan balance is increased in steps, with the lender determining that work has been done before advancing the money.

When lots are to be sold, they must be released from the subdivision loan to allow for construction financing of the house. Generally, these release provisions require more money for a given lot than the proportionate amount of subdivision loan. For example, if there is a $3 million loan on a 100 unit development, or $30,000 per lot, the loan may require that $40,000 be paid to release a lot from the subdivision loan. That way the lender forces the subdivider to stay with the project and earn a profit only after the development has proven its success through a near sell-out.

To keep loans and costs under control, the subdivider will build in phases, having enough inventory of lots to satisfy demand, but not too much, because of cost considerations.

City inspectors will examine the work at various stages to give approvals. Approvals are generally to determine compliance with city building codes for safety reasons, but they are also to protect the public. Installation of pipes must be done properly to insure they won't leak natural gas, that they will carry sewage without leaking, and that water pipes won't get contaminated. Inspectors also assure that electrical connections are safe.

Generally, lots are considered a sale of real estate and may be sold by licensed real estate brokers and salespeople. If lots are offered for sale out of state, the subdividers must file a statement with the *Office of Interstate Land Sales Registration.* This is to make buyers aware of risks, especially buyers who might not visit the property. This is often the case for resort property.

DEVELOPMENT

Development is not only getting the land prepared for use, it is also building on it and often financing a permanent loan. Matters to be considered when developing real estate are the site, market, financial feasibility, lease, financing, and construction.

Zoning

Zoning is a municipality's right to regulate land use. Legally, it comes under the heading of *police power,* the right to regulate for the health, safety, and welfare of the general public. Every major city in the United States except Houston, Texas, has zoning. Houston does enforce private deed restrictions.

Each parcel of land is in a zoning category. Major categories include residential (with subcategories for density), commercial, and industrial uses. Often the city has a *master plan* to guide future land use. Properties may be zoned agricultural when the land is first annexed by the city. The zoning may then be changed (rezoned) upon the owner's application to the city's planning staff and elected officials.

A property owner may appeal a minor infringement of a zoning provision by requesting a *variance.* This may be used to relieve hardship such as that caused by building setback requirements on small or oddly shaped lots.

When a provision of a zoning law is changed or a physical change occurs, a structure may no longer comply with current zoning. For example, if a street is widened, the buildings facing it may no longer comply with the setback requirement. They become *nonconforming.* Past uses are allowed to continue as long as the property remains intact. If the property is damaged or remodeled, however, its replacement must conform to the new zoning law.

Other Land Use Controls

Other land use controls include private deed restrictions and subdivision restrictions. A subdivider or a condominium complex may initially file a subdivision plan that limits land use. Even if this is approved, enforcement may require court action.

> **Trick Question**
>
> A common trick question on a licensing exam asks the purpose of zoning. An erroneous answer is "to enhance or sustain property values." The correct reason is "to promote the health, welfare, and safety of a community." Most laypersons don't know the correct reason. They would answer "to protect property value," which is a wrong answer on exams.

Site Zoning and Adjacent Land Uses

A site must have proper zoning before development can begin. A building permit is required. Frequently, sites suitable for residential development may be zoned agricultural, or those intended for retail use are not zoned for commercial. Then, there must be an exploration of the attitudes of the local residents, municipal zoning staff, and the various governmental bodies that approve a proposed project, such as the Planning and Zoning Commission (P&Z) and City Council.

Proximity of Competition and Traffic-Generating Businesses

Conditions outside the site affect the success of the proposed development. Adjacent uses may generate desirable or undesirable traffic depending upon the type of business. For example, fast food restaurants typically find that shopping centers, high schools, and theaters are desired traffic generators. A residential development wants quiet neighbors. Both would tend to feel that salvage yards, factories, and auto dealerships generate undesired traffic.

Size and Shape

Most developments need some street frontage. Rectangular sites are preferred for some uses because they make the most efficient layout for building, parking, and access.

Traffic Access, Count, and Curb Cuts

A motorist must be able to enter and exit a site safely without causing traffic congestion. Although many developers prefer at least two curb cuts into a subdivision, some cities will not allow more than one curb cut for a small development having less than 150 feet of frontage.

Visibility

Some uses, such as retail properties, require good visibility to attract customers. Poor visibility of a site is related to accessibility: a site that is difficult to see is also difficult to enter. A person traveling at 35 miles per hour should be able to see the property (or its sign) in time to enter the parking area safely. Corner lots generally provide better visibility and access than interior lots. Sign requirements vary by city; one should check with the planning or zoning department.

Topography, Drainage, and Utilities

A flat site is preferred for commercial development, whereas many homebuyers prefer land with scenic qualities—a lake, stream, or hillside. The contour of the land affects the site development and building design. A level or gently sloping (less than three percent) grade is easier to build on; slopes of more than five percent require special consideration. Low-lying areas with poor drainage may require storm sewers, water retention ponds, or structural building considerations. The unavailability of gas, water, sewer, or electric service at or near the site is a major problem.

Parking and Landscaping Requirements

For a commercial development, the parking area must support the property's image of attractiveness and convenience. Therefore, the layout, dimensions and arrangements, grading, paving, landscaping, and lighting must all be considered in the site planning process. Many cities impose off-street parking requirements.

FINANCIAL FEASIBILITY

There are two methods of trying to assure the financial feasibility of a proposed development. One is to determine the cost (or operating expenses plus debt service) that a completed building will incur. The price (or rental rate) is established to exceed that sum by a reasonable amount, depending on the equity capital required. This method is appropriate when attempting to build a unique structure with special market acceptance. However, when building into a competitive market, it is preferable to determine the going price (or rental rate) in the market. Then construction and related building costs must be kept within a strict budget so that market rents will be ample to pay for debt service plus operating expenses and to provide an adequate return on equity. Costs to be included are:

- Land
- Land acquisition fees
- Interest during construction
- Construction loan fees
- Permanent loan fees
- Permits
- Building and paving
- Site work
- Utility hookups
- Legal and accounting
- Builder's and developer's profit

LEASING

The lease is the key to developing real estate for a nonspeculative rental project. It allows property development and reduces risk for all parties who may be concerned—the landlord, tenant, lender, developer, builder, and investor.

The strength of a lease provides the basis for issuing credit. An institutional lender will provide a commitment for permanent financing based on the terms of a lease and credit standing of a tenant. The lease should specify, in sufficient detail, all aspects of the development. The permanent lender is obligated to fund the project upon completion of all terms described in the lease. The commitment need not be funded if the lease terms are not met, for example, if the completed building differs from the lease specifications or doesn't meet building codes, or there are other unfulfilled lease provisions. A permanent financing commitment paves the way to construction financing. When permanent financing is arranged, the construction money is available to begin the project.

FINANCING

The two major sources of real estate financing are equity and debt. Equity funds are the owner's money in the property. Debt funds are obtained from third parties, predominantly from financial institutions.

Equity Financing

When the prospective owner's cash resources are insufficient to provide needed equity, other sources may be available. A partnership is one option. Joint ventures are similar to partnership agreements, but are for owning and operating one specific business; they terminate upon disposition of that business. Many developers prefer to use their own efforts, called "sweat equity," instead of money.

Debt Financing

Under normal circumstances, financing for a newly built real estate development is obtained from two different lenders: a long-term (permanent) lender and an interim (construction) lender. The permanent lender issues a takeout commitment that promises to repay the construction loan upon project completion. Only with a takeout commitment (or acceptable substitute) can a construction loan be arranged. The borrower prefers a permanent loan commitment from a long-term lender rather than a short-term lender.

Financing alternatives include a permanent loan (with or without participation) or a miniperm or balloon loan. Each may be structured to pay interest only for a specified period followed by amortization at a specified rate, with level payments, or an adjustable-rate mortgage (ARM). The ARM rates on commercial properties are usually indexed to short-term rates such as the prime or U.S. Treasury bill rate. Rate adjustments may have annual caps on rate changes and a ceiling on the rate over the life of the ARM.

Participation mortgages provide an inflation hedge to the lender as compensation for increasing market interest rates. The borrower gives up a portion of annual earnings and pays annual debt service plus a percentage of gross income or net operating income above a specified level. Lenders prefer to use gross income because amounts are easier to validate than net income or cash flow.

Lenders

Savings and loan associations, commercial banks, real estate investment trusts (REITs), and mortgage bankers are the primary market (original sources) of loans for housing.

Mortgage REITs finance every step of a real estate development. REITs may originate loans or purchase them from other originators. *Pension fund* investment managers typically place a portion of their assets in bank trust departments, trust companies, and life insurance companies. These assets are placed in commingled funds for investment in real estate equities and mortgages.

Insurance companies are interested in investing in commercial real estate mortgages, often through mortgage bankers who have correspondents. *Finance companies* are construction and gap financing lenders for commercial real estate. Underwriting criteria are competitive with commercial banks. Loan structures are also similar. Finance companies will accept more risk in higher L/V ratios than commercial banks with corresponding higher commitment fees, points, and interest rates.

THE LOAN PROCESS

The loan process includes four steps. In *loan submission,* a comprehensive package of the borrower's financial condition and the project's characteristics, explaining feasibility, is presented to the lender. *Loan processing* includes analysis of the loan submission to verify borrower and project information submitted. *Loan underwriting* includes analysis of the borrower's creditworthiness and project feasibility to reach a lending decision. Negotiations between borrower and lender are needed to reach an agreement regarding loan terms. *Loan closing* activates the borrower and lender decisions and establishes contractual arrangements.

CONSTRUCTION FINANCING

The construction loan is secured by the land and subsequent improvements. The lender expects the loan to be repaid in a lump sum when construction is completed. For commercial property with key tenants committed and a permanent loan package arranged, the developer-borrower can shop for an interim (construction) loan. Institutional loans, other third party loans, and joint ventures are three options usually available for construction financing.

The lender usually contracts with an independent architect or engineer and legal counsel to conduct the inspections. On each disbursement, county records are examined to be sure that no new liens have been filed. Typically, funds are disbursed to cover costs up to five to ten percent less than work in place plus stored materials.

Ten percent of the total construction loan amount is typically retained after construction completion, disbursed when the lender has verified that all subcontractors, materialmen, and the general contractor have been paid and no mechanics' liens have been filed on the property. With the final disbursement, the construction lender expects the property to be occupied and the permanent loan to be funded, paying off the interim loan.

These steps do not eliminate construction and economic risks. Construction risks include stoppages due to labor problems, unfavorable weather, unanticipated increase in construction costs, errors in plans and specifications, poor construction management, misappropriation of construction funds, and developer insolvency. Economic risks include market change that reduces demand, neighborhood decline, poor space design and construction defects, and poor marketing analysis leading to incorrect pricing and marketing strategy.

PHYSICAL DEVELOPMENT

To physically develop the property, the landowner must act as a builder or hire a construction firm. If the landowner is the general contractor and developer, construction knowledge is essential. Since many landowners do not have the required time or expertise, most projects are built by firms in the construction business.

CONSTRUCTION ARRANGEMENTS

A contractor will agree to do the specified work by a certain date for a fixed price or a certain markup above cost. Generally, contracts are complex and consist of several documents including:

1. Invitation to bid
2. Instructions to bidders
3. General legal conditions of the contract
4. Supplementary operating conditions of the contract
5. Technical specifications
6. Drawings and blueprints
7. Bid bond
8. Contract agreement(s)
    a. Owner-architect
    b. Owner-contractor
    c. Contractor-subcontractor
9. Performance bond

The construction requirements and specifications of the project are exposed for bids; generally, the contract is awarded to the low bidder unless there are extenuating circumstances.

There are several cost estimation services available to help estimate the cost of alternative construction features; three are the *Boeckh Building-Cost Manual,* the *F. W. Dodge Construction Cost Manuals,* and the *Marshall Valuation Service.*

To reduce the chances for conflict among owners, architects, engineers, and contractors, standardized contract documents are commonly used. The American Institute of Architects offers a variety of forms that are frequently used. The National Society of Professional Engineers, the American Public Works Association, and the Associated General Contractors of America also provide standardized contract forms.

TYPES OF CONTRACTS

Although there are many variations, most contracts fall into one of the following three categories and are awarded through competitive bidding, negotiation, or a joint venture.

A *fixed-price contract* requires the contractor to complete the project for a specific dollar amount. With a *guaranteed-maximum-price contract,* the owner pays for only the actual cost incurred (including a contractor's fee) within the maximum price guarantee. A *cost-plus contract* reimburses the contractor for actual costs incurred plus an additional percentage or fixed amount.

SELECTING A CONTRACTOR

There are several questions to be raised in selecting a contractor. Does the contractor have prior experience building similar projects? Can a list of these projects be provided, along with original cost estimates and the actual cost of each? Does the contractor have experience building in the local area?

Building regulations, labor supplies, and the source of materials vary with each city; a contractor not familiar with local conditions may have conflicts, time delays, or greater expenses.

Also, it is easier to check the references of a local contractor with lenders, government agencies, trade unions, and building supply houses.

Will the contractor provide a financial statement to indicate that (a) the firm has sufficient funds to be responsible for specific performance, (b) the firm is able to handle the cash flow requirements, and (c) there are no excessive short-term debts that might create problems after the project is started? Who are the subcontractors to be used for each major activity area? What is their reputation for quality and reliability and their relationship with the general contractor? The general contractor is only as good as the subcontractors used.

CONDOMINIUMS AND COOPERATIVES

Developers or converters of a proposed condominium or cooperative, or any other shared housing arrangement, in most states must file a statement indicating their plan. The developer must provide disclosure of the plans. This includes an architect's or engineer's report, past or expected expenses, management arrangements, and other legal and financial arrangements. It should include the conditions, covenants, and restrictions (CCRs) of use. These are often written in a document called the Declaration of Restrictions.

Condominium and co-op converters must follow specific rules in their community, depending on the converter's plans to allow for eviction of existing nonbuyers or a noneviction plan.

Housing offerings are printed in red on at least part of the first page (red herrings) upon preliminary review. Then final plans are printed in black ink.

ACQUISITION DUE DILIGENCE

Due diligence relates to the process of inquiring about the property before buying. Because buyers and brokers are not engineers or geologists, it is in their best interest to hire qualified personnel to inspect the property.

Every house should be inspected for wood-destroying insects by a qualified, licensed pest control operator. Termites, carpenter ants, old house borers, and other pests should be discovered and the damage assessed. A qualified inspector will check the plumbing, electrical, and mechanical systems. Finally, a structural engineer may be helpful to determine soundness.

Commercial properties may need those types of inspectors plus a high-level engineer. Environmental inspections may include searching for contaminants such as lead-based paints, the presence of underground storage tanks, groundwater contaminants from adjacent properties, asbestos, and pollution from the subject property into the air and water.

Due diligence for income property also includes a review of market conditions including proposed competition, tenant leases, and financing availability.

ENVIRONMENTAL CONSIDERATIONS

Increased environmental sensitivity has raised the liability of those involved in real estate. Every gas station could have a leaking tank, pump, or pipe that contaminates groundwater. Dry cleaners and car washes may use chemicals that, if released, are harmful. Even seemingly innocent property uses such as golf courses and agricultural land are suspect because of fertilizer and pesticide applications.

The Federal Home Loan Mortgage Corporation proposed guidelines of due diligence for environmental standards of single-family homes that had been previously defined as non-contaminating. A house may host radon (suspected of causing lung cancer), have had pesticides liberally applied to the lawn, or have urea formaldehyde or asbestos insulation. It may be near a landfill or have been built on contaminated land. Due diligence for the investigation of homes is being proposed. It would require an inspection of records of prior ownership and land use and an inspection of the property.

Liability

Superfund is the common name for the 1980 law that affects many issues of real estate contamination. Its real name is the Comprehensive Environmental Response, Compensation, and Liability Act (CERCLA). It is enforced by the Environmental Protection Agency (EPA). Superfund, reauthorized in 1986 as SARA, includes stronger clean-up standards, disclosure requirements, and funding. *Superfund* requires the clean-up but pays only in extreme situations, and it can render property worthless. It imposes liability on those involved with hazardous materials, liability that is *strict, joint and several, and retroactive.* Liability is created by Superfund in any connection with the property: as an owner, operator, generator, or transporter. *Strict* means that it doesn't matter whether such a person acted knowingly or reasonably—that person bears liability. The absence of negligence or other wrongdoing is not a defense. There are some legal defenses, but they are limited.

Joint and several liability means that every responsible party is liable for the full cost. The government or Superfund claimant may find anyone with a "deep pocket" to pay costs, and it doesn't have to sort out who was responsible for how much damage. Huge clean-up costs could conceivably be assessed against someone who took title for an instant during a closing to facilitate a transaction but otherwise had no involvement with the property. The unlucky party would be responsible for the full amount but could seek reimbursement from other responsible parties.

Retroactive liability means that it reaches back to prior owners and operators. This overrides "as is" clauses in sales contracts. It also precedes any mortgage lien.

Activities

Activities that may cause contamination include manufacturing, assembly, repair, laboratory, storage, machine shop services, and cleaning. A partial list of types of businesses that may be perpetrators includes any that use, manufacture, or supply products to these uses: electronics, leathers, paints, pesticides, petroleum, pharmaceutical, plastics, refining, smelting, or textiles. In addition to those activities, events that may cause serious problems include fires, explosions, spills, and tank, pump, and pipe leaks.

Contaminated Items

Contamination can be found in buildings, soil, and groundwater. Within a building, asbestos may have been used for insulation and pipe wrapping. There is a serious problem when the asbestos is friable, which means it crumbles. Interior components of a building may have thick residues of chemicals that were used—these may have migrated through concrete floors into the soil. Radon is a naturally occurring radioactive material found in air and water. Radon has been linked to lung cancer.

Contamination that spreads through groundwater can cause problems to adjacent property. An extensive site assessment might be needed to determine the extent of damage. Buyers of homes built before 1978 are to be advised that the house may contain lead-based paint.

Detection

A proper *site assessment* by a qualified professional is the best approach when considering a purchase. More and more lenders are requiring one as a requisite for a mortgage loan on commercial or industrial property. Contaminants can render a property worthless, and a foreclosure makes the lender an owner who is liable. Site assessment costs range from a few thousand to hundreds of thousands and include drilling tests, electrical conducting, ground penetration, and laboratory analysis. An aerial photograph may help identify suspect points including ditches, landfills, lagoons, pits, ponds, tanks, and waste piles.

Other methods of due diligence to determine detection include reviewing public records and newspapers and checking with the local fire department and state or federal environmental or pollution agencies.

A Phase I assessment should be performed by a buyer as part of the due diligence acquisition process; it will help avoid clean-up liability under Superfund. Phase II is to estimate the cost of clean-up, and Phase III is to perform the work.

QUESTIONS ON CHAPTER 22

1. Subdividing is a process of
   (A) passive investment in land
   (B) clearing land of unwanted debris
   (C) preparing land to be built on
   (D) building on land

2. A developer is
   (A) the same as a subdivider
   (B) one who subdivides, builds, and generally arranges financing
   (C) a builder and tenant
   (D) a broker who builds

3. Land appraisal is most frequently done by which approach?
   (A) income
   (B) replacement cost
   (C) direct sales comparison
   (D) all of the above

4. Construction loan funds are
   (A) provided in steps as construction progresses
   (B) all loaned at the start of a project
   (C) all loaned at the completion of a project
   (D) always the same dollar amount as permanent loan funds

5. Land for commercial uses ideally is on a (an)
   (A) slope or grade
   (B) heavily travelled street or road
   (C) in a residential neighborhood
   (D) interstate freeway

6. Banned health hazard products include all except
   (A) radon
   (B) asbestos
   (C) radioactive materials
   (D) Drano

7. Construction cost sources include all except
   (A) Wiley
   (B) Dodge
   (C) Boechk
   (D) Marshall

8. In most states, a condominium converter must
   (A) not renew leases of tenants
   (B) file a plan and provide disclosure
   (C) make a special deal with each tenant over age 65
   (D) not evict anyone under any circumstances

9. Before breaking ground, a ________ must be obtained from the city.
   (A) construction loan
   (B) red herring
   (C) occupancy plan
   (D) building permit

10. The final stage of building inspection is the granting of a
    (A) prospectus
    (B) certificate of occupancy
    (C) inspection certificate
    (D) conforming use certificate

11. A nonconforming use
    (A) is allowed to continue unless the building is damaged
    (B) is grandfathered
    (C) was conforming before a zoning change
    (D) all of the above

12. Private restrictions on the use of land may be created by
    (A) private-land use controls
    (B) written agreement
    (C) general plan restrictions in subdivisions
    (D) all of the above

13. Conditions, covenants, and restrictions (CCRs) are usually recorded by a document called
    (A) a plat
    (B) subdivision rules
    (C) Declaration of Restrictions
    (D) common area designations

14. If CCRs and zoning conflict with each other,
    (A) the more stringent rules apply
    (B) either can be followed
    (C) courts or mediation will be needed to settle the matter
    (D) the developer will decide what to do

15. A *turnkey* project is one that is
    (A) under development
    (B) under contract
    (C) complete and ready for use or occupancy
    (D) under lease

16. A development has restrictive covenants. These may not regulate
    (A) exterior paint color
    (B) fences
    (C) satellite dishes
    (D) race of owner

17. Zoning ordinances typically regulate the
    (A) number of occupants allowed per dwelling unit
    (B) uses of each parcel of land
    (C) maximum rent that may be charged
    (D) building code

18. Tony's Pizzeria has signed a purchase contract with X Oil Company to purchase a parcel of property. X Oil Company has failed to disclose to Tony's that toxic substances were dumped on the property over a ten-year period. When this omission is discovered by Tony's, the contract is
    (A) valid
    (B) voidable
    (C) void
    (D) unenforceable

19. If an area is rezoned industrial and a commercial establishment may continue its operation in that area, this is an example of
    (A) variance
    (B) nonconforming use
    (C) conditional use permit
    (D) spot zoning

20. You want to build your house 10 feet closer to the road than the zoning ordinance permits. You will likely try to get
    (A) a zoning ordinance change
    (B) a variance
    (C) a deed restriction amendment
    (D) a nonconforming use

21. Which of the following is true of high concentrations of radon gas?
    (A) It has a very noticeable odor.
    (B) It has been linked to lung cancer in long-term exposure.
    (C) It is released by plastic building materials.
    (D) It enters the house through openings in the attic.

22. Which of the following is *not* governed by zoning?
    (A) use of a residence partly for a business
    (B) racial considerations in a sale
    (C) maximum number of apartment units per acre
    (D) building height

23. The Comprehensive Environmental Response, Compensation, and Liability Act (CERCLA) is enforced by the
    (A) Department of Homeland Security
    (B) Department of Housing and Urban Development (HUD)
    (C) Federal Emergency Management Agency (FEMA)
    (D) Environmental Protection Agency (EPA)

24. A buyer of a house built before 1978 must be advised of the possibility that it may contain
    (A) asbestos
    (B) radon
    (C) lead-based paint
    (D) termites

25. EPA stands for
    (A) earnings per annum
    (B) effort put away
    (C) extra-pretty agent
    (D) Environmental Protection Agency

26. When the zoning of a property changes, uses that were once compliant with zoning are
    (A) terminated because they are no longer compliant
    (B) allowed to continue as nonconforming uses
    (C) granted a variance
    (D) allowed to continue only if changes are made to comply with the new zoning

27. The presence of wetlands
    (A) lowers a property's value
    (B) causes the property to be useless
    (C) enhances the property's value because the wetlands attract wildlife
    (D) should be disclosed to potential buyers

28. A condominium owner would likely pay all *except*
    (A) association fees
    (B) maintenance fees
    (C) property taxes
    (D) stock transfers

29. Every major city in the United States except Houston applies
    (A) zoning
    (B) Fair Housing Act
    (C) Truth in Lending Act
    (D) Real Estate Settlement Procedures Act

30. Deed restrictions are typically enforced by
    (A) the landowner
    (B) a court
    (C) the police force
    (D) the registrar of deeds

31. During a property inspection, the inspector learned that the building had been used as a facility to repair automobile brakes. This should suggest the possible presence of
    (A) asbestos
    (B) radon
    (C) hydrocarbons
    (D) lead-based paint

32. A builder of new homes seeks to have an 18-foot setback, whereas zoning requires a 20-foot setback. The builder should request
    (A) spot zoning
    (B) a nonconforming use
    (C) a variance
    (D) that authorities look the other way

33. The prospective buyer wants to convert the living room into a commercial beauty parlor. The buyer's sales agent should
    (A) advise against having a business in your own home
    (B) determine whether the use is permitted under the zoning law
    (C) suggest that the buyer pay more for the house
    (D) help the buyer find customers for the shop

34. The city rezoned a block of industrial property to become retail. Martin owns an industrial building that has been leased as a factory for the past five years. Martin now
    (A) can continue to use the property as a factory though it becomes a nonconforming use
    (B) must stop using the property as a factory when the lease expires
    (C) is exempt from zoning because of the age of the building
    (D) may terminate the lease

35. Sources of groundwater contamination include all *except*
   (A) leaking petroleum tanks
   (B) disposal of containers from pesticides used on a farm
   (C) radiation from a nuclear power plant
   (D) satellite transmissions

36. Superfund imposes
   (A) partial liability for cleanup
   (B) waste
   (C) forfeiture of land
   (D) strict liability for damage

37. Which of the following pairs does *not* make sense?
   (A) grantor—gives deed
   (B) landlord—gives tenancy
   (C) mortgage lender—gives loan
   (D) police power—gives deed restriction

38. Zoning is authorized by what power of government?
   (A) condemnation
   (B) eminent domain
   (C) regulation
   (D) police power

39. A gas produced by radioactive materials naturally found in the earth is
   (A) carbon dioxide
   (B) carbon monoxide
   (C) nitrogen
   (D) radon

40. Which of the following is *not* an environmental hazard?
   (A) leaking underground storage tanks
   (B) radon gas
   (C) xenon lights
   (D) lead-based paint

41. The right to zone the use of land is an exercise of
   (A) the police power of government
   (B) eminent domain
   (C) municipal entitlement
   (D) condemnation

42. A property that complied with zoning when built but no longer complies because of a zoning change is
   (A) nonconforming
   (B) illegal
   (C) a variance
   (D) not usable

43. Building permits are issued to
   (A) assure the safety of property
   (B) provide visually appealing designs
   (C) protect workers from injury
   (D) maintain economic balance

44. The purpose of a certificate of occupancy is to ensure
   (A) safety for occupants
   (B) adequacy of rent
   (C) harmony of architectural style
   (D) payment of taxes

45. The purpose of zoning is
   (A) to elevate property values
   (B) to protect health, safety, and welfare
   (C) to allow utility lines to be built
   (D) to compensate landowners for a taking

46. Zoning and building codes are an exercise of the government right of
   (A) federal restrictions
   (B) police power
   (C) hazardous duties
   (D) civil rights

47. A variance is often granted because of a
    (A) minor failure to meet the zoning code
    (B) need to violate the zoning code flagrantly
    (C) need to obtain an occupancy permit
    (D) neighbor trespass

48. Generally, a nonconforming use may continue unless it is
    (A) seriously damaged by wind
    (B) seriously damaged by fire
    (C) remodeled
    (D) any of the above

49. Zoning is a _________ law.
    (A) federal
    (B) state
    (C) municipal
    (D) any of the above

50. The document that allows humans to occupy a structure is a(n)
    (A) occupancy permit or certificate of occupancy
    (B) inspection certificate
    (C) fire certificate
    (D) building permit

ANSWERS

1. **C**	11. **D**	21. **B**	31. **A**	41. **A**
2. **B**	12. **D**	22. **B**	32. **C**	42. **A**
3. **C**	13. **C**	23. **D**	33. **B**	43. **A**
4. **A**	14. **A**	24. **C**	34. **A**	44. **A**
5. **B**	15. **C**	25. **D**	35. **D**	45. **B**
6. **D**	16. **D**	26. **B**	36. **D**	46. **B**
7. **A**	17. **B**	27. **D**	37. **D**	47. **A**
8. **B**	18. **B**	28. **D**	38. **D**	48. **D**
9. **D**	19. **B**	29. **A**	39. **D**	49. **C**
10. **B**	20. **B**	30. **B**	40. **C**	50. **A**

PART SIX
Real Estate Licensing Examinations

Examination Preparation 23

STUDYING FOR THE EXAM

You can't expect to walk into the examination room cold and pass with flying colors. You have to be ready for the exam. This means more than just doing a lot of reading and trying to cram your head full of facts. You have to be psychologically ready and properly prepared to take the examination so that you can *do your best* on it. To do this, you have to acquire the following things:

1. Knowledge
2. A positive attitude
3. Information about exam taking
4. Confidence
5. Rest

Let's take these things one by one.

1. **KNOWLEDGE.** Of course you have to have the knowledge to get through the examination; this means knowing the subject matter well. Begin by studying this book and your state's license law. Also, if your state provides special examination preparation materials, study those, too. You can't expect to acquire all the requisite knowledge if you start just a short time before the examination. You should follow a *reasonable* study schedule so that you will be at maximum readiness the day of the examination (see Chapter 1).
2. **A POSITIVE ATTITUDE.** Your attitude should be one of confidence; after all, if you have worked hard to learn all the real estate concepts, then you DO know the required material. The exam simply is the place where you get your chance to demonstrate that you are qualified for a real estate license.
3. **INFORMATION ABOUT EXAM TAKING.** Some people panic at the prospect of taking an exam and end up doing badly as a result. This sort of thing does happen; but if you look at the examination in the proper perspective, it won't happen to you. If you are concerned about taking an examination after having been out of school for such a long time, the thing to do is to practice: take the model examinations in Chapter 24. Set up actual "examination conditions." Give yourself a time limit, and go right through each practice exam just as if it were the real thing. Use all the examination-taking techniques you'll read about in this chapter. Each model exam has an answer key at the end of the book so that you can grade your own performance. To guide your future study, note the areas where you did well and those where you did poorly. Do this for *each* model examination. Take the first exam and grade it. Then do more studying in your weaker areas before you take the next one.

In this way, you'll accomplish three important objectives. First, you'll get used to the examination situation. Second, you'll gain practice in answering the kinds of questions featured on most real estate license examinations. Third, you'll be able to keep track of the subject areas where you need more study and practice.

4. **CONFIDENCE.** Here are the areas of subject matter that examinees worry most about: (a) arithmetic, (b) contracts, and (c) for broker examinees, closing statements. We have included detailed sections on all of these as well as on all other real estate subjects. The arithmetic you need to know is exactly the same sort you had to do for homework in the sixth and seventh grades. Contracts inspire awe because they're supposed to be the province of lawyers. By now you know that a contract is just a piece of paper on which people put down facts concerning an agreement they've made. In regard to a contract, all you have to be sure of is to write what you are supposed to and to be clear in what you say.

   Closing statements concern broker examinees because they look so mysterious. Actually, they're nothing but a record of where people's money goes in a real estate transaction. Once again, the important thing is accuracy. Also, if you think a bit about most of the items involved, it is easy to see if they should be paid by (or to) the seller or the buyer. All that's left is to line up all the numbers neatly so that they can be added easily. The tricky part about closing statements usually is the prorating of various items. As Chapter 18 explains, there is nothing mysterious about prorating. However, the calculations are long and clumsy, so there's room for error to sneak in if you're not careful.

   Throughout this book, an important aim has been to explain everything thoroughly. If you understand the subjects discussed, you should have no confidence problem: you *know* the basic principles of real estate.

5. **REST.** Don't stay up studying the night before the exam! No one is at his best when he's tired or short of sleep. The day before the exam should be one of rest and relaxation.

   You can, however, design a little last-minute study session that actually will help you, without disturbing your relaxation. Prepare a brief outline of the *important* points on which you know you need special effort. If contracts bug you, make a list of all the contracts and a brief description of each. Study this material for one-half hour in the morning, one-half hour in the afternoon, and one-half hour in the evening, but finish your studying *at least* two hours before bedtime. In that way, you'll be strengthening your weaknesses. However, you won't be spending so much time on studying that it interferes with the really important task at hand: getting yourself relaxed. The morning of the examination, if you have time, you can spend another half hour going over those notes one last time.

TWO IMPORTANT RULES

Your objective when you take any examination is to get as high a score as possible. The way to do that is to provide as many right answers as you can. This leads us directly to the most important rule of all in taking examinations.

ALWAYS ANSWER EVERY QUESTION. Even if you have to guess, write down an answer. If you leave the question blank, you can be certain that it will be counted as wrong because a blank answer is a wrong answer. If you guess, there is always the chance that the guess will be correct. When boiled down, the situation is as simple as this: you know you will get the question wrong if you leave it blank, but if you guess at an answer you may guess the right one.

The second rule is SKIP THE HARD QUESTIONS AT FIRST AND SAVE THEM FOR LAST. Many people start at the beginning of an examination and work their way through, one question at a time, until they run out of time. If they encounter a hard question, they battle with it for a long time and won't go on until they've licked it (or it has licked them). These people are lucky to get to the end of the exam before they run out of time. If they don't get to the end, they never even see some of the questions—and they might have been able to answer enough of those questions to earn a passing grade on the exam. On all of the national examinations, all questions are worth the same. Most independent states score their examinations that way too. So what sense is there in spending 15 minutes, half an hour, or more on a single question when you can use that same time to read and answer a couple of *dozen* questions correctly? (The materials you get from your state's licensing authority will tell you what kinds of questions to expect and how they are counted.)

When you take the exam, start at the beginning. Read the instructions, and then read the first question. Answer it if you can, and then go on to number 2. Answer that one if you can, and then go on to the next and the next and the next. When you come across a hard question that you have to think about or that will take a lot of time to answer even if you know exactly how to get the answer (some arithmetic problems can be like that), skip it for the time being. Come back to it later after you have gone all the way through the exam and answered all the questions that you can do fairly quickly. Be sure to make some kind of note for yourself of the questions you skip, so you'll remember to go back to them!

STRATEGIES FOR HANDLING COMMON TYPES OF QUESTIONS

In order to know how to attack your state's particular examination, you have to be familiar with the special strategies for handling specific kinds of questions.

Multiple-Choice Questions

This is a multiple-choice question:

The title of this chapter is
(A) "How to Hammer a Nail"
(B) "Examination Preparation"
(C) "I Was an Elephant for the FBI"
(D) none of the above

The national examinations are composed entirely of multiple-choice questions. The essence of this kind of question is that you must choose one of the answers as the correct one.

Always remember that with a multiple-choice question *you are looking at the right answer*! You just have to pick it out. Your strategy for answering such questions is as follows:

1. Read the question *completely* and *carefully*. Many people miss little key words such as *if, not, but,* and *except* that can completely change or limit the meaning of a sentence. Read all of the answer choices. You may find that more than one is correct. In that case, choice D is the correct answer because it will state "both A and B" or "all of the above" or similar. However, if you stop reading at choice A because you find it to be correct, you will select the wrong answer. Sometimes by reading beyond choice A, you will find a better answer. Remember that the test is looking for the "best" answer.

   What happens when you are sure that choices A and B are both correct but are not sure whether or not choice C is correct? You may be offered "both A and B" as choice D. In that case, you should pick choice D. However, if choice D is "all of the above," then choice C must be correct. In that case, choice D is still the best choice.
2. If the answer is obvious to you, mark it properly and go on to the next question. Then forget about the question you have just answered. Almost always the answer you pick first is the right one, and you will very rarely improve an answer by coming back and stewing over it later.
3. If the answer is not obvious, begin the elimination strategy. First look for all the choices that you know are wrong. Eliminate those. You may not be sure of the right answer, but that doesn't mean you can't find some answers you know are wrong. Then guess an answer from the choices that are left.

   Remember to make your guesses educated. Usually you will be able to guide your guesswork by using some of your knowledge. You may be able to decide that a certain answer is much more likely to be right than the others. Always be sure to eliminate answers you *know* are wrong in order to limit your choices and thus improve your chances of guessing the right one. However, if it looks as though a long time will be needed to reach even an elimination answer, skip the question and come back to it later if you have the time.

USING COMPUTER ANSWER FORMS

Figure 23-1 is an example of a computer examination form. Forms like these used to be the standard for all kinds of examinations, but their use is declining in favor of touch-screen computers. Even so, you can make a few copies of the form to mark your answers on for the model exams in this book—or you can simply circle your answers directly on the book pages.

If you have to use one of these answer forms, you will be given the form and a *test booklet.* The examination questions and instructions will be written in the test booklet; you will "code" your answers onto the answer sheet. Usually you will be asked to make no marks at all on the test booklet; you should put your name and any other identification asked for in the appropriate spaces on the answer sheet. Remember one critical rule: *Do not make any marks at all on either the test or the answer sheet unless and until you are instructed to.* Normally the test booklet will contain very detailed and specific instructions on how to enter the necessary

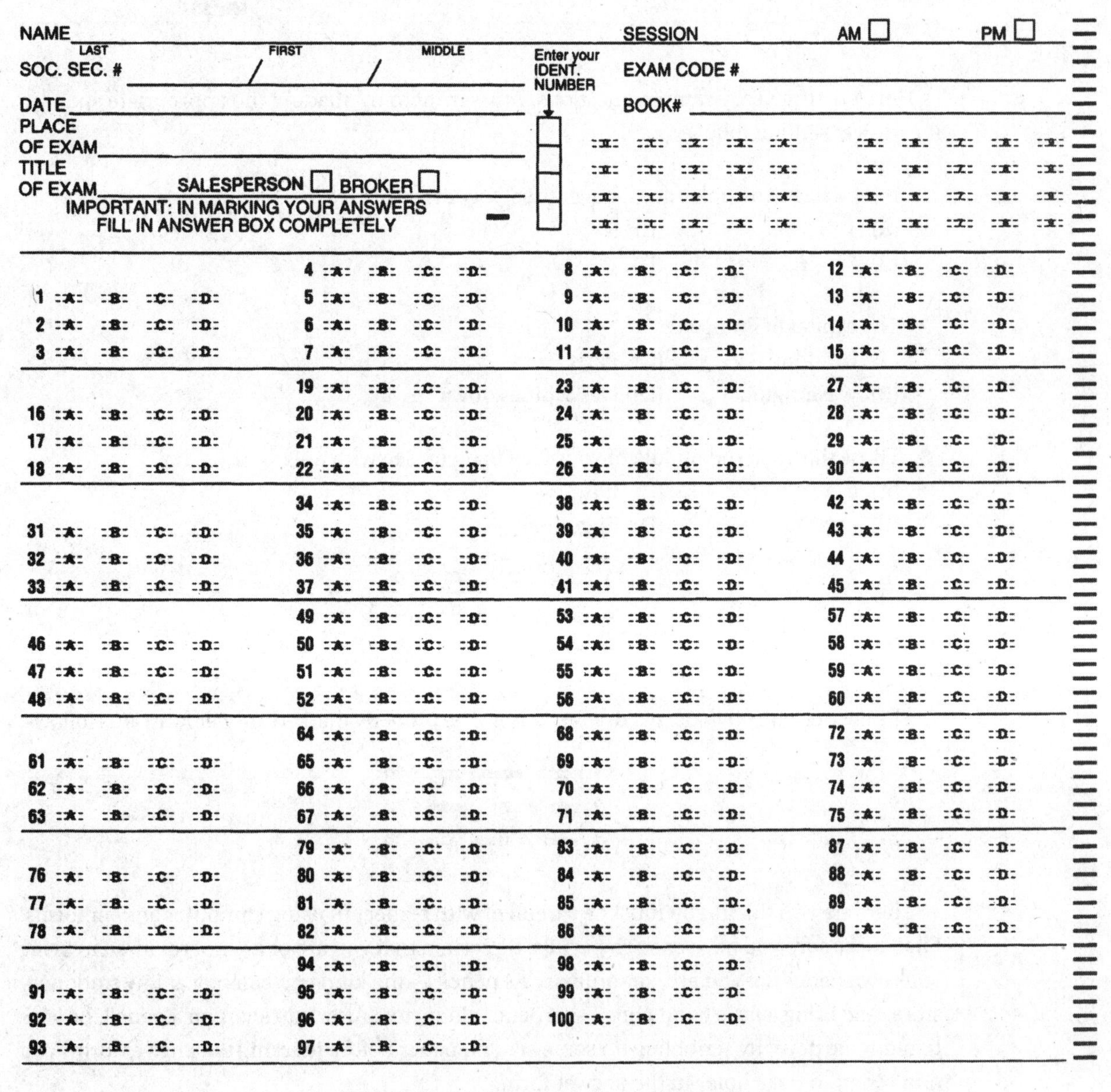

NAME ______ LAST ______ FIRST ______ MIDDLE ______ SESSION ______ AM ☐ PM ☐

SOC. SEC. # ______ / ______ / ______ Enter your IDENT. NUMBER EXAM CODE # ______

DATE ______ BOOK# ______

PLACE OF EXAM ______

TITLE OF EXAM ______ SALESPERSON ☐ BROKER ☐

IMPORTANT: IN MARKING YOUR ANSWERS FILL IN ANSWER BOX COMPLETELY

	4 A B C D	8 A B C D	12 A B C D
1 A B C D	5 A B C D	9 A B C D	13 A B C D
2 A B C D	6 A B C D	10 A B C D	14 A B C D
3 A B C D	7 A B C D	11 A B C D	15 A B C D
	19 A B C D	23 A B C D	27 A B C D
16 A B C D	20 A B C D	24 A B C D	28 A B C D
17 A B C D	21 A B C D	25 A B C D	29 A B C D
18 A B C D	22 A B C D	26 A B C D	30 A B C D
	34 A B C D	38 A B C D	42 A B C D
31 A B C D	35 A B C D	39 A B C D	43 A B C D
32 A B C D	36 A B C D	40 A B C D	44 A B C D
33 A B C D	37 A B C D	41 A B C D	45 A B C D
	49 A B C D	53 A B C D	57 A B C D
46 A B C D	50 A B C D	54 A B C D	58 A B C D
47 A B C D	51 A B C D	55 A B C D	59 A B C D
48 A B C D	52 A B C D	56 A B C D	60 A B C D
	64 A B C D	68 A B C D	72 A B C D
61 A B C D	65 A B C D	69 A B C D	73 A B C D
62 A B C D	66 A B C D	70 A B C D	74 A B C D
63 A B C D	67 A B C D	71 A B C D	75 A B C D
	79 A B C D	83 A B C D	87 A B C D
76 A B C D	80 A B C D	84 A B C D	88 A B C D
77 A B C D	81 A B C D	85 A B C D	89 A B C D
78 A B C D	82 A B C D	86 A B C D	90 A B C D
	94 A B C D	98 A B C D	
91 A B C D	95 A B C D	99 A B C D	
92 A B C D	96 A B C D	100 A B C D	
93 A B C D	97 A B C D		

Figure 23-1. Computer Examination Form

information on the answer sheet; in addition, the people administering the examination will give an oral and visual presentation on the same subject.

Using these answer sheets is really quite simple. Below is an example of a set of answer spaces for three different questions:

1 ::A:: ::B:: ::C:: ::D::
2 ::A:: ::B:: ::C:: ::D::
3 ::A:: ::B:: ::C:: ::D::

For each question, you have four answers to choose from: A, B, C, and D. (For the appraiser exams, there are five answer choices.) When you have picked your answer, enter it on the examination form by *shading* the appropriate space with your pencil. Suppose, for example, you choose C. Notice the picture below on the right; the space for choice C is shaded completely. You need to shade all your answers as completely as this.

Here are three easy sample questions. Answer them by shading the appropriate space in the answer set that follows.

1. If John has two apples and I have three, together we have
   (A) 4 (C) 6
   (B) 5 (D) 7

2. The capital of France is
   (A) London (C) Paris
   (B) Washington (D) Des Moines, Iowa

3. Those things in the middle of your face that you see with are
   (A) eyes (C) nose
   (B) ears (D) fingernails

1 ::A:: ::B:: ::C:: ::D::
2 ::A:: ::B:: ::C:: ::D::
3 ::A:: ::B:: ::C:: ::D::

The answers are obvious: 1 is B, 2 is C, 3 is A. The properly marked answer form is as follows:

1 ::A:: ▬ ::C:: ::D::
2 ::A:: ::B:: ▬ ::D::
3 ▬ ::B:: ::C:: ::D::

There are two things you must be careful of with respect to using computer answer forms. First, make sure you have enough pencils. Be certain that you are or are not required to bring your own pencils; if you are, get ordinary #2 pencils—the kind you can buy at any stationery store. And bring a little hand pencil sharpener along, too. After you sharpen a pencil, be sure to blunt the point by scribbling on some scratch paper before marking the answer form. You don't want to poke holes in the answer form.

Second, our strategy of skipping the hard questions and saving them for last means that you have to *make absolutely certain that you are putting the answer in the right space for that question!* Be especially careful here. The computer is very stupid! If you put answer marks in the wrong spaces, it will count them wrong.

By the way, the computer form used on the exam you take may be slightly different from the sample form shown here. However, it will be used in just the same way we use ours.

ONSITE SCORING SYSTEMS (KEYPADS AND TOUCH SCREENS)

Pencil-and-paper examinations are being replaced by various kinds of computerized exams. One form uses a "keypad" attached to a small terminal (screen) that displays the questions and the answer choices. You enter your choice of answer on the keypad, which also has keys allowing you to "erase" an answer, "scroll" back and forth in the exam, skip a question, and so on. Other forms use what appears to be a typical desktop personal computer, complete with a normal keyboard. The most common kind of computerized testing system uses a touch screen. It looks like a normal computer monitor, but to register your responses you tap the screen at appropriate places. If you've ever used the do-it-yourself checkout line at a supermarket or a do-it-yourself photo machine at a drugstore, you've used a touch screen.

Whatever the system used, you usually can get some literature ahead of time that explains how to use the system. Naturally, it is important to know as much about it as you can in advance. Also, before you take the exam, the administrators will provide a detailed explanation and demonstration of how to use the equipment. Whatever the system, usually the instructions for using it (entering answers, erasing, scrolling, etc.) either appear on the screen (usually at the bottom or top) or can be accessed easily.

There are two major advantages to these devices. First, the machines make it easier to skip around within the exam without losing track of where you are and without any risk that you will put your answers in the wrong spaces. Second, most of the machines allow for instant scoring of your exam. When you are done, you "lock" the machine by pressing a special key. After that time, you no longer can enter or change answers. At that point, the administrators of the exam can instantly provide your score since your answers are entered into the computer already. Usually you will be given a sheet that will tell you which questions you answered correctly and which you missed. In some examining centers if the exam isn't over yet, you may be allowed to check your answers against the exam you just took and so tell exactly where you went wrong on the questions you missed.

24 Model Examinations

The following are nine exams for you to take. Each has 100 questions. The first exam is on federal laws and regulations, including Fair Housing laws, federal taxation, TRID (TILA-RESPA Integrated Disclosure), and environmental matters. This is the only model exam that focuses on one broad issue.

Exams 2 through 8 each have questions on a wide variety of topics. Applicants for broker and sales exams should take all of these exams. Only appraiser applicants should take Exam 9.

After taking each exam, score yourself. For wrong answers, determine why you were wrong. You'll learn more from wrong answers than from right ones. Proceed through all the exams until you can comfortably pass each one. Good luck!

ANSWER SHEET
Model Examination 1

Federal Law and Regulation

1. A B C D
2. A B C D
3. A B C D
4. A B C D
5. A B C D
6. A B C D
7. A B C D
8. A B C D
9. A B C D
10. A B C D
11. A B C D
12. A B C D
13. A B C D
14. A B C D
15. A B C D
16. A B C D
17. A B C D
18. A B C D
19. A B C D
20. A B C D
21. A B C D
22. A B C D
23. A B C D
24. A B C D
25. A B C D
26. A B C D
27. A B C D
28. A B C D
29. A B C D
30. A B C D
31. A B C D
32. A B C D
33. A B C D
34. A B C D
35. A B C D
36. A B C D
37. A B C D
38. A B C D
39. A B C D
40. A B C D
41. A B C D
42. A B C D
43. A B C D
44. A B C D
45. A B C D
46. A B C D
47. A B C D
48. A B C D
49. A B C D
50. A B C D
51. A B C D
52. A B C D
53. A B C D
54. A B C D
55. A B C D
56. A B C D
57. A B C D
58. A B C D
59. A B C D
60. A B C D
61. A B C D
62. A B C D
63. A B C D
64. A B C D
65. A B C D
66. A B C D
67. A B C D
68. A B C D
69. A B C D
70. A B C D
71. A B C D
72. A B C D
73. A B C D
74. A B C D
75. A B C D
76. A B C D
77. A B C D
78. A B C D
79. A B C D
80. A B C D
81. A B C D
82. A B C D
83. A B C D
84. A B C D
85. A B C D
86. A B C D
87. A B C D
88. A B C D
89. A B C D
90. A B C D
91. A B C D
92. A B C D
93. A B C D
94. A B C D
95. A B C D
96. A B C D
97. A B C D
98. A B C D
99. A B C D
100. A B C D

Model Examination 1

FEDERAL LAW AND REGULATION

See page 497 for the answer key.

1. A tax deduction for business property that requires no cash outlay is
   (A) interest expense
   (B) vacancy loss expense
   (C) depreciation expense
   (D) management expense

2. The two most important income tax deductions for homeowners are
   (A) repairs and maintenance
   (B) mortgage interest and repairs
   (C) property taxes up to $10,000 and mortgage interest
   (D) mortgage payments and sales taxes

3. The initial cost of an asset, increased by capital expenditures and reduced by depreciation expenses, from which gain or loss is measured is
   (A) basic adjustment
   (B) basal metabolism
   (C) adjustment of base
   (D) adjusted tax basis

4. The federal income tax exemption on the gain from the sale of a personal residence held long enough by a married couple is up to
   (A) $25,000
   (B) $125,000
   (C) $250,000
   (D) $500,000

5. The federal income tax exemption on the gain from the sale of a personal residence held long enough by a single individual is up to
   (A) $25,000
   (B) $125,000
   (C) $250,000
   (D) $500,000

6. Cash or other unlike property included to equalize values in a Section 1031 exchange is called
   (A) slipper
   (B) equity
   (C) parity
   (D) boot

7. A Real Estate Investment Trust (REIT) must
   (A) be closely held (five or fewer owners)
   (B) distribute nearly all of its taxable income
   (C) own stocks or bonds
   (D) be the same as a limited partnership

8. Federal income tax rates are generally
   (A) regressive
   (B) proportional
   (C) progressive
   (D) none of the above

9. Federal income tax deductions for a principal residence include
   (A) mortgage interest on up to $750,000 of debt
   (B) real estate taxes up to $10,000
   (C) uninsured casualty losses, over a certain amount
   (D) all of the above

10. A homeowner, on his or her federal income tax return, may not deduct
    (A) fixing-up expenses
    (B) the cost of a new porch
    (C) repainting
    (D) any of the above

11. The report of a sale to the IRS is normally made on Form 1099 by the
    (A) buyer
    (B) seller
    (C) broker
    (D) closing agent

12. Harold bought 20 acres for $125,000 and sold the land three years later for $230,000. On his income tax return, he reports
    (A) a short-term capital gain of $105,000 in the year of sale
    (B) a long-term capital gain of $105,000 in the year of sale
    (C) ordinary income of $105,000 in the year of sale
    (D) annual income of $35,000

13. Indications of independent contractor status include
   (A) freedom of hours to work
   (B) use of own vehicle
   (C) payment of own professional dues
   (D) all of the above

14. A REIT is like
   (A) tenancy in common
   (B) a joint tenancy
   (C) community property
   (D) a mutual fund

15. Depreciation expense claimed for income tax purposes
   (A) reduces cash flow
   (B) reduces taxable income
   (C) increases potential gross income
   (D) increases cash operating expenses

16. Instead of selling business or investment property and paying a tax on the gain, one may prefer
   (A) a capital gain
   (B) Section 1231
   (C) a tax-free exchange
   (D) ordinary income

17. Taxable gain on the sale of commercial real estate is calculated by
   (A) sales price minus the basis
   (B) adjusted sales price minus the basis
   (C) realized selling price minus the adjusted tax basis
   (D) contract sales price minus the original tax basis

18. When a salesperson's agreement with a broker is to be an independent contractor, the broker
   (A) withholds federal taxes and social security tax
   (B) withholds only social security
   (C) withholds taxes only when they exceed $600 in a year
   (D) does not withhold federal taxes or social security

19. In the purchase of a home, which is frequently tax deductible?
   (A) attorney fees
   (B) recording fees
   (C) title insurance
   (D) points paid for a mortgage that are customary

20. Owners of improved rental property can generally deduct on their federal income tax returns in the year incurred
    (A) a new roof
    (B) parking lot repaving
    (C) mortgage principal payments
    (D) toilet repairs on multiple units

21. To qualify as a Section 1031 exchange, the property must be
    (A) a personal residence
    (B) used in a trade or business or be held as an investment
    (C) land
    (D) mortgaged

22. Mr. and Mrs. Martin jointly bought a principal residence in 2020 for $325,000 and sold it in 2025 for $500,000. What portion of their $175,000 gain is subject to capital gains tax?
    (A) $0
    (B) $325,000
    (C) $500,000
    (D) $175,000

23. Under Internal Revenue Code Section 1031, tax on the profit of an exchange is
    (A) reduced
    (B) deferred
    (C) eliminated
    (D) mitigated

24. In a Subchapter S Corporation, the corporation itself
    (A) pays no federal income tax
    (B) pays no property tax on land it owns
    (C) gets a discount on state income tax
    (D) is taxed just like a C Corporation

25. Double taxation is generally not avoided when using which form of ownership?
    (A) S Corporation
    (B) general partnership
    (C) tenancy in common
    (D) corporation

26. The 1968 Fair Housing Act, as amended, prohibits housing discrimination on the basis of
    (A) race and color
    (B) race, color, and religion
    (C) race, color, religion, and national origin
    (D) race, color, religion, national origin, sex, familial status, or handicap

27. Court action may be taken by an individual under the Fair Housing Act
(A) only if a complaint is filed with HUD
(B) if action is taken within one year of the alleged discriminatory act
(C) if the alleged discriminatory act occurred on public property
(D) if the alleged discriminatory act caused damage of at least $500

28. Which of the following is exempt from the federal Fair Housing Act of 1968?
(A) salesperson selling lots in a subdivision
(B) salesperson renting apartments in a building with 10 units
(C) owner renting vacant single-family unit
(D) none of the above

29. If a landlord refuses to rent to a minority prospective tenant with a poor credit record,
(A) federal Fair Housing laws are violated
(B) the minority applicant should report this violation
(C) there is no violation
(D) the landlord may face a significant fine

30. Sullivan wants to rent an apartment but is confined to a wheelchair. At Sullivan's request, the landlord must
(A) build a ramp at the landlord's expense
(B) build a ramp at Sullivan's expense
(C) do nothing
(D) reduce the rent to Sullivan's affordability level

31. A seller asks his broker about the race of a potential buyer. The broker
(A) must answer the question if a single agent
(B) violates Fair Housing laws if he answers the question
(C) may answer the question because of the constitutional right to free speech
(D) may tell the seller but no one else

32. Which amendment to the Constitution provides support by banning racial discrimination?
(A) Second
(B) Third
(C) Twelfth
(D) Thirteenth

33. Broker Bob is showing houses to Joe Chen, who is of Chinese ancestry. Bob shows Joe houses in a neighborhood of predominantly Chinese-American residents because Bob thinks Joe would feel more at home there. This is
(A) illegal steering under Fair Housing laws
(B) superior and tactful service
(C) done for Joe's benefit
(D) effective sales technique

34. Where the dominant population base of an area consists of minorities,
    (A) brokers need not be concerned with Fair Housing laws
    (B) brokers must still observe Fair Housing laws
    (C) there are no reporting requirements
    (D) violation of Fair Housing laws is not possible

35. The Civil Rights Act of 1968 prohibits
    (A) integration
    (B) greenlining
    (C) discrimination in housing
    (D) school busing

36. "Familial status" refers to
    (A) families with two or more children
    (B) families with children under age 18
    (C) the earning capacity of the family
    (D) TRID

37. Broker Barb is acting as a buyer's agent for Joe. Joe asks whether any people of a certain race live in a specific neighborhood. How should Barb respond?
    (A) She should advise Joe that, under fair housing laws, she cannot provide that information.
    (B) She should advise Joe of the race of all the neighbors she is aware of.
    (C) She should advise Joe that he should contact the demographic agency to obtain the information.
    (D) She should advise Joe to contact the school district to obtain the information.

38. A broker has contacted owners in a certain subdivision and advised them that several members of a particular race have bought homes in the area. The broker has offered to list properties at a reduced rate and encouraged the owners to sell quickly. This is
    (A) steering
    (B) profiling
    (C) blockbusting
    (D) scaring

39. Ellen owns a two-family house and lives in one side. She wants to advertise the other side as nonsmoking. Can she legally do so?
    (A) Yes, because the right to smoke is not protected by law.
    (B) Yes, because she owns the house.
    (C) No, because a property owner cannot advertise discriminatory practices.
    (D) No, because smoking is perfectly legal.

40. No federal Fair Housing laws are violated if a landlord refuses to rent to
    (A) families with children
    (B) tenants of Indian descent
    (C) deaf persons
    (D) college students

41. The Civil Rights Act of 1866 prohibits
   (A) zoning
   (B) redlining
   (C) open housing
   (D) racial discrimination

42. A minority group is moving into an area immediately adjacent to an old subdivision. ABC Realty offers to list homes in the subdivision at a lower-than-usual commission rate if the owners list within 45 days. There is no mention of race, and the broker acts in good faith. Which of the following is true?
   (A) Such practice is not illegal.
   (B) Brokers cannot lower their standard rate of commission.
   (C) The broker's license can be revoked for this action.
   (D) This is blockbusting.

43. A couple of Greek ancestry looking for a house ask a real estate agent to show them houses in Greek neighborhoods. Which of the following is correct?
   (A) The agent must accommodate this request because the buyers requested it.
   (B) The agent should tell the buyers that he will show them houses in a number of different neighborhoods.
   (C) The agent should refuse to show any houses to the buyers.
   (D) The agent should agree and then show the buyers houses in other neighborhoods.

44. "Whites only" is the advertisement for a house that's for sale. This is
   (A) legal if the house is owner occupied
   (B) illegal unless a real estate agent is used
   (C) legal because it's a one-family house
   (D) illegal

45. The Aggie Oaks Apartments advertising states that families with children are welcome. The management has allocated a certain section of the complex for residents with children, and those tenants must pay an additional security deposit for each child under twelve. Is this discriminatory?
   (A) No, because families with children are welcome in the complex.
   (B) Yes, because it is based on marital status.
   (C) Yes, because it is based on familial status.
   (D) No, because children are not a named protected class.

46. Agent Curt has been asked by his seller not to advertise in a Spanish-language newspaper. Curt and his office would not violate the law if
   (A) the paper does not serve the market area of his office
   (B) he doesn't like the paper's editorial positions on many issues
   (C) his seller directed him in writing
   (D) his office has never used the paper because of the high rates charged by the paper

MODEL EXAMINATION 1

47. When discussing the possibility of listing their house, the owners tell the salesperson, "We want to sell only to buyers of a certain national origin." The salesperson should
   (A) accept that condition and write it into the listing to comply with the principal's request
   (B) report the owners to the Department of Housing and Urban Development
   (C) accept the condition because it is only a listing but be prepared to show the property to others
   (D) explain that this would be a violation of the federal Fair Housing Act; if the owners persist, refuse the listing

48. A blind man with a service dog seeks to rent an apartment in a building that does not allow pets. Under the federal Fair Housing Act, the apartment building owner
   (A) can reject the applicant as a tenant
   (B) can increase the security deposit compared with what is charged other tenants
   (C) must lease, though the lease can require the tenant to repair any damage caused by the animal
   (D) cannot treat the tenant any differently than any other prospective tenant

49. Which of the following phrases does HUD consider discriminatory advertising?
   (A) "Senior housing"
   (B) "Friendly neighbors"
   (C) "Exclusive neighborhood"
   (D) "No one excluded"

50. An ad reads: "Home for sale only to a family." This ad violates
   (A) Truth in Lending and Fair Housing Act
   (B) TRID and Fair Housing Act
   (C) Fair Housing Act and Americans with Disabilities Act
   (D) HUD and EPA

51. A homeowner wishes to lease his house for a year while he takes a sabbatical. He instructs the licensed agent to lease only to a white family. The agent must
   (A) be loyal and comply
   (B) ask for the request in writing
   (C) refuse to follow the order because it violates the Fair Housing Act
   (D) ask the broker what the firm's policy is

52. An apartment complex is intended for the elderly only. What percentage of units must have at least one person over age 55?
   (A) 100%
   (B) 80%
   (C) 50%
   (D) 0%

53. Under the federal Fair Housing Act, which of the following is protected as a "handicap"?
(A) inadequate income to pay rent
(B) use of a controlled substance for more than a year
(C) inability to get an FHA mortgage
(D) a mental handicap that limits a major life activity

54. A builder wishes to erect a 150-unit two-story building without an elevator. Under the federal Fair Housing Act, which units must be accessible to the handicapped?
(A) all units
(B) ground-floor units only
(C) second-floor units only
(D) 20% units set-aside

55. Allen, who is not a real estate licensee, wants to sell his brother's house to a member of their same religious affiliation. Allen can
(A) do anything provided he has his brother's oral permission
(B) charge a commission for the sale of the home
(C) advertise that the house is for sale only to members of that religion
(D) discuss the matter with his brother

56. Houses located near a college offer rooms or apartments for rent. Which of the following is violating the federal Fair Housing Act?
(A) owners who rent out two rooms in their house but not to a Greek couple
(B) owners who live in their house, rent out three rooms, and require a $500 greater deposit from college students
(C) owners who live elsewhere and refuse to rent their house to single individuals
(D) owners who rent a room in the house they live in, as well as two apartments in the house they own next door, but won't rent to a married couple with young children

57. A broker advertises a home for rent as follows: "4BR, 2BA, nr. schools & shopping. Elderly married couples only." Which law has been broken?
(A) Uniform Residential Appraisal Report
(B) Americans with Disabilities
(C) Truth in Lending
(D) Federal Fair Housing Act

58. A lawyer wishes to buy a 40-year-old office building. To comply with the Americans with Disabilities Act, the lawyer must
(A) widen hallways
(B) install elevators
(C) install a wheelchair ramp for at least one entrance
(D) remove carpet from hallways

59. Under the federal Fair Housing Act, which of the following prospects can a landlord legally refuse to rent to?
    (A) an unmarried couple
    (B) a married couple with children
    (C) unemployed retirees
    (D) a person unable to afford the rent

60. A church owns a retirement facility and requires that tenants be members of the church. This is
    (A) a violation of Fair Housing laws
    (B) acceptable under all tenant requirements
    (C) acceptable provided there is no discrimination based on race, color, sex, national origin, familial status, or handicap
    (D) not permitted as an exception of Fair Housing laws

61. Which of the following is not protected under the federal Fair Housing Act?
    (A) race
    (B) religion
    (C) national origin
    (D) sexual preference

62. Which of the following is *not* discriminatory under Fair Housing laws?
    (A) steering
    (B) blockbusting
    (C) refusing to rent to a single woman
    (D) owning two units, living in one, and refusing to rent the other to families with children

63. Which of the following neighborhood characteristics may an appraiser not describe in an appraisal?
    (A) economic characteristics
    (B) homogeneity of property
    (C) racial composition
    (D) proximity to transportation

64. Which government agency investigates alleged violations of Fair Housing laws?
    (A) FHA
    (B) EEOC
    (C) HUD
    (D) FNMA

65. The Comprehensive Environmental Response, Compensation, and Liability Act (CERCLA) is enforced by
(A) Department of Homeland Security
(B) Department of Housing and Urban Development (HUD)
(C) Federal Emergency Management Agency (FEMA)
(D) Environmental Protection Agency (EPA)

66. When land is partially in a designated wetland area,
(A) the value is enhanced by the abundance of wildlife
(B) the land is worthless
(C) Superfund will buy it
(D) this is a material fact to be disclosed to a potential buyer

67. A buyer of a house built before 1978 must be advised of the possibility that it may contain
(A) asbestos
(B) radon
(C) lead-based paint
(D) termites

68. EPA stands for
(A) earnings per annum
(B) effort put away
(C) extra-pretty agent
(D) Environmental Protection Agency

69. The federal agency that enforces Superfund is
(A) FEMA
(B) HUD
(C) EPA
(D) DOD

70. During a property inspection, the inspector learns that the building was used as a facility to repair automobile brakes. This should suggest the possible presence of
(A) asbestos
(B) radon
(C) hydrocarbons
(D) lead-based paint

71. Sources of groundwater contamination include all *except*
(A) leaking petroleum tanks
(B) disposal of containers from pesticides used on a farm
(C) radiation from a nuclear power plant
(D) satellite transmissions

72. Superfund imposes
    (A) partial liability for cleanup
    (B) waste
    (C) forfeiture of land
    (D) strict liability for damage

73. A gas produced by radioactive materials naturally found in the earth is
    (A) carbon dioxide
    (B) carbon monoxide
    (C) nitrogen
    (D) radon

74. Which of the following is *not* an environmental hazard?
    (A) leaking underground storage tanks
    (B) radon gas
    (C) xenon lights
    (D) lead-based paint

75. Which of the following is true of high concentrations of radon gas?
    (A) It has a very noticeable odor.
    (B) It has been linked to lung cancer in long-term exposure.
    (C) It is released by plastic building materials.
    (D) It enters the house through openings in the attic.

76. Which of the following must be included as a finance charge for real property credit transactions?
    (A) fee for title insurance
    (B) fee for deed preparation
    (C) monthly payment, in dollars
    (D) none of the above

77. The annual percentage rate (APR) means the
    (A) true interest rate charged
    (B) total dollar amount of finance charges
    (C) monthly payment, in dollars
    (D) percentage of loan paid off each year

78. TRID applies to a
    (A) first mortgage loan used to purchase a dwelling
    (B) second mortgage used to improve a dwelling
    (C) home-improvement loan more than one year old
    (D) none of the above

79. Under TRID, two of the most critical facts that must be disclosed to buyers or borrowers are
   (A) duration of the contract and discount rate
   (B) finance charges and closing costs
   (C) carrying charges and advertising expense
   (D) installment payments and cancellation rights

80. A purpose of TRID is to provide the
   (A) broker's address
   (B) salesperson's name
   (C) total price
   (D) opportunity for borrowers to shop for a loan

81. Under TRID, the lender must provide the
   (A) Loan Estimate
   (B) simple rate of interest
   (C) second mortgage source
   (D) Cost of Funds Index

82. TRID combines the requirements of the Truth in Lending Act and
   (A) USPAP
   (B) NAR
   (C) RESPA
   (D) FNMA

83. TRID requires a form be prepared before closing called
   (A) Recording Costs
   (B) Attorney Fees
   (C) Title Costs
   (D) Closing Disclosure

84. TRID requires the proposed lender to give the borrower a
   (A) Loan Estimate
   (B) Closing Disclosure
   (C) both A and B
   (D) neither A nor B

85. TRID applies to
   (A) all real estate financing
   (B) contracts for deed
   (C) second mortgages only
   (D) financing of one- to four-family residences

86. Mortgage lenders may not discriminate on the basis of age, but they need not lend to
    (A) parents of more than two children
    (B) childless couples
    (C) part-time workers
    (D) minors

87. Under TRID, lenders must
    (A) provide a Loan Estimate
    (B) limit the number of discount points to not more than 2
    (C) let the buyer void the transaction up to ten days after closing
    (D) not charge more than 1% for lender fees

88. TRID applies to
    (A) the first mortgage loan on one's proposed home
    (B) seller financing
    (C) an investor purchase
    (D) all real estate loans

89. The purpose of TRID is to give information about settlement costs to
    (A) loan applicants
    (B) low-income buyers only
    (C) sellers only
    (D) brokers only

90. If the following transactions involve a new first mortgage, to which would TRID apply?
    (A) summer house to be occupied by purchaser
    (B) house purchased by an investor to remodel and resell
    (C) 50-unit apartment building
    (D) 50-acre farm with home for use by tenant farmer

91. TRID requires what information?
    (A) buyer's closing costs only
    (B) seller's closing costs only
    (C) Loan Estimate and Closing Disclosure
    (D) mortgage delinquency data

92. Which of the following is *not* a consideration of TRID?
    (A) Loan Estimate
    (B) Closing Disclosure
    (C) payment of loan fees
    (D) commission rates of real estate brokers

93. A lender refuses to make mortgage loans on properties in a certain area of a city regardless of the qualifications of the borrower. This is
(A) discrimination
(B) price fixing
(C) blockbusting
(D) redlining

94. An FDIC-insured lender is considering a loan application to remodel an existing house. The applicant is a member of a minority. Under the federal Fair Housing Act, the lender must make the loan unless
(A) the home is in a minority neighborhood
(B) the home is more than 50 years old
(C) the applicant is not economically qualified
(D) the property has lead-based paint

95. Which of the following terms can be included in an ad with no further disclosure required?
(A) "Take eight years to pay"
(B) "Only $10,000 down"
(C) "Payments of $500 per month"
(D) "6% annual percentage rate"

96. A lender is required by TRID to provide to an applicant for a real estate mortgage
(A) good title
(B) his FICO score
(C) a Loan Estimate
(D) an amortization schedule

97. The law that requires a lender to disclose to a potential borrower the closing costs is called
(A) Equal Credit Opportunity Act
(B) Interstate Land Sales Full Disclosure
(C) Real Estate Settlement Procedures Act
(D) TRID

98. A broker is advertising property for a homebuilder. Which phrase is *not* allowed?
(A) "New homes for $350,000"
(B) "Veterans may apply for loans with nothing down"
(C) "No down payment required"
(D) "Near schools and shopping"

99. TRID requires which of the following be included in a disclosure statement?
    (A) the broker's license number
    (B) the house style
    (C) the inspector's report
    (D) any mortgage prepayment penalties

100. Violations of TRID can bring
    (A) fines
    (B) imprisonment
    (C) both fines and imprisonment
    (D) license suspension

ANSWER KEY
Model Examination 1

1. **C**	26. **D**	51. **C**	76. **D**
2. **C**	27. **B**	52. **B**	77. **A**
3. **D**	28. **C**	53. **D**	78. **A**
4. **D**	29. **C**	54. **B**	79. **B**
5. **C**	30. **B**	55. **D**	80. **D**
6. **D**	31. **B**	56. **D**	81. **A**
7. **B**	32. **D**	57. **D**	82. **C**
8. **C**	33. **A**	58. **C**	83. **D**
9. **D**	34. **B**	59. **D**	84. **C**
10. **D**	35. **C**	60. **C**	85. **D**
11. **D**	36. **B**	61. **D**	86. **D**
12. **B**	37. **A**	62. **D**	87. **A**
13. **D**	38. **C**	63. **C**	88. **A**
14. **D**	39. **A**	64. **C**	89. **A**
15. **B**	40. **D**	65. **D**	90. **A**
16. **C**	41. **D**	66. **D**	91. **C**
17. **C**	42. **A**	67. **C**	92. **D**
18. **D**	43. **B**	68. **D**	93. **D**
19. **D**	44. **D**	69. **C**	94. **C**
20. **D**	45. **C**	70. **A**	95. **D**
21. **B**	46. **D**	71. **D**	96. **C**
22. **A**	47. **D**	72. **D**	97. **D**
23. **B**	48. **C**	73. **D**	98. **C**
24. **A**	49. **C**	74. **C**	99. **D**
25. **D**	50. **C**	75. **B**	100. **C**

ANSWER SHEET
Model Examination 2

Broker and Salesperson

1. (A) (B) (C) (D)
2. (A) (B) (C) (D)
3. (A) (B) (C) (D)
4. (A) (B) (C) (D)
5. (A) (B) (C) (D)
6. (A) (B) (C) (D)
7. (A) (B) (C) (D)
8. (A) (B) (C) (D)
9. (A) (B) (C) (D)
10. (A) (B) (C) (D)
11. (A) (B) (C) (D)
12. (A) (B) (C) (D)
13. (A) (B) (C) (D)
14. (A) (B) (C) (D)
15. (A) (B) (C) (D)
16. (A) (B) (C) (D)
17. (A) (B) (C) (D)
18. (A) (B) (C) (D)
19. (A) (B) (C) (D)
20. (A) (B) (C) (D)
21. (A) (B) (C) (D)
22. (A) (B) (C) (D)
23. (A) (B) (C) (D)
24. (A) (B) (C) (D)
25. (A) (B) (C) (D)
26. (A) (B) (C) (D)
27. (A) (B) (C) (D)
28. (A) (B) (C) (D)
29. (A) (B) (C) (D)
30. (A) (B) (C) (D)
31. (A) (B) (C) (D)
32. (A) (B) (C) (D)
33. (A) (B) (C) (D)
34. (A) (B) (C) (D)
35. (A) (B) (C) (D)
36. (A) (B) (C) (D)
37. (A) (B) (C) (D)
38. (A) (B) (C) (D)
39. (A) (B) (C) (D)
40. (A) (B) (C) (D)
41. (A) (B) (C) (D)
42. (A) (B) (C) (D)
43. (A) (B) (C) (D)
44. (A) (B) (C) (D)
45. (A) (B) (C) (D)
46. (A) (B) (C) (D)
47. (A) (B) (C) (D)
48. (A) (B) (C) (D)
49. (A) (B) (C) (D)
50. (A) (B) (C) (D)
51. (A) (B) (C) (D)
52. (A) (B) (C) (D)
53. (A) (B) (C) (D)
54. (A) (B) (C) (D)
55. (A) (B) (C) (D)
56. (A) (B) (C) (D)
57. (A) (B) (C) (D)
58. (A) (B) (C) (D)
59. (A) (B) (C) (D)
60. (A) (B) (C) (D)
61. (A) (B) (C) (D)
62. (A) (B) (C) (D)
63. (A) (B) (C) (D)
64. (A) (B) (C) (D)
65. (A) (B) (C) (D)
66. (A) (B) (C) (D)
67. (A) (B) (C) (D)
68. (A) (B) (C) (D)
69. (A) (B) (C) (D)
70. (A) (B) (C) (D)
71. (A) (B) (C) (D)
72. (A) (B) (C) (D)
73. (A) (B) (C) (D)
74. (A) (B) (C) (D)
75. (A) (B) (C) (D)
76. (A) (B) (C) (D)
77. (A) (B) (C) (D)
78. (A) (B) (C) (D)
79. (A) (B) (C) (D)
80. (A) (B) (C) (D)
81. (A) (B) (C) (D)
82. (A) (B) (C) (D)
83. (A) (B) (C) (D)
84. (A) (B) (C) (D)
85. (A) (B) (C) (D)
86. (A) (B) (C) (D)
87. (A) (B) (C) (D)
88. (A) (B) (C) (D)
89. (A) (B) (C) (D)
90. (A) (B) (C) (D)
91. (A) (B) (C) (D)
92. (A) (B) (C) (D)
93. (A) (B) (C) (D)
94. (A) (B) (C) (D)
95. (A) (B) (C) (D)
96. (A) (B) (C) (D)
97. (A) (B) (C) (D)
98. (A) (B) (C) (D)
99. (A) (B) (C) (D)
100. (A) (B) (C) (D)

Model Examination 2

BROKER AND SALESPERSON

See page 516 for the answer key. Math problems are worked out on page 517.

1. When estimating the market value of the subject house, which is 20 years old, which of the following would be of greatest importance?
   (A) replacement cost new
   (B) reproduction cost new
   (C) market value of recently sold comparable properties
   (D) market value less depreciation

2. To appraise a house valued at over $1,000,000 in a federally related transaction, the person performing the appraisal would need to
   (A) be state certified
   (B) be federally licensed
   (C) be MAI designated
   (D) have 10 years' experience in appraisal

3. The subject house being appraised has a swimming pool and a garage. The appraiser estimates that a pool contributes $20,000 to the value and a garage contributes $15,000. A recent sale in the neighborhood for $300,000 is similar except that it has a garage but no pool. The adjusted value of the comparable house is
   (A) $300,000
   (B) $315,000
   (C) $320,000
   (D) $335,000

4. An appraiser opines that a commercial property, if fully leased, would produce $20,000 per month in rent. A vacancy allowance of 9% is applied. By using a capitalization rate of 12%, the market value of the property is estimated at
   (A) $1,820,000
   (B) $1,666,666
   (C) $1,516,667
   (D) $2,400,000

5. A main goal of an appraiser is
   (A) measuring obsolescence
   (B) rendering a market value opinion
   (C) narrative reporting
   (D) meeting sales contract amount

6. A major factor in appraising a business opportunity is the
   (A) net income
   (B) number of employees
   (C) number of customers
   (D) owner's expertise

7. Prices of comparable sales are adjusted
   (A) to the subject property
   (B) upward if a feature of the comparable is inferior to the subject
   (C) downward if a feature of the comparable is superior to the subject
   (D) all of the above

8. All other things being equal, the cost per square foot of a two-story house compared to a one-story house having the same square footage
   (A) is less
   (B) is more
   (C) is the same
   (D) depends on height

9. Ellen bought a half-acre lot at $10 per square foot and built a house 50′ × 50′ for $75 per square foot. Her total cost
   (A) was $405,300
   (B) was $623,100
   (C) was $236,550
   (D) cannot be determined from information given

10. Wear and tear is a form of loss in value called
   (A) physical deterioration
   (B) functional obsolescence
   (C) deferred maintenance
   (D) economic obsolescence

11. In which appraisal approach does the appraiser estimate the land value separately from improvements?
   (A) cost
   (B) sales comparison
   (C) income
   (D) depreciation

12. Which is *not* an appraisal approach?
    (A) sales comparison approach
    (B) income capitalization approach
    (C) cost approach
    (D) depreciation approach

13. An appraiser should
    (A) raise or lower the appraised value to the contract amount
    (B) determine the price
    (C) be impartial
    (D) be a member of the broker's firm

14. The best way for a broker acting as a dual agent to minimize his or her risk is to
    (A) hire an attorney
    (B) obtain written permission from both parties to act for both sides
    (C) read the state's real estate law
    (D) provide full disclosure to all parties

15. When representing a buyer in a real estate transaction, a broker must disclose to the seller all of the following *except*
    (A) the buyer's motivation
    (B) the broker's relationship with the buyer
    (C) any agreement to compensate the broker out of the listing broker's commission
    (D) the broker's agreement with the buyer

16. Broker Herman must have an escrow account
    (A) before the state licensing agency will issue his broker's license
    (B) when he holds earnest money
    (C) before his agents can work
    (D) when his personal account is overdrawn

17. Which of the following best describes the role of a broker who seeks a buyer?
    (A) general contractor
    (B) independent contractor
    (C) special agent
    (D) attorney-in-fact

18. How is a buyer's broker typically paid?
    (A) by the buyer at closing
    (B) by a finder's fee paid outside of closing
    (C) by the buyer prior to closing
    (D) by the listing broker through a commission split

19. The rate of commission on a real estate sale is
    (A) typically 6% of the selling price
    (B) mandated at various rates by brokerage law
    (C) paid to the salesperson, who gives 50% to the broker
    (D) negotiable between the broker and (typically) seller

20. As a general rule, escrow money received by the broker
    (A) is given to the seller on signing the contract
    (B) may be spent by the broker for repairs to the property
    (C) is deposited in the broker's trust account soon after receipt
    (D) is shared equally by the broker and salesperson

21. An unlicensed employee of a broker who prepares an advertisement to sell a house
    (A) is free to do so
    (B) must have the broker's prior written approval
    (C) must become licensed
    (D) must accurately describe the house

22. An owner who sells property that he developed
    (A) may engage an unlicensed sales staff who receive commissions
    (B) may engage an unlicensed sales staff who receive salaries only
    (C) must acquire a broker's license
    (D) must acquire a salesperson's license

23. In a no brokerage agency relationship, the seller or buyer
    (A) is responsible for the broker's acts
    (B) is not responsible for the broker's acts
    (C) is not considered a customer of the broker
    (D) is an agent of the broker

24. In a no brokerage agency arrangement, parties give up their rights to a broker's
    (A) undivided loyalty
    (B) honest dealings
    (C) use of skill and care
    (D) accounting for funds

25. Some ways to terminate a brokerage relationship are
    (A) agreement between the parties
    (B) fulfillment of the contract
    (C) expiration of the term
    (D) all of the above

26. A broker with no agency relationship may assist
    (A) buyer
    (B) seller
    (C) both buyer and seller
    (D) all of the above

27. *Prima facie evidence* refers to evidence that is good and sufficient on its face to
    (A) sell real estate
    (B) buy real estate
    (C) establish a given fact
    (D) file a lawsuit

28. Brokerage representation may legally include
    (A) a broker working for the seller
    (B) a broker working as a single agent for buyer or seller (but not both in the same transaction)
    (C) broker representing neither buyer nor seller but facilitating the transaction
    (D) all of the above

29. A salesperson is ____________ of the broker.
    (A) an agent
    (B) an employee
    (C) a paid contractor
    (D) a REALTOR®

30. Termination of a brokerage relationship may occur as a result of
    (A) bankruptcy of the principal or customer
    (B) death of either party
    (C) destruction of the property
    (D) all of the above

31. When a broker is an agent of the owner, the salesperson is
    (A) a REALTOR®
    (B) an independent contractor
    (C) a subagent
    (D) an employee

32. A single agent must
    (A) deal honestly and fairly
    (B) be loyal
    (C) provide full disclosure
    (D) all of the above

33. A person who wishes to sell his or her own house
    (A) must employ a broker
    (B) must employ a salesperson
    (C) must employ a REALTOR®
    (D) may employ any of the above but is not required to do so

34. Which of the following, when performed for others, are included as real estate services for the purposes of determining licensure requirements?
    (A) buying and selling property
    (B) renting and leasing property
    (C) auction or exchange of property
    (D) all of the above

35. *Dual agency* is defined as an arrangement in which
   (A) the same broker represents both the buyer and the seller
   (B) the broker is the seller
   (C) the salesperson and the broker split the commission
   (D) the listing broker and selling broker cooperate with each other

36. When a broker serves in a single agent capacity, his obligations to the customer include
   (A) loyalty
   (B) confidentiality
   (C) obedience
   (D) all of the above

37. If a broker uses a title company to hold escrow deposits, the broker must turn over the funds
   (A) on the same day
   (B) in the same time frame as would be required if the broker maintained the account
   (C) when convenient
   (D) at the end of each month

38. If a buyer and seller reach oral agreement over the price and terms of a contract but seek a broker's help to complete written contract forms, the relationship with the broker is likely to be
   (A) single agent
   (B) transaction agent
   (C) no brokerage relationship
   (D) any of the above

39. A single agent relationship creates a ______________ with a seller.
   (A) fiduciary relationship
   (B) transaction agency
   (C) no business relationship
   (D) subagency

40. When a real estate attorney wants to earn a commission for performing real estate services, he or she must
   (A) become a REALTOR®
   (B) obtain a real estate license
   (C) join a franchise
   (D) be a member of the bar association

41. It is a good practice for a broker to prequalify
   (A) a prospective buyer
   (B) a seller
   (C) a sales agent
   (D) all clients

42. Herb is 17 when he signs a contract to lease a small house. After the property manager accepts his offer, the lease contract is
(A) void
(B) voidable by Herb
(C) voidable by either party
(D) illegal

43. Carol is buying a house and has 30 days in which to obtain a financing commitment as specified in the sales contract. The contract states that Carol must notify the seller within that period if she cannot arrange financing or risk default if she does not close. The contract states that "time is of the essence" applies to financing approval. Which is true?
(A) Carol can arrange financing at her convenience.
(B) The financing approval is automatically extended.
(C) The lender must check Carol's credit references.
(D) Carol must adhere to the stated time frame, which will end at 11:59 P.M. on the 30th day.

44. Harry gives Sam a $5,000 payment and a written agreement stating that, on July 1 of the following year, Harry will purchase the property from Sam for $150,000 cash, or Sam may keep the money. This transaction is
(A) a lease with an option to purchase
(B) a purchase contract with a delayed settlement
(C) an option agreement
(D) a limited partnership

45. Sally has shown Rebecca 11 homes, and Rebecca has selected one and made an offer. An hour later, Rebecca changes her mind. Which of the following is true?
(A) Rebecca has already committed to the first seller and cannot withdraw.
(B) Rebecca may withdraw the offer but will forfeit the earnest money.
(C) The broker earned a commission.
(D) Rebecca may withdraw before the seller accepts the offer.

46. A bank acquired property through foreclosure. Upon sale of the property, the bank covenants and warrants the property's title only against defects occurring during the bank's ownership. This conveyance is a(n)
(A) administrator's deed
(B) general warranty deed
(C) special warranty deed
(D) covenant deed

47. A deed is the instrument used to convey ownership of
(A) personal property
(B) short-term use of real estate
(C) real property
(D) any of the above

48. Which of the following is a voluntary alienation of real property?
(A) escheat
(B) condemnation
(C) foreclosure
(D) a deed of trust

49. Developers subdivide land and assign lot and block numbers to identify lots in the subdivision. This is a legal description known as
(A) a block party
(B) recorded plat description
(C) developer plat
(D) lot survey method

50. A township is
(A) 36 square miles
(B) one square mile
(C) a small city
(D) 640 acres

51. What is the type of legal description that uses imaginary lines of meridians and base lines?
(A) metes and bounds
(B) recorded plat
(C) lot and block
(D) rectangular survey

52. Broker Barb is acting as a buyer's agent for Joe. Joe asks whether any people of a certain race live in a specific neighborhood. How should Barb respond?
(A) She should advise Joe that, under Fair Housing laws, she cannot provide that information.
(B) She should advise Joe of the race of all the neighbors she is aware of.
(C) She should advise Joe that he should contact the demographic agency to obtain the information.
(D) She should advise Joe to contact the school district to obtain the information.

53. A broker has contacted owners in a certain subdivision and advised them that several members of a particular race have bought homes in the area. The broker has offered to list properties at a reduced rate and encouraged the owners to sell quickly. This is
(A) steering
(B) profiling
(C) blockbusting
(D) scaring

54. Mortgage lenders may not discriminate on the basis of age, but they need not lend to
(A) parents of more than two children
(B) childless couples
(C) part-time workers
(D) minors

55. Ellen owns a two-family house and lives in one side. She wants to advertise the other side as nonsmoking. Can she legally do so?
(A) Yes, because the right to smoke is not protected by law.
(B) Yes, because she owns the house.
(C) No, because a property owner cannot advertise discriminatory practices.
(D) No, because smoking is perfectly legal.

56. A tenant applicant confined to a wheelchair is interested in renting a patio home. A request is made to the landlord to allow the tenant to build an access ramp. Which of the following is true?
(A) The tenant's application may be rejected.
(B) The landlord must allow the tenant to make the modification at the tenant's expense.
(C) The landlord is required to pay for modifications to the property to accommodate the access.
(D) The landlord may collect an additional deposit to assure compliance.

57. No federal Fair Housing laws are violated if a landlord refuses to rent to
(A) families with children
(B) tenants of Indian descent
(C) deaf persons
(D) college students

58. The federal ban on discrimination based on familial status is intended to provide equal access to rentals for
(A) unmarried couples under 25
(B) people with children
(C) single tenants
(D) the elderly

59. The TILA-RESPA Integrated Disclosure (TRID) law applies to
(A) all real estate financing
(B) contracts for deed
(C) second mortgages only
(D) financing of one- to four-family residences

60. The Civil Rights Act of 1866 prohibits
(A) zoning
(B) redlining
(C) open housing
(D) racial discrimination

61. If a landlord refuses to rent to a minority prospective tenant with a poor credit record,
   (A) federal Fair Housing laws are violated
   (B) the minority applicant should report this violation
   (C) there is no violation
   (D) the landlord may face a significant fine

62. Sullivan wants to rent an apartment but is confined to a wheelchair. At Sullivan's request, the landlord
   (A) must build a ramp at the landlord's expense
   (B) must build a ramp at Sullivan's expense
   (C) need do nothing
   (D) must reduce the rent to Sullivan's affordability level

63. A seller asks his broker about the race of a potential buyer. The broker
   (A) must answer the question if a single agent
   (B) violates Fair Housing laws if he answers the question
   (C) may answer the question because of the Constitutional right to free speech
   (D) may tell the seller but no one else

64. Broker Bob is showing houses to Joe Chen, who is of Chinese ancestry. Bob shows Joe houses in a neighborhood of predominantly Chinese-American residences because Bob thinks Joe would feel more at home there. This is
   (A) illegal steering under Fair Housing laws
   (B) superior and tactful service
   (C) done for Joe's benefit
   (D) effective sales technique

65. Where the dominant population base of an area consists of minorities,
   (A) brokers need not be concerned with Fair Housing laws
   (B) brokers must still observe Fair Housing laws
   (C) there are no reporting requirements
   (D) violation of Fair Housing laws is not possible

66. The Civil Rights Act of 1968 prohibits
   (A) integration
   (B) greenlining
   (C) discrimination in housing
   (D) school busing

67. *Familial status* refers to
   (A) families with two or more children
   (B) families with children under age 18
   (C) the earning capacity of the family
   (D) TRID

68. Casualty insurance is
(A) a financing expense
(B) a capital expenditure
(C) a fixed expense
(D) a variable expense

69. A retail tenant in a mall pays monthly rent and additional amounts for property maintenance, taxes, utilities, and cleaning services. This is known as a
(A) net lease
(B) percentage lease
(C) ground lease
(D) gross lease

70. Broker Morris has a written agreement with Sarah Lee as the only broker to find a one-story house with three bedrooms and two baths in Lancaster. What type of agreement does Broker Morris have?
(A) an open buyer's agent agreement
(B) a dual agent agreement
(C) an exclusive buyer's agent agreement
(D) an agency agreement

71. Which of the following would terminate a listing agreement?
(A) death of the broker
(B) revocation of the salesperson's license
(C) an offer to purchase
(D) expiration of the salesperson's license

72. The Floyds' home was listed with a broker, but they located a buyer and sold the property themselves. However, they were still obligated to pay a brokerage fee. What kind of listing had they agreed to?
(A) a net listing
(B) an open listing
(C) an exclusive right to sell
(D) a multiple listing

73. If the agent has not located a buyer by the end of the listing period, what happens?
(A) The listing is automatically extended for an additional 30 days.
(B) The listing price is reduced by 10%.
(C) The agency relationship terminates.
(D) The broker can charge for advertising.

74. A contract in which an owner promises to pay compensation to a broker for providing a buyer but may give listings to other brokers is known as
(A) an exclusive right to sell
(B) a net listing
(C) an open listing
(D) multiple listing

75. Stephen, a salesperson associated with Broker Bob, lists a house for sale for $320,000, with 6% commission due at closing. Three weeks later, the owner accepts an offer of $300,000, brought in by Stephen. Bob's practice is that 45% of commissions goes to the office and the remainder to the salesperson. How much will Stephen make on the sale?
(A) $18,000
(B) $10,560
(C) $9,500
(D) $9,900

76. Before Mary listed her house, she reserved the right to sell the property to anyone she found without paying a broker. What kind of listing is this?
(A) exclusive right to sell
(B) exclusive agency
(C) multiple
(D) open

77. Roger has contracted with three brokers to sell his ranch. He has agreed to pay a commission to the broker who brings a ready, willing, and able buyer and who meets his terms. This is a(n)
(A) exclusive listing
(B) exclusive right to sell
(C) open listing
(D) multiple listing

78. The listing broker will receive a commission fee of 6% on the first $250,000 of the sale price and 4% on any amount above $250,000. If the property sells for $450,000, how much is the total commission?
(A) $15,000
(B) $18,000
(C) $27,000
(D) $23,000

79. Which of the following is included in an agent's responsibilities to the party the agent is *not* representing?
(A) The agent must disclose all material facts to the other party.
(B) The agent owes no loyalty to the other party.
(C) The agent must provide all information about the principal.
(D) The agent must advise the other party.

80. Broker Barb listed Tony's house and subsequently showed it to her sister, who made an offer to purchase. What is Barb's responsibility of disclosure?
(A) No disclosure is required.
(B) Barb must disclose her family relationship to the purchaser.
(C) No disclosure is required if the property is sold at the list price.
(D) It is a violation of fiduciary duty for Barb to show the property to a relative.

81. The seller has agreed to pay two points to the lending institution to help the buyers obtain a mortgage loan. The house was listed for $320,000 and is being sold for $300,000. The buyers will pay 20% in cash and borrow the rest. How much will the seller owe to the lender for points?
(A) $6,400
(B) $4,800
(C) $5,120
(D) $6,000

82. The lender's underwriting criteria specify a maximum housing expense to income ratio of 28%. If the applicant proves annual earnings of $75,000 in the previous year, and that salary is continuing, the maximum monthly PITI would be
(A) $2,100
(B) $1,750
(C) $2,800
(D) $2,300

83. Which of the following is an advantage of a biweekly mortgage payment plan?
(A) The borrower pays less per month in return for a longer term.
(B) It is equivalent to 12 monthly payments each year.
(C) There are lower interest rates.
(D) The loan is paid off sooner than it would be with 12 monthly payments.

84. Common covenants found in a mortgage include all of the following *except*
(A) keep the property in good repair
(B) pay all current real estate taxes
(C) provide unlimited access to the mortgagee
(D) not damage any improvements while securing the loan

85. The function of the Federal Housing Administration (FHA) is to
(A) lend money
(B) insure loans
(C) guarantee loans
(D) buy loans

86. The buyer who agrees to pay an existing loan but does not take personal responsibility
(A) assumes the loan
(B) takes the property subject to the loan
(C) subordinates the loan
(D) forecloses on the loan

87. In general, the lien with first claim on the real estate is the
(A) largest one
(B) one that is signed
(C) one that was recorded first
(D) one that is foreclosed

88. When a borrower is required to maintain an escrow account with the lending institution, money in that account may be used to pay the homeowner's
(A) REALTOR®
(B) utility bills
(C) property taxes or insurance
(D) life insurance

89. After Harvey bought his apartment, he started receiving his own property tax bills. This indicates that he has bought into a
(A) condominium
(B) cooperative
(C) leasehold
(D) syndicate

90. Mr. and Mrs. Buyer asked Broker Bob about the best way to take title to the property they had agreed to buy. What should Bob do?
(A) advise the Buyers to ask their parents
(B) advise the Buyers to contact an attorney
(C) have his accountant advise the Buyers of the best way to take title
(D) advise the Buyers to contact the county recorder

91. Helen and David, who are not married, want to buy a house together. To ensure that if one dies, the other will automatically become the full owner, their deed must state that they are
(A) tenants in common
(B) tenants in severalty
(C) tenants by the entirety
(D) joint tenants

92. Real property is
(A) the land
(B) the land, the improvements, and the bundle of rights
(C) the land and the right to possess it
(D) the land, buildings, and easements

93. A house has a built-in bookcase and elaborate chandelier. The buyer is also getting a dinette set and a table lamp. Which is correct?
(A) The built-in bookcase and chandelier are included in the real estate contract. The lamp and dinette set are chattels and are sold separately.
(B) All the items must be included in the real estate contract.
(C) The chandelier is real estate; the other items must be sold on a bill of sale.
(D) All of the items are real estate because they are sold with the house.

94. Man-made additions to land are known as
(A) chattels
(B) trade fixtures
(C) easements
(D) improvements

95. Land can be owned apart from the buildings on it under the arrangement known as a
(A) net lease
(B) life estate
(C) remainder
(D) ground lease

96. No one seems to own the vacant land next to Mary's house, so she has used it for a pasture for many years. She may have a good chance of obtaining ownership by
(A) poaching
(B) foreclosure
(C) adverse possession
(D) eminent domain

97. An abstract of title contains
(A) the summary of a title search
(B) an attorney's opinion of title
(C) a registrar's certificate of title
(D) a quiet title lawsuit

98. In order to reach the beach, the Smiths have a permanent right-of-way across their neighbor's beachfront property. The Smiths own
(A) an easement
(B) a license
(C) a deed restriction
(D) a lien

99. Zoning ordinances typically regulate the
(A) number of occupants allowed per dwelling unit
(B) uses of each parcel of land
(C) maximum rent that may be charged
(D) building code

100. The business relationship between broker and salesperson for income tax purposes is usually
(A) employer-employee
(B) independent contractor
(C) single agent
(D) self-employment

ANSWER KEY
Model Test 2

1. **C**	26. **D**	51. **D**	76. **B**
2. **A**	27. **C**	52. **A**	77. **C**
3. **C**	28. **D**	53. **C**	78. **D**
4. **A**	29. **A**	54. **D**	79. **B**
5. **B**	30. **D**	55. **A**	80. **B**
6. **A**	31. **C**	56. **B**	81. **B**
7. **D**	32. **D**	57. **D**	82. **B**
8. **A**	33. **D**	58. **B**	83. **D**
9. **A**	34. **D**	59. **D**	84. **C**
10. **A**	35. **A**	60. **D**	85. **B**
11. **A**	36. **D**	61. **C**	86. **B**
12. **D**	37. **B**	62. **B**	87. **C**
13. **C**	38. **C**	63. **B**	88. **C**
14. **B**	39. **A**	64. **A**	89. **A**
15. **A**	40. **B**	65. **B**	90. **B**
16. **B**	41. **A**	66. **C**	91. **D**
17. **C**	42. **B**	67. **B**	92. **B**
18. **D**	43. **D**	68. **C**	93. **A**
19. **D**	44. **C**	69. **A**	94. **D**
20. **C**	45. **D**	70. **C**	95. **D**
21. **B**	46. **C**	71. **A**	96. **C**
22. **B**	47. **C**	72. **C**	97. **A**
23. **B**	48. **D**	73. **C**	98. **A**
24. **A**	49. **B**	74. **C**	99. **B**
25. **D**	50. **A**	75. **D**	100. **B**

ANSWER EXPLANATIONS—ARITHMETIC QUESTIONS

NOTE: The section contains explanations for only the math questions in the model exam.

3. Comparable sold for $300,000 without a pool. Add $20,000 for a pool = $320,000.

4.	$20,000 monthly rent × 12	$240,000 annual
	Vacancy rate at 9%	– $21,600
	Net rental income	$218,400
	Divide by 0.12 capitalization rate	$1,820,000
9.	Lot = 43,560 × $10 × ½	$217,800
	House = 50 × 50 = 2,500 × $75	$187,500
	Total	$405,300
75.	Sales price	$300,000
	Commission rate	× .06
	Total commission	$18,000
	Salesperson rate	× .55
	Stephen's commission	$ 9,900
78.	6% of $250,000 =	$15,000
	4% of $200,000 =	$8,000
	Total	$23,000
81.	$300,000 × (1.00 – .20 = .80)	$240,000 mortgage
	2 points = 2% of mortgage	× .02
	Points	$ 4,800
82.	$75,000 × .28	= $21,000
	Divide by 12 months	= $1,750

MODEL EXAMINATION 2

ANSWER SHEET
Model Examination 3

Broker and Salesperson

1. Ⓐ Ⓑ Ⓒ Ⓓ	**26.** Ⓐ Ⓑ Ⓒ Ⓓ	**51.** Ⓐ Ⓑ Ⓒ Ⓓ	**76.** Ⓐ Ⓑ Ⓒ Ⓓ
2. Ⓐ Ⓑ Ⓒ Ⓓ	**27.** Ⓐ Ⓑ Ⓒ Ⓓ	**52.** Ⓐ Ⓑ Ⓒ Ⓓ	**77.** Ⓐ Ⓑ Ⓒ Ⓓ
3. Ⓐ Ⓑ Ⓒ Ⓓ	**28.** Ⓐ Ⓑ Ⓒ Ⓓ	**53.** Ⓐ Ⓑ Ⓒ Ⓓ	**78.** Ⓐ Ⓑ Ⓒ Ⓓ
4. Ⓐ Ⓑ Ⓒ Ⓓ	**29.** Ⓐ Ⓑ Ⓒ Ⓓ	**54.** Ⓐ Ⓑ Ⓒ Ⓓ	**79.** Ⓐ Ⓑ Ⓒ Ⓓ
5. Ⓐ Ⓑ Ⓒ Ⓓ	**30.** Ⓐ Ⓑ Ⓒ Ⓓ	**55.** Ⓐ Ⓑ Ⓒ Ⓓ	**80.** Ⓐ Ⓑ Ⓒ Ⓓ
6. Ⓐ Ⓑ Ⓒ Ⓓ	**31.** Ⓐ Ⓑ Ⓒ Ⓓ	**56.** Ⓐ Ⓑ Ⓒ Ⓓ	**81.** Ⓐ Ⓑ Ⓒ Ⓓ
7. Ⓐ Ⓑ Ⓒ Ⓓ	**32.** Ⓐ Ⓑ Ⓒ Ⓓ	**57.** Ⓐ Ⓑ Ⓒ Ⓓ	**82.** Ⓐ Ⓑ Ⓒ Ⓓ
8. Ⓐ Ⓑ Ⓒ Ⓓ	**33.** Ⓐ Ⓑ Ⓒ Ⓓ	**58.** Ⓐ Ⓑ Ⓒ Ⓓ	**83.** Ⓐ Ⓑ Ⓒ Ⓓ
9. Ⓐ Ⓑ Ⓒ Ⓓ	**34.** Ⓐ Ⓑ Ⓒ Ⓓ	**59.** Ⓐ Ⓑ Ⓒ Ⓓ	**84.** Ⓐ Ⓑ Ⓒ Ⓓ
10. Ⓐ Ⓑ Ⓒ Ⓓ	**35.** Ⓐ Ⓑ Ⓒ Ⓓ	**60.** Ⓐ Ⓑ Ⓒ Ⓓ	**85.** Ⓐ Ⓑ Ⓒ Ⓓ
11. Ⓐ Ⓑ Ⓒ Ⓓ	**36.** Ⓐ Ⓑ Ⓒ Ⓓ	**61.** Ⓐ Ⓑ Ⓒ Ⓓ	**86.** Ⓐ Ⓑ Ⓒ Ⓓ
12. Ⓐ Ⓑ Ⓒ Ⓓ	**37.** Ⓐ Ⓑ Ⓒ Ⓓ	**62.** Ⓐ Ⓑ Ⓒ Ⓓ	**87.** Ⓐ Ⓑ Ⓒ Ⓓ
13. Ⓐ Ⓑ Ⓒ Ⓓ	**38.** Ⓐ Ⓑ Ⓒ Ⓓ	**63.** Ⓐ Ⓑ Ⓒ Ⓓ	**88.** Ⓐ Ⓑ Ⓒ Ⓓ
14. Ⓐ Ⓑ Ⓒ Ⓓ	**39.** Ⓐ Ⓑ Ⓒ Ⓓ	**64.** Ⓐ Ⓑ Ⓒ Ⓓ	**89.** Ⓐ Ⓑ Ⓒ Ⓓ
15. Ⓐ Ⓑ Ⓒ Ⓓ	**40.** Ⓐ Ⓑ Ⓒ Ⓓ	**65.** Ⓐ Ⓑ Ⓒ Ⓓ	**90.** Ⓐ Ⓑ Ⓒ Ⓓ
16. Ⓐ Ⓑ Ⓒ Ⓓ	**41.** Ⓐ Ⓑ Ⓒ Ⓓ	**66.** Ⓐ Ⓑ Ⓒ Ⓓ	**91.** Ⓐ Ⓑ Ⓒ Ⓓ
17. Ⓐ Ⓑ Ⓒ Ⓓ	**42.** Ⓐ Ⓑ Ⓒ Ⓓ	**67.** Ⓐ Ⓑ Ⓒ Ⓓ	**92.** Ⓐ Ⓑ Ⓒ Ⓓ
18. Ⓐ Ⓑ Ⓒ Ⓓ	**43.** Ⓐ Ⓑ Ⓒ Ⓓ	**68.** Ⓐ Ⓑ Ⓒ Ⓓ	**93.** Ⓐ Ⓑ Ⓒ Ⓓ
19. Ⓐ Ⓑ Ⓒ Ⓓ	**44.** Ⓐ Ⓑ Ⓒ Ⓓ	**69.** Ⓐ Ⓑ Ⓒ Ⓓ	**94.** Ⓐ Ⓑ Ⓒ Ⓓ
20. Ⓐ Ⓑ Ⓒ Ⓓ	**45.** Ⓐ Ⓑ Ⓒ Ⓓ	**70.** Ⓐ Ⓑ Ⓒ Ⓓ	**95.** Ⓐ Ⓑ Ⓒ Ⓓ
21. Ⓐ Ⓑ Ⓒ Ⓓ	**46.** Ⓐ Ⓑ Ⓒ Ⓓ	**71.** Ⓐ Ⓑ Ⓒ Ⓓ	**96.** Ⓐ Ⓑ Ⓒ Ⓓ
22. Ⓐ Ⓑ Ⓒ Ⓓ	**47.** Ⓐ Ⓑ Ⓒ Ⓓ	**72.** Ⓐ Ⓑ Ⓒ Ⓓ	**97.** Ⓐ Ⓑ Ⓒ Ⓓ
23. Ⓐ Ⓑ Ⓒ Ⓓ	**48.** Ⓐ Ⓑ Ⓒ Ⓓ	**73.** Ⓐ Ⓑ Ⓒ Ⓓ	**98.** Ⓐ Ⓑ Ⓒ Ⓓ
24. Ⓐ Ⓑ Ⓒ Ⓓ	**49.** Ⓐ Ⓑ Ⓒ Ⓓ	**74.** Ⓐ Ⓑ Ⓒ Ⓓ	**99.** Ⓐ Ⓑ Ⓒ Ⓓ
25. Ⓐ Ⓑ Ⓒ Ⓓ	**50.** Ⓐ Ⓑ Ⓒ Ⓓ	**75.** Ⓐ Ⓑ Ⓒ Ⓓ	**100.** Ⓐ Ⓑ Ⓒ Ⓓ

Model Examination 3

BROKER AND SALESPERSON

See page 538 for the answer key. Math problems are worked out on pages 539–540.

1. If, upon receipt of an offer to purchase under certain terms, the seller makes a counteroffer, the prospective purchaser is
   (A) bound by his original offer
   (B) bound to accept the counteroffer
   (C) bound by the agent's decision
   (D) relieved of his original offer

2. A broker who makes profitable investments with earnest money deposits
   (A) must share 50% of the profits with the owners of the money he used
   (B) has done something illegal
   (C) may keep all the profits
   (D) must turn over the profits to the state's licensing authorities

3. A real estate broker must comply with
   (A) agency law
   (B) her state's real estate licensing law
   (C) both A and B
   (D) neither A nor B

4. Smith makes an offer on Jones's property and states that the offer will remain open for 3 days. The day after Smith has made the offer, he decides to withdraw it since Jones has neither rejected nor accepted it.
   (A) Smith cannot do this.
   (B) Smith can do this only if Jones was planning to reject the offer anyway.
   (C) Smith must give Jones at least half the remaining time to make a decision.
   (D) Smith may withdraw the offer.

5. In order to sell property belonging to a trust for which she is trustee, the trustee must have
   (A) a broker's license
   (B) a salesperson's license
   (C) a trustor's license
   (D) none of the above

6. Appraised valuation is $250,000. Tax is based on 20% of appraised valuation. City tax is 50 mills; county tax is 40 mills. Which of the following is true?
   (A) City tax is $2,500.
   (B) County tax is $20,000.
   (C) City and county tax add up to $6,000.
   (D) none of the above

7. Under TRID, the lender must advise the borrower of the loan's
   (A) owner
   (B) Loan Estimate
   (C) seller
   (D) closing date

8. In the absence of an agreement to the contrary, the mortgage normally having priority will be the one that
   (A) is for the greatest amount
   (B) is a permanent mortgage
   (C) was recorded first
   (D) is a construction loan mortgage

9. A development in which a person owns her dwelling unit and, in common with other owners in the same project, also owns common property is
   (A) a leased fee
   (B) a condominium
   (C) a homestead
   (D) none of the above

10. Which of the following is true?
   (A) A condominium owner need not pay condominium fees that he feels are too high.
   (B) Condominium units are attached housing units.
   (C) Condominium is an ownership form, not an architectural style.
   (D) Condominiums cannot be rented.

11. The money for making FHA loans is provided by
   (A) qualified lending institutions
   (B) the Department of Housing and Urban Development
   (C) the Federal Housing Administration
   (D) the Federal Savings and Loan Insurance Corporation

12. License law forbids
   (A) soliciting for listings before one is licensed
   (B) collecting a commission from more than one party to a transaction
   (C) showing property to other licensees
   (D) the purchase by a broker of property that she has listed

13. The term REALTOR®
   (A) is a registered trademark
   (B) refers to anyone who has a real estate license
   (C) refers to anyone who sells real estate
   (D) none of the above

14. Complaints about housing discrimination may be brought
   (A) to the Secretary of Housing and Urban Development
   (B) directly to court
   (C) either A or B
   (D) neither A nor B

15. Which of the following is (are) ALWAYS exempt from real property taxes?
   (A) income-producing property owned by a church
   (B) government-owned property
   (C) most restaurants
   (D) a private, for-profit school

16. Tenancy in severalty refers to
   (A) ownership by a married couple
   (B) ownership by one person
   (C) ownership by people who are related, but not married
   (D) ownership by unrelated people

17. For every contract of sale there must be
   (A) an offer and acceptance
   (B) a mortgage loan
   (C) a broker
   (D) good consideration

18. Real estate brokers can lose their licenses for which of the following?
   (A) using moneys received as commissions to pay office help
   (B) representing the buyer in a transaction
   (C) refusing a listing
   (D) paying a commission to a nonlicensed person

19. A broker must keep all earnest money deposits in
   (A) her office safe
   (B) her business checking account
   (C) an escrow or trust account
   (D) a savings account

20. Eminent domain is
    (A) the right of the government to take private property for public use
    (B) the extent to which state boundaries reach out to sea
    (C) an ancient form of ownership not common today
    (D) the right of the federal government to pass laws that supersede state law

21. An 8-year-old building was worth $190,000 after depreciating at a rate of 3% per year. What was its original value?
    (A) $195,700
    (B) $235,600
    (C) $250,000
    (D) $277,778

22. Flaherty's mortgage loan is for $200,000 and carries an annual interest rate of 6%. Monthly payments are $1,200. How much will the principal be reduced by the *second* monthly payment?
    (A) $200.00
    (B) $201.00
    (C) $1,499.00
    (D) $1,500.00

23. Bernard signs a listing that guarantees the listing broker a commission payment if the sale is effected by any licensed agent. This is
    (A) an open listing
    (B) an exclusive agency listing
    (C) an exclusive right to sell listing
    (D) a net listing

24. Macrae agrees with Ortez to list Ortez's property once Macrae is issued his real estate license. When the license is issued, Ortez signs a listing with Macrae.
    (A) The listing is valid.
    (B) The listing is invalid.
    (C) The listing is valid only at the listing price.
    (D) Macrae can collect only half the normal commission.

25. A licensee can lose his license for
    (A) selling properties quickly, at low prices
    (B) buying property for his own use from his principal
    (C) splitting a commission with another participating broker
    (D) none of the above

26. A type of mortgage in which the lender makes periodic payments to the borrower, who is required to be age 62 or older in the FHA program, is called
    (A) opposite
    (B) accelerate
    (C) reverse
    (D) deficit

27. A conventional mortgage is
    (A) amortizing
    (B) guaranteed by FHA
    (C) not guaranteed by a government agency
    (D) approved by the VA

28. Among other things, the agent is obligated to
    (A) be loyal to the principal
    (B) be honest with the principal
    (C) both A and B
    (D) neither A nor B

29. Which of the following is *not* required to have a real estate license?
    (A) the resident manager of an apartment project
    (B) the resident manager of an apartment project who, for a fee, sells a house across the street
    (C) a student who sells houses as part of a research project
    (D) all of the above

30. Recordation of a deed is the responsibility of the
    (A) grantor
    (B) grantee
    (C) both A and B
    (D) neither A nor B

Questions 31–34 refer to the Far Hills Estates diagram below.

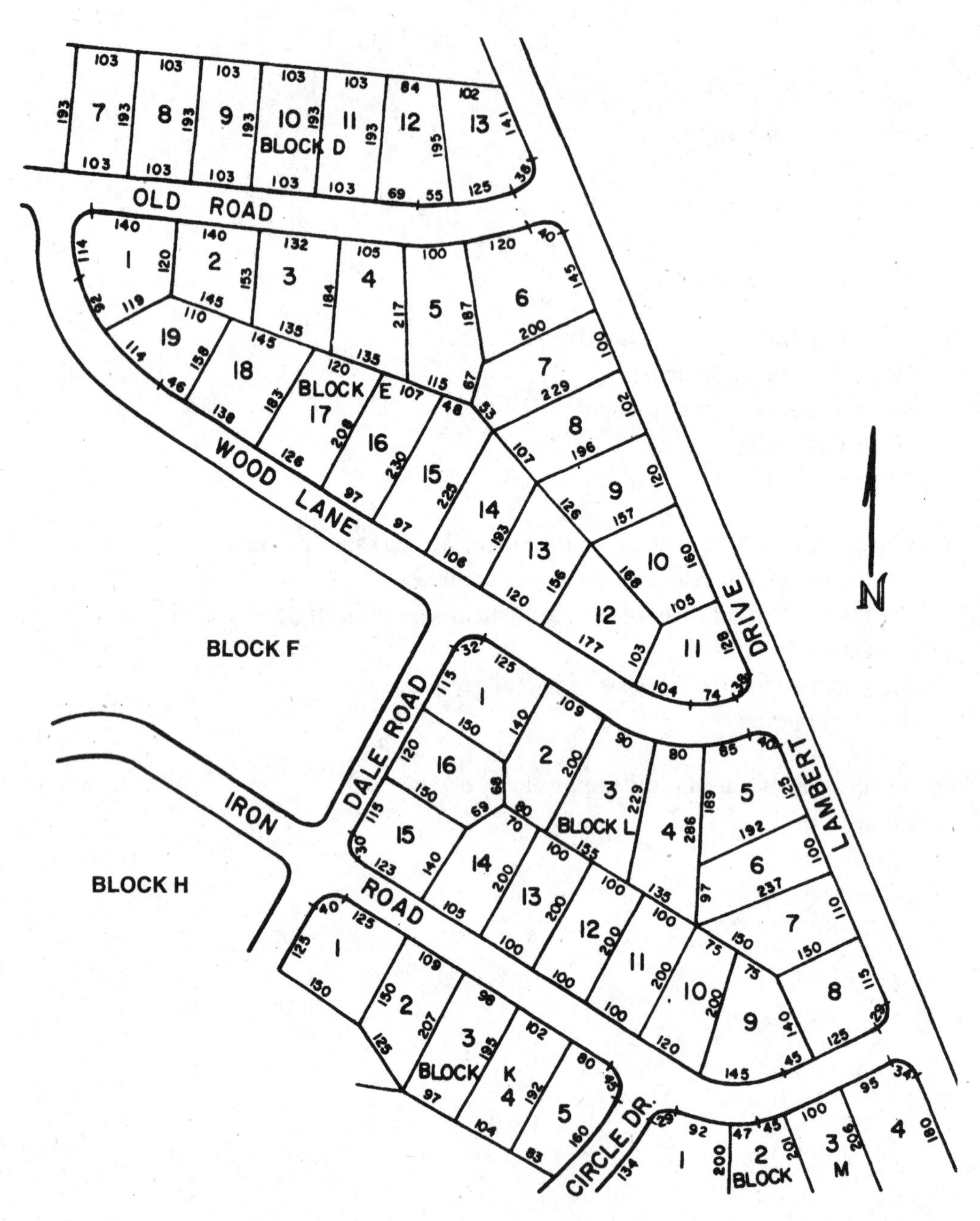

Far Hills Estates

31. Which of following statements is (are) true?
    (A) Four lots in Block E have frontage on two streets.
    (B) Iron Road has more lots fronting on it than any of the other streets on the plat.
    (C) both A and B
    (D) neither A nor B

32. Which lot has the greatest footage on Wood Lane?
(A) Lot 11, Block E
(B) Lot 12, Block E
(C) Lot 1, Block L
(D) Lot 19, Block E

33. Which lot has the greatest depth?
(A) Lot 4, Block L
(B) Lot 7, Block E
(C) Lot 5, Block E
(D) Lot 16, Block E

34. Which of following statements is (are) true?
(A) All the lots on the westerly side of Dale Road should appear in Block F.
(B) There is no indication of where to find a plat of the easterly side of Lambert Drive.
(C) both A and B
(D) neither A nor B

---

35. The part of conveyance that defines or limits the quantity of the estate granted is
(A) habendum
(B) premises
(C) equity
(D) consideration

36. The Statute of Frauds
(A) requires certain contracts to be in writing to be enforceable
(B) requires a license to operate as broker or salesperson
(C) regulates escrow accounts
(D) regulates the estate owning real estate

37. A warranty deed protects the grantee against a loss by
(A) casualty
(B) defective title
(C) both A and B
(D) neither A nor B

38. Brown's building rents for $8,500 per month. The building's market value is $470,000. Taxes are 75 mills, based on 40% of market value. What percentage of the building's income must be paid out in tax?
(A) 6.9%
(B) 10.7%
(C) 13.8%
(D) 34.5%

39. Henderson bought a home for $420,000. Four years later, he sold it for $599,000. What is the average annual rate of appreciation?
    (A) 9.92%
    (B) 10.65%
    (C) 13.12%
    (D) 42.62%

40. The income approach to appraisal would be most suitable for
    (A) a newly opened subdivision
    (B) commercial and investment property
    (C) property heavily mortgaged
    (D) property heavily insured

41. Real estate is defined as
    (A) land and buildings
    (B) land and all permanent attachments to it
    (C) land and everything growing on it
    (D) land only

42. A second mortgage is
    (A) a lien on real estate that has a prior mortgage on it
    (B) the first mortgage recorded
    (C) always made by the seller
    (D) smaller in amount than a first mortgage

43. Depreciation can be caused by
    (A) physical deterioration
    (B) functional obsolescence
    (C) economic obsolescence
    (D) all of the above

44. TRID applies to
    (A) all real estate salespersons and brokers
    (B) all retail stores
    (C) all wholesale establishments
    (D) none of the above

45. Perkins bought 11 lots for $21,000 each. He keeps four and sells the remaining lots for a total of $28,000 more than he originally paid for all of them. What was the average sale price of each lot that he sold?
    (A) $21,000
    (B) $33,000
    (C) $37,000
    (D) $49,000

46. One discount point is equal to
    (A) 1% of the sales price
    (B) 1% of the interest rate
    (C) 1% of the loan amount
    (D) none of the above

47. Which of the following is *not* realty?
    (A) fee simple estate
    (B) leasehold for indefinite duration
    (C) lumber
    (D) life estate

48. A lease can state that the rent is to be paid in
    (A) labor
    (B) crops
    (C) cash
    (D) any of the above

49. A mortgaged property can be
    (A) sold without the consent of the mortgagee
    (B) conveyed by the grantor making a deed to the grantee
    (C) both A and B
    (D) neither A nor B

Questions 50 and 51 refer to the diagram below of a house and a lot.

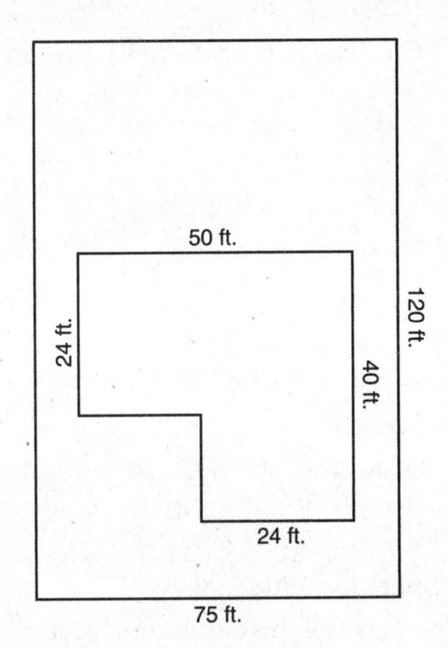

50. How many square feet are there in the house?
    (A) 384
    (B) 1,200
    (C) 1,584
    (D) 2,000

51. What percentage of the area of the lot is taken up by the house?
    (A) 13½%
    (B) 17.6%
    (C) 22.2%
    (D) 24.0%

---

52. TRID requires
    (A) borrowing from relatives
    (B) borrowing by a business
    (C) Closing Disclosure
    (D) inheritances

53. A contract that gives someone the right but not the obligation to buy at a specified price within a specified time is
    (A) a contract of sale
    (B) an option
    (C) an agreement of sale
    (D) none of the above

54. The prohibitions of the 1968 Fair Housing Act apply to
    (A) multifamily dwellings of five or more units
    (B) multifamily dwellings of four or fewer units if the owner occupies one of the units
    (C) single-family dwellings
    (D) none of the above

55. Consideration that is of value only to the person who receives it is called
    (A) good consideration
    (B) near consideration
    (C) valuable consideration
    (D) no consideration

56. An easement in gross
    (A) covers an entire property and all parts of it
    (B) extends to one person only
    (C) extends to the general public
    (D) occurs when the public has used private land without hindrance for a certain period of time

57. A condominium homeowner's association may
    (A) require that certain owners sell their units and leave
    (B) assess fees for the upkeep of common property
    (C) practice discrimination if it is done subtly
    (D) none of the above

58. If government takes private property it must
   (A) make just compensation for the property taken
   (B) require the property for a public use
   (C) both A and B
   (D) neither A nor B

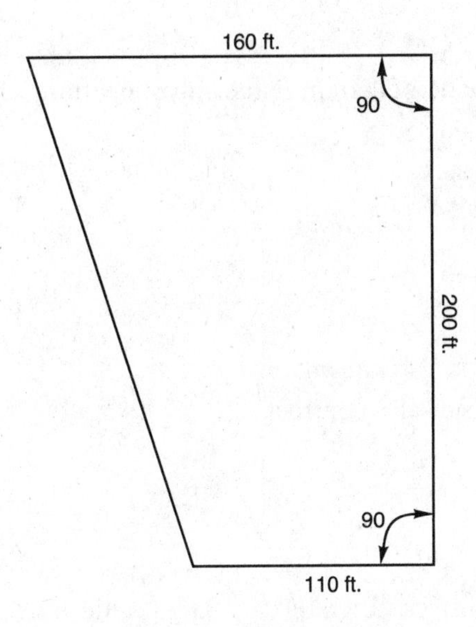

59. The lot shown above sold for $78,300. What was the price per square foot?
   (A) $4.17
   (B) $3.56
   (C) $2.90
   (D) $2.45

60. A salesperson receives 60% of the total commission on a sale for $315,000. The salesperson received $15,120. What was the rate of commission?
   (A) 4.8%
   (B) 6%
   (C) 6¾%
   (D) 8%

61. A millage rate of 84.5 mills is the same as
   (A) $84.50 per $1,000
   (B) 8.45%
   (C) both A and B
   (D) neither A nor B

62. A single-family house privately owned by an individual owning fewer than three such houses may be sold or rented without being subject to the provisions of the Fair Housing Act unless
    (A) a broker is used
    (B) discriminatory advertising is used
    (C) both A and B
    (D) neither A nor B

63. The transfer, by a tenant, of certain rights and obligations of an existing lease to another tenant is called
    (A) assignment of lease
    (B) a release
    (C) subletting
    (D) an eviction

64. A "contract for deed" is also known as
    (A) an installment land sales contract
    (B) a land contract
    (C) both A and B
    (D) neither A nor B

65. Nadel buys a new house costing $188,000. Land value is 15% of total price. The house contains 1,733 sq. ft. What is the cost per square foot of the house alone (not including the land)?
    (A) $16.44
    (B) $92.21
    (C) $109.64
    (D) $127.20

66. Salesman Jones works for a broker. Jones receives 42% of all commissions he brings in. Jones sells a home for $140,700, at a 6½% commission. What is the *broker's* share of the proceeds?
    (A) $3,841
    (B) $5,304
    (C) $9,146
    (D) $1,407

67. Rent for a 90 ft. by 60 ft. office is $6,300 per month. What is the annual rent per square foot?
    (A) $14
    (B) $11
    (C) $7.56
    (D) $140

68. An estoppel certificate is required when the
    (A) mortgage is sold to an investor
    (B) property is sold
    (C) property is being foreclosed
    (D) mortgage is assumed

69. An owner who seeks a mortgage loan and offers three properties as security will give
    (A) a blanket mortgage
    (B) an FHA mortgage
    (C) a conventional mortgage
    (D) a chattel mortgage

70. Ferraro bought a lot for $20,000 in 2003 and spent $130,000 to build a house on it in 2005. Today the house has increased 40% in value while the lot has increased 800%. What is the combined value of house and lot today?
    (A) $180,000
    (B) $182,000
    (C) $210,000
    (D) $362,000

71. A contract that has no force or effect is said to be
    (A) voidable
    (B) void
    (C) avoided
    (D) voidiated

72. Which of the following statements is (are) false?
    (A) FHA loans are insured loans.
    (B) VA loans are guaranteed loans.
    (C) both A and B
    (D) neither A nor B

73. To be enforceable, a sales agreement must have
    (A) the signature of the wife of a married seller
    (B) an earnest money deposit
    (C) competent parties
    (D) witnesses

74. A broker listed a house for $350,000 at a 6% commission. The eventual sales price was $315,000. How much less was the broker's commission than it would have been if the house had sold at the listed price?
    (A) $210
    (B) $2,100
    (C) $18,900
    (D) $35,000

75. How many square feet are there in 1¼ acres?
    (A) 10,890
    (B) 44,649
    (C) 54,450
    (D) 152,460

76. Under TRID, borrowers must receive a
    (A) Loan Estimate
    (B) Closing Disclosure
    (C) both A and B
    (D) neither A nor B

77. An item of personalty that is affixed to realty so as to be used as a part of it is
    (A) a fixture
    (B) a chattel
    (C) personal property
    (D) encumbered

78. The requirement that all parties to a contract have an understanding of the conditions and stipulations of the agreement is
    (A) consenting realty
    (B) a proper offer
    (C) good and valuable consideration
    (D) mutual agreement

79. A deed of conveyance must be signed by
    (A) the grantee and the grantor
    (B) only by the grantor
    (C) both A and B
    (D) neither A nor B

80. In the application of the income approach to appraising, which of the following statements is true?
    (A) The higher the capitalization rate, the lower the appraised value.
    (B) The higher the capitalization rate, the higher the appraised value.
    (C) The present value is equal to future income.
    (D) none of the above

81. A real estate broker may not
    (A) identify local schools, whether public or private
    (B) mention ethnic festivities held in the city
    (C) state the racial composition of the neighborhood
    (D) map religious facilities in the area

82. A prospective tenant in a 50-unit apartment complex required a wheelchair. The tenant wanted to rent a unit, install grab bars in the bathroom, and lower door handles. Under the federal Fair Housing Act, the landlord
   (A) must allow and pay for these changes
   (B) must allow the tenant to make these changes at the tenant's expense and can require the tenant to restore the property at the end of the lease
   (C) can refuse to rent the unit because the unit is not handicapped accessible
   (D) must allow the tenant's modifications and accept the property as modified when the lease ends

83. A salesperson was taking a listing on a million-dollar-plus house. The owner told the salesperson that she would be unwilling to sell to a minority buyer. The salesperson should
   (A) honor the principal's request
   (B) smile and take the listing but be prepared to disobey
   (C) decline because following the seller's instruction would violate the federal Fair Housing Act
   (D) report the owner to HUD

84. An American citizen of Mexican ancestry asks to be shown a house in a certain neighborhood. How should the broker respond?
   (A) "I'd be pleased to show you a house for sale anywhere."
   (B) "You'll be happier in a house elsewhere."
   (C) "Neighbors there will not be friendly."
   (D) "The seller does not want me to show that house."

85. The presence of wetlands
   (A) lowers property value
   (B) causes the property to be useless
   (C) enhances the property value because it attracts wildlife
   (D) should be disclosed to potential buyers

86. For tax purposes, the typical employment agreement between salespersons and a sponsoring broker is arranged as
   (A) employment
   (B) outside salesperson
   (C) independent contractor
   (D) dependent contractor

87. Which of the following is generally not tax deductible for homeowners who itemize tax deductions?
   (A) interest expense
   (B) real estate taxes
   (C) discount points
   (D) utilities

88. An independent contractor relationship requires the broker to
    (A) fund a retirement plan for salespersons
    (B) insist that salespersons work eight hours a day
    (C) withhold income taxes from commissions earned
    (D) allow salespersons freedom from obligations as an employee

89. The Civil Rights Act of 1866 prohibits discrimination based on
    (A) race
    (B) sex
    (C) religion
    (D) national origin

90. Steering is
    (A) helping a buyer find a house in a neighborhood of people of similar ethnicity
    (B) helping a buyer find a house with a suitable floor plan
    (C) helping a buyer to qualify for a mortgage
    (D) encouraging a buyer to look at homes in neighborhoods with specific ethnic characteristics and discouraging the buyer from looking in other neighborhoods

91. The owner of a house lists it for sale with ABC Realty, asking the firm not to sell to a minority prospect. ABC Realty, to comply with the federal Fair Housing Act, should
    (A) refuse to accept the listing
    (B) disregard the request
    (C) agree to the request but show the house to minority prospects if requested
    (D) report the owner to HUD

92. Zoning is authorized by what power of government?
    (A) condemnation
    (B) eminent domain
    (C) regulation
    (D) police power

93. The Residential Lead-Based Paint Hazard Reduction Act of 1992 requires disclosure of
    (A) lead-based paint
    (B) radon in air
    (C) asbestos
    (D) radon in water

94. A broker shows an Asian family houses only in Asian neighborhoods. This is
    (A) legal steering
    (B) blockbusting
    (C) effective marketing
    (D) illegal steering

95. A property was used for retail sales, but the zoning has been changed to residential. The property can continue as retail use but is
    (A) limited
    (B) biased
    (C) nonconforming
    (D) spot zoning

96. A handy person is in the business of buying rundown homes, fixing them up, and reselling them at a profit. He buys a single-family home for $110,000. Under TRID, disclosure of discount points
    (A) does not apply to a business loan
    (B) is not required for loans over $100,000
    (C) is required for all single-family homes
    (D) is required only when more than 2 points is charged

97. When property to be sold is likely to contain hazardous materials, the licensed salesperson should
    (A) disclose what he or she knows to interested buyers
    (B) let buyers find out on their own
    (C) comply with the owner's instructions
    (D) refer all questions to the owner

98. Federal Fair Housing laws protect all of the following *except*
    (A) racial minorities
    (B) families with children
    (C) pregnant women
    (D) graduate students

99. A blind man with a service dog seeks to rent an apartment in a building that does not allow pets. Under the federal Fair Housing Act, the apartment building owner
    (A) can reject the applicant as a tenant
    (B) can increase the security deposit compared with what is charged other tenants
    (C) must lease, although the lease can require the tenant to repair any damage caused by the animal
    (D) cannot treat the tenant any differently than any other prospective tenant

100. Which of the following phrases does HUD consider discriminatory advertising?
    (A) "Senior housing"
    (B) "Friendly neighbors"
    (C) "Exclusive neighborhood"
    (D) "No one excluded"

ANSWER KEY
Model Test 3

1. **D**
2. **B**
3. **C**
4. **D**
5. **D**
6. **A**
7. **B**
8. **C**
9. **B**
10. **C**
11. **A**
12. **A**
13. **A**
14. **C**
15. **B**
16. **B**
17. **A**
18. **D**
19. **C**
20. **A**
21. **C**
22. **B**
23. **B**
24. **B**
25. **D**
26. **C**
27. **C**
28. **C**
29. **A**
30. **B**
31. **B**
32. **A**
33. **A**
34. **B**
35. **A**
36. **A**
37. **B**
38. **C**
39. **B**
40. **B**
41. **B**
42. **A**
43. **D**
44. **D**
45. **C**
46. **C**
47. **C**
48. **D**
49. **C**
50. **C**
51. **B**
52. **C**
53. **B**
54. **A**
55. **A**
56. **B**
57. **B**
58. **C**
59. **C**
60. **D**
61. **C**
62. **C**
63. **C**
64. **C**
65. **B**
66. **B**
67. **A**
68. **A**
69. **A**
70. **D**
71. **B**
72. **D**
73. **C**
74. **B**
75. **C**
76. **C**
77. **A**
78. **D**
79. **B**
80. **A**
81. **C**
82. **B**
83. **C**
84. **A**
85. **D**
86. **C**
87. **D**
88. **D**
89. **A**
90. **D**
91. **A**
92. **D**
93. **A**
94. **D**
95. **C**
96. **A**
97. **A**
98. **D**
99. **C**
100. **C**

ANSWER EXPLANATIONS—ARITHMETIC QUESTIONS

NOTE: The section contains explanations for only the arithmetic questions in the model exam.

6. If the tax is based on 20% of the appraised value of $250,000, then the tax is based on

$$0.2 \times \$250{,}000 = \$50{,}000$$

City tax is 50 mills, or 5%, so

$$\$50{,}000 \times 0.05 = \$2{,}500 \text{ city tax}$$

County tax is 40 mills, or 4%, so

$$\$50{,}000 \times 0.04 = \$2{,}000 \text{ county tax}$$

21. After depreciating for 8 years at 3% per year, the building has lost 24% of original value ($8 \times 0.03 = 0.24$). This means that it is now worth the remaining 76% of its original value. If that sum is $190,000, then the original value is

$$\$190{,}000 \div 0.76 = \$250{,}000$$

22. Six percent of $200,000 is 0.06 × $200,000 = $12,000. Interest for a single month is $12,000 ÷ 12, or $1,000. Therefore, of the $1,200 *first* payment, $1,000 goes to interest, leaving $200 to reduce the principal. This means that Flaherty pays interest on $199,800 the second month. By figuring the same way, we get a monthly interest charge of $999 for the second month on that amount; this leaves $201 by which the second monthly payment reduces the principal.

38. Annual rent is $8,500 × 12 = $102,000. Taxes are based on 40% of appraisal of $470,000, or 0.4 × $470,000 = $188,000. The tax rate is 75 mills, or 7½%, so the tax is

$$0.075 \times \$188{,}000 = \$14{,}100$$

This represents 13.8% of the annual rent ($14,100 ÷ $102,000).

39. Total appreciation is $599,000 − $420,000 = $179,000. Total percent appreciation is $179,000 ÷ $420,000 = 42.619%. Average over 4 years is

$$42.619\% \div 4 = 10.65\%$$

45. The 11 lots originally cost a total of 11 × $21,000 = $231,000; Perkins sells 7 lots for a total of $28,000 more than that, or $231,000 + $28,000 = $259,000. The average sale price is

$$\$259{,}000 \div 7 = \$37{,}000$$

50. The house is a combination of two rectangles: one is 50 ft. × 24 ft. (1,200 sq. ft.) and the other, smaller part jutting out to the bottom is 24 ft. × 16 ft. (384 sq. ft.). You get the 16 ft. measurement by subtracting the 24 ft. of the short side of the house from the 40 ft. of the long side.

$$1{,}200 \text{ sq. ft.} + 384 \text{ sq. ft.} = 1{,}584 \text{ sq. ft.}$$

51. The lot is 120 ft. × 75 ft. = 9,000 sq. ft. The area taken by the house is

$$1{,}584 \text{ sq. ft.} \div 9{,}000 \text{ sq. ft.} = 0.176 = 17.6\%$$

59. This figure is a trapezoid. The top (T) and bottom (B) are of unequal lengths, while the height (H) does not change. The formula for this figure is

$$\text{Area} = \tfrac{1}{2} \times (T + B) \times H$$

The area of the lot, then, is

$$\tfrac{1}{2} \times (110 \text{ ft.} + 160 \text{ ft.}) \times 200 \text{ ft.} = \tfrac{1}{2} \times 270 \text{ ft.} \times 200 \text{ ft.} = 27{,}000 \text{ sq. ft.}$$

If the lot is sold for $78,300.00, the price per square foot is

$$\$78{,}300.00 \div 27{,}000.00 = \$2.90$$

60. The salesperson's $15,120 was 60% of total commission, so

$$\$15{,}120 \div 0.6 = \$25{,}200 \text{ total commission}$$

Since the sale price was $315,000, the *rate* of commission was

$$\$25{,}200 \div \$315{,}000 = 0.08 = 8\%$$

65. First find the value of the house alone. Land is 15% of the total value, which means that the house is 85% of the total (100% − 15% = 85%). Then the house is worth

$$0.85 \times \$188{,}000 = \$159{,}800$$

If the house has 1,733 sq. ft., the cost per square foot is

$$\$159{,}800 \div 1{,}733 \text{ sq. ft.} = \$92.21$$

66. First find the amount of the total commission:

$$0.065 \times \$140{,}700 = \$9{,}145.50$$

If Jones can keep 42% of this, then the broker must get to keep the remaining 58%. Therefore, the broker's share is

$$0.58 \times \$9{,}145.50 = \$5{,}304.39$$

67. Area = 90 ft. × 60 ft. = 5,400 sq. ft., and $6,300 × 12 mo. = $75,600 per year. Annual rent per square foot is

$$\$75{,}600 \div 5{,}400 \text{ sq. ft.} = \$14.00$$

70. The lot value has increased by 800%. Don't forget that the lot was worth 100% of its value in 2003, so if it has gone up by 800%, then today it is worth 900%, or 9 times, its original value.

$$9 \times \$20{,}000 = \$180{,}000 \text{ lot value today}$$

Similarly, since the house is worth 40% more, today it is worth 140% of its original value:

$$1.4 \times \$130{,}000 = \$182{,}000 \text{ house value today}$$

House and lot together are worth $180,000 + $182,000 = $362,000 today.

74. There are two ways of answering this question.

(A) You can figure the commissions on $350,000 and $315,000 and subtract.

$$0.06 \times \$350{,}000 = \$21{,}000$$

$$0.06 \times \$315{,}000 = \$18{,}900$$

$$\$210{,}000 - \$189{,}000 = \$2{,}100, \text{ the correct answer}$$

(B) You can subtract $315,000 from $350,000 and get $35,000, which is the difference in price. The broker will lose the commission on that amount, so

$$0.06 \times \$35{,}000 = \$2{,}100, \text{ also the correct answer}$$

75. One acre contains 43,560 sq. ft., so

$$1\tfrac{1}{4} \times 43{,}560 \text{ sq. ft.} = 54{,}450 \text{ sq. ft.}$$

ANSWER SHEET
Model Examination 4

Broker and Salesperson

1. A B C D
2. A B C D
3. A B C D
4. A B C D
5. A B C D
6. A B C D
7. A B C D
8. A B C D
9. A B C D
10. A B C D
11. A B C D
12. A B C D
13. A B C D
14. A B C D
15. A B C D
16. A B C D
17. A B C D
18. A B C D
19. A B C D
20. A B C D
21. A B C D
22. A B C D
23. A B C D
24. A B C D
25. A B C D
26. A B C D
27. A B C D
28. A B C D
29. A B C D
30. A B C D
31. A B C D
32. A B C D
33. A B C D
34. A B C D
35. A B C D
36. A B C D
37. A B C D
38. A B C D
39. A B C D
40. A B C D
41. A B C D
42. A B C D
43. A B C D
44. A B C D
45. A B C D
46. A B C D
47. A B C D
48. A B C D
49. A B C D
50. A B C D
51. A B C D
52. A B C D
53. A B C D
54. A B C D
55. A B C D
56. A B C D
57. A B C D
58. A B C D
59. A B C D
60. A B C D
61. A B C D
62. A B C D
63. A B C D
64. A B C D
65. A B C D
66. A B C D
67. A B C D
68. A B C D
69. A B C D
70. A B C D
71. A B C D
72. A B C D
73. A B C D
74. A B C D
75. A B C D
76. A B C D
77. A B C D
78. A B C D
79. A B C D
80. A B C D
81. A B C D
82. A B C D
83. A B C D
84. A B C D
85. A B C D
86. A B C D
87. A B C D
88. A B C D
89. A B C D
90. A B C D
91. A B C D
92. A B C D
93. A B C D
94. A B C D
95. A B C D
96. A B C D
97. A B C D
98. A B C D
99. A B C D
100. A B C D

Model Examination 4

BROKER AND SALESPERSON

See page 558 for the answer key. Math problems are worked out on page 559.

NOTE: In this model examination, all the math questions (questions 91–100) are grouped at the end, but you may expect that in most states, the math questions will be scattered throughout the examination.

1. Jackson's will left to Mrs. Jackson the right to use, occupy, and enjoy Jackson's real estate until her death. At that time, the real estate will become the property of their children. Mrs. Jackson is a
   (A) remainderman
   (B) life tenant
   (C) joint tenant
   (D) tenant in common

2. To prove his right to a commission, the broker must show
   (A) that he was licensed throughout the transaction
   (B) that he had a contract of employment
   (C) that he was the "efficient and procuring cause" of the sale
   (D) all of the above

3. Contracts made by a minor are
   (A) enforceable at all times
   (B) void
   (C) voidable by either party
   (D) voidable only by the minor

4. Mr. Beans owns land worth $25,000; Mr. Pork owns a house worth $150,000, subject to a $100,000 mortgage, which Beans will assume. For a fair trade,
   (A) Pork should pay $25,000 cash in addition
   (B) Beans should pay $25,000 cash in addition
   (C) they may trade properties evenly
   (D) none of the above

5. From the standpoint of the grantor, which of the following types of deed creates the least liability?
   (A) special warranty
   (B) general warranty
   (C) bargain and sale
   (D) quitclaim

6. The lending of money at a rate of interest above the legal rate is
   (A) speculating
   (B) usury
   (C) both A and B
   (D) neither A nor B

7. A licensee's license must be
   (A) carried in her wallet at all times
   (B) posted in a public place in the broker's office
   (C) kept on the wall at the licensee's home
   (D) kept on the wall at the Real Estate Commission

8. The parties to a deed are the
   (A) vendor and vendee
   (B) grantor and grantee
   (C) offeror and offeree
   (D) acceptor and acceptee

9. A person must be licensed if he is to sell
   (A) his home
   (B) property belonging to an estate for which he is executor
   (C) property belonging to other clients who pay him a commission
   (D) property that he has inherited

10. Which of the following is *not* an appraisal approach?
   (A) cost
   (B) sales comparison
   (C) income
   (D) trade

11. The most comprehensive ownership of land at law is known as
   (A) estate for years
   (B) life estate
   (C) fee simple
   (D) defeasible title

12. Each of the following pairs of words or phrases describes the same extent of land *except*
    (A) 1 acre—43,560 square feet
    (B) 1 mile—5,280 linear feet
    (C) 1 square mile—460 acres
    (D) 1 section—640 acres

13. In order to sell property, one must
    (A) have a realtor's license
    (B) belong to a realtor's association
    (C) hire only realtors
    (D) none of the above

14. What are the broker's responsibilities under the 1968 Fair Housing Act?
    (A) to show all houses to all prospects
    (B) to treat all prospects equally
    (C) both A and B
    (D) neither A nor B

15. In order to be recorded, a deed must be in writing and must
    (A) be signed by the grantee
    (B) state the actual purchase price
    (C) be acknowledged
    (D) be free of all liens

16. Which of the following is *not* a good reason why real property taxes are so popular?
    (A) Real estate ownership is easy to hide.
    (B) Real estate is valuable.
    (C) Real estate is easy to find.
    (D) Real estate can be foreclosed to provide payment of unpaid tax levies.

17. The purpose of TRID is to
    (A) set the maximum interest rates that may be charged
    (B) let borrowers know the cost of credit
    (C) both A and B
    (D) neither A nor B

18. Which of the following contractual arrangements would be unenforceable?
    (A) A agrees to buy B's house.
    (B) A agrees with B that B shall steal money from C.
    (C) A agrees to find a buyer for B's car.
    (D) A agrees with B that B shall make restitution to C for money stolen by B.

19. The two parties to a lease contract are the
    (A) landlord and the serf
    (B) rentor and the rentee
    (C) lessor and the lessee
    (D) grantor and the grantee

20. When a person deliberately lies in order to mislead a fellow party to a contract, this act is
    (A) fraud
    (B) misrepresentation
    (C) legal if no third parties are involved
    (D) all right if the lie is not written into the contract

21. A contract that transfers possession but not ownership of property is
    (A) a special warranty deed
    (B) an option
    (C) an easement
    (D) a lease

22. Egbert owns a building in life estate. Upon Egbert's death, ownership of the building will go to Ethel. Ethel is a
    (A) remainderman
    (B) life tenant
    (C) common tenant
    (D) reversionary interest

23. A person owning an undivided interest in land with at least one other, and having the right of survivorship, is said to be a
    (A) tenant in common
    (B) tenant at will
    (C) joint tenant
    (D) tenant at sufferance

24. A person who has some rights to use land, but not all possessory rights, is said to have
    (A) an interest in land
    (B) an estate in land
    (C) a life estate in land
    (D) a tenancy in common

25. Which of the following is *not* corporeal property?
    (A) fee simple real estate
    (B) leasehold
    (C) easement
    (D) fixture

26. A listing contract that says the broker will receive a commission no matter who sells the property is called
    (A) an open listing
    (B) a net listing
    (C) an exclusive agency listing
    (D) an exclusive right to sell listing

27. Which of the following best describes an installment or land contract?
   (A) a contract to buy land only
   (B) a mortgage on land
   (C) a means of conveying title immediately while the purchaser pays for the property
   (D) a method of selling real estate whereby the purchaser pays for the property in regular installments while the seller retains title to the property until final payment

28. Ms. Maloney has a 3-month option on 20 acres at $2,000 per acre. She may
   (A) buy the property for $40,000
   (B) sell the option to another
   (C) not buy the land
   (D) all of the above

29. The mortgagor's right to reestablish ownership after delinquency is known as
   (A) reestablishment
   (B) satisfaction
   (C) equity of redemption
   (D) acceleration

30. A broker must place funds belonging to others in
   (A) his office safe, to which only he knows the combination
   (B) a safety deposit box
   (C) an account maintained by the Real Estate Commission
   (D) a trust, or escrow, account

31. A licensee's license can be revoked for
   (A) closing a deal
   (B) intentionally misleading someone into signing a contract that she ordinarily would not sign
   (C) submitting a ridiculous offer to a seller
   (D) all of the above

32. Capitalization is a process used to
   (A) convert income stream into a lump-sum capital value
   (B) determine cost
   (C) establish depreciation
   (D) determine potential future value

33. The number of square feet in 1 acre is
   (A) 64,000
   (B) 460
   (C) 440
   (D) 43,560

34. When changed surroundings cause an existing house to lose value, there is
   (A) physical deterioration
   (B) economic obsolescence
   (C) functional obsolescence
   (D) all of the above

35. An appraisal is
   (A) a forecast of value
   (B) an estimate of value
   (C) a prediction of value
   (D) a precise estimation of value

36. A lien for unpaid property taxes
   (A) can be sold at auction
   (B) takes priority over all other liens
   (C) cannot exist unless taxes are at least 36 months overdue
   (D) is a form of adverse possession

37. The prohibitions of the 1968 Fair Housing Act apply to privately owned housing when
   (A) a broker or other person engaged in selling or renting dwellings is used
   (B) discriminatory advertising is used
   (C) both A and B
   (D) neither A nor B

38. Under TRID, the lender must provide a Closing Disclosure form to the borrower at least __________ before closing.
   (A) 1 day
   (B) 3 days
   (C) 6 days
   (D) 1 week

39. A agrees to trade his car to B in exchange for a vacant lot that B owns.
   (A) This is a valid contractual agreement.
   (B) This is not a contract because no money changes hands.
   (C) This is not a contract because "unlike" items can't be traded.
   (D) This is not a valid contract because the car is titled in A's name.

40. A percentage lease requires the tenant to pay
   (A) a percentage of taxes and insurance
   (B) a percentage of net income as rent
   (C) a percentage of sales as rent
   (D) none of the above

41. A seller of real estate is also known as the
   (A) vendee
   (B) grantor
   (C) vendor
   (D) grantee

42. An estate at will is
(A) a limited partnership
(B) a tenancy of uncertain duration
(C) an inheritance by will
(D) a life tenancy

43. A hands B a deed with the intent to pass title and orally requests B not to record the deed until A dies. When is the deed valid?
(A) immediately
(B) when B records the deed
(C) when A dies
(D) never

44. A quitclaim deed conveys only the interest of the
(A) guaranteed
(B) property
(C) claimant
(D) grantor

45. A gross lease requires the tenant to pay rent based on
(A) gross sales
(B) net sales
(C) gross profit
(D) none of the above

46. The recording of a deed
(A) passes the title
(B) insures the title
(C) guarantees the title
(D) gives constructive notice of ownership

47. The law that requires most real estate contracts to be written to be enforceable is the
(A) Statute of Limitations
(B) Statute of Frauds
(C) Statute of Written Real Estate Agreements
(D) Law of Property

48. In the absence of an agreement to the contrary, the mortgage normally having priority will be the one that
(A) is for the greatest amount
(B) is a permanent mortgage
(C) was recorded first
(D) is a construction loan mortgage

49. A net listing is one
   (A) that requires the broker to seek a net price for the property
   (B) that is legal in all states
   (C) that most ethical brokers prefer to use
   (D) in which the broker's commission is the amount by which the sale price exceeds the agreed-upon net price the seller desires

50. Under TRID, a lender must provide
   (A) a Loan Estimate
   (B) a first mortgage loan used to purchase one's dwelling
   (C) a home-improvement loan more than 1 year old
   (D) none of the above

51. Which of the following does not terminate an agency relationship?
   (A) making an offer
   (B) death of either party
   (C) resignation of agent
   (D) destruction of subject matter

52. The 1968 Fair Housing Act protects against housing discrimination on the basis of
   (A) race and color
   (B) race, color, and religion
   (C) race, color, religion, and national origin
   (D) race, color, religion, national origin, sex, familial status, and handicap

53. A person who dies leaving no will is said to have died
   (A) intestate
   (B) without heirs
   (C) unbequeathed
   (D) unwillingly

54. From the point of view of the grantee, the safest kind of deed that can be received is a
   (A) general warranty deed
   (B) special warranty deed
   (C) quitclaim or release deed
   (D) trustee's deed

55. What is *not* an essential element of a valid contract?
   (A) offer and acceptance
   (B) capacity of participants
   (C) lack of ambiguity
   (D) legal objective

56. A freehold estate is
   (A) an estate acquired without paying anything
   (B) any leasehold
   (C) any estate wherein one may use the property as one wishes
   (D) an estate of uncertain duration

57. In order to do business, a licensee must
 (A) make proper application for a license
 (B) pass a licensing examination
 (C) have a license issued by the appropriate state agency
 (D) all of the above

58. When a loan is assumed on property that is sold,
 (A) the original borrower is relieved of further responsibility
 (B) the purchaser becomes liable for the debt
 (C) the purchaser must obtain a certificate of eligibility
 (D) all of the above

59. An estoppel certificate is often required when a
 (A) mortgage is sold to an investor
 (B) property is sold
 (C) property is being foreclosed
 (D) mortgage is assumed

60. Which of the following will not terminate a lease?
 (A) performance
 (B) breach
 (C) surrender
 (D) vacancy

61. The adjustment process in the direct sales comparison approach involves the principle of
 (A) contribution
 (B) diminishing returns
 (C) variable proportions
 (D) anticipation

62. A contract in which A agrees to allow B to use A's real estate in return for periodic payments of money by B is a
 (A) deed
 (B) contract of sale
 (C) lease
 (D) mortgage

63. Dower rights assure that
 (A) a husband receives a certain portion of his deceased wife's estate
 (B) wives and, in some states, children receive a certain portion of a deceased husband's or father's estate
 (C) a homeowner cannot lose his entire investment in his home
 (D) husbands and wives share equally in property acquired during marriage

64. When an individual holds property past the expiration of a lease without the landlord's consent, the leasehold estate she has is a
   (A) tenancy at sufferance
   (B) freehold estate
   (C) common of pasturage
   (D) holdover tenancy

65. A broker's unlicensed secretary
   (A) may sell property providing he does it under the broker's direct supervision
   (B) may sell or negotiate deals so long as he does not leave the office
   (C) may refer interested clients to the broker or her employed licensees
   (D) all of the above

66. An ARM is an
   (A) alternative reliable mode
   (B) automobile repair mechanic
   (C) adjustable-rate mortgage
   (D) authentic riparian model

67. Which of the following is *not* required of an agent with respect to his principal?
   (A) to be loyal
   (B) to act in person
   (C) to account for the agent's own personal finances
   (D) to act in the principal's best interests

68. Tenancy in severalty refers to
   (A) ownership by one person only
   (B) ownership by two persons only
   (C) ownership by at least three persons
   (D) a special form of joint ownership available only to married couples

69. A person who has permission to use land, but has no other rights, has
   (A) tenancy at sufferance
   (B) tenancy in common
   (C) license
   (D) fee simple estate

70. When tastes and standards cause an existing house to lose value, there is
   (A) physical deterioration
   (B) economic obsolescence
   (C) functional obsolescence
   (D) all of the above

71. A rule-of-thumb method for determining the price a wage earner can afford to pay for a home is to multiply her annual income by
   (A) 1½
   (B) 2½
   (C) 4
   (D) 6

72. A real estate license, once received,
   (A) remains in effect indefinitely
   (B) is good for a limited period of time
   (C) must be filed in county records
   (D) may be inherited by the holder's spouse

73. Which of the following is most accurately described as personal property?
   (A) a fixture
   (B) a chattel
   (C) an improvement
   (D) realty

74. A real estate broker is a
   (A) general agent
   (B) special agent
   (C) secret agent
   (D) travel agent

75. A broker may use escrow moneys held on behalf of others
   (A) for collateral for business loans
   (B) for collateral for personal loans
   (C) for salary advances to licensees in his employ
   (D) none of the above

76. Which of the following forms of deeds has one or more guarantees of title?
   (A) quitclaim
   (B) executor's
   (C) warranty
   (D) special form

77. The highest price a buyer is willing, but not compelled, to pay and the lowest price a seller is willing, but not compelled, to accept is
   (A) estimated value
   (B) economic value
   (C) marginal value
   (D) market value

78. To be valid, a deed need *not* necessarily be
(A) signed
(B) written
(C) sealed
(D) delivered

79. A person who receives title to land by virtue of having used and occupied it for a certain period of time, without actually paying the previous owner for it, receives title by
(A) will
(B) descent
(C) alienation
(D) adverse possession

80. A conventional mortgage is
(A) amortizing
(B) guaranteed by the FHA
(C) not guaranteed by a government agency
(D) approved by the VA

81. The party to whom a deed conveys real estate is the
(A) grantee
(B) grantor
(C) beneficiary
(D) recipient

82. Which is *not* considered a permanent attachment to land?
(A) trees
(B) fixtures
(C) chattels
(D) anything built on the land

83. The main appeal of VA mortgages to borrowers lies in
(A) low interest rates
(B) minimum down payments
(C) unlimited mortgage ceiling
(D) easy availability

84. For which reason is a deed recorded?
(A) to insure certain title
(B) to give notice to the world
(C) to meet a state requirement
(D) to save title insurance cost

85. Ownership of real property is transferred
(A) when the grantor signs the deed
(B) when the grantor's signature has been notarized
(C) when the deed is delivered
(D) when the correct documentary stamps are put on the deed and canceled

86. A valid contract of purchase or sale of real property must be signed by the
(A) broker
(B) agent and seller
(C) seller only
(D) buyer and seller

87. A sublease is a
(A) lease made by a lessor
(B) lease made by a lessee and a third party
(C) lease for basement space
(D) condition of property

88. "Hand money" paid upon the signing of an agreement of sale is called
(A) an option
(B) a recognizance
(C) earnest money
(D) a freehold estate

89. Market value appraisals assume that
(A) the purchaser pays all cash (no mortgage financing)
(B) FHA or VA financing is employed
(C) the appraiser can determine the types of financing involved
(D) the financing, if any, is on terms generally available in that area

90. A licensed salesperson
(A) must work under the supervision of a broker
(B) can collect commission payments only from his broker
(C) must have his license held by his employing broker
(D) all of the above

91. If the property is assessed at $180,000 and the millage rate is 32.5, the tax is
(A) $58.50
(B) $585
(C) $5,850
(D) $58,500

92. Smith sold three lots for a total of $90,000. The first lot sold for 1½ times the price of the second lot. The second lot sold for twice the price of the third lot. How much did the first lot sell for?
(A) $15,000
(B) $30,000
(C) $45,000
(D) $90,000

93. Aaronson owns a 44-acre tract of land. In order to develop it, he must set aside 10% of the area for parks and must use 6.6 acres for streets, drainage, and other uses. If the minimum permissible lot size is 7,500 sq. ft., what is the maximum possible number of lots Aaronson can lay out?
(A) 191
(B) 180
(C) 293
(D) 81

94. Velez borrowed $7,500 for 4 years, paying interest every quarter. The total amount of interest she paid was $2,700. What was the annual interest rate?
(A) 9%
(B) 10%
(C) 27%
(D) 36%

95. Jim sold a parcel of land for $164,450. He made a profit of 43%. What was his purchase price?
(A) $82,755.00
(B) $93,736.50
(C) $115,000.00
(D) $164,450.00

96. Anne referred a customer to Connie, who sold the customer a house for $139,500. Connie paid Anne a referral fee of 12% of Connie's commission. If Anne received $1,171.80, what was the rate of the commission on Connie's sale?
(A) 5%
(B) 6%
(C) 6½%
(D) 7%

97. What monthly rent must Sam get on his land in order to earn an annual return of 11% on the $210,000 that he paid for the land? Assume that the tenant pays all property taxes.
(A) $275
(B) $1,552
(C) $1,925
(D) $23,100

98. Smith bought three lots for $10,000 each. If they increase in value by 25% each year *compounded*, in how many years will she be able to sell one of the lots for as much as she originally paid for all three?
    (A) 4
    (B) 5
    (C) 6
    (D) 7

99. How many acres are contained in a rectangular tract of land measuring 1,700 ft. by 2,100 ft.?
    (A) 81.96
    (B) 92.91
    (C) 112.01
    (D) 115.00

100. A backyard measures 100 ft. by 80 ft. The house is 40 ft. wide. A fence 4 ft. high is to be built around the yard with the width of the house as part of the barrier. If fence fabric is $1.80 per square yard, how much will the necessary fence fabric cost?
    (A) $64
    (B) $256
    (C) $288
    (D) $2,304

ANSWER KEY
Model Test 4

1. **B**	26. **D**	51. **A**	76. **C**
2. **D**	27. **D**	52. **D**	77. **D**
3. **D**	28. **D**	53. **A**	78. **C**
4. **B**	29. **C**	54. **A**	79. **D**
5. **D**	30. **D**	55. **C**	80. **C**
6. **B**	31. **B**	56. **D**	81. **A**
7. **B**	32. **A**	57. **D**	82. **C**
8. **B**	33. **D**	58. **B**	83. **B**
9. **C**	34. **B**	59. **A**	84. **B**
10. **D**	35. **B**	60. **D**	85. **C**
11. **C**	36. **B**	61. **A**	86. **D**
12. **C**	37. **C**	62. **C**	87. **B**
13. **D**	38. **B**	63. **B**	88. **C**
14. **B**	39. **A**	64. **A**	89. **D**
15. **C**	40. **C**	65. **C**	90. **D**
16. **A**	41. **C**	66. **C**	91. **C**
17. **B**	42. **B**	67. **C**	92. **C**
18. **B**	43. **A**	68. **A**	93. **A**
19. **C**	44. **D**	69. **C**	94. **A**
20. **A**	45. **D**	70. **C**	95. **C**
21. **D**	46. **D**	71. **B**	96. **D**
22. **A**	47. **B**	72. **B**	97. **C**
23. **C**	48. **C**	73. **B**	98. **B**
24. **A**	49. **D**	74. **B**	99. **A**
25. **C**	50. **A**	75. **D**	100. **B**

ANSWER EXPLANATIONS—ARITHMETIC QUESTIONS

NOTE: The section contains explanations for only the arithmetic questions in the model exam.

91. The millage rate of 32.5 is 3.25%, so tax on \$180,000 is

$$0.0325 \times \$180,000 = \$5,850$$

92. Let the price of the *third* lot be P. The price of the second lot, then, is $2 \times P$. The price of the first lot is $1\frac{1}{2} \times 2 \times P$. All three cost \$90,000, so

$$P + (2 \times P) + (1\frac{1}{2} \times 2 \times P) = \$90,000, \text{ or } 6 \times P = \$90,000$$

Therefore, $P = \$90,000 \div 6 = \$15,000$. The price of the second lot is twice that, or \$30,000. The third lot is 1½ times the second lot, or \$45,000.

93. If Aaronson uses 10% of the 44 acres for parks, he will use 4.4 acres. Add that to the 6.6 other acres that must be used for streets, etc., and you get a total of 11 acres not devoted to lots. This leaves 33 acres, or 33 × 43,560 sq. ft. = 1,437,480 sq. ft. for lots. If each lot must be at least 7,500 sq. ft., then the maximum number of lots is

$$1,437,480 \text{ sq. ft.} \div 7,500 \text{ sq. ft.} = 191.664, \text{ or 191 full lots}$$

94. Velez paid \$2,700 interest over 4 years; this comes to \$675 per year (\$2,700 ÷ 4). The annual interest rate was

$$\$675 \div \$7,500 = 0.09 = 9\%$$

95. If Jim made a profit of 43%, then he sold the land for 143% of its purchase price. The purchase price was

$$\$164,450 \div 1.43 = \$115,000$$

96. Anne's \$1,171.80 was 12% of Connie's commission, which must have been \$1,171.80 ÷ 0.12 = \$9,765. If the sale price was \$139,500, the rate of commission was

$$\$9,765 \div \$139,500 = 0.07 = 7\%$$

97. Eleven percent of \$210,000 is \$23,100.

$$\$23,100 \div 12 = \$1,925 \text{ monthly}$$

98. How many years will be required for \$10,000 to compound to \$30,000 at 25% per year?

One year:	1.25 × \$10,000 = \$12,500
Two years:	1.25 × \$12,500 = \$15,625
Three years:	1.25 × \$15,625 = \$19,531
Four years:	1.25 × \$19,531 = \$24,414
Five years:	1.25 × \$24,414 = \$30,518

It will take about 5 years.

99. Area = 1,700 ft. × 2,100 ft. = 3,570,000 sq. ft.

$$3,570,000 \text{ sq. ft.} \div 43,560 \text{ sq. ft.} = 81.96 \text{ acres}$$

100. The backyard is 100 ft. + 80 ft. + 100 ft. + 80 ft. = 360 ft. around. From this we subtract the 40 ft. taken up by the house to get 320 lineal ft. of fence needed. If the fence is 4 ft. high, it will need 320 ft. × 4 ft. = 1,280 sq. ft. of fence fabric; 1,280 sq. ft. is 142.222 sq. yd.

$$142.222 \text{ sq. yd.} \times \$1.80 = \$256.00 \text{ total cost}$$

ANSWER SHEET
Model Examination 5

Broker and Salesperson

1. Ⓐ Ⓑ Ⓒ Ⓓ	26. Ⓐ Ⓑ Ⓒ Ⓓ	51. Ⓐ Ⓑ Ⓒ Ⓓ	76. Ⓐ Ⓑ Ⓒ Ⓓ
2. Ⓐ Ⓑ Ⓒ Ⓓ	27. Ⓐ Ⓑ Ⓒ Ⓓ	52. Ⓐ Ⓑ Ⓒ Ⓓ	77. Ⓐ Ⓑ Ⓒ Ⓓ
3. Ⓐ Ⓑ Ⓒ Ⓓ	28. Ⓐ Ⓑ Ⓒ Ⓓ	53. Ⓐ Ⓑ Ⓒ Ⓓ	78. Ⓐ Ⓑ Ⓒ Ⓓ
4. Ⓐ Ⓑ Ⓒ Ⓓ	29. Ⓐ Ⓑ Ⓒ Ⓓ	54. Ⓐ Ⓑ Ⓒ Ⓓ	79. Ⓐ Ⓑ Ⓒ Ⓓ
5. Ⓐ Ⓑ Ⓒ Ⓓ	30. Ⓐ Ⓑ Ⓒ Ⓓ	55. Ⓐ Ⓑ Ⓒ Ⓓ	80. Ⓐ Ⓑ Ⓒ Ⓓ
6. Ⓐ Ⓑ Ⓒ Ⓓ	31. Ⓐ Ⓑ Ⓒ Ⓓ	56. Ⓐ Ⓑ Ⓒ Ⓓ	81. Ⓐ Ⓑ Ⓒ Ⓓ
7. Ⓐ Ⓑ Ⓒ Ⓓ	32. Ⓐ Ⓑ Ⓒ Ⓓ	57. Ⓐ Ⓑ Ⓒ Ⓓ	82. Ⓐ Ⓑ Ⓒ Ⓓ
8. Ⓐ Ⓑ Ⓒ Ⓓ	33. Ⓐ Ⓑ Ⓒ Ⓓ	58. Ⓐ Ⓑ Ⓒ Ⓓ	83. Ⓐ Ⓑ Ⓒ Ⓓ
9. Ⓐ Ⓑ Ⓒ Ⓓ	34. Ⓐ Ⓑ Ⓒ Ⓓ	59. Ⓐ Ⓑ Ⓒ Ⓓ	84. Ⓐ Ⓑ Ⓒ Ⓓ
10. Ⓐ Ⓑ Ⓒ Ⓓ	35. Ⓐ Ⓑ Ⓒ Ⓓ	60. Ⓐ Ⓑ Ⓒ Ⓓ	85. Ⓐ Ⓑ Ⓒ Ⓓ
11. Ⓐ Ⓑ Ⓒ Ⓓ	36. Ⓐ Ⓑ Ⓒ Ⓓ	61. Ⓐ Ⓑ Ⓒ Ⓓ	86. Ⓐ Ⓑ Ⓒ Ⓓ
12. Ⓐ Ⓑ Ⓒ Ⓓ	37. Ⓐ Ⓑ Ⓒ Ⓓ	62. Ⓐ Ⓑ Ⓒ Ⓓ	87. Ⓐ Ⓑ Ⓒ Ⓓ
13. Ⓐ Ⓑ Ⓒ Ⓓ	38. Ⓐ Ⓑ Ⓒ Ⓓ	63. Ⓐ Ⓑ Ⓒ Ⓓ	88. Ⓐ Ⓑ Ⓒ Ⓓ
14. Ⓐ Ⓑ Ⓒ Ⓓ	39. Ⓐ Ⓑ Ⓒ Ⓓ	64. Ⓐ Ⓑ Ⓒ Ⓓ	89. Ⓐ Ⓑ Ⓒ Ⓓ
15. Ⓐ Ⓑ Ⓒ Ⓓ	40. Ⓐ Ⓑ Ⓒ Ⓓ	65. Ⓐ Ⓑ Ⓒ Ⓓ	90. Ⓐ Ⓑ Ⓒ Ⓓ
16. Ⓐ Ⓑ Ⓒ Ⓓ	41. Ⓐ Ⓑ Ⓒ Ⓓ	66. Ⓐ Ⓑ Ⓒ Ⓓ	91. Ⓐ Ⓑ Ⓒ Ⓓ
17. Ⓐ Ⓑ Ⓒ Ⓓ	42. Ⓐ Ⓑ Ⓒ Ⓓ	67. Ⓐ Ⓑ Ⓒ Ⓓ	92. Ⓐ Ⓑ Ⓒ Ⓓ
18. Ⓐ Ⓑ Ⓒ Ⓓ	43. Ⓐ Ⓑ Ⓒ Ⓓ	68. Ⓐ Ⓑ Ⓒ Ⓓ	93. Ⓐ Ⓑ Ⓒ Ⓓ
19. Ⓐ Ⓑ Ⓒ Ⓓ	44. Ⓐ Ⓑ Ⓒ Ⓓ	69. Ⓐ Ⓑ Ⓒ Ⓓ	94. Ⓐ Ⓑ Ⓒ Ⓓ
20. Ⓐ Ⓑ Ⓒ Ⓓ	45. Ⓐ Ⓑ Ⓒ Ⓓ	70. Ⓐ Ⓑ Ⓒ Ⓓ	95. Ⓐ Ⓑ Ⓒ Ⓓ
21. Ⓐ Ⓑ Ⓒ Ⓓ	46. Ⓐ Ⓑ Ⓒ Ⓓ	71. Ⓐ Ⓑ Ⓒ Ⓓ	96. Ⓐ Ⓑ Ⓒ Ⓓ
22. Ⓐ Ⓑ Ⓒ Ⓓ	47. Ⓐ Ⓑ Ⓒ Ⓓ	72. Ⓐ Ⓑ Ⓒ Ⓓ	97. Ⓐ Ⓑ Ⓒ Ⓓ
23. Ⓐ Ⓑ Ⓒ Ⓓ	48. Ⓐ Ⓑ Ⓒ Ⓓ	73. Ⓐ Ⓑ Ⓒ Ⓓ	98. Ⓐ Ⓑ Ⓒ Ⓓ
24. Ⓐ Ⓑ Ⓒ Ⓓ	49. Ⓐ Ⓑ Ⓒ Ⓓ	74. Ⓐ Ⓑ Ⓒ Ⓓ	99. Ⓐ Ⓑ Ⓒ Ⓓ
25. Ⓐ Ⓑ Ⓒ Ⓓ	50. Ⓐ Ⓑ Ⓒ Ⓓ	75. Ⓐ Ⓑ Ⓒ Ⓓ	100. Ⓐ Ⓑ Ⓒ Ⓓ

Model Examination 5

BROKER AND SALESPERSON

See page 578 for the answer key. Math problems are worked out on pages 579–580.

1. An appraisal is
   (A) a forecast of value
   (B) a prediction of value
   (C) an appraiser's opinion of value
   (D) a statement of exact value

2. The number of square feet in 1 acre is
   (A) 640
   (B) 45,630
   (C) 43,560
   (D) 53,460

3. How is the *gross rent multiplier* calculated?
   (A) market value/market rental
   (B) monthly payment/market rental
   (C) market rental/market price
   (D) sales price/market price

4. A perfectly rectangular tract of land contains exactly 10.6 acres. The measurement on one side is 181 ft. To the nearest foot, how deep is the tract?
   (A) 255 ft.
   (B) 385 ft.
   (C) 1,817 ft.
   (D) 2,551 ft.

5. What is the maximum number of 8,000-sq.-ft. lots that can be platted from a 17.1-acre tract if 19% of the land must be used for streets and parks?
   (A) 17
   (B) 62
   (C) 75
   (D) 93

6. Laws that set minimum construction standards are
   (A) building codes
   (B) zoning codes
   (C) environmental laws
   (D) condemnation laws

7. Which of the following is *not* an appraisal approach?
   (A) cost
   (B) trade
   (C) sales comparison
   (D) income

8. Value is determined by
   (A) supply and demand
   (B) asking prices
   (C) interest rates
   (D) active brokers

9. Permission for land use not normally permitted by the zoning classification of the property is
   (A) a differential
   (B) a zone change
   (C) a variance
   (D) an egregement

10. Smith sells a tract of land 400 ft. by 665 ft. for $17,100. To the nearest dollar, what is the price per acre?
    (A) $2,800
    (B) $2,950
    (C) $3,117
    (D) $3,228

11. In Green's city, the property tax rate is 71 mills, based upon 30% of the market value of the property. Green just paid $170,000 for a home. What will the tax be?
    (A) $3,621
    (B) $4,470
    (C) $7,317
    (D) $12,070

12. When changes in taste cause kitchen fixtures to become less desirable, there is
    (A) physical deterioration
    (B) functional obsolescence
    (C) economic obsolescence
    (D) all of the above

13. The income approach is generally most suitable for appraising
    (A) commercial and investment property
    (B) single-family homes
    (C) heavily mortgaged property
    (D) heavily insured property

Questions 14–16 concern the following situation:

Mr. Jones died, leaving Mrs. Jones the right to use, occupy, and enjoy his real estate until her death, at which time their son Willis would receive a fee simple estate in the real estate.

14. Mrs. Jones is a
    (A) remainderman
    (B) life tenant
    (C) joint tenant
    (D) tenant in common

15. Willis is a
    (A) remainderman
    (B) life tenant
    (C) joint tenant
    (D) tenant in common

16. Mrs. Jones has a
    (A) fee simple estate in joint tenancy with Willis
    (B) fee simple estate as tenant in common with Willis
    (C) life estate
    (D) reversionary interest

---

17. A building 40 ft. by 22 ft. has exterior walls 10 ft. high. There are 13 windows, 4 sq. ft. each, and a door measuring 48 sq. ft. One gallon of paint covers 400 sq. ft. If the door and the windows are not to be painted, how many gallons of paint are needed to give the walls *two* full coats?
    (A) 2.85
    (B) 5.7
    (C) 3.1
    (D) 6.2

18. Bernini borrows $12,000. He pays $210 each quarter in interest. What is the annual interest rate on this loan?
    (A) 1.75%
    (B) 5.75%
    (C) 7%
    (D) 8.5%

19. A freehold estate is
    (A) an estate acquired without paying anything for it
    (B) an estate always acquired by adverse possession
    (C) an estate of uncertain duration
    (D) any leasehold estate

20. A person who appears to own a piece of real estate, but actually does not, has
    (A) good title
    (B) recorded title
    (C) constructive notice
    (D) color of title

21. A real estate broker is
    (A) a general agent
    (B) an attorney-in-fact
    (C) a special agent
    (D) an agent provocateur

22. Contracts made by a minor are
    (A) void
    (B) voidable by the minor
    (C) voidable by all parties
    (D) voidable by adult parties

23. The parties to a deed are
    (A) vendor and vendee
    (B) testator and testatee
    (C) grantor and grantee
    (D) offeror and offeree

24. Ms. Levy has the choice of renting a house for $1,800 per month, including utilities, or for $1,400 per month if she pays the utility bills. Average utility bills are $3,840 per year, but Ms. Levy feels she can reduce this by 35%. How much *per month* does she expect to save by paying the utility bills herself?
    (A) $0
    (B) $120
    (C) $144
    (D) $192

25. Salesman O'Hara gets 50% of the first $50,000 of commissions he brings in and 60% of all above that amount in 1 year. Last year he sold $2,750,000 of real estate, all at 7% commission. How much did O'Hara get to keep?
    (A) $192,500
    (B) $115,500
    (C) $110,500
    (D) $85,500

26. A quitclaim deed conveys only the interest of the
   (A) grantor
   (B) claimant
   (C) quittor
   (D) property

27. Which of the following will *not* terminate a lease?
   (A) breach
   (B) performance
   (C) vacancy
   (D) surrender

28. One discount point is equal to
   (A) 1% of the sale price
   (B) 1% of the loan amount
   (C) 1% of the interest rate
   (D) none of the above

29. Which of the following is *not* covered by title insurance?
   (A) forged deed
   (B) deed by incompetent
   (C) tornado damage
   (D) undisclosed heirs

30. Ownership of realty by one person is
   (A) tenancy in severalty
   (B) conjoint tenancy
   (C) tenancy by the entirety
   (D) tenancy sole

31. Green made an offer to purchase real estate from Blue. The offer gave Blue 7 days to consider it. Two days later, without hearing anything from Blue, Green found another property that he liked better. Green then wanted to withdraw his offer to Blue.
   (A) Green had to wait until the 7 days had passed.
   (B) Green was required to notify Blue of his desire to withdraw the offer and to give Blue "reasonable time" to accept or reject it.
   (C) Green could withdraw the offer immediately.
   (D) none of the above

32. The secondary mortgage market is the market
   (A) for second mortgages
   (B) in which existing mortgages are bought and sold
   (C) in which junior mortgages are originated
   (D) for older, low-interest loan assumptions

33. A landowner leases her land to a lessee, who in turn leases the land to a sublessee. Who holds the *sandwich lease*?
(A) landowner
(B) lessee
(C) sublessee
(D) none of the above

34. The covenant whereby a person warrants that he is the possessor and owner of property being conveyed is the covenant of
(A) seizin
(B) habendum
(C) possession
(D) further assurance

35. Title to land passes
(A) on the date shown on the deed
(B) upon recordation of the deed
(C) when the deed is signed
(D) upon delivery of the deed

36. Provisions for defeat of the mortgage are found in the __________ clause.
(A) alienation
(B) acceleration
(C) foreclosure
(D) defeasance

37. All mortgages are
(A) due on sale
(B) security for a debt
(C) recorded
(D) none of the above

38. A fixture is
(A) anything that cannot be removed from the real estate without leaving a hole
(B) anything the property owner says is a fixture
(C) anything necessary to the proper and efficient use of the real estate
(D) none of the above

39. Byrd lives in an apartment owned by Lyon. The lease has expired, but Byrd has stayed on with Lyon's permission and continues to pay rent to Lyon. This is an example of tenancy
(A) by remainder
(B) at will
(C) at suffrance
(D) in common

40. Which of the following is *not* real estate?
(A) a flagpole affixed to a house
(B) a tomato crop not ready to harvest
(C) a greenhouse
(D) all of the above *are* real estate

41. The law that requires that all transfers of ownership rights to real estate be in writing is the
(A) Statute of Frauds
(B) Statute of Limitations
(C) Parol Evidence Rule
(D) Statute of Liberties

42. After paying a 6½% commission, Jackson received net proceeds of $336,132.50 on the sale of his home. What was the sale price?
(A) $21,848.60
(B) $357,981.10
(C) $359,500.00
(D) $375,000.00

43. When a valid lease of real estate exists and the rent is paid on time, which of the following is false?
(A) The lessor cannot move in and use the property until the lease expires.
(B) The lessee possesses the right of occupancy.
(C) The lessee holds the fee to the real estate.
(D) The lessor has a reversionary interest in the real estate.

44. The clause in a mortgage note that allows the lender to demand immediate payment in full of the remaining balance, if payment of the note is not made as contracted, is the __________ clause.
(A) alienation
(B) acceleration
(C) foreclosure
(D) amortization

45. All of the following should be recorded *except*
(A) an easement
(B) a 20-year lease
(C) a mortgage
(D) a 6-month lease

46. In a typical mortgage loan transaction, the mortgagee is the
(A) lender
(B) borrower
(C) appraiser
(D) closing agent

47. Which of the following has access to documents that have been recorded in the public records?
(A) prospective buyers
(B) prospective lenders
(C) appraisers
(D) all of the above

48. Ms. Anderson sold a tract of land. After paying a 9% commission and paying 3% of the selling price in taxes, fees, and closing costs, she ended up with $176,000. What was the sale price of the land?
(A) $181,280
(B) $191,840
(C) $197,120
(D) $200,000

49. Farley's warehouse measures 52 ft. by 36 ft. The walls are 9 in. thick, and there are 13 support pillars inside the building, each 6 in. by 6 in. What is the area, in square feet, of the *net* interior floor space?
(A) 1,739
(B) 1,742 1/4
(C) 1,803 5/16
(D) 1,806 9/16

50. Title that establishes ownership of real estate in a reasonably clear manner upon examination of the public records is __________ title.
(A) owner's
(B) marketable
(C) equitable
(D) quiet

51. A certain property has no road frontage; however, there exists a recorded right of access across a neighboring parcel of land. This right is
(A) a conditional title
(B) an entroversion
(C) an easement
(D) a defeasance

52. An estate that may be terminated by any party at any time is an estate
(A) for years
(B) at will
(C) in possession
(D) in termination

53. If the commission in a listing contract is arranged so that the broker receives all of the purchase price in excess of a certain figure, the listing is a(n) __________ listing.
(A) open
(B) net
(C) gross
(D) multiple

54. Which of the following does *not* have the right of survivorship?
(A) tenant in common
(B) joint tenant
(C) tenant by the entirety
(D) all *have* the right of survivorship

55. In a VA loan transaction, to whom is the discount (if any) paid?
(A) the VA
(B) the lender
(C) the broker
(D) the buyer

56. What agency administers VA loan guarantees?
(A) Veterans Authority
(B) Virginia Association of Realtors
(C) Department of Veterans Affairs
(D) Veterinarians Association

57. Because his business was robbed, Montez can't afford rent on a building he leased for his business. Which of the following is true?
(A) The lease is no longer enforceable.
(B) Montez does not have to pay the rent any longer.
(C) Montez may deduct the loss from the rent.
(D) Montez must continue to pay the rent until the lease expires.

58. Which of the following is neither real estate nor a fixture?
(A) ceiling
(B) chandelier on ceiling
(C) lightbulb in chandelier
(D) light switch

59. Real estate brokers and salespeople should be familiar with
(A) common law
(B) agency law
(C) license law
(D) all of the above

60. Which of the following is *not* always a right possessed by an owner of a fee simple estate?
    (A) possession
    (B) easement
    (C) occupancy
    (D) disposition

61. Which of the following would *not* be considered commercial real estate?
    (A) office building
    (B) condominium home
    (C) shopping center
    (D) doctors' office complex

62. Smith wishes to develop a subdivision containing 55 lots averaging 10,500 sq. ft. each. An average of 1,380 sq. ft. of street, sidewalk, and other space must be provided for each lot. What is the minimum number of acres that Smith will need?
    (A) 12
    (B) 14
    (C) 15
    (D) 17

63. Which of the following instructions from a seller may *not* be complied with by a real estate broker?
    (A) Don't put up a for sale sign in the yard.
    (B) Don't present offers for less than the listing price.
    (C) Don't show the property to persons who do not speak English.
    (D) Don't show the property on religious holidays observed by seller.

64. Real estate brokers' commissions usually are set by
    (A) the local Board of Realtors
    (B) the laws of the state
    (C) agreement with other brokerage firms
    (D) agreement between property owner and broker

65. When a licensed salesperson advertises property for sale,
    (A) the name of the broker must be mentioned
    (B) only the salesperson's name must be mentioned
    (C) the name of the property owner must be revealed
    (D) the price must not be mentioned

66. Which form of listing does *not* allow the seller to sell her listed property by herself without paying the listing broker a commission?
    (A) open listing
    (B) exclusive right to sell listing
    (C) multiple listing
    (D) exclusive agency listing

67. A broker has a listing the terms of which allow him to earn a commission only if he finds a buyer before the seller or another broker does. This is an example of
(A) an open listing
(B) an exclusive right to sell listing
(C) a net listing
(D) an exclusive agency listing

68. Part of an addition that Sam builds to his home turns out to be on land belonging to June. This is an example of
(A) accretion
(B) riparian rights
(C) easement
(D) encroachment

69. Tina has the legal right to use Fred's driveway for her lifetime. Tina has
(A) a life estate
(B) an easement in gross
(C) a license to use
(D) a riparian right

70. Simpson wishes to install wall-to-wall carpeting in his home, which measures 42 × 28 ft. The outside walls are 6 in. thick, and the interior walls are 4 in. thick. There are 135 linear ft. of wall inside the home. The two bathrooms (each with interior measurements of 9 ft. × 6 ft.) and the kitchen (12 ft. × 12 ft.) are not to be carpeted. If installed carpeting costs $13.95 per square yard, how much will the job cost?
(A) $1,255.50
(B) $1,222.95
(C) $1,506.60
(D) $1,715.85

71. Ms. Farrell contracts to buy a house for $330,000. She pays $9,000 in earnest money and applies for a 75% loan. How much more money will she need to make up the purchase price of the home?
(A) $80,250
(B) $73,500
(C) $64,500
(D) $29,250

72. Leo allows Lemmie to stay in Leo's house while Leo is on vacation, but Lemmie must leave when Leo returns. Lemmie has
(A) a short-term lease
(B) a proprietary right
(C) a license
(D) an easement

73. An estate that has an indefinite duration is
    (A) a freehold estate
    (B) a renewable leasehold
    (C) an estate at suffrance
    (D) a nonfreehold estate

74. How many 100 ft. × 100 ft. lots can be made out of a 3-acre plot of land?
    (A) 6
    (B) 11
    (C) 13
    (D) 43

75. A lot sold for $225 a front foot. If the lot was 96 ft. deep and had an area of 6,336 sq. ft., how much did the lot sell for?
    (A) $11,770
    (B) $14,256
    (C) $14,850
    (D) $21,600

76. A "legal description" of land must
    (A) be sufficient to identify the property
    (B) carry measurements to the nearest inch
    (C) show the area of the land being described
    (D) all of the above

77. Increase in the size of a plot of land because of soil deposited by the flow of a stream is called
    (A) accretion
    (B) riparian right
    (C) depletion
    (D) amortization

78. If an offer is *rescinded,* it is
    (A) revised
    (B) altered
    (C) terminated
    (D) accepted

79. Which of the following is *not* necessarily required in a deed?
    (A) acknowledgment
    (B) grantee's signature
    (C) date
    (D) description of real estate

80. Government's right to regulate land use derives from
    (A) escheat
    (B) just compensation
    (C) police power
    (D) eminent domain

81. TRID requires what information be provided to the borrower?
    (A) Loan Estimate
    (B) Closing Disclosure
    (C) both Loan Estimate and Closing Disclosure
    (D) neither Loan Estimate nor Closing Disclosure

82. If the following transactions involve a new first mortgage, to which would TRID apply?
    (A) summer house to be occupied by purchaser
    (B) house purchased by an investor to remodel and resell
    (C) 50-unit apartment building
    (D) 50-acre farm with home for use by tenant farmer

83. A home seller agrees to take back a $25,000 second mortgage note on the sale. On the closing statement, this is a
    (A) credit to the buyer, debit to the seller
    (B) credit to the buyer, credit to the seller
    (C) debit to the buyer, credit to the seller
    (D) debit to the buyer, debit to the seller

84. TRID is an acronym for
    (A) TILA-RESPA Integrated Disclosure
    (B) thanks regarding interest development
    (C) troubled real estate interest deficits
    (D) third realty investment disclosures

85. Under TRID,
    (A) borrowers can shop for loans
    (B) discount points may be not more than 2
    (C) the buyer may void the transaction up to ten days after closing
    (D) lenders may not charge more than 1% for lender fees

86. Which of the following is *not* a consideration of TRID?
    (A) Closing Disclosures
    (B) Loan Estimates
    (C) Borrower Disclosures
    (D) commissions of real estate brokers

87. When the zoning of a property changes, uses that were once compliant with zoning are
    (A) terminated because they are no longer compliant
    (B) allowed to continue as nonconforming uses
    (C) granted a variance
    (D) allowed to continue only if changes are made to comply with the new zoning

88. Which of the following terms can be included in a real estate ad with no further disclosure required?
    (A) "Take eight years to pay"
    (B) "Only $10,000 down"
    (C) "Payments of $500 per month"
    (D) "6% annual percentage rate"

89. A lender is required by TRID to provide to an applicant for a real estate mortgage
    (A) good title
    (B) his FICO score
    (C) a Loan Estimate
    (D) an amortization schedule

90. The law that requires a lender to disclose to a potential borrower the true expenses of borrowing is the
    (A) Equal Credit Opportunity Act
    (B) Interstate Land Sales Full Disclosure
    (C) Real Estate Settlement Procedures Act
    (D) TRID

91. A broker is advertising property for a homebuilder. Which phrase is *not* acceptable?
    (A) "New homes for $350,000"
    (B) "Veterans may apply for loans with nothing down"
    (C) "No down payment required for veterans"
    (D) "Near schools and shopping"

92. What is the fee to retire a loan early?
    (A) the amount of attorney fees for mortgage preparation
    (B) the mortgage lender's fee for title insurance
    (C) a limitation on how often the mortgage can be sold
    (D) mortgage prepayment penalty

93. A handy person is in the business of buying rundown homes, fixing them up, and reselling them at a profit. He buys a single-family home for $110,000. Under the Truth in Lending Act, the disclosure of discount points
    (A) does not apply to a business loan
    (B) is not required for loans over $100,000
    (C) is required for all single-family homes
    (D) is required only when more than 2 points is charged

94. Under TRID, willful violation of disclosure requirements can result in
    (A) fines
    (B) imprisonment
    (C) both fines and imprisonment
    (D) license suspension

95. The Comprehensive Environmental Response, Compensation, and Liability Act (CERCLA) is enforced by the
    (A) Department of Homeland Security
    (B) Department of Housing and Urban Development (HUD)
    (C) Federal Emergency Management Agency (FEMA)
    (D) Environmental Protection Agency (EPA)

96. When land is partially in a designated wetland area,
    (A) the value of the land is enhanced by the abundance of wildlife
    (B) the land is worthless
    (C) Superfund will buy it
    (D) this is a material fact to be disclosed to a potential buyer

97. A buyer of a house built before 1978 must be advised of the possibility that it may contain
    (A) asbestos
    (B) radon
    (C) lead-based paint
    (D) termites

98. EPA stands for
    (A) earnings per annum
    (B) effort put away
    (C) extra-pretty agent
    (D) Environmental Protection Agency

99. Townhouses intended to be sold in a group mainly to investors and to be used by transients for temporary vacation homes are pooled into a rental group to be professionally managed. They are often sold as
    (A) securities
    (B) tenancy in common
    (C) joint tenants
    (D) fee simple

100. A property was used for retail sales, but the zoning has been changed to residential. The property can continue its retail use but is
    (A) limited
    (B) biased
    (C) nonconforming
    (D) spot zoning

ANSWER KEY
Model Test 5

1. C	26. A	51. C	76. A
2. C	27. C	52. B	77. A
3. A	28. B	53. B	78. C
4. D	29. C	54. A	79. B
5. C	30. A	55. B	80. C
6. A	31. C	56. C	81. C
7. B	32. B	57. D	82. A
8. A	33. B	58. C	83. A
9. C	34. A	59. D	84. A
10. A	35. D	60. B	85. A
11. A	36. D	61. B	86. D
12. B	37. B	62. C	87. B
13. A	38. D	63. C	88. D
14. B	39. B	64. D	89. C
15. A	40. B	65. A	90. D
16. C	41. A	66. B	91. C
17. B	42. C	67. A	92. D
18. C	43. C	68. D	93. A
19. C	44. B	69. B	94. C
20. D	45. D	70. A	95. D
21. C	46. A	71. B	96. D
22. B	47. D	72. C	97. C
23. C	48. D	73. A	98. D
24. D	49. A	74. C	99. A
25. C	50. B	75. C	100. C

ANSWER EXPLANATIONS—ARITHMETIC QUESTIONS

NOTE: The section contains explanations for only the arithmetic questions in the model exam.

4. The number of square feet in 10.6 acres is

   $$10.6 \times 43{,}560 \text{ sq. ft.} = 461{,}736 \text{ sq. ft.}$$

   If one side of the tract is 181 ft., the other must be

   $$461{,}736 \text{ sq. ft.} \div 181 \text{ ft.} = 2{,}551.0276 = 2{,}551 \text{ ft., rounded off}$$

5. The number of square feet in 17.1 acres is

   $$17.1 \text{ acres} \times 43{,}560 \text{ sq. ft.} = 744{,}876 \text{ sq. ft.}$$

   If 19% must be set aside, 81% is left for lots, or

   $$0.81 \times 744{,}876 \text{ sq. ft.} = 603{,}349.56 \text{ sq. ft.}$$

   If each lot must have at least 8,000 sq. ft., the maximum number of lots is 603,349.56 sq. ft. ÷ 8,000 sq. ft. = 75.4, or a maximum of 75 full lots.

10. Area = 400 ft. × 665 ft. = 266,000 sq. ft.
    266,000 sq. ft. ÷ 43,560 sq. ft. = 6.1065 acres
    $17,100.00 ÷ 6.1065 acres = $2,800.29, or $2,800 per acre

11. Taxable value = $170,000 × 0.30 = $51,000
    Since 71 mills is the same as a tax rate of 7.1%, the tax is

    $$0.071 \times \$51{,}000 = \$3{,}621$$

17. A building that is 40 ft. × 22 ft. is 124 ft. around (40 + 22 + 40 + 22). The total wall area is 124 ft. × 10 ft., or 1,240 sq. ft. From this we subtract the 52 sq. ft. of windows (4 sq. ft. × 13) and the 48 sq. ft. of door to get 1,140 sq. ft. of wall area needing paint. One gallon of paint covers 400 sq. ft. For one coat of paint we need 1,140 sq. ft. ÷ 400 sq. ft. = 2.85 gal., so two coats will require twice as much, or 5.7 gal.

18. Annual interest = $210 per quarter × 4 = $840. The interest rate is

    $$\$840 \div \$12{,}000 = 0.07 = 7\%$$

24. Ms. Levy will pay $400 less rent per month if she pays the utility bills herself. Average bills are $3,840 per year, which is $320 per month ($3,840 ÷ 12). Levy thinks she can reduce this by 35%, which means that she expects to pay only 65% of the $320 monthly average, or 0.65 × $320 = $208. She saves $400 on rent and expects to pay $208 of that for utilities, so she expects to save $192 per month.

25. Total commissions brought in by O'Hara = 7% × $2,250,000 = $192,500. He gets 50% of the first $50,000 (or $25,000) *plus* 60% of the rest. In his case, the rest is $142,500.

    $$0.6 \times \$142{,}500 = \$85{,}500$$

    To this we add his $25,000 share of the first $50,000 to get a total income for him of $110,500.

42. If Jackson paid a 6½% commission, then the $336,132.50 he has left represents the remaining 93½% of the sale price. The sale price then must be

    $$\$336{,}132.50 \div 0.935 = \$359{,}500$$

48. Ms. Anderson paid out a total of 12% of the selling price in fees, so her $176,000 is the 88% of the sale price that she has left. Therefore, the sale price was

$$\$176{,}000 \div 0.88 = \$200{,}000$$

49. The 9-in.-thick walls are ¾ ft. thick. Therefore, the interior dimensions of the floor are 50½ ft. × 34½ ft. (remember, the walls are at *both* ends):

$$50.5 \times 34.5 = 1{,}742.25 \text{ sq. ft. gross}$$

from which we subtract the area taken up by the pillars. They are each ½ ft. × ½ ft., or ¼ sq. ft. Their total area is

$$13 \times \tfrac{1}{4} \text{ sq. ft.} = 3\tfrac{1}{4} \text{ sq. ft.}$$

Subtracting this amount from 1,742.25 leaves a net of 1,739 sq. ft.

62. The lots contain 10,500 sq. ft. each and must each have 1,380 sq. ft. additional amenity. Therefore, Smith will need a minimum of 11,880 sq. ft. for each lot (10,500 sq. ft. + 1,380 sq. ft.). Since he wants 55 lots, he will need at least 55 × 11,880 sq. ft. = 653,400 sq. ft.; 653,400 sq. ft. ÷ 43,560 sq. ft. = 15 acres exactly.

70. Interior gross dimensions are 41 ft. × 27 ft. = 1,107 sq. ft. (Remember to subtract 6 in. of outside wall from each end.) Interior wall area is ⅓ ft. × 135 ft. = 45 sq. ft. Each bathroom is 54 sq. ft. The kitchen is 144 sq. ft. This is a total area of 297 sq. ft. *not* to be carpeted, leaving 810 sq. ft. because 1 sq. yd. contains 9 sq. ft. (3 ft. × 3 ft.). The total number of square yards to be carpeted is 810 sq. ft. ÷ 9 sq. ft. = 90 sq. yd. At $13.95 per square yard, the total cost is $1,255.50 (90 sq. yd. × $13.95).

71. To determine how much Ms. Farrell still needs, we subtract from the $330,000 purchase price the $9,000 earnest money payment and the 75% loan. The loan is 75% of $330,000, or $330,000 × 0.75 = $247,500. $330,000 less $9,000 is $321,000. Subtracting the $247,500 loan leaves $73,500 needed.

74. A lot 100 ft. × 100 ft. contains 100 × 100 = 10,000 sq. ft. The 3-acre plot contains 43,560 × 3 = 130,680 sq. ft.; 130,680 ÷ 10,000 = 13.068, or 13 full lots.

75. The width of the lot is 6,336 ÷ 96 = 66 ft. At $225 a front ft. the lot would sell for 66 × $225 = $14,850.

ANSWER SHEET
Model Examination 6

Broker and Salesperson

1. A B C D
2. A B C D
3. A B C D
4. A B C D
5. A B C D
6. A B C D
7. A B C D
8. A B C D
9. A B C D
10. A B C D
11. A B C D
12. A B C D
13. A B C D
14. A B C D
15. A B C D
16. A B C D
17. A B C D
18. A B C D
19. A B C D
20. A B C D
21. A B C D
22. A B C D
23. A B C D
24. A B C D
25. A B C D
26. A B C D
27. A B C D
28. A B C D
29. A B C D
30. A B C D
31. A B C D
32. A B C D
33. A B C D
34. A B C D
35. A B C D
36. A B C D
37. A B C D
38. A B C D
39. A B C D
40. A B C D
41. A B C D
42. A B C D
43. A B C D
44. A B C D
45. A B C D
46. A B C D
47. A B C D
48. A B C D
49. A B C D
50. A B C D
51. A B C D
52. A B C D
53. A B C D
54. A B C D
55. A B C D
56. A B C D
57. A B C D
58. A B C D
59. A B C D
60. A B C D
61. A B C D
62. A B C D
63. A B C D
64. A B C D
65. A B C D
66. A B C D
67. A B C D
68. A B C D
69. A B C D
70. A B C D
71. A B C D
72. A B C D
73. A B C D
74. A B C D
75. A B C D
76. A B C D
77. A B C D
78. A B C D
79. A B C D
80. A B C D
81. A B C D
82. A B C D
83. A B C D
84. A B C D
85. A B C D
86. A B C D
87. A B C D
88. A B C D
89. A B C D
90. A B C D
91. A B C D
92. A B C D
93. A B C D
94. A B C D
95. A B C D
96. A B C D
97. A B C D
98. A B C D
99. A B C D
100. A B C D

Model Examination 6

BROKER AND SALESPERSON

See page 598 for the answer key. Math problems are worked out on pages 599–600.

1. Will openly occupied Ward's land for a period of time, without interference from Ward, and then received fee simple title to the land. This was an example of
   (A) an easement in gross
   (B) adverse possession
   (C) estoppel
   (D) subrogation

2. Max has Beth's power of attorney. Max is called
   (A) an agent in place
   (B) a real estate broker
   (C) a lawyer
   (D) an attorney-in-fact

3. The right by which the state takes title to real estate for which no legal owner can be found is
   (A) police power
   (B) tenancy
   (C) eminent domain
   (D) escheat

4. Which of the following types of deeds has no warranty?
   (A) executor's
   (B) quitclaim
   (C) general warranty
   (D) trustee's

5. Louise does work on Joe's house, but is not paid. She may file a
   (A) mechanic's lien
   (B) satisfaction piece
   (C) notice of foreclosure
   (D) sheriff's auction

6. What is the cash-on-cash ratio for a property having a cash flow of $8,360 and an initial investor's cash equity of $76,000?
   (A) 0.10%
   (B) 0.105%
   (C) 0.11%
   (D) 0.12%

7. What percent of 1 square mile is 96 acres?
   (A) 21%
   (B) 19.6%
   (C) 15%
   (D) 12%

8. Which claim is paid first at a foreclosure?
   (A) tax lien
   (B) first mortgage
   (C) second mortgage
   (D) mechanic's lien

9. In an agency relationship, the principal is the
   (A) seller
   (B) buyer
   (C) broker
   (D) employer

10. In a lease contract, the landlord is the
    (A) leasee
    (B) leasor
    (C) lessee
    (D) lessor

11. If consideration has value only to the person receiving it, it is
    (A) personal consideration
    (B) good consideration
    (C) valuable consideration
    (D) good and valuable consideration

12. A lease that requires the landlord to pay operating expenses of the property is a __________ lease.
    (A) gross
    (B) flat
    (C) net
    (D) step

13. Taxes for the calendar year are due on June 1 of the year in which they are assessed. Tax rate is 41 mills. Assessed value is $53,250. If a sale of the property is closed on August 15, then at settlement (using the statutory year)
    (A) seller owes buyer $818.72
    (B) buyer owes seller $818.72
    (C) buyer owes seller $1,364.53
    (D) seller owes buyer $1,364.53

14. A properly done appraisal is
    (A) an authentication of value
    (B) an estimate of value
    (C) a prediction of value
    (D) a statement of exact value

15. If Murphy defaults on the payments to his mortgagee, under which clause may the lender demand immediate payment in full of the entire remaining loan balance?
    (A) defeasance
    (B) subrogation
    (C) acceleration
    (D) due-on-sale

16. Which of the following does *not* have the right of survivorship?
    (A) joint tenant
    (B) tenant in severalty
    (C) tenant by the entirety
    (D) all of the above

17. An easement can be
    (A) the right to cross someone else's land to get to the road
    (B) a variety of fraud exclusive to real estate
    (C) the right to inherit if there is no will
    (D) none of the above

18. A broker's license is revoked. The broker's salespeople
    (A) also have their licenses revoked
    (B) may continue to operate the broker's business
    (C) may, upon proper application, transfer their licenses to another broker
    (D) must place their licenses on inactive status for a year

19. In an exclusive right to sell listing,
    (A) only one broker is authorized to act as the seller's agent
    (B) if the seller finds the buyer, the seller still must pay the broker a commission
    (C) the broker must use her best efforts to solicit offers
    (D) all of the above

20. Among the broker's functions, he may
    (A) accept an offer on behalf of the seller
    (B) solicit offers for the seller's listed property
    (C) try to negotiate better terms before submitting an offer
    (D) work to get the buyer the best possible price

21. Which of the following is an "improvement" to land?
    (A) rezoning
    (B) driveway
    (C) orchard
    (D) all of the above

22. Which of the following is a nonfreehold estate?
    (A) fee simple estate
    (B) leasehold estate
    (C) life estate
    (D) dower estate

23. The rent for Sam's store is based at least in part upon Sam's gross business revenues. Sam's lease is a __________ lease.
    (A) reappraisal
    (B) net
    (C) step
    (D) percentage

24. What is usury?
    (A) collecting more interest than that allowed by law
    (B) building a structure that extends over someone else's land
    (C) selling property for less than the asking price
    (D) selling real estate without a license

25. Joe wants to build a patio 20 yd. by 20 ft. and 6 in. thick. How many cubic yards of concrete will he need?
    (A) 22.2
    (B) 40.0
    (C) 20.0
    (D) 400

26. The number of square feet in 1 acre is
    (A) 640
    (B) 43,560
    (C) 45,360
    (D) 53,460

27. How is the *gross rent multiplier* calculated?
    (A) market value/market rental
    (B) monthly payment/market rental
    (C) market rental/market price
    (D) sales price/market price

28. Minimum allowable construction standards are established by
   (A) building codes
   (B) zoning codes
   (C) environmental laws
   (D) condemnation laws

29. An allowed land use that is *not* normally permitted by the property's zoning classification is
   (A) a dispensation
   (B) a zone change
   (C) a variance
   (D) a restrictive covenant

30. Peeling paint and loose floorboards are examples of
   (A) physical deterioration
   (B) functional obsolescence
   (C) economic obsolescence
   (D) both A and B

31. The sales comparison approach is generally most suitable for appraising
   (A) commercial and investment property
   (B) single-family homes
   (C) heavily mortgaged property
   (D) heavily insured property

32. A feature of a freehold estate is that it
   (A) is acquired without paying anything for it
   (B) is always acquired by adverse possession
   (C) has uncertain duration
   (D) is any leasehold estate

33. A person who has the appearance of owning land, but does not own it, has
   (A) good title
   (B) recorded title
   (C) constructive notice
   (D) color of title

34. A contract entered into by a minor can be
   (A) void
   (B) voidable by the minor
   (C) voidable by all parties
   (D) voidable by adult parties

35. Which of the following will *not* terminate a lease?
   (A) surrender
   (B) performance
   (C) vacancy
   (D) breach

36. In mortgage lending, one discount point is equal to
    (A) 1% of the sale price
    (B) 1% of the loan amount
    (C) 1% of the interest rate
    (D) 1% of the commission

37. Hooper's offer to purchase Looper's real estate stated that the offer would become void if not accepted within 5 days. The day after the offer was made, having heard nothing from Looper, Hooper wanted to withdraw his offer. Which of the following is true?
    (A) Hooper had to wait until the 5 days had passed.
    (B) Hooper was required to notify Looper of his desire to withdraw the offer and to give Looper "reasonable time" to accept or reject it.
    (C) Hooper could withdraw the offer immediately.
    (D) none of the above

38. In a real estate transaction, title to land passes
    (A) on the date shown on the deed
    (B) upon recordation of the deed
    (C) upon notarization of the deed
    (D) upon delivery of the deed

39. A fixture is
    (A) anything that cannot be removed from the real estate without leaving a hole larger than 6 in.
    (B) anything the seller says is a fixture
    (C) anything needed for the proper use of the real estate
    (D) none of the above

40. The law that requires that all transfers of ownership rights to real estate be in writing is the
    (A) Contract Act
    (B) Statute of Limitations
    (C) Statute of Recordation
    (D) Statute of Frauds

41. All of the following should be recorded *except*
    (A) a deed of trust
    (B) a 20-year lease
    (C) an executor's deed
    (D) a contract of sale

42. In a typical mortgage loan transaction, the mortgagor is the
    (A) lender
    (B) borrower
    (C) appraiser
    (D) closing agent

43. A salesperson sold a property for $106,000. If the broker received 50% of the commission and the salesperson's 60% share of that amount was $2,544, what was the commission rate charged on the sale?
(A) 7%
(B) 7.5%
(C) 7.75%
(D) 8%

44. An estate that any party can terminate at any time is an estate
(A) for years
(B) at will
(C) in possession
(D) leasehold entailed

45. A listing contract says that the seller receives $125,000 and the broker receives all of the purchase price over $125,000 as his commission. This is a(n) ________ listing.
(A) open
(B) net
(C) multiple
(D) exclusive

46. Jones leases a building that is damaged by a storm. Under common law,
(A) the lease is no longer enforceable
(B) Jones needn't pay rent until the building is repaired
(C) Jones must repair the building but may deduct the cost from rent
(D) Jones must continue to pay rent until the lease expires

47. In each locality, real estate brokerage commissions are determined by
(A) the local Board of Realtors
(B) the laws of the state
(C) the board of estimate
(D) agreement between property owner and broker

48. A lot sold for $980 a front foot. If the lot was 132 ft. deep and had an area of 6,468 sq. ft., how much did the lot sell for?
(A) $64,680
(B) $57,710
(C) $48,000
(D) $39,000

49. Which of the following is required for "mutual agreement" in a contract?
(A) offer and acceptance
(B) proper consideration
(C) description of the land
(D) legal form

50. In most states, a 6-month lease
    (A) must be in writing
    (B) need not be written
    (C) may not be in writing
    (D) ought not to be written

51. The expenses of settlement are
    (A) paid by the broker
    (B) paid by the seller
    (C) negotiated between buyer and seller
    (D) paid by the buyer

52. In most states, a primary source of revenue for local government is
    (A) income taxes
    (B) sales taxes
    (C) property taxes
    (D) severance taxes

53. The purpose of real estate licensing laws is to protect
    (A) salespersons
    (B) lawyers and legislators
    (C) developers
    (D) the general public

54. A broker in a no-agency broker relationship
    (A) may offer both buyer and seller confidential information about the other
    (B) does not have an agency relationship with either buyer or seller
    (C) deals only with transactions that do not involve leasing
    (D) all of the above

55. Salesperson Jones pockets an earnest money deposit and is found out. Jones's broker must
    (A) report the incident to the state attorney general
    (B) repay the money if Jones cannot do so
    (C) pay Jones's legal fees, if Jones is charged with a crime
    (D) both A and B

56. Listed real estate should be advertised in the name of the
    (A) owner
    (B) salesperson
    (C) broker
    (D) closing agent

57. A 2.6-acre lot sold for $188,000. What was the price per square foot?
    (A) $1.19
    (B) $1.55
    (C) $1.66
    (D) $1.82

58. Hamilton's broker license is revoked by the real estate commission. Hamilton may
   (A) apply for a salesperson's license
   (B) appeal the revocation to the courts
   (C) wait 6 months and apply for reinstatement
   (D) continue to operate his business until his existing listings are sold or have expired

59. Commissions from the sale of real estate
   (A) must be divided equally between broker and salesperson
   (B) must be divided equally among all participating brokers
   (C) are not taxable income
   (D) none of the above

60. Which is the superior lien?
   (A) tax lien
   (B) first mortgage
   (C) junior mortgage
   (D) mechanic's lien

61. A tenancy in severalty exists when
   (A) one person owns real estate
   (B) husband and wife own real estate together
   (C) any related persons own real estate together
   (D) none of the above

62. Which of the following is *not* a test of a fixture?
   (A) manner of attachment
   (B) intent of the person who put it there
   (C) cost of the item
   (D) custom in the community

63. During the life of the life tenant, the remainderman has
   (A) an easement in gross
   (B) a reversionary interest
   (C) a fee simple estate
   (D) a renewing leasehold estate

64. At settlement, the seller paid the buyer $351.00 toward the annual tax bill of $972.00. If taxes are assessed for the calendar year and must be paid by July 15, what was the date of the settlement?
   (A) May 10
   (B) August 20
   (C) July 15
   (D) can't be determined

65. When parties to a contract agree to amend the contract, the document they prepare and sign is called a
(A) deed amendment
(B) satisfaction piece
(C) novation
(D) relinquishment

66. What kind of estate is received if the deed states that the grantor grants to the grantee "and his heirs and assigns forever"?
(A) fee simple
(B) leasehold
(C) life estate
(D) nonfreehold

67. A deed that conveys only the interest of the grantor is a
(A) general warranty deed
(B) bargain and sale deed
(C) quitclaim deed
(D) special warranty deed

68. A real estate license should be
(A) carried in the licensee's wallet
(B) posted at the Real Estate Commission
(C) posted in a public place in the broker's office
(D) posted in the licensee's home

69. The usefulness of the cost approach in appraisal may be limited if the subject property is
(A) a new structure
(B) functionally obsolescent
(C) in an inactive market
(D) proposed construction

70. A tenant paid rent at the beginning of the month; the property sold in the middle of the month. To prorate rent at settlement, you would
(A) credit the buyer
(B) debit the seller
(C) both A and B
(D) neither A nor B

71. Which of the following is considered a brokerage commission for the purposes of TRID?
(A) title insurance fee
(B) deed preparation fee
(C) monthly payment
(D) none of the above

72. A person who is employed by a broker to rent property but not to sell it
   (A) must be licensed
   (B) need not be licensed
   (C) must have a special rent-only license
   (D) must be a licensed broker

73. The process whereby a person may have her real estate sold to pay a debt or claim is called
   (A) lien
   (B) foreclosure
   (C) covenant
   (D) defeasance

74. A person who owns an undivided interest in real estate with at least one other person and has the right of survivorship is called a
   (A) tenant in severalty
   (B) life tenant
   (C) tenant in common
   (D) joint tenant

75. Which is *not* considered real estate?
   (A) fixtures
   (B) trees
   (C) chattels
   (D) sidewalk

76. A person who dies and leaves no will is said to have died
   (A) without heirs
   (B) intestate
   (C) unbequeathed
   (D) unherited

77. A broker must keep funds entrusted to her but belonging to others in
   (A) an office safe to which only she has the combination
   (B) a savings account in the name of the person whose money it is
   (C) a trust or escrow account
   (D) a special account managed by the state

78. When a party to a contract has deliberately lied in order to mislead the other party(ies) into agreeing to the contract, this action is an example of
   (A) fraud
   (B) misrepresentation
   (C) duress
   (D) defeasance

79. If a party to a contract acts so as to make performance under the contract impossible, this action is an example of
   (A) discharge of contract
   (B) performance of contract
   (C) abandonment of contract
   (D) breach of contract

80. A minor may be bound by the courts to contracts for
   (A) personal property
   (B) realty only
   (C) necessaries
   (D) rent

81. Edgar purchases a condo costing $232,500. The loan requires a 15% down payment, an origination fee of 0.75%, 2.25 points discount, and a 0.5% PMI fee. How much money does Edgar need for these expenses?
   (A) $23,250.00
   (B) $28,995.75
   (C) $41,791.88
   (D) $43,012.50

82. A makes an offer to buy B's real estate. B makes a counteroffer to A. Which of the following is true?
   (A) A is bound by his original offer.
   (B) A must accept the counteroffer.
   (C) A's original offer no longer exists.
   (D) A may not counteroffer back to B.

83. A contract of sale of real estate must be signed by
   (A) buyer and seller
   (B) broker, buyer, and seller
   (C) buyer only
   (D) broker only

84. In a contract of sale, the seller is the
   (A) grantor
   (B) vendee
   (C) grantee
   (D) vendor

85. Which of the following statements is true?
   (A) To be valid, a contract of sale must be signed by the broker if a broker assists in the transaction.
   (B) A corporation may be a party to a contract of sale.
   (C) A contract of sale need not be written if it is closed within 1 year.
   (D) A person who is an attorney-in-fact must also be an attorney-at-law.

86. Which of the following must be included in a deed?
   (A) proper description of the real estate
   (B) street address, if the real estate is a house
   (C) area of the land ("more or less")
   (D) all of the above

87. A deed is recorded
   (A) to give public notice
   (B) to insure title
   (C) to satisfy the law
   (D) to avoid extra taxes

88. The part of a deed that defines or limits the quantity of estate granted is the
   (A) habendum
   (B) premises
   (C) equity
   (D) consideration

89. A final payment larger than the intermediate payments to a note is called
   (A) an escalator
   (B) an amortization
   (C) a balloon
   (D) a reappraisal

90. The interest rate on a loan is 5.125%. The interest for the month of July was $891.20. What was the loan balance at the beginning of July?
   (A) $200,000
   (B) $207,480
   (C) $208,671
   (D) $221,223

91. A conventional mortgage loan is
   (A) self-amortizing
   (B) not government insured/guaranteed
   (C) approved by the FHA
   (D) uninsurable

92. The mortgage with the highest priority is usually the one
   (A) with the highest unpaid balance
   (B) with the highest original loan amount
   (C) with the highest interest rate
   (D) that was recorded first

93. The seller of real estate takes a note secured by a mortgage on the real estate as partial payment. The mortgage is
   (A) sale financed
   (B) a purchase-money mortgage
   (C) a secondary mortgage
   (D) a first lien

94. An ad valorem tax is based on
   (A) the taxpayer's income
   (B) the sale price of the article taxed
   (C) the size and/or weight of the article
   (D) the value of the article taxed

95. A person who is too young to be held to a contract is called
   (A) minority impaired
   (B) youthful
   (C) unavailable in law
   (D) incompetent

96. A contract in which rights to use and occupy real estate are transferred for a specified period of time is
   (A) a deed
   (B) an easement
   (C) a life estate
   (D) a lease

97. Whenever all parties agree to the terms of a contract, there has been
   (A) mutual agreement
   (B) legality of object
   (C) consideration
   (D) competency

98. Recording a deed
   (A) passes title
   (B) gives constructive notice
   (C) insures title
   (D) removes liens

99. Multiple listing
    (A) causes lost commissions
    (B) is a listing-sharing organization of brokers
    (C) is illegal in some states
    (D) is a violation of antitrust law

100. A contract that gives a person the right to buy during a specified time, but carries no obligation to do so, is
    (A) a sale contract
    (B) a land contract
    (C) an option
    (D) a bargain and sale

ANSWER KEY
Model Test 6

1. **B**	26. **B**	51. **C**	76. **B**
2. **D**	27. **A**	52. **C**	77. **C**
3. **D**	28. **A**	53. **D**	78. **A**
4. **B**	29. **C**	54. **B**	79. **D**
5. **A**	30. **A**	55. **B**	80. **C**
6. **C**	31. **B**	56. **C**	81. **C**
7. **C**	32. **C**	57. **C**	82. **C**
8. **A**	33. **D**	58. **B**	83. **A**
9. **D**	34. **B**	59. **D**	84. **D**
10. **D**	35. **C**	60. **A**	85. **B**
11. **B**	36. **B**	61. **A**	86. **A**
12. **A**	37. **C**	62. **C**	87. **A**
13. **B**	38. **D**	63. **B**	88. **A**
14. **B**	39. **D**	64. **A**	89. **C**
15. **C**	40. **D**	65. **C**	90. **C**
16. **B**	41. **D**	66. **A**	91. **B**
17. **A**	42. **B**	67. **C**	92. **D**
18. **C**	43. **D**	68. **C**	93. **B**
19. **D**	44. **B**	69. **B**	94. **D**
20. **B**	45. **B**	70. **C**	95. **D**
21. **B**	46. **D**	71. **D**	96. **D**
22. **B**	47. **D**	72. **A**	97. **A**
23. **D**	48. **C**	73. **B**	98. **B**
24. **A**	49. **A**	74. **D**	99. **B**
25. **A**	50. **B**	75. **C**	100. **C**

ANSWER EXPLANATIONS—ARITHMETIC QUESTIONS

NOTE: The section contains explanations for only the arithmetic questions in the model exam.

6. Cash-on-cash ratio is annual cash return received, divided by the cash spent (original equity) to acquire the investment.

$$\$8{,}360 \div \$76{,}000 = 0.11$$

7. Because 1 sq. mi. contains 640 acres,

$$96 \div 640 = 0.15, \text{ or } 15\%$$

13. The sale closed August 15, but the taxes were due on June 1. Therefore, the taxes have already been paid by the seller. So *buyer pays seller* at closing to reimburse the taxes already paid.
Assessed value is $53,250; tax rate is 41 mills (which is 4.1%, or 0.041). So the year's tax is $53,250 \times 0.041 = \$2,183.25$.
Seller owned property for exactly 7.5 mo. (January through July, and half of August). So buyer owes seller the taxes for the remaining 4.5 mo. of the year that the buyer will own the property. One month's taxes amount to $\$2,183.25 \div 12 = \$181.9375$; $4.5 \times \$181.9375 = \$818.719$, or (rounded) $818.72 that buyer owes seller.

25. Because 20 yd. is $20 \times 3 = 60$ ft., the area of the patio is $60 \times 20 = 1,200$ sq. ft. The patio is 6 in., or ½ ft., thick. Then $1,200 \times 0.5 = 600$ cu. ft., the *volume* of the patio. One cubic yard is $3 \times 3 \times 3 = 27$ cu. ft. Therefore, Joe will need

$$600 \div 27 = 22.22, \text{ or (rounded) } 22.2 \text{ cu. yd. of concrete}$$

43. First we find the amount of the total commission. Salesperson's share of $2,544 was 60% of the total the firm received. $\$2,544 \div 0.6 = \$4,240$ for the firm's share, which was 50% of the total. $\$4,240 \div 0.5 = \$8,480$ total commission.
Because the house sold for $106,000

$$\$8{,}480 \div \$106{,}000 = 0.08, \text{ or } 8\%$$

48. We have to determine the number of front feet in the lot and multiply that number by $980 to determine the price the lot sold for.

$$6{,}468 \div 132 = 49 \text{ front ft. } 49 \times \$980 = \$48{,}020, \text{ or } \$48{,}000 \text{ rounded}$$

57. $\$188,000 \div 2.6$ acres $= \$72,307.69$ per acre

$$\$72{,}307.69 \div 43{,}560 = \$1.65997, \text{ or } \$1.66 \text{ rounded}$$

64. We have to determine how long the seller owned the property. We know he owned it at the beginning of the year. We calculate time from his tax payment and then figure from January 1 to determine the date of settlement. Remember: the settlement date is considered a day that the *seller* owned the property. The total tax bill is $972.
$\$972 \div 12 = \$81$ per month

$$\$81 \div 30 = \$2.70 \text{ per day}$$

First we determine how many *whole months* the seller owned the property by dividing his share of taxes ($351) by the monthly tax share of $81; $\$351 \div \$81 = 4.33333$. Thus, the seller owned the property for 4 whole months (January through April) and part of May, the month in which settlement must have occurred.

Now we subtract the 4 whole months worth of taxes from the seller's share; $4 \times \$81 = \$324$ and $\$351 - \$324 = \$27$. So the seller also owned the property for \$27 "worth" of May. $\$27 \div \$2.70 = 10$ of May's 31 days (statutory year). Thus, the closing had to be on May 10.

81. We have to calculate the down payment, the origination fee, the discount, and the PMI fee and then add all of these to determine the answer.

Down payment is 15%; $\$232{,}500 \times 0.15 = \$34{,}875$ down payment

Origination fee, discount, and PMI fee all are calculated as part of the *loan amount* (not the sales price). The loan will be the sales price less the down payment, or $\$232{,}500 - \$34{,}875 = \$197{,}625$ loan amount

Origination fee (0.75%) is $\$197{,}625 \times 0.0075 = \$1{,}482.19$ rounded

Discount (2.25%) is $\$197{,}625 \times 0.0225 = \$4{,}446.56$ (rounded)

PMI fee (0.5%, or ½%) is $\$197{,}625 \times 0.005 = \$988.13$ (rounded)

$$\$34{,}875 + \$1{,}482.19 + \$4{,}446.56 + \$988.13 = \$41{,}791.88$$

90. We have to calculate the monthly interest rate, because we are given a monthly interest payment; $5.125\% \div 12 = 0.42708\% = 0.0042708$. The interest payment of \$891.20 is 0.42708% of the loan amount.

$$\$891.20 \div 0.0042708 = \$208{,}671 \text{ rounded}$$

ANSWER SHEET
Model Examination 7

Broker and Salesperson

1. Ⓐ Ⓑ Ⓒ Ⓓ
2. Ⓐ Ⓑ Ⓒ Ⓓ
3. Ⓐ Ⓑ Ⓒ Ⓓ
4. Ⓐ Ⓑ Ⓒ Ⓓ
5. Ⓐ Ⓑ Ⓒ Ⓓ
6. Ⓐ Ⓑ Ⓒ Ⓓ
7. Ⓐ Ⓑ Ⓒ Ⓓ
8. Ⓐ Ⓑ Ⓒ Ⓓ
9. Ⓐ Ⓑ Ⓒ Ⓓ
10. Ⓐ Ⓑ Ⓒ Ⓓ
11. Ⓐ Ⓑ Ⓒ Ⓓ
12. Ⓐ Ⓑ Ⓒ Ⓓ
13. Ⓐ Ⓑ Ⓒ Ⓓ
14. Ⓐ Ⓑ Ⓒ Ⓓ
15. Ⓐ Ⓑ Ⓒ Ⓓ
16. Ⓐ Ⓑ Ⓒ Ⓓ
17. Ⓐ Ⓑ Ⓒ Ⓓ
18. Ⓐ Ⓑ Ⓒ Ⓓ
19. Ⓐ Ⓑ Ⓒ Ⓓ
20. Ⓐ Ⓑ Ⓒ Ⓓ
21. Ⓐ Ⓑ Ⓒ Ⓓ
22. Ⓐ Ⓑ Ⓒ Ⓓ
23. Ⓐ Ⓑ Ⓒ Ⓓ
24. Ⓐ Ⓑ Ⓒ Ⓓ
25. Ⓐ Ⓑ Ⓒ Ⓓ
26. Ⓐ Ⓑ Ⓒ Ⓓ
27. Ⓐ Ⓑ Ⓒ Ⓓ
28. Ⓐ Ⓑ Ⓒ Ⓓ
29. Ⓐ Ⓑ Ⓒ Ⓓ
30. Ⓐ Ⓑ Ⓒ Ⓓ
31. Ⓐ Ⓑ Ⓒ Ⓓ
32. Ⓐ Ⓑ Ⓒ Ⓓ
33. Ⓐ Ⓑ Ⓒ Ⓓ
34. Ⓐ Ⓑ Ⓒ Ⓓ
35. Ⓐ Ⓑ Ⓒ Ⓓ
36. Ⓐ Ⓑ Ⓒ Ⓓ
37. Ⓐ Ⓑ Ⓒ Ⓓ
38. Ⓐ Ⓑ Ⓒ Ⓓ
39. Ⓐ Ⓑ Ⓒ Ⓓ
40. Ⓐ Ⓑ Ⓒ Ⓓ
41. Ⓐ Ⓑ Ⓒ Ⓓ
42. Ⓐ Ⓑ Ⓒ Ⓓ
43. Ⓐ Ⓑ Ⓒ Ⓓ
44. Ⓐ Ⓑ Ⓒ Ⓓ
45. Ⓐ Ⓑ Ⓒ Ⓓ
46. Ⓐ Ⓑ Ⓒ Ⓓ
47. Ⓐ Ⓑ Ⓒ Ⓓ
48. Ⓐ Ⓑ Ⓒ Ⓓ
49. Ⓐ Ⓑ Ⓒ Ⓓ
50. Ⓐ Ⓑ Ⓒ Ⓓ
51. Ⓐ Ⓑ Ⓒ Ⓓ
52. Ⓐ Ⓑ Ⓒ Ⓓ
53. Ⓐ Ⓑ Ⓒ Ⓓ
54. Ⓐ Ⓑ Ⓒ Ⓓ
55. Ⓐ Ⓑ Ⓒ Ⓓ
56. Ⓐ Ⓑ Ⓒ Ⓓ
57. Ⓐ Ⓑ Ⓒ Ⓓ
58. Ⓐ Ⓑ Ⓒ Ⓓ
59. Ⓐ Ⓑ Ⓒ Ⓓ
60. Ⓐ Ⓑ Ⓒ Ⓓ
61. Ⓐ Ⓑ Ⓒ Ⓓ
62. Ⓐ Ⓑ Ⓒ Ⓓ
63. Ⓐ Ⓑ Ⓒ Ⓓ
64. Ⓐ Ⓑ Ⓒ Ⓓ
65. Ⓐ Ⓑ Ⓒ Ⓓ
66. Ⓐ Ⓑ Ⓒ Ⓓ
67. Ⓐ Ⓑ Ⓒ Ⓓ
68. Ⓐ Ⓑ Ⓒ Ⓓ
69. Ⓐ Ⓑ Ⓒ Ⓓ
70. Ⓐ Ⓑ Ⓒ Ⓓ
71. Ⓐ Ⓑ Ⓒ Ⓓ
72. Ⓐ Ⓑ Ⓒ Ⓓ
73. Ⓐ Ⓑ Ⓒ Ⓓ
74. Ⓐ Ⓑ Ⓒ Ⓓ
75. Ⓐ Ⓑ Ⓒ Ⓓ
76. Ⓐ Ⓑ Ⓒ Ⓓ
77. Ⓐ Ⓑ Ⓒ Ⓓ
78. Ⓐ Ⓑ Ⓒ Ⓓ
79. Ⓐ Ⓑ Ⓒ Ⓓ
80. Ⓐ Ⓑ Ⓒ Ⓓ
81. Ⓐ Ⓑ Ⓒ Ⓓ
82. Ⓐ Ⓑ Ⓒ Ⓓ
83. Ⓐ Ⓑ Ⓒ Ⓓ
84. Ⓐ Ⓑ Ⓒ Ⓓ
85. Ⓐ Ⓑ Ⓒ Ⓓ
86. Ⓐ Ⓑ Ⓒ Ⓓ
87. Ⓐ Ⓑ Ⓒ Ⓓ
88. Ⓐ Ⓑ Ⓒ Ⓓ
89. Ⓐ Ⓑ Ⓒ Ⓓ
90. Ⓐ Ⓑ Ⓒ Ⓓ
91. Ⓐ Ⓑ Ⓒ Ⓓ
92. Ⓐ Ⓑ Ⓒ Ⓓ
93. Ⓐ Ⓑ Ⓒ Ⓓ
94. Ⓐ Ⓑ Ⓒ Ⓓ
95. Ⓐ Ⓑ Ⓒ Ⓓ
96. Ⓐ Ⓑ Ⓒ Ⓓ
97. Ⓐ Ⓑ Ⓒ Ⓓ
98. Ⓐ Ⓑ Ⓒ Ⓓ
99. Ⓐ Ⓑ Ⓒ Ⓓ
100. Ⓐ Ⓑ Ⓒ Ⓓ

Model Examination 7

BROKER AND SALESPERSON

See page 619 for the answer key. Math problems are worked out on page 620.

1. A buyer purchased a house with cash but neither took possession nor recorded the deed. What error has the buyer made, either by his actions or by his failure to act?
   (A) given constructive notice of ownership
   (B) acquired ownership through adverse possession
   (C) hidden from tax authorities
   (D) failed to give constructive notice of ownership

2. Allen has given Betty the right to park in his driveway for three weeks. This right is best called
   (A) license
   (B) easement
   (C) right-of-way
   (D) lease

3. What type of rights does *not* refer to the use of water?
   (A) riparian
   (B) usufructory
   (C) dower
   (D) littoral

4. The most extensive form of ownership in real estate is
   (A) a paid-up fee
   (B) life estate
   (C) fee tail
   (D) fee simple

5. Your grandfather died intestate, and you received his house. Title was acquired
   (A) by will
   (B) by estovers
   (C) by descent
   (D) by mitigation

6. When title to property is taken as tenants in common,
   (A) all parties have an equal interest
   (B) each tenant in common may sell his interest
   (C) each tenant in common must pay rent to the landlord
   (D) when a tenant dies, the share is distributed equally to the surviving tenants in common

7. The owner of a life estate
   (A) may possess the property for his or her lifetime
   (B) pays no property taxes
   (C) may not receive rental income
   (D) may not improve the property

8. A person who inherits real estate by will is a
   (A) bequest
   (B) life estate
   (C) devisee
   (D) executor

9. A hunting license represents
   (A) permission to go onto another's land
   (B) an easement
   (C) a fee tail
   (D) an estate

10. Farmer Brown plowed land and harvested crops on an open field next to land he owned. He did this openly for the statutory period of time. He then claimed ownership based on
    (A) escheat
    (B) estovers
    (C) eminent domain
    (D) adverse possession

11. Three people bought property with the understanding that if one of them died, his heirs would inherit that interest. Most likely they took title under which type of estate?
    (A) joint tenancy
    (B) tenancy in common
    (C) estate in common
    (D) tenancy by entireties

12. Recording a deed provides __________ to the public.
    (A) actual notice
    (B) real notice
    (C) constructive notice
    (D) available notice

13. When a husband and wife take ownership as tenants in common, they
    (A) can determine who inherits their share
    (B) enjoy the rights of survivorship
    (C) are the same as tenants by the entireties
    (D) have created community property

14. Title insurance is normally
    (A) required to protect the lender and is optional for the property's equity buyer
    (B) required to protect the property buyer and is optional for the lender
    (C) paid annually
    (D) used in only a few states

15. A broker with a listing brings a contract from a ready, willing, and able buyer, at the listed price, which the seller is considering. Generally, the broker has earned his commission
    (A) now
    (B) when the seller accepts
    (C) when contract contingencies are met
    (D) after closing

16. The fiduciary has a relationship with the client that includes
    (A) hostility
    (B) trust and confidence
    (C) peace
    (D) dishonesty

17. A broker has a 90-day exclusive right to sell listing. With ten days left on the listing, the broker shows the property to Mr. B. The owner and Mr. B. wait twelve days to sign a contract in an attempt to avoid paying the broker's commission. The broker learns about the sale and should
    (A) receive a commission by a broker protection provision
    (B) go away because the listing expired before the sale contract
    (C) complain to the Board of Realtors
    (D) collect by alienation

18. A broker was responsible for the sale of a house under a legal written listing. The owner refused to pay the commission. The broker's best course of action is to
    (A) collect from the buyer
    (B) reverse the transaction
    (C) place a mechanic's lien on the property
    (D) sue the seller to collect

19. A salesman is working as a buyer's broker. The salesman should
    (A) show only properties for which a commission is assured
    (B) encourage the buyer to pay the asking price
    (C) counsel the buyer about his needs and about the real estate market
    (D) wait for the best deal to come along

20. A real estate salesperson is responsible primarily to
    (A) the buyer
    (B) the seller
    (C) the real estate commission
    (D) the broker who holds his license

21. A property owner requests that the listing broker not let anyone show the house on Sunday. Because Sunday is a prime showing day, the broker should
    (A) comply personally but not consider himself responsible for the actions of other brokers
    (B) comply with the owner's wishes
    (C) refuse to accept the listing
    (D) list the house and show it on Sunday only to very serious prospects

22. When answering questions a buyer raises about home defects, a salesperson should
    (A) not disclose any defects
    (B) answer honestly to the best of his knowledge
    (C) plead ignorance of any defects
    (D) tell only the inspector

23. Broker Bob has a listing that requires a 6% brokerage commission. Broker Bob offers half the commission to a selling broker. Broker Sam brings a buyer who purchases at $200,000. The seller must pay Broker Bob
    (A) $3,000
    (B) $6,000
    (C) $9,000
    (D) $12,000

24. A broker who has a dozen salespersons working under his license asks salespersons to encourage clients to use a certain lender. The broker has an ownership interest in that lender. The salespersons should
    (A) be loyal to their broker
    (B) disclose the broker's interest in that lender when discussing lenders with clients
    (C) disregard instructions from the broker on lending matters
    (D) find the lenders with the lowest interest rate

25. A corporation has more than 100 stockholders. Its only assets are four shopping centers. To sell stock interstate, the salesperson must be licensed to sell
    (A) real estate
    (B) securities
    (C) timeshares
    (D) property management

26. A broker received two earnest money checks on two houses in one day. The broker should
    (A) hold the checks if requested by the prospective buyers
    (B) deposit the checks into the broker's trust or escrow account
    (C) split the earnest money with the listing broker
    (D) hold the checks uncashed until the closings

27. A homeowner sent a letter to six local real estate brokerage firms offering to pay a 10% commission to any firm that produced a buyer within 30 days. This created
    (A) an exclusive right to sell with all six firms
    (B) an open listing with the six firms
    (C) an intense competition
    (D) an invitation to talk with the owner

28. The lead broker should provide information about the fiduciary obligations of the broker's firm to all of the following parties *except* the
    (A) broker's salesperson
    (B) buyer's broker
    (C) subagents of the broker
    (D) buyer's lenders

29. A broker has a listing on a house. A buyer tells the broker that he will pay more than the listing price if the owner requests it. The broker
    (A) must tell the owner
    (B) must not tell the owner
    (C) must ask the buyer his top price
    (D) must step aside and let the owner negotiate

30. A broker learns that a 30-foot utility easement extends across the front of a vacant lot and that a 50-foot setback is required by zoning. The prospective buyer is considering building a home. The broker should
    (A) say nothing about either the easement or the setback
    (B) disclose both the easement and the setback
    (C) refer questions to the seller
    (D) disclose only the setback because the utility easement is within it

31. A real estate salesperson resigned from one broker and was soon employed by another. The listings the salesperson brought to the first broker
    (A) remain with the first broker
    (B) can be transferred to the new broker after 10 days
    (C) can be terminated by the client
    (D) can be kept by a salesperson if he or she becomes a broker

32. When a buyer asks about proximity to a floodplain, the salesperson should
    (A) say nothing
    (B) ask the owner
    (C) ask a neighbor about flooding
    (D) suggest the buyer obtain a current flood map from FEMA

33. On Tuesday, prospective buyers offered, in writing, a contract for $500,000 to purchase a house including the grand piano. On Wednesday, the owners counteroffered in writing with $500,000 without the piano. On Thursday, the owners changed their minds and were willing to accept $500,000, including the piano, stated in writing. On Friday, the buyers
    (A) are bound by their original offer, which was accepted on Thursday
    (B) are relieved of their original offer and not legally obligated to buy because their offer was rejected
    (C) must make a response to the seller
    (D) have violated the law by requesting that the piano be included

34. To be a dual agent, in states that allow it, the licensee must
    (A) receive compensation from the seller only
    (B) receive compensation from the buyer only
    (C) disclose the relationship and gain consent from both parties
    (D) be sure that the husband and wife both approve of the transaction

35. A contract is orally explained to an adult who is illiterate. The buyer signs with an X. The contract is
    (A) void
    (B) voidable by the illiterate
    (C) unenforceable
    (D) valid

36. For a contract to be binding, there must be
    (A) a meeting of the minds
    (B) consideration
    (C) legal purpose
    (D) all of the above

37. An enforceable contract may not include
    (A) affection
    (B) money
    (C) duress
    (D) love

38. Why do property owners sometimes ask surveyors to plant stakes in the ground?
    (A) to help locate boundaries
    (B) to make property marketable
    (C) to provide exact dimensions
    (D) to assist tax assessor

39. What is the best method to determine whether there is an encroachment?
    (A) get a survey
    (B) read a title policy
    (C) check the deed
    (D) get an appraisal

40. When a deed is recorded at the county courthouse, it provides what protection to the buyer?
    (A) guarantee of ownership
    (B) title insurance
    (C) constructive notice
    (D) chain of title

41. In the owner's absence, a relative with the owner's written general power of attorney
    (A) may sell the property
    (B) can sell anything except real estate
    (C) needs a court order to sell real estate
    (D) needs a specific power of attorney

42. A buyer defaulted on a contract of sale for which the buyer had paid a $5,000 earnest money deposit. What happens to the deposit?
    (A) The seller may keep it.
    (B) It belongs to the broker when a broker is involved.
    (C) The buyer can stop payment on the check if it has not been cashed.
    (D) The seller and buyer split it.

43. Which of the following is true about a contract of sale?
    (A) It conveys legal title when signed.
    (B) It binds both parties to complete the sale when conditions are met.
    (C) It requires specific performance of the buyer.
    (D) Upon signing, the broker has the right to collect a commission.

44. When the furniture in a house is sold with the house, the furniture is usually included in a
    (A) quitclaim deed
    (B) bill of sale
    (C) depreciation schedule
    (D) Section 1031 exchange

45. Upon reaching a written agreement in a contract of sale, the buyer acquires
    (A) fee simple ownership
    (B) leasehold interest
    (C) equitable title
    (D) deed to secure ownership

46. A farm is leased to a tenant farmer. The owner of the farm sells it with growing crops. The growing crops belong to the
    (A) tenant farmer
    (B) buyer who bought lock, stock, and barrel
    (C) seller
    (D) lender

47. Which parties determine the amount of earnest money in a real estate contract?
    (A) broker and buyer
    (B) local Board of Realtors
    (C) buyer and seller
    (D) seller and broker

48. When a contract of sale includes a due diligence contingency for the buyer's inspection,
    (A) the buyer should provide a punch list for the seller to fix
    (B) the buyer may terminate the contract if the results of an inspection are negative
    (C) the broker should get the repairs done
    (D) the seller must repair all problems regardless of the cost

49. A deed without any warranty is called a
    (A) general deed
    (B) special deed
    (C) limited deed
    (D) quitclaim deed

50. A restrictive covenant provides a(n)
    (A) limitation on ownership or use
    (B) exclusive use
    (C) limited warranty deed
    (D) government limitation

51. In a gross lease, who pays for operating expenses?
    (A) lessor
    (B) lessee
    (C) sandwich lease
    (D) mortgagee

52. When discussing the possibility of listing their house, the owners tell the salesperson, "We want to sell only to buyers of a certain national origin." The salesperson should
    (A) accept that condition and write it into the listing to comply with the principal's request
    (B) report the owners to the Department of Housing and Urban Development
    (C) because it is only a listing, accept the condition but be prepared to show the property to others
    (D) explain that this would be a violation of the federal Fair Housing Act; if the owners, persist, refuse the listing

53. An ad reads: "Home for sale only to a family." This ad violates
(A) TRID and Fair Housing Act
(B) TRID and Civil Rights Act of 1866
(C) Fair Housing Act and Americans with Disabilities Act
(D) HUD and EPA

54. An apartment complex is intended for the elderly only. What percentage of units must have at least one person over age 55?
(A) 100%
(B) 80%
(C) 50%
(D) 0%

55. A builder wishes to erect a 150-unit two-story building without an elevator. Under the federal Fair Housing Act, which units must be accessible to the handicapped?
(A) all units
(B) ground-floor units only
(C) second-floor units only
(D) 20% units set-aside

56. A broker advertises a home for rent as follows: "4BR, 2BA, nr. schools & shopping. Elderly married couples only." Which law has been broken?
(A) Uniform Residential Appraisal Report
(B) Americans with Disabilities
(C) TRID
(D) Federal Fair Housing Act

57. A lawyer wishes to buy a 40-year-old office building. To comply with the Americans with Disabilities Act, the lawyer must
(A) widen hallways
(B) install elevators
(C) install a wheelchair ramp for at least one entrance
(D) remove carpet from hallways

58. Under the federal Fair Housing Act, which of the following is a legal reason for refusing to rent to a member of a minority? The prospects are
(A) an unmarried couple
(B) a married couple with children
(C) unemployed retirees
(D) unable to afford the rent

59. A church owns a retirement facility and requires that tenants be members of the church. This is
    (A) a violation of Fair Housing laws
    (B) acceptable under all tenant requirements
    (C) acceptable provided there is no discrimination based on race, color, sex, national origin, familial status, or handicap
    (D) not permitted as an exception to Fair Housing laws

60. Which of the following is *not* protected under the federal Fair Housing Act?
    (A) race
    (B) religion
    (C) national origin
    (D) sexual preference

61. Which of the following neighborhood characteristics may an appraiser *not* describe in an appraisal?
    (A) economic characteristics
    (B) homogeneity of property
    (C) racial composition
    (D) proximity to transportation

62. An American citizen of Mexican ancestry asks to be shown a house in a certain neighborhood. How should the broker respond?
    (A) "I'd be pleased to show you a house for sale anywhere."
    (B) "You'll be happier in a house elsewhere."
    (C) "Neighbors there will not be friendly."
    (D) "The seller does not want me to show that house."

63. The law that requires a lender to disclose to a potential borrower the loan expenses of borrowing is
    (A) Equal Credit Opportunity Act
    (B) Interstate Land Sales Full Disclosure
    (C) Real Estate Settlement Procedures Act
    (D) TRID

64. A handy person is in the business of buying rundown homes, fixing them up, and reselling them at a profit. He buys a single-family home for $110,000. Under TRID, disclosure of discount points
    (A) does not apply to a business loan
    (B) is not required for loans over $100,000
    (C) is required for all single-family homes
    (D) is required only when more than 2 points are charged

65. The Comprehensive Environmental Response, Compensation, and Liability Act (CERCLA) is enforced by the
    (A) Department of Homeland Security
    (B) Department of Housing and Urban Development (HUD)
    (C) Federal Emergency Management Agency (FEMA)
    (D) Environmental Protection Agency (EPA)

66. When the zoning of a property changes, uses that were once compliant with zoning are
    (A) terminated because they are no longer compliant
    (B) allowed to continue as nonconforming uses
    (C) granted a variance
    (D) allowed to continue only if changes are made to comply with the new zoning

67. When a landfill is located near a subdivision and houses clearly lose value, the depreciation is called
    (A) physical deterioration
    (B) functional obsolescence
    (C) economic or external obsolescence
    (D) not in my back yard (NIMBY)

68. In a limited partnership, the limited partners have
    (A) limited liability
    (B) no input on everyday decisions
    (C) no ownership
    (D) only A and B

69. A builder of new homes seeks to have an 18-foot setback, whereas zoning requires a 20-foot setback. The builder should request
    (A) spot zoning
    (B) a nonconforming use
    (C) a variance
    (D) that authorities look the other way

70. The city rezoned a block of industrial property to become retail. Martin owns an industrial building that has been leased as a factory for the past five years. Martin now
    (A) can continue to use the property as a factory although it becomes a nonconforming use
    (B) must stop using the property as a factory when the lease expires
    (C) is exempt from zoning because of the age of the building
    (D) may terminate the lease

71. When property to be sold is likely to contain hazardous materials, the licensed salesperson should
    (A) disclose what he or she knows to interested buyers
    (B) let buyers find out on their own
    (C) comply with the owner's instructions
    (D) refer all questions to the owner

72. Which of the following pairs does not make sense?
   (A) grantor—gives deed
   (B) landlord—gives tenancy
   (C) mortgagee—gives loan
   (D) police power—gives deed restriction

73. Which of the following is true of high concentrations of radon gas?
   (A) It has a very noticeable odor.
   (B) It has been linked to lung cancer in long-term exposure.
   (C) It is released by plastic building materials.
   (D) It enters the house through openings in the attic.

74. Which of the following entities does not acquire first mortgages on individual homes?
   (A) FNMA
   (B) FHLMC
   (C) GNMA
   (D) FDIC

75. A 30-year mortgage for $104,400 at 7¾% requires monthly payments of $747.96. If the mortgage runs to maturity, how much total interest will be paid, rounded to the nearest dollar?
   (A) $104,400
   (B) $22,439
   (C) $164,866
   (D) $269,266

76. Lender *A* has a first mortgage but allows Lender *B*'s loan to have priority. This is an example of
   (A) home equity swap
   (B) upside-down loan
   (C) subordination
   (D) deficiency judgment

77. Which agency is not active in the secondary mortgage market?
   (A) VA
   (B) FNMA
   (C) GNMA
   (D) FHLMC

78. A homebuyer agrees to assume the seller's mortgage. The seller wants to be released from liability. The document used is called
   (A) an assumption of mortgage
   (B) a novation
   (C) a quitclaim deed
   (D) a foreclosure deed

79. A home equity line of credit is
   (A) secured
   (B) unsecured
   (C) for a 15-, 20-, or 30-year term
   (D) less than a first mortgage

80. What is the maximum prepayment penalty in a VA or an FHA mortgage?
   (A) 0%
   (B) 1%
   (C) 2%
   (D) no maximum percentage

81. The Federal National Mortgage Association (FNMA)
   (A) originates new mortgages to qualified individual applicants
   (B) buys mortgages in the secondary mortgage market
   (C) is no longer a large player in the mortgage market
   (D) is owned by Freddie Mac

82. Which type of loan is *not* likely to be provided as a second mortgage?
   (A) construction
   (B) home equity
   (C) wraparound
   (D) mechanic's lien

83. What type of loan calls for periodic changes in interest rates based on the interest rate of Treasury bills?
   (A) accelerated rate
   (B) amortized rate
   (C) adjustable rate
   (D) admonished rate

84. A homebuyer purchases a house subject to the existing mortgage on the home. The seller is
   (A) released from the debt
   (B) required to pay off the debt immediately
   (C) foolish for doing such a transaction
   (D) still obligated under the existing mortgage

85. A two-story house has all the bedrooms upstairs, and the only bathroom is on the first level. This is an example of
   (A) physical obsolescence
   (B) functional obsolescence
   (C) environmental obsolescence
   (D) physical depreciation

86. Clay (terra cotta) pipe and cast-iron pipe used for sewer lines have mostly been replaced in new installations by plastic (PVC) pipe. This is an example of
    (A) physical deterioration
    (B) functional obsolescence
    (C) economic obsolescence
    (D) material changes

87. Which type of depreciation is nearly always incurable?
    (A) physical depreciation
    (B) functional depreciation
    (C) functional obsolescence
    (D) external obsolescence

88. A real estate speculator purchases three adjacent lots for $10,000 each. He combines them into one large lot worth $50,000. This is an example of
    (A) combinatory increment
    (B) sum of parts
    (C) accretion
    (D) plottage

89. When appraising a vacant building that was previously used as a church, the most applicable approach would most likely be the
    (A) cost approach
    (B) income approach
    (C) capitalization
    (D) faith based

90. When applying the cost approach, the appraiser considers all of the following *except*
    (A) land at market value
    (B) replacement cost or reproduction cost
    (C) depreciation
    (D) gross rental income

91. A competitive market analysis considers
    (A) comparable sales
    (B) highest and best use
    (C) gross income
    (D) leased sales

92. Faded and peeling paint is referred to in an appraisal as
    (A) deferred maintenance
    (B) preventive maintenance
    (C) lack of care
    (D) property waste

93. An appraisal approach that places emphasis on market rental income minus operating expenses is the
(A) market approach
(B) direct sales comparison
(C) cost approach
(D) income capitalization

94. From all potential possible uses of a property, the highest and best use of land if vacant
(A) attracts the greatest gross income
(B) attracts the greatest net income
(C) provides the highest return to the land
(D) has the greatest floor area coverage ratio

95. The direct sales comparison approach is most applicable for a
(A) retail shopping center
(B) warehouse
(C) church
(D) ten-year-old single-family house in an active housing market

96. The income approach to value
(A) is the best approach for appraising a single-family home
(B) is the best way to determine the highest and best use of vacant land
(C) is used to estimate the value of investment property
(D) is one of a dozen acceptable approaches to appraising real estate

97. To qualify as a Section 1031 exchange, the property must be
(A) a personal residence
(B) used in a trade or business or held as an investment
(C) land
(D) mortgaged

98. A house is appraised at $250,000. The assessment ratio is 60%. The tax rate is 30 mills. Taxes for the year are
(A) $750
(B) $4,500
(C) $5,000
(D) $7,500

99. Mr. and Mrs. Martin jointly bought a principal residence in 2004 for $325,000 and sold it in 2020 for $500,000. What portion of their $175,000 gain is subject to capital gains tax?
    (A) $0
    (B) $325,000
    (C) $500,000
    (D) $175,000

100. In a Subchapter S Corporation, the corporation itself
    (A) pays no federal income tax
    (B) pays no property tax on land it owns
    (C) gets a discount on state income tax
    (D) is taxed just like a C Corporation

ANSWER KEY
Model Test 7

1. **D**
2. **A**
3. **C**
4. **D**
5. **C**
6. **B**
7. **A**
8. **C**
9. **A**
10. **D**
11. **B**
12. **C**
13. **A**
14. **A**
15. **A**
16. **B**
17. **A**
18. **D**
19. **C**
20. **D**
21. **B**
22. **B**
23. **D**
24. **B**
25. **B**
26. **B**
27. **D**
28. **D**
29. **A**
30. **B**
31. **A**
32. **D**
33. **B**
34. **C**
35. **D**
36. **D**
37. **C**
38. **A**
39. **A**
40. **C**
41. **A**
42. **A**
43. **B**
44. **B**
45. **C**
46. **A**
47. **C**
48. **B**
49. **D**
50. **A**
51. **A**
52. **D**
53. **C**
54. **B**
55. **B**
56. **D**
57. **C**
58. **D**
59. **C**
60. **D**
61. **C**
62. **A**
63. **D**
64. **A**
65. **D**
66. **B**
67. **C**
68. **D**
69. **C**
70. **A**
71. **A**
72. **D**
73. **B**
74. **D**
75. **C**
76. **C**
77. **A**
78. **B**
79. **A**
80. **A**
81. **B**
82. **A**
83. **C**
84. **D**
85. **B**
86. **B**
87. **D**
88. **D**
89. **A**
90. **D**
91. **A**
92. **A**
93. **D**
94. **C**
95. **D**
96. **C**
97. **B**
98. **B**
99. **A**
100. **A**

ANSWER EXPLANATIONS—ARITHMETIC QUESTIONS

NOTE: The section contains explanations for only the arithmetic questions in the model exam.

23. $200,000 × .06 = $12,000

75. 30 years × 12 months = 360 payments
    360 × $747.96 = $269,265.60 total paid
    $269,266 – $104,400 original cost = $164,866 interest paid

98. $250,000 × 60% ratio = $150,000; $150,000 × 30 mills = $4,500

99. No tax on gains up to $500,000 for a married couple. Answer A is correct.

ANSWER SHEET
Model Examination 8

Broker and Salesperson

1. Ⓐ Ⓑ Ⓒ Ⓓ	26. Ⓐ Ⓑ Ⓒ Ⓓ	51. Ⓐ Ⓑ Ⓒ Ⓓ	76. Ⓐ Ⓑ Ⓒ Ⓓ
2. Ⓐ Ⓑ Ⓒ Ⓓ	27. Ⓐ Ⓑ Ⓒ Ⓓ	52. Ⓐ Ⓑ Ⓒ Ⓓ	77. Ⓐ Ⓑ Ⓒ Ⓓ
3. Ⓐ Ⓑ Ⓒ Ⓓ	28. Ⓐ Ⓑ Ⓒ Ⓓ	53. Ⓐ Ⓑ Ⓒ Ⓓ	78. Ⓐ Ⓑ Ⓒ Ⓓ
4. Ⓐ Ⓑ Ⓒ Ⓓ	29. Ⓐ Ⓑ Ⓒ Ⓓ	54. Ⓐ Ⓑ Ⓒ Ⓓ	79. Ⓐ Ⓑ Ⓒ Ⓓ
5. Ⓐ Ⓑ Ⓒ Ⓓ	30. Ⓐ Ⓑ Ⓒ Ⓓ	55. Ⓐ Ⓑ Ⓒ Ⓓ	80. Ⓐ Ⓑ Ⓒ Ⓓ
6. Ⓐ Ⓑ Ⓒ Ⓓ	31. Ⓐ Ⓑ Ⓒ Ⓓ	56. Ⓐ Ⓑ Ⓒ Ⓓ	81. Ⓐ Ⓑ Ⓒ Ⓓ
7. Ⓐ Ⓑ Ⓒ Ⓓ	32. Ⓐ Ⓑ Ⓒ Ⓓ	57. Ⓐ Ⓑ Ⓒ Ⓓ	82. Ⓐ Ⓑ Ⓒ Ⓓ
8. Ⓐ Ⓑ Ⓒ Ⓓ	33. Ⓐ Ⓑ Ⓒ Ⓓ	58. Ⓐ Ⓑ Ⓒ Ⓓ	83. Ⓐ Ⓑ Ⓒ Ⓓ
9. Ⓐ Ⓑ Ⓒ Ⓓ	34. Ⓐ Ⓑ Ⓒ Ⓓ	59. Ⓐ Ⓑ Ⓒ Ⓓ	84. Ⓐ Ⓑ Ⓒ Ⓓ
10. Ⓐ Ⓑ Ⓒ Ⓓ	35. Ⓐ Ⓑ Ⓒ Ⓓ	60. Ⓐ Ⓑ Ⓒ Ⓓ	85. Ⓐ Ⓑ Ⓒ Ⓓ
11. Ⓐ Ⓑ Ⓒ Ⓓ	36. Ⓐ Ⓑ Ⓒ Ⓓ	61. Ⓐ Ⓑ Ⓒ Ⓓ	86. Ⓐ Ⓑ Ⓒ Ⓓ
12. Ⓐ Ⓑ Ⓒ Ⓓ	37. Ⓐ Ⓑ Ⓒ Ⓓ	62. Ⓐ Ⓑ Ⓒ Ⓓ	87. Ⓐ Ⓑ Ⓒ Ⓓ
13. Ⓐ Ⓑ Ⓒ Ⓓ	38. Ⓐ Ⓑ Ⓒ Ⓓ	63. Ⓐ Ⓑ Ⓒ Ⓓ	88. Ⓐ Ⓑ Ⓒ Ⓓ
14. Ⓐ Ⓑ Ⓒ Ⓓ	39. Ⓐ Ⓑ Ⓒ Ⓓ	64. Ⓐ Ⓑ Ⓒ Ⓓ	89. Ⓐ Ⓑ Ⓒ Ⓓ
15. Ⓐ Ⓑ Ⓒ Ⓓ	40. Ⓐ Ⓑ Ⓒ Ⓓ	65. Ⓐ Ⓑ Ⓒ Ⓓ	90. Ⓐ Ⓑ Ⓒ Ⓓ
16. Ⓐ Ⓑ Ⓒ Ⓓ	41. Ⓐ Ⓑ Ⓒ Ⓓ	66. Ⓐ Ⓑ Ⓒ Ⓓ	91. Ⓐ Ⓑ Ⓒ Ⓓ
17. Ⓐ Ⓑ Ⓒ Ⓓ	42. Ⓐ Ⓑ Ⓒ Ⓓ	67. Ⓐ Ⓑ Ⓒ Ⓓ	92. Ⓐ Ⓑ Ⓒ Ⓓ
18. Ⓐ Ⓑ Ⓒ Ⓓ	43. Ⓐ Ⓑ Ⓒ Ⓓ	68. Ⓐ Ⓑ Ⓒ Ⓓ	93. Ⓐ Ⓑ Ⓒ Ⓓ
19. Ⓐ Ⓑ Ⓒ Ⓓ	44. Ⓐ Ⓑ Ⓒ Ⓓ	69. Ⓐ Ⓑ Ⓒ Ⓓ	94. Ⓐ Ⓑ Ⓒ Ⓓ
20. Ⓐ Ⓑ Ⓒ Ⓓ	45. Ⓐ Ⓑ Ⓒ Ⓓ	70. Ⓐ Ⓑ Ⓒ Ⓓ	95. Ⓐ Ⓑ Ⓒ Ⓓ
21. Ⓐ Ⓑ Ⓒ Ⓓ	46. Ⓐ Ⓑ Ⓒ Ⓓ	71. Ⓐ Ⓑ Ⓒ Ⓓ	96. Ⓐ Ⓑ Ⓒ Ⓓ
22. Ⓐ Ⓑ Ⓒ Ⓓ	47. Ⓐ Ⓑ Ⓒ Ⓓ	72. Ⓐ Ⓑ Ⓒ Ⓓ	97. Ⓐ Ⓑ Ⓒ Ⓓ
23. Ⓐ Ⓑ Ⓒ Ⓓ	48. Ⓐ Ⓑ Ⓒ Ⓓ	73. Ⓐ Ⓑ Ⓒ Ⓓ	98. Ⓐ Ⓑ Ⓒ Ⓓ
24. Ⓐ Ⓑ Ⓒ Ⓓ	49. Ⓐ Ⓑ Ⓒ Ⓓ	74. Ⓐ Ⓑ Ⓒ Ⓓ	99. Ⓐ Ⓑ Ⓒ Ⓓ
25. Ⓐ Ⓑ Ⓒ Ⓓ	50. Ⓐ Ⓑ Ⓒ Ⓓ	75. Ⓐ Ⓑ Ⓒ Ⓓ	100. Ⓐ Ⓑ Ⓒ Ⓓ

Model Examination 8

BROKER AND SALESPERSON

See page 639 for the answer key.

1. Which of the following is *not* real estate?
   (A) a load of new shingles waiting to be installed
   (B) a built-in dishwasher
   (C) an underground storage tank used for heating oil
   (D) planted shrubs

2. A speculator combined two lots at a cost of $10,000 each, resulting in one lot worth $30,000. This is an example of
   (A) competition
   (B) capitalism
   (C) plottage value
   (D) unjust enrichment

3. A hurricane causes a violent shift in surface soil, removing the soil from the subject property and distributing it downstream. This is called
   (A) erosion
   (B) riparian
   (C) avulsion
   (D) decomposition

4. Which of the following is *not* true of a life estate?
   (A) The life tenant may take estovers.
   (B) The life tenant may not commit waste.
   (C) The life tenant may live on the property.
   (D) The life tenant must obtain permission from the remainderman before making any substantive changes.

5. Two unrelated people owned property as joint tenants. One died. His share goes to
   (A) his heirs at law
   (B) his heirs under his will
   (C) the remaining joint tenant
   (D) his wife if he is married

6. Five individuals own a small shopping center as tenants in common. One wishes to sell his interest. He
   (A) may do so
   (B) must offer it to the remaining owners at the same terms as to outsiders
   (C) cannot sell because of survivorship
   (D) can sell only if given written permission by the other owners

7. A city wants to buy land to build a convention center. As to the use of eminent domain,
   (A) it is not allowed because a convention center is not a necessity
   (B) it can be applied with market value paid for the land
   (C) it will be too costly to enforce
   (D) only federal and state governments have that power, not a city

8. Which of the following is true about eminent domain?
   (A) The process used is called condemnation.
   (B) Governments can use it whenever they want.
   (C) Governments can pay whatever amount they want.
   (D) Constitutionally, it comes under the First Amendment.

9. The right of __________ gives the state ownership of property left by one who died without heirs and without a will.
   (A) estovers
   (B) intestate
   (C) emblements
   (D) escheat

10. A lender's title insurance policy protects the lender against
    (A) unpaid interest
    (B) unpaid principal
    (C) undiscovered existing title defects
    (D) home equity loan defaults

11. Swimming pools and tennis courts in a condominium development are known as
    (A) amenities
    (B) PUDs
    (C) associations
    (D) common areas

12. A title insurance policy generally protects the insured party or parties from
    (A) valid matters to which the policy has not taken an exception
    (B) death or illness of the seller
    (C) a change in property use
    (D) an unrecorded deed to which the policy has taken an exception

13. Sarah and Rachel bought a house as tenants in common. If Sarah dies,
  (A) Rachel inherits Sarah's interest
  (B) Rachel keeps her own part and Sarah's heirs get Sarah's interest
  (C) the result is the same as it would have been if they had been joint tenants
  (D) the rules of community property are applied

14. Brokers who violate the Sherman Antitrust Act may be slapped with a fine of up to $100,000 plus a maximum prison term of _________ years.
  (A) one
  (B) two
  (C) three
  (D) ten

15. A seller's agent should do all of the following *except*
  (A) tell everything about the offer
  (B) present all offers
  (C) explain the advantages of an offer
  (D) discuss the seller's pending divorce with the buyer

16. A broker acting as a seller's agent
  (A) can raise but not reduce the listing price without the seller's approval
  (B) must split his commission with a salesperson sponsored by a different broker
  (C) must present all offers to the principal unless instructed by the seller not to do so
  (D) keep as confidential whatever information the principal requests

17. The rate of brokerage commissions for the sale of a house is determined by
  (A) standard practice in the community
  (B) realtor guidelines in the state
  (C) national norms
  (D) negotiations between the owner and the broker

18. After an offer was accepted, the salesperson received $5,000 of earnest money. What should he do with the check?
  (A) cash it immediately
  (B) deposit it in his personal checking account
  (C) deposit it in his broker's escrow account
  (D) keep the amount that equals his commission and deposit any excess into an escrow account

19. Typically, the listing broker has a fiduciary relationship with the
  (A) sponsored salesperson
  (B) tenant
  (C) client
  (D) Board of Realtors

20. When a salesperson shows a house that is listed with her firm to a buyer she represents, there is a(n)
   (A) dual agency
   (B) general agency
   (C) multiple list
   (D) exclusive right to sell

21. A salesperson has listed a house for sale. A potential buyer is owed what duty?
   (A) loyalty
   (B) honesty
   (C) full disclosure
   (D) none

22. A broker who represents a buyer with a contract of sale has learned that the seller inherited the property and has two siblings who may have an interest. The buyer's broker should tell the listing broker that
   (A) the deal is off
   (B) he should suggest that the seller get legal advice, possibly to get quitclaim deeds from the siblings
   (C) his buyer will walk away from this fishy deal
   (D) a significant price reduction is in order for the risk assumed

23. A broker is representing a buyer and showed him a lot in a new subdivision. The broker did not ask the developer but, knowing that the developer had put in streets in other subdivisions, told the buyer that the developer would do the same in this one. The buyer relied on the broker's statement. The developer did not put in streets in this subdivision. Who is liable for damages?
   (A) no one because nothing was put in writing
   (B) the developer for not doing here what he had done elsewhere
   (C) the broker because he misrepresented what would happen
   (D) both the broker and the developer

24. The owner of a shopping center signed a management agreement with a real estate broker. Is there an agency relationship?
   (A) yes
   (B) no
   (C) maybe
   (D) never

25. At 2 P.M., licensed broker John receives an offer to purchase a home that Kevin owns and John listed. John arranges with Kevin to meet at 8 P.M. Another broker who cooperates with John shows the same property at 4 P.M. A third broker, also cooperating, calls John with an interested purchaser to show the property at 6 P.M. John should
   (A) meet with Kevin at 8 P.M. and present all offers
   (B) meet with Kevin at 8 P.M. and present the highest offer
   (C) meet with Kevin at 8 P.M. and present the first offer
   (D) not allow the property to be shown after the first offer at 2 P.M.

26. Prospective buyers are orally told that a house is on a sewer system. They negotiate a contract. The day before closing, the buyers discover that it is not on a sewer system. They refuse to close. Which of the following is correct?
   (A) The buyers must close because they signed the contract.
   (B) The seller can keep the earnest money.
   (C) The broker can collect a commission.
   (D) The buyers are within their rights to walk away because of the misrepresentation.

27. The day before closing, the listing broker discovers that the roof is leaking and has damaged a bedroom ceiling. The broker should
   (A) keep quiet
   (B) tell the selling broker and sellers of this issue
   (C) repair the roof
   (D) read the seller's disclosure and decide what to do

28. A prospective buyer asks the listing broker whether the property is in or near a floodplain. The broker should suggest that the buyer ask
   (A) the owner
   (B) the lender
   (C) the U.S. Army Corps of Engineers
   (D) the Federal Emergency Management Agency, which prepares flood maps

29. A real estate broker pays its salespersons 50% of commissions up to $100,000 in commissions received in a year and 60% of commissions in excess of $100,000. Salesperson Sam has generated $250,000 of total gross commissions. He should earn
   (A) $106,000
   (B) $125,000
   (C) $140,000
   (D) $150,000

30. Owners tell the listing agent that the roof leaks. To avoid misrepresentation, the listing agent must
   (A) disclose the leak to prospective buyers
   (B) suggest that buyers get an inspection
   (C) keep silent if requested by the owners
   (D) refer questions about the roof to the owners

31. To advertise a property through social media, the broker must get written authorization from
    (A) Angie's List or similar media
    (B) the seller
    (C) the state real estate commission
    (D) Facebook or Google

32. A broker has a general liability insurance policy. This should protect against
    (A) property damage from a fire in the broker's office
    (B) an injury when a chair in the broker's office breaks when an obese client sits down
    (C) severance pay to terminated employees
    (D) a car accident caused by the broker's salesperson

33. If a potential buyer expresses concern about the possible presence of radon gas in a house being considered, the salesperson's best response is to
    (A) ask the sellers if radon gas is present
    (B) offer less money for the house
    (C) explain that radon gas is not harmful
    (D) suggest that the house be tested for the presence of radon gas

34. A salesperson working for a broker may
    (A) be classified as an independent contractor for tax purposes
    (B) be paid the minimum wage rate
    (C) receive a commission from another broker
    (D) advertise in her own name, without including the broker, provided she pays for the ad herself

35. A buyer in a contract of sale is diagnosed with dementia. His legal caretaker seeks to set the contract aside, declaring it void. Most likely
    (A) it can be declared void because of the buyer's incompetence
    (B) the deal is done
    (C) the seller may keep the earnest money
    (D) the broker has earned a commission

36. Which of the following is a unilateral contract?
    (A) lease
    (B) option
    (C) contract for deed
    (D) exclusive right to sell listing

37. A broker who represented a buyer told the buyer where the property boundaries were but did not disclose that they were uncertain. After closing, the buyer discovered that his new fence, although within the boundaries the broker showed him, was actually encroaching on his neighbor's land. The broker
   (A) may be liable because he did not tell the buyer that the boundaries he described were uncertain
   (B) is not liable because he is not a surveyor
   (C) did adequate due diligence
   (D) is not responsible for anything dealing with boundaries

38. Which of the following is an example of encroachment?
   (A) legal access road connecting a public road with a property and running across another property
   (B) electric power lines across a property
   (C) a portable storage building that straddles a property line
   (D) growing crops planted by a tenant farmer

39. *Annexation* is the way in which
   (A) something is attached to real estate
   (B) a city acquires more land
   (C) both A and B
   (D) neither A nor B

40. A licensed salesperson presented the owner with an offer for $150,000. The owner counteroffered at $160,000. While the potential buyer is considering the counteroffer, the owner sends an e-mail rescinding that counteroffer. Which of the following is correct?
   (A) The buyer gets 48 hours to consider the counteroffer.
   (B) There is no contract.
   (C) The buyer must provide a reply.
   (D) The broker has earned a commission.

41. There is a clause in a contract of sale allowing an inspection for wood-destroying insects. Termites are discovered. Who pays to correct the problem?
   (A) the buyer
   (B) the seller
   (C) the buyer and seller each pay half
   (D) the person named in the contract as the responsible party

42. A real estate sales contract
   (A) binds the parties to complete the contract, subject to resolving contingencies
   (B) transfers title upon signing
   (C) entitles the broker to keep the earnest money no matter what happens thereafter
   (D) gives the seller the right to demand closing

43. The owner of a rented retail building received an offer to buy, but the lease contains a clause that gives the tenant the right to buy the property first and with the same terms. This clause is called
(A) right of first refusal
(B) option
(C) Section 1031 exchange
(D) tenant at sufferance

44. Neil enters into an oral contract to purchase real estate and provides a $1,500 check as earnest money. The contract is
(A) unenforceable
(B) binding on the buyer
(C) voidable by the seller
(D) negotiable

45. When an offer includes the phrase "time is of the essence," it means that the offer must be accepted or rejected
(A) within a week
(B) within a day
(C) upon delivery
(D) as soon as possible

46. To be binding, a real estate sales agreement must
(A) include a title search
(B) include a survey
(C) include an appraisal
(D) be signed by the seller

47. To be valid, a deed must be signed by
(A) the grantor
(B) the grantee
(C) the mortgagor
(D) the mortgagee

48. Typically, a deed restriction serves to
(A) prevent a lease
(B) prevent a mortgage
(C) prevent a marriage
(D) control the future use of the property

49. Several restrictions on use of a property are noted in prior deeds. These are called
(A) covenants not to compete
(B) restrictive covenants
(C) land use planning
(D) zoning categories

50. Which of the following phrases does HUD consider discriminatory advertising?
   (A) "Senior housing"
   (B) "Friendly neighbors"
   (C) "Exclusive neighborhood"
   (D) "No one excluded"

51. A homeowner wishes to lease his house for a year while he takes a sabbatical. He instructs the licensed agent to lease to a white family only. The agent must
   (A) be loyal and comply
   (B) ask for the request in writing
   (C) refuse to follow the order because it violates the Fair Housing Act
   (D) ask the broker what the firm's policy is

52. A real estate broker may not
   (A) identify local schools, whether public or private
   (B) mention ethnic festivities held in the city
   (C) state the racial composition of the neighborhood
   (D) map religious facilities in the area

53. An FDIC-insured lender is considering a loan application to remodel an existing house. The applicant is a member of a minority. Under the federal Fair Housing Act, the lender must make the loan unless
   (A) the home is in a minority neighborhood
   (B) the home is more than 50 years old
   (C) the applicant is not economically qualified
   (D) the property has lead-based paint

54. A prospective tenant in a 50-unit apartment complex requires a wheelchair. The tenant wants to rent a unit, install grab bars in the bathroom, and lower door handles. Under the federal Fair Housing Act, the landlord
   (A) must allow and pay for these changes
   (B) must allow the tenant to make these changes at the tenant's expense and can require the tenant to restore the property at the end of the lease
   (C) can refuse to rent the unit because the unit is not handicapped accessible
   (D) must allow the tenant's modifications and accept the property as modified when the lease ends

55. A salesperson was taking a listing on a million-dollar-plus house. The owner told the salesperson that she would be unwilling to sell to a minority buyer. The salesperson should
   (A) honor the principal's request
   (B) smile and take the listing but be prepared to disobey
   (C) decline because following the seller's instruction would violate the federal Fair Housing Act
   (D) report the owner to HUD

56. A broker shows an Asian family houses only in Asian neighborhoods. This is
   (A) legal steering
   (B) blockbusting
   (C) effective marketing
   (D) illegal steering

57. A lender refuses to make mortgage loans on properties in a certain area of a city regardless of the qualifications of the borrower. This is
   (A) discrimination
   (B) price fixing
   (C) blockbusting
   (D) redlining

58. Which government agency investigates alleged violations of Fair Housing laws?
   (A) FHA
   (B) EEOC
   (C) HUD
   (D) FNMA

59. Which of the following terms can be included in a real estate ad with no further disclosure required?
   (A) "Take eight years to pay"
   (B) "Only $10,000 down"
   (C) "Payments of $500 per month"
   (D) "6% annual percentage rate"

60. TRID requires which of the following in a Closing Disclosure statement?
   (A) the amount of attorney fees for estate planning
   (B) the mortgage lender's fee for advertising
   (C) a limitation on how often the mortgage can be sold
   (D) any mortgage prepayment penalties

61. A buyer of a house built before 1978 must be advised of the possibility that it contains
   (A) asbestos
   (B) radon
   (C) lead-based paint
   (D) termites

62. A property was used for retail sales, but the zoning has been changed to residential. The property can continue as retail use but is
   (A) limited
   (B) biased
   (C) nonconforming
   (D) spot zoning

63. The presence of wetlands
 (A) lowers property value
 (B) causes the property to be useless
 (C) enhances the property value because it attracts wildlife
 (D) should be disclosed to potential buyers

64. Every major city in the United States except Houston applies
 (A) zoning
 (B) Fair Housing Act
 (C) TRID
 (D) Real Estate Settlement Procedures Act

65. An individual who attends a sales presentation is offered use of a certain unit, along with common areas of a project, for a certain week each year. This is known as a
 (A) cooperative
 (B) condominium
 (C) rental pool
 (D) timeshare

66. During a property inspection, the inspector learned that the building had been used as a facility to repair automobile brakes. This should suggest the possible presence of
 (A) asbestos
 (B) radon
 (C) hydrocarbons
 (D) lead-based paint

67. A subdivision is in an area zoned residential. A homeowner wishes to open a tailor shop in his basement. He requires a
 (A) conditional use permit
 (B) variance
 (C) occupancy permit
 (D) nonconforming use

68. Superfund imposes
 (A) partial liability for cleanup
 (B) waste
 (C) forfeiture of land
 (D) strict liability for damage

69. A gas produced by radioactive materials naturally found in the earth is
 (A) carbon dioxide
 (B) carbon monoxide
 (C) nitrogen
 (D) radon

70. Which of the following is *not* governed by zoning?
   (A) use of a residence partly for a business
   (B) racial considerations in a sale
   (C) maximum number of apartment units per acre
   (D) building height

71. All of the following are categorized as a leasehold estate *except*
   (A) periodic estate
   (B) estate for years
   (C) estate from year to year
   (D) vacation timeshare

72. A real estate broker is asked to help lease a shopping center that has 20 tenants. The broker suggests that a standard lease form be used for all tenants. In trying to sign tenants, many want terms different from those on the standard form. The broker should
   (A) modify each lease to serve the tenant
   (B) ask the owner how to proceed
   (C) seek legal assistance from a broker with a law degree
   (D) tell the prospective tenants to go elsewhere

73. Landlords may raise rents in a rent-controlled building annually by 6% of the cost of newly installed improvements. An apartment currently rents for $500 per month. The landlord spends $2,000 on a security system and wants to raise rent by $12 per month. By how much would this exceed an eligible increase?
   (A) $0
   (B) $2
   (C) $5
   (D) $10

74. A lease has two years left. The lessee may withhold rent if
   (A) the property is sold
   (B) the lessee moves out of town
   (C) the property is condemned for health reasons as a result of the lessor's neglect
   (D) the lessee's family outgrows the space

75. A developer secures a long-term land lease from the landowner. The developer intends to build an office tower. The developer has what interest in the land?
   (A) leasehold
   (B) leased fee
   (C) sandwich lease
   (D) sale/leaseback

76. Which of the following does not fall under the law of agency?
    (A) unlicensed assistant to a real estate broker
    (B) broker with a listing
    (C) salesperson sponsored by a broker
    (D) property manager under an agreement

77. Which of the following is true of both an exclusive right to sell listing and an exclusive agency listing?
    (A) Only one broker may represent the seller.
    (B) The owner may sell without paying a commission.
    (C) Both are open listings.
    (D) Neither can be entered in the MLS.

78. A property is titled in the name of a person who is now deceased. She left four children, two of whom will sign the listing. Which of the following is correct?
    (A) The listing is binding on all the children.
    (B) Two signatures is 50%, which is inadequate. The listing needs three signatures.
    (C) The broker doesn't know who owns the property.
    (D) All siblings are bound by the consent of any one of them.

79. A broker has an exclusive right to sell listing on a new subdivision. Potential buyers, who are individuals without agents, want to make offers. The broker prepares contracts for these buyers to sign. The broker
    (A) can represent both parties
    (B) must deal honestly with all parties
    (C) owes loyalty to the buyers
    (D) cannot serve two masters in the same contract

80. A property manager's compensation is most often based on a
    (A) percentage of gross income
    (B) percentage of net operating income
    (C) percentage of operating expense
    (D) percentage of expenses saved

81. If an owner reserves the right to give a listing to more than one broker, she would use
    (A) multiple listing
    (B) exclusive listing
    (C) open listing
    (D) net listing

82. A broker lists a property, but the owner places unusually severe restrictions on showing it. The broker may
    (A) cancel the listing contract
    (B) not cancel the listing contract
    (C) seek a release from the owner
    (D) complain to the state real estate commission

83. A real estate salesperson prepares an online ad that says, "This property has the most spectacular view in the city." This is an example of
(A) deception
(B) misrepresentation
(C) fraud
(D) puffing

84. When stating the seller's listing price and terms to a prospective buyer, the seller's broker can disclose
(A) the lowest amount the seller will accept
(B) the price and terms stated in the listing
(C) the lowest amount the broker thinks the property will appraise for
(D) the highest amount the broker thinks the property will appraise for

85. When a licensed salesperson prepares a listing that is signed by the property owner, the listing is a contract between
(A) salesperson's broker and owner
(B) salesperson and buyer
(C) salesperson and owner
(D) salesperson and broker

86. An owner sold a lakefront lot using a *contract for deed*, taking 10% down and with payments to be made over the next 10 years. When does the buyer get a deed?
(A) upon final payment
(B) upon closing
(C) when the contract expires
(D) when the seller dies

87. A seller provided an option to the buyer for 90 days with an option fee of $5,000 and a property price of $100,000. During the option period, the buyer decides not to buy. What does the buyer need to do?
(A) nothing
(B) notify the seller of his intentions
(C) reconsider the decision once more
(D) buy as required by the option

88. Mr. X, a property owner, sells an option to buy land for $1,000 per acre for 1½ years. Mr. X dies a month later. What happens?
(A) The option remains valid.
(B) The option is void upon Mr. X's death.
(C) The option expires one year after Mr. X's death.
(D) The option is void but can be renewed by Mr. X's heirs.

89. To obtain a VA loan, the borrower is *not* required to
   (A) have served in the U.S. military
   (B) pay 10% down
   (C) find a VA-approved lender
   (D) get any insurance coverage

90. What is the term for making a loan with an illegally high interest rate?
   (A) usufructory
   (B) usury
   (C) encroachment
   (D) leverage

91. An owner financed the sale of his home and received a better interest rate than banks offered on deposits. This loan is known as
   (A) shrewd
   (B) purchase-money mortgage
   (C) wraparound mortgage
   (D) risky mortgage

92. A mortgage has a rate that may change every six months depending on changes in the rate of Treasury bills. This is
   (A) an indexed loan
   (B) an adjustable-rate mortgage
   (C) a capped loan
   (D) an adjustable-payment mortgage

93. Which of the following types of first mortgage loan is typically described as "conventional"?
   (A) 80% loan from mortgage banker
   (B) 97% FHA loan
   (C) 100% VA loan
   (D) contract for deed

94. When a house is sold and an existing mortgage is assumed,
   (A) the buyer becomes liable for the full loan
   (B) the seller is relieved of the debt
   (C) the lender must approve the sale
   (D) the down payment is zero

95. After paying off a mortgage loan, the property owner should be sure that which of the following instruments is recorded?
   (A) relief of debt
   (B) novation
   (C) satisfaction of mortgage
   (D) burned mortgage

96. Under TRID, lenders must
   (A) provide a Loan Estimate
   (B) limit the number of discount points to not more than 2
   (C) let the buyer void the transaction up to ten days after closing
   (D) not charge more than 1% for lender fees

97. The sale of a condominium with a mandatory rental pool for units not being used by owners is a sale of
   (A) real estate
   (B) securities
   (C) condominiums
   (D) timeshares

98. Federal Fair Housing laws protect all *except*
   (A) racial minorities
   (B) families with children
   (C) pregnant women
   (D) graduate students

99. A broker is advertising property for a homebuilder. Which phrase is *not* appropriate to use?
   (A) "New homes for $350,000"
   (B) "Veterans may apply for loans with nothing down"
   (C) "No down payment required for veterans"
   (D) "Near schools and shopping"

100. Under TRID, willful violation of disclosure requirements can result in
   (A) fines
   (B) imprisonment
   (C) both fines and imprisonment
   (D) license suspension

ANSWER KEY
Model Test 8

1 **A**	26. **D**	51. **C**	76. **A**
2. **C**	27. **B**	52. **C**	77. **A**
3. **C**	28. **D**	53. **C**	78. **C**
4. **D**	29. **C**	54. **B**	79. **B**
5. **C**	30. **A**	55. **C**	80. **A**
6. **A**	31. **B**	56. **D**	81. **C**
7. **B**	32. **B**	57. **D**	82. **C**
8. **A**	33. **D**	58. **C**	83. **D**
9. **D**	34. **A**	59. **D**	84. **B**
10. **C**	35. **A**	60. **D**	85. **A**
11. **D**	36. **B**	61. **C**	86. **A**
12. **A**	37. **A**	62. **C**	87. **A**
13. **B**	38. **C**	63. **D**	88. **A**
14. **C**	39. **C**	64. **A**	89. **B**
15. **D**	40. **B**	65. **D**	90. **B**
16. **C**	41. **D**	66. **A**	91. **B**
17. **D**	42. **A**	67. **A**	92. **B**
18. **C**	43. **A**	68. **D**	93. **A**
19. **C**	44. **A**	69. **D**	94. **A**
20. **A**	45. **D**	70. **B**	95. **C**
21. **B**	46. **D**	71. **D**	96. **A**
22. **B**	47. **A**	72. **B**	97. **B**
23. **C**	48. **D**	73. **B**	98. **D**
24. **A**	49. **B**	74. **C**	99. **C**
25. **A**	50. **C**	75. **A**	100. **C**

ANSWER SHEET
Model Examination 9

Appraiser

1. Ⓐ Ⓑ Ⓒ Ⓓ Ⓔ	26. Ⓐ Ⓑ Ⓒ Ⓓ Ⓔ	51. Ⓐ Ⓑ Ⓒ Ⓓ Ⓔ	76. Ⓐ Ⓑ Ⓒ Ⓓ Ⓔ
2. Ⓐ Ⓑ Ⓒ Ⓓ Ⓔ	27. Ⓐ Ⓑ Ⓒ Ⓓ Ⓔ	52. Ⓐ Ⓑ Ⓒ Ⓓ Ⓔ	77. Ⓐ Ⓑ Ⓒ Ⓓ Ⓔ
3. Ⓐ Ⓑ Ⓒ Ⓓ Ⓔ	28. Ⓐ Ⓑ Ⓒ Ⓓ Ⓔ	53. Ⓐ Ⓑ Ⓒ Ⓓ Ⓔ	78. Ⓐ Ⓑ Ⓒ Ⓓ Ⓔ
4. Ⓐ Ⓑ Ⓒ Ⓓ Ⓔ	29. Ⓐ Ⓑ Ⓒ Ⓓ Ⓔ	54. Ⓐ Ⓑ Ⓒ Ⓓ Ⓔ	79. Ⓐ Ⓑ Ⓒ Ⓓ Ⓔ
5. Ⓐ Ⓑ Ⓒ Ⓓ Ⓔ	30. Ⓐ Ⓑ Ⓒ Ⓓ Ⓔ	55. Ⓐ Ⓑ Ⓒ Ⓓ Ⓔ	80. Ⓐ Ⓑ Ⓒ Ⓓ Ⓔ
6. Ⓐ Ⓑ Ⓒ Ⓓ Ⓔ	31. Ⓐ Ⓑ Ⓒ Ⓓ Ⓔ	56. Ⓐ Ⓑ Ⓒ Ⓓ Ⓔ	81. Ⓐ Ⓑ Ⓒ Ⓓ Ⓔ
7. Ⓐ Ⓑ Ⓒ Ⓓ Ⓔ	32. Ⓐ Ⓑ Ⓒ Ⓓ Ⓔ	57. Ⓐ Ⓑ Ⓒ Ⓓ Ⓔ	82. Ⓐ Ⓑ Ⓒ Ⓓ Ⓔ
8. Ⓐ Ⓑ Ⓒ Ⓓ Ⓔ	33. Ⓐ Ⓑ Ⓒ Ⓓ Ⓔ	58. Ⓐ Ⓑ Ⓒ Ⓓ Ⓔ	83. Ⓐ Ⓑ Ⓒ Ⓓ Ⓔ
9. Ⓐ Ⓑ Ⓒ Ⓓ Ⓔ	34. Ⓐ Ⓑ Ⓒ Ⓓ Ⓔ	59. Ⓐ Ⓑ Ⓒ Ⓓ Ⓔ	84. Ⓐ Ⓑ Ⓒ Ⓓ Ⓔ
10. Ⓐ Ⓑ Ⓒ Ⓓ Ⓔ	35. Ⓐ Ⓑ Ⓒ Ⓓ Ⓔ	60. Ⓐ Ⓑ Ⓒ Ⓓ Ⓔ	85. Ⓐ Ⓑ Ⓒ Ⓓ Ⓔ
11. Ⓐ Ⓑ Ⓒ Ⓓ Ⓔ	36. Ⓐ Ⓑ Ⓒ Ⓓ Ⓔ	61. Ⓐ Ⓑ Ⓒ Ⓓ Ⓔ	86. Ⓐ Ⓑ Ⓒ Ⓓ Ⓔ
12. Ⓐ Ⓑ Ⓒ Ⓓ Ⓔ	37. Ⓐ Ⓑ Ⓒ Ⓓ Ⓔ	62. Ⓐ Ⓑ Ⓒ Ⓓ Ⓔ	87. Ⓐ Ⓑ Ⓒ Ⓓ Ⓔ
13. Ⓐ Ⓑ Ⓒ Ⓓ Ⓔ	38. Ⓐ Ⓑ Ⓒ Ⓓ Ⓔ	63. Ⓐ Ⓑ Ⓒ Ⓓ Ⓔ	88. Ⓐ Ⓑ Ⓒ Ⓓ Ⓔ
14. Ⓐ Ⓑ Ⓒ Ⓓ Ⓔ	39. Ⓐ Ⓑ Ⓒ Ⓓ Ⓔ	64. Ⓐ Ⓑ Ⓒ Ⓓ Ⓔ	89. Ⓐ Ⓑ Ⓒ Ⓓ Ⓔ
15. Ⓐ Ⓑ Ⓒ Ⓓ Ⓔ	40. Ⓐ Ⓑ Ⓒ Ⓓ Ⓔ	65. Ⓐ Ⓑ Ⓒ Ⓓ Ⓔ	90. Ⓐ Ⓑ Ⓒ Ⓓ Ⓔ
16. Ⓐ Ⓑ Ⓒ Ⓓ Ⓔ	41. Ⓐ Ⓑ Ⓒ Ⓓ Ⓔ	66. Ⓐ Ⓑ Ⓒ Ⓓ Ⓔ	91. Ⓐ Ⓑ Ⓒ Ⓓ Ⓔ
17. Ⓐ Ⓑ Ⓒ Ⓓ Ⓔ	42. Ⓐ Ⓑ Ⓒ Ⓓ Ⓔ	67. Ⓐ Ⓑ Ⓒ Ⓓ Ⓔ	92. Ⓐ Ⓑ Ⓒ Ⓓ Ⓔ
18. Ⓐ Ⓑ Ⓒ Ⓓ Ⓔ	43. Ⓐ Ⓑ Ⓒ Ⓓ Ⓔ	68. Ⓐ Ⓑ Ⓒ Ⓓ Ⓔ	93. Ⓐ Ⓑ Ⓒ Ⓓ Ⓔ
19. Ⓐ Ⓑ Ⓒ Ⓓ Ⓔ	44. Ⓐ Ⓑ Ⓒ Ⓓ Ⓔ	69. Ⓐ Ⓑ Ⓒ Ⓓ Ⓔ	94. Ⓐ Ⓑ Ⓒ Ⓓ Ⓔ
20. Ⓐ Ⓑ Ⓒ Ⓓ Ⓔ	45. Ⓐ Ⓑ Ⓒ Ⓓ Ⓔ	70. Ⓐ Ⓑ Ⓒ Ⓓ Ⓔ	95. Ⓐ Ⓑ Ⓒ Ⓓ Ⓔ
21. Ⓐ Ⓑ Ⓒ Ⓓ Ⓔ	46. Ⓐ Ⓑ Ⓒ Ⓓ Ⓔ	71. Ⓐ Ⓑ Ⓒ Ⓓ Ⓔ	96. Ⓐ Ⓑ Ⓒ Ⓓ Ⓔ
22. Ⓐ Ⓑ Ⓒ Ⓓ Ⓔ	47. Ⓐ Ⓑ Ⓒ Ⓓ Ⓔ	72. Ⓐ Ⓑ Ⓒ Ⓓ Ⓔ	97. Ⓐ Ⓑ Ⓒ Ⓓ Ⓔ
23. Ⓐ Ⓑ Ⓒ Ⓓ Ⓔ	48. Ⓐ Ⓑ Ⓒ Ⓓ Ⓔ	73. Ⓐ Ⓑ Ⓒ Ⓓ Ⓔ	98. Ⓐ Ⓑ Ⓒ Ⓓ Ⓔ
24. Ⓐ Ⓑ Ⓒ Ⓓ Ⓔ	49. Ⓐ Ⓑ Ⓒ Ⓓ Ⓔ	74. Ⓐ Ⓑ Ⓒ Ⓓ Ⓔ	99. Ⓐ Ⓑ Ⓒ Ⓓ Ⓔ
25. Ⓐ Ⓑ Ⓒ Ⓓ Ⓔ	50. Ⓐ Ⓑ Ⓒ Ⓓ Ⓔ	75. Ⓐ Ⓑ Ⓒ Ⓓ Ⓔ	100. Ⓐ Ⓑ Ⓒ Ⓓ Ⓔ

Model Examination 9

APPRAISER

See page 661 for the answer key. Math problems are worked out on pages 662–663.

NOTE: The last 25 questions are for general appraisal certification only.

1. Natural or man-made features that affect a neighborhood and its geographic location are __________ influences.
   (A) social
   (B) economic
   (C) government
   (D) environmental
   (E) legal

2. The four broad forces influencing value are
   (A) utility, transferability, demand, and supply (scarcity)
   (B) governmental, economic, social, and political
   (C) supply, demand, location, and popular taste
   (D) governmental, economic, social, and physical
   (E) police power, eminent domain, taxation, and escheat

3. The federal government is active in which of the following areas?
   (A) housing and urban development
   (B) environmental protection
   (C) monetary and fiscal policy
   (D) secondary mortgage market encouragement
   (E) all of the above

4. Population increases __________ demand for housing.
   (A) depress the
   (B) are capable of diminishing the
   (C) have no effect on the
   (D) can create a
   (E) are determined by the

5. In eminent domain, "just compensation" means
   (A) the fair market value of the property
   (B) the current market value plus compensation for anticipated future benefits
   (C) the market value when the property was bought by the current owner
   (D) the insurable value
   (E) none of the above

6. The right of government or quasi-government units to take private property for public use upon the payment of just compensation is
   (A) escheat
   (B) condemnation
   (C) eminent domain
   (D) estoppel
   (E) an unconstitutional practice

7. __________ is the exercise of the right of government to take private property for public use.
   (A) Condemnation
   (B) Certified appraisal
   (C) Immediate justice
   (D) Land removal
   (E) both A and B

8. The fullest and most common type of estate in realty is
   (A) fee tail
   (B) mortgage
   (C) supra leasehold
   (D) community property
   (E) fee simple

9. An estate in severalty is ownership by
   (A) more than two parties
   (B) two parties
   (C) one or two parties
   (D) one party
   (E) two or more parties with right of survivorship

10. Which of the following is evidence of real estate ownership?
    (A) title
    (B) defaulted mortgage
    (C) estate
    (D) fee simple
    (E) tenancy

11. The concept of value in exchange assumes that
    (A) supply does not affect market value
    (B) a commodity is traded in a marketplace, and for that reason prices may be measured objectively
    (C) value can be determined without exposure in a marketplace
    (D) the real estate appraiser can be held liable if the property does not sell in a reasonable time for the appraised value
    (E) the property value will support a reasonable mortgage debt

12. The definition of market value in appraisal does *not* include
    (A) payment in cash or equivalent
    (B) exposure in the market
    (C) informed parties
    (D) topography of the property
    (E) probable price

13. "Value in exchange" is most closely related to
    (A) value in use
    (B) market value
    (C) investment value
    (D) sentimental value
    (E) none of the above

14. Which statement is FALSE?
    (A) Value in use may be lower than value in exchange.
    (B) Market price is the amount actually paid for a good or service, a historical fact from a particular transaction.
    (C) Cost represents the amount paid for the construction of a building or the amount paid for its acquisition.
    (D) Value in use may be far higher than value in exchange.
    (E) The appraiser must be professionally designated for the value to be accurate.

15. The three common types of legal description are
    (A) metes and bounds; recorded lot, block, and tract; and government rectangular survey
    (B) government rectangular survey, private survey, and house number
    (C) metes and bounds, acreage blocks, and government rectangular survey
    (D) land survey, building survey, and depreciation
    (E) short legal, average legal, and long legal

16. Valuation of property for real estate tax purposes results in
    (A) assessed value
    (B) appraisal
    (C) a special assessment
    (D) a millage rate
    (E) none of the above

17. __________ can render existing supply obsolete and less valuable.
    (A) Inflation
    (B) Interest rates
    (C) Employment
    (D) Income levels
    (E) Changes in tastes and standards

18. One implication of competition and excess profit is that
    (A) a certain optimum combination of land, labor, capital, and entrepreneurship exists
    (B) an estimate of value should be based on future expectations
    (C) abnormally high profits cannot be expected to continue indefinitely
    (D) maximum value accrues to real estate when social and economic homogeneity is present in a neighborhood
    (E) the direct sales comparison approach becomes irrelevant

19. Because real estate markets deal with different desires and needs, they are said to be
    (A) fractionated
    (B) structured
    (C) segmented
    (D) submarketed
    (E) spacious

20. Demand for real estate clearly exists when
    (A) population and employment are on the rise
    (B) there is desire or need for space, plus available mortgage financing
    (C) there is desire or need for space, plus ability to pay
    (D) purchasing power increases
    (E) farm prices rise

21. A mortgage is
    (A) a gift from the mortgagee
    (B) a transfer of real estate to a financial institution
    (C) any loan
    (D) a pledge of real estate as collateral for a loan
    (E) a cloud on the title to real estate

22. The portion of the loan payment for recapture of the investment capital in a mortgage is
    (A) principal, interest, taxes, and insurance
    (B) principal reduction or amortization
    (C) interest and principal payment
    (D) negative amortization
    (E) mortgage banking

23. An appraisal is
   (A) an establishment of value
   (B) a prediction of sales price
   (C) a mathematically precise forecast of value
   (D) an estimate of rental levels
   (E) an estimate of value

24. An appraiser
   (A) determines value
   (B) determines price
   (C) estimates value
   (D) measures price
   (E) forecasts price

25. An appraiser
   (A) determines rent rates
   (B) sets value
   (C) suggests financing
   (D) estimates value
   (E) inspects electrical, mechanical, and plumbing for working condition

26. The __________ establishes the market conditions prevailing when the appraisal is made.
   (A) statement of limitations
   (B) statement of conclusions
   (C) introductory note
   (D) transmittal letter
   (E) date of the value estimate

27. Analysis of a location involves which three levels?
   (A) general, specific, detailed
   (B) country, state, community
   (C) residential, commercial, industrial
   (D) country, state, county
   (E) region, neighborhood, site

28. Market data are used
   (A) in the direct sales comparison approach
   (B) in the income approach
   (C) in the cost approach
   (D) in statistical analysis
   (E) all of the above

29. The cost approach is most applicable when the subject property
   (A) has new improvements that represent the highest and best use
   (B) is in an active local market for similar properties
   (C) produces a positive cash flow
   (D) is in an area where many comparable properties have recently been sold
   (E) exhibits a great deal of functional obsolescence

30. In preparing an appraisal report, your analysis concludes that one of the approaches to value is not applicable to this particular case. You should therefore
    (A) omit the approach altogether
    (B) base the approach on hypothetical data
    (C) state that the approach is not relevant
    (D) state that the approach is not applicable, explain the reasons for this contention, and provide supporting data
    (E) find another approach so as to include three approaches

31. Reconciliation involves
    (A) averaging the unadjusted sales prices of comparables
    (B) recalculating all data
    (C) averaging all estimates derived, weighting each according to its importance
    (D) placing primary emphasis on the estimate deemed most reliable
    (E) averaging estimates from the three approaches, giving each equal weight

32. Cost and market value are more likely to be almost the same when properties are
    (A) new
    (B) old
    (C) depreciated
    (D) syndicated
    (E) appraised

33. In appraisal, reconciliation is
    (A) an estimate of value
    (B) one of the three approaches used in estimating value
    (C) a process of reevaluation that leads to the final value estimate
    (D) an assurance of checkbook accuracy
    (E) a process that is similar to correlation in statistics

34. The essential elements of an appraisal report
    (A) usually follow the scope of work
    (B) vary with the type of report
    (C) depend on the client's needs
    (D) depend on the fee charged
    (E) depend on the number of comparables located

35. A restricted appraisal report
    (A) contains many items that are not included in the usual report
    (B) is an appraisal report prepared for a fussy client
    (C) is generally the least detailed report
    (D) may be based on hypothetical or assumed data
    (E) should be prepared on a nontypical property to show appraisal expertise

36. A parcel of land that is improved to the point of being ready to be built upon is called
(A) realty
(B) land
(C) a site
(D) terrain
(E) a location

37. Before performing a site inspection, an appraiser should
(A) gather basic information about the site
(B) know the reasons for the site inspection
(C) have a map or drawing of the site
(D) have writing materials for taking notes
(E) all of the above

38. Environmental hazards that an appraiser must be conscious of include all of the following *except*
(A) asbestos
(B) radon
(C) Drano
(D) urea formaldehyde
(E) dry cleaning chemicals

39. The term that denotes the attractiveness and usefulness of a property is
(A) price estimate
(B) highest and best use
(C) location
(D) functional utility
(E) value in use

40. Combining two or more sites in order to develop one site with a greater value than the individual sites have separately provides
(A) assemblage
(B) plottage
(C) surplus land
(D) excess land
(E) highest and best use of land

41. In the cost approach, the valuation of land involves the principle of
(A) conformity
(B) contribution
(C) highest and best use
(D) marginal productivity
(E) variable proportions

42. The highest and best use of land is the reasonable use
    (A) to which land is currently being put
    (B) to which land can be put without adverse effect over the short term
    (C) to which land can most profitably be put over the short term
    (D) to which land can most profitably be put over the long term
    (E) whereby land will become agriculturally productive

43. The sales comparison approach should be used
    (A) without exception
    (B) on residential properties only
    (C) on residential and income properties
    (D) in all cases where comparable sales are available
    (E) on vacant land

44. The sales comparison approach involves
    (A) analyzing sales
    (B) comparing properties that have recently been sold to a subject property
    (C) analyzing market rentals
    (D) both A and B
    (E) none of the above

45. If the monthly rental for the subject property is $575 and the gross rent multiplier is 127, what estimate of value is indicated?
    (A) $74,750
    (B) $75,000
    (C) $72,925
    (D) $73,125
    (E) $73,000

46. The sales comparison approach
    (A) uses the replacement cost of improvements, added to a comparable vacant land value
    (B) uses data on recent sales of comparable properties
    (C) uses the income-producing capability of the property
    (D) involves multiple listing service data
    (E) requires at least three comparable sales for application

47. What is the indicated value of a property that rents for $750 per month, using a monthly gross rent multiplier of 110, if the expenses attributable to the property are $125 per month?
    (A) $75,670
    (B) $82,500
    (C) $68,750
    (D) $61,125
    (E) $13,750

48. A 7-year-old residence is currently valued at $72,000. What was its original value if it has appreciated by 60% since it was built?
(A) $27,000
(B) $37,800
(C) $45,000
(D) $115,200
(E) none of the above

49. For residential property, market value appraisals assume that
(A) the purchaser pays all cash; no money is borrowed
(B) an FHA or VA mortgage is used
(C) a purchase money mortgage is considered
(D) the value estimate is based on no special financing
(E) the seller pays no more than five points

50. In considering comparable sales in direct sales comparison appraisal,
(A) the seller's motivation is significant
(B) the date of sale is significant
(C) the proximity of the properties is most important
(D) both A and B
(E) none of the above

Questions 51–53 are based on the following information:

Seven comparables have been found for a single-family dwelling being appraised and gross rent multipliers calculated from the sales prices as follows:

Comparable	GRM
1	125
2	125
3	127
4	127
5	127
6	128
7	130

51. What is the mean GRM?
(A) 125
(B) 126
(C) 127
(D) 128
(E) 130

52. What is the median GRM?
   (A) 125
   (B) 126
   (C) 127
   (D) 128
   (E) 130

53. What is the mode for the GRMs?
   (A) 125
   (B) 126
   (C) 127
   (D) 128
   (E) 130

54. Land prices are analyzed and adjusted in the sales comparison process. These adjustments may involve
   (A) adding or subtracting lump-sum amounts
   (B) adding or subtracting percentages
   (C) accumulating percentages
   (D) using several units of comparison
   (E) any of the above

55. The adjustment process in the direct sales comparison technique involves
   (A) identifying the similarities between properties and adjusting the sales prices
   (B) analyzing the comparables and adjusting the prices for similarities
   (C) identifying the significant differences between the subject and comparable properties and adjusting the sales prices of the comparables for differences
   (D) locating competitive properties and ranking them in the order of desirability
   (E) adjusting the subject to be like the comparables

56. In the cost approach, the site is valued as if it were
   (A) vacant and available for development to its highest and best use
   (B) improved and suited for its intended use or development
   (C) developed and operating
   (D) attractively landscaped
   (E) lacking nearby utilities

57. The direct sales comparison approach is especially suitable for appraising land zoned for which of the following types of use?
   (A) single-family residential
   (B) commercial
   (C) industrial
   (D) multifamily residential
   (E) any of the above

58. All of the following are accepted classifications of accrued depreciation for appraisal purposes *except*
   (A) functional obsolescence
   (B) economic obsolescence
   (C) accounting allocation
   (D) physical deterioration
   (E) external obsolescence

59. Accrued depreciation can be defined in appraisal terms as
   (A) an increase in value from inflationary gains
   (B) a total reduction in value from all causes
   (C) diminished utility from trade imbalance
   (D) competitive pressures
   (E) functional form changes

60. The cost estimates used in appraisal typically reflect
   (A) wholesale costs
   (B) current cost levels
   (C) typical costs to build a building like this subject
   (D) the actual historic cost of the building being appraised
   (E) both B and C

61. Cost indexes are used to
   (A) derive units of comparison
   (B) catalog building components
   (C) estimate operating expenses
   (D) update past costs into current costs
   (E) estimate the local consumer price index

62. The most detailed, time-consuming, and costly method of estimating cost new is
   (A) unit of comparison
   (B) quantity survey
   (C) trade breakdown
   (D) unit-in-place
   (E) segregated cost

63. In calculating depreciation, a limitation of the age-life method is that it
   (A) tends to ignore physical deterioration
   (B) cannot be used with estimates of cost new
   (C) is based on replacement cost rather than reproduction cost
   (D) tends to ignore functional and economic obsolescence
   (E) is difficult to compute

64. Two baths are required to serve a three-bedroom house if
    (A) the appraiser believes they are necessary
    (B) the selling price is more than $75,000
    (C) two baths are demanded by a typical purchaser of a three-bedroom house in the given market
    (D) four or more persons will live in the house
    (E) one of the bedrooms is at the opposite end of the house from the other two

65. Which is the most precise yet least used cost estimation method?
    (A) index method
    (B) quantity survey method
    (C) comparative square foot method
    (D) unit-in-place method
    (E) segregated cost method

66. A defect is considered curable if
    (A) major structural alterations are not required
    (B) the cost of the cure represents more than 25% of the remaining utility
    (C) the cost to cure the condition is less than or equal to the anticipated addition to value
    (D) the cost to cure increases the remaining economic life of the improvement
    (E) the cure is in conformity with building codes

67. External or economic obsolescence is normally *not*
    (A) curable by the owner, the landlord, or the tenant
    (B) discovered during the neighborhood analysis portion of appraisal
    (C) evident in both land and buildings
    (D) caused by forces outside the property
    (E) also called environmental obsolescence

68. A house built in 1985 and appraised in 2020 would have
    (A) a useful life of 35 years
    (B) a chronological age of 35 years
    (C) an effective age of 35 years
    (D) an economic age of 35 years
    (E) a 35-year-old replacement cost

69. In one step of the land residual technique, the building capitalization rate is applied to the known building value to estimate the
    (A) highest and best use of the site
    (B) cost of the building
    (C) income needed to support the land
    (D) net operating income needed to support the building
    (E) land value

70. An example of partial interest is
   (A) a life estate
   (B) a leasehold estate
   (C) a lengthy attention span
   (D) a short attention span
   (E) both A and B

71. A condominium is
   (A) a type of mortgage on real estate
   (B) a building like a cooperative
   (C) a legal concept of ownership
   (D) a zero-lot-line house
   (E) a style of housing

72. Clarity and accuracy contribute to the
   (A) quality of an appraisal
   (B) quantity of evidence
   (C) appraisal fee
   (D) final value estimate
   (E) appearance of the report

73. If an appraiser is asked to undertake an assignment on a property type with which she has no previous experience and is otherwise unfamiliar, she should
   (A) refuse the assignment
   (B) accept the assignment, thereby expanding her abilities, but reduce the fee
   (C) ask the client to hire another appraiser to help her
   (D) associate herself with another appraiser experienced in this type of assignment and inform the client of this fact
   (E) get the necessary computer software to solve the problem

74. Real estate markets are composed of
   (A) buyers only
   (B) sellers only
   (C) types of property
   (D) buyers and sellers
   (E) appraisers and counselors

75. Real estate supply factors include
   (A) the current housing supply
   (B) new construction activity
   (C) tax depreciation allowances
   (D) both A and B
   (E) none of the above

The remaining 25 questions are for general appraisal certification only.

76. The mortgage constant is a function of all of the following *except*
    (A) the term of the loan
    (B) the borrower's income level
    (C) the interest rate
    (D) the term of amortization
    (E) It is a function of all of the above.

77. The benefit(s) forgone from a project that cannot be built is (are) called
    (A) anticipation
    (B) substitution
    (C) opportunity cost
    (D) surplus productivity
    (E) marginal cost

78. The debt coverage ratio
    (A) is federally approved for residential valuation
    (B) can be applied uniformly to all properties
    (C) is the most equitable method of determining a cap rate
    (D) indicates the safety of the loan
    (E) is generally less than 1.00

79. The ____________ is a key safety factor for the lender.
    (A) maturity term
    (B) sales price
    (C) loan-to-value ratio
    (D) age of the improvements
    (E) land value

80. A parcel of land is sold for $115,000. If it appreciated at 12% compounded per year and the seller held it for four years, how much did the seller pay for it (disregard taxes, selling costs, etc.)?
    (A) $60,000
    (B) $101,000
    (C) $73,000
    (D) $62,500
    (E) $81,700

81. A 30-year-old building with an effective age of 20 years has a total life expectancy of 50 years. How much depreciation has occurred?
    (A) 10%
    (B) 20%
    (C) 40%
    (D) 60%
    (E) none of the above

82. A plot of land, 100 ft. × 200 ft., along an interstate freeway is situated 12 miles north of the central business district of a city with a population approaching 1 million. Which of the following would be the highest and best use of the land?
(A) service station
(B) convenience store
(C) two-story office building
(D) medical office building
(E) There is not enough information to make a determination.

83. A $2 million shopping center is purchased with a 75% loan-to-value ratio mortgage, payable monthly over 25 years at 12% interest. It generates $205,000 annual net operating income. The land is considered to be worth $500,000; the balance of cost is represented by land improvements. What is the improvement ratio?
(A) 25%
(B) 50%
(C) 75%
(D) 100%
(E) none of the above

84. The adjustment for below-market-rate financing may provide an estimate of
(A) market price
(B) cash equivalence
(C) property price
(D) creative financing
(E) replacement cost

85. Land or site appraisals assist in
(A) the sale and purchase of land
(B) land development
(C) ad valorem and income tax situations
(D) all of the above
(E) none of the above

86. Overall rate of return is
(A) the annual net operating income divided by the purchase price
(B) annual income and all gains or losses prorated to an effective annual amount minus an internal rate of return
(C) a percentage or decimal rate that, when divided into a periodic income amount, offers a lump-sum capital value for the income
(D) an income that is unchanging with time
(E) none of the above

87. All of the following are used in the valuation of income-producing property *except*
    (A) rental rates
    (B) operating expenses
    (C) income taxes
    (D) net leasable area
    (E) vacancy and collection allowance

88. A $1 million property will have a 75% loan at an 8% annual mortgage constant. What must the net operating income be to produce a 12% cash-on-cash return?
    (A) $75,000
    (B) $127,500
    (C) $90,000
    (D) $150,000
    (E) $750,000

89. If a particular buyer requires a recapture of the building portion of the purchase price in 25 years, what is the indicated recapture rate for the building, assuming straight-line capture?
    (A) 0.25%
    (B) 2%
    (C) 4%
    (D) 20%
    (E) 25%

90. In a high-rise 100-unit apartment building, there is a basement laundry area that brings in $100 monthly from the concessionaire. The laundry income is
    (A) included, as miscellaneous income, in potential gross income
    (B) included, as other income, in effective gross income
    (C) deducted from effective gross income
    (D) added to before-tax cash flow
    (E) distributed to the maintenance workers

91. In an appraisal of income property, which of the following items should be excluded from the expense statement?
    (A) mortgage loan interest payments
    (B) ordinary and necessary current expenses
    (C) projected replacement reserve
    (D) management fees
    (E) advertising expenses

Questions 92–95 are based on the following information:

A building contains 25 one-bedroom units and 75 two-bedroom units. The one-bedroom units rent for $550 monthly; two-bedrooms are $675. The vacancy rate is 7%; operating expenses are estimated at 40% of effective gross income. There is $1,000 annual income from vending machines.

92. Potential gross income is
    (A) $773,500
    (B) $772,500
    (C) $719,425
    (D) $64,375
    (E) none of the above

93. Effective gross income is
    (A) $773,500
    (B) $772,500
    (C) $719,425
    (D) $64,375
    (E) none of the above

94. Net operating income is
    (A) $64,375
    (B) $286,970
    (C) $431,655
    (D) $719,425
    (E) none of the above

95. Operating expenses are
    (A) $1,000
    (B) $54,075
    (C) $287,770
    (D) $719,425
    (E) none of the above

96. Which approach involves an investigation of the rent schedules of the subject property and the comparables?
    (A) cost approach
    (B) just compensation evaluation
    (C) income approach
    (D) comparable rent approach
    (E) all of the above

97. Property free and clear of indebtedness is offered for sale. The property is net-net leased (the tenant pays all operating expenses) for $90,000 per year under a 30-year lease. Rent is payable annually at the end of each year. The building cost $850,000, including a developer's profit and risk allowance. The land costs $100,000, and its value is well established by comparable sales. It is estimated that the building will be worth only about half its current cost at the end of the lease term and that the land will remain constant in value over that period. A 10% discount rate is considered appropriate. Using the building residual technique, estimate the value of the building.
    (A) between $250,000 and $300,000
    (B) between $300,000 and $750,000
    (C) between $750,000 and $1,000,000
    (D) between $1,000,000 and $1,500,000
    (E) not within any of the above ranges

98. Which of the following statements is true of a *gross lease*?
    (A) The tenant pays all operating expenses.
    (B) The landlord pays all operating expenses.
    (C) This lease is used only for commercial properties.
    (D) Rent rises with the cost of living.
    (E) This lease must be drafted by an attorney.

99. Land purchased for $100,000 cash appreciates at the rate of 15%, compounded annually. About how much is the land worth after 5 years? Disregard taxes, insurance, and selling expenses.
    (A) $200,000
    (B) $175,000
    (C) $115,000
    (D) $75,000
    (E) $15,000

100. An allowance for vacancy and collection loss is estimated as a percentage of
    (A) net operating income
    (B) before-tax cash flow
    (C) effective gross income
    (D) potential gross income
    (E) after-tax cash flow

ANSWER KEY
Model Test 9

1. **D**	26. **E**	51. **C**	76. **B**
2. **D**	27. **E**	52. **C**	77. **C**
3. **E**	28. **E**	53. **C**	78. **D**
4. **D**	29. **A**	54. **E**	79. **C**
5. **A**	30. **D**	55. **C**	80. **C**
6. **C**	31. **D**	56. **A**	81. **C**
7. **A**	32. **A**	57. **E**	82. **E**
8. **E**	33. **C**	58. **C**	83. **C**
9. **D**	34. **A**	59. **B**	84. **B**
10. **A**	35. **C**	60. **E**	85. **D**
11. **B**	36. **C**	61. **D**	86. **A**
12. **D**	37. **E**	62. **B**	87. **C**
13. **B**	38. **C**	63. **D**	88. **C**
14. **E**	39. **D**	64. **C**	89. **C**
15. **A**	40. **B**	65. **B**	90. **B**
16. **A**	41. **C**	66. **C**	91. **A**
17. **E**	42. **D**	67. **A**	92. **B**
18. **C**	43. **D**	68. **B**	93. **C**
19. **C**	44. **D**	69. **D**	94. **C**
20. **C**	45. **E**	70. **E**	95. **C**
21. **D**	46. **B**	71. **C**	96. **C**
22. **B**	47. **B**	72. **A**	97. **C**
23. **E**	48. **C**	73. **D**	98. **B**
24. **C**	49. **D**	74. **D**	99. **A**
25. **D**	50. **D**	75. **D**	100. **D**

ANSWER EXPLANATIONS—ARITHMETIC QUESTIONS

NOTE: The section contains explanations for only the arithmetic questions in the model exam.

45. Value = rent × GRM:

$575 × 127 = $73,025, or $73,000 rounded

47. When using GRM, you ignore expenses, so don't pay any attention to the $125 monthly expenses. Just use the monthly rent of $750. Value = rent × GRM:

$750 × 110 = $82,500

48. The residence is now worth 160% of what is was worth when it was built:

$72,000 ÷ 1.6 = $45,000

51. To find the mean, add the seven GRMs, divide by 7:

(125 + 125 + 127 + 127 + 127 + 128 + 130) ÷ 7 = 889 ÷ 7 = 127

52. The GRMs are already arranged in order; the median is the one in the middle (the fourth): 127.

53. The mode is the value that appears most often. Since 127 appears three times, more than any other GRM value, the mode is 127.

80. Purchase price will be $115,000/$(1.12)^4$. $(1.12)^4$ = 1.12 × 1.12 × 1.12 × 1.12 = 1.574.

$115,000 ÷ 1.574 = $73,084 = $73,000 rounded

81. Effective depreciation is figured for 20 out of a total of 50 years, or 20/50 = 0.4 = 40%.

83. You are given more information than you need. Improvement ratio is value of improvements/total property value. Since total value is $2,000,000 and land is worth $500,000, improvements are worth $1,500,000.

$1,500,000 ÷ $2,000,000 = 0.75 = 75%

88. Loan is 75% of value: $1,000,000 × 0.75 = $750,000. 12% constant:

$750,000 × 0.12 = $90,000.

There is $250,000 of equity, for which 15% cash-on-cash is desired:

$250,000 × 0.15 = $37,500.

NOI will have to be mortgage payment + desired cash return:

$37,500 + $90,000 = $127,500

89. Recapture rate is 100%/25 years = 4% per year.

92. Monthly rents: 25 × $550 = $13,750; 75 × $675 = $50,625; $13,750 + $50,625 = $64,375.

Potential annual gross income = $64,375 × 12 = $772,500

93. Effective gross income is potential gross income plus miscellaneous income, less vacancy.
Vacancy is 7% of potential gross rent: $772,500 × 0.07 = $54,075 vacancy

$$\$772{,}500 + \$1{,}000 - \$54{,}075 = \$719{,}425$$

94. NOI is effective gross less operating expenses, which are 40% of effective gross.
NOI is 60% of effective gross:

$$\$719{,}425 \times 0.6 = \$431{,}655$$

95. Operating expenses are 40% of NOI:

$$\$719{,}425 \times 0.4 = \$287{,}770$$

97. Land value is $100,000; 10% of that ($10,000) is return to land, so

$$\$90{,}000 - \$10{,}000 = \$80{,}000$$

is the return to the building. This implies a value of a little less than $800,000 for the income over 30 years ($80,000/0.1) *plus* the present value of a reversion of about $400,000, which wouldn't be very much. The range $750,000 to $1,000,000 clearly is correct.

99. Value will be $(1.15)^5 \times \$100{,}000 = 2.011 \times \$100{,}000 = \$201{,}100 = \$200{,}000$ rounded.

Index

B

E

H

I

L

M

T

W

Y

Z